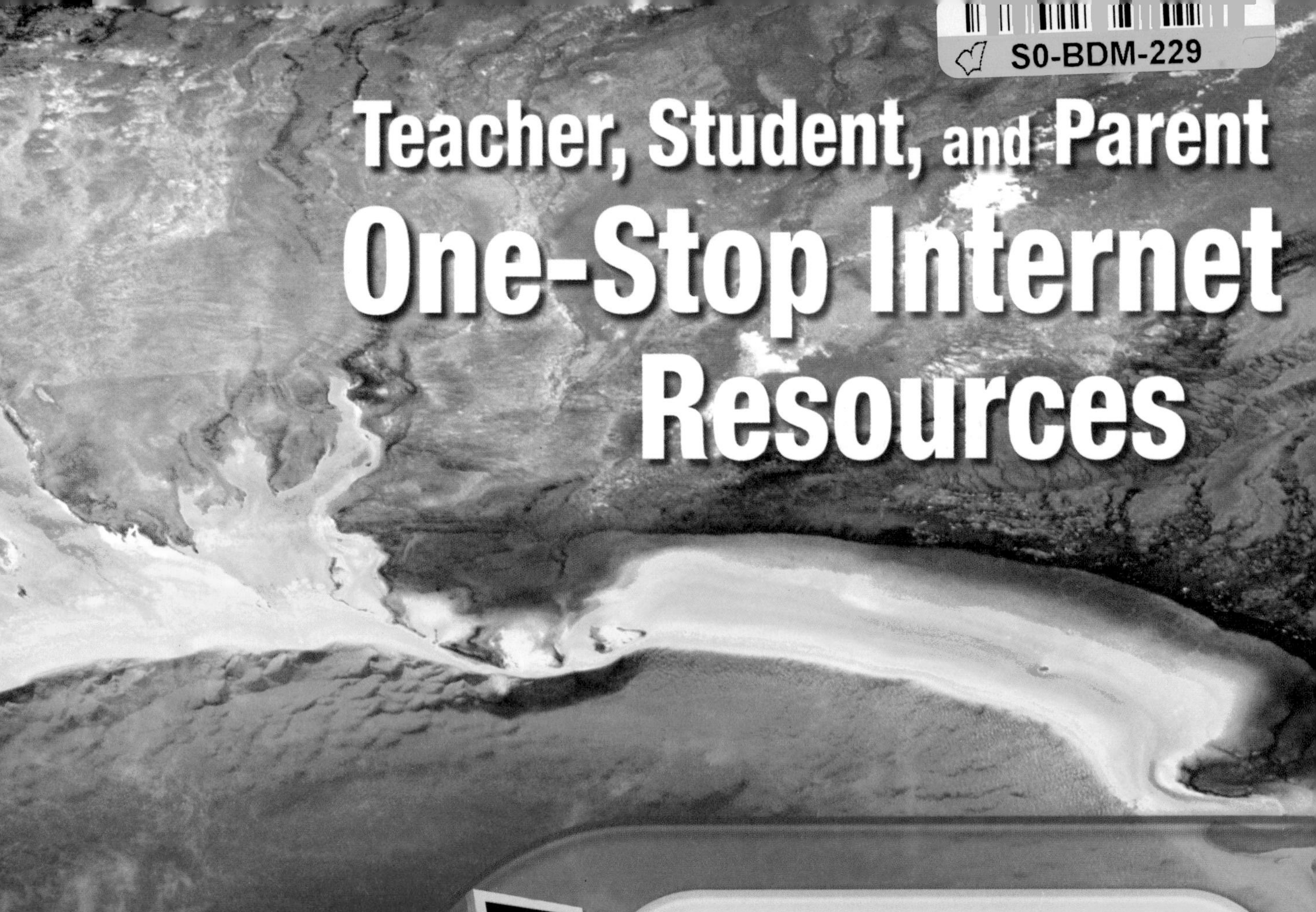

Teacher, Student, and Parent One-Stop Internet Resources

Log on to earth.msscience.com

ONLINE STUDY TOOLS

- Section Self-Check Quizzes
- Interactive Tutor
- Chapter Review Tests
- Standardized Test Practice
- Vocabulary PuzzleMaker

ONLINE RESEARCH

- WebQuest Projects
- Prescreened Web Links
- Career Links
- Microscopy Image Links
- Internet Labs

INTERACTIVE ONLINE STUDENT EDITION

- Complete Interactive Student Edition available at mhln.com

FOR TEACHERS

- Teacher Bulletin Board
- Teaching Today—Professional Development

SAFETY SYMBOLS

SAFETY SYMBOLS	HAZARD	EXAMPLES	PRECAUTION	REMEDY
DISPOSAL	Special disposal procedures need to be followed.	certain chemicals, living organisms	Do not dispose of these materials in the sink or trash can.	Dispose of wastes as directed by your teacher.
BIOLOGICAL	Organisms or other biological materials that might be harmful to humans	bacteria, fungi, blood, unpreserved tissues, plant materials	Avoid skin contact with these materials. Wear mask or gloves.	Notify your teacher if you suspect contact with material. Wash hands thoroughly.
EXTREME TEMPERATURE	Objects that can burn skin by being too cold or too hot	boiling liquids, hot plates, dry ice, liquid nitrogen	Use proper protection when handling.	Go to your teacher for first aid.
SHARP OBJECT	Use of tools or glassware that can easily puncture or slice skin	razor blades, pins, scalpels, pointed tools, dissecting probes, broken glass	Practice common-sense behavior and follow guidelines for use of the tool.	Go to your teacher for first aid.
FUME	Possible danger to respiratory tract from fumes	ammonia, acetone, nail polish remover, heated sulfur, moth balls	Make sure there is good ventilation. Never smell fumes directly. Wear a mask.	Leave foul area and notify your teacher immediately.
ELECTRICAL	Possible danger from electrical shock or burn	improper grounding, liquid spills, short circuits, exposed wires	Double-check setup with teacher. Check condition of wires and apparatus.	Do not attempt to fix electrical problems. Notify your teacher immediately.
IRRITANT	Substances that can irritate the skin or mucous membranes of the respiratory tract	pollen, moth balls, steel wool, fiberglass, potassium permanganate	Wear dust mask and gloves. Practice extra care when handling these materials.	Go to your teacher for first aid.
CHEMICAL	Chemicals can react with and destroy tissue and other materials	bleaches such as hydrogen peroxide; acids such as sulfuric acid, hydrochloric acid; bases such as ammonia, sodium hydroxide	Wear goggles, gloves, and an apron.	Immediately flush the affected area with water and notify your teacher.
TOXIC	Substance may be poisonous if touched, inhaled, or swallowed.	mercury, many metal compounds, iodine, poinsettia plant parts	Follow your teacher's instructions.	Always wash hands thoroughly after use. Go to your teacher for first aid.
FLAMMABLE	Flammable chemicals may be ignited by open flame, spark, or exposed heat.	alcohol, kerosene, potassium permanganate	Avoid open flames and heat when using flammable chemicals.	Notify your teacher immediately. Use fire safety equipment if applicable.
OPEN FLAME	Open flame in use, may cause fire.	hair, clothing, paper, synthetic materials	Tie back hair and loose clothing. Follow teacher's instruction on lighting and extinguishing flames.	Notify your teacher immediately. Use fire safety equipment if applicable.

Eye Safety
Proper eye protection should be worn at all times by anyone performing or observing science activities.

Clothing Protection
This symbol appears when substances could stain or burn clothing.

Animal Safety
This symbol appears when safety of animals and students must be ensured.

Handwashing
After the lab, wash hands with soap and water before removing goggles.

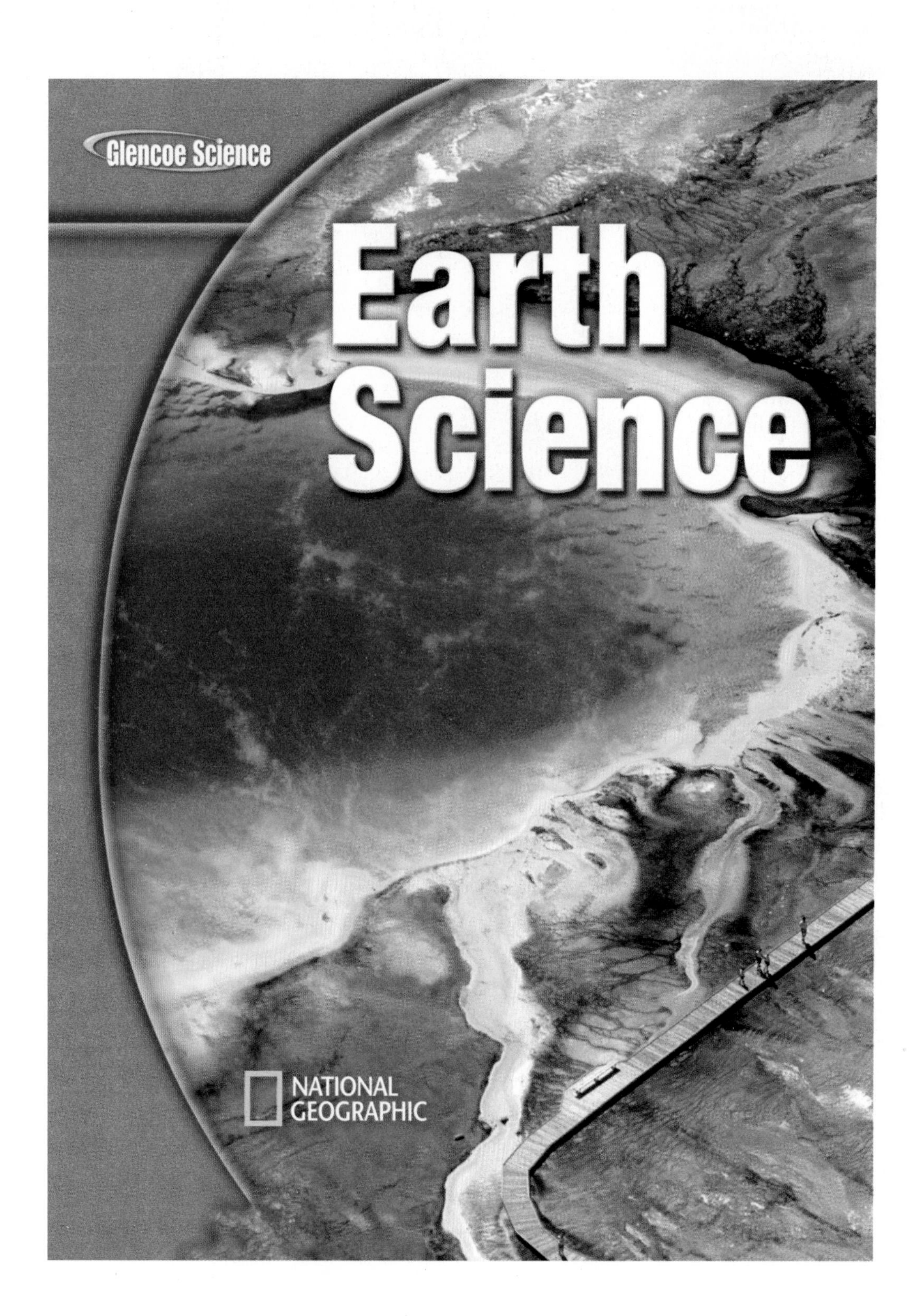

New York, New York Columbus, Ohio Chicago, Illinois Woodland Hills, California

Glencoe Science

Earth Science

Visitors to Yellowstone National Park stroll along a boardwalk near the edge of Grand Prismatic Spring, a natural hot spring. The temperature at the center of the pool is 93°C. The different colors are caused by bacteria which live in different temperatures of water.

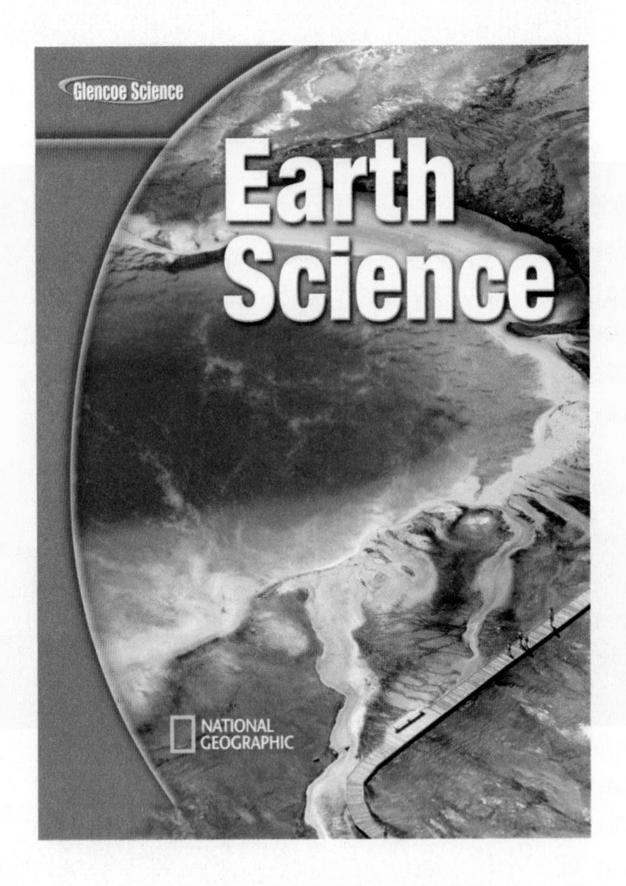

Glencoe

The McGraw·Hill Companies

Send all inquiries to:
Glencoe/McGraw-Hill
8787 Orion Place
Columbus, OH 43240-4027

ISBN: 978-0-07-877802-5
MHID: 0-07-877802-6

Printed in the United States of America.

4 5 6 7 8 9 10 027/055 12 11 10 09

Contents In Brief

Authors

Education Division
Washington, D.C.

Ralph M. Feather Jr., PhD
Assistant Professor
Department of Educational Studies
and Secondary Education
Bloomsburg University
Bloomsburg, PA

Susan Leach Snyder
Retired Teacher, Consultant
Jones Middle School
Upper Arlington, OH

Dinah Zike
Educational Consultant
Dinah-Might Activities, Inc.
San Antonio, TX

Science Consultants

William C. Keel, PhD
Department of Physics and Astronomy
University of Alabama
Tuscaloosa, AL

Robert Nierste
Science Department Head
Hendrick Middle School, Plano ISD
Plano, TX

Series Consultants

MATH

Michael Hopper, DEng
Manager of Aircraft Certification
L-3 Communications
Greenville, TX

Teri Willard, EdD
Mathematics Curriculum Writer
Belgrade, MT

READING

Elizabeth Babich
Special Education Teacher
Mashpee Public Schools
Mashpee, MA

Barry Barto
Special Education Teacher
John F. Kennedy Elementary
Manistee, MI

Carol A. Senf, PhD
School of Literature,
Communication, and Culture
Georgia Institute of Technology
Atlanta, GA

Rachel Swaters-Kissinger
Science Teacher
John Boise Middle School
Warsaw, MO

SAFETY

Aileen Duc, PhD
Science 8 Teacher
Hendrick Middle School, Plano ISD
Plano, TX

Sandra West, PhD
Department of Biology
Texas State University-San Marcos
San Marcos, TX

ACTIVITY TESTERS

Nerma Coats Henderson
Pickerington Lakeview Jr. High School
Pickerington, OH

Mary Helen Mariscal-Cholka
William D. Slider Middle School
El Paso, TX

Science Kit and Boreal Laboratories
Tonawanda, NY

Series Reviewers

Deidre Adams
West Vigo Middle School
West Terre Haute, IN

Sharla Adams
IPC Teacher
Allen High School
Allen, TX

Maureen Barrett
Thomas E. Harrington Middle School
Mt. Laurel, NJ

John Barry
Seeger Jr.-Sr. High School
West Lebanon, IN

Desiree Bishop
Environmental Studies Center
Mobile County Public Schools
Mobile, AL

William Blair
Retired Teacher
J. Marshall Middle School
Billerica, MA

Tom Bright
Concord High School
Charlotte, NC

Lois Burdette
Green Bank Elementary-Middle School
Green Bank, WV

Marcia Chackan
Pine Crest School
Boca Raton, FL

Obioma Chukwu
J.H. Rose High School
Greenville, NC

Nerma Coats Henderson
Pickerington Lakeview Jr. High School
Pickerington, OH

Karen Curry
East Wake Middle School
Raleigh, NC

Inga Dainton
Merrilvills High School
Merrilville, IN

Joanne Davis
Murphy High School
Murphy, NC

Robin Dillon
Hanover Central High School
Cedar Lake, IN

Anthony J. DiSipio, Jr.
8th Grade Science
Octorana Middle School
Atglen, PA

Annette D'Urso Garcia
Kearney Middle School
Commerce City, CO

Dwight Dutton
East Chapel Hill High School
Chapel Hill, NC

Carolyn Elliott
South Iredell High School
Statesville, NC

Sueanne Esposito
Tipton High School
Tipton, IN

Sandra Everhart
Dauphin/Enterprise Jr. High Schools
Enterprise, AL

Mary Ferneau
Westview Middle School
Goose Creek, SC

Cory Fish
Burkholder Middle School
Henderson, NV

Linda V. Forsyth
Retired Teacher
Merrill Middle School
Denver, CO

George Gabb
Great Bridge Middle School
Chesapeake Public Schools
Chesapeake, VA

Judith Helton
RS Middle School
Rutherfordton, NC

Lynne Huskey
Chase Middle School
Forest City, NC

Maria E. Kelly
Principal
Nativity School
Catholic Diocese of Arlington
Burke, VA

Michael Mansour
Board Member
National Middle Level Science
Teacher's Association
John Page Middle School
Madison Heights, MI

Mary Helen Mariscal-Cholka
William D. Slider Middle School
El Paso, TX

Michelle Mazeika
Whiting Middle School
Whiting, IN

Joe McConnell
Speedway Jr. High School
Indianapolis, IN

Sharon Mitchell
William D. Slider Middle School
El Paso, TX

Amy Morgan
Berry Middle School
Hoover, AL

Norma Neely, EdD
Associate Director for Regional
Projects
Texas Rural Systemic Initiative
Austin, TX

Annette Parrott
Lakeside High School
Atlanta, GA

Nora M. Prestinari Burchett
Saint Luke School
McLean, VA

Mark Sailer
Pioneer Jr.-Sr. High School
Royal Center, IN

Vicki Snell
Monroe Central Jr.-Sr. High School
Parker City, IN

Joanne Stickney
Monticello Middle School
Monticello, NY

Dee Stout
Penn State University
University Park, PA

Darcy Vetro-Ravndal
Hillsborough High School
Tampa, FL

Karen Watkins
Perry Meridian Middle School
Indianapolis, IN

Clabe Webb
Permian High School
Ector County ISD
Odessa, TX

Alison Welch
William D. Slider Middle School
El Paso, TX

Kim Wimpey
North Gwinnett High School
Suwanee, GA

Michael Wolter
Muncie Central High School
Muncie, IN

Kate Ziegler
Durant Road Middle School
Raleigh, NC

Teacher Advisory Board

The Teacher Advisory Board gave the authors, editorial staff, and design team feedback on the content and design of the Student Edition. They provided valuable input in the development of the 2008 edition of ***Glencoe Earth Science.***

John Gonzales
Challenger Middle School
Tucson, AZ

Rachel Shively
Aptakisic Jr. High School
Buffalo Grove, IL

Roger Pratt
Manistique High School
Manistique, MI

Kirtina Hile
Northmor Jr. High/High School
Galion, OH

Marie Renner
Diley Middle School
Pickerington, OH

Nelson Farrier
Hamlin Middle School
Springfield, OR

Jeff Remington
Palmyra Middle School
Palmyra, PA

Erin Peters
Williamsburg Middle School
Arlington, VA

Rubidel Peoples
Meacham Middle School
Fort Worth, TX

Kristi Ramsey
Navasota Jr. High School
Navasota, TX

Student Advisory Board

The Student Advisory Board gave the authors, editorial staff, and design team feedback on the design of the Student Edition. We thank these students for their hard work and creative suggestions in making the 2008 edition of ***Glencoe Earth Science*** student friendly.

Jack Andrews
Reynoldsburg Jr. High School
Reynoldsburg, OH

Peter Arnold
Hastings Middle School
Upper Arlington, OH

Emily Barbe
Perry Middle School
Worthington, OH

Kirsty Bateman
Hilliard Heritage Middle School
Hilliard, OH

Andre Brown
Spanish Emersion Academy
Columbus, OH

Chris Dundon
Heritage Middle School
Westerville, OH

Ryan Manafee
Monroe Middle School
Columbus, OH

Addison Owen
Davis Middle School
Dublin, OH

Teriana Patrick
Eastmoor Middle School
Columbus, OH

Ashley Ruz
Karrer Middle School
Dublin, OH

The Glencoe middle school science Student Advisory Board taking a timeout at COSI, a science museum in Columbus, Ohio.

HOW TO...

Use Your Science Book

Why do I need my science book?

Have you ever been in class and not understood all of what was presented? Or, you understood everything in class, but at home, got stuck on how to answer a question? Maybe you just wondered when you were ever going to use this stuff?

These next few pages are designed to help you understand everything your science book can be used for . . . besides a paperweight!

Before You Read

- **Chapter Opener** Science is occurring all around you, and the opening photo of each chapter will preview the science you will be learning about. The **Chapter Preview** will give you an idea of what you will be learning about, and you can try the **Launch Lab** to help get your brain headed in the right direction. The **Foldables** exercise is a fun way to keep you organized.
- **Section Opener** Chapters are divided into two to four sections. The **As You Read** in the margin of the first page of each section will let you know what is most important in the section. It is divided into four parts. **What You'll Learn** will tell you the major topics you will be covering. **Why It's Important** will remind you why you are studying this in the first place! The **Review Vocabulary** word is a word you already know, either from your science studies or your prior knowledge. The **New Vocabulary** words are words that you need to learn to understand this section. These words will be in **boldfaced** print and highlighted in the section. Make a note to yourself to recognize these words as you are reading the section.

As You Read

- **Headings** Each section has a title in large red letters, and is further divided into blue titles and small red titles at the beginnings of some paragraphs. To help you study, make an outline of the headings and subheadings.
- **Margins** In the margins of your text, you will find many helpful resources. The **Science Online** exercises and **Integrate** activities help you explore the topics you are studying. **MiniLabs** reinforce the science concepts you have learned.
- **Building Skills** You also will find an **Applying Math** or **Applying Science** activity in each chapter. This gives you extra practice using your new knowledge, and helps prepare you for standardized tests.
- **Student Resources** At the end of the book you will find **Student Resources** to help you throughout your studies. These include **Science, Technology,** and **Math Skill Handbooks,** an **English/Spanish Glossary,** and an **Index.** Also, use your **Foldables** as a resource. It will help you organize information, and review before a test.
- **In Class** Remember, you can always ask your teacher to explain anything you don't understand.

FOLDABLES™ Study Organizer

Science Vocabulary Make the following Foldable to help you understand the vocabulary terms in this chapter.

STEP 1 **Fold** a vertical sheet of notebook paper from side to side.

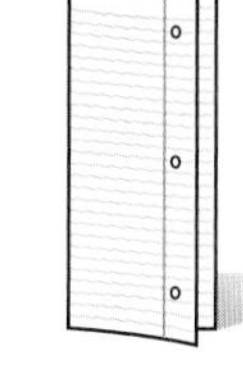

STEP 2 **Cut** along every third line of only the top layer to form tabs.

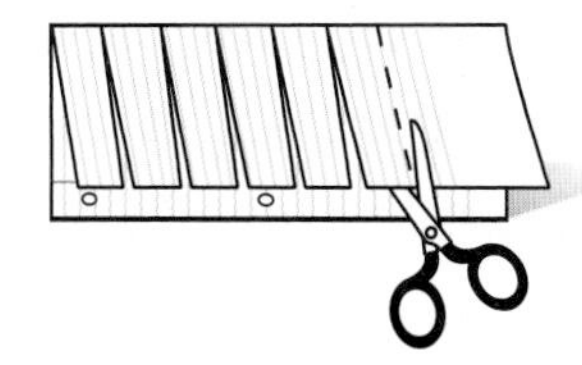

STEP 3 **Label** each tab with a vocabulary word from the chapter.

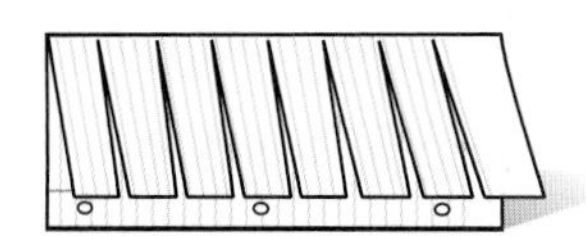

Build Vocabulary As you read the chapter, list the vocabulary words on the tabs. As you learn the definitions, write them under the tab for each vocabulary word.

Look For...
FOLDABLES™
At the beginning of every section.

In Lab

Working in the laboratory is one of the best ways to understand the concepts you are studying. Your book will be your guide through your laboratory experiences, and help you begin to think like a scientist. In it, you not only will find the steps necessary to follow the investigations, but you also will find helpful tips to make the most of your time.

- Each lab provides you with a **Real-World Question** to remind you that science is something you use every day, not just in class. This may lead to many more questions about how things happen in your world.
- Remember, experiments do not always produce the result you expect. Scientists have made many discoveries based on investigations with unexpected results. You can try the experiment again to make sure your results were accurate, or perhaps form a new hypothesis to test.
- Keeping a **Science Journal** is how scientists keep accurate records of observations and data. In your journal, you also can write any questions that may arise during your investigation. This is a great method of reminding yourself to find the answers later.

Look For...
Launch Labs start every chapter.
MiniLabs in the margin of each chapter.
Two Full-Period Labs in every chapter.
EXTRA Try at Home Labs at the end of your book.
the Web site with laboratory demonstrations.

Before a Test

Admit it! You don't like to take tests! However, there *are* ways to review that make them less painful. Your book will help you be more successful taking tests if you use the resources provided to you.

- Review all of the **New Vocabulary** words and be sure you understand their definitions.
- Review the notes you've taken on your **Foldables,** in class, and in lab. Write down any question that you still need answered.
- Review the **Summaries** and **Self Check questions** at the end of each section.
- Study the concepts presented in the chapter by reading the **Study Guide** and answering the questions in the **Chapter Review.**

a or b?

?

T or F?

Look For...

- **Reading Checks** and **caption questions** throughout the text.
- the **summaries** and **self check questions** at the end of each section.
- the **Study Guide** and **Review** at the end of each chapter.
- the **Standardized Test Practice** after each chapter.

Let's Get Started

To help you find the information you need quickly, use the Scavenger Hunt below to learn where things are located in Chapter 1.

1. What is the title of this chapter?
2. What will you learn in Section 1?
3. Sometimes you may ask, "Why am I learning this?" State a reason why the concepts from Section 2 are important.
4. What is the main topic presented in Section 2?
5. How many reading checks are in Section 1?
6. What is the Web address where you can find extra information?
7. What is the main heading above the sixth paragraph in Section 2?
8. There is an integration with another subject mentioned in one of the margins of the chapter. What subject is it?
9. List the new vocabulary words presented in Section 2.
10. List the safety symbols presented in the first Lab.
11. Where would you find a Self Check to be sure you understand the section?
12. Suppose you're doing the Self Check and you have a question about concept mapping. Where could you find help?
13. On what pages are the Chapter Study Guide and Chapter Review?
14. Look in the Table of Contents to find out on which page Section 2 of the chapter begins.
15. You complete the Chapter Review to study for your chapter test. Where could you find another quiz for more practice?

unit 1 Earth Materials—2

chapter 1 The Nature of Science—4

chapter 2 Matter—32

chapter 3 Minerals—60

chapter 4 Rocks—88

In each chapter, look for these opportunities for review and assessment:

- Reading Checks
- Caption Questions
- Section Review
- Chapter Study Guide
- Chapter Review
- Standardized Test Practice
- Online practice at earth.msscience.com

Get Ready to Read Strategies

Contents

Earth's Energy and Mineral Resources—118

unit 2 The Changing Surface of Earth—150

Views of Earth—152

In each chapter, look for these opportunities for review and assessment:
- Reading Checks
- Caption Questions
- Section Review
- Chapter Study Guide
- Chapter Review
- Standardized Test Practice
- Online practice at earth.msscience.com

Contents

In each chapter, look for these opportunities for review and assessment:
- Reading Checks
- Caption Questions
- Section Review
- Chapter Study Guide
- Chapter Review
- Standardized Test Practice
- Online practice at earth.msscience.com

Earth's Air and Water—422

Contents

chapter 17

Climate—482

Get Ready to Read Strategies

chapter 18

Ocean Motion—512

Oceanography—540

unit 6 You and the Environment—570

chapter 20 Our Impact on Land—572

chapter 21 Our Impact on Water and Air—598

In each chapter, look for these opportunities for review and assessment:

- Reading Checks
- Caption Questions
- Section Review
- Chapter Study Guide
- Chapter Review
- Standardized Test Practice
- Online practice at earth.msscience.com

unit 7 Astronomy—624

chapter 22 Exploring Space—626

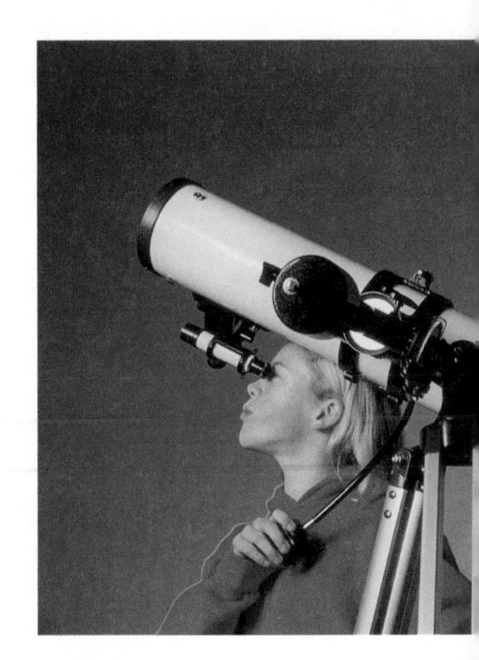

Contents

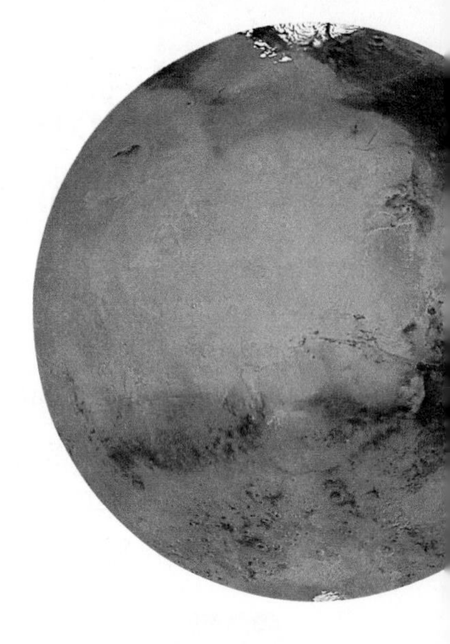

Chapter 23 The Sun-Earth-Moon System—658

Chapter 24 The Solar System—688

Chapter 25 Stars and Galaxies—722

Get Ready to Read Strategies

Student Resources—754

Science Skill Handbook—756

In each chapter, look for these opportunities for review and assessment:
- Reading Checks
- Caption Questions
- Section Review
- Chapter Study Guide
- Chapter Review
- Standardized Test Practice
- Online practice at earth.msscience.com

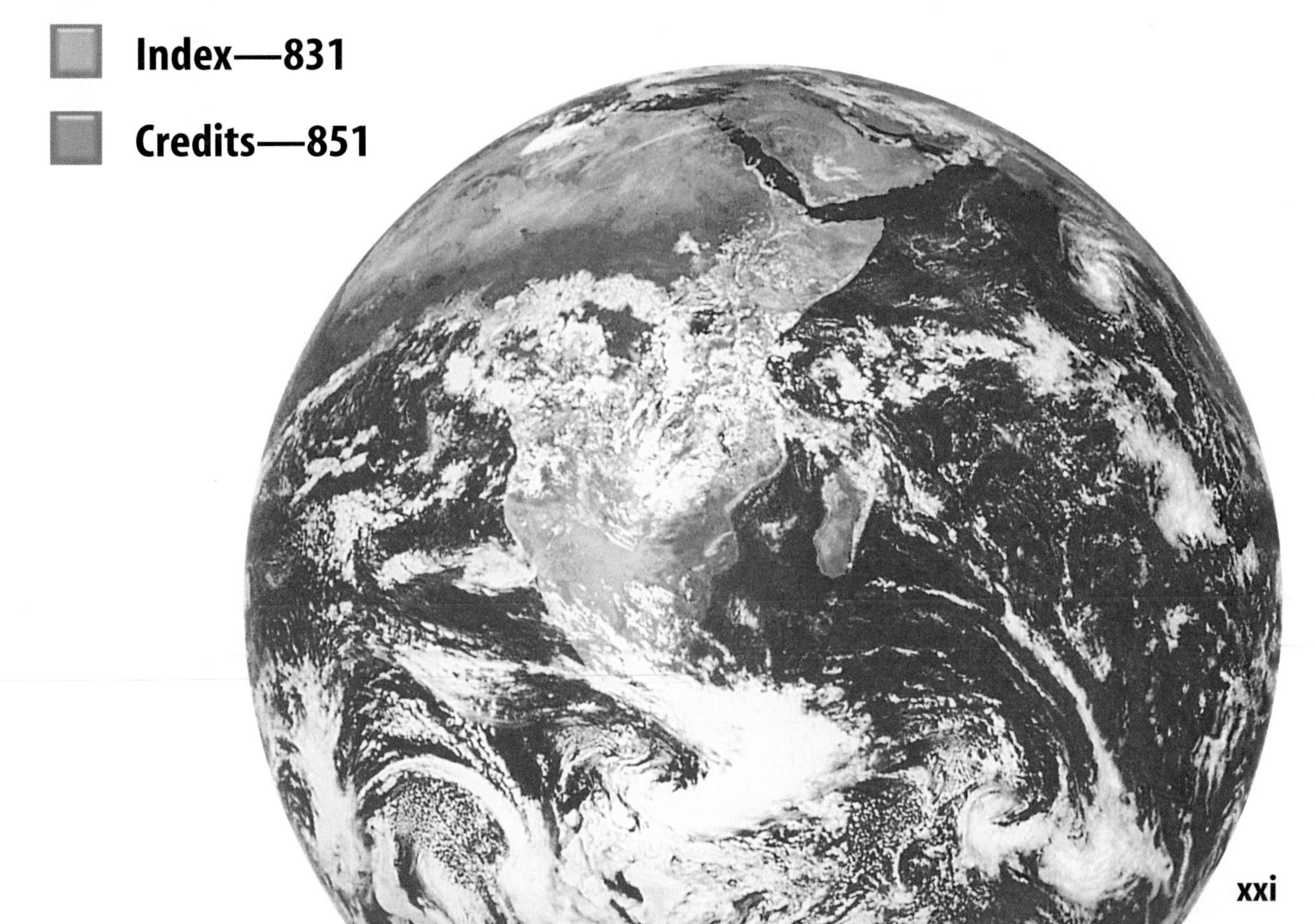

Cross-Curricular Readings

NATIONAL GEOGRAPHIC Unit Openers

NATIONAL GEOGRAPHIC VISUALIZING

TIME SCIENCE AND Society

TIME SCIENCE AND HISTORY

Oops! Accidents in SCIENCE

Science and Language Arts

SCIENCE Stats

DVD available as a video lab

Mini LAB

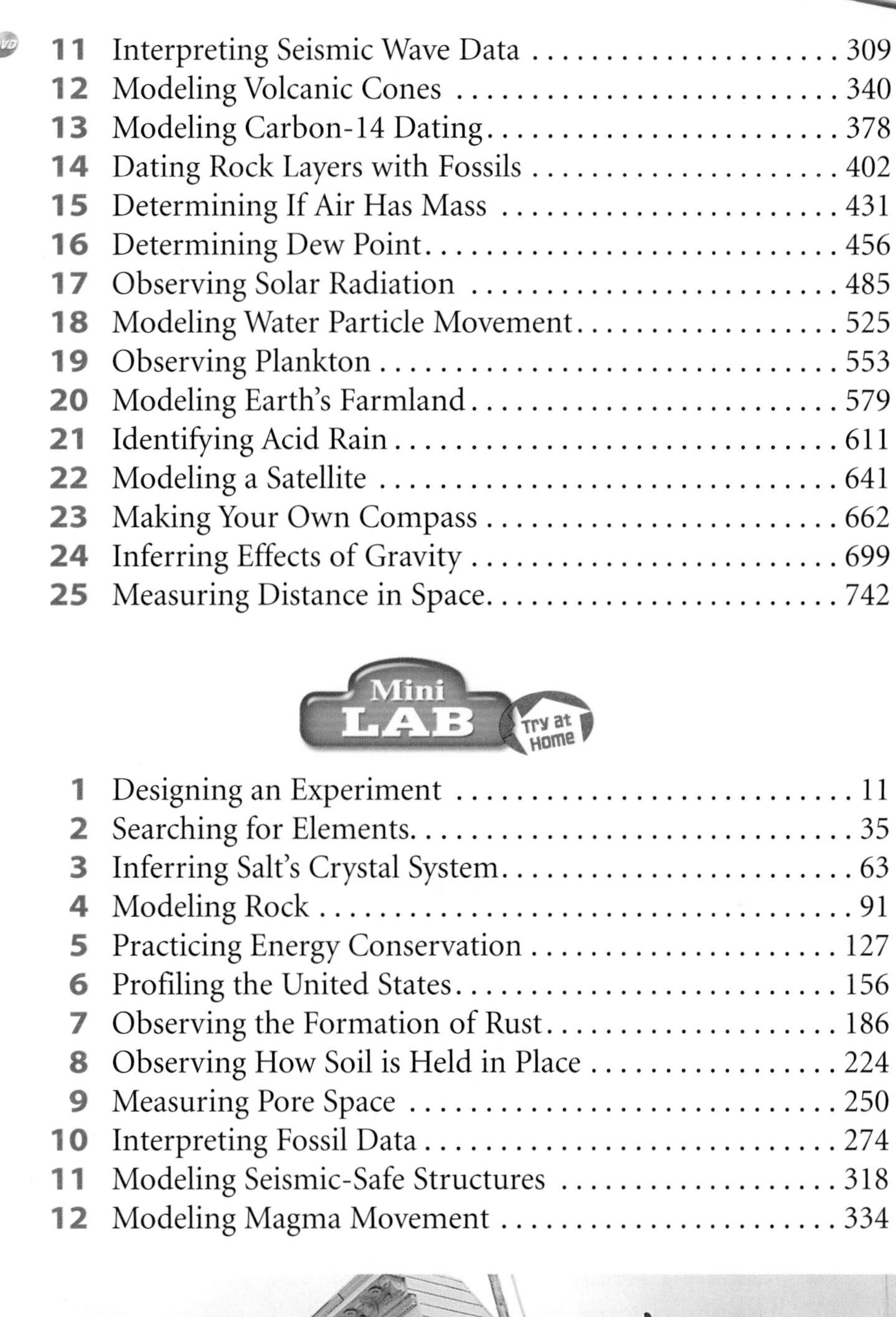

Mini LAB Try at Home

DVD available as a video lab

Content Details

One-Page Labs

Two-Page Labs

Design Your Own Labs

Model and Invent Labs

Use the Internet Labs

Share your data with other students at **earth.msscience.com/internet_lab**.

Content Details

Activities

Applying Math

Applying Science

INTEGRATE

Science Online

Standardized Test Practice

unit 1

Earth Materials

How Are Rocks & Fluorescent Lights Connected?

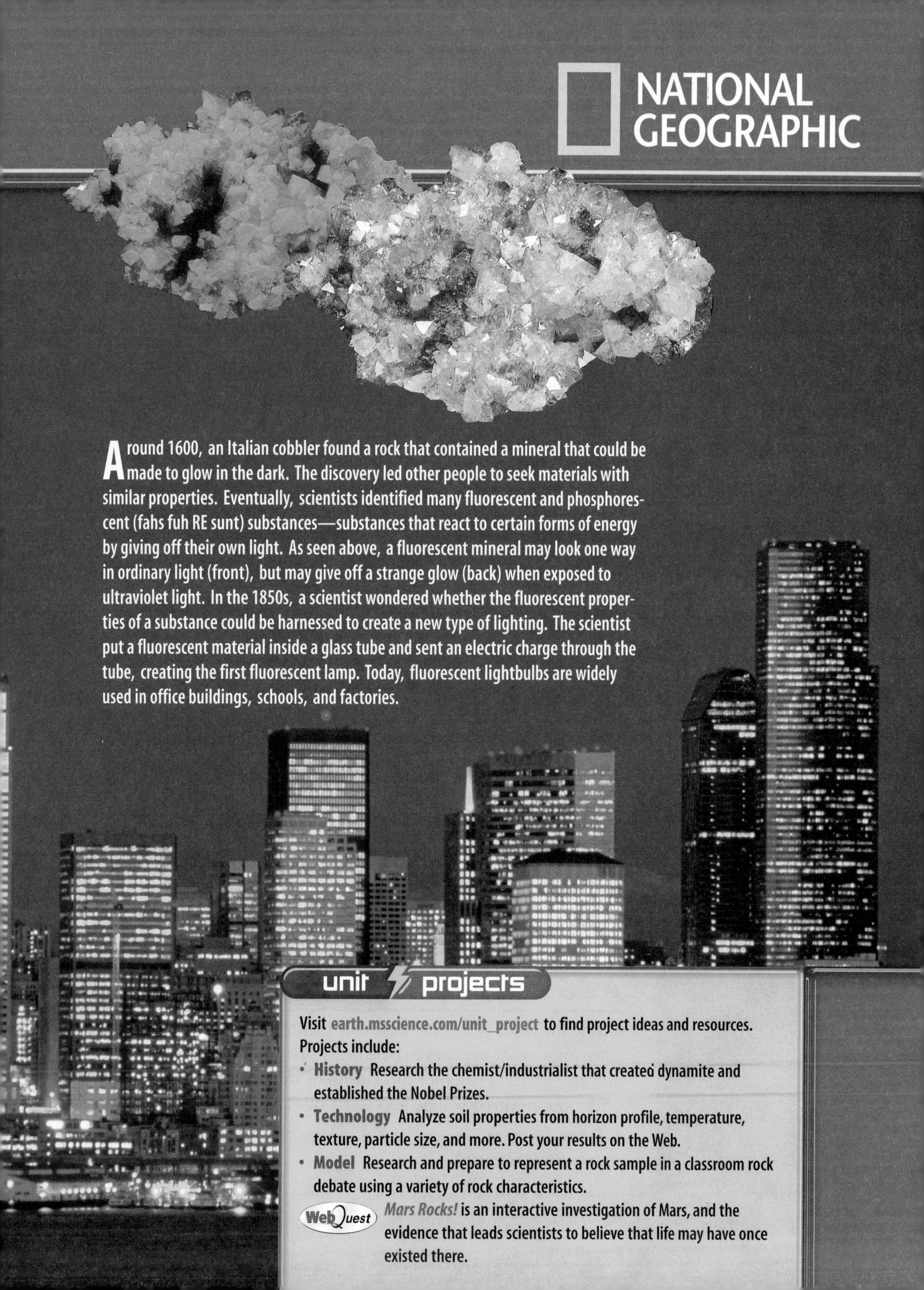

Around 1600, an Italian cobbler found a rock that contained a mineral that could be made to glow in the dark. The discovery led other people to seek materials with similar properties. Eventually, scientists identified many fluorescent and phosphorescent (fahs fuh RE sunt) substances—substances that react to certain forms of energy by giving off their own light. As seen above, a fluorescent mineral may look one way in ordinary light (front), but may give off a strange glow (back) when exposed to ultraviolet light. In the 1850s, a scientist wondered whether the fluorescent properties of a substance could be harnessed to create a new type of lighting. The scientist put a fluorescent material inside a glass tube and sent an electric charge through the tube, creating the first fluorescent lamp. Today, fluorescent lightbulbs are widely used in office buildings, schools, and factories.

unit projects

Visit earth.msscience.com/unit_project to find project ideas and resources. Projects include:

- **History** Research the chemist/industrialist that created dynamite and established the Nobel Prizes.
- **Technology** Analyze soil properties from horizon profile, temperature, texture, particle size, and more. Post your results on the Web.
- **Model** Research and prepare to represent a rock sample in a classroom rock debate using a variety of rock characteristics.

WebQuest *Mars Rocks!* is an interactive investigation of Mars, and the evidence that leads scientists to believe that life may have once existed there.

chapter 1

The Nature of Science

The BIG Idea

Science is a process of observing, studying, and thinking about things in order to gain knowledge.

SECTION 1
Science All Around
Main Idea Testing, or experimenting, is an important part of science.

SECTION 2
Scientific Enterprise
Main Idea Scientific knowledge has changed and continues to change as new advances are made.

A 66-million-year-old heart?

Inside the chest of a small dinosaur nicknamed Willo is something amazing—what appears to be a heart preserved as stone. Scientists still are debating whether this clump of stone is a preserved heart, and more research will be necessary. But that's the nature of science.

Science Journal How do you think scientists could learn more about the clump of stone that could be a heart?

Start-Up Activities

Measure in SI

Big and *small* are words people use a lot. But, the meaning of these words depends on your experiences and what you are describing. Early in human history, people developed ways to measure things. In the following lab, try some of these measuring devices.

1. Using only your hands and fingers as measuring devices, measure the length and width of the cover of this book.
2. Compare your measurements with those of other students.
3. Using a metric ruler, repeat the measurement process.
4. Again, compare your measurements with the measurements of other students in the classroom.
5. **Think Critically** Infer and describe several advantages of using standardized measuring devices.

Preview this chapter's content and activities at earth.msscience.com

Science Vocabulary Make the following Foldable to help you understand the vocabulary terms in this chapter.

STEP 1 **Fold** a vertical sheet of notebook paper from side to side.

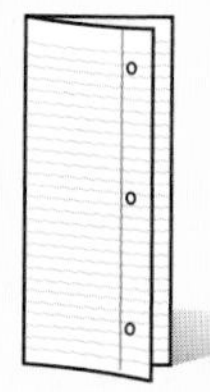

STEP 2 **Cut** along every third line of only the top layer to form tabs.

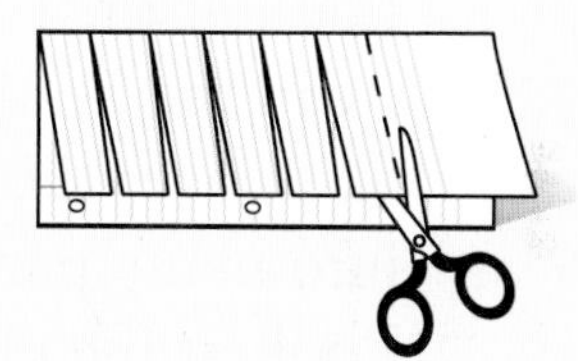

STEP 3 **Label** each tab with a vocabulary word from the chapter.

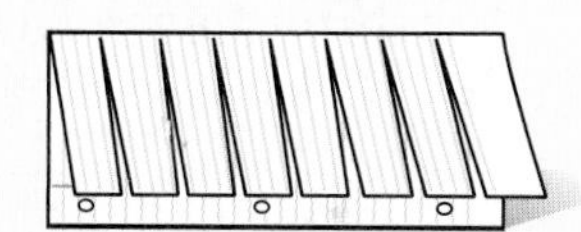

Build Vocabulary As you read the chapter, list the vocabulary words on the tabs. As you learn the definitions, write them under the tab for each vocabulary word. Exchange your Vocabulary Foldable with a classmate and quiz each other to see how many vocabulary words you can define without looking under the tabs.

Get Ready to Read

Preview

① Learn It! If you know what to expect before reading, it will be easier to understand ideas and relationships presented in the text. Follow these steps to preview your reading assignments.

1. Look at the title and any illustrations that are included.
2. Read the headings, subheadings, and anything in bold letters.
3. Skim over the passage to see how it is organized. Is it divided into many parts?
4. Look at the graphics—pictures, maps, or diagrams. Read their titles, labels, and captions.
5. Set a purpose for your reading. Are you reading to learn something new? Are you reading to find specific information?

② Practice It! Take some time to preview this chapter. Skim all the main headings and subheadings. With a partner, discuss your answers to these questions.

- Which part of this chapter looks most interesting to you?
- Are there any words in the headings that are unfamiliar to you?
- Choose one of the lesson review questions to discuss with a partner.

③ Apply It! Now that you have skimmed the chapter, write a short paragraph describing one thing you want to learn from this chapter.

Reading Tip

As you preview this chapter, be sure to scan the illustrations, tables, and graphs. Skim the captions.

Target Your Reading

Use this to focus on the main ideas as you read the chapter.

1. **Before you read** the chapter, respond to the statements below on your worksheet or on a numbered sheet of paper.
   - Write an **A** if you **agree** with the statement.
   - Write a **D** if you **disagree** with the statement.

2. **After you read** the chapter, look back to this page to see if you've changed your mind about any of the statements.
   - If any of your answers changed, explain why.
   - Change any false statements into true statements.
   - Use your revised statements as a study guide.

Before You Read A or D		Statement	After You Read A or D
	1	Science can be described as a process of observing, studying, and thinking about things.	
	2	A hypothesis can be a possible solution to a problem or a temporary assumption that explains something.	
	3	The different factors that can change, or vary, in an experiment are called variables.	
	4	Very few experiments require a control, or standard, to which results can be compared.	
	5	For an experimental result to be considered reliable, it must be confirmed by many tests.	
	6	A scientific problem requires variables that can be observed, measured, and tested.	
	7	A scientific theory is an explanation backed by results obtained from one test or experiment.	
	8	Usually, a scientific law explains why something happens in a given situation.	

Science Online
Print out a worksheet of this page at earth.msscience.com

section 1

Science All Around

as you read

What You'll Learn

- **Describe** scientific methods.
- **Define** science and Earth science.
- **Distinguish** among independent variables, dependent variables, constants, and controls.

Why It's Important

Scientific methods are used every day when you solve problems.

Review Vocabulary

analyze: to examine methodically

New Vocabulary

- hypothesis
- scientific methods
- science
- Earth science
- variable
- independent variable
- constant
- dependent variable
- control
- technology

Mysteries and Problems

Scientists are often much like detectives trying to solve a mystery. One such mystery occurred in 1996 when Japanese scientists were looking through historical records. They reported finding accounts of a tsunami that had smashed the coast of the island of Honshu on January 27, 1700. That led to the question: What had triggered these huge ocean waves?

The Search for Answers The scientists suspected that an earthquake along the coast of North America was to blame. From the coast of British Columbia to northern California is an area called the Cascadia subduction zone, shown in **Figure 1.** A subduction zone is where one section of Earth's outer, rigid layer, called a plate, is sinking beneath another plate. In areas like this, earthquakes are common. However, one problem remained. Based on the size of the tsunami, the earthquake had to have been an extremely powerful one, sending waves rolling all the way across the Pacific Ocean. That would be a much stronger earthquake than any known to have occurred in the area. Could evidence be found for such a large earthquake?

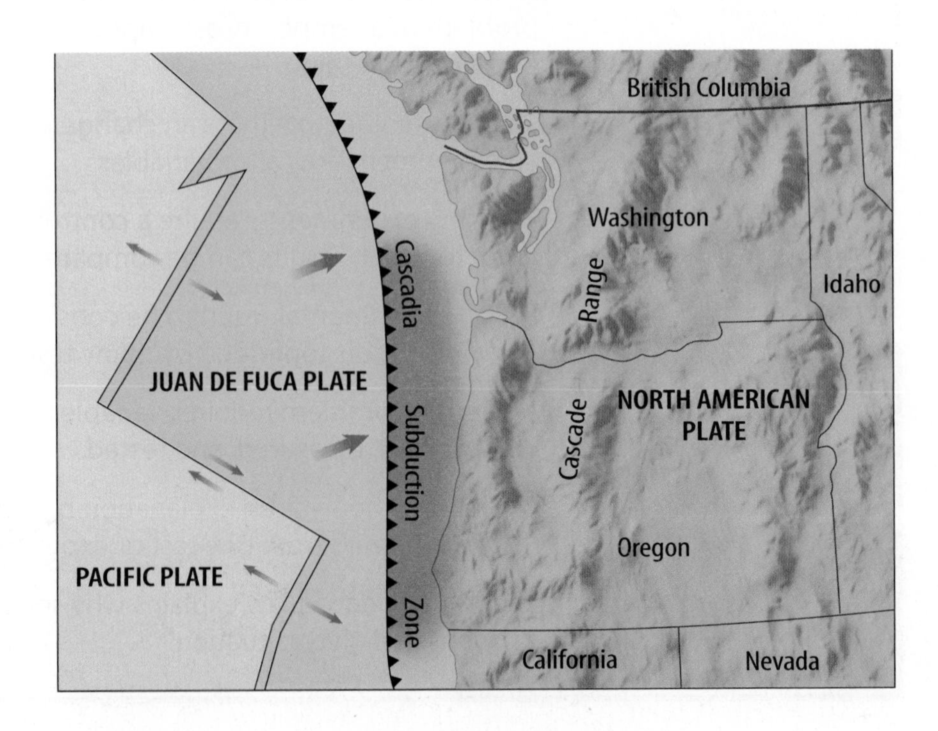

Figure 1 Along the Cascadia subduction zone, the Juan de Fuca Plate is sinking under the North American Plate.

Gathering Evidence Evidence of a large earthquake in the distant past did seem to exist along the coasts of Washington and Oregon. Much of the coast in that area had sunk, submerging coastal forests and killing thousands of trees. However, dating the earthquake to a specific year would be difficult.

Figure 2 You can see the growth rings in this tree trunk. **Determine** *How much time does each ring represent?*

A Possible Solution One scientist, whose field of study was tree rings, thought he knew how the earthquake could be dated. He made an educated guess, called a **hypothesis,** that tree rings in the drowned trees could be used to determine when the earthquake occurred.

Reading Check *What is a hypothesis?*

The hypothesis was based on what scientists know about tree growth. Each year, a living tree makes a new ring of tissue in its trunk, called an annual growth ring. You can see the annual rings in the cross section of a tree trunk shown in **Figure 2.** Two groups of scientists analyzed the rings in drowned trees along the coast, like the remains of cedar trees shown in **Figure 3.** Their data showed that the trees had died or were damaged after August 1699 but before the spring growing season of 1700. That evidence put the date of the earthquake in the same time period as the tsunami on Honshu.

Importance of Solving the Mystery In addition to solving the mystery of what caused the tsunami, the tree rings also provided a warning for people living in the Pacific Northwest. Earthquakes much stronger than any that have occurred in modern times are possible. Scientists warn that it's only a matter of time until another huge quake occurs.

Figure 3 Growth rings from these and other trees linked a huge earthquake along the coast of Washington to a tsunami in Japan that occurred more than 300 years ago.

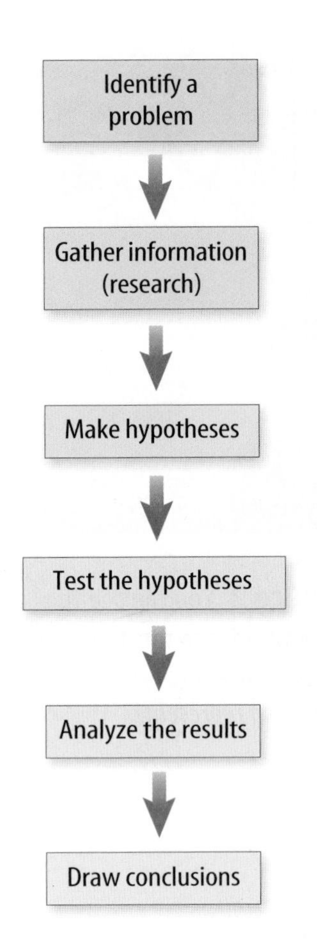

Figure 4 By using scientific methods, you can solve many problems.

Scientific Methods

When scientists try to solve a mystery like what caused the tsunami in Japan in 1700, they perform problem-solving procedures called **scientific methods.** As shown in **Figure 4,** some of the scientific methods they use include identifying a problem, gathering information (researching), developing hypotheses, testing the hypotheses, analyzing the results, and drawing conclusions. When you use methods like these, you are solving problems in a scientific way.

Science

Science means "having knowledge." **Science** is a process of observing, studying, and thinking about things in your world to gain knowledge. Many observations can't be explained easily. When people can't explain things, they ask questions. For example, you might observe that the sky appears to be blue during the day but often appears to be red at sunset and sunrise. You might ask yourself why this happens. You might visit or see a picture of Devils Postpile in California, shown in **Figure 5,** and notice that the dark rock is divided into long, thin, six-sided columns. Many fallen columns lie at the base of this mass of rock. You might wonder how and when this strange-looking rock formed. You also might wonder why rocks can be smooth or rough, shiny or dull, and can be so many different colors. Science involves trying to answer questions and solve problems to better understand the world. Every time you attempt to find out how and why things look and behave the way they do, you are performing science.

Reading Check *What is science?*

Figure 5 The columns in Devils Postpile rise between 12 m and 18 m from the valley floor. This unusual formation was created when hot lava cooled and cracked.

Figure 6 Earth science includes the study of climate, volcanoes, space, and much more.
Identify *Which Earth science topics are represented here?*

Earth Science Science is divided into different areas of study. The kind of science you will learn about this year is Earth science. **Earth science** is the study of Earth and space. Some Earth science topics include rocks, minerals, soil, volcanoes, earthquakes, maps, fossils, mountains, climates, weather, ocean water, and objects in space. Some of these topics are represented in **Figure 6.** Much of the information you'll learn about has been discovered through the ages by people who conducted scientific tests or investigations. However, many unanswered questions remain and much more is waiting to be discovered.

Reading Check *What topics do Earth scientists study?*

Working in the Lab

Testing, or experimenting, is an important part of science, and if you really want to learn from an investigation, the experiment must be carefully designed. Suppose that after listening to advertisements for several dishwashing liquids, you want to know which brand of dishwashing liquid cleans dishes the best. To find the answer, you would need to do some library or Internet research on dishwashing liquids. After researching, several thoughts might go through your mind. For example, you might hypothesize that brand X will clean dishes better than any other brand. You also might consider that there might be no difference in how well the different liquids clean.

Next, you would design an experiment that tests the validity of your hypotheses. You would need to think about which dishwashing liquids you would test, the amount of each dishwashing liquid you would use, the temperature of the water, the number of dishes you would wash, the kind and amount of grease you would put on the dishes, and the brand of paper towels you would use. All these factors can affect the outcome.

Topic: Earth Science
Visit earth.msscience.com for Web links to information about the different areas of Earth science.

Activity Prepare a collage that illustrates what you learn.

Figure 7 Wiping each dish in the same manner with a different paper towel is an important constant.
Explain *why it is necessary to have a constant in your experiment.*

Variables and Constants The different factors that can change in an experiment are **variables.** However, you want to design your experiment so you test only one variable at a time. The variable you want to test is the brand of dishwashing liquid. This is called the **independent variable**—the variable that you change. **Constants** are the variables that do not change in an experiment. Constants in this experiment would be the amount of dishwashing liquid used, the amount of water, the water temperature, the number of dishes, the kind and amount of grease applied to each dish, the brand of paper towels that were used, and the manner in which each dish was wiped. For example, you might use 20 equally greasy dishes that are identical in size, soaked in 20 L of hot water (30°C) to which 10 mL of dishwashing liquid have been added. You might rub each dish with a different dry paper towel of the same brand after it has soaked for 20 min and air dried, as the student in **Figure 7** is doing. If grease does not appear on the towel, you would consider the dish to be clean. The amount of grease on the towel is a measure of how clean each dish is and is called the dependent variable. A **dependent variable** is the variable being measured.

Soil Experiment Suppose you wanted to design an experiment to find out what kind of soil is best for growing cactus plants. What would be your variables and constants in the experiment?

Controls Many experiments also need a control. A **control** is a standard to which your results can be compared. The control in your experiment is the same number of greasy dishes, placed in 20 L of hot water except that no dishwashing liquid is added to the water. These dishes also are allowed to soak for 20 min and air dry. Then they are wiped with paper towels in the same manner as the other dishes were wiped.

Reading Check *Why is a control used in an experiment?*

Repeating Experiments For your results to be valid or reliable, your tests should be repeated many times to see whether you can confirm your original results. For example, you might design your experiment so you repeat the procedures five times for each different dishwashing liquid and control. Also, the number of samples being tested should be large. That is why 20 plates would be chosen for each test of each dishwashing liquid. The control group also would have 20 plates. By repeating an experiment five times, you can be more confident that your conclusions are accurate because your total sample for each dishwashing liquid would be 100 plates. If something in an experiment occurs just once, you can't base a scientific conclusion on it. However, if you can show that brand X cleans best in 100 trials under the same conditions, then you have a conclusion you can feel confident about.

Testing After you have decided how you will conduct an experiment, you can begin testing. During the experiment, you should observe what happens and carefully record your data in a table, like the one shown in **Figure 8.** Your final step is to draw your conclusions. You analyze your results and try to understand what they mean.

When you are making and recording observations, be sure to include any unexpected results. Many discoveries have been made when experiments produced unexpected results.

Designing an Experiment

Procedure

1. Design an experiment to test the question: *Which flashlight battery lasts the longest?*
2. In your design, be sure to include detailed steps of your experiment.
3. Identify the independent variable, constants, dependent variable, and control.

Analysis

1. List the equipment you would need to do your experiment.
2. Explain why you should repeat the experiment.

Try at Home

Figure 8 Arranging your data in a table makes the information easier to understand and analyze.

Technology

Science doesn't just add to the understanding of your natural surroundings, it also allows people to make discoveries that help others. Science makes the discoveries, and technology puts the discoveries to use. **Technology** is the use of scientific discoveries for practical purposes.

When people first picked up stones to use as tools or weapons, the age of technology had started. The discovery of fire and its ability to change clay into pottery or rocks into metals made the world you live in possible. Think back to the Launch Lab at the beginning of this chapter. Measuring devices like the metric ruler you used are examples of technology.

Everywhere you look, you can see ways that science and technology have shaped your world. Look at **Figure 9** to see how many examples of technology you can identify in each of the pictures. **Figure 10** shows a time line of some important examples of technology used in Earth science. Notice how different cultures have added to discoveries and inventions over the centuries.

Figure 9 Examples of technology are all around you.
Identify *What are some ways these examples affect your life?*

VISUALIZING THE HISTORY OF EARTH SCIENCE TECHNOLOGY

NATIONAL GEOGRAPHIC

Figure 10

For thousands of years, discoveries made by people of many cultures have advanced the study of Earth. This time line shows milestones and inventions that have shaped the development of Earth science technology and led to a greater understanding of the planet and its place in the universe.

10,000 B.C.: First pottery (Japan)

7000 B.C.: Copper metalworking (Turkey)

3500 B.C.: Bronze tools and weapons (Mesopotamia)

1000 B.C.

80 B.C.: Astronomical calendar (Greece)

A.D. 100

A.D. 132 (CHINA) This early seismograph helped detect earthquakes.

650: Windmill (Persia)

900: Terraced field for soil conservation (Peru)

1000 (NORWAY) Streamers tied through holes in Viking wind vanes indicated wind direction and strength.

Streamer holes

1090: Magnetic compass (Arabia, China)

1538: Diving bell (Spain)

1592: Thermometer (Italy)

1814 (GERMANY) A spectroscope allowed scientists to determine which elements are present in an object or substance that is giving off light.

1880: Modern seismograph (England)

1926: Liquid-fuel rocket (United States)

1943 (FRANCE) Breathing from tanks of compressed air lets divers move underwater without being tethered to an air source at the surface.

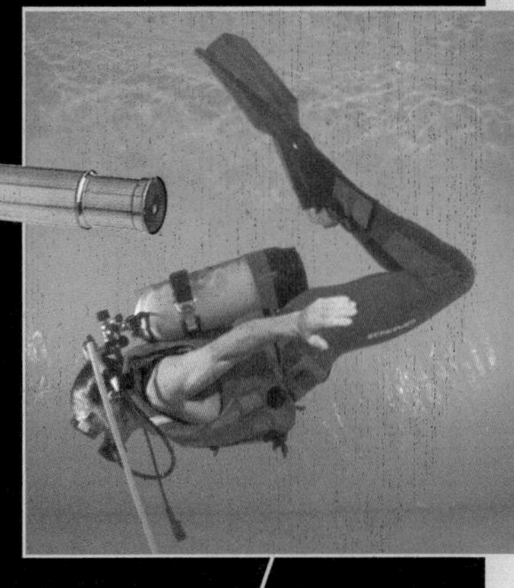

1957: Space satellite (former U.S.S.R.)

1998–2006: *INTERNATIONAL SPACE STATION* With participants from 16 countries, the *International Space Station* is helping scientists better understand Earth and beyond.

2000

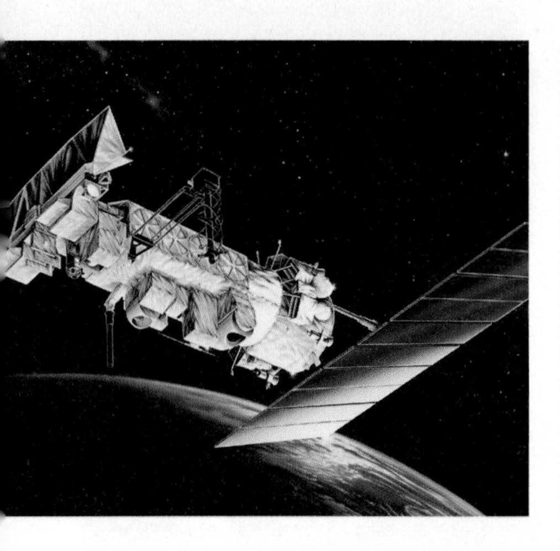

Figure 11 Weather satellites help forecasters predict future storms.
Predict *How might this same technology be used to protect endangered species?*

Using Technology Most people immediately think of complex and exotic inventions when the word *technology* is mentioned. However, the use of scientific knowledge has resulted in such common yet important things as paper, can openers, buckets, aspirin, rubber boots, locks and keys, microfiber clothing, ironing boards, bandages, and scissors. It also has resulted in robots that check underwater oil rigs for leaks and others that manufacture cars. Technology also includes calculators and computers that process information.

Transferable Technology Technology is a natural outcome of using scientific knowledge to solve problems and make people's lives easier and better. The wonderful thing about technology is that it is transferable, which means that it can be applied to new situations. For example, many types of technology that are now common were originally developed for use in outer space.

Scientists developed robotic parts, new fibers, and microminiaturized instruments for spacecraft and satellites. After these materials were developed, many were modified for use here on Earth. Technology that once was developed by the military, such as radar and sonar, has applications in the study of space, weather, Earth's structures, and medicine.

Earth scientists rely on information from weather satellites like the one in **Figure 11** to gather weather data. But biologists also use satellites to track animals. A tiny radio transmitter attached to an animal sends signals up to a satellite. The satellite then sends data on the animal's location to a ground station. Some researchers use the data to track bird migration.

section 1 review

Summary

Mysteries and Problems

- Scientists develop and test hypotheses to explain phenomena they observe.

Scientific Methods

- Scientific methods consist of a series of problem-solving procedures.

Working in the Lab

- Scientists experiment with independent and dependent variables, constants, and controls.

Technology

- Technology uses scientific discoveries for practical purposes.

Self Check

1. **Define** the term *hypothesis.* Why must it be testable?
2. **Define** the term *Earth science.*
3. **Explain** why it is important that scientists perform an experiment more than one time.
4. **Think Critically** Why is it important to use constants in an experiment?

Applying Skills

5. **Communicate** In your Science Journal, write a paragraph about how you would try to describe a modern device such as a TV, microwave oven, or computer to someone living in 1800.

Science Online earth.msscience.com/self_check_quiz

section 2

Scientific Enterprise

A Work in Progress

Throughout time, people have been both frightened by and curious about their surroundings. Storms, erupting volcanoes, comets, seasonal changes, and other natural phenomena fascinated people thousands of years ago, and they fascinate people today. As shown in **Figure 12,** early people relied on mythology to explain what they observed. They believed that mythological gods were responsible for creating storms, causing volcanoes to erupt, causing earthquakes, bringing the seasons, and making comets appear in the sky.

as you read

What You'll Learn

- **Explain** why science is always changing.
- **Compare and contrast** scientific theories and scientific laws.
- **Discuss** the limits of science.

Why It's Important

Science helps you understand the world around you.

Review Vocabulary

observation: act of using the senses to gather information

New Vocabulary

- scientific theory
- scientific law
- ethics
- bias

Recording Observations Some early civilizations went so far as to record what they saw. They developed calendars that described natural recurring phenomena. Six thousand years ago, Egyptian farmers observed that the Nile River flooded their lands every summer. Their crops had to be planted at the right time in order to make use of this water. The farmers noticed that shortly before flood time, the brightest star in the sky, Sirius, appeared at dawn in the east. The Egyptians developed a calendar based on the appearance of this star, which occurred about every 365 days.

Later, civilizations created instruments to measure with. As you saw in the Launch Lab, instruments allow for precise measurements. As instruments became better, accuracy of observations improved. While observations were being made, people tried to reason why things happened the way they did. They made inferences, or conclusions, to help explain things. Some people developed hypotheses that they tested. Their experimental conclusions allowed them to learn even more.

Figure 12 Early Scandinavian and Germanic peoples believed that a god named Thor controlled the weather. In this drawing, Thor is creating a storm. Lightning flashed whenever he threw his heavy hammer.

The History of Meteorology

Today, scientists know what they know because of all the knowledge that has been collected over time. The history of meteorology, which is the study of weather, illustrates how an understanding of one area of Earth science has developed over time.

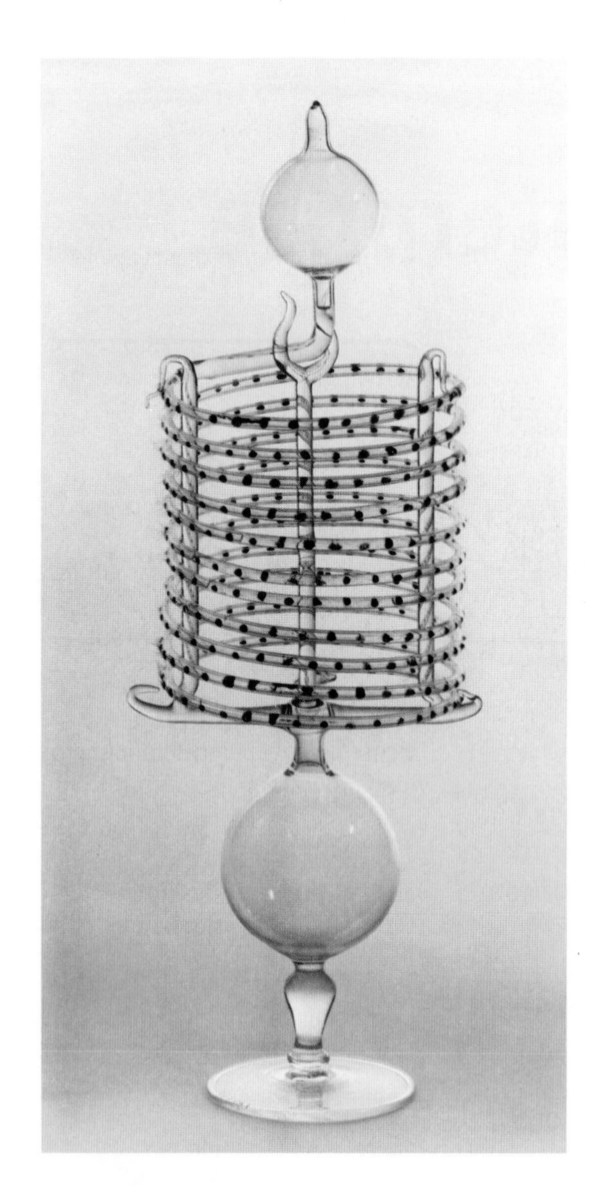

Figure 13 This photo shows a replica of a 1660 Italian alcohol thermometer.

Weather Instruments As you have read, ancient peoples believed that their gods controlled weather. However, even early civilizations observed and recorded some weather information. The rain gauge was probably the first weather instrument. The earliest reference to the use of a rain gauge to record the amount of rainfall appears in a book by the ruler of India from 321 B.C. to 296 B.C.

It wasn't until the 1600s that scientists in Italy began to use instruments extensively to study weather. These instruments included the barometer—to measure air pressure; the thermometer—to measure temperature, shown in **Figure 13;** the hygrometer—to measure water vapor in the air; and the anemometer—to measure wind speed. With these instruments, the scientists set up weather stations across Italy.

What instruments were used extensively in Italy in the 1600s to study weather?

Weather Prediction in the United States Benjamin Franklin was the first American to suggest that weather could be predicted. Franklin read accounts of storms from newspapers across the country. From these articles, Franklin concluded that severe storms generally move across the country from west to east. He also concluded that observers could monitor a storm and notify those ahead of its path that it was coming. Franklin's ideas were put to practical use shortly after the telegraph was invented in 1837.

By 1849, an organized system of weather observation sites was set up and weather reports from volunteer weather observers were sent by telegraph to the Smithsonian Institution. In 1850, Joseph Henry, secretary of the Smithsonian Institution in the United States, began drawing maps from the weather data he received. A large weather map was displayed at the Smithsonian and a weather report was sent to the *Washington Evening Post* to be published in the newspaper.

National Weather Service By the late 1800s, the United States Weather Bureau was functioning with more than 350 observing sites across the country. By 1923, weather forecasts were being carried by 140 radio stations across the United States. In 1970, the bureau's name was changed to the National Weather Service and it became part of the National Oceanic and Atmospheric Administration (NOAA).

Today's weather is forecast using orbiting satellites, weather balloons, radar, and other sophisticated technology. Each day about 60,000 reports from weather stations, ships, aircraft, and radar transmitters are gathered and filed. **Figure 14** shows instruments used to gather data at a weather station. All the information gathered is compiled into a report that is distributed to radio stations, television networks, and other news media.

Today, if you want to know about the weather anywhere in the world—at any time of day or night—you could watch a television weather channel, listen to a radio news station, or check an internet site. If you live in an area that has tornadoes, hurricanes, or other severe weather conditions, you know it is important to have weather watches and warnings available to your community.

Science Online

Topic: Weather Forecasting

Visit earth.msscience.com for Web links to information about weather forecasting.

Activity Prepare a detailed forecast for an imaginary snowstorm using information based on the research you have conducted.

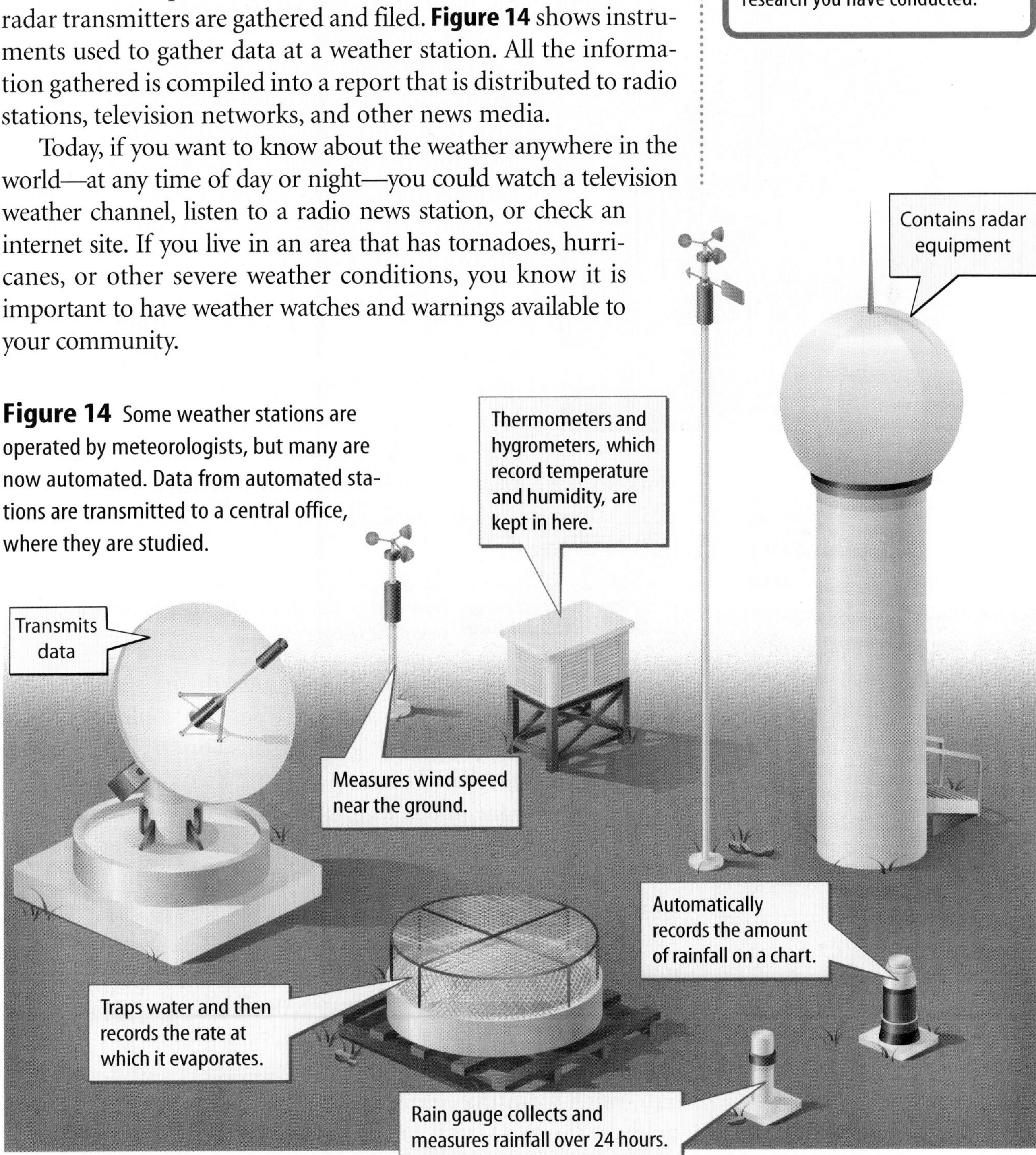
Contains radar equipment
Thermometers and hygrometers, which record temperature and humidity, are kept in here.
Transmits data
Measures wind speed near the ground.
Automatically records the amount of rainfall on a chart.
Traps water and then records the rate at which it evaporates.
Rain gauge collects and measures rainfall over 24 hours.

Figure 14 Some weather stations are operated by meteorologists, but many are now automated. Data from automated stations are transmitted to a central office, where they are studied.

Figure 15 The Global Positioning System (GPS) can pinpoint a person's location on Earth. A radio receiver gets signals from several orbiting *Navstar* satellites like this one. By comparing how far the receiver is from each satellite, the receiver's position can be determined and displayed.

Continuing Research

Scientific knowledge continues to change as scientists develop better instruments and testing procedures. As it changes, scientists have a greater understanding of nature. As you saw in **Figure 14,** scientists use a variety of technologies to study weather. Scientists have similar technologies to study Earth's interior, the oceans, environmental problems, and space. How could the technology shown in **Figure 15** be used by Earth scientists?

It is impossible to predict the types of instruments scientists will have in the future. But it is easy to predict that as research continues and instruments improve, knowledge will grow. Perhaps one day you will make a scientific breakthrough that changes people's understanding of the world.

Scientific Theories As you learned earlier, scientists test hypotheses. If data gathered over a long period of time support a hypothesis, scientists become convinced that the hypothesis is useful. They use results from many scientists' work to develop a scientific theory. A **scientific theory** is an explanation or model backed by results obtained from many tests or experiments.

Reading Check *How can a scientific hypothesis become a scientific theory?*

Examine how one hypothesis became a theory. Comets once were believed to be the forecasters of disaster. People often were terrified yet fascinated by the ghostly balls appearing in the sky. Slowly over the years, comets lost much of their mystery. However, from the 1800s until 1949, most scientists hypothesized that comets were made of many particles of different kinds of materials swarming in a cluster. Based on this hypothesis, a comet was described as a swirling cloud of dust.

In 1949, American astronomer Fred L. Whipple proposed a hypothesis that a comet was more like a dirty snowball—that the nucleus of a comet contains practically all of a comet's mass and consists of ice and dust. If a comet's orbit brings it close to the Sun, the heat vaporizes some of the ice, releasing dust and gas, which form the comet's tail. Dr. Whipple's hypothesis was published in the March 1950 *Astrophysical Journal.*

Hypothesis Supported Before it became an accepted theory, Dr. Whipple's hypothesis was subjected to many years of tests and observations. Some of the most important were the 1986 observations of Halley's comet, shown in **Figure 16.** A group of astronomers from the University of Arizona, headed by Dr. Susan Wyckoff, studied the composition of the comet. Dr. Wyckoff observed the comet many times, using giant telescopes in Arizona and Chile in South America. At other times, she studied the observations of other astronomers, including those who studied data collected by the *Giotto* and other spacecrafts. All these observations and data supported Dr. Whipple's original hypothesis. With so much support, Dr. Whipple's hypothesis has become an accepted scientific theory.

INTEGRATE Physics

Scientific Laws A **scientific law** is a rule that describes the behavior of something in nature. Usually, a scientific law describes what will happen in a given situation but doesn't explain why it happens. An example of a scientific law is Newton's first law of motion. According to this law, an object, such as a marble or a spacecraft, will continue in motion or remain at rest until it's acted upon by an outside force. According to Newton's second law of motion, when a force acts on an object, the object will change speed, direction, or both. Finally, according to Newton's third law, for every action, there is an equal and opposite reaction. This law explains how rockets that are used to launch space probes to study Halley's comet and other objects in space work. When a rocket forces burning gases out of its engines, the gases push back on the rocket with a force of equal strength and propel the rocket forward.

Observing a Scientific Law

Procedure

1. Cut one end from a **shoe box.**
2. Put the box on the floor. Place a **rubber ball** in the closed end of the box.
3. Pushing on the closed end of the box, move the box rapidly across the floor. Then suddenly stop pushing.

Analysis

1. What happened when the box stopped?
2. How does Newton's first law of motion explain this?

Figure 16 The view of Halley's comet from the *Giotto* spacecraft allowed scientists to determine the size of the icy nucleus, and that the nucleus was covered by a black crust of dust. Jets of gas blasted out from holes in the crust to form the comet's tail.

Figure 17 Ethical questions can't be solved by using scientific methods.

These animals live on the African plains.
Form an Opinion *Should they be hunted as trophies?*

Disease-carrying mosquitoes can live in this swamp.
Debate *Should swamps be drained, even if other species lose their habitat?*

Helmets reduce serious head injuries.
Think Critically *Should the government require motorcycle riders to wear helmets?*

Limits of Science

Will science always provide answers to all your questions? No, science doesn't have answers to all the questions and problems in the universe. Science is limited in what it can explain. For a question or problem to be scientifically studied, there must be variables that can be observed, measured, and tested. Problems that deal with ethics and belief systems cannot be answered using these methods. **Ethics** deals with moral values about what is good or bad. Belief systems deal with religious and/or other beliefs. Examples of ethical and belief-system questions that science cannot answer are: Do humans have more value on Earth than other life-forms?, Should the federal government regulate car emissions?, and Should animals be used in medical experiments? Look at **Figure 17.** What's your opinion?

Why can't science be used to answer ethical questions?

Science Ethics The question of whether or not to use humans in medical research studies is matter of ethics. As a class, discuss and list some pros and cons of using humans as test subjects. Explain why there is no right or wrong answer to this question.

Doing Science Right

Although ethical questions cannot be answered by science, there are ethical ways of doing science. The correct approach to doing science is to perform experiments in a way that honestly tests hypotheses and draws conclusions in an unbiased way.

Being Objective When you do scientific experiments, be sure that you design your experiments in such a way that you objectively test your hypotheses. If you don't, your **bias,** or personal opinion, can affect your observations. For example, in the 1940s, Soviet scientist Trofim Lysenko believed that individuals of the same species would not compete with one another. His ideas were based on the political beliefs held in the Soviet Union at that time. Based on his personal opinion, Lysenko ordered 300,000 tree seedlings planted in groups in a reforestation project. He believed that the trees in each group would aid one another in competing against other plant species. However, the area where the trees were planted was extremely dry, and all of the trees were competing for water and nutrients. As a result, many trees died. Lysenko's personal opinion and lack of knowledge turned out to be a costly experiment for the Soviet government.

Figure 18 These seedlings are crowded into a single pot.
Predict *How many do you think could survive?*

Suppose you wanted to grow as many plants as possible in a single flowerpot. Would you assume that all of the plants in the pot shown in **Figure 18** could survive, or would you set up an experiment to objectively test this hypothesis? Unless you test various numbers of plants in pots under the same conditions, you could not make a valid conclusion.

Applying Science

How can bias affect your observations?

Do you think bias can affect a person's observations? With the help of her classmates, Sharon performed an experiment to find out.

Identifying the Problem

Sharon showed ten friends a photograph of an uncut amethyst and asked them to rank the quality of color from 1 to 10. She then wrote the words *Prize Amethyst* on top of the photo and asked ten more friends to rank the quality of color.

Rankings Without Hint	
5	7
4	5
6	4
5	6
5	3
Average: 5.0	

Rankings With Hint	
7	8
8	9
9	8
10	8
7	9
Average: 8.3	

Solving the Problem

1. Examine the tables. Do you think the hint affected the way Sharon's classmates rated the amethyst? What effect did the hint have on them?
2. Do you think bias could affect the results of a scientific experiment? Explain. How could this bias be prevented?

Figure 19 Scientists take detailed notes of procedures and observations when they do science experiments.
Explain *why you should do the same thing.*

Being Ethical and Open People who perform science in ethical and unbiased ways keep detailed notes of their procedures, like the scientists shown in **Figure 19.** Their conclusions are based on precise measurements and tests. They communicate their discoveries by publishing their research in journals or presenting reports at scientific meetings. This allows other scientists to examine and evaluate their work. Scientific knowledge advances when people work together. Much of the science you know today has come about because of the collaboration of investigations done by many different people over many years.

The opposite of ethical behavior in science is fraud. Scientific fraud involves dishonest acts or statements. Fraud could include such things as making up data, changing the results of experiments, or taking credit for work done by others.

section 2 review

Summary

A Work in Progress

- Early people used mythology to explain what they observed.

Continuing Research

- After data are gathered over a long period of time to test a hypothesis, the information might be developed into a scientific theory.
- A scientific law is a rule that describes the behavior of something.

Limits of Science

- Science is limited to what it can explain.
- Scientists need to remain open and unbiased in their research.

Self Check

1. **Explain** why science is always changing.
2. **List** ways a hypothesis can be supported.
3. **Compare and contrast** scientific theory and scientific law.
4. **Determine** What kinds of questions can't be answered by science?
5. **Think Critically** When reading science articles, why should you look for the authors' biases?

Applying Skills

6. **Draw Conclusions** Describe what would have happened if the 1986 observations of Halley's comet had not supported Dr. Whipple's original hypothesis.

Science Online earth.msscience.com/self_check_quiz

Understanding Science Articles

Scientists conduct investigations to learn things about our world. It is important for researchers to share what they learn so other researchers can repeat and expand upon their results. One important way that scientific results are shared is by publishing them in journals and magazines.

Real-World Question

What information about Earth science and scientific methods can you learn by reading an appropriate magazine article?

Goals

- Obtain a recent magazine article concerning a research topic in Earth science.
- Identify aspects of science and scientific methods in the article.

Materials

magazine articles about Earth science topics

Procedure

1. Locate a recent magazine article about a topic in Earth science research.
2. Read the article, paying attention to details that are related to science, research, and scientific methods.
3. What branch of Earth science does the article discuss?
4. **Describe** what the article is about. Does it describe a particular event or discuss more general research?
5. Are the names of any scientists mentioned? If so, what were their roles?
6. Are particular hypotheses being tested? If so, is the research project complete or is it still continuing?

7. **Describe** how the research is conducted. What is being measured? What observations are recorded?

Conclude and Apply

1. **Explain** Are data available that do or do not support the hypotheses?
2. **Infer** What do other scientists think about the research?
3. Are references provided that tell you where you can find more information about this particular research or the more general topic? If not, what are some sources where you might locate more information?

Communicating Your Data

Prepare an oral report on the article you read. Present your report to the class. **For more help, refer to the** Science Skill Handbook.

Testing Variables of a Pendulum

Goals

- **Manipulate** variables of a pendulum.
- **Draw** conclusions from experimentation with pendulums.

Materials

string (60 cm)
metal washers (5)
watch with a second hand
metric ruler
paper clip
protractor

Safety Precautions

Real-World Question

A pendulum is an old, but accurate, timekeeping device. It works because of two natural phenomena—gravity and inertia—that are important in the study of Earth science. Gravity makes all objects fall toward Earth's surface. Inertia makes matter remain at rest or in motion unless acted upon by an external force. In the following lab, you will test some variables that might affect the swing of a pendulum. How do the length of a pendulum, the attached mass, and the angle of the release of the mass affect the swing of a pendulum?

Procedure

1. Copy the three data tables into your Science Journal.
2. Bend the paper clip into an S shape and tie it to one end of the string.
3. Hang one washer from the paper clip.

Table 1 Length of the Pendulum

Length of String (cm)	Swings Per Minute		
	Trial 1	Trial 2	Average
10			
20			
30	Do not write in this book.		
40			
50			

Table 2 Amount of Mass on the Pendulum

Units of Mass	Swings Per Minute		
	Trial 1	Trial 2	Average
1			
2			
3	Do not write in this book.		
4			
5			

Table 3 Angle of the Release of the Mass

Angle of Release	Swings Per Minute		
	Trial 1	Trial 2	Average
90°			
80°			
70°	Do not write in this book.		
60°			
50°			

4. **Measure** 10 cm of string from the washer and hold the string at that distance with one hand.
5. Use your other hand to pull back the end of the pendulum with the washer so it is parallel with the ground. Let go of the washer.
6. Count the number of complete swings the pendulum makes in 1 min. Record this number in **Table 1.**
7. Repeat steps 5 and 6 and record the number of swings in **Table 1** under "Trial 2."
8. Average the results of steps 6 and 7 and record the average swings per minute in **Table 1.**
9. Repeat steps 4 through 8, using string lengths of 20 cm, 30 cm, 40 cm, and 50 cm. Record your data in **Table 1.**
10. Copy the data with the string length of 50 cm in **Table 2.**
11. Repeat steps 5 through 8 using a 50 cm length of string and two, three, four, and five washers. Record these data in **Table 2.**
12. Use 50 cm of string and one washer for the third set of tests.
13. Use the protractor to measure a 90° drop of the mass. Repeat this procedure, calculate the average, and record the data in **Table 3.**
14. Repeat procedures 12 and 13, using angles of 80°, 70°, 60°, and 50°.

Conclude and Apply

1. **Explain** When you tested the effect of the angle of the drop of the pendulum on the swings per minute, which variables did you keep constant?
2. **Infer** which of the variables you tested affects the swing of a pendulum.
3. **Predict** Suppose you have a pendulum clock that indicates an earlier time than it really is. (This means it has too few swings per minute.) What could you do to the clock to make it keep better time?

Communicating Your Data

Graph the data from your tables. Title and label the graphs. Use different colored pencils for each graph. **Compare** your graphs with the graphs of other members of your class.

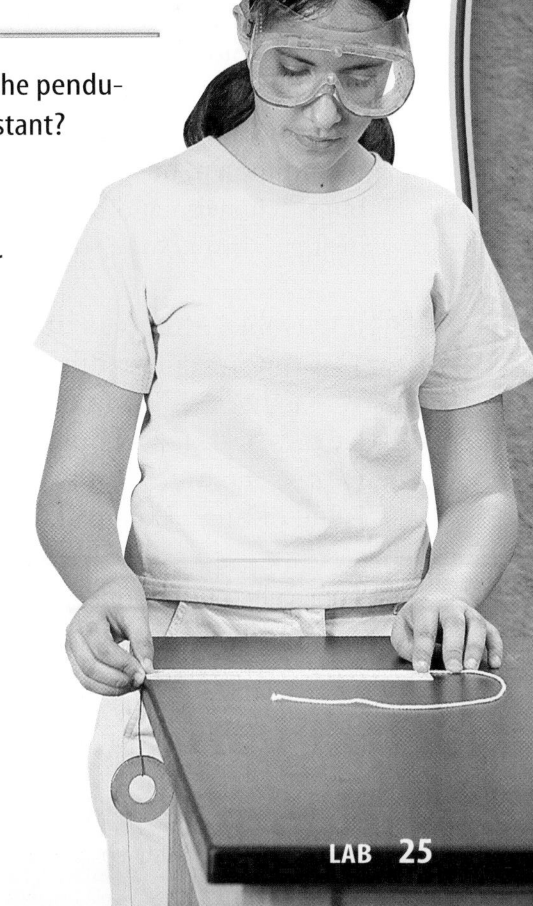

Science and Language Arts

"The Microscope"

by Maxine Kumin

Maxine Kumin

Anton Leeuwenhoek was Dutch.
He sold pincushions, cloth, and such.
The waiting townsfolk fumed and fussed
As Anton's dry goods gathered dust.

He worked, instead of tending store,
At grinding special lenses for
A microscope. Some of the things
He looked at were: mosquitoes' wings,
the hairs of sheep, the legs of lice,
the skin of people, dogs, and mice;
ox eyes, spiders' spinning gear,
fishes' scales, a little smear
of his own blood, and best of all,
the unknown, busy, very small
bugs that swim and bump and hop
inside a simple water drop.

Impossible! Most Dutchmen said.
This Anton's crazy in the head!
We ought to ship him off to Spain.
He says he's seen a housefly's brain.
He says the water that we drink
Is full of bugs. He's mad, we think!

They called him *dumkopf,* which means "dope."
That's how we got the microscope.

Understanding Literature

Rhyming Couplets A couplet is a poetic convention in which every two lines rhyme. Some of the most famous poems that use rhyming couplets describe heroic deeds and often are epic tales. An epic tale is a long story that describes a journey of exploration. Why do you think the poet used rhyming couplets in the poem you just read?

Respond to the Reading

1. Do you think Anton was a scientist by trade? Why or why not?
2. What did Anton find inside a water drop?
3. **Linking Science and Writing** Write a heroic verse or poem that rhymes. Pick a scientific method or discovery from your textbook.

Scientific instruments can increase scientific knowledge. For example, microscopes have changed the scale at which humans can make observations. Electron microscopes allow observers to obtain images at magnifications of 10,000× to 1,000,000×. In a microscope with a magnification of 10,000×, a 0.001-mm object will appear as a 1-cm image. Microscopes are used in Earth science to observe the arrangement and composition of minerals in rocks. These give clues about the conditions that formed the rock.

Reviewing Main Ideas

Section 1 Science All Around

1. Scientific methods include identifying a problem or question, gathering information, developing hypotheses, designing an experiment to test the hypotheses, performing the experiment, collecting and analyzing data, and forming conclusions.
2. Science experiments should be repeated to see whether results are consistent.
3. In an experiment, the independent variable is the variable being tested. Constants are variables that do not change. The variable being measured is the dependent variable. A control is a standard to which things can be compared.
4. Technology is the use of scientific discoveries.

Section 2 Scientific Enterprise

1. Today, everything known in science results from knowledge that has been collected over time. Science has changed and will continue to change because of continuing research and improvements in instruments and testing procedures.
2. Scientific theories are explanations or models that are supported by repeated experimentation.
3. Scientific laws are rules that describe the behavior of something in nature. They do not explain why something happens.
4. Problems that deal with ethics and belief systems cannot be answered using scientific methods.

Visualizing Main Ideas

Copy and complete the following concept map about variables and constants.

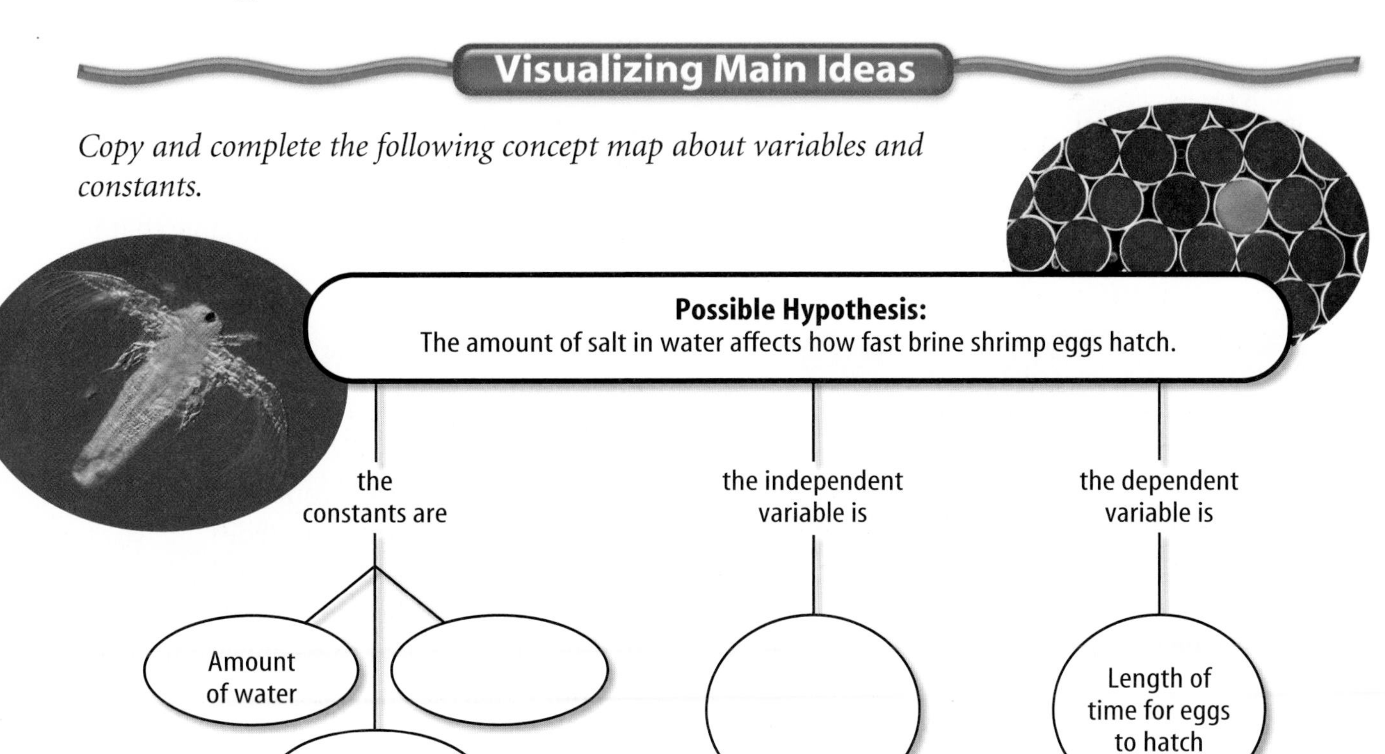

Using Vocabulary

bias p. 21	independent variable p. 10
constant p. 10	science p. 8
control p. 10	scientific law p. 19
dependent variable p. 10	scientific methods p. 8
Earth science p. 9	scientific theory p. 18
ethics p. 20	technology p. 12
hypothesis p. 7	variable p. 10

Use what you know about the vocabulary words to explain the differences between the words in the following sets. Then explain how the words are related.

1. constant—control
2. dependent variable—independent variable
3. scientific law—scientific theory
4. science—technology
5. hypothesis—scientific theory
6. science—Earth science
7. independent variable—constant
8. variable—control
9. Earth science—technology
10. ethics—bias

Checking Concepts

Choose the word or phrase that best answers the question.

11. Which word means an educated guess?
    A) theory
    B) hypothesis
    C) variable
    D) law

12. The idea that a comet is like a dirty snowball is which of the following?
    A) hypothesis **C)** law
    B) variable **D)** theory

13. Which of the following is the first step in using scientific methods?
    A) develop hypotheses
    B) make conclusions
    C) test hypotheses
    D) identify a problem

14. The statement that an object at rest will remain at rest unless acted upon by a force is an example of which of the following?
    A) hypothesis **C)** law
    B) variable **D)** theory

15. Which of the following questions could NOT be answered using scientific methods?
    A) Should lying be illegal?
    B) Does sulfur affect the growth of grass?
    C) How do waves cause erosion?
    D) Does land heat up faster than water?

16. Which of the following describes variables that stay the same in an experiment?
    A) dependent variables
    B) independent variables
    C) constants
    D) controls

17. Which of the following is a variable that is being tested in a science experiment?
    A) dependent variable
    B) independent variable
    C) constant
    D) control

18. What should you do if your data are different from what you expected?
    A) Conclude that you made a mistake in the way you collected the data.
    B) Change your data to be consistent with your expectation.
    C) Conclude that you made a mistake when you recorded your data.
    D) Conclude that your expectation might have been wrong.

Thinking Critically

19. **Recognize Cause and Effect** Suppose you had two plants—a cactus and a palm. You planted them in soil and watered them daily. After two weeks, the cactus was dead. What scientific methods could you use to find out why the cactus died?

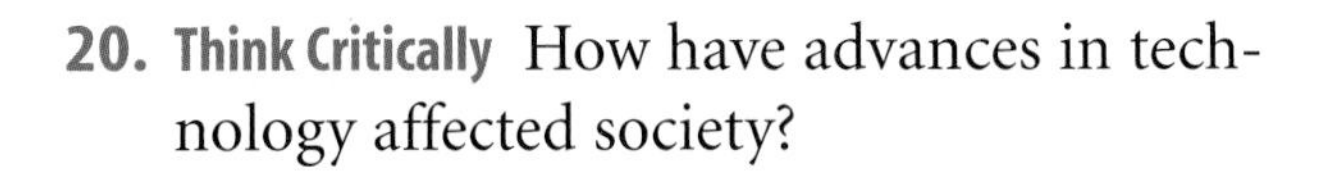

20. **Think Critically** How have advances in technology affected society?

21. **Explain** what is meant by the statement, *Technology is transferable.*

22. **Evaluate** Why don't all hypotheses become theories?

23. **Identify** some scientific methods you use every day to answer questions or solve problems?

24. **Identify and Manipulate Variables and Controls** How would you set up a simple experiment to test whether salt-crystal growth is affected by temperature?

25. **Form Hypotheses** You observe two beakers containing clear liquid and ice cubes. In the first beaker, the ice cubes are floating. In the second, the ice cubes are on the bottom of the beaker. Write a hypothesis to explain the difference in your observations about the two beakers.

26. **Recognize Cause and Effect** Explain why scientific methods cannot be used to answer ethical questions.

27. **Draw Conclusions** A laboratory tests a hypothesis through an experiment and publishes its findings that confirm the hypothesis is true. Ten other laboratories attempt to duplicate the findings, but none are able to prove the hypothesis true. Give a possible explanation why the labs' results did not agree.

Performance Activities

28. **Poster** Research an example of Earth science technology that is not shown in Figure 10. Create a poster that explains the contribution this technology made to the understanding of Earth science.

Applying Math

Use the table below to answer questions 29–30.

Color and Heat Absorption

Color	Beginning Temperature (°C)	Temperature (°C) after 10 minutes
Red	24°	26°
Black	24°	28°
Blue	24°	27°
White	24°	25°
Green	24°	27°

29. **A Color Experiment** A friend tells you that dark colors absorb more heat than light colors do. You conduct an experiment to determine which color of fabric absorbs the most heat. Analyze your data below. Was your friend correct? Explain.

30. **Variables** Identify the independent variables and the dependent variables of the experiment.

chapter 

Standardized Test Practice

Part 1 Multiple Choice

Record your answers on the answer sheet provided by your teacher or on a sheet of paper.

Use the illustration below to answer question 1.

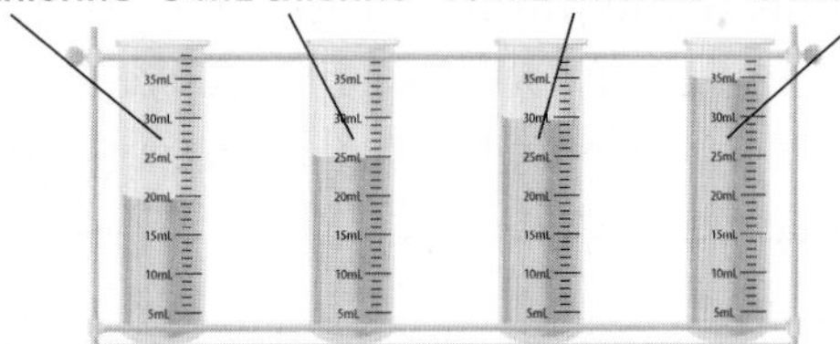

1. The test tubes were left at room temperature for a week to see if algae would grow. Which variable is being investigated?
   A. the volume of water used
   B. the temperature of the test tube's contents
   C. the amount of chlorine present
   D. the amount of algae present

2. Which of the following is the study of Earth and space?
   A. life science
   B. Earth science
   C. physical science
   D. chemical science

3. Which of these is a factor to which experimental results can be compared?
   A. independent variable
   B. dependent variable
   C. control
   D. constant

4. What is the use of scientific discoveries for practical purposes?
   A. bias
   B. scientific methods
   C. science
   D. technology

5. Which of the following is an explanation or model that is supported by many experiments and observations?
   A. hypothesis **C.** theory
   B. law **D.** estimate

6. Which is a rule that describes the behavior of something in nature?
   A. hypothesis **C.** estimate
   B. law **D.** theory

Use the illustrations below to answer questions 7–9.

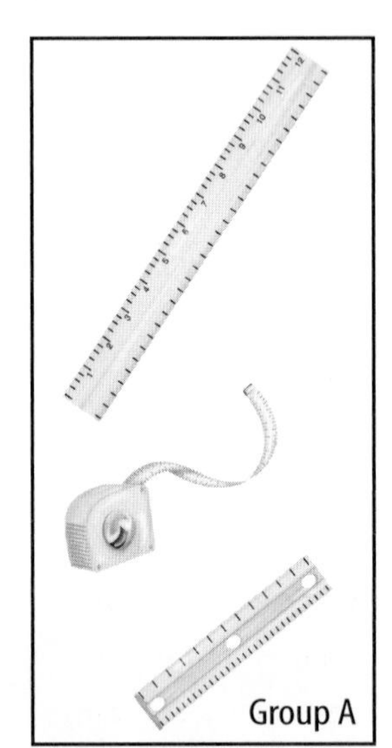

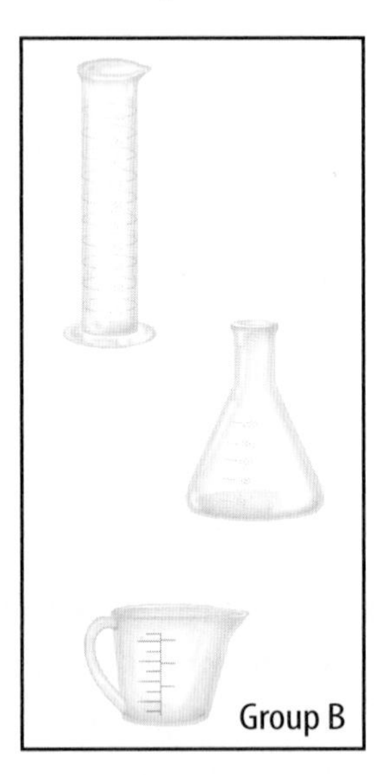

7. Which quality can be measured using the tools in group A?
   A. distance **C.** volume
   B. weight **D.** mass

8. Which quality can be measured using the tools in group B?
   A. distance **C.** volume
   B. weight **D.** mass

9. Which of the following belongs in group B above?
   A. spring scale **C.** beaker
   B. thermometer **D.** stopwatch

Part 2 | Short Response/Grid In

Record your answers on the answer sheet provided by your teacher or on a sheet of paper.

10. What's the difference between an independent variable and a dependent variable?

11. Which types of questions can be answered using scientific methods?

12. What can scientists do to ensure that they perform experiments objectively?

13. Why is it a good idea to repeat an experiment?

14. Would a scientist be convinced that his or her results were accurate after one trial? Why or why not?

Use the illustration below to answer questions 15–17.

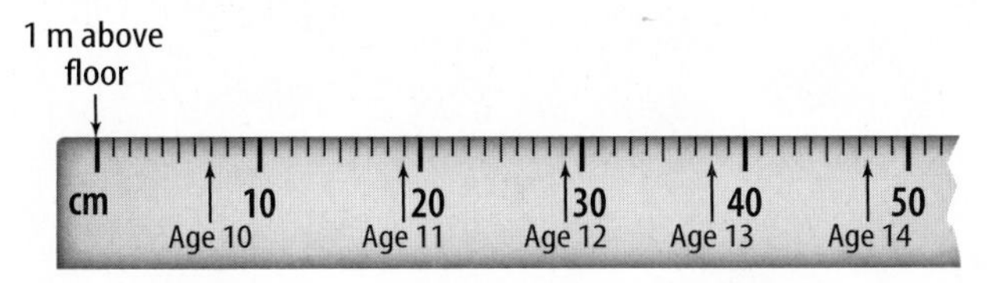

15. Alicia taped the meterstick shown above to her bedroom wall and recorded her height on her birthday for five consecutive years. The bottom of the meterstick was 1 m above the floor. How many centimeters tall was Alicia when she was 10 years old?

16. How many centimeters did she grow between her 10th birthday and her 14th birthday?

17. What was the maximum amount that Alicia grew in any one year?

18. Describe the steps of the scientific method. Give an example experiment using the steps.

19. Define the terms *scientific theory* and *scientific law*. Give an example of each.

Part 3 | Open Ended

Record your answers on a sheet of paper.

Use the graph below to answer questions 20 and 21.

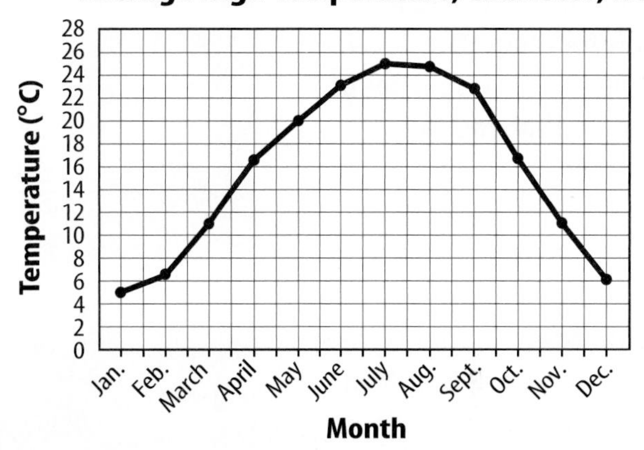

20. Describe how the average high temperature changes through the year in Charlotte, North Carolina. Which month is warmest? Which is coldest? How much is the difference?

21. How do you think the average temperature data for Charlotte, N.C. were obtained? What measurements were recorded? What calculations were performed?

22. Why is it important to take good notes and record all data when performing an experiment?

23. Why does science lead to better technology and technology lead to better science?

Test-Taking Tip

Your Experiences Remember to recall any hands-on experience as you read the question. Base your answer on the information given on the test.

Question 21 Recall the procedure from the pendulum lab. Remember how the record of your data helped you complete the lab.

chapter 2

Matter

The BIG Idea

The structures of different types of atoms and how they join together determine all the properties of matter that you can observe.

SECTION 1
Atoms

Main Idea Matter is made up of tiny particles called atoms.

SECTION 2
Combinations of Atoms

Main Idea Compounds and mixtures are formed from combinations of atoms.

SECTION 3
Properties of Matter

Main Idea The physical properties of a substance are the characteristics that can be observed without changing the substance into a new substance.

Matter, Matter Everywhere!

Everything in this scene is matter, which can exist as solid, liquid, or gas. You can see solids and liquids, but not gas. Only one thing occurs naturally on Earth in all three forms. Can you guess what it is?

Science Journal What is matter made of and how can it take such varied forms? Write what you know now in your Science Journal, and compare it with what you learn after you read the chapter.

Start-Up Activities

Change the State of Water

On Earth, water is unique because it is found as a solid, liquid, or gas. Water is invisible as a gas, but you know it is there when fog forms over a lake or a puddle of water dries up. The following lab will help you visualize how matter can change states.

1. Pour 500 mL of water into a 1,000-mL glass beaker.
2. Mark the level of water in the beaker with the bottom edge of a piece of tape.
3. Place the beaker on a hot plate.
4. With the help of an adult, heat the water until it boils for 5 min. Let the water cool.
5. With the help of an adult, compare the level of the water to the bottom edge of the tape.
6. **Think Critically** Did the amount of water in the beaker change? In your Science Journal, explain what happened to the water.

Matter Make the following Foldable to help you understand the vocabulary terms in this chapter.

STEP 1 **Fold** a vertical sheet of notebook paper from side to side.

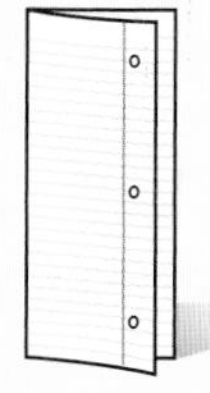

STEP 2 **Cut** along every third line of only the top layer to form tabs.

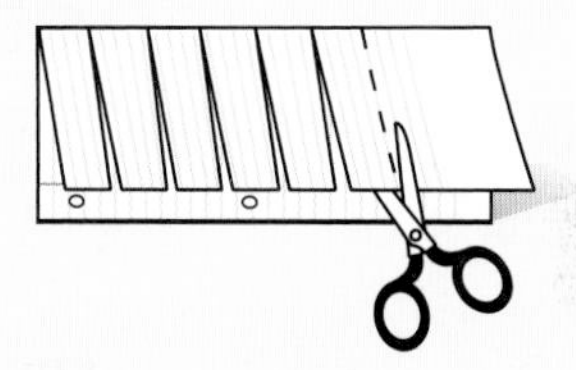

STEP 3 **Label** each tab with vocabulary words.

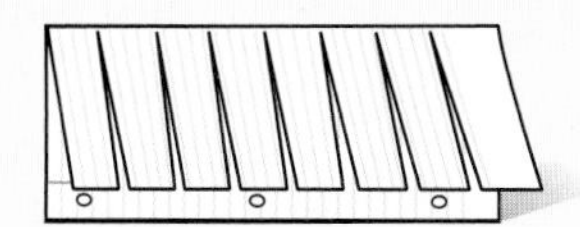

Build Vocabulary As you read the chapter, list the vocabulary words on the tabs. As you learn the definitions, write them under the tab for each vocabulary word.

Preview this chapter's content and activities at earth.msscience.com

Get Ready to Read

Identify the Main Idea

1 Learn It! Main ideas are the most important ideas in a paragraph, section, or chapter. Supporting details are facts or examples that explain the main idea. Understanding the main idea allows you to grasp the whole picture.

2 Practice It! Read the following paragraph. Draw a graphic organizer like the one below to show the main idea and supporting details.

> On Earth, matter occurs in four physical states. These four states are solid, liquid, gas, and plasma. You might have had solid toast and liquid milk or juice for breakfast this morning. You breathe air, which is a gas. A lightning bolt during a storm is an example of matter in its plasma state.
>
> *—from page 47*

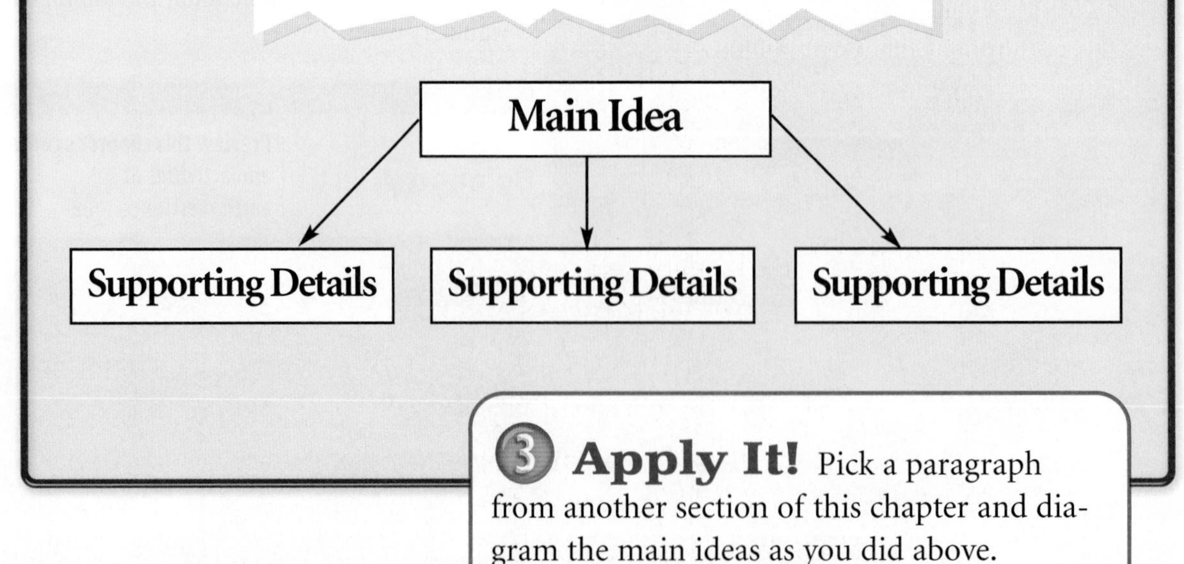

3 Apply It! Pick a paragraph from another section of this chapter and diagram the main ideas as you did above.

Target Your Reading

Use this to focus on the main ideas as you read the chapter.

1. **Before you read** the chapter, respond to the statements below on your worksheet or on a numbered sheet of paper.
   - Write an **A** if you **agree** with the statement.
   - Write a **D** if you **disagree** with the statement.

2. **After you read** the chapter, look back to this page to see if you've changed your mind about any of the statements.
   - If any of your answers changed, explain why.
   - Change any false statements into true statements.
   - Use your revised statements as a study guide.

Reading Tip

The main idea is often the first sentence in a paragraph but not always.

Before You Read A or D		Statement	After You Read A or D
	1	Matter is anything that has mass and takes up space.	
	2	Everything that exists is matter (including heat, light, and ideas).	
	3	Liquids and gases are not matter.	
	4	Electrons are atomic particles with a negative charge that exist outside the atomic nucleus.	
	5	Atoms are the same as molecules.	
	6	If the ends of a pair of magnets have opposite poles, the two ends will repel each other.	
	7	The components of a mixture can be separated by physical means.	
	8	In solids, the individual particles vibrate, but they don't switch positions with each other.	
	9	Anything that melts, freezes, or boils has water in it.	
	10	Gases are less dense than solids because the spacing between the particles is greater.	

Science Online
Print out a worksheet of this page at earth.msscience.com

section 1

Atoms

as you read

What You'll Learn

- **Identify** the states of matter.
- **Describe** the internal structure of an atom.
- **Compare** isotopes of an element.

Why It's Important

Nearly everything around you—air, water, food, and clothes—is made of atoms.

Review Vocabulary

mass: amount of matter in an object

New Vocabulary

- matter
- atom
- element
- proton
- neutron
- electron
- atomic number
- mass number
- isotope

The Building Blocks of Matter

What do the objects you see, the air you breathe, and the food you eat have in common? They are matter. **Matter** is anything that has mass and takes up space. Heat and light are not matter, because they have no mass and do not take up space. Glance around the room. If all the objects you see are matter, why do they look so different from one another?

Atoms Matter, in its various forms, surrounds you. You can't see all matter as clearly as you see water, which is a transparent liquid, or rocks, which are colorful solids. You can't see air, for example, because air is colorless gases. The forms or properties of one type of matter differ from those of another type because matter is made up of tiny particles called **atoms.** The structures of different types of atoms and how they join together determine all the properties of matter that you can observe. **Figure 1** illustrates how small objects, like atoms, can be put together in different ways.

This figure shows only two types of atoms represented by the two colors. In reality, there are over 90 types of atoms having different sizes, making great variety possible.

Figure 1 Like atoms, the same few blocks can combine in many ways.
Infer *How could this model help explain the variety of matter?*

The Structure of Matter Matter is joined together much like the blocks shown in **Figure 1.** The building blocks of matter are atoms. The types of atoms in matter and how they attach to each other give matter its properties.

Elements When atoms combine, they form many different types of matter. Your body contains several types of atoms combined in different ways. These atoms form the proteins, DNA, tissues, and other matter that make you the person you are. Most other objects that you see also are made of several different types of atoms. However, some substances are made of only one type of atom. **Elements** are substances that are made of only one type of atom and cannot be broken down into simpler substances by normal chemical or physical means.

Elements combine to make a variety of items you depend on every day. They also combine to make up the minerals that compose Earth's crust. Minerals usually are combinations of atoms that occur in nature as solid crystals and are usually found as mixtures in ores. Some minerals, however, are made up of only one element. These minerals, which include copper and silver, are called native elements. **Table 1** shows some common elements and their uses. A table of the elements, called the periodic table of the elements, is included on the inside back cover of this book.

Searching for Elements

Procedure

1. Obtain a copy of the **periodic table of the elements** and familiarize yourself with the elements.
2. Search your house for items made of various elements.
3. Use a **highlighter** to highlight the elements you discover on your copy of the periodic table.

Analysis

1. Were certain types of elements more common?
2. Infer why you did not find many of the elements.

Try at Home

Table 1 Some Common Uses of Elements

Element	Phosphorus	Silver	Copper	Carbon
Native state of the element	Phosphorus	Silver	Copper	Graphite
Uses of the element	Fertilizer	Tableware	Wire	Ski wax

Figure 2 This model airplane is a small-scale version of a large object.

Modeling the Atom

How can you study things that are too small to be seen with the unaided eye? When something is too large or too small to observe directly, models can be used. The model airplane, shown in **Figure 2,** is a small version of a larger object. A model also can describe tiny objects, such as atoms, that otherwise are difficult or impossible to see.

The History of the Atomic Model More than 2,300 years ago, the Greek philosopher Democritus (dih MAH kruh tuss) proposed that matter is composed of small particles. He called these particles atoms and said that different types of matter were composed of different types of atoms. More than 2,000 years later, John Dalton expanded on these ideas. He theorized that all atoms of an element contain the same type of atom.

Protons and Neutrons In the early 1900s, additional work led to the development of the current model of the atom, shown in **Figure 3.** Three basic particles make up an atom—protons, neutrons (NOO trahnz), and electrons. **Protons** are particles that have a positive electric charge. **Neutrons** have no electric charge. Both particles are located in the nucleus—the center of an atom. With no negative charge to balance the positive charge of the protons, the charge of the nucleus is positive.

Electrons Particles with a negative charge are called **electrons,** and they exist outside of the nucleus. In 1913, Niels Bohr, a Danish scientist, proposed that an atom's electrons travel in orbitlike paths around the nucleus. He also proposed that electrons in an atom have energy that depends on their distance from the nucleus. Electrons in paths that are closer to the nucleus have lower energy, and electrons farther from the nucleus have higher energy.

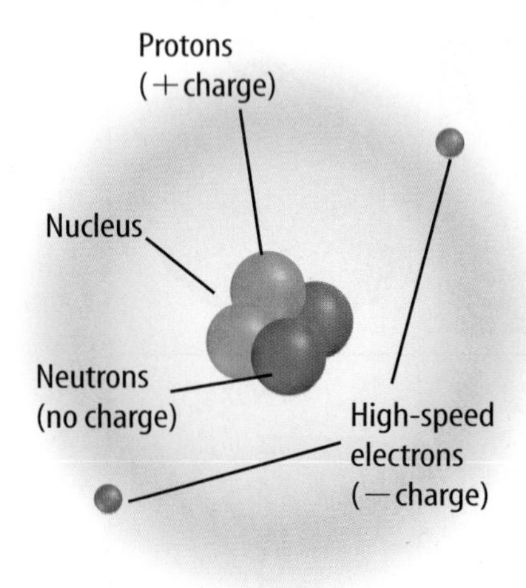

Figure 3 This model of a helium atom shows two protons and two neutrons in the nucleus, and two electrons in the electron cloud.

The Current Atomic Model Over the next several decades, research showed that electrons can be grouped into energy levels, each holding only a specific number of electrons. Also, electrons do not travel in orbitlike paths. Instead, scientists use a model that resembles a cloud surrounding the nucleus. Electrons can be anywhere within the cloud, but evidence suggests that they are located near the nucleus most of the time. To understand how this might work, imagine a beehive. The hive represents the nucleus of an atom. The bees swarming around the hive are like electrons moving around the nucleus. As they swarm, you can't predict their exact location, but they usually stay close to the hive.

Counting Atomic Particles

You now know where protons, neutrons, and electrons are located, but how many of each are in an atom? The number of protons in an atom depends on the element. All atoms of the same element have the same number of protons. For example, all iron atoms—whether in train tracks or breakfast cereal—contain 26 protons, and all atoms with 26 protons are iron atoms. The number of protons in an atom is equal to the **atomic number** of the element. This number can be found above the element symbol on the periodic table that is printed in the back of this book. Notice that as you go from left to right on the periodic table, the atomic number of the element increases by one.

Isotopes Some isotopes of elements are radioactive. Physicians can introduce these isotopes into a patient's circulatory system. The low-level radiation they emit allows the isotopes to be tracked as they move throughout the patient's body. Explain how this would be helpful in diagnosing a disease.

How many electrons? In a neutral atom, the number of protons is equal to the number of electrons. This makes the overall charge of the atom zero. Therefore, for a neutral atom:

Atomic number = number of protons = number of electrons

Atoms of an element can lose or gain electrons and still be the same element. When this happens, the atom is no longer neutral. Atoms with fewer electrons than protons have a positive charge, and atoms with more electrons than protons have a negative charge.

How many neutrons? Unlike protons, atoms of the same element can have different numbers of neutrons. The number of neutrons in an atom isn't found on the periodic table. Instead, you need to be given the atom's mass number. The **mass number** of an atom is equal to the number of protons plus the number of neutrons. The number of neutrons is determined by subtracting the atomic number from the mass number. For example, if the mass number of nitrogen is 14, subtracting its atomic number, seven, tells you that nitrogen has seven neutrons. In **Figure 4,** the number of neutrons can be determined by counting the blue spheres and the number of protons by counting orange spheres. Atoms of the same element that have different numbers of neutrons are called **isotopes. Table 2** lists useful isotopes of some elements.

Figure 4 This radioactive carbon atom is found in organic material.
Determine *this atom's mass number.*

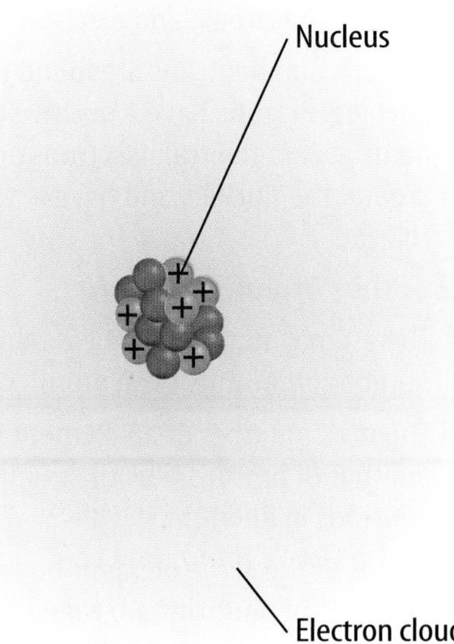

How are isotopes of the same element different?

Table 2 Some Useful Isotopes

Isotope	Number of Protons	Number of Neutrons	Number of Electrons	Atomic Number	Mass Number
Hydrogen-1	1	0	1	1	1
Hydrogen-2	1	1	1	1	2
Hydrogen-3	1	2	1	1	3
Carbon-12	6	6	6	6	12
Carbon-14	6	8	6	6	14

Uses of Isotopes Scientists have found uses for isotopes that benefit humans. For example, medical doctors use radioactive isotopes to treat certain types of cancer, such as prostate cancer. Geologists use isotopes to date some rocks and fossils.

section 1 review

Summary

The Building Blocks of Matter

- Matter has mass and takes up space.
- Matter is made of particles called atoms.
- Elements are substances that are made of only one type of atom.

Modeling the Atom

- The current model of an atom includes protons, neutrons, and electrons.
- Protons and neutrons are found in the nucleus. Protons have a positive charge and neutrons are neutral. Electrons are located around the nucleus and have a negative charge.

Counting Atomic Particles

- The atomic number of an element equals the number of protons in an atom.
- The mass number of an element equals the number of protons plus the number of neutrons in an atom.
- Atoms of the same element having different numbers of neutrons are called isotopes.

Self Check

1. **Explain** how the air you breathe fits the definition of matter.
2. **Explain** why it is helpful to have a model of an atom.
3. **Determine** the charge of an atom that has five protons and five electons.
4. **Explain** how isotopes can be used to benefit humans.
5. **Think Critically** Oxygen-16 and oxygen-17 are isotopes of oxygen. The numbers 16 and 17 represent their mass numbers, respectively. If the element oxygen has an atomic number of 8, how many protons and neutrons are in these two isotopes?

Applying Math

6. **Use Numbers** If a sodium atom has 11 protons and 12 neutrons, what is its mass number?
7. **Simple Equations** The mass number of a nitrogen atom is 14. Find its atomic number in the periodic table shown on the inside back cover of this book. Then determine the number of neutrons in its nucleus.

section 2

Combinations of Atoms

Interactions of Atoms

When you take a shower, eat your lunch, or do your homework on the computer, you probably don't think about elements. But everything you touch, eat, or use is made from them. Elements are all around you and in you.

There are about 90 naturally occurring elements on Earth. When you think about the variety of matter in the universe, you might find it difficult to believe that most of it consists of combinations of these same elements. How could so few elements produce so many different things? This happens because elements can combine in countless ways. For example, the same oxygen atoms that you breathe also might be found in many other objects, as shown in **Figure 5.** As you can see, each combination of atoms is unique. How do these combinations form and what holds them together?

as you read

What You'll Learn

- **Describe** ways atoms combine to form compounds.
- **List** differences between compounds and mixtures.

Why It's Important

On Earth, most matter exists as compounds or mixtures.

Review Vocabulary
force: a push or a pull

New Vocabulary
- compound
- molecule
- ion
- mixture
- heterogeneous mixture
- homogeneous mixture
- solution

Solid limestone has oxygen within its structure.

This canister contains pure oxygen gas.

Oxygen also is present in the juices of these apples.

Figure 5 Oxygen is a common element found in many different solids, liquids, and gases.
Infer *How can the same element, made from the same type of atoms, be found in so many different materials?*

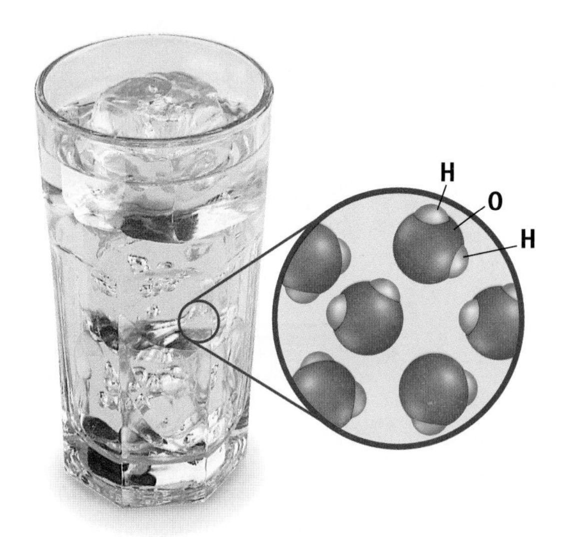

Figure 6 The water you drink is a compound consisting of hydrogen and oxygen atoms.
Identify *how hydrogen and oxygen are united to form water.*

Compounds When the atoms of more than one element combine, they form a compound. A **compound** contains atoms of more than one type of element that are chemically bonded together. Water, shown in **Figure 6,** is a compound in which two hydrogen atoms are bonded to each oxygen atom. Table salt—sodium chloride—is a compound consisting of sodium atoms bonded to chlorine atoms. Compounds are represented by chemical formulas that show the ratios and types of atoms in the compound. For example, the chemical formula for sodium chloride is NaCl. The formula for water is H_2O.

Reading Check *What atoms form the compound water?*

The properties of compounds often are very different from the properties of the elements that combine to form them. Sodium is a soft, silvery metal, and chlorine is a greenish, poisonous gas, but the compound they form is the white, crystalline table salt you use to season food. Under normal conditions on Earth, the hydrogen and oxygen that form water are gases. Water can be solid ice, liquid water, or gas. Which form do you think is most common for water at Earth's south pole?

Chemical Properties A property that describes a change that occurs when one substance reacts with another is called a chemical property. For example, one chemical property of water is that it changes to hydrogen gas and oxygen gas when an electric current passes through it. The chemical properties of a substance depend on what elements are in that substance and how they are arranged. Iron atoms in the mineral biotite will react with water and oxygen to form iron oxide, or rust, but iron mixed with chromium and nickel in stainless steel resists rusting.

Topic: Periodic Table
Visit earth.msscience.com for Web links to information about the periodic table and chemical bonding.

Activity Research five elements that you are unfamiliar with and make a table showing their names, atomic number, properties, and how they are used.

Bonding

The forces that hold the atoms together in compounds are called chemical bonds. These bonds form when atoms share or exchange electrons. However, only those electrons having the highest energies in the electron cloud can form bonds. As you read in the last section, these are found farthest from the nucleus. An atom can have only eight electrons in this highest energy level. If more electrons exist, they must form a new, higher energy level. If an atom has exactly eight electrons in its outermost level, it is unlikely to form bonds. If an atom has fewer than eight electrons in its outermost level, it is unstable and is more likely to combine with other atoms.

Covalent Bonds Atoms can combine to form compounds in two different ways. One way is by sharing the electrons in their outermost energy levels. The type of bond that forms by sharing outer electrons is a covalent bond. A group of atoms connected by covalent bonds is called a **molecule.** For example, two atoms of hydrogen can share electrons with one atom of oxygen to form a molecule of water, as shown in **Figure 7.** Each of the hydrogen atoms has one electron in its outermost level, and the oxygen has six electrons in its outermost level. This arrangement causes hydrogen and oxygen atoms to bond together. Each of the hydrogen atoms becomes stable by sharing one electron with the oxygen atom, and the oxygen atom becomes stable by sharing two electrons with the two hydrogen atoms.

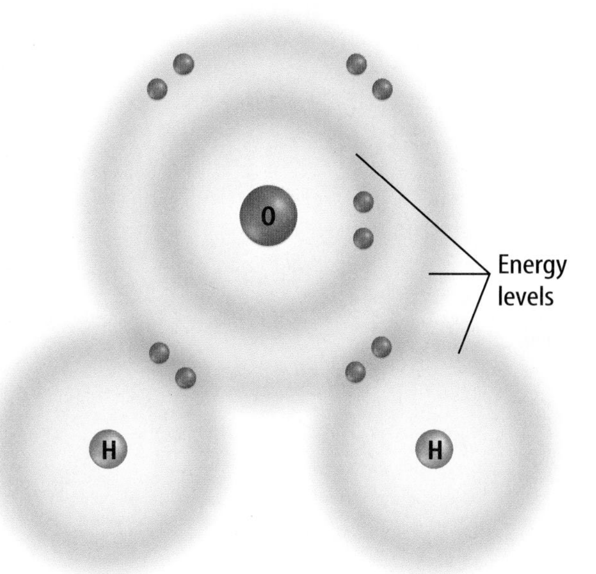

Figure 7 A molecule of water consists of two atoms of hydrogen that share outer electrons with one atom of oxygen.

Ionic Bonds In addition to sharing electrons, atoms also combine if they become positively or negatively charged. This type of bond is called an ionic bond. Atoms can be neutral, or under certain conditions, atoms can lose or gain electrons. When an atom loses electrons, it has more protons than electrons, so the atom is positively charged. When an atom gains electrons, it has more electrons than protons, so the atom is negatively charged. Electrically charged atoms are called **ions.**

Ions are attracted to each other when they have opposite charges. This is similar to the way magnets behave. If the ends of a pair of magnets have the same type of pole, they repel each other. Conversely, if the ends have opposite poles, they attract one another. Ions form electrically neutral compounds when they join. The mineral halite, commonly used as table salt, forms in this way. A sodium (Na) atom loses an outer electron and becomes a positively charged ion. As shown in **Figure 8,** if the sodium ion comes close to a negatively charged chlorine (Cl) ion, they attract each other and form the salt you use on french fries or popcorn.

Figure 8 Table salt forms when a sodium ion and a chlorine ion are attracted to one another.
Draw Conclusions *What kind of bond holds ions together?*

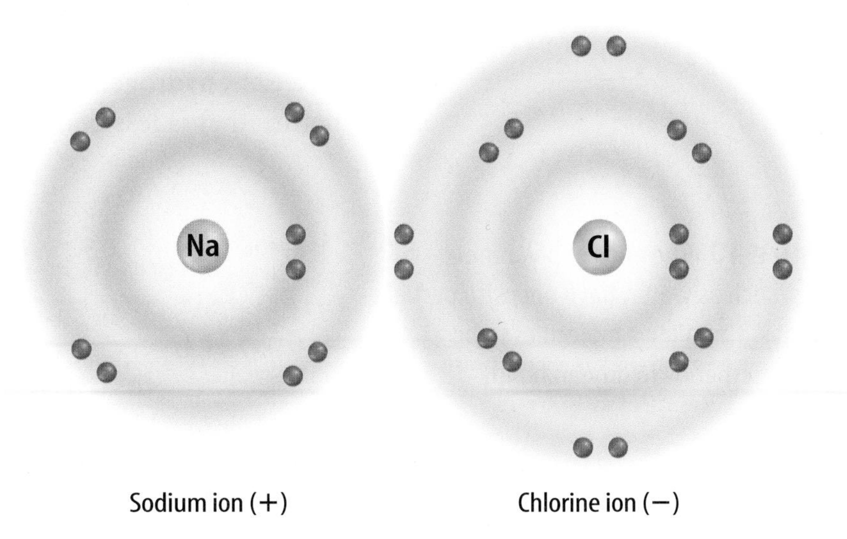

Figure 9 This machine is making cable from spools of wire. Electrons move freely along this wire, passing from one copper ion to another.
Identify *What type of bond holds copper atoms together?*

Metallic Bonds Metallic bonds are found in metals such as copper, gold, aluminum, and silver. In this type of bond, electrons are free to move from one positively charged ion to another. This free movement of electrons is responsible for key characteristics of metals. The movement of electrons, or conductivity, allows metals like copper, shown in **Figure 9,** to pass an electric current easily.

Hydrogen Bonds Some bonds, called hydrogen bonds, can form without the interactions of electrons. The arrangement of hydrogen and oxygen atoms in water molecules causes them to be polar molecules. A polar molecule has a positive end and a negative end. This happens because the atoms do not share electrons equally. When hydrogen and oxygen atoms form a molecule with covalent bonds, the hydrogen atoms produce an area of partial positive charge and the oxygen atom produces an area of partial negative charge. The positive end of one molecule is attracted to the negative end of another molecule, as shown in **Figure 10,** and a weak hydrogen bond is formed. The different parts of the water molecule are slightly charged, but as a whole, the molecule has no charge. This type of bond is easily broken, indicating that the charges are weak.

Hydrogen bonds are responsible for several properties of water, some of which are unique. Cohesion is the attraction between water molecules that allows them to form raindrops and to form beads on flat surfaces. Hydrogen bonds cause water to exist as a liquid, rather than a gas, at room temperature. As water freezes, hydrogen bonds force water molecules apart, into a structure that is less dense than liquid water.

Figure 10 The ends of polar molecules, such as water, have opposite partial charges. This allows molecules to be held together by hydrogen bonds.

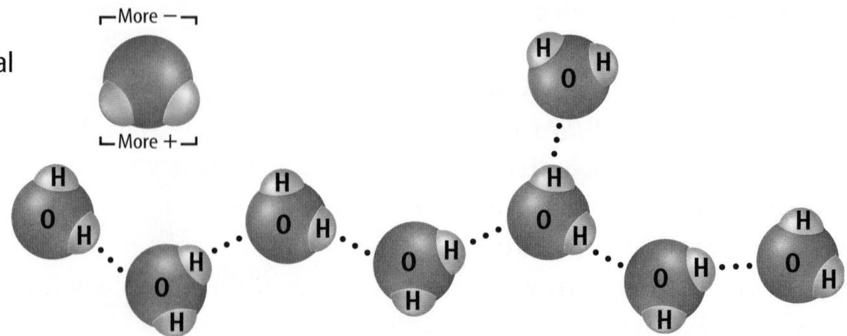

Figure 11 This rock contains a variety of mineral compounds that, together, form a mixture.

Mixtures

Sometimes compounds and elements mix together but do not combine chemically. A **mixture** is composed of two or more substances that are not chemically combined. There are two different types of mixtures—heterogeneous and homogeneous. The components of a **heterogeneous mixture** are not mixed evenly and each component retains its own properties. Maybe you've seen a rock like the one in **Figure 11.** Several different minerals are mixed together, but if you were to examine the minerals separately, you would find that they have the same properties and appearance as they have in the rock.

The components of a **homogeneous mixture** are evenly mixed throughout. You can't see the individual components. Another name for a homogeneous mixture is a **solution.** The properties of the components of this type of mixture often are different from the properties of the mixture. Ocean water is an example of a liquid solution that consists of salts mixed with liquid water.

Reading Check *What is a solution?*

Separating Mixtures

The components of a mixture can be separated by physical means. For example, you can sit at your desk and pick out the separate items in your backpack, or you can let the water evaporate from a saltwater mixture and the salt will remain.

Separating the components of a mixture is a relatively easy task compared to separating those of a compound. The substances in a compound cannot be separated by physical means. An existing compound can be changed to one or more new substances by chemically breaking down the original compound. For example, a drop of dilute hydrochloric acid (HCl) can be placed on calcium carbonate ($CaCO_3$) and carbon dioxide (CO_2) is released. Can you think of an example of a substance that could be chemically broken down?

Classifying Forms of Matter

Procedure

1. Make a chart with the columns *Mixtures, Compounds,* and *Elements.*
2. Classify each of these items into the proper column on your chart: **air, sand, hydrogen, muddy water, sugar, ice, sugar water, water, salt, oxygen, copper.**
3. Make a solution using two or more of the items listed above.

Analysis

1. How does a solution differ from other types of mixtures?
2. How does an element differ from a compound?

Figure 12 The ocean is a mixture of many different forms of matter. The ocean water itself is a solution, a homogeneous mixture.

Exploring Matter

Air, sweetened tea, salt water, and the contents of your backpack are examples of mixtures. The combination of rocks, fish, and coral shown in **Figure 12** also is a mixture. In each case, the materials within the mixture are not chemically combined. The individual components are made of compounds or elements. The atoms in these compounds lost their individual properties when they combined.

Seashells and coral reefs contain calcium carbonate, which has the formula $CaCO^3$. Properties of $CaCO^3$ differ greatly from those of its elements, calcium, carbon, and oxygen. For example, calcium is a soft, silvery metal, oxygen is a gas, and carbon can be a black solid. In contrast, calcium carbonate is hard and white. For example, it also is found in limestone and marble.

section 2 review

Summary

Interactions of Atoms

- A compound contains atoms of more than one type of element that are chemically bonded.

Bonding

- Atoms share electrons in covalent bonds.
- Atoms lose or gain electrons in ionic bonds.
- In metallic bonds electrons move freely from one metal ion to another.
- Hydrogen bonds can form between polar molecules.

Mixtures

- A combination of two or more substances that are not chemically combined is a mixture.
- Components of a heterogeneous mixture are not mixed evenly.
- Components of a homogeneous mixture or solution are evenly mixed.

Self Check

1. **Explain** how atoms or ions combine to form compounds.
2. **Classify** sweetened tea as a solution or a compound.
3. **Infer** Why do metals transmit electricity so well?
4. **Identify** What does the formula tell you about a chemical compound?
5. **Think Critically** How can you determine whether salt water is a solution or a compound?

Applying Skills

6. **Infer** You have seen how the Na^+ ion attracts the Cl^- ion forming the compound sodium chloride, NaCl. What compound would form from Ca^{+2} and Cl^-?
7. **Design** How would you separate a mixture of sugar and sand? Devise an experiment to do this. Discuss your procedure with your teacher. Perform the experiment and write the results.

Scales of Measurement

How would you describe some of the objects in your classroom? Perhaps your desktop is about one-half the size of a door. Measuring physical properties in a laboratory experiment will help you make better observations.

Real-World Question

How are physical properties of objects measured?

Goals

- **Measure** various physical properties in SI.
- **Determine** sources of error.

Materials

triple beam balance
100-mL graduated cylinder
metersticks (2)
non-mercury thermometers (3)
stick or dowel
rock sample
string
globe
water

Safety Precautions

WARNING: *Never "shake down" lab thermometers.*

Procedure

1. Go to every station and determine the measurement requested. Record your observations in a data table and list sources of error.
   - **a.** Use a balance to determine the mass, to the nearest 0.1 g, of the rock sample.
   - **b.** Use a graduated cylinder to measure the water volume to the nearest 0.5 mL.
   - **c.** Use three thermometers to determine the average temperature, to the nearest 0.5°C, at a selected location in the room.
   - **d.** Use a meterstick to measure the length, to the nearest 0.1 cm, of the stick or dowel.
   - **e.** Use a meterstick and string to measure the circumference of the globe. Be accurate to the nearest 0.1 cm.

Measurement and Error

Sample at Station	Value of Measurement	Causes of Error
a.	mass = ___ g	Do not
b.	volume = ___ mL	write
c. (location)	average temp. = ___ °C	in this
d.	length= ___ cm	book.
e.	circumference= ___ cm	

Conclude and Apply

1. **Compare** your results with those provided by your teacher.
2. **Calculate** your percentage of error in each case. Use this formula.

$$\%\text{ error} = \frac{\text{your val.} - \text{teacher's val.}}{\text{teacher's val.}} \times 100$$

3. Being within five to seven percent of the correct value is considered good. If your error exceeds ten percent, what could you do to improve your results and reduce error?

Communicating Your Data

Compare your conclusions with those of other students in your class. **For more help, refer to the** Science Skill Handbook.

section 3

Properties of Matter

as you read

What You'll Learn

- **Describe** the physical properties of matter.
- **Identify** what causes matter to change state.
- **List** the four states of matter.

Why It's Important

You can recognize many substances by their physical properties.

Review Vocabulary
energy: the ability to cause change

New Vocabulary
- density

Physical Properties of Matter

In addition to the chemical properties of matter that you have already investigated in this chapter, matter also has other properties that can be described. You might describe a pair of blue jeans as soft, blue, and about 80 cm long. A sandwich could have two slices of bread, lettuce, tomato, cheese, and turkey. These descriptions can be made without altering the sandwich or the blue jeans in any way. The properties that you can observe without changing a substance into a new substance are physical properties.

One physical property that you will use to describe matter is density. **Density** is a measure of the mass of an object divided by its volume. Generally, this measurement is given in grams per cubic centimeter (g/cm^3). For example, the average density of liquid water is about 1 g/cm^3. So 1 cm^3 of pure water has a mass of about 1 g.

An object that's more dense than water will sink in water. On the other hand, an object that's not as dense as water will float in water. When oil spills occur on the ocean, as shown in **Figure 13,** the oil floats on the surface of the water and washes up on beaches. Because the oil floats, even a small spill can spread out and cover large areas.

Figure 13 Oil spills on the ocean spread across the surface of the water.
Infer *How does the density of oil compare to the density of water?*

States of Matter

On Earth, matter occurs in four physical states. These four states are solid, liquid, gas, and plasma. You might have had solid toast and liquid milk or juice for breakfast this morning. You breathe air, which is a gas. A lightning bolt during a storm is an example of matter in its plasma state. What are the differences among these four states of matter?

Solids The reason some matter is solid is that its particles are in fixed positions relative to each other. The individual particles vibrate, but they don't switch positions with each other. Solids have a definite shape and take up a definite volume.

Suppose you have a puzzle that is completely assembled. The pieces are connected so one piece cannot switch positions with another piece. However, the pieces can move a little, but stay attached to one another. The puzzle pieces in this model represent particles of a substance in a solid state. Such particles are strongly attracted to each other and resist being separated.

Applying Math — Solve One-Step Equations

CALCULATING DENSITY You want to find the density of a small cube of an unknown material. It measures 1 cm × 1 cm × 2 cm. It has a mass of 8 g.

Solution

1. *This is what you know:*
   - mass: $m = 8$ g
   - volume: $v = 1\text{ cm} \times 1\text{ cm} \times 2\text{ cm} = 2\text{ cm}^3$
   - $d = m/v$
2. *This is what you need to find out:* density: d
3. *This is the procedure you need to use:*
   - substitute: $d = 8\text{ g}/2\text{ cm}^3$
   - Divide to solve for d: $d = 4\text{ g/cm}^3$
   - The density is 4 g/cm^3
4. *Check your answer:* Multiply by the volume. You should get the given mass.

Practice Problems

1. You discover a gold bar while exploring an old shipwreck. It measures 10 cm × 5 cm × 2 cm. It has a mass of 1,930 g. Find the density of gold.
2. A bar of soap measures 8 cm × 5 cm × 2 cm. Its mass is 90 g. Calculate its density. Predict whether this soap will float.

Science Online — For more practice, visit earth.msscience.com/math_practice

Topic: States of Matter
Visit earth.msscience.com for Web links to information about the four states of matter.

Activity Search for information on a family of elements called the halogens. Use the information you find to write a "Wanted Poster" on one halogen. Include items such as physical description, distinguishing characteristics, and known associates.

Liquids Particles in a liquid are attracted to each other, but are not in fixed positions as they are in the solid shown in **Figure 15.** This is because liquid particles have more energy than solid particles. This energy allows them to move around and change positions with each other.

When you eat breakfast, you might have several liquids at the table such as syrup, juice, and milk. These are substances in the liquid state, even though one flows more freely than the others at room temperature. The particles in a liquid can change positions to fit the shape of the container they are held in. You can pour any liquid into any container, and it will flow until it matches the shape of its new container.

Gases The particles that make up gases have enough energy to overcome any attractions between them. This allows them to move freely and independently. Unlike liquids and solids, gases spread out and fill the container in which they are placed. Air fresheners work in a similar way. If an air freshener is placed in a corner, it isn't long before the particles from the air freshener have spread throughout the room. Look at the hot-air balloon shown in **Figure 15C.** The particles in the balloon are evenly spaced throughout the balloon. The balloon floats in the sky, because the hot air inside the balloon is less dense than the colder air around it.

Reading Check *How do air fresheners work?*

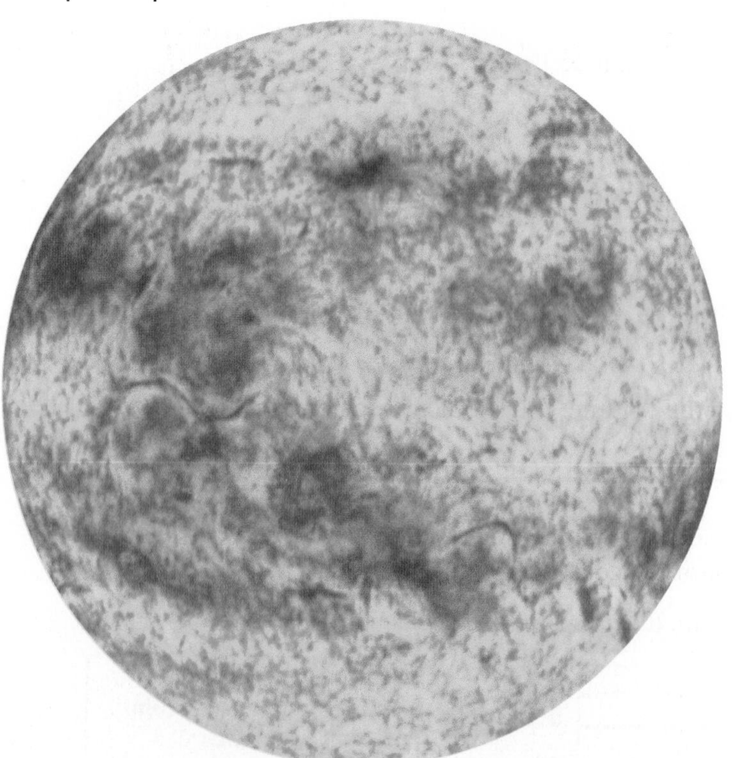

Figure 14 The Sun is an example of a plasma.

Plasma Although it is probably unfamiliar to most people, plasma is the most common state of matter in the universe. This state is associated with high temperatures. Can you name something that is in the plasma state? Stars like the Sun, shown in **Figure 14,** are composed of matter in the plasma state. Plasma also exists in Jupiter's magnetic field. On Earth, plasma is found in lightning bolts, as shown in **Figure 15D.** Plasma is composed of ions and electrons. It forms when high temperatures cause some of the electrons normally found in an atom's electron cloud to escape and move outside of the electron cloud.

NATIONAL GEOGRAPHIC VISUALIZING STATES OF MATTER

Figure 15

Matter on Earth exists naturally in four different states—solid, liquid, gas, and plasma—as shown here. The state of a sample of matter depends upon the amount of energy its atoms or molecules possess. The more energy that matter contains, the more freely its atoms or molecules move, because they are able to overcome the attractive forces that tend to hold them together.

D PLASMA Electrically charged particles in lightning are free moving.

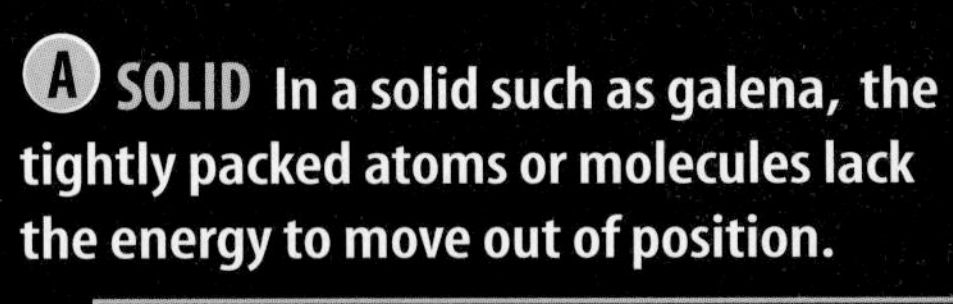

A SOLID In a solid such as galena, the tightly packed atoms or molecules lack the energy to move out of position.

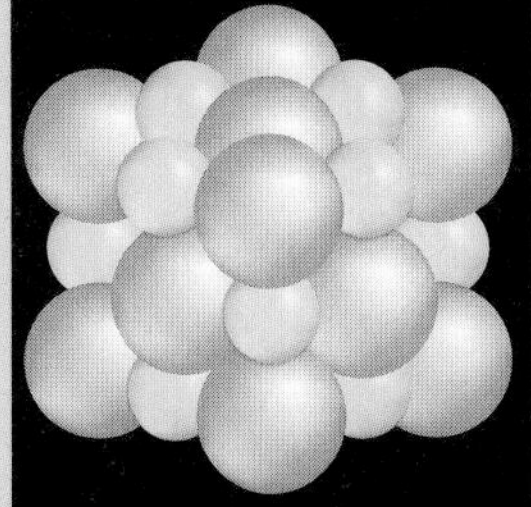

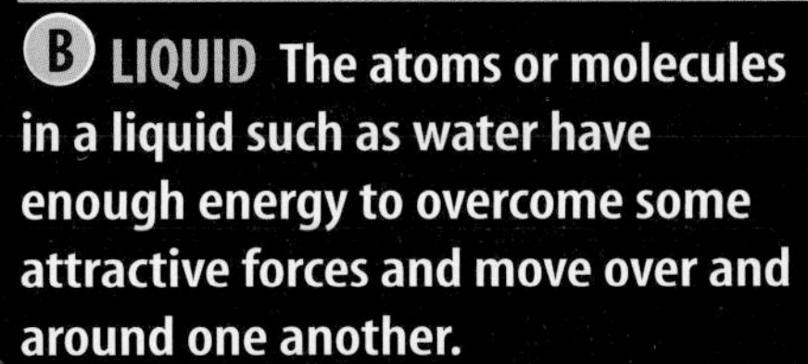

B LIQUID The atoms or molecules in a liquid such as water have enough energy to overcome some attractive forces and move over and around one another.

C GAS In air and other gases, atoms or molecules have sufficient energy to separate from each other completely and move in all directions.

Figure 16 A solid metal can be changed to a liquid by adding thermal energy to its molecules. **Describe** *what is happening to the molecules during this change.*

Changing the State of Matter

Matter is changed from a liquid to a solid at its freezing point and from a liquid to a gas at its boiling point. You may know the freezing and boiling points of water. Water begins to change from a liquid to a solid at its freezing point of 0°C. It boils at 100°C. Water is the only substance that occurs naturally on Earth as a solid, liquid, and gas. Other substances don't naturally occur in these three states on Earth because of the limited temperature range Earth experiences. For example, temperatures on Earth do not get cold enough for solid carbon dioxide to exist naturally. However, it can be manufactured.

The attraction between particles of a substance and their rate of movement are factors that determine the state of matter. When thermal energy is added to ice, the rate of movement of its molecules increases. This allows the molecules to move more freely and causes the ice to melt. As **Figure 16** shows, even solid metal can melt when enough thermal energy is added.

Changes in state also occur because of increases or decreases in pressure. For example, decreasing pressure lowers the boiling points of liquids. At high elevations, water boils at a lower temperature because the air pressure has decreased. Also, many solids melt at lower temperatures when pressure is decreased. Intense pressure keeps material in Earth's mantle in a solid or semi-solid state even though the temperature is high. A decrease in pressure can cause some material to melt and form magma. However, extreme changes in pressure usually are required to alter the melting points of solids by only a few degrees.

Figure 17 If ice were more dense than water, lakes would freeze solid from the bottom up. **Infer** *What effect might this have on the fish?*

Changes in Physical Properties

Chemical properties of matter don't change when the matter changes state, but some of its physical properties change. For example, the density of water changes as water changes state. Ice floats in liquid water, as seen in **Figure 17,** because it is less dense than liquid water. This is unique, because most materials are denser in their solid state than in their liquid state.

Reading Check *Why does ice float in water?*

Some physical properties of substances don't change when they change state. For example, water is colorless and transparent in each of its states.

Matter on Mars

Matter in one state often can be changed to another state by adding or removing thermal energy. Changes in thermal energy might explain why Mars appears to have had considerable water on its surface in the past but now has little or no water on its surface. Recent images of Mars reveal that there might still be some groundwater that occasionally reaches the surface, as shown in **Figure 18.** But what could explain the huge water-carved channels that formed long ago? Much of the liquid water on Mars might have changed state as the planet cooled to its current temperature. Scientists believe that some of Mars's liquid water soaked into the ground and froze, forming permafrost. Some of the water might have frozen to form the polar ice caps. Even more of the water might have evaporated into the atmosphere and escaped to space.

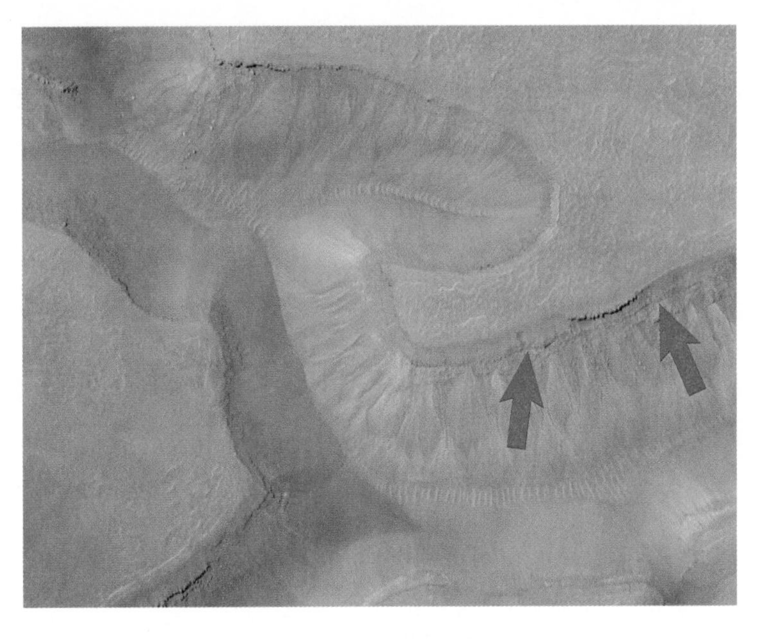

Figure 18 Groundwater might reach the surface of Mars along the edge of this large channel.

section 3 review

Summary

Physical Properties of Matter

- Density is the mass of an object divided by its volume.

States of Matter

- Solids have a definite shape and volume.
- Liquids take the shape of their containers.
- Gases spread out and fill their containers.
- Plasma occurs at high temperatures and has such high energy that some electrons may escape their electron clouds.

Changing the State of Matter

- Both temperature and pressure can cause changes in the state of matter.

Changes in Physical Properties

- Chemical properties do not change when matter changes state, but physical properties can change.

Self Check

1. **List** the four states of matter in order from lowest to highest particle movement.
2. **Explain** how temperature can bring about changes in the state of matter.
3. **Explain** why an ice cube will melt if compressed, even though the temperature remains the same.
4. **Think Critically** Suppose you blow up a balloon and then place it in a freezer. Later, you find that the balloon has shrunk and has bits of ice in it. Explain.

Applying Skills

5. **Classify** Assign each of the following items to one of the four states of matter and describe their characteristics: groundwater, lightning, lava, snow, textbook, ice cap, notebook, apple juice, eraser, glass, cotton, helium, iron oxide, lake, limestone, and water vapor.
6. **Infer** You have probably noticed that some liquids, such as honey and molasses, flow slowly at room temperature. How does heating affect flow rate?

Design Your Own

DETERMINING DENSITY

Goals

- **List** some ways that the density of an object can be measured.
- **Design** an experiment that compares the densities of several materials.

Possible Materials

pan
triple-beam balance
100-mL beaker
250-mL graduated cylinder
water
chalk
piece of quartz
piece of clay
small wooden block
small metal block
small cork
rock
ruler

Safety Precautions

WARNING: *Be wary of sharp edges on some of the materials and take care not to break the beaker or graduated cylinder. Wash hands thoroughly with soap and water when finished.*

Real-World Question

Which has a greater density—a rock or a piece of wood? Is cork more dense than clay? Density is the ratio of an object's mass to its volume.

Form a Hypothesis

State a hypothesis about what process you can use to measure and compare the densities of several materials.

Test Your Hypothesis

Make a Plan

1. As a group, agree upon and write the hypothesis statement.
2. As a group, list the steps that you need to take to test your hypothesis. Be specific, describing exactly what you will do at each step. List your materials.
3. Working as a group, use the equation: density = mass/volume. Devise a method of determining the mass and volume of each material to be tested.
4. **Design** a data table in your Science Journal so that it is ready to use as your group collects data.

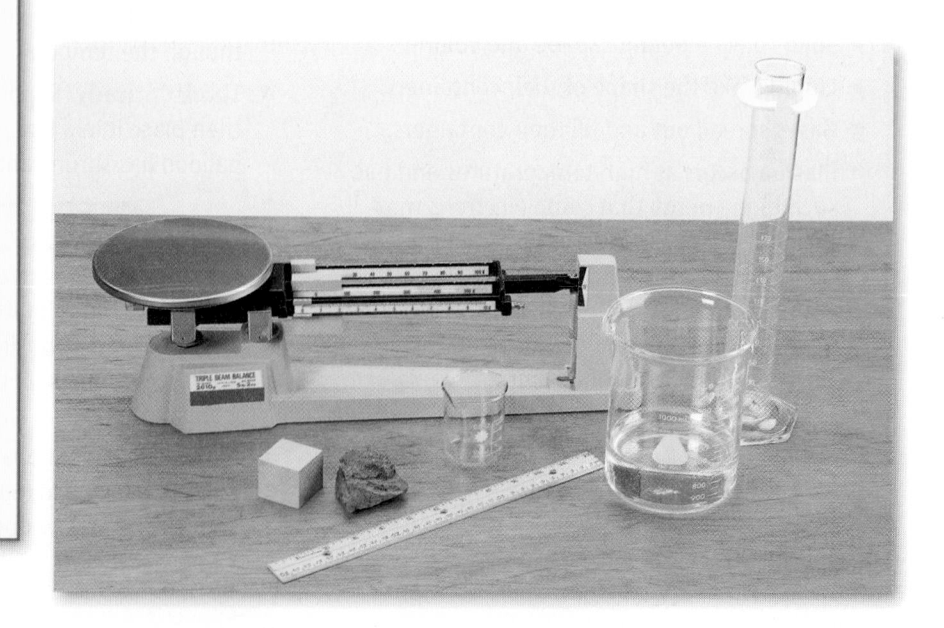

5. Read over your entire experiment to make sure that all steps are in a logical order.
6. Should you run the process more than once for any of the materials?
7. **Identify** any constants, variables, and controls of the experiment.

Follow Your Plan

1. Make sure your teacher approves your plan before you start.
2. Carry out the experiment as planned.
3. While the experiment is going on, write any observations that you make and complete the data table in your Science Journal.

Analyze Your Data

1. **Observe** Do you observe anything about the way objects with greater density feel compared with objects of lower density?
2. **Predict** Which of those objects you measured directly would float in water? Which would sink?
3. **Predict** how your volume measurements might be affected by using a tool to push a floating object under water. Explain how this error might increase or decrease the density you obtained.

Conclude and Apply

1. **Form Hypotheses** Based on your results, would you hypothesize that a cork is more dense, the same density, or less dense than water?
2. **Draw Conclusions** Without measuring the density of an object that floats, conclude how you know that it has a density of less than 1.0 g/cm^3.
3. **Predict** Would the density of the clay be affected if you were to break it into smaller pieces?
4. **Explain** why ships float, even though they are made mostly of steel that has a density much greater than that of water.

Communicating Your Data

Write an informational pamphlet on different methods for determining the density of objects. Include equations and a step-by-step procedure.

Science Stats

Amazing Atoms

Did you know . . .

. . . Uranium has the greatest mass of the abundant natural elements. One atom of uranium has a mass number that is more than 235 times greater than the mass number of one hydrogen atom, the element with the least mass. However, the diameter of a uranium atom is only about three times the size of a hydrogen atom, similar to the difference between a baseball and a volleyball.

. . . The melting point of cesium is 28.4°C. It would melt in your hand if you held it. You would not want to hold cesium, though, because it would react strongly with your skin. In fact, the metal might even catch fire.

. . . The diameter of an atom is about 100,000 times as great as the diameter of its nucleus. Suppose that when you sit in a chair, you represent the nucleus of an atom. The nearest electron in your atom would be about 120 km away—nearly half the distance across the Florida peninsula.

. . . More than ninety elements occur naturally. However, about 98 percent of Earth's crust consists of only the eight elements shown here.

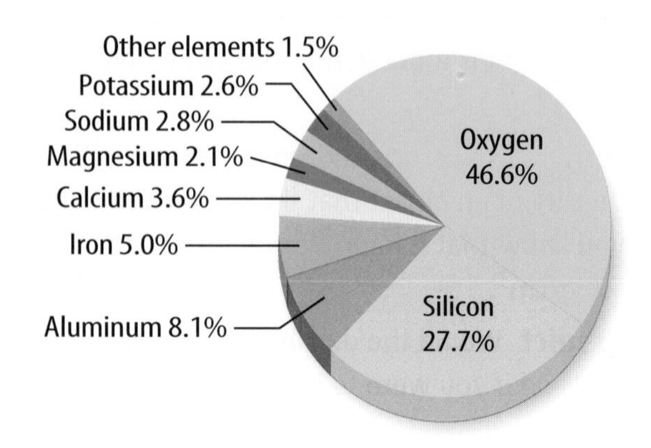

Applying Math Looking at the circle graph, which is the third most abundant element in Earth's crust?

Find Out About It

Visit earth.msscience.com/science_stats to find out more about atoms and isotopes. What is a radioactive isotope of an element? How are isotopes used in science?

Reviewing Main Ideas

Section 1 Atoms

1. Matter is anything that has mass and takes up space.
2. The nucleus of an atom contains protons with a positive charge and neutrons with no charge. Electrons, which have a negative charge, surround the nucleus.
3. Isotopes are atoms of the same element that have different numbers of neutrons.

Section 2 Combinations of Atoms

1. Atoms join to form compounds and molecules. A compound is a substance made of two or more elements. The chemical and physical properties of a compound differ from those of the elements of which it is composed.
2. A mixture is a substance in which the components are not chemically combined.

Section 3 Properties of Matter

1. Physical properties can be observed and measured without causing a chemical change in a substance. Chemical properties can be observed only when one substance reacts with another substance.
2. Atoms or molecules in a solid are in fixed positions relative to one another. In a liquid, the atoms or molecules are close together but are freer to change positions. Atoms or molecules in a gas move freely to fill any container.
3. Because of Earth's narrow temperature range, water is the only substance known that occurs naturally as a solid, liquid, and gas.
4. One physical property that is used to describe matter is density. Density is a ratio of the mass of an object to its volume. A material that is less dense will float in a material that is more dense.

Visualizing Main Ideas

Copy and complete the following concept map. Use the terms: liquids, plasma, matter, *and* solids.

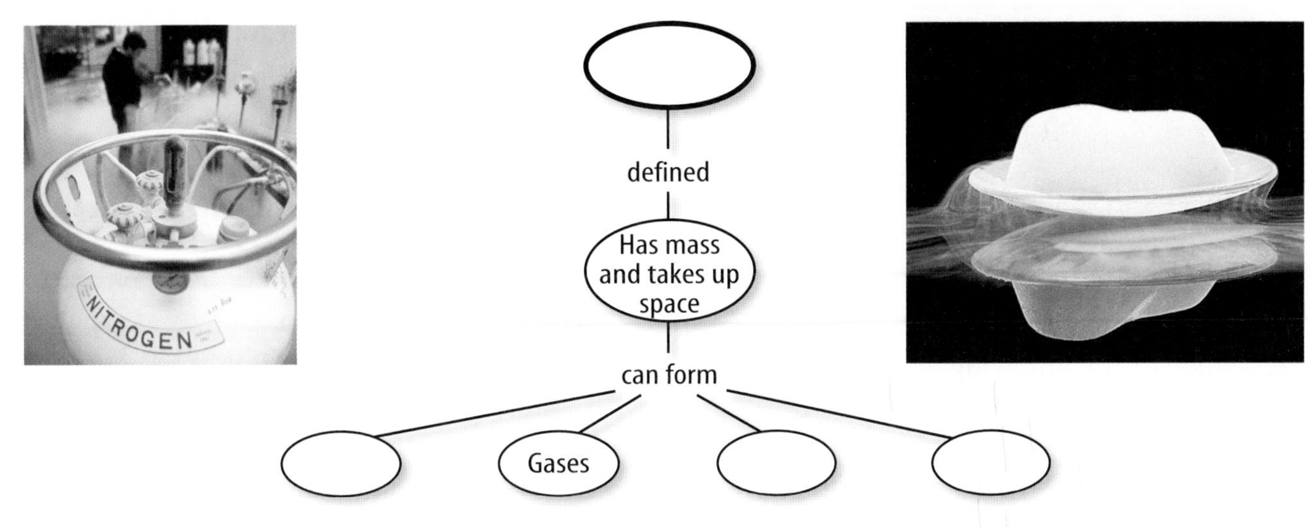

Using Vocabulary

Explain the difference between the vocabulary words in each of the following sets.

1. atom—element
2. mass number—atomic number
3. solution—heterogeneous mixture
4. matter—compound—element
5. heterogeneous mixture—homogeneous mixture
6. proton—neutron—electron
7. isotope—atom
8. atom—ion
9. mixture—compound
10. neutron—mass number

Checking Concepts

Choose the word or phrase that best answers the question.

11. Which of the following contains only one type of atom?
    A) compound **C)** element
    B) mixture **D)** solution

12. Which of the following has a positive electric charge?
    A) electron **C)** neutron
    B) proton **D)** atom

13. In an atom, what forms a cloud around the nucleus?
    A) electrons **C)** neutrons
    B) protons **D)** positive ions

14. A carbon atom has a mass number of 12. How many protons and how many neutrons does it have?
    A) 6, 6 **C)** 6, 12
    B) 12, 12 **D)** 12, 6

15. On Earth, oxygen usually exists as which of the following?
    A) solid **C)** liquid
    B) gas **D)** plasma

Use the illustration below to answer question 16.

Scandium	Titanium	Vanadium	Chromium	Manganese
21	22	23	24	25
Sc	Ti	V	Cr	Mn
44.956	47.88	50.942	51.996	54.938

16. In the section of the periodic table shown above, which element has 24 protons?
    A) titanium **C)** chromium
    B) manganese **D)** vanadium

17. An isotope known as iodine-131 has 53 protons. How many neutrons does it have?
    A) 78 **C)** 68
    B) 53 **D)** 184

18. Which of the following are electrically charged?
    A) molecule **C)** isotope
    B) solution **D)** ion

19. Which of the following is not a physical property of water?
    A) transparent
    B) colorless
    C) higher density in the liquid state than in the solid state
    D) changes to hydrogen and oxygen when electricity passes through it

Science Online earth.msscience.com/vocabulary_puzzlemaker

Thinking Critically

20. **Infer** If an atom has no electric charge, what can be said about the number of protons and electrons it contains?

21. **Identify** Carbon has six protons and nitrogen has seven protons. Which has the greatest number of neutrons—carbon-13, carbon-14, or nitrogen-14?

22. **Explain** Would isotopes of the same element have the same number of electrons?

23. **Infer** If a sodium atom loses an electron and becomes a sodium ion with a charge of 1+, what would happen if a calcium atom loses two electrons?

24. **Predict** You are told that an unknown liquid has a density of 0.79 g/cm^3 and will not mix evenly with water. Predict what will happen if you pour some of this liquid into a glass of water, stir, and wait five minutes.

Use the table below to answer question 25.

Atomic Number v. Mass Number

Element	Atomic Number	Mass Number
Fluorine	9	19
Lithium	3	7
Carbon	6	12
Nitrogen	7	14
Beryllium	4	9
Boron	5	11

25. **Make and Use Graphs** Use the table above to make a line graph. For each isotope, plot the mass number along the *y*-axis and the atomic number along the *x*-axis. What is the relationship between mass number and atomic number?

Performance Activities

26. **Classify** Use the periodic table of the elements, located on the inside back cover, to classify the following substances as elements or compounds: iron, aluminum, carbon dioxide, gold, water, and sugar.

Applying Math

27. **Will it float?** You have a heavy piece of wood that measures 2 cm × 10 cm × 5 cm. You find its mass is 89 g. Will this piece of wood float?

Use the graph below to answer question 28.

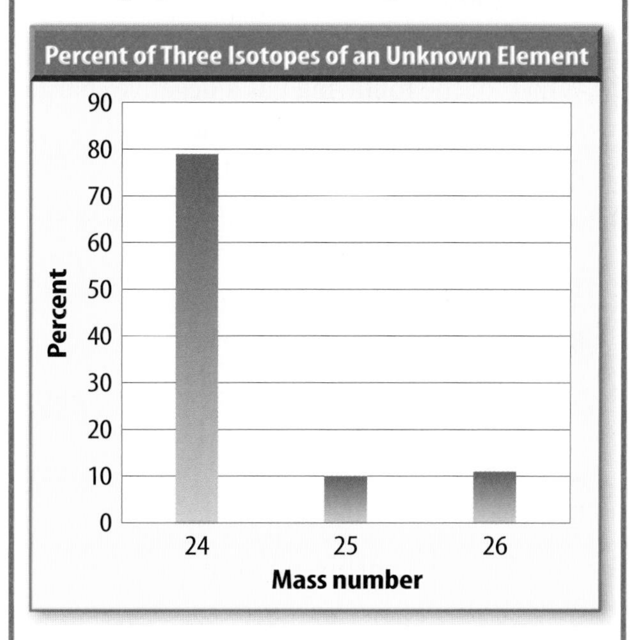

28. **Interpret Graphs** Many elements occur in nature as mixtures of several isotopes. The graph above shows the distribution in nature of three isotopes of an element. Assuming that the most abundant isotope has an equal number of protons and neutrons, use the periodic table on the inside back cover of this book to identify this element and name its three isotopes.

Standardized Test Practice

Part 1 Multiple Choice

Record your answers on the answer sheet provided by your teacher or on a sheet of paper.

Use the figure below to answer question 1.

1. Which of the following terms best describes the snack shown above?
   A. heterogeneous mixture
   B. homogeneous mixture
   C. solution
   D. compound

2. Which of the following has a negative charge?
   A. electron **C.** nucleus
   B. proton **D.** neutron

3. In which type of bond do atoms share electrons?
   A. metallic **C.** ionic
   B. hydrogen **D.** covalent

4. Which state of matter consists of ions and electrons?
   A. solid **C.** gas
   B. plasma **D.** liquid

5. Which particle orbits an atom's nucleus?
   A. isotope **C.** proton
   B. neutron **D.** electron

6. Which of the following terms best describes seawater?
   A. solution **C.** isotope
   B. ion **D.** element

Test-Taking Tip

Instructions Listen carefully to the instructions from the teacher and read the directions and each question carefully.

7. Which of the events described below is an example of a change of state?
   A. river water flowing into an ocean
   B. air being heated in a hot air balloon to make it rise
   C. ice being crushed for snow cones
   D. a puddle of water evaporating after a rain

8. Which of the following particles always are present in equal number in a neutral atom?
   A. protons, neutrons
   B. electrons, neutrons
   C. protons, electrons
   D. electrons, ions

9. In which state of matter do atoms vibrate but remain in fixed positions?
   A. solid **C.** plasma
   B. gas **D.** liquid

This block was taken from the periodic table. Use the illustration below to answer questions 10–12.

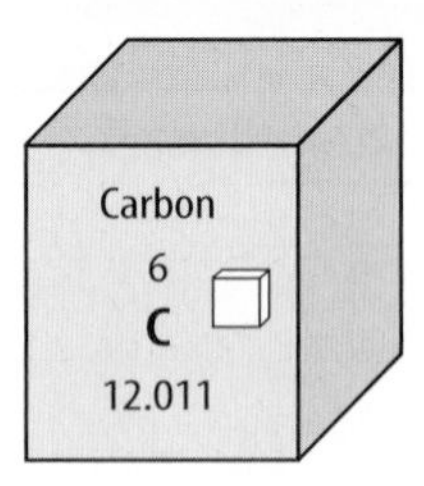

10. What is the chemical symbol for carbon?
    A. C **C.** S
    B. H **D.** Sn

11. In which state does pure carbon exist at room temperature?
    A. gas **C.** solid
    B. liquid **D.** plasma

12. What is carbon's atomic number?
    A. 10 **C.** 12.011
    B. 13 **D.** 6

Part 2 Short Response/Grid In

Record your answers on the answer sheet provided by your teacher or on a sheet of paper.

13. What is an atom?

14. How is an element different from a compound?

15. How do homogeneous mixtures differ from heterogeneous mixtures?

16. Why does ice float in liquid water?

17. What liquid formed channels on the surface of Mars long ago? What property of liquids made this possible?

Use the table below to answer questions 18–21.

Density of Some Metals	
Metal	**Density (g/cm^3)**
copper	8.9
silver	10.5
lead	11.3
gold	19.3
platinum	21.5
aluminum	2.7

18. How much more dense is platinum than gold, in grams per cubic centimeter?

19. What is the mass in grams of one cubic centimeter of pure gold? *Hint: density = mass ÷ volume*

20. How many cubic centimeters of space are taken up by 10.5 g of silver? How many are taken up by the same mass of gold?

21. An aluminum lid has a mass of 6.5 g. It has a volume of 2.4 cm^3. Calculate the density of aluminum in grams per cubic centimeter.

Part 3 Open Ended

Record your answers on a sheet of paper.

Use the illustration below to answer question 22.

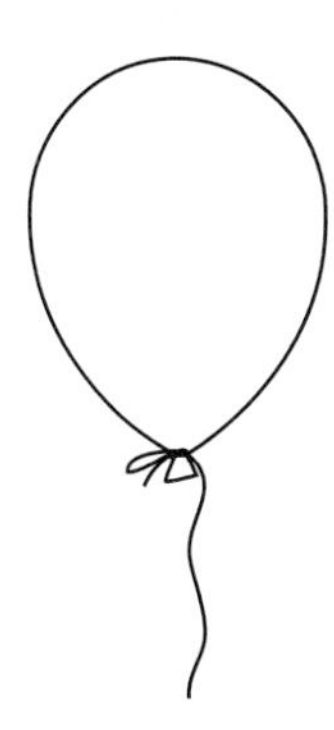

22. A balloon contains helium gas. How are the helium atoms distributed in the balloon? Do the atoms move? If so, how? Copy the sketch above on your paper and draw the helium atoms inside it.

23. Compare and contrast protons, neutrons, and electrons.

24. What is an isotope? Why are some isotopes useful to society?

25. What is the difference between chemical properties and physical properties? List one example of each type.

26. How is atomic number different from mass number?

27. Explain what happens to water molecules when ice melts.

28. Compare the covalent, ionic, metallic, and hydrogen bonds. Explain how these bonds form and describe their properties.

29. Explain, using examples, how the properties of compounds differ from those of atoms that combine to form them.

30. How does the current atomic model describe the movement and location of electrons?

chapter 3

Minerals

The BIG Idea

Minerals compose much of Earth's crust and can be identified by their physical properties.

SECTION 1
Minerals

Main Idea Minerals are formed by natural processes, are inorganic, have definite chemical compositions, and are crystalline solids.

SECTION 2
Mineral Identification

Main Idea Each mineral is identified by its physical properties.

SECTION 3
Uses of Minerals

Main Idea Minerals are important because some are rare, have special properties, or contain materials that have many uses.

Nature's Beautiful Creation

Although cut by gemologists to enhance their beauty, these gorgeous diamonds formed naturally—deep within Earth. One requirement for a substance to be a mineral is that it must occur in nature. Human-made diamonds serve their purpose in industry but are not considered minerals.

Science Journal Write two questions you would ask a gemologist about the minerals that he or she works with.

Start-Up Activities

Distinguish Rocks from Minerals

When examining rocks, you'll notice that many of them are made of more than one material. Some rocks are made of many different crystals of mostly the same mineral. A mineral, however, will appear more like a pure substance and will tend to look the same throughout. Can you tell a rock from a mineral?

1. Use a magnifying lens to observe a quartz crystal, salt grains, and samples of sandstone, granite, calcite, mica, and schist (SHIHST).
2. Draw a sketch of each sample.
3. Infer which samples are made of one type of material and should be classified as minerals.
4. Infer which samples should be classified as rocks.
5. **Think Critically** In your Science Journal, compile a list of descriptions for the minerals you examined and a second list of descriptions for the rocks. Compare and contrast your observations of minerals and rocks.

Minerals Make the following Foldable to help you better understand minerals.

STEP 1 **Fold** a vertical sheet of notebook paper from side to side.

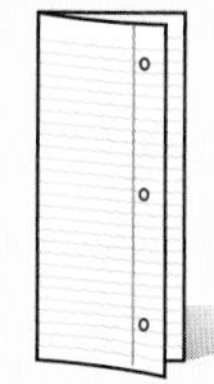

STEP 2 **Cut** along every third line of only the top layer to form tabs.

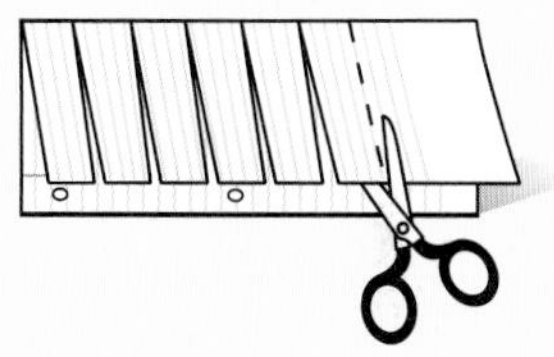

STEP 3 **Label** each tab with a question.

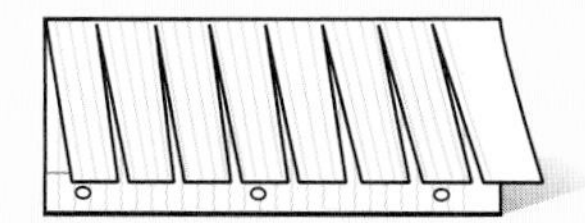

Ask Questions Before you read the chapter, write questions you have about minerals on the front of the tabs. As you read the chapter, add more questions and write answers under the appropriate tabs.

Preview this chapter's content and activities at earth.msscience.com

Get Ready to Read

New Vocabulary

1 Learn It! What should you do if you find a word you don't know or understand? Here are some suggested strategies:

1. Use context clues (from the sentence or the paragraph) to help you define it.
2. Look for prefixes, suffixes, or root words that you already know.
3. Write it down and ask for help with the meaning.
4. Guess at its meaning.
5. Look it up in the glossary or a dictionary.

2 Practice It! Look at the word ***natural*** in the following passage. See how context clues can help you understand its meaning.

Context Clue
describes types of processes

Context Clue
occur with no input from humans

Context Clue
examples of two substances to compare and contrast

. . . all minerals are formed by natural processes. These are processes that occur on or inside Earth with no input from humans. For example, salt formed by the natural evaporation of seawater is the mineral halite, but salt formed by evaporation of saltwater solutions in laboratories is not a mineral.

—from page 62

3 Apply It! Make a vocabulary bookmark with a strip of paper. As you read, keep track of words you do not know or want to learn more about.

Read a paragraph containing a vocabulary word from beginning to end. Then, go back to determine the meaning of the word.

Target Your Reading

Use this to focus on the main ideas as you read the chapter.

1. **Before you read** the chapter, respond to the statements below on your worksheet or on a numbered sheet of paper.
   - Write an **A** if you **agree** with the statement.
   - Write a **D** if you **disagree** with the statement.

2. **After you read** the chapter, look back to this page to see if you've changed your mind about any of the statements.
   - If any of your answers changed, explain why.
   - Change any false statements into true statements.
   - Use your revised statements as a study guide.

Before You Read A or D	Statement	After You Read A or D
	1 All minerals are solids, but not all solids are minerals.	
	2 The word *crystalline* means that atoms are arranged in a repeating pattern.	
	3 The two most abundant elements in Earth's crust are silicon and carbon.	
	4 Like vitamins, minerals are organic substances, which means they contain carbon.	
	5 Color is always the best physical property to use when attempting to identify minerals.	
	6 A mineral's hardness is a measure of how easily it can be scratched.	
	7 Most gems or gemstones are special varieties of particular minerals.	
	8 Synthetic, or human-made, diamonds are minerals.	
	9 A mineral or rock is called an ore only if it contains a substance that can be mined for a profit.	

Science Online

Print out a worksheet of this page at earth.msscience.com

section 1

Minerals

as you read

What You'll Learn

- **Describe** characteristics that all minerals share.
- **Explain** how minerals form.

Why It's Important

You use minerals and products made from them every day.

Review Vocabulary
atoms: tiny particles that make up matter; composed of protons, electrons, and neutrons

New Vocabulary
- mineral
- crystal
- magma
- silicate

What is a mineral?

How important are minerals to you? Very important? You actually own or encounter many things made from minerals every day. Ceramic, metallic, and even some paper items are examples of products that are derived from or include minerals. **Figure 1** shows just a few of these things. Metal bicycle racks, bricks, and the glass in windows would not exist if it weren't for minerals. A **mineral** is a naturally occurring, inorganic solid with a definite chemical composition and an orderly arrangement of atoms. About 4,000 different minerals are found on Earth, but they all share these four characteristics.

Mineral Characteristics First, all minerals are formed by natural processes. These are processes that occur on or inside Earth with no input from humans. For example, salt formed by the natural evaporation of seawater is the mineral halite, but salt formed by evaporation of saltwater solutions in laboratories is not a mineral. Second, minerals are inorganic. This means that they aren't made by life processes. Third, every mineral is an element or compound with a definite chemical composition. For example, halite's composition, NaCl, gives it a distinctive taste that adds flavor to many foods. Fourth, minerals are crystalline solids. All solids have a definite volume and shape. Gases and liquids like air and water have no definite shape, and they aren't crystalline. Only a solid can be a mineral, but not all solids are minerals.

Figure 1 You probably use minerals or materials made from minerals every day without thinking about it.
Infer *How many objects in this picture might be made from minerals?*

Atom Patterns The word *crystalline* means that atoms are arranged in a pattern that is repeated over and over again. For example, graphite's atoms are arranged in layers. Opal, on the other hand, is not a mineral in the strictest sense because its atoms are not all arranged in a definite, repeating pattern, even though it is a naturally occurring, inorganic solid.

Figure 2 More than 200 years ago, the smooth, flat surfaces on crystals led scientists to infer that minerals had an orderly structure inside.

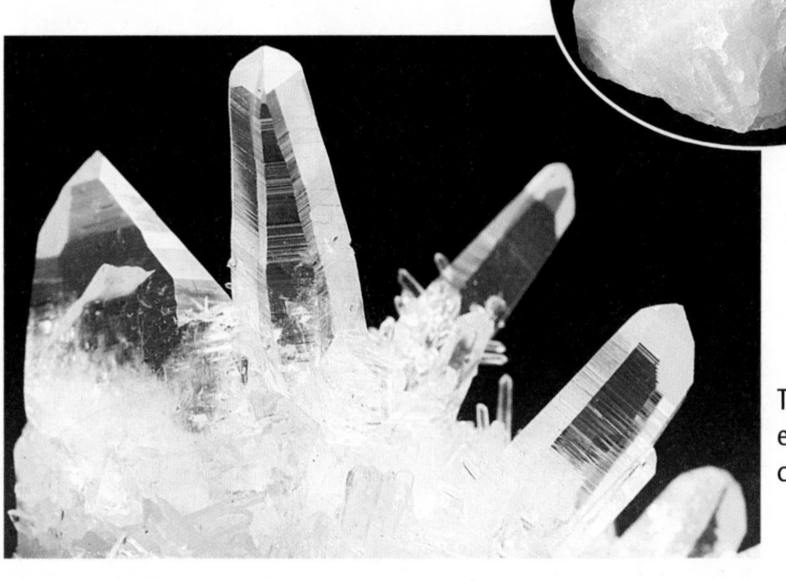

Even though this rose quartz looks uneven on the outside, its atoms have an orderly arrangement on the inside.

The well-formed crystal shapes exhibited by these clear quartz crystals suggest an orderly structure.

The Structure of Minerals

Do you have a favorite mineral sample or gemstone? If so, perhaps it contains well-formed crystals. A **crystal** is a solid in which the atoms are arranged in orderly, repeating patterns. You can see evidence for this orderly arrangement of atoms when you observe the smooth, flat outside surfaces of crystals. A crystal system is a group of crystals that have similar atomic arrangements and therefore similar external crystal shapes.

Reading Check *What is a crystal?*

Crystals Not all mineral crystals have smooth surfaces and regular shapes like the clear quartz crystals in **Figure 2.** The rose quartz in the smaller photo of **Figure 2** has atoms arranged in repeating patterns, but you can't see the crystal shape on the outside of the mineral. This is because the rose quartz crystals developed in a tight space, while the clear quartz crystals developed freely in an open space. The six-sided, or hexagonal crystal shape of the clear quartz crystals in **Figure 2,** and other forms of quartz can be seen in some samples of the mineral. **Figure 3** illustrates the six major crystal systems, which classify minerals according to their crystal structures. The hexagonal system to which quartz belongs is one example of a crystal system.

Crystals form by many processes. Next, you'll learn about two of these processes—crystals that form from magma and crystals that form from solutions of salts.

Inferring Salt's Crystal System

Procedure

1. Use a **magnifying lens** to observe grains of common **table salt** on a dark sheet of **construction paper.** Sketch the shape of a salt grain. **WARNING:** *Do not taste or eat mineral samples. Keep hands away from your face.*
2. Compare the shapes of the salt crystals with the shapes of crystals shown in **Figure 3.**

Analysis

1. Which characteristics do all the grains have in common?
2. Research another mineral with the same crystal system as salt. What is this crystal system called?

Try at Home

NATIONAL GEOGRAPHIC VISUALIZING CRYSTAL SYSTEMS

Figure 3

A crystal's shape depends on how its atoms are arranged. Crystal shapes can be organized into groups known as crystal systems—shown here in 3-D with geometric models (in blue). Knowing a mineral's crystal system helps researchers understand its atomic structure and physical properties.

▲ CUBIC Fluorite is an example of a mineral that forms cubic crystals. Minerals in the cubic crystal system are equal in size along all three principal dimensions.

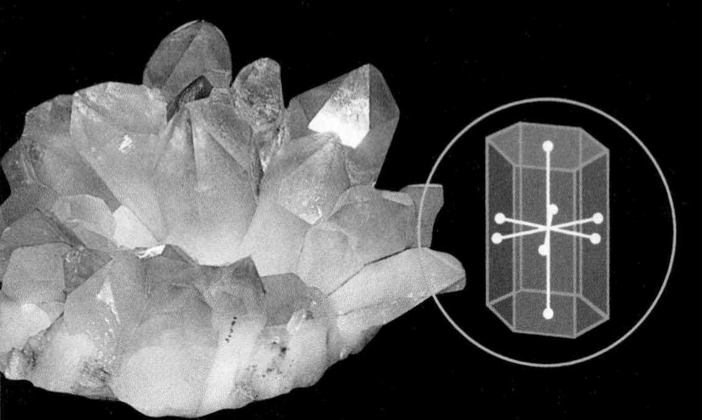

▲ HEXAGONAL (hek SA guh nul) In hexagonal crystals, horizontal distances between opposite crystal surfaces are equal. These crystal surfaces intersect to form 60° or 120° angles. The vertical length is longer or shorter than the horizontal lengths.

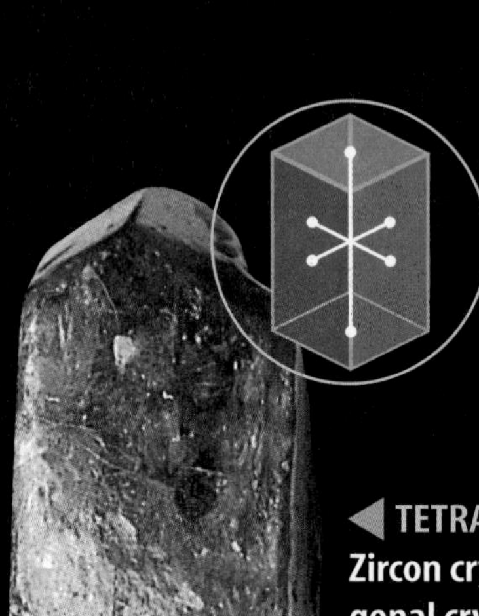

◀ TETRAGONAL (te TRA guh nul) Zircon crystals are tetragonal. Tetragonal crystals are much like cubic crystals, except that one of the principal dimensions is longer or shorter than the other two dimensions.

▲ ORTHORHOMBIC (awr thuh RAHM bihk) Minerals with orthorhombic structure, such as barite, have dimensions that are unequal in length, resulting in crystals with a brick-like shape.

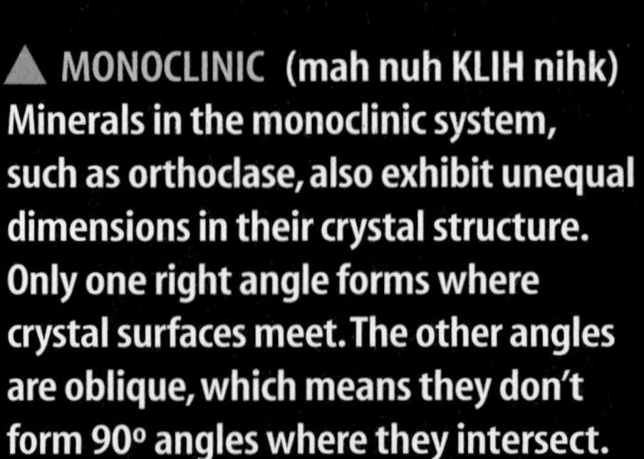

▲ MONOCLINIC (mah nuh KLIH nihk) Minerals in the monoclinic system, such as orthoclase, also exhibit unequal dimensions in their crystal structure. Only one right angle forms where crystal surfaces meet. The other angles are oblique, which means they don't form 90° angles where they intersect.

▲ TRICLINIC (tri KLIH nihk) The triclinic crystal system includes minerals exhibiting the least symmetry. Triclinic crystals, such as rhodonite (ROH dun ite), are unequal in all dimensions, and all angles where crystal surfaces meet are oblique.

Figure 4 Minerals form by many natural processes.

A This rock formed as magma cooled slowly, allowing large mineral grains to form.

Labradorite

B Some minerals form when salt water evaporates, such as these white crystals of halite in Death Valley, California.

Crystals from Magma Natural processes form minerals in many ways. For example, hot melted rock material, called **magma,** cools when it reaches Earth's surface, or even if it's trapped below the surface. As magma cools, its atoms lose heat energy, move closer together, and begin to combine into compounds. During this process, atoms of the different compounds arrange themselves into orderly, repeating patterns. The type and amount of elements present in a magma partly determine which minerals will form. Also, the size of the crystals that form depends partly on how rapidly the magma cools.

When magma cools slowly, the crystals that form are generally large enough to see with the unaided eye, as shown in **Figure 4A.** This is because the atoms have enough time to move together and form into larger crystals. When magma cools rapidly, the crystals that form will be small. In such cases, you can't easily see individual mineral crystals.

Crystals from Solution Crystals also can form from minerals dissolved in water. When water evaporates, as in a dry climate, ions that are left behind can come together to form crystals like the halite crystals in **Figure 4B.** Or, if too much of a substance is dissolved in water, ions can come together and crystals of that substance can begin to form in the solution. Minerals can form from a solution in this way without the need for evaporation.

Crystal Formation Evaporites commonly form in dry climates. Research the changes that take place when a saline lake or shallow sea evaporates and halite or gypsum forms.

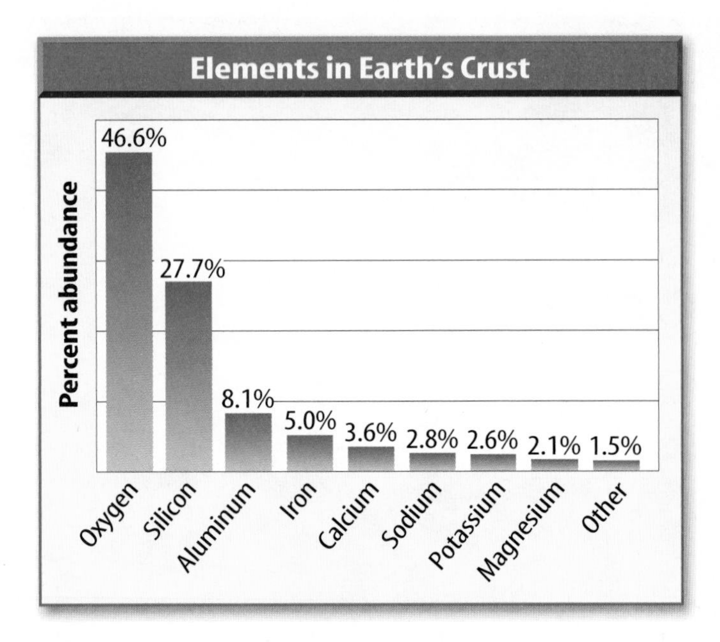

Figure 5 Most of Earth's crust is composed of eight elements.

Mineral Compositions and Groups

Ninety elements occur naturally in Earth's crust. Approximately 98 percent (by weight) of the crust is made of only eight of these elements, as shown in **Figure 5.** Of the thousands of known minerals, only a few dozen are common, and these are mostly composed of the eight most common elements in Earth's crust.

Most of the common rock-forming minerals belong to a group called the silicates. **Silicates** (SIH luh kayts) are minerals that contain silicon (Si) and oxygen (O) and usually one or more other elements. Silicon and oxygen are the two most abundant elements in Earth's crust. These two elements alone combine to form the basic building blocks of most of the minerals in Earth's crust and mantle. Feldspar and quartz, which are silicates, and calcite, which is a carbonate, are examples of common, rock-forming minerals. Other mineral groups also are defined according to their compositions.

section 1 review

Summary

What is a mineral?

- Many products used by humans are made from minerals.
- Minerals are defined by four main characteristics.

The Structure of Minerals

- The crystal shape of a mineral reflects the way in which its atoms are arranged.
- Minerals are classified according to the types of atoms in their structures and the way that the atoms are arranged.

Mineral Compositions and Groups

- Only eight elements form approximately 98 percent (by weight) of Earth's crust.
- The majority of Earth's crust is composed of silicate minerals.

Self Check

1. **List** four characteristics that all minerals share.
2. **Describe** two ways that minerals can form from solution.
3. **Explain** whether diamonds made in the laboratory are considered to be minerals.
4. **Describe** how crystals of minerals are classified.
5. **Think Critically** The mineral dolomite, a rock-forming mineral, contains oxygen, carbon, magnesium, and calcium. Is dolomite a silicate? Explain.

Applying Skills

6. **Graph** Make a graph of your own design that shows the relative percentages of the eight most common elements in Earth's crust. Then determine the approximate percentage of the crust that is made up of iron and aluminum. If one is available, you may use an electronic spreadsheet program to make your graph and perform the calculation.

Science Online earth.msscience.com/self_check_quiz

Crystal Formation

In this lab, you'll have a chance to learn how crystals form from solutions.

Real-World Question

How do crystals form from solution?

Goals

- **Compare and contrast** the crystals that form from salt and sugar solutions.
- **Observe** crystals and infer how they formed.

Materials

250-mL beakers (2)
cardboard
large paper clip
table salt
flat wooden stick
granulated sugar
cotton string
hot plate
magnifying lens
thermal mitt
shallow pan
spoon

Safety Precautions

WARNING: *Never taste or eat any lab materials.*

Procedure

1. Gently mix separate solutions of salt in water and sugar in water in the two beakers. Keep stirring the solutions as you add salt or sugar to the water. Stop mixing when no more salt or sugar will dissolve in the solutions. Label each beaker.
2. Place the sugar solution beaker on a hot plate. Use the hot plate to heat the sugar solution gently. **WARNING:** *Do not touch the hot beaker without protecting your hands.*
3. Tie one end of the thread to the middle of the wooden stick. Tie a large paper clip to the free end of the string for weight. Place the stick across the opening of the sugar beaker so the thread dangles in the sugar solution.
4. Remove the beaker from the hot plate and cover it with cardboard. Place it in a location where it won't be disturbed.
5. Pour a thin layer of the salt solution into the shallow pan.
6. Leave the beaker and the shallow pan undisturbed for at least one week.
7. After one week, examine each solution with a magnifying lens to see whether crystals have formed.

Conclude and Apply

1. **Compare and contrast** the crystals that formed from the salt and the sugar solutions. How do they compare with samples of table salt and sugar?
2. **Describe** what happened to the saltwater solution in the shallow pan.
3. Did this same process occur in the sugar solution? Explain.

Communicating Your Data

Make a poster that describes your methods of growing salt and sugar crystals. Present your results to your class.

section 2

Mineral Identification

as you read

What You'll Learn

- **Describe** physical properties used to identify minerals.
- **Identify** minerals using physical properties such as hardness and streak.

Why It's Important

Identifying minerals helps you recognize valuable mineral resources.

Review Vocabulary

physical property: any characteristic of a material that you can observe without changing the identity of the material

New Vocabulary

- hardness
- luster
- specific gravity
- streak
- cleavage
- fracture

Physical Properties

Why can you recognize a classmate when you see him or her in a crowd away from school? A person's height or the shape of his or her face helps you tell that person from the rest of your class. Height and facial shape are two properties unique to individuals. Individual minerals also have unique properties that distinguish them.

Mineral Appearance Just like height and facial characteristics help you recognize someone, mineral properties can help you recognize and distinguish minerals. Color and appearance are two obvious clues that can be used to identify minerals.

However, these clues alone aren't enough to recognize most minerals. The minerals pyrite and gold are gold in color and can appear similar, as shown in **Figure 6.** As a matter of fact, pyrite often is called fool's gold. Gold is worth a lot of money, whereas pyrite has little value. You need to look at other properties of minerals to tell them apart. Some other properties to study include how hard a mineral is, how it breaks, and its color when crushed into a powder. Every property you observe in a mineral is a clue to its identity.

Figure 6 The general appearance of a mineral often is not enough to identify it.

Using only color, observers can be fooled when trying to distinguish between pyrite and gold.

The mineral azurite is identified readily by its striking blue color.

Hardness A measure of how easily a mineral can be scratched is its **hardness.** The mineral talc is so soft you can scratch it loose with your fingernail. Talcum powder is made from this soft mineral. Diamonds, on the other hand, are the hardest mineral. Some diamonds are used as cutting tools, as shown in **Figure 7.** A diamond can be scratched only by another diamond. Diamonds can be broken, however.

Reading Check *Why is hardness sometimes referred to as scratchability?*

Sometimes the concept of hardness is confused with whether or not a mineral will break. It is important to understand that even though a diamond is extremely hard, it can shatter if given a hard enough blow in the right direction along the crystal.

Figure 7 Some saw blades have diamonds embedded in them to help slice through materials, such as this limestone. Blades are kept cool by running water over them.

Mohs Scale In 1824, the Austrian scientist Friedrich Mohs developed a list of common minerals to compare their hardnesses. This list is called Mohs scale of hardness, as seen in **Table 1.** The scale lists the hardness of ten minerals. Talc, the softest mineral, has a hardness value of one, and diamond, the hardest mineral, has a value of ten.

Here's how the scale works. Imagine that you have a clear or whitish-colored mineral that you know is either fluorite or quartz. You try to scratch it with your fingernail and then with an iron nail. You can't scratch it with your fingernail but you can scratch it with the iron nail. Because the hardness of your fingernail is 2.5 and that of the iron nail is 4.5, you can determine the unknown mineral's hardness to be somewhere around 3 or 4. Because it is known that quartz has a hardness of 7 and fluorite has a hardness of 4, the mystery mineral must be fluorite.

Some minerals have a hardness range rather than a single hardness value. This is because atoms are arranged differently in different directions in their crystal structures.

Table 1 Mineral Hardness

Mohs Scale	Hardness	Hardness of Common Objects	
Talc (softest)	1		
Gypsum	2	fingernail	(2.5)
Calcite	3	piece of copper	(2.5 to 3.0)
Fluorite	4	iron nail	(4.5)
Apatite	5	glass	(5.5)
Feldspar	6	steel file	(6.5)
Quartz	7	streak plate	(7.0)
Topaz	8		
Corundum	9		
Diamond (hardest)	10		

Figure 8 Luster is an important physical property that is used to distinguish minerals. Graphite has a metallic luster. Fluorite has a nonmetallic, glassy luster.

Luster The way a mineral reflects light is known as **luster.** Luster can be metallic or nonmetallic. Minerals with a metallic luster, like the graphite shown in **Figure 8,** shine like metal. Metallic luster can be compared to the shine of a metal belt buckle, the shiny chrome trim on some cars, or the shine of metallic cooking utensils. When a mineral does not shine like metal, its luster is nonmetallic. Examples of terms for nonmetallic luster include dull, pearly, silky, and glassy. Common examples of minerals with glassy luster are quartz, calcite, halite, and fluorite.

Specific Gravity Minerals also can be distinguished by comparing the weights of equal-sized samples. The **specific gravity** of a mineral is the ratio of its weight compared with the weight of an equal volume of water. Like hardness, specific gravity is expressed as a number. If you were to research the specific gravities of gold and pyrite, you'd find that gold's specific gravity is about 19, and pyrite's is 5. This means that gold is about 19 times heavier than water and pyrite is 5 times heavier than water. You could experience this by comparing equal-sized samples of gold and pyrite in your hands—the pyrite would feel much lighter. The term *heft* is sometimes used to describe how heavy a mineral sample feels.

Applying Science

How can you identify minerals?

Properties of Minerals

Mineral	Hardness	Streak
Copper	2.5–3	copper-red
Galena	2.5	dark gray
Gold	2.5–3	yellow
Hematite	5.5–6.5	red to brown
Magnetite	6–6.5	black
Silver	2.5–3	silver-white

You have learned that minerals are identified by their physical properties, such as streak, hardness, cleavage, and color. Use your knowledge of mineral properties and your ability to read a table to solve the following problems.

Identifying the Problem

The table includes hardnesses and streak colors for several minerals. How can you use these data to distinguish minerals?

Solving the Problem

1. What test would you perform to distinguish hematite from copper? How would you carry out this test?
2. How could you distinguish copper from galena? What tool would you use?
3. What would you do if two minerals had the same hardness and the same streak color?

Streak When a mineral is rubbed across a piece of unglazed porcelain tile, as in **Figure 9,** a streak of powdered mineral is left behind. **Streak** is the color of a mineral when it is in a powdered form. The streak test works only for minerals that are softer than the streak plate. Gold and pyrite can be distinguished by a streak test. Gold has a yellow streak and pyrite has a greenish-black or brownish-black streak.

Some soft minerals will leave a streak even on paper. The last time you used a pencil to write on paper, you left a streak of the mineral graphite. One reason that graphite is used in pencil lead is because it is soft enough to leave a streak on paper.

Figure 9 Streak is more useful for mineral identification than is mineral color. Hematite, for example, can be dark red, gray, or silver in color. However, its streak is always dark reddish-brown.

Why do gold and pyrite leave a streak, but quartz does not?

Cleavage and Fracture The way a mineral breaks is another clue to its identity. Minerals that break along smooth, flat surfaces have **cleavage** (KLEE vihj). Cleavage, like hardness, is determined partly by the arrangement of the mineral's atoms. Mica is a mineral that has one perfect cleavage. **Figure 10** shows how mica breaks along smooth, flat planes. If you were to take a layer cake and separate its layers, you would show that the cake has cleavage. Not all minerals have cleavage. Minerals that break with uneven, rough, or jagged surfaces have **fracture.** Quartz is a mineral with fracture. If you were to grab a chunk out of the side of that cake, it would be like breaking a mineral that has fracture.

Figure 10 Weak or fewer bonds within the structures of mica and halite allow them to be broken along smooth, flat cleavage planes.
Infer *If you broke quartz, would it look the same?*

Observing Mineral Properties

Procedure

1. Obtain samples of some of the following clear minerals: **gypsum, muscovite mica, halite,** and **calcite.**
2. Place each sample over the print on this page and observe the letters.

Analysis

1. Which mineral can be identified by observing the print's double image?
2. What other special property is used to identify this mineral?

Figure 11 Some minerals are natural magnets, such as this lodestone, which is a variety of magnetite.

Other Properties Some minerals have unique properties. Magnetite, as you can guess by its name, is attracted to magnets. Lodestone, a form of magnetite, will pick up iron filings like a magnet, as shown in **Figure 11.** Light forms two separate rays when it passes through calcite, causing you to see a double image when viewed through transparent specimens. Calcite also can be identified because it fizzes when hydrochloric acid is put on it.

Now you know that you sometimes need more information than color and appearance to identify a mineral. You also might need to test its streak, hardness, luster, and cleavage or fracture. Although the overall appearance of a mineral can be different from sample to sample, its physical properties remain the same.

section 2 review

Summary

Physical Properties

- Minerals are identified by observing their physical properties.
- Hardness is a measure of how easily a mineral can be scratched.
- Luster describes how a mineral reflects light.
- Specific gravity is the ratio of the weight of a mineral sample compared to the weight of an equal volume of water.
- Streak is the color of a powdered mineral.
- Minerals with cleavage break along smooth, flat surfaces in one or more directions.
- Fracture describes any uneven manner in which a mineral breaks.
- Some minerals react readily with acid, form a double image, or are magnetic.

Self Check

1. **Compare and contrast** a mineral fragment that has one cleavage direction with one that has only fracture.
2. **Explain** how an unglazed porcelain tile can be used to identify a mineral.
3. **Explain** why streak often is more useful for mineral identification than color.
4. **Determine** What hardness does a mineral have if it does not scratch glass but it scratches an iron nail?
5. **Think Critically** What does the presence of cleavage planes within a mineral tell you about the chemical bonds that hold the mineral together?

Applying Skills

6. **Draw Conclusions** A large piece of the mineral halite is broken repeatedly into several perfect cubes. How can this be explained?

Science Online earth.msscience.com/self_check_quiz

section 3

Uses of Minerals

Gems

Walking past the window of a jewelry store, you notice a large selection of beautiful jewelry—a watch sparkling with diamonds, a necklace holding a brilliant red ruby, and a gold ring. For thousands of years, people have worn and prized minerals in their jewelry. What makes some minerals special? What unusual properties do they have that make them so valuable?

Properties of Gems As you can see in **Figure 12, gems** or gemstones are highly prized minerals because they are rare and beautiful. Most gems are special varieties of a particular mineral. They are clearer, brighter, or more colorful than common samples of that mineral. The difference between a gem and the common form of the same mineral can be slight. Amethyst is a gem form of quartz that contains just traces of iron in its structure. This small amount of iron gives amethyst a desirable purple color. Sometimes a gem has a crystal structure that allows it to be cut and polished to a higher quality than that of a nongem mineral. **Table 2** lists popular gems and some locations where they have been collected.

as you read

What You'll Learn

- **Describe** characteristics of gems that make them more valuable than other minerals.
- **Identify** useful elements that are contained in minerals.

Why It's Important

Minerals are necessary materials for decorative items and many manufactured products.

Review Vocabulary

metal: element that typically is a shiny, malleable solid that conducts heat and electricity well

New Vocabulary

- gem
- ore

Figure 12 It is easy to see why gems are prized for their beauty and rarity. Shown here is The Imperial State Crown, made for Queen Victoria of England in 1838. It contains thousands of jewels, including diamonds, rubies, sapphires, and emeralds.

Table 2 Minerals and Their Gems

Fun Facts	Mineral	Gem Example	Some Important Locations
Beryl is named for the element beryllium, which it contains. Some crystals reach several meters in length.	Beryl 	Emerald 	Colombia, Brazil, South Africa, North Carolina
A red spinel in the British crown jewels has a mass of 352 carats. A carat is 0.2 g.	Spinel 	Ruby spinel 	Sri Lanka, Thailand, Myanmar (Burma)
Purplish-blue examples of zoisite were discovered in 1967 near Arusha, Tanzania.	Zoisite 	Tanzanite 	Tanzania
The most valuable examples are yellow, pink, and blue varieties.	Topaz (uncut) 	Topaz (gem) 	Siberia, Germany, Japan, Mexico, Brazil, Colorado, Utah, Texas, California, Maine, Virginia, South Carolina

Fun Facts	Mineral	Gem Example	Some Important Locations
Olivine composes a large part of Earth's upper mantle. It is also present in moon rocks.	Olivine	Peridot	Myanmar (Burma), Zebirget (Saint John's Island, located in the Red Sea), Arizona, New Mexico
Garnet is a common mineral found in a wide variety of rock types. The red color of the variety almandine is caused by iron in its crystal structure.	Garnet	Almandine	Ural Mountains, Italy, Madagascar, Czech Republic, India, Sri Lanka, Brazil, North Carolina, Arizona, New Mexico
Quartz makes up about 30 percent of Earth's continental crust.	Quartz	Amethyst	Colorless varieties in Hot Springs, Arkansas; Amethyst in Brazil, Uruguay, Madagascar, Montana, North Carolina, California, Maine
The blue color of sapphire is caused by iron or titanium in corundum. Chromium in corundum produces the red color of ruby.	Corundum	Blue sapphire	Thailand, Cambodia, Sri Lanka, Kashmir

Important Gems

All gems are prized, but some are truly spectacular and have played an important role in history. For example, the Cullinan diamond, found in South Africa in 1905, was the largest uncut diamond ever discovered. Its mass was 3,106.75 carats (about 621 g). The Cullinan diamond was cut into 9 main stones and 96 smaller ones. The largest of these is called the Cullinan 1 or Great Star of Africa. Its mass is 530.20 carats (about 106 g), and it is now part of the British monarchy's crown jewels, shown in **Figure 13A.**

Another well-known diamond is the blue Hope diamond, shown in **Figure 13B.** This is perhaps the most notorious of all diamonds. It was purchased by Henry Philip Hope around 1830, after whom it is named. Because his entire family as well as a later owner suffered misfortune, the Hope diamond has gained a reputation for bringing its owner bad luck. The Hope diamond's mass is 45.52 carats (about 9 g). Currently it is displayed in the Smithsonian Institution in Washington, D.C.

Topic: Gemstone Data

Visit earth.msscience.com for Web links to information about gems at the Smithsonian Museum of Natural History.

Activity List three important examples of gems other than those described on this page. Prepare a data table with the heads *Gem Name/Type, Weight (carats/grams), Mineral,* and *Location.* Fill in the table entries for the gemstones you selected.

Useful Gems

In addition to their beauty, some gems serve useful purposes. You learned earlier that diamonds have a hardness of 10 on Mohs scale. They can scratch almost any material—a property that makes them useful as industrial abrasives and cutting tools. Other useful gems include rubies, which are used to produce specific types of laser light. Quartz crystals are used in electronics and as timepieces. When subjected to an electric field, quartz vibrates steadily, which helps control frequencies in electronic devices and allows for accurate timekeeping.

Most industrial diamonds and other gems are synthetic, which means that humans make them. However, the study of natural gems led to their synthesis, allowing the synthetic varieties to be used by humans readily.

Figure 13 These gems are among the most famous examples of precious stones.

A The Great Star of Africa is part of a sceptre in the collection of British crown jewels.

B Beginning in 1668, the Hope diamond was part of the French crown jewels. Then known as the French Blue, it was stolen in 1792 and later surfaced in London, England in 1812.

Bauxite

Figure 14 Bauxite, an ore of aluminum, is processed to make pure aluminum metal for useful products.

Useful Elements in Minerals

Gemstones are perhaps the best-known use of minerals, but they are not the most important. Look around your home. How many things made from minerals can you name? Can you find anything made from iron?

Ores Iron, used in everything from frying pans to ships, is obtained from its ore, hematite. A mineral or rock is an **ore** if it contains a useful substance that can be mined at a profit. Magnetite is another mineral that contains iron.

When is a mineral also an ore?

INTEGRATE Chemistry Aluminum sometimes is refined, or purified, from the ore bauxite, shown in **Figure 14.** In the process of refining aluminum, aluminum oxide powder is separated from unwanted materials that are present in the original bauxite. After this, the aluminum oxide powder is converted to molten aluminum by a process called smelting.

During smelting, a substance is melted to separate it from any unwanted materials that may remain. Aluminum can be made into useful products like bicycles, soft-drink cans, foil, and lightweight parts for airplanes and cars. The plane flown by the Wright brothers during the first flight at Kitty Hawk had an engine made partly of aluminum.

Historical Mineralogy An early scientific description of minerals was published by Georgius Agricola in 1556. Use print and online resources to research the mining techniques discussed by Agricola in his work *De Re Metallica.*

Figure 15 The mineral sphalerite (greenish when nearly pure) is an important source of zinc. Iron often is coated with zinc to prevent rust in a process called galvanization.

Vein Minerals Under certain conditions, metallic elements can dissolve in fluids. These fluids then travel through weaknesses in rocks and form mineral deposits. Weaknesses in rocks include natural fractures or cracks, faults, and surfaces between layered rock formations. Mineral deposits left behind that fill in the open spaces created by the weaknesses are called vein mineral deposits.

Reading Check *How do fluids move through rocks?*

Sometimes vein mineral deposits fill in the empty spaces after rocks collapse. An example of a mineral that can form in this way is shown in **Figure 15.** This is the shiny mineral sphalerite, a source of the element zinc, which is used in batteries. Sphalerite sometimes fills spaces in collapsed limestone.

Minerals Containing Titanium You might own golf clubs with titanium shafts or a racing bicycle containing titanium. Perhaps you know someone who has a titanium hip or knee replacement. Titanium is a durable, lightweight, metallic element derived from minerals that contain this metal in their crystal structures. Two minerals that are sources of the element titanium are ilmenite (IHL muh nite) and rutile (rew TEEL), shown in **Figure 16.** Ilmenite and rutile are common in rocks that form when magma cools and solidifies. They also occur as vein mineral deposits and in beach sands.

Figure 16 Rutile and ilmenite are common ore minerals of the element titanium.

Rutile

Ilmenite

Uses for Titanium Titanium is used in automobile body parts, such as connecting rods, valves, and suspension springs. Low density and durability make it useful in the manufacture of aircraft, eyeglass frames, and sports equipment such as tennis rackets and bicycles. Wheelchairs used by people who want to race or play basketball often are made from titanium, as shown in **Figure 17.** Titanium is one of many examples of useful materials that come from minerals and that enrich humans' lives.

Figure 17 Wheelchairs used for racing and playing basketball often have parts made from titanium.

section 3 review

Summary

Gems

- Gems are highly prized mineral specimens often used as decorative pieces in jewelry or other items.
- Some gems, especially synthetic ones, have industrial uses.

Useful Elements in Minerals

- Economically important quantities of useful elements or compounds are present in ores.
- Ores generally must be processed to extract the desired material.
- Iron, aluminum, zinc, and titanium are common metals that are extracted from minerals.

Self Check

1. **Explain** why the Cullinan diamond is an important gem.
2. **Identify** Examine **Table 2.** What do rubies and sapphires have in common?
3. **Describe** how vein minerals form.
4. **Explain** why bauxite is considered to be a useful rock.
5. **Think Critically** Titanium is nontoxic. Why is this important in the manufacture of artificial body parts?

Applying Skills

6. **Use Percentages** Earth's average continental crust contains 5 percent iron and 0.007 percent zinc. How many times more iron than zinc is present in average continental crust?

Mineral Identification

Goals

- **Hypothesize** which properties of each mineral are most useful for identification purposes.
- **Test** your hypothesis as you attempt to identify unknown mineral samples.

Materials

mineral samples
magnifying lens
pan balance
graduated cylinder
water
piece of copper
**copper penny*
glass plate
small iron nail
steel file
streak plate
5% HCl with dropper
Mohs scale of hardness
Minerals Appendix
**minerals field guide*
safety goggles
**Alternate materials*

Safety Precautions

WARNING: *If an HCl spill occurs, notify your teacher and rinse with cool water until you are told to stop. Do not taste, eat, or drink any lab materials.*

Real-World Question

Although certain minerals can be identified by observing only one property, others require testing several properties to identify them. How can you identify unknown minerals?

Procedure

1. Copy the data table into your Science Journal. Obtain a set of unknown minerals.
2. Observe a numbered mineral specimen carefully. Write a star in the table entry that represents what you hypothesize is an important physical property. Choose one or two properties that you think will help most in identifying the sample.
3. Perform tests to observe your chosen properties first.
   a. To estimate hardness:
      - Rub the sample firmly against objects of known hardness and observe whether it leaves a scratch on the objects.
      - Estimate a hardness range based on which items the mineral scratches.
   b. To estimate specific gravity: Perform a density measurement.
      - Use the pan balance to determine the sample's mass, in grams.

- Measure its volume using a graduated cylinder partially filled with water. The amount of water displaced by the immersed sample, in mL, is an estimate of its volume in cm^3.
- Divide mass by volume to determine density. This number, without units, is comparable to specific gravity.

4. With the help of the Mineral Appendix or a field guide, attempt to identify the sample using the properties from step 2. Perform more physical property observations until you can identify the sample. Repeat steps 2 through 4 for each unknown.

Physical Properties of Minerals

Sample Number	Hardness	Cleavage or Fracture	Color	Specific Gravity	Luster and Streak	Crystal Shape	Other Properties	Mineral Name
1								
2			Do not write in this book.					
etc.								

Analyze Your Data

1. Which properties were most useful in identifying your samples? Which properties were least useful?
2. **Compare** the properties that worked best for you with those that worked best for other students.

Conclude and Apply

1. **Determine** two properties that distinguish clear, transparent quartz from clear, transparent calcite. Explain your choice of properties.
2. Which physical properties would be easiest to determine if you found a mineral specimen in the field?

Communicating Your Data

For three minerals, list physical properties that were important for their identification. **For more help, refer to the** Science Skill Handbook.

TIME SCIENCE AND HISTORY

SCIENCE CAN CHANGE THE COURSE OF HISTORY!

Dr. Dorothy Crowfoot Hodgkin

Like X rays, electrons are diffracted by crystalline substances, revealing information about their internal structures and symmetry. This electron diffraction pattern of titanium was obtained with an electron beam focused along a specific direction in the crystal.

Trailblazing scientist and humanitarian

What contributions did Dorothy Crowfoot Hodgkin make to science?

Dr. Hodgkin used a method called X-ray crystallography (kris tuh LAH gruh fee) to figure out the structures of crystalline substances, including vitamin B^{12}, vitamin D, penicillin, and insulin.

What's X-ray crystallography?

Scientists expose a crystalline sample to X rays. As X rays travel through a crystal, the crystal diffracts, or scatters, the X rays into a regular pattern. Like an individual's fingerprints, each crystalline substance has a unique diffraction pattern. Crystallography has applications in the life, Earth, and physical sciences. For example, geologists use X-ray crystallography to identify and study minerals found in rocks.

1910–1994

What were some obstacles Hodgkin overcame?

During the 1930s, there were few women scientists. Hodgkin was not even allowed to attend meetings of the chemistry faculty where she taught because she was a woman. Eventually, she won over her colleagues with her intelligence and tenacity.

How does Hodgkin's research help people today?

Dr. Hodgkin's discovery of the structure of insulin helped scientists learn how to control diabetes, a disease that affects more than 15 million Americans. Diabetics' bodies are unable to process sugar efficiently. Diabetes can be fatal. Fortunately, Dr. Hodgkin's research with insulin has saved many lives.

Research Look in reference books or go to the Glencoe Science Web site for information on how X-ray crystallography is used to study minerals. Write your findings and share them with your class.

For more information, visit earth.msscience.com/time

Reviewing Main Ideas

Section 1 Minerals

1. Much of what you use each day is made at least in some part from minerals.
2. All minerals are formed by natural processes and are inorganic solids with definite chemical compositions and orderly arrangements of atoms.
3. Minerals have crystal structures in one of six major crystal systems.

Section 2 Mineral Identification

1. Hardness is a measure of how easily a mineral can be scratched.
2. Luster describes how light reflects from a mineral's surface.
3. Streak is the color of the powder left by a mineral on an unglazed porcelain tile.
4. Minerals that break along smooth, flat surfaces have cleavage. When minerals break with rough or jagged surfaces, they are displaying fracture.
5. Some minerals have special properties that aid in identifying them. For example, magnetite is identified by its attraction to a magnet.

Section 3 Uses of Minerals

1. Gems are minerals that are more rare and beautiful than common minerals.
2. Minerals are useful for their physical properties and for the elements they contain.

Visualizing Main Ideas

Copy and complete the following concept map about minerals. Use the following words and phrases: the way a mineral breaks, the way a mineral reflects light, ore, a rare and beautiful mineral, how easily a mineral is scratched, streak, *and* a useful substance mined for profit.

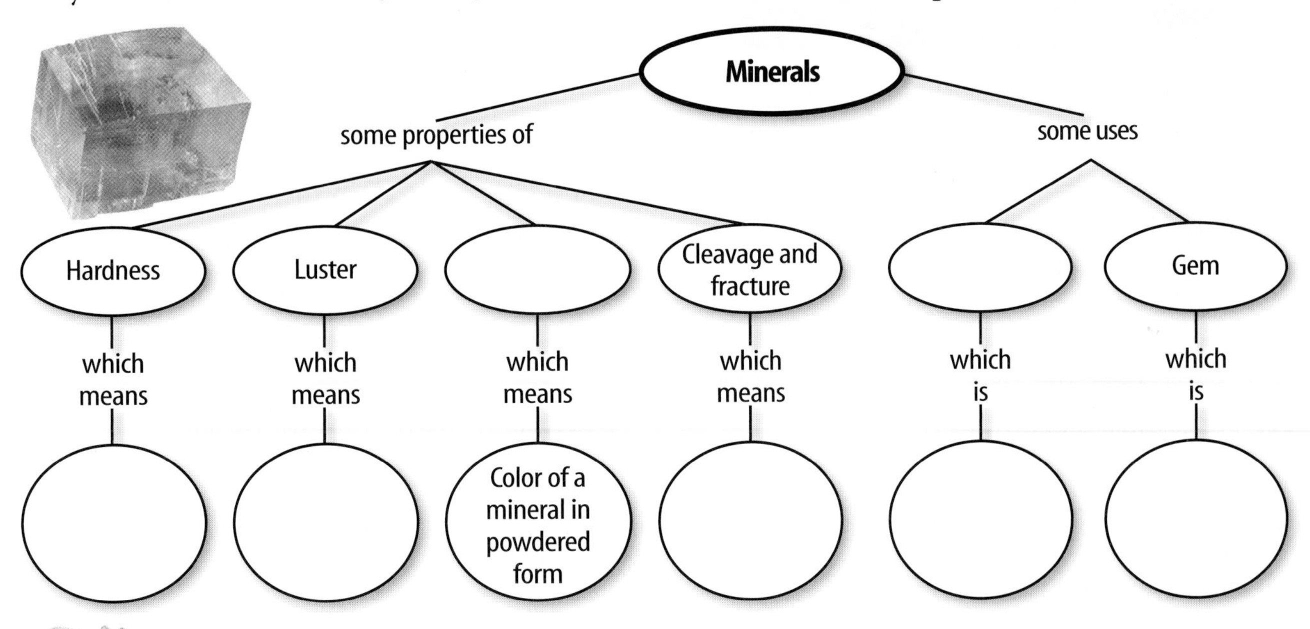

Using Vocabulary

cleavage p. 71	magma p. 65
crystal p. 63	mineral p. 62
fracture p. 71	ore p. 77
gem p. 73	silicate p. 66
hardness p. 69	specific gravity p. 70
luster p. 70	streak p. 71

Explain the difference between the vocabulary words in each of the following sets.

1. cleavage—fracture
2. crystal—mineral
3. luster—streak
4. magma—crystal
5. hardness—specific gravity
6. ore—mineral
7. crystal—luster
8. mineral—silicate
9. gem—crystal
10. streak—specific gravity

Checking Concepts

Choose the word or phrase that best answers the question.

11. Which is a characteristic of a mineral?
    A) It can be a liquid.
    B) It is organic.
    C) It has no crystal structure.
    D) It is inorganic.

12. What must all silicates contain?
    A) magnesium
    B) silicon and oxygen
    C) silicon and aluminum
    D) oxygen and carbon

13. What is the measure of how easily a mineral can be scratched?
    A) luster
    B) hardness
    C) cleavage
    D) fracture

Use the photo below to answer question 14.

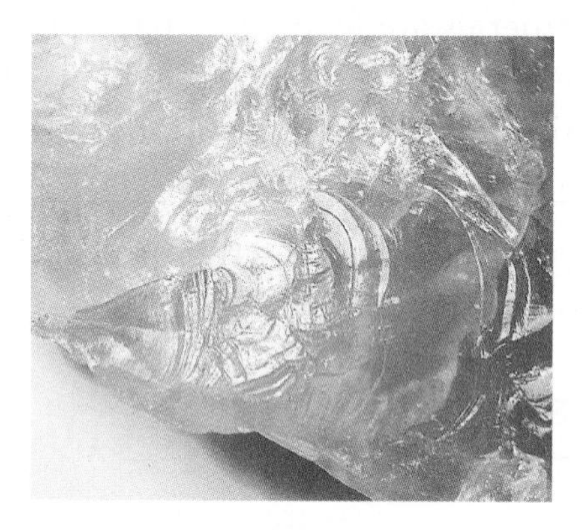

14. Examine the photo of quartz above. In what way does quartz break?
    A) cleavage
    B) fracture
    C) luster
    D) flat planes

15. Which of the following must crystalline solids have?
    A) carbonates
    B) cubic structures
    C) orderly arrangement of atoms
    D) cleavage

16. What is the color of a powdered mineral formed when rubbing it against an unglazed porcelain tile?
    A) luster
    B) density
    C) hardness
    D) streak

17. Which is hardest on Mohs scale?
    A) talc
    B) quartz
    C) diamond
    D) feldspar

Science Online earth.msscience.com/vocabulary_puzzlemaker

Thinking Critically

18. **Classify** Water is an inorganic substance that is formed by natural processes on Earth. It has a unique composition. Sometimes water is a mineral and other times it is not. Explain.

19. **Determine** how many sides a perfect salt crystal has.

20. **Apply** Suppose you let a sugar solution evaporate, leaving sugar crystals behind. Are these crystals minerals? Explain.

21. **Predict** Will a diamond leave a streak on a streak plate? Explain.

22. **Collect Data** Make an outline of how at least seven physical properties can be used to identify unknown minerals.

23. **Explain** how you would use **Table 1** to determine the hardness of any mineral.

24. **Concept Map** Copy and complete the concept map below, which includes two crystal systems and two examples from each system. Use the following words and phrases: *hexagonal, corundum, halite, fluorite,* and *quartz.*

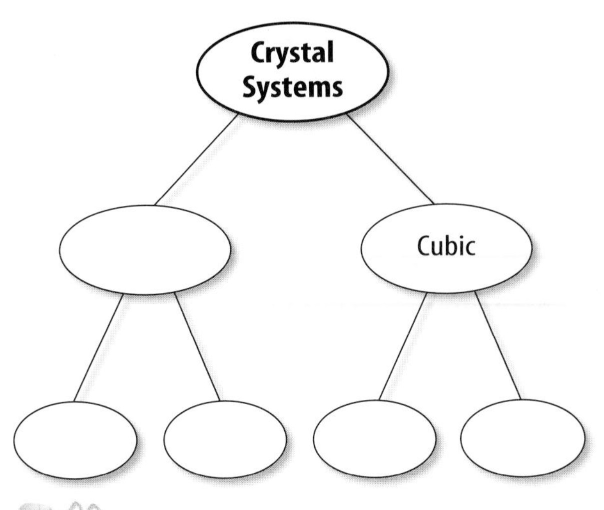

Performance Activities

25. **Display** Make a display that shows the six crystal systems of minerals. Research the crystal systems of minerals and give three examples for each crystal system. Indicate whether any of the minerals are found in your state. Describe any important uses of these minerals. Present your display to the class.

Applying Math

26. **Mineral Volume** Recall that 1 mL = 1 cm^3. Suppose that the volume of water in a graduated cylinder is 107.5 mL. A specimen of quartz, tied to a piece of string, is immersed in the water. The new water level reads 186 mL. What is the volume, in cm^3, of the piece of quartz?

Use the graph below to answer questions 27 and 28.

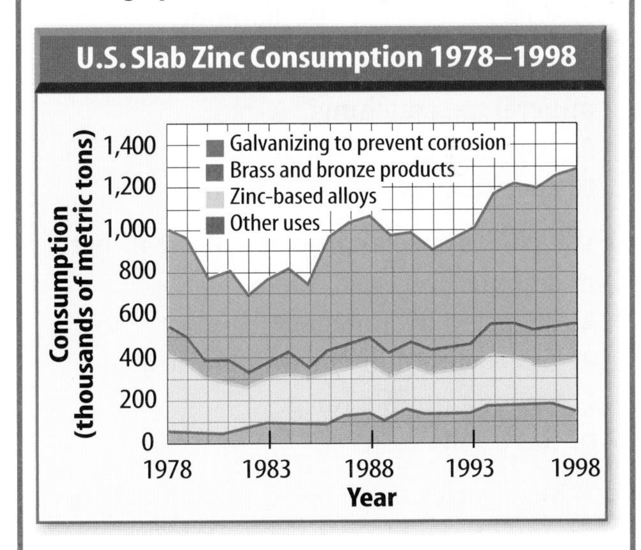

27. **Zinc Use** According to the graph above, what was the main use of zinc consumed in the United States between 1978 and 1998?

28. **Metal Products** According to the graph, approximately how many thousand metric tons of zinc were used to make brass and bronze products in 1998?

Part 1 Multiple Choice

Record your answers on the answer sheet provided by your teacher or on a sheet of paper.

Use the photo below to answer question 1.

1. To which crystal system does the crystal shown above belong?
   A. hexagonal **C.** triclinic
   B. cubic **D.** monoclinic

2. Which of the following is a common rock-forming mineral?
   A. azurite **C.** quartz
   B. gold **D.** diamond

3. Which term refers to the resistance of a mineral to scratching?
   A. hardness **C.** luster
   B. specific gravity **D.** fracture

4. Which is a special property of the mineral magnetite?
   A. attracted by a magnet
   B. fizzes with dilute hydrochloric acid
   C. forms a double image
   D. has a salty taste

5. Which causes some minerals to break along smooth, flat surfaces?
   A. streak **C.** luster
   B. cleavage **D.** fracture

Test-Taking Tip

If you are taking a timed test, keep track of time during the test. If you find that you're spending too much time on a multiple-choice question, mark your best guess and move on.

6. Which of these forms in cracks or along faults?
   A. bauxite
   B. silicates
   C. vein minerals
   D. rock-forming minerals

7. Which is the most abundant element in Earth's crust?
   A. silicon **C.** iron
   B. manganese **D.** oxygen

Use the table below to answer questions 8–10.

Mineral	Hardness
Talc	1
Gypsum	2
Calcite	3
Fluorite	4
Apatite	5
Feldspar	6
Quartz	7
Topaz	8
Corundum	9
Diamond	10

8. Which mineral in the table is softest?
   A. diamond **C.** talc
   B. feldspar **D.** gypsum

9. Which mineral will scratch feldspar but not topaz?
   A. quartz **C.** apatite
   B. calcite **D.** diamond

10. After whom is the scale shown above named?
    A. Neil Armstrong
    B. Friedrich Mohs
    C. Alfred Wegener
    D. Isaac Newton

Part 2 Short Response/Grid In

Record your answers on the answer sheet provided by your teacher or on a sheet of paper.

11. What is the definition of a mineral?
12. Why are gems valuable?
13. Explain the difference between fracture and cleavage.
14. Why is mineral color sometimes not helpful for identifying minerals?

Use the conversion factor and table below to answer questions 15–17.

1.0 carat = 0.2 grams

Diamond	Carats	Grams
Uncle Sam: largest diamond found in United States	40.4	?
Punch Jones: second largest U.S. diamond; named after boy who discovered it	?	6.89
Theresa: discovered in Wisconsin in 1888	21.5	4.3
2001 diamond production from western Australia	21,679,930	?

15. How many grams is the *Uncle Sam* diamond?
16. How many carats is the *Punch Jones* diamond?
17. How many grams of diamond were produced in western Australia in 2001?
18. What is the source of most of the diamonds that are used for industrial purposes?
19. Explain how minerals are useful to society. Describe some of their uses.

Part 3 Open Ended

Record your answers on a sheet of paper.

Use the photo below to answer question 20.

20. The mineral crystals in the rock above formed when magma cooled and are visible with the unaided eye. Hypothesize about how fast the magma cooled.
21. What is a crystal system? Why is it useful to classify mineral crystals this way?
22. How can a mineral be identified using its physical properties?
23. What is a crystal? Do all crystals have smooth crystal faces? Explain.
24. Are gases that are given off by volcanoes minerals? Why or why not?
25. What is the most abundant mineral group in Earth's crust? What elements always are found in the minerals included in this group?
26. Several layers are peeled from a piece of muscovite mica? What property of minerals does this illustrate? Describe this property in mica.

Rocks

The BIG Idea

Rocks continuously change as they are subjected to the processes of the rock cycle.

SECTION 1
The Rock Cycle

Main Idea Rocks are solid mixtures of minerals or other natural materials that change slowly through time.

SECTION 2
Igneous Rocks

Main Idea Igneous rocks are formed from molten or liquid rock material called magma.

SECTION 3
Metamorphic Rocks

Main Idea Metamorphic rocks form when solid rocks are squeezed, heated, or exposed to fluids, changing them into new rocks.

SECTION 4
Sedimentary Rocks

Main Idea Sedimentary rocks form when sediment is compacted and cemented together, or when minerals form from solutions.

How did it get there?

The giant rocky peak of El Capitan towers majestically in Yosemite National Park. Surrounded by flat landscape, it seems out of place. How did this expanse of granite rock come to be?

Science Journal Are you a rock collector? If so, write two sentences about your favorite rock. If not, describe the rocks you see in the photo in enough detail that a non-sighted person could visualize them.

Start-Up Activities

Observe and Describe Rocks

Some rocks are made of small mineral grains that lock together, like pieces of a puzzle. Others are grains of sand tightly held together or solidified lava that once flowed from a volcano. If you examine rocks closely, you sometimes can tell what they are made of.

1. Collect three different rock samples near your home or school.
2. Draw a picture of the details you see in each rock.
3. Use a magnifying lens to look for different types of materials within the same rock.
4. Describe the characteristics of each rock. Compare your drawings and descriptions with photos, drawings, and descriptions in a rocks and minerals field guide.
5. Use the field guide to try to identify each rock.
6. **Think Critically** Decide whether you think your rocks are mixtures. If so, infer or suggest what these mixtures might contain. Write your explanations in your Science Journal.

Major Rock Types Make the following Foldable to help you organize facts about types of rocks.

STEP 1 **Fold** a sheet of paper in half lengthwise. Make the back edge about 5 cm longer than the front edge.

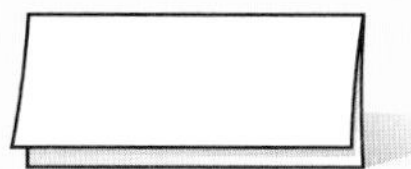

STEP 2 **Turn** the paper so the fold is on the bottom. Then fold it into thirds.

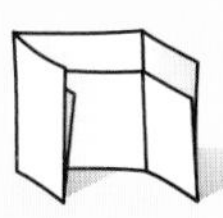

STEP 3 **Unfold and cut** only the top layer along both folds to make three tabs.

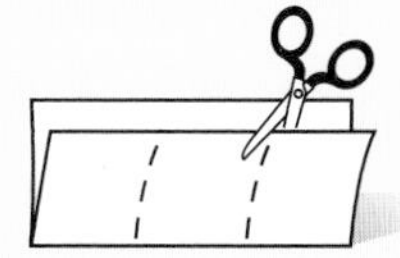

STEP 4 **Label** the Foldable as shown.

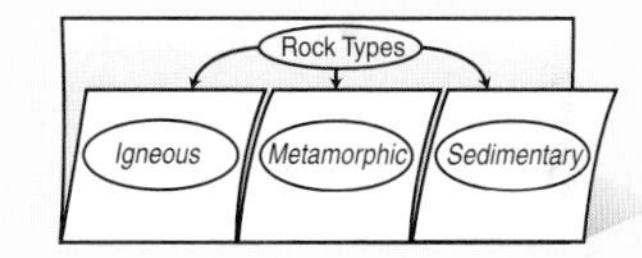

Make an Organizational Study Fold As you read the chapter, write and illustrate what you learn about the three main types of rocks in your study fold.

Preview this chapter's content and activities at earth.msscience.com

Get Ready to Read

Monitor

1 Learn It! An important strategy to help you improve your reading is monitoring, or finding your reading strengths and weaknesses. As you read, monitor yourself to make sure the text makes sense. Discover different monitoring techniques you can use at different times, depending on the type of test and situation.

2 Practice It! The paragraph below appears in Section 1. Read the passage and answer the questions that follow. Discuss your answers with other students to see how they monitor their reading.

> . . . rocks change by many processes. For example, a sedimentary rock can change by heat and pressure to form a metamorphic rock. The metamorphic rock then can melt and later cool to form an igneous rock. The igneous rock then could be broken into fragments by weathering and erode away. The fragments might later compact and cement together to form another sedimentary rock.
>
> *—from page 91*

- What questions do you still have after reading?
- Do you understand all of the words in the passage?
- Did you have to stop reading often? Is the reading level appropriate for you?

3 Apply It! Identify one paragraph that is difficult to understand. Discuss it with a partner to improve your understanding.

Monitor your reading by slowing down or speeding up depending on your understanding of the text.

Target Your Reading

Use this to focus on the main ideas as you read the chapter.

1. **Before you read** the chapter, respond to the statements below on your worksheet or on a numbered sheet of paper.
   - Write an **A** if you **agree** with the statement.
   - Write a **D** if you **disagree** with the statement.

2. **After you read** the chapter, look back to this page to see if you've changed your mind about any of the statements.
   - If any of your answers changed, explain why.
   - Change any false statements into true statements.
   - Use your revised statements as a study guide.

Before You Read A or D		Statement	After You Read A or D
	1	The three major types of rock are igneous, sedimentary, and metamorphic rocks.	
	2	During the rock cycle, any given rock can change into any of the three major rock types.	
	3	When magma reaches Earth's surface and flows from volcanoes, it is called lava.	
	4	The pressure exerted by rocks produces all the heat used to form magma.	
	5	All igneous rock is formed from lava that cooled on Earth's surface.	
	6	Before any rock is transformed into a metamorphic rock, some of the minerals must be melted.	
	7	Metamorphic rock can form only under intense heat and pressure.	
	8	Sandstone, limestone, chalk, rock salt, and coal are all examples of sedimentary rocks.	
	9	Sedimentary rocks can be made of just about any material found in nature.	

Science Online
Print out a worksheet of this page at earth.msscience.com

section 1

The Rock Cycle

as you read

What You'll Learn

- **Distinguish** between a rock and a mineral.
- **Describe** the rock cycle and some changes that a rock could undergo.

Why It's Important

Rocks exist everywhere, from under deep oceans and in high mountain ranges, to the landscape beneath your feet.

Review Vocabulary

mineral: a naturally occurring, inorganic solid with a definite chemical composition and an orderly arrangement of atoms

New Vocabulary

- rock
- rock cycle

What is a rock?

Imagine you and some friends are exploring a creek. Your eye catches a glint from a piece of rock at the edge of the water. As you wander over to pick up the rock, you notice that it is made of different-colored materials. Some of the colors reflect light, while others are dull. You put the rock in your pocket for closer inspection in science lab.

Common Rocks The next time you walk past a large building or monument, stop and take a close look at it. Chances are that it is made out of common rock. In fact, most rock used for building stone contains one or more common minerals, called rock-forming minerals, such as quartz, feldspar, mica, or calcite. When you look closely, the sparkles you see are individual crystals of minerals. A **rock** is a mixture of such minerals, rock fragments, volcanic glass, organic matter, or other natural materials. **Figure 1** shows minerals mixed together to form the rock granite. You might even find granite near your home.

Figure 1 Mount Rushmore, in South Dakota, is made of granite. Granite is a mixture of feldspar, quartz, mica, hornblende, and other minerals.

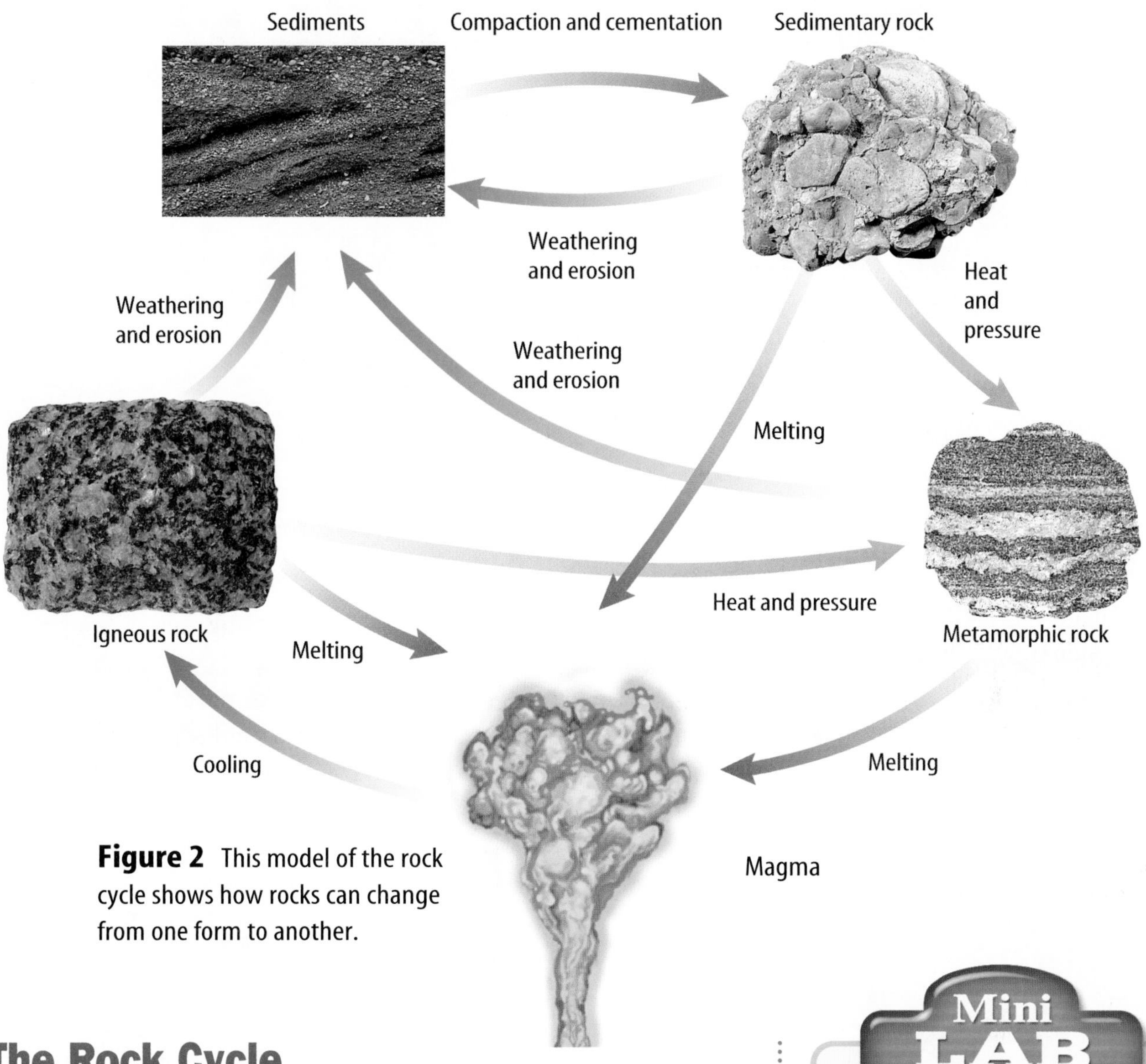

Figure 2 This model of the rock cycle shows how rocks can change from one form to another.

The Rock Cycle

To show how rocks slowly change through time, scientists have created a model called the **rock cycle,** shown in **Figure 2.** It illustrates the processes that create and change rocks. The rock cycle shows the three types of rock—igneous, metamorphic, and sedimentary—and the processes that form them.

Look at the rock cycle and notice that rocks change by many processes. For example, a sedimentary rock can change by heat and pressure to form a metamorphic rock. The metamorphic rock then can melt and later cool to form an igneous rock. The igneous rock then could be broken into fragments by weathering and erode away. The fragments might later compact and cement together to form another sedimentary rock. Any given rock can change into any of the three major rock types. A rock even can transform into another rock of the same type.

Reading Check *What is illustrated by the rock cycle?*

Mini LAB

Modeling Rock

Procedure

1. Mix about 10 mL of **white glue** with about 7 g of **dirt** or **sand** in a **small paper cup.**
2. Stir the mixture and then allow it to harden overnight.
3. Tear away the paper cup carefully from your mixture.

Analysis

1. Which rock type is similar to your hardened mixture?
2. Which part of the rock cycle did you model?

Try at Home

NATIONAL GEOGRAPHIC

VISUALIZING THE ROCK CYCLE

Figure 3

Rocks continuously form and transform in a process that geologists call the rock cycle. For example, molten rock—from volcanoes such as Washington's Mount Rainier, background—cools and solidifies to form igneous rock. It slowly breaks down when exposed to air and water to form sediments. These sediments are compacted or cemented into sedimentary rock. Heat and pressure might transform sedimentary rock into metamorphic rock. When metamorphic rock melts and hardens, igneous rock forms again. There is no distinct beginning, nor is there an end, to the rock cycle.

▲ The black sand beach of this Polynesian island is sediment weathered and eroded from the igneous rock of a volcano nearby.

▲ This alluvial fan on the edge of Death Valley, California, was formed when gravel, sand, and finer sediments were deposited by a stream emerging from a mountain canyon.

▲ Layers of shale and chalk form Kansas's Monument Rocks. They are remnants of sediments deposited on the floor of the ancient sea that once covered much of this region.

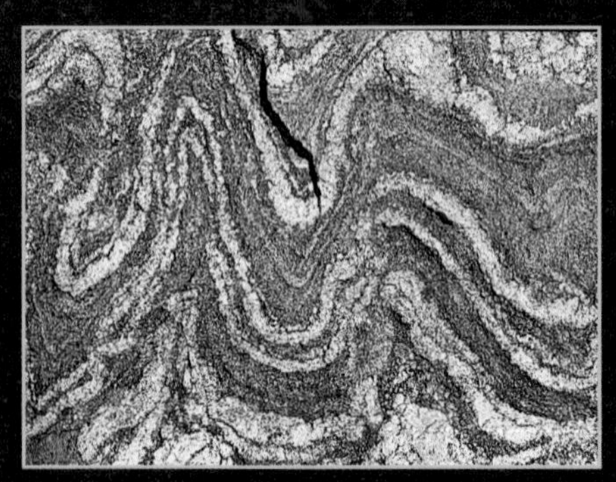

▲ Heat and pressure deep below Earth's surface can change rock into metamorphic rock, like this banded gneiss.

Matter and the Rock Cycle

The rock cycle, illustrated in **Figure 3,** shows how rock can be weathered to small rock and mineral grains. This material then can be eroded and carried away by wind, water, or ice. When you think of erosion, it might seem that the material is somehow destroyed and lost from the cycle. This is not the case. The chemical elements that make up minerals and rocks are not destroyed. This fact illustrates the principle of conservation of matter. The changes that take place in the rock cycle never destroy or create matter. The elements are just redistributed in other forms.

What is the principle of conservation of matter?

Figure 4 The rock formations at Siccar Point, Scotland, show that rocks undergo constant change.

Discovering the Rock Cycle James Hutton, a Scottish physician and naturalist, first recognized in 1788 that rocks undergo profound changes. Hutton noticed, among other things, that some layers of solid rock in Siccar Point, shown in **Figure 4,** had been altered since they formed. Instead of showing a continuous pattern of horizontal layering, some of the rock layers at Siccar Point are tilted and partly eroded. However, the younger rocks above them are nearly horizontal.

Hutton published these and other observations, which proved that rocks are subject to constant change. Hutton's early recognition of the rock cycle continues to influence geologists.

section 1 review

Summary

What is a rock?

- Rocks are mixtures of minerals, rock fragments, organic matter, volcanic glass, and other materials found in nature.

The Rock Cycle

- The three major types of rock are igneous, metamorphic, and sedimentary.
- Rock cycle processes do not create or destroy matter.
- Processes that are part of the rock cycle change rocks slowly over time.
- In the late eighteenth century, James Hutton recognized some rock cycle processes by observing rocks in the field.
- Some of Hutton's ideas continue to influence geologic thinking today.

Self Check

1. **Explain** how rocks differ from minerals.
2. **Compare and contrast** igneous and metamorphic rock formation.
3. **Describe** the major processes of the rock cycle.
4. **Explain** one way that the rock cycle can illustrate the principle of conservation of matter.
5. **Think Critically** How would you define magma based on the illustration in **Figure 2?** How would you define sediment and sedimentary rock?

Applying Skills

6. **Communicate** Review the model of the rock cycle in **Figure 2.** In your Science Journal, write a story or poem that explains what can happen to a sedimentary rock as it changes throughout the rock cycle.

section 2

Igneous Rocks

as you read

What You'll Learn

- **Recognize** magma and lava as the materials that cool to form igneous rocks.
- **Contrast** the formation of intrusive and extrusive igneous rocks.
- **Contrast** granitic and basaltic igneous rocks.

Why It's Important

Igneous rocks are the most abundant kind of rock in Earth's crust. They contain many valuable resources.

Review Vocabulary

element: substance made of one type of atom that cannot be broken down by ordinary chemical or physical means

New Vocabulary

- igneous rock
- lava
- intrusive
- extrusive
- basaltic
- granitic

Formation of Igneous Rocks

Perhaps you've heard of recent volcanic eruptions in the news. When some volcanoes erupt, they eject a flow of molten rock material, as shown in **Figure 5.** Molten rock material, called magma, flows when it is hot and becomes solid when it cools. When hot magma cools and hardens, it forms **igneous** (IHG nee us) **rock.** Why do volcanoes erupt, and where does the molten material come from?

Magma In certain places within Earth, the temperature and pressure are just right for rocks to melt and form magma. Most magmas come from deep below Earth's surface. Magma is located at depths ranging from near the surface to about 150 km below the surface. Temperatures of magmas range from about 650°C to 1,200°C, depending on their chemical compositions and pressures exerted on them.

The heat that melts rocks comes from sources within Earth's interior. One source is the decay of radioactive elements within Earth. Some heat is left over from the formation of the planet, which originally was molten. Radioactive decay of elements contained in rocks balances some heat loss as Earth continues to cool.

Because magma is less dense than surrounding solid rock, it is forced upward toward the surface, as shown in **Figure 6.** When magma reaches Earth's surface and flows from volcanoes, it is called **lava.**

Figure 5 Some lava is highly fluid and free-flowing, as shown by this spectacular lava fall in Volcano National Park, East Rift, Kilauea, Hawaii.

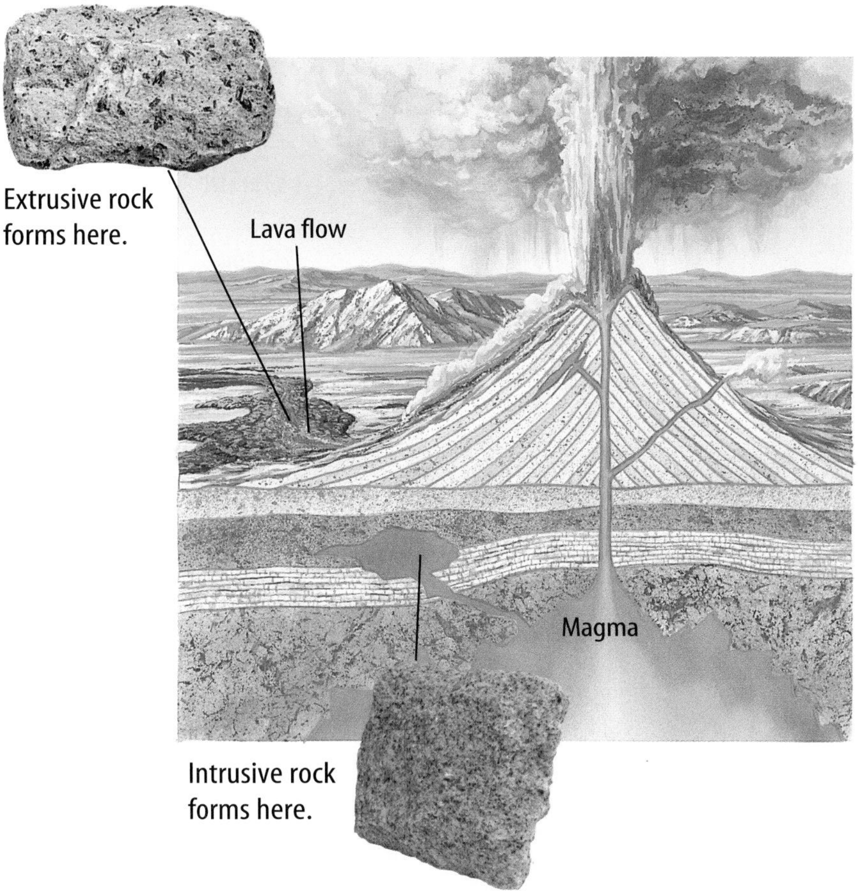

Extrusive rock forms here.

Lava flow

Magma

Intrusive rock forms here.

Figure 6 Intrusive rocks form from magma trapped below Earth's surface. Extrusive rocks form from lava flowing at the surface.

Intrusive Rocks Magma is melted rock material composed of common elements and fluids. As magma cools, atoms and compounds in the liquid rearrange themselves into new crystals called mineral grains. Rocks form as these mineral grains grow together. Rocks that form from magma below the surface, as illustrated in **Figure 6,** are called **intrusive** igneous rocks. Intrusive rocks are found at the surface only after the layers of rock and soil that once covered them have been removed by erosion. Erosion occurs when the rocks are pushed up by forces within Earth. Because intrusive rocks form at depth and they are surrounded by other rocks, it takes a long time for them to cool. Slowly cooled magma produces individual mineral grains that are large enough to be observed with the unaided eye.

Extrusive Rocks **Extrusive** igneous rocks are formed as lava cools on the surface of Earth. When lava flows on the surface, as illustrated in **Figure 6,** it is exposed to air and water. Lava, such as the basaltic lava shown in **Figure 5,** cools quickly under these conditions. The quick cooling rate keeps mineral grains from growing large, because the atoms in the liquid don't have the time to arrange into large crystals. Therefore, extrusive igneous rocks are fine grained.

What controls the grain size of an igneous rock?

Table 1 Common Igneous Rocks			
Magma Type	Basaltic	Andesitic	Granitic
Intrusive	Gabbro	Diorite	Granite
Extrusive	Basalt Scoria	Andesite	Rhyolite Pumice Obsidian

Volcanic Glass Pumice, obsidian, and scoria are examples of volcanic glass. These rocks cooled so quickly that few or no mineral grains formed. Most of the atoms in these rocks are not arranged in orderly patterns, and few crystals are present.

In the case of pumice and scoria, gases become trapped in the gooey molten material as it cools. Some of these gases eventually escape, but holes are left behind where the rock formed around the pockets of gas.

Topic: Rock Formation
Visit earth.msscience.com for Web links to information about intrusive and extrusive rocks.

Activity List several geographic settings where intrusive or extrusive rocks are found. Select one setting for intrusive rocks, and one for extrusive rocks. Describe how igneous rocks form in the two settings, and locate an example of each on a map.

Classifying Igneous Rocks

Igneous rocks are intrusive or extrusive depending on how they are formed. A way to further classify these rocks is by the magma from which they form. As shown in **Table 1,** an igneous rock can form from basaltic, andesitic, or granitic magma. The type of magma that cools to form an igneous rock determines important chemical and physical properties of that rock. These include mineral composition, density, color, and melting temperature.

Reading Check *Name two ways igneous rocks are classified.*

Basaltic Rocks **Basaltic** (buh SAWL tihk) igneous rocks are dense, dark-colored rocks. They form from magma that is rich in iron and magnesium and poor in silica, which is the compound SiO_2. The presence of iron and magnesium in minerals in basalt gives basalt its dark color. Basaltic lava is fluid and flows freely from volcanoes in Hawaii, such as Kilauea. How does this explain the black beach sand common in Hawaii?

Granitic Rocks **Granitic** igneous rocks are light-colored rocks of lower density than basaltic rocks. Granitic magma is thick and stiff and contains lots of silica but lesser amounts of iron and magnesium. Because granitic magma is stiff, it can build up a great deal of gas pressure, which is released explosively during violent volcanic eruptions.

Andesitic Rocks Andesitic igneous rocks have mineral compositions between those of basaltic and granitic rocks. Many volcanoes around the rim of the Pacific Ocean formed from andesitic magmas. Like volcanoes that erupt granitic magma, these volcanoes also can erupt violently.

Take another look an **Table 1.** Basalt forms at the surface of Earth because it is an extrusive rock. Granite forms below Earth's surface from magma with a high concentration of silica. When you identify an igneous rock, you can infer how it formed and the type of magma that it formed from.

Melting Rock Inside Earth, materials contained in rocks can melt. In your Science Journal, describe what is happening to the atoms and molecules to cause this change of state.

section 2 review

Summary

Formation of Igneous Rocks

- When molten rock material, called magma, cools and hardens, igneous rock forms.
- Intrusive igneous rocks form as magma cools and hardens slowly, beneath Earth's surface.
- Extrusive igneous rocks form as lava cools and hardens rapidly, at or above Earth's surface.

Classifying Igneous Rocks

- Igneous rocks are further classified according to their mineral compositions.
- The violent nature of some volcanic eruptions is partly explained by the composition of the magma that feeds them.

Self Check

1. **Explain** why some types of magma form igneous rocks that are dark colored and dense.
2. **Identify** the property of magma that causes it to be forced upward toward Earth's surface.
3. **Explain** The texture of obsidian is best described as glassy. Why does obsidian contain few or no mineral grains?
4. **Think Critically** Study the photos in **Table 1.** How are granite and rhyolite similar? How are they different?

Applying Skills

5. **Make and Use Graphs** Four elements make up most of the rocks in Earth's crust. They are: *oxygen—46.6 percent, aluminum—8.1 percent, silicon—27.7 percent,* and *iron—5.0 percent.* Make a bar graph of these data. What might you infer from the low amount of iron?

Igneous Rock Clues

You've learned how color often is used to estimate the composition of an igneous rock. The texture of an igneous rock describes its overall appearance, including mineral grain sizes and the presence or absence of bubble holes, for example. In most cases, grain size relates to how quickly the magma or lava cooled. Crystals you can see without a magnifying lens indicate slower cooling. Smaller, fine-grained crystals indicate quicker cooling, possibly due to volcanic activity. Rocks with glassy textures cooled so quickly that there was no time to form mineral grains.

Real-World Question

What does an igneous rock's texture and color indicate about its formation history?

Goals

- **Classify** different samples of igneous rocks by color and infer their composition.
- **Observe** the textures of igneous rocks and infer how they formed.

Materials

rhyolite
basalt
vesicular basalt
pumice
granite
obsidian
gabbro
magnifying lens

Safety Precautions

WARNING: *Some rock samples might have sharp edges. Always use caution while handling samples.*

Procedure

1. **Arrange** rocks according to color (light or dark). Record your observations in your Science Journal.
2. **Arrange** rocks according to similar texture. Consider grain sizes and shapes, presence of holes, etc. Use your magnifying lens to see small features more clearly. Record your observations.

Conclude and Apply

1. **Infer** which rocks are granitic based on color.
2. **Infer** which rocks cooled quickly. What observations led you to this inference?
3. **Identify** any samples that suggest gases were escaping from them as they cooled.
4. **Describe** Which samples have a glassy appearance? How did these rocks form?
5. **Infer** which samples are not volcanic. Explain.

Communicating Your Data

Research the compositions of each of your samples. Did the colors of any samples lead you to infer the wrong compositions? Communicate to your class what you learned.

section 3

Metamorphic Rocks

Formation of Metamorphic Rocks

Have you ever packed your lunch in the morning and not been able to recognize it at lunchtime? You might have packed a sandwich, banana, and a large bottle of water. You know you didn't smash your lunch on the way to school. However, you didn't think about how the heavy water bottle would damage your food if the bottle was allowed to rest on the food all day. The heat in your locker and the pressure from the heavy water bottle changed your sandwich. Like your lunch, rocks can be affected by changes in temperature and pressure.

Metamorphic Rocks Rocks that have changed because of changes in temperature and pressure or the presence of hot, watery fluids are called **metamorphic rocks.** Changes that occur can be in the form of the rock, shown in **Figure 7,** the composition of the rock, or both. Metamorphic rocks can form from igneous, sedimentary, or other metamorphic rocks. What Earth processes can change these rocks?

as you read

***What* You'll Learn**

- **Describe** the conditions in Earth that cause metamorphic rocks to form.
- **Classify** metamorphic rocks as foliated or nonfoliated.

***Why* It's Important**

Metamorphic rocks are useful because of their unique properties.

Review Vocabulary
pressure: the amount of force exerted per unit of area

New Vocabulary
- metamorphic rock
- foliated
- nonfoliated

Figure 7 The mineral grains in granite are flattened and aligned when heat and pressure are applied to them. As a result, gneiss is formed. **Describe** *other conditions that can cause metamorphic rocks to form.*

Topic: Shale Metamorphism

Visit earth.msscience.com for Web links to information about the metamorphism of shale. Communicate to your class what you learn.

Activity Make a table with headings that are major rock types that form from shale metamorphism. Under each rock heading, make a list of minerals that can occur in the rock.

Heat and Pressure Rocks beneath Earth's surface are under great pressure from rock layers above them. Temperature also increases with depth in Earth. In some places, the heat and pressure are just right to cause rocks to melt and magma to form. In other areas where melting doesn't occur, some mineral grains can change by dissolving and recrystallizing—especially in the presence of fluids. Sometimes, under these conditions, minerals exchange atoms with surrounding minerals and new, bigger minerals form.

Depending upon the amount of pressure and temperature applied, one type of rock can change into several different metamorphic rocks, and each type of metamorphic rock can come from several kinds of parent rocks. For example, the sedimentary rock shale will change into slate. As increasing pressure and temperature are applied, the slate can change into phyllite, then schist, and eventually gneiss. Schist also can form when basalt is metamorphosed, or changed, and gneiss can come from granite.

How can one type of rock change into several different metamorphic rocks?

Hot Fluids Did you know that fluids can move through rock? These fluids, which are mostly water with dissolved elements and compounds, can react chemically with a rock and change its composition, especially when the fluids are hot. That's what happens when rock surrounding a hot magma body reacts with hot fluids from the magma, as shown in **Figure 8.** Most fluids that transform rocks during metamorphic processes are hot and mainly are comprised of water and carbon dioxide.

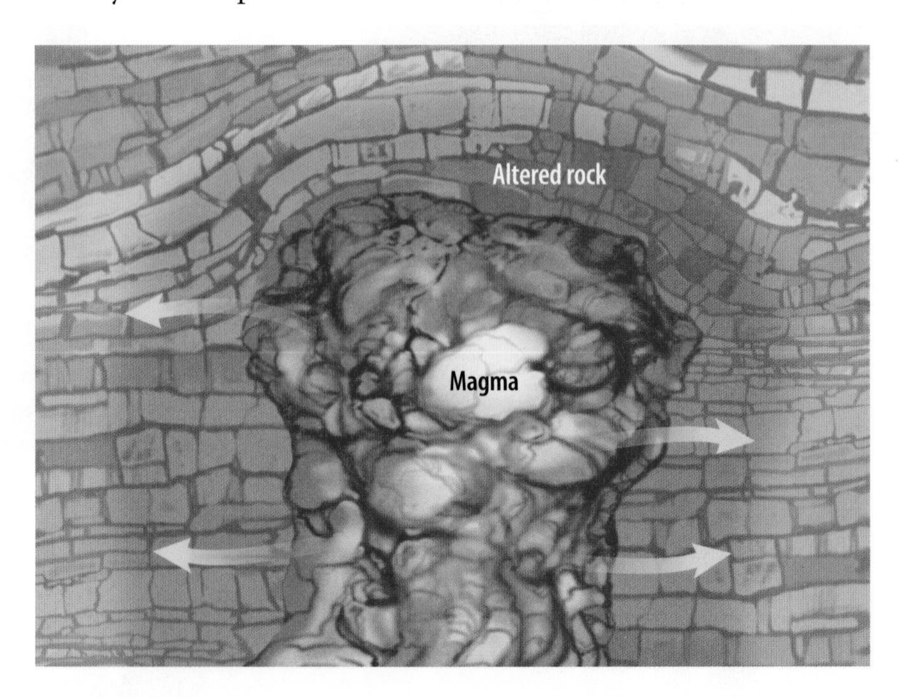

Figure 8 In the presence of hot, water-rich fluids, solid rock can change in mineral composition without having to melt.

Classifying Metamorphic Rocks

Metamorphic rocks form from igneous, sedimentary, or other metamorphic rocks. Heat, pressure, and hot fluids trigger the changes. Each resulting rock can be classified according to its composition and texture.

Foliated Rocks When mineral grains line up in parallel layers, the metamorphic rock is said to have a **foliated** texture. Two examples of foliated rocks are slate and gneiss. Slate forms from the sedimentary rock shale. The minerals in shale arrange into layers when they are exposed to heat and pressure. As **Figure 9** shows, slate separates easily along these foliation layers.

The minerals in slate are pressed together so tightly that water can't pass between them easily. Because it's watertight, slate is ideal for paving around pools and patios. The naturally flat nature of slate and the fact that it splits easily make it useful for roofing and tiling many surfaces.

Gneiss (NISE), another foliated rock, forms when granite and other rocks are changed. Foliation in gneiss shows up as alternating light and dark bands. Movement of atoms has separated the dark minerals, such as biotite mica, from the light minerals, which are mainly quartz and feldspar.

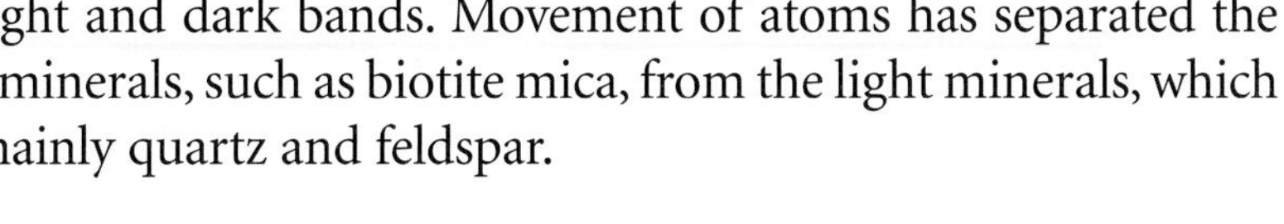

What type of metamorphic rock is composed of mineral grains arranged in parallel layers?

Figure 9 Slate often is used as a building or landscaping material. **Identify** *the properties that make slate so useful for these purposes.*

Figure 10 This exhibit in Vermont shows the beauty of carved marble.

Nonfoliated Rocks In some metamorphic rocks, layering does not occur. The mineral grains grow and rearrange, but they don't form layers. This process produces a **nonfoliated** texture.

Sandstone is a sedimentary rock that's often composed mostly of quartz grains. When sandstone is heated under a lot of pressure, the grains of quartz grow in size and become interlocking, like the pieces of a jigsaw puzzle. The resulting rock is called quartzite.

Marble is another nonfoliated metamorphic rock. Marble forms from the sedimentary rock limestone, which is composed of the mineral calcite. Usually, marble contains several other minerals besides calcite. For example, hornblende and serpentine give marble a black or greenish tone, whereas hematite makes it red. As **Figure 10** shows, marble is a popular material for artists to sculpt because it is not as hard as other rocks.

So far, you've investigated only a portion of the rock cycle. You still haven't observed how sedimentary rocks are formed and how igneous and metamorphic rocks evolve from them. The next section will complete your investigation of the rock cycle.

section 3 review

Summary

Formation of Metamorphic Rocks

- Changes in pressure, temperature, or the presence of fluids can cause metamorphic rocks to form.
- Rock, altered by metamorphic processes at high temperatures and pressures, changes in the solid state without melting.
- Hot fluids that move through and react with preexisting rock are composed mainly of water and carbon dioxide.
- One source of hot, watery fluids is magma bodies close to the changing rock.
- Any parent rock type—igneous, metamorphic, or sedimentary—can become a metamorphic rock.

Classifying Metamorphic Rocks

- Texture and mineral composition determine how a metamorphic rock is classified.
- Physical properties of metamorphic rocks, such as the watertight nature of slate, make them useful for many purposes.

Self Check

1. **Explain** what role fluids play in rock metamorphism.
2. **Describe** how metamorphic rocks are classified. What are the characteristics of rocks in each of these classifications?
3. **Identify** Give an example of a foliated and a nonfoliated metamorphic rock. Name one of their possible parent rocks.
4. **Think Critically** Marble is a common material used to make sculptures, but not just because it's a beautiful stone. What properties of marble make it useful for this purpose?

Applying Skills

5. **Concept Map** Put the following events in an events-chain concept map that explains how a metamorphic rock might form from an igneous rock. *Hint: Start with "Igneous Rock Forms."* Use each event just once.

   Events: *sedimentary rock forms, weathering occurs, heat and pressure are applied, igneous rock forms, metamorphic rock forms, erosion occurs, sediments are formed, deposition occurs*

Science Online earth.msscience.com/self_check_quiz

section 4

Sedimentary Rocks

Formation of Sedimentary Rocks

Igneous rocks are the most common rocks on Earth, but because most of them exist below the surface, you might not have seen too many of them. That's because 75 percent of the rocks exposed at the surface are sedimentary rocks.

Sediments are loose materials such as rock fragments, mineral grains, and bits of shell that have been moved by wind, water, ice, or gravity. If you look at the model of the rock cycle, you will see that sediments come from already-existing rocks that are weathered and eroded. **Sedimentary rock** forms when sediments are pressed and cemented together, or when minerals form from solutions.

Stacked Rocks Sedimentary rocks often form as layers. The older layers are on the bottom because they were deposited first. Sedimentary rock layers are a lot like the books and papers in your locker. Last week's homework is on the bottom, and today's notes will be deposited on top of the stack. However, if you disturb the stack, the order in which the books and papers are stacked will change, as shown in **Figure 11.** Sometimes, forces within Earth overturn layers of rock, and the oldest are no longer on the bottom.

as you read

What You'll Learn

- **Explain** how sedimentary rocks form from sediments.
- **Classify** sedimentary rocks as detrital, chemical, or organic in origin.
- **Summarize** the rock cycle.

Why It's Important

Some sedimentary rocks, like coal, are important sources of energy.

Review Vocabulary
weathering: surface processes that work to break down rock mechanically or chemically

New Vocabulary
- sediment
- sedimentary rock
- compaction
- cementation

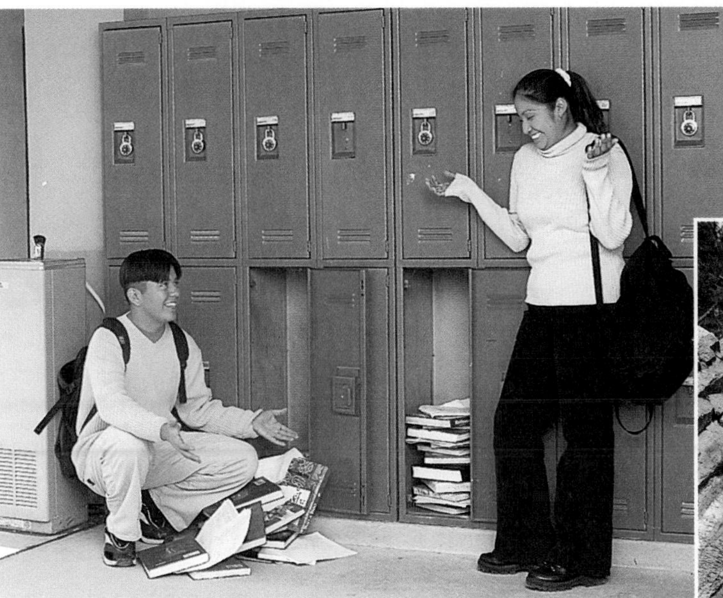

Figure 11 Like sedimentary rock layers, the oldest paper is at the bottom of the stack. If the stack is disturbed, then it is no longer in order.

Classifying Sediments

Procedure

WARNING: *Use care when handling sharp objects.*

1. Collect different samples of **sediment.**
2. Spread them on a sheet of **paper.**
3. Use **Table 2** to determine the size range of gravel-sized sediment.
4. Use **tweezers or a dissecting probe** and a **magnifying lens** to separate the gravel-sized sediments.
5. Separate the gravel into piles—rounded or angular.

Analysis

1. Describe the grains in both piles.
2. Determine what rock could form from each type of sediment you have.

Classifying Sedimentary Rocks

Sedimentary rocks can be made of just about any material found in nature. Sediments come from weathered and eroded igneous, metamorphic, and sedimentary rocks. Sediments also come from the remains of some organisms. The composition of a sedimentary rock depends upon the composition of the sediments from which it formed.

Like igneous and metamorphic rocks, sedimentary rocks are classified by their composition and by the manner in which they formed. Sedimentary rocks usually are classified as detrital, chemical, or organic.

Detrital Sedimentary Rocks

The word *detrital* (dih TRI tul) comes from the Latin word *detritus,* which means "to wear away." Detrital sedimentary rocks, such as those shown in **Table 2,** are made from the broken fragments of other rocks. These loose sediments are compacted and cemented together to form solid rock.

Weathering and Erosion When rock is exposed to air, water, or ice, it is unstable and breaks down chemically and mechanically. This process, which breaks rocks into smaller pieces, is called weathering. **Table 2** shows how these pieces are classified by size. The movement of weathered material is called erosion.

Compaction Erosion moves sediments to a new location, where they then are deposited. Here, layer upon layer of sediment builds up. Pressure from the upper layers pushes down on the lower layers. If the sediments are small, they can stick together and form solid rock. This process, shown in **Figure 12,** is called **compaction.**

Reading Check *How do rocks form through compaction?*

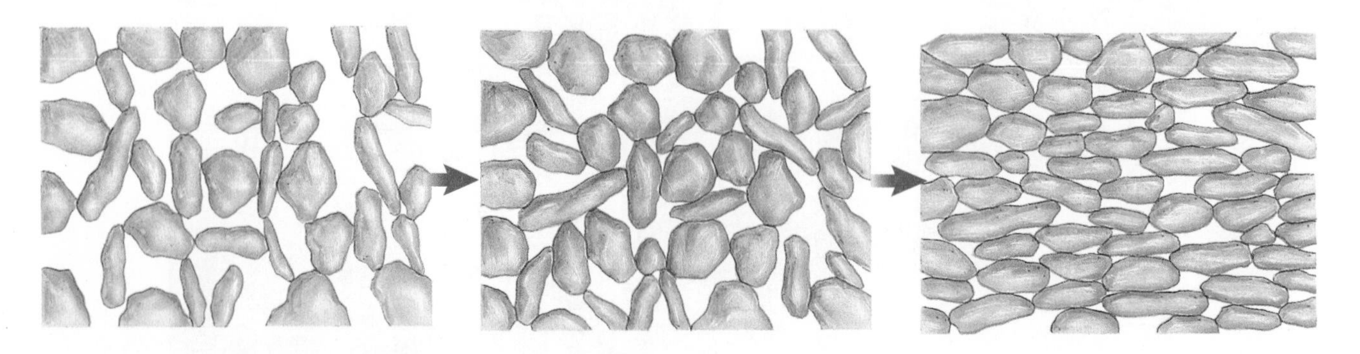

Figure 12 During compaction, pore space between sediments decreases, causing them to become packed together more tightly.

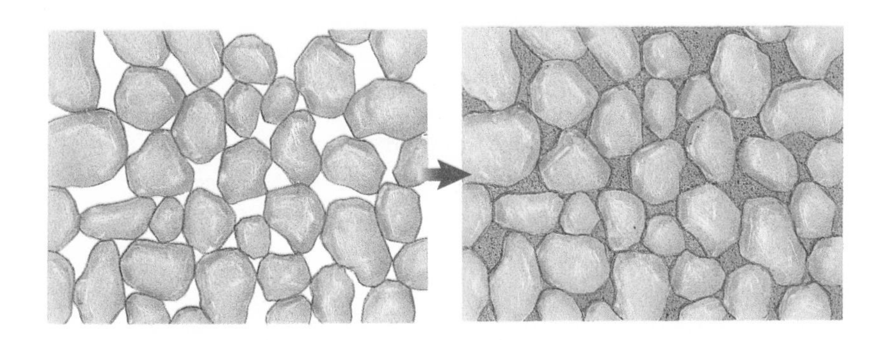

Figure 13 Sediments are cemented together as minerals crystallize between grains.

Cementation If sediments are large, like sand and pebbles, pressure alone can't make them stick together. Large sediments have to be cemented together. As water moves through soil and rock, it picks up materials released from minerals during weathering. The resulting solution of water and dissolved materials moves through open spaces between sediments. **Cementation,** which is shown in **Figure 13,** occurs when minerals such as quartz, calcite, and hematite are deposited between the pieces of sediment. These minerals, acting as natural cements, hold the sediment together like glue, making a detrital sedimentary rock.

Shape and Size of Sediments Detrital rocks have granular textures, much like granulated sugar. They are named according to the shapes and sizes of the sediments that form them. For example, conglomerate and breccia both form from large sediments, as shown in **Table 2.** If the sediments are rounded, the rock is called conglomerate. If the sediments have sharp angles, the rock is called breccia. The roundness of sediment particles depends on how far they have been moved by wind or water.

Table 2 Sediment Sizes and Detrital Rocks

Sediment	Clay	Silt	Sand	Gravel
Size Range	<0.004 mm	0.004–0.063 mm	0.063–2 mm	>2 mm
Example	Shale	Siltstone	Sandstone	Conglomerate (shown) or Breccia

Figure 14 Although concrete strongly resembles conglomerate, concrete is not a rock because it does not occur in nature.

Materials Found in Sedimentary Rocks The gravel-sized sediments in conglomerate and breccia can consist of any type of rock or mineral. Often, they are composed of chunks of the minerals quartz and feldspar. They also can be pieces of rocks such as gneiss, granite, or limestone. The cement that holds the sediments together usually is made of quartz or calcite.

Have you ever looked at the concrete in sidewalks, driveways, and stepping stones? The concrete in **Figure 14** is made of gravel and sand grains that have been cemented together. Although the structure is similar to that of naturally occurring conglomerate, it cannot be considered a rock.

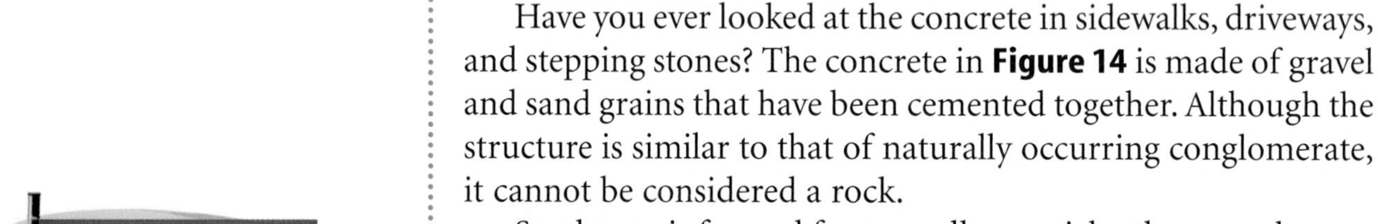

Sedimentary Petrology Research the work done by sedimentary petrologists. Include examples of careers in academia and in industry.

Sandstone is formed from smaller particles than conglomerates and breccias. Its sand-sized sediments can be just about any mineral, but they are usually grains of minerals such as quartz and feldspar that are resistant to weathering. Siltstone is similar to sandstone except it is made of smaller, silt-sized particles. Shale is a detrital sedimentary rock that is made mainly of clay-sized particles. Clay-sized sediments are compacted together by pressure from overlying layers.

Chemical Sedimentary Rocks

Chemical sedimentary rocks form when dissolved minerals come out of solution. You can show that salt is deposited in the bottom of a glass or pan when saltwater solution evaporates. In a similar way, minerals collect when seas or lakes evaporate. The deposits of minerals that come out of solution form sediments and rocks. For example, the sediment making up New Mexico's White Sands desert consists of pieces of a chemical sedimentary rock called rock gypsum. Chemical sedimentary rocks are different. They are not made from pieces of preexisting rocks.

Reading Check *How do chemical sedimentary rocks form?*

Limestone Calcium carbonate is carried in solution in ocean water. When calcium carbonate ($CaCO_3$) comes out of solution as calcite and its many crystals grow together, limestone forms. Limestone also can contain other minerals and sediments, but it must be at least 50 percent calcite. Limestone usually is deposited on the bottom of lakes or shallow seas. Large areas of the central United States have limestone bedrock because seas covered much of the country for millions of years. It is hard to imagine Kansas being covered by ocean water, but it has happened several times throughout geological history.

Rock Salt When water that is rich in dissolved salt evaporates, it often deposits the mineral halite. Halite forms rock salt, shown in **Figure 15.** Rock salt deposits can range in thickness from a few meters to more than 400 m. Companies mine these deposits because rock salt is an important resource. It's used in the manufacturing of glass, paper, soap, and dairy products. The halite in rock salt is processed and used as table salt.

Figure 15 Rock salt is extracted from this mine in Germany. The same salt can be processed and used to season your favorite foods.

Organic Sedimentary Rocks

Rocks made of the remains of once-living things are called organic sedimentary rocks. One of the most common organic sedimentary rocks is fossil-rich limestone. Like chemical limestone, fossil-rich limestone is made of the mineral calcite. However, fossil-rich limestone mostly contains remains of once-living ocean organisms instead of only calcite that formed directly from ocean water.

Animals such as mussels, clams, corals, and snails make their shells from $CaCO_3$ that eventually becomes calcite. When they die, their shells accumulate on the ocean floor. When these shells are cemented together, fossil-rich limestone forms. If a rock is made completely of shell fragments that you can see, the rock is called coquina (koh KEE nuh).

Chalk Chalk is another organic sedimentary rock that is made of microscopic shells. When you write with naturally occurring chalk, you're crushing and smearing the calcite-shell remains of once-living ocean organisms.

Coal Another useful organic sedimentary rock is coal, shown in **Figure 16.** Coal forms when pieces of dead plants are buried under other sediments in swamps. These plant materials are chemically changed by microorganisms. The resulting sediments are compacted over millions of years to form coal, an important source of energy. Much of the coal in North America and Europe formed during a period of geologic time that is so named because of this important reason. The Carboniferous Period, which spans from approximately 360 to 286 million years ago, was named in Europe. So much coal formed during this interval of time that coal's composition—primarily carbon—was the basis for naming a geologic period.

Applying Math — Calculate Thickness

COAL FORMATION It took 300 million years for a layer of plant matter about 0.9 m thick to produce a bed of bituminous coal 0.3 m thick. Estimate the thickness of plant matter that produced a bed of coal 0.15 m thick.

Solution

1. *This is what you know:*
   - original thickness of plant matter = 0.9 m
   - original coal thickness = 0.3 m
   - new coal thickness = 0.15 m
2. *This is what you need to know:* thickness of plant matter needed to form 0.15 m of coal
3. *This is the equation you need to use:* (thickness of plant matter)/(new coal thickness) = (original thickness of plant matter)/(original coal thickness)
4. *Substitute the known values:* (? m plant matter)/(0.15 m coal) = (0.9 m plant matter)/(0.3 m coal)
5. *Solve the equation:* (? m plant matter) = (0.9 m plant matter)(0.15 m coal)/(0.3 m coal) = 0.45 m plant matter
6. *Check your answer:* Multiply your answer by the original coal thickness. Divide by the original plant matter thickness to get the new coal thickness.

Practice Problems

1. Estimate the thickness of plant matter that produced a bed of coal 0.6 m thick.
2. About how much coal would have been produced from a layer of plant matter 0.50 m thick?

For more practice, visit earth.msscience.com/math_practice

Figure 16 This coal layer in Alaska is easily identified by its jet-black color, as compared with other sedimentary layers.

Another Look at the Rock Cycle

You have seen that the rock cycle has no beginning and no end. Rocks change continually from one form to another. Sediments can become so deeply buried that they eventually become metamorphic or igneous rocks. These reformed rocks later can be uplifted and exposed to the surface—possibly as mountains to be worn away again by erosion.

All of the rocks that you've learned about in this chapter formed through some process within the rock cycle. All of the rocks around you, including those used to build houses and monuments, are part of the rock cycle. Slowly, they are all changing, because the rock cycle is a continuous, dynamic process.

section 4 review

Summary

Formation of Sedimentary Rocks

- Sedimentary rocks form as layers, with older layers near the bottom of an undisturbed stack.

Classifying Sedimentary Rocks

- To classify a sedimentary rock, determine its composition and texture.

Detrital Sedimentary Rocks

- Rock and mineral fragments make up detrital rocks.

Chemical Sedimentary Rocks

- Chemical sedimentary rocks form from solutions of dissolved minerals.

Organic Sedimentary Rocks

- The remains of once-living organisms make up organic sedimentary rocks.

Self Check

1. **Identify** where sediments come from.
2. **Explain** how compaction is important in the formation of coal.
3. **Compare and contrast** detrital and chemical sedimentary rock.
4. **List** chemical sedimentary rocks that are essential to your health or that are used to make life more convenient. How is each used?
5. **Think Critically** Explain how pieces of granite and slate could both be found in the same conglomerate. How would the granite and slate pieces be held together?

Applying Math

6. **Calculate Ratios** Use information in **Table 2** to estimate how many times larger the largest grains of silt and sand are compared to the largest clay grains.

Sedimentary Rocks

Goals
- **Observe** sedimentary rock characteristics.
- **Compare and contrast** sedimentary rock textures.
- **Classify** sedimentary rocks as detrital, chemical, or organic.

Materials
unknown sedimentary rock samples
marking pen
5% hydrochloric acid (HCl) solution
dropper
paper towels
water
magnifying lens
metric ruler

Safety Precautions

WARNING: *HCl is an acid and can cause burns. Wear goggles and a lab apron. Rinse spills with water and wash hands afterward.*

Sedimentary rocks are formed by compaction and cementation of sediment. Because sediment is found in all shapes and sizes, do you think these characteristics could be used to classify detrital sedimentary rocks? Sedimentary rocks also can be classified as chemical or organic.

Real-World Question

How are rock characteristics used to classify sedimentary rocks as detrital, chemical, or organic?

Procedure

1. Make a Sedimentary Rock Samples chart in your Science Journal similar to the one shown on the next page.
2. **Determine** the sizes of sediments in each sample, using a magnifying lens and a metric ruler. Using **Table 2,** classify any grains of sediment in the rocks as gravel, sand, silt, or clay. In general, the sediment is silt if it is gritty and just barely visible, and clay if it is smooth and if individual grains are not visible.
3. Place a few drops of 5% HCl solution on each rock sample. Bubbling on a rock indicates the presence of calcite.
4. **Examine** each sample for fossils and describe any that are present.
5. **Determine** whether each sample has a granular or nongranular texture.

Sedimentary Rock Samples					
Sample	Observations	Minerals or Fossils Present	Sediment Size	Detrital, Chemical, or Organic	Rock Name
A					
B		Do not write in this book.			
C					
D					
E					

Analyze Your Data

1. **Classify** your samples as detrital, chemical, or organic.
2. **Identify** each rock sample.

Conclude and Apply

1. **Explain** why you tested the rocks with acid. What minerals react with acid?
2. **Compare and contrast** sedimentary rocks that have a granular texture with sedimentary rocks that have a nongranular texture.

Communicating Your Data

Compare your conclusions with those of other students in your class. **For more help, refer to the** Science Skill Handbook.

Australia's controversial rock star

One of the most famous rocks in the world is causing serious problems for Australians

Uluru (yew LEW rew), also known as Ayers Rock, is one of the most popular tourist destinations in Australia. This sandstone skyscraper is more than 8 km around, over 300 m high, and extends as much as 4.8 km below the surface. One writer describes it as an iceberg in the desert. Geologists hypothesize that the mighty Uluru rock began forming 550 million years ago during Precambrian time. That's when large mountain ranges started to form in Central Australia.

For more than 25,000 years, this geological wonder has played an important role in the lives of the Aboriginal peoples, the Anangu (a NA noo). These native Australians are the original owners of the rock and have spiritual explanations for its many caves, holes, and scars.

Athlete Nova Benis-Kneebone had the honor of receiving the Olympic torch near the sacred Uluru and carried it partway to the Olympic stadium.

Tourists Take Over

In the 1980s, some 100,000 tourists visited—and many climbed—Uluru. In 2000, the rock attracted about 400,000 tourists. The Anangu take offense at anyone climbing their sacred rock. However, if climbing the rock were outlawed, tourism would be seriously hurt. That would mean less income for Australians.

To respect the Anangu's wishes, the Australian government returned Ayers Rock to the Anangu in 1985 and agreed to call it by its traditional name. The Anangu leased back the rock to the Australian government until the year 2084, when its management will return to the Anangu. Until then, the Anangu will collect 25 percent of the money people pay to visit the rock.

The Aboriginal people encourage tourists to respect their beliefs. They offer a walking tour around the rock, and they show videos about Aboriginal traditions. The Anangu sell T-shirts that say "I *didn't* climb Uluru." They hope visitors to Uluru will wear the T-shirt with pride and respect.

Write Research a natural landmark or large natural land or water formation in your area. What is the geology behind it? When was it formed? How was it formed? Write a folktale that explains its formation. Share your folktale with the class.

For more information, visit earth.msscience.com/time

Reviewing Main Ideas

Section 1 The Rock Cycle

1. A rock is a mixture of one or more minerals, rock fragments, organic matter, or volcanic glass.
2. The rock cycle includes all processes by which rocks form.

Section 2 Igneous Rocks

1. Magma and lava are molten materials that harden to form igneous rocks.
2. Intrusive igneous rocks form when magma cools slowly below Earth's surface. Extrusive igneous rocks form when lava cools rapidly at the surface.
3. The compositions of most igneous rocks range from granitic to andesitic to basaltic.

Section 3 Metamorphic Rocks

1. Heat, pressure, and fluids can cause metamorphic rocks to form.
2. Slate and gneiss are examples of foliated metamorphic rocks. Quartzite and marble are examples of nonfoliated metamorphic rocks.

Section 4 Sedimentary Rocks

1. Detrital sedimentary rocks form when fragments of rocks and minerals are compacted and cemented together.
2. Chemical sedimentary rocks come out of solution or are left behind by evaporation.
3. Organic sedimentary rocks contain the remains of once-living organisms.

Visualizing Main Ideas

Copy and complete the following concept map on rocks. Use the following terms: organic, metamorphic, foliated, extrusive, igneous, *and* chemical.

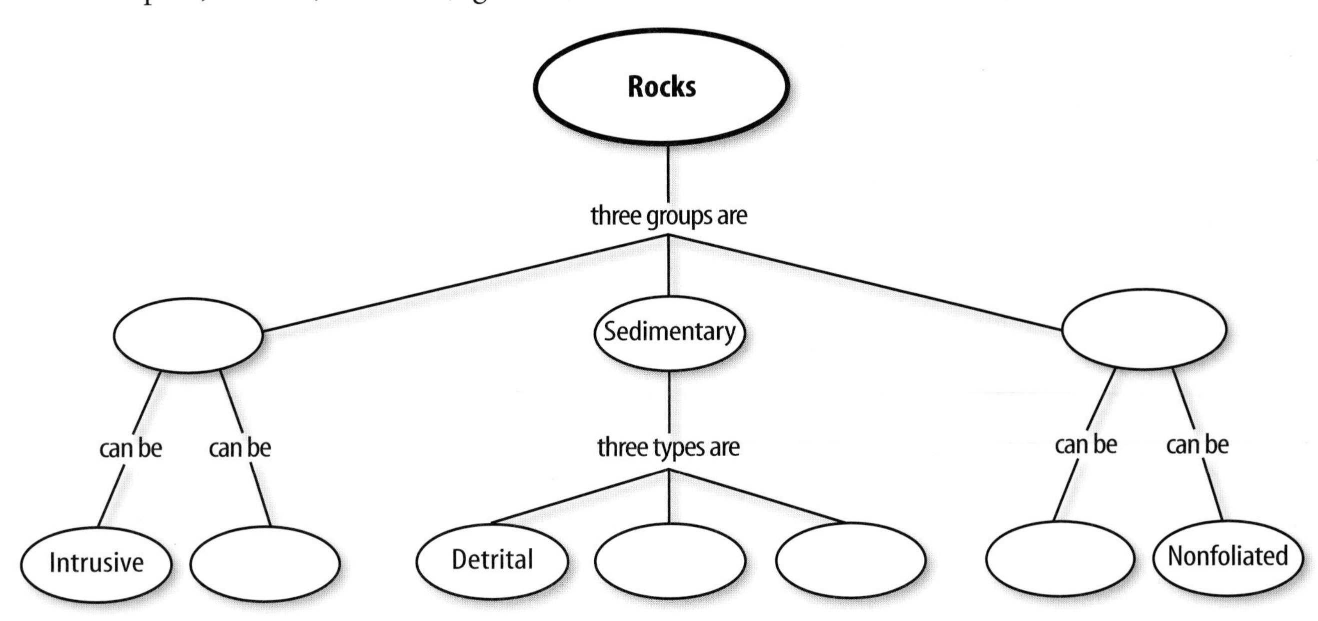

Using Vocabulary

basaltic p. 97
cementation p. 105
compaction p. 104
extrusive p. 95
foliated p. 101
granitic p. 97
igneous rock p. 94
intrusive p. 95
lava p. 94
metamorphic rock p. 99
nonfoliated p. 102
rock p. 90
rock cycle p. 91
sediment p. 103
sedimentary rock p. 103

Explain the difference between the vocabulary words in each of the following sets.

1. foliated—nonfoliated
2. cementation—compaction
3. sediment—lava
4. extrusive—intrusive
5. rock—rock cycle
6. metamorphic rock—igneous rock—sedimentary rock
7. sediment—sedimentary rock
8. lava—igneous rock
9. rock—sediment
10. basaltic—granitic

Checking Concepts

Choose the word or phrase that best answers the question.

11. Why does magma tend to rise toward Earth's surface?
    A) It is more dense than surrounding rocks.
    B) It is more massive than surrounding rocks.
    C) It is cooler than surrounding rocks.
    D) It is less dense than surrounding rocks.

12. During metamorphism of granite into gneiss, what happens to minerals?
    A) They partly melt.
    B) They become new sediments.
    C) They grow smaller.
    D) They align into layers.

13. Which rock has large mineral grains?
    A) granite **C)** obsidian
    B) basalt **D)** pumice

14. Which type of rock is shown in this photo?
    A) foliated
    B) nonfoliated
    C) intrusive
    D) extrusive

15. What do igneous rocks form from?
    A) sediments **C)** gravel
    B) mud **D)** magma

16. What sedimentary rock is made of large, angular pieces of sediments?
    A) conglomerate **C)** limestone
    B) breccia **D)** chalk

17. Which of the following is an example of a detrital sedimentary rock?
    A) limestone **C)** breccia
    B) evaporite **D)** chalk

18. What is molten material at Earth's surface called?
    A) limestone **C)** breccia
    B) lava **D)** granite

19. Which of these is an organic sedimentary rock?
    A) coquina **C)** rock salt
    B) sandstone **D)** conglomerate

Science Online earth.msscience.com/vocabulary_puzzlemaker

Thinking Critically

20. **Infer** Granite, pumice, and scoria are igneous rocks. Why doesn't granite have airholes like the other two?
21. **Infer** why marble rarely contains fossils.
22. **Predict** Would you expect quartzite or sandstone to break more easily? Explain your answer.
23. **Compare and contrast** basaltic and granitic magmas.
24. **Form Hypotheses** A geologist was studying rocks in a mountain range. She found a layer of sedimentary rock that had formed in the ocean. Hypothesize how this could happen.

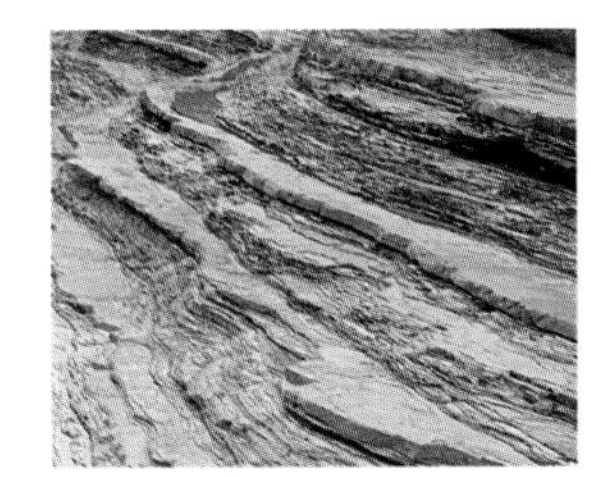

25. **Concept Map** Copy and complete the concept map shown below. Use the following terms and phrases: *magma, sediments, igneous rock, sedimentary rock, metamorphic rock*. Add and label any missing arrows.

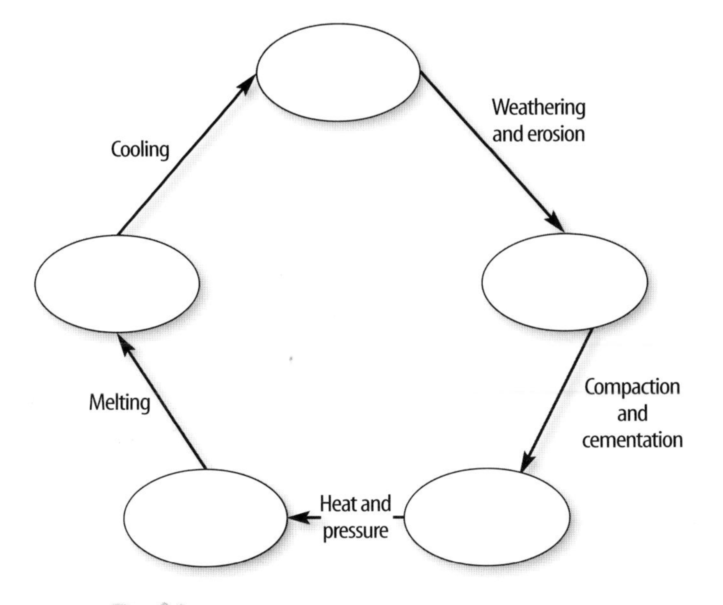

Performance Activities

26. **Poster** Collect a group of rocks. Make a poster that shows the classifications of rocks, and glue your rocks to the poster under the proper headings. Describe your rocks and explain where you found them.

Applying Math

27. **Grain Size** Assume that the conglomerate shown on the second page of the "Sedimentary Rocks" lab is one-half of its actual size. Determine the average length of the gravel in the rock.
28. **Plant Matter** Suppose that a 4-m layer of plant matter was compacted to form a coal layer 1 m thick. By what percent has the thickness of organic material been reduced?

Use the graph below to answer questions 29 and 30.

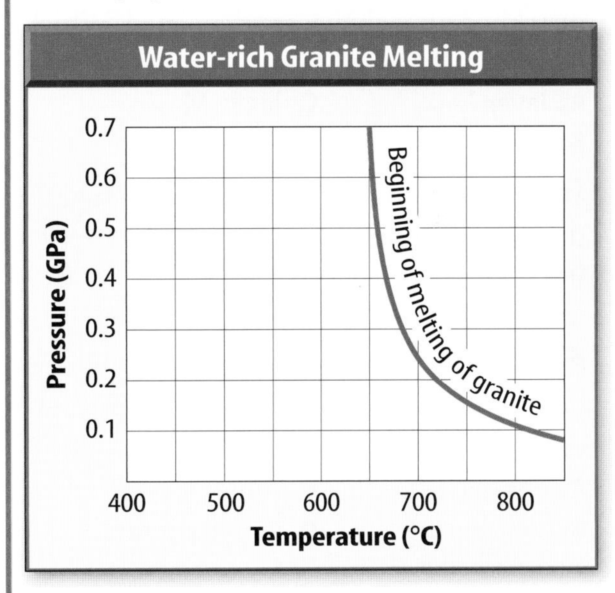

29. **Melting Granite** Determine the melting temperature of a water-rich granite at a pressure of 0.2 GPa.

    Pressure conversions:
    1 GPa, or gigapascal, = 10,000 bars
    1 bar = 0.9869 atmospheres
30. **Melting Pressure** At about what pressure will a water-rich granite melt at 680°C?

Part 1 Multiple Choice

Record your answers on the answer sheet provided by your teacher or on a sheet of paper.

Use the illustration below to answer question 1.

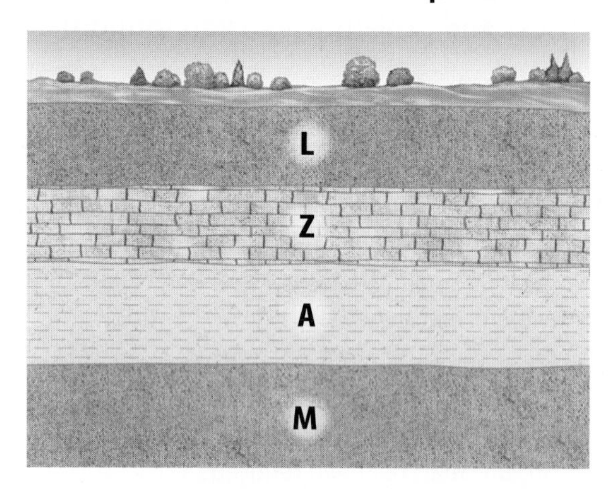

1. These layers of sedimentary rock were not disturbed after they were deposited. Which layer was deposited first?
   A. layer L **C.** layer M
   B. layer Z **D.** layer A

2. Who realized that rocks undergo changes through long periods of time after observing rocks at Siccar Point, Scotland?
   A. James Hutton **C.** Galileo Galilei
   B. Neil Armstrong **D.** Albert Einstein

3. During which process do minerals precipitate in the spaces between sediment grains?
   A. compaction **C.** cementation
   B. weathering **D.** conglomerate

4. Which rock often is sculpted to create statues?
   A. shale **C.** coquina
   B. marble **D.** conglomerate

Test-Taking Tip

Careful Reading Read each question carefully for full understanding.

5. Which of the following rocks is a metamorphic rock?
   A. shale **C.** slate
   B. granite **D.** pumice

6. Which rock consists mostly of pieces of seashell?
   A. sandstone **C.** pumice
   B. coquina **D.** granite

Use the diagram below to answer questions 7–9.

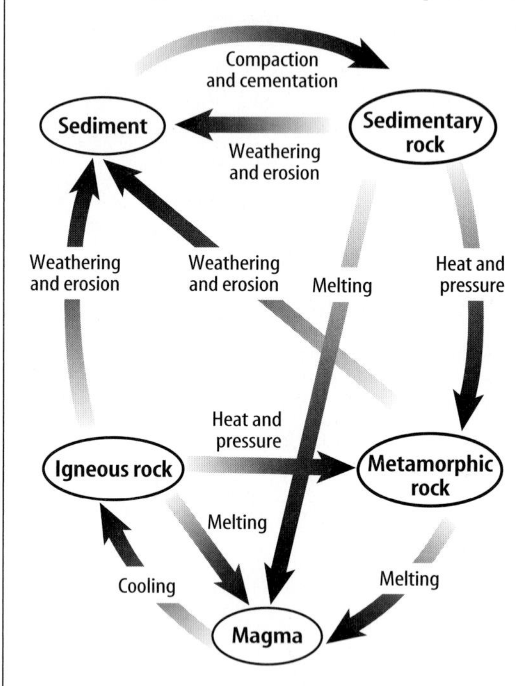

7. Which process in the rock cycle causes magma to form?
   A. melting **C.** weathering
   B. erosion **D.** cooling

8. What forms when rocks are weathered and eroded?
   A. igneous rock **C.** sedimentary rock
   B. sediment **D.** metamorphic rock

9. Which type of rock forms because of high heat and pressure without melting?
   A. igneous rock **C.** sedimentary rock
   B. intrusive rock **D.** metamorphic rock

Part 2 Short Response/Grid In

Record your answers on the answer sheet provided by your teacher or on a sheet of paper.

10. What is a rock? How is a rock different from a mineral?

11. Explain why some igneous rocks are coarse and others are fine.

12. What is foliation? How does it form?

13. How do chemical sedimentary rocks, such as rock salt, form?

14. Why do some rocks contain fossils?

15. How is the formation of chemical sedimentary rocks similar to the formation of cement in detrital sedimentary rocks?

Use the graph below to answer questions 16–17.

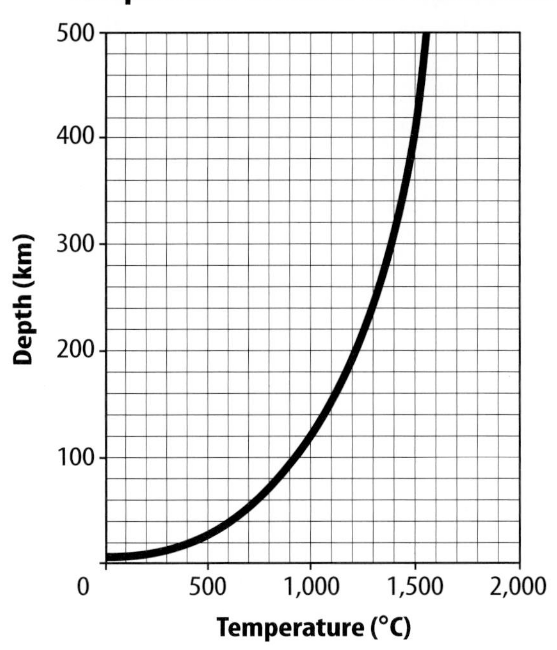

16. According to the graph, about how deep below a continent does the temperature reach 1,000°C?

17. In general, what happens to temperature as depth below Earth's surface increases?

Part 3 Open Ended

Record your answers on a sheet of paper.

Use the table below to answer questions 18 and 19.

Magma Type	Basaltic	Andesitic	Granitic
Intrusive	Do not write in this book.		
Extrusive			

18. Copy the table on your paper. Then, fill in the empty squares with a correct rock name.

19. Explain how igneous rocks are classified.

20. Explain how loose sediment can become sedimentary rock.

21. Why does pressure increase with depth in Earth? How does higher pressure affect rocks?

22. Why is slate sometimes used as shingles for roofs? What other rocks are used for important purposes in society?

23. How are organic sedimentary rocks different from other rocks? List an example of an organic sedimentary rock.

24. Why is the rock cycle called a cycle?

25. A geologist found a sequence of rocks in which 200-million-year-old shales were on top of 100-million-year-old sandstones. Hypothesize how this could happen.

26. Explain why coquina could be classified in more than one way.

chapter 5

Earth's Energy and Mineral Resources

The BIG Idea

Earth's resources provide materials and energy for everyday living.

SECTION 1
Nonrenewable Energy Resources

Main Idea Nonrenewable energy resources, including fossil fuels, are used faster than they can be replaced.

SECTION 2
Renewable Energy Resources

Main Idea Energy resources that can be replaced or restored within a relatively short amount of time are called renewable energy resources.

SECTION 3
Mineral Resources

Main Idea People use a variety of Earth's mineral resources to meet a diverse range of needs.

Where do we find energy?

Much of the energy consumed in the world comes from oil and gas. Other sources of energy come from moving water, wind, and the Sun's rays. In this chapter you'll learn about many types of energy resources and the importance of conserving these resources.

Science Journal Write three ways electricity is generated at a power plant.

Start-Up Activities

Finding Energy Reserves

The physical properties of Earth materials determine how easily liquids and gases move through them. Geologists use these properties, in part, to predict where reserves of energy resources like petroleum or natural gas can be found.

1. Obtain a sample of sandstone and a sample of shale from your teacher.
2. Make sure that your samples can be placed on a tabletop so that the sides facing up are reasonably flat and horizontal.
3. Place the two samples side by side in a shallow baking pan.
4. Using a dropper, place three drops of cooking oil on each sample.
5. For ten minutes, observe what happens to the oil on the samples.
6. **Think Critically** Write your observations in your Science Journal. Infer which rock type might be a good reservoir for petroleum.

Preview this chapter's content and activities at earth.msscience.com

Energy Resources Make the following Foldable to help you identify energy resources.

STEP 1 **Fold** a sheet of paper in half lengthwise. Make the back edge about 1.25 cm longer than the front edge.

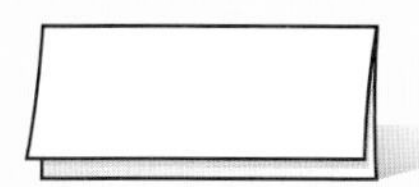

STEP 2 **Turn** lengthwise and **fold** into thirds.

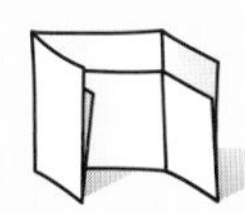

STEP 3 **Unfold and cut** only the top layer along both folds to make three tabs.

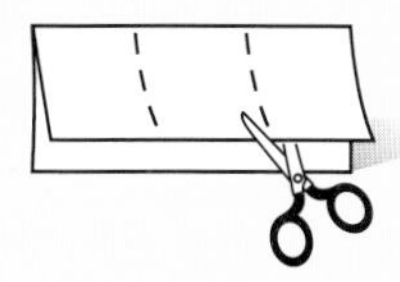

STEP 4 **Label** each tab as shown.

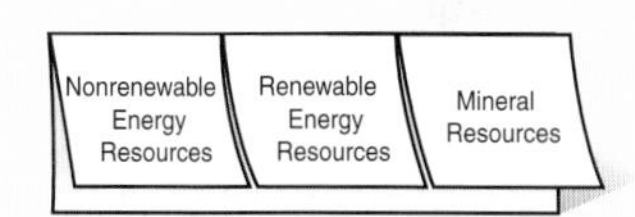

Find Main Ideas As you read the chapter, list examples on the front of the tabs and write about each type of resource under the tabs.

Get Ready to Read

Visualize

1 Learn It! Visualize by forming mental images of the text as you read. Imagine how the text descriptions look, sound, feel, smell, or taste. Look for any pictures or diagrams on the page that may help you add to your understanding.

2 Practice It! Read the following paragraph. As you read, use the underlined details to form a picture in your mind.

> Oil and natural gas are often <u>found in layers of rock</u> that have become <u>tilted or folded</u>. Because they are less dense than water, <u>oil and gas are forced upward</u>. Rock layers that are impermeable, such as shale, stop this upward movement. When this happens, a folded shale layer can <u>trap the oil and natural gas below it</u>.
>
> *—from page 123*

Based on the description above, try to visualize oil and gas accumulation in folded rock layers. Now look at the illustration on page 123.

- How closely does it match your mental picture?
- Reread the passage and look at the picture again. Did your ideas change?
- Compare your image with what others in your class visualized.

3 Apply It! Read the chapter and list three subjects you were able to visualize. Make a rough sketch showing what you visualized.

Forming your own mental images will help you remember what you read.

Target Your Reading

Use this to focus on the main ideas as you read the chapter.

1. **Before you read** the chapter, respond to the statements below on your worksheet or on a numbered sheet of paper.
   - Write an **A** if you **agree** with the statement.
   - Write a **D** if you **disagree** with the statement.

2. **After you read** the chapter, look back to this page to see if you've changed your mind about any of the statements.
   - If any of your answers changed, explain why.
   - Change any false statements into true statements.
   - Use your revised statements as a study guide.

Before You Read A or D	Statement	After You Read A or D
	1 Fossil fuels formed millions of years ago from the remains of plants and animals.	
	2 Coal is classified as metamorphic rock.	
	3 Nonrenewable energy resources are being used faster than natural Earth processes can replace them.	
	4 Nuclear energy produced by the splitting of heavy elements is a nonrenewable energy resource.	
	5 Solar energy is an example of a renewable energy resource.	
	6 Most places have winds strong enough to commercially generate electricity.	
	7 Economic profitability does not have to be considered when describing a mineral deposit as an ore.	
	8 Most mineral resources are renewable resources.	
	9 The recycling process often uses less energy than it takes to obtain new material.	

Science Online
Print out a worksheet of this page at earth.msscience.com

section 1

Nonrenewable Energy Resources

as you read

What You'll Learn

- **Identify** examples of nonrenewable energy resources.
- **Describe** the advantages and disadvantages of using fossil fuels.
- **Explain** the advantages and disadvantages of using nuclear energy.

Why It's Important

Nonrenewable resources should be conserved to ensure their presence for future generations.

Review Vocabulary

fuel: a material that provides useful energy

New Vocabulary

- fossil fuel
- coal
- oil
- natural gas
- reserve
- nuclear energy

Energy

The world's population relies on energy of all kinds. Energy is the ability to cause change. Some energy resources on Earth are being used faster than natural Earth processes can replace them. These resources are referred to as nonrenewable energy resources. Most of the energy resources used to generate electricity are nonrenewable.

Fossil Fuels

Nonrenewable energy resources include fossil fuels. **Fossil fuels** are fuels such as coal, oil, and natural gas that form from the remains of plants and other organisms that were buried and altered over millions of years. Coal is a sedimentary rock formed from the compacted and transformed remains of ancient plant matter. Oil is a liquid hydrocarbon that often is referred to as petroleum. Hydrocarbons are compounds that contain hydrogen and carbon atoms. Other naturally occurring hydrocarbons occur in the gas or semisolid states. Fossil fuels are processed to make gasoline for cars, to heat homes, and for many other uses, as shown in **Table 1.**

Table 1 Uses of Fossil Fuels

Fuel	Uses
Coal	■ To generate electricity
Oil	■ To produce gasoline and other fuels ■ As lubricants ■ To make plastics, home shingles, and other products
Natural Gas	■ To heat buildings ■ As a source of sulfur

Figure 1 This coal layer is located in Castle Gate, Utah. **Analyze and Conclude** *Using the map and legend below, can you determine what type of coal it is?*

Coal The most abundant fossil fuel in the world is coal, shown in **Figure 1.** If the consumption of coal continues at the current rate, it is estimated that the coal supply will last for about another 250 years.

Coal is a rock that contains at least 50 percent plant remains. Coal begins to form when plants die in a swampy area. The dead plants are covered by more plants, water, and sediment, preventing atmospheric oxygen from coming into contact with the plant matter. The lack of atmospheric oxygen prevents the plant matter from decaying rapidly. Bacterial growth within the plant material causes a gradual breakdown of molecules in the plant tissue, leaving carbon and some impurities behind. This is the material that eventually will become coal after millions of years. Bacteria also cause the release of methane gas, carbon dioxide, ammonia, and water as the original plant matter breaks down.

What happens to begin the formation of coal in a swampy area?

Synthetic Fuels Unlike gasoline, which is refined from petroleum, other fuels called synthetic fuels are extracted from solid organic material. Synthetic fuels can be created from coal—a sedimentary rock containing hydrocarbons. The hydrocarbons are extracted from coal to form liquid and gaseous synthetic fuels. Liquid synthetic fuels can be processed to produce gasoline for automobiles and fuel oil for home heating. Gaseous synthetic fuels are used to generate electricity and heat buildings.

Coal Formation The coal found in the eastern and midwestern United States formed from plants that lived in great swamps about 300 million years ago during the Pennsylvanian Period of geologic time. Research the Pennsylvanian Period to find out what types of plants lived in these swamps. Describe the plants in your Science Journal.

Stages of Coal Formation As decaying plant material loses gas and moisture, the concentration of carbon increases. The first step in this process, shown in **Figure 2,** results in the formation of peat. Peat is a layer of organic sediment. When peat burns, it releases large amounts of smoke because it has a high concentration of water and impurities.

As peat is buried under more sediment, it changes into lignite, which is a soft, brown coal with much less moisture. Heat and pressure produced by burial force water out of peat and concentrate carbon in the lignite. Lignite releases more energy and less smoke than peat when it is burned.

As the layers are buried deeper, bituminous coal, or soft coal, forms. Bituminous coal is compact, black, and brittle. It provides lots of heat energy when burned. Bituminous coal contains various levels of sulfur, which can pollute the environment.

If enough heat and pressure are applied to buried layers of bituminous coal, anthracite coal forms. Anthracite coal contains the highest amount of carbon of all forms of coal. Therefore, anthracite coal is the cleanest burning of all coals.

Figure 2 Coal is formed in four basic stages.

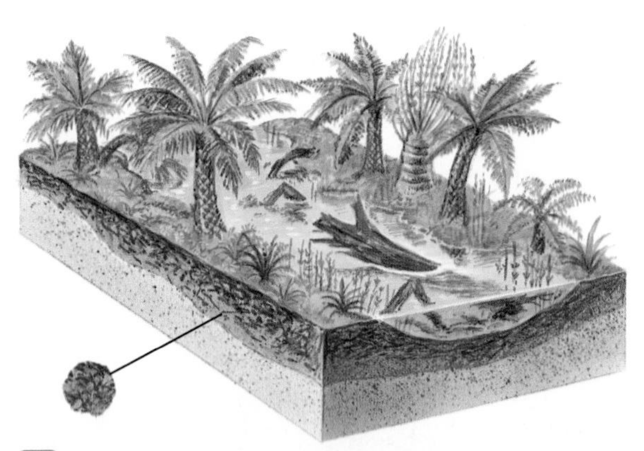

A Dead plant material accumulates in swamps and eventually forms a layer of peat.

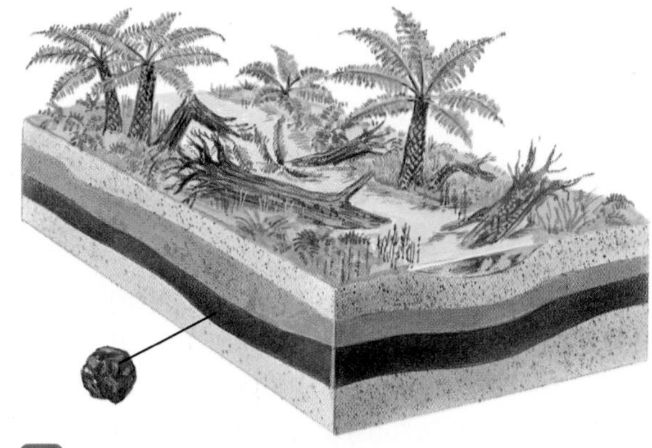

B Over time, heat and pressure cause the peat to change into lignite coal.

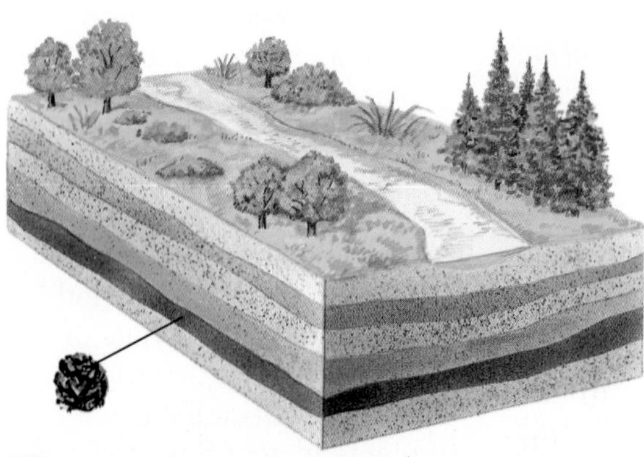

C As the lignite coal becomes buried by more sediments, heat and pressure change it into bituminous coal.

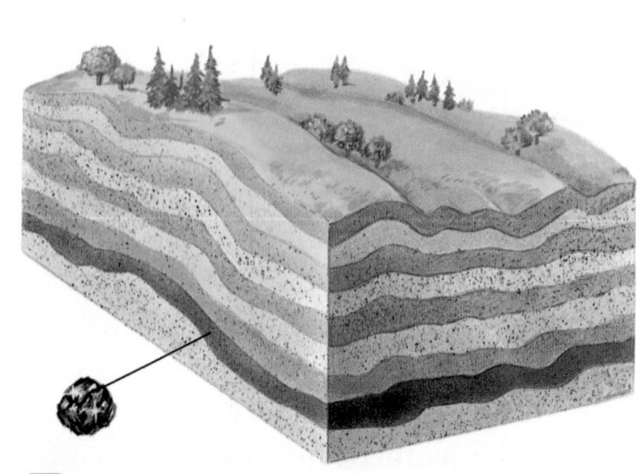

D When bituminous coal is heated and squeezed during metamorphism, anthracite coal forms.

Oil and Natural Gas Coal isn't the only fossil fuel used to obtain energy. Two other fossil fuels that provide large quantities of the energy used today are oil and natural gas. **Oil** is a thick, black liquid formed from the buried remains of microscopic marine organisms. **Natural gas** forms under similar conditions and often with oil, but it forms in a gaseous state. Oil and natural gas are hydrocarbons. However, natural gas is composed of hydrocarbon molecules that are lighter than those in oil.

Residents of the United States burn vast quantities of oil and natural gas for daily energy requirements. As shown in **Figure 3,** Americans obtain most of their energy from these sources. Natural gas is used mostly for heating and cooking. Oil is used in many ways, including as heating oil, gasoline, lubricants, and in the manufacture of plastics and other important compounds.

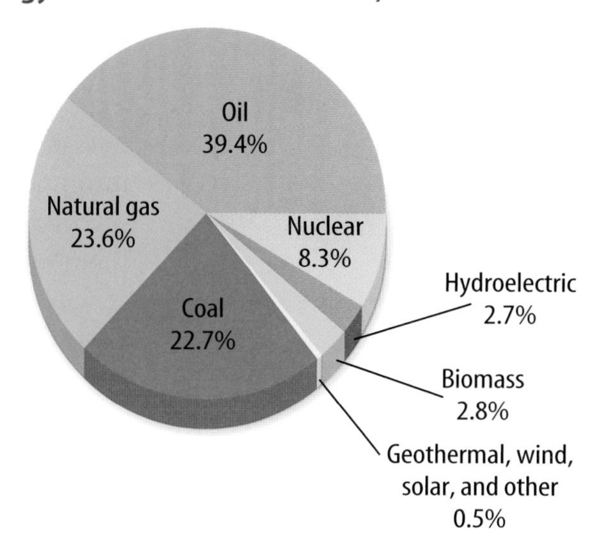

Figure 3 This circle graph shows the percentages of energy that the United States derives from various energy resources.
Calculate *What percentage is from nonrenewable energy resources?*

Formation of Oil and Natural Gas Most geologists agree that petroleum forms over millions of years from the remains of tiny marine organisms in ocean sediment. The process begins when marine organisms called plankton die and fall to the seafloor. Similar to the way that coal is buried, sediment is deposited over them. The temperature rises with depth in Earth, and increased heat eventually causes the dead plankton to change to oil and gas after they have been buried deeply by sediment.

Oil and natural gas often are found in layers of rock that have become tilted or folded. Because they are less dense than water, oil and natural gas are forced upward. Rock layers that are impermeable, such as shale, stop this upward movement. When this happens, a folded shale layer can trap the oil and natural gas below it. Such a trap for oil and gas is shown in **Figure 4.** The rock layer beneath the shale in which the petroleum and natural gas accumulate is called a reservoir rock.

Figure 4 Oil and natural gas are fossil fuels formed by the burial of marine organisms. These fuels can be trapped and accumulate beneath Earth's surface.

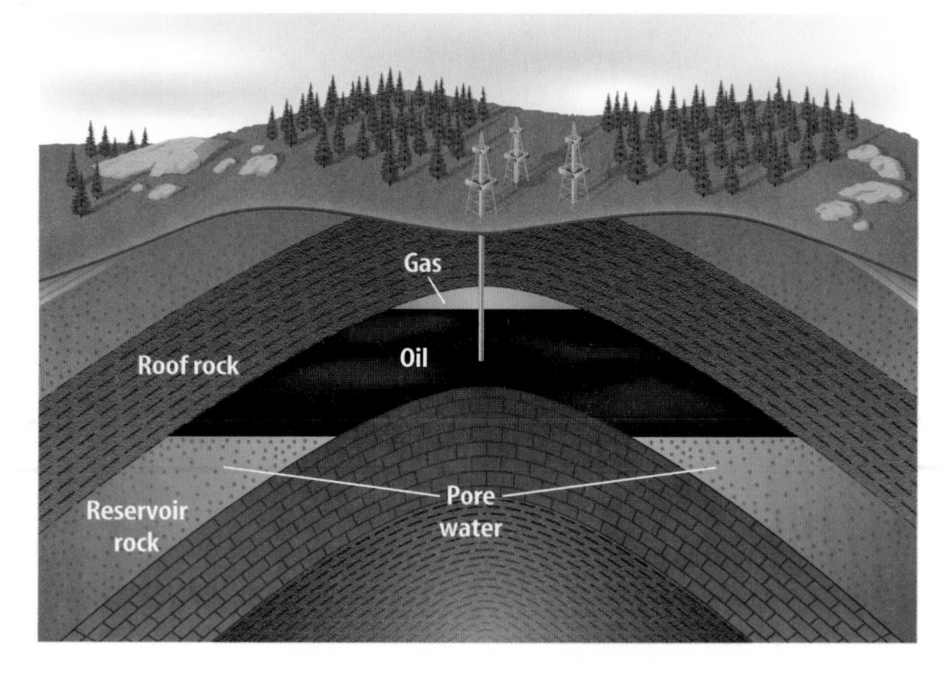

Removing Fossil Fuels from the Ground

Coal is removed from the ground using one of several methods of excavation. The two most common methods are strip mining, also called open-pit mining, and underground mining, shown in **Figure 5.** Oil and natural gas are removed by pumping them out of the ground.

Coal Mining During strip mining, as shown in **Figure 5,** layers of soil and rock above coal are removed and piled to one side. The exposed coal then is removed and loaded into trucks or trains and transported elsewhere. After the coal has been removed, mining companies often return the soil and rock to the open pit and cover it with topsoil. Trees and grass are planted in a process called land reclamation. If possible, animals native to the area are reintroduced. Strip mining is used only when the coal deposits are close to the surface.

In one method of underground coal mining, tunnels are dug and pillars of rock are left to support the rocks surrounding the tunnels. Two types of underground coal mines are drift mines and slope mines. Drift mining, shown in the **Figure 5** inset photo, is the removal of coal that is not close to Earth's surface through a horizontal opening in the side of a hill or mountain. In slope mining, an angled opening and air shaft are made in the side of a mountain to remove coal.

Figure 5 Coal is a fossil fuel that can be removed from Earth in many different ways.

During strip mining, coal is accessed by removing the soil and rock above it.

During drift mining, tunnels are made into Earth.
Explain *how you think the coal is removed from these tunnels.*

Drilling for Oil and Gas Oil and natural gas are fossil fuels that can be pumped from underground deposits. Geologists and engineers drill wells through rocks where these resources might be trapped, as shown in **Figure 6.** As the well is being drilled, it is lined with pipe to prevent it from caving in. When the drill bit reaches the rock layer containing oil, drilling is stopped. Equipment is installed to control the flow of oil. The surrounding rock then is fractured to allow oil and gas to flow into the well. The oil and gas are pumped to the surface.

How are oil and natural gas brought to Earth's surface?

Figure 6 Oil and natural gas are recovered from Earth by drilling deep wells.

Fossil Fuel Reserves

The amount of a fossil fuel that can be extracted at a profit using current technology is known as a **reserve.** This is not the same as a fossil fuel resource. A fossil fuel resource has fossil fuels that are concentrated enough that they can be extracted from Earth in useful amounts. However, a resource is not classified as a reserve unless the fuel can be extracted economically. What might cause a known fossil fuel resource to become classified as a reserve?

Methane Hydrates You have learned that current reserves of coal will last about 250 years. Enough natural gas is located in the United States to last about 60 more years. However, recent studies indicate that a new source of methane, which is the main component of natural gas, might be located beneath the seafloor. Icelike substances known as methane hydrates could provide tremendous reserves of methane.

Methane hydrates are stable molecules found hundreds of meters below sea level in ocean floor sediment. They form under conditions of relatively low temperatures and high pressures. The hydrocarbons are trapped within the cagelike structure of ice, as described in **Figure 7.** Scientists estimate that more carbon is contained in methane hydrates than in all current fossil fuel deposits combined. Large accumulations of methane hydrates are estimated to exist off the eastern coast of the United States. Can you imagine what it would mean to the world's energy supply if relatively clean-burning methane could be extracted economically from methane hydrates?

Topic: Methane Hydrates
Visit earth.msscience.com for Web links to information about methane hydrates.

Activity Identify which oceans might contain significant amounts of methane hydrates.

NATIONAL GEOGRAPHIC

VISUALIZING METHANE HYDRATES

Figure 7

Reserves of fossil fuels—such as oil, coal, and natural gas—are limited and will one day be used up. Methane hydrates could be an alternative energy source. This icelike substance, background, has been discovered in ocean floor sediments and in permafrost regions worldwide. If scientists can harness this energy, the world's gas supply could be met for years to come.

Methane hydrates are highly flammable compounds made up of methane—the main component of natural gas—trapped in a cage of frozen water. Methane hydrates represent an enormous source of potential energy. However, they contain a greenhouse gas that might intensify global warming. More research is needed to determine how to safely extract them from the seafloor.

In the photo above, a Russian submersible explores a site in the North Atlantic that contains methane hydrate deposits.

Conserving Fossil Fuels Do you sometimes forget to turn off the lights when you walk out of a room? Wasteful habits might mean that electricity to run homes and industries will not always be as plentiful and cheap as it is today. Fossil fuels take millions of years to form and are used much faster than Earth processes can replenish them.

Today, coal provides about 25 percent of the energy that is used worldwide and 22 percent of the energy used in the United States. Oil and natural gas provide almost 61 percent of the world's energy and about 65 percent of the U.S. energy supply. At the rate these fuels are being used, they could run out someday. How can this be avoided?

By remembering to turn off lights and appliances, you can avoid wasting fossil fuels. Another way to conserve fossil fuels is to make sure doors and windows are shut tightly during cold weather so heat doesn't leak out of your home. If you have air-conditioning, run it as little as possible. Ask the adults you live with if more insulation could be added to your home or if an insulated jacket could be put on the water heater.

Energy from Atoms

Most electricity in the United States is generated in power plants that use fossil fuels. However, alternate sources of energy exist. **Nuclear energy** is an alternate energy source produced from atomic reactions. When the nucleus of a heavy element is split, lighter elements form and energy is released. This energy can be used to light a home or power the submarines shown in **Figure 8.**

The splitting of heavy elements to produce energy is called nuclear fission. Nuclear fission is carried out in nuclear power plants using a type of uranium as fuel.

Practicing Energy Conservation

Procedure

1. Have an adult help you find the **electric meter** for your home and record the reading in your **Science Journal.**
2. Do this for several days, taking your meter readings at about the same time each day.
3. List things you and your family can do to reduce your electricity use.
4. Encourage your family to try some of the listed ideas for several days.

Analysis

1. Keep taking meter readings and infer whether the changes make any difference.
2. Have you and your family helped conserve energy?

Try at Home

Figure 8 Atoms can be a source of energy.

These submarines are powered by nuclear fission.

During nuclear fission, energy is given off when a heavy atom, like uranium, splits into lighter atoms.

Heavy atom

Lighter atoms

+ Energy

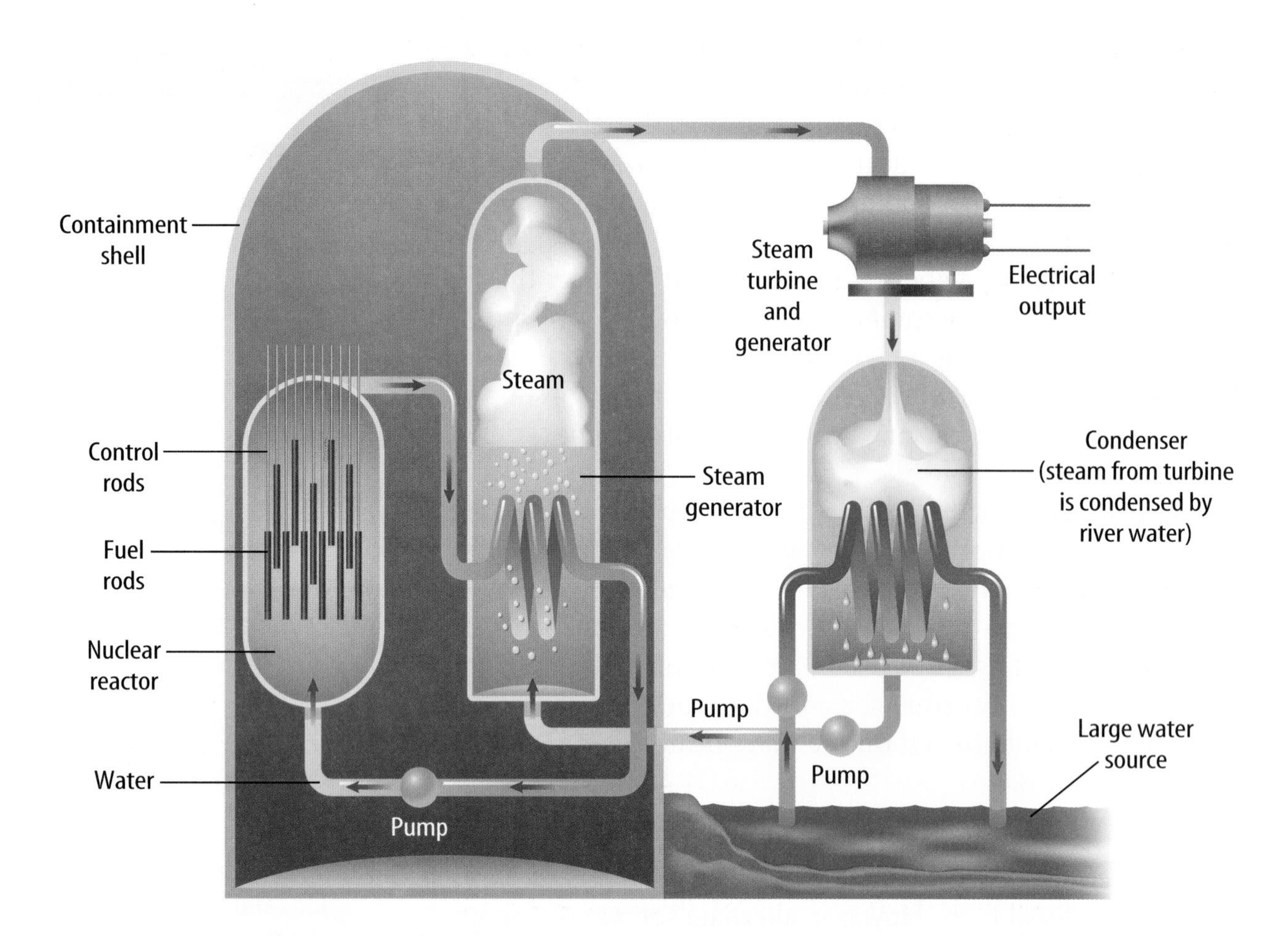

Figure 9 Heat released in nuclear reactors produces steam, which in turn is used to produce electricity. This is an example of transforming nuclear energy into electrical energy.
Infer *Why do you think nuclear power plants are located near rivers and lakes?*

Electricity from Nuclear Energy A nuclear power plant, shown in **Figure 9,** has a large chamber called a nuclear reactor. Within the nuclear reactor, uranium fuel rods sit in a pool of cooling water. Neutrons are fired into the fuel rods. When the uranium-235 atoms are hit, they break apart and fire out neutrons that hit other atoms, beginning a chain reaction. As each atom splits, it not only fires neutrons but also releases heat that is used to boil water to make steam. The steam drives a turbine, which turns a generator that produces electricity.

Reading Check *How is nuclear energy used to produce electricity?*

Nuclear energy from fission is considered to be a nonrenewable energy resource because it uses uranium-235 as fuel. A limited amount of uranium-235 is available for use. Another problem with nuclear energy is the waste material that it produces. Nuclear waste from power plants consists of highly radioactive elements formed by the fission process. Some of this waste will remain radioactive for thousands of years. The Environmental Protection Agency (EPA) has determined that nuclear waste must be stored safely and contained for at least 10,000 years before reentering the environment.

Fusion Environmental problems related to nuclear power could be eliminated if usable energy could be obtained from fusion. The Sun is a natural fusion power plant that provides energy for Earth and the solar system. Someday fusion also might provide energy for your home.

During fusion, materials of low mass are fused together to form a substance of higher mass. No fuel problem exists if the low-mass material is a commonly occurring substance. Also, if the end product is not radioactive, storing nuclear waste is not a problem. In fact, fusion of hydrogen into helium would satisfy both of these conditions. However, technologies do not currently exist to enable humans to fuse hydrogen into helium at reasonably low temperatures in a controlled manner. But research is being conducted, as shown in **Figure 10.** If this is accomplished, nuclear energy could be considered as an alternative energy resource along with renewable resources. You will learn the importance of renewable energy resources in the next section.

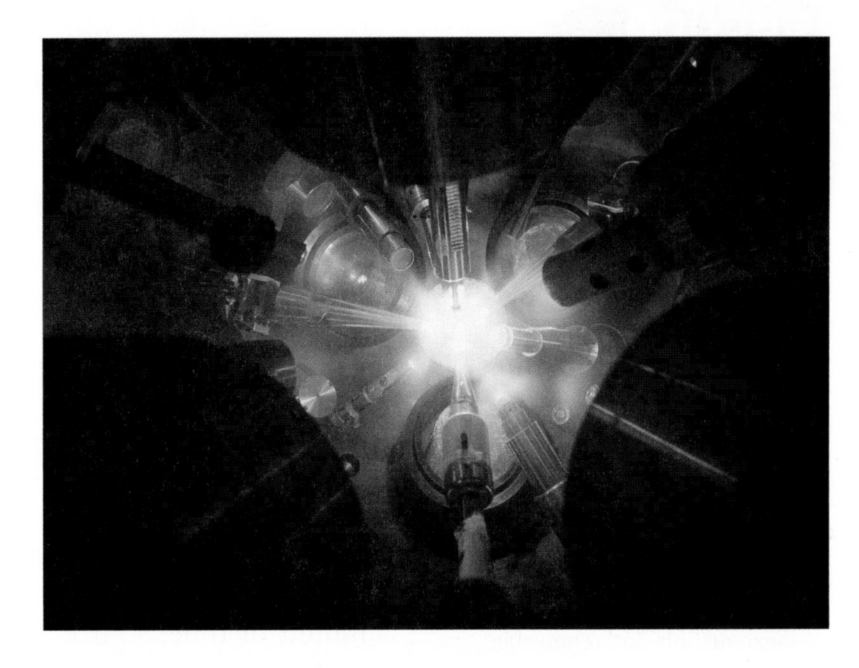

Figure 10 Lasers are used in research facilities to help people understand and control fusion.

section 1 review

Summary

Fossil Fuels

- Coal, natural gas, and oil are all nonrenewable energy sources.
- Synthetic fuels are human-made fuels that can be derived from coal.
- The four stages of coal formation are peat, lignite, bituminous coal, and anthracite coal.
- Oil and gas are made from the decay of ancient marine organisms.
- Strip mining and underground mining are two common methods that are used to extract coal reserves.

Energy from Atoms

- Energy is released during a fission reaction when a heavy atom is split into lighter atoms.
- Fusion occurs when two atoms come together to form a single atom.

Self Check

1. **Explain** why coal, oil, and natural gas are fossil fuels.
2. **Explain** why fossil fuels are considered to be nonrenewable energy resources.
3. **Describe** two disadvantages of nuclear energy.
4. **Think Critically** Why are you likely to find natural gas and oil deposits in the same location, but less likely to find coal and petroleum deposits at the same location?

Applying Math

5. **Design a Graph** Current energy consumption by source in the U.S. is as follows: oil, 39%; natural gas, 24%; coal, 23%; nuclear energy, 8%; renewable resources, 6%. Design a bar graph to show the energy consumption by source in the U.S. Display the sources from greatest to least.

section 2

Renewable Energy Resources

as you read

What You'll Learn

- **Compare and contrast** inexhaustible and renewable energy resources.
- **Explain** why inexhaustible and renewable resources are used less than nonrenewable resources.

Why It's Important

As fossil fuel reserves continue to diminish, alternate energy resources will be needed.

Review Vocabulary
energy: the ability to cause change

New Vocabulary
- solar energy
- wind farm
- hydroelectric energy
- geothermal energy
- biomass energy

Renewable Energy Resources

How soon the world runs out of fossil fuels depends on how they are used and conserved. A renewable resource can be replaced or restored as it is used or within a relatively short amount of time. Sources of renewable energy include the Sun, wind, water, and geothermal energy.

Energy from the Sun When you sit in the Sun, walk into the wind, or sail against an ocean current, you are experiencing the power of solar energy. **Solar energy** is energy from the Sun. You already know that the Sun's energy heats Earth, and it causes circulation in Earth's atmosphere and oceans. Global winds and ocean currents are examples of nature's use of solar energy. Thus, solar energy is used indirectly when the wind and some types of moving water are used to do work.

People can use solar energy in a passive way or in an active way. South-facing windows on buildings act as passive solar collectors, warming exposed rooms. Solar cells actively collect energy from the Sun and transform it into electricity. Solar cells were invented to generate electricity for satellites. Now they also are used to power calculators, streetlights, and experimental cars. Some people have installed solar energy cells on their roofs, as shown in **Figure 11.**

Figure 11 Solar panels, such as on this home in Laguna Niguel, California, can be used to collect renewable solar energy to power appliances and heat water.

Figure 12 Wind farms are used to produce electricity.
Evaluate *Some people might argue that windmills produce visual pollution. Why do you think this is?*

Disadvantages of Solar Energy Solar energy is clean and renewable, but it does have some disadvantages. Solar cells work less efficiently on cloudy days and cannot work at all at night. Some systems use batteries to store solar energy for use at night or on cloudy days, but it is difficult to store large amounts of energy in batteries. Worn out batteries also must be discarded. This can pollute the environment if not done properly.

Energy from Wind What is better to do on a warm, windy day than fly a kite? A strong wind can lift a kite high in the sky and whip it around. The pull of the wind is so great that you wonder if it will whip the kite right out of your hands. Wind is a source of energy. It was and still is used to power sailing ships. Windmills have used wind energy to grind corn and pump water. Today, windmills can be used to generate electricity. When a large number of windmills are placed in one area for the purpose of generating electricity, the area is called a **wind farm,** as shown in **Figure 12.**

Wind energy has advantages and disadvantages. Wind is nonpolluting and free. It does little harm to the environment and produces no waste. However, only a few regions of the world have winds strong enough to generate electricity. Also, wind isn't steady. Sometimes it blows too hard and at other times it is too weak or stops entirely. For an area to use wind energy consistently, the area must have a persistent wind that blows at an appropriate speed.

Why are some regions better suited for wind farms than others?

Physicists The optimal speed of wind needed to rotate blades on a windmill is something a physicist would study. They can calculate the energy produced based on the speed at which the blades turn. Some areas in the country are better suited for wind farms than others. Find out which areas utilize wind farms and report in your Science Journal how much electric-ity is produced and what it is used for. What kinds of organizations would a physicist work for in these locations?

Energy from Water For a long time, waterwheels steadily spun next to streams and rivers. The energy in the flowing water powered the wheels that ground grain or cut lumber. More than a pretty picture, using a waterwheel in this way is an example of microhydropower. Microhydropower has been used throughout the world to do work.

Running water also can be used to generate electricity. Electricity produced by waterpower is called **hydroelectric energy.** To generate electricity from water running in a river, a large concrete dam is built to retain water, as illustrated in **Figure 13.** A lake forms behind the dam. As water is released, its force turns turbines at the base of the dam. The turbines then turn generators that make electricity.

At first it might appear that hydroelectric energy doesn't create any environmental problems and that the water is used with little additional cost. However, when dams are built, upstream lakes fill with sediment and downstream erosion increases. Land above the dam is flooded, and wildlife habitats are damaged.

Energy from Earth Erupting volcanoes and geysers like Old Faithful are examples of geothermal energy in action. The energy that causes volcanoes to erupt or water to shoot up as a geyser also can be used to generate electricity. Energy obtained by using hot magma or hot, dry rocks inside Earth is called **geothermal energy.**

Bodies of magma can heat large reservoirs of groundwater. Geothermal power plants use steam from the reservoirs to produce electricity, as shown in **Figure 14.** In a developing method, water becomes steam when it is pumped through broken, hot, dry rocks. The steam then is used to turn turbines that run generators to make electricity. The advantage of using hot, dry rocks is that they are found just about everywhere. Geothermal energy presently is being used in Hawaii and in parts of the western United States.

Figure 13 Hydroelectric power is important in many regions of the United States. Hoover Dam was built on the Colorado River to supply electricity for a large area.

Power lines
Power plant
Lake
Intake pipe
Generator
Turbine
Discharge pipe

The power of running water is converted to usable energy in a hydroelectric power plant.

Figure 14 Geothermal energy is used to supply electricity to industries and homes.

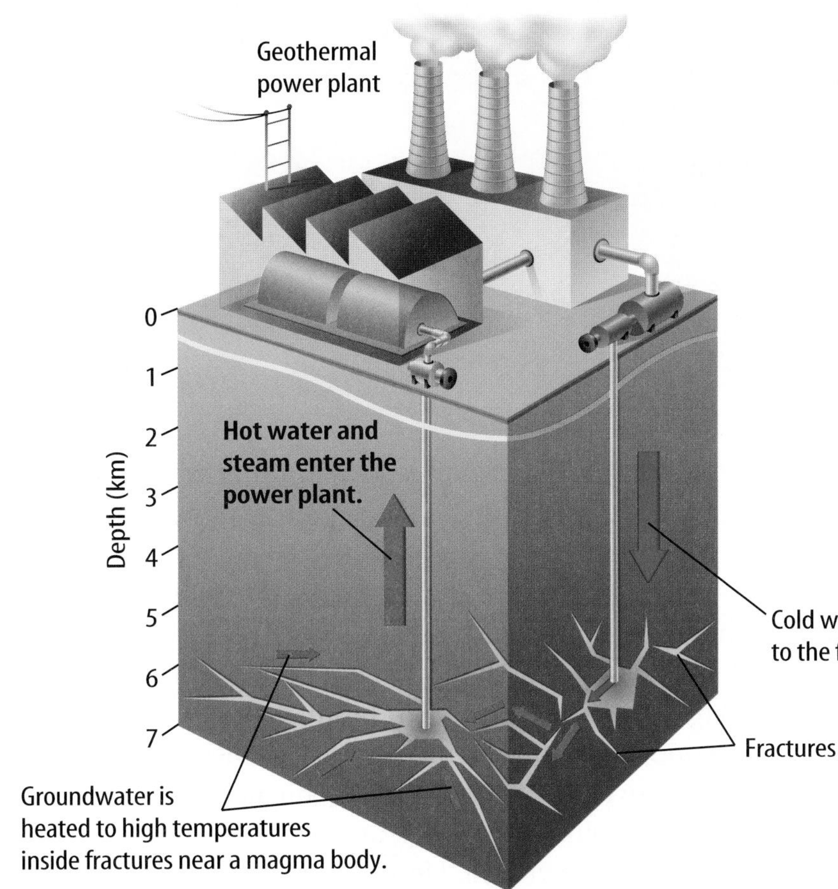

Geothermal power plant
0
1
2
3
4
5
6
7
Depth (km)
Hot water and steam enter the power plant.
Cold water is returned to the fractured hot rock.
Fractures
Groundwater is heated to high temperatures inside fractures near a magma body.

What by-product is produced in this geothermal plant in California? Is it considered a pollutant?

Disadvantages of Geothermal Energy Geothermal energy sounds like the perfect solution—harness the heat energy already in the Earth. However, often the hot, dry rocks or magma are located deep within Earth. More energy would be used to get to the rocks or magma than would be taken from it. Also, digging for the energy destroys habitat in much the same way that drilling for fossil fuels does.

Other Renewable Energy Resources

Some other energy resources are not inexhaustible, but if used responsibly, can be replaced within a human lifetime. These are replaced either by nature or humans. For example, trees can be cut down and others can be planted in their place.

Biomass Energy A major renewable energy resource is biomass materials. **Biomass energy** is energy derived from burning organic material such as wood, alcohol, and garbage. The term *biomass* is derived from the words *biological* and *mass.*

Science Online

Topic: Biomass Energy
Visit earth.msscience.com for Web links to information about biomass energy.

Activity List three new technologies that turn biomass into useable energy. Give two examples of each type of biomass and its energy technology.

Figure 15 These campers are using wood, a renewable energy resource, to produce heat and light.
Discuss *Why do you think wood is the most commonly used biomass fuel?*

Energy from Wood If you've ever sat around a campfire, like the campers shown in **Figure 15,** or close to a wood-burning fireplace to keep warm, you have used energy from wood. The burning wood is releasing stored solar energy as heat energy. Humans have long used wood as an energy resource. Much of the world still cooks with wood. In fact, firewood is used more widely today than any other type of biomass fuel.

Using wood as a biomass fuel has its problems. Gases and small particles are released when wood is burned. These materials can pollute the air. When trees are cut down for firewood, natural habitats are destroyed. However, if proper conservation methods are employed or if tree farms are maintained specifically for use as fuel, energy from wood can be a part of future energy resources.

Figure 16 Gasohol sometimes is used to reduce dependence on fossil fuels.

Energy from Alcohol Biomass fuel can be burned directly, such as when wood or peat is used for heating and cooking. However, it also can be transformed into other materials that might provide cleaner, more efficient fuels.

For example, during distillation, biomass fuel, such as corn, is changed to an alcohol such as ethanol. Ethanol then can be mixed with another fuel. When the other fuel is gasoline, the mixture is called gasohol. Gasohol can be used in the same way as gasoline, as shown in **Figure 16,** but it cuts down on the amount of fossil fuel needed to produce gasoline. Fluid biomass fuels are more efficient and have more uses than solid biomass fuels do.

The problem with this process is that presently, growing the corn and distilling the ethanol often uses more energy from burning fossil fuels than the amount of energy that is derived from burning ethanol. At present, biomass fuel is best used locally.

Reading Check ***What are the drawbacks of biomass fuels?***

Energy from Garbage Every day humans throw away a tremendous amount of burnable garbage. As much as two thirds of what is thrown away could be burned. If more garbage were used for fuel, as shown in **Figure 17,** human dependence on fossil fuels would decrease. Burning garbage is a cheap source of energy and also helps reduce the amount of material that must be dumped into landfills.

Compared to other nations, the United States lags in the use of municipal waste as a renewable energy resource. For example, in some countries in Western Europe, as much as half of the waste generated is used for biomass fuel. When the garbage is burned, heat is produced, which turns water to steam. The steam turns turbines that run generators to produce electricity.

Unfortunately, some problems can be associated with using energy from garbage. Burning municipal waste can produce toxic ash residue and air pollution. Substances such as heavy metals could find their way into the smoke from garbage and thus into the atmosphere.

Figure 17 Garbage can be burned to produce electricity at trash-burning power plants such as this one in Virginia.

section 2 review

Summary

Renewable Energy Resources

- Solar cells are used to collect the Sun's energy.
- Wind energy produces no waste or pollution, however only a few areas are conducive for creating significant energy supplies.
- Dams are used to help provide running water, which is used to produce electricity.
- Energy obtained by using heat from inside Earth is called geothermal energy.

Other Renewable Energy Resources

- Biomass energy is produced when organic material such as wood, alcohol, or garbage is burned.
- Trash-burning power plants convert waste into electricty by burning garbage.

Self Check

1. **List** three advantages and disadvantages of using solar energy, wind energy, and hydroelectric energy.
2. **Explain** the difference between nonrenewable and renewable energy resources. Give two examples of each.
3. **Describe** how geothermal energy is used to create electricity.
4. **Infer** why nonrenewable resources are used more than renewable resources.
5. **Think Critically** How could forests be classified as a renewable and a nonrenewable resource?

Applying Skills

6. **Use a Spreadsheet** Make a table of energy resources. Include an example of how each resource is used. Then describe how you could reduce the use of energy resources at home.

Soaking Up Solar Energy

Winter clothing tends to be darker in color than summer clothing. The color of the material used in the clothing affects its ability to absorb energy. In this lab, you will use different colors of soil to study this effect.

Real-World Question

How does color affect the absorption of energy?

Goals

- **Determine** whether color has an effect on the absorption of solar energy.
- **Relate** the concept of whether color affects absorption to other applications.

Materials

dry, black soil
dry, brown soil
dry, sandy, white soil
thermometers (3)
ring stand
graph paper
colored pencils (3)
metric ruler
clear-glass or plastic dishes (3)
200-watt gooseneck lamp
**200-watt lamp with reflector and clamp*
watch or clock with second hand
**stopwatch*
**Alternate materials*

Safety Precautions

WARNING: *Handle glass with care so as not to break it. Wear thermal mitts when handling the light source.*

Procedure

1. Fill each dish with a different color of soil to a depth of 2.5 cm.
2. Arrange the dishes close together on your desk and place a thermometer in each dish. Be sure to cover the thermometer bulb in each dish completely with the soil.
3. Position the lamp over all three dishes.
4. **Design** a data table for your observations similar to the sample table above. You will need to read the temperature of each dish every 30 s for 20 min after the light is turned on.
5. Turn on the light and begin your experiment.
6. Use the data to construct a graph. Time should be plotted on the horizontal axis and temperature on the vertical axis. Use a different colored pencil to plot the data for each type of soil, or use a computer to design a graph that illustrates your data.

Time and Temperature

Time (min)	Temperature Dish A (°C)	Temperature Dish B (°C)	Temperature Dish C (°C)
0.0			
0.5	Do not write in this book.		
1.0			
1.5			

Conclude and Apply

1. **Observe** which soil had the greatest temperature change. The least?
2. **Explain** why the curves on the graph flatten.
3. **Infer** Why do flat-plate solar collectors have black plates behind the water pipes?
4. **Explain** how the color of a material affects its ability to absorb energy.
5. **Infer** Why is most winter clothing darker in color than summer clothing?

section 3

Mineral Resources

Metallic Mineral Resources

If your room at home is anything like the one shown in **Figure 18,** you will find many metal items. Metals are obtained from Earth materials called metallic mineral resources. A **mineral resource** is a deposit of useful minerals. See how many metals you can find. Is there anything in your room that contains iron? What about the metal in the frame of your bed? Is it made of iron? If so, the iron might have come from the mineral hematite. What about the framing around the windows in your room? Is it aluminum? Aluminum, like that in a soft-drink can, comes from a mixture of minerals known as bauxite. Many minerals contain these and other useful elements. Which minerals are mined as sources for the materials you use every day?

Ores Deposits in which a mineral or minerals exist in large enough amounts to be mined at a profit are called **ores.** Generally, the term ore is used for metallic deposits, but this is not always the case. The hematite that was mentioned earlier as an iron ore and the bauxite that was mentioned earlier as an aluminum ore are metallic ores.

Reading Check *What is an ore?*

as you read

What You'll Learn

- **Explain** the conditions needed for a mineral to be classified as an ore.
- **Describe** how market conditions can cause a mineral to lose its value as an ore.
- **Compare and contrast** metallic and nonmetallic mineral resources.

Why It's Important

Many products you use are made from mineral resources.

Review Vocabulary
metal: a solid material that is generally hard, shiny, pliable and a good electrical conductor

New Vocabulary
- mineral resource
- ore
- recycling

Copper in wires found in electrical equipment comes from the mineral chalcopyrite.
Many bed frames contain iron, which is extracted from minerals such as hematite.
Aluminum comes from a mixture of minerals called bauxite.
Stainless steel contains chromium, which comes from the mineral chromite.

Figure 18 Many items in your home are made from metals obtained from metallic mineral resources.

Figure 19 Iron ores are smelted to produce nearly pure iron.
List *three examples of what this iron could be used for.*

Economic Effects When is a mineral deposit considered an ore? The mineral in question must be in demand. Enough of it must be present in the deposit to make it worth removing. Some mining operations are profitable only if a large amount of the mineral is needed. It also must be fairly easy to separate the mineral from the material in which it is found. If any one of these conditions isn't met, the deposit might not be considered an ore.

Supply and demand is an important part of life. You might have noticed that when the supply of fresh fruit is down, the price you pay for it at the store goes up. Economic factors largely determine what an ore is.

Refining Ore The process of extracting a useful substance from an ore involves two operations—concentrating and refining. After a metallic ore is mined from Earth's crust, it is crushed and the waste rock is removed. The waste rock that must be removed before a mineral can be used is called gangue (GANG).

INTEGRATE Chemistry Refining produces a pure or nearly pure substance from ore. For example, iron can be concentrated from the ore hematite, which is composed of iron oxide. The concentrated ore then is refined to be as close to pure iron as possible. One method of refining is smelting, illustrated in **Figure 19.** Smelting is a chemical process that removes unwanted elements from the metal that is being processed. During one smelting process, a concentrated ore of iron is heated with a specific chemical. The chemical combines with oxygen in the iron oxide, resulting in pure iron. Note that one resource, fossil fuel, is burned to produce the heat that is needed to obtain the finished product of another resource, in this case iron.

Nonmetallic Mineral Resources

Any mineral resources not used as fuels or as sources of metals are nonmetallic mineral resources. These resources are mined for the nonmetallic elements contained in them and for the specific physical and chemical properties they have. Generally, nonmetallic mineral resources can be divided into two different groups—industrial minerals and building materials. Some materials, such as limestone, belong to both groups of nonmetallic mineral resources, and others are specific to one group or the other.

Industrial Minerals Many useful chemicals are obtained from industrial minerals. Sandstone is a source of silica (SiO_2), which is a compound that is used to make glass. Some industrial minerals are processed to make fertilizers for farms and gardens. For example, sylvite, a mineral that forms when seawater evaporates, is used to make potassium fertilizer.

Many people enjoy a little sprinkle of salt on french fries and pretzels. Table salt is a product derived from halite, a nonmetallic mineral resource. Halite also is used to help melt ice on roads and sidewalks during winter and to help soften water.

Other industrial minerals are useful because of their characteristic physical properties. For example, abrasives are made from deposits of corundum and garnet. Both of these minerals are hard and able to scratch most other materials they come into contact with. Small particles of garnet can be glued onto a sheet of heavy paper to make abrasive sandpaper. **Figure 20** illustrates just a few ways in which nonmetallic mineral resources help make your life more convenient.

Observing the Effects of Insulation

Procedure

1. Pour **warm water** into a **thermos bottle.** Cap it and set it aside.
2. Pour **cold water** with **ice** into a **glass** surrounded by a **thermal cup holder.**
3. Pour warm water—the same temperature as in step 1—into an **uncovered cup.** Pour cold water with ice into a glass container that is not surrounded by a thermal cup holder.
4. After 24 h, measure the temperature of each of the liquids.

Analysis

1. Infer how the insulation affected the temperature of each liquid.
2. Relate the usefulness of insulation in a thermos bottle to the usefulness of fiberglass insulation in a home.

Figure 20 You benefit from the use of industrial minerals every day.

Road salt melts ice on streets.

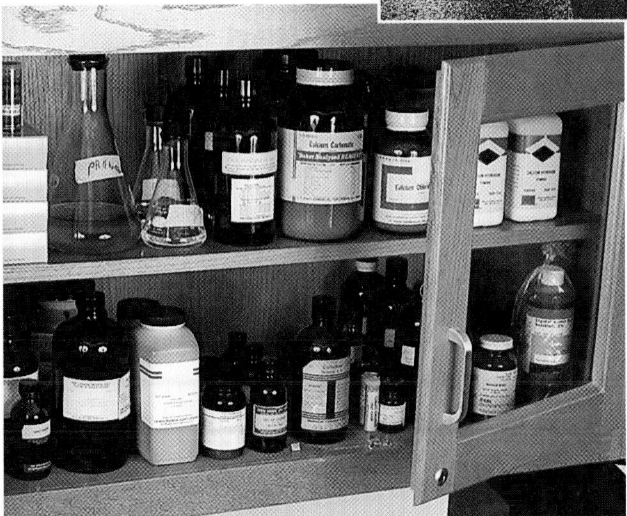

Many important chemicals are made from industrial minerals.

An industrial mineral called trona is important for making glass.

Building Materials One of the most important nonmetallic mineral resources is aggregate. Aggregate is composed of crushed stone or a mixture of gravel and sand and has many uses in the building industry. For example, aggregates can be mixed with cement and water to form concrete. Quality concrete is vital to the building industry. Limestone also has industrial uses. It is used as paving stone and as part of concrete mixtures. Have you ever seen the crushed rock in a walking path or driveway? The individual pieces might be crushed limestone. Gypsum, a mineral that forms when seawater evaporates, is soft and lightweight and is used in the production of plaster and wallboard. If you handle a piece of broken plaster or wallboard, note its appearance, which is similar to the mineral gypsum.

Rock also is used as building stone. You might know of buildings in your region that are made from granite, limestone, or sandstone. These rocks and others are quarried and cut into blocks and sheets. The pieces then can be used to construct buildings. Some rock also is used to sculpt statues and other pieces of art.

Reading Check *What are some important nonmetallic mineral resources?*

Applying Science

Why should you recycle?

Recycling in the United States has become a way of life. In 2000, 88 percent of Americans participated in recycling. Recycling is important because it saves precious raw materials and energy. Recycling aluminum saves 95 percent of the energy required to obtain it from its ore. Recycling steel saves up to 74 percent in energy costs, and recycling glass saves up to 22 percent.

Identifying the Problem

The following table includes materials that currently are being recycled and rates of recycling for the years 1995, 1997, and 2001. Examine the table to determine materials for which recycling increased or decreased between 1995 and 2001.

Recycling Rates in the United States

Material	1995 (%)	1997 (%)	2001 (%)
Glass	24.5	24.3	27.2
Steel	36.5	38.4	43.5 (est.)
Aluminum	34.6	31.2	33.0
Plastics	5.3	5.2	7.0

Solving the Problem

1. Has the recycling of materials increased or decreased over time? Which materials are recycled most? Which materials are recycled least? Discuss why some materials might be recycled more than others.
2. How can recycling benefit society? Explain your answer.

Recycling Mineral Resources

Mineral resources are nonrenewable. You've learned that nonrenewable energy resources are being used faster than natural Earth processes can replace them. Most mineral resources take millions of years to form. Have you ever thrown away an empty soft-drink can? Many people do. These cans become solid waste. Wouldn't it be better if these cans and other items made from mineral resources were recycled into new items?

Recycling is using old materials to make new ones. Recycling has many advantages. It reduces the demand for new mineral resources. The recycling process often uses less energy than it takes to obtain new material. Because supplies of some minerals might become limited in the future, recycling could be required to meet needs for certain materials, as shown in **Figure 21.**

Recycling also can be a profitable experience. Some companies purchase scrap metal and empty soft-drink cans for the aluminum and tin content. The seller receives a small amount of money for turning in the material. Schools and other groups earn money by recycling soft-drink cans.

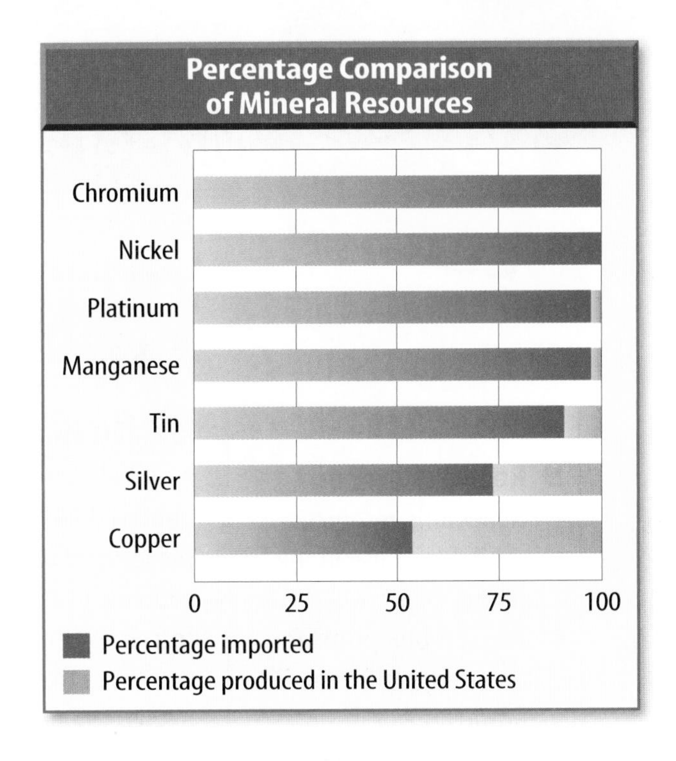

Figure 21 The United States produces only a small percentage of the metallic resources it consumes.

section 3 review

Summary

Metallic Mineral Resources

- Minerals found in rocks that can be mined for an economic profit are called ores.

Nonmetallic Mineral Resources

- Nonmetallic mineral resources can be classified into groups: industrial minerals and building materials.
- Sedimentary rocks such as limestone and sandstone can be used as building materials to make things like buildings and statues.

Recycling Mineral Resources

- Recycling materials helps to preserve Earth's resources by reusing old or used materials without extracting new resources from Earth.
- The recycling process may use fewer resources than it takes to obtain new material.

Self Check

1. **Explain** how metals obtained from metallic mineral resources are used in your home and school. Which of these products could be recycled easily?
2. **List** two industrial uses for nonmetallic mineral resources.
3. **Explain** how supply and demand of a material can cause a mineral to become an ore.
4. **Think Critically** Gangue is waste rock remaining after a mineral ore is removed. Why is gangue sometimes reprocessed?

Applying Skills

5. **Classify** the following mineral resources as metallic or nonmetallic: *hematite, limestone, bauxite, sandstone, garnet,* and *chalcopyrite.* Explain why you classified each one as you did.

Model and Invent

Home Sweet Home

Goals

- **Research** various renewable resources available to use in the home.
- **Design** blueprints for an energy-efficient home and/or design and build a model of an energy-efficient home.

Possible Materials

paper
ruler
pencils
cardboard
glue
aluminum foil

Real-World Question

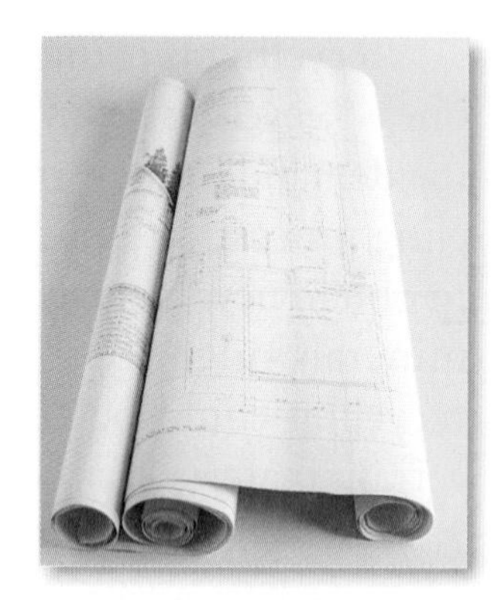

As fossil fuel supplies continue to be depleted, an increasing U. S. population has recognized the need for alternative energy sources. United States residents might be forced to consider using renewable energy resources to meet some of their energy needs. The need for energy-efficient housing is more relevant now than ever before. A designer of energy-efficient homes considers proper design and structure, a well chosen building site with wise material selection, and selection of efficient energy generation systems to power the home. Energy-efficient housing uses less energy and produces fewer pollutants. What does the floor plan, building plan, or a model of an energy efficient home look like? How and where should your house be designed and built to efficiently use the alternative energy resources you've chosen?

Using Scientific Methods

Make a Model

Plan

1. **Research** current information about energy-efficient homes.
2. **Research** renewable energy resources such as wind, hydroelectric power, or solar power, as well as energy conservation. Decide which energy resources are most efficient for your home design.
3. Decide where your house should be built to use energy efficiently.
4. Decide how your house will be laid out and draw mock blueprints for your home. Highlight energy issues such as where solar panels can be placed.
5. Build a model of your energy-efficient home.

Do

1. Ask your peers for input on your home. As you research, become an expert in one area of alternative energy generation and share your information with your classmates.
2. **Compare** your home's design to energy-efficient homes you learn about through your research.

Test Your Model

1. Think about how most of the energy in a home is used. Remember as you plan your home that energy-efficient homes not only generate energy—they also use it more efficiently.
2. Carefully consider where your home should be built. For instance, if you plan to use wind power, will your house be built in an area that receives adequate wind?
3. Be sure to plan for backup energy generation. For instance, if you plan to use mostly solar energy, what will you do if it's a cloudy day?

Analyze Your Data

Devise a budget for building your home. Could your energy-efficient home be built at a reasonable price? Could anyone afford to build it?

Conclude and Apply

Create a list of pro and con statements about the use of energy-efficient homes. Why aren't renewable energy sources widely used in homes today?

Present your model to the class. Explain which energy resources you chose to use in your home and why. Have an open house. Take prospective home owners/classmates on a tour of your home and sell it.

Oops! Accidents in SCIENCE

SOMETIMES GREAT DISCOVERIES HAPPEN BY ACCIDENT!

BLACK GOLD!

What if you went out to your backyard, started digging a hole, and all of the sudden oil spurted out of the ground? Dollar signs might flash before your eyes.

It wasn't quite that exciting for Charles Tripp. Tripp, a Canadian, is credited with being the first person to strike oil. And he wasn't even looking for what has become known as "black gold."

In 1851, Tripp built a factory in Ontario, Canada, not far from Lake Erie. He used a natural, black, thick, sticky substance that could be found nearby to make asphalt for paving roads and to construct buildings.

In 1855, Tripp dug a well looking for freshwater for his factory. After digging just 2 m or so, he unexpectedly came upon liquid. It wasn't clear, clean, and delicious; it was smelly thick, and black. You guessed it—oil! Tripp didn't understand the importance of his find. Two years after his accidental discovery, Tripp sold his company to James Williams. In 1858, Williams continued to search for water for the factory, but, as luck would have it, diggers kept finding oil.

Some people used TNT to search for oil. This photo was taken in 1943.

Some people argue that the first oil well in North America was in Titusville, Pennsylvania, when Edwin Drake hit oil in 1859. However, most historians agree that Williams was first in 1858. But they also agree that it was Edwin Drake's discovery that led to the growth of the oil industry. So, Drake and Williams can share the credit!

The Titusville, Pennsylvania, oil well drilled by Edwin Drake. This photo was taken in 1864.

Today, many oil companies are drilling beneath the sea for oil.

Make a Graph Research the leading oil-producing nations and make a bar graph of the top five producers. Research how prices of crude oil affect the U.S. and world economies. Share your findings with your class.

For more information, visit earth.msscience.com/oops

Reviewing Main Ideas

Section 1 Nonrenewable Energy Resources

1. Fossil fuels are considered to be nonrenewable energy resources.
2. The higher the concentration of carbon in coal is, the cleaner it burns.
3. Oil and natural gas form from altered and buried marine organisms and often are found near one another.
4. Nuclear energy is obtained from the fission of heavy isotopes.

Section 2 Renewable Energy Resources

1. Some energy resources—solar energy, wind energy, hydroelectric energy, and geothermal energy—are classified as renewable energy resources.
2. Renewable energy resources are replaced within a relatively short period of time.
3. Biomass energy is derived from organic material such as wood and corn.

Section 3 Mineral Resources

1. Metallic mineral resources provide metals.
2. Ores are mineral resources that can be mined at a profit.
3. Smelting is a chemical process that removes unwanted elements from a metal that is being processed.
4. Nonmetallic mineral resources are classified as industrial minerals or building materials.

Visualizing Main Ideas

Copy and complete the following table that lists advantages and disadvantages of energy resources.

Energy Resources

Resource	Advantages	Disadvantages
Fossil fuels		
Nuclear energy		
Solar energy	Do not write in this book.	
Wind energy		
Geothermal energy		
Biomass fuel		

Using Vocabulary

biomass energy p. 133	nuclear energy p. 127
coal p. 121	oil p. 123
fossil fuel p. 120	ore p. 137
geothermal energy p. 132	recycling p. 141
hydroelectric energy p. 132	reserve p. 125
mineral resource p. 137	solar energy p. 130
natural gas p. 123	wind farm p. 131

Each phrase below describes a vocabulary word from the list. Write the word that matches the phrase describing it.

1. mineral resource mined at a profit
2. fuel that is composed mainly of the remains of dead plants
3. method of conservation in which items are processed to be used again
4. renewable energy resource that is used to power the *Hubble Space Telescope*
5. energy resource that is based on fission
6. liquid from remains of marine organisms

Checking Concepts

Choose the word or phrase that best answers the question.

7. Which has the highest content of carbon?
   A) peat **C)** bituminous coal
   B) lignite **D)** anthracite coal

8. Which is the first step in coal formation?
   A) peat **C)** bituminous coal
   B) lignite **D)** anthracite

9. Which of the following is an example of a fossil fuel?
   A) wind **C)** natural gas
   B) water **D)** uranium-235

10. What is the waste material that must be separated from an ore?
    A) smelter **C)** mineral resource
    B) gangue **D)** petroleum

11. What common rock structure can trap oil and natural gas under it?
    A) folded rock **C)** porous rock
    B) sandstone rock **D)** permeable rock

Use the figure below to answer question 12.

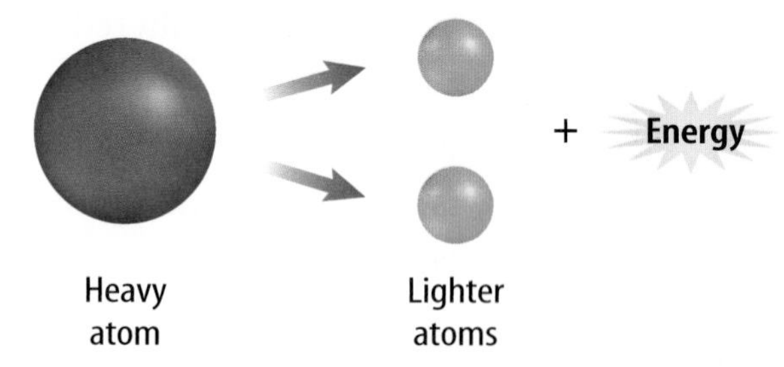

12. What other particles are released in the reaction above?
    A) protons **C)** uranium atoms
    B) neutrons **D)** heavy atoms

13. What is a region where many windmills are located in order to generate electricity from wind called?
    A) wind farm
    B) hydroelectric dam
    C) oil well
    D) steam-driven turbine

14. Which of the following is a deposit of hematite that can be mined at a profit?
    A) ore **C)** gangue
    B) anthracite **D)** energy resource

15. What is an important use of petroleum?
    A) making plaster **C)** as abrasives
    B) making glass **D)** making gasoline

16. Which of the following is a nonrenewable energy resource?
    A) water **C)** geothermal
    B) wind **D)** petroleum

Science Online earth.msscience.com/vocabulary_puzzlemaker

Thinking Critically

17. **Describe** the major problems associated with generating electricity using nuclear power plants.
18. **Explain** why wind is considered to be a renewable energy resource.
19. **Determine** which type of energy resources are considered to be biomass fuels. List three biomass fuels.
20. **Discuss** two conditions which could occur to cause gangue to be reclassified as an ore.
21. **Predict** If a well were drilled into a rock layer containing petroleum, natural gas, and water, which substance would be encountered first? Illustrate your answer with a labeled diagram.
22. **Compare and contrast** solar energy and wind energy by creating a table.
23. **Concept Map** Copy and complete the following concept map about mineral resources.

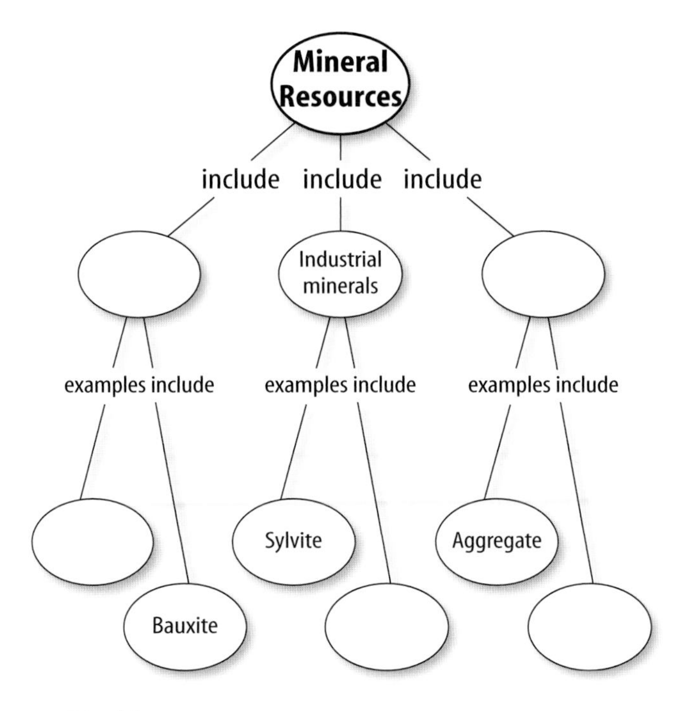

Performance Activities

24. **Make Models** Make a blueprint of a house that has been built to use passive solar energy. Which side of the house will face the sun?
25. **Letter** Write a letter to the Department of Energy asking how usable energy might be obtained from methane hydrates in the future. Also inquire about methods to extract methane hydrates.

Applying Math

Use the table below to answer questions 26–28.

Big Canyon Mine

Ore Mineral	Metal	Percent Composition	Value (dollars/kg)
Bauxite	Aluminum	5	1.00
Hematite	Iron	2	4.00
Chalcopyrite	Copper	1	6.00
Galena	Lead	7	1.00

26. **Ore Composition** If 100 kg of rock are extracted from this mine, what percentage of the rock is gangue?
    A. 15%
    B. 26%
    C. 74%
    D. 85%
27. **Total Composition** Graph the total composition of the extracted rock using a circle graph. Label each component clearly. Provide a title for your graph.
28. **Economic Geology** Of that 100 kg of extracted rock, determine how many kilograms of each ore mineral is extracted. List the total dollar value for each metal after the gangue has been eliminated and the ore mineral extracted.

Part 1 Multiple Choice

Record your answers on the answer sheet provided by your teacher or on a sheet of paper.

1. Which is a sedimentary rock formed from decayed plant matter?
   A. biomass **C.** natural gas
   B. coal **D.** oil

Use the graph below to answer questions 2 and 3.

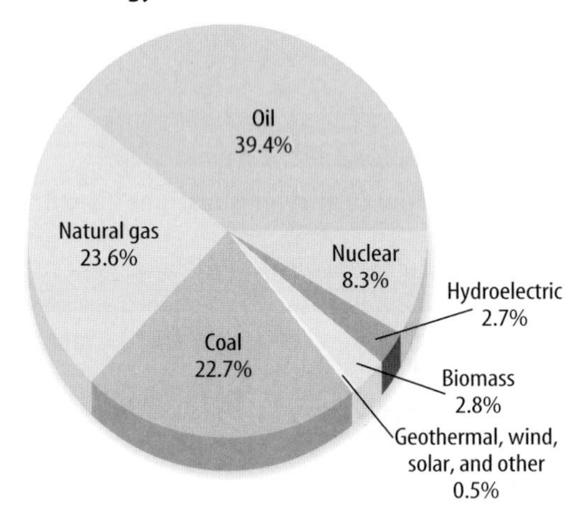

2. What percentage of the energy used in the United States comes from fossils fuels?
   A. 6% **C.** 47%
   B. 24% **D.** 86%

3. What percentage of our energy sources would be lost if coal runs out?
   A. 6% **C.** 23%
   B. 8% **D.** 39%

4. Which is a new potential source of methane?
   A. coal **C.** hydrocarbons
   B. hydrates **D.** petroleum

5. Which is a renewable resource?
   A. coal
   B. nuclear
   C. oil
   D. solar

Use the illustration below to answer questions 6 and 7.

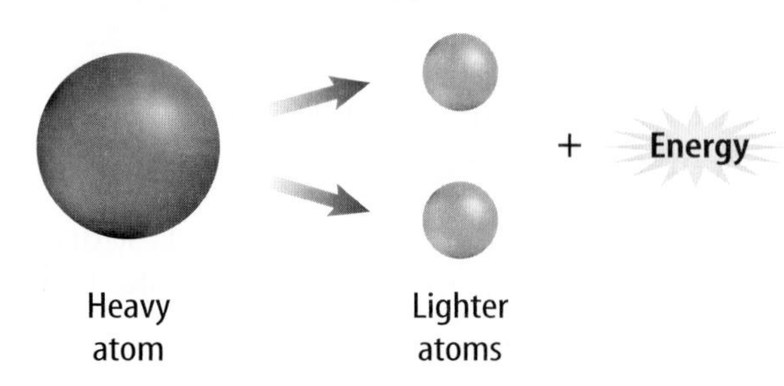

6. Which type of energy source is shown in this diagram?
   A. coal **C.** fusion
   B. fission **D.** natural gas

7. What is produced that drives a turbine, which turns a generator?
   A. atoms **C.** steam
   B. neutrons **D.** waste

8. Which type of energy uses magma or hot dry rocks to generate electricity?
   A. geothermal
   B. hydroelectric
   C. nuclear
   D. solar

9. What is combined to make gasohol?
   A. ethanol and gasoline
   B. oil and gasoline
   C. oil and petroleum
   D. wood and gasoline

10. Which helps to reduce the demand for new mineral resources?
    A. generating **C.** recycling
    B. mining **D.** refining

Test-Taking Tip

Circle Graphs If the question asks about the sum of multiple segments of a circle graph, do your addition on scratch paper and double-check your math before selecting an answer.

Part 2 Short Response/Grid In

Record your answers on the answer sheet provided by your teacher or on a sheet of paper.

11. What are two advantages of burning garbage for fuel?
12. What conditions are necessary for a wind farm?
13. Contrast methods used to remove coal with methods used to remove oil and natural gas.
14. How are bacteria involved in the formation of coal?
15. Why are methane hydrates so difficult to extract from the seafloor?
16. List two advantages of using fusion as an energy source.
17. Compare and contrast a mineral resource and an ore. How could a mineral resource become an ore? Is it possible for an ore to become just a mineral resource? Explain your answers.

Use the photo below to answer question 18.

18. What type of nonmetallic mineral resource is being used in this picture? List two other uses for this nonmetallic mineral.

Part 3 Open Ended

Record your answers on a sheet of paper.

19. Contrast the amount of heat released and smoke produced when burning peat, lignite, bituminous coal, and anthracite coal.
20. What are some household ways to help conserve fossil fuels?

Use the illustration below to answer question 21.

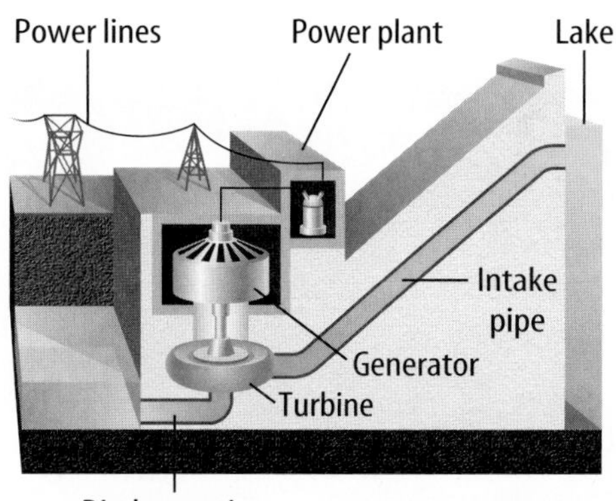

21. How is energy to run the turbine being produced? Discuss environmental issues associated with this energy source.
22. Design a 4-part, time-lapse illustration to show the path of iron from the hematite mine to pure iron.
23. How do population growth and technology affect the use of nonrenewable resources?
24. Some sources describe the Sun, wind, water, and geothermal energy as renewable energy resources. What might be some limitations to these resources?
25. Are mineral resources considered to be renewable or nonrenewable? Explain your answer.

unit 2 The Changing Surface of Earth

How Are Rivers & Writing Connected?

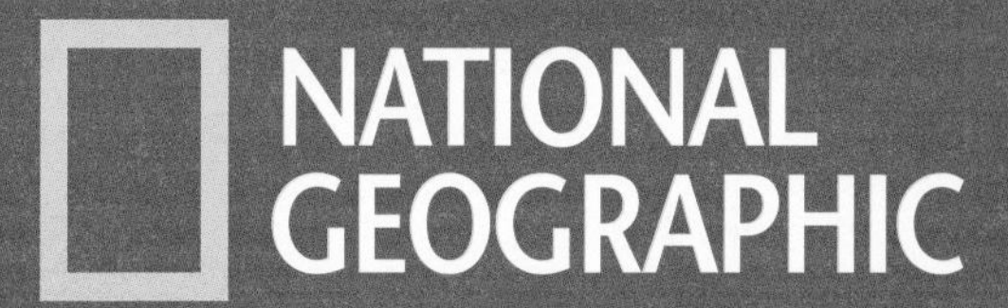

Rivers change the surface of Earth by moving material from place to place. In the process, rivers help build up fertile soil. Thousands of years ago, the Egyptian civilization took root in the extremely fertile soil along the Nile River (seen here as it looks today). Agriculture flourished, society grew more complex, and people needed a way to keep track of everything from harvests to history. Around 3100 B.C., the Egyptians developed one of the first systems of writing—a type of picture writing called hieroglyphics (left). Later peoples probably borrowed from the Egyptian system to create their own writing systems. By about 1000 B.C., the Phoenicians had developed an alphabet, with symbols that stood for individual sounds. The Phoenician alphabet was adopted and modified by other peoples, eventually giving rise to the alphabet we use today.

unit projects

Visit earth.msscience.com/unit_project to find project ideas and resources. Projects include:

- **History** Discover the characteristics of a variety of landforms and then design a tourist brochure about your specific feature.
- **Career** Explore a geology-related career, learn the tools of the trade, and present your career at a class job fair.
- **Model** Create a picture book with adventurous characters to demonstrate your new knowledge of Earth's changing surface.

WebQuest Investigate the *Barrier Islands*. Form an opinion as to whether developers should build in this environmentally fragile ecosystem.

chapter 6

Views of Earth

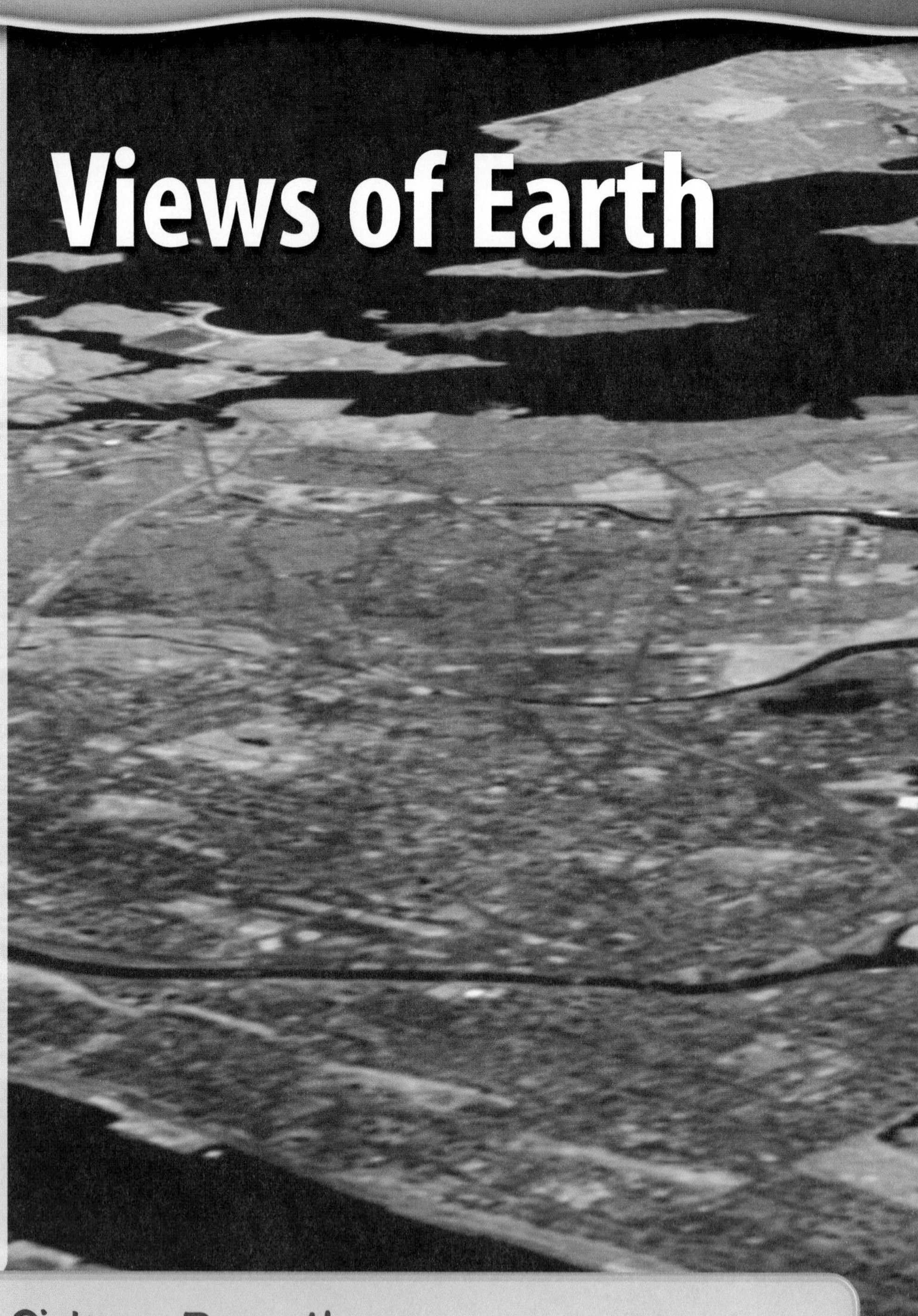

The BIG Idea

Different viewpoints are used when observing various features of Earth's surface.

SECTION 1
Landforms

Main Idea Three of Earth's basic landforms are plains, plateaus, and mountains.

SECTION 2
Viewpoints

Main Idea Latitude and longitude lines are used to identify exact locations on Earth's surface and are also useful when determining the time of day or night.

SECTION 3
Maps

Main Idea Maps are used to model Earth's surface and differ depending on their intended uses.

Pictures From Above

Remote sensing from satellites is a powerful way to learn about Earth's landforms, weather, and vegetation. In this image, vegetation shows up as green, uncovered land is red, water is blue, and human-made structures appear gray.

Science Journal Assume that you want to build a home at a location shown somewhere in this photograph. Describe where you would build your new home and why you would build at your chosen location.

Start-Up Activities

Describe Landforms

Pictures of Earth from space are acquired by instruments attached to satellites. Scientists use these images to make maps because they show features of Earth's surface, such as mountains and rivers.

1. Using a globe, atlas, or a world map, locate the following features and describe their positions on Earth relative to other major features.
   a. Andes mountains
   b. Amazon, Ganges, and Mississippi Rivers
   c. Indian Ocean, the Sea of Japan, and the Baltic Sea
   d. Australia, South America, and North America
2. Provide any other details that would help someone else find them.
3. **Think Critically** Choose one country on the globe or map and describe its major physical features in your Science Journal.

Preview this chapter's content and activities at earth.msscience.com

Views of Earth Make the following Foldable to help identify what you already know, what you want to know, and what you learned about the views of Earth.

STEP 1 **Fold** a vertical sheet of paper from side to side. Make the front edge about 1.25 cm shorter than the back edge.

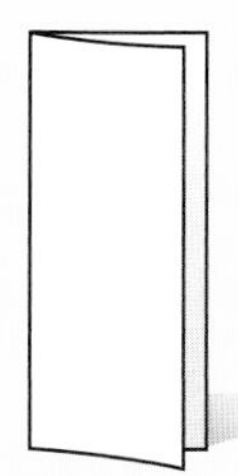

STEP 2 **Turn** lengthwise and **fold** into thirds.

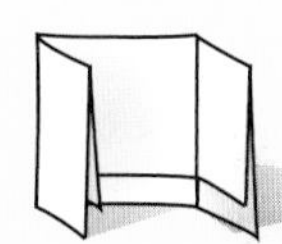

STEP 3 **Unfold and cut** only the top layer along both folds to make three tabs.

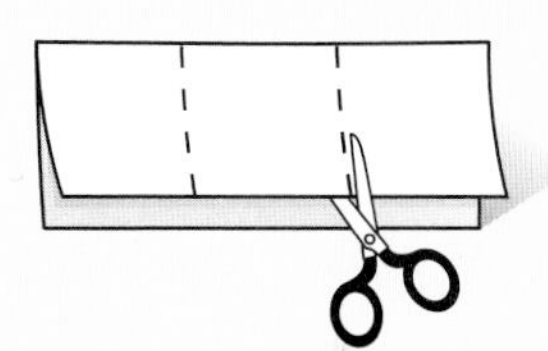

STEP 4 **Label** each tab.

Identify Questions Before you read the chapter, write what you already know about the views of Earth under the left tab of your Foldable, and write questions about what you want to know under the center tab. After you read the chapter, list what you learned under the right tab.

Get Ready to Read

Questioning

1 Learn It! Asking questions helps you to understand what you read. As you read, think about the questions you'd like answered. Often you can find the answer in the next paragraph or section. Learn to ask good questions by asking who, what, when, where, why, and how.

2 Practice It! Read the following passage from Section 1.

> Mountains with snowcapped peaks often are shrouded in clouds and tower above the surrounding land. If you climb them, the views are spectacular. The world's highest mountain peak is Mount Everest in the Himalaya—more than 8,800 m above see level. By contrast, the highest mountain peaks in the United States reach just over 6,000 m. Mountains also vary in how they are formed. The four main types of mountains are folded, upwarped, fault-block, and volcanic.
>
> *—from page 157*

Here are some questions you might ask about this paragraph:

- What are the four main types of mountains?
- What causes these four types of mountains to be different?
- Where is the world's highest mountain peak?

3 Apply It! As you read the chapter, look for answers to section headings that are in the form of questions.

Test yourself. Create questions and then read to find answers to your own questions.

Target Your Reading

Use this to focus on the main ideas as you read the chapter.

1. **Before you read** the chapter, respond to the statements below on your worksheet or on a numbered sheet of paper.
   - Write an **A** if you **agree** with the statement.
   - Write a **D** if you **disagree** with the statement.

2. **After you read** the chapter, look back to this page to see if you've changed your mind about any of the statements.
   - If any of your answers changed, explain why.
   - Change any false statements into true statements.
   - Use your revised statements as a study guide.

Before You Read A or D		Statement	After You Read A or D
	1	Plateaus are flat, raised landforms made of nearly horizontal rocks with a steep-sloped boundary.	
	2	Folded mountains are formed by tremendous forces inside Earth squeezing horizontal rock layers.	
	3	Volcanic mountains are cone-shaped structures that formed when molten rock rose to the surface.	
	4	Latitude lines run north to south.	
	5	Latitude lines are also called meridians.	
	6	A map scale is used to measure the weight of heavy maps.	
	7	A map legend is a historic map.	
	8	Contour lines run up and down on hillsides.	
	9	Contour intervals indicate horizontal distance on topographic maps.	
	10	Geologic cross sections can be used to visualize the slope of rock layers beneath Earth's surface.	

Science Online
Print out a worksheet of this page at earth.msscience.com

section 1

Landforms

as you read

What You'll Learn

- **Discuss** differences between plains and plateaus.
- **Describe** folded, upwarped, fault-block, and volcanic mountains.

Why It's Important

Landforms influence how people can use land.

Review Vocabulary
landform: a natural feature of a land surface

New Vocabulary
- plain
- plateau
- folded mountain
- upwarped mountain
- fault-block mountain
- volcanic mountain

Plains

Earth offers abundant variety—from tropics to tundras, deserts to rain forests, and freshwater mountain streams to saltwater tidal marshes. Some of Earth's most stunning features are its landforms, which can provide beautiful vistas, such as vast, flat, fertile plains; deep gorges that cut through steep walls of rock; and towering, snowcapped peaks. **Figure 1** shows the three basic types of landforms—plains, plateaus, and mountains.

Even if you haven't ever visited mountains, you might have seen hundreds of pictures of them in your lifetime. Plains are more common than mountains, but they are more difficult to visualize. **Plains** are large, flat areas, often found in the interior regions of continents. The flat land of plains is ideal for agriculture. Plains often have thick, fertile soils and abundant, grassy meadows suitable for grazing animals. Plains also are home to a variety of wildlife, including foxes, ground squirrels, and snakes. When plains are found near the ocean, they're called coastal plains. Together, interior plains and coastal plains make up half of all the land in the United States.

Figure 1 Three basic types of landforms are plains, plateaus, and mountains.

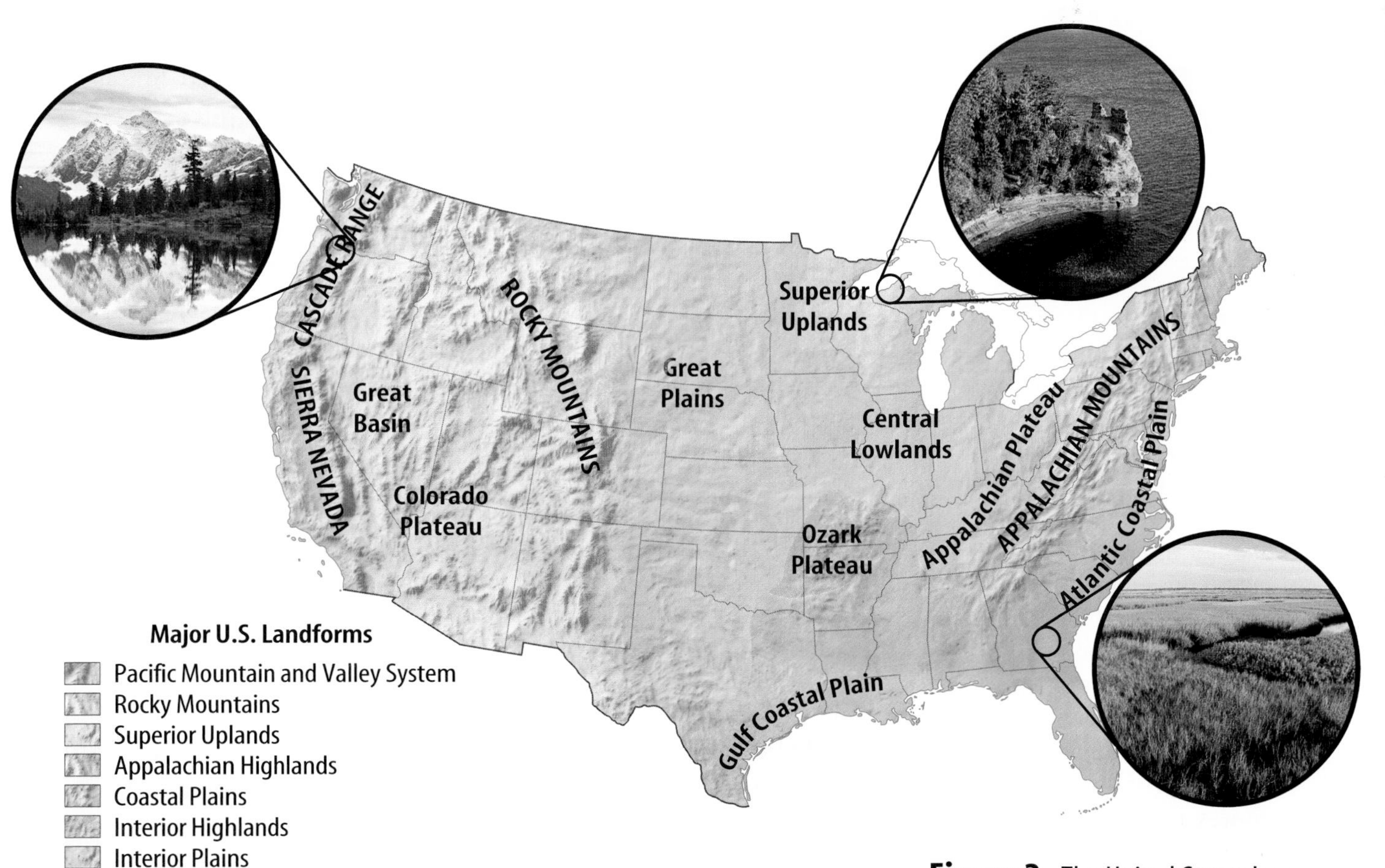

Figure 2 The United States has eight major landform regions, which include plains, mountains, and plateaus.
Describe *the region that you live in.*

Coastal Plains A coastal plain often is called a lowland because it is lower in elevation, or distance above sea level, than the land around it. You can think of the coastal plains as being the exposed portion of a continental shelf. The continental shelf is the part of a continent that extends into the ocean. The Atlantic Coastal Plain is a good example of this type of landform. It stretches along the east coast of the United States from New Jersey to Florida. This area has low rolling hills, swamps, and marshes. A marsh is a grassy wetland that usually is flooded with water.

The Atlantic Coastal Plain, shown in **Figure 2,** began forming about 70 million years ago as sediment began accumulating on the ocean floor. Sea level eventually dropped, and the seafloor was exposed. As a result, the coastal plain was born. The size of the coastal plain varies over time. That's because sea level rises and falls. During the last ice age, the coastal plain was larger than it is now because so much of Earth's water was contained in glaciers.

The Gulf Coastal Plain includes the lowlands in the southern United States that surround the Gulf of Mexico. Much of this plain was formed from sediment deposited in deltas by the many rivers that enter the Gulf of Mexico.

Reading Check *How are coastal plains formed?*

Profiling the United States

Procedure

1. Place the bottom edge of a piece of **paper** across the middle of **Figure 2,** extending from the west coast to the east coast.
2. Mark where different landforms are located along this edge.
3. Use a **map of the United States** and the **descriptions of the landforms in Section 1** to help you draw a profile, or side view, of the United States. Use steep, jagged lines to represent mountains. Low, flat lines can represent plains.

Analysis

1. Describe how your profile changed shape as you moved from west to east.
2. Describe how the shape of your profile would be different if you oriented your paper north to south.

Try at Home

Interior Plains The central portion of the United States is comprised largely of interior plains. Shown in **Figure 3,** you'll find them between the Rocky Mountains, the Appalachian Mountains, and the Gulf Coastal Plain. They include the Central Lowlands around the Missouri and Mississippi Rivers and the rolling hills of the Great Lakes area.

A large part of the interior plains is known as the Great Plains. This area lies between the Mississippi River and the Rocky Mountains. It is a flat, grassy, dry area with few trees. The Great Plains also are referred to as the high plains because of their elevation, which ranges from 350 m above sea level at the eastern border to 1,500 m in the west. The Great Plains consist of nearly horizontal layers of sedimentary rocks.

Plateaus

At somewhat higher elevations, you will find plateaus (pla TOHZ). **Plateaus** are flat, raised areas of land made up of nearly horizontal rocks that have been uplifted by forces within Earth. They are different from plains in that their edges rise steeply from the land around them. Because of this uplifting, it is common for plateaus, such as the Colorado Plateau, to be cut through by deep river valleys and canyons. The Colorado River, as shown in **Figure 3,** has cut deeply into the rock layers of the plateau, forming the Grand Canyon. Because the Colorado Plateau is located mostly in what is now a dry region, only a few rivers have developed on its surface. If you hiked around on this plateau, you would encounter a high, rugged environment.

Figure 3 Plains and plateaus are fairly flat, but plateaus have higher elevation.

Mountains

Mountains with snowcapped peaks often are shrouded in clouds and tower high above the surrounding land. If you climb them, the views are spectacular. The world's highest mountain peak is Mount Everest in the Himalaya—more than 8,800 m above sea level. By contrast, the highest mountain peaks in the United States reach just over 6,000 m. Mountains also vary in how they are formed. The four main types of mountains are folded, upwarped, fault-block, and volcanic.

Reading Check *What is the highest mountain peak on Earth?*

Topic: Landforms
Visit earth.msscience.com for Web links to information about some ways landforms affect economic development.

Activity Create four colorful postcards with captions explaining how landforms have affected economic development in your area.

Folded Mountains The Appalachian Mountains and the Rocky Mountains in Canada, shown in **Figure 4,** are comprised of folded rock layers. In **folded mountains,** the rock layers are folded like a rug that has been pushed up against a wall.

INTEGRATE Physics To form folded mountains, tremendous forces inside Earth squeeze horizontal rock layers, causing them to fold. The Appalachian Mountains formed between 480 million and 250 million years ago and are among the oldest and longest mountain ranges in North America. The Appalachians once were higher than the Rocky Mountains, but weathering and erosion have worn them down. They now are less than 2,000 m above sea level. The Ouachita (WAH shuh tah) Mountains of Arkansas are extensions of the same mountain range.

Figure 4 Folded mountains form when rock layers are squeezed from opposite sides. These mountains in Banff National Park, Canada, consist of folded rock layers.

Figure 5 The southern Rocky Mountains are upwarped mountains that formed when crust was pushed up by forces inside Earth.

Upwarped Mountains The Adirondack Mountains in New York, the southern Rocky Mountains in Colorado and New Mexico, and the Black Hills in South Dakota are upwarped mountains. **Figure 5** shows a mountain range in Colorado. Notice the high peaks and sharp ridges that are common to this type of mountain. **Upwarped mountains** form when blocks of Earth's crust are pushed up by forces inside Earth. Over time, the soil and sedimentary rocks at the top of Earth's crust erode, exposing the hard, crystalline rock underneath. As these rocks erode, they form the peaks and ridges.

Fault-Block Mountains **Fault-block mountains** are made of huge, tilted blocks of rock that are separated from surrounding rock by faults. These faults are large fractures in rock along which mostly vertical movement has occurred. The Grand Tetons of Wyoming, shown in **Figure 6,** and the Sierra Nevada in California, are examples of fault-block mountains. As **Figure 6** shows, when these mountains formed, one block was pushed up, while the adjacent block dropped down. This mountain-building process produces majestic peaks and steep slopes.

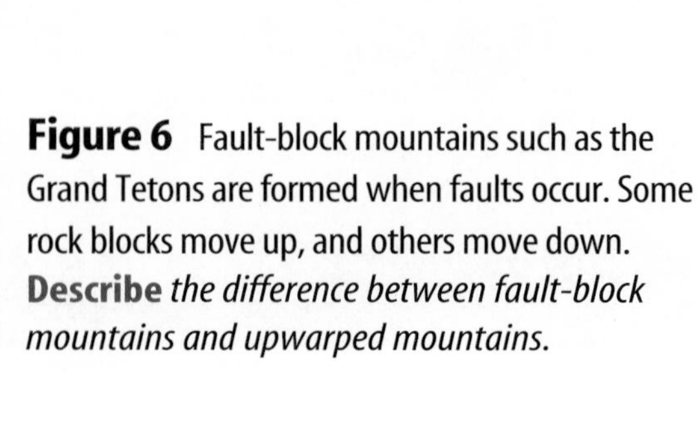

Figure 6 Fault-block mountains such as the Grand Tetons are formed when faults occur. Some rock blocks move up, and others move down. **Describe** *the difference between fault-block mountains and upwarped mountains.*

Figure 7 Mount Shasta is a volcanic mountain made up of layers of lava flows and ash.

Volcanic Mountains **Volcanic mountains,** like the one shown in **Figure 7,** begin to form when molten material reaches the surface through a weak area of the crust. The deposited materials pile up, layer upon layer, until a cone-shaped structure forms. Two volcanic mountains in the United States are Mount St. Helens in Washington and Mount Shasta in California. The Hawaiian Islands are the peaks of huge volcanoes that sit on the ocean floor. Measured from the base, Mauna Loa in Hawaii would be higher than Mount Everest.

Plains, plateaus, and mountains offer different kinds of landforms to explore. They range from low, coastal plains and high, desert plateaus to mountain ranges thousands of meters high.

section 1 review

Summary

Plains and Plateaus

- Plains are large, flat landforms that are usually found in the interior region of a continent.
- Plateaus are flat, raised landforms made of nearly horizontal, uplifted rocks.

Mountains

- Folded mountains form when horizontal rock layers are squeezed from opposite sides.
- Upwarped mountains form when blocks of Earth's crust are pushed up by forces inside Earth.
- Fault-block mountains form from huge, tilted blocks of rock that are separated by faults.
- Volcanic mountains form when molten rock forms cone-shaped structures at Earth's surface.

Self Check

1. **Describe** the eight major landform regions in the United States that are mentioned in this chapter.
2. **Compare and contrast** volcanic mountains, folded mountains, and upwarped mountains using a three-circle Venn diagram.
3. **Think Critically** If you wanted to know whether a particular mountain was formed by movement along a fault, what would you look for? Support your reasoning.

Applying Skills

4. **Concept Map** Make an events-chain concept map to explain how interior plains and coastal plains form.

section 2

Viewpoints

as you read

What You'll Learn

- **Define** latitude and longitude.
- **Explain** how latitude and longitude are used to identify locations on Earth.
- **Determine** the time and date in different time zones.

Why It's Important

Latitude and longitude allow you to locate places on Earth.

Review Vocabulary

pole: either end of an axis of a sphere

New Vocabulary

- equator
- latitude
- prime meridian
- longitude

Latitude and Longitude

During hurricane season, meteorologists track storms as they form in the Atlantic Ocean. To identify the exact location of a storm, latitude and longitude lines are used. These lines form an imaginary grid system that allows people to locate any place on Earth accurately.

Latitude Look at **Figure 8.** The **equator** is an imaginary line around Earth exactly halfway between the north and south poles. It separates Earth into two equal halves called the northern hemisphere and the southern hemisphere. Lines running parallel to the equator are called lines of **latitude,** or parallels. Latitude is the distance, measured in degrees, either north or south of the equator. Because they are parallel, lines of latitude do not intersect, or cross, one another.

The equator is at 0° latitude, and the poles are each at 90° latitude. Locations north and south of the equator are referred to by degrees north latitude and degrees south latitude, respectively. Each degree is further divided into segments called minutes and seconds. There are 60 minutes in one degree and 60 seconds in one minute.

Figure 8 Latitude and longitude are measurements that are used to indicate locations on Earth's surface.

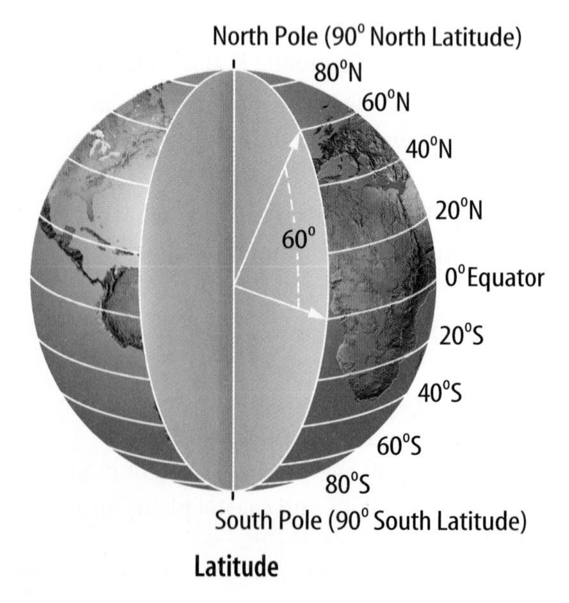

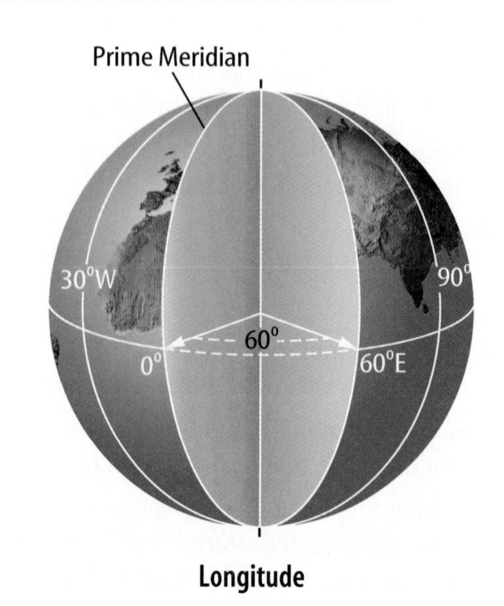

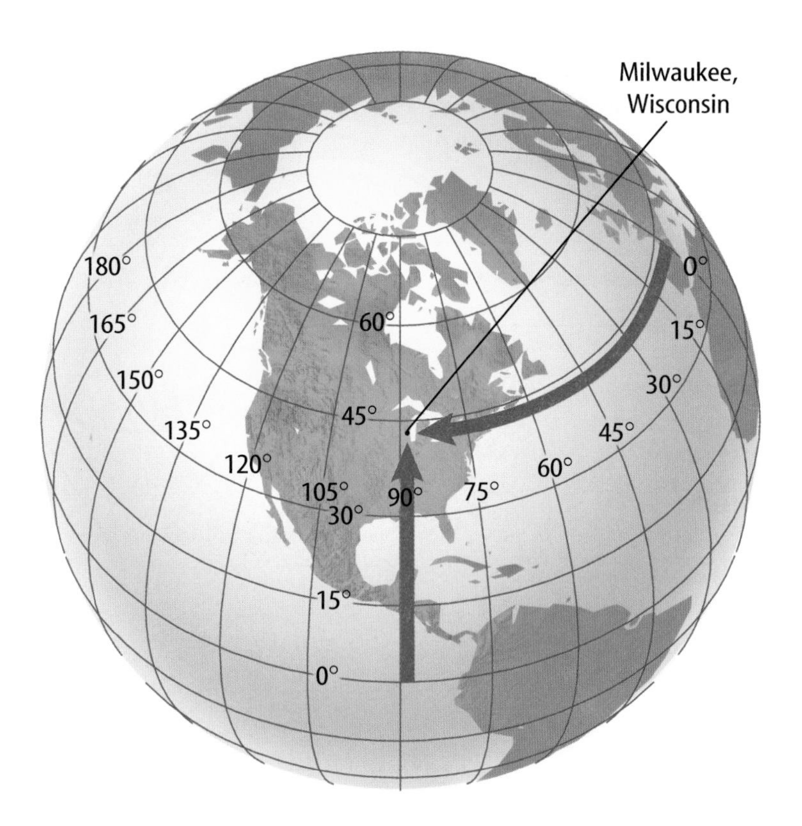

Figure 9 The city of Milwaukee, Wisconsin is located at about 43°N, 88°W. **Explain** *the difference between latitude and longitude.*

Longitude The vertical lines, seen in **Figure 8,** have two names—meridians and lines of longitude. Longitude lines are different from latitude lines in many important ways. Just as the equator is used as a reference point for lines of latitude, there's a reference point for lines of longitude—the **prime meridian.** This imaginary line represents 0° longitude. In 1884, astronomers decided the prime meridian should go through the Greenwich (GREN ihtch) Observatory near London, England. The prime meridian had to be agreed upon, because no natural point of reference exists.

Longitude refers to distances in degrees east or west of the prime meridian. Points west of the prime meridian have west longitude measured from 0° to 180°, and points east of the prime meridian have east longitude, measured similarly.

Prime Meridian The prime meridian does not circle Earth as the equator does. Rather, it runs from the north pole through Greenwich, England, to the south pole. The line of longitude on the opposite side of Earth from the prime meridian is the 180° meridian. East lines of longitude meet west lines of longitude at the 180° meridian. You can locate places accurately using latitude and longitude as shown in **Figure 9.** Note that latitude position always comes first when a location is given.

What line of longitude is found opposite the prime meridian?

Interpreting Latitude and Longitude

Procedure

1. Find the equator and prime meridian on a **world map.**
2. Move your finger to latitudes north of the equator, then south of the equator. Move your finger to longitudes west of the prime meridian, then east of the prime meridian.

Analysis

1. Identify the cities that have the following coordinates:
   a. 56°N, 38°E
   b. 34°S, 18°E
   c. 23°N, 82°W
2. Determine the latitude and longitude of the following cities:
   a. London, England
   b. Melbourne, Australia
   c. Buenos Aires, Argentina

Figure 10 The United States has six time zones.

A Washington, D.C., lies in the eastern time zone. Students there would be going to sleep at 9:00 P.M.

B But students in Seattle, Washington, which lies in the Pacific time zone, are eating dinner. **Determine** *what time it would be in Seattle when the students in Washington, D.C., are sleeping at 9:00 P.M.*

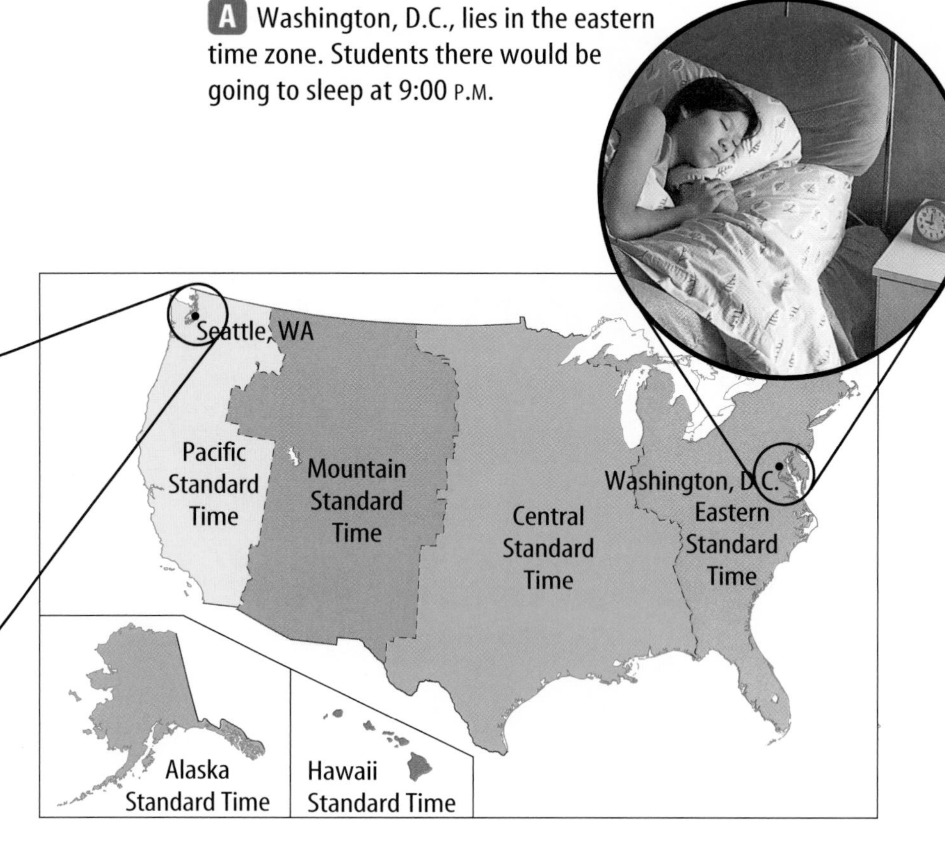

Time Zones

What time it is depends on where you are on Earth. Time is measured by tracking Earth's movement in relation to the Sun. Each day has 24 h, so Earth is divided into 24 time zones. Each time zone is about 15° of longitude wide and is 1 h different from the zones on each side of it. The United States has six different time zones. As you can see in **Figure 10,** people in different parts of the country don't experience dusk simultaneously. Because Earth rotates, the eastern states end a day while the western states are still in sunlight.

Reading Check *What is the basis for dividing Earth into 24 time zones?*

Time zones do not follow lines of longitude strictly. Time zone boundaries are adjusted in local areas. For example, if a city were split by a time zone boundary, the results would be confusing. In such a situation, the time zone boundary is moved outside the city.

Calendar Dates

In each time zone, one day ends and the next day begins at midnight. If it is 11:59 P.M. Tuesday, then 2 min later it will be 12:01 A.M. Wednesday in that particular time zone.

International Travel If you travel east or west across three or more time zones, you could suffer from jet lag. Jet lag occurs when your internal time clock does not match the new time zone. Jet lag can disrupt the daily rhythms of sleeping and eating. Have you or any of your classmates ever traveled to a foreign country and suffered from jet lag?

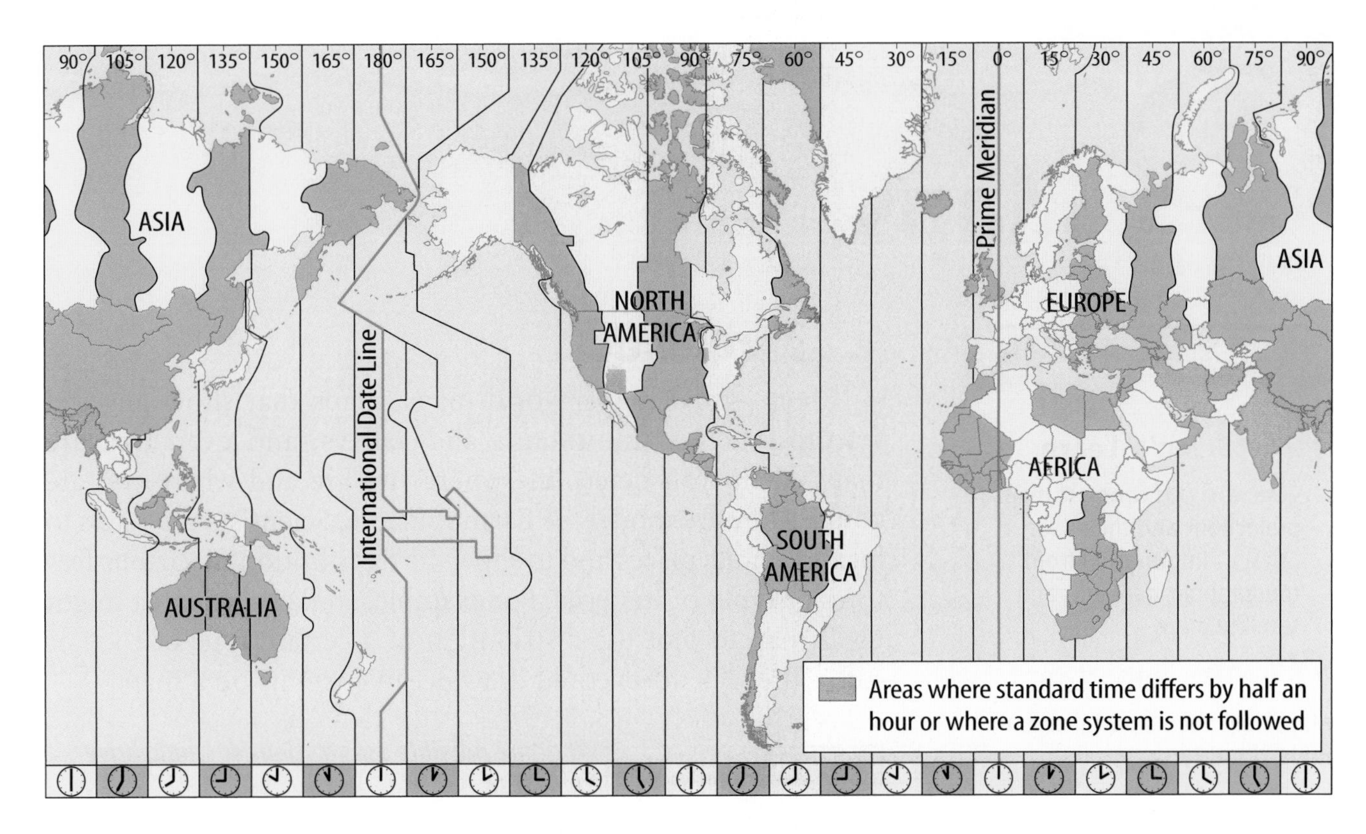

Figure 11 Lines of longitude roughly determine the locations of time zone boundaries. These boundaries are adjusted locally to avoid splitting cities and other political subdivisions, such as counties, into different time zones.

International Date Line You gain or lose time when you enter a new time zone. If you travel far enough, you can gain or lose a whole day. The International Date Line, shown on **Figure 11,** is the transition line for calendar days. If you were traveling west across the International Date Line, located near the 180° meridian, you would move your calendar forward one day. Traveling east, you would move your calendar back one day.

section 2 review

Summary

Latitude and Longitude

- The equator is the imaginary line that wraps around Earth at 0° latitude.
- Latitude is the distance in degrees north or south of the equator.
- The prime meridian is the imaginary line that represents 0° longitude and runs north to south through Greenwich, England.
- Longitude is the distance in degrees east or west of the prime meridian.

Time Zones and Calendar Dates

- Earth is divided into 24 one-hour time zones.
- The International Date Line is the transition line for calendar days.

Self Check

1. **Explain** how lines of latitude and longitude help people find locations on Earth.
2. **Determine** the latitude and longitude of New Orleans, Louisiana.
3. **Calculate** what time it would be in Los Angeles if it were 7:00 P.M. in New York City.
4. **Think Critically** How could you leave home on Monday to go sailing on the ocean, sail for 1 h on Sunday, and return home on Monday?

Applying Math

5. **Use Fractions** If you started at the prime meridian and traveled east one-fourth of the way around Earth, what line of longitude would you reach?

section 3

Maps

as you read

What You'll Learn

- **Compare and contrast** map projections and their uses.
- **Analyze** information from topographic, geologic, and satellite maps.

Why It's Important

Maps help people navigate and understand Earth.

Review Vocabulary

globe: a spherical representation of Earth

New Vocabulary

- conic projection
- topographic map
- contour line
- map scale
- map legend

Map Projections

Maps—road maps, world maps, maps that show physical features such as mountains and valleys, and even treasure maps—help you determine where you are and where you are going. They are models of Earth's surface. Scientists use maps to locate various places and to show the distribution of various features or types of material. For example, an Earth scientist might use a map to plot the distribution of a certain type of rock or soil. Other scientists could draw ocean currents on a map.

Reading Check *What are possible uses a scientist would have for maps?*

Many maps are made as projections. A map projection is made when points and lines on a globe's surface are transferred onto paper, as shown in **Figure 12.** Map projections can be made in several different ways, but all types of projections distort the shapes of landmasses or their areas. Antarctica, for instance, might look smaller or larger than it is as a result of the projection that is used for a particular map.

Figure 12 Lines of longitude are drawn parallel to one another in Mercator projections.
Describe *what happens near the poles in Mercator projections.*

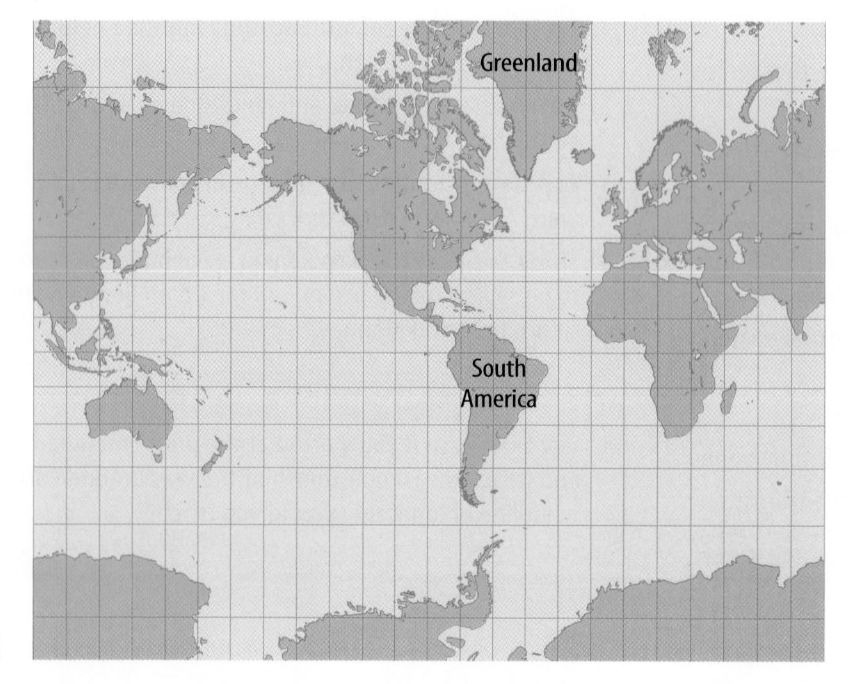

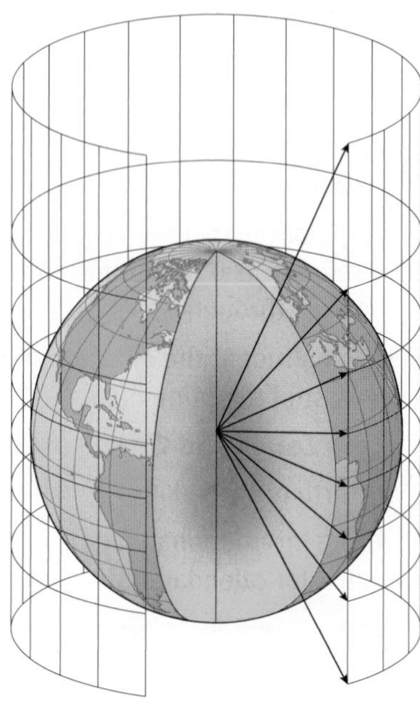

Figure 13 Robinson projections show little distortion in continent shapes and sizes.

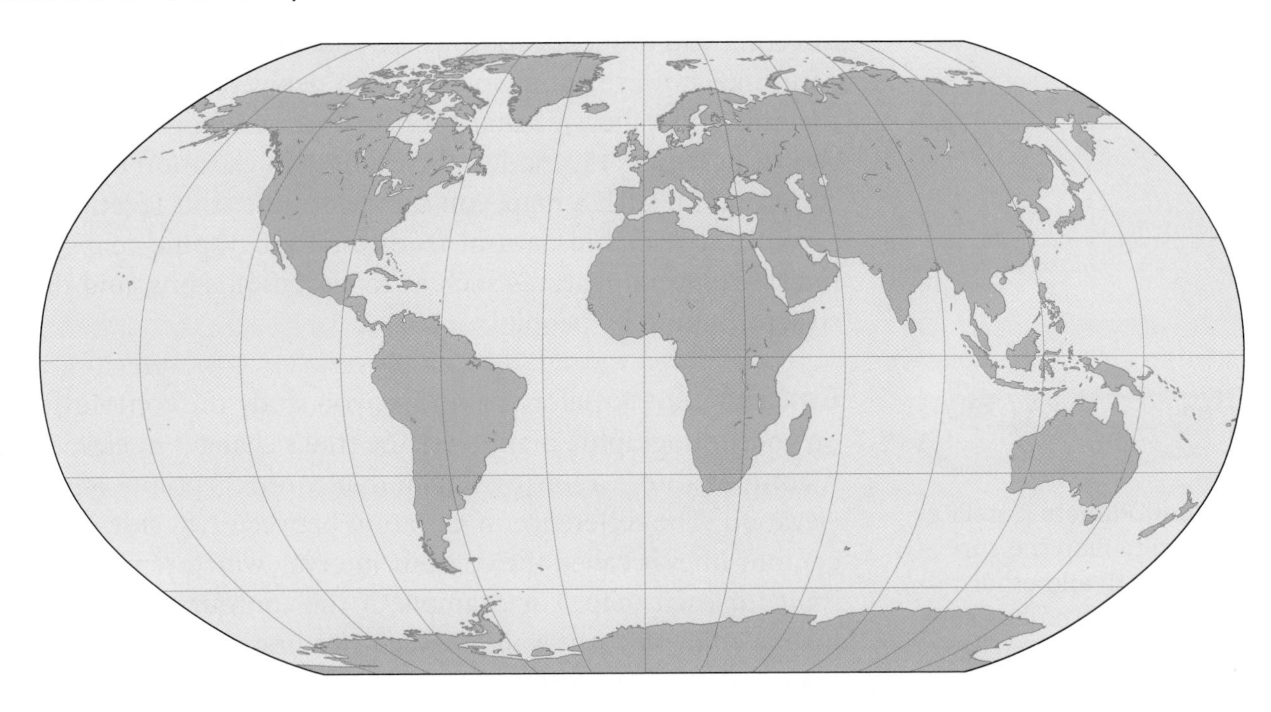

Mercator Projection Mercator (mer KAY ter) projections are used mainly on ships. They project correct shapes of continents, but the areas are distorted. Lines of longitude are projected onto the map parallel to each other. As you learned earlier, only latitude lines are parallel. Longitude lines meet at the poles. When longitude lines are projected as parallel, areas near the poles appear bigger than they are. Greenland, in the Mercator projection in **Figure 12,** appears to be larger than South America, but Greenland is actually smaller.

Robinson Projection A Robinson projection shows accurate continent shapes and more accurate land areas. As shown in **Figure 13,** lines of latitude remain parallel, and lines of longitude are curved as they are on a globe. This results in less distortion near the poles.

Conic Projection When you look at a road map or a weather map, you are using a conic (KAH nihk) projection. Conic projections, like the one shown in **Figure 14,** often are used to produce maps of small areas. These maps are well suited for middle latitude regions but are not as useful for mapping polar or equatorial regions. **Conic projections** are made by projecting points and lines from a globe onto a cone.

Reading Check *How are conic projections made?*

Figure 14 Small areas are mapped accurately using conic projections.

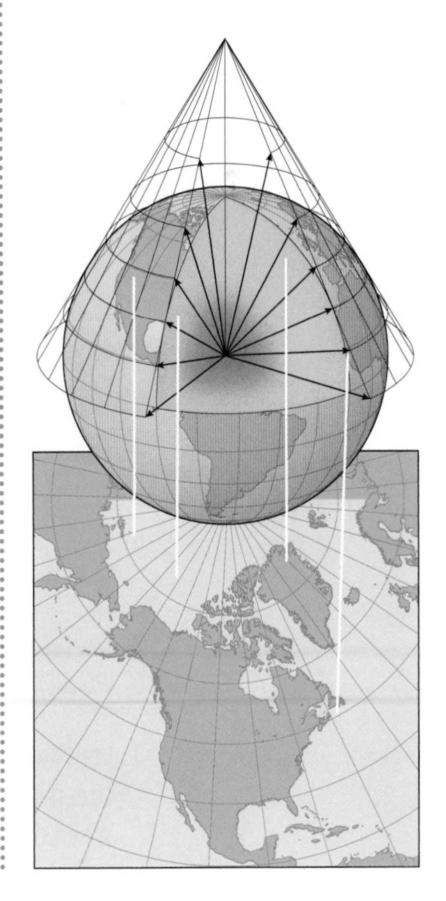

Topographic Maps

For nature hiking, a conic map projection can be helpful by directing you to the location where you will start your hike. On your hike, however, you would need a detailed map identifying the hills and valleys of that specific area. A **topographic map,** shown in **Figure 15,** models the changes in elevation of Earth's surface. With such a map, you can determine your location relative to identifiable natural features. Topographic maps also indicate cultural features such as roads, cities, dams, and other structures built by people.

Mapping Planets Satellites are used to map the surface of Earth and other planets. Space probes have made topographic maps of Venus and Mars. Satellites and probes send a radar beam or laser pulses to the surface and measure how long it takes for the beam or pulses to return to the space vehicle.

Contour Lines Before your hike, you study the contour lines on your topographic map to see the trail's changes in elevation. A **contour line** is a line on a map that connects points of equal elevation. The difference in elevation between two side-by-side contour lines is called the contour interval, which remains constant for each map. For example, if the contour interval on a map is 10 m and you walk between two lines anywhere on that map, you will have walked up or down 10 m.

In mountainous areas, the contour lines are close together. This situation models a steep slope. However, if the change in elevation is slight, the contour lines will be far apart. Often large contour intervals are used for mountainous terrain, and small contour intervals are used for fairly flat areas. Why? **Table 1** gives additional tips for examining contour lines.

Index Contours Some contour lines, called index contours, are marked with their elevation. If the contour interval is 5 m, you can determine the elevation of other lines around the index contour by adding or subtracting 5 m from the elevation shown on the index contour.

Table 1 Contour Rules
1. **Contour lines close around hills and basins.** To decide whether you're looking at a hill or basin, you can read the elevation numbers or look for hachures (ha SHOORZ). These are short lines drawn at right angles to the contour line. They show depressions by pointing toward lower elevations.
2. **Contour lines never cross.** If they did, it would mean that the spot where they cross would have two different elevations.
3. **Contour lines form Vs that point upstream when they cross streams.** This is because streams flow in depressions that are beneath the elevation of the surrounding land surface. When the contour lines cross the depression, they appear as Vs pointing upstream on the map.

NATIONAL GEOGRAPHIC

VISUALIZING TOPOGRAPHIC MAPS

Figure 15

Planning a hike? A topographic map will show you changes in elevation. With such a map, you can see at a glance how steep a mountain trail is, as well as its location relative to rivers, lakes, roads, and cities nearby. The steps in creating a topographic map are shown here.

Ⓐ To create a topographic map of Old Rag Mountain in Shenandoah National Park, Virginia, mapmakers first measure the elevation of the mountain at various points.

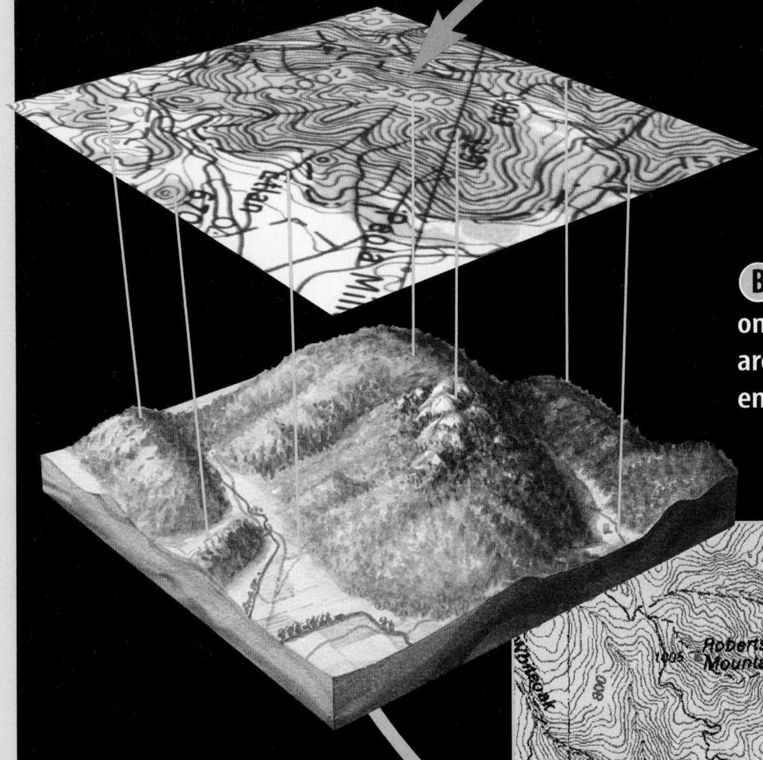

Ⓑ These points are then projected onto paper. Points at the same elevation are connected, forming contour lines that encircle the mountain.

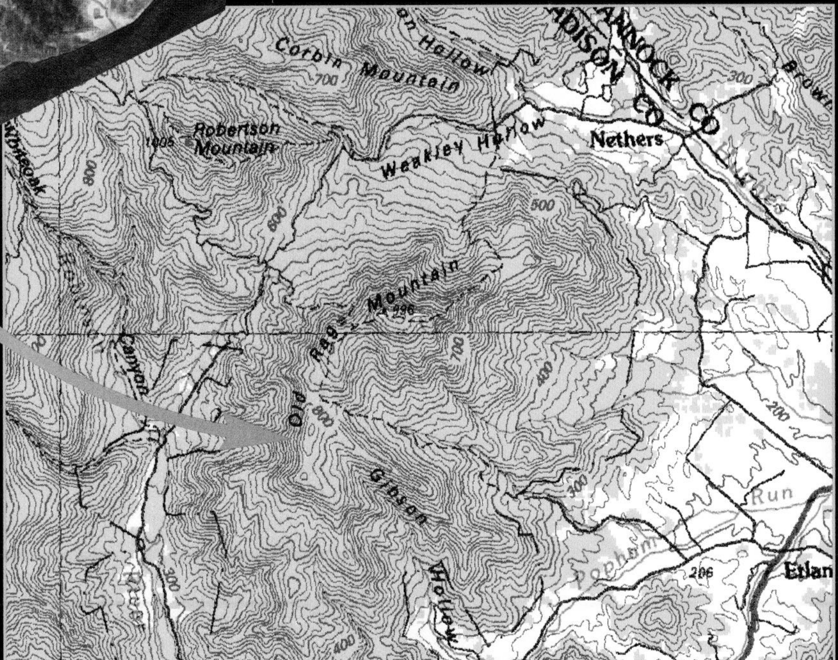

Ⓒ Where contour lines on a topographic map are close together, elevation is changing rapidly—and the trail is very steep!

Map Scale When planning your hike, you'll want to determine the distance to your destination before you leave. Because maps are small models of Earth's surface, distances and sizes of things shown on a map are proportional to the real thing on Earth. Therefore, real distances can be found by using a scale.

The **map scale** is the relationship between the distances on the map and distances on Earth's surface. Scale often is represented as a ratio. For example, a topographic map of the Grand Canyon might have a scale that reads 1:80,000. This means that one unit on the map represents 80,000 units on land. If the unit you wanted to use was a centimeter, then 1 cm on the map would equal 80,000 cm on land. The unit of distance could be feet or millimeters or any other measure of distance. However, the units of measure on each side of the ratio must always be the same. A map scale also can be shown in the form of a small bar that is divided into sections and scaled down to match real distances on Earth.

Topic: Map Technology
Visit earth.msscience.com for Web links to information about modern cartography and technology.

Activity Become a cartographer by developing a plan that uses map technology. For example, plan a traveling vacation or design a home or recreation area. Make a map of your trip or new environment.

Map Legend Topographic maps and most other maps have a legend. A **map legend** explains what the symbols used on the map mean. Some frequently used symbols for topographic maps are shown in the appendix at the back of the book.

Map Series Topographic maps are made to cover different amounts of Earth's surface. A map series includes maps that have the same dimensions of latitude and longitude. For example, one map series includes maps that are 7.5 minutes of latitude by 7.5 minutes of longitude. Other map series include maps covering larger areas of Earth's surface.

Geologic Maps

One of the more important tools to Earth scientists is the geologic map. Geologic maps show the arrangement and types of rocks at Earth's surface. Using geologic maps and data collected from rock exposures, a geologist can infer how rock layers might look below Earth's surface. The block diagram in **Figure 16** is a 3-D model that illustrates a solid section of Earth. The top surface of the block is the geologic map. Side views of the block are called cross sections, which are derived from the surface map. Developing geologic maps and cross sections is extremely important for the exploration and extraction of natural resources. What can a scientist do to determine whether a cross section accurately represents the underground features?

Figure 16 Geologists use block diagrams to understand Earth's subsurface. The different colors represent different rock layers.

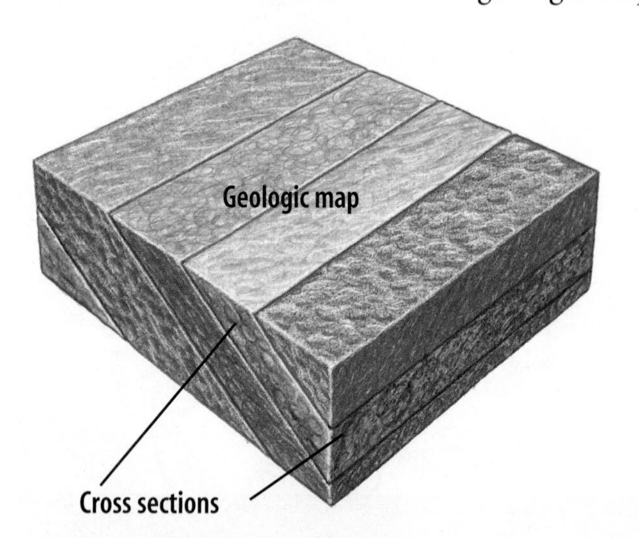

Three-Dimensional Maps Topographic maps and geologic maps are two-dimensional models that are used to study features of Earth's surface. To visualize Earth three dimensionally, scientists often rely on computers. Using computers, information is digitized to create a three-dimensional view of features such as rock layers or river systems. Digitizing is a process by which points are plotted on a coordinate grid.

Map Uses As you have learned, Earth can be viewed in many different ways. Maps are chosen depending upon the situation. If you wanted to determine New Zealand's location relative to Canada and you didn't have a globe, you probably would examine a Mercator projection. In your search, you would use lines of latitude and longitude, and a map scale. If you wanted to travel across the country, you would rely on a road map, or conic projection. You also would use a map legend to help locate features along the way. To climb the highest peak in your region, you would take along a topographic map.

Applying Science

How can you create a cross section from a geologic map?

Earth scientists are interested in knowing the types of rocks and their configurations underground. To help them visualize this, they use geologic maps. Geologic maps offer a two-dimensional view of the three-dimensional situation found under Earth's surface. You don't have to be a professional geologist to understand a geologic map. Use your ability to create graphs to interpret this geologic map.

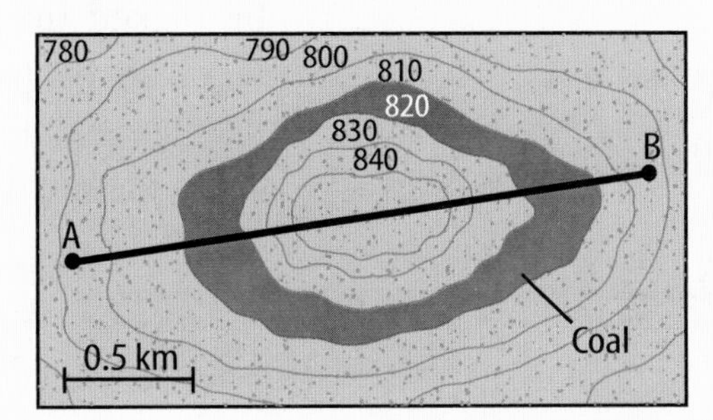

Identifying the Problem

Above is a simple geologic map showing where a coal seam is found on Earth's surface. Place a straight edge of paper along the line marked A–B and mark the points where it meets a contour. Make a different color mark where it meets the exposure of coal. Make a graph on which the various elevations (in meters) are marked on the *y*-axis. Lay your marked edge of paper along the *x*-axis and transfer the points directly above onto the proper elevation line. Now connect the dots to draw in the land's surface and connect the marks you made for the coal seam separately.

Solving the Problem

1. What type of topography does the map represent?
2. At what elevation is the coal seam?
3. Does this seam tilt, or is it horizontal? Explain how you know.

Figure 17 Hurricane Isabel's wind lashed the North Carolina and Virginia coasts on September 18, 2003.
Analyze *this satellite photo of Hurricane Isabel approaching the North Carolina Outer Banks. How many states do you think might be affected by this weather system?*

Remote Sensing

Scientists use remote-sensing techniques to collect much of the data used for making maps. Remote sensing is a way of collecting information about Earth from a distance, often using satellites.

Landsat One way that Earth's surface has been studied is with data collected from Landsat satellites, as shown in **Figure 17.** These satellites take pictures of Earth's surface using different wavelengths of light. The images can be used to make maps of snow cover over the United States or to evaluate the impact of forest fires, such as those that occurred in the western United States during the summer of 2000. The newest Landsat satellite, *Landsat 7,* can acquire detailed images by detecting light reflected off landforms on Earth.

Global Positioning System The Global Positioning System, or GPS, is a satellite-based, radio-navigation system that allows users to determine their exact position anywhere on Earth. Twenty-four satellites orbit 20,200 km above the planet. Each satellite sends a position signal and a time signal. The satellites are arranged in their orbits so that signals from at least six can be picked up at any given moment by someone using a GPS receiver. By processing the signals, the receiver calculates the user's exact location. GPS technology is used to navigate, to create detailed maps, and to track wildlife.

section 3 review

Summary

Map Projections
- A map projection is the projection of points and lines of a globe's surface onto paper.

Topographic Maps
- Topographic maps show the changes in elevation of Earth's surface by using contour lines.

Geologic Maps
- Geologic maps show the arrangement and types of rocks at Earth's surface.

Remote Sensing
- Remote sensing is a way of collecting information about Earth from a distance, often by using satellites.
- Distant planets can be mapped using satellites.

Self Check

1. **Compare and contrast** Mercator and conic projections.
2. **Explain** why Greenland appears larger on a Mercator projection than it does on a Robinson projection.
3. **Describe** why contour lines never cross.
4. **Explain** whether a topographic map or a geologic map would be most useful for drilling a water well.
5. **Think Critically** Review the satellite photograph at the beginning of this chapter. Is most of the city near or far from the water? Why is it located there?

Applying Skills

6. **Make Models** Architects make detailed maps called scale drawings to help them plan their work. Make a scale drawing of your classroom.

earth.msscience.com/self_check_quiz

Making a Topographic Map

Have you ever wondered how topographic maps are made? Today, radar and remote-sensing devices aboard satellites collect data, and computers and graphic systems make the maps. In the past, surveyors and aerial photographers collected data. Then, maps were hand drawn by cartographers, or mapmakers. In this lab, you can practice cartography.

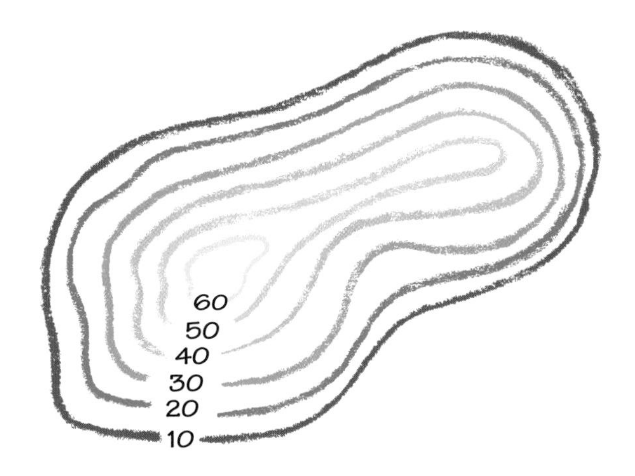

Real-World Question

How is a topographic map made?

Goals

- **Draw** a topographic map.
- **Compare and contrast** contour intervals.

Materials

plastic model of a landform
water tinted with food coloring
transparency
clear-plastic storage box with lid
beaker
metric ruler
tape
transparency marker

Procedure

1. Using the ruler and the transparency marker, make marks up the side of the storage box that are 2 cm apart.
2. Secure the transparency to the outside of the box lid with tape.
3. Place the plastic model in the box. The bottom of the box will be zero elevation.
4. Using the beaker, pour water into the box to a height of 2 cm. Place the lid on the box.
5. Use the transparency marker to trace the top of the water line on the transparency.
6. Using the scale 2 cm = 10 m, mark the elevation on the line.
7. Repeat the process of adding 2 cm of water and tracing until the landform is mapped.
8. Transfer the tracing of the landform onto a sheet of white paper.

Conclude and Apply

1. **Identify** the contour interval of this topographic map.
2. **Evaluate** how the distance between contour lines on the map shows the steepness of the slope on the landform model.
3. **Determine** the total elevation of the landform you have selected.
4. **Describe** how elevation was represented on your map.
5. **Explain** how elevations are shown on topographic maps.
6. Must all topographic maps have a contour line that represents 0 m of elevation? Explain.

Model and Invent

Constructing Landforms

Goals

- **Research** how contour lines show relief on a topographic map.
- **Determine** what scale you can best use to model a landscape of your choice.
- Working cooperatively with your classmates, model a landscape in three dimensions from the information given on a topographic map.

Possible Materials

U.S. Geological Survey 7.5-minute quadrangle maps
sandbox sand
rolls of brown paper towels
spray bottle filled with water
ruler

Real-World Question

Most maps perform well in helping you get from place to place. A road map, for example, will allow you to choose the shortest route from one place to another. If you are hiking, though, distance might not be so important. You might want to choose a route that avoids steep terrain. In this case you need a map that shows the highs and lows of Earth's surface, called relief. Topographic maps use contour lines to show the landscape in three dimensions. Among their many uses, such maps allow hikers to choose routes that maximize the scenery and minimize the physical exertion. What does a landscape depicted on a two-dimensional topographic map look like in three dimensions? How can you model a landscape?

Make a Model

1. **Choose** a topographic map showing a landscape easily modeled using sand. Check to see what contour interval is used on the map. Use the index contours to find the difference between the lowest and the highest elevations shown on the landscape. Check the distance scale to determine how much area the landscape covers.
2. **Determine** the scale you will use to convert the elevations shown on your map to heights on your model. Make sure the scale is proportional to the distances on your map.
3. **Plan** a model of the landscape in sand by sketching the main features and their scaled heights onto paper. Note the degree of steepness found on all sides of the features.

4. **Prepare** a document that shows the scale you plan to use for your model and the calculations you used to derive that scale. Remember to use the same scale for distance as you use for height. If your landscape is fairly flat, you can exaggerate the vertical scale by a factor of two or three. Be sure your paper is neat, is easy to follow, and includes all units. Present the document to your teacher for approval.

Test Your Model

1. Using the sand, spray bottle, and ruler, create a scale model of your landscape on the brown paper towels.
2. **Check** your topographic map to be sure your model includes the landscape features at their proper heights and proper degrees of steepness.

Analyze Your Data

1. **Determine** if your model accurately represents the landscape depicted on your topographic map. Discuss the strengths and weaknesses of your model.
2. **Explain** why it was important to use the same scale for height and distance. If you exaggerated the height, why was it important to indicate the exaggeration on your model?

Conclude and Apply

1. **Infer** why the mapmakers chose the contour interval used on your topographic map?
2. **Predict** the contour intervals mapmakers might choose for topographic maps of the world's tallest mountains—the Himalaya—and for topographic maps of Kansas, which is fairly flat.

Communicating Your Data

Prepare a vacation getaway commercial to advertise the topographical features of your model landscape. Be sure to discuss the landscape elevation and features, scale, and similarities to actual landforms.

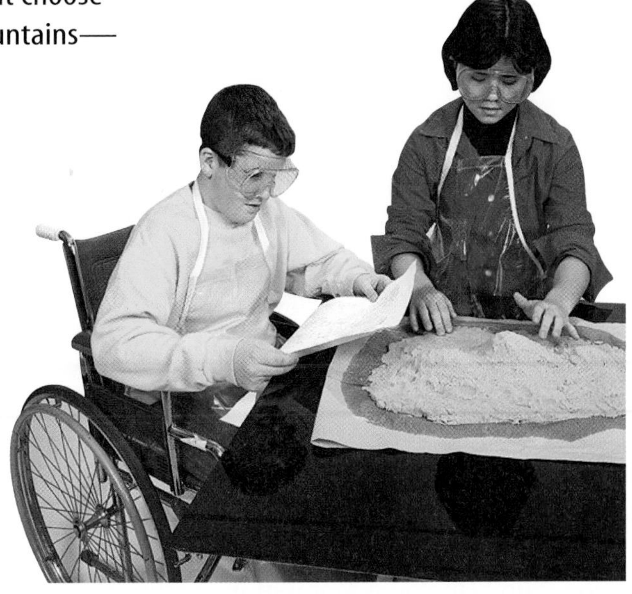

TIME SCIENCE AND HISTORY

SCIENCE CAN CHANGE THE COURSE OF HISTORY!

New York Harbor in 1849

Rich Midwest farmland

Georgia peaches

Alaska pipeline

Maine fishing and lobster industry

LOCATION, LOCATION

Why is New York City at the mouth of the Hudson River and not 300 km inland? Why are there more farms in Iowa than in Alaska? What's the reason for growing lots of peaches in Georgia but not in California's Death Valley? It's all about location. The landforms, climate, soil, and resources in an area determine where cities and farms grow and what people connected with them do.

LANDFORMS ARE KEY

When many American cities were founded hundreds of years ago, waterways were the best means of transportation. Old cities such as New York City and Boston are located on deep harbors where ships could land with people and goods. Rivers also were major highways centuries ago. They still are.

Topography and soil also play a role in where activities such as farming take root. States such as Iowa and Illinois have many farms because they have flat land and fertile soil. Growing crops is more difficult in mountainous areas or where soil is stony and poor.

CLIMATE AND SOIL

Climate limits the locations of cities and farms, as well. The fertile soil and warm, moist climate of Georgia make it a perfect place to grow peaches. California's Death Valley can't support such crops because it's a hot, dry desert.

RESOURCES RULE

The location of an important natural resource can change the rules. A gold deposit or an oil field can cause a town to grow in a place where the topography, soil, and climate are not favorable. For example, thousands of people now live in parts of Alaska only because of the great supply of oil there. Maine has a harsh climate and poor soil. But people settled along its coast because they could catch lobsters and fish in the nearby North Atlantic.

The rules that govern where towns grow and where people live are different now than they used to be. Often information, not goods, moves from place to place on computers that can be anywhere. But as long as people farm, use minerals, and transport goods from place to place, the natural environment and natural resources will always help determine where people are and what they do.

Research **Why was your community built where it is? Research its history. What types of economic activity were important when it was founded? Did topography, climate, or resources determine its location? Design a Moment in History to share your information.**

For more information, visit earth.msscience.com/time

Reviewing Main Ideas

Section 1 Landforms

1. The three main types of landforms are plains, plateaus, and mountains.
2. Plains are large, flat areas. Plateaus are relatively flat, raised areas of land made up of nearly horizontal rocks that have been uplifted. Mountains rise high above the surrounding land.

Section 2 Viewpoints

1. Latitude and longitude form an imaginary grid system that enables points on Earth to be located exactly.
2. Latitude is the distance in degrees north or south of the equator. Longitude is the distance in degrees east or west of the prime meridian.
3. Earth is divided into 24 time zones. Each time zone represents a 1-h difference. The International Date Line separates different calendar days.

Section 3 Maps

1. Mercator, Robinson, and conic projections are made by transferring points and lines on a globe's surface onto paper.
2. Topographic maps show the elevation of Earth's surface. Geologic maps show the types of rocks that make up Earth's surface.
3. Remote sensing is a way of collecting information from a distance. Satellites are important remote-sensing devices.

Visualizing Main Ideas

Copy and complete the following concept map on landforms.

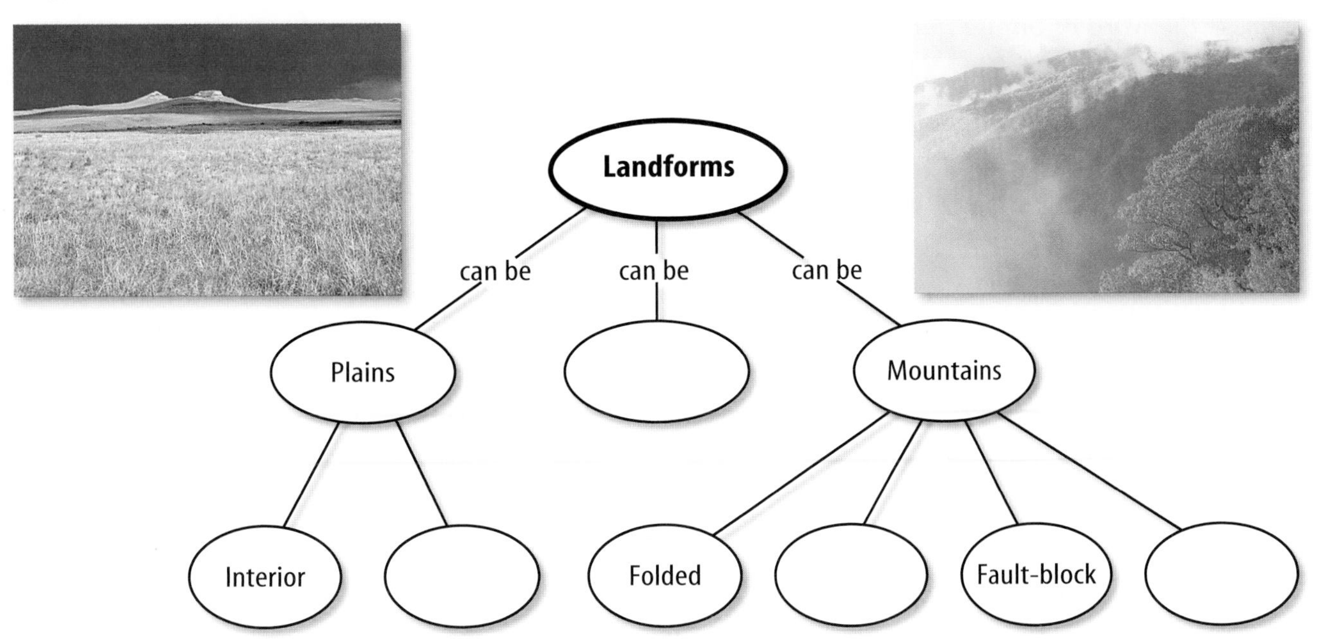

Using Vocabulary

conic projection p. 165	map scale p. 168
contour line p. 166	plain p. 154
equator p. 160	plateau p. 156
fault-block mountain p. 158	prime meridian p. 161
folded mountain p. 157	topographic map p. 166
latitude p. 160	upwarped mountain p. 158
longitude p. 161	volcanic mountain p. 159
map legend p. 168	

For each set of terms below, choose the one term that does not belong and explain why it does not belong.

1. upwarped mountain—equator—volcanic mountain
2. plain—plateau—prime meridian
3. topographic map—contour line—volcanic mountain
4. prime meridian—equator—folded mountain
5. fault-block mountain—upwarped mountain—plateau

Checking Concepts

Choose the word or phrase that best answers the question.

6. What makes up about 50 percent of all land areas in the United States?
   A) plateaus **C)** mountains
   B) plains **D)** volcanoes
7. Which type of map shows changes in elevation at Earth's surface?
   A) conic **C)** Robinson
   B) topographic **D)** Mercator
8. How many degrees apart are the 24 time zones?
   A) 10° **C)** 15°
   B) 34° **D)** 25°

Use the photo below to answer question 9.

9. What kind of mountains are the Grand Tetons of Wyoming?
   A) fault-block **C)** upwarped
   B) volcanic **D)** folded
10. Landsat satellites collect data by using
   A) sonar. **C)** sound waves.
   B) echolocation. **D)** light waves.
11. Which type of map is most distorted at the poles?
   A) conic **C)** Robinson
   B) topographic **D)** Mercator
12. Where is the north pole located?
   A) 0°N **C)** 50°N
   B) 180°N **D)** 90°N
13. What is measured with respect to sea level?
   A) contour interval
   B) elevation
   C) conic projection
   D) sonar
14. What kind of map shows rock types making up Earth's surface?
   A) topographic **C)** geologic
   B) Robinson **D)** Mercator
15. Which major U.S. landform includes the Grand Canyon?
   A) Great Plains **C)** Colorado Plateau
   B) Great Basin **D)** Gulf Coastal Plain

Science Online earth.msscience.com/vocabulary_puzzlemaker

Thinking Critically

16. **Explain** how a topographic map of the Atlantic Coastal Plain differs from a topographic map of the Rocky Mountains.

17. **Determine** If you left Korea early Wednesday morning and flew to Hawaii, on what day of the week would you arrive?

Use the illustration below to answer question 18.

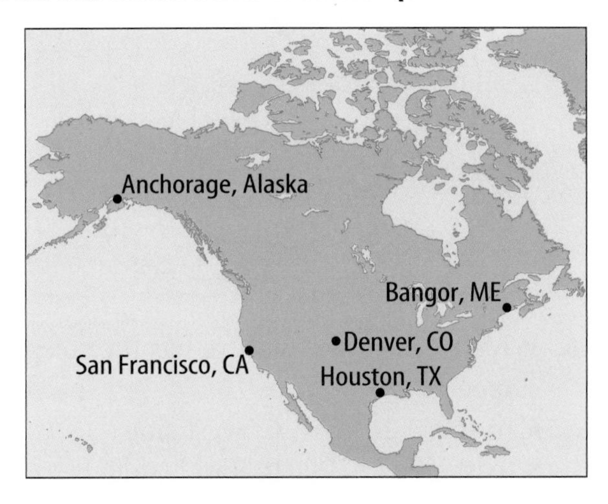

18. **Determine** Using the map above, arrange these cities in order from the city with the earliest time to the one with the latest time on a given day: Anchorage, Alaska; San Francisco, California; Bangor, Maine; Denver, Colorado; Houston, Texas.

19. **Describe** how a map with a scale of 1:50,000 is different from a map with a scale of 1:24,000.

20. **Compare and contrast** Mercator, Robinson, and conic map projections.

21. **Form Hypotheses** You are visiting a mountain in the northwest part of the United States. The mountain has steep sides and is not part of a mountain range. A crater can be seen at the top of the mountain. Hypothesize about what type of mountain you are visiting.

22. **Concept Map** Copy and complete the following concept map about parts of a topographic map.

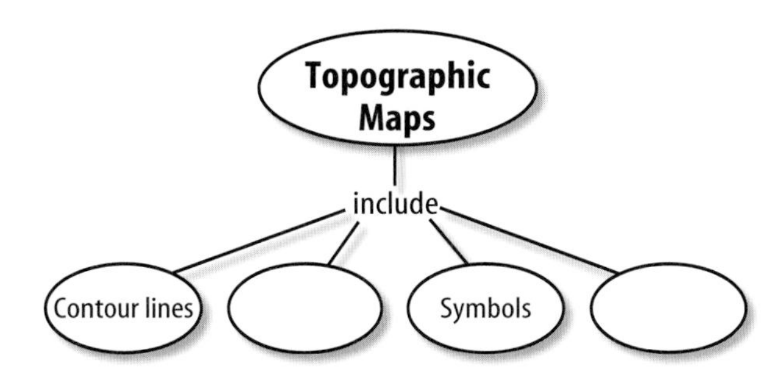

Performance Activities

23. **Poem** Create a poem about one type of landform. Include characteristics of the landform in your poem. How can the shape of your poem add meaning to your poem? Display your poem with those of your classmates.

Applying Math

24. **Calculate** If you were flying directly south from the north pole and reached 70° north latitude, how many more degrees of latitude would you pass over before reaching the south pole? Illustrate and label a diagram to support your answer.

Use the map below to answer question 25.

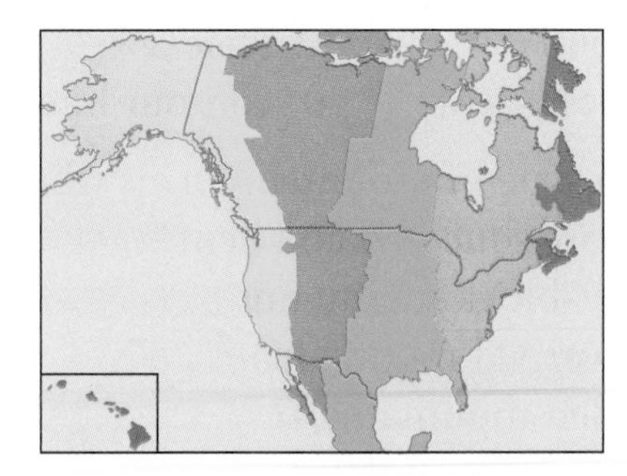

25. **Calculate** If it is 2:00 P.M. in Orlando, Florida, what time is it in Los Angeles, California? In Anchorage, Alaska? In your hometown?

chapter 6 Standardized Test Practice

Part 1 Multiple Choice

Record your answers on the answer sheet provided by your teacher or on a sheet of paper.

Use the map below to answer question 1.

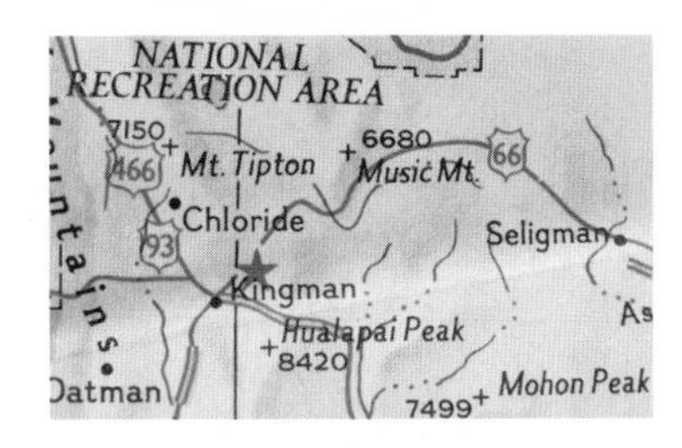

1. Which of the following is shown above?
   A. cross section **C.** topographic map
   B. geologic map **D.** road map

2. Which landform is a relatively flat area that has high elevation?
   A. mountain **C.** coastal plain
   B. interior plain **D.** plateau

3. Which of the following can provide detailed information about your position on Earth's surface?
   A. prime meridian
   B. global positioning system
   C. International Date Line
   D. LandSat 7

4. What connects points of equal elevation on a map?
   A. legend **C.** scale
   B. series **D.** contour line

5. Which type of mountain forms when rock layers are squeezed and bent?
   A. fault-block mountains
   B. upwarped mountains
   C. folded mountains
   D. volcanic mountains

Test-Taking Tip

Read Carefully Read all choices before answering the questions.

Use the illustration below to answer questions 6–8. The numbers on the drawing represent meters above sea level.

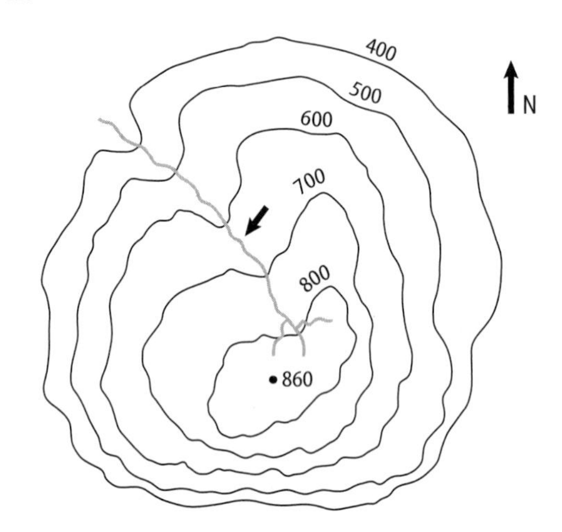

6. Which side of the feature has the steepest slope?
   A. north side **C.** west side
   B. east side **D.** south side

7. What is the highest elevation on the feature?
   A. 800 meters **C.** 860 meters
   B. 960 meters **D.** 700 meters

8. What does the line that is marked by the arrow represent?
   A. a contour line **C.** a high ridge
   B. a stream **D.** a glacier

9. Which type of map is made by projecting points and lines from a globe onto a cone?
   A. Mercator projection
   B. conic projection
   C. Robinson projection
   D. geologic map

10. Which are useful for measuring position north or south of the equator?
    A. lines of latitude
    B. lines of longitude
    C. index contours
    D. map legends

Part 2 Short Response/Grid In

Record your answers on the answer sheet provided by your teacher or on a separate sheet of paper.

11. How do volcanic mountains form?

12. What is a time zone? How are time zones determined around the world? Why are they needed?

13. Which type of map would you use to find the location of a layer of coal at Earth's surface? Why?

14. How are plateaus similar to plains? How are they different?

15. Why are lines of latitude sometimes called parallels?

16. List the locations on Earth that represent 0° latitude, 90°N latitude, and 90°S latitude.

Use the scale at the bottom of the map to answer questions 17–19.

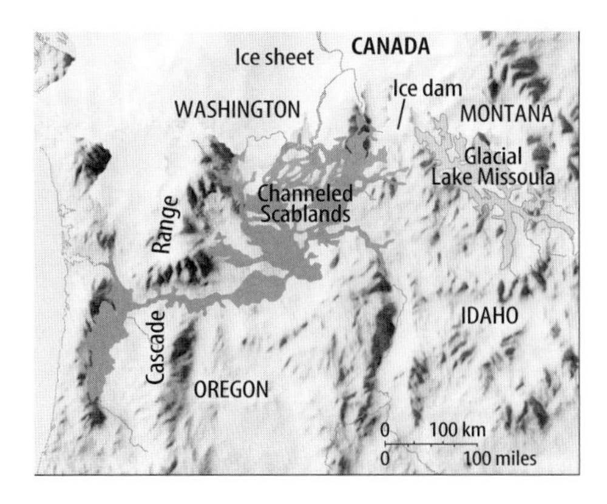

17. How many centimeters on Earth's surface are represented by one centimeter on the map?

18. How many meters on Earth's surface does one centimeter represent?

19. How many kilometers on Earth's surface does one centimeter represent?

Part 3 Open Ended

Record your answers on a sheet of paper.

Use the map below to answer questions 20–21.

20. Where do mountains, plateaus, and plains occur in the United States? Describe these regions and how they were formed.

21. Take the role of a fur trader, pioneer, or explorer. Write three journal entries that give a general description of how the landforms change across the United States.

22. Compare and contrast interior plains and coastal plains.

23. Why is remote sensing important to society? What types of information are obtained?

24. Why are computers often used to make maps?

25. Why might you experience jet lag if you travel across the United States on a plane? Predict which direction it would be more difficult to adjust to jet lag—when traveling from Hawaii to New York or New York to Hawaii. Support your answer with an example of each situation.

chapter 7

Weathering and Soil

The BIG Idea

Soil is a natural resource that must be monitored, managed, and protected.

SECTION 1
Weathering

Main Idea Weathering processes weaken and break apart rock material into smaller pieces.

SECTION 2
The Nature of Soil

Main Idea Soil is a mixture of weathered rock, decayed organic matter, mineral fragments, water, and air.

SECTION 3
Soil Erosion

Main Idea Soil erosion is harmful because plants do not grow as well when topsoil has been removed.

What's a tor?

A tor, shown in the photo, is a pile of boulders left on the land. Tors form because of weathering, which is a natural process that breaks down rock. Weathering weakened the rock that used to be around the boulders. This weakened rock then was eroded away, and the boulders are all that remain.

Science Journal Write a poem about a tor. Use words in your poem that rhyme with the word *tor*.

Start-Up Activities

Stalactites and Stalagmites

During weathering, minerals can be dissolved by acidic water. If this water seeps into a cave, minerals might precipitate. In this lab, you will model the formation of stalactites and stalagmites.

1. Pour 700 mL of water into two 1,000-mL beakers and place the beakers on a large piece of cardboard. Stir Epsom salt into each beaker until no more will dissolve.
2. Add two drops of yellow food coloring to each beaker and stir.
3. Measure and cut three 75-cm lengths of cotton string. Hold the three pieces of string in one hand and twist the ends of all three pieces to form a loose braid of string.
4. Tie each end of the braid to a large steel nut.
5. Soak the braid of string in one of the beakers until it is wet with the solution. Drop one nut into one beaker and the other nut into the second beaker. Allow the string to sag between the beakers. Observe for several days.
6. **Think Critically** Record your observations in your Science Journal. How does this activity model the formation of stalactites and stalagmites?

Weathering and Soil Make the following Foldable to help you understand the vocabulary terms in this chapter.

STEP 1 **Fold** a vertical sheet of notebook paper from side to side.

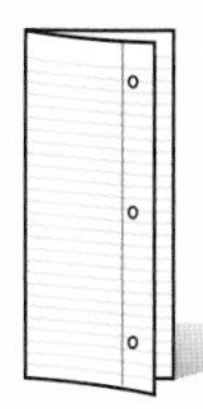

STEP 2 **Cut** along every third line of only the top layer to form tabs.

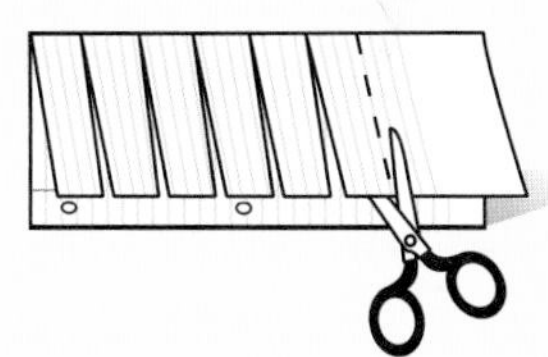

STEP 3 **Label** each tab.

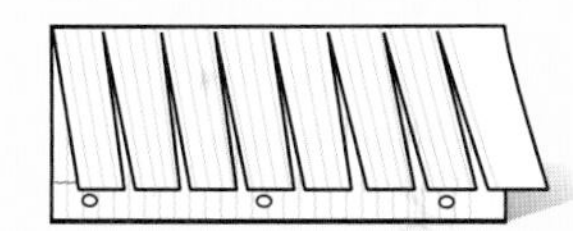

Build Vocabulary As you read the chapter, list the vocabulary words about weathering and soil on the tabs. As you learn the definitions, write them under the tab for each vocabulary word.

Preview this chapter's content and activities at earth.msscience.com

Get Ready to Read

Make Predictions

1 Learn It! A prediction is an educated guess based on what you already know. One way to predict while reading is to guess what you believe the author will tell you next. As you are reading, each new topic should make sense because it is related to the previous paragraph or passage.

2 Practice It! Read the excerpt below from section. Based on what you have read, make predictions about what you will read in the rest of the lesson. After you read section, go back to your predictions to see if they were correct.

Think about how you would describe different climates in different regions on Earth.

Predict some different types of soils for different places. What factors might produce different types of soils?

Determine how soil temperature and moisture content could affect the quality of soils.

Different regions on Earth have **different climates.** Deserts are dry, prairies are semidry, and temperate forests are mild and moist. These places also have **different types of soils. Soil temperature and moisture content** affect the quality of soils. Soils in deserts contain little organic material and are thinner than soils in wetter climates. Prairie soils have thick, dark A horizons because the grasses that grow there contribute lots of organic matter.

—from page 193

3 Apply It! Before you read, skim the questions in the Chapter Review. Choose three questions and predict the answers.

As you read, check the predictions you made to see if they were correct.

Target Your Reading

Use this to focus on the main ideas as you read the chapter.

1. **Before you read** the chapter, respond to the statements below on your worksheet or on a numbered sheet of paper.
   - Write an **A** if you **agree** with the statement.
   - Write a **D** if you **disagree** with the statement.

2. **After you read** the chapter, look back to this page to see if you've changed your mind about any of the statements.
   - If any of your answers changed, explain why.
   - Change any false statements into true statements.
   - Use your revised statements as a study guide.

Before You Read A or D		Statement	After You Read A or D
	1	Weathering breaks rocks into smaller and smaller pieces, such as sand, silt, or clay.	
	2	Exposure to atmospheric water and gases causes rocks to change chemically.	
	3	Soil is a mixture of weathered rock, decayed organic matter, mineral fragments, water, and air.	
	4	Because of weathering, new soil is usually produced rapidly in all regions on Earth.	
	5	The different layers of soil are called horizons.	
	6	Climate does not affect the type of soil produced in Earth's different regions.	
	7	Most plants grow well when topsoil erodes.	
	8	In tropical, deforested areas, soil is useful to farmers for only a few years before the topsoil is gone.	
	9	Contour farming is a practice of planting crops in large, circular mounds.	

Science Online
Print out a worksheet of this page at earth.msscience.com

section 1

Weathering

as you read

What You'll Learn

- **Explain** how mechanical weathering and chemical weathering differ.
- **Describe** how weathering affects Earth's surface.
- **Explain** how climate affects weathering.

Why It's Important

Through time, weathering turns mountains into sediment.

Review Vocabulary
surface area: the area of a rock or other object that is exposed to the surroundings

New Vocabulary
- weathering
- mechanical weathering
- ice wedging
- chemical weathering
- oxidation
- climate

Weathering and Its Effects

Can you believe that tiny moss plants, a burrowing vole shrew, and even oxygen in the air can affect solid rock? These things weaken and break apart rock at Earth's surface. Surface processes that work to break down rock are called **weathering.**

Weathering breaks rock into smaller and smaller pieces, such as sand, silt, and clay. These particles are called sediment. The terms *sand, silt,* and *clay* are used to describe specific particle sizes, which contribute to soil texture. Sand grains are larger than silt, and silt is larger than clay.

Soil texture influences virtually all mechanical and chemical processes in the soil, including the ability to hold moisture and nutrients.

Over millions of years, weathering has changed Earth's surface. The process continues today. Weathering wears mountains down to hills, as shown in **Figure 1.** Rocks at the top of mountains are broken down by weathering, and the sediment is moved downhill by gravity, water, and ice. Weathering also produces strange rock formations like those shown at the beginning of this chapter. Two different types of weathering—mechanical weathering and chemical weathering—work together to shape Earth's surface.

Figure 1 Over long periods of time, weathering wears mountains down to rolling hills.
Explain *how this occurs.*

Figure 2 Growing tree roots can be agents of mechanical weathering.

Tree roots can crack a sidewalk.

Tree roots also can grow into cracks and break rock apart.

Mechanical Weathering

Mechanical weathering occurs when rocks are broken apart by physical processes. This means that the overall chemical makeup of the rock stays the same. Each fragment has characteristics similar to the original rock. Growing plants, burrowing animals, and expanding ice are some of the things that can mechanically weather rock. These physical processes produce enough force to break rocks into smaller pieces.

Reading Check *What can cause mechanical weathering?*

Plants and Animals Water and nutrients that collect in the cracks of rocks result in conditions in which plants can grow. As the roots grow, they enlarge the cracks. You've seen this kind of mechanical weathering if you've ever tripped on a crack in a sidewalk near a tree, as shown in **Figure 2.** Sometimes tree roots wedge rock apart, also shown in **Figure 2.**

Burrowing animals also cause mechanical weathering, as shown in **Figure 3.** As these animals burrow, they loosen sediment and push it to the surface. Once the sediment is brought to the surface, other weathering processes act on it.

Figure 3 Small animals mechanically weather rock when they burrow by breaking apart sediment.

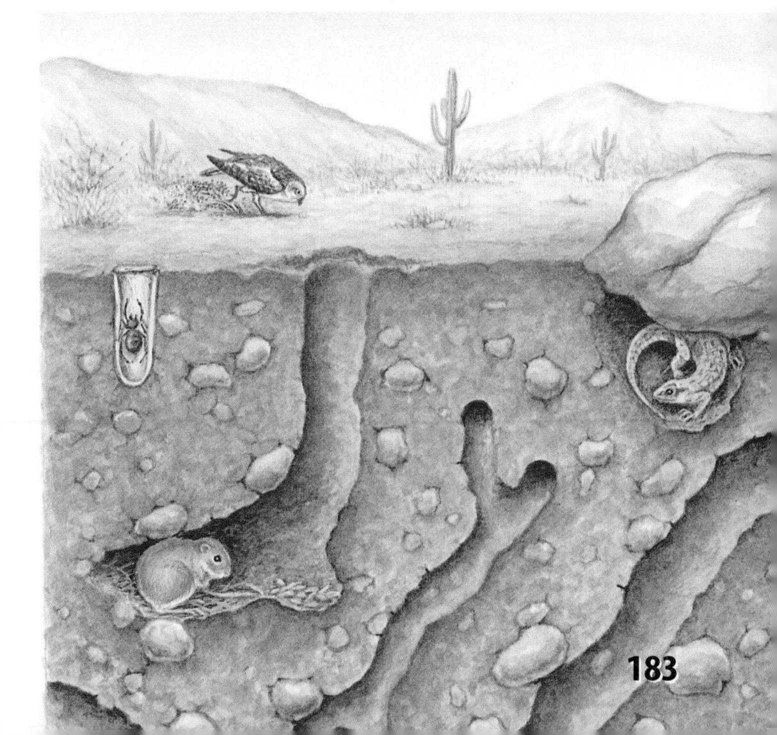

Figure 4 When water enters cracks in rock and freezes, it expands, causing the cracks to enlarge and the rock to break apart.

Figure 5 As rock is broken apart by mechanical weathering, the amount of rock surface exposed to air and water increases. The background squares show the total number of surfaces exposed.

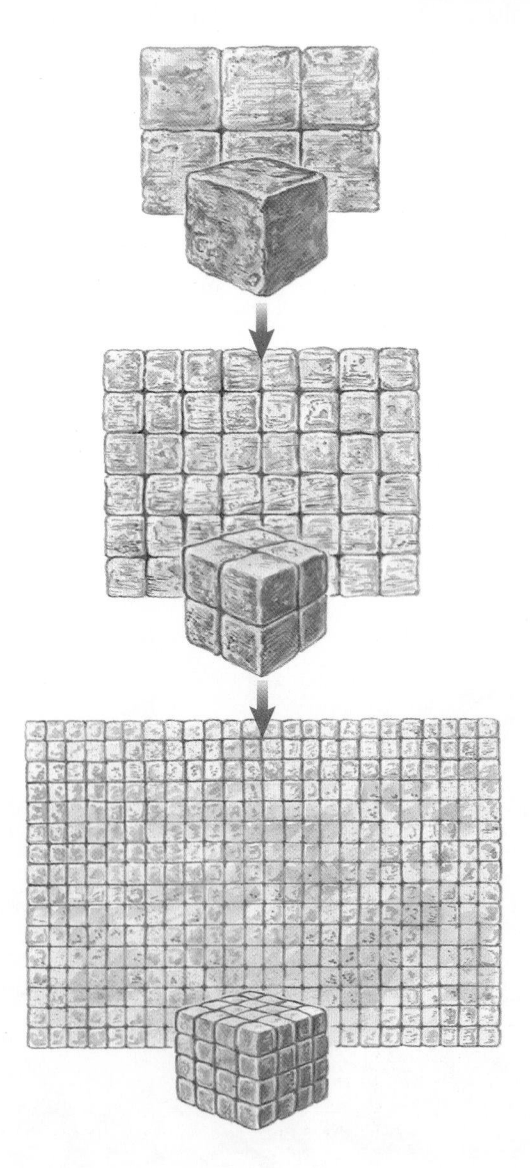

Ice Wedging A mechanical weathering process called ice wedging is shown in **Figure 4. Ice wedging** occurs in temperate and cold climates where water enters cracks in rocks and freezes. Because water expands when it turns to ice, pressure builds up in the cracks. This pressure can extend the cracks and break apart rock. The ice then melts, allowing more water to enter the crack, where it freezes and breaks the rock even more. Ice wedging is most noticeable in the mountains, where warm days and cold nights are common. It is one process that wears down mountain peaks. This cycle of freezing and thawing not only breaks up rocks, but also can break up roads and highways. When water enters cracks in road pavement and freezes, it forces the pavement apart. This causes potholes to form in roads.

Surface Area Mechanical weathering by plants, animals, and ice wedging reduces rocks to smaller pieces. These small pieces have more surface area than the original rock body, as shown in **Figure 5.** As the amount of surface area increases, more rock is exposed to water and oxygen. This speeds up a different type of weathering called chemical weathering, which continues to reduce the particle size of sediments from a coarse to a finer texture.

Chemical Weathering

The second type of weathering, **chemical weathering,** occurs when chemical reactions dissolve or alter the minerals in rocks or change them into different minerals. This type of weathering occurs at or near Earth's surface and changes the chemical composition of the rock, which can weaken the rock.

Topic: Chemical Weathering

Visit earth.msscience.com for Web links to information about chemical weathering.

Activity List different types of chemical weathering. Next to each type, write an effect that you have observed.

Natural Acids Naturally formed acids can weather rocks. When water reacts with carbon dioxide in the air or soil, a weak acid, called carbonic acid, forms. Carbonic acid reacts with minerals such as calcite, which is the main mineral that makes up limestone. This reaction causes the calcite to dissolve. Over many thousands of years, carbonic acid has weathered so much limestone that caves have formed, as shown in **Figure 6**.

Chemical weathering also occurs when naturally formed acids come in contact with other rocks. Over a long period of time, the mineral feldspar, which is found in granite, some types of sandstone, and other rocks, is broken down into a clay mineral called kaolinite (KAY oh luh nite). Kaolinite clay is common in some soils. Clay is an end product of weathering.

Reading Check *How does kaolinite clay form?*

Plant Acids Some roots and decaying plants give off acids that also dissolve minerals in rock. When the minerals dissolve, the rock is weakened. Eventually, the rock breaks into smaller pieces. As the rock weathers, nutrients become available to plants.

Carbon dioxide + Water = Carbonic acid
Carbonic acid dissolves limestone.

Figure 6 Caves form when slightly acidic groundwater dissolves limestone.
Explain *why the groundwater is acidic.*

Observing Chemical and Mechanical Weathering

Procedure

1. Collect and rinse two handfuls of **common rock or shells.**
2. Place equal amounts of rock into two **plastic bottles.**
3. Fill one bottle with **water** to cover the rock and seal with a **lid**.
4. Cover the rock in the second bottle with **lemon juice** and seal.
5. Shake both bottles for ten minutes.
6. Tilt the bottles so you can observe the liquids in each.

Analysis

1. Describe the appearance of each liquid.
2. Explain any differences.

pH Scale The strength of acids and bases is measured on the pH scale with a range of 0 to 14. On this scale, 0 is extremely acidic, 14 is extremely basic or alkaline, and 7 is neutral. Most minerals are more soluble in acidic soils than in neutral or slightly alkaline soils. Different plants grow best at different pH values. For example, peanuts grow best in soils that have a pH of 5.3 to 6.6, while alfalfa grows best in soils having a pH of 6.2 to 7.8.

Oxygen Oxygen also causes chemical weathering. **Oxidation** (ahk sih DAY shun) occurs when some materials are exposed to oxygen and water. For example, when minerals containing iron are exposed to water and the oxygen in air, the iron in the mineral reacts to form a new material that resembles rust. One common example of this type of weathering is the alteration of the iron-bearing mineral magnetite to a rustlike material called limonite, as shown in **Figure 7.** Oxidation of minerals gives some rock layers a red color.

Reading Check *How does oxygen cause weathering?*

Effects of Climate

Climate affects soil temperature and moisture and also affects the rate of mechanical and chemical weathering. **Climate** is the pattern of weather that occurs in a particular area over many years. In cold climates, where freezing and thawing are frequent, mechanical weathering rapidly breaks down rock through the process of ice wedging. Chemical weathering is more rapid in warm, wet climates. High temperatures tend to increase the rate of chemical reactions. Thus, chemical weathering tends to occur quickly in tropical areas. Lack of moisture in deserts and low temperatures in polar regions slow down chemical weathering.

Figure 7 Iron-containing minerals like the magnetite shown here can weather to form a rustlike material called limonite.
Explain *how this is similar to rust forming on your bicycle chain.*

Marble statue

Granite statue

Figure 8 Different types of rock weather at different rates. In humid climates, marble statues weather rapidly and become discolored. Granite statues weather more slowly.

Effects of Rock Type Rock type also can affect the rate of weathering in a particular climate. In wet climates, for example, marble weathers more rapidly than granite, as shown in **Figure 8.**

The weathering of rocks and the process of soil formation alter rock minerals so that soil minerals are mostly inherited from the parent rock type. Weathering begins the process of forming soil from rock and sediment and also affects particle size and soil texture. Recall that sand, silt, and clay simply describe the different particle sizes of the soil's mineral content.

section 1 review

Summary

Weathering and Its Effects

- Weathering includes processes that break down rock.
- Weathering affects Earth's landforms.

Mechanical Weathering

- During mechanical weathering, rock is broken apart, but it is not changed chemically.
- Plant roots, burrowing animals, and expanding ice all weather rock.

Chemical Weathering

- During chemical weathering, minerals in rock dissolve or change to other minerals.
- Agents of chemical weathering include natural acids and oxygen.

Self Check

1. **Describe** how weathering reduces the height of mountains through millions of years.
2. **Explain** how both tree roots and prairie dogs mechanically weather rock.
3. **Summarize** the effects of carbonic acid on limestone.
4. **Describe** how climate affects weathering.
5. **Think Critically** Why does limestone often form cliffs in dry climates but rarely form cliffs in wet climates?

Applying Skills

6. **Venn Diagram** Make a Venn diagram to compare and contrast mechanical weathering and chemical weathering. Include the causes of mechanical and chemical weathering in your diagram.

section 2

The Nature of Soil

as you read

What You'll Learn

- **Explain** how soil forms.
- **Describe** soil characteristics.
- **Describe** factors that affect the development of soil.

Why It's Important

Much of the food that you eat is grown in soil.

Review Vocabulary
profile: a vertical slice through rock, sediment, or soil

New Vocabulary
- soil
- humus
- horizon
- soil profile
- litter
- leaching

Formation of Soil

The word *ped* is from a Greek word that means "ground" and from a Latin word that means "foot." The pedal under your foot, when you're bicycling, is named from the word *ped.* The part of Earth under your feet, when you're walking on the ground, is the pedosphere, or soil. Soil science is called pedology.

What is soil and where does it come from? A layer of rock and mineral fragments produced by weathering covers the surface of Earth. As you learned in Section 1, weathering gradually breaks rocks into smaller and smaller fragments. However, these fragments do not become high-quality soil until plants and animals live in them. Plants and animals add organic matter, the remains of once-living organisms, to the rock fragments. Organic matter can include leaves, twigs, roots, and dead worms and insects. **Soil** is a mixture of weathered rock, decayed organic matter, mineral fragments, water, and air.

Soil can take thousands of years to form and ranges from 60 m thick in some areas to just a few centimeters thick in others. Climate, slope, types of rock, types of vegetation, and length of time that rock has been weathering all affect the formation of soil, as shown in **Figure 9.** For example, different kinds of soils develop in tropical regions than in polar regions. Soils that develop on steep slopes are different from soils that develop on flat land. **Figure 10** illustrates how soil develops from rock.

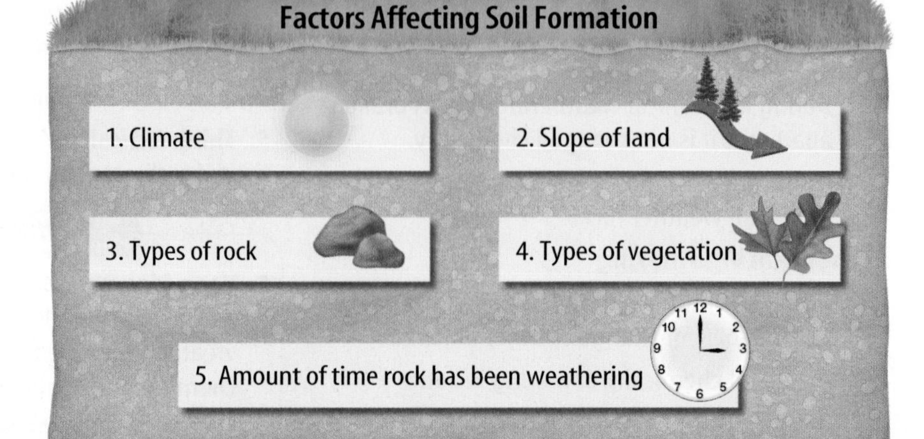

Factors Affecting Soil Formation
1. Climate
2. Slope of land
3. Types of rock
4. Types of vegetation
5. Amount of time rock has been weathering

Figure 9 Five different factors affect soil formation.
Explain *how time influences the development of soils.*

NATIONAL GEOGRAPHIC

VISUALIZING SOIL FORMATION

Figure 10

It may take thousands of years to form, but soil is constantly evolving from solid rock, as this series of illustrations shows. Soil is a mixture of weathered rock, mineral fragments, and organic material—the remains of dead plants and animals—along with water and air.

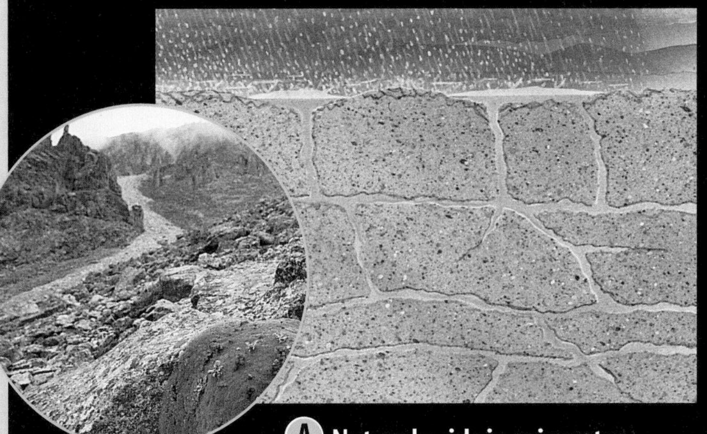

A Natural acids in rainwater weather the surface of exposed bedrock. Water can also freeze in cracks, causing rocks to fracture and break apart. The inset photo shows weathered rock in the Tien Shan Mountains of Central Asia.

B Plants take root in the cracks and among bits of weathered rock—shown in the inset photo above. As they grow, plants, along with other natural forces, continue the process of breaking down rocks, and a thin layer of soil begins to form.

C Like the grub in the inset photo, insects, worms, and other living things take up residence among plant roots. Their wastes, along with dead plant material, add organic matter to the soil.

D As organic matter increases and underlying bedrock continues to break down, the soil layer thickens. Rich topsoil supports trees and other plants with large root systems.

Comparing Components of Soil

Procedure

1. Complete a safety worksheet.
2. Collect a sample of **soil.**
3. Observe it closely with a **magnifying lens.**
4. Record evidence of plant and animal components and their activities.

Analysis

1. Describe the different particles found in your sample. Did you find any remains of organisms?
2. Explain how living organisms might affect the soil.
3. Compare and contrast your sample with those other students have collected.

Composition of Soil

Soil is made up of rock and mineral fragments, organic matter, air, and water. The rock and mineral fragments come from rocks that have been weathered. Most of these fragments are small particles of sediment such as clay, silt, and sand.

Most organic matter in soil comes from plants. Plant leaves, stems, and roots all contribute organic matter to soil. Animals and microorganisms provide additional organic matter when they die. After plant and animal material gets into soil, fungi and bacteria cause it to decay. The decayed organic matter turns into a dark-colored material called **humus** (HYEW mus). Humus serves as a source of nutrients for plants. As worms, insects, and rodents burrow throughout soil, they mix the humus with the fragments of rock. Good-quality surface soil has approximately equal amounts of humus and weathered rock material.

Water Infiltration Soil has many small spaces between individual soil particles that are filled with water or air. When soil is moist, the spaces hold the water that plants need to grow. During a drought, the spaces are almost entirely filled with air. When water soaks into the ground, it infiltrates the pores. Infiltration rate is determined by calculating the time it takes for water sitting on soil to drop a fixed distance. This rate changes as the soil pore spaces fill with water.

Figure 11 This soil, which developed beneath a grassy prairie, has three main horizons.
Describe *how the A horizon is different from the other two horizons.*

Soil Profile

You have seen layers of soil if you've ever dug a deep hole or driven along a road that has been cut into a hillside. You probably observed that most plant roots grow in the top layer of soil. The top layer typically is darker than the soil layers below it. These different layers of soil are called **horizons.** All the horizons of a soil form a **soil profile.** Most soils have three horizons—labeled A, B, and C, as shown in **Figure 11.**

A Horizon The A horizon is the top layer of soil. In a forest, the A horizon might be covered with litter. **Litter** consists of leaves, twigs, and other organic material that can be changed to humus by decomposing organisms. Litter helps prevent erosion and evaporation of water from soil. The A horizon also is known as topsoil. Topsoil has more humus and fewer rock and mineral particles than the other layers in a soil profile. The A horizon generally is dark and fertile. The dark color of the soil is caused by the humus, which provides nutrients for plant growth.

Since dark color absorbs solar energy more readily, soil color can greatly affect soil temperature. Darker color also may indicate a higher content of soil moisture. Soil moisture and soil temperature are important in determining seed germination for plants and the vitality of decomposing organisms.

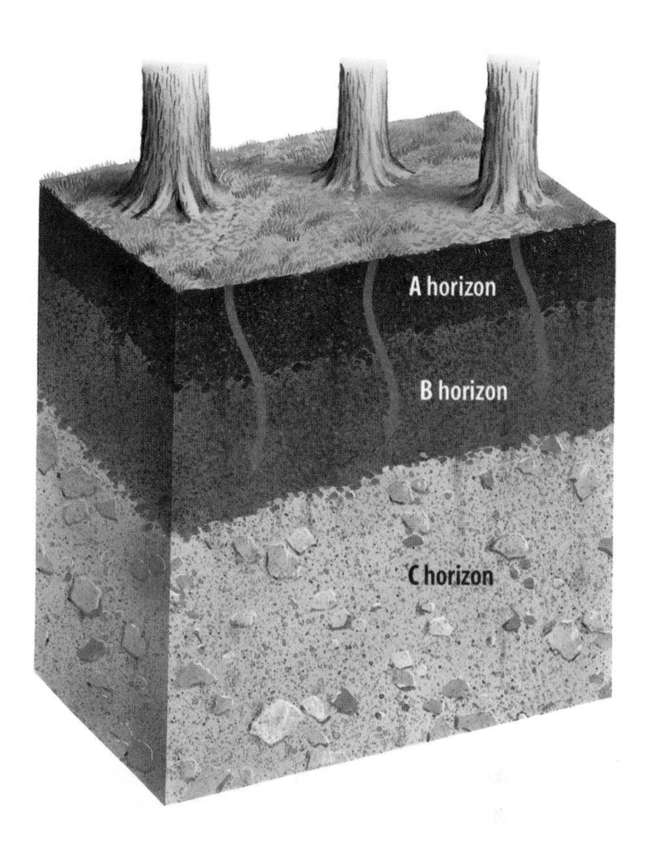

Figure 12 Leaching removes material from the upper layer of soil. Much of this material then is deposited in the B horizon.

B Horizon The layer below the A horizon is the B horizon, also known as subsoil. Because less organic matter is added to this horizon, it is lighter in color than the A horizon and contains less humus. As a result, the B horizon is less fertile. The B horizon contains material moved down from the A horizon by the process of leaching.

Leaching is the removal of minerals that have been dissolved in water. The process of leaching resembles making coffee in a drip coffeemaker. In soil, water seeps through the A horizon and reacts with humus and carbon dioxide to form acid. The acid dissolves some of the minerals in the A horizon and carries the material into the B horizon, as shown in **Figure 12.**

How does leaching transport material from the A horizon to the B horizon?

C Horizon The C horizon consists of partially weathered rock and is the bottom horizon in a soil profile. It is often the thickest soil horizon. This horizon does not contain much organic matter and is not strongly affected by leaching. It usually is composed of coarser sediment than the soil horizons above it. What would you find if you dug to the bottom of the C horizon? As you might have guessed, you would find rock—the rock that gave rise to the soil horizons above it. This rock is called the parent material of the soil. The C horizon is the soil layer that is most like the parent material.

Soil Fertility Plants need a variety of nutrients for growth. They need things like nitrogen, phosphorus, potassium, sulfur, calcium, and magnesium called macronutrients They get these nutrients from the minerals and organic material in soil. Fertile soil supplies the nutrients that plants need in the proper amounts. Soil fertility usually is determined in a laboratory by a soil chemist. However, fertility sometimes can be inferred by looking at plants. Do research to discover more important plant nutrients.

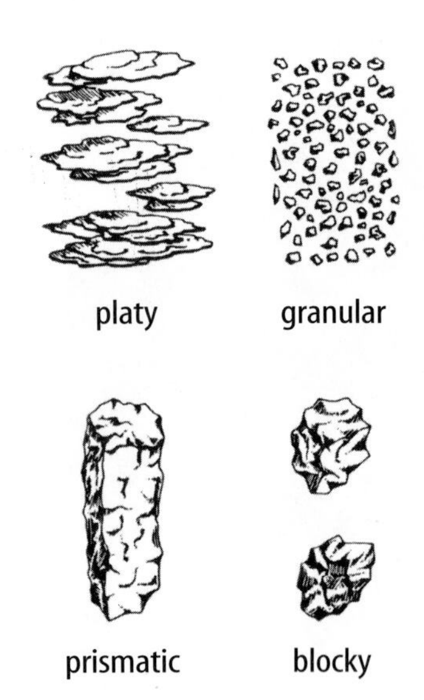

Figure 13 Four major classes characterize soil structure.

Soil Structure Individual soil particles clump together. Examine soil closely and you will see natural clumps called *peds.* Soil structure affects pore space and will affect a plant's ability to penetrate roots. **Figure 13** shows four classes of soil structure. Granular structures are common in surface soils with high organic content that glues minerals together. Earthworms, frost, and rodents mix the soil, keeping the peds small, which provides good porosity and movement of air and water. Platy structures are often found in subsurface soils that have been leached or compacted by animals or machinery. Blocky structures are common in subsoils or surface soils with high clay content, which shrinks and swells, producing cracks. Prismatic structures, found in B horizons, are very dense and difficult for plant roots to penetrate. Vertical cracks result from freezing and thawing, wetting and drying, and downward movement of water and roots. Soil consistency refers to the ability of peds and soil particles to stick together and hold their shapes.

Applying Math — Calculate Percentages

SOIL TEXTURE Some soil is coarse, some is fine. This property of soil is called soil texture. The texture of soil often is determined by finding the percentages of sand, silt, and clay. Calculate the percentage of clay shown by the circle graph.

20 g Sand particles
15 g Clay particles
15 g Silt particles

Solution

1 *This is what you know:*
- sand weight: 20 g
- clay weight: 15 g
- silt weight: 15 g

2 *This is what you need to find:*
- total weight of the sample
- percentage of clay particles

3 *This is the procedure you need to use:*
- Add all the masses to determine the total sample mass: 20 g sand + 15 g silt + 15 g clay = 50 g sample
- Divide the clay mass by the sample mass; multiply by 100: 15 g clay/50 g sample × 100 = 30% clay in the sample

Practice Problems

1. Calculate the percentage of sand in the sample.
2. Calculate the percentage of silt in the sample.

For more practice, visit earth.msscience.com/math_practice

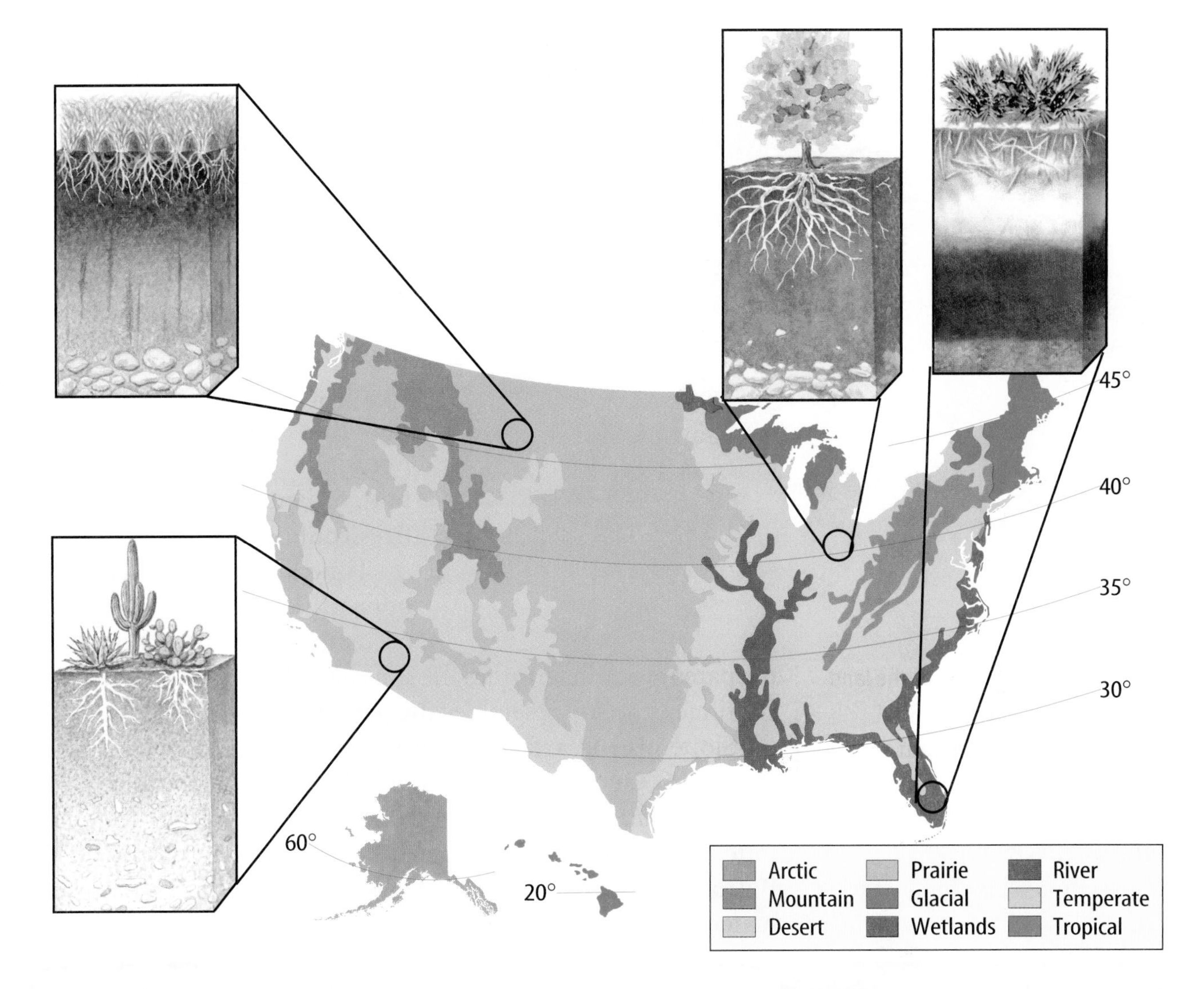

Figure 14 The United States has many different soil types. They vary in color, depth, texture, and fertility.
Identify *the soil type in your region.*

Soil Types

If you travel across the country, you will notice that not all soils are the same. Some are thick and red. Some are brown with hard rock nodules, and some have thick, black A horizons. They vary in color, depth, texture, fertility, pH, temperature, and moisture content. Many soils exist, as shown in **Figure 14.**

Soil Types Reflect Climate Different regions on Earth have different climates. Deserts are dry, prairies are semidry, and temperate forests are mild and moist. These places also have different types of soils. Soil temperature and moisture content affect the quality of soils. Soils in deserts contain little organic material and also are thinner than soils in wetter climates. Prairie soils have thick, dark A horizons because the grasses that grow there contribute lots of organic matter. Temperate forest soils have less organic matter and thinner A horizons than prairie soils do. Other regions such as tundra and tropical areas also have distinct soils.

Other Factors Parent rock material affects soils that develop from it. Clay soils develop on rocks like basalt, because minerals in the rock weather to form clay. Rock type also affects vegetation, because different rocks provide different amounts of nutrients.

Soil pH, controls many chemical and biological activities that take place in soil. Activities of organisms, acid rain, or land management practices could affect soil quality.

Time also affects soil development. If weathering has been occurring for only a short time, the parent rock determines the soil characteristics. As weathering continues, the soil resembles the parent rock less and less.

Figure 15 The slope of the land affects soil development. Thin, poorly developed soils form on steep slopes, but valleys often have thick, well-developed soils. **Infer** *why this is so.*

Slope also is a factor affecting soil profiles, as shown in **Figure 15.** On steep slopes, soils often are poorly developed, because material moves downhill before it can be weathered much. In bottomlands, sediment and water are plentiful. Bottomland soils are often thick, dark, and full of organic material.

section 2 review

Summary

Formation of Soil

- Soil is a mixture of rock and mineral fragments, decayed organic matter, water, and air.

Composition of Soil

- Organic matter gradually changes to humus.
- Soil moisture is important for plant growth.

Soil Profile

- The layers in a soil profile are called horizons.
- Most soils have an A, B, and C horizon.

Soil Types

- Many different types of soils occur in the United States.
- Climate and other factors determine the type of soil that develops.

Self Check

1. **List** the five factors that affect soil development.
2. **Explain** how soil forms.
3. **Explain** why A horizons often are darker than B horizons or C horizons.
4. **Describe** how leaching affects soil.
5. **Think Critically** Why is a soil profile in a tropical rain forest different from one in a desert? A prairie?

Applying Skills

6. **Use Statistics** A farmer collected five soil samples from a field and tested their acidity, or pH. His data were the following: 7.5, 8.2, 7.7, 8.1, and 8.0. Calculate the mean of these data. Also, determine the range and median.

Science Online earth.msscience.com/self_check_quiz

Soil Texture

Soils have different amounts of different sizes of particles. When you determine how much sand, silt, and clay a soil contains, you describe the soil's texture.

Real-World Question

What is the texture of your soil?

Goals

- **Estimate** soil texture by making a ribbon.

Materials

soil sample (100 g)
water bottle

Safety Precautions

Procedure

1. Take some soil and make it into a ball. Work the soil with your fingers. Slowly add water to the soil until it is moist.
2. After your ball of soil is moist, try to form a thin ribbon of soil. Use the following descriptions to categorize your soil:
   a. If you can form a long, thin ribbon, you have a clay soil.
   b. If you formed a long ribbon but it breaks easily, you have a clay loam soil.
   c. If you had difficulty forming a long ribbon, you have loam soil.
3. Now make your soil classification more detailed by selecting one of these descriptions:
   a. If the soil feels smooth, add the word *silty* to your soil name.
   b. If the soil feels slightly gritty, don't add any word to your soil name.
   c. If the soil feels very gritty, add the word *sandy* before your soil name.

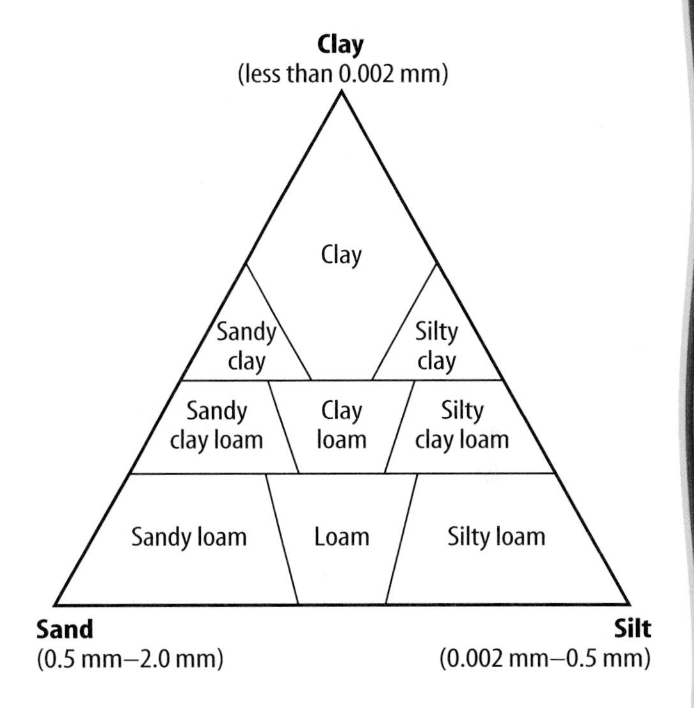

Conclude and Apply

1. **Classify** Which texture class name did you assign to your soil?
2. **Observe** Find your soil texture class name on the triangle above. Notice that the corners of the triangle are labeled *sand, silt,* and *clay.*
3. **Determine** Is your soil texture class close to one of the three corners or near the middle of the diagram? If your soil texture class is close to a corner, which one?
4. **Describe** Does your soil contain mostly sand, silt, or clay, or does it have nearly equal amounts of each? *Hint: If your soil name is close to a corner, it has mostly that size of sediment. If your soil name is in the middle of the triangle, it has nearly equal amounts of each sediment size.*

section 3

Soil Erosion

as you read

What You'll Learn

- **Explain** why soil is important.
- **Evaluate** ways that human activity has affected Earth's soil.
- **Describe** ways to reduce soil erosion.

Why It's Important

If topsoil is eroded, soil becomes less fertile.

Review Vocabulary

erosion: the picking up and moving of sediment or soil

New Vocabulary

- no-till farming
- contour farming
- terracing

Soil—An Important Resource

While picnicking at a local park, a flash of lightning and a clap of thunder tell you that a storm is upon you. Watching the pounding rain from the park shelter, you notice that the water flowing off of the ball diamond is muddy, not clear. The flowing water is carrying away some of the sediment that used to be on the field. This process is called soil erosion. Soil erosion is harmful because plants do not grow as well when topsoil has been removed.

Causes and Effects of Soil Erosion

Soil erodes when it is moved from the place where it formed. Erosion occurs as water flows over Earth's surface or when wind picks up and transports sediment. Generally, erosion is more severe on steep slopes than on gentle slopes. It's also more severe in areas where there is little vegetation. Under normal conditions, a balance between soil production and soil erosion often is maintained. This means that soil forms at about the same rate as it erodes. However, humans sometimes cause erosion to occur faster than new soil can form. One example is when people remove ground cover. Ground cover is vegetation that covers the soil and protects it from erosion. When vegetation is cleared, as shown in **Figure 16,** soil erosion often increases.

Figure 16 Removing vegetation can increase soil erosion.

Trees protect the soil from erosion in forested regions.

When forest is removed, soil erodes rapidly.

Figure 17 Tropical rain forests often are cleared by burning. **Explain** *how this can increase soil erosion.*

Agricultural Cultivation Soil erosion is a serious problem for agriculture. Topsoil contains many nutrients, holds water well, and has a porous structure that is good for plant growth. If topsoil is eroded, the quality of the soil is reduced. For example, plants need nutrients to grow. Each year, nutrients are both added to the soil and removed from the soil. The difference between the amount of nutrients added and the amount of nutrients removed is called the nutrient balance. If topsoil erodes rapidly, the nutrient balance might be negative. Farmers might have to use more fertilizer to compensate for the nutrient loss. In addition, the remaining soil might not have the same open structure and water-holding ability that topsoil does.

Forest Harvesting When forests are removed, soil is exposed and erosion increases. This creates severe problems in many parts of the world, but tropical regions are especially at risk. Each year, thousands of square kilometers of tropical rain forest are cleared for lumber, farming, and grazing, as shown in **Figure 17.** Soils in tropical rain forests appear rich in nutrients but are almost infertile below the first few centimeters. The soil is useful to farmers for only a few years before the topsoil is gone. Farmers then clear new land, repeating the process and increasing the damage to the soil.

Overgrazing In most places, land can be grazed with little damage to soil. However, overgrazing can increase soil erosion. In some arid regions of the world, sheep and cattle raised for food are grazed on grasses until almost no ground cover remains to protect the soil. When natural vegetation is removed from land that receives little rain, plants are slow to grow back. Without protection, soil is carried away by wind, and the moisture in the soil evaporates.

Soil Scientist Elvia Niebla is a soil scientist at the U.S. Environmental Protection Agency (EPA). Soil scientists at the EPA work to reduce soil erosion and pollution. Niebla's research even helped keep hamburgers safe to eat. How? In a report for the EPA, she explained how meat can be contaminated when cattle graze on polluted soil.

Topic: Land Use

Visit earth.msscience.com for Web links to information about how land use affects Earth's soil and about measures taken to reduce the impact.

Activity Debate with classmates about the best ways to protect rich farmland. Consider advantages and disadvantages of each method.

Excess Sediment If soil erosion is severe, sediment can damage the environment. Severe erosion sometimes occurs where land is exposed. Examples might include strip-mined areas or large construction sites. Eroded soil is moved to a new location where it is deposited. If the sediment is deposited in a stream, as shown in **Figure 18,** the stream channel might fill.

Figure 18 Erosion from exposed land can cause streams to fill with excessive amounts of sediment. **Explain** *how this could damage streams.*

Preventing Soil Erosion

Each year more than 1.5 billion metric tons of soil are eroded in the United States. Soil is a natural resource that must be managed and protected. People can do several things to conserve soil.

Manage Crops All over the world, farmers work to slow soil erosion. They plant shelter belts of trees to break the force of the wind and plant crops to cover the ground after the main harvest. In dry areas, instead of plowing under crops, many farmers graze animals on the vegetation. Proper grazing management can maintain vegetation and reduce soil erosion.

In recent years, many farmers have begun to practice no-till farming. Normally, farmers till or plow their fields one or more times each year. Using **no-till farming,** seen in **Figure 19,** farmers leave plant stalks in the field over the winter months. At the next planting, they seed crops without destroying these stalks and without plowing the soil. Farm machinery makes a narrow slot in the soil, and the seed is planted in this slot. No-till farming provides cover for the soil year-round, which reduces water runoff and soil erosion. One study showed that no-till farming can leave as much as 80% of the soil covered by plant residue. The leftover stalks also keep weeds from growing in the fields.

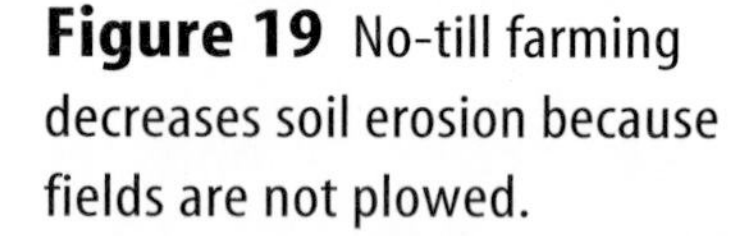

Figure 19 No-till farming decreases soil erosion because fields are not plowed.

Reading Check *How can farmers reduce soil erosion?*

Reduce Erosion on Slopes On gentle slopes, planting along the natural contours of the land, called **contour farming,** reduces soil erosion. This practice, shown in **Figure 20,** slows the flow of water down the slope and helps prevent the formation of gullies.

Where slopes are steep, terracing often is used. **Terracing** (TER uh sing) is a method in which steep-sided, level topped areas are built onto the sides of steep hills and mountains so that crops can be grown. These terraces reduce runoff by creating flat areas and shorter sections of slope. In the Philippines, Japan, China, and Peru, terraces have been used for centuries.

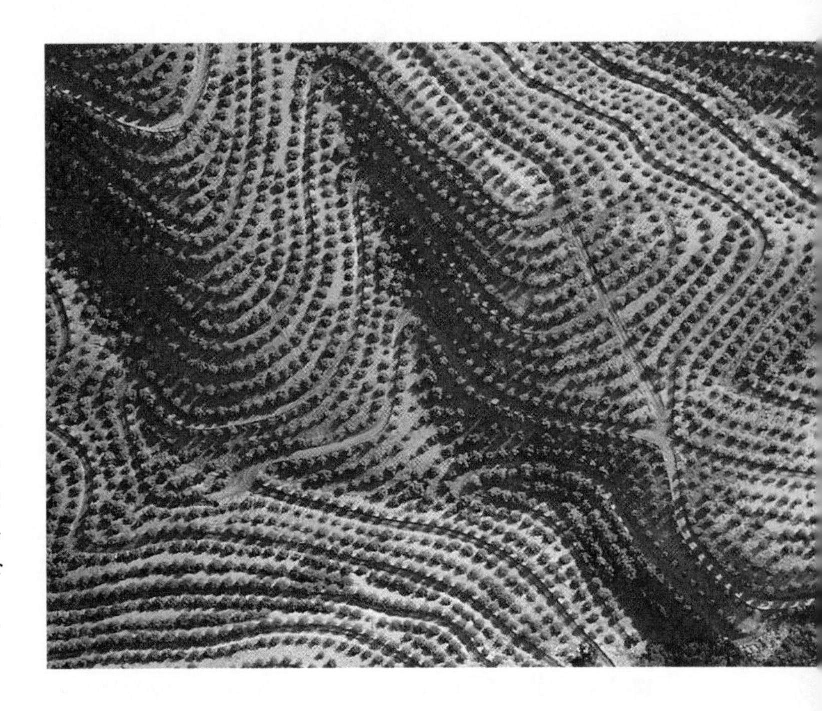

Figure 20 This orchard was planted along the natural contours of the land.
Summarize *the benefits of using contour farming on slopes.*

Reduce Erosion of Exposed Soil A variety of methods are used to control erosion where soil is exposed. During the construction process water is sometimes sprayed onto bare soil to prevent erosion by wind. When construction is complete, topsoil is added in areas where it was removed and trees are planted. At strip mines, water flow can be controlled so that most of the eroded soil is kept from leaving the mine. After mining is complete, the land is reclaimed. This means that steep slopes are flattened and vegetation is planted.

section 3 review

Summary

Soil—An Important Resource

- Soil erosion is a serious problem because topsoil is removed from the land.

Causes and Effects of Soil Erosion

- Soil erosion occurs rapidly on steep slopes and areas that are not covered by vegetation.
- The quality of farmland is reduced when soil erosion occurs.

Preventing Soil Erosion

- Farmers reduce erosion by planting shelter belts, using no-till farming, and planting cover crops after harvesting.
- Contour farming and terracing are used to control erosion on slopes.

Self Check

1. **Explain** why soil is important.
2. **Explain** how soil erosion damages soil.
3. **Describe** no-till farming.
4. **Explain** how overgrazing increases soil erosion.
5. **Think Critically** How does contour farming help water soak into the ground?

Applying Skills

6. **Communicate** Do research to learn about the different methods that builders use to reduce soil erosion during construction. Write a newspaper article describing how soil erosion at large construction sites is being controlled in your area.

Design Your Own

WEATHERING CHALK

Goals

- **Design** experiments to evaluate the effects of acidity, surface area, and temperature on the rate of chemical weathering of chalk.
- **Describe** factors that affect chemical weathering.
- **Explain** how the chemical weathering of chalk is similar to the chemical weathering of rocks.

Possible Materials

pieces of chalk (6)
small beakers (2)
metric ruler
water
white vinegar (100 mL)
hot plate
computer probe for temperature
**thermometer*
**Alternate materials*

Safety Precautions

Wear safety goggles when pouring vinegar. Be careful when using a hot plate and heated solutions.

WARNING: *If mixing liquids, always add acid to water.*

Real-World Question

Chalk is a type of limestone made of the shells of microscopic organisms. The famous White Cliffs of Dover, England, are made up of chalk. This lab will help you understand how chalk can be chemically weathered. How can you simulate chemical weathering of chalk?

Form a Hypothesis

How do you think acidity, surface area, and temperature affect the rate of chemical weathering of chalk? What happens to chalk in water? What happens to chalk in acid (vinegar)? How will the size of the chalk pieces affect the rate of weathering? What will happen if you heat the acid? Make hypotheses to support your ideas.

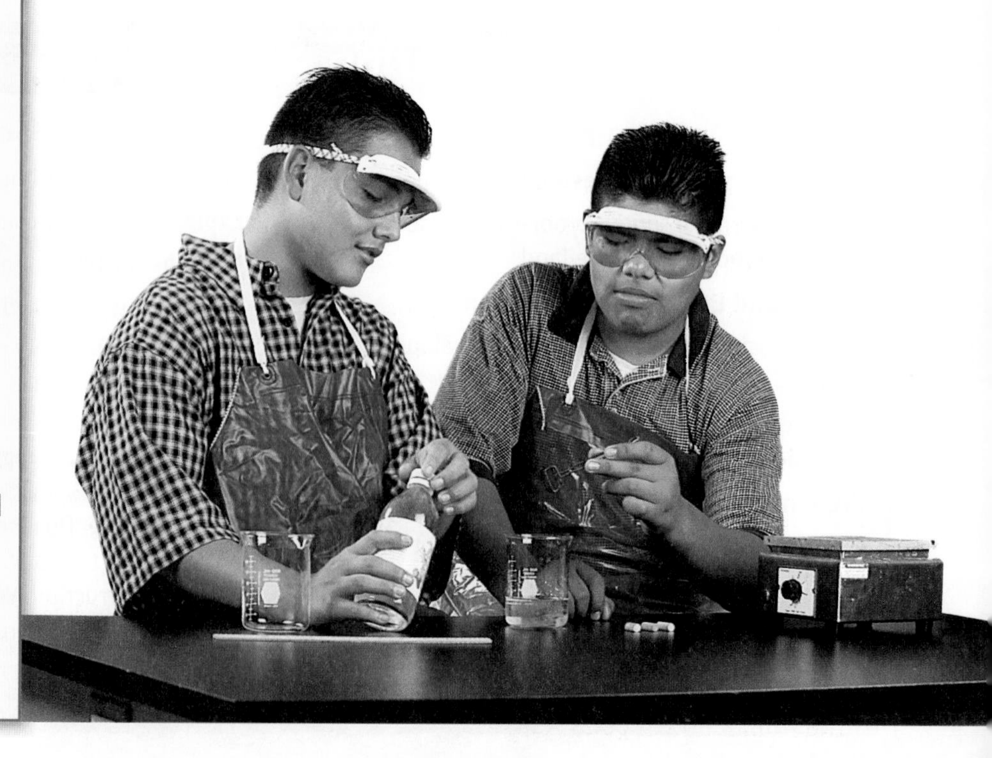

Test Your Hypothesis

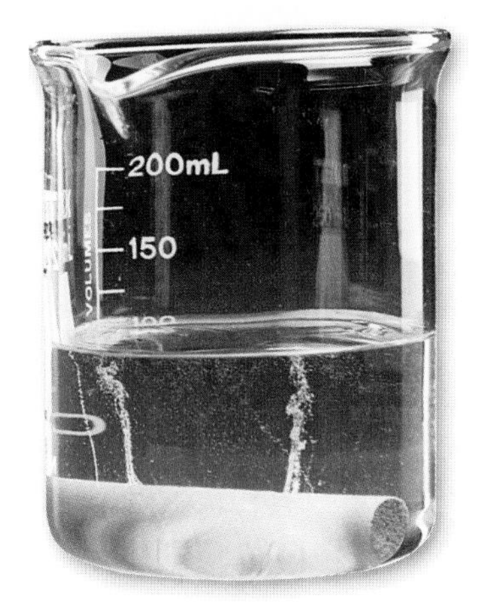

Make a Plan

1. **Develop** hypotheses about the effects of acidity, surface area, and temperature on the rate of chemical weathering.
2. **Decide** how to test your first hypothesis. List the steps needed to test the hypothesis.
3. Repeat step 2 for your other two hypotheses.
4. **Design** data tables in your Science Journal. Make one for acidity, one for surface area, and one for temperature.
5. **Identify** what remains constant in your experiment and what varies. Change only one variable in each procedure.
6. **Summarize** your data in a graph. Decide from reading the Science Skill Handbook which type of graph to use.

Follow Your Plan

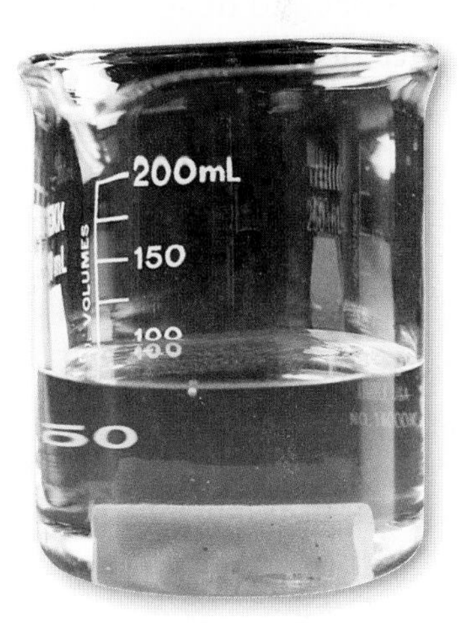

1. Make sure your teacher approves your plan before you start.
2. Carry out the three experiments as planned.
3. While you are conducting the experiments, record your observations and complete the data tables in your Science Journal.
4. Graph your data to show how each variable affected the rate of weathering.

Analyze Your Data

1. **Analyze** your graph to find out which substance—water or acid—weathered the chalk more quickly. Was your hypothesis supported by your data?
2. **Infer** from your data whether the amount of surface area makes a difference in the rate of chemical weathering. Explain.

Conclude and Apply

1. **Explain** how the chalk was chemically weathered.
2. How does heat affect the rate of chemical weathering?
3. What does this imply about weathering in the tropics and in polar regions?

Compare your results with those of your classmates. How were your data similar? How were they different? **For more help, refer to the Science Skill Handbook.**

Science and Language Arts

Landscape, History, and the Pueblo Imagination

by Leslie Marmon Silko

Leslie Marmon Silko, a woman of Pueblo, Hispanic, and American heritage, explains what ancient Pueblo people believed about the circle of life on Earth.

You see that after a thing is dead, it dries up. It might take weeks or years, but eventually if you touch the thing, it crumbles under your fingers. It goes back to dust. The soul of the thing has long since departed. With the plants and wild game the soul may have already been borne back into bones and blood or thick green stalk and leaves. Nothing is wasted. What cannot be eaten by people or in some way used must then be left where other living creatures may benefit. What domestic animals or wild scavengers can't eat will be fed to the plants. The plants feed on the dust of these few remains.

. . . Corn cobs and husks, the rinds and stalks and animal bones were not regarded by the ancient people as filth or garbage. The remains were merely resting at a mid-point in their journey back to dust. . . .

The dead become dust . . . . The ancient Pueblo people called the earth the Mother Creator of all things in this world. Her sister, the Corn mother, occasionally merges with her because all . . . green life rises out of the depths of the earth.

Rocks and clay . . . become what they once were. Dust.

A rock shares this fate with us and with animals and plants as well.

Understanding Literature

Repetition The recurrence of sounds, words, or phrases is called repetition. What is Silko's purpose of the repeated use of the word *dust?*

Respond to the Reading

1. What one word is repeated throughout this passage?
2. What effect does the repetition of this word have on the reader?
3. **Linking Science and Writing** Using repetition, write a one-page paper on how to practice a type of soil conservation.

This chapter discusses how weathered rocks and mineral fragments combine with organic matter to make soil. Silko's writing explains how the ancient Pueblo people understood that all living matter returns to the earth, or becomes dust. Lines such as "green life rises out of the depths of the earth," show that the Pueblo people understood that the earth, or rocks and mineral fragments, must combine with living matter in order to make soil and support plant life.

Reviewing Main Ideas

Section 1 Weathering

1. Weathering helps to shape Earth's surface.
2. Mechanical weathering breaks apart rock without changing its chemical composition. Plant roots, animals, and ice wedging are agents of mechanical weathering.
3. Chemical weathering changes the chemical composition of rocks. Natural acids and oxygen in the air can cause chemical weathering.

Section 2 The Nature of Soil

1. Soil is a mixture of rock and mineral fragments, organic matter, air, and water.
2. A soil profile contains different layers that are called horizons.
3. Climate, parent rock, slope of the land, type of vegetation, and the time that rock has been weathering are factors that affect the development of soil.

Section 3 Soil Erosion

1. Soil is eroded when it is moved to a new location by wind or water.
2. Human activities can increase the rate of soil erosion.
3. Windbreaks, no-till farming, contour farming, and terracing reduce soil erosion on farm fields.

Visualizing Main Ideas

Copy and complete the following concept map about weathering.

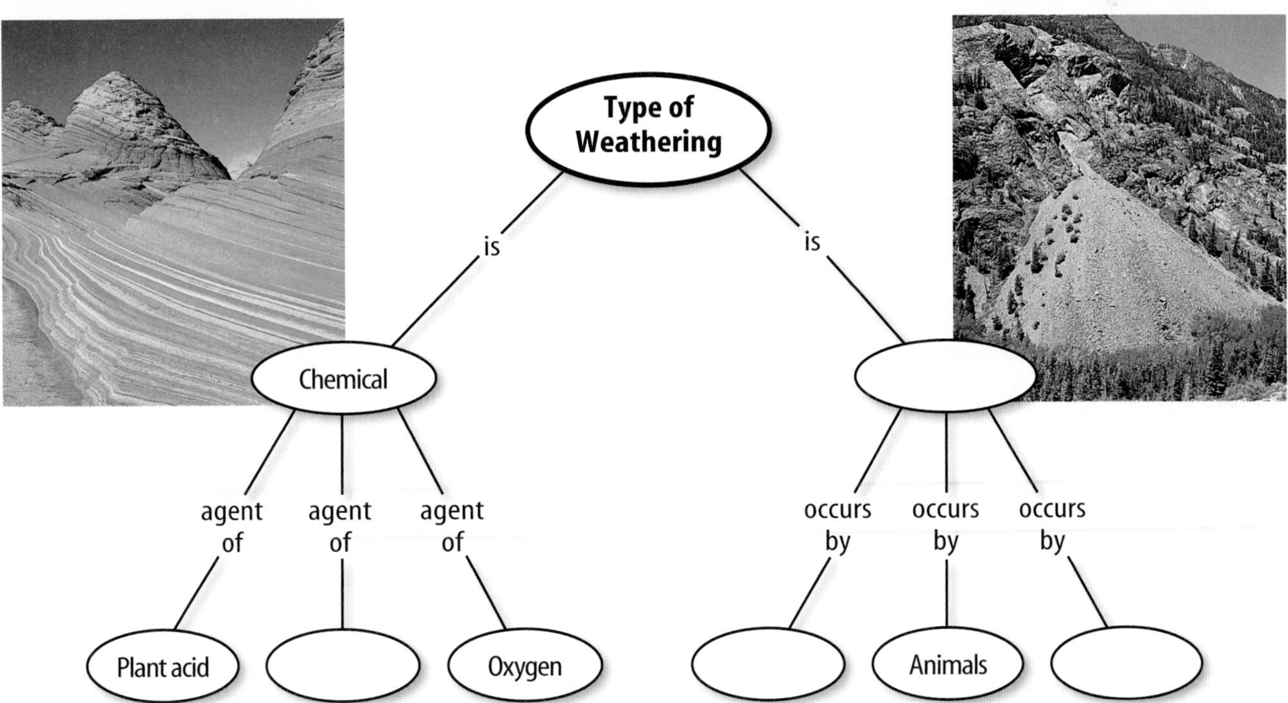

chapter 7 Review

Using Vocabulary

chemical weathering p. 185	mechanical weathering p. 183
climate p. 186	no-till farming p. 198
contour farming p. 199	oxidation p. 186
horizon p. 190	soil p. 188
humus p. 190	soil profile p. 190
ice wedging p. 184	terracing p. 199
leaching p. 191	weathering p. 182
litter p. 191	

Fill in the blanks with the correct vocabulary word or words.

1. ________ changes the composition of rock.
2. ________ forms from organic matter such as leaves and roots.
3. The horizons of a soil make up the ________.
4. ________ transports material to the B horizon.
5. ________ occurs when many materials containing iron are exposed to oxygen and water.
6. ________ means that crops are planted along the natural contours of the land.
7. ________ is the pattern of weather that occurs in a particular area for many years.

Checking Concepts

Choose the word or phrase that best answers the question.

8. Which of the following can be caused by acids produced by plant roots?
   A) soil erosion
   B) oxidation
   C) mechanical weathering
   D) chemical weathering

Use the graph below to answer question 9.

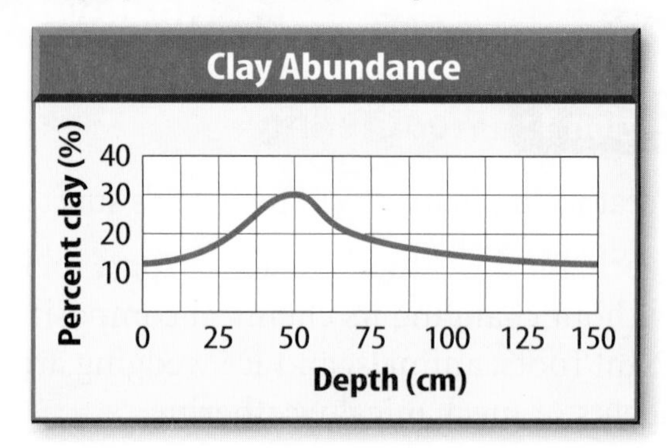

9. The above graph shows the percentage of clay in a soil profile at varying depths. Which depth has the highest amount of clay?
   A) 25 cm **C)** 50 cm
   B) 150 cm **D)** 100 cm
10. Which of the following is an agent of mechanical weathering?
    A) animal burrowing
    B) carbonic acid
    C) leaching
    D) oxidation
11. In which region is chemical weathering most rapid?
    A) cold, dry **C)** warm, moist
    B) cold, moist **D)** warm, dry
12. What is a mixture of rock and mineral fragments, organic matter, air, and water called?
    A) soil **C)** horizon
    B) limestone **D)** clay
13. What is organic matter in soil?
    A) leaching **C)** horizon
    B) humus **D)** profile
14. What is done to reduce soil erosion on steep slopes?
    A) no-till farming
    B) contour farming
    C) terracing
    D) grazing

Science Online earth.msscience.com/vocabulary_puzzlemaker

Thinking Critically

15. **Predict** which type of weathering—mechanical or chemical—you would expect to have a greater effect in a polar region. Explain.

16. **Recognize Cause and Effect** How does soil erosion reduce the quality of soil?

17. **Concept Map** Copy and complete the concept map about layers in soil.

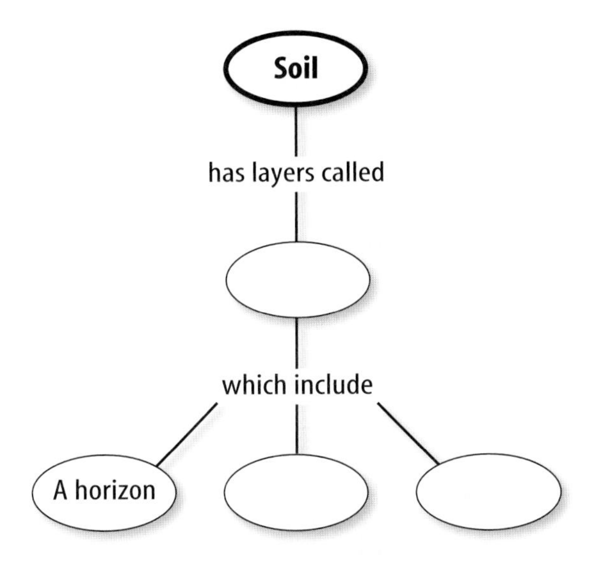

18. **Recognize Cause and Effect** Why do rows of trees along the edges of farm fields reduce wind erosion of soil?

19. **Form a Hypothesis** A pile of boulders lies at the base of a high-mountain cliff. Form a hypothesis explaining how the pile of rock might have formed.

20. **Test a Hypothesis** How would you test your hypothesis from question 19?

21. **Identify a Question** Many scientists are conducting research to learn more about how soil erosion occurs and how it can be reduced. Write a question about soil erosion that you would like to research. With your teacher's help, carry out an investigation to answer your question.

Performance Activities

22. **Design a Landscape** Find a slope in your area that might benefit from erosion maintenance. Develop a plan for reducing erosion on this slope. Make a map showing your plan.

23. **Describing Peds** Natural clumps of soil are called peds. Collect a large sample of topsoil. Describe the shape of the peds. Sketch the peds in your Science Journal.

Applying Math

Use the illustration below to answer questions 24–26.

24. **Fertilizer Nutrients** A bag of fertilizer is labeled to list the nutrients as three numbers. The numbers represent the percentages of nitrogen, phosphate, and potash in that order. What are the percentages of these nutrients for a fertilizer with the following information on the label: 5-10-10?

25. **Fertilizer Ratio** The fertilizer ratio tells you the proportions of the different nutrients in a fertilizer. To find the fertilizer ratio, divide each nutrient value by the lowest value. Calculate the fertilizer ratio for the fertilizer in question 24.

26. **Relative Amounts of Nutrients** Which nutrient is least abundant in the fertilizer? Which nutrients are most abundant? How many times more potash does the fertilizer contain than nitrogen?

Part 1 Multiple Choice

Record your answers on the answer sheet provided by your teacher or on a separate sheet of paper.

Use the photo below to answer question 1.

1. Which method for reducing soil erosion is shown on the hillsides above?
   A. no-till farming **C.** contour farming
   B. terracing **D.** shelter belts

2. Which of the following terms might describe a soil's texture?
   A. red **C.** porous
   B. coarse **D.** wet

3. Which soil horizon often has a dark color because of the presence of humus?
   A. E horizon **C.** B horizon
   B. C horizon **D.** A horizon

4. Which of the following is an agent of chemical weathering?
   A. ice wedging
   B. burrowing animals
   C. carbonic acid
   D. growing tree roots

Test-Taking Tip

Come Back To It Never skip a question. If you are unsure of an answer, mark your best guess on another sheet of paper and mark the question in your test booklet to remind you to come back to it at the end of the test.

5. Which of the following might damage a soil's structure?
   A. a gentle rain **C.** earthworms
   B. organic matter **D.** compaction

6. In which of the following types of rock are caves most likely to form?
   A. limestone **C.** granite
   B. sandstone **D.** basalt

7. Which of the following is most likely to cause erosion of farmland during a severe drought?
   A. water runoff
   B. soil creeping downhill
   C. wind
   D. ice

Use the table below to answer questions 8–10.

Texture Data for a Soil Profile			
Horizon	Percent		
	Sand	Silt	Clay
A	16.2	54.4	29.4
B	10.5	50.2	39.3
C	31.4	48.4	20.2
R (bedrock)	31.7	50.1	18.2

8. According to the table, which horizon in this soil has the lowest percentage of sand?
   A. A horizon **C.** C horizon
   B. B horizon **D.** R horizon

9. Which of the following is the R horizon?
   A. topsoil **C.** bedrock
   B. humus **D.** gravel

10. Which of the following is the best description of the soil represented by the table?
    A. sandy **C.** clayey
    B. silty **D.** organic

Part 2 Short Response/Grid In

Record your answers on the answer sheet provided by your teacher or on a sheet of paper.

Use the illustration below to answer question 11.

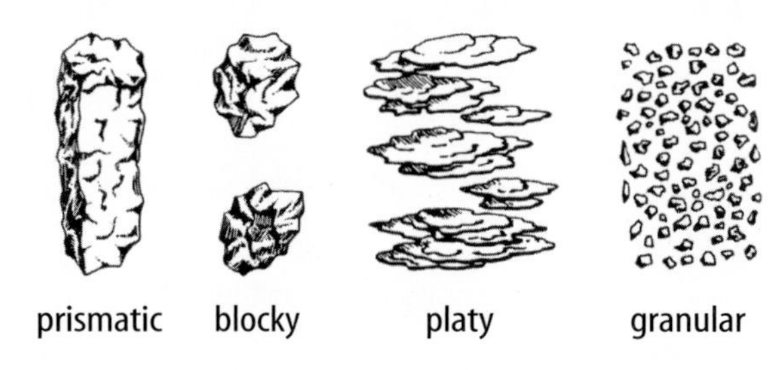

11. Natural clumps of soil are called peds. Compare and contrast the different types of peds in the sketch. Explain how the names of the peds describe their shape.

12. Explain how caves form. What role does carbonic acid have?

Use the diagram below to answer questions 13–15.

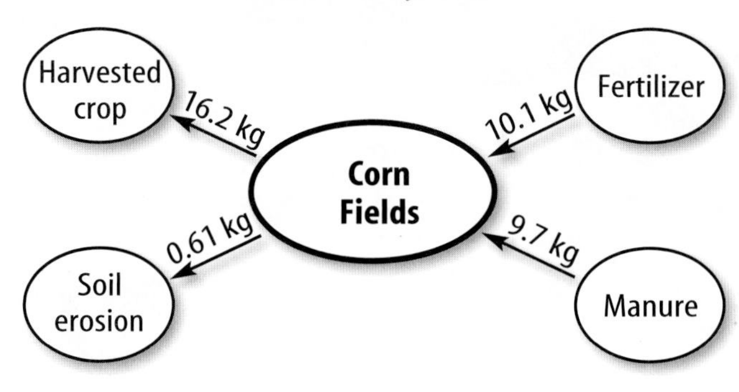

13. According to the diagram above, what was the total amount of phosphorus added to each acre?

14. What is the total amount of phorphorus lost from each acre?

15. What is the difference between the amount of phosphorus added and the amount of phosphorus lost?

Part 3 Open Ended

Record your answers on a sheet of paper.

16. Describe ways that humans affect Earth's soil. How can damage to soil be reduced?

17. How does weathering change Earth's surface?

18. How does no-till farming reduce soil erosion?

19. How does time affect soil development?

20. How does humus form? What does it form from?

Use the graph below to answer questions 21–23.

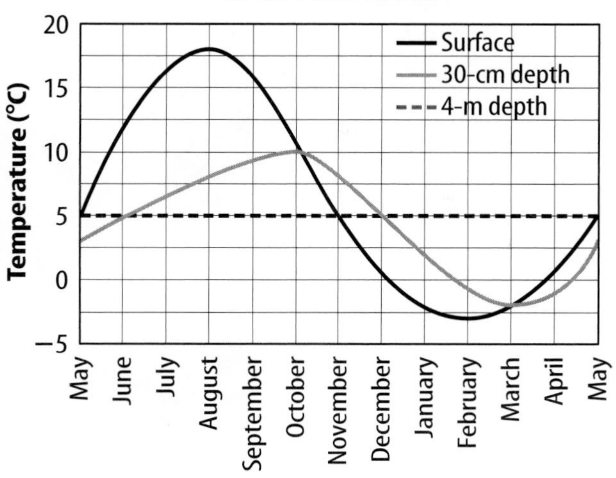

21. During which month was the surface soil warmest? During which month was it coldest? Explain.

22. During which month was the soil at 30 cm warmest? During which month was it coldest? How is this different than surface soil?

23. Why didn't soil temperature vary at a depth of 4 m?

chapter

Erosional Forces

The BIG Idea

Erosion is a process that wears away surface materials and moves them from one place to another.

SECTION 1
Erosion by Gravity

Main Idea Mass movement is a type of erosion that happens as gravity moves materials downslope.

SECTION 2
Glaciers

Main Idea As glaciers pass over land, they erode it, changing features on Earth's surface.

SECTION 3
Wind

Main Idea When air moves, it picks up loose material and transports it to other places.

For Sale—Ocean View

This home, once perched on a hillside overlooking the water, has been destroyed by landslides and flooding. In this chapter you will learn how large amounts of soil, such as the soil that once supported this house, can move from one place to another.

Science Journal Name three major landforms around the world and explain what erosional forces helped shape them.

Start-Up Activities

Demonstrate Sediment Movement

Can you think of ways to move something without touching it? In nature, sediment is moved from one location to another by a variety of forces. What are some of these forces? In this lab, you will investigate to find out the answers to these questions.

WARNING: *Do not pour sand or gravel down the drain.*

1. Place a small pile of a sand-and-gravel mixture into a large shoe-box lid.
2. Move the sediment pile to the other end of the lid without touching the particles with your hands. You can touch and manipulate the box lid.
3. Try to move the mixture in a number of different ways.
4. **Think Critically** In your Science Journal, describe the methods you used to move the sediment. Which method was most effective? Explain how your methods compare with forces of nature that move sediment.

Erosion and Deposition Make the following Foldable to help you identify the examples of erosion and deposition.

STEP 1 **Fold** one piece of paper widthwise into thirds.

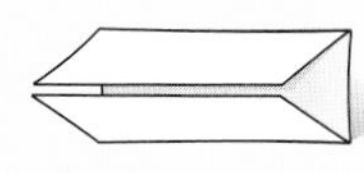

STEP 2 **Fold** the paper lengthwise into fourths.

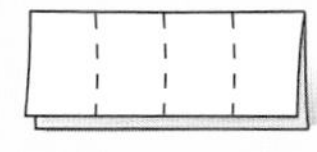

STEP 3 **Unfold,** lay the paper lengthwise, and draw lines along the folds.

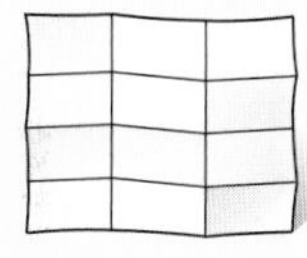

STEP 4 **Label** your table as shown.

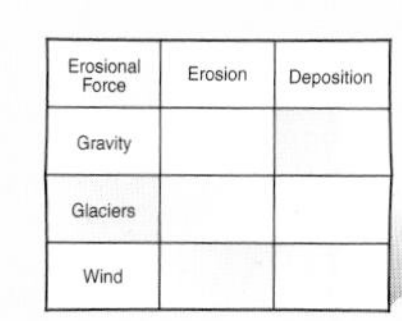

Make a Table As you read the chapter, complete the table, listing specific examples of erosion and deposition for each erosional force.

Preview this chapter's content and activities at earth.msscience.com

Get Ready to Read

Identify Cause and Effect

1 Learn It! A cause is the reason something happens. The result of what happens is called an effect. Learning to identify causes and effects helps you understand why things happen. By using graphic organizers, you can sort and analyze causes and effects as you read.

2 Practice It! Read the following paragraph. Then use the graphic organizer below to show what happened when ice freezes in the cracks of rocks.

> ...Rockfalls happen when blocks of rock break loose from a steep slope and tumble through the air. As they fall, these rocks crash into other rocks and knock them loose. More and more rocks break loose and tumble to the bottom. The fall of a single, large rock down a steep slope can cause serious damage to structures at the bottom. During the winter, when ice freezes in the cracks of rocks, the cracks expand and extend. In the spring, the pieces of rock break loose and fall down the mountainside...
>
> —*from page 212*

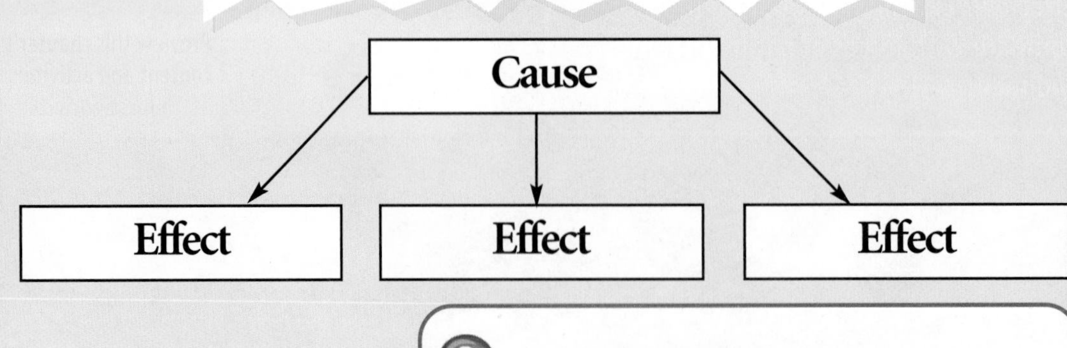

3 Apply It! As you read the chapter, be aware of causes and effects of gravity and ice. Find five causes and their effects.

Graphic organizers such as the Cause-Effect organizer help you organize what you are reading so you can remember it later.

Target Your Reading

Use this to focus on the main ideas as you read the chapter.

1. **Before you read** the chapter, respond to the statements below on your worksheet or on a numbered sheet of paper.
   - Write an **A** if you **agree** with the statement.
   - Write a **D** if you **disagree** with the statement.

2. **After you read** the chapter, look back to this page to see if you've changed your mind about any of the statements.
   - If any of your answers changed, explain why.
   - Change any false statements into true statements.
   - Use your revised statements as a study guide.

Before You Read A or D		Statement	After You Read A or D
	1	Gravity, water, wind, and glaciers are common agents of erosion.	
	2	Deposition occurs when agents of erosion lose energy and drop the sediments they are carrying.	
	3	Mass movement is the slow process of changing rock into soil.	
	4	The two broad categories of glaciers are called continental glaciers and valley glaciers.	
	5	During the most recent ice age, continental glaciers covered the entire Earth.	
	6	Valley glaciers carve deep, V-shaped valleys.	
	7	Abrasion can be caused by windblown sediment striking and wearing away the surface of rock.	
	8	Most dunes move, or migrate away from the direction of the wind.	
	9	During sandstorms, large sand grains are often carried high into the atmosphere.	

Science Online
Print out a worksheet of this page at earth.msscience.com

section 1

Erosion by Gravity

as you read

What You'll Learn

- **Explain** the differences between erosion and deposition.
- **Compare** and contrast slumps, creep, rockfalls, rock slides, and mudflows.
- **Explain** why building on steep slopes might not be wise.

Why It's Important

Many natural features throughout the world were shaped by erosion.

Review Vocabulary

sediment: loose materials, such as mineral grains and rock fragments, that have been moved by erosional forces

New Vocabulary

- erosion
- deposition
- mass movement
- slump
- creep

Erosion and Deposition

Do you live in an area where landslides occur? As **Figure 1** shows, large piles of sediment and rock can move downhill with devastating results. Such events often are triggered by heavy rainfall. The muddy debris at the lower end of the slide comes from material that once was further up the hillside. The displaced soil and rock debris is a product of erosion (ih ROH zhun). **Erosion** is a process that wears away surface materials and moves them from one place to another.

What wears away sediments? How were you able to move the pile of sediments in the Launch Lab? If you happened to tilt the pan, you took advantage of an important erosional force—gravity. Gravity is the force of attraction that pulls all objects toward Earth's center. Other causes of erosion, also called agents of erosion, are water, wind, and glaciers.

Water and wind erode materials only when they have enough energy of motion to do work. For example, air can't move much sediment on a calm day, but a strong wind can move dust and even larger particles. Glacial erosion works differently by slowly moving sediment that is trapped in solid ice. As the ice melts, sediment is deposited, or dropped. Sometimes sediment is carried farther by moving meltwater.

Figure 1 The jumbled sediment at the base of a landslide is material that once was located farther uphill.
Define *the force that moves materials toward the center of Earth.*

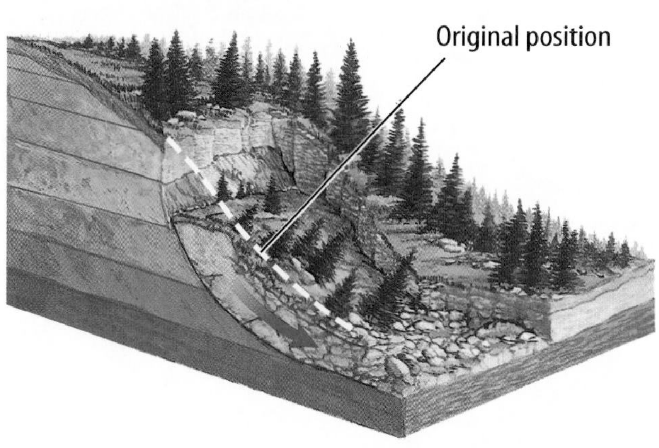

Figure 2 Slump occurs when material slips downslope as one large mass.
Infer *What might have caused this slump to happen?*

Dropping Sediments Agents of erosion drop the sediments they are carrying as they lose energy. This is called **deposition.** When sediments are eroded, they are not lost from Earth—they are just relocated.

Mass Movement

The greater an object's mass is, the greater its gravitational force is. Earth has such a great mass that gravity is a major force of erosion and deposition. Rocks and other materials, especially on steep slopes, are pulled toward the center of Earth by gravity.

A **mass movement** is any type of erosion that happens as gravity moves materials downslope. Some mass movements are so slow that you hardly notice they're happening. Others happen quickly—possibly causing catastrophes. Common types of mass movement include slump, creep, rockfalls, rock slides, and mudflows. Landslides are mass movements that can be one of these types or a combination of these types of mass movement.

Reading Check *What is a mass movement?*

Slump When a mass of material slips down along a curved surface, the mass movement is called **slump.** Often, when a slope becomes too steep, the base material no longer can support the rock and sediment above it. The soil and rock slip downslope as one large mass or break into several sections.

Sometimes a slump happens when water moves to the base of a slipping mass of sediment. This water weakens the slipping mass and can cause movement of material downhill. Or, if a strong rock layer lies on top of a weaker layer—commonly clay—the clay can weaken further under the weight of the rock. The clay no longer can support the strong rock on the hillside. As shown in **Figure 2,** a curved scar is left where the slumped materials originally rested.

Modeling Slump

Procedure

WARNING: *Do not pour lab materials down the drain.*

1. Place one end of a **baking pan** on **two bricks** and position the other end over a sink with a sealed drain.
2. Fill the bottom half of the pan with **gelatin powder** and the top half of the pan with **aquarium gravel.** Place a large, **flat rock** on the gravel.
3. Using a **watering can,** sprinkle water on the materials in the pan for several minutes. Record your observations in your **Science Journal.**

Analysis

1. What happened to the different sediments in the pan?
2. Explain how your experiment models slump.

Figure 3 Over time, creep has caused these tree trunks to lean downhill. The trees then curved back toward the Sun.

Creep The next time you travel, look along the roadway or trail for slopes where trees and fence posts lean downhill. Leaning trees and human-built structures show another mass movement called creep. **Creep** occurs when sediments slowly shift their positions downhill, as **Figure 3** illustrates. Creep is common in areas of frequent freezing and thawing.

Rockfalls and Rock Slides Signs along mountainous roadways warn of another type of mass movement called rockfalls. Rockfalls happen when blocks of rock break loose from a steep slope and tumble through the air. As they fall, these rocks crash into other rocks and knock them loose. More and more rocks break loose and tumble to the bottom. The fall of a single, large rock down a steep slope can cause serious damage to structures at the bottom. During the winter, when ice freezes in the cracks of rocks, the cracks expand and extend. In the spring, the pieces of rock break loose and fall down the mountainside, as shown in **Figure 4.**

Rock slides occur when layers of rock—usually steep layers—slip downslope suddenly. Rock slides, like rockfalls, are fast and can be destructive in populated areas. They commonly occur in mountainous areas or in areas with steep cliffs, also as shown in **Figure 5.** Rock slides happen most often after heavy rains or during earthquakes, but they can happen on any rocky slope at any time without warning.

Figure 4 Rockfalls, such as this one, occur as material free falls through the air.

Figure 5 Rock slides are common in regions where layers of rock are steep.

Mudflows What would happen if you took a long trip and forgot to turn off the sprinkler in your hillside garden before you left? If the soil is usually dry, the sprinkler water could change your yard into a muddy mass of material much like chocolate pudding. Part of your garden might slide downhill. You would have made a mudflow, a thick mixture of sediments and water flowing down a slope. The mudflow in **Figure 6** caused a lot of destruction.

Mudflows usually occur in areas that have thick layers of loose sediments. They often happen after vegetation has been removed by fire. When heavy rains fall on these areas, water mixes with sediment, causing it to become thick and pasty. Gravity causes this mass to flow downhill. When a mudflow finally reaches the bottom of a slope, it loses its energy of motion and deposits all the sediment and everything else it has been carrying. These deposits often form a mass that spreads out in a fan shape. Why might mudflows cause more damage than floodwaters?

Figure 6 Mudflows, such as these in the town of Sarno, Italy, have enough energy to move almost anything in their paths. **Explain** *how mudflows differ from slumps, creep, and rock slides.*

What conditions are favorable for triggering mudflows?

Mudflows, rock slides, rockfalls, creep, and slump are similar in some ways. They all are most likely to occur on steep slopes, and they all depend on gravity to make them happen. Also, all types of mass movement occur more often after a heavy rain. The water adds mass and creates fluid pressure between grains and layers of sediment. This makes the sediment expand—possibly weakening it.

Driving Force The force that drives most types of erosion is gravity. Water at an elevation has potential, or stored energy. When water drops in elevation this energy changes to kinetic energy, or energy of motion. Water may then become a powerful agent of erosion. Find out how water has shaped the region in which you live.

Consequences of Erosion

People like to have a great view and live in scenic areas away from noise and traffic. To live this way, they might build or move into houses and apartments on the sides of hills and mountains. When you consider gravity as an agent of erosion, do you think steep slopes are safe places to live?

Building on Steep Slopes When people build homes on steep slopes, they constantly must battle naturally occurring erosion. Sometimes builders or residents make a slope steeper or remove vegetation. This speeds up the erosion process and creates additional problems. Some steep slopes are prone to slumps because of weak sediment layers underneath.

Figure 7 Some slopes are stabilized by building walls made from concrete or stone.

Making Steep Slopes Safe Plants can be beautiful or weedlike—but they all have root structures that hold soil in place. One of the best ways to reduce erosion is to plant vegetation. Deep tree roots and fibrous grass roots bind soil together, reducing the risk of mass movement. Plants also absorb large amounts of water. Drainage pipes or tiles inserted into slopes can prevent water from building up, too. These materials help increase the stability of a slope by allowing excess water to flow out of a hillside more easily.

Walls made of concrete or boulders also can reduce erosion by holding soil in place, as shown in **Figure 7.** However, preventing mass movements on a slope is difficult because rain or earthquakes can weaken all types of Earth materials, eventually causing them to move downhill.

Reading Check *What can be done to slow erosion on steep slopes?*

People who live in areas with erosion problems spend a lot of time and money trying to preserve their land. Sometimes they're successful in slowing down erosion, but they never can eliminate erosion and the danger of mass movement. Eventually, gravity wins. Sediment moves from place to place, constantly reducing elevation and changing the shape of the land.

section 1 review

Summary

Erosion and Deposition

- Gravity is the force that pulls all objects toward Earth's center.
- Water and wind erode materials only when they have enough energy of motion to do work.
- Agents of erosion drop sediment as they lose energy.

Mass Movement

- The greater an object's mass is, the greater its gravitational force is.
- Gravity is a major force of erosion and deposition.
- Common types of mass movement include slump, creep, rockfalls, rock slides, and mudflows.

Self Check

1. **Define** the term *erosion* and name the forces that cause it.
2. **Explain** how deposition changes the surface of Earth.
3. **Describe** the characteristics that all types of mass movements have in common.
4. **Describe** ways to help slow erosion on steep slopes.
5. **Think Critically** When people build houses and roads, they often pile up dirt or cut into the sides of hills. Predict how this might affect sediment on a slope. Explain how to control the effects of such activities.

Applying Skills

6. **Compare and Contrast** What are the similarities and differences between rock falls and rock slides?

Science Online earth.msscience.com/self_check_quiz

section 2

Glaciers

How Glaciers Form and Move

If you've ever gone sledding, snowboarding, or skiing, you might have noticed that after awhile, the snow starts to pack down under your weight. A snowy hillside can become icy if it is well traveled. In much the same way, glaciers form in regions where snow accumulates. Some areas of the world, as shown in **Figure 8,** are so cold that snow remains on the ground year-round. When snow doesn't melt, it piles up. As it accumulates slowly, the increasing weight of the snow becomes great enough to compress the lower layers into ice. Eventually, there can be enough pressure on the ice so that it becomes plasticlike. The mass slowly begins to flow in a thick, plasticlike lower layer, and ice slowly moves away from its source. A large mass of ice and snow moving on land under its own weight is a **glacier.**

Ice Eroding Rock

Glaciers are agents of erosion. As glaciers pass over land, they erode it, changing features on the surface. Glaciers then carry eroded material along and deposit it somewhere else. Glacial erosion and deposition change large areas of Earth's surface. How is it possible that something as fragile as snow or ice can push aside trees, drag rocks along, and slowly change the surface of Earth?

as you read

What **You'll Learn**

- **Explain** how glaciers move.
- **Describe** evidence of glacial erosion and deposition.
- **Compare and contrast** till and outwash.

Why **It's Important**

Glacial erosion and deposition create many landforms on Earth.

Review Vocabulary
plasticlike: not completely solid or liquid; capable of being molded or changing form

New Vocabulary
- glacier
- plucking
- till
- moraine
- outwash

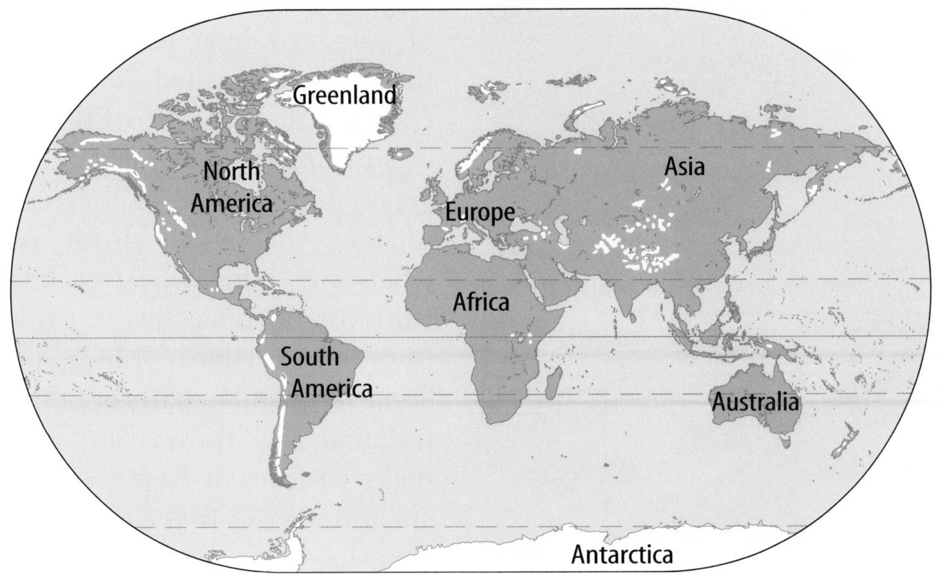

Figure 8 The white regions on this map show areas that are glaciated today.

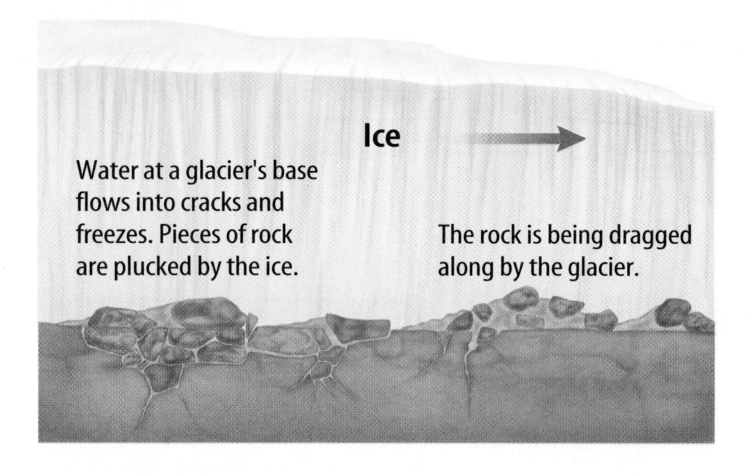

Ice
Water at a glacier's base flows into cracks and freezes. Pieces of rock are plucked by the ice.
The rock is being dragged along by the glacier.

Figure 9 Plucking is a process that occurs when a moving glacier picks up loosened rock particles.

Plucking Glaciers weather and erode solid rock. When glacial ice melts, water flows into cracks in rocks. Later, the water refreezes in these cracks, expands, and fractures the rock. Pieces of rock then are lifted out by the ice, as shown in **Figure 9.** This process, called **plucking,** results in boulders, gravel, and sand being added to the bottom and sides of a glacier.

Reading Check *What is plucking?*

Transporting and Scouring As it moves forward over land, a glacier can transport huge volumes of sediment and rock. Plucked rock fragments and sand at its base scour and scrape the soil and bedrock like sandpaper against wood, eroding the ground below even more. When bedrock is gouged deeply by rock fragments being dragged along, marks such as those in **Figure 10** are left behind. These marks, called grooves, are deep, long, parallel scars on rocks. Shallower marks are called striations (stri AY shunz). Grooves and striations indicate the direction in which the glacier moved.

Ice Depositing Sediment

When glaciers begin to melt, they are unable to carry much sediment. The sediment drops, or is deposited, on the land. When a glacier melts and begins to shrink back, it is said to retreat. As it retreats, a jumble of boulders, sand, clay, and silt is left behind. This mixture of different-sized sediments is called **till.** Till deposits can cover huge areas of land. Thousands of years ago, huge ice sheets in the northern United States left enough till behind to fill valleys completely and make these areas appear flat. Till areas include the wide swath of what are now wheat farms running northwestward from Iowa to northern Montana. Some farmland in parts of Ohio, Indiana, and Illinois and the rocky pastures of New England are also regions that contain till deposits.

Figure 10 When glaciers melt, striations or grooves can be found on the rocks beneath. These glacial grooves on Kelley's Island, Ohio, give evidence of past glacial erosion and movement.

Figure 11 Moraines are forming along the sides of this glacier. Unlike the moraines that form at the ends of glaciers, these moraines form as rock and sediment fall from nearby slopes.

Moraine Deposits Till also is deposited at the end of a glacier when it is not moving forward. Unlike the till that is left behind as a sheet of sediment over the land, this type of deposit doesn't cover such a wide area. Rocks and soil are moved to the end of the glacier, much like items on a grocery store conveyor belt. Because of this, a big ridge of material piles up that looks as though it has been pushed along by a bulldozer. Such a ridge is called a **moraine.** Moraines also are deposited along the sides of a glacier, as shown in **Figure 11.**

Outwash Deposits When glacial ice starts to melt, the meltwater can deposit sediment that is different from till. Material deposited by the meltwater from a glacier, most often beyond the end of the glacier, is called **outwash.** Meltwater carries sediments and deposits them in layers. Heavier sediments drop first, so bigger pieces of rock are deposited closer to the glacier. The outwash from a glacier also can form into a fan-shaped deposit when the stream of meltwater deposits sand and gravel in front of the glacier.

What is outwash?

Eskers Another type of outwash deposit looks like a long, winding ridge. This deposit forms in a melting glacier when meltwater forms a river within the ice, as shown in the diagram in **Figure 12.** This river carries sand and gravel and deposits them within its channel. When the glacier melts, a winding ridge of sand and gravel, called an esker (ES kur), is left behind. An esker is shown in the photograph in **Figure 12.**

Figure 12 Eskers are glacial deposits formed by meltwater.

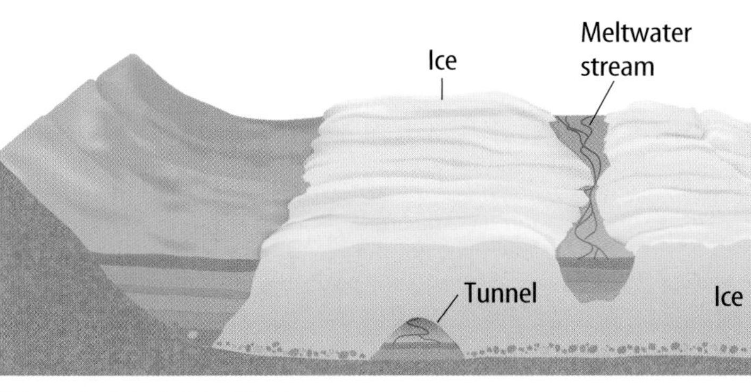

Eskers form when sediment deposited in ice tunnels or by streams on top of the ice is left behind on Earth's surface.

The snake-like shape of this esker in North Dakota is characteristic of this type of glacial deposit.

Continental Glaciers

The two types of glaciers are continental glaciers and valley glaciers. Today, continental glaciers, like the one in **Figure 13** cover ten percent of Earth, mostly near the poles in Antarctica and Greenland. These continental glaciers are huge masses of ice and snow. Continental glaciers are thicker than some mountain ranges. Glaciers make it impossible to see most of the land features in Antarctica and Greenland.

In what regions on Earth would you expect to find continental glaciers?

Figure 13 Continental glaciers and valley glaciers are agents of erosion and deposition. This continental glacier covers a large area in Antarctica.

Climate Changes In the past, continental glaciers covered as much as 28 percent of Earth. **Figure 14** shows how much of North America was covered by glaciers during the most recent ice advance. These periods of widespread glaciation are known as ice ages. Extensive glaciers have covered large portions of Earth many times over the last 2 million to 3 million years. During this time, glaciers advanced and retreated many times over much of North America. The average air temperature on Earth was about 5°C lower during these ice ages than it is today. The last major advance of ice reached its maximum extent about 18,000 years ago. After this last advance of glaciers, the ends of the ice sheets began to recede, or move back, by melting.

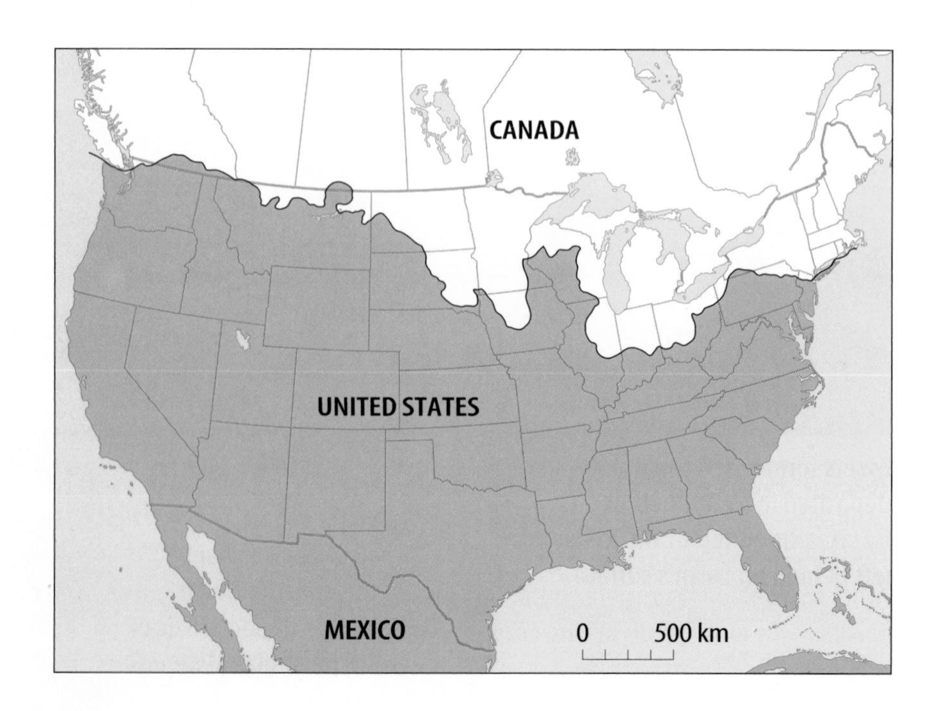

Figure 14 This map shows how much of North America was covered by continental glaciers about 18,000 years ago.
Observe *Was your location covered? If so, what evidence of glaciers does your area show?*

Valley Glaciers

Valley glaciers occur even in today's warmer global climate. In the high mountains where the average temperature is low enough to prevent snow from melting during the summer, valley glaciers grow and creep along. **Figure 15** shows valley glaciers in Africa.

Figure 15 Valley glaciers, like these on Mount Kilimanjaro in north Tanzania, Africa, form between mountain peaks that lie above the snow line, where snow lasts all year.

Evidence of Valley Glaciers If you visit the mountains, you can tell whether valley glaciers ever existed there. You might look for striations, then search for evidence of plucking. Glacial plucking often occurs near the top of a mountain where a glacier is mainly in contact with solid rock. Valley glaciers erode bowl-shaped basins, called cirques (SURKS), into the sides of mountains. If two valley glaciers side by side erode a mountain, a long ridge called an arête (ah RAYT) forms between them. If valley glaciers erode a mountain from several directions, a sharpened peak called a horn might form. **Figure 16** shows some features formed by valley glaciers.

Valley glaciers flow down mountain slopes and along valleys, eroding as they go. Valleys that have been eroded by glaciers have a different shape from those eroded by streams. Stream-eroded valleys are normally V-shaped. Glacially eroded valleys are U-shaped because a glacier plucks and scrapes soil and rock from the sides as well as from the bottom. A large U-shaped valley and smaller hanging valleys are illustrated in **Figure 16.**

Figure 16 Valley glaciers transform the mountains over which they pass.

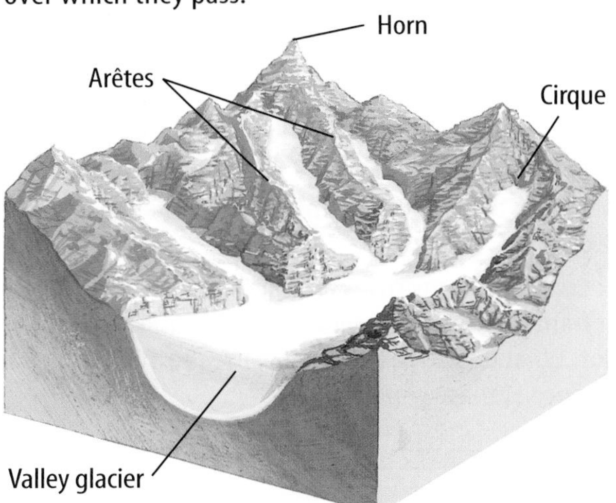

Bowl-shaped basins called cirques form by erosion at the start of a valley glacier. Arêtes form where two adjacent valley glaciers meet and erode a long, sharp ridge. Horns are sharpened peaks formed by glacial action in three or more cirques.

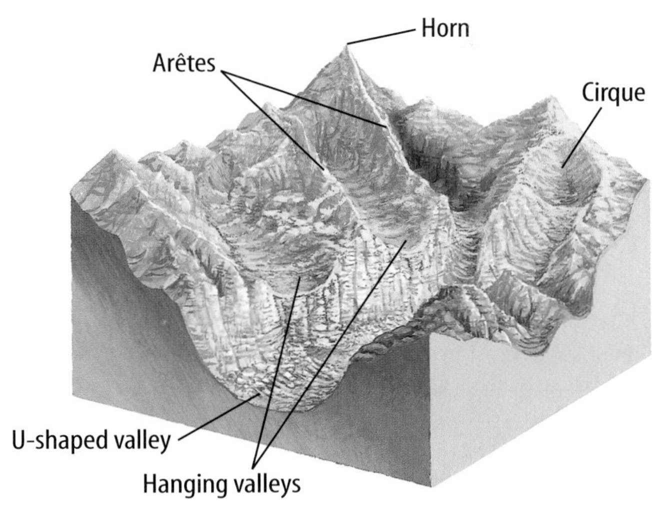

U-shaped valleys result when valley glaciers move through regions once occupied by streams. A tributary glacial valley whose mouth is high above the floor of the main valley is called a hanging valley. The discordance between the different valley floors is due to the greater erosive power of the trunk glacier.

Figure 17 Sand and gravel deposits left by glaciers are important starting materials for the construction of roadways and buildings.

Topic: Glacial Deposits
Visit earth.msscience.com for Web links to information about uses of glacial deposits.

Activity List various uses of glacial deposits and name the methods of removing this material.

Importance of Glaciers

Glaciers have had a profound effect on Earth's surface. They have eroded mountaintops and transformed valleys. Vast areas of the continents have sediments that were deposited by great ice sheets. Today, glaciers in polar regions and in mountains continue to change the surface features of Earth.

In addition to changing the appearance of Earth's surface, glaciers leave behind sediments that are economically important, as illustrated in **Figure 17.** The sand and gravel deposits from glacial outwash and eskers are important resources. These deposits are excellent starting materials for the construction of roads and buildings.

section 2 review

Summary

How Glaciers Move and Form

- Glaciers form in regions where snow accumulates and remains year round.
- The weight of snow compresses the lower layers into ice and causes the ice to become plasticlike.

Ice Eroding Rock

- Glaciers are agents of erosion.
- Glacial erosion and deposition change large areas of the Earth's surface.

Ice Depositing Sediment

- Glaciers melt and retreat, leaving behind sediment.
- Forms of glacial deposits include till, moraines, outwash deposits and eskers.

Continental and Valley Glaciers

- Continental and valley glaciers are the two types of glaciers.

Self Check

1. **Describe** how glaciers move.
2. **Identify** two common ways in which a glacier can cause erosion.
3. **Determine** Till and outwash are glacial deposits. Explain how till and outwash are different.
4. **Discuss** How do moraines form? What are moraines made of?
5. **Think Critically** Many rivers and lakes that receive water from glacial meltwater often appear milky blue in color. What do you think might cause the milk appearance of these waters?

Applying Skills

6. **Recognize Cause and Effect** Since 1900, the Alps have lost 50 percent of their ice caps, and New Zealand's glaciers have shrunk by 26 percent. Describe what you think some causes and effects of this glacial melting have been.

GLACIAL GROOVING

Throughout the world's mountainous regions, 200,000 valley glaciers are moving in response to gravity.

Real-World Question

How is the land affected when a valley glacier moves downslope?

Goals

- **Compare** stream and glacial valleys.

Materials

sand
large plastic or metal tray
**stream table*
ice block
books (2 or 3)
**wood block*
metric ruler
overhead light source with reflector

**Alternate materials*

Safety Precautions

WARNINGS: *Do not pour sand down the drain. Make sure source is plugged into a GFI electrical outlet. Do not touch light source—it may be hot.*

Procedure

1. Set up the large tray of sand as shown above. Place books under one end of the tray to make a slope.
2. Cut a narrow riverlike channel through the sand. Measure and record its width and depth in a table similar to the one shown. Draw a sketch that includes these measurements.
3. Position the overhead light source to shine on the channel as shown.
4. Force the ice block into the channel at the upper end of the tray.
5. Gently push the ice along the channel until it's halfway between the top and bottom of the tray, and directly under the light.

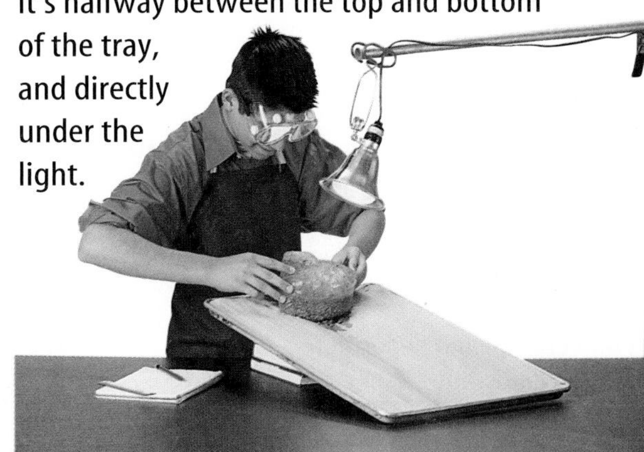

6. Turn on the light and allow the ice to melt. Record what happens.
7. Record the width and depth of the ice channel in the table. Make a scale drawing.

Conclude and Apply

1. **Explain** how you can determine the direction that a glacier traveled from the location of deposits.
2. **Explain** how you can determine the direction of glacial movement from sediments deposited by meltwater.
3. **Describe** how valley glaciers affect the surface over which they move.

Glacier Data

Sample Data	Width (cm)	Depth (cm)	Observations
Original channel	1–2	3	Stream channel looked V-shaped
Glacier channel		Do not write in this book.	
Meltwater channel			

section 3

Wind

as you read

What You'll Learn

- **Explain** how wind causes deflation and abrasion.
- **Recognize** how loess and dunes form.

Why It's Important

Wind erosion and deposition change landscapes, especially in dry climates.

Review Vocabulary
friction: force that opposes the motion of an object when the object is in contact with another object or surface

New Vocabulary
- deflation
- abrasion
- loess
- dune

Wind Erosion

When air moves, it picks up loose material and transports it to other places. Air differs from other erosional forces because it usually cannot pick up heavy sediments. Unlike rivers that move in confined places like channels and valleys, wind carries and deposits sediments over large areas. For example, wind is capable of picking up and carrying dust particles from fields or volcanic ash high into the atmosphere and depositing them thousands of kilometers away.

Deflation Wind erodes Earth's surface by deflation (dih FLAY shun) and abrasion (uh BRAY zhun). When wind erodes by **deflation,** it blows across loose sediment, removing small particles such as silt and sand. The heavier, coarser material is left behind.

Abrasion When windblown sediment strikes rock, the surface of the rock gets scraped and worn away by a process called **abrasion.** Abrasion, shown in **Figure 18,** is similar to sandblasting. Workers use machines that spray a mixture of sand and water under high pressure against a building. The friction wears away dirt from stone, concrete, or brick walls. It also polishes the walls of buildings by breaking away small pieces and leaving an even, smooth finish. Wind acts like a sandblasting machine, bouncing and blowing sand grains along. These sand grains strike against rocks and break off small fragments. The rocks become pitted and are worn down gradually.

Figure 18 The odd shape of this boulder was produced by wind abrasion.

Reading Check *How is wind abrasion similar to sandblasting?*

Deflation and abrasion happen to all land surfaces but occur mostly in deserts, beaches, and plowed fields. These areas have fewer plants to hold the sediments in place. When winds blow over them, they can be eroded rapidly. Grassland or pasture land have many plants that hold the soil in place, therefore there is little soil erosion caused by the wind.

Sandstorms Even when the wind blows strongly, it seldom carries sand grains higher than 0.5 m from the ground. However, sandstorms do occur. When the wind blows forcefully in the sandy parts of deserts, sand grains bounce along and hit other sand grains, causing more and more grains to rise into the air. These windblown sand grains form a low cloud just above the ground. Most sandstorms occur in deserts, but they can occur in other arid regions.

Dust Storms When soil is moist, it stays packed on the ground, but when it dries out, it can be eroded by wind. Soil is composed largely of silt- and clay-sized particles. Because these small particles weigh less than sand-sized particles of the same material, wind can move them high into the air.

Silt and clay particles are small and stick together. A faster wind is needed to lift these fine particles of soil than is needed to lift grains of sand. However, after they are airborne, the wind can carry them long distances. Where the land is dry, dust storms can cover hundreds of kilometers. These storms blow topsoil from open fields, overgrazed areas, and places where vegetation has disappeared. In the 1930s, silt and dust that was picked up in Kansas fell in New England and in the North Atlantic Ocean. Dust blown from the Sahara has been traced as far away as the West Indies—a distance of at least 6,000 km.

Dust Bowl Poor agricultural practices and a long period of sustained drought caused the Dust Bowl of the 1930s. Research how this affected the livelihood of the people of the southern plains.

Applying Science

What factors affect wind erosion?

Many factors compound the effects of wind erosion. But can anything be done to minimize erosion?

Identifying the Problem

Wind velocity and duration, the size of sediment particles, the size of the area subjected to the wind, and the amount of vegetation present all affect how much soil is eroded by wind. The table shows different combinations of these factors. It also includes an erosion rating that depends upon what factors pertain to an area.

Factors That Affect Wind Erosion

Factor	Descriptions				
Wind velocity	high	high	low	low	low
Duration of wind	long	long	short	long	long
Particle size	coarse	medium	coarse	coarse	medium
Surface area	large	large	small	small	large
Amount of vegetation	high	low	high	high	high
Erosion rating	some	a lot	a little	some	?

Solving the Problem

1. Looking at the table, can you figure out which factors increase and which factors decrease the amount of erosion?
2. From what you've discovered, can you estimate the missing erosion rating?

Observing How Soil Is Held in Place

Procedure

1. Obtain a piece of **sod** (a chunk of soil about 5 cm thick with grass growing from it).
2. Carefully remove soil from the sod roots by hand. Examine the roots with a **magnifying lens.**
3. Wash hands thoroughly with soap and water.

Analysis

1. Draw several of these roots in your **Science Journal.**
2. What characteristics of grass roots help hold soil in place and thus reduce erosion?

Try at Home

Reducing Wind Erosion

As you've learned, wind erosion is most common where there are no plants to protect the soil. Therefore, one of the best ways to slow or stop wind erosion is to plant vegetation. This practice helps conserve soil and protect valuable farmland.

Windbreaks People in many countries plant vegetation to reduce wind erosion. For centuries, farmers have planted trees along their fields to act as windbreaks that prevent soil erosion. As the wind hits the trees, its energy of motion is reduced. It no longer is able to lift particles.

In one study, a thin belt of cottonwood trees reduced the effect of a 25-km/h wind to about 66 percent of its normal speed, or to about 16.5 km/h. Tree belts also trap snow and hold it on land. This increases the moisture level of the soil, which helps prevent further erosion.

Roots Along many seacoasts and deserts, vegetation is planted to reduce erosion. Plants with fibrous root systems, such as grasses, work best at stopping wind erosion. Grass roots are shallow and slender with many fibers. They twist and turn between particles in the soil and hold it in place.

Planting vegetation is a good way to reduce the effects of deflation and abrasion. Even so, if the wind is strong and the soil is dry, nothing can stop erosion completely. **Figure 19** shows a project designed to decrease wind erosion.

Topic: Conservation Practices

Visit earth.msscience.com for Web links to collect data on various methods of protecting soil from wind erosion.

Activity List methods that conserve soil. What is the most commonly used method by farmers?

Figure 19 Rows of grasses and rocks were installed on these dunes in Qinghai, China, to reduce wind erosion.

Deposition by Wind

Sediments blown away by wind eventually are deposited. Over time, these windblown deposits develop into landforms, such as dunes and accumulations of loess.

Loess Some examples of large deposits of windblown sediments are found near the Mississippi and Missouri Rivers. These wind deposits of fine-grained sediments known as **loess** (LES) are shown in **Figure 20.** Strong winds that blew across glacial outwash areas carried the sediments and deposited them. The sediments settled on hilltops and in valleys. Once there, the particles packed together, creating a thick, unlayered, yellowish-brown-colored deposit. Loess is as fine as talcum powder. Many farmlands of the midwestern United States have fertile soils that developed from loess deposits.

Figure 20 This sediment deposit is composed partially of windblown loess.

Dunes Do you notice what happens when wind blows sediments against an obstacle such as a rock or a clump of vegetation? The wind sweeps around or over the obstacle. Like a river, air drops sediment when its energy decreases. Sediment starts to build up behind the obstacle. The sediment itself then becomes an obstacle, trapping even more material. If the wind blows long enough, the mound will become a dune, as shown in **Figure 21.** A **dune** (DOON) is a mound of sediments drifted by the wind.

Reading Check *What is a dune?*

Dunes are common in desert regions. You also can see sand dunes along the shores of oceans, seas, or lakes. If dry sediments exist in an area where prevailing winds or sea breezes blow daily, dunes build up. Sand or other sediment will continue to build up and form a dune until the sand runs out or the obstruction is removed. Some desert sand dunes can grow to 100 m high, but most are much shorter.

Figure 21 Loose sediment of any type can form a dune if enough of it is present and an obstacle lies in the path of the wind.

Moving Dunes A sand dune has two sides. The side facing the wind has a gentler slope. The side away from the wind is steeper. Examining the shape of a dune tells you the direction from which the wind usually blows.

Unless sand dunes are planted with grasses, most dunes move, or migrate away from the direction of the wind. This process is shown in **Figure 22.** Some dunes are known as traveling dunes because they move rapidly across desert areas. As they lose sand on one side, they build it up on the other.

NATIONAL GEOGRAPHIC

VISUALIZING HOW DUNES FORM AND MIGRATE

Figure 22

Sand blown loose from dry desert soil often builds up into dunes. A dune may begin to form when windblown sand is deposited in the sheltered area behind an obstacle, such as a rock outcrop. The sand pile grows as more grains accumulate. As shown in the diagram at right, dunes are mobile, gradually moved along by wind.

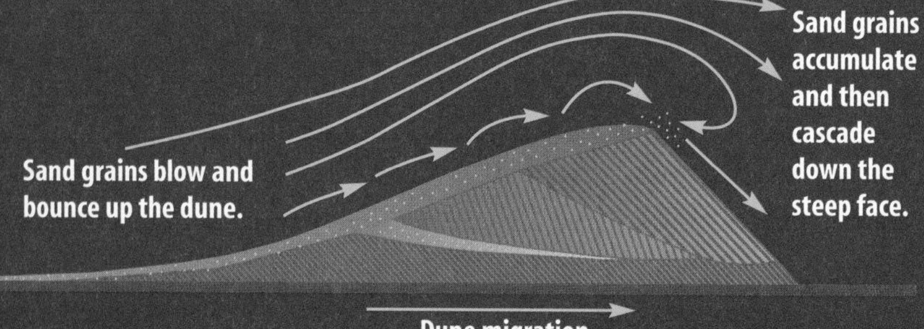

Sand grains accumulate and then cascade down the steep face.
Sand grains blow and bounce up the dune.
Dune migration

▲ A dune migrates as sand blows up its sloping side and then cascades down the steeper side. Gradually, a dune moves forward—in the same direction that the wind is blowing—as sand, lost from one side, piles up on the other side.

▲ Dunes are made of sediments eroded from local materials. Although many dunes are composed of quartz and feldspar, the brilliant white dunes in White Sands National Park, New Mexico, are made of gypsum.

▲ Deserts may expand when humans move into the transition zone between habitable land and desert. Here, villagers in Mauritania in northwestern Africa shovel the sand that encroaches on their schoolhouse daily.

◀ The dunes at left are coastal dunes from the Laguna Madre region of South Texas on the Gulf of Mexico. Note the vegetation in the photo, which has served as an obstacle to trap sand.

Dune Shape The shape of a dune depends on the amount of sand or other sediment available, the wind speed and direction, and the amount of vegetation present. One common dune shape is a crescent-shaped dune known as a barchan (BAR kun) dune. The open side of a barchan dune faces the direction that the wind is blowing. When viewed from above, the points of the crescent are directed downwind. This type of dune forms on hard surfaces where the sand supply is limited.

Another common type of dune, called a transverse dune, forms where sand is abundant. Transverse dunes are so named because the long directions of these dunes are perpendicular to the general wind direction. In regions where the wind direction changes, star dunes, shown in **Figure 23,** form pointed structures. Other dune forms also exist, some of which show a combination of features.

Figure 23 Star dunes form in areas where the wind blows from several different directions.

Shifting Sediments When dunes and loess form, the landscape changes. Wind, like gravity, running water, and glaciers, shapes the land. New landforms created by these agents of erosion are themselves being eroded. Erosion and deposition are part of a cycle of change that constantly shapes and reshapes the land.

section 3 review

Summary

Wind Erosion

- Air movement picks up loose material and transports it to other places.
- Deflation and abrasion happen mainly in deserts, beaches, and plowed fields.

Reducing Wind Erosion

- Planting vegetation can reduce wind erosion.
- Farmers use windbreaks to protect their crop fields from wind erosion.

Deposition by Wind

- Windblown deposits develop into landforms, such as dunes and accumulation of loess.
- Many farmlands of the midwestern United States have fertile soils that developed from loess deposits.

Self Check

1. **Compare and contrast** abrasion and deflation. Describe how they affect the surface of Earth.
2. **Explain** the differences between dust storms and sandstorms. Describe how energy of motion affects the deposition of sand and dust by these storms.
3. **Think Critically** You notice that sand is piling up behind a fence outside your apartment building. Explain why this occurs.

Applying Math

4. **Solve One-Step Equations** Between 1972 and 1992, the Sahara increased by nearly 700 km^2 in Mali and the Sudan. Calculate the average number of square kilometers the desert increased each year between 1972 and 1992.

Design Your Own

Blowing in the Wind

Goals

- **Observe** the effects of soil moisture and wind speed on wind erosion.
- **Design** and carry out experiments that test the effects of soil moisture and wind speed on wind erosion.

Possible Materials

flat pans (4)
fine sand (400 mL)
gravel (400 mL)
hair dryer
sprinkling can
water
28-cm × 35-cm cardboard sheets (4)
tape
mixing bowl
metric ruler
wind speed indicator

Safety Precautions

Wear your safety goggles at all times when using the hair dryer on sediments. Make sure the dryer is plugged into a GFI electrical outlet.

Real-World Question

Have you ever played a sport outside and suddenly had the wind blow dust into your eyes? What did you do? Turn your back? Cover your eyes? How does wind pick up sediment? Why does wind pick up some sediments and leave others on the ground? What factors affect wind erosion?

Form a Hypothesis

How does moisture in sediment affect the ability of wind to erode sediments? Does the speed of the wind limit the size of sediments it can transport? Form a hypothesis about how sediment moisture affects wind erosion. Form another hypothesis about how wind speed affects the size of the sediment the wind can transport.

Test Your Hypothesis

Make a Plan

1. As a group, agree upon and write your hypothesis statements.
2. **List** the steps needed to test your first hypothesis. Plan specific steps and vary only one factor at a time. Then, list the steps needed to test your second hypothesis. Test only one factor at a time.

3. Mix the sediments in the pans. Plan how you will fold cardboard sheets and attach them to the pans to keep sediments contained.
4. **Design** data tables in your Science Journal. Use them as your group collects data.
5. **Identify** all constants, variables, and controls of the experiment. One example of a control is a pan of sediment not subjected to any wind.

Follow Your Plan

1. Make sure your teacher approves your plan before you start.
2. Carry out the experiments as planned.
3. While doing the experiments, write any observations that you or other members of your group make. Summarize your data in the data tables you designed in your Science Journal.

Sediment Movement

Sediment	Wind Speed	Sediment Moved
Fine sand (dry)	low	
	high	
Fine sand (wet)	low	Do not write in this book.
	high	
Gravel (dry)	low	
	high	
Gravel (wet)	low	
	high	
Fine sand and gravel (dry)	low	
	high	
Fine sand and gravel (wet)	low	
	high	

Analyze Your Data

1. **Compare** your results with those of other groups. Explain what might have caused any differences among the groups.
2. **Explain** the relationship that exists between the speed of the wind and the size of the sediments it transports.

Conclude and Apply

1. How does energy of motion of the wind influence sediment transport? What is the general relationship between wind speed and erosion?
2. **Explain** the relationship between the sediment moisture and the amount of sediment moved by the wind.

Design a table that summarizes the results of your experiment, and use it to explain your interpretations to others in the class.

SCIENCE Stats

Losing Against Erosion

Did you know...

...Glaciers, one of nature's most powerful erosional forces, can move more than 30 m per day. In one week, a fast-moving glacier can travel the length of almost two football fields. Glaciers such as these are unusual—most move less than 10 cm per day.

...Some sand dunes migrate as much as 30 m per year. In a coastal region of France, traveling dunes have buried farms and villages. The dunes were halted by anti-erosion practices, such as planting grass in the sand and growing a barrier of trees between the dunes and farmland.

Applying Math If a sand dune is traveling at 30 m per year, how many meters does it travel in one month?

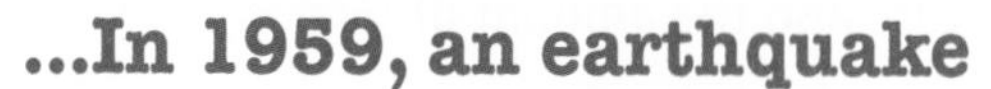

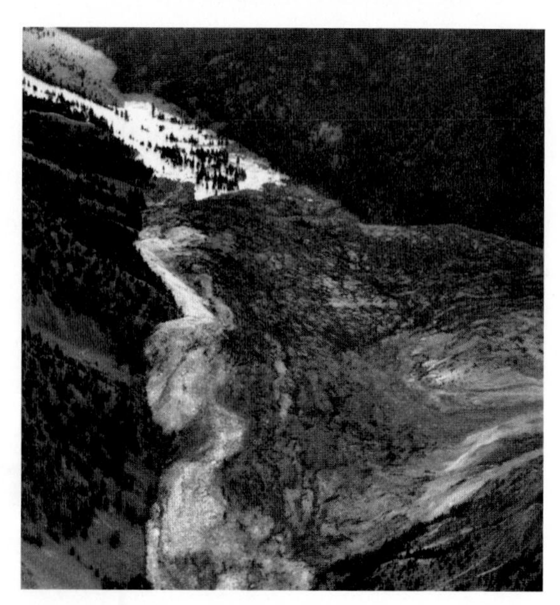

...In 1959, an earthquake triggered a mass movement in Madison River Canyon, Montana. About 21 million km^3 of rock and soil slid down the canyon at an estimated 160 km/h. This type of mass movement of earth is called a rock slide.

Find Out About It

Visit earth.msscience.com/science_stats **to learn about landslides. When is a landslide called a mudflow? In which U.S. states are mudflows most likely to occur?**

Reviewing Main Ideas

Section 1 Erosion by Gravity

1. Erosion is the process that picks up and transports sediment.
2. Deposition occurs when an agent of erosion loses its energy and can no longer carry its load of sediment.
3. Slump, creep, rock slides, and mudflows are all mass movements caused by gravity.

Section 2 Glaciers

1. Glaciers are powerful agents of erosion. As water freezes and thaws in cracks, it breaks off pieces of surrounding rock. These pieces then are incorporated into glacial ice by plucking.
2. As sediment embedded in the base of a glacier moves across the land, grooves and striations form. Glaciers deposit two kinds of material—till and outwash.

Section 3 Wind

1. Deflation occurs when wind erodes only fine-grained sediments, leaving coarse sediments behind.
2. The pitting and polishing of rocks and grains of sediment by windblown sediment is called abrasion.
3. Wind deposits include loess and dunes. Loess consists of fine-grained particles such as silt and clay. Dunes form when windblown sediments accumulate behind an obstacle.

Visualizing Main Ideas

Copy and complete the following concept map on erosional forces. Use the following terms and phrases: striations, leaning trees and structures, curved scar on slope, deflation, *and* mudflows.

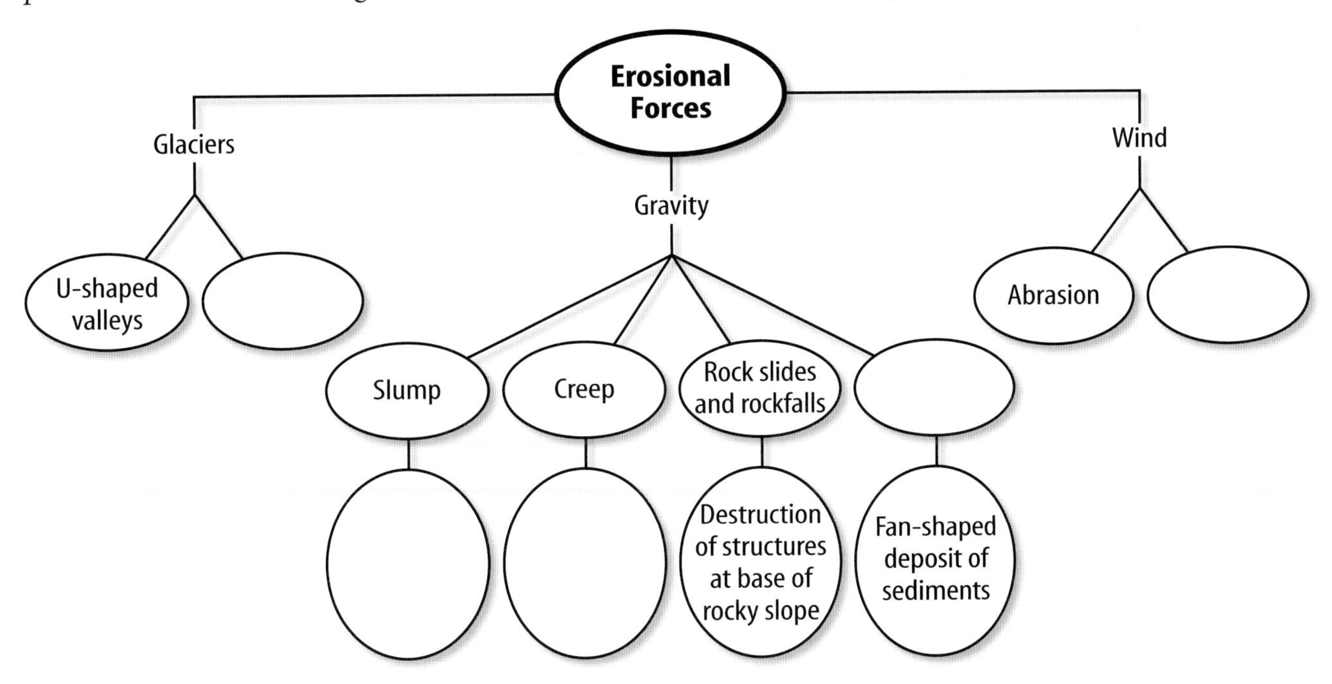

Using Vocabulary

abrasion p. 222
creep p. 212
deflation p. 222
deposition p. 211
dune p. 225
erosion p. 210
glacier p. 215
loess p. 225
mass movement p. 211
moraine p. 217
outwash p. 217
plucking p. 216
slump p. 211
till p. 216

Each phrase below describes a vocabulary word from the list. In your Science Journal, write the term that matches each description.

1. loess, dunes, and moraines are examples
2. slowest mass movement
3. ice picking up pieces of rock
4. much like sandblasting
5. gravity transport of material downslope
6. sand and gravel deposited by meltwater
7. glacial deposit composed of sediment with many sizes and shapes

Checking Concepts

Choose the word or phrase that best answers the question.

8. Which term is an example of a feature created by deposition?
   A) cirque **C)** striation
   B) abrasion **D)** dune

9. The best plants for reducing wind erosion have what type of root system?
   A) taproot **C)** fibrous
   B) striated **D)** sheet

10. What does a valley glacier create at the point where it starts?
    A) esker **C)** till
    B) moraine **D)** cirque

Use the photo below to answer question 11.

11. Which of the following is suggested by leaning trees curving upright on a hillside?
    A) abrasion **C)** slump
    B) creep **D)** mudflow

12. What shape do glacier-created valleys have?
    A) V-shape **C)** U-shape
    B) L-shape **D)** S-shape

13. Which is formed by glacial erosion?
    A) eskers **C)** moraines
    B) arêtes **D)** warmer climate

14. What type of wind erosion leaves pebbles and boulders behind?
    A) deflation **C)** abrasion
    B) loess **D)** sandblasting

15. What is a ridge formed by deposition of till called?
    A) striation **C)** cirque
    B) esker **D)** moraine

16. What is the material called that is deposited by meltwater beyond the end of a glacier?
    A) esker
    B) cirque
    C) outwash
    D) moraine

Science Online earth.msscience.com/vocabulary_puzzlemaker

Thinking Critically

17. **Explain** how striations can give information about the direction that a glacier moved.

18. **Describe** how effective a retaining wall made of fine wire mesh would be against erosion.

19. **Determine** what can be done to prevent the migration of beach dunes.

20. **Recognize Cause and Effect** A researcher finds evidence of movement of ice within a glacier. Explain how this movement could occur.

21. **Think Critically** The end of a valley glacier is at a lower elevation than its point of origin is. How does this help explain melting at its end while snow and ice still are accumulating where it originated?

22. **Make and Use Tables** Make a table to contrast continental and valley glaciers.

23. **Concept Map** Copy and complete the events-chain concept map below to show how a sand dune forms. Use the terms and phrases: *sand accumulates, dune, dry sand,* and *obstruction traps.*

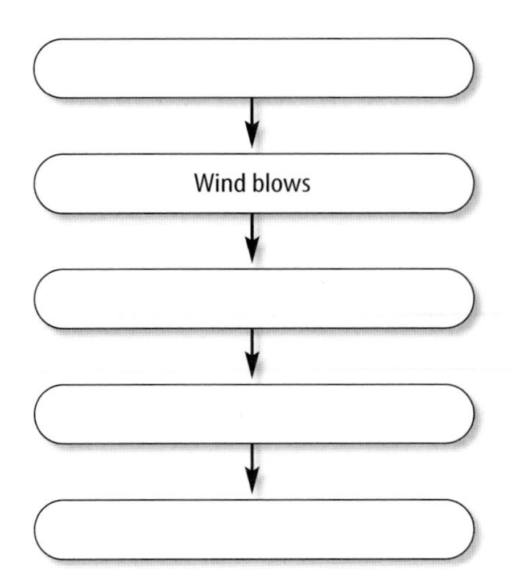

24. **Form a Hypothesis** Hypothesize why silt in loess deposits is transported farther than sand in dune deposits.

25. **Test a Hypothesis** Explain how to test the effect of glacial thickness on a glacier's ability to erode.

26. **Classify** the following as erosional or depositional features: loess, cirque, U-shaped valley, sand dune, abraded rock, striation, and moraine.

Performance Activities

27. **Poster** Make a poster with magazine photos showing glacial features in North America. Add a map to locate each feature.

28. **Design an experiment** to see how the amount of moisture added to sediments affects mass movement. Keep all variables constant except the amount of moisture in the sediment. Try your experiment.

Applying Math

29. **Slope** Gravity is a very powerful erosional force. This means the steeper a slope is, the more soil will move. A person can calculate how steep a slope is by using the height (rise) divided by the length (run). This answer is then multiplied by 100 to get percent slope. If you had a slope 15 m high and 50 m long, what would be the percent slope?

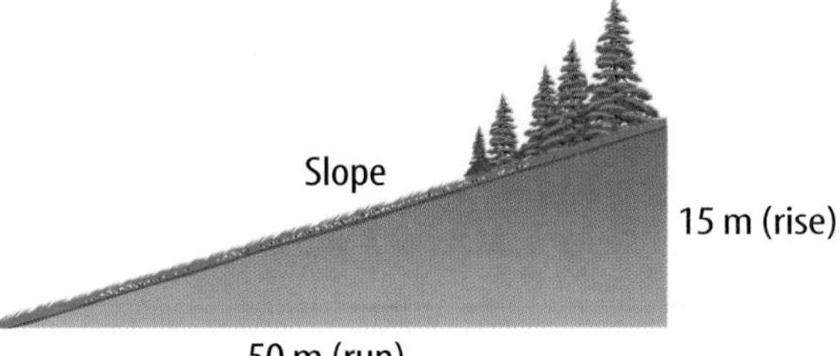

30. **Traveling Sand** A sand dune can travel up to 30 m per year. How far does the sand dune move per day?

Standardized Test Practice

Part 1 Multiple Choice

Record your answers on the answer sheet provided by your teacher or on a sheet of paper.

Use the illustration below to answer question 1.

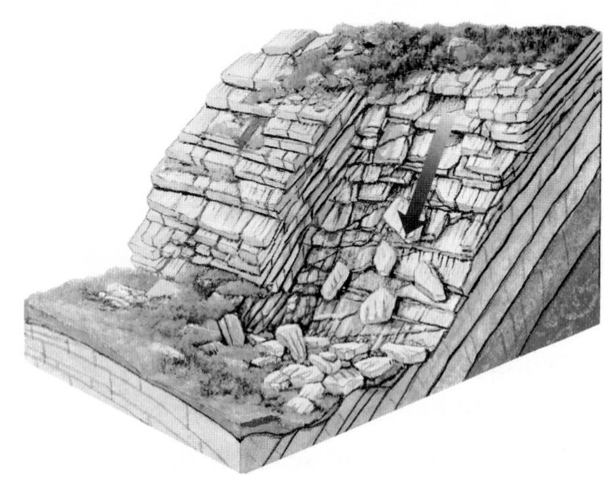

1. Which type of mass movement is shown above?
   A. slump
   B. creep
   C. rock slide
   D. mudflow

2. Which term refers to sediment that is deposited by glacier ice?
   A. outwash
   B. till
   C. loess
   D. esker

3. During which process does wind pick up fine sediment?
   A. abrasion
   B. mass movement
   C. deflation
   D. deposition

4. On which of the following continents do continental glaciers exist today?
   A. Antarctica
   B. Africa
   C. Australia
   D. Europe

Test-Taking Tip

Take Your Time Stay focused during the test and don't rush, even if you notice that other students are finishing the test early.

5. Which forms when a rock in glacier ice slides over Earth's surface?
   A. moraine
   B. esker
   C. horn
   D. groove

6. What causes sediment and rock to move to lower elevations through time?
   A. sunlight
   B. plant roots
   C. gravity
   D. dust storms

7. Which can reduce wind erosion?
   A. windbreaks
   B. dunes
   C. eskers
   D. horns

8. Which consists of fine-grained, wind-blown sediment?
   A. moraine
   B. loess
   C. till
   D. rock fall

Use the diagram below to answer questions 9–11.

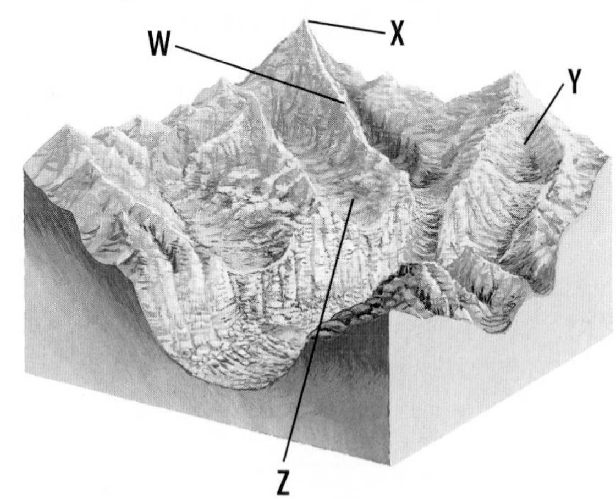

9. Which term describes point X?
   A. horn
   B. arête
   C. cirque
   D. hanging valley

10. Which term describes point Z?
    A. horn
    B. arête
    C. cirque
    D. hanging valley

11. Which agent of erosion created the landscape in the diagram?
    A. wind
    B. water
    C. gravity
    D. ice

Part 2 Short Response/Grid In

Record your answers on the answer sheet provided by your teacher or on a separate sheet of paper.

12. Give three examples of erosion? How does it affect Earth's surface?

13. What is deposition? Give three examples of how it changes Earth's surface.

14. How is a rock fall different from a rock slide? Use a labeled diagram to support your answer.

15. Explain how a glacier can erode the land, and then describe three forms of glacial deposition.

The graph below shows data about how much water flows through a stream. The stream is fed by glacial meltwater. Use the graph to answer questions 16–18.

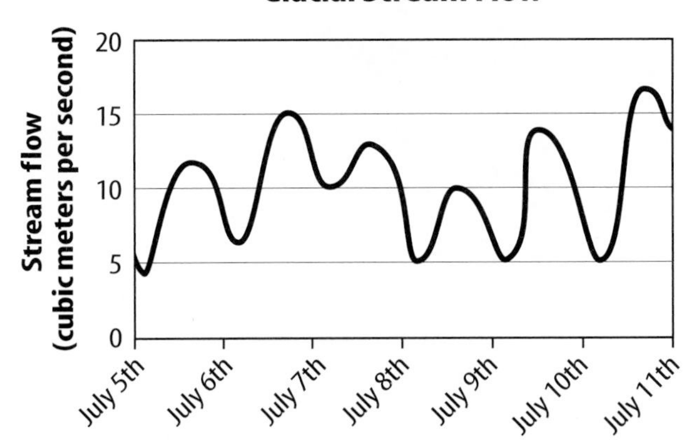

16. What were the lowest and highest amounts of stream flow on July 6th?

17. What were the lowest and highest amounts of stream flow on July 8th?

18. Notice that these data were obtained in July. Explain why the amount of stream flow from a glacier would vary each day. Give three examples to support your reasoning.

Part 3 Open Ended

Record your answers on a sheet of paper.

Use the diagram below to answer questions 19–22.

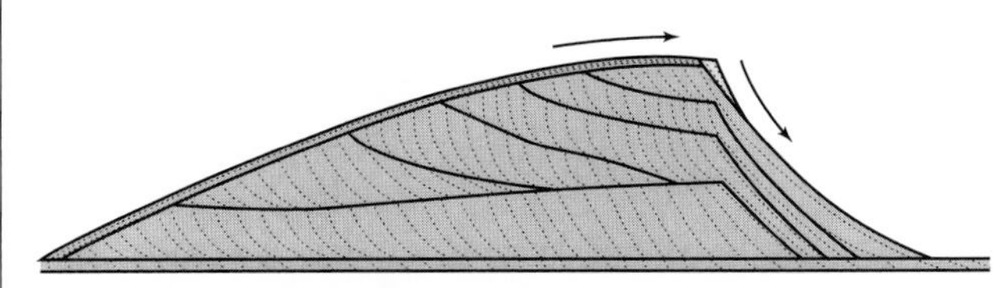

19. Describe the process that causes sand to move up the less steep side of the dune. Use a labeled diagram to support your answer.

20. Why does sand move down the steeper side of the dune? Use a labeled diagram to support your answer.

21. Design three time-lapse illustrations to show how sand dunes move across land.

22. Describe three ways to slow down the movement of sand dunes.

23. What types of damage are caused by landslides? How are people affected by landslides physically and economically?

24. Give three ways that damage from landslides can be reduced.

25. What is a dust storm? Where would you expect dust storms to occur? Give two safety suggestions for people caught in a dust storm.

26. Describe two ways that glacier ice can move across the surface.

27. How is creep different from most other types of mass movement? Explain the forces that cause creep, as well as the effect of creep.

28. Create a chart to show how continental glaciers are different from valley glaciers. Include their causes, physical features, and geological effects on the land.

chapter 9

Water Erosion and Deposition

The BIG Idea

Surface water reshapes Earth through erosion and deposition processes.

SECTION 1
Surface Water

Main Idea As water moves across Earth's surface, it erodes soil and rock from one location and deposits the sediment in another.

SECTION 2
Groundwater

Main Idea Water that soaks into the ground becomes part of a system that can include unique forms of erosion and deposition.

SECTION 3
Ocean Shoreline

Main Idea Waves, currents, and tides reshape shorelines through erosion and deposition processes.

Nature's Sculptor

Bryce Canyon National Park in Utah is home to the Hoodoos—tall, column-like formations. They were made by one of the most powerful forces on Earth—moving water. In this chapter you will learn how moving water shapes Earth's surface.

Science Journal What might have formed the narrowing of each Hoodoo? What will happen if this narrowing continues?

Start-Up Activities

Model How Erosion Works

Moving water has great energy. Sometimes rainwater falls softly and soaks slowly into soil. Other times it rushes down a slope with tremendous force and carries away valuable topsoil. What determines whether rain soaks into the ground or runs off and wears away the surface?

1. Place an aluminum pie pan on your desktop.
2. Put a pile of dry soil about 7 cm high into the pan.
3. Slowly drip water from a dropper onto the pile and observe what happens next.
4. Drip the water faster and continue to observe what happens.
5. Repeat steps 1 through 4, but this time change the slope of the hill by increasing the central pile. Start again with dry soil.
6. **Think Critically** In your Science Journal, write about the effect the water had on the different slopes.

Characteristics of Surface Water, Groundwater, and Shoreline Water Make the following Foldable to help you identify the main concepts relating to surface water, groundwater, and shoreline water.

STEP 1 **Fold** the top of a vertical piece of paper down and the bottom up to divide the paper into thirds.

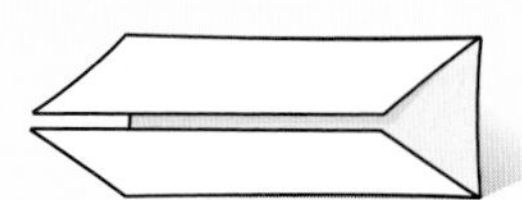

STEP 2 **Turn** the paper horizontally; **unfold and label** the three columns as shown.

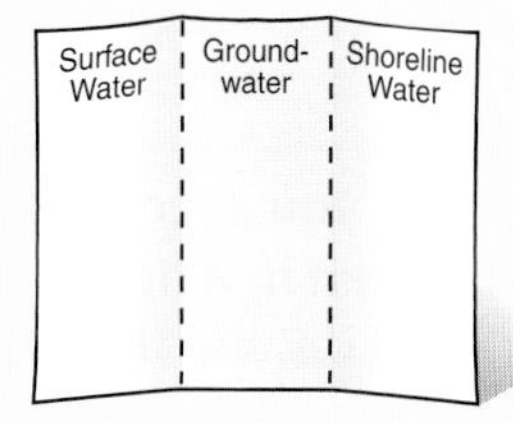

Read for Main Ideas As you read the chapter, list the concepts relating to surface water, groundwater, shoreline water.

Preview this chapter's content and activities at earth.msscience.com

Get Ready to Read

Make Connections

1 Learn It! Make connections between what you read and what you already know. Connections can be based on personal experiences (text-to-self), what you have read before (text-to-text), or events in other places (text-to-world).

As you read, ask connecting questions. Are you reminded of a personal experience? Have you read about the topic before? Did you think of a person, a place, or an event in another part of the world?

2 Practice It! Read the excerpt below and make connections to your own knowledge and experience.

Have you ever seen or been in a flood? What do you think causes floods?

What have you read about floods in other chapters? What types of weather events cause floods?

How can people protect themselves from floods? How do people unintentionally contribute to the damage caused by floods?

Sometimes heavy rains or sudden melting of snow can cause large amounts of water to enter a river system. What happens when a river system has too much water in it? The water needs to go somewhere, and out and over the banks is the only place it can go. A river that overflows its banks can bring disaster by flooding homes or washing away bridges or crops.

—from page 246

3 Apply It! As you read this chapter, choose five words or phrases that make a connection to something you already know.

Reading Tip

Make connections with memorable events, places, or people in your life. The better the connection, the more likely you will remember.

Target Your Reading

Use this to focus on the main ideas as you read the chapter.

1. **Before you read** the chapter, respond to the statements below on your worksheet or on a numbered sheet of paper.
   - Write an **A** if you **agree** with the statement.
   - Write a **D** if you **disagree** with the statement.

2. **After you read** the chapter, look back to this page to see if you've changed your mind about any of the statements.
   - If any of your answers changed, explain why.
   - Change any false statements into true statements.
   - Use your revised statements as a study guide.

Before You Read A or D		Statement	After You Read A or D
	1	Water often causes sheet erosion when it flows in thin, broad sheets.	
	2	Streams are classified as young, mature, or old.	
	3	The stages of development of a stream depend only on the actual age of the stream.	
	4	The largest drainage basin in the United States is the Grand Canyon.	
	5	Water that soaks into the ground and collects in tiny pores in underlying rock is called groundwater.	
	6	The water table below Earth's surface always remains at the same depth.	
	7	The three major forces at work on a shoreline are waves, currents, and tides.	
	8	A longshore current runs perpendicular to the shoreline.	
	9	All sand is made of the mineral quartz.	

Science Online
Print out a worksheet of this page at earth.msscience.com

section 1

Surface Water

as you read

What You'll Learn

- **Identify** the causes of runoff.
- **Compare** rill, gully, sheet, and stream erosion.
- **Identify** three different stages of stream development.
- **Explain** how alluvial fans and deltas form.

Why It's Important

Runoff and streams shape Earth's surface.

Review Vocabulary
erosion: transport of surface materials by agents such as gravity, wind, water, or glaciers

New Vocabulary
- runoff
- channel
- sheet erosion
- drainage basin
- meander

Runoff

Picture this. You pour a glass of milk, and it overflows, spilling onto the table. You grab a towel to clean up the mess, but the milk is already running through a crack in the table, over the edge, and onto the floor. This is similar to what happens to rainwater when it falls to Earth. Some rainwater soaks into the ground and some evaporates, turning into a gas. The rainwater that doesn't soak into the ground or evaporate runs over the ground. Eventually, it enters streams, lakes, or the ocean. Water that doesn't soak into the ground or evaporate but instead flows across Earth's surface is called **runoff.** If you've ever spilled milk while pouring it, you've experienced something similar to runoff.

Factors Affecting Runoff What determines whether rain soaks into the ground or runs off? The amount of rain and the length of time it falls are two factors that affect runoff. Light rain falling over several hours probably will have time to soak into the ground. Heavy rain falling in less than an hour or so will run off because it cannot soak in fast enough, or it can't soak in because the ground cannot hold any more water.

Figure 1 In areas with gentle slopes and vegetation, little runoff and erosion take place. Lack of vegetation has led to severe soil erosion in some areas.

Other Factors Another factor that affects the amount of runoff is the steepness, or slope, of the land. Gravity, the attractive force between all objects, causes water to move down slopes. Water moves rapidly down steep slopes so it has little chance to soak into the ground. Water moves more slowly down gentle slopes and across flat areas. Slower movement allows water more time to soak into the ground.

Vegetation, such as grass and trees, also affects the amount of runoff. Just like milk running off the table, water will run off smooth surfaces that have little or no vegetation. Imagine a tablecloth on the table. What would happen to the milk then? Runoff slows down when it flows around plants. Slower-moving water has a greater chance to sink into the ground. By slowing down runoff, plants and their roots help prevent soil from being carried away. Large amounts of soil may be carried away in areas that lack vegetation, as shown in **Figure 1.**

INTEGRATE Physics

Effects of Gravity When you lie on the ground and feel as if you are being held in place, you are experiencing the effects of gravity. Gravity is the attracting force all objects have for one another. The greater the mass of an object is, the greater its force of gravity is. Because Earth has a much greater mass than any of the objects on it, Earth's gravitational force pulls objects toward its center. Water runs downhill because of Earth's gravitational pull. When water begins to run down a slope, it picks up speed. As its speed increases, so does its energy. Fast-moving water, shown in **Figure 2,** carries more soil than slow-moving water does.

Conservation Farmers sometimes have to farm on some kind of slope. The steeper the slope, the more erosion will occur. Not only is slope an important factor but other factors have to be considered as well. The Natural Resources Conservation Service, a government agency, studies these factors to determine soil loss from a given area. Find out what other factors this agency uses to determine soil loss.

Figure 2 During floods, the high volume of fast-moving water erodes large amounts of soil.

Figure 3 Heavy rains can remove large amounts of sediment, forming deep gullies in the side of a slope.

Water Erosion

Suppose you and several friends walk the same way to school each day through a field or an empty lot. You always walk in the same footsteps as you did the day before. After a few weeks, you've worn a path through the field. When water travels down the same slope time after time, it also wears a path. The movement of soil and rock from one place to another is called erosion.

Rill and Gully Erosion You may have noticed a groove or small ditch on the side of a slope that was left behind by running water. This is evidence of rill erosion. Rill erosion begins when a small stream forms during a heavy rain. As this stream flows along, it has enough energy to erode and carry away soil. Water moving down the same path creates a groove, called a **channel,** on the slope where the water eroded the soil. If water frequently flows in the same channel, rill erosion may change over time into another type of erosion called gully erosion.

During gully erosion, a rill channel becomes broader and deeper. **Figure 3** shows gullies that were formed when water carried away large amounts of soil.

Sheet Erosion Water often erodes without being in a channel. Rainwater that begins to run off during a rainstorm often flows as thin, broad sheets before forming rills and streams. For example, when it rains over an area, the rainwater accumulates until it eventually begins moving down a slope as a sheet, like the water flowing off the hood of the car in **Figure 4.** Water also can flow as sheets if it breaks out of its channel.

Figure 4 When water accumulates, it can flow in sheets like the water seen flowing over the hood of this car.

Floodwaters spilling out of a river can flow as sheets over the surrounding flatlands. Streams flowing out of mountains fan out and may flow as sheets away from the foot of the mountain. **Sheet erosion** occurs when water that is flowing as sheets picks up and carries away sediments.

Stream Erosion Sometimes water continues to flow along a low place it has formed. As the water in a stream moves along, it picks up sediments from the bottom and sides of its channel. By this process, a stream channel becomes deeper and wider.

The sediment that a stream carries is called its load. Water picks up and carries some of the lightweight sediments, called the suspended load. Larger, heavy particles called the bed load just roll along the bottom of the stream channel, as shown in **Figure 5.** Water can even dissolve some rocks and carry them away in solution. The different-sized sediments scrape against the bottom and sides of the channel like a piece of sandpaper. Gradually, these sediments can wear away the rock by a process called abrasion.

Figure 5 This cross section of a stream channel shows the location of the suspended load and the bed load.
Describe *how the stream carries dissolved material.*

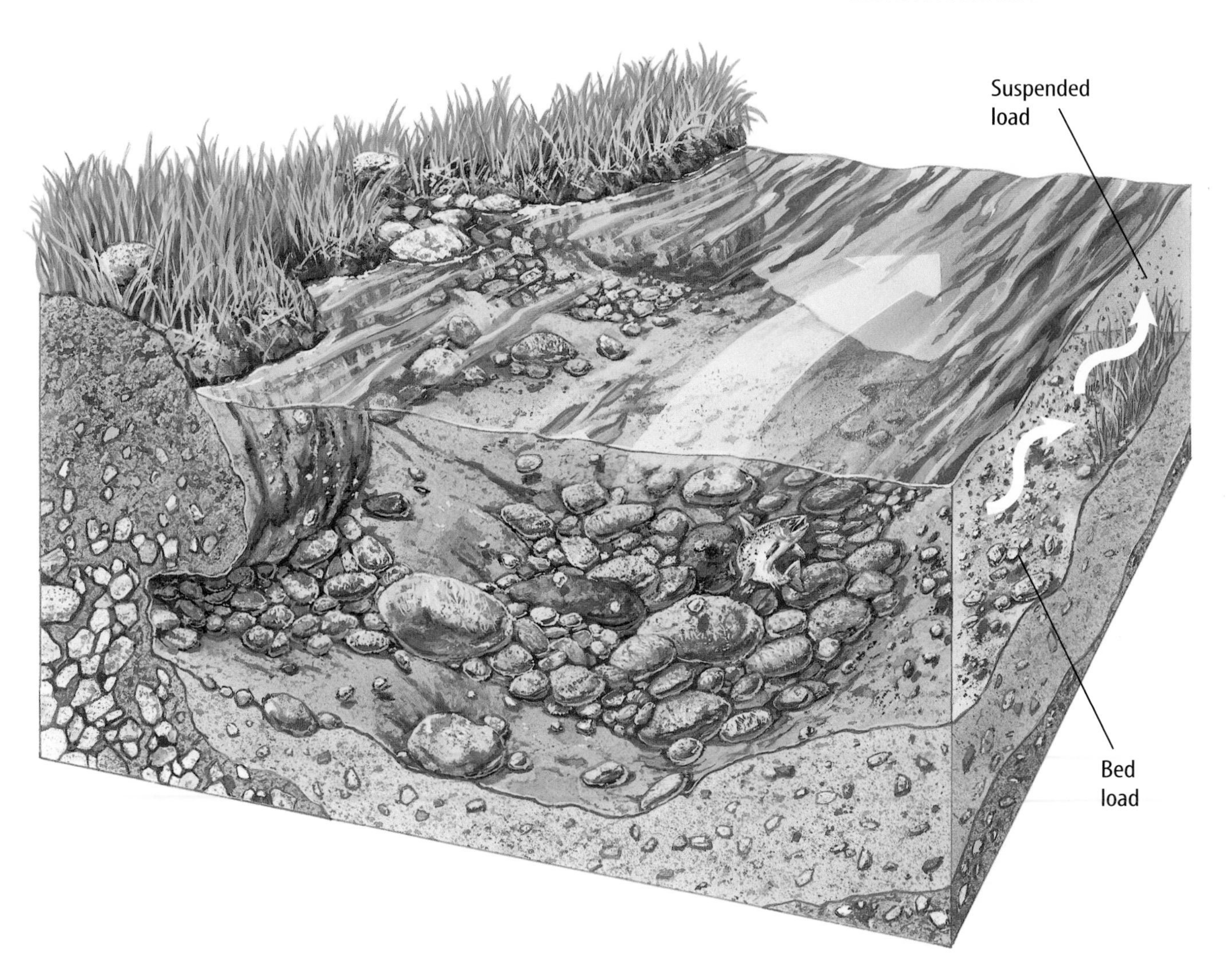

Topic: Drainage Basins

Visit earth.msscience.com for Web links to information about drainage basins in your region.

Activity Locate the drainage basin in which you live. Make a poster showing the shape, or boundary, of the drainage basin. Also, include the main river, or trunk, and the major tributaries, or branches.

River System Development

Have you spent time near a river or stream in your community? Each day, probably millions of liters of water flow through that stream. Where does all the water come from? Where is it flowing to?

River Systems Streams are parts of river systems. The water comes from rills, gullies, and smaller streams located upstream. Just as the tree in **Figure 6** is a system containing twigs, branches, and a trunk, a river system also has many parts. Runoff enters small streams, which join together to form larger streams. Larger streams come together to form rivers. Rivers grow and carry more water as more streams join.

Drainage Basins A **drainage basin** is the area of land from which a stream or river collects runoff. Compare a drainage basin to a bathtub. Water that collects in a bathtub flows toward one location—the drain. Likewise, all of the water in a river system eventually flows to one location—the main river, or trunk. The largest drainage basin in the United States is the Mississippi River drainage basin shown in **Figure 6.**

Reading Check *What is a drainage basin?*

Figure 6 River systems can be compared with the structure of a tree.

The system of twigs, branches, and trunk that make up a tree is similar to the system of streams and rivers that make up a river system.

CANADA
Columbia River Drainage Basin
Mississippi River Drainage Basin
Colorado River Drainage Basin
MEXICO

A large number of the streams and rivers in the United States are part of the Mississippi River drainage basin, or watershed.
State *what river represents the trunk of this system.*

Stages of Stream Development

Streams come in a variety of forms. Some are narrow and swift moving, and others are wide and slow moving. Streams differ because they are in different stages of development. These stages depend on the slope of the ground over which the stream flows. Streams are classified as young, mature, or old. **Figure 8** shows how the stages come together to form a river system.

The names of the stages of development aren't always related to the actual age of a river. The New River in West Virginia is one of the oldest rivers in North America. However, it has a steep valley and flows swiftly. As a result, it is classified as a young stream.

Figure 7 A meander is a broad bend in a river or stream. As time passes, erosion of the outer bank increases the bend.

Young Streams A stream that flows swiftly through a steep valley is a young stream. A young stream may have whitewater rapids and waterfalls. Water flowing through a steep channel with a rough bottom has a high level of energy and erodes the stream bottom faster than its sides.

Mature Streams The next stage in the development of a stream is the mature stage. A mature stream flows more smoothly through its valley. Over time, most of the rocks in the streambed that cause waterfalls and rapids are eroded by running water and the sediments it carries.

Erosion is no longer concentrated on the bottom in a mature stream. A mature stream starts to erode more along its sides, and curves develop. These curves form because the speed of the water changes throughout the width of the channel.

Water in a shallow area of a stream moves slower because it drags along the bottom. In the deeper part of the channel, the water flows faster. If the deep part of the channel is next to one side of the river, water will erode that side and form a slight curve. Over time, the curve grows to become a broad arc called a **meander** (mee AN dur), as shown in **Figure 7.**

The broad, flat valley floor formed by a meandering stream is called a floodplain. When a stream floods, it often will cover part or all of the floodplain.

NATIONAL GEOGRAPHIC VISUALIZING STREAM DEVELOPMENT

Figure 8

Although no two streams are exactly alike, all go through three main stages—young, mature, and old—as they flow from higher to lower ground. A young stream, below, surging over steep terrain, moves rapidly. In a less steep landscape, right, a mature stream flows more smoothly. On nearly level ground, the stream—considered old—winds leisurely through its valley. The various stages of a stream's development are illustrated here.

A A young stream begins at a source—here, a melting mountain glacier. From its source, the stream flows swiftly downhill, cutting a narrow valley.

B A mature stream flows smoothly through its valley. Mature streams often develop broad curves called meanders.

C Old streams flow through broad, flat floodplains. Near its mouth, the stream gradually drops its load of silt. This sediment forms a delta, an area of flat, fertile land extending into the ocean.

Oxbow lake

Topic: Classification of Rivers

Visit earth.msscience.com for Web links to information about major rivers in the United States. Classify two of these streams as young, mature, or old.

Activity Write a paragraph about why you think the Colorado River is a young, mature, or old stream.

Old Streams The last stage in the development of a stream is the old stage. An old stream flows smoothly through a broad, flat floodplain that it has deposited. South of St. Louis, Missouri, the lower Mississippi River is in the old stage.

Major river systems, such as the Mississippi River, usually contain streams in all stages of development. In the upstream portion of a river system, you find whitewater streams moving swiftly down mountains and hills. At the bottom of mountains and hills, you find streams that start to meander and are in the mature stage of development. These streams meet at the trunk of the drainage basin and form a major river.

Reading Check *How do old streams differ from young streams?*

Too Much Water

Sometimes heavy rains or a sudden melting of snow can cause large amounts of water to enter a river system. What happens when a river system has too much water in it? The water needs to go somewhere, and out and over the banks is the only choice. A river that overflows its banks can bring disaster by flooding homes or washing away bridges or crops.

Figure 9 Flooding causes problems for people who live along major rivers. Floodwater broke through a levee during the Mississippi River flooding in 1993.

Dams and levees are built in an attempt to prevent this type of flooding. A dam is built to control the water flow downstream. It may be built of soil, sand, or steel and concrete. Levees are mounds of earth that are built along the sides of a river. Dams and levees are built to prevent rivers from overflowing their banks. Unfortunately, they do not stop the water when flooding is great. This was the case in 1993 when heavy rains caused the Mississippi River to flood parts of nine midwestern states. Flooding resulted in billions of dollars in property damage. **Figure 9** shows some of the damage caused by this flood.

As you have seen, floods can cause great amounts of damage. But at certain times in Earth's past, great floods have completely changed the surface of Earth in a large region. Such floods are called catastrophic floods.

Figure 10 The Channeled Scablands formed when Lake Missoula drained catastrophically.

These channels were formed by the floodwaters.

Catastrophic Floods During Earth's long history, many catastrophic floods have dramatically changed the face of the surrounding area. One catastrophic flood formed the Channeled Scablands in eastern Washington State, shown here in **Figure 10.** A vast lake named Lake Missoula covered much of western Montana. A natural dam of ice formed this lake. As the dam melted or was eroded away, tremendous amounts of water suddenly escaped through what is now the state of Idaho into Washington. In a short period of time, the floodwater removed overlying soil and carved channels into the underlying rock, some as deep as 50 m. Flooding occurred several more times as the lake refilled with water and the dam broke loose again. Scientists say the last such flood occurred about 13,000 years ago.

Deposition by Surface Water

You know how hard it is to carry a heavy object for a long time without putting it down. As water moves throughout a river system, it loses some of its energy of motion. The water can no longer carry some of its sediment. As a result, it drops, or is deposited, to the bottom of the stream.

Some stream sediment is carried only a short distance. In fact, sediment often is deposited within the stream channel itself. Other stream sediment is carried great distances before being deposited. Sediment picked up when rill and gully erosion occur is an example of this. Water usually has a lot of energy as it moves down a steep slope. When water begins flowing on a level surface, it slows, loses energy, and deposits its sediment. Water also loses energy and deposits sediment when it empties into an ocean or lake.

Observing Runoff Collection

Procedure

1. Put a plastic **rain gauge** into a narrow **drinking glass** and place the glass in the **sink.**
2. Fill a plastic **sprinkling can** with **water.**
3. Hold the sprinkling can one-half meter above the sink for 30 s.
4. Record the amount of water in the rain gauge.
5. After emptying the rain gauge, place a **plastic funnel** into the rain gauge and sprinkle again for 30 s.
6. Record the amount of water in the gauge.

Analysis

Explain how a small amount of rain falling on a drainage basin can have a big effect on a river or stream.

Agriculture
is important
on the Nile Delta.

Figure 11 This satellite image of the Nile River Delta in Egypt shows the typical triangular shape. The green color shows areas of vegetation.

Deltas and Fans Sediment that is deposited as water empties into an ocean or lake forms a triangular, or fan-shaped, deposit called a delta, shown in **Figure 11.** When the river waters empty from a mountain valley onto an open plain, the deposit is called an alluvial (uh LEW vee ul) fan. The Mississippi River exemplifies the topics presented in this section. Runoff causes rill and gully erosion. Sediment is picked up and carried into the larger streams that flow into the Mississippi River. As the Mississippi River flows, it cuts into its banks and picks up more sediment. Where the land is flat, the river deposits some of its sediment in its own channel. As the Mississippi enters the Gulf of Mexico, it slows, dropping much of its sediment and forming the Mississippi River delta.

section 1 review

Summary

Runoff

- Rainwater that doesn't soak into the ground or evaporate becomes runoff.
- Slope of land and vegetation affect runoff.

Water Erosion

- Water flowing over the same slope causes rills and gullies to form.

River System Development

- A drainage basin is an area of land from which a stream or river collects runoff.

Self Check

1. **Explain** how the slope of an area affects runoff.
2. **Compare and contrast** rill and gully erosion.
3. **Describe** the three stages of stream development.
4. **Think Critically** How is a stream's rate of flow related to the amount of erosion it causes? How is it related to the size of the sediments it deposits?

Applying Skills

5. **Compare and contrast** the formation of deltas and alluvial fans.

Science Online earth.msscience.com/self_check_quiz

section 2

Groundwater

Groundwater Systems

What would have happened if the spilled milk in Section 1 ran off the table onto a carpeted floor? It probably would have quickly soaked into the carpet. Water that falls on Earth can soak into the ground just like the milk into the carpet.

Water that soaks into the ground becomes part of a system, just as water that stays above ground becomes part of a river system. Soil is made up of many small rock and mineral fragments. These fragments are all touching one another, as shown in **Figure 12,** but some empty space remains between them. Holes, cracks, and crevices exist in the rock underlying the soil. Water that soaks into the ground collects in these pores and empty spaces and becomes part of what is called **groundwater.**

How much of Earth's water do you think is held in the small openings in rock? Scientists estimate that 14 percent of all freshwater on Earth exists as groundwater. This is almost 30 times more water than is contained in all of Earth's lakes and rivers.

as you read

What You'll Learn

- **Recognize** the importance of groundwater.
- **Describe** the effect that soil and rock permeability have on groundwater movement.
- **Explain** how groundwater dissolves and deposits minerals.

Why It's Important

The groundwater system is an important source of your drinking water.

Review Vocabulary
pore: a small, or minute, opening in rock or soil

New Vocabulary
- groundwater
- permeable
- impermeable
- aquifer
- water table
- spring
- geyser
- cave

Figure 12 Soil has many small, connected pores that are filled with water when soil is wet.

Measuring Pore Space

Procedure

1. Use two identical, **clear-plastic containers.**
2. Put 3 cm of **sand** in one container and 3 cm of **gravel** in the other.
3. Pour **water** slowly into the containers and stop when the water just covers the top of the sediment.
4. Record the volume of water used in each.

Analysis

Which substance has more pore space—sand or gravel?

Try at Home

Permeability A groundwater system is similar to a river system. However, instead of having channels that connect different parts of the drainage basin, the groundwater system has connecting pores. Soil and rock are **permeable** (PUR mee uh bul) if the pore spaces are connected and water can pass through them. Sandstone is an example of a permeable rock.

Soil or rock that has many large, connected pores is permeable. Water can pass through it easily. However, if a rock or sediment has few pore spaces or they are not well connected, then the flow of groundwater is blocked. These materials are **impermeable,** which means that water cannot pass through them. Granite has few or no pore spaces at all. Clay has many small pore spaces, but the spaces are not well connected.

Reading Check *How does water move through permeable rock?*

Groundwater Movement How deep into Earth's crust does groundwater go? **Figure 13** shows a model of a groundwater system. Groundwater keeps going deeper until it reaches a layer of impermeable rock. When this happens, the water stops moving down. As a result, water begins filling up the pores in the rocks above. A layer of permeable rock that lets water move freely is an **aquifer** (AK wuh fur). The area where all of the pores in the rock are filled with water is the zone of saturation. The upper surface of this zone is the **water table.**

Figure 13 A stream's surface level is the water table. Below that is the zone of saturation.

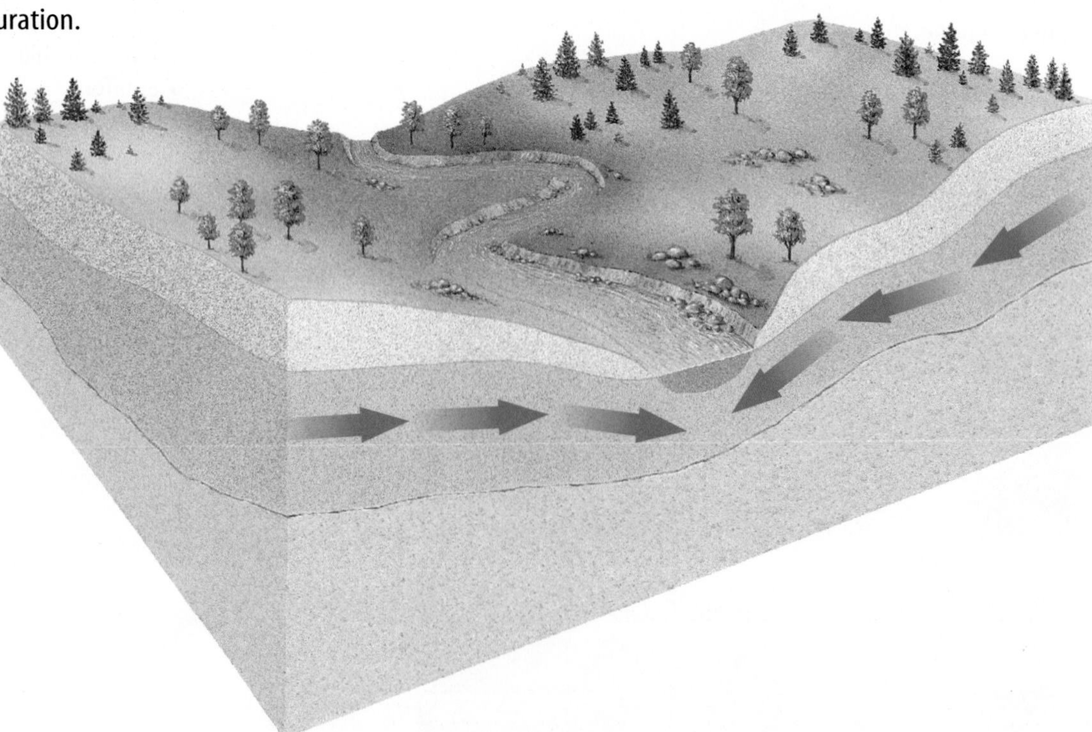

Water Table

Why are the zone of saturation and the water table so important? An average United States resident uses about 626 L of water per day. That's enough to fill nearly two thousand soft drink cans. Many people get their water from groundwater through wells that have been drilled into the zone of saturation. However, the supply of groundwater is limited. During a drought, the water table drops. This is why you should conserve water.

Applying Math — Calculate Rate of Flow

GROUNDWATER FLOW You and your family are hiking and the temperature is hot. You feel as if you can't walk one step farther. Luckily, relief is in sight. On the side of a nearby hill you see a stream, and you rush to splash some water on your face. Although you probably feel that it's taking you forever to reach the stream, your pace is quick when compared to how long it takes groundwater to flow through the aquifer that feeds the stream. The following problem will give you some idea of just how slowly groundwater flows through an aquifer.

The groundwater flows at a rate of 0.6 m/day. You've run 200 m to get some water from a stream. How long does it take the groundwater in the aquifer to travel the same distance?

Solution

1 *This is what you know:*

- the distance that the groundwater has to travel: $d = 200$ m
- the rate that groundwater flows through the aquifer: $r = 0.6$ m/day

2 *This is what you want to find:* time = t

3 *This is the equation you use:* $r \times t = d$ (rate × time = distance)

4 *Solve the equation for* t *and then substitute known values:*

$$t = \frac{d}{r} = \frac{(200 \text{ m})}{(0.6 \text{ m/day})} = 333.33 \text{ days}$$

Practice Problems

1. The groundwater in an aquifer flows at a rate of 0.5 m/day. How far does the groundwater move in a year?
2. How long does it take groundwater in the above aquifer to move 100 m?

For more practice, visit earth.msscience.com/math_practice

Figure 14 The years on the pole show how much the ground level dropped in the San Joaquin Valley, California, between 1925 and 1977.

Figure 15 The pressure of water in a sloping aquifer keeps an artesian well flowing.
Describe *what limits how high water can flow in an artesian well.*

Wells A good well extends deep into the zone of saturation, past the top of the water table. Groundwater flows into the well, and a pump brings it to the surface. Because the water table sometimes drops during very dry seasons, even a good well can go dry. Then time is needed for the water table to rise, either from rainfall or through groundwater flowing from other areas of the aquifer.

Where groundwater is the main source of drinking water, the number of wells and how much water is pumped out are important. If a large factory were built in such a town, the demand on the groundwater supply would be even greater. Even in times of normal rainfall, the wells could go dry if water were taken out at a rate greater than the rate at which it can be replaced.

In areas where too much water is pumped out, the land level can sink from the weight of the sediments above the now-empty pore spaces. **Figure 14** shows what occurred when too much groundwater was removed in a region of California.

One type of well doesn't need a pump to bring water to the surface. An artesian well is a well in which water rises to the surface under pressure. Artesian wells are less common than other types of wells because of the special conditions they require.

As shown in **Figure 15,** the aquifer for an artesian well needs to be located between two impermeable layers that are sloping. Water enters at the high part of the sloping aquifer. The weight of the water in the higher part of the aquifer puts pressure on the water in the lower part. If a well is drilled into the lower part of the aquifer, the pressurized water will flow to the surface. Sometimes, the pressure is great enough to force the water into the air, forming a fountain.

Reading Check *How does water move through permeable rock?*

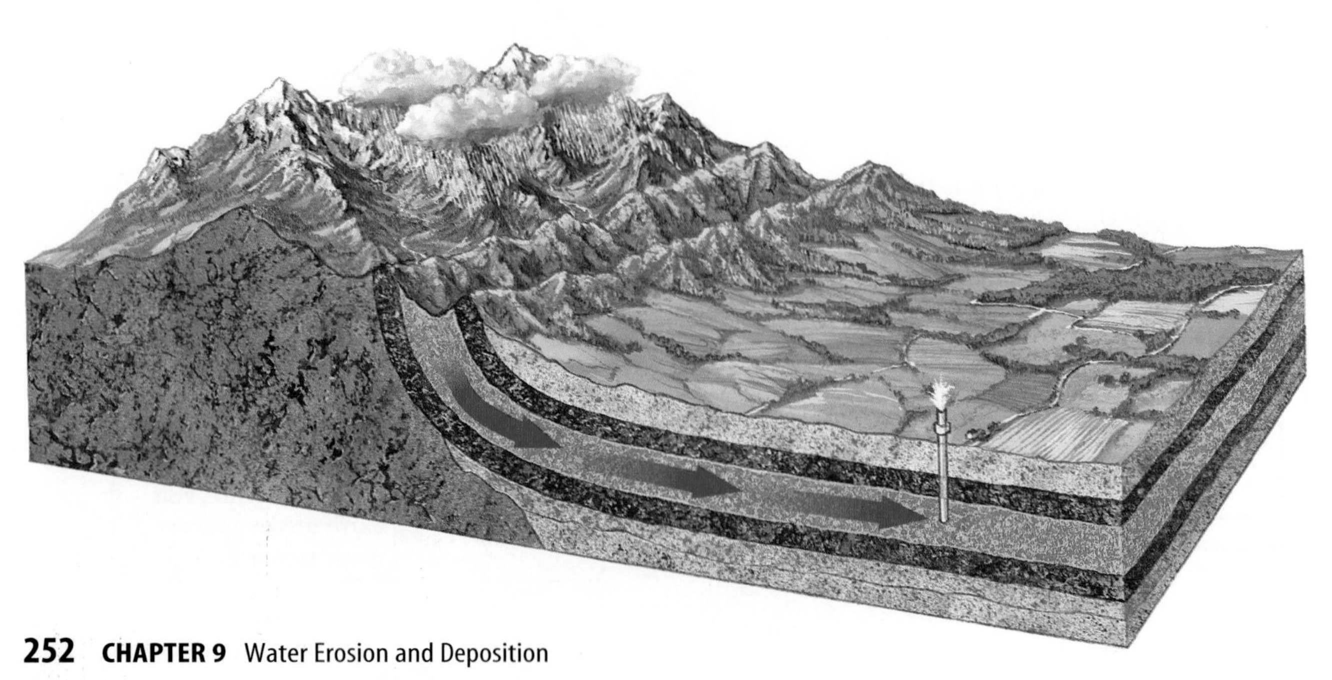

Springs In some places, the water table is so close to Earth's surface that water flows out and forms a **spring.** Springs are found on hillsides or other places where the water table meets a sloping surface. Springs often are used as a source of freshwater.

The water from most springs is a constant, cool temperature because soil and rock are good insulators and protect the groundwater from changes in temperature on Earth's surface. However, in some places, magma rises to within a few kilometers of Earth's surface and heats the surrounding rock. Groundwater that comes in contact with these hot rocks is heated and can come to the surface as a hot spring.

Acid Rain Effects Acid rain occurs when gases released by burning oil and coal mix with water in the air. Infer what effect acid rain can have on a statue made of limestone.

Geysers When water is put into a teakettle to boil, it heats slowly at first. Then some steam starts to come out of the cap on the spout, and suddenly the water starts boiling. The teakettle starts whistling as steam is forced through the cap. A similar process can occur with groundwater. One of the places where groundwater is heated is in Yellowstone National Park in Wyoming. Yellowstone has hot springs and geysers. A **geyser** is a hot spring that erupts periodically, shooting water and steam into the air. Groundwater is heated to high temperatures, causing it to expand underground. This expansion forces some of the water out of the ground, taking the pressure off the remaining water. The remaining water boils quickly, with much of it turning to steam. The steam shoots out of the opening like steam out of a teakettle, forcing the remaining water out with it. Yellowstone's famous geyser, Old Faithful, pictured in **Figure 16,** shoots between 14,000 and 32,000 L of water and steam into the air about once every 80 min.

Figure 16 Yellowstone's famous geyser, Old Faithful, used to erupt once about every 76 min. An earthquake on January 9, 1998, slowed Old Faithful's "clock" by 4 min to an average of one eruption about every 80 min. The average height of the geyser's water is 40.5 m.

The Work of Groundwater

Although water is the most powerful agent of erosion on Earth's surface, it also can have a great effect underground. Water mixes with carbon dioxide gas to form a weak acid called carbonic acid. Some of this carbon dioxide is absorbed from the air by rainwater or surface water. Most carbon dioxide is absorbed by groundwater moving through soil. One type of rock that is dissolved easily by this acid is limestone. Acidic groundwater moves through natural cracks and pores in limestone, dissolving the rock. Gradually, the cracks in the limestone enlarge until an underground opening called a **cave** is formed.

Cave Formation You've probably seen a picture of the inside of a cave, like the one shown in **Figure 17,** or perhaps you've visited one. Groundwater not only dissolves limestone to make caves, but it also can make deposits on the insides of caves.

Water often drips slowly from cracks in the cave walls and ceilings. This water contains calcium ions dissolved from the limestone. If the water evaporates while hanging from the ceiling of a cave, a deposit of calcium carbonate is left behind. Stalactites form when this happens over and over. Where drops of water fall to the floor of the cave, a stalagmite forms. The words *stalactite* and *stalagmite* come from Greek words that mean "to drip."

Sinkholes If underground rock is dissolved near the surface, a sinkhole may form. A sinkhole is a depression on the surface of the ground that forms when the roof of a cave collapses or when material near the surface dissolves. Sinkholes are common features in places like Florida and Kentucky that have lots of limestone and enough rainfall to keep the groundwater system supplied with water. Sinkholes can cause property damage if they form in a populated area.

In summary, when rain falls and becomes groundwater, it might dissolve limestone and form a cave, erupt from a geyser, or be pumped from a well to be used at your home.

Figure 17 Water dissolves rock to form caves and also deposits material to form spectacular formations, such as these in Carlsbad Caverns in New Mexico.

section 2 review

Summary

Groundwater Systems

- Water that soaks into the ground and collects in pore spaces is called groundwater.
- 14 percent of all freshwater on Earth exists as groundwater.
- Groundwater systems have connecting pores.
- The zone of saturation is the area where all pores in the rock are filled with water.

Water Table

- The supply of groundwater is limited.

Self Check

1. **Describe** how the permeability of soil and rocks affects the flow of groundwater.
2. **Describe** why a well might go dry.
3. **Explain** how caves form.
4. **Think Critically** Why would water in wells, geysers, and hot springs contain dissolved materials?

Applying Skills

5. **Compare and contrast** wells, geysers, and hot springs.

Science Online earth.msscience.com/self_check_quiz

section 3

Ocean Shoreline

The Shore

Picture yourself sitting on a beautiful, sandy beach like the one shown in **Figure 18.** Nearby, palm trees sway in the breeze. Children play in the quiet waves lapping at the water's edge. It's hard to imagine a place more peaceful. Now, picture yourself sitting along another shore. You're on a high cliff watching waves crash onto boulders far below. Both of these places are shorelines. An ocean shoreline is where land meets the ocean.

The two shorelines just described are different even though both experience surface waves, tides, and currents. These actions cause shorelines to change constantly. Sometimes you can see these changes from hour to hour. Why are shorelines so different? You'll understand why they look different when you learn about the forces that shape shorelines.

as you read

What You'll Learn

- **Identify** the different causes of shoreline erosion.
- **Compare and contrast** different types of shorelines.
- **Describe** some origins of sand.

Why It's Important

Constantly changing shorelines impact the people who live and work by them.

Review Vocabulary

tides: the alternating rise and fall of sea level caused by the gravitational attraction of the Moon and the Sun

New Vocabulary

- longshore current
- beach

Figure 18 Waves, tides, and currents cause shorelines to change constantly. Waves approaching the shoreline at an angle create a longshore current.
Describe *the effects longshore currents have on a shoreline.*

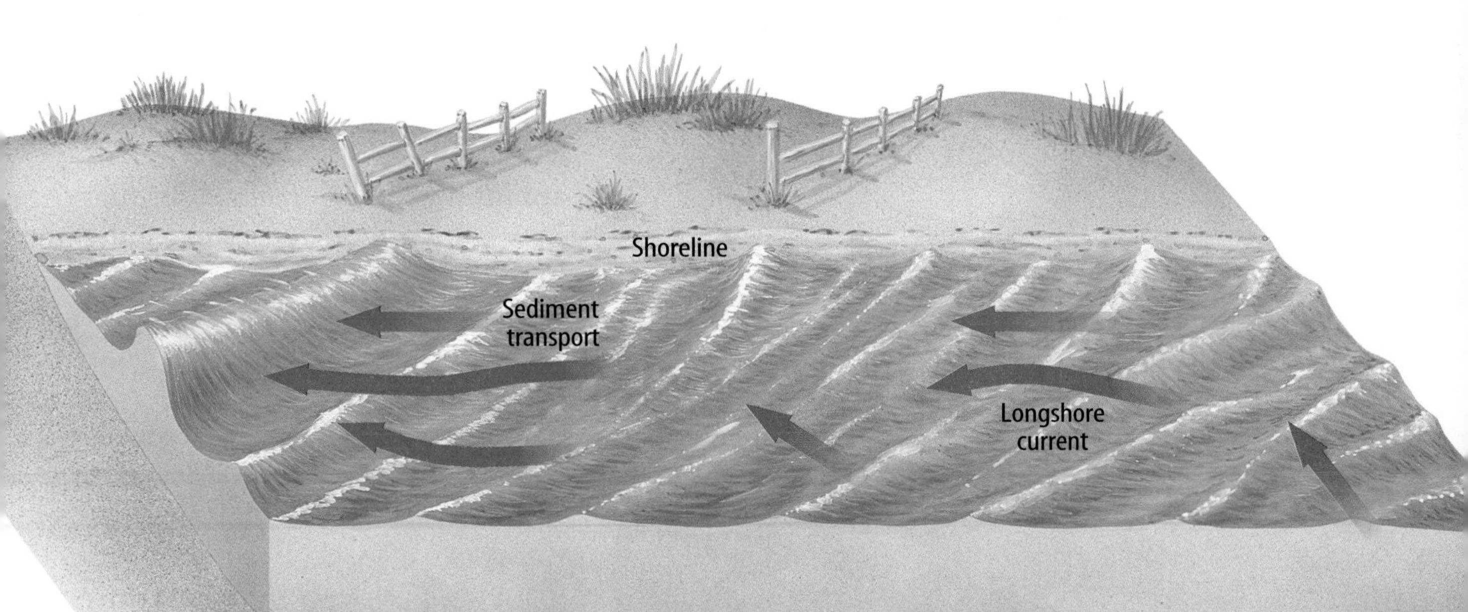

Shoreline Forces When waves constantly pound against the shore, they break rocks into ever-smaller pieces. Currents move many metric tons of sediment along the shoreline. The sediment grains grind against each other like sandpaper. The tide goes out carrying sediment to deeper water. When the tide returns, it brings new sediment with it. These forces are always at work, slowly changing the shape of the shoreline. Water is always in motion along the shore.

The three major forces at work on the shoreline are waves, currents, and tides. Winds blowing across the water make waves. Waves, crashing against a shoreline, are a powerful force. They can erode and move large amounts of material in a short time. Waves usually collide with a shore at slight angles. This creates a **longshore current** of water that runs parallel to the shoreline. Longshore currents, shown in **Figure 18,** carry many metric tons of loose sediments and act like rivers of sand in the ocean.

Reading Check *How does a longshore current form?*

Tides create currents that move at right angles to the shore. These are called tidal currents. Outgoing tides carry sediments away from the shore, and incoming tides bring new sediments toward the shore. Tides work with waves to shape shorelines. You've seen the forces that affect all shorelines. Now you will see the differences that make one shore a flat, sandy beach and another shore a steep, rocky cliff.

Figure 19 Along a rocky shoreline, the force of pounding waves breaks rock fragments loose, then grinds them into smaller and smaller pieces.

Rocky Shorelines

Rocks and cliffs are the most common features along rocky shorelines like the one in **Figure 19.** Waves crash against the rocks and cliffs. Sediments in the water grind against the cliffs, slowly wearing the rock away. Then rock fragments broken from the cliffs are ground up by the endless motion of waves. They are transported as sediment by longshore currents.

Softer rocks become eroded before harder rocks do, leaving islands of harder rocks. This takes thousands of years, but remember that the ocean never stops. In a single day, about 14,000 waves crash onto shore.

This quartz sand from a Texas beach is clear and glassy.

Some Hawaiian beaches are composed of black basalt sand.

Figure 20 Beach sand varies in size, color, and composition.

Sandy Beaches

Smooth, gently sloping shorelines are different from steep, rocky shorelines. Beaches are the main feature here. **Beaches** are deposits of sediment that are parallel to the shore.

Beaches are made up of different materials. Some are made of rock fragments from the shoreline. Many beaches are made of grains of quartz, and others are made of seashell fragments. These fragments range in size from stones larger than your hand to fine sand. Sand grains range from 0.06 mm to 2 mm in diameter. Why do many beaches have particles of this size? Waves break rocks and seashells down to sand-sized particles like those shown in **Figure 20.** The constant wave motion bumps sand grains together. This bumping not only breaks particles into smaller pieces but also smooths off their jagged corners, making them more rounded.

Reading Check *How do waves affect beach particles?*

Sand in some places is made of other things. For example, Hawaii's black sands are made of basalt, and its green sands are made of the mineral olivine. Jamaica's white sands are made of coral and shell fragments.

Figure 21 Shorelines change constantly. Human development is often at risk from shoreline erosion.

Sand Erosion and Deposition

Longshore currents carry sand along beaches to form features such as barrier islands, spits, and sandbars. Storms and wind also move sand. Thus, beaches are fragile, short-term land features that are damaged easily by storms and human activities such as some types of construction. Communities in widely separated places such as Long Island, New York; Malibu, California; and Padre Island, Texas, have problems because of beach erosion.

Barrier Islands Barrier islands are sand deposits that lay parallel to the shore but are separated from the mainland. These islands start as underwater sand ridges formed by breaking waves. Hurricanes and storms add sediment to them, raising some to sea level. When a barrier island becomes large enough, the wind blows the loose sand into dunes, keeping the new island above sea level. As with all seashore features, barrier islands are short term, lasting from a few years to a few centuries.

The forces that build barrier islands also can erode them. Storms and waves carry sediments away. Beachfront development, as in **Figure 21,** can be affected by shoreline erosion.

section 3 review

Summary

The Shore

- An ocean shoreline is where land meets the ocean.
- The forces that shape shorelines are waves, currents, and tides.

Rocky Shorelines

- Rocks and cliffs are the most common features along rocky shorelines.

Sandy Beaches

- Beaches are made up of different materials. Some are made of rock fragments and others are made of seashell fragments.

Sand Erosion and Deposition

- Longshore currents, storms, and wind move sand.
- Beaches are fragile, short-term land features.

Self Check

1. **Identify** major forces that cause shoreline erosion.
2. **Compare and contrast** the features you would find along a steep, rocky shoreline with the features you would find along a gently sloping, sandy shoreline.
3. **Explain** how the type of shoreline could affect the types of sediments you might find there.
4. **List** several materials that beach sand might be composed of. Where do these materials come from?
5. **Think Critically** How would erosion and deposition of sediment along a shoreline be affected if the longshore current was blocked by a wall built out into the water?

Applying Math

6. **Solve One-Step Equations** If 14,000 waves crash onto a shore daily, how many waves crash onto it in a year? How many crashed onto it since you were born?

Classifying Types of Sand

Sand is made of different kinds of grains, but did you realize that the slope of a beach is related to the size of its grains? The coarser the grain size is, the steeper the beach is. The composition of sand also is important. Many sands are mined because they have economic value.

Real-World Question

What characteristics can be used to classify different types of beach sand?

Goals

- **Observe** differences in sand.
- **Identify** characteristics of beach sand.
- **Infer** sediment sources.

Materials

samples of different sands (3)
magnifying lens
**stereomicroscope*
magnet
**Alternate materials*

Safety Precautions 

Procedure

1. **Design** a five-column data table to compare the three sand samples. Use column one for the samples and the others for the characteristics you will be examining.

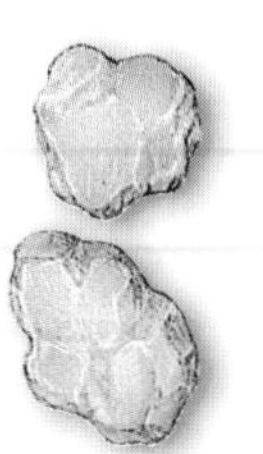
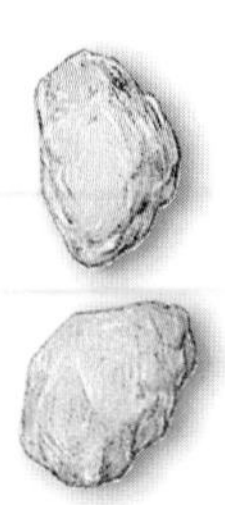

Angular | Sub-angular | Sub-rounded | Rounded

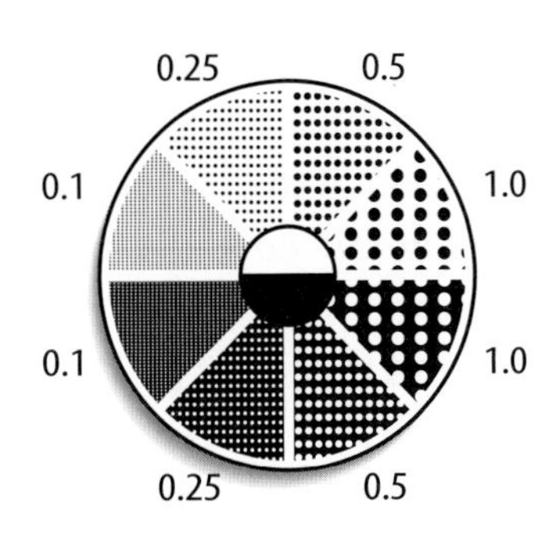

Sand gauge
(measurements in mm)

2. **Use the diagram** in the previous column to determine the average roundness of each sample.
3. **Identify** the grain size of your samples by using the sand gauge above. To determine the grain size, place sand grains in the middle of the circle of the sand gauge. Use the upper half of the circle for dark-colored particles and the bottom half for light-colored particles.
4. **Decide** on two other characteristics to examine that will help you classify your samples.

Conclude and Apply

1. **Compare and contrast** some characteristics of beach sand.
2. **Describe** why there are variations in the characteristics of different sand samples.
3. **Explain** what your observations tell you about the sources of the three samples.

Communicating Your Data

Compare your results with those of other students.

Water Speed and Erosion

Goals

- **Assemble** an apparatus for measuring the effect of water speed on erosion.
- **Observe and measure** the ability of water traveling at different speeds to erode sand.

Materials

paint roller pan
**disposable wallpaper trays*
sand
1-L beaker
rubber tubing (20 cm)
metric ruler
water
stopwatch
fine-mesh screen
wood block
**Alternate materials*

Safety Precautions

Wash your hands after you handle the sand. Immediately clean up any water that spills on the floor.

Real-World Question

What would it be like to make a raft and use it to float on a river? Would it be easy? Would you feel like Tom Sawyer? Probably not. You'd be at the mercy of the current. Strong currents create fast rivers. But does fast moving water affect more than just floating rafts and other objects? How does the speed of a stream or river affect its ability to erode?

Procedure

1. Copy the data table on the following page.
2. Place the screen in the sink. Pour moist sand into your pan and smooth out the sand. Set one end of the pan on the wood block and hang the other end over the screen in the sink. Excess water will flow onto the screen in the sink.
3. Attach one end of the hose to the faucet and place the other end in the beaker. Turn on the water so that it trickles into the beaker. Time how long it takes for the trickle of water to fill the beaker to the 1-L mark. Divide 1 L by your time in seconds to calculate the water speed. Record the speed in your data table.
4. Without altering the water speed, hold the hose over the end of the pan that is resting on the wood block. Allow the water to flow into the sand for 2 min. At the end of 2 min, turn off the water.
5. **Measure** the depth and length of the eroded channel formed by the water. Count the number of branches formed on the channel. Record your measurements and observations in your data table.

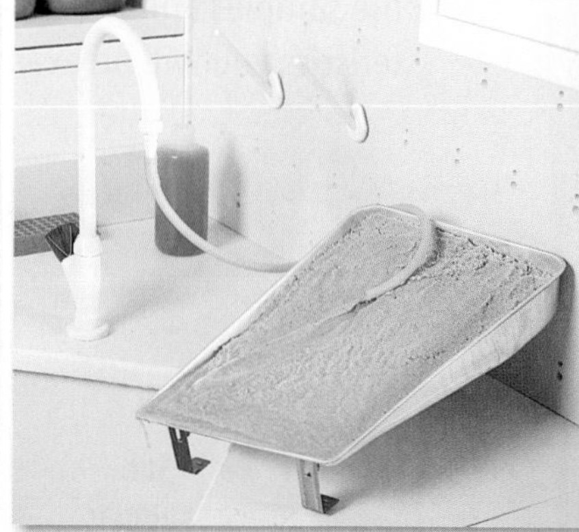

Water Speed and Erosion

Water Speed (Liters per Second)	Depth of Channel	Length of Channel	Number of Channel Branches
	Do not write in this book.		

6. Empty the excess water from the tray and smooth out the sand. Repeat steps 3 through 5 two more times, increasing your water speed each time.

Conclude and Apply

1. **Identify** the constants and variables in your experiment.
2. **Observe** Which water speed created the deepest and longest channel?
3. **Observe** Which water speed created the greatest number of branches?
4. **Infer** the effect that water speed has on erosion.
5. **Predict** how your results would have differed if one end of the pan had been raised higher.
6. **Infer** how streams and rivers can shape Earth's surface.

Communicating Your Data

Write a pamphlet for people buying homes near rivers or streams that outlines the different effects that water erosion could have on their property.

Is there hope for America's coastlines or is beach erosion a "shore" thing?

Sands in Time

Water levels are rising along the coastline of the United States. Serious storms and the building of homes and businesses along the shore are leading to the erosion of anywhere from 70 percent to 90 percent of the U.S. coastline. A report from the Federal Emergency Management Agency (FEMA) confirms this. The report says that one meter of United States beaches will be eaten away each year for the next 60 years. Since 1965, the federal government has spent millions of dollars replenishing more than 1,300 eroding sandy shores around the country. And still beaches continue to disappear.

The slowly eroding beaches are upsetting to residents and officials of many communities, who depend on their shore to earn money from visitors. Some city and state governments are turning to beach nourishment—a process in which sand is taken from the seafloor and dumped on beaches. The process is expensive, however. The state of Delaware, for example, is spending 7,000,000 dollars to bring in sand for its beaches.

This beach house will collapse as its underpinnings are eroded.

Other methods of saving eroding beaches are being tried. In places along the Great Lakes shores and coastal shores, one company has installed fabrics underwater to slow currents. By slowing currents, sand is naturally deposited and kept in place.

Another shore-saving device is a synthetic barrier that is shaped like a plastic snowflake. A string of these barriers is secured just offshore. They absorb the energy of incoming waves. Reducing wave energy can prevent sand from being eroded from the beach. New sand also might accumulate because the barriers slow down the currents that flow along the shore.

Many people believe that communities along the shore must restrict the beachfront building of homes, hotels, and stores. Since some estimates claim that by the year 2025, nearly 75 percent of the U.S. population will live in coastal areas, it's a tough solution. Says one geologist, "We can retreat now and save our beaches or we can retreat later and probably ruin the beaches in the process."

Debate Using the facts in this article and other research you have done in your school media center or through Web links at msscience.com, make a list of methods that could be used to save beaches. Debate the issue with your classmates.

For more information, visit earth.msscience.com/time

Reviewing Main Ideas

Section 1 Surface Water

1. Rainwater that does not soak into the ground is pulled down the slope by gravity. This water is called runoff.
2. Runoff can erode sediment. Factors such as steepness of slope and number and type of plants affect the amount of erosion. Rill, gully, and sheet erosion are types of surface water erosion caused by runoff.
3. Runoff generally flows into streams that merge with larger rivers until emptying into a lake or ocean. Major river systems usually contain several different types of streams.
4. Young streams flow through steep valleys and have rapids and waterfalls. Mature streams flow through gentler terrain and have less energy. Old streams are often wide and meander across their floodplains.

Section 2 Groundwater

1. When water soaks into the ground, it becomes part of a vast groundwater system.
2. Although rock may seem solid, many types are filled with connected spaces called pores. Such rocks are permeable and can contain large amounts of groundwater.

Section 3 Ocean Shoreline

1. Ocean shorelines are always changing.
2. Waves and currents have tremendous amounts of energy which break up rocks into tiny fragments called sediment. Over time, the deposition and relocation of sediment can change beaches, sandbars, and barrier islands.

Visualizing Main Ideas

Copy and complete the following concept map on caves.

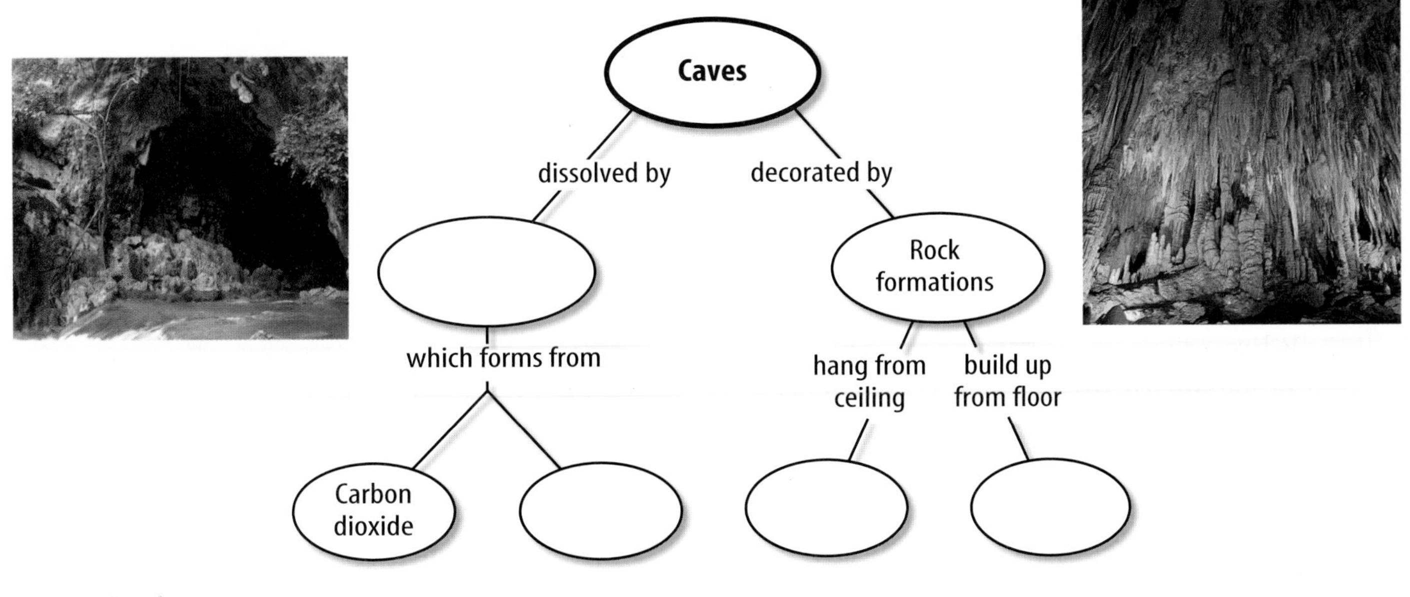

chapter 9 Review

Using Vocabulary

aquifer p. 250
beach p. 257
cave p. 253
channel p. 240
drainage basin p. 242
geyser p. 253
groundwater p. 240
impermeable p. 250
longshore current p. 256
meander p. 243
permeable p. 250
runoff p. 238
sheet erosion p. 241
spring p. 253
water table p. 250

Explain the difference between the vocabulary words in each of the following sets.

1. runoff—sheet erosion
2. channel—drainage basin
3. aquifer—cave
4. spring—geyser
5. permeable—impermeable
6. sheet erosion—meander
7. groundwater—water table
8. permeable—aquifer
9. longshore current—beach
10. meander—channel

Checking Concepts

Choose the word or phrase that best answers the question.

11. Where are beaches most common?
    A) rocky shorelines
    B) flat shorelines
    C) aquifers
    D) young streams

12. What is the network formed by a river and all the smaller streams that contribute to it?
    A) groundwater system
    B) zone of saturation
    C) river system
    D) water table

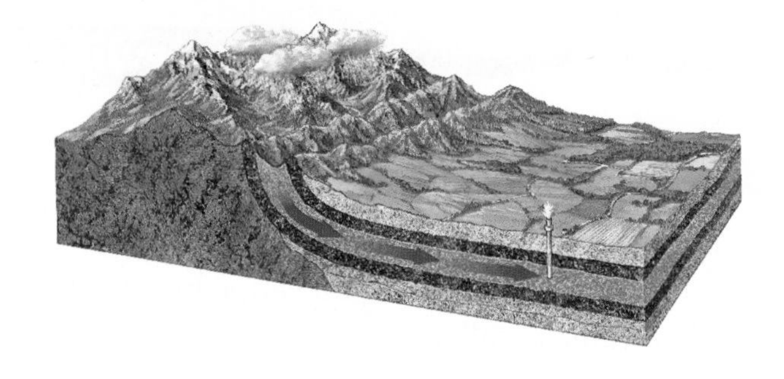

13. Why does water rise in an artesian well?
    A) a pump **C)** heat
    B) erosion **D)** pressure

14. Which term describes rock through which fluids can flow easily?
    A) impermeable **C)** saturated
    B) meanders **D)** permeable

15. Identify an example of a structure created by deposition.
    A) beach **C)** cave
    B) rill **D)** geyser

16. Which stage of development are mountain streams in?
    A) young **C)** old
    B) mature **D)** meandering

17. What forms as a result of the water table meeting Earth's surface?
    A) meander **C)** aquifer
    B) spring **D)** stalactite

18. What contains heated groundwater that reaches Earth's surface?
    A) water table **C)** aquifer
    B) cave **D)** hot spring

19. What is a layer of permeable rock that water flows through?
    A) an aquifer **C)** a water table
    B) a pore **D)** impermeable

20. Name the deposit that forms when a mountain river runs onto a plain.
    A) subsidence **C)** infiltration
    B) an alluvial fan **D)** water diversion

Science Online earth.msscience.com/vocabulary_puzzlemaker

Thinking Critically

21. **Concept Map** Copy and complete the concept map below using the following terms: *developed meanders, gentle curves, gentle gradient, old, rapids, steep gradient, wide floodplain,* and *young.*

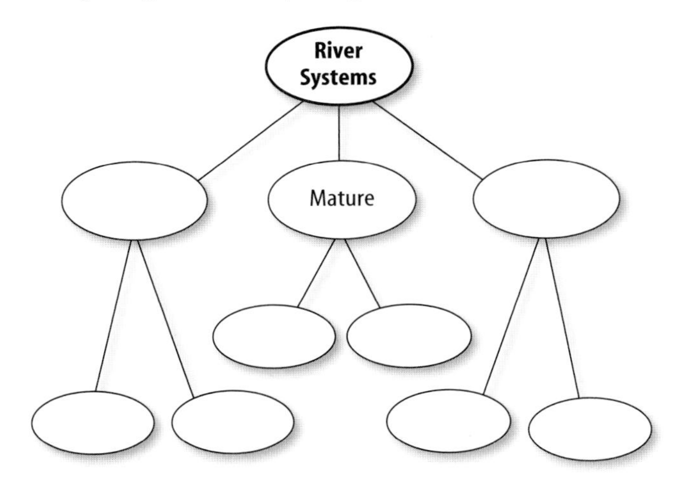

22. **Describe** what determines whether a stream erodes its bottom or its sides.

23. **Interpret Data** The rate of water flowing out of the Brahmaputra River in India, the La Plata River in South America, and the Mississippi River in North America are given in the table below. Infer which river carries the most sediment.

River Flow Rates

River	Flow (m^3/s)
Brahmaputra River, India	19,800
La Plata River, South America	79,300
Mississippi River, North America	175,000

24. **Explain** why the Mississippi River has meanders along its course.

25. **Outline** Make an outline that explains the three stages of stream development.

26. **Form Hypotheses** Hypothesize why most of the silt in the Mississippi delta is found farther out to sea than the sand-sized particles are.

27. **Infer** Along what kind of shoreline would you find barrier islands?

28. **Explain** why you might be concerned if developers of a new housing project started drilling wells near your well.

29. **Use Variables, Constants, and Controls** Explain how you could test the effect of slope on the amount of runoff produced.

Performance Activities

30. **Poster** Research a beach that interests you. Make a poster that shows different features you would find at a beach.

Applying Math

Use the illustration below to answer questions 31–32.

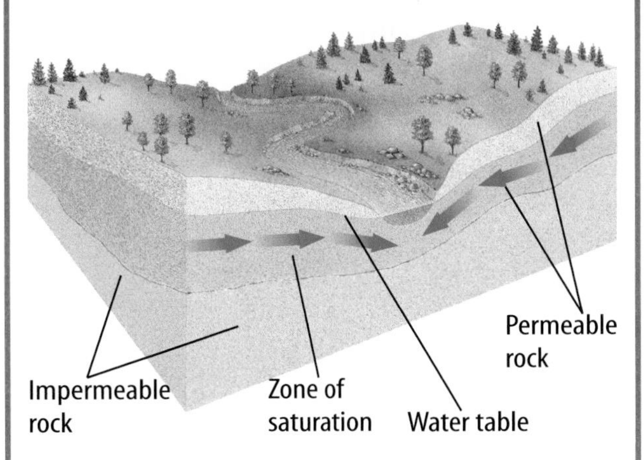

31. **Flow Distance** The groundwater in an aquifer flows at a rate of 0.2 m/day. How far does the groundwater move in one week?

32. **Flow Time** If groundwater in an aquifer flows at a rate of 0.4 m/day, how long does it take groundwater to move 24 m?

Part 1 Multiple Choice

Record your answers on the answer sheet provided by your teacher or on a sheet of paper.

1. Which is erosion over a large, flat area?
   - **A.** gully
   - **B.** rill
   - **C.** runoff
   - **D.** sheet
2. Which is erosion where a rill becomes broader and deeper?
   - **A.** gully
   - **B.** rill
   - **C.** runoff
   - **D.** sheet
3. Which is the area of land from which a stream collects runoff?
   - **A.** drainage basin
   - **B.** gully
   - **C.** runoff
   - **D.** stream channel
4. Which type of soil or rock allows water to pass through them?
   - **A.** impermeable
   - **B.** nonporous
   - **C.** permeable
   - **D.** underground

Refer to the figure below to answer question 5.

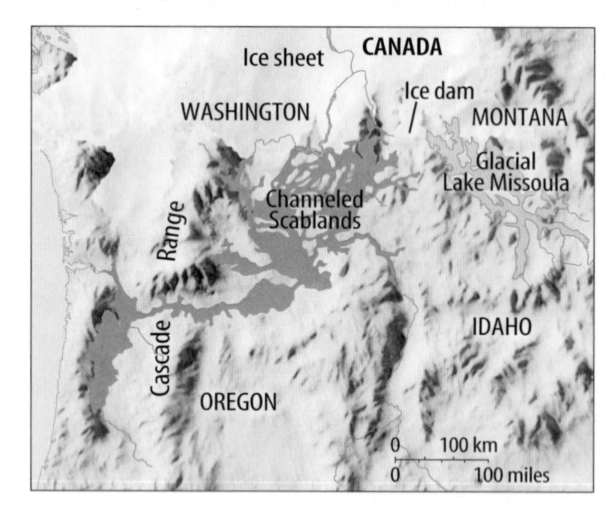

5. Which formed the Channeled Scablands?
   - **A.** deposition
   - **B.** floodwaters
   - **C.** rill erosion
   - **D.** sheet erosion
6. Which dissolves limestone to form caves?
   - **A.** carbonic acid
   - **B.** hydrochloric acid
   - **C.** stalactites
   - **D.** stalagmites
7. Which forms on the ceilings of caves as water drips through cracks?
   - **A.** aquifer
   - **B.** geyser
   - **C.** stalactite
   - **D.** stalagmite
8. Which are piles of sand found on barrier islands?
   - **A.** deltas
   - **B.** dunes
   - **C.** geysers
   - **D.** streams
9. Which creates springs and geysers?
   - **A.** groundwater
   - **B.** gullies
   - **C.** rills
   - **D.** runoff

Refer to the figure below to answer question 10 and 11.

10. Which feature is shown?
    - **A.** artesian well
    - **B.** aquifer
    - **C.** geyser
    - **D.** waterfall
11. Which provides the water?
    - **A.** groundwater
    - **B.** runoff
    - **C.** stream
    - **D.** surface water
12. Which is a layer of permeable rock through which water moves freely?
    - **A.** aquifer
    - **B.** clay
    - **C.** geyser
    - **D.** granite

Test-Taking Tip

Correct Answer Bubbles For each question, double check that you are filling in the correct answer bubble for the question number you are working on.

Part 2 | Short Response/Grid In

Record your answers on the answer sheet provided by your teacher or on a sheet of paper.

13. How does gravity affect water erosion?

14. Describe the different types of load in a stream.

15. What will happen to homes and businesses located in a floodplain?

16. Explain how the water table and the zone of saturation are related.

17. How would harmful chemicals in the soil enter into the groundwater system?

Refer to the picture below to answer questions 18 and 19.

18. What type of erosion is shown?

19. What led to the erosion shown?

20. What factors affect runoff?

21. What are the characteristics of a mature stream?

Part 3 | Open Ended

Record your answers on a sheet of paper.

22. Compare and contrast rocky shorelines and sandy beaches.

23. What are the Outer Banks in North Carolina and how were they formed?

24. What can humans do to try to control flood waters?

25. Explain how waves produce longshore currents.

Refer to the figure below to answer question 26.

26. What stream formation is shown in the diagram above? Explain how differing speeds of water in a stream can cause this type of stream formation.

27. Explain how a drainage basin works. Compare and contrast a drainage basin to the gutter drainage system of a rooftop.

28. Compare and contrast a pumped well and an artesian well.

29. Compare and contrast alluvial fans and deltas.

How Are Volcanoes & Fish Connected?

It's hard to know exactly what happened four and a half billion years ago, when Earth was very young. But it's likely that Earth was much more volcanically active than it is today. Along with lava and ash, volcanoes emit gases—including water vapor. Some scientists think that ancient volcanoes spewed tremendous amounts of water vapor into the early atmosphere. When the water vapor cooled, it would have condensed to form liquid water. Then the water would have fallen to the surface and collected in low areas, creating the oceans. Scientists hypothesize that roughly three and a half billion to four billion years ago, the first living things developed in the oceans. According to this hypothesis, these early life-forms gradually gave rise to more and more complex organisms—including the multitudes of fish that swim through the world's waters.

unit projects

Visit earth.msscience.com/unit_project to find project ideas and resources. Projects include:

- **History** Create a time line of volcano trivia with facts such as location, greatest magnitude, most destructive, and first volcano recorded. Can volcanoes be predicted?
- **Careers** Study the specialized skills of various careers as you design and prepare a city for a natural disaster.
- **Model** Research, design, construct, test, evaluate, and present your home seismograph in a 5-minute infomercial.

WebQuest *Volcanoes and the Ring of Fire* is an online study of plate tectonics. Design a chart of recent volcano activity, and use it to produce a map of the Ring of Fire with the names and ages of each volcano.

Plate Tectonics

The BIG Idea

The combination of ideas from continental drift, seafloor spreading, and many other discoveries led to the theory of plate tectonics.

SECTION 1
Continental Drift

Main Idea The continental drift hypothesis states that continents have moved slowly to their current locations.

SECTION 2
Seafloor Spreading

Main Idea New discoveries led to the theory of seafloor spreading as an explanation for continental drift.

SECTION 3
Theory of Plate Tectonics

Main Idea The theory of plate tectonics explains the formation of many of Earth's features and geologic events.

Will this continent split?

Ol Doinyo Lengai is an active volcano in the East African Rift Valley, a place where Earth's crust is being pulled apart. If the pulling continues over millions of years, Africa will separate into two landmasses. In this chapter, you'll learn about rift valleys and other clues that the continents move over time.

Science Journal Pretend you're a journalist with an audience that assumes the continents have never moved. Write about the kinds of evidence you'll need to convince people otherwise.

Start-Up Activities

Reassemble an Image

Can you imagine a giant landmass that broke into many separate continents and Earth scientists working to reconstruct Earth's past? Do this lab to learn about clues that can be used to reassemble a supercontinent.

1. Collect interesting photographs from an old magazine.
2. You and a partner each select one photo, but don't show them to each other. Then each of you cut your photos into pieces no smaller than about 5 cm or 6 cm.
3. Trade your cut-up photo for your partner's.
4. Observe the pieces, and reassemble the photograph your partner has cut up.
5. **Think Critically** Write a paragraph describing the characteristics of the cut-up photograph that helped you put the image back together. Think of other examples in which characteristics of objects are used to match them up with other objects.

Preview this chapter's content and activities at earth.msscience.com

Plate Tectonics Make the following Foldable to help identify what you already know, what you want to know, and what you learned about plate tectonics.

STEP 1 **Fold** a vertical sheet of paper from side to side. Make the front edge about 1.25 cm shorter than the back edge.

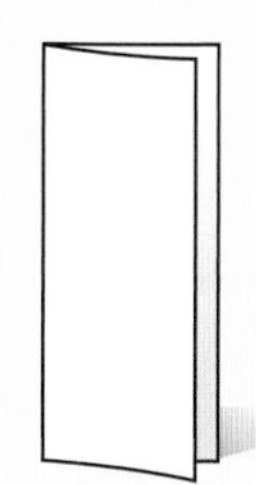

STEP 2 **Turn** lengthwise and fold into thirds.

STEP 3 **Unfold and cut** only the layer along both folds to make three tabs.

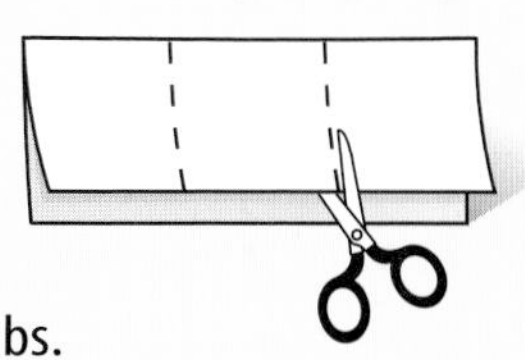

STEP 4 **Label** each tab.

Identify Questions Before you read the chapter, write what you already know about plate tectonics under the left tab of your Foldable, and write questions about what you'd like to know under the center tab. After you read the chapter, list what you learned under the right tab.

Get Ready to Read

Summarize

1 Learn It! Summarizing helps you organize information, focus on main ideas, and reduce the amount of information to remember. To summarize, restate the important facts in a short sentence or paragraph. Be brief and do not include too many details.

2 Practice It! Read the text on pages 272–274 labeled *Evidence for Continental Drift.* Then read the summary below and look at the important facts from that passage.

Summary

The puzzle-like fit of the continents, along with fossil, climate, and rock clues, were the main types of evidence supporting Wegener's hypothesis of continental drift.

Important Facts

- The edges of some continents look as though they could fit together like a puzzle.
- Fossils of the freshwater reptile *Mesosaurus* have been found in South America and Africa. The fossil plant *Glossopteris* has been found in Africa, Australia, India, South America, and Antarctica.
- Fossils of warm-weather plants were found on an island in the Arctic Ocean. Evidence of glaciers was found in temperate and tropical areas.
- Rocks and rock structures found in parts of the Appalachian Mountains of the eastern United States are similar to those found in Greenland and western Europe.

3 Apply It! Practice summarizing as you read this chapter. Stop after each section and write a brief summary.

Reread your summary to make sure you didn't change the author's original meaning or ideas.

Target Your Reading

Use this to focus on the main ideas as you read the chapter.

1. **Before you read** the chapter, respond to the statements below on your worksheet or on a numbered sheet of paper.
   - Write an **A** if you **agree** with the statement.
   - Write a **D** if you **disagree** with the statement.

2. **After you read** the chapter, look back to this page to see if you've changed your mind about any of the statements.
   - If any of your answers changed, explain why.
   - Change any false statements into true statements.
   - Use your revised statements as a study guide.

Before You Read A or D		Statement	After You Read A or D
	1	Fossils of tropical plants are never found in Antarctica.	
	2	Because of all the evidence that Alfred Wegener collected, scientists initially accepted his hypothesis of continental drift.	
	3	Wegener's continental drift hypothesis explains how, when, and why the continents drifted apart.	
	4	Earthquakes and volcanic eruptions often occur underwater along mid-ocean ridges.	
	5	Seafloor spreading provided part of the explanation of how continents could move.	
	6	Earth's broken crust rides on several large plates that move on a plastic-like layer of Earth's mantle.	
	7	The San Andreas Fault is part of a plate boundary.	
	8	When two continental plates move toward each other, one continent sinks beneath the other.	
	9	Scientists have proposed several explanations of how heat moves in Earth's interior.	

Science Online
Print out a worksheet of this page at earth.msscience.com

section 1

Continental Drift

as you read

What You'll Learn

- **Describe** the hypothesis of continental drift.
- **Identify** evidence supporting continental drift.

Why It's Important

The hypothesis of continental drift led to plate tectonics—a theory that explains many processes in Earth.

Review Vocabulary
continent: one of the six or seven great divisions of land on the globe

New Vocabulary
- continental drift
- Pangaea

Evidence for Continental Drift

If you look at a map of Earth's surface, you can see that the edges of some continents look as though they could fit together like a puzzle. Other people also have noticed this fact. For example, Dutch mapmaker Abraham Ortelius noted the fit between the coastlines of South America and Africa more than 400 years ago.

Pangaea German meteorologist Alfred Wegener (VEG nur) thought that the fit of the continents wasn't just a coincidence. He suggested that all the continents were joined together at some time in the past. In a 1912 lecture, he proposed the hypothesis of continental drift. According to the hypothesis of **continental drift,** continents have moved slowly to their current locations. Wegener suggested that all continents once were connected as one large landmass, shown in **Figure 1,** that broke apart about 200 million years ago. He called this large landmass **Pangaea** (pan JEE uh), which means "all land."

Reading Check *Who proposed continental drift?*

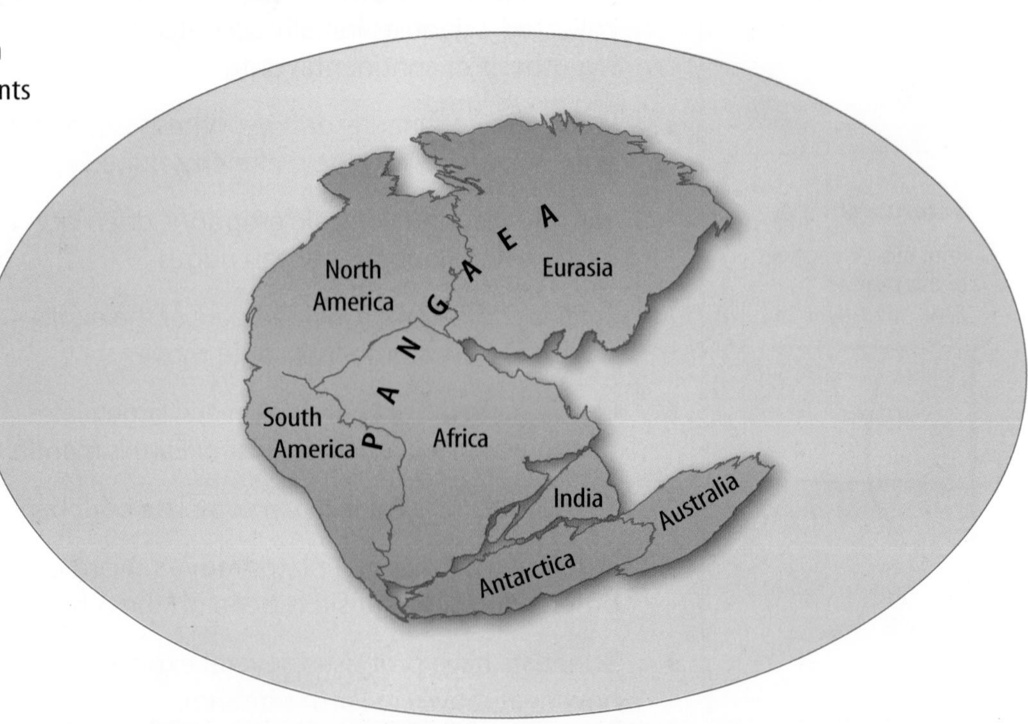

Figure 1 This illustration represents how the continents once were joined to form Pangaea. This fitting together of continents according to shape is not the only evidence supporting the past existence of Pangaea.

A Controversial Idea Wegener's ideas about continental drift were controversial. It wasn't until long after Wegener's death in 1930 that his basic hypothesis was accepted. The evidence Wegener presented hadn't been enough to convince many people during his lifetime. He was unable to explain exactly how the continents drifted apart. He proposed that the continents plowed through the ocean floor, driven by the spin of Earth. Physicists and geologists of the time strongly disagreed with Wegener's explanation. They pointed out that continental drift would not be necessary to explain many of Wegener's observations. Other important observations that came later eventually supported Wegener's earlier evidence.

Topic: Continental Drift
Visit earth.msscience.com for Web links to information about the continental drift hypothesis.

Activity Research and write a brief report about the initial reactions, from the public and scientific communities, toward Wegener's continental drift hypothesis.

Fossil Clues Besides the puzzlelike fit of the continents, fossils provided support for continental drift. Fossils of the reptile *Mesosaurus* have been found in South America and Africa, as shown in **Figure 2.** This swimming reptile lived in freshwater and on land. How could fossils of *Mesosaurus* be found on land areas separated by a large ocean of salt water? It probably couldn't swim between the continents. Wegener hypothesized that this reptile lived on both continents when they were joined.

Reading Check *How do* Mesosaurus *fossils support the past existence of Pangaea?*

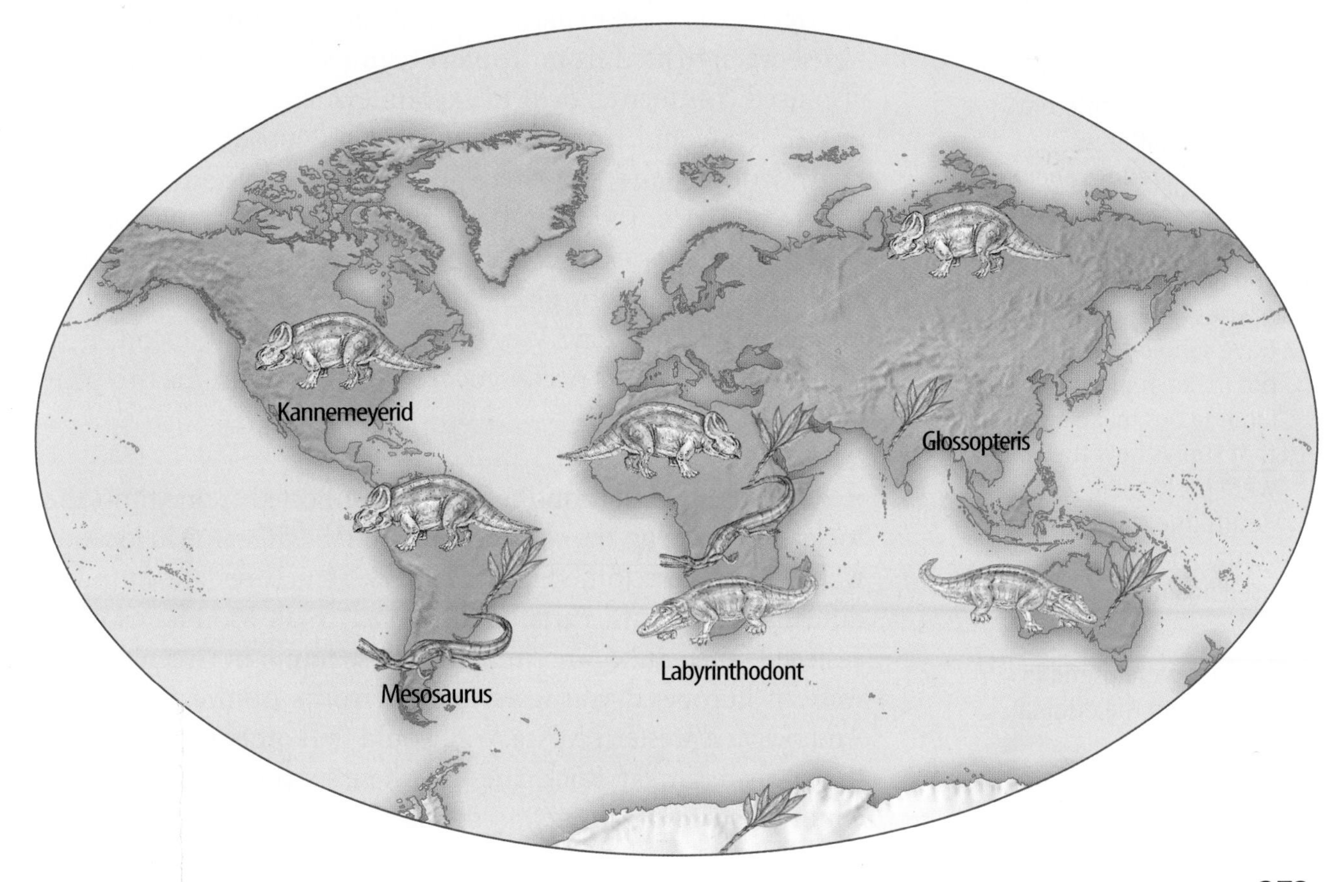

Figure 2 Fossil remains of plants and animals that lived in Pangaea have been found on more than one continent.
Evaluate *How do the locations of* Glossopteris, Mesosaurus, Kannemeyerid, Labyrinthodont, *and other fossils support Wegener's hypothesis of continental drift?*

Figure 3 This fossil plant, *Glossopteris,* grew in a temperate climate.

Interpreting Fossil Data

Procedure

1. Build a three-layer landmass using **clay** or **modeling dough.**
2. Mold the clay into mountain ranges.
3. Place similar **"fossils"** into the clay at various locations around the landmass.
4. Form five continents from the one landmass. Also, form two smaller landmasses out of different clay with different mountain ranges and fossils.
5. Place the five continents and two smaller landmasses around the room.
6. Have someone who did not make or place the landmasses make a model that shows how they once were positioned.
7. Return the clay to its container so it can be used again.

Analysis

What clues were useful in reconstructing the original landmass?

Try at Home

A Widespread Plant Another fossil that supports the hypothesis of continental drift is *Glossopteris* (glahs AHP tur us). **Figure 3** shows this fossil plant, which has been found in Africa, Australia, India, South America, and Antarctica. The presence of *Glossopteris* in so many areas also supported Wegener's idea that all of these regions once were connected and had similar climates.

Climate Clues Wegener used continental drift to explain evidence of changing climates. For example, fossils of warm-weather plants were found on the island of Spitsbergen in the Arctic Ocean. To explain this, Wegener hypothesized that Spitsbergen drifted from tropical regions to the arctic. Wegener also used continental drift to explain evidence of glaciers found in temperate and tropical areas. Glacial deposits and rock surfaces scoured and polished by glaciers are found in South America, Africa, India, and Australia. This shows that parts of these continents were covered with glaciers in the past. How could you explain why glacial deposits are found in areas where no glaciers exist today? Wegener thought that these continents were connected and partly covered with ice near Earth's south pole long ago.

Rock Clues If the continents were connected at one time, then rocks that make up the continents should be the same in locations where they were joined. Similar rock structures are found on different continents. Parts of the Appalachian Mountains of the eastern United States are similar to those found in Greenland and western Europe. If you were to study rocks from eastern South America and western Africa, you would find other rock structures that also are similar. Rock clues like these support the idea that the continents were connected in the past.

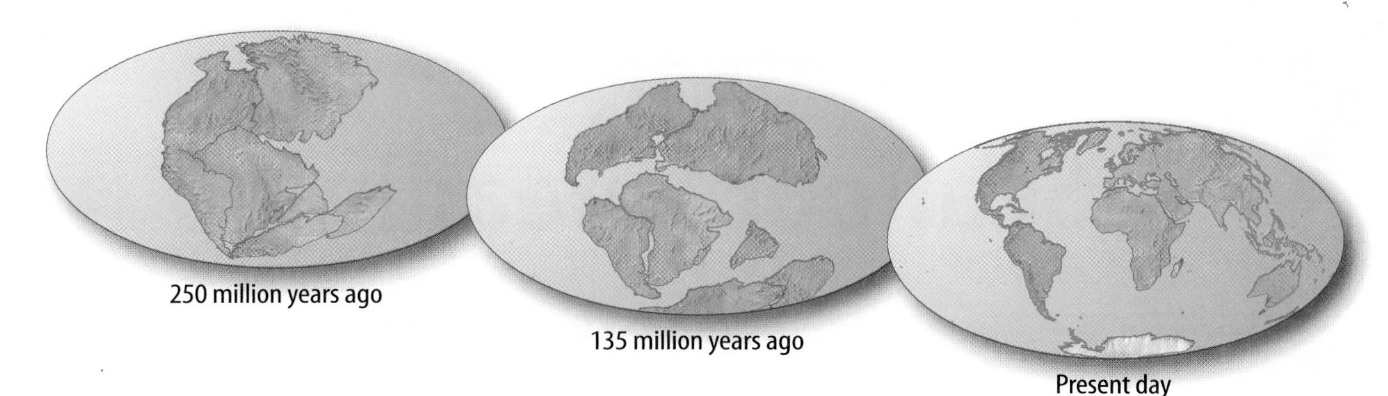

Figure 4 These computer models show the probable course the continents have taken. On the far left is their position 250 million years ago. In the middle is their position 135 million years ago. At right is their current position.

How could continents drift?

Although Wegener provided evidence to support his hypothesis of continental drift, he couldn't explain how, when, or why these changes, shown in **Figure 4,** took place. The idea suggested that lower-density, continental material somehow had to plow through higher-density, ocean-floor material. The force behind this plowing was thought to be the spin of Earth on its axis—a notion that was quickly rejected by physicists. Because other scientists could not provide explanations either, Wegener's idea of continental drift was initially rejected. The idea was so radically different at that time that most people closed their minds to it.

Rock, fossil, and climate clues were the main types of evidence for continental drift. After Wegener's death, more clues were found, largely because of advances in technology, and new ideas that related to continental drift were developed. You'll learn about a new idea, seafloor spreading, in the next section.

section 1 review

Summary

Evidence for Continental Drift

- Alfred Wegener proposed in his hypothesis of continental drift that all continents were once connected as one large landmass called Pangaea.
- Evidence of continental drift came from fossils, signs of climate change, and rock structures from different continents.

How could continents drift?

- During his lifetime, Wegener was unable to explain how, when, or why the continents drifted.
- After his death, advances in technology permitted new ideas to be developed to help explain his hypothesis.

Self Check

1. **Explain** how Wegener used climate clues to support his hypothesis of continental drift.
2. **Describe** how rock clues were used to support the hypothesis of continental drift.
3. **Summarize** the ways that fossils helped support the hypothesis of continental drift.
4. **Think Critically** Why would you expect to see similar rocks and rock structures on two landmasses that were connected at one time.

Applying Skills

5. **Compare and contrast** the locations of fossils of the temperate plant *Glossopteris,* as shown in **Figure 2,** with the climate that exists at each location today.

section 2

Seafloor Spreading

as you read

What You'll Learn

- **Explain** seafloor spreading.
- **Recognize** how age and magnetic clues support seafloor spreading.

Why It's Important

Seafloor spreading helps explain how continents moved apart.

Review Vocabulary
seafloor: portion of Earth's crust that lies beneath ocean waters

New Vocabulary
- seafloor spreading

Mapping the Ocean Floor

If you were to lower a rope from a boat until it reached the seafloor, you could record the depth of the ocean at that particular point. In how many different locations would you have to do this to create an accurate map of the seafloor? This is exactly how it was done until World War I, when the use of sound waves was introduced by German scientists to detect submarines. During the 1940s and 1950s, scientists began using sound waves on moving ships to map large areas of the ocean floor in detail. Sound waves echo off the ocean bottom—the longer the sound waves take to return to the ship, the deeper the water is.

Using sound waves, researchers discovered an underwater system of ridges, or mountains, and valleys like those found on the continents. In some of these underwater ridges are rather long rift valleys where volcanic eruptions and earthquakes occur from time to time. Some of these volcanoes actually are visible above the ocean surface. In the Atlantic, the Pacific, and in other oceans around the world, a system of ridges, called the mid-ocean ridges, is present. These underwater mountain ranges, shown in **Figure 5,** stretch along the center of much of Earth's ocean floor. This discovery raised the curiosity of many scientists. What formed these mid-ocean ridges?

Reading Check *How were mid-ocean ridges discovered?*

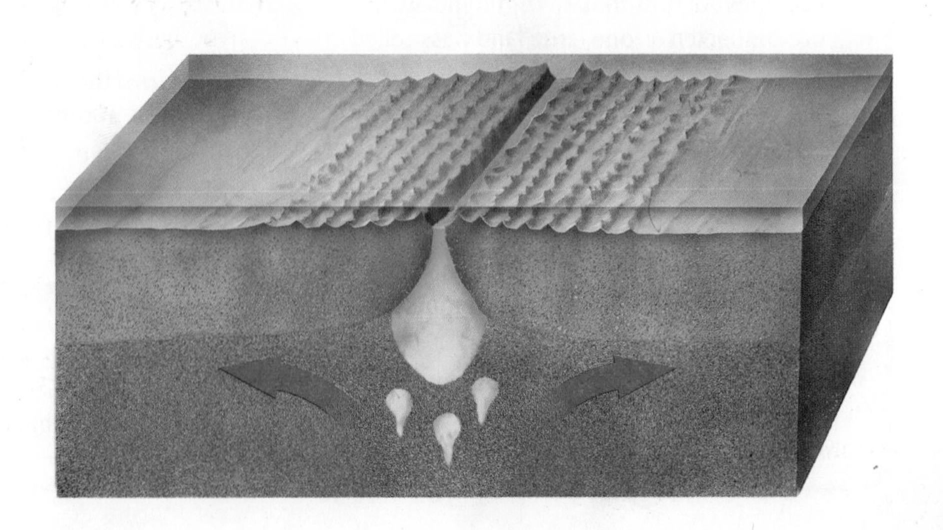

Figure 5 As the seafloor spreads apart at a mid-ocean ridge, new seafloor is created. The older seafloor moves away from the ridge in opposite directions.

The Seafloor Moves In the early 1960s, Princeton University scientist Harry Hess suggested an explanation. His now-famous theory is known as **seafloor spreading.** Hess proposed that hot, less dense material below Earth's crust rises toward the surface at the mid-ocean ridges. Then, it flows sideways, carrying the seafloor away from the ridge in both directions, as seen in **Figure 5.**

As the seafloor spreads apart, magma is forced upward and flows from the cracks. It becomes solid as it cools and forms new seafloor. As new seafloor moves away from the mid-ocean ridge, it cools, contracts, and becomes denser. This denser, colder seafloor sinks, helping to form the ridge. The theory of seafloor spreading was later supported by the following observations.

How does new seafloor form at mid-ocean ridges?

Figure 6 Many new discoveries have been made on the seafloor. These giant tube worms inhabit areas near hot water vents along mid-ocean ridges.

Evidence for Spreading

In 1968, scientists aboard the research ship *Glomar Challenger* began gathering information about the rocks on the seafloor. *Glomar Challenger* was equipped with a drilling rig that allowed scientists to drill into the seafloor to obtain rock samples. Scientists found that the youngest rocks are located at the mid-ocean ridges. The ages of the rocks become increasingly older in samples obtained farther from the ridges, adding to the evidence for seafloor spreading.

Using submersibles along mid-ocean ridges, new seafloor features and life-forms also were discovered there, as shown in **Figure 6.** As molten material is forced upward along the ridges, it brings heat and chemicals that support exotic life-forms in deep, ocean water. Among these are giant clams, mussels, and tube worms.

INTEGRATE Physics

Magnetic Clues Earth's magnetic field has a north and a south pole. Magnetic lines, or directions, of force leave Earth near the south pole and enter Earth near the north pole. During a magnetic reversal, the lines of magnetic force run the opposite way. Scientists have determined that Earth's magnetic field has reversed itself many times in the past. These reversals occur over intervals of thousands or even millions of years. The reversals are recorded in rocks forming along mid-ocean ridges.

Curie Point Find out what the Curie point is and describe in your Science Journal what happens to iron-bearing minerals when they are heated to the Curie point. Explain how this is important to studies of seafloor spreading.

Normal magnetic polarity
Reverse magnetic polarity

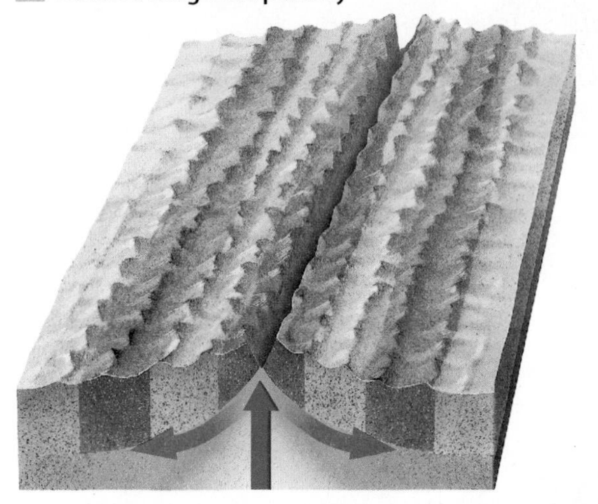

Figure 7 Changes in Earth's magnetic field are preserved in rock that forms on both sides of mid-ocean ridges.
Explain *why this is considered to be evidence of seafloor spreading.*

Magnetic Time Scale Iron-bearing minerals, such as magnetite, that are found in the rocks of the seafloor can record Earth's magnetic field direction when they form. Whenever Earth's magnetic field reverses, newly forming iron minerals will record the magnetic reversal.

Using a sensing device called a magnetometer (mag nuh TAH muh tur) to detect magnetic fields, scientists found that rocks on the ocean floor show many periods of magnetic reversal. The magnetic alignment in the rocks reverses back and forth over time in strips parallel to the mid-ocean ridges, as shown in **Figure 7.** A strong magnetic reading is recorded when the polarity of a rock is the same as the polarity of Earth's magnetic field today. Because of this, normal polarities in rocks show up as large peaks. This discovery provided strong support that seafloor spreading was indeed occurring. The magnetic reversals showed that new rock was being formed at the mid-ocean ridges. This helped explain how the crust could move—something that the continental drift hypothesis could not do.

section 2 review

Summary

Mapping the Ocean Floor

- Mid-ocean ridges, along the center of the ocean floor, have been found by using sound waves, the same method once used to detect submarines during World War I.
- Harry Hess suggested, in his seafloor spreading hypothesis, that the seafloor moves.

Evidence for Spreading

- Scientists aboard *Glomar Challenger* provided evidence of spreading by discovering that the youngest rocks are located at ridges and become increasingly older farther from the ridges.
- Magnetic alignment of rocks, in alternating strips that run parallel to ridges, indicates reversals in Earth's magnetic field and provides further evidence of seafloor spreading.

Self Check

1. **Summarize** What properties of iron-bearing minerals on the seafloor support the theory of seafloor spreading?
2. **Explain** how the ages of the rocks on the ocean floor support the theory of seafloor spreading.
3. **Summarize** How did Harry Hess's hypothesis explain seafloor movement?
4. **Explain** why some partly molten material rises toward Earth's surface.
5. **Think Critically** The ideas of Hess, Wegener, and others emphasize that Earth is a dynamic planet. How is seafloor spreading different from continental drift?

Applying Skills

6. **Solve One-Step Equations** North America is moving about 1.25 cm per year away from a ridge in the middle of the Atlantic Ocean. Using this rate, how much farther apart will North America and the ridge be in 200 million years?

Science Online earth.msscience.com/self_check_quiz

Seafloor Spreading Rates

How did scientists use their knowledge of seafloor spreading and magnetic field reversals to reconstruct Pangaea? Try this lab to see how you can determine where a continent may have been located in the past.

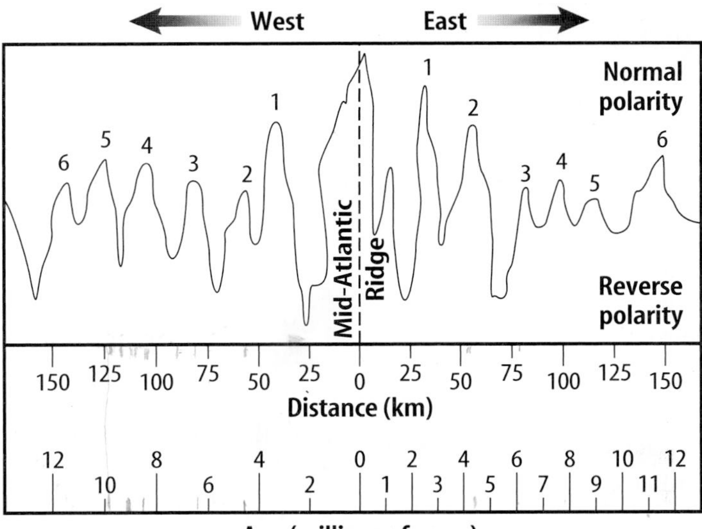

Real-World Question

Can you use clues, such as magnetic field reversals on Earth, to help reconstruct Pangaea?

Goals

- **Interpret** data about magnetic field reversals. Use these magnetic clues to reconstruct Pangaea.

Materials

metric ruler
pencil

Procedure

1. Study the magnetic field graph above. You will be working only with normal polarity readings, which are the peaks above the baseline in the top half of the graph.
2. Place the long edge of a ruler vertically on the graph. Slide the ruler so that it lines up with the center of peak 1 west of the Mid-Atlantic Ridge.
3. **Determine** and record the distance and age that line up with the center of peak 1 west. Repeat this process for peak 1 east of the ridge.
4. **Calculate** the average distance and age for this pair of peaks.
5. Repeat steps 2 through 4 for the remaining pairs of normal-polarity peaks.
6. **Calculate** the rate of movement in cm per year for the six pairs of peaks. Use the formula rate = distance/time. Convert kilometers to centimeters. For example, to calculate a rate using normal-polarity peak 5, west of the ridge:

$$\text{rate} = \frac{125 \text{ km}}{10 \text{ million years}} = \frac{12.5 \text{ km}}{\text{million years}} = \frac{1{,}250{,}000 \text{ cm}}{1{,}000{,}000 \text{ years}} = 1.25 \text{ cm/year}$$

Conclude and Apply

1. **Compare** the age of igneous rock found near the mid-ocean ridge with that of igneous rock found farther away from the ridge.
2. If the distance from a point on the coast of Africa to the Mid-Atlantic Ridge is approximately 2,400 km, calculate how long ago that point in Africa was at or near the Mid-Atlantic Ridge.
3. How could you use this method to reconstruct Pangaea?

section 3

Theory of Plate Tectonics

as you read

What You'll Learn

- **Compare and contrast** different types of plate boundaries.
- **Explain** how heat inside Earth causes plate tectonics.
- **Recognize** features caused by plate tectonics.

Why It's Important

Plate tectonics explains how many of Earth's features form.

Review Vocabulary

converge: to come together
diverge: to move apart
transform: to convert or change

New Vocabulary

- plate tectonics
- plate
- lithosphere
- asthenosphere
- convection current

Plate Tectonics

The idea of seafloor spreading showed that more than just continents were moving, as Wegener had thought. It was now clear to scientists that sections of the seafloor and continents move in relation to one another.

Plate Movements In the 1960s, scientists developed a new theory that combined continental drift and seafloor spreading. According to the theory of **plate tectonics,** Earth's crust and part of the upper mantle are broken into sections. These sections, called **plates,** move on a plasticlike layer of the mantle. The plates can be thought of as rafts that float and move on this layer.

Composition of Earth's Plates Plates are made of the crust and a part of the upper mantle, as shown in **Figure 8.** These two parts combined are the **lithosphere** (LIH thuh sfihr). This rigid layer is about 100 km thick and generally is less dense than material underneath. The plasticlike layer below the lithosphere is called the **asthenosphere** (as THE nuh sfihr). The rigid plates of the lithosphere float and move around on the asthenosphere.

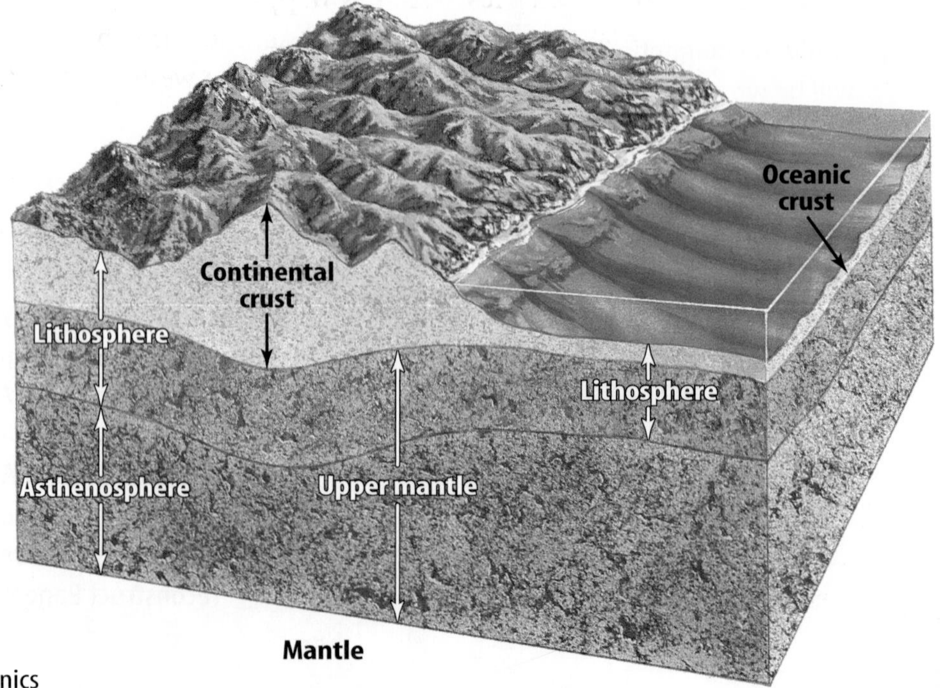

Figure 8 Plates of the lithosphere are composed of oceanic crust, continental crust, and rigid upper mantle.

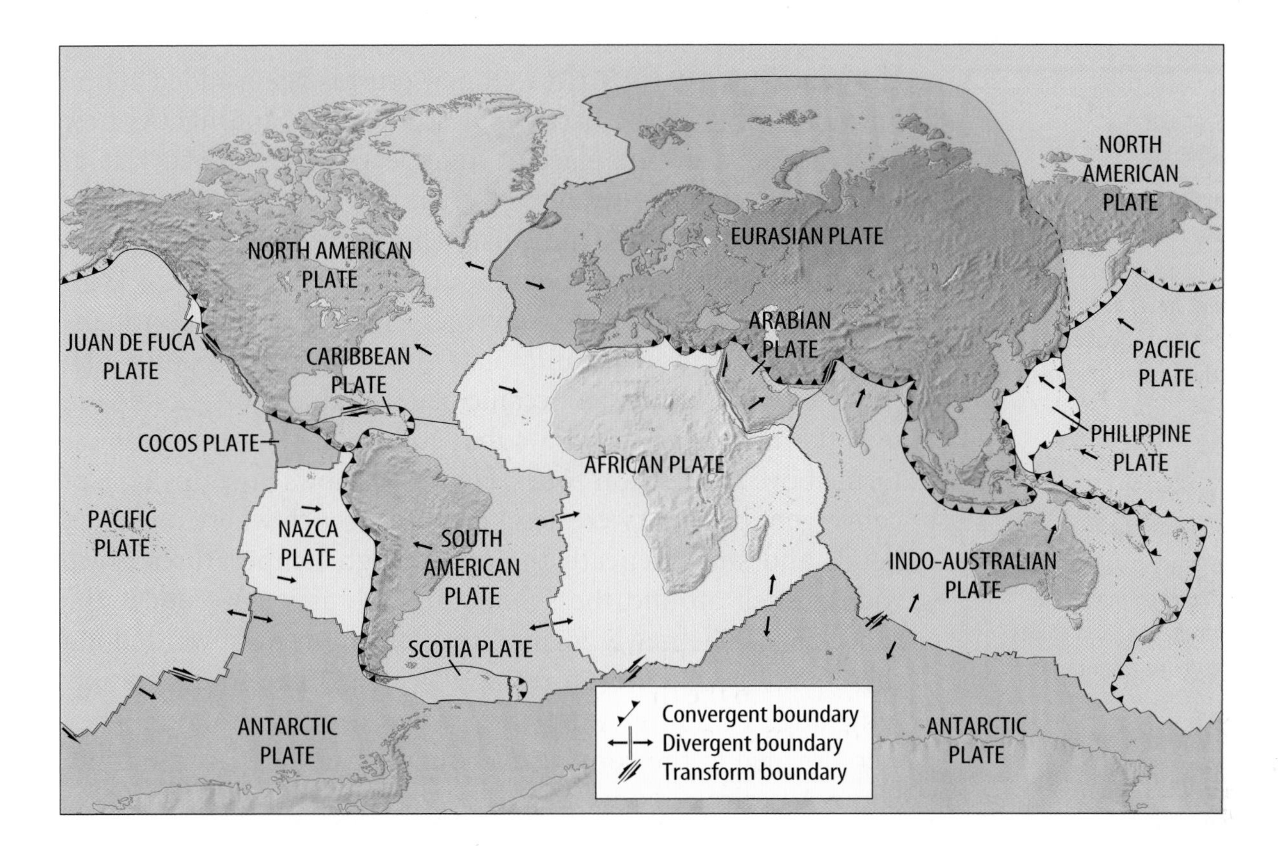

Figure 9 This diagram shows the major plates of the lithosphere, their direction of movement, and the type of boundary between them.
Analyze and Conclude *Based on what is shown in this figure, what is happening where the Nazca Plate meets the Pacific Plate?*

Plate Boundaries

When plates move, they can interact in several ways. They can move toward each other and converge, or collide. They also can pull apart or slide alongside one another. When the plates interact, the result of their movement is seen at the plate boundaries, as in **Figure 9.**

Reading Check *What are the general ways that plates interact?*

Movement along any plate boundary means that changes must happen at other boundaries. What is happening to the Atlantic Ocean floor between the North American and African Plates? Compare this with what is happening along the western margin of South America.

Plates Moving Apart The boundary between two plates that are moving apart is called a divergent boundary. You learned about divergent boundaries when you read about seafloor spreading. In the Atlantic Ocean, the North American Plate is moving away from the Eurasian and the African Plates, as shown in **Figure 9.** That divergent boundary is called the Mid-Atlantic Ridge. The Great Rift Valley in eastern Africa might become a divergent plate boundary. There, a valley has formed where a continental plate is being pulled apart. **Figure 10** shows a side view of what a rift valley might look like and illustrates how the hot material rises up where plates separate.

Science Online

Topic: Earthquakes and Volcanoes

Visit earth.msscience.com for Web links to recent news or magazine articles about earthquakes and volcanic activity related to plate tectonics.

Activity Prepare a group demonstration about recent volcanic and earthquake events. Divide tasks among group members. Find and copy maps, diagrams, photographs, and charts to highlight your presentation. Emphasize the locations of events and the relationship to plate tectonics.

Plates Moving Together If new crust is being added at one location, why doesn't Earth's surface keep expanding? As new crust is added in one place, it disappears below the surface at another. The disappearance of crust can occur when seafloor cools, becomes denser, and sinks. This occurs where two plates move together at a convergent boundary.

When an oceanic plate converges with a less dense continental plate, the denser oceanic plate sinks under the continental plate. The area where an oceanic plate subducts, or goes down, into the mantle is called a subduction zone. Some volcanoes form above subduction zones. **Figure 10** shows how this type of convergent boundary creates a deep-sea trench where one plate bends and sinks beneath the other. High temperatures cause rock to melt around the subducting slab as it goes under the other plate. The newly formed magma is forced upward along these plate boundaries, forming volcanoes. The Andes mountain range of South America contains many volcanoes. They were formed at the convergent boundary of the Nazca and the South American Plates.

Applying Science

How well do the continents fit together?

Recall the Launch Lab you performed at the beginning of this chapter. While you were trying to fit pieces of a cut-up photograph together, what clues did you use?

Identifying the Problem

Take a copy of a map of the world and cut out each continent. Lay them on a tabletop and try to fit them together, using techniques you used in the Launch Lab. You will find that the pieces of your Earth puzzle—the continents—do not fit together well. Yet, several of the areas on some continents fit together extremely well.

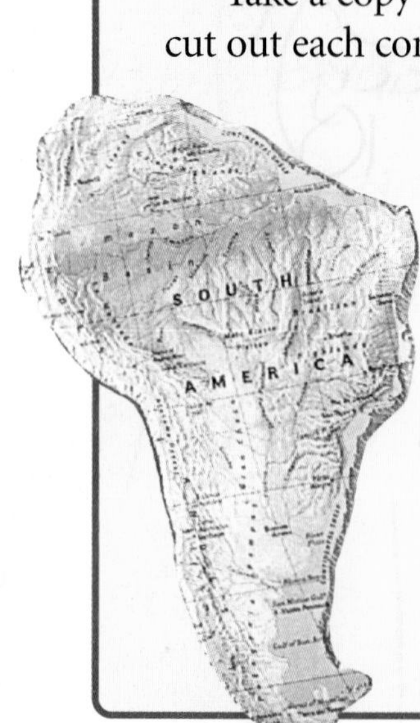

Take out another world map—one that shows the continental shelves as well as the continents. Copy it and cut out the continents, this time including the continental shelves.

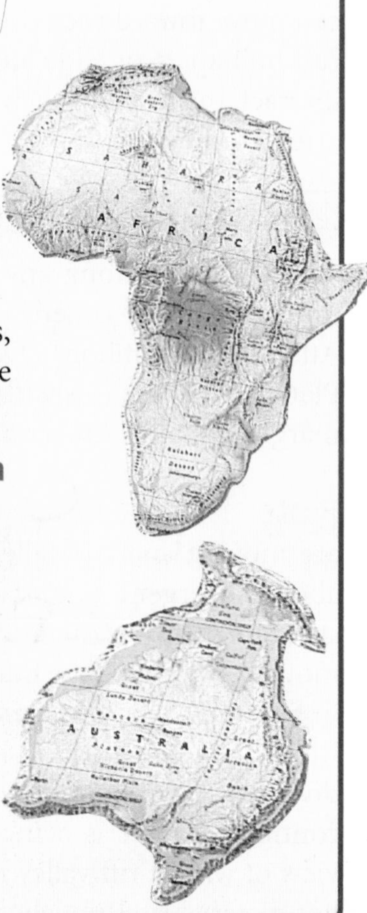

Solving the Problem

1. Does including the continental shelves solve the problem of fitting the continents together?
2. Why should continental shelves be included with maps of the continents?

NATIONAL GEOGRAPHIC

VISUALIZING PLATE BOUNDARIES

Figure 10

By diverging at some boundaries and converging at others, Earth's plates are continually—but gradually—reshaping the landscape around you. The Mid-Atlantic Ridge, for example, was formed when the North and South American Plates pulled apart from the Eurasian and African Plates (see globe). Some features that occur along plate boundaries—rift valleys, volcanoes, and mountain ranges—are shown on the right and below.

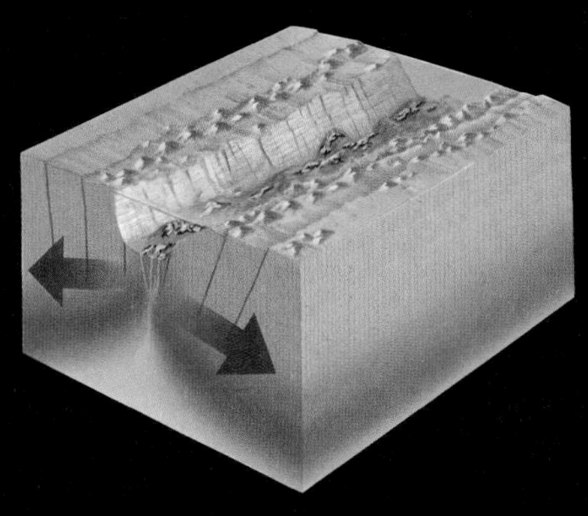

A RIFT VALLEY When continental plates pull apart, they can form rift valleys. The African continent is separating now along the East African Rift Valley.

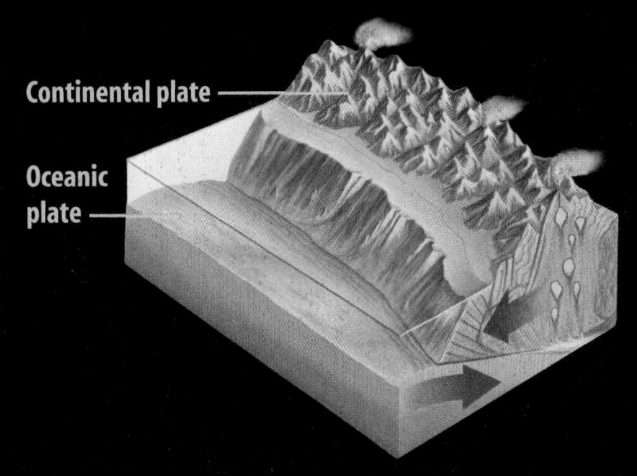

SUBDUCTION Where oceanic and continental plates collide, the oceanic plate plunges beneath the less dense continental plate. As the plate descends, molten rock (yellow) forms and rises toward the surface, creating volcanoes.

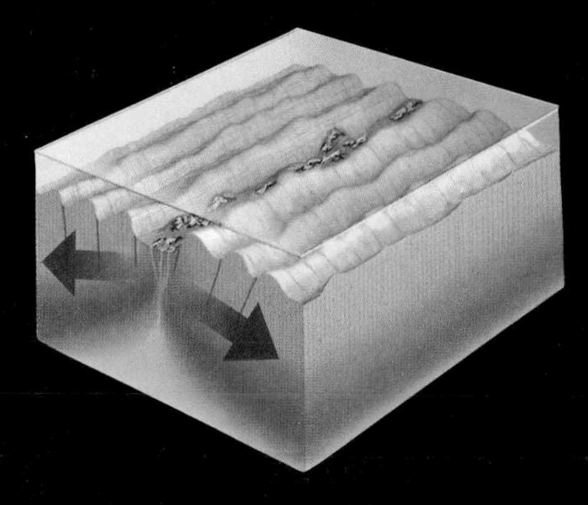

SEAFLOOR SPREADING A mid-ocean ridge, like the Mid-Atlantic Ridge, forms where oceanic plates continue to separate. As rising magma (yellow) cools, it forms new oceanic crust.

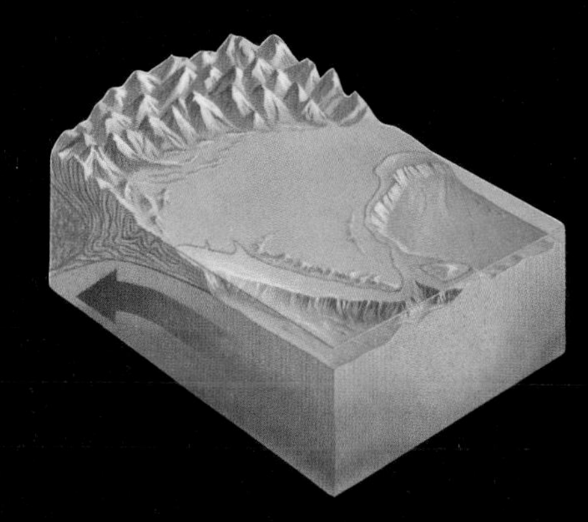

CONTINENTAL COLLISION Where two continental plates collide, they push up the crust to form mountain ranges such as the Himalaya.

Where Plates Collide A subduction zone also can form where two oceanic plates converge. In this case, the colder, older, denser oceanic plate bends and sinks down into the mantle. The Mariana Islands in the western Pacific are a chain of volcanic islands formed where two oceanic plates collide.

Usually, no subduction occurs when two continental plates collide, as shown in **Figure 10.** Because both of these plates are less dense than the material in the asthenosphere, the two plates collide and crumple up, forming mountain ranges. Earthquakes are common at these convergent boundaries. However, volcanoes do not form because there is no, or little, subduction. The Himalaya in Asia are forming where the Indo-Australian Plate collides with the Eurasian Plate.

Where Plates Slide Past Each Other The third type of plate boundary is called a transform boundary. Transform boundaries occur where two plates slide past one another. They move in opposite directions or in the same direction at different rates. When one plate slips past another suddenly, earthquakes occur. The Pacific Plate is sliding past the North American Plate, forming the famous San Andreas Fault in California, as seen in **Figure 11.** The San Andreas Fault is part of a transform plate boundary. It has been the site of many earthquakes.

Figure 11 The San Andreas Fault in California occurs along the transform plate boundary where the Pacific Plate is sliding past the North American Plate.

Overall, the two plates are moving in roughly the same direction. **Explain** *Why, then, do the red arrows show movement in opposite directions?*

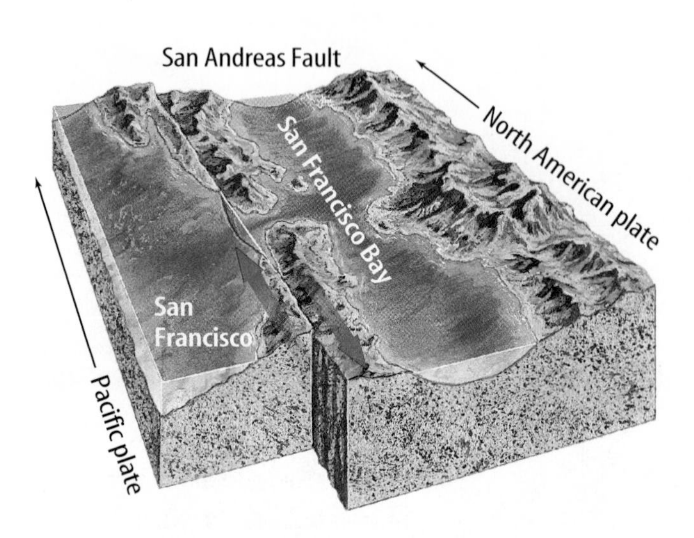

This photograph shows an aerial view of the San Andreas Fault.

Causes of Plate Tectonics

Many new discoveries have been made about Earth's crust since Wegener's day, but one question still remains. What causes the plates to move? Scientists now think they have a good idea. They think that plates move by the same basic process that occurs when you heat soup.

Convection Inside Earth Soup that is cooking in a pan on the stove contains currents caused by an unequal distribution of heat in the pan. Hot, less dense soup is forced upward by the surrounding, cooler, denser soup. As the hot soup reaches the surface, it cools and sinks back down into the pan. This entire cycle of heating, rising, cooling, and sinking is called a **convection current.** A version of this same process, occurring in the mantle, is thought to be the force behind plate tectonics. Scientists suggest that differences in density cause hot, plasticlike rock to be forced upward toward the surface.

Moving Mantle Material Wegener wasn't able to come up with an explanation for why plates move. Today, researchers who study the movement of heat in Earth's interior have proposed several possible explanations. All of the hypotheses use convection in one way or another. It is, therefore, the transfer of heat inside Earth that provides the energy to move plates and causes many of Earth's surface features. One hypothesis is shown in **Figure 12.** It relates plate motion directly to the movement of convection currents. According to this hypothesis, convection currents cause the movements of plates.

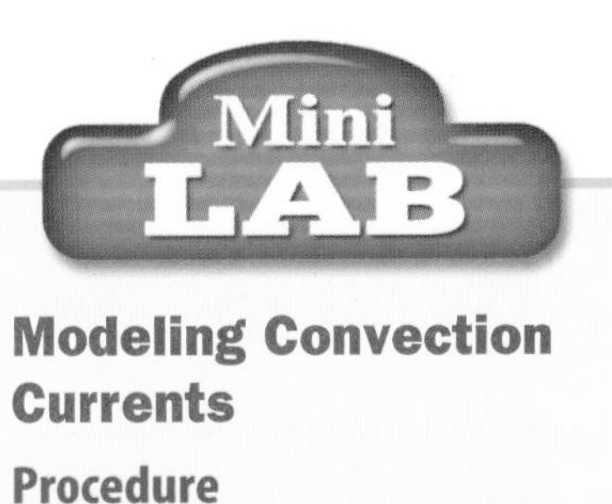

Modeling Convection Currents

Procedure

1. Pour **water** into **a clear, colorless casserole dish** until it is 5 cm from the top.
2. Center the dish on a **hot plate** and heat it. **WARNING:** *Wear* ***thermal mitts*** *to protect your hands.*
3. Add a few drops of **food coloring** to the water above the center of the hot plate.
4. Looking from the side of the dish, observe what happens in the water.
5. Illustrate your observations in your **Science Journal.**

Analysis

1. Determine whether any currents form in the water.
2. Infer what causes the currents to form.

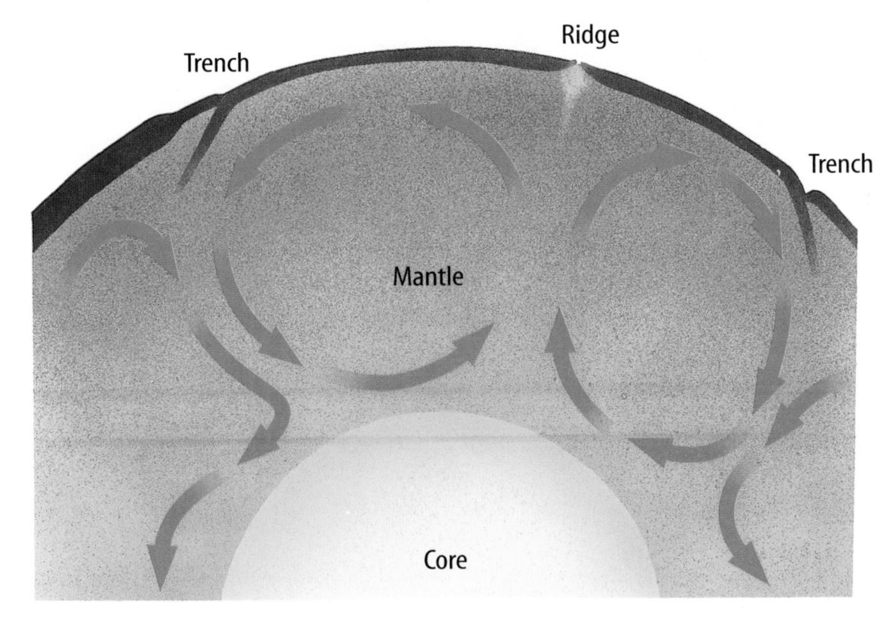

Figure 12 In one hypothesis, convection currents occur throughout the mantle. Such convection currents (see arrows) are the driving force of plate tectonics.

Features Caused by Plate Tectonics

Earth is a dynamic planet with a hot interior. This heat leads to convection, which powers the movement of plates. As the plates move, they interact. The interaction of plates produces forces that build mountains, create ocean basins, and cause volcanoes. When rocks in Earth's crust break and move, energy is released in the form of seismic waves. Humans feel this release as earthquakes. You can see some of the effects of plate tectonics in mountainous regions, where volcanoes erupt, or where landscapes have changed from past earthquake or volcanic activity.

Reading Check ***What happens when seismic energy is released as rocks in Earth's crust break and move?***

Figure 13 Fault-block mountains can form when Earth's crust is stretched by tectonic forces. The arrows indicate the directions of moving blocks.
Name *the type of force that occurs when Earth's crust is pulled in opposite directions.*

Normal Faults and Rift Valleys Tension forces, which are forces that pull apart, can stretch Earth's crust. This causes large blocks of crust to break and tilt or slide down the broken surfaces of crust. When rocks break and move along surfaces, a fault forms. Faults interrupt rock layers by moving them out of place. Entire mountain ranges can form in the process, called fault-block mountains, as shown in **Figure 13.** Generally, the faults that form from pull-apart forces are normal faults—faults in which the rock layers above the fault move down when compared with rock layers below the fault.

Rift valleys and mid-ocean ridges can form where Earth's crust separates. Examples of rift valleys are the Great Rift Valley in Africa, and the valleys that occur in the middle of mid-ocean ridges. Examples of mid-ocean ridges include the Mid-Atlantic Ridge and the East Pacific Rise.

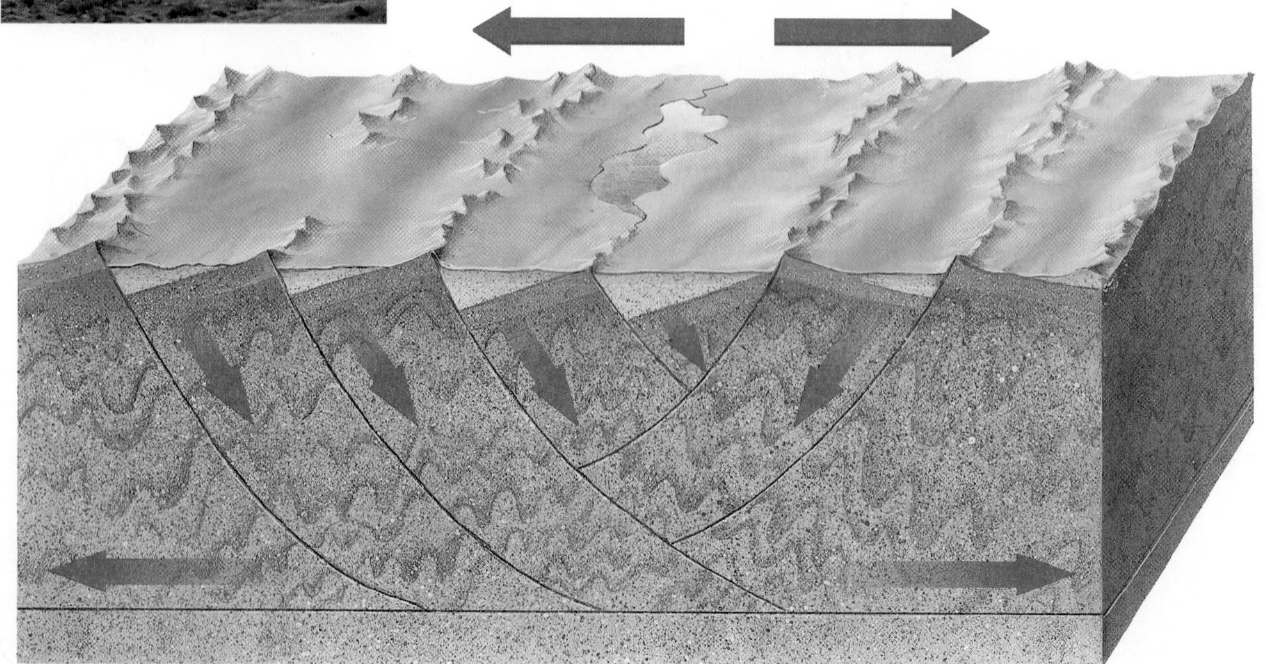

Mountains and Volcanoes Compression forces squeeze objects together. Where plates come together, compression forces produce several effects. As continental plates collide, the forces that are generated cause massive folding and faulting of rock layers into mountain ranges such as the Himalaya, shown in **Figure 14,** or the Appalachian Mountains. The type of faulting produced is generally reverse faulting. Along a reverse fault, the rock layers above the fault surface move up relative to the rock layers below the fault.

Reading Check *What features occur where plates converge?*

As you learned earlier, when two oceanic plates converge, the denser plate is forced beneath the other plate. Curved chains of volcanic islands called island arcs form above the sinking plate. If an oceanic plate converges with a continental plate, the denser oceanic plate slides under the continental plate. Folding and faulting at the continental plate margin can thicken the continental crust to produce mountain ranges. Volcanoes also typically are formed at this type of convergent boundary.

Volcanologist This person's job is to study volcanoes in order to predict eruptions. Early warning of volcanic eruptions gives nearby residents time to evacuate. Volcanologists also educate the public about the hazards of volcanic eruptions and tell people who live near volcanoes what they can do to be safe in the event of an eruption. Volcanologists travel all over the world to study new volcanic sites.

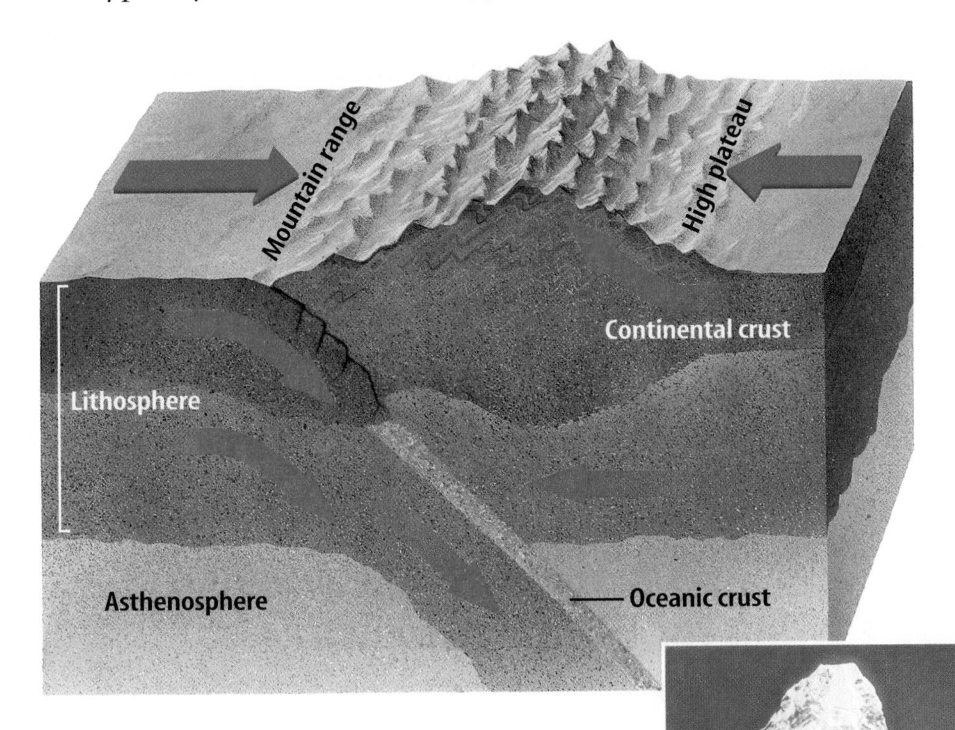

Figure 14 The Himalaya still are forming today as the Indo-Australian Plate collides with the Eurasian Plate.

Figure 15 Most of the movement along a strike-slip fault is parallel to Earth's surface. When movement occurs, human-built structures along a strike-slip fault are offset, as shown here in this road.

Strike-Slip Faults At transform boundaries, two plates slide past one another without converging or diverging. The plates stick and then slide, mostly in a horizontal direction, along large strike-slip faults. In a strike-slip fault, rocks on opposite sides of the fault move in opposite directions, or in the same direction at different rates. This type of fault movement is shown in **Figure 15.** One such example is the San Andreas Fault. When plates move suddenly, vibrations are generated inside Earth that are felt as an earthquake.

Earthquakes, volcanoes, and mountain ranges are evidence of plate motion. Plate tectonics explains how activity inside Earth can affect Earth's crust differently in different locations. You've seen how plates have moved since Pangaea separated. Is it possible to measure how far plates move each year?

Testing for Plate Tectonics

Until recently, the only tests scientists could use to check for plate movement were indirect. They could study the magnetic characteristics of rocks on the seafloor. They could study volcanoes and earthquakes. These methods supported the theory that the plates have moved and still are moving. However, they did not provide proof—only support—of the idea.

New methods had to be discovered to be able to measure the small amounts of movement of Earth's plates. One method, shown in **Figure 16,** uses lasers and a satellite. Now, scientists can measure exact movements of Earth's plates of as little as 1 cm per year.

Direction of Forces In which directions do forces act at convergent, divergent, and transform boundaries? Demonstrate these forces using wooden blocks or your hands.

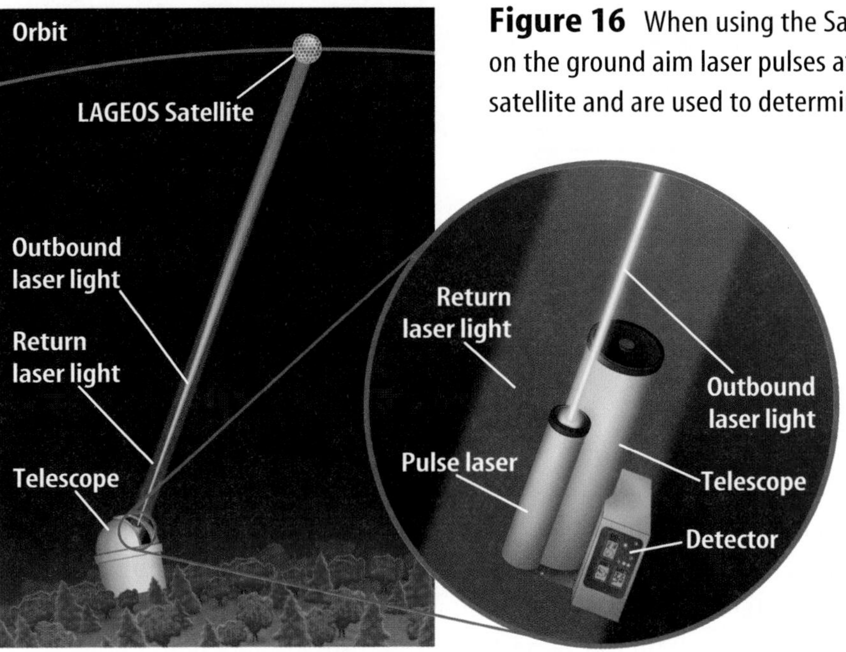

Figure 16 When using the Satellite Laser Ranging System, scientists on the ground aim laser pulses at a satellite. The pulses reflect off the satellite and are used to determine a precise location on the ground.

Current Data Satellite Laser Ranging System data show that Hawaii is moving toward Japan at a rate of about 8.3 cm per year. Maryland is moving away from England at a rate of 1.7 cm per year. Using such methods, scientists have observed that the plates move at rates ranging from about 1 cm to 12 cm per year.

section 3 review

Summary

Plate Tectonics

- The theory of plate tectonics states that sections of the seafloor and continents move as plates on a plasticlike layer of the mantle.

Plate Boundaries

- The boundary between two plates moving apart is called a divergent boundary.
- Plates move together at a convergent boundary.
- Transform boundaries occur where two plates slide past one another.

Causes of Plate Tectonics

- Convection currents are thought to cause the movement of Earth's plates.

Features Caused by Plate Tectonics

- Tension forces cause normal faults, rift valleys, and mid-ocean ridges at divergent boundaries.
- At convergent boundaries, compression forces cause folding, reverse faults, and mountains.
- At transform boundaries, two plates slide past one another along strike-slip faults.

Self Check

1. **Describe** what occurs at plate boundaries that are associated with seafloor spreading.
2. **Describe** three types of plate boundaries where volcanic eruptions can occur.
3. **Explain** how convection currents are related to plate tectonics.
4. **Think Critically** Using **Figure 9** and a world map, determine what natural disasters might occur in Iceland. Also determine what disasters might occur in Tibet. Explain why some Icelandic disasters are not expected to occur in Tibet.

Applying Skills

5. **Predict** Plate tectonic activity causes many events that can be dangerous to humans. One of these events is a seismic sea wave, or tsunami. Learn how scientists predict the arrival time of a tsunami in a coastal area.
6. **Use a Word Processor** Write three separate descriptions of the three basic types of plate boundaries—divergent boundaries, convergent boundaries, and transform boundaries. Then draw a sketch of an example of each boundary next to your description.

Use the Internet

Predicting Tectonic Activity

Goals

- **Research** the locations of earthquakes and volcanic eruptions around the world.
- **Plot** earthquake epicenters and the locations of volcanic eruptions.
- **Predict** locations that are tectonically active based on a plot of the locations of earthquake epicenters and active volcanoes.

Data Source

Science Online

Visit **earth.msscience.com/internet_lab** for more information about earthquake and volcano sites, and data from other students.

Real-World Question

The movement of plates on Earth causes forces that build up energy in rocks. The release of this energy can produce vibrations in Earth that you know as earthquakes. Earthquakes occur every day. Many of them are too small to be felt by humans, but each event tells scientists something more about the planet. Active volcanoes can do the same and often form at plate boundaries.

Can you predict tectonically active areas by plotting locations of earthquake epicenters and volcanic eruptions?

Think about where earthquakes and volcanoes have occurred in the past. Make a hypothesis about whether the locations of earthquake epicenters and active volcanoes can be used to predict tectonically active areas.

Make a Plan

1. Make a data table in your Science Journal like the one shown.
2. Collect data for earthquake epicenters and volcanic eruptions for at least the past two weeks. Your data should include the longitude and latitude for each location. For help, refer to the data sources given on the opposite page.

Locations of Epicenters and Eruptions

Earthquake Epicenter/ Volcanic Eruption	Longitude	Latitude
	Do not write in this book.	

Follow Your Plan

1. Make sure your teacher approves your plan before you start.
2. **Plot** the locations of earthquake epicenters and volcanic eruptions on a map of the world. Use an overlay of tissue paper or plastic.
3. After you have collected the necessary data, predict where the tectonically active areas on Earth are.
4. **Compare and contrast** the areas that you predicted to be tectonically active with the plate boundary map shown in **Figure 9.**

Analyze Your Data

1. What areas on Earth do you predict to be the locations of tectonic activity?
2. How close did your prediction come to the actual location of tectonically active areas?

Conclude and Apply

1. How could you make your predictions closer to the locations of actual tectonic activity?
2. Would data from a longer period of time help? Explain.
3. What types of plate boundaries were close to your locations of earthquake epicenters? Volcanic eruptions?
4. **Explain** which types of plate boundaries produce volcanic eruptions. Be specific.

Communicating Your Data

Find this lab using the link below. Post your data in the table provided. **Compare** your data to those of other students. Combine your data with those of other students and **plot** these combined data on a map to recognize the relationship between plate boundaries, volcanic eruptions, and earthquake epicenters.

Science Online
earth.msscience.com/internet_lab

Science and Language Arts

Listening In

by Gordon Judge

I'm just a bit of seafloor on this mighty solid sphere.
With no mind to be broadened, I'm quite content
down here.
The mantle churns below me, and the sea's in turmoil, too;
But nothing much disturbs me, I'm rock solid through
and through.

I do pick up occasional low-frequency vibrations –
(I think, although I can't be sure, they're sperm whales'
conversations).
I know I shouldn't listen in, but what else can I do?
It seems they are all studying for degrees from the OU.

They've mentioned me in passing, as their minds begin
improving:

I think I've heard them say
"The theory says the sea-
floor's moving...".
They call it "Plate Tectonics", this
new theory in their noddle.
If they would only ask me, I
could tell them it's all
twaddle....

But, how can I be moving, when I know full well myself
That I'm quite firmly anchored to a continental shelf?
"Well, the continent is moving, too; you're *pushing* it,
you see,"
I hear those OU whales intone, hydro-acoustically....

Well, thank you very much, OU. You've upset my
composure.
Next time you send your student whales to look at
my exposure
I'll tell them it's a load of tosh: it's *they* who move,
not me,
Those arty-smarty blobs of blubber, clogging up the sea!

Understanding Literature

Point of View Point of view refers to the perspective from which an author writes. This poem begins, "I'm just a bit of seafloor...." Right away, you know that the poem, or story, is being told from the point of view of the speaker, or the "first person." What effect does the first-person narration have on the story?

Respond to the Reading

1. Who is narrating the poem?
2. Why might the narrator think he or she hasn't moved?
3. **Linking Science and Writing** Using the first-person point of view, write an account from the point of view of a living or nonliving thing.

Volcanoes can occur where two plates move toward each other. When an oceanic plate and a continental plate collide, a volcano will form. Subduction zones occur when one plate sinks under another plate. Rocks melt in the zones where these plates converge, causing magma to move upward and form volcanic mountains.

Reviewing Main Ideas

Section 1 Continental Drift

1. Alfred Wegener suggested that the continents were joined together at some point in the past in a large landmass he called Pangaea. Wegener proposed that continents have moved slowly, over millions of years, to their current locations.
2. The puzzlelike fit of the continents, fossils, climatic evidence, and similar rock structures support Wegener's idea of continental drift. However, Wegener could not explain what process could cause the movement of the landmasses.

Section 2 Seafloor Spreading

1. Detailed mapping of the ocean floor in the 1950s showed underwater mountains and rift valleys.
2. In the 1960s, Harry Hess suggested seafloor spreading as an explanation for the formation of mid-ocean ridges.
3. The theory of seafloor spreading is supported by magnetic evidence in rocks and by the ages of rocks on the ocean floor.

Section 3 Theory of Plate Tectonics

1. In the 1960s, scientists combined the ideas of continental drift and seafloor spreading to develop the theory of plate tectonics. The theory states that the surface of Earth is broken into sections called plates that move around on the asthenosphere.
2. Currents in Earth's mantle called convection currents transfer heat in Earth's interior. It is thought that this transfer of heat energy moves plates.
3. Earth is a dynamic planet. As the plates move, they interact, resulting in many of the features of Earth's surface.

Visualizing Main Ideas

Copy and complete the concept map below about continental drift, seafloor spreading, and plate tectonics.

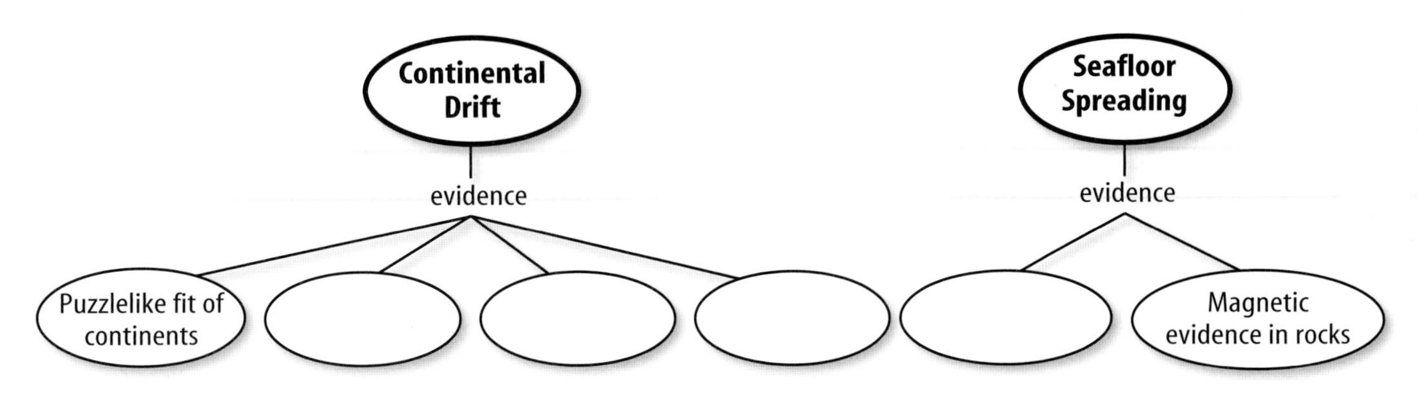

Using Vocabulary

asthenosphere p. 280	Pangaea p. 272
continental drift p. 272	plate p. 280
convection current p. 285	plate tectonics p. 280
lithosphere p. 280	seafloor spreading p. 277

Each phrase below describes a vocabulary term from the list. Write the term that matches the phrase describing it.

1. plasticlike layer below the lithosphere
2. idea that continents move slowly across Earth's surface
3. large, ancient landmass that consisted of all the continents on Earth
4. composed of oceanic or continental crust and upper mantle
5. explains locations of mountains, trenches, and volcanoes
6. theory proposed by Harry Hess that includes processes along mid-ocean ridges

Checking Concepts

Choose the word or phrase that best answers the question.

7. Which layer of Earth contains the asthenosphere?
   A) crust **C)** outer core
   B) mantle **D)** inner core
8. What type of plate boundary is the San Andreas Fault part of?
   A) divergent **C)** convergent
   B) subduction **D)** transform
9. What hypothesis states that continents slowly moved to their present positions on Earth?
   A) subduction **C)** continental drift
   B) erosion **D)** seafloor spreading

Use the illustration below to answer question 10.

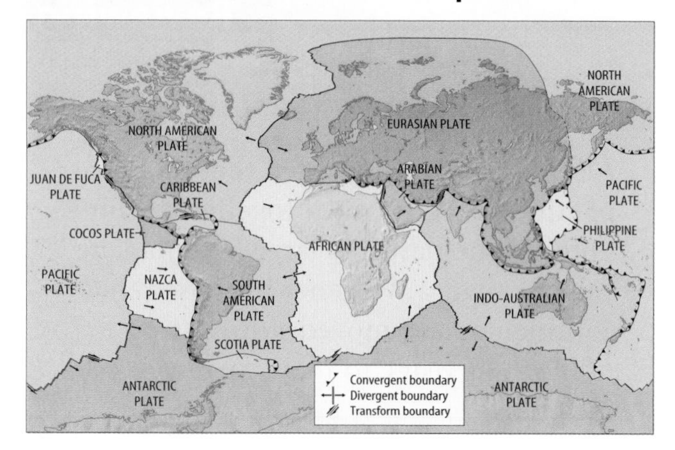

10. Which plate is subducting beneath the South American Plate?
    A) Nazca **C)** North American
    B) African **D)** Indo-Australian
11. Which of the following features are evidence that many continents were at one time near Earth's south pole?
    A) glacial deposits **C)** volcanoes
    B) earthquakes **D)** mid-ocean ridges
12. What evidence in rocks supports the theory of seafloor spreading?
    A) plate movement
    B) magnetic reversals
    C) subduction
    D) convergence
13. Which type of plate boundary is the Mid-Atlantic Ridge a part of?
    A) convergent **C)** transform
    B) divergent **D)** subduction
14. What theory states that plates move around on the asthenosphere?
    A) continental drift
    B) seafloor spreading
    C) subduction
    D) plate tectonics

Science Online earth.msscience.com/vocabulary_puzzlemaker

Thinking Critically

15. **Infer** Why do many earthquakes but few volcanic eruptions occur in the Himalaya?
16. **Explain** Glacial deposits often form at high latitudes near the poles. Explain why glacial deposits have been found in Africa.
17. **Describe** how magnetism is used to support the theory of seafloor spreading.
18. **Explain** why volcanoes do not form along the San Andreas Fault.
19. **Explain** why the fossil of an ocean fish found on two different continents would not be good evidence of continental drift.
20. **Form Hypotheses** Mount St. Helens in the Cascade Range is a volcano. Use **Figure 9** and a U.S. map to hypothesize how it might have formed.
21. **Concept Map** Make an events-chain concept map that describes seafloor spreading along a divergent plate boundary. Choose from the following phrases: *magma cools to form new seafloor, convection currents circulate hot material along divergent boundary,* and *older seafloor is forced apart.*

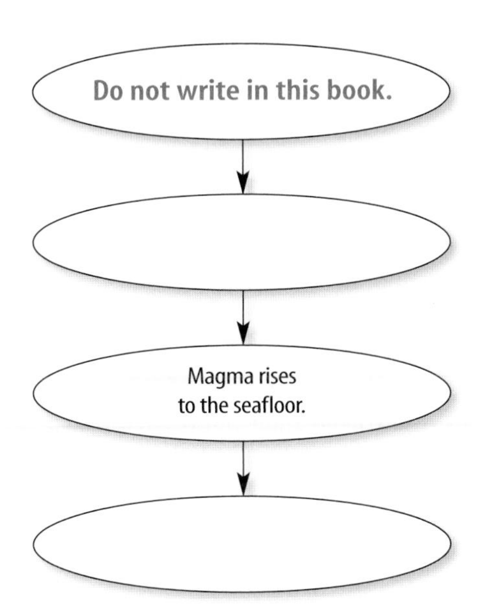

Performance Activities

22. **Observe and Infer** In the MiniLab called "Modeling Convection Currents," you observed convection currents produced in water as it was heated. Repeat the experiment, placing sequins, pieces of wood, or pieces of rubber bands into the water. How do their movements support your observations and inferences from the MiniLab?

Applying Math

23. **A Growing Rift** Movement along the African Rift Valley is about 2.1 cm per year. If plates continue to move apart at this rate, how much larger will the rift be (in meters) in 1,000 years? In 15,500 years?

Use the illustration below to answer questions 24 and 25.

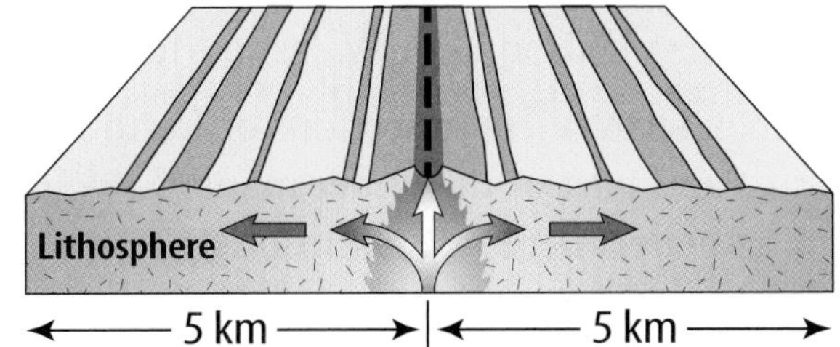

24. **New Seafloor** 10 km of new seafloor has been created in 50,000 years, with 5 km on each side of a mid-ocean ridge. What is the rate of movement, in km per year, of each plate? In cm per year?
25. **Use a Ratio** If 10 km of seafloor were created in 50,000 years, how many kilometers of seafloor were created in 10,000 years? How many years will it take to create a total of 30 km of seafloor?

Standardized Test Practice

Part 1 Multiple Choice

Record your answers on the answer sheet provided by your teacher or on a sheet of paper.

Use the illustration below to answer question 1.

1. What is the name of the ancient supercontinent shown above?
   A. Pangaea **C.** Laurasia
   B. Gondwanaland **D.** North America

2. Who developed the continental drift hypothesis?
   A. Harry Hess **C.** Alfred Wegener
   B. J. Tuzo Wilson **D.** W. Jason Morgan

3. Which term refers to sections of Earth's crust and part of the upper mantle?
   A. asthenosphere **C.** lithosphere
   B. plate **D.** core

4. About how fast do plates move?
   A. a few millimeters each year
   B. a few centimeters each year
   C. a few meters each year
   D. a few kilometers each year

Test-Taking Tip

Marking Answers Be sure to ask if it is okay to mark in the test booklet when taking the test, but make sure you mark all answers on your answer sheet.

5. Where do Earth's plates slide past each other?
   A. convergent boundaries
   B. divergent boundaries
   C. transform boundaries
   D. subduction zones

Study the diagram below before answering questions 6 and 7.

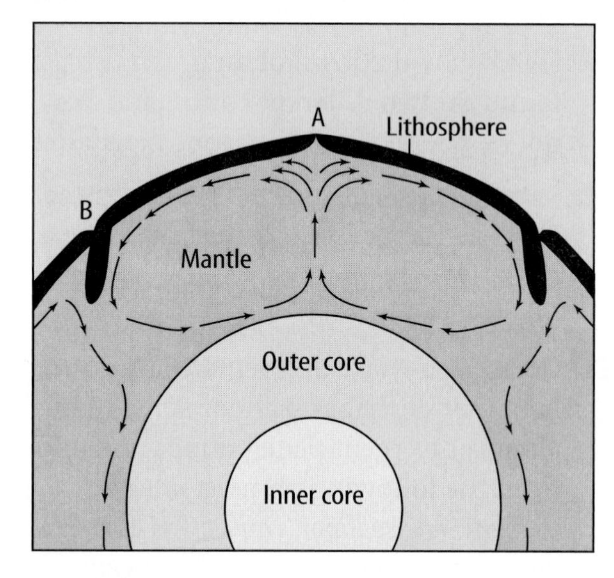

6. Suppose that the arrows in the diagram represent patterns of convection in Earth's mantle. Which type of plate boundary is most likely to occur along the region labeled "A"?
   A. transform
   B. reverse
   C. convergent
   D. divergent

7. Which statement is true of the region marked "B" on the diagram?
   A. Plates move past each other sideways.
   B. Plates move apart and volcanoes form.
   C. Plates move toward each other and volcanoes form.
   D. Plates are not moving.

Part 2 Short Response/Grid In

Record your answers on the answer sheet provided by your teacher or on a sheet of paper.

8. What is an ocean trench? Where do they occur?

9. How do island arcs form?

10. Why do earthquakes occur along the San Andreas Fault?

11. Describe a mid-ocean ridge.

12. Why do plates sometimes sink into the mantle?

Use the graph below to answer questions 13–15.

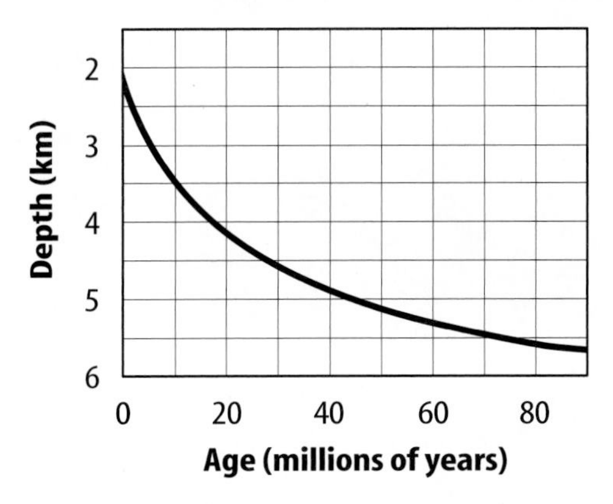

13. Use the graph to estimate the average depth below the ocean of ocean crust that has just formed.

14. Estimate the average depth of ocean crust that is 60 million years old.

15. Describe how the depth of ocean crust is related to the age of ocean crust.

16. On average, about how fast do plates move?

17. What layer in Earth's mantle do plates slide over?

18. Describe how scientists make maps of the ocean floor.

Part 3 Open Ended

Record your answers on a sheet of paper.

Use the illustration below to answer question 19.

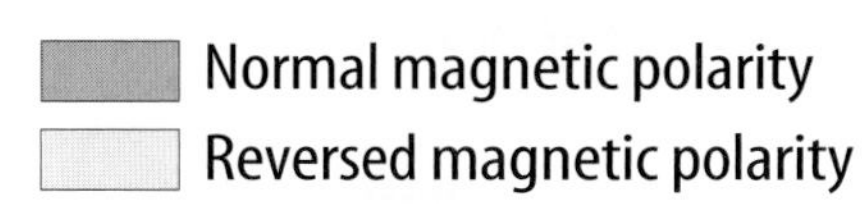

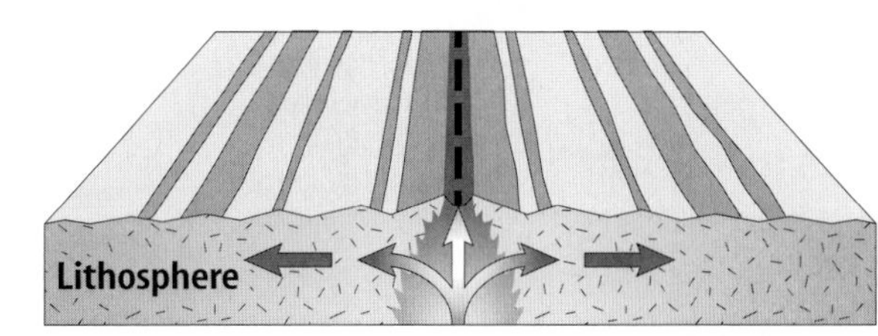

19. Examine the diagram above. Explain how the magnetic stripes form in rock that makes up the ocean crust.

20. What causes convection in Earth's mantle?

21. Explain the theory of plate tectonics.

22. What happened to the continents that made up Pangaea after it started to break up?

23. How does Earth's lithosphere differ from Earth's asthenosphere?

24. What types of life have been discovered near mid-ocean ridges?

25. What are the three types of motion that occur at plate boundaries? Describe each motion.

26. What forms when continents collide? Describe the process.

27. What occurs at the center of a mid-ocean ridge? What might you find there?

28. What evidence do we have that supports the hypothesis of continental drift?

29. Who proposed the first theories about plate tectonics? Explain why other scientists questioned these theories.

chapter 11

Earthquakes

The BIG Idea

Earthquakes are natural hazards that result from movement of Earth's plates.

SECTION 1
Forces Inside Earth

Main Idea Most earthquakes occur at plate boundaries when rocks break and move along faults.

SECTION 2
Features of Earthquakes

Main Idea Seismic waves provide data that can be interpreted to determine earthquake locations and features of Earth's interior.

SECTION 3
People and Earthquakes

Main Idea The effects of an earthquake depend on its size and the geology and types of structures in a region.

Was anyone hurt?

On October 17, 1989, the Loma Prieta earthquake rocked San Francisco, CA, leaving 62 dead and many more injured. Seismologists try to predict when and where earthquakes will occur so they can warn people of possible danger.

Science Journal Write *three* things that you would ask a scientist studying earthquakes.

Start-Up Activities

Why do earthquakes occur?

The bedrock beneath the soil can break to form cracks and move, forming faults. When blocks of bedrock move past each other along a fault, they cause the ground to shake. Why doesn't a block of bedrock move all the time, causing constant earthquakes? You'll find out during this activity.

1. Tape a sheet of medium-grain sandpaper to the tabletop.
2. Tape a second sheet of sandpaper to the cover of a textbook.
3. Place the book on the table so that both sheets of sandpaper meet.
4. Tie two large, thick rubber bands together and loop one of the rubber bands around the edge of the book so that it is not touching the sandpaper.
5. Pull on the free rubber band until the book moves. Record your observations.
6. **Think Critically** Write a paragraph that describes how the book moved. Using this model, predict why blocks of bedrock don't move all the time.

Earthquakes and Earth's Crust Make the following Foldable to help you understand the cause-and-effect relationship between earthquakes and movement in Earth's crust.

STEP 1 **Fold** a sheet of paper in half lengthwise.

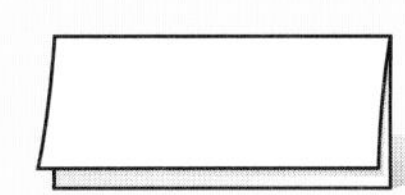

STEP 2 **Fold** paper down 2.5 cm from the top. (Hint: From the tip of your index finger to your middle knuckle is about 2.5 cm.)

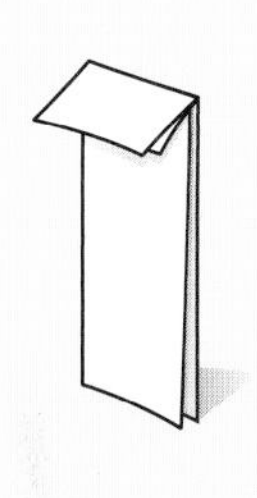

STEP 3 **Open and draw** lines along the 2.5 cm fold. **Label** as shown.

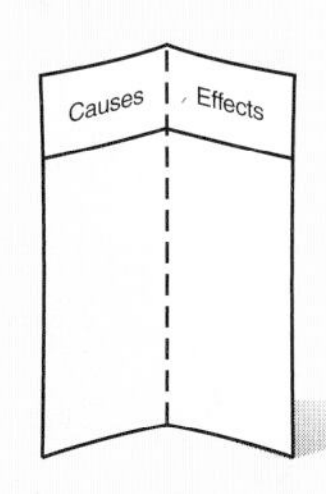

Read and Write As you read the chapter, write the causes and effects of earthquakes on your Foldable.

Preview this chapter's content and activities at earth.msscience.com

Get Ready to Read

Compare and Contrast

1 Learn It! Good readers compare and contrast information as they read. This means they look for similarities and differences to help them to remember important ideas. Look for signal words in the text to let you know when the author is comparing or contrasting.

Compare and Contrast Signal Words	
Compare	**Contrast**
as	but
like	or
likewise	unlike
similarly	however
at the same time	although
in a similar way	on the other hand

2 Practice It! Read the excerpt below and notice how the author uses contrast signal words to describe the differences between normal faults, reverse faults, and strike-slip faults.

Have you ever tried to drink a thick milkshake from a cup? Sometimes the milkshake is so thick that it won't flow. How do you make the milkshake flow? You shake it. Something **similar** can happen to very wet soil during an earthquake. Wet soil can be strong most of the time, **but** the shaking from an earthquake can cause the soil to act more **like** a liquid. This is called liquefaction.

3 Apply It! Compare and contrast normal faults, reverse faults, and strike-slip faults on pages 302 and 303.

Reading Tip

As you read, use other skills, such as summarizing and connecting, to help you understand comparisons and contrasts.

Target Your Reading

Use this to focus on the main ideas as you read the chapter.

1. **Before you read** the chapter, respond to the statements below on your worksheet or on a numbered sheet of paper.
   - Write an **A** if you **agree** with the statement.
   - Write a **D** if you **disagree** with the statement.

2. **After you read** the chapter, look back to this page to see if you've changed your mind about any of the statements.
   - If any of your answers changed, explain why.
   - Change any false statements into true statements.
   - Use your revised statements as a study guide.

Before You Read A or D		Statement	After You Read A or D
	1	Movement of Earth's plates can cause large sections of rock to bend, compress, or stretch.	
	2	A fault can be a large break, or crack, in Earth's crust even though there has never been movement along that break.	
	3	Earthquakes occur when rocks break and move along a fault and vibrations are created.	
	4	The shaking, or vibrations, that people feel during an earthquake are called seismic waves.	
	5	All seismic waves travel through Earth at the same speed.	
	6	The Richter magnitude scale is used to describe the strength of an earthquake.	
	7	Most earthquakes have magnitudes too low to be felt by humans.	
	8	Scientists can predict when and where an earthquake will occur.	

Science Online
Print out a worksheet of this page at earth.msscience.com

section 1

Forces Inside Earth

as you read

What You'll Learn

- **Explain** how earthquakes result from the buildup of energy in rocks.
- **Describe** how compression, tension, and shear forces make rocks move along faults.
- **Distinguish** among normal, reverse, and strike-slip faults.

Why It's Important

Earthquakes cause billions of dollars in property damage and kill an average of 10,000 people every year.

Review Vocabulary

plate: a large section of Earth's crust and rigid upper mantle that moves around on the asthenosphere

New Vocabulary

- fault
- earthquake
- normal fault
- reverse fault
- strike-slip fault

Earthquake Causes

Recall the last time you used a rubber band. Rubber bands stretch when you pull them. Because they are elastic, they return to their original shape once the force is released. However, if you stretch a rubber band too far, it will break. A wooden craft stick behaves in a similar way. When a force is first applied to the stick, it will bend and change shape. The energy needed to bend the stick is stored inside the stick as potential energy. If the force keeping the stick bent is removed, the stick will return to its original shape, and the stored energy will be released as energy of motion.

Fault Formation There is a limit to how far a wooden craft stick can bend. This is called its elastic limit. Once its elastic limit is passed, the stick remains bent or breaks, as shown in **Figure 1.** Rocks behave in a similar way. Up to a point, applied forces cause rocks to bend and stretch, undergoing what is called elastic deformation. Once the elastic limit is passed, the rocks may break. When rocks break, they move along surfaces called **faults.** A tremendous amount of force is required to overcome the strength of rocks and to cause movement along a fault. Rock along one side of a fault can move up, down, or sideways in relation to rock along the other side of the fault.

When a force is applied, the stick will bend and change shape.
When the elastic limit is passed, the stick will break.

Figure 1 The bending and breaking of wooden craft sticks are similar to how rocks bend and break.

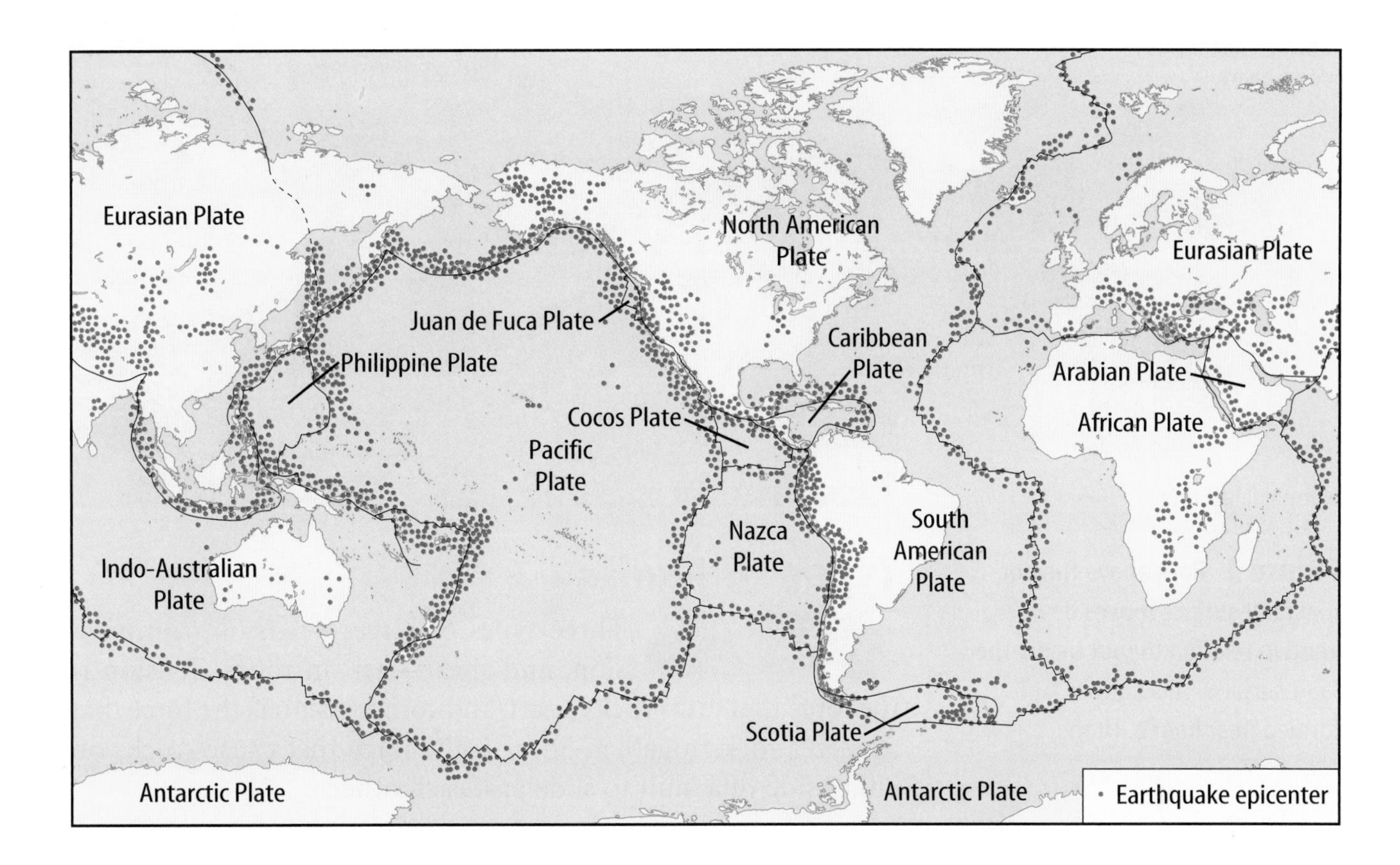

Figure 2 The dots represent the epicenters of major earthquakes over a ten-year period. Note that most earthquakes occur near plate boundaries.
Form a hypothesis *to explain why earthquakes rarely occur in the middle of a plate.*

What causes faults? What produces the forces that cause rocks to break and faults to form? The surface of Earth is in constant motion because of forces inside the planet. These forces cause sections of Earth's surface, called plates, to move. This movement puts stress on the rocks near the plate edges. To relieve this stress, the rocks tend to bend, compress, or stretch. If the force is great enough, the rocks will break. An **earthquake** is the vibrations produced by the breaking of rock. **Figure 2** shows how the locations of earthquakes outline the plates that make up Earth's surface.

Reading Check *Why do most earthquakes occur near plate boundaries?*

How Earthquakes Occur As rocks move past each other along a fault, their rough surfaces catch, temporarily halting movement along the fault. However, forces keep driving the rocks to move. This action builds up stress at the points where the rocks are stuck. The stress causes the rocks to bend and change shape. When the rocks are stressed beyond their elastic limit, they can break, move along the fault, and return to their original shapes. An earthquake results. Earthquakes range from unnoticeable vibrations to devastating waves of energy. Regardless of their intensity, most earthquakes result from rocks moving over, under, or past each other along fault surfaces.

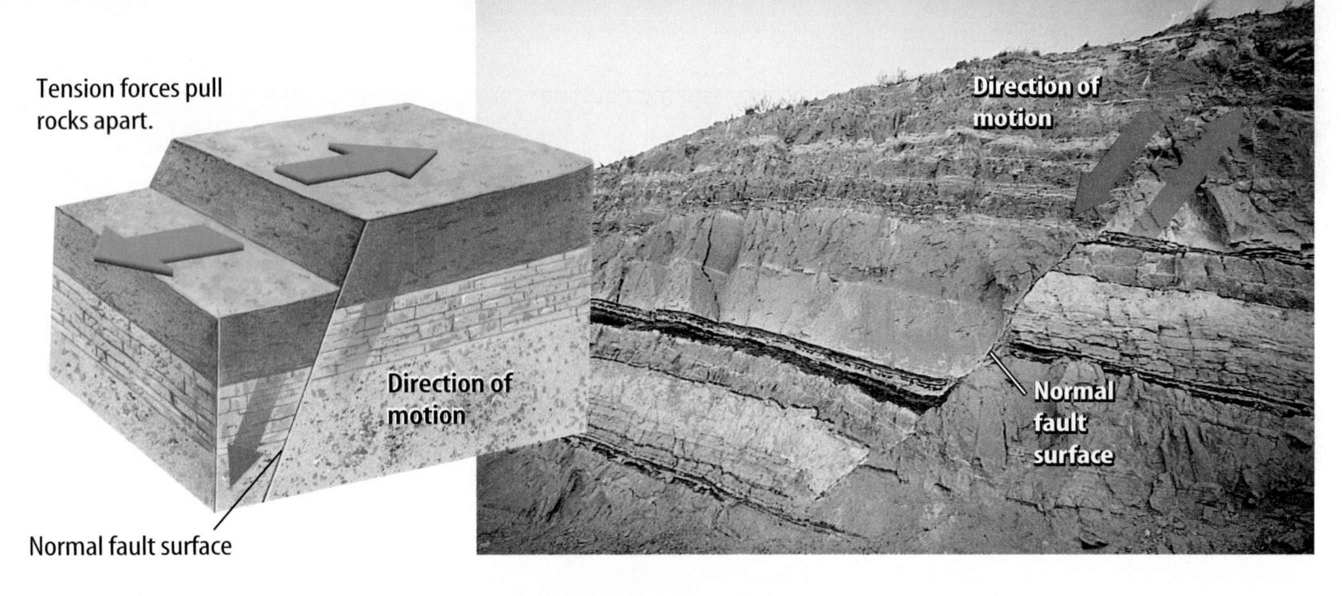

Figure 3 Rock above the normal fault surface moves downward in relation to rock below the fault surface. This normal fault formed near Kanab, Utah.

Types of Faults

INTEGRATE Physics Three types of forces—tension, compression, and shear—act on rocks. Tension is the force that pulls rocks apart, and compression is the force that squeezes rocks together. Shear is the force that causes rocks on either side of a fault to slide past each other.

Normal Faults Tensional forces inside Earth cause rocks to be pulled apart. When rocks are stretched by these forces, a normal fault can form. Along a **normal fault,** rock above the fault surface moves downward in relation to rock below the fault surface. The motion along a normal fault is shown in **Figure 3.** Notice the normal fault shown in the photograph above.

Reverse Faults Reverse faults result from compression forces that squeeze rock. **Figure 4** shows the motion along a reverse fault. If rock breaks from forces pushing from opposite directions, rock above a **reverse fault** surface is forced up and over the rock below the fault surface. The photo below shows a large reverse fault in California.

Figure 4 The rock above the reverse fault surface moves upward in relation to the rock below the fault surface.

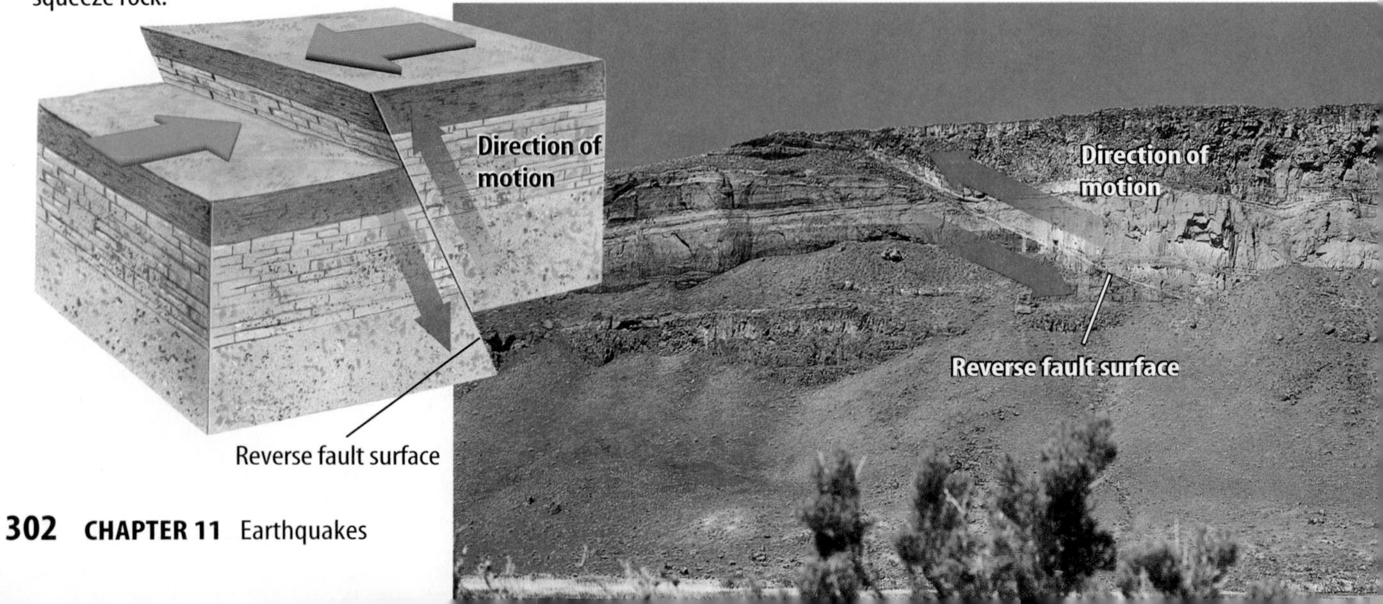

Figure 5 Shear forces push on rock in opposite—but not directly opposite—horizontal directions. When they are strong enough, these forces split rock and create strike-slip faults.

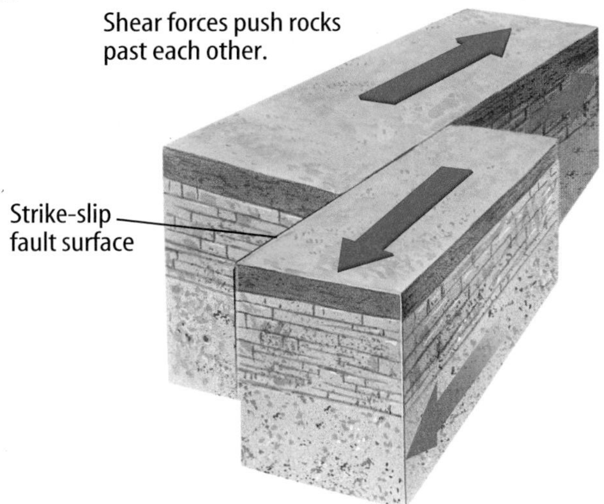

Strike-Slip Faults At a **strike-slip fault,** shown in **Figure 5,** rocks on either side of the fault are moving past each other without much upward or downward movement. The photo above shows the largest fault in California—the San Andreas Fault—which stretches more than 1,100 km through the state. The San Andreas Fault is the boundary between two of Earth's plates that are moving sideways past each other.

Reading Check *What is a strike-slip fault?*

section 1 review

Summary

Earthquake Causes

- Faults form when stressed rocks break along surfaces.
- Stresses on rock are created by plate movements.
- When rocks break along a fault, vibrations are created. This is an earthquake.

Types of Faults

- Normal faults can form when rocks undergo tension.
- Compression forces produce reverse faults.
- Strike-slip faults result when rocks move past each other without much upward or downward movement.

Self Check

1. **Infer** The Himalaya in Tibet formed when two of Earth's plates collided. What types of faults would you expect to find in these mountains? Why?
2. **State** In what direction do rocks move above a normal fault surface? What force causes this?
3. **Describe** how compression forces make rocks move along a reverse fault.
4. **Think Critically** Why is it easier to predict where an earthquake will occur than it is to predict when it will occur?

Applying Skills

5. **Infer** Why do the chances of an earthquake increase rather than decrease as time passes since the last earthquake?

section 2

Features of Earthquakes

as you read

What You'll Learn

- **Explain** how earthquake energy travels in seismic waves.
- **Distinguish** among primary, secondary, and surface waves.
- **Describe** the structure of Earth's interior.

Why It's Important

Seismic waves are responsible for most damage caused by earthquakes.

Review Vocabulary
wave: rhythmic movement that carries energy through matter and space

New Vocabulary
- seismic wave
- focus
- primary wave
- secondary wave
- surface wave
- epicenter
- seismograph

Seismic Waves

When two people hold opposite ends of a rope and shake one end, as shown in **Figure 6,** they send energy through the rope in the form of waves. Like the waves that travel through the rope, **seismic** (SIZE mihk) **waves** generated by an earthquake travel through Earth. During a strong earthquake, the ground moves forward and backward, heaves up and down, and shifts from side to side. The surface of the ground can ripple like waves do in water. Imagine trying to stand on ground that had waves traveling through it. This is what you might experience during a strong earthquake.

Origin of Seismic Waves You learned earlier that rocks move past each other along faults, creating stress at points where the rocks' irregular surfaces catch each other. The stress continues to build up until the elastic limit is exceeded and energy is released in the form of seismic waves. The point where this energy release first occurs is the **focus** (plural, *foci*) of the earthquake. The foci of most earthquakes are within 65 km of Earth's surface. A few have been recorded as deep as 700 km. Seismic waves are produced and travel outward from the earthquake focus.

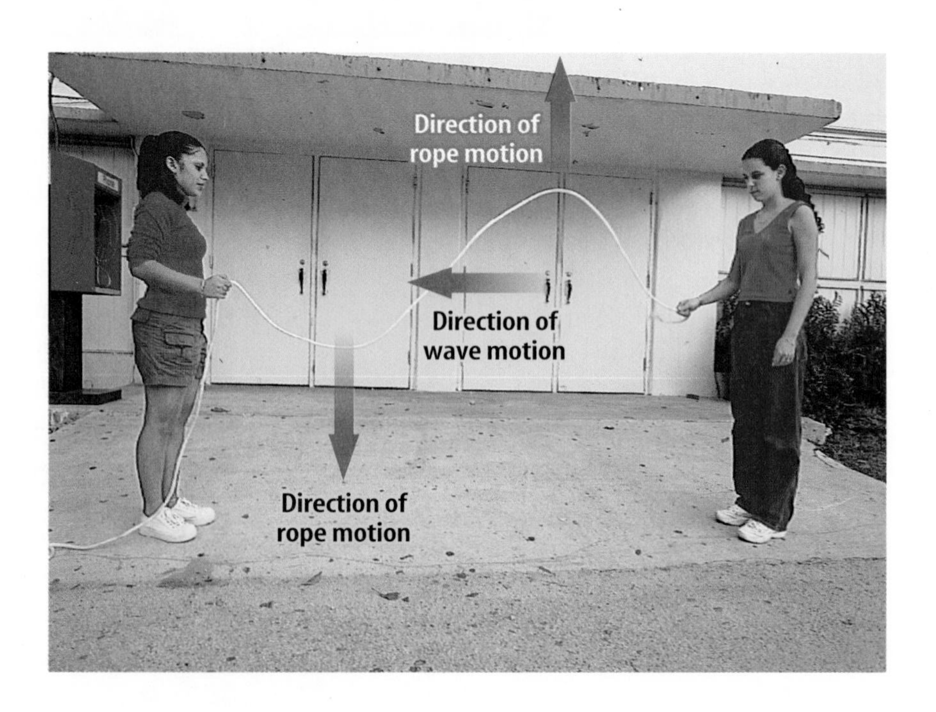

Figure 6 Some seismic waves are similar to the wave that is traveling through the rope. Note that the rope moves perpendicular to the wave direction.

Primary Waves When earthquakes occur, three different types of seismic waves are produced. All of the waves are generated at the same time, but each behaves differently within Earth. **Primary waves** (P-waves) cause particles in rocks to move back and forth in the same direction that the wave is traveling. If you squeeze one end of a coiled spring and then release it, you cause it to compress and then stretch as the wave travels through the spring, as shown in **Figure 7.** Particles in rocks also compress and then stretch apart, transmitting primary waves through the rock.

Figure 7 Primary waves move through Earth the same way that a wave travels through a coiled spring.

Secondary and Surface Waves **Secondary waves** (S-waves) move through Earth by causing particles in rocks to move at right angles to the direction of wave travel. The wave traveling through the rope shown in **Figure 6** is an example of a secondary wave.

Surface waves cause most of the destruction resulting from earthquakes. **Surface waves** move rock particles in a backward, rolling motion and a side-to-side, swaying motion, as shown in **Figure 8.** Many buildings are unable to withstand intense shaking because they are made with stiff materials. The buildings fall apart when surface waves cause different parts of the building to move in different directions.

Reading Check *Why do surface waves damage buildings?*

Surface waves are produced when earthquake energy reaches the surface of Earth. Surface waves travel outward from the epicenter. The earthquake **epicenter** (EH pih sen tur) is the point on Earth's surface directly above the earthquake focus. Find the focus and epicenter in **Figure 9.**

Sound Waves When sound is produced, waves move through air or some other material. Research sound waves to find out which type of seismic wave they are similar to.

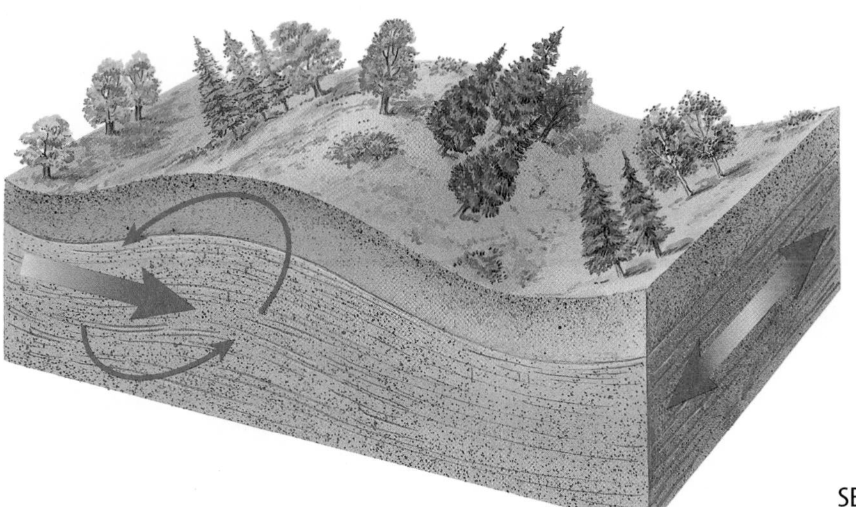

Figure 8 Surface waves move rock particles in a backward, rolling motion and a side-to-side, swaying motion.
Compare and contrast *surface waves and secondary waves.*

NATIONAL GEOGRAPHIC VISUALIZING SEISMIC WAVES

Figure 9

As the plates that form Earth's lithosphere move, great stress is placed on rocks. They bend, stretch, and compress. Occasionally, rocks break, producing earthquakes that generate seismic waves. As shown here, different kinds of seismic waves—each with distinctive characteristics—move outward from the focus of the earthquake.

C The point on Earth's surface directly above an earthquake's focus is known as the epicenter. Surface waves spread out from the epicenter like ripples in a pond.

D The amplitudes, or heights, of surface waves are greater than those of primary and secondary waves. Surface waves cause the most damage during an earthquake.

B Primary waves and secondary waves originate at the focus and travel outward in all directions. Primary waves travel about twice as fast as secondary waves.

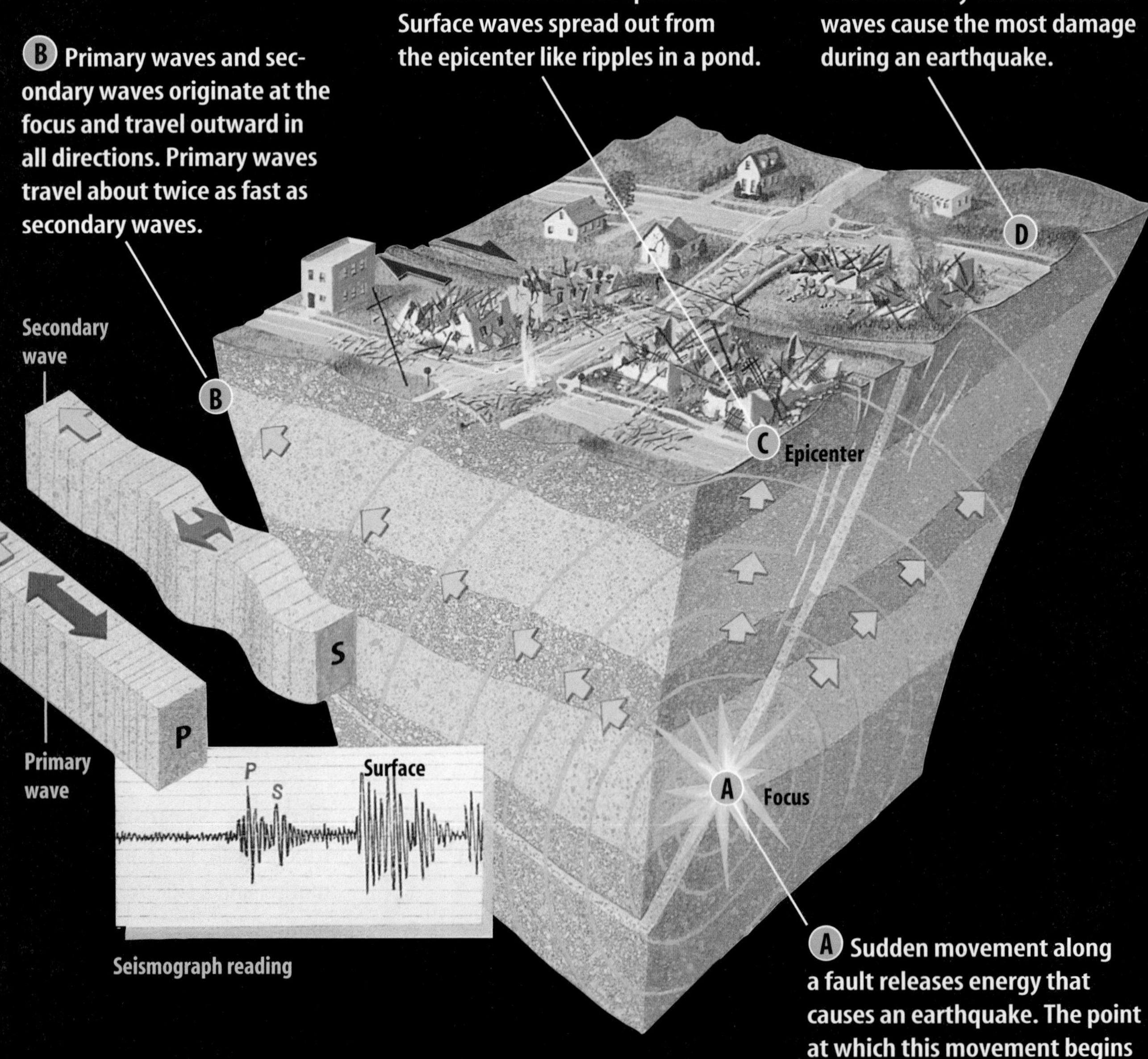

A Sudden movement along a fault releases energy that causes an earthquake. The point at which this movement begins is called the earthquake's focus.

Locating an Epicenter

Different seismic waves travel through Earth at different speeds. Primary waves are the fastest, secondary waves are slower, and surface waves are the slowest. Can you think of a way this information could be used to determine how far away an earthquake epicenter is? Think of the last time you saw two people running in a race. You probably noticed that the faster person got further ahead as the race continued. Like runners in a race, seismic waves travel at different speeds.

Scientists have learned how to use the different speeds of seismic waves to determine the distance to an earthquake epicenter. When an epicenter is far from a location, the primary wave has more time to put distance between it and the secondary and surface waves, just like the fastest runner in a race.

Measuring Seismic Waves Seismic waves from earthquakes are measured with an instrument known as a **seismograph.** Seismographs register the waves and record the time that each arrived. Seismographs consist of a rotating drum of paper and a pendulum with an attached pen. When seismic waves reach the seismograph, the drum vibrates but the pendulum remains at rest. The stationary pen traces a record of the vibrations on the moving drum of paper. The paper record of the seismic event is called a seismogram. **Figure 10** shows two types of seismographs that measure either vertical or horizontal ground movement, depending on the orientation of the drum.

Topic: Earthquake Data
Visit earth.msscience.com for Web links to the National Earthquake Information Center and the World Data Center for Seismology.

Activity List the locations and distances of each reference that seigmograph stations used to determine the epicenter of the most recent earthquake.

Figure 10 Seismographs differ according to whether they are intended to measure horizontal or vertical seismic motions.
Infer *why one seismograph can't measure both horizontal and vertical motions.*

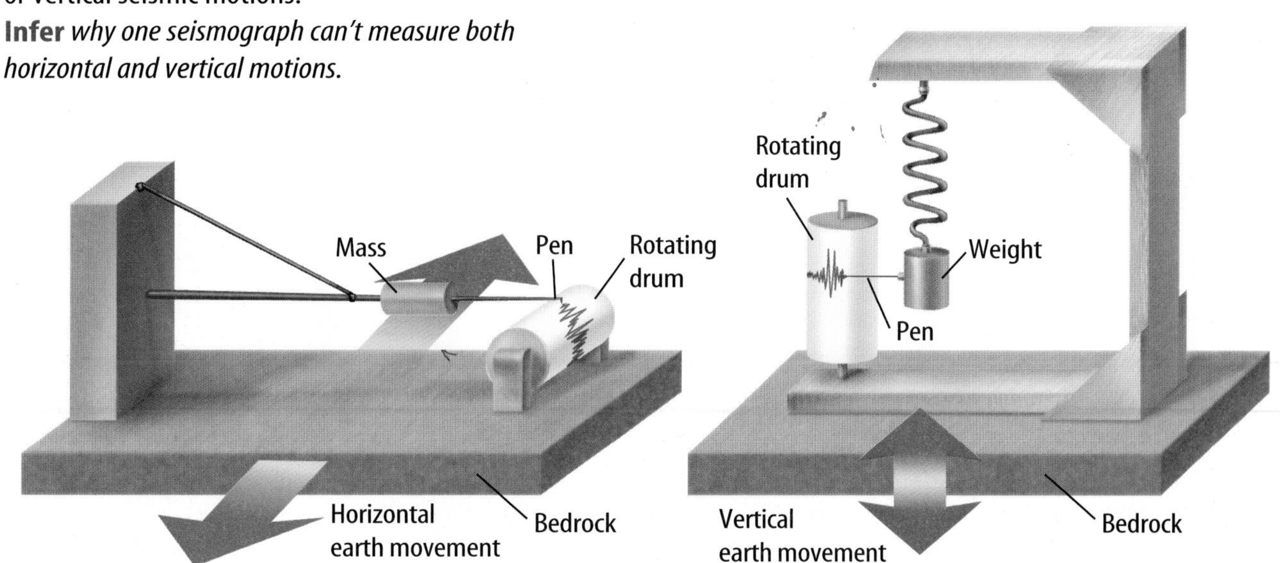

Seismograph Stations Each type of seismic wave reaches a seismograph station at a different time based on its speed. Primary waves arrive first at seismograph stations, and secondary waves, which travel slower, arrive second. Because surface waves travel slowest, they arrive at seismograph stations last. This difference in arrival times is used to calculate the distance from the seismograph station to the earthquake epicenter, as shown in **Figure 11.** If a seismograph station is located 4,000 km from an earthquake epicenter, primary waves will reach the station about 6 minutes before secondary waves.

If seismic waves reach three or more seismograph stations, the location of the epicenter can be determined. To locate an epicenter, scientists draw circles around each station on a map. The radius of each circle equals that station's distance from the earthquake epicenter. The point where all three circles intersect, shown in **Figure 12,** is the location of the earthquake epicenter.

Seismologists usually describe earthquakes based on their distances from the seismograph. Local events occur less than 100 km away. Regional events occur 100 km to 1,400 km away. Teleseismic events are those that occur at distances greater than 1,400 km.

Figure 11 Primary waves arrive at a seismograph station before secondary waves do.
Use Graphs *If primary waves reach a seismograph station two minutes before secondary waves, how far is the station from the epicenter?*

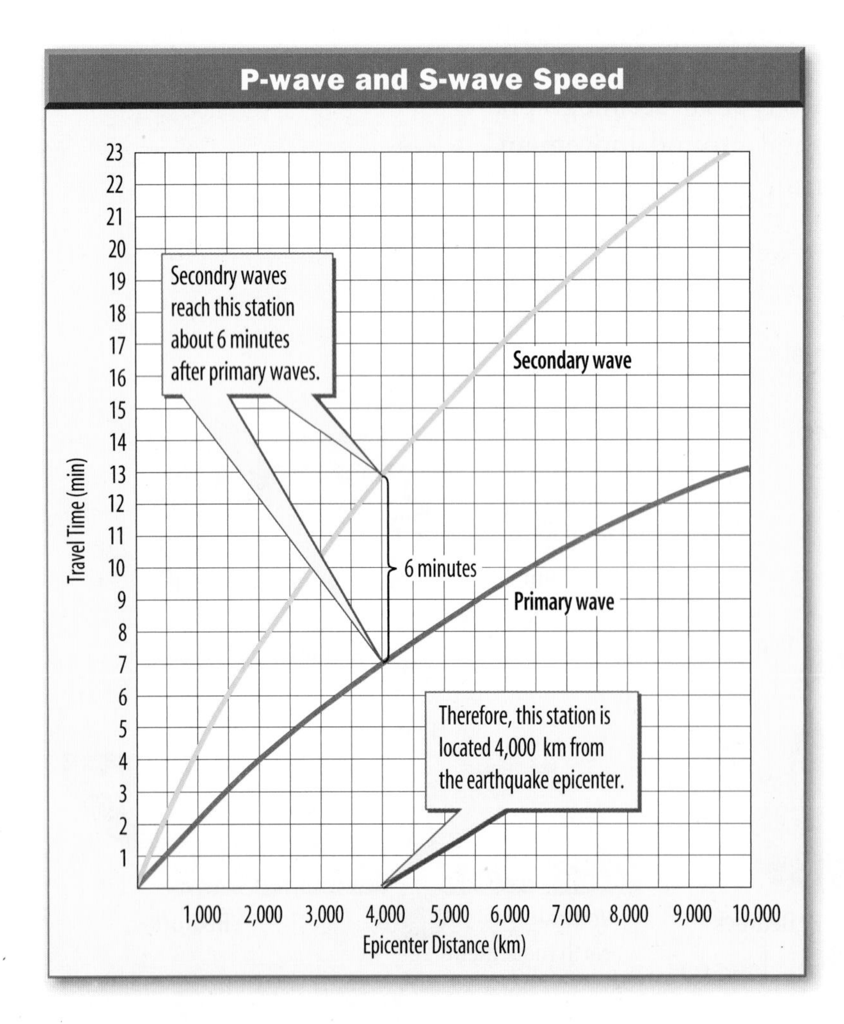

Figure 12 The radius of each circle is equal to the distance from the epicenter to each seismograph station. The intersection of the three circles is the location of the epicenter.

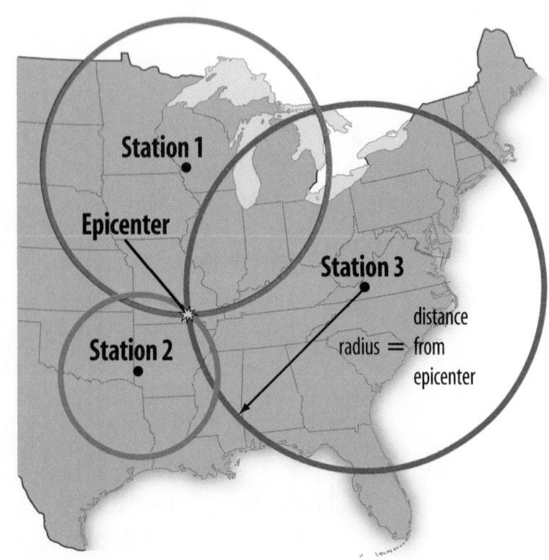

Basic Structure of Earth

Figure 13 shows Earth's internal structure. At the very center of Earth is a solid, dense inner core made mostly of iron with smaller amounts of nickel, oxygen, silicon, and sulfur. Pressure from the layers above causes the inner core to be solid. Above the solid inner core lies the liquid outer core, which also is made mainly of iron.

Reading Check *How do the inner and outer cores differ?*

Earth's mantle is the largest layer, lying directly above the outer core. It is made mostly of silicon, oxygen, magnesium, and iron. The mantle often is divided into an upper part and a lower part based on changing seismic wave speeds. A portion of the upper mantle, called the asthenosphere (as THE nuh sfihr), consists of weak rock that can flow slowly.

Earth's Crust The outermost layer of Earth is the crust. Together, the crust and a part of the mantle just beneath it make up Earth's lithosphere (LIH thuh sfihr). The lithosphere is broken into a number of plates that move over the asthenosphere beneath it.

The thickness of Earth's crust varies. It is more than 60 km thick in some mountainous regions and less than 5 km thick under some parts of the oceans. Compared to the mantle, the crust contains more silicon and aluminum and less magnesium and iron. Earth's crust generally is less dense than the mantle beneath it.

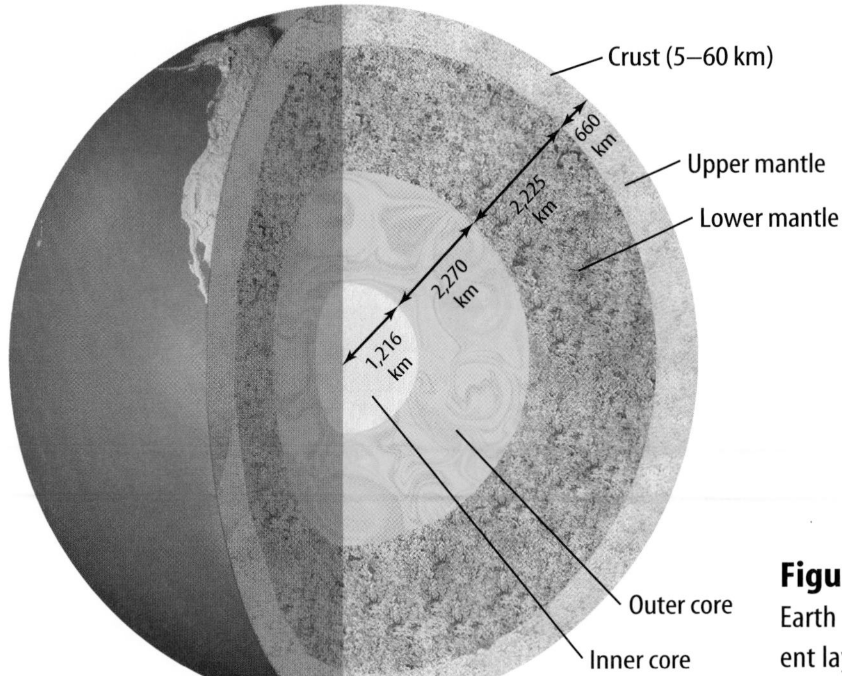

Figure 13 The internal structure of Earth shows that it is made of different layers.

Interpreting Seismic Wave Data

Procedure

Copy the table below into your **Science Journal.** Use the **graph** in **Figure 11** to determine the difference in arrival times for primary and secondary waves at the distances listed in the data table below. Two examples are provided for you.

Wave Data

Distance (km)	Difference in Arrival Time
1,500	2 min, 50 s
2,250	Do not write in this book.
2,750	
3,000	
4,000	5 min, 55 s
7,000	
9,000	

Analysis

1. What happens to the difference in arrival times as the distance from the earthquake increases?
2. If the difference in arrival times at a seismograph station is 6 min, 30 s, how far away is the epicenter?

Mapping Earth's Internal Structure As shown in **Figure 14,** the speeds and paths of seismic waves change as they travel through materials with different densities. By studying seismic waves that have traveled through Earth, scientists have identified different layers with different densities. In general, the densities increase with depth as pressures increase. Studying seismic waves has allowed scientists to map Earth's internal structure without being there.

Early in the twentieth century, scientists discovered that large areas of Earth don't receive seismic waves from an earthquake. In the area on Earth between 105° and 140° from the earthquake focus, no waves are detected. This area, called the shadow zone, is shown in **Figure 14.** Secondary waves are not transmitted through a liquid, so they stop when they hit the liquid outer core. Primary waves are slowed and bent but not stopped by the liquid outer core. Because of this, scientists concluded that the outer core and mantle are made of different materials. Primary waves speed up again as they travel through the solid inner core. The bending of primary waves and the stopping of secondary waves create the shadow zone.

Reading Check *Why do seismic waves change speed as they travel through Earth?*

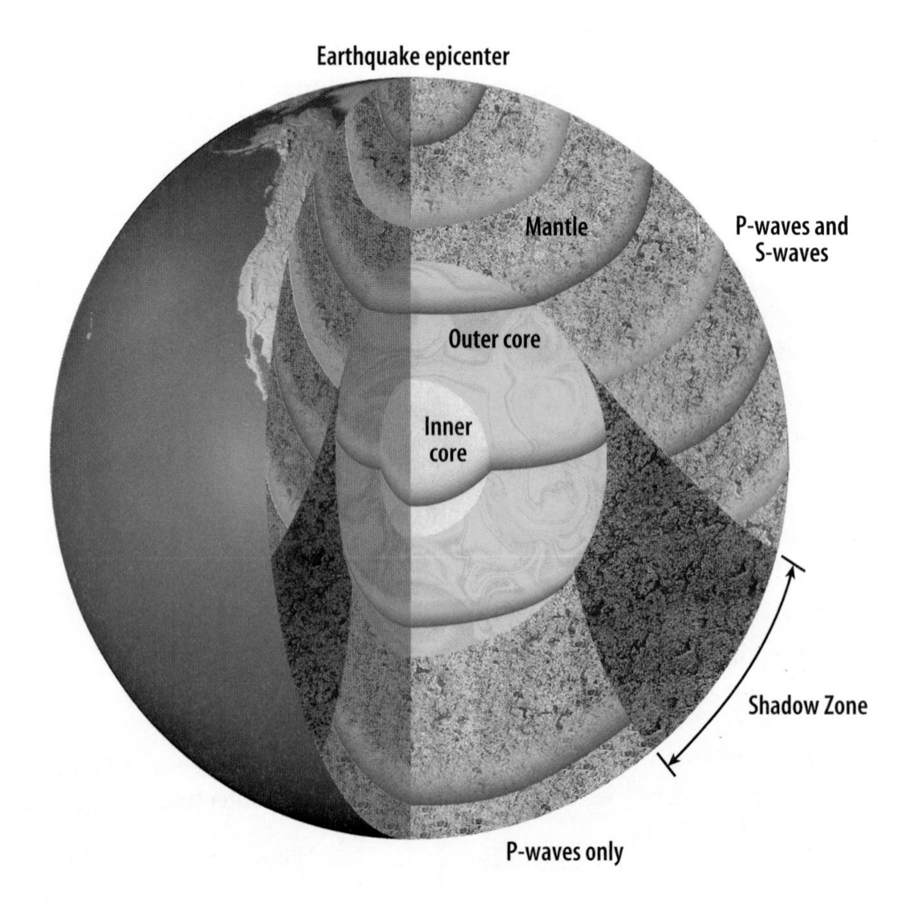

Figure 14 Seismic waves bend and change speed as the density of rock changes. Primary waves bend when they contact the outer core, and secondary waves are stopped completely. This creates a shadow zone where no seismic waves are received.

Layer Boundaries **Figure 15** shows how seismic waves change speed as they pass through layers of Earth. Seismic waves speed up when they pass through the bottom of the crust and enter the upper mantle, shown on the far left of the graph. This boundary between the crust and upper mantle is called the Mohorovicic discontinuity (moh huh ROH vee chihch • dis kahn tuh NEW uh tee), or Moho.

The mantle is divided into layers based on changes in seismic wave speeds. For example, primary and secondary waves slow down again when they reach the asthenosphere. Then they generally speed up as they move through a more solid region of the mantle below the asthenosphere.

The core is divided into two layers based on how seismic waves travel through it. Secondary waves do not travel through the liquid core, as you can see in the graph. Primary waves slow down when they reach the outer core, but they speed up again upon reaching the solid inner core.

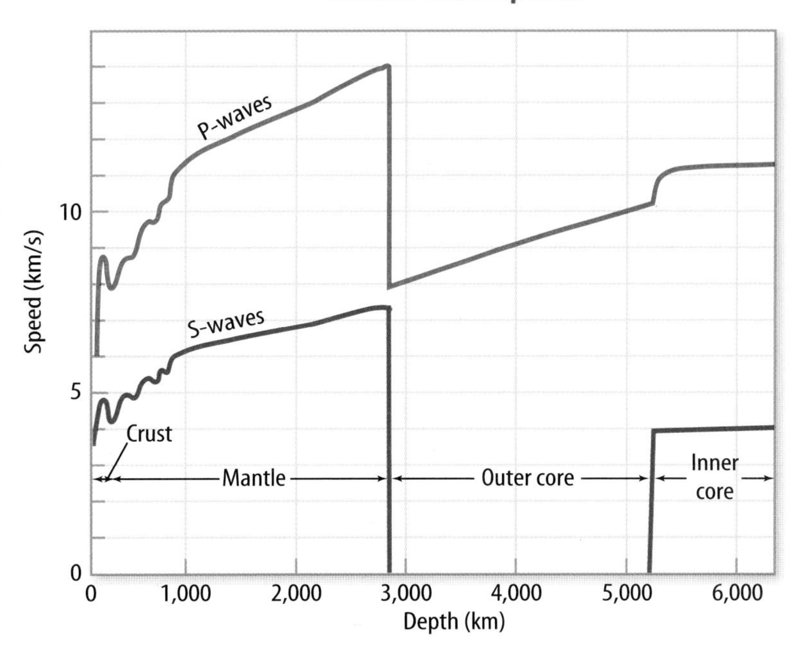

Figure 15 Changes in the speeds of seismic waves allowed scientists to detect boundaries between Earth's layers. S waves in the inner core form when P waves strike its surface.

section 2 review

Summary

Seismic Waves

- Stress builds up at the points where the surfaces of the fault touch.
- When the elastic limit of the rock is exceeded, it moves, producing seismic waves.
- There are three types of seismic waves—primary, secondary, and surface.

Locating an Epicenter

- A seismograph measures seismic waves.
- Three seismograph stations are needed to determine the location of an epicenter.

Basic Structure of Earth

- The inner core, the outer core, the lower mantle, the upper mantle, and the crust make up Earth.

Self Check

1. **Compare and contrast** the movement of rocks by primary waves, secondary waves, and surface waves.
2. **Explain** why surface waves cause the most damage to property.
3. **Describe** what makes up most of Earth's inner core.
4. **Explain** why three seismograph stations are needed to determine the location of an epicenter.
5. **Think Critically** Why do some seismograph stations receive both primary and secondary waves from an earthquake but other stations don't?

Applying Skills

6. **Simple Equations** Primary waves travel about 6 km/s through Earth's crust. The distance from Los Angeles to Phoenix is about 600 km. How long would it take primary waves to travel between the two cities?

Epicenter Location

In this lab you can plot the distance of seismograph stations from the epicenters of earthquakes and determine the location of earthquake epicenters.

Real-World Question

How can plotting the distance of several seismograph stations from an earthquake epicenter allow you to determine the locations of the epicenter?

Goals

- **Plot** the distances from several seismograph stations based on primary and secondary wave arrival times.
- **Interpret** the location of earthquake epicenters from these plots.

Materials

string globe
metric ruler chalk

Earthquake Data

Location of Seismograph	Wave	Wave Arrival Times Earthquake A	Earthquake B
New York, New York	P	2:24:05 P.M.	1:19:42 P.M.
	S	2:29:15 P.M.	1:25:27 P.M.
Seattle, Washington	P	2:24:40 P.M.	1:14:37 P.M.
	S	2:30:10 P.M.	1:16:57 P.M.
Rio de Janeiro, Brazil	P	2:29:10 P.M.	—
	S	2:37:50 P.M.	—
Paris, France	P	2:30:30 P.M.	1:24:57 P.M.
	S	2:40:10 P.M.	1:34:27 P.M.
Tokyo, Japan	P	—	1:24:27 P.M.
	S	—	1:33:27 P.M.

Procedure

1. Determine the difference in arrival time between the primary and secondary waves at each station for each earthquake listed in the table.
2. After you determine the arrival time differences for each seismograph station, use the graph in **Figure 11** to determine the distance in kilometers of each seismograph from the epicenter of each earthquake. Record these data in a data table. For example, the difference in arrival times in Paris for earthquake B is 9 min, 30 s. On the graph, the primary and secondary waves are separated along the vertical axis by 9 min, 30 s at a distance of 8,975 km.
3. Using the string, measure the circumference of the globe. Determine a scale of centimeters of string to kilometers on Earth's surface. (Earth's circumference is 40,000 km.)
4. For each earthquake, place one end of the string at each seismic station location on the globe. Use the chalk to draw a circle with a radius equal to the distance to the earthquake's epicenter.
5. **Identify** the epicenter for each earthquake.

Conclude and Apply

1. How is the distance of a seismograph from the earthquake related to the arrival times of the waves?
2. **Identify** the location of the epicenter for each earthquake.
3. How many stations were needed to locate each epicenter accurately?
4. **Explain** why some seismographs didn't receive seismic waves from some quakes.

section 3

People and Earthquakes

Earthquake Activity

Imagine waking up in the middle of the night with your bed shaking, windows shattering, and furniture crashing together. That's what many people in Northridge, California, experienced at 4:30 A.M. on January 17, 1994. The ground beneath Northridge shook violently—it was an earthquake.

Although the earthquake lasted only 15 s, it killed 51 people, injured more than 9,000 people, and caused $44 billion in damage. More than 22,000 people were left homeless. **Figure 16** shows some of the damage caused by the Northridge earthquake and a seismogram made by that quake.

Earthquakes are natural geological events that provide information about Earth. Unfortunately, they also cause billions of dollars in property damage and kill an average of 10,000 people every year. With so many lives lost and such destruction, it is important for scientists to learn as much as possible about earthquakes to try to reduce their impact on society.

as you read

What **You'll Learn**

- **Explain** where most earthquakes in the United States occur.
- **Describe** how scientists measure earthquakes.
- **List** ways to make your classroom and home more earthquake-safe.

Why **It's Important**

Earthquake preparation can save lives and reduce damage.

Review Vocabulary
crest: the highest point of a wave

New Vocabulary
- magnitude
- liquefaction
- tsunami

Figure 16 The 1994 Northridge, California, earthquake was a costly disaster. Several major highways were damaged and 51 lives were lost.

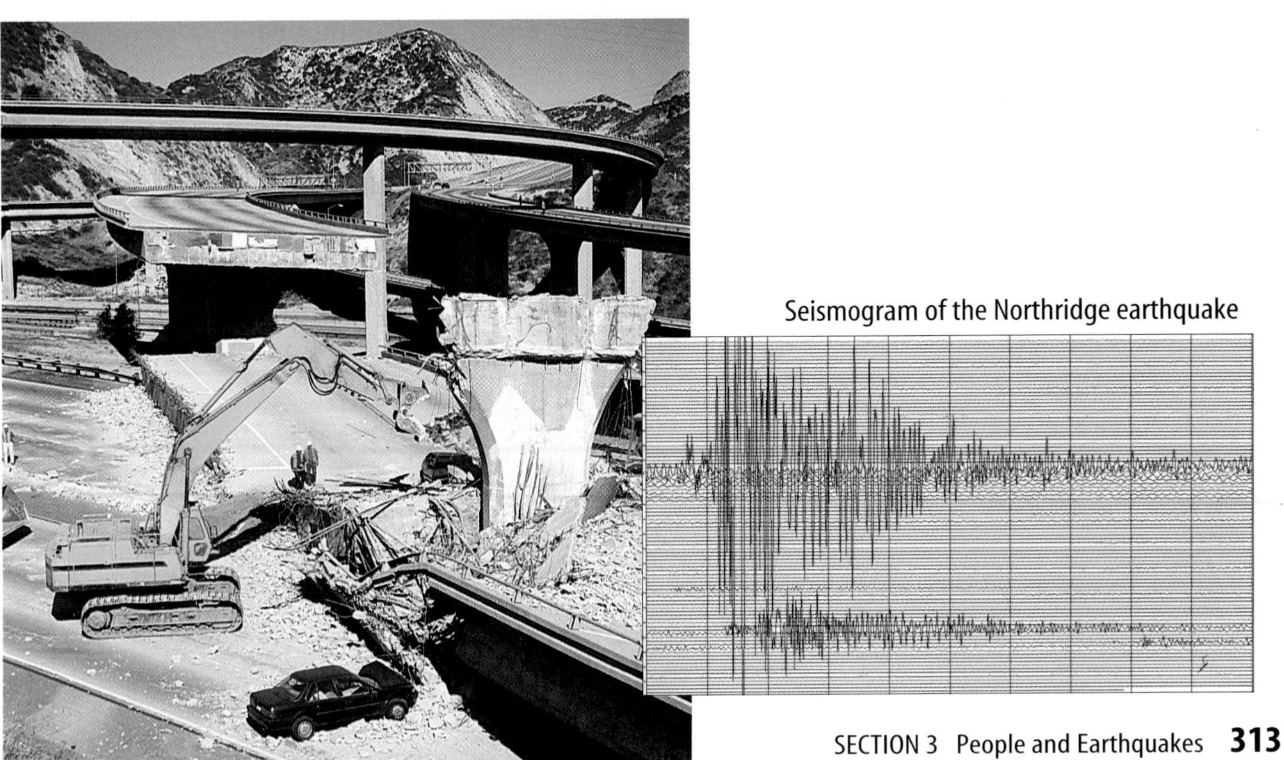

Figure 17 The 1999 earthquake in Turkey released about 32 times more energy than the 1994 Northridge earthquake did.

Studying Earthquakes Scientists who study earthquakes and seismic waves are seismologists. As you learned earlier, the instrument that is used to record primary, secondary, and surface waves from earthquakes all over the world is called a seismograph. Seismologists can use records from seismographs, called seismograms, to learn more than just where the epicenter of an earthquake is located.

Measuring Earthquake Magnitude The height of the lines traced on the paper of a seismograph is a measure of the energy that is released, or the **magnitude,** of the earthquake. The Richter magnitude scale is used to describe the strength of an earthquake and is based on the height of the lines on the seismogram. The Richter scale has no upper limit. However, scientists think that a value of about 9.5 would be the maximum strength an earthquake could register. For each increase of 1.0 on the Richter scale, the height of the line on a seismogram is ten times greater. However, about 32 times as much energy is released for every increase of 1.0 on the scale. For example, an earthquake with a magnitude of 8.5 releases about 32 times more energy than an earthquake with a magnitude of 7.5.

Table 1 Large-Magnitude Earthquakes

Year	Location	Magnitude	Deaths
1556	Shensi, China	?	830,000
1886	Charleston, SC	?	60
1906	San Francisco, CA	8.3	700 to 800
1923	Tokyo, Japan	9.2	143,000
1960	Chile	9.5	490 to 2,290
1975	Laoning Province, China	7.5	few
1976	Tangshan, China	8.2	242,000
1990	Iran	7.7	50,000
1994	Northridge, CA	6.8	51
2001	India	7.7	>20,000
2003	Bam, Iran	6.6	30,000

Past Earthquakes Damage from the 7.8-magnitude earthquake in Turkey in 1999 is shown in **Figure 17. Table 1** is a list of some large-magnitude earthquakes that have occurred around the world and the damage they have caused. Most of the earthquakes you hear about are large ones that cause great damage. However, of all the earthquakes detected throughout the world each year, most have magnitudes too low to be felt by humans. Scientists record thousands of earthquakes every day with magnitudes of less than 3.0. Each year, about 55,000 earthquakes are felt but cause little or no damage. These minor earthquakes have magnitudes that range from approximately 3.0 to 4.9 on the Richter scale.

Describing Earthquake Intensity Earthquakes also can be described by the amount of damage they cause. The modified Mercalli intensity scale describes the intensity of an earthquake using the amount of structural and geologic damage in a specific location. The amount of damage done depends on the strength of the earthquake, the nature of surface material, the design of structures, and the distance from the epicenter.

Under ideal conditions, only a few people would feel an intensity-I earthquake, and it would cause no damage. An intensity-IV earthquake would be felt by everyone indoors during the day but would be felt by only a few people outdoors. Pictures might fall off walls and books might fall from shelves. However, an intensity-IX earthquake would cause considerable damage to buildings and would cause cracks in the ground. An intensity-XII earthquake would cause total destruction of buildings, and objects such as cars would be thrown upward into the air. The 1994 6.8-magnitude earthquake in Northridge, California, was listed at an intensity of IX because of the damage it caused.

Magnetism In 1975, Chinese scientists successfully predicted an earthquake by measuring a slow tilt of Earth's surface and small changes in Earth's magnetism. Many lives were saved as a result of this prediction. Research the jobs that seismologists do and the types of organizations that they work for. Find out why most earthquakes have not been predicted.

Liquefaction Have you ever tried to drink a thick milkshake from a cup? Sometimes the milkshake is so thick that it won't flow. How do you make the milkshake flow? You shake it. Something similar can happen to very wet soil during an earthquake. Wet soil can be strong most of the time, but the shaking from an earthquake can cause it to act more like a liquid. This is called **liquefaction.** When liquefaction occurs in soil under buildings, the buildings can sink into the soil and collapse, as shown in **Figure 18.** People living in earthquake regions should avoid building on loose soils.

Figure 18 San Francisco's Marina district suffered extensive damage from liquefaction in the 1989 Loma Prieta earthquake because it is built on a landfilled marsh.

Topic: Tsunamis

Visit earth.msscience.com for Web links to information about tsunamis.

Activity Make a table that displays the location, date, earthquake magnitude, and maximum runup of the five most recent tsunamis.

Tsunamis Most earthquake damage occurs when surface waves cause buildings, bridges, and roads to collapse. People living near the seashore, however, have another problem. An earthquake under the ocean causes a sudden movement of the ocean floor. The movement pushes against the water, causing a powerful wave that can travel thousands of kilometers in all directions.

Ocean waves caused by earthquakes are called seismic sea waves, or **tsunamis** (soo NAH meez). Far from shore, a wave caused by an earthquake is so long that a large ship might ride over it without anyone noticing. But when one of these waves breaks on a shore, as shown in **Figure 19,** it forms a towering crest that can reach 30 m in height.

Tsunami Warnings Just before a tsunami crashes onto shore, the water along a shoreline might move rapidly toward the sea, exposing a large portion of land that normally is underwater. This should be taken as a warning sign that a tsunami could strike soon. You should head for higher ground immediately.

Because of the number of earthquakes that occur around the Pacific Ocean, the threat of tsunamis is constant. To protect lives and property, a warning system has been set up in coastal areas and for the Pacific Islands to alert people if a tsunami is likely to occur. The Pacific Tsunami Warning Center, located near Hilo, Hawaii, provides warning information including predicted tsunami arrival times at coastal areas.

However, even tsunami warnings can't prevent all loss of life. In the 1960 tsunami that struck Hawaii, 61 people died when they ignored the warning to move away from coastal areas.

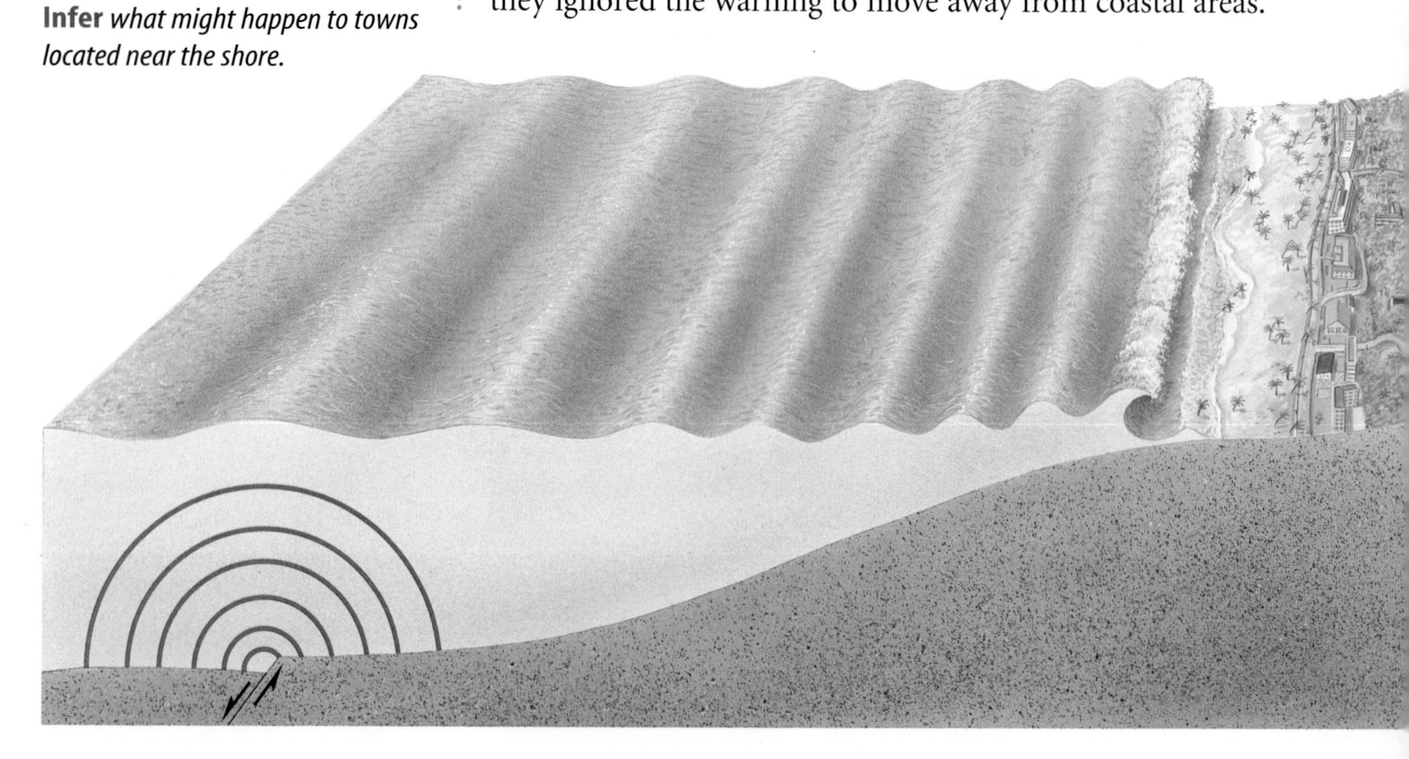

Figure 19 A tsunami begins over the earthquake focus.
Infer *what might happen to towns located near the shore.*

Earthquake Safety

Although earthquakes cannot be predicted reliably, **Figure 20** shows where earthquakes are most likely to occur in the United States. Knowing where earthquakes are likely to occur helps in long-term planning. Cities can take action to reduce damage and loss of life. Many buildings withstood the 1989 Loma Prieta earthquake because they were built with the expectation that such an earthquake would occur someday.

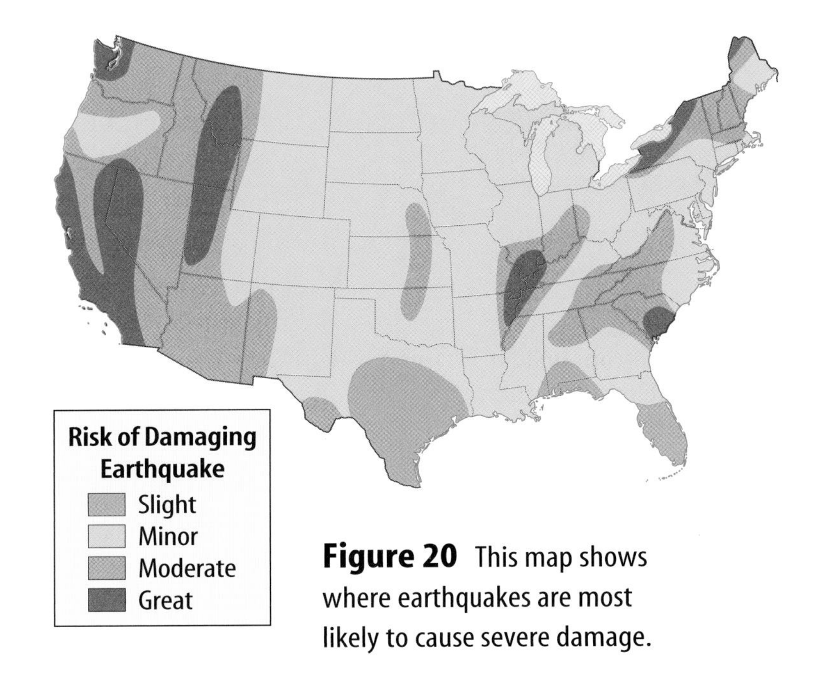

Figure 20 This map shows where earthquakes are most likely to cause severe damage.

Applying Math — Find a Ratio

EARTHQUAKE ENERGY An increase of one magnitude on the Richter scale for an earthquake means that 32 times more energy is released. How many times greater is the energy released by a magnitude-6 earthquake than the energy released by a magnitude-3 earthquake?

Solution

1 *This is what you know:*

- magnitude-6 earthquake, magnitude-3 earthquake
- energy increases 32 times per magnitude number

2 *This is what you need to find out:*

How many times greater is the energy of the larger earthquake than the energy of the smaller earthquake?

3 *This is the procedure you need to use:*

- Find the difference in magnitudes: $6 - 3 = 3$.
- This is the number of times 32 is multiplied times itself. $32 \times 32 \times 32 = 32{,}768$.
- The magnitude-6 earthquake releases 32,768 times more energy than the magnitude-3 earthquake.

4 *Check your answer:*

Count how many times you need to divide 32,768 by 32 to obtain 1. You should get 3.

Practice Problems

1. How many times greater is the energy released by a magnitude-7 earthquake than the energy released by a magnitude-2 earthquake?
2. How many times greater is the energy released by a magnitude-5 earthquake than the energy released by a magnitude-3 earthquake?

Science Online — For more practice, visit earth.msscience.com/math_practice

Modeling Seismic-Safe Structures

Procedure

1. On a **tabletop,** build a structure out of **building blocks** by simply placing one block on top of another.
2. Build a second structure by wrapping sections of three blocks together with **rubber bands.** Then, wrap larger rubber bands around the entire completed structure.
3. Set the second structure on the tabletop next to the first one and pound on the side of the table with a slow, steady rhythm.

Analysis

1. Which of your two structures was better able to withstand the "earthquake" caused by pounding on the table?
2. How might the idea of wrapping the blocks with rubber bands be used in construction of supports for elevated highways?

Try at Home

Quake-Resistant Structures During earthquakes, buildings, bridges, and highways can be damaged or destroyed. Most loss of life during an earthquake occurs when people are trapped in or on these crumbling structures. What can be done to reduce loss of life?

Seismic-safe structures stand up to vibrations that occur during an earthquake. **Figure 21** shows how buildings can be built to resist earthquake damage. Today in California, some new buildings are supported by flexible, circular moorings placed under the buildings. The moorings are made of steel plates filled with alternating layers of rubber and steel. The rubber acts like a cushion to absorb earthquake waves. Tests have shown that buildings supported in this way should be able to withstand an earthquake measuring up to 8.3 on the Richter scale without major damage.

In older buildings, workers often install steel rods to reinforce building walls. Such measures protect buildings in areas that are likely to experience earthquakes.

Reading Check *What are seismic-safe structures?*

Figure 21 The rubber portions of this building's moorings absorb most of the wave motion of an earthquake. The building itself only sways gently.
Infer *what purpose the rubber serves.*

Before an Earthquake To make your home as earthquake-safe as possible, certain steps can be taken. To reduce the danger of injuries from falling objects, move heavy objects from high shelves to lower shelves. Learn how to turn off the gas, water, and electricity in your home. To reduce the chance of fire from broken gas lines, make sure that water heaters and other gas appliances are held securely in place as shown in **Figure 22.** A newer method that is being used to minimize the danger of fire involves placing sensors on gas lines. The sensors automatically shut off the gas when earthquake vibrations are detected.

Figure 22 Sturdy metal straps on this gas water heater help reduce the danger of fires from broken gas lines during an earthquake.

During an Earthquake If you're indoors, move away from windows and any objects that could fall on you. Seek shelter in a doorway or under a sturdy table or desk. If you're outdoors, stay in the open—away from power lines or anything that might fall. Stay away from chimneys or other parts of buildings that could fall on you.

After an Earthquake If water and gas lines are damaged, the valves should be shut off by an adult. If you smell gas, leave the building immediately and call authorities from a phone away from the leak area. Stay away from damaged buildings. Be careful around broken glass and rubble, and wear boots or sturdy shoes to keep from cutting your feet. Finally, stay away from beaches. Tsunamis sometimes hit after the ground has stopped shaking.

section 3 review

Summary

Earthquake Activity

- The height of the lines traced on a seismogram can be used to determine an earthquake's magnitude.
- The intensity of an earthquake is determined by examining the amount of damage caused by the earthquake.

Earthquake Safety

- Knowing where large earthquakes are likely to occur helps people plan how to reduce damage.
- If you're ever in an earthquake, move away from windows or any object that might fall on you. Seek shelter in a doorway or under a sturdy table or desk.

Self Check

1. **Explain** how you can determine if you live in an area where an earthquake is likely to occur.
2. **Compare and contrast** the Richter and the Mercalli scales.
3. **Explain** what causes a tsunami.
4. **Describe** three ways an earthquake causes damage.
5. **Think Critically** How are shock absorbers on a car similar to the circular moorings used in modern earthquake-safe buildings? How do they absorb shock?

Applying Skills

6. **Infer** Seismographs around the world record the occurrence of thousands of earthquakes every day. Why are so few earthquakes in the news?

Earthquake Depths

Goals
- **Observe** any connection between earthquake-focus depth and epicenter location using the data provided on the next page.
- **Describe** any observed relationship between earthquake-focus depth and the movement of plates at Earth's surface.

Materials
graph paper
pencil

Real-World Question

You learned in this chapter that Earth's crust is broken into sections called plates. Stresses caused by movement of plates generate energy within rocks that must be released. When this release is sudden and rocks break, an earthquake occurs. Can a study of the foci of earthquakes tell you about plate movement in a particular region?

Analyze Your Data

1. Use graph paper and the data table on the right to make a graph plotting the depths of earthquake foci and the distances from the coast of a continent for each earthquake epicenter.
2. Use the graph below as a reference to draw your own graph. Place *Distance from the coast* and units on the *x*-axis. Begin labeling at the far left with 100 km west. To the right of it should be 0 km, then 100 km east, 200 km east, 300 km east, and so on through 700 km east. What point on your graph represents the coast?
3. Label the *y*-axis *Depth below Earth's surface.* Label the top of the graph *0 km* to represent Earth's surface. Label the bottom of the *y*-axis *–800 km.*
4. **Plot** the focus depths against the distance and direction from the coast for each earthquake in the table below.

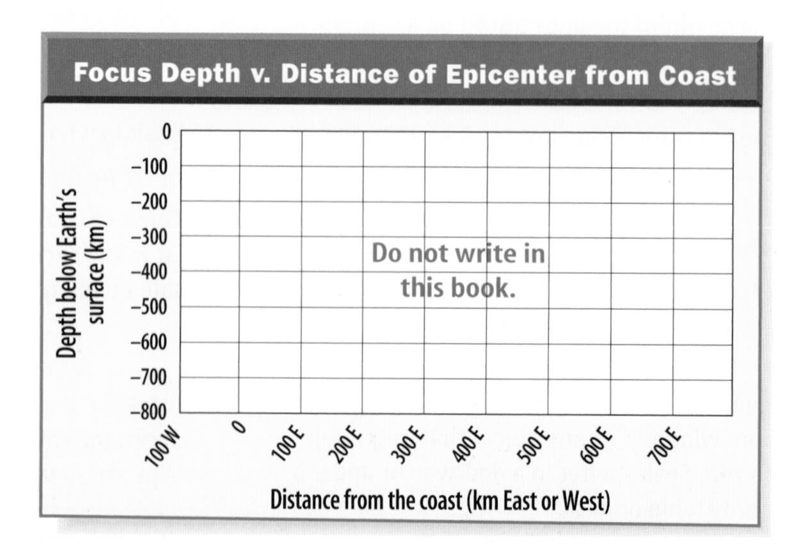

Conclude and Apply

1. **Describe** any observed relationship between the location of earthquake epicenters and the depth of foci.
2. **Explain** why none of the plotted earthquakes occurred below 700 km.
3. Based on your graph, form a hypothesis to explain what is happening to the plates at Earth's surface in the vicinity of the plotted earthquake foci. In what direction are the plates moving relative to each other?
4. **Infer** what process is causing the earthquakes you plotted on your graph.
5. **Infer** whether these earthquakes are occurring along the eastern side of a continent or along the western side of a continent.
6. **Draw and label** a cross section of the Earth beneath this coast. Label the eastern plate, the western plate, and use arrows to show the directions the plates are moving.
7. **Form a hypothesis** to predict which continent these data might apply to. Apply what you have learned in this lab and the information in **Figure 2.** Explain your answer.

Focus and Epicenter Data

Earthquake	Focus Depth (km)	Distance of Epicenter from Coast (km)
A	−55	0
B	−295	100 east
C	−390	455 east
D	−60	75 east
E	−130	255 east
F	−195	65 east
G	−695	400 east
H	−20	40 west
I	−505	695 east
J	−520	390 east
K	−385	335 east
L	−45	95 east
M	−305	495 east
N	−480	285 east
O	−665	545 east
P	−85	90 west
Q	−525	205 east
R	−85	25 west
S	−445	595 east
T	−635	665 east
U	−55	95 west
V	−70	100 west

Communicating Your Data

Compare your graph with those of other members of your class. **For more help, refer to the Science Skill Handbook.**

SCIENCE Stats

Moving Earth!

Did you know...

... Tsunamis can travel as fast as commercial jets and can reach heights of 30 m. A wave that tall would knock over this lighthouse. Since 1945, more people have been killed by tsunamis than by the ground shaking from earthquakes.

... The most powerful earthquake to hit the United States in recorded history shook Alaska in 1964. At 8.5 on the Richter scale, the quake shook all of Alaska for nearly 5 min, which is a long time for an earthquake. Nearly 320 km of roads near Anchorage suffered damage, and almost half of the 204 bridges had to be rebuilt.

Applying Math How many 3.0-magnitude earthquakes would it take to equal the energy released by one 8.0-magnitude earthquake?

... Snakes can sense the vibrations made by a small rodent up to 23 m away. Does this mean that they can detect vibrations prior to major earthquakes? Unusual animal behavior was observed just before a 1969 earthquake in China— an event that was successfully predicted.

Write About It

Visit earth.msscience.com/science_stats to research the history and effects of earthquakes in the United States. In a paragraph, describe how the San Francisco earthquake of 1906 affected earthquake research.

Reviewing Main Ideas

Section 1 Forces Inside Earth

1. Plate movements can cause rocks to bend and stretch. Rocks can break if the forces on them are beyond their elastic limit.
2. Earthquakes are vibrations produced when rocks break along a fault.
3. Normal faults form when rocks are under tension. Reverse faults form under compression and shearing forces produce strike-slip faults.

Section 2 Features of Earthquakes

1. Primary waves stretch and compress rock particles. Secondary waves move particles at right angles to the direction of wave travel.
2. Surface waves move rock particles in a backward, rolling motion and a side-to-side swaying motion.
3. Earthquake epicenters are located by recording seismic waves.
4. The boundaries between Earth's internal layers are determined by observing the speeds and paths of seismic waves.

Section 3 People and Earthquakes

1. A seismograph measures the magnitude of an earthquake.
2. The magnitude of an earthquake is related to the energy released by the earthquake.

Visualizing Main Ideas

Copy and complete the following concept map on earthquake damage.

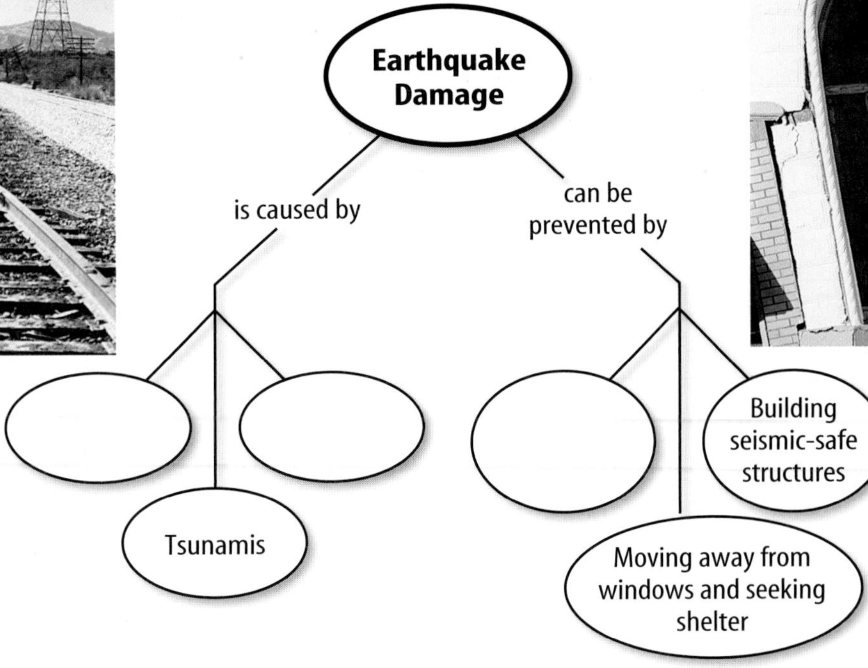

Earthquake Damage
is caused by
can be prevented by
Tsunamis
Building seismic-safe structures
Moving away from windows and seeking shelter

chapter 11 Review

Using Vocabulary

earthquake p. 301	reverse fault p. 302
epicenter p. 305	secondary wave p. 305
fault p. 300	seismic wave p. 304
focus p. 304	seismograph p. 307
liquefaction p. 315	strike-slip fault p. 303
magnitude p. 314	surface wave p. 305
normal fault p. 302	tsunami p. 316
primary wave p. 305	

Fill in the blanks with the correct words.

1. _________ causes most of the damage in earthquakes because of the side to side swaying motion that many buildings are unable to withstand.
2. At a(n) _________, rocks move past each other without much upward or downward movement.
3. The point on Earth's surface directly above the earthquake focus is the _________.
4. The measure of the energy released during an earthquake is its _________.
5. An earthquake under the ocean can cause a(n) _________ that travels thousands of kilometers.

Checking Concepts

Choose the word or phrase that best answers the question.

6. Earthquakes can occur when which of the following is passed?
   A) tension limit **C)** elastic limit
   B) seismic unit **D)** shear limit
7. When the rock above the fault surface moves down relative to the rock below the fault surface, what kind of fault forms?
   A) normal **C)** reverse
   B) strike-slip **D)** shear

Use the illustration below to answer question 8.

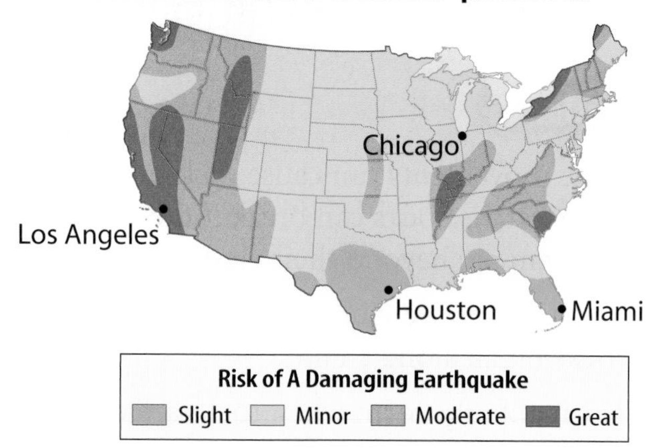

8. Using the figure above, which city should be most prepared for an earthquake?
   A) Miami, FL **C)** Chicago, IL
   B) Houston, TX **D)** Los Angeles, CA
9. From which of the following do primary and secondary waves move outward?
   A) epicenter **C)** Moho
   B) focus **D)** tsunami
10. What kind of earthquake waves stretch and compress rocks?
    A) surface **C)** secondary
    B) primary **D)** shear
11. What are the slowest seismic waves?
    A) surface **C)** secondary
    B) primary **D)** pressure
12. What is the fewest number of seismograph stations that are needed to locate the epicenter of an earthquake?
    A) two **C)** four
    B) three **D)** five
13. What happens to primary waves when they pass from liquids into solids?
    A) slow down **C)** stay the same
    B) speed up **D)** stop
14. What part of a seismograph does not move during an earthquake?
    A) sheet of paper **C)** drum
    B) fixed frame **D)** pendulum

Science Online earth.msscience.com/vocabulary_puzzlemaker

Thinking Critically

15. **Infer** The 1960 earthquake in the Pacific Ocean off the coast of Chile caused damage and loss of life in Chile, Hawaii, Japan, and other areas along the Pacific Ocean border. How could this earthquake do so much damage to areas thousands of kilometers from its epicenter?

16. **Explain** why a person who is standing outside in an open field is relatively safe during a strong earthquake.

17. **Describe** how a part of the seismograph remains at rest during an earthquake.

18. **Explain** why it is incorrect to call a tsunami a tidal wave.

19. **Predict** which is likely to be more stable during an earthquake—a single-story wood-frame house or a brick building. Explain.

20. **Measure in SI** Use an atlas and a metric ruler to answer the following question. Primary waves travel at about 6 km/s in continental crust. How long would it take a primary wave to travel from San Francisco, California, to Reno, Nevada?

Use the table below to answer question 21.

Seismograph Station Data

Station	Latitude	Longitude	Distance from Earthquake
1	45° N	120° W	1,300 km
2	35° N	105° W	1,200 km
3	40° N	115° W	790 km

21. **Use Tables** Use a map of the United States that has a distance scale, a compass for drawing circles, and the table above to determine the location of the earthquake epicenter.

Performance Activities

22. **Model** Use layers of different colors of clay to illustrate the three different kinds of faults. Label each model, explaining the forces involved and the rock movement.

Applying Math

23. **Earthquake Magnitude** An increase of one on the Richter scale corresponds to an increase of 10 in the size of the largest wave on a seismogram. How many times larger is the largest wave of a Richter magnitude-6 earthquake than a Richter magnitude-3 earthquake?

24. **Tsunami Speed** An underwater earthquake produces a tsunami 1,500 km away from Hawaii. If the tsunami travels at 600 km/h, how long will it take to reach Hawaii?

Use the graph below to answer question 25.

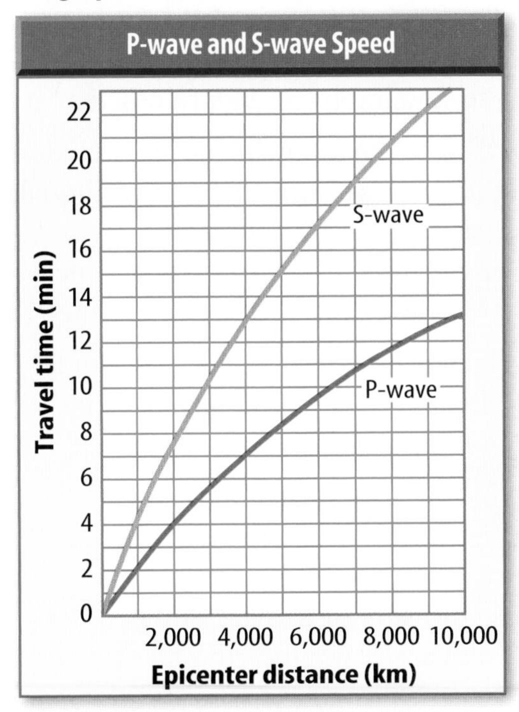

25. **Earthquake Waves** The graph above shows a P-wave and an S-wave plotted on a time-distance graph. According to the graph, which wave moves at the greater speed?

Part 1 Multiple Choice

Record your answers on the answer sheet provided by your teacher or on a sheet of paper.

Use the photo below to answer question 1.

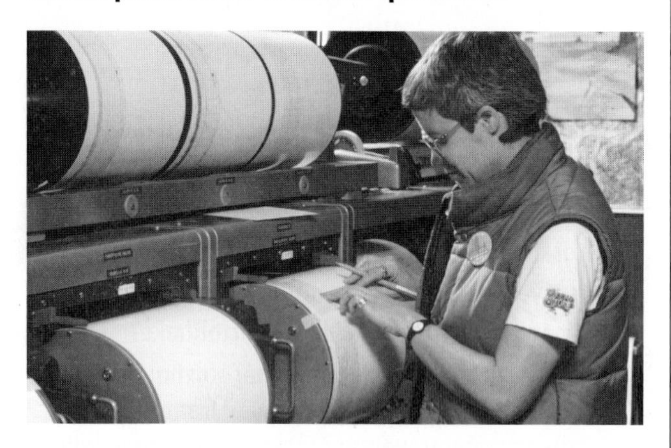

1. The instrument above records seismic waves from an earthquake. Which of the following is the name of this instrument?
   A. seismogram **C.** tiltmeter
   B. seismograph **D.** strainmeter

2. Which of the following terms is used to indicate the region where no earthquake waves reach Earth's surface?
   A. light zone **C.** shadow zone
   B. waveless zone **D.** seismic zone

3. Which is used to measure magnitude?
   A. Richter scale **C.** shadow zone
   B. Mercalli scale **D.** seismic gap

4. What is earthquake intensity?
   A. a measure of energy released
   B. a measure of seismic risk
   C. a measure of damage done
   D. a measure of an earthquake's focus

Test-Taking Tip

Eliminate Incorrect Answers If you don't know the answer to a multiple choice question, try to eliminate as many incorrect answers as possible.

5. Which of the following describes liquefaction?
   A. the stopping of S-waves by Earth's molten outer core
   B. ice melting during an earthquake to cause flooding
   C. seismic waves shaking sediment, causing it to become more liquid like
   D. rivers diverted by the motion of earthquake flooding

6. Which of the following describes the motion of secondary waves?
   A. a backward rolling and a side-to-side swaying motion
   B. a back-and-forth motion that is parallel to the direction of travel
   C. vibration in directions that are perpendicular to the direction of wave travel
   D. a forward rolling motion

Use the illustration below to answer question 7.

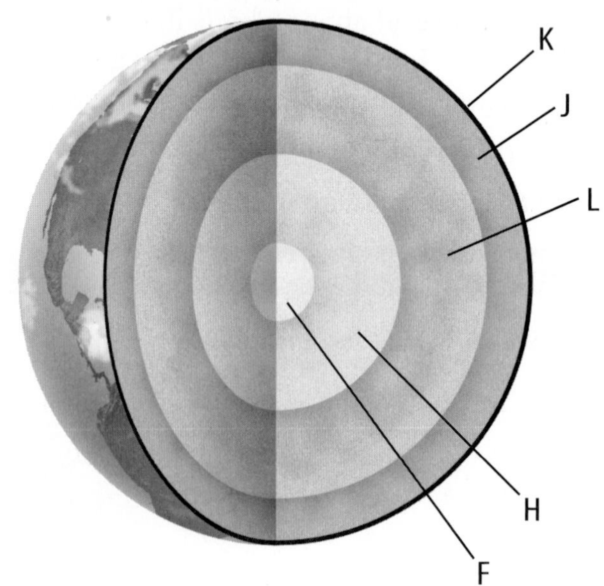

7. Which letter corresponds to the lower mantle?
   A. F **C.** L
   B. H **D.** J

Part 2 | Short Response/Grid In

Record your answers on the answer sheet provided by your teacher or on a sheet of paper.

8. Explain how earthquakes occur. Include a description of how energy builds up in rocks and is later released.

Use the illustration below to answer questions 9–10.

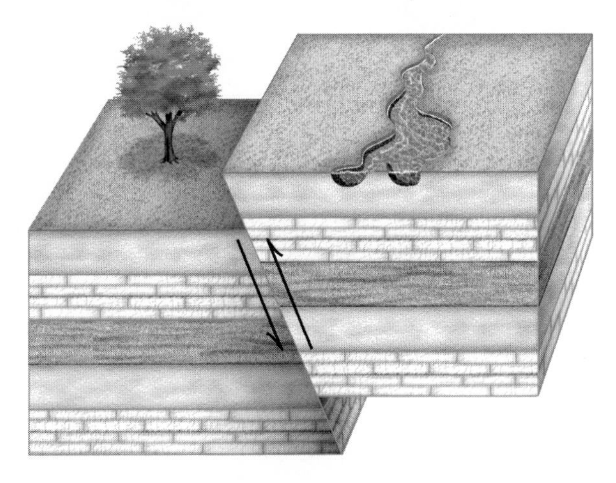

9. What type of fault is shown?
10. What type of force caused this fault to form?
11. Where is the fault plane?
12. Explain why most earthquakes occur along plate boundaries.
13. An earthquake that occurred in San Fernando, California, in 1971 caused $500 million in damage. An earthquake that occurred in Whittier, California, in 1987 caused $358 million in damage. How much more damage was caused by the San Fernando earthquake?
14. What factors affect damage done by an earthquake?
15. Explain the relationship between worldwide earthquake distribution and tectonic boundaries.

Part 3 | Open Ended

Record your answers on a sheet of paper.

16. The three types of faults are normal faults, reverse faults, and strike-slip faults. Draw each type of fault including arrows that show which way the rocks move.
17. Describe how a person should prepare for an earthquake and how a person should react if an earthquake occurs.

Use the map below to answer question 18.

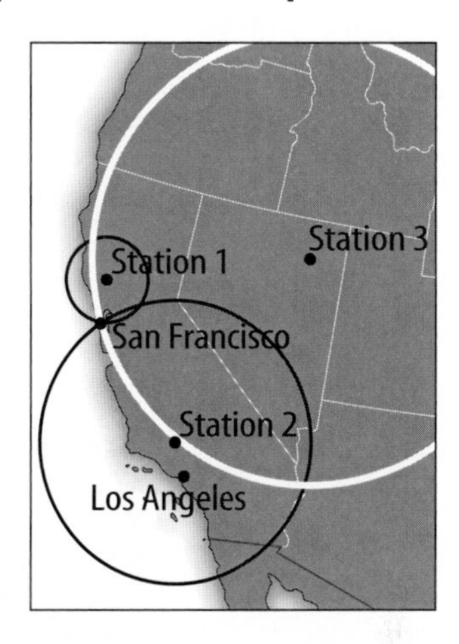

18. The map above shows three circles drawn around three different seismograph stations. The circles have radii equal to the distance between the seismograph station and the earthquake's epicenter. Which labeled point on the map represents the earthquake's epicenter? How do you know?
19. Compare and contrast the three types of seismic waves.
20. Draw three diagrams to show how each type of seismic wave moves through rocks.
21. Describe what happens to S-waves when they contact Earth's outer core. Decribe what happens to P-waves when they reach Earth's outer core. What is the *shadow zone?*

chapter 12

Volcanoes

The BIG Idea

Volcanic eruptions are caused by magma reaching Earth's surface.

SECTION 1
Volcanoes and Earth's Moving Plates

Main Idea Most volcanic activity occurs along plate boundaries and at locations called hot spots.

SECTION 2
Types of Volcanoes

Main Idea The composition of magma controls volcanic eruptions and determines the types of volcanic features.

SECTION 3
Igneous Rock Features

Main Idea Magma often solidifies underground and forms igneous rock features that may become exposed by erosion.

Beautiful but Dangerous?

In late October, 2002, earthquakes triggered this vigorous eruption from Mt. Etna, a volcano on the Italian island of Sicily. Strombolian eruptions and lava fountains spewed gas, bombs, blocks, and liquid lava. Ash settled as far away as Libya.

Science Journal Do all volcanoes begin with violent, explosive eruptions? Write about your current beliefs, then do some research and write about your discoveries.

Start-Up Activities

Map a Volcano

You've seen pictures of volcanoes from the ground, but what would a volcano look like on a map? Volcanoes can be represented on maps that show the elevation of the land, topographic maps.

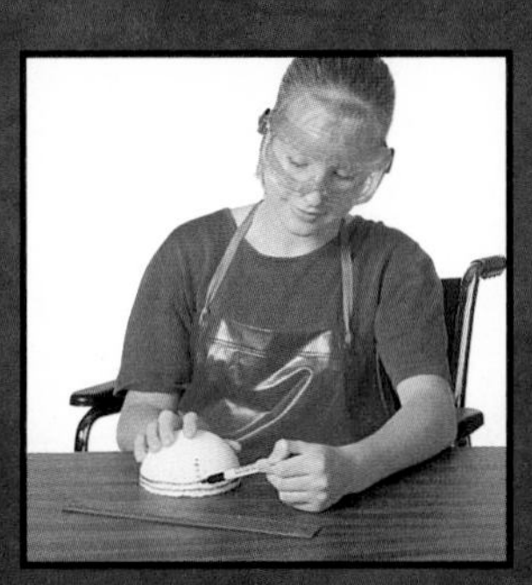

1. Obtain half of a foam ball from your teacher and place it on the top of a table with the flat side down.
2. Using a metric ruler and a permanent marker, mark 1-cm intervals on the foam ball. Start at the base of the ball and mark up at several places around the ball.
3. Connect the marks of equal elevation by drawing a line around the ball at the 1-cm mark, at the 2-cm mark, etc.
4. Look directly down on the top of the ball. Make a drawing of what you see in your Science Journal.
5. **Think Critically** In your Science Journal, write a paragraph that explains how your drawing shows the volcano's general shape.

Volcanoes Make the following Foldable to compare and contrast the characteristics of explosive and quiet volcanic eruptions.

STEP 1 **Fold** one sheet of paper lengthwise.

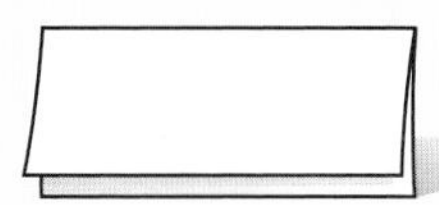

STEP 2 **Fold** into thirds.

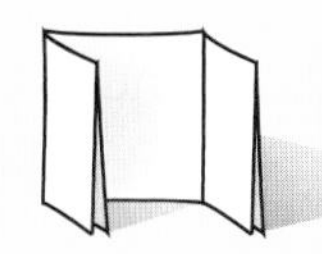

STEP 3 **Unfold and draw** overlapping ovals. **Cut** the top sheet along the folds.

STEP 4 **Label** the ovals *Explosive Eruptions*, *Both*, and *Quiet Eruptions,* as shown.

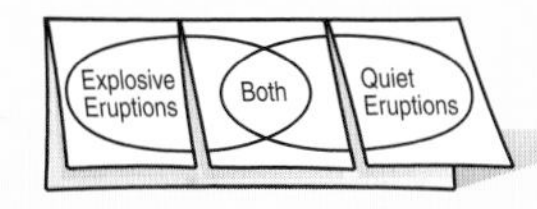

Construct a Venn Diagram As you read the chapter, list the characteristics unique to explosive eruptions under the left tab, those unique to quiet eruptions under the right tab, and those characteristics common to both under the middle tab.

Preview this chapter's content and activities at earth.msscience.com

Get Ready to Read

Make Inferences

1 Learn It! When you make inferences, you draw conclusions that are not directly stated in the text. This means you "read between the lines." You interpret clues and draw upon prior knowledge. Authors rely on a reader's ability to infer because all the details are not always given.

2 Practice It! Read the excerpt below and pay attention to highlighted words as you make inferences. Use this Think-Through chart to help you make inferences.

When **sulfurous gases** from volcanoes mix with water vapor in the atmosphere, **acid rain forms**. The vegetation, lakes, and streams around Soufrière Hills volcano were impacted significantly by acid rain. As the vegetation died, shown in **Figure 3,** the organisms that lived in the forest were **forced to leave or also died**.

—*from page 331*

Text	Question	Inferences
sulfurous gases	What are sulfurous gases?	Gases containing sulfur?
acid rain forms	How does acid rain form?	The gas chemically combines with water vapor and then it precipitates?
forced to leave or also died	Why were the organisms forced to leave?	They were dependent on the vegetation?

3 Apply It! As you read this chapter, practice your skill at making inferences by making connections and asking questions.

Sometimes you make inferences by using other reading skills, such as questioning and predicting.

Target Your Reading

Use this to focus on the main ideas as you read the chapter.

① **Before you read** the chapter, respond to the statements below on your worksheet or on a numbered sheet of paper.
- Write an **A** if you **agree** with the statement.
- Write a **D** if you **disagree** with the statement.

② **After you read** the chapter, look back to this page to see if you've changed your mind about any of the statements.
- If any of your answers changed, explain why.
- Change any false statements into true statements.
- Use your revised statements as a study guide.

Before You Read A or D	Statement	After You Read A or D
	1 Some deep, underground rocks are so hot that a drop in pressure can cause them to form magma.	
	2 Deep in Earth's interior, most of Earth's mantle is molten, liquid magma.	
	3 Magma is forced quickly toward Earth's surface because it is more dense than the rock around it.	
	4 Most volcanic eruptions occur near plate boundaries or at locations called hot spots.	
	5 Magma that is deep underground can contain water vapor and other gases.	
	6 Water vapor in magma usually produces volcanoes that erupt quietly with lava that flows smoothly.	
	7 Some volcanoes can form without lava flows.	
	8 Most of the magma that forms underground never reaches Earth's surface to form volcanoes.	
	9 When a volcano stops erupting, the magma inside the vent sinks deep into Earth, forming a bottomless pit.	

Science Online
Print out a worksheet of this page at earth.msscience.com

section 1

Volcanoes and Earth's Moving Plates

as you read

What You'll Learn

- **Describe** how volcanoes can affect people.
- **List** conditions that cause volcanoes to form.
- **Identify** the relationship between volcanoes and Earth's moving plates.

Why It's Important

Volcanoes can be dangerous to people and their communities.

Review Vocabulary
lava: molten rock material flowing from volcanoes onto Earth's surface

New Vocabulary
- volcano
- vent
- crater
- hot spot

What are volcanoes?

A **volcano** is an opening in Earth that erupts gases, ash, and lava. Volcanic mountains form when layers of lava, ash, and other material build up around these openings. Can you name any volcanoes? Did you know that Earth has more than 600 active volcanoes?

Most Active Volcanoes Kilauea (kee low AY ah), located in Hawaii, is the world's most active volcano. For centuries, this volcano has been erupting, but not explosively. In May of 1990, most of the town of Kalapana Gardens was destroyed, but no one was hurt because the lava moved slowly and people could escape. The most recent series of eruptions from Kilauea began in January 1983 and still continues.

The island country of Iceland is also famous for its active volcanoes. It sits on an area where Earth's plates move apart and is known as the land of fire and ice. The February 26, 2000, eruption of Hekla, in Iceland, is shown in **Figure 1.**

Figure 1 This photo of the February 26, 2000, eruption of Hekla shows why Iceland is known as the land of fire and ice.

Figure 2 This town on Montserrat was devastated by the eruption of Soufrière Hills volcano.

Effects of Eruptions

When volcanoes erupt, they often have direct, dramatic effects on the lives of people and their property. Lava flows destroy everything in their path. Falling volcanic ash can collapse buildings, block roads, and in some cases cause lung disease in people and animals. Sometimes, volcanic ash and debris rush down the side of the volcano. This is called a pyroclastic flow. The temperatures inside the flow can be high enough to ignite wood. When big eruptions occur, people often are forced to abandon their land and homes. People who live farther away from volcanoes are more likely to survive, but cities, towns, crops, and buildings in the area can be damaged by falling debris.

Human and Environmental Impacts The eruption of Soufrière (sew FREE er) Hills volcano in Montserrat, which began in July of 1995, was one of the largest recent volcanic eruptions near North America. Geologists knew it was about to erupt, and the people who lived near it were evacuated. On June 25, 1997, large pyroclastic flows swept down the volcano. As shown in **Figure 2,** they buried cities and towns that were in their path. The eruption killed 20 people who ignored the evacuation order.

INTEGRATE Earth Science When sulfurous gases from volcanoes mix with water vapor in the atmosphere, acid rain forms. The vegetation, lakes, and streams around Soufrière Hills volcano were impacted significantly by acid rain. As the vegetation died, shown in **Figure 3,** the organisms that lived in the forest were forced to leave or also died.

Figure 3 The vegetation near the volcano on Chances Peak, on the island of Montserrat in the West Indies, was destroyed by acid rain, heat, and ash.

Volcanologists
Volcanologists research many aspects of volcanoes, including space volcanoes. Io, a moon of Jupiter, has many active volcanoes. Become an amateur volcanologist; choose an aspect of volcanology that interests you, research your topic, and then create an exciting news broadcast to share the information with the class.

How do volcanoes form?

What happens inside Earth to create volcanoes? Why are some areas of Earth more likely to have volcanoes than others? Deep inside Earth, heat and pressure changes cause rock to melt, forming liquid rock or magma. Some deep rocks already are melted. Others are hot enough that a small rise in temperature or drop in pressure can cause them to melt and form magma. What makes magma come to the surface?

Magma Forced Upward Magma is less dense than the rock around it, so it is forced slowly toward Earth's surface. You can see this process if you turn a bottle of cold syrup upside down. Watch the dense syrup force the less dense air bubbles slowly toward the top.

Reading Check *Why is magma forced toward Earth's surface?*

After many thousands or even millions of years, magma reaches Earth's surface and flows out through an opening called a **vent.** As lava flows out, it cools quickly and becomes solid, forming layers of igneous rock around the vent. The steep-walled depression around a volcano's vent is the **crater. Figure 4** shows magma being forced out of a volcano.

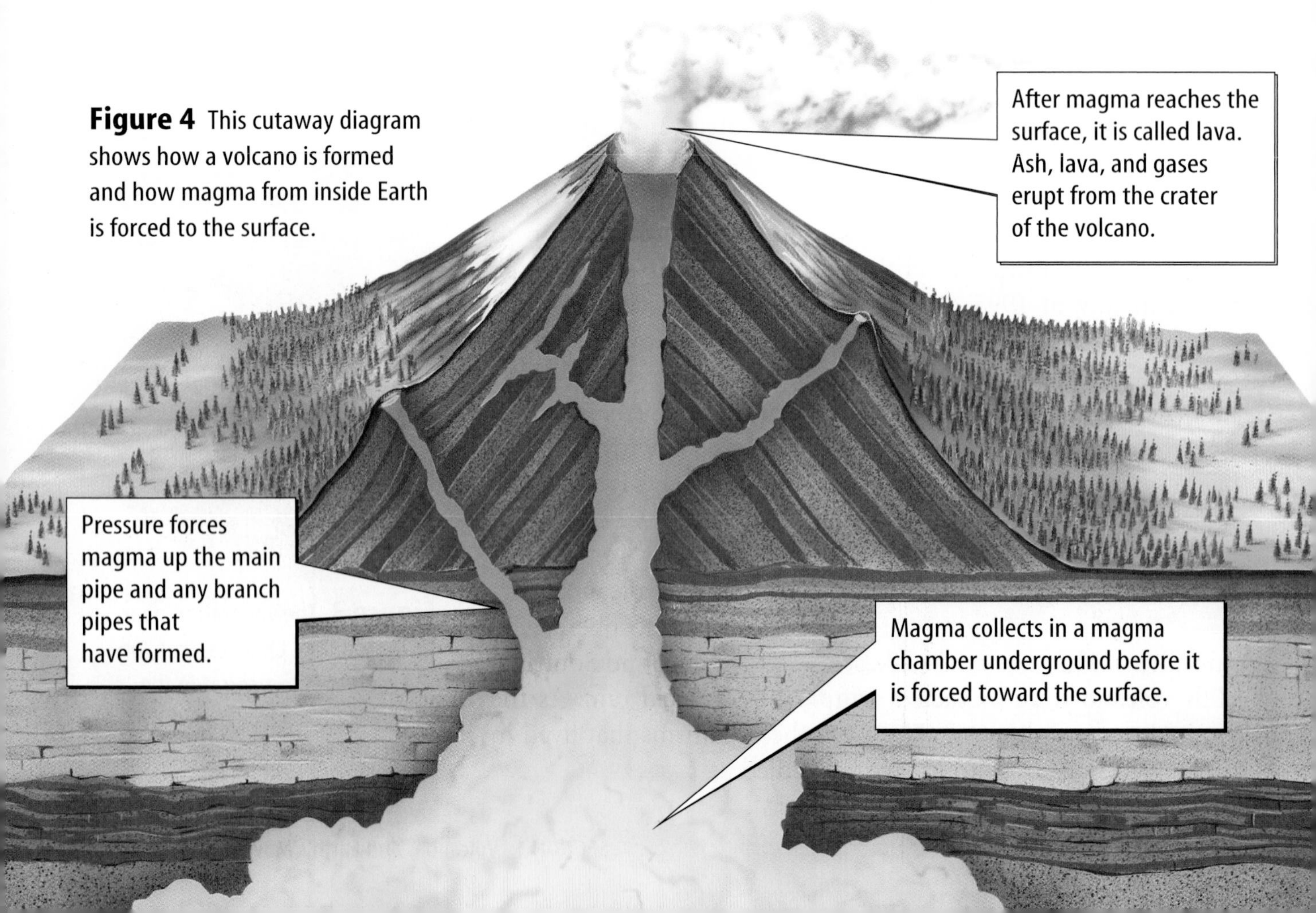

After magma reaches the surface, it is called lava. Ash, lava, and gases erupt from the crater of the volcano.
Pressure forces magma up the main pipe and any branch pipes that have formed.
Magma collects in a magma chamber underground before it is forced toward the surface.

Figure 4 This cutaway diagram shows how a volcano is formed and how magma from inside Earth is forced to the surface.

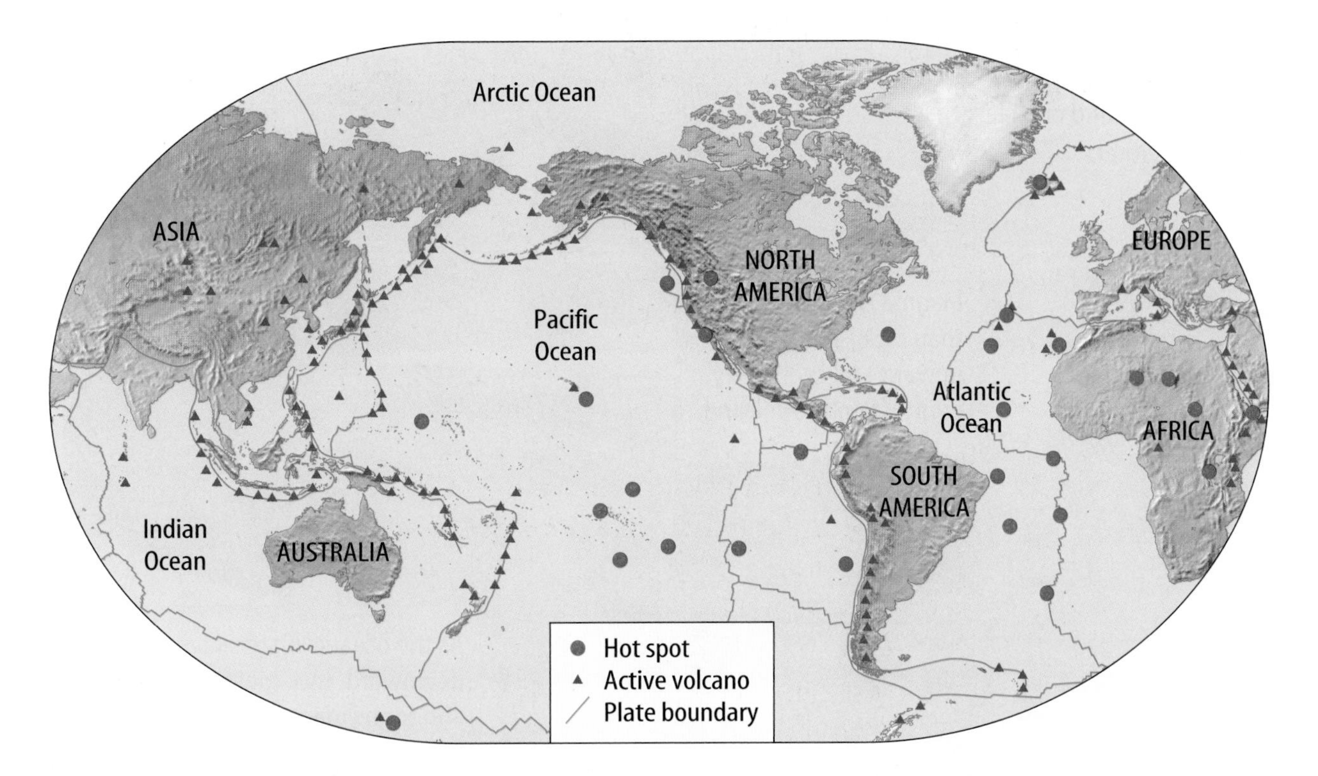

Figure 5 This map shows the locations of volcanoes, hot spots, and plate boundaries around the world. The Ring of Fire is a belt of active volcanoes that circles the Pacific Ocean.

Where do volcanoes occur?

Volcanoes often form in places where plates are moving apart, where plates are moving together, and at locations called hot spots. You can find locations of active volcanoes at plate boundaries and at hot spots on the map in **Figure 5.** Many examples can be found of volcanoes around the world that form at these three different kinds of areas. You'll explore volcanoes in Iceland, on the island of Montserrat, and in Hawaii.

Divergent Plate Boundaries Iceland is a large island in the North Atlantic Ocean. It is near the Arctic Circle and therefore has some glaciers. Iceland has volcanic activity because it is part of the Mid-Atlantic Ridge.

The Mid-Atlantic Ridge is a divergent plate boundary, which is an area where Earth's plates are moving apart. When plates separate, they form long, deep cracks called rifts. Lava flows from these rifts and is cooled quickly by seawater. **Figure 6** shows how magma rises at rifts to form new volcanic rock. As more lava flows and hardens, it builds up on the seafloor. Sometimes, the volcanoes and rift eruptions rise above sea level, forming islands such as Iceland. In 1963, the new island Surtsey was formed during a volcanic eruption.

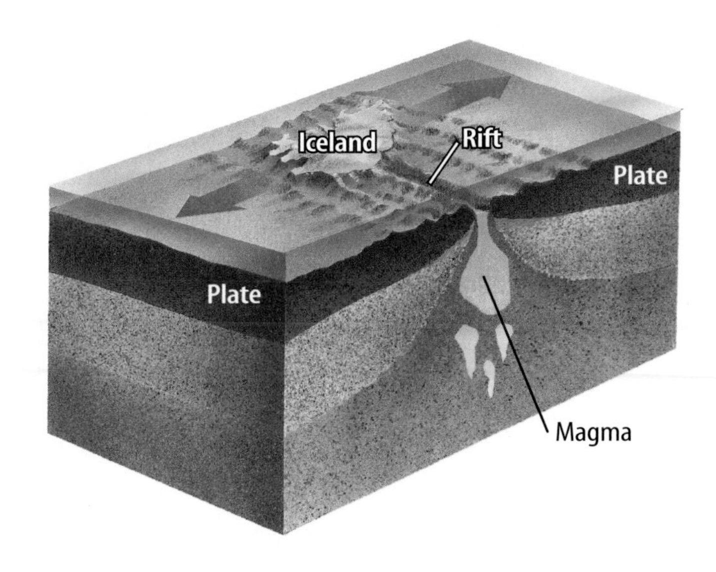

Figure 6 This diagram shows how volcanic activity occurs where Earth's plates move apart.

Figure 7 Volcanoes can form where plates collide and one plate slides below the other.

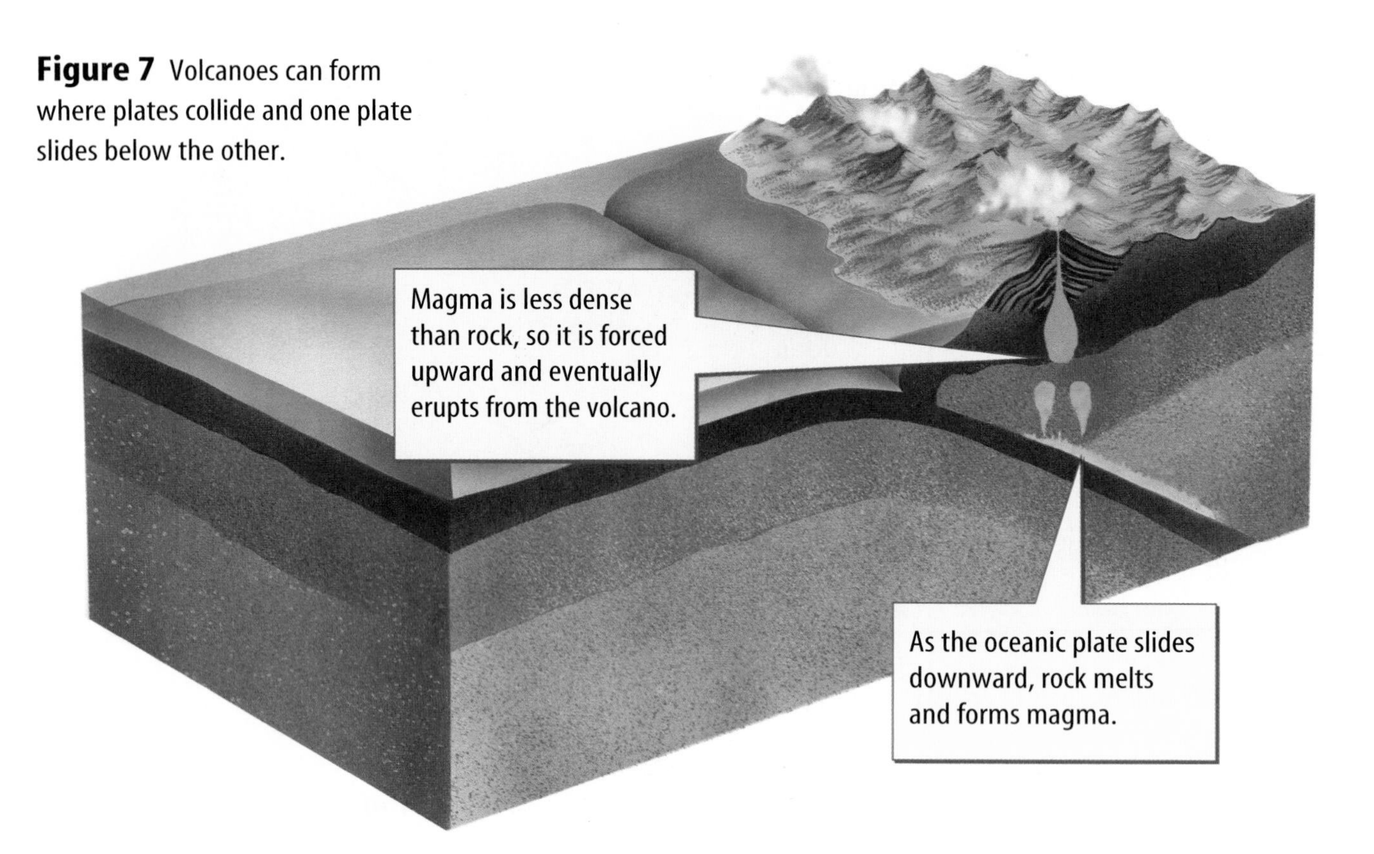
Magma is less dense than rock, so it is forced upward and eventually erupts from the volcano.
As the oceanic plate slides downward, rock melts and forms magma.

Mini LAB

Modeling Magma Movement

Procedure

1. Pour **water** into a **clear-plastic cup.**
2. Pour a small amount of **olive oil** into a separate plastic cup.
3. Extract a small amount of oil with a **dropper.**
4. Submerge the dropper tip into the water cup and slowly squeeze oil drops into the water.

Analysis

1. Describe what happened to the oil.
2. How do your observations compare with the movement of magma within Earth's crust?

Try at Home

Convergent Plate Boundaries Places where Earth's plates move together are called convergent plate boundaries. They include areas where an oceanic plate slides below a continental plate as in **Figure 7,** and where one oceanic plate slides below another oceanic plate. The Andes in South America began forming when an oceanic plate started sliding below a continental plate. Volcanoes that form on convergent plate boundaries tend to erupt more violently than other volcanoes do.

Magma forms when the plate sliding below another plate gets deep enough and hot enough to melt partially. The magma then is forced upward to the surface, forming volcanoes like Soufrière Hills on the island of Montserrat.

Hot Spots The Hawaiian Islands are forming as a result of volcanic activity. However, unlike Iceland, they haven't formed at a plate boundary. The Hawaiian Islands are in the middle of the Pacific Plate, far from its edges. What process could be forming them?

It is thought that some areas at the boundary between Earth's mantle and core are unusually hot. Hot rock at these areas is forced toward the crust where it melts partially to form a **hot spot.** The Hawaiian Islands sit on top of a hot spot under the Pacific Plate. Magma has broken through the crust to form several volcanoes. The volcanoes that rise above the water form the Hawaiian Islands, shown in **Figure 8.**

Figure 8 This satellite photo shows five of the Hawaiian Islands, which actually are volcanoes. **Explain** *why they are in a relatively straight line.*

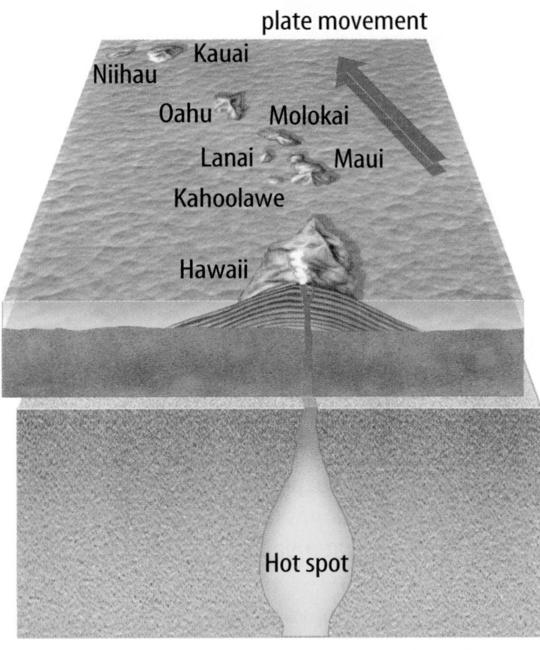

This illustration shows that the Hawaiian Islands were formed over a hot spot.

The Hawaiian Islands As you can see in **Figure 8,** the Hawaiian Islands are all in a line. This is because the Pacific Plate is moving over a stationary hot spot. Kauai, the oldest Hawaiian island, was once located where the big island, Hawaii, is situated today. As the plate moved, Kauai moved away from the hot spot and became dormant. As the Pacific Plate continued to move, the islands of Oahu, Molokai, Maui, and Hawaii were formed. The Hawaiian Islands formed over a period of about 5 million years.

section 1 review

Summary

What are volcanoes?

- A volcano is an opening in Earth's surface that erupts gases, ash, and lava.

Effects of Eruptions

- Direct effects of volcanic eruptions can be caused by lava flows, pyroclastic flows, and falling ash.
- Volcanic eruptions also produce indirect effects, such as acid rain.

How do volcanoes form?

- Volcanoes form when magma is forced up and flows onto Earth's surface as lava.
- A crater is a steep-walled depression around a volcano's vent.

Where do volcanoes occur?

- Volcanoes form where one plate sinks beneath another plate, where two plates are moving apart, and at hot spots.

Self Check

1. **Explain** why volcanoes are commonly found at the edges of Earth's moving plates.
2. **Describe** what effects pyroclastic flows have on people.
3. **Explain** why lava cools rapidly along a mid-ocean ridge. How might underwater lava differ from surface lava?
4. **Describe** the processes that cause Soufrière Hills volcano to erupt.
5. **Think Critically** If the Pacific Plate stopped moving, what might happen to the Big Island of Hawaii?

Applying Skills

6. **Concept Map** Make a concept map that shows how the Hawaiian Islands formed. Use the following phrases: *volcano forms, plate moves, volcano becomes dormant,* and *new volcano forms.* Draw and label an illustration of this process.

section 2

Types of Volcanoes

as you read

What You'll Learn

- **Explain** how the explosiveness of a volcanic eruption is related to the silica and water vapor content of its magma.
- **List** three forms of volcanoes.

Why It's Important

If you know the type of volcano, you can predict how it will erupt.

Review Vocabulary
magma: hot, melted rock material beneath Earth's surface

New Vocabulary
- shield volcano
- tephra
- cinder cone volcano
- composite volcano

What controls eruptions?

Some volcanic eruptions are explosive, like those from Soufrière Hills volcano, Mount Pinatubo, and Mount St. Helens. In others, the lava quietly flows from a vent, as in the Kilauea eruptions. What causes these differences?

Two important factors control whether an eruption will be explosive or quiet. One factor is the amount of water vapor and other gases that are trapped in the magma. The second factor is how much silica is present in the magma. Silica is a compound composed of the elements silicon and oxygen.

Trapped Gases When you shake a soft-drink container and then quickly open it, the pressure from the gas in the drink is released suddenly, spraying the drink all over. In the same way, gases such as water vapor and carbon dioxide are trapped in magma by the pressure of the surrounding magma and rock. As magma nears the surface, it is under less pressure. This allows the gas to escape from the magma. Gas escapes easily from some magma during quiet eruptions. However, gas that builds up to high pressures eventually causes explosive eruptions such as the one shown in **Figure 9.**

Figure 9 Mount St. Helens erupted on May 18, 1980.

8:32 A.M.

38 seconds later

Water Vapor The magma at some convergent plate boundaries contains a lot of water vapor. This is because oceanic plate material and some of its water slide under other plate material at some convergent plate boundaries. The trapped water vapor in the magma can cause explosive eruptions.

Composition of Magma

The second major factor that affects the nature of the eruption is the composition of the magma. Magma can be divided into two major types—silica poor and silica rich.

Quiet Eruptions Magma that is relatively low in silica is called basaltic magma. It is fluid and produces quiet, nonexplosive eruptions such as those at Kilauea. This type of lava pours from volcanic vents and runs down the sides of a volcano. As this *pahoehoe* (pa-HOY-hoy) lava cools, it forms a ropelike structure. If the same lava flows at a lower temperature, a stiff, slowly moving *aa* (AH-ah) lava forms. In fact, you can walk right up to some aa lava flows on Kilauea.

Figure 10 shows some different types of lava. These quiet eruptions form volcanoes over hot spots such as the Hawaiian volcanoes. Basaltic magmas also flow from rift zones, which are long, deep cracks in Earth's surface. Many lava flows in Iceland are of this type. Because basaltic magma is fluid when it is forced upward in a vent, trapped gases can escape easily in a nonexplosive manner, sometimes forming lava fountains. Lavas that flow underwater form pillow lava formations. They consist of rock structures shaped like tubes, balloons, or pillows.

Topic: Kilauea Volcano
Visit earth.msscience.com for Web links to information about Kilauea volcano in Hawaii.

Activity On a map of the Hawaiian Islands, identify the oldest and most recent islands. Next, indicate where Kilauea volcano is located on the Big Island. Do you see a directional pattern? Indicate on your Hawaiian map where you believe the next Hawaiian island will form.

42 seconds later

53 seconds later

NATIONAL GEOGRAPHIC VISUALIZING LAVA

Figure 10

Lava rarely travels faster than a few kilometers an hour. Therefore, it poses little danger to people. However, homes and property can be damaged. On land, there are two main types of lava flows—aa and pahoehoe. When lava comes out of cracks in the ocean floor, it is called pillow lava. The lava cooling here came from a volcanic eruption on the island of Hawaii.

Aa flows, like this one on Mount Etna in Italy, carry sharp angular chunks of rock called scoria. Aa flows move slowly and are intensely hot.

Pillow lava occurs where lava oozes out of cracks in the ocean floor. It forms pillow-shaped lumps as it cools. Pillow lava is the most common type of lava on Earth.

Pahoehoe flows, like this one near Kilauea's Mauna Ulu Crater in Hawaii, are more fluid than aa flows. They develop a smooth skin and form ropelike patterns when they cool.

Figure 11 Magmas that are rich in silica produce violent eruptions.

Violent eruptions, such as this one in Alaska, often produce a lot of volcanic ash.

Magnification: 450×

This color enhanced view of volcanic ash, from a 10 million year old volcano in Nebraska, shows the glass particles that make up ash.

Explosive Magma Silica-rich, or granitic, magma produces explosive eruptions such as those at Soufrière Hills volcano. This magma sometimes forms where Earth's plates are moving together and one plate slides under another. As the plate that is sliding under the other goes deeper, some rock is melted. The magma is forced upward by denser surrounding rock, comes in contact with the crust, and becomes enriched in silica. Silica-rich granitic magma is thick, and gas gets trapped inside, causing pressure to build up. When an explosive eruption occurs, as shown in **Figure 11,** the gases expand rapidly, often carrying pieces of lava in the explosion.

Reading Check ***What type of magmas produce violent eruptions?***

Some magmas have an andesitic composition. Andesitic magma is more silica rich than basaltic magma is, but it is less silica rich than granitic magma. It often forms at convergent plate boundaries where one plate slides under the other. Because of their higher silica content, they also erupt more violently than basaltic magmas. One of the biggest eruptions in recorded history, Krakatau, was primarily andesitic in composition. The word *andesitic* comes from the Andes, which are mountains located along the western edge of South America, where andesite rock is common. Many of the volcanoes encircling the Pacific Ocean also are made of andesite.

Volcanic Ash When volcanoes erupt, ash often is spread over a great distance. People who live near volcanoes must be careful not to inhale too much of the ash particles because the particles can cause respiratory problems. In your Science Journal, describe what people can do to prevent exposure to volcanic ash.

Modeling Volcanic Cones

Procedure

1. Pour **dry sand** or **sugar** onto one spot on a **paper plate. WARNING:** *Do not taste, eat, or drink any materials used in the lab.*
2. Mix a batch of **plaster of paris** and pour it onto one spot on another paper plate.
3. Allow the plaster of paris to dry. Use a **protractor** to measure the slope angles of the sides of the volcanoes.

Analysis

What form of volcano is represented by the model with steeper sides?

Forms of Volcanoes

A volcano's form depends on whether it is the result of a quiet or an explosive eruption and the type of lava it is made of—basaltic, granitic, or andesitic (intermediate). The three basic types of volcanoes are shield volcanoes, cinder cone volcanoes, and composite volcanoes.

Shield Volcano Quiet eruptions of basaltic lava spread out in flat layers. The buildup of these layers forms a broad volcano with gently sloping sides called a **shield volcano,** as seen in **Figure 12.** The Hawaiian Islands are examples of shield volcanoes. Basaltic lava also can flow onto Earth's surface through large cracks called fissures. This type of eruption forms flood basalts, not volcanoes, and accounts for the greatest volume of erupted volcanic material. The basaltic lava flows over Earth's surface, covering large areas with thick deposits of basaltic igneous rock when it cools. The Columbia Plateau located in the northwestern United States was formed in this way. Much of the new seafloor that originates at mid-ocean ridges forms as underwater flood basalts.

Cinder Cone Volcano Explosive eruptions throw lava and rock high into the air. Bits of rock or solidified lava dropped from the air are called **tephra** (TEH fruh). Tephra varies in size from volcanic ash, to cinders, to larger rocks called bombs and blocks. When tephra falls to the ground, it forms a steep-sided, loosely packed **cinder cone volcano,** as seen in **Figure 13.**

Figure 12 A shield volcano like Mauna Loa, shown here, is formed when lava flows from one or more vents without erupting violently.

Paricutín On February 20, 1943, a Mexican farmer learned about cinder cones when he went to his cornfield. He noticed that a hole in his cornfield that had been there for as long as he could remember was giving off smoke. Throughout the night, hot glowing cinders were thrown high into the air. In just a few days, a cinder cone several hundred meters high covered his cornfield. This is the volcano named Paricutín.

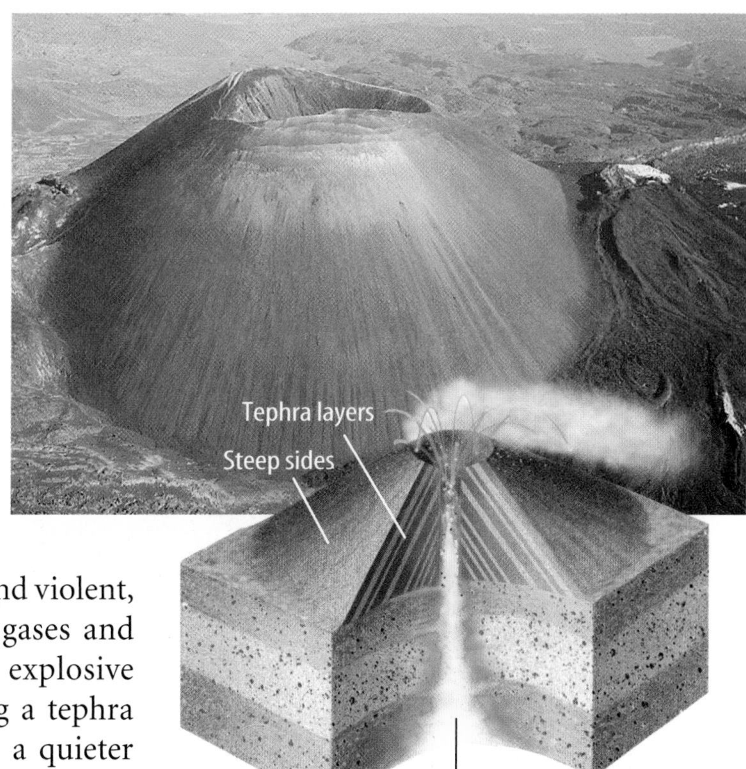

Figure 13 Paricutín is a large, cinder cone volcano located in Mexico.

Composite Volcano Some volcanic eruptions can vary between quiet and violent, depending on the amount of trapped gases and how rich in silica the magma is. An explosive period can release gas and ash, forming a tephra layer. Then, the eruption can switch to a quieter period, erupting lava over the top of the tephra layer. When this cycle of lava and tephra is repeated over and over in alternating layers, a **composite volcano** is formed. Composite volcanoes, shown in **Figure 14,** are found mostly where Earth's plates come together and one plate slides below the other. Soufrière Hills volcano is an example. As you can see in **Table 1** on the next page, many things affect eruptions and the form of a volcano.

Figure 14 Mount Rainier in the state of Washington is an example of a composite volcano.

Violent Eruptions Soufrière Hills volcano formed as ocean floor of the North American Plate and the South American Plate slid beneath the Caribbean Plate, causing magma to form. Successive eruptions of lava and tephra produced the majestic composite volcanoes that tower above the surrounding landscape on Montserrat and other islands in the Lesser Antilles. Before the 1995 eruption, silica-rich magma rose and was trapped beneath the surface. As the magma was forced toward Earth's surface, the pressure on the underlying magma was released. This started a series of eruptions that were still continuing in the year 2003.

Table 1 Thirteen Selected Eruptions

Volcano and Location	Year	Type	Eruptive Force	Magma Content		Ability of Magma to Flow	Products of Eruption
				Silica	H_2O		
Mount Etna, Sicily	1669	composite	moderate	high	low	medium	lava, ash
Tambora, Indonesia	1815	cinder cone	high	high	high	low	cinders, ash
Krakatau, Indonesia	1883	composite	high	high	high	low	cinders, ash
Mount Pelée, Martinique	1902	cinder cone	high	high	high	low	gas, ash
Vesuvius, Italy	1906	composite	moderate	high	low	medium	lava, ash
Mount Katmai, Alaska	1912	composite	high	high	high	low	lava, ash, gas
Paricutín, Mexico	1943	cinder cone	moderate	high	low	medium	ash, cinders
Surtsey, Iceland	1963	shield	moderate	low	low	high	lava, ash
Mount St. Helens, Washington	1980	composite	high	high	high	low	gas, ash
Kilauea, Hawaii	1983	shield	low	low	low	high	lava
Mount Pinatubo, Philippines	1991	composite	high	high	high	low	gas, ash
Soufrière Hills, Montserrat	1995	composite	high	high	high	low	gas, ash, rocks
Popocatépetl, Mexico	2000	composite	moderate	high	low	medium	gas, ash

Figure 15 Not much was left after Krakatau erupted in 1883.

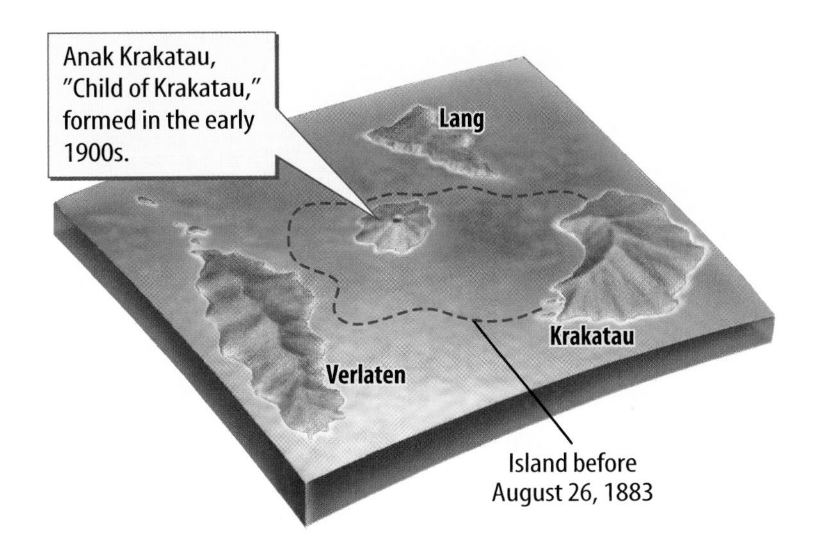

Krakatau One of the most violent eruptions in recent times occurred on an island in the Sunda Straits near Indonesia in August of 1883. Krakatau, a volcano on the island, erupted with such force that the island disappeared, as shown in **Figure 15A.** Most of the island collapsed into the emptied magma chamber. The noise of the eruption was so loud that it woke people in Australia and was heard as far away as 4,653 km from the island. Ash from the eruption fell in Singapore, which is 840 km to the north, and the area around the volcano was in complete darkness for 24 h. More than 36,000 people were killed, most by the giant tsunami waves created by the eruption. Global temperatures were lowered as much as 1.2°C by particles blown into the atmosphere and didn't return to normal until 1888.

section 2 review

Summary

What controls eruptions?

- The amount of water vapor and other gases control the type of eruption and the amount of silica present in the magma.

Composition of Magma

- Magma can be divided into two major types—silica rich and silica poor.

Forms of Volcanoes

- A shield volcano is a broad, gently sloping volcano formed by quiet eruptions of basaltic lava.
- A cinder cone volcano is a steep-sided, loosely packed volcano formed from tephra.
- Composite volcanoes are formed by alternating explosive and quiet eruptions that produce layers of tephra and lava.

Self Check

1. **Define** the term *tephra,* and where it can be found.
2. **Describe** the differences between basaltic and granitic magma.
3. **Identify** the specific water vapor and silica conditions that cause differences in eruptions.
4. **Describe** how the Hawaiian Islands formed.
5. **Think Critically** In 1883, Krakatau in Indonesia erupted. Infer which kind of lava Krakatau erupted—lava rich in silica or lava low in silica. Support your inference using data in **Table 1.**

Applying Skills

6. **Compare and contrast** Kilauea and Mount Pinatubo using information from **Table 1.**

Identifying Types of Volcanoes

You have learned that certain properties of magma are related to the type of eruption and the form of the volcano that will develop. Do this lab to see how to make and use a table that relates the properties of magma to the form of volcano that develops.

Real-World Question

Are the silica and water content of a magma related to the form of volcano that develops?

Goals

- **Determine** any relationship between the ability of magma to flow and eruptive force.
- **Determine** any relationship between magma composition and eruptive force.

Materials

Table 1 (thirteen selected eruptions)
paper
pencil

Procedure

1. Copy the graph shown above.
2. Using the information from **Table 1,** plot the magma content for each of the volcanoes listed by writing the name of the basic type of volcano in the correct spot on the graph.

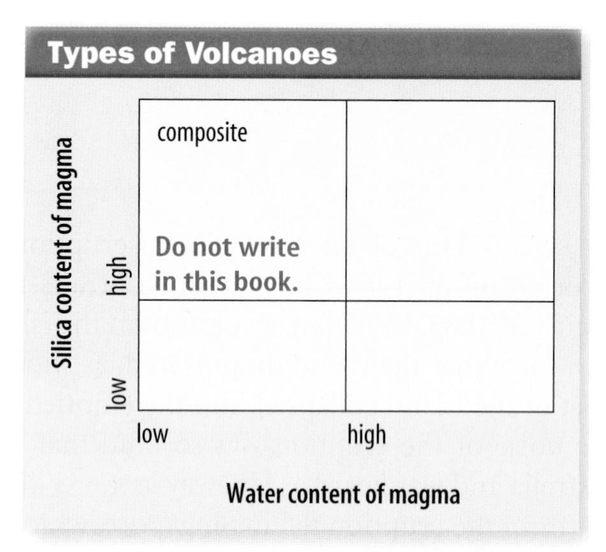

Conclude and Apply

1. What relationship appears to exist between the ability of the magma to flow and the eruptive force of the volcano?
2. Which would be more liquidlike: magma that flows easily or magma that flows with difficulty?
3. What relationship appears to exist between the silica or water content of the magma and the nature of the material ejected from the volcano?
4. How is the ability of a magma to flow related to its silica content?
5. **Infer** which of the two variables, silica or water content, appears to have the greater effect on the eruptive force of the volcano.
6. **Describe** the relationship that appears to exist between the silica and water content of the magma and the type of volcano that is produced.

Communicating Your Data

Create a flowchart that shows the relationship between magma composition and the type of volocano formed. **For more help, refer to the** Science Skill Handbook.

section 3

Igneous Rock Features

Intrusive Features

You can observe volcanic eruptions because they occur at Earth's surface. However, far more activity occurs underground. In fact, most magma never reaches Earth's surface to form volcanoes or to flow as flood basalts. This magma cools slowly underground and produces underground rock bodies that could become exposed later at Earth's surface by erosion. These rock bodies are called intrusive igneous rock features. There are several different types of intrusive features. Some of the most common are batholiths, sills, dikes, and volcanic necks. What do intrusive igneous rock bodies look like? You can see illustrations of these features in **Figure 16.**

as you read

What You'll Learn

- **Describe** intrusive igneous rock features and how they form.
- **Explain** how a volcanic neck and a caldera form.

Why It's Important

Many features formed underground by igneous activity are exposed at Earth's surface by erosion.

Review Vocabulary
intrude: to enter by force; cut in
extrude: to force or push out

New Vocabulary
- batholith
- dike
- sill
- volcanic neck
- caldera

Figure 16 This diagram shows intrusive and other features associated with volcanic activity.
Identify *which features shown are formed above ground. Which are formed by intrusive activities?*

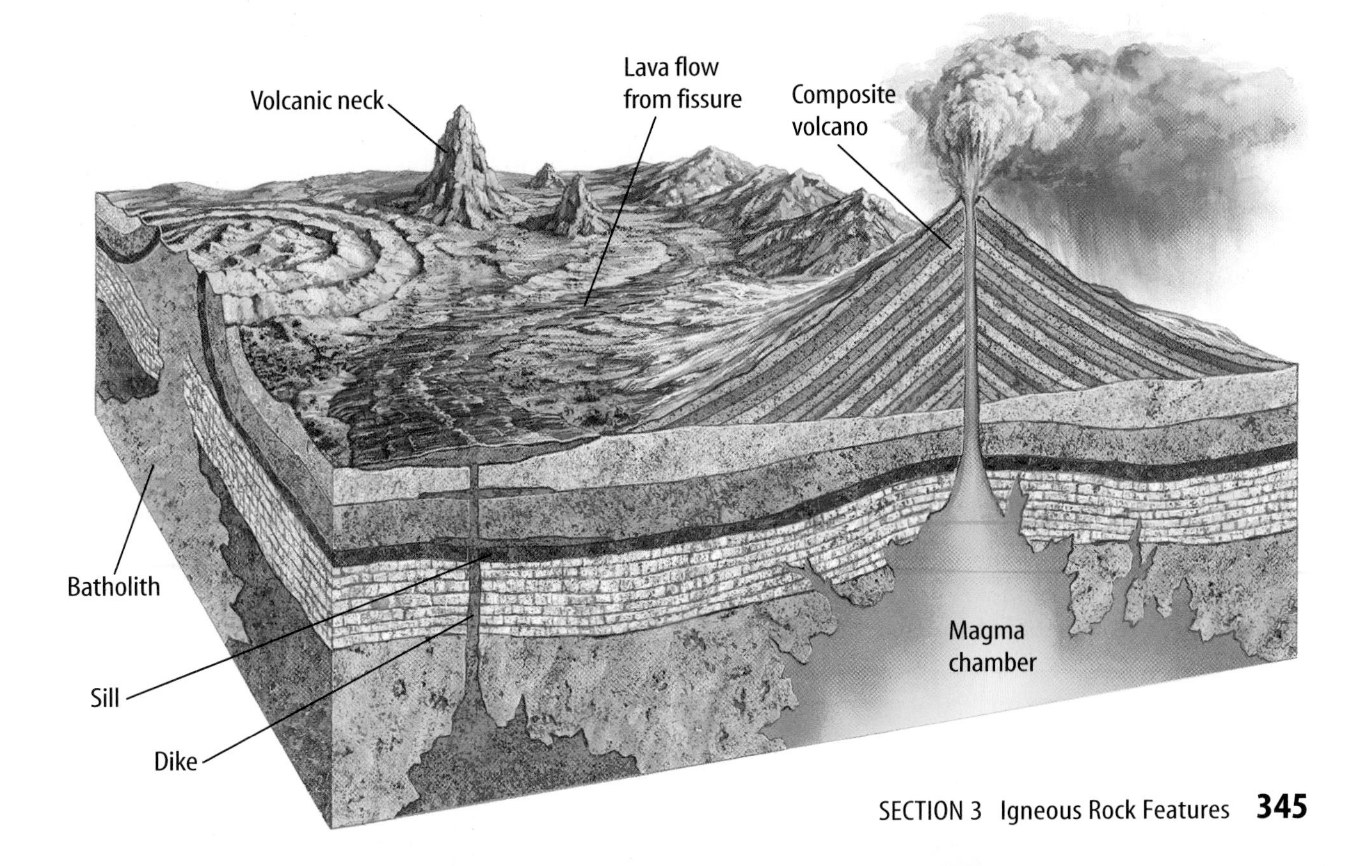

Batholiths The largest intrusive igneous rock bodies are **batholiths.** They can be many hundreds of kilometers in width and length and several kilometers thick. Batholiths form when magma bodies that are being forced upward from inside Earth cool slowly and solidify before reaching the surface. However, not all of them remain hidden inside Earth. Some batholiths have been exposed at Earth's surface by many millions of years of erosion. The granite domes of Yosemite National Park are the remains of a huge batholith that stretches across much of the length of California.

Applying Math — Calculate Percent

CLASSIFYING IGNEOUS ROCKS Igneous rocks are classified into three types depending on the amount of silica they contain. Basaltic rocks contain approximately 45 percent to 52 percent silica. Andesitic, or intermediate, rocks contain about 52 percent to 66 percent silica, and granitic rocks have more than 66 percent silica. The lighter the color is, the higher the silica content is. A 900-kg block of igneous rock contains 630 kg of silica. Calculate the percent of silica in the rock to classify it.

Solution

1. *This is what you know:*
   - rock = 900 kg
   - silica = 630 kg
2. *This is what you need to find:* The percentage of silica: x
3. *This is the equation you need to use:* Mass of silica / mass of rock = x / 100
4. *Solve the equation for* x*:*
   - $x = (630 \text{ kg}/900 \text{ kg}) \times 100$
   - x = 70 percent, therefore, the rock is granitic.

Check your answer by dividing it by 100, then multiplying by 900. Did you get the given amount of silica?

Practice Problems

1. A 250-kg boulder of basalt contains 125 kg of silica. Use the classification system to determine whether basalt is light or dark.
2. Andesite is an intermediate, medium-colored rock with a silica content ranging from 52 percent to 66 percent. About how many kilograms of silica would you predict to be in a 68-kg boulder of andesite?

Science Online — For more practice, visit earth.msscience.com/math_practice

Dikes and Sills Magma sometimes squeezes into cracks in rock below the surface. This is like squeezing toothpaste into the spaces between your teeth. Magma that is forced into a crack that cuts across rock layers and hardens is called a **dike.** Magma that is forced into a crack parallel to rock layers and hardens is called a **sill.** These features are shown in **Figure 17.** Most dikes and sills run from a few meters to hundreds of meters long.

Other Features

When a volcano stops erupting, the magma hardens inside the vent. Erosion, usually by water and wind, begins to wear away the volcano. The cone is much softer than the solid igneous rock in the vent. Thus, the cone erodes first, leaving behind the solid igneous core as a **volcanic neck.** Ship Rock in New Mexico, shown in **Figure 17,** is a good example of a volcanic neck.

Topic: Igneous Rock Features

Visit earth.msscience.com for Web links to information about igneous rock features.

Activity Create a collage for artistic competition by using a variety of pictures of igneous rock features. For extra challenge, research Devils Tower, Wyoming. Develop your own hypothesis for its formation, and present your ideas as a panel discussion with other classmates.

Dike
Sill
Sill
Dike
A sill is formed when magma is forced between parallel rock layers.
The dikes near Ship Rock were formed when magma squeezed into vertical cracks cutting across rock layers.

Figure 17 Igneous features can form in many different sizes and shapes.

Calderas Sometimes after an eruption, the top of a volcano can collapse, as seen in **Figure 18.** This produces a large depression called a **caldera.** Crater Lake in Oregon, shown in **Figure 19,** is a caldera that filled with water and is now a lake. Crater Lake formed after the violent eruption and destruction of Mount Mazama about 7,000 years ago.

Figure 18 Calderas form when the top of a volcano collapses.

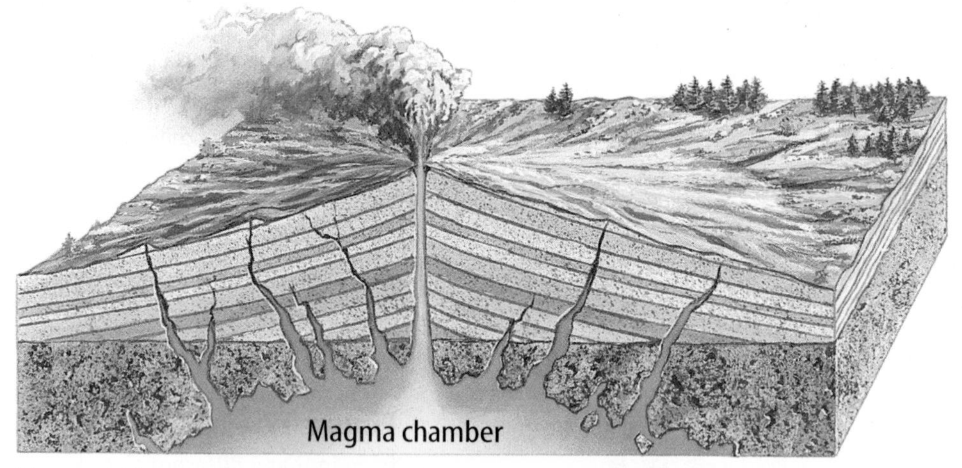

Magma is forced upward, causing volcanic activity to occur.

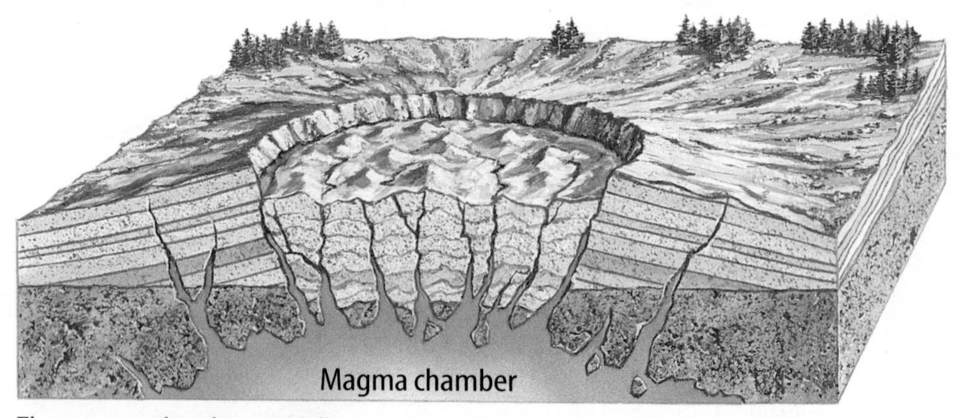

The magma chamber partially empties, causing rock to collapse into the emptied chamber below the surface. This forms a circular-shaped caldera.

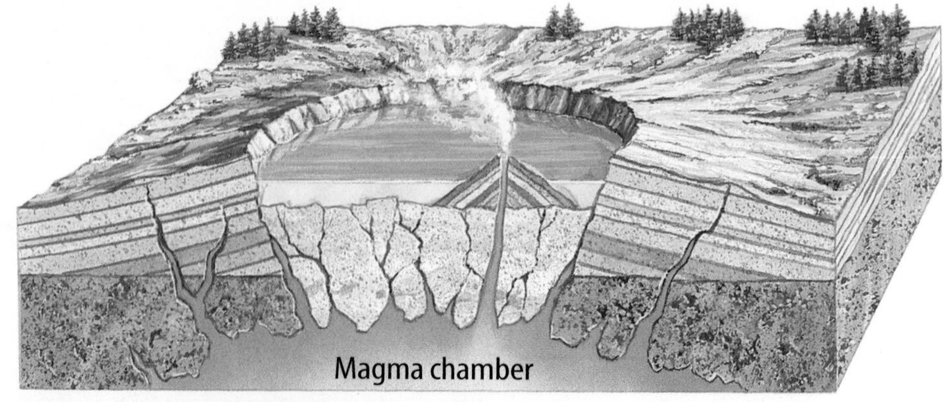

Crater Lake in Oregon formed when water collected in the circular space left when surface material collapsed.

Figure 19 Wizard Island in Crater Lake is a cinder cone volcano that erupted after the formation of the caldera.
Explain *what causes a caldera to form.*

Igneous Features Exposed You have learned in this chapter that Earth's surface is built up and worn down continually. The surface of Earth is built up by volcanoes. Also, igneous rock is formed when magma hardens below ground. Eventually, the processes of weathering and erosion wear down rock at the surface, exposing features like batholiths, dikes, and sills.

section 3 review

Summary

Intrusive Features

- Intrusive igneous rock features are formed from magma that is forced upward toward Earth's crust, then slowly cools and solidifies underground before reaching the surface.
- Batholiths, dikes, and sills are a few examples of intrusive igneous rock features.

Other Features

- A volcanic neck is the solid igneous core of a volcano left behind after the softer cone has been eroded.
- A caldera is a large, circular-shaped depression that forms when the top of a volcano collapses.

Self Check

1. **Compare and contrast** a caldera and a crater.
2. **Illustrate** how a sill forms. How is it different from a dike?
3. **Describe** a batholith and explain how it forms.
4. **Think Critically** Why are the large, granite dome features of Yosemite National Park in California considered to be intrusive volcanic features when they are exposed at the surface?

Applying Math

5. **Calculate** Basaltic rocks contain approximately 45 percent to 52 percent silica. About how many kilograms of silica would you predict to be in a 68-kg boulder of basalt?

Design Your Own

How do calderas form?

Goals

- **Design** a volcano setup that will demonstrate how a caldera could form.
- **Observe** what happens during trials with your volcano setup.
- **Describe** what you observe.

Possible Materials

small box
small balloon
paper
newspaper
flour
plastic tubing
clamp for tubing
tape
scissors

Safety Precautions

Real-World Question

A caldera is a depression that forms when the top of a volcano collapses after an eruption. What might cause the top of a volcano to collapse?

Form a Hypothesis

Based on your reading about volcanoes, state a hypothesis about what would happen if the magma inside the magma chamber of a volcano were suddenly removed.

Test Your Hypothesis

Make a Plan

1. As a group, agree upon the hypothesis and identify which results will support the hypothesis.
2. **Design** a volcano that allows you to test your hypothesis. What materials will you use to build your volcano?
3. What will you remove from inside your volcano to represent the loss of magma? How will you remove it?
4. Where will you place your volcano? What will you do to minimize messes?
5. **Identify** all constants, variables, and controls of the experiment.

Follow Your Plan

1. Make sure your teacher approves your plan before you start.
2. **Construct** your volcano with any features that will be required to test your hypothesis.
3. **Conduct** one or more appropriate trials to test your hypothesis. Record any observations that you make and any other data that are appropriate to test your hypothesis.

Analyze Your Data

1. **Describe** in words or with a drawing what your volcano looked like before you began.
2. **Observe** what happened to your volcano during the experiment that you conducted? Did its appearance change?
3. **Describe** in words or with a drawing what your volcano looked like after the trial.
4. **Observe** What other observations did you make?
5. **Describe** any other data that you recorded.

Conclude and Apply

1. **Draw Conclusions** Did your observations support your hypothesis? Explain.
2. **Explain** how your demonstration was similar to what might happen to a real volcano. How was it different?

Communicating Your Data

Make a 4-sequence time-lapse diagram with labels and descriptions of how a caldera forms. Use your visual aid to describe caldera formation to students in another class.

Oops! Accidents in SCIENCE

SOMETIMES GREAT DISCOVERIES HAPPEN BY ACCIDENT!

Buried in Ash

A long-forgotten city is accidentally found after 2,000 years

In the heat of the Italian Sun, a farmer digs a new well for water. He thrusts his shovel into the ground one more time. But instead of hitting water, the shovel strikes something hard; a slab of smooth white marble.

Under the ground lay the ancient city of Herculaneum (her kew LAY nee um). The city, and its neighbor Pompeii (pom PAY) had been buried for more than 1,600 years. On August 24, 79 A.D., Mount Vesuvius, a nearby volcano, erupted and buried both cities with pumice, rocks, mud, and ash.

Back in Time

The Sun shone over the peaceful town of Herculaneum on that August morning almost 2,000 years ago. But at about 1 P.M., that peace was shattered forever.

With massive force, the peak of Vesuvius exploded, sending six cubic kilometers of ash and pumice into the sky. Hours later, a fiery surge made its way from the volcano to the city. These pyroclastic flows continued as more buildings were crushed and buried by falling ash and pumice. Within six hours, much of the city was totally buried under the flows. After six surges from Vesuvius, the deadly eruption ceased. But the city had disappeared under approximately 21 m of ash, rock, and mud.

Excavated ruins with Mount Vesuvius in the background

A City Vanishes

More than 3,600 people were killed in the natural disaster. Scientists believe that many died trying to protect their faces from the pyroclastic surges that filled the air with hot ash. Those able to escape returned to find no trace of their city. Over hundreds of years, grass and fields covered Herculaneum, erasing it from human memory.

Archaeologists have unearthed perfectly preserved mosaics and a library with ancient scrolls in excellent condition. Archaeologists found skeletons and voids that were filled with plaster to form casts of people who died the day Vesuvius erupted. Visitors to the site can see a Roman woman, a teen-aged girl, and a soldier with his sword still in his hand.

Much of Herculaneum still lies buried beneath thick layers of volcanic ash, and archaeologists still are digging to expose more of the ruins. Their work is helping scientists better understand everyday life in an ancient Italian town. But, if it weren't for a farmer's search for water, Herculaneum might not have been discovered at all!

Research the history of your town. Ask your local librarian to help "unearth" maps, drawings, or photos that let you travel back in time! Design a two-layer map that shows the past and the present.

For more information, visit earth.msscience.com/oops

Reviewing Main Ideas

Section 1 Volcanoes and Earth's Moving Plates

1. Volcanoes can be dangerous to people because they can cause deaths and destroy property.
2. Rocks in the crust and mantle melt to form magma, which is forced toward Earth's surface. When the magma flows through vents, it's called lava and forms volcanoes.
3. Volcanoes can form over hot spots or when Earth's plates pull apart or come together.

Section 2 Types of Volcanoes

1. The three types of volcanoes are composite volcanoes, cinder cone volcanoes, and shield volcanoes.
2. Shield volcanoes produce quiet eruptions. Cinder cone and composite volcanoes can produce explosive eruptions.
3. Some lavas are thin and flow easily, producing quiet eruptions. Other lavas are thick and stiff, producing violent eruptions.

Section 3 Igneous Rock Features

1. Intrusive igneous rock bodies such as batholiths, dikes, and sills form when magma solidifies underground.
2. Batholiths are the most massive igneous rock bodies. Dikes and sills form when magma squeezes into cracks.
3. A caldera forms when the top of a volcano collapses, forming a large depression.

Visualizing Main Ideas

Copy and complete the following concept map on types of volcanic eruptions.

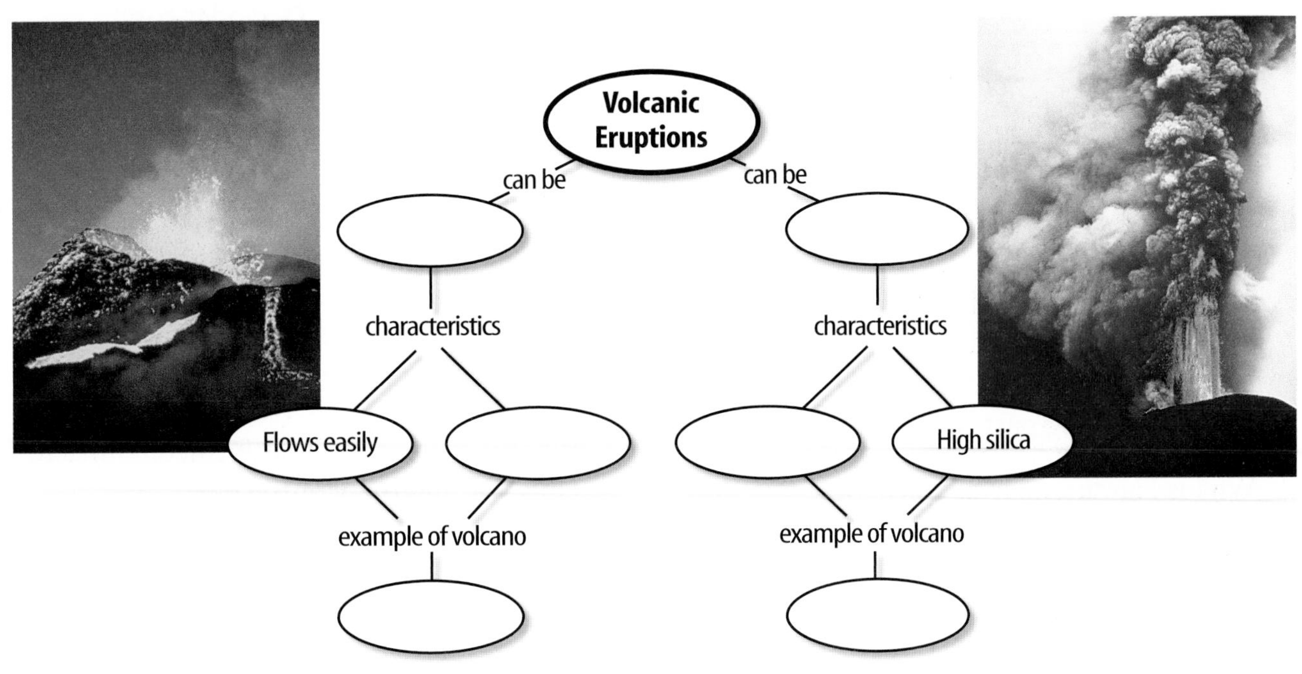

chapter 12 Review

Using Vocabulary

batholith p. 346
caldera p. 348
cinder cone volcano p. 340
composite volcano p. 341
crater p. 332
dike p. 347
hot spot p. 334
shield volcano p. 340
sill p. 347
tephra p. 340
vent p. 332
volcanic neck p. 347
volcano p. 330

Fill in the blanks with the correct vocabulary word or words.

1. A broad volcano with gently sloping sides is called a(n) ________.
2. Bits of rock or solidified lava dropped from the air after a volcanic eruption are ________.
3. Magma squeezed into a horizontal crack between rock layers is a(n) ________.
4. The steep-walled depression around a volcano's vent is called a(n) ________.
5. Magma squeezed into a vertical crack across rock layers is called a(n) ________.

Checking Concepts

Choose the word or phrase that best answers the question.

6. What type of boundary is associated with composite volcanoes?
   A) plates moving apart
   B) plates sticking and slipping
   C) plates moving together
   D) plates sliding past each other
7. Why is Hawaii made of volcanoes?
   A) Plates are moving apart.
   B) A hot spot exists.
   C) Plates are moving together.
   D) Rift zones exist.
8. What kind of magmas produce violent volcanic eruptions?
   A) those rich in silica
   B) those that are fluid
   C) those forming shield volcanoes
   D) those rich in iron
9. Magma that is low in silica generally produces what kind of eruptions?
   A) thick **C)** quiet
   B) caldera **D)** explosive

Use the photo below to answer question 10.

10. Which type of volcano, shown above, is made entirely of tephra?
    A) shield **C)** cinder cone
    B) caldera **D)** composite
11. What kind of volcano is Kilauea?
    A) shield **C)** cinder cone
    B) composite **D)** caldera cone
12. What is the largest intrusive igneous rock body?
    A) dike **C)** sill
    B) volcanic neck **D)** batholith
13. What is the process that formed Soufrière Hills volcano on Montserrat?
    A) plates sticking and slipping
    B) caldera formation
    C) plates sliding sideways
    D) plates moving together

Science Online earth.msscience.com/vocabulary_puzzlemaker

Thinking Critically

14. **Explain** how glaciers and volcanoes can exist on Iceland.

15. **Describe** what kind of eruption is produced when basaltic lava that is low in silica flows from a volcano.

16. **Explain** how volcanoes are related to earthquakes.

17. **Infer** Misti is a volcano in Peru. Peru is on the western edge of South America. How might this volcano have formed?

18. **Describe** the layers of a composite volcano. Which layers represent violent eruptions?

19. **Classify** the volcano Fuji, which has steep sides and is made of layers of silica-rich lava and ash.

Use the map below to answer question 20.

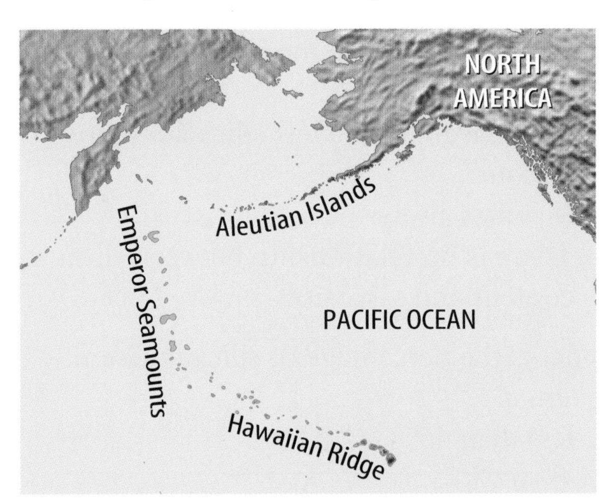

20. **Interpret Scientific Illustrations** Look at the map above. The Hawaiian Islands and Emperor Seamounts were formed when the Pacific Plate moved over a fixed hot spot. If the Emperor chain trends in a direction different from the Hawaiian Islands, what can you infer about the Pacific Plate?

21. **Concept Map** Make a network-tree concept map about where volcanoes can occur. Include the following words and phrases: *hot spots, divergent plate boundaries, convergent plate boundaries, volcanoes, can occur, examples, Iceland, Soufrière Hills,* and *Hawaiian Islands.*

Performance Activities

22. **Poster** Make a Venn diagram of the three basic types of volcanoes. Label them and indicate what cone formation, lava composition, eruption, and geologic location are expected of each type of volcano.

Applying Math

23. **Sea Level** The base of the volcano Mauna Loa is about 5,000 m below sea level. The total height of the volcano is 9,170 m. What percentage of the volcano is above sea level? Below sea level?

Use the table below to answer questions 24 and 25.

Volcano	Year of Eruption	Amount of Material Ejected
Tambora	1815	131 km^3
Katmai	1912	30 km^3
Novarupta	1912	15 km^3
Mt. St. Helens	1980	1.3 km^3
Pinatubo	1991	5.5 km^3

24. **Ejected Material** How many times greater was the volume of ejected material from Tambora, as compared to Mt. St. Helens?

25. **Graph** Design a bar graph to show the amount of ejected material from the volcanoes. Present the information from least to greatest volume.

Standardized Test Practice

Part 1 Multiple Choice

Record your answers on the answer sheet provided by your teacher or on a sheet of paper.

Use the photo below to answer question 1.

1. Which of the following terms best describes the rock in the photo above?
   A. aa
   B. pahoehoe
   C. pillow lava
   D. ash

2. Which of the following is made of layers of ash and cooled lava flows?
   A. shield volcano
   B. plateau basalts
   C. composite volcano
   D. cinder cone volcano

3. Which of the following volcanoes is located in the United States?
   A. Hekla
   B. Paricutin
   C. Mount Vesuvius
   D. Mount St. Helens

4. Which of the following igneous features is parallel to the rock layers that it intrudes?
   A. batholith
   B. volcanic neck
   C. sill
   D. dike

5. Which of the following forms when the top of a volcano collapses into a partially emptied magma chamber?
   A. fissure
   B. crater
   C. caldera
   D. volcanic neck

Test-Taking Tip

Relax Stay calm during the test. If you feel yourself getting nervous, close your eyes and take five slow, deep breaths.

Use the graph below to answer questions 6 and 7.

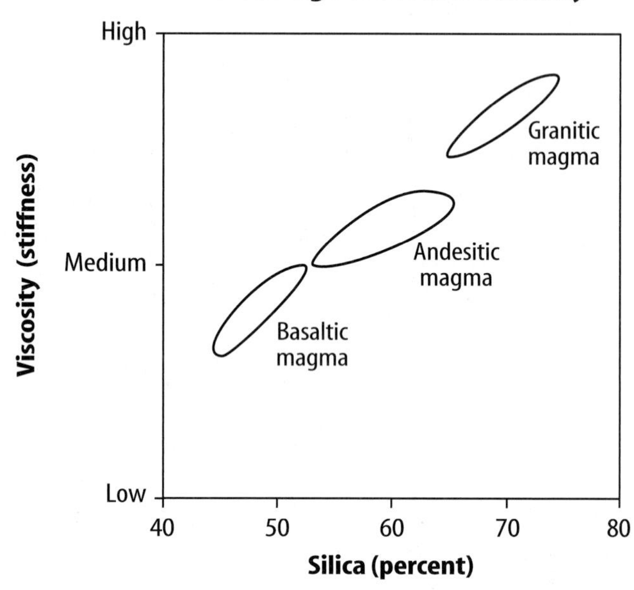

6. What relationship can be inferred from the graph?
   A. Magmas that have more silica are more viscous.
   B. Magmas that have less silica are more viscous.
   C. Magmas always have low viscosity.
   D. There is no relationship between silica content and viscosity.

7. What is the percentage of silica in Granitic magma?
   A. less than 45%
   B. 45–52%
   C. 53–65%
   D. greater than 65%

8. Which of the following is the finest type of tephra?
   A. volcanic ash
   B. volcanic bombs
   C. volcanic cinders
   D. volcanic blocks

Part 2 Short Response/Grid In

Record your answers on the answer sheet provided by your teacher or on a sheet of paper.

9. What is a hot spot? Why do volcanoes often form at hot spots?

10. Why are the Hawaiian Islands in a line?

11. How is a dike different from a sill? Support your answer with a Venn diagram.

Use the table below to answer questions 12–14.

Eruption	Volume Percent Water Vapor
1	58.7
2	60.1
3	61.4
4	59.3
5	59.6

12. Calculate the mean, median, and range of the water vapor data in the table? Describe how this information would be helpful to a volcanologist.

13. Using the mean value that you calculated in question 12, what percentage of the volcanic gas consists of gases other than water vapor?

14. The water vapor content of Kilauea is above average when compared to other volcanoes. How might these data help to explain why lava fountains often occur on Kilauea?

15. What is the difference between magma and lava?

16. Explain how igneous rock forms from lava.

Part 3 Open Ended

Record your answers on a sheet of paper.

17. Explain how volcanic necks, such as Ship Rock, form. Support your answer with a labeled diagram.

18. How does tephra form?

19. Why do some volcanoes occur where one plate sinks beneath another plate? Support your answer with a labeled diagram.

20. How can pillow-shaped bodies form from lava?

Use the map below to answer questions 21 and 22.

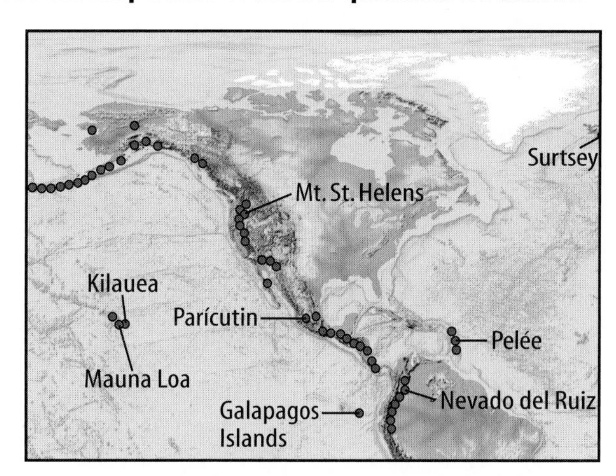

21. What kind of magma was required to create the Columbia Plateau as compared to Mt. St. Helens, only 200 miles away? What would you predict would be the percentage of water vapor in the two types of magma?

22. Where do volcanoes occur in the United States?

23. How do volcanoes affect people and their property? List four safety precautions for people living in volcanic areas.

How Are Canals & the Paleozoic Era Connected?

Before the invention of the locomotive, canals, such as the one at upper right, were an important means of transportation. In the 1790s, an engineer traveled around England to study new canals. The engineer noticed something odd: All across the country, certain types of rocks seemed to lie in predictable layers, or strata. And the same strata always had the same kinds of fossils in them. Since each layer of sedimentary rock typically forms on top of the previous one, scientists realized that the strata recorded the history of life on Earth. By the mid-1800s, the known rock strata had been organized into a system that we now know as the geologic time scale. In this system, Earth's history is divided into units called eras, which in turn are divided into periods. Many of the rock layers in the Grand Canyon (background) date from the Paleozoic, or "ancient life," Era.

unit projects

Visit earth.msscience.com/unit_project to find project ideas and resources. Projects include:

- **History** Discover some of Earth's inhabitants of different time periods using the fossil record. Create a drawing of a scene in Earth's history.
- **Technology** Choose an extinct animal to investigate. How has technology allowed paleontologists to learn about how it lived?
- **Model** As a group, design a wall mural or diorama depicting the layers of the geologic time scale, or a particular scene of interest from an era.

WebQuest Use online resources to form your own opinion concerning plate tectonics. Investigate the *Fossils of Antarctica* and what they could tell us about its ancient climate and location.

chapter 13

Clues to Earth's Past

The BIG Idea

Fossils, along with the relative ages and absolute ages of rocks, provide evidence of past life, climates, and environments on Earth.

SECTION 1
Fossils
Main Idea Fossils are used as evidence of past life on Earth.

SECTION 2
Relative Ages of Rocks
Main Idea Relative ages of rocks can be determined by examining their locations within a sequence of rock layers.

SECTION 3
Absolute Ages of Rocks
Main Idea Absolute ages of rocks can be determined by using properties of the atoms that make up materials.

Reading the Past

The pages of Earth's history, much like the pages of human history, can be read if you look in the right place. Unlike the pages of a book, the pages of Earth's past are written in stone. In this chapter you will learn how to read the pages of Earth's history to understand what the planet was like in the distant past.

Science Journal List three fossils that you would expect to find a million years from now in the place you live today.

Start-Up Activities

Clues to Life's Past

Fossil formation begins when dead plants or animals are buried in sediment. In time, if conditions are right, the sediment hardens into sedimentary rock. Parts of the organism are preserved along with the impressions of parts that don't survive. Any evidence of once-living things contained in the rock record is a fossil.

1. Fill a small jar (about 500 mL) one-third full of plaster of paris. Add water until the jar is half full.
2. Drop in a few small shells.
3. Cover the jar and shake it to model a swift, muddy stream.
4. Now model the stream flowing into a lake by uncovering the jar and pouring the contents into a paper or plastic bowl. Let the mixture sit for an hour.
5. Crack open the hardened plaster to locate the model fossils.
6. **Think Critically** Remove the shells from the plaster and study the impressions they made. In your Science Journal, list what the impressions would tell you if found in a rock.

FOLDABLES™ Study Organizer

Age of Rocks Make the following Foldable to help you understand how scientists determine the age of a rock.

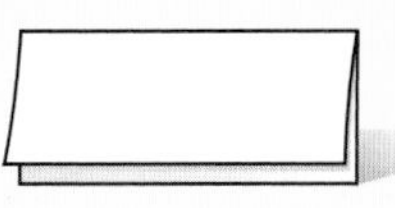

STEP 1 **Fold** a sheet of paper in half lengthwise.

STEP 2 **Fold** paper down 2.5 cm from the top. (Hint: From the tip of your index finger to your middle knuckle is about 2.5 cm.)

STEP 3 **Open and draw** lines along the 2.5-cm fold. **Label** as shown.

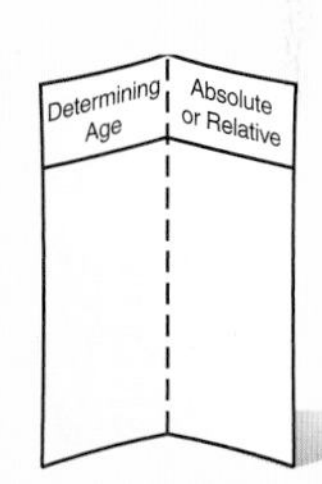

Summarize in a Table As you read the chapter, in the left column, list four different ways in which one could determine the age of a rock. In the right column, note whether each method gives an absolute or a relative age.

Preview this chapter's content and activities at earth.msscience.com

Get Ready to Read

Take Notes

1 Learn It! The best way for you to remember information is to write it down, or take notes. Good note-taking is useful for studying and research. When you are taking notes, it is helpful to

- phrase the information in your own words;
- restate ideas in short, memorable phrases;
- stay focused on main ideas and only the most important supporting details.

2 Practice It! Make note-taking easier by using a chart to help you organize information clearly. Write the main ideas in the left column. Then write at least three supporting details in the right column. Read the text from Section 1 of this chapter under the heading *Conditions Needed for Fossil Formation,* page 363. Then take notes using a chart, such as the one below.

Main Idea	Supporting Details
	1. 2. 3. 4. 5.
	1. 2. 3. 4. 5.

3 Apply It! As you read this chapter, make a chart of the main ideas. Next to each main idea, list at least three supporting details.

Reading Tip

Read one or two paragraphs first and take notes after you read. You are likely to take down too much information if you take notes as you read.

Target Your Reading

Use this to focus on the main ideas as you read the chapter.

1. **Before you read** the chapter, respond to the statements below on your worksheet or on a numbered sheet of paper.
   - Write an **A** if you **agree** with the statement.
   - Write a **D** if you **disagree** with the statement.

2. **After you read** the chapter, look back to this page to see if you've changed your mind about any of the statements.
   - If any of your answers changed, explain why.
   - Change any false statements into true statements.
   - Use your revised statements as a study guide.

Before You Read A or D		Statement	After You Read A or D
	1	All fossils are made from the hard parts of animals.	
	2	Fossils can be used as evidence to show that past climates and environments have changed.	
	3	A trace fossil is the outline, or copy, of a fossil.	
	4	Sediment typically accumulates in horizontal beds, which can later form layers of sedimentary rock.	
	5	The relative age of a rock layer indicates whether the layer is older or younger when compared to other rock layers.	
	6	The principle of superposition refers to a high concentration of fossils within a small area.	
	7	Most sequences of rock layers are complete.	
	8	Geologists often can match up, or correlate, layers of rock over great distances.	
	9	The absolute age of a material refers to the actual age, in years, of the material.	

Science Online
Print out a worksheet of this page at earth.msscience.com

section 1

Fossils

as you read

What You'll Learn

- **List** the conditions necessary for fossils to form.
- **Describe** several processes of fossil formation.
- **Explain** how fossil correlation is used to determine rock ages.
- **Determine** how fossils can be used to explain changes in Earth's surface, life forms, and environments.

Why It's Important

Fossils help scientists find oil and other sources of energy necessary for society.

Review Vocabulary
paleontologist: a scientist who studies fossils

New Vocabulary
- fossil
- permineralized remains
- carbon film
- mold
- cast
- index fossil

Traces of the Distant Past

A giant crocodile lurks in the shallow water of a river. A herd of *Triceratops* emerges from the edge of the forest and cautiously moves toward the river. The dinosaurs are thirsty, but danger waits for them in the water. A large bull *Triceratops* moves into the river. The others follow.

Does this scene sound familiar to you? It's likely that you've read about dinosaurs and other past inhabitants of Earth. But how do you know that they really existed or what they were like? What evidence do humans have of past life on Earth? The answer is fossils. Paleontologists, scientists who study fossils, can learn about extinct animals from their fossil remains, as shown in **Figure 1.**

Figure 1 Scientists can learn how dinosaurs looked and moved using fossil remains. A skeleton can then be reassembled and displayed in a museum.

Formation of Fossils

Fossils are the remains, imprints, or traces of prehistoric organisms. Fossils have helped scientists determine approximately when life first appeared, when plants and animals first lived on land, and when organisms became extinct. Fossils are evidence of not only when and where organisms once lived, but also how they lived.

For the most part, the remains of dead plants and animals disappear quickly. Scavengers eat and scatter the remains of dead organisms. Fungi and bacteria invade, causing the remains to rot and disappear. If you've ever left a banana on the counter too long, you've seen this process begin. In time, compounds within the banana cause it to break down chemically and soften. Microorganisms, such as bacteria, cause it to decay. What keeps some plants and animals from disappearing before they become fossils? Which organisms are more likely to become fossils?

Figure 2 These fossil shark teeth are hard parts. Soft parts of animals do not become fossilized as easily.

Conditions Needed for Fossil Formation Whether or not a dead organism becomes a fossil depends upon how well it is protected from scavengers and agents of physical destruction, such as waves and currents. One way a dead organism can be protected is for sediment to bury the body quickly. If a fish dies and sinks to the bottom of a lake, sediment carried into the lake by a stream can cover the fish rapidly. As a result, no waves or scavengers can get to it and tear it apart. The body parts then might be fossilized and included in a sedimentary rock like shale. However, quick burial alone isn't always enough to make a fossil.

Organisms have a better chance of becoming fossils if they have hard parts such as bones, shells, or teeth. One reason is that scavengers are less likely to eat these hard parts. Hard parts also decay more slowly than soft parts do. Most fossils are the hard parts of organisms, such as the fossil teeth in **Figure 2.**

Predicting Fossil Preservation

Procedure

1. Take a brief walk outside and observe your neighborhood.
2. Look around and notice what kinds of plants and animals live nearby.

Analysis

1. Predict what remains from your time might be preserved far into the future.
2. Explain what conditions would need to exist for these remains to be fossilized.

Try at Home

Types of Preservation

Perhaps you've seen skeletal remains of *Tyrannosaurus rex* towering above you in a museum. You also have some idea of what this dinosaur looked like because you've seen illustrations. Artists who draw *Tyrannosaurus rex* and other dinosaurs base their illustrations on fossil bones. What preserves fossil bones?

Figure 3 Opal and various minerals have replaced original materials and filled the hollow spaces in this permineralized dinosaur bone.
Explain *why this fossil retained the shape of the original bone.*

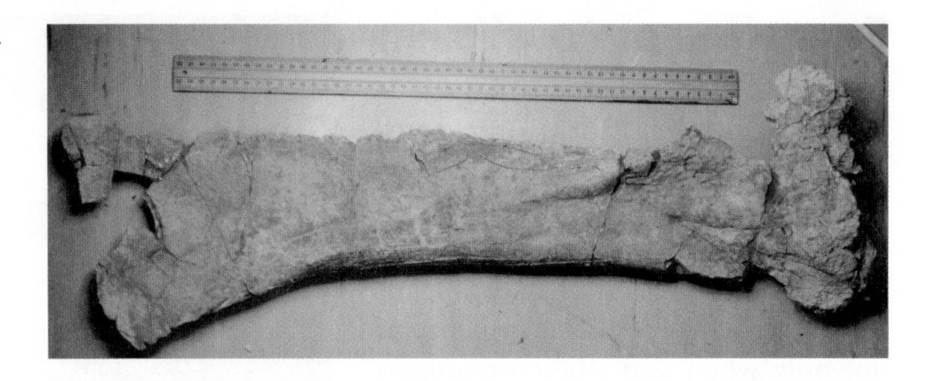

Mineral Replacement Most hard parts of organisms such as bones, teeth, and shells have tiny spaces within them. In life, these spaces can be filled with cells, blood vessels, nerves, or air. When the organism dies and the soft materials inside the hard parts decay, the tiny spaces become empty. If the hard part is buried, groundwater can seep in and deposit minerals in the spaces. **Permineralized remains** are fossils in which the spaces inside are filled with minerals from groundwater. In permineralized remains, some original material from the fossil organism's body might be preserved—encased within the minerals from groundwater. It is from these original materials that DNA, the chemical that contains an organism's genetic code, can sometimes be recovered.

Sometimes minerals replace the hard parts of fossil organisms. For example, a solution of water and dissolved silica (the compound SiO_2) might flow into and through the shell of a dead organism. If the water dissolves the shell and leaves silica in its place, the original shell is replaced.

Often people learn about past forms of life from bones, wood, and other remains that became permineralized or replaced with minerals from groundwater, as shown in **Figure 3,** but many other types of fossils can be found.

Figure 4 Graptolites lived hundreds of millions of years ago and drifted on currents in the oceans. These organisms often are preserved as carbon films.

Carbon Films The tissues of organisms are made of compounds that contain carbon. Sometimes fossils contain only carbon. Fossils usually form when sediments bury a dead organism. As sediment piles up, the organism's remains are subjected to pressure and heat. These conditions force gases and liquids from the body. A thin film of carbon residue is left, forming a silhouette of the original organism called a **carbon film.** **Figure 4** shows the carbonized remains of graptolites, which were small marine animals. Graptolites have been found in rocks as old as 500 million years.

Coal In swampy regions, large volumes of plant matter accumulate. Over millions of years, these deposits become completely carbonized, forming coal. Coal is an important fuel source, but since the structure of the original plant is usually lost, it cannot reveal as much about the past as other kinds of fossils.

Reading Check *In what sort of environment does coal form?*

Coal Mining Many of the first coal mines in the United States were located in eastern states like Pennsylvania and West Virginia. In your Science Journal, discuss how the environments of the past relate to people's lives today.

Molds and Casts In nature, impressions form when seashells or other hard parts of organisms fall into a soft sediment such as mud. The object and sediment are then buried by more sediment. Compaction, together with cementation, which is the deposition of minerals from water into the pore spaces between sediment particles, turns the sediment into rock. Other open pores in the rock then let water and air reach the shell or hard part. The hard part might decay or dissolve, leaving behind a cavity in the rock called a **mold.** Later, mineral-rich water or other sediment might enter the cavity, form new rock, and produce a copy or **cast** of the original object, as shown in **Figure 5.**

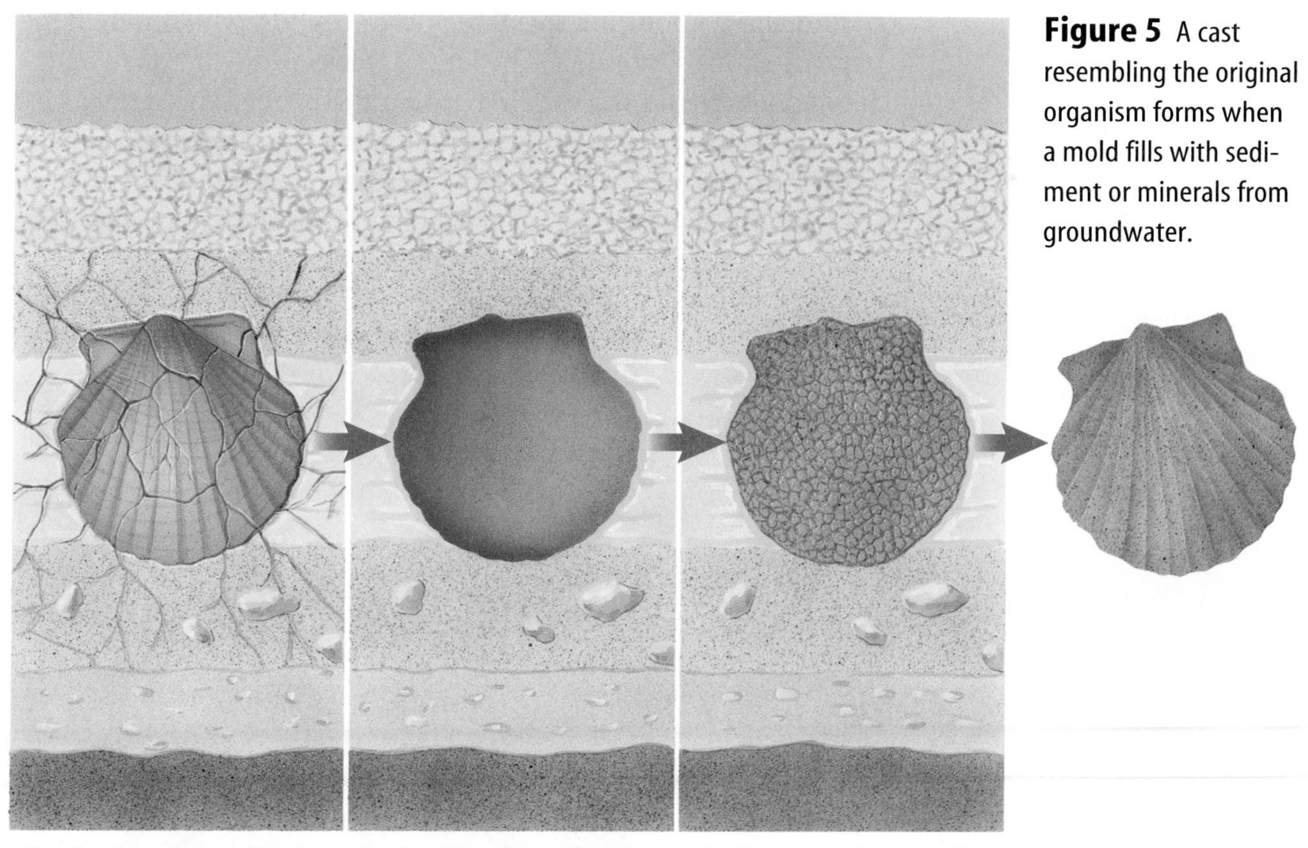

The fossil begins to dissolve as water moves through spaces in the rock layers.

The fossil has been dissolved away. The harder rock once surrounding it forms a mold.

Sediment washes into the mold and is deposited, or mineral crystals form.

A cast results.

Figure 5 A cast resembling the original organism forms when a mold fills with sediment or minerals from groundwater.

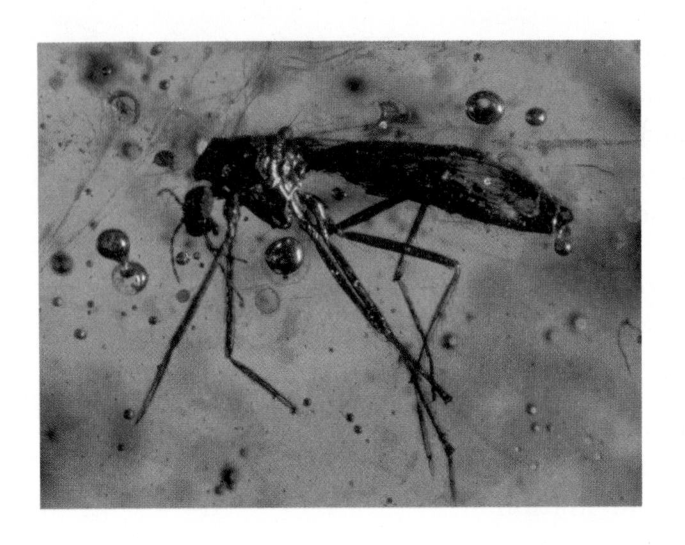

Figure 6 The original soft parts of this mosquito have been preserved in amber for millions of years.

Original Remains Sometimes conditions allow original soft parts of organisms to be preserved for thousands or millions of years. For example, insects can be trapped in amber, a hardened form of sticky tree resin. The amber surrounds and protects the original material of the insect's exoskeleton from destruction, as shown in **Figure 6.** Some organisms, such as the mammoth, have been found preserved in frozen ground in Siberia. Original remains also have been found in natural tar deposits, such as the La Brea tar pits in California.

Trace Fossils Do you have a handprint in plaster that you made when you were in kindergarten? If so, it's a record that tells something about you. From it, others can guess your size and maybe your weight at that age. Animals walking on Earth long ago left similar tracks, such as those in **Figure 7.** Trace fossils are fossilized tracks and other evidence of the activity of organisms. In some cases, tracks can tell you more about how an organism lived than any other type of fossil. For example, from a set of tracks at Davenport Ranch, Texas, you might be able to learn something about the social life of sauropods, which were large, plant-eating dinosaurs. The largest tracks of the herd are on the outer edges and the smallest are on the inside. These tracks led some scientists to hypothesize that adult sauropods surrounded their young as they traveled—perhaps to protect them from predators. A nearby set of tracks might mean that another type of dinosaur, an allosaur, was stalking the herd.

Figure 7 Tracks made in soft mud, and now preserved in solid rock, can provide information about animal size, speed, and behavior.

The dinosaur track below is from the Glen Rose Formation in north-central Texas.

The tracks to the right are located on a Navajo reservation in Arizona.

Trails and Burrows Other trace fossils include trails and burrows made by worms and other animals. These, too, tell something about how these animals lived. For example, by examining fossil burrows you can sometimes tell how firm the sediment the animals lived in was. As you can see, fossils can tell a great deal about the organisms that have inhabited Earth.

Reading Check ***How are trace fossils different from fossils that are the remains of an organism's body?***

Index Fossils

One thing you can learn by studying fossils is that species of organisms have changed over time. Some species of organisms inhabited Earth for long periods of time without changing. Other species changed a lot in comparatively short amounts of time. It is these organisms that scientists use as index fossils.

Index fossils are the remains of species that existed on Earth for relatively short periods of time, were abundant, and were widespread geographically. Because the organisms that became index fossils lived only during specific intervals of geologic time, geologists can estimate the ages of rock layers based on the particular index fossils they contain. However, not all rocks contain index fossils. Another way to approximate the age of a rock layer is to compare the spans of time, or ranges, over which more than one fossil appears. The estimated age is the time interval where fossil ranges overlap, as shown in **Figure 8.**

Figure 8 The fossils in a sequence of sedimentary rock can be used to estimate the ages of each layer. The chart shows when each organism inhabited Earth.
Explain *why it is possible to say that the middle layer of rock was deposited between 440 million and 410 million years ago.*

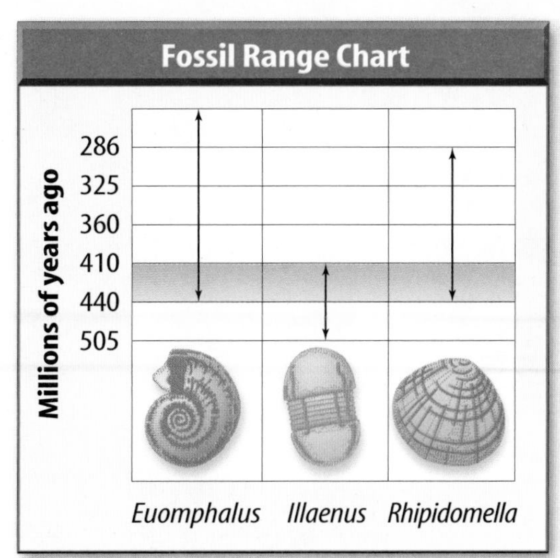

Ancient Ecology
Ecology is the study of how organisms interact with each other and with their environment. Some paleontologists study the ecology of ancient organisms. Discuss the kinds of information you could use to determine how ancient organisms interacted with their environment.

Fossils and Ancient Environments

Scientists can use fossils to determine what the environment of an area was like long ago. Using fossils, you might be able to find out whether an area was land or whether it was covered by an ocean at a particular time. If the region was covered by ocean, it might even be possible to learn the depth of the water. What clues about the depth of water do you think fossils could provide?

Fossils also are used to determine the past climate of a region. For example, rocks in parts of the eastern United States contain fossils of tropical plants. The environment of this part of the United States today isn't tropical. However, because of the fossils, scientists know that it was tropical when these plants were living. **Figure 9** shows that North America was located near the equator when these fossils formed.

Figure 9 The equator passed through North America 310 million years ago. At this time, warm, shallow seas and coal swamps covered much of the continent, and ferns like the *Neuropteris,* below, were common.

Deep ocean
Exposed land
Shallow seas
Deep ocean
Exposed land
Shallow seas
Equator
Exposed land
Shallow seas
Coal swamps
AFRICA
Deep ocean

Figure 10 The crinoid on the left lived in warm, shallow seas that once covered part of North America. Crinoids like the one on the right typically live in warm, shallow waters in the Pacific Ocean.

Shallow Seas How would you explain the presence of fossilized crinoids—animals that lived in shallow seas—in rocks found in what is today a desert? **Figure 10** shows a fossil crinoid and a living crinoid. When the fossil crinoids were alive, a shallow sea covered much of western and central North America. The crinoid hard parts were included in rocks that formed from the sediments at the bottom of this sea. Fossils provide information about past life on Earth and also about the history of the rock layers that contain them. Fossils can provide information about the ages of rocks and the climate and type of environment that existed when the rocks formed.

section 1 review

Summary

Formation of Fossils

- Fossils are the remains, imprints, or traces of past organisms.
- Fossilization is most likely if the organism had hard parts and was buried quickly.

Fossil Preservation

- Permineralized remains have open spaces filled with minerals from groundwater.
- Thin carbon films remain in the shapes of dead organisms.
- Hard parts dissolve to leave molds.
- Trace fossils are evidence of past activity.

Index Fossils

- Index fossils are from species that were abundant briefly, but over wide areas.
- Scientists can estimate the ages of rocks containing index fossils.

Fossils and Ancient Environments

- Fossils tell us about the environment in which the organisms lived.

Self Check

1. **Describe** the typical conditions necessary for fossil formation.
2. **Explain** how a fossil mold is different from a fossil cast.
3. **Discuss** how the characteristics of an index fossil are useful to geologists.
4. **Describe** how carbon films form.
5. **Think Critically** What can you say about the ages of two widely separated layers of rock that contain the same type of fossil?

Applying Skills

6. **Communicate** what you learn about fossils. Visit a museum that has fossils on display. Make an illustration of each fossil in your Science Journal. Write a brief description, noting key facts about each fossil and how each fossil might have formed.
7. **Compare and contrast** original remains with other kinds of fossils. What kinds of information would only be available from original remains? Are there any limitations to the use of original remains?

section 2

Relative Ages of Rocks

as you read

What You'll Learn

- **Describe** methods used to assign relative ages to rock layers.
- **Interpret** gaps in the rock record.
- **Give** an example of how rock layers can be correlated with other rock layers.

Why It's Important

Being able to determine the age of rock layers is important in trying to understand a history of Earth.

Review Vocabulary

sedimentary rock: rock formed when sediments are cemented and compacted or when minerals are precipitated from solution

New Vocabulary

- principle of superposition
- relative age
- unconformity

Superposition

Imagine that you are walking to your favorite store and you happen to notice an interesting car go by. You're not sure what kind it is, but you remember that you read an article about it. You decide to look it up. At home you have a stack of magazines from the past year, as seen in **Figure 11.**

You know that the article you're thinking of came out in the January edition, so it must be near the bottom of the pile. As you dig downward, you find magazines from March, then February. January must be next. How did you know that the January issue of the magazine would be on the bottom? To find the older edition under newer ones, you applied the principle of superposition.

Oldest Rocks on the Bottom According to the **principle of superposition,** in undisturbed layers of rock, the oldest rocks are on the bottom and the rocks become progressively younger toward the top. Why is this the case?

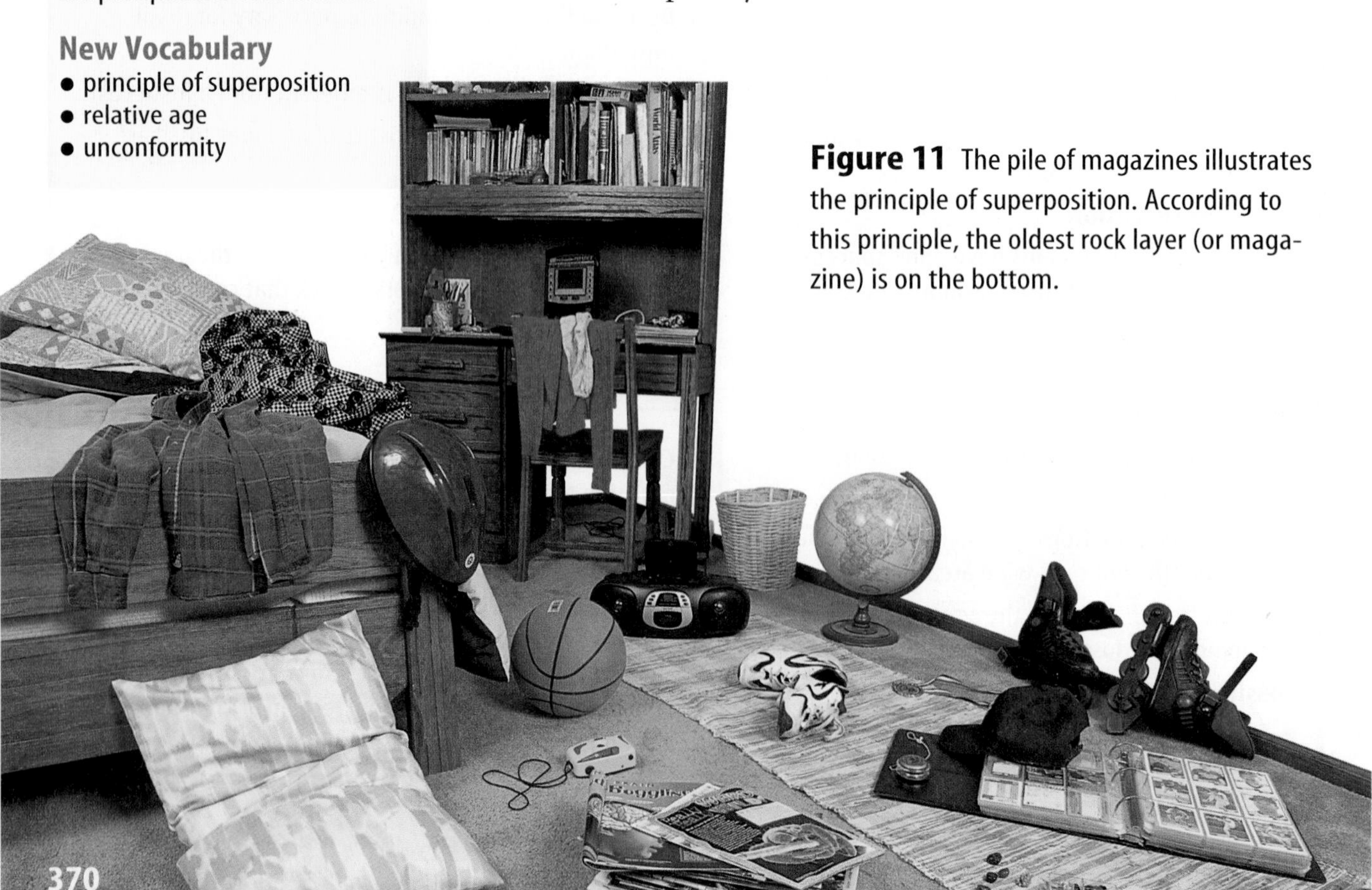

Figure 11 The pile of magazines illustrates the principle of superposition. According to this principle, the oldest rock layer (or magazine) is on the bottom.

Rock Layers Sediment accumulates in horizontal beds, forming layers of sedimentary rock. The first layer to form is on the bottom. The next layer forms on top of the previous one. Because of this, the oldest rocks are at the bottom. However, forces generated by mountain formation sometimes can turn layers over. When layers have been turned upside down, it's necessary to use other clues in the rock layers to determine their original positions and relative ages.

Topic: Relative Dating

Visit earth.msscience.com for Web links to information about relative dating of rocks and other materials.

Activity Imagine yourself at an archaeological dig. You have found a rare artifact and want to know its age. Make a list of clues you might look for to provide a relative date and explain how each would allow you to approximate the artifact's age.

Relative Ages

Now you want to look for another magazine. You're not sure how old it is, but you know it arrived after the January issue. You can find it in the stack by using the principle of relative age.

The **relative age** of something is its age in comparison to the ages of other things. Geologists determine the relative ages of rocks and other structures by examining their places in a sequence. For example, if layers of sedimentary rock are offset by a fault, which is a break in Earth's surface, you know that the layers had to be there before a fault could cut through them. The relative age of the rocks is older than the relative age of the fault. Relative age determination doesn't tell you anything about the age of rock layers in actual years. You don't know if a layer is 100 million or 10,000 years old. You only know that it's younger than the layers below it and older than the fault cutting through it.

Other Clues Help Determination of relative age is easy if the rocks haven't been faulted or turned upside down. For example, look at **Figure 12.** Which layer is the oldest? In cases where rock layers have been disturbed you might have to look for fossils and other clues to date the rocks. If you find a fossil in the top layer that's older than a fossil in a lower layer, you can hypothesize that layers have been turned upside down by folding during mountain building.

Figure 12 In a stack of undisturbed sedimentary rocks, the oldest rocks are at the bottom. This stack of rocks can be folded by forces within Earth.
Explain *how you can tell if an older rock is above a younger one.*

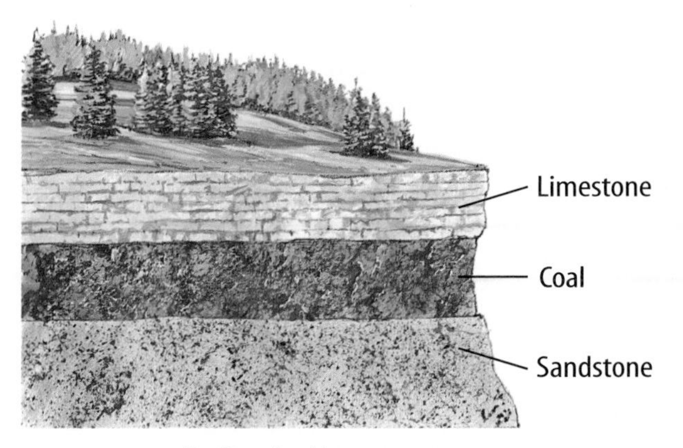

Undisturbed Layers

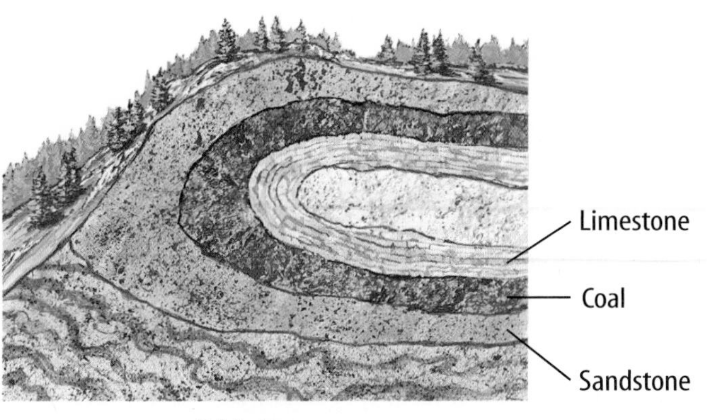

Folded Layers

Figure 13 An angular unconformity results when horizontal layers cover tilted, eroded layers.

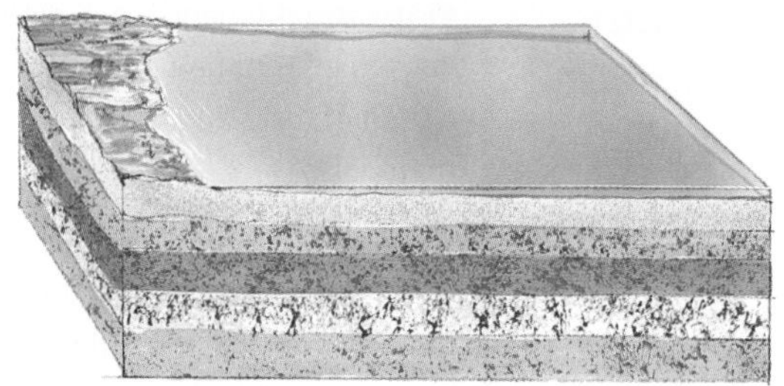

A Sedimentary rocks are deposited originally as horizontal layers.

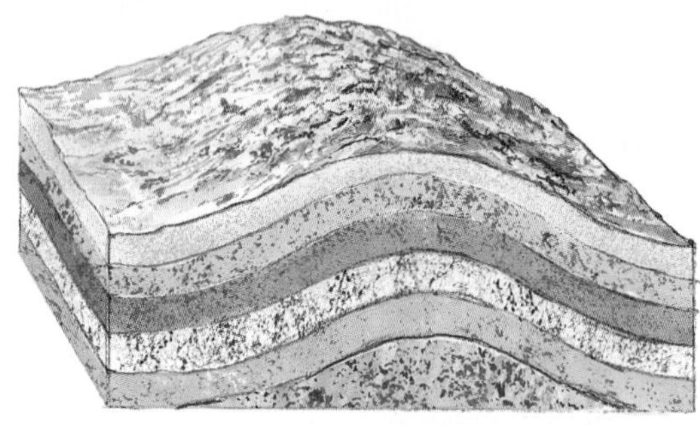

B The horizontal rock layers are tilted as forces within Earth deform them.

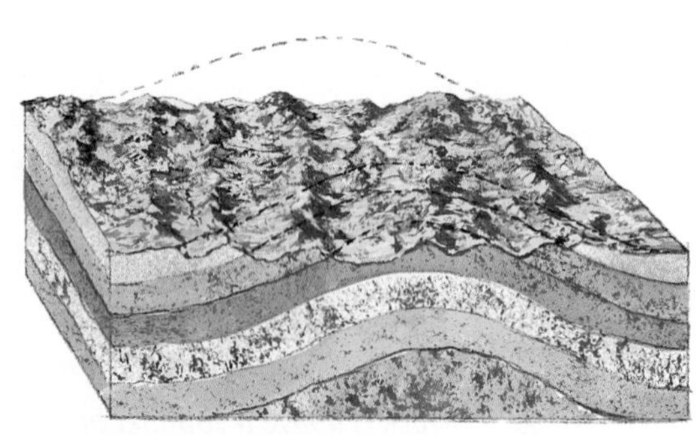

C The tilted layers erode.

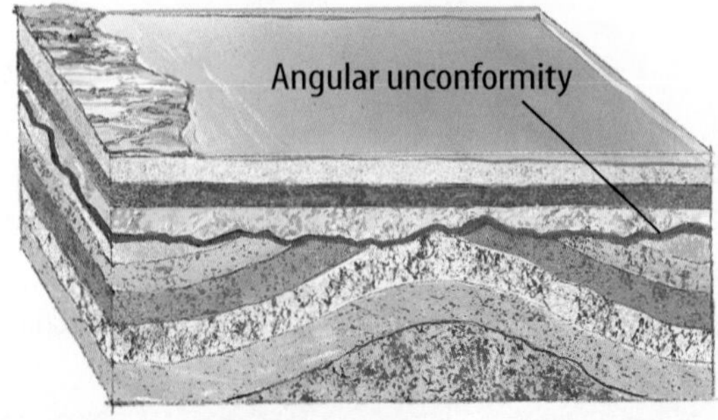

D An angular unconformity results when new layers form on the tilted layers as deposition resumes.

Unconformities

A sequence of rock is a record of past events. But most rock sequences are incomplete—layers are missing. These gaps in rock sequences are called **unconformities** (un kun FOR muh teez). Unconformities develop when agents of erosion such as running water or glaciers remove rock layers by washing or scraping them away.

Reading Check *How do unconformities form?*

Angular Unconformities Horizontal layers of sedimentary rock often are tilted and uplifted. Erosion and weathering then wear down these tilted rock layers. Eventually, younger sediment layers are deposited horizontally on top of the tilted and eroded layers. Geologists call such an unconformity an angular unconformity. **Figure 13** shows how angular unconformities develop.

Disconformity Suppose you're looking at a stack of sedimentary rock layers. They look complete, but layers are missing. If you look closely, you might find an old surface of erosion. This records a time when the rocks were exposed and eroded. Later, younger rocks formed above the erosion surface when deposition of sediment began again. Even though all the layers are parallel, the rock record still has a gap. This type of unconformity is called a disconformity. A disconformity also forms when a period of time passes without any new deposition occurring to form new layers of rock.

Nonconformity Another type of unconformity, called a nonconformity, occurs when metamorphic or igneous rocks are uplifted and eroded. Sedimentary rocks are then deposited on top of this erosion surface. The surface between the two rock types is a nonconformity. Sometimes rock fragments from below are incorporated into sediments deposited above the nonconformity. All types of unconformities are shown in **Figure 14.**

NATIONAL GEOGRAPHIC

VISUALIZING UNCONFORMITIES

Figure 14

An unconformity is a gap in the rock record caused by erosion or a pause in deposition. There are three major kinds of unconformities—nonconformity, angular unconformity, and disconformity.

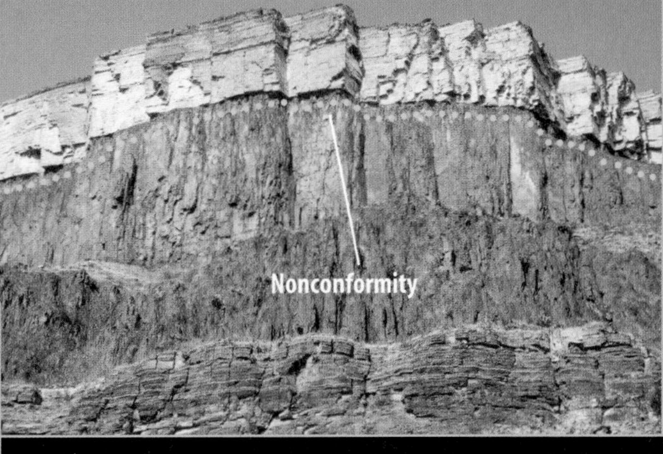

▲ In a nonconformity, horizontal layers of sedimentary rock overlie older igneous or metamorphic rocks. A nonconformity in Big Bend National Park, Texas, is shown above.

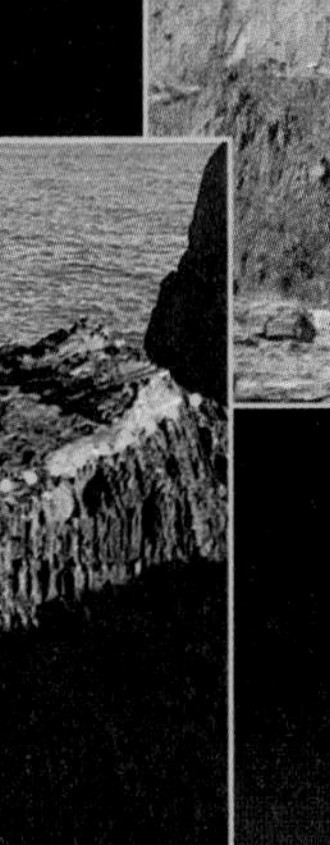

▲ An angular unconformity develops when new horizontal layers of sedimentary rock form on top of older sedimentary rock layers that have been folded by compression. An example of an angular unconformity at Siccar Point in southeastern Scotland is shown above.

▼ A disconformity develops when horizontal rock layers are exposed and eroded, and new horizontal layers of rock are deposited on the eroded surface. The disconformity shown below is in the Grand Canyon.

Science Online

Topic: Correlating with Index Fossils

Visit earth.msscience.com for Web links to information about using index fossils to match up layers of rock.

Activity Make a chart that shows the rock layers of both the Grand Canyon and Capitol Reef National Park in Utah. For each layer that appears in both parks, list an index fossil you could find to correlate the layers.

Matching Up Rock Layers

Suppose you're studying a layer of sandstone in Bryce Canyon in Utah. Later, when you visit Canyonlands National Park, Utah, you notice that a layer of sandstone there looks just like the sandstone in Bryce Canyon, 250 km away. Above the sandstone in the Canyonlands is a layer of limestone and then another sandstone layer. You return to Bryce Canyon and find the same sequence—sandstone, limestone, and sandstone. What do you infer? It's likely that you're looking at the same layers of rocks in two different locations. **Figure 15** shows that these rocks are parts of huge deposits that covered this whole area of the western United States. Geologists often can match up, or correlate, layers of rocks over great distances.

Evidence Used for Correlation It's not always easy to say that a rock layer exposed in one area is the same as a rock layer exposed in another area. Sometimes it's possible to walk along the layer for kilometers and prove that it's continuous. In other cases, such as at the Canyonlands area and Bryce Canyon as seen in **Figure 16,** the rock layers are exposed only where rivers have cut through overlying layers of rock and sediment. How can you show that the limestone sandwiched between the two layers of sandstone in Canyonlands is likely the same limestone as at Bryce Canyon? One way is to use fossil evidence. If the same types of fossils were found in the limestone layer in both places, it's a good indication that the limestone at each location is the same age, and, therefore, one continuous deposit.

Reading Check *How do fossils help show that rocks at different locations belong to the same rock layer?*

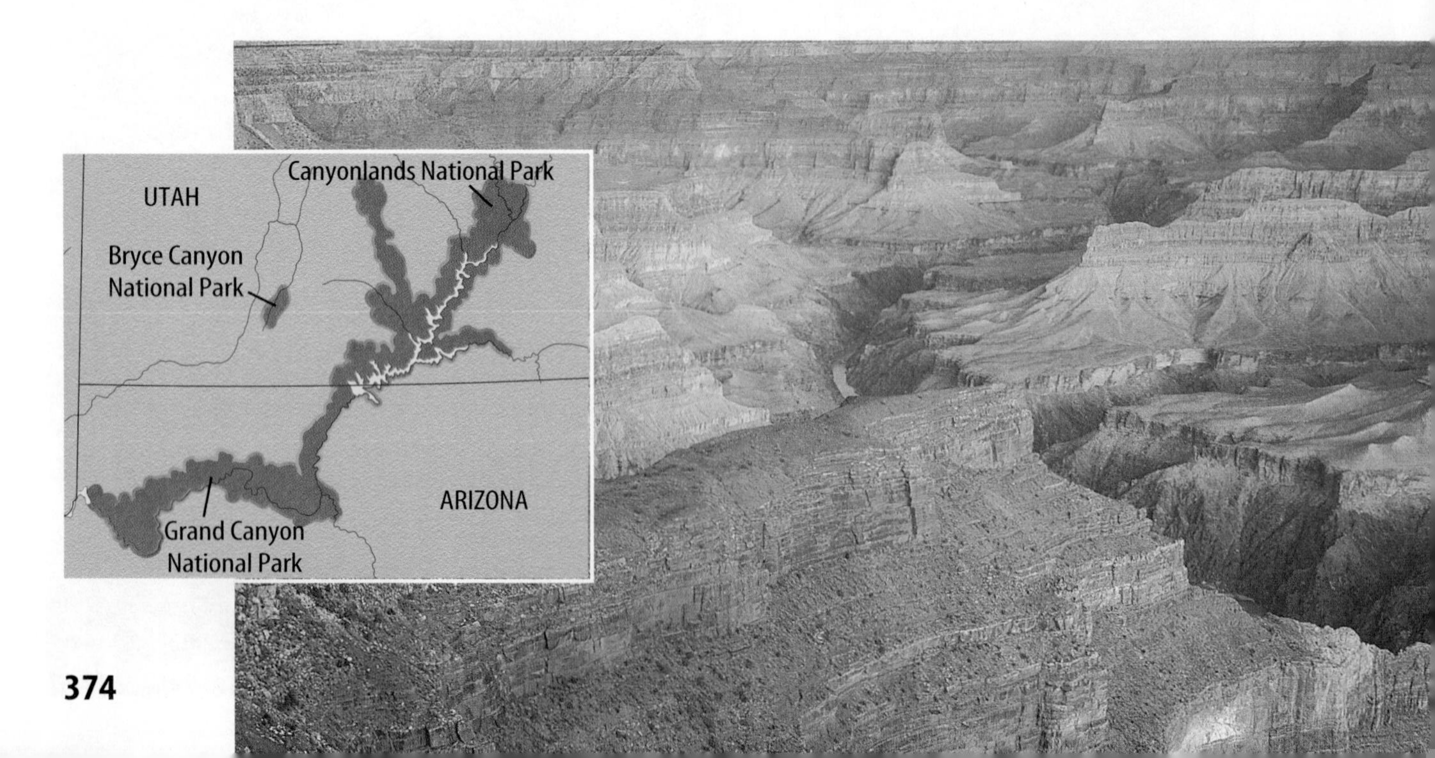

Figure 15 These rock layers, exposed at Hopi Point in Grand Canyon National Park, Arizona, can be correlated, or matched up, with rocks from across large areas of the western United States.

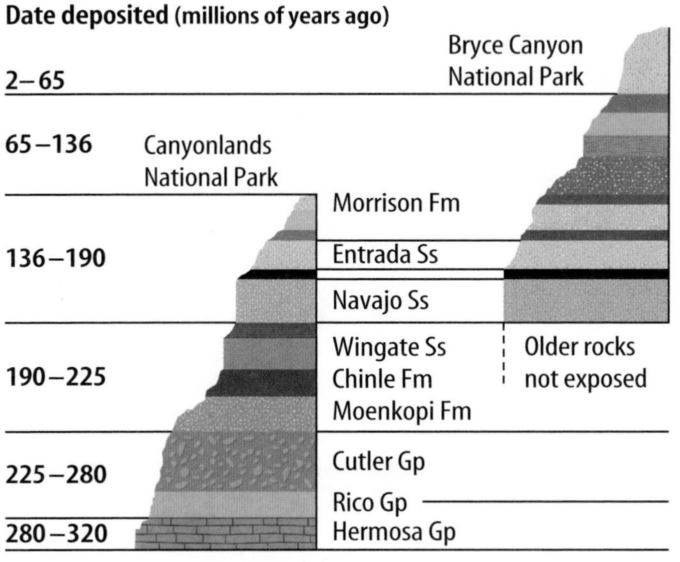

Figure 16 Geologists have named the many rock layers, or formations, in Canyonlands and in Bryce Canyon, Utah. They also have correlated some formations between the two canyons.
List *the labeled layers present at both canyons.*

Can layers of rock be correlated in other ways? Sometimes determining relative ages isn't enough and other dating methods must be used. In Section 3, you'll see how the numerical ages of rocks can be determined and how geologists have used this information to estimate the age of Earth.

section 2 review

Summary

Superposition

- Superposition states that in undisturbed rock, the oldest layers are on the bottom.

Relative Ages

- Rock layers can be ranked by relative age.

Unconformities

- Angular unconformities are new layers deposited over tilted and eroded rock layers.
- Disconformities are gaps in the rock record.
- Nonconformities divide uplifted igneous or metamorphic rock from new sedimentary rock.

Matching Up Rock Layers

- Rocks from different areas may be correlated if they are part of the same layer.

Self Check

1. **Discuss** how to find the oldest paper in a stack of papers.
2. **Explain** the concept of relative age.
3. **Illustrate** a disconformity.
4. **Describe** one way to correlate similar rock layers.
5. **Think Critically** Explain the relationship between the concept of relative age and the principle of superposition.

Applying Skills

6. **Interpret data** to determine the oldest rock bed. A sandstone contains a 400-million-year-old fossil. A shale has fossils that are over 500 million years old. A limestone, below the sandstone, contains fossils between 400 million and 500 million years old. Which rock bed is oldest? Explain.

Relative Ages

Which of your two friends is older? To answer this question, you'd need to know their relative ages. You wouldn't need to know the exact age of either of your friends—just who was born first. The same is sometimes true for rock layers.

Real-World Question

Can you determine the relative ages of rock layers?

Goals

- **Interpret** illustrations of rock layers and other geological structures and determine the relative order of events.

Materials

paper   pencil

Procedure

1. **Analyze Figures A and B.**
2. Make a sketch of **Figure A.** On it, identify the relative age of each rock layer, igneous intrusion, fault, and unconformity. For example, the shale layer is the oldest, so mark it with a 1. Mark the next-oldest feature with a 2, and so on.
3. Repeat step 2 for **Figure B.**

Conclude and Apply

Figure A

1. **Identify** the type of unconformity shown. Is it possible that there were originally more layers of rock than are shown?
2. **Describe** how the rocks above the fault moved in relation to rocks below the fault.
3. **Hypothesize** how the hill on the left side of the figure formed.

A

B

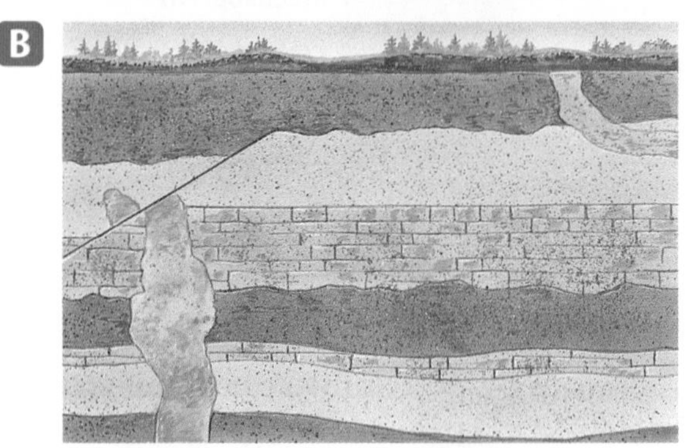

Granite   Limestone

Sandstone   Shale

Figure B

4. Is it possible to conclude if the igneous intrusion on the left is older or younger than the unconformity nearest the surface?
5. **Describe** the relative ages of the two igneous intrusions. How did you know?
6. **Hypothesize** which two layers of rock might have been much thicker in the past.

Compare your results with other students' results. **For more help, refer to the Science Skill Handbook.**

section 3

Absolute Ages of Rocks

Absolute Ages

As you sort through your stack of magazines looking for that article about the car you saw, you decide that you need to restack them into a neat pile. By now, they're in a jumble and no longer in order of their relative age, as shown in **Figure 17.** How can you stack them so the oldest are on the bottom and the newest are on top? Fortunately, magazine dates are printed on the cover. Thus, stacking magazines in order is a simple process. Unfortunately, rocks don't have their ages stamped on them. Or do they? **Absolute age** is the age, in years, of a rock or other object. Geologists determine absolute ages by using properties of the atoms that make up materials.

as you read

What You'll Learn

- **Identify** how absolute age differs from relative age.
- **Describe** how the half-lives of isotopes are used to determine a rock's age.

Why It's Important

Events in Earth's history can be better understood if their absolute ages are known.

Review Vocabulary
isotopes: atoms of the same element that have different numbers of neutrons

New Vocabulary
- absolute age
- radioactive decay
- half-life
- radiometric dating
- uniformitarianism

Radioactive Decay

INTEGRATE Physics

Atoms consist of a dense central region called the nucleus, which is surrounded by a cloud of negatively charged particles called electrons. The nucleus is made up of protons, which have a positive charge, and neutrons, which have no electric charge. The number of protons determines the identity of the element, and the number of neutrons determines the form of the element, or isotope. For example, every atom with a single proton is a hydrogen atom. Hydrogen atoms can have no neutrons, a single neutron, or two neutrons. This means that there are three isotopes of hydrogen.

Reading Check *What particles make up an atom's nucleus?*

Some isotopes are unstable and break down into other isotopes and particles. Sometimes a lot of energy is given off during this process. The process of breaking down is called **radioactive decay.** In the case of hydrogen, atoms with one proton and two neutrons are unstable and tend to break down. Many other elements have stable and unstable isotopes.

Figure 17 The magazines that have been shuffled through no longer illustrate the principle of superposition.

Modeling Carbon-14 Dating

Procedure

1. Count out 80 **red jelly beans.**
2. Remove half the red jelly beans and replace them with **green jelly beans.**
3. Continue replacing half the red jelly beans with green jelly beans until only 5 red jelly beans remain. Count the number of times you replace half the red jelly beans.

Analysis

1. How did this activity model the decay of carbon-14 atoms?
2. How many half lives of carbon-14 did you model during this activity?
3. If the atoms in a bone experienced the same number of half lives as your jelly beans, how old would the bone be?

Alpha and Beta Decay In some isotopes, a neutron breaks down into a proton and an electron. This type of radioactive decay is called beta decay because the electron leaves the atom as a beta particle. The nucleus loses a neutron but gains a proton. When the number of protons in an atom is changed, a new element forms. Other isotopes give off two protons and two neutrons in the form of an alpha particle. Alpha and beta decay are shown in **Figure 18.**

Half-Life In radioactive decay reactions, the parent isotope undergoes radioactive decay. The daughter product is produced by radioactive decay. Each radioactive parent isotope decays to its daughter product at a certain rate. Based on this decay rate, it takes a certain period of time for one half of the parent isotope to decay to its daughter product. The **half-life** of an isotope is the time it takes for half of the atoms in the isotope to decay. For example, the half-life of carbon-14 is 5,730 years. So it will take 5,730 years for half of the carbon-14 atoms in an object to change into nitrogen-14 atoms. You might guess that in another 5,730 years, all of the remaining carbon-14 atoms will decay to nitrogen-14. However, this is not the case. Only half of the atoms of carbon-14 remaining after the first 5,730 years will decay during the second 5,730 years. So, after two half-lives, one fourth of the original carbon-14 atoms still remain. Half of them will decay during another 5,730 years. After three half-lives, one eighth of the original carbon-14 atoms still remain. After many half-lives, such a small amount of the parent isotope remains that it might not be measurable.

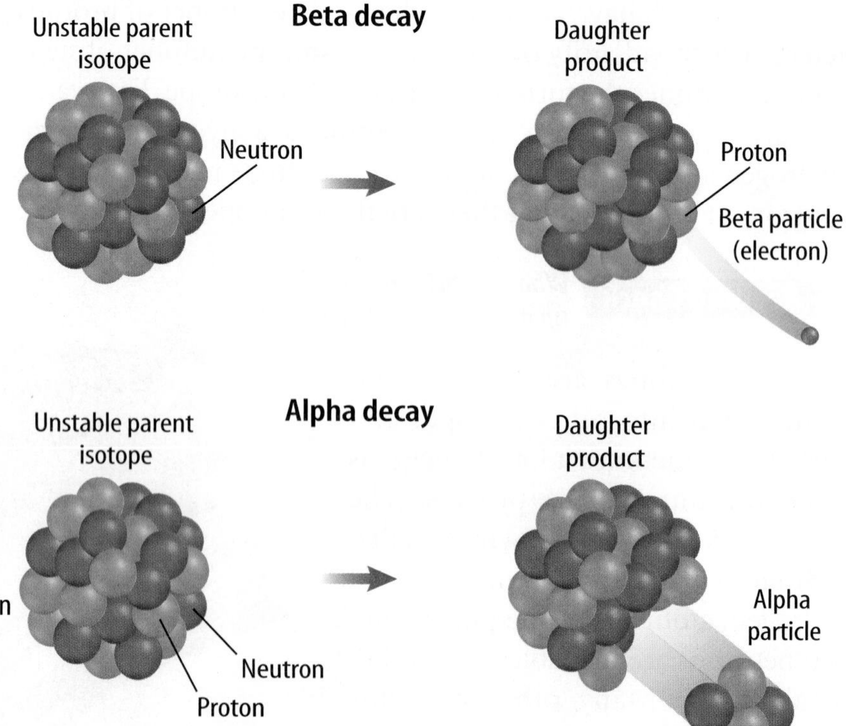

Figure 18 In beta decay, a neutron changes into a proton by giving off an electron. This electron has a lot of energy and is called a beta particle.

In the process of alpha decay, an unstable parent isotope nucleus gives off an alpha particle and changes into a new daughter product. Alpha particles contain two neutrons and two protons.

Radiometric Ages

Decay of radioactive isotopes is like a clock keeping track of time that has passed since rocks have formed. As time passes, the amount of parent isotope in a rock decreases as the amount of daughter product increases, as in **Figure 19.** By measuring the ratio of parent isotope to daughter product in a mineral and by knowing the half-life of the parent, in many cases you can calculate the absolute age of a rock. This process is called **radiometric dating.**

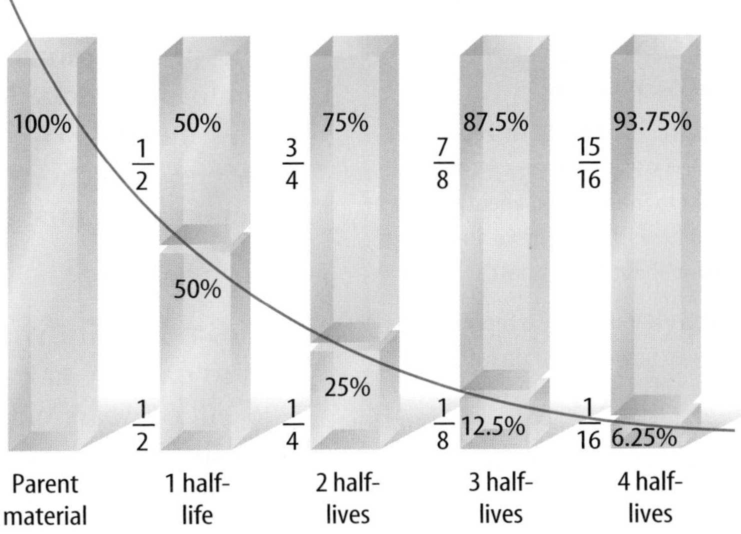

Figure 19 During each half-life, one half of the parent material decays to the daughter product. **Explain** *how one uses both parent and daughter material to estimate age.*

A scientist must decide which parent isotope to use when measuring the age of a rock. If the object to be dated seems old, then the geologist will use an isotope with a long half-life. The half-life for the decay of potassium-40 to argon-40 is 1.25 billion years. As a result, this isotope can be used to date rocks that are many millions of years old. To avoid error, conditions must be met for the ratios to give a correct indication of age. For example, the rock being studied must still retain all of the argon-40 that was produced by the decay of potassium-40. Also, it cannot contain any contamination of daughter product from other sources. Potassium-argon dating is good for rocks containing potassium, but what about other things?

Radiocarbon Dating Carbon-14 is useful for dating bones, wood, and charcoal up to 75,000 years old. Living things take in carbon from the environment to build their bodies. Most of that carbon is carbon-12, but some is carbon-14, and the ratio of these two isotopes in the environment is always the same. After the organism dies, the carbon-14 slowly decays. By determining the amounts of the isotopes in a sample, scientists can evaluate how much the isotope ratio in the sample differs from that in the environment. For example, during much of human history, people built campfires. The wood from these fires often is preserved as charcoal. Scientists can determine the amount of carbon-14 remaining in a sample of charcoal by measuring the amount of radiation emitted by the carbon-14 isotope in labs like the one in **Figure 20.** Once they know the amount of carbon-14 in a charcoal sample, scientists can determine the age of the wood used to make the fire.

Figure 20 Radiometric ages are determined in labs like this one.

Topic: Isotopes in Ice Cores
Visit earth.msscience.com for Web links to information about ice cores and how isotopes in ice are used to learn about Earth's past.

Activity Prepare a report that shows how isotopes in ice cores can tell us about past Earth environments. Include how these findings can help us understand today's climate.

Age Determinations Aside from carbon-14 dating, rocks that can be radiometrically dated are mostly igneous and metamorphic rocks. Most sedimentary rocks cannot be dated by this method. This is because many sedimentary rocks are made up of particles eroded from older rocks. Dating these pieces only gives the age of the preexisting rock from which it came.

The Oldest Known Rocks Radiometric dating has been used to date the oldest rocks on Earth. These rocks are about 3.96 billion years old. By determining the age of meteorites, and using other evidence, scientists have estimated the age of Earth to be about 4.5 billion years. Earth rocks greater than 3.96 billion years old probably were eroded or changed by heat and pressure.

Why can't most sedimentary rocks be dated radiometrically?

Applying Science

When did the Iceman die?

Carbon-14 dating has been used to date charcoal, wood, bones, mummies from Egypt and Peru, the Dead Sea Scrolls, and the Italian Iceman. The Iceman was found in 1991 in the Italian Alps, near the Austrian border. Based on carbon-14 analysis, scientists determined that the Iceman is 5,300 years old. Determine approximately in what year the Iceman died.

Half-Life of Carbon-14

Percent Carbon-14	Years Passed
100	0
50	5,730
25	11,460
12.5	17,190
6.25	22,920
3.125	

Reconstruction of Iceman

Identifying the Problem

The half-life chart shows the decay of carbon-14 over time. Half-life is the time it takes for half of a sample to decay. Fill in the years passed when only 3.125 percent of carbon-14 remain. Is there a point at which no carbon-14 would be present? Explain.

Solving the Problem

1. Estimate, using the data table, how much carbon-14 still was present in the Iceman's body that allowed scientists to determine his age.
2. If you had an artifact that originally contained 10.0 g of carbon-14, how many grams would remain after 17,190 years?

Uniformitarianism

Can you imagine trying to determine the age of Earth without some of the information you know today? Before the discovery of radiometric dating, many people estimated that Earth is only a few thousand years old. But in the 1700s, Scottish scientist James Hutton estimated that Earth is much older. He used the principle of **uniformitarianism.** This principle states that Earth processes occurring today are similar to those that occurred in the past. Hutton's principle is often paraphrased as "the present is the key to the past."

Hutton observed that the processes that changed the landscape around him were slow, and he inferred that they were just as slow throughout Earth's history. Hutton hypothesized that it took much longer than a few thousand years to form the layers of rock around him and to erode mountains that once stood kilometers high. **Figure 21** shows Hutton's native Scotland, a region shaped by millions of years of geologic processes.

Today, scientists recognize that Earth has been shaped by two types of change: slow, everyday processes that take place over millions of years, and violent, unusual events such as the collision of a comet or asteroid about 65 million years ago that might have caused the extinction of the dinosaurs.

Figure 21 The rugged highlands of Scotland were shaped by erosion and uplift.

section 3 review

Summary

Absolute Ages

- The absolute age is the actual age of an object.

Radioactive Decay

- Some isotopes are unstable and decay into other isotopes and particles.
- Decay is measured in half-lives, the time it takes for half of a given isotope to decay.

Radiometric Ages

- By measuring the ratio of parent isotope to daughter product, one can determine the absolute age of a rock.
- Living organisms less than 75,000 years old can be dated using carbon-14.

Uniformitarianism

- Processes observable today are the same as the processes that took place in the past.

Self Check

1. **Evaluate** the age of rocks. You find three undisturbed rock layers. The middle layer is 120 million years old. What can you say about the ages of the layers above and below it?
2. **Determine** the age of a fossil if it had only one eighth of its original carbon-14 content remaining.
3. **Explain** the concept of uniformitarianism.
4. **Describe** how radioactive isotopes decay.
5. **Think Critically** Why can't scientists use carbon-14 to determine the age of an igneous rock?

Applying Math

6. **Make and use a table** that shows the amount of parent material of a radioactive element that is left after four half-lives if the original parent material had a mass of 100 g.

Model and Invent

Trace Fossils

Goals
- **Construct** a model of trace fossils.
- **Describe** the information that you can learn from looking at your model.

Possible Materials
construction paper
wire
plastic (a fairly rigid type)
scissors
plaster of paris
toothpicks
sturdy cardboard
clay
pipe cleaners
glue

Safety Precautions

Real-World Question

Trace fossils can tell you a lot about the activities of organisms that left them. They can tell you how an organism fed or what kind of home it had. How can you model trace fossils that can provide information about the behavior of organisms? What materials can you use to model trace fossils? What types of behavior could you show with your trace fossil model?

Make a Model

1. **Decide** how you are going to make your model. What materials will you need?
2. **Decide** what types of activities you will demonstrate with your model. Were the organisms feeding? Resting? Traveling? Were they predators? Prey? How will your model indicate the activities you chose?
3. What is the setting of your model? Are you modeling the organism's home? Feeding areas? Is your model on land or water? How can the setting affect the way you build your model?
4. Will you only show trace fossils from a single species or multiple species? If you include more than one species, how will you provide evidence of any interaction between the species?

Check the Model Plans

1. Compare your plans with those of others in your class. Did other groups mention details that you had forgotten to think about? Are there any changes you would like to make to your plan before you continue?
2. Make sure your teacher approves your plan before you continue.

Test Your Model

1. Following your plan, construct your model of trace fossils.
2. Have you included evidence of all the behaviors you intended to model?

Analyze Your Data

1. **Evaluate** Now that your model is complete, do you think that it adequately shows the behaviors you planned to demonstrate? Is there anything that you think you might want to do differently if you were going to make the model again?
2. **Describe** how using different kinds of materials might have affected your model. Can you think of other materials that would have allowed you to show more detail than you did?

Conclude and Apply

1. **Compare and contrast** your model of trace fossils with trace fossils left by real organisms. Is one more easily interpreted than the other? Explain.
2. **List** behaviors that might not leave any trace fossils. Explain.

Communicating Your Data

Ask other students in your class or another class to look at your model and describe what information they can learn from the trace fossils. Did their interpretations agree with what you intended to show?

Oops! Accidents in SCIENCE

SOMETIMES GREAT DISCOVERIES HAPPEN BY ACCIDENT!

The World's Oldest Fish Story

A catch-of-the-day set science on its ears

Camouflage marks

First dorsal fin

Second dorsal fin

Pectoral fin

Anal fin

Pelvic fin

Some scientists call the coelacanth "Old Four Legs." It got its nickname because the fish has paired fins that look something like legs.

On a December day in 1938, just before Christmas, Marjorie Courtenay-Latimer went to say hello to her friends on board a fishing boat that had just returned to port in South Africa. Courtenay-Latimer, who worked at a museum, often went aboard her friends' ship to check out the catch. On this visit, she received a surprise Christmas present—an odd-looking fish. As soon as the woman spotted its strange blue fins among the piles of sharks and rays, she knew it was special.

Courtenay-Latimer took the fish back to her museum to study it. "It was the most beautiful fish I had ever seen, five feet long, and a pale mauve blue with iridescent silver markings," she later wrote. Courtenay-Latimer sketched it and sent the drawing to a friend of hers, J. L. B. Smith.

Smith was a chemistry teacher who was passionate about fish. After a time, he realized it was a coelacanth (SEE luh kanth). Fish experts knew that coelacanths had first appeared on Earth 400 million years ago. But the experts thought the fish were extinct. People had found fossils of coelacanths, but no one had seen one alive. It was assumed that the last coelacanth species had died out 65 million years ago. They were wrong. The ship's crew had caught one by accident.

Smith figured there might be more living coelacanths. So he decided to offer a reward for anyone who could find a living specimen. After 14 years of silence, a report came in that a coelacanth had been caught off the east coast of Africa.

Today, scientists know that there are at least several hundred coelacanths living in the Indian Ocean, just east of central Africa. Many of these fish live near the Comoros Islands. The coelacanths live in underwater caves during the day but move out at night to feed. The rare fish are now a protected species. With any luck, they will survive for another hundred million years.

Write a short essay describing the discovery of the coelacanths and describe the reaction of scientists to this discovery.

For more information, visit earth.msscience.com/oops

Reviewing Main Ideas

Section 1 Fossils

1. Fossils are more likely to form if hard parts of the dead organisms are buried quickly.
2. Some fossils form when original materials that made up the organisms are replaced with minerals. Other fossils form when remains are subjected to heat and pressure, leaving only a carbon film behind. Some fossils are the tracks or traces left by ancient organisms.

Section 2 Relative Ages of Rocks

1. The principle of superposition states that, in undisturbed layers, older rocks lie underneath younger rocks.
2. Unconformities, or gaps in the rock record, are due to erosion or periods of time during which no deposition occurred.
3. Rock layers can be correlated using rock types and fossils.

Section 3 Absolute Ages of Rocks

1. Absolute dating provides an age in years for the rocks.
2. The half-life of a radioactive isotope is the time it takes for half of the atoms of the isotope to decay into another isotope.

Visualizing Main Ideas

Copy and complete the following concept map on fossils.

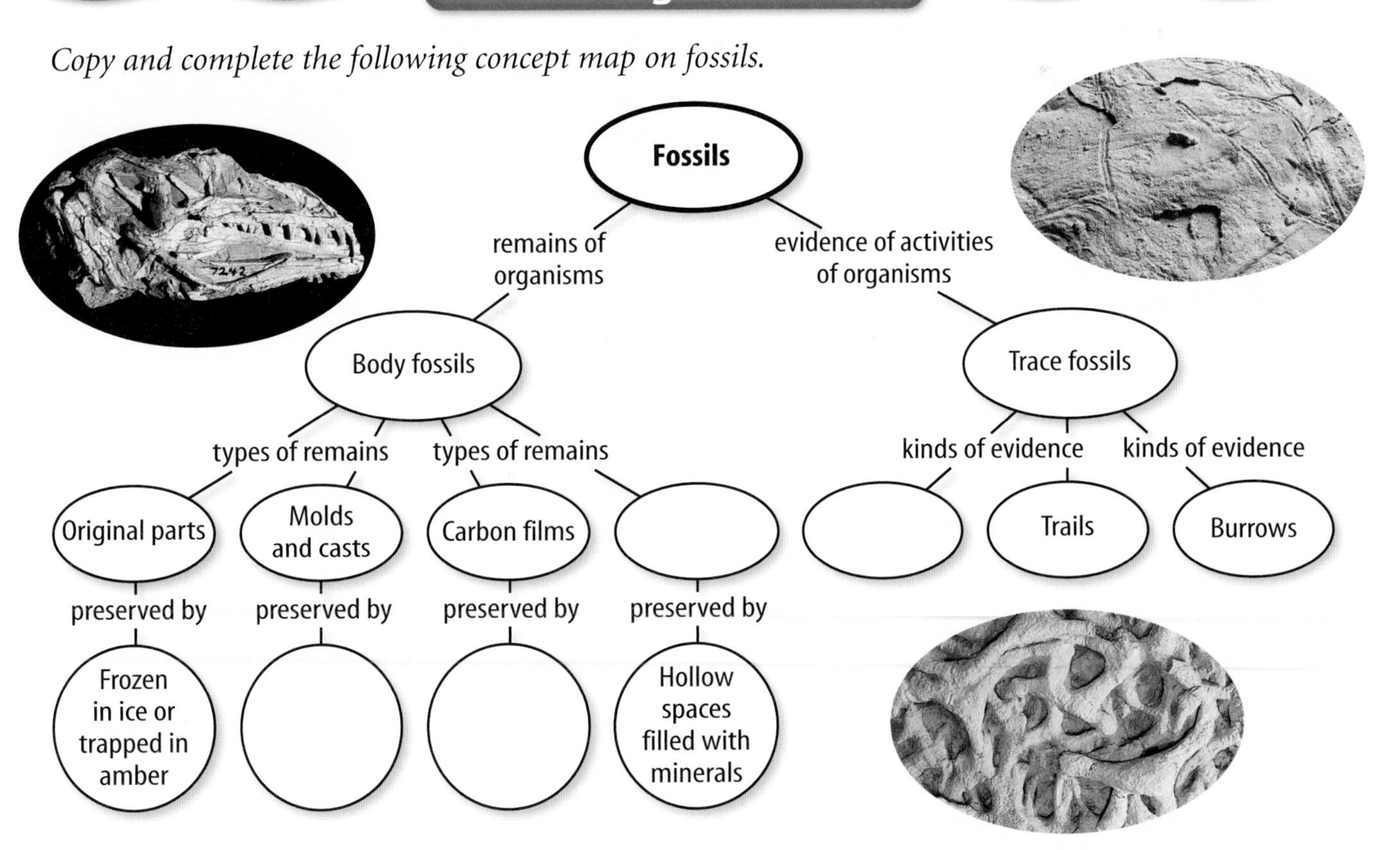

chapter 13 Review

Using Vocabulary

absolute age p. 377	principle of superposition p. 370
carbon film p. 364	radioactive decay p. 377
cast p. 365	radiometric dating p. 379
fossil p. 363	relative age p. 371
half-life p. 378	unconformity p. 372
index fossil p. 367	uniformitarianism p. 381
mold p. 365	
permineralized remains p. 364	

Write an original sentence using the vocabulary word to which each phrase refers.

1. thin film of carbon preserved as a fossil
2. older rocks lie under younger rocks
3. processes occur today as they did in the past
4. gap in the rock record
5. time needed for half the atoms to decay
6. fossil organism that lived for a short time
7. gives the age of rocks in years
8. minerals fill spaces inside fossil
9. a copy of a fossil produced by filling a mold with sediment or crystals

Checking Concepts

Choose the word or phrase that best answers the question.

10. What is any evidence of ancient life called?
    A) half-life
    B) fossil
    C) unconformity
    D) disconformity

11. Which of the following conditions makes fossil formation more likely?
    A) buried slowly
    B) attacked by scavengers
    C) made of hard parts
    D) composed of soft parts

12. What are cavities left in rocks when a shell or bone dissolves called?
    A) casts
    B) molds
    C) original remains
    D) carbon films

13. To say "the present is the key to the past" is a way to describe which of the following principles?
    A) superposition
    B) succession
    C) radioactivity
    D) uniformitarianism

14. A fault can be useful in determining which of the following for a group of rocks?
    A) absolute age
    B) index age
    C) radiometric age
    D) relative age

15. Which of the following is an unconformity between parallel rock layers?
    A) angular unconformity
    B) fault
    C) disconformity
    D) nonconformity

Use the illustration below to answer question 16.

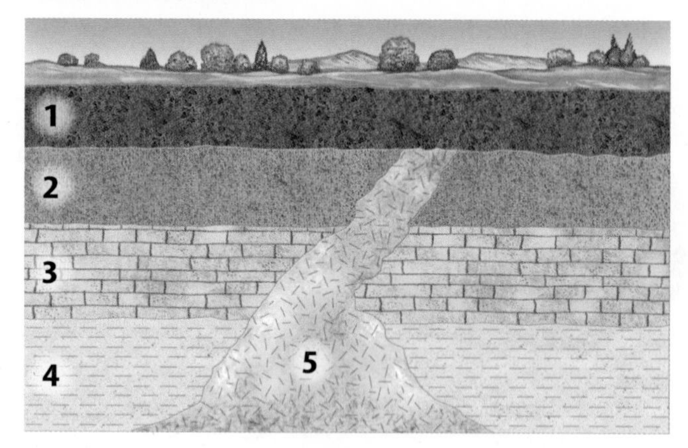

16. Which of the following puts the layers in order from oldest to youngest?
    A) 5-4-3-2-1
    B) 1-2-3-4-5
    C) 2-3-4-5-1
    D) 4-3-2-5-1

17. Which process forms new elements?
    A) superposition
    B) uniformitarianism
    C) permineralization
    D) radioactive decay

Science Online earth.msscience.com/vocabulary_puzzlemaker

Thinking Critically

18. **Explain** why the fossil record of life on Earth is incomplete. Give some reasons why.

19. **Infer** Suppose a lava flow was found between two sedimentary rock layers. How could you use the lava flow to learn about the ages of the sedimentary rock layers? *(Hint: Most lava contains radioactive isotopes.)*

20. **Infer** Suppose you're correlating rock layers in the western United States. You find a layer of volcanic ash deposits. How can this layer help you in your correlation over a large area?

21. **Recognize Cause and Effect** Explain how some woolly mammoths could have been preserved intact in frozen ground. What conditions must have persisted since the deaths of these animals?

22. **Classify** each of the following fossils in the correct category in the table below: *dinosaur footprint, worm burrow, dinosaur skull, insect in amber, fossil woodpecker hole,* and *fish tooth.*

Types of Fossils

Trace Fossils	Body Fossils
Do not write	in this book.

23. **Compare and contrast** the three different kinds of unconformities. Draw sketches of each that illustrate the features that identify them.

24. **Describe** how relative and absolute ages differ. How might both be used to establish ages in a series of rock layers?

25. **Discuss** uniformitarianism in the following scenario. You find a shell on the beach, and a friend remembers seeing a similar fossil while hiking in the mountains. What does this suggest about the past environment of the mountain?

Performance Activities

26. **Illustrate** Create a model that allows you to explain how to establish the relative ages of rock layers.

27. **Use a Classification System** Start your own fossil collection. Label each find as to type, approximate age, and the place where it was found. Most state geological surveys can provide you with reference materials on local fossils.

Applying Math

28. **Calculate** how many half-lives have passed in a rock containing one-eighth the original radioactive material and seven-eighths of the daughter product.

Use the graphs below to answer question 29.

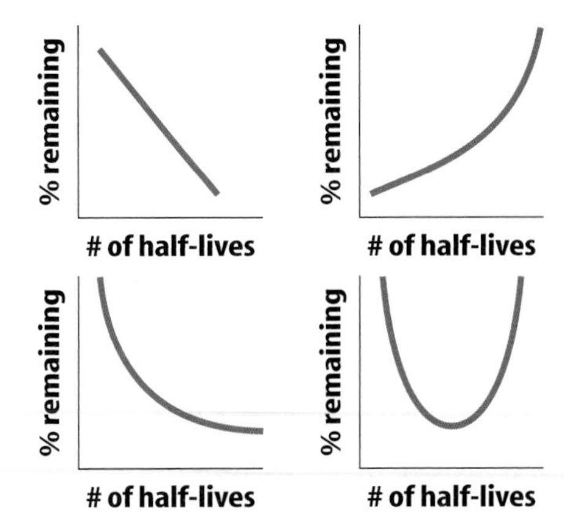

29. **Interpret Data** Which of the above curves best illustrates radioactive decay?

Part 1 Multiple Choice

Record your answers on the answer sheet provided by your teacher or on a sheet of paper.

Use the photo below to answer question 1.

1. Which type of fossil preservation is shown above?
A. trace fossil
B. original remains
C. carbon film
D. permineralized remains

2. Which principle states that the oldest rock layer is found at the bottom in an undisturbed stack of rock layers?
A. half-life
B. absolute dating
C. superposition
D. uniformitarianism

3. Which type of scientist studies fossils?
A. meteorologist
B. chemist
C. astronomer
D. paleontologist

4. Which are the remains of species that existed on Earth for relatively short periods of time, were abundant, and were widespread geographically?
A. trace fossils
B. index fossils
C. carbon films
D. body fossils

5. Which term means matching up rock layers in different places?
A. superposition
B. correlation
C. uniformitarianism
D. absolute dating

6. Which of the following is least likely to be found as a fossil?
A. clam shell
B. shark tooth
C. snail shell
D. jellyfish imprint

7. Which type of fossil preservation is a thin carbon silhouette of the original organism?
A. cast
B. carbon film
C. mold
D. permineralized remains

8. Which isotope is useful for dating wood and charcoal that is less than about 75,000 years old?
A. carbon-14
B. potassium-40
C. uranium-238
D. argon-40

Use the diagram below to answer questions 9–11.

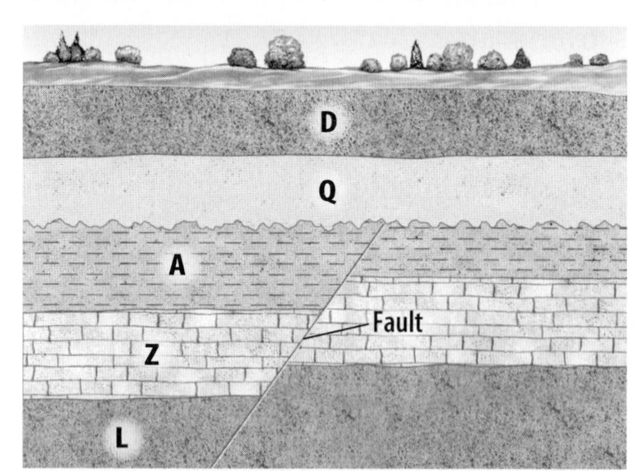

9. Which sequence of letters describes the rock layers in the diagram from oldest to youngest?
A. D, Q, A, Z, L
B. L, Z, A, Q, D
C. Z, L, A, D, Q
D. Q, D, L, Z, A

10. What does the wavy line between layers A and Q represent?
A. a disconformity
B. a fault
C. a nonconformity
D. an angular unconformity

11. Which of the following correctly describes the relative age of the fault?
A. younger than A, but older than Q
B. younger than Z, but older than L
C. younger than Q, but older than A
D. younger than D, but older than Q

Part 2 | Short Response/Grid In

Record your answers on the answer sheet provided by your teacher or on a sheet of paper.

12. What is a fossil?

13. How is a fossil cast different from a fossil mold?

14. Describe the principle of uniformitarianism.

15. Explain how the original remains of an insect can be preserved as a fossil in amber.

16. Why do scientists hypothesize that Earth is about 4.5 billion years old?

17. Describe the process of radioactive decay. Use the terms *isotope, nucleus,* and *half-life* in your answer.

Use the table below to answer questions 18–20.

Number of Half-lives	Parent Isotope Remaining (%)
1	100
2	X
3	25
4	12.5
5	Y

18. What value should replace the letter X in the table above?

19. What value should replace the letter Y in the table above?

20. Explain the relationship between the number of half-lives that have elapsed and the amount of parent isotope remaining.

21. Compare and contrast the three types of unconformities.

22. Why are index fossils useful for estimating the age of rock layers?

Part 3 | Open Ended

Record your answers on a sheet of paper.

23. Why are fossils important? What information do they provide?

24. List three different types of trace fossils. Explain how each type forms.

Examine the graph below and answer questions 25–27.

Relationship Between Sediment Burial Rate and Potential for Remains to Become Fossils

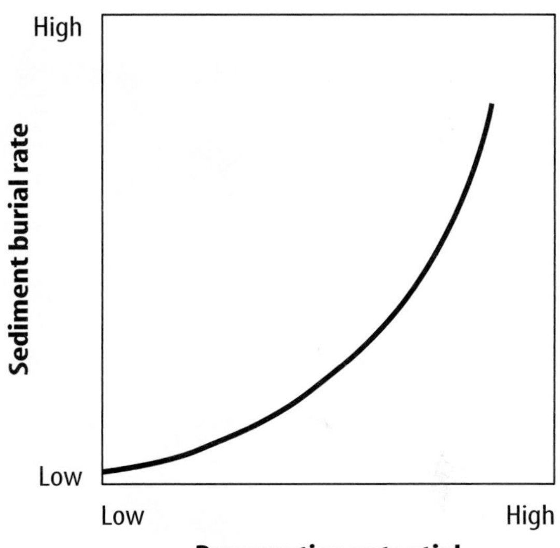

25. How does the potential for remains to be preserved change as the rate of burial by sediment increases?

26. Why do you think this relationship exists?

27. What other factors affect the potential for the remains of organisms to become fossils?

28. How could a fossil of an organism that lived in ocean water millions of years ago be found in the middle of North America?

Test-Taking Tip

Check It Again Double check your answers before turning in the test.

chapter

14

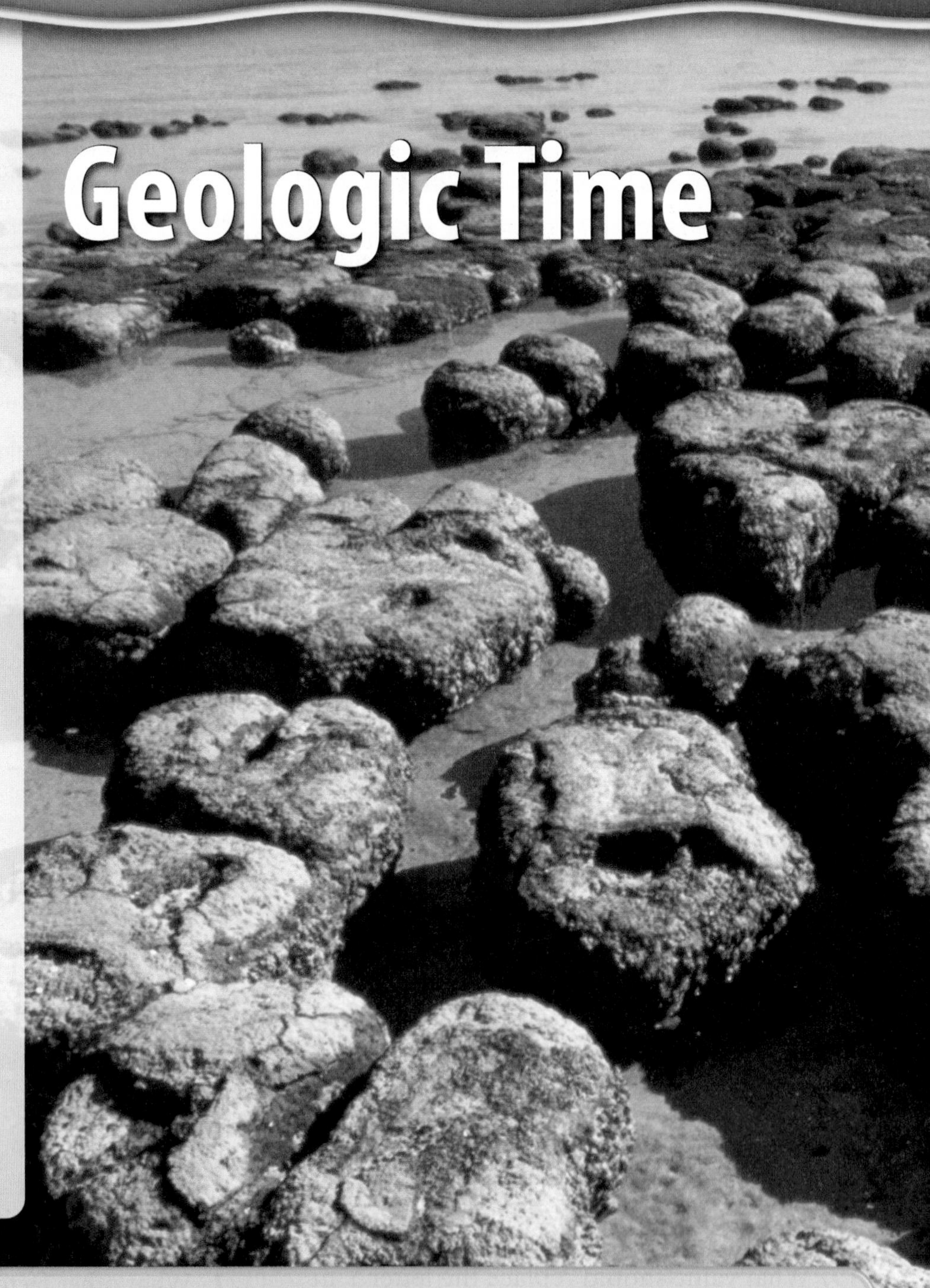

Geologic Time

The BIG Idea

Scientists use units of geologic time to interpret the history of life on Earth.

SECTION 1
Life and Geologic Time

Main Idea Fossils provide evidence that life on Earth has evolved throughout geologic time.

SECTION 2
Early Earth History

Main Idea Primitive life forms existed on Earth during Precambrian time and the Paleozoic Era.

SECTION 3
Middle and Recent Earth History

Main Idea Life forms continued to evolve through the Mesozoic Era and the current Cenozoic Era.

Looking at the Past

The stromatolites in the picture hardly have changed since they first appeared 3.5 billion years ago. Looking at these organisms today allows us to imagine what the early Earth might have looked like. In this chapter, you will see how much Earth has changed over time, even as some parts remain the same.

Science Journal Describe how an animal or plant might change if Earth becomes hotter in the next million years.

Start-Up Activities

Survival Through Time

Environments include the living and nonliving things that surround and affect organisms. Whether or not an organism survives in its environment depends upon its characteristics. Only if an organism survives until adulthood can it reproduce and pass on its characteristics to its offspring. In this lab, you will use a model to find out how one characteristic can determine whether individuals can survive in an environment.

1. Cut 15 pieces each of green, orange, and blue yarn into 3-cm lengths.
2. Scatter them on a sheet of green construction paper.
3. Have your partner use a pair of tweezers to pick up as many pieces as possible in 15 s.
4. **Think Critically** In your Science Journal, discuss which colors your partner selected. Which color was least selected? Suppose that the construction paper represents grass, the yarn pieces represent insects, and the tweezers represent an insect-eating bird. Which color of insect do you predict would survive to adulthood?

Geological Time Make the following Foldable to help you identify the major events in each era of geologic time.

STEP 1 **Fold** the top of a vertical piece of paper down and the bottom up to divide the paper into thirds.

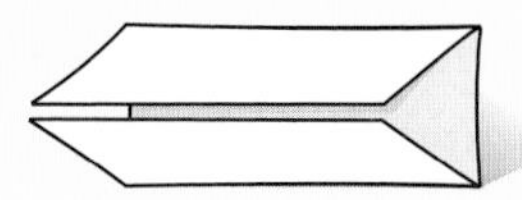

STEP 2 **Turn** the paper horizontally; **unfold and label** the three columns as shown.

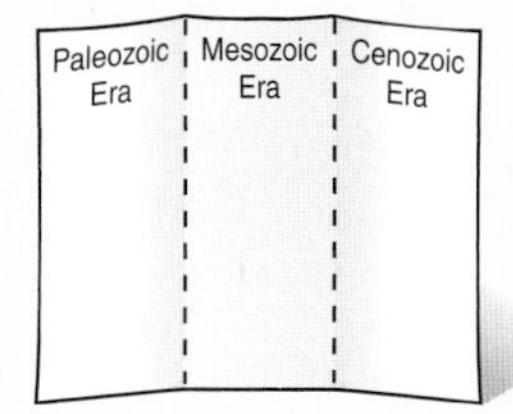

Read for Main Ideas As you read the chapter, list at least three major events that occurred in each era. Keep the events in chronological order. For each event, note the period in which it took place.

Preview this chapter's content and activities at earth.msscience.com

Get Ready to Read

Questions and Answers

1 Learn It! Knowing how to find answers to questions will help you on reviews and tests. Some answers can be found in the textbook, while other answers require you to go beyond the textbook. These answers might be based on knowledge you already have or things you have experienced.

2 Practice It! Read the excerpt below. Answer the following questions and then discuss them with a partner.

> The Paleozoic Era, or era of ancient life, began about 544 million years ago and ended about 248 million years ago. Traces of life are much easier to find in Paleozoic rocks than in Precambrian rocks.
>
> *—from page 402*

- What does the term *Paleozoic* mean?
- What does the term *Precambrian* mean?
- Why are traces of life easier to find in Paleozoic rocks than in Precambrian rocks?

3 Apply It! Look at some questions in the chapter review. Which questions can be answered directly from the text? Which require you to go beyond the text?

As you read, keep track of questions you answer in the chapter. This will help you remember what you read.

Target Your Reading

Use this to focus on the main ideas as you read the chapter.

1. **Before you read** the chapter, respond to the statements below on your worksheet or on a numbered sheet of paper.
   - Write an **A** if you **agree** with the statement.
   - Write a **D** if you **disagree** with the statement.

2. **After you read** the chapter, look back to this page to see if you've changed your mind about any of the statements.
   - If any of your answers changed, explain why.
   - Change any false statements into true statements.
   - Use your revised statements as a study guide.

Before You Read A or D		Statement	After You Read A or D
	1	The fossil record shows that species have changed over geologic time.	
	2	Geologic time units are based on life-forms that lived only during certain periods of time.	
	3	Eras are longer than eons.	
	4	Precambrian time is the shortest part of Earth's history.	
	5	No life-forms existed on Earth during Precambrian time.	
	6	Oxygen gas has always been a major component of Earth's atmosphere throughout geologic time.	
	7	All dinosaurs were large, slow-moving, cold-blooded reptiles.	
	8	Dinosaurs lived during the Mesozoic Era.	
	9	Many scientists hypothesize that a comet or asteroid collision with Earth, ended Mesozoic Era.	

Science Online
Print out a worksheet of this page at earth.msscience.com

section 1

Life and Geologic Time

as you read

What You'll Learn

- **Explain** how geologic time can be divided into units.
- **Relate** changes of Earth's organisms to divisions on the geologic time scale.
- **Describe** how plate tectonics affects species.

Why It's Important

The life and landscape around you are the products of change through geologic time.

Review Vocabulary
fossils: remains, traces, or imprints of prehistoric organisms

New Vocabulary
- geologic time scale
- eon
- era
- period
- epoch
- organic evolution
- species
- natural selection
- trilobite
- Pangaea

Geologic Time

A group of students is searching for fossils. By looking in rocks that are hundreds of millions of years old, they hope to find many examples of trilobites (TRI loh bites) so that they can help piece together a puzzle. That puzzle is to find out what caused the extinction of these organisms. **Figure 1** shows some examples of what they are finding. The fossils are small, and their bodies are divided into segments. Some of them seem to have eyes. Could these interesting fossils be trilobites?

Trilobites are small, hard-shelled organisms that crawled on the seafloor and sometimes swam through the water. Most ranged in size from 2 cm to 7 cm in length and from 1 cm to 3 cm in width. They are considered to be index fossils because they lived over vast regions of the world during specific periods of geologic time.

The Geologic Time Scale The appearance or disappearance of types of organisms throughout Earth's history marks important occurrences in geologic time. Paleontologists have been able to divide Earth's history into time units based on the life-forms that lived only during certain periods. This division of Earth's history makes up the **geologic time scale.** However, sometimes fossils are not present, so certain divisions of the geologic time scale are based on other criteria.

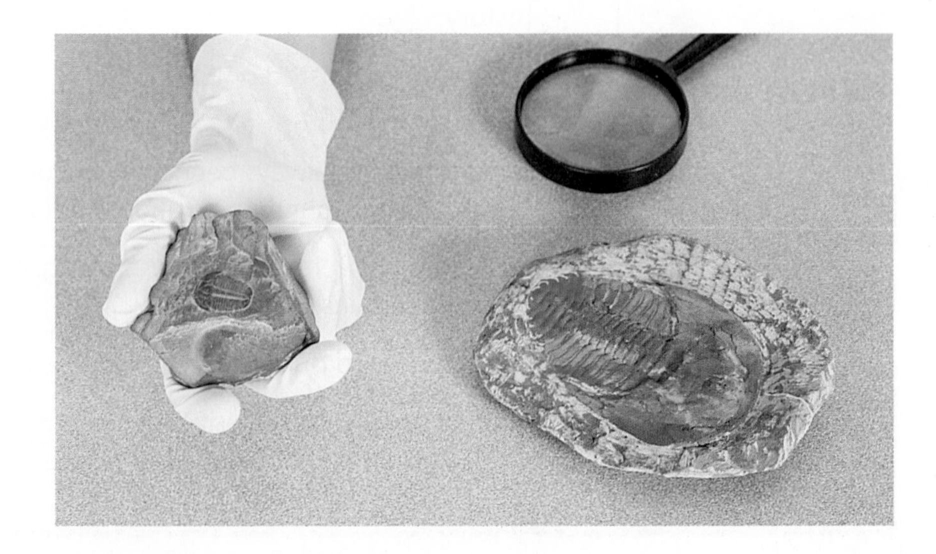

Figure 1 Many sedimentary rocks in the United States are rich in invertebrate fossils such as these trilobites.

Major Subdivisions of Geologic Time The oldest rocks on Earth contain no fossils. Then, for many millions of years after the first appearance of fossils, the fossil record remained sparse. Later in Earth's history came an explosion in the abundance and diversity of organisms. These organisms left a rich fossil record. As shown in **Figure 2,** four major subdivisions of geologic time are used—eons, eras, periods, and epochs. The longest subdivisions—**eons**—are based upon the abundance of certain fossils.

Reading Check *What are the major subdivisions of geologic time?*

Next to eons, the longest subdivisions are the **eras,** which are marked by major, striking, and worldwide changes in the types of fossils present. For example, at the end of the Mesozoic Era, many kinds of invertebrates, birds, mammals, and reptiles became extinct.

Eras are subdivided into periods. **Periods** are units of geologic time characterized by the types of life existing worldwide at the time. Periods can be divided into smaller units of time called **epochs.** Epochs also are characterized by differences in life-forms, but some of these differences can vary from continent to continent. Epochs of periods in the Cenozoic Era have been given specific names. Epochs of other periods usually are referred to simply as early, middle, or late. Epochs are further subdivided into units of shorter duration.

Dividing Geologic Time There is a limit to how finely geologic time can be subdivided. It depends upon the kind of rock record that is being studied. Sometimes it is possible to distinguish layers of rock that formed during a single year or season. In other cases, thick stacks of rock that have no fossils provide little information that could help in subdividing geologic time.

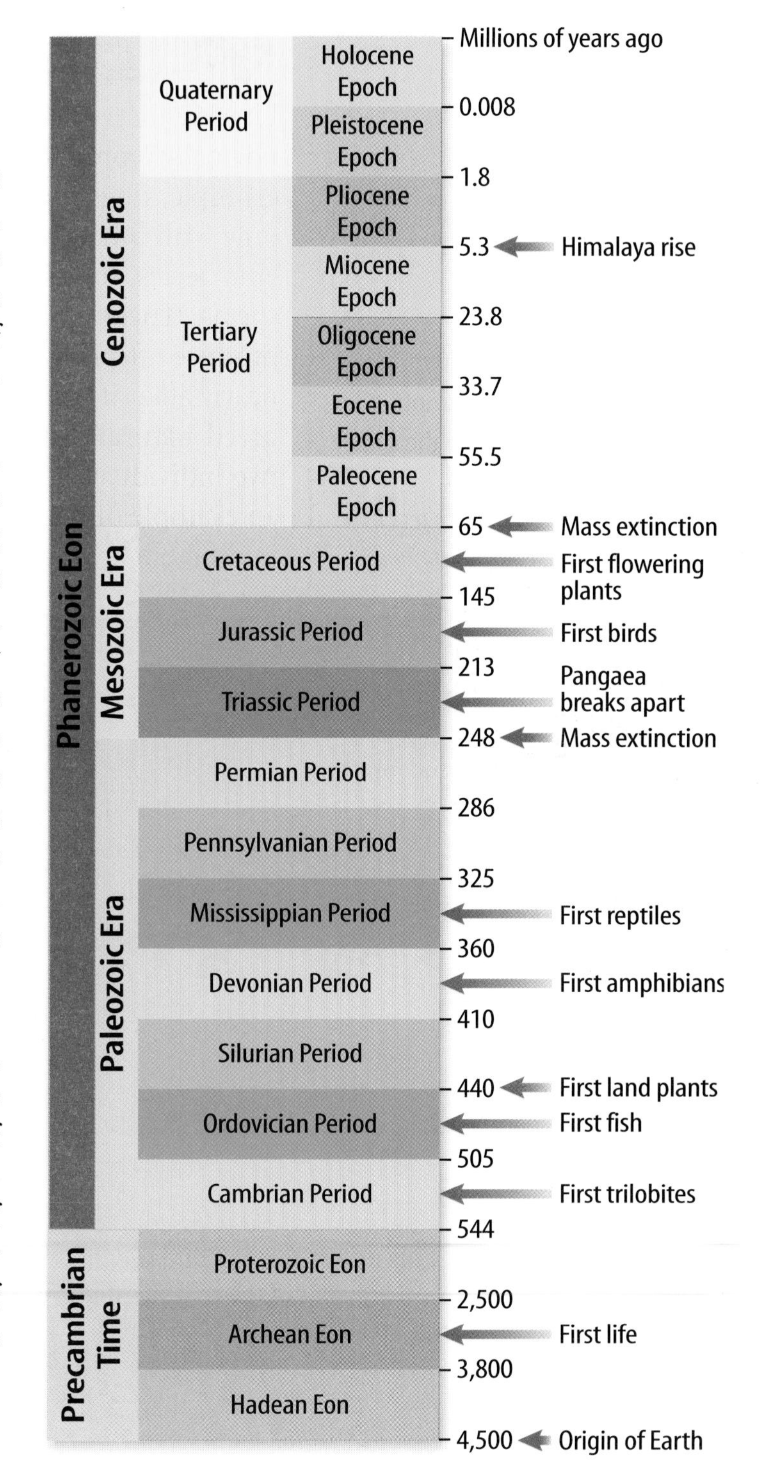

Figure 2 Scientists have divided the geologic time scale into subunits based upon the appearance and disappearance of types of organisms.
Explain *how the even blocks in this chart can be misleading.*

Organic Evolution

The fossil record shows that species have changed over geologic time. This change through time is known as **organic evolution.** According to most theories about organic evolution, environmental changes can affect an organism's survival. Those organisms that are not adapted to changes are less likely to survive or reproduce. Over time, the elimination of individuals that are not adapted can cause changes to species of organisms.

INTEGRATE Life Science

Species Many ways of defining the term species (SPEE sheez) have been proposed. Life scientists often define a **species** as a group of organisms that normally reproduces only with other members of their group. For example, dogs are a species because dogs mate and reproduce only with other dogs. In some rare cases, members of two different species, such as lions and tigers, can mate and produce offspring. These offspring, however, are usually sterile and cannot produce offspring of their own. Even though two organisms look nearly alike, if the populations they each come from do not interbreed naturally and produce offspring that can reproduce, the two individuals do not belong to the same species. **Figure 3** shows an example of two species that look similar to each other but live in different areas and do not mate naturally with each other.

Figure 3 Just because two organisms look alike does not mean that they belong to the same species.
Describe *an experiment to test if these lizards are separate species.*

The coast horned lizard lives along the coast of central and southern California.

The desert horned lizard lives in arid regions of the southwestern United States.

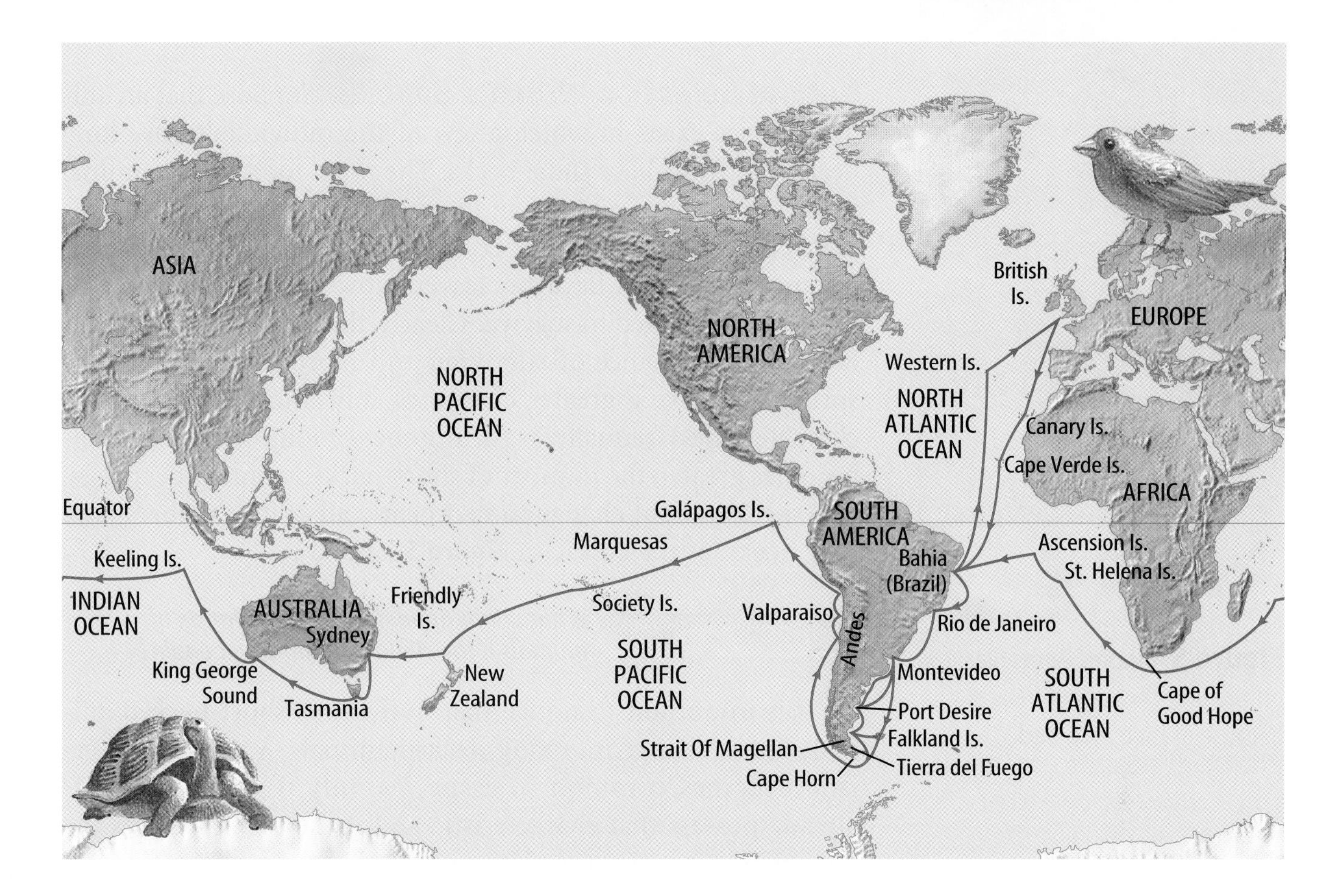

Natural Selection Charles Darwin was a naturalist who sailed around the world from 1831 to 1836 to study biology and geology. **Figure 4** shows a map of his journey. With some of the information about the plants and animals he observed on this trip in mind, he later published a book about the theory of evolution by natural selection.

In his book, he proposed that **natural selection** is a process by which organisms with characteristics that are suited to a certain environment have a better chance of surviving and reproducing than organisms that do not have these characteristics. Darwin knew that many organisms are capable of producing more offspring than can survive. This means that organisms compete with each other for resources necessary for life, such as food and living space. He also knew that individual organisms within the same species could be different, or show variations, and that these differences could help or hurt the individual organism's chance of surviving.

Some organisms that were well suited to their environment lived longer and had a better chance of producing offspring. Organisms that were poorly adapted to their environment produced few or no offspring. Because many characteristics are inherited, the characteristics of organisms that are better adapted to the environment get passed on to offspring more often. According to Darwin, this can cause a species to change over time.

Figure 4 Charles Darwin sailed around the world between 1831 and 1836 aboard the HMS *Beagle* as a naturalist. On his journey he saw an abundance of evidence for natural selection, especially on the Galápagos Islands off the western coast of South America.

Figure 5 Giraffes can eat leaves off the branches of tall trees because of their long necks.

Figure 6 Cat breeders have succeeded in producing a great variety of cats by using the principle of artificial selection.

Natural Selection Within a Species Suppose that an animal species exists in which a few of the individuals have long necks, but most have short necks. The main food for the animal is the leafy foliage on trees in the area. What happens if the climate changes and the area becomes dry? The lower branches of the trees might not have any leaves. Now which of the animals will be better suited to survive? Clearly, the long-necked animals have a better chance of surviving and reproducing. Their offspring will have a greater chance of inheriting the important characteristic. Gradually, as the number of long-necked animals becomes greater, the number of short-necked animals decreases. The species might change so that nearly all of its members have long necks, as the giraffe in **Figure 5** has.

Reading Check *What might happen to the population of animals if the climate became wet again?*

It is important to notice that individual, short-necked animals didn't change into long-necked animals. A new characteristic becomes common in a species only if some members already possess that characteristic and if the trait increases the animal's chance of survival. If no animal in the species possessed a long neck in the first place, a long-necked species could not have evolved by means of natural selection.

Artificial Selection Humans have long used the principle of artificial selection when breeding domestic animals. By carefully choosing individuals with desired characteristics, animal breeders have created many breeds of cats, dogs, cattle, and chickens. **Figure 6** shows the great variety of cats produced by artificial selection.

The Evolution of New Species Natural selection explains how characteristics change and how new species arise. For example, if the short-necked animals migrated to a different location, they might have survived. They could have continued to reproduce in the new location, eventually developing enough different characteristics from the long-necked animals that they might not be able to breed with each other. At this point, at least one new species would have evolved.

Trilobites

Remember the trilobites? The term *trilobite* comes from the structure of the hard outer skeleton or exoskeleton. The exoskeleton of a **trilobite** consists of three lobes that run the length of the body. As shown in **Figure 7,** the trilobite's body also has a head (cephalon), a segmented middle section (thorax), and a tail (pygidium).

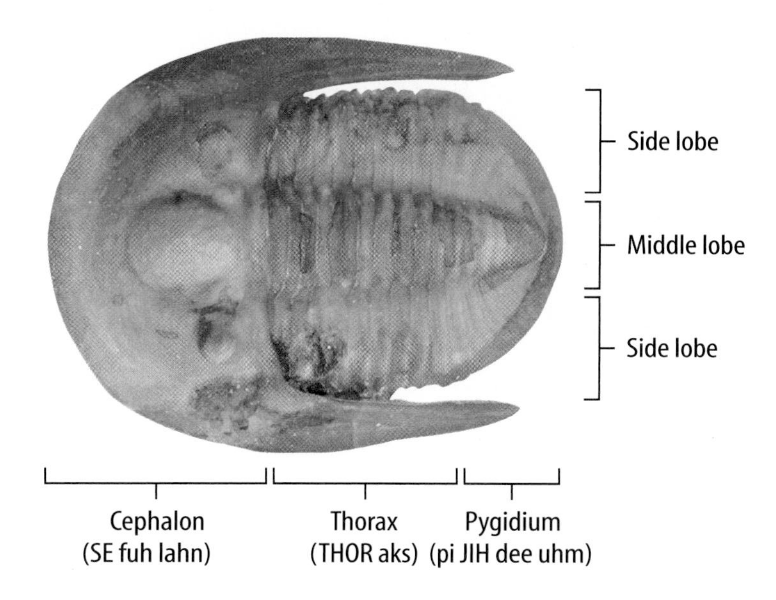

Figure 7 The trilobite's body was divided into three lobes that run the length of the body—two side lobes and one middle lobe.

Changing Characteristics of Trilobites Trilobites inhabited Earth's oceans for more than 200 million years. Throughout the Paleozoic Era, some species of trilobites became extinct and other new species evolved. Species of trilobites that lived during one period of the Paleozoic Era showed different characteristics than species from other periods of this era. As **Figure 8** shows, paleontologists can use these different characteristics to demonstrate changes in trilobites through geologic time. These changes can tell you about how different trilobites from different periods lived and responded to changes in their environments.

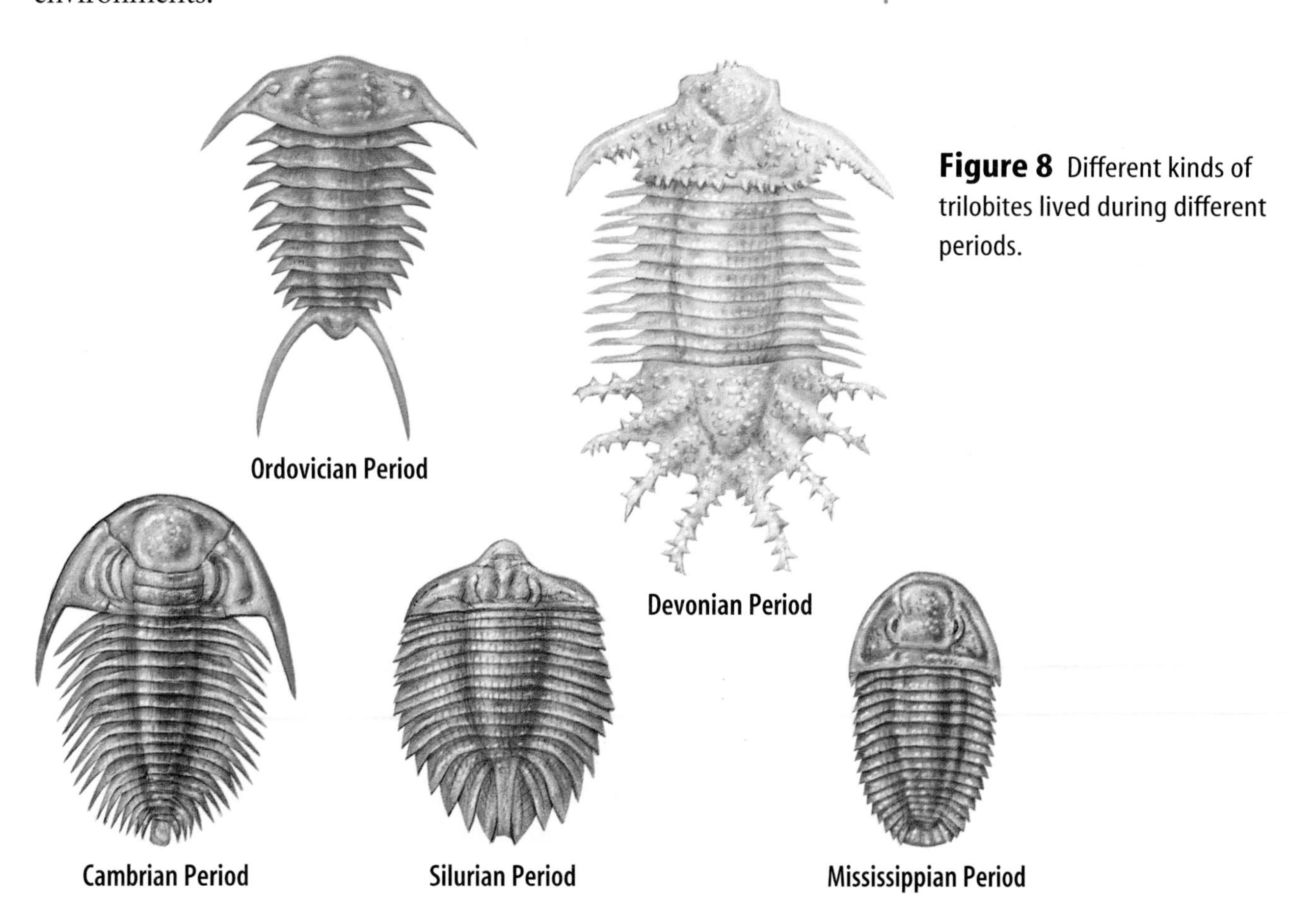

Figure 8 Different kinds of trilobites lived during different periods.

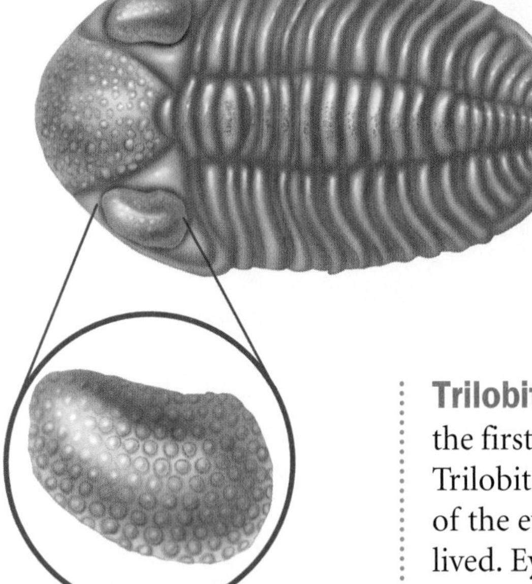

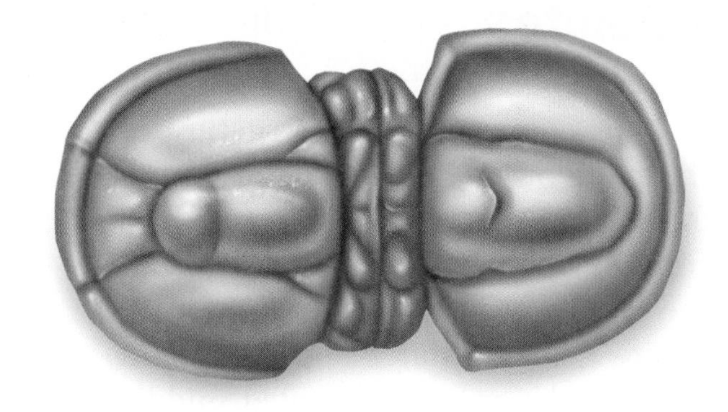

Figure 9 Trilobites had many different types of eyes. Some had eyes that contained hundreds of small circular lenses, somewhat like an insect. The blind trilobite (right) had no eyes.

Trilobite Eyes Trilobites, shown in **Figure 9,** might have been the first organisms that could view the world with complex eyes. Trilobite eyes show the result of natural selection. The position of the eyes on an organism gives clues about where it must have lived. Eyes that are located toward the front of the head indicate an organism that was adapted for active swimming. If the eyes are located toward the back of the head, the organism could have been a bottom dweller. In most species of trilobites, the eyes were located midway on the head—a compromise for an organism that was adapted for crawling on the seafloor and swimming in the water.

Over time, the eyes in trilobites changed. In many trilobite species, the eyes became progressively smaller until they completely disappeared. Blind trilobites, such as the one on the right in **Figure 9,** might have burrowed into sediments on the seafloor or lived deeper than light could penetrate. In other species, however, the eyes became more complex. One kind of trilobite, *Aeglina,* developed large compound eyes that had numerous individual lenses. Some trilobites developed stalks that held the eyes upward. Where would this be useful?

Trilobite Bodies The trilobite body and tail also underwent significant changes in form through time, as you can see in **Figure 8** on the previous page. A special case is *Olenellus,* shown in **Figure 10.** This trilobite, which lived during the Early Cambrian Period, had an extremely segmented body—perhaps more so than any other known species of trilobite. It is thought that *Olenellus,* and other species that have so many body segments, are primitive trilobites.

Fossils Show Changes Trilobite exoskeletons changed as trilobites adapted to changing environments. Species that could not adapt became extinct. What processes on Earth caused environments to change so drastically that species adapted or became extinct?

Figure 10 *Olenellus* is one of the most primitive trilobite species.

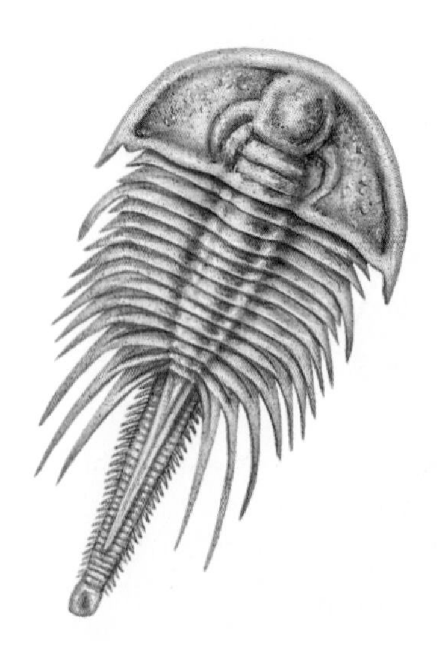

Plate Tectonics and Earth History

Plate tectonics is one possible answer to the riddle of trilobite extinction. Earth's moving plates caused continents to collide and separate many times. Continental collisions formed mountains and closed seas caught between continents. Continental separations created wider, deeper seas between continents. By the end of the Paleozoic Era, sea levels had dropped and the continents had come together to form one giant landmass, the supercontinent **Pangaea** (pan JEE uh). Because trilobites lived in the oceans, their environment was changed or destroyed. **Figure 11** shows the arrangement of continents at the end of the Paleozoic Era. What effect might these changes have had on the trilobite populations?

Figure 11 The amount of shallow water environment was reduced when Pangaea formed. **Describe** *how this change affected organisms that lived along the coasts of continents.*

Not all scientists accept the above explanation for the extinctions at the end of the Paleozoic Era, and other possibilities—such as climate change—have been proposed. As in all scientific debates, you must consider the evidence carefully and come to conclusions based on the evidence.

section 1 review

Summary

Geologic Time

- Earth's history is divided into eons, eras, periods, and epochs, based on fossils.

Organic Evolution

- The fossil record indicates that species have changed over time.
- Charles Darwin proposed natural selection to explain change in species.
- In natural selection, organisms best suited to their environments survive and produce the most offspring.

Trilobites

- Trilobites were abundant in the Paleozoic fossil record and can be used as index fossils.

Plate Tectonics and Earth History

- Continents moving through time have influenced the environments of past organisms.

Self Check

1. **Discuss** how fossils relate to the geologic time scale.
2. **Infer** how plate tectonics might lead to extinction.
3. **Infer** how the eyes of a trilobite show how it lived.
4. **Explain** how paleontologists use trilobite fossils as index fossils for various geologic time periods.
5. **Think Critically** Aside from moving plates, what other factors could cause an organism's environment to change? How would this affect species?

Applying Skills

6. **Recognize Cause and Effect** Answer the questions below.
   a. How does natural selection cause evolutionary change to take place?
   b. How could the evolution of a characteristic within one species affect the evolution of a characteristic within another species? Give an example.

section 2

Early Earth History

as you read

What You'll Learn

- **Identify** characteristic Precambrian and Paleozoic life-forms.
- **Draw** conclusions about how species adapted to changing environments in Precambrian time and the Paleozoic Era.
- **Describe** changes in Earth and its life-forms at the end of the Paleozoic Era.

Why It's Important

The Precambrian includes most of Earth's history.

Review Vocabulary

life: state of being in which one grows, reproduces, and maintains a constant internal environment

New Vocabulary

- Precambrian time
- cyanobacteria
- Paleozoic Era

Precambrian Time

It may seem strange, but **Figure 12** is probably an accurate picture of Earth's first billion years. Over the next 3 billion years, simple life-forms began to colonize the oceans.

Look again at the geologic time scale shown in **Figure 2.** **Precambrian** (pree KAM bree un) **time** is the longest part of Earth's history and includes the Hadean, Archean, and Proterozoic Eons. Precambrian time lasted from about 4.5 billion years ago to about 544 million years ago. The oldest rocks that have been found on Earth are about 4 billion years old. However, rocks older than about 3.5 billion years are rare. This probably is due to remelting and erosion.

Although the Precambrian was the longest interval of geologic time, relatively little is known about the organisms that lived during this time. One reason is that many Precambrian rocks have been so deeply buried that they have been changed by heat and pressure. Many fossils can't withstand these conditions. In addition, most Precambrian organisms didn't have hard parts that otherwise would have increased their chances to be preserved as fossils.

Figure 12 During the early Precambrian, Earth was a lifeless planet with many volcanoes.

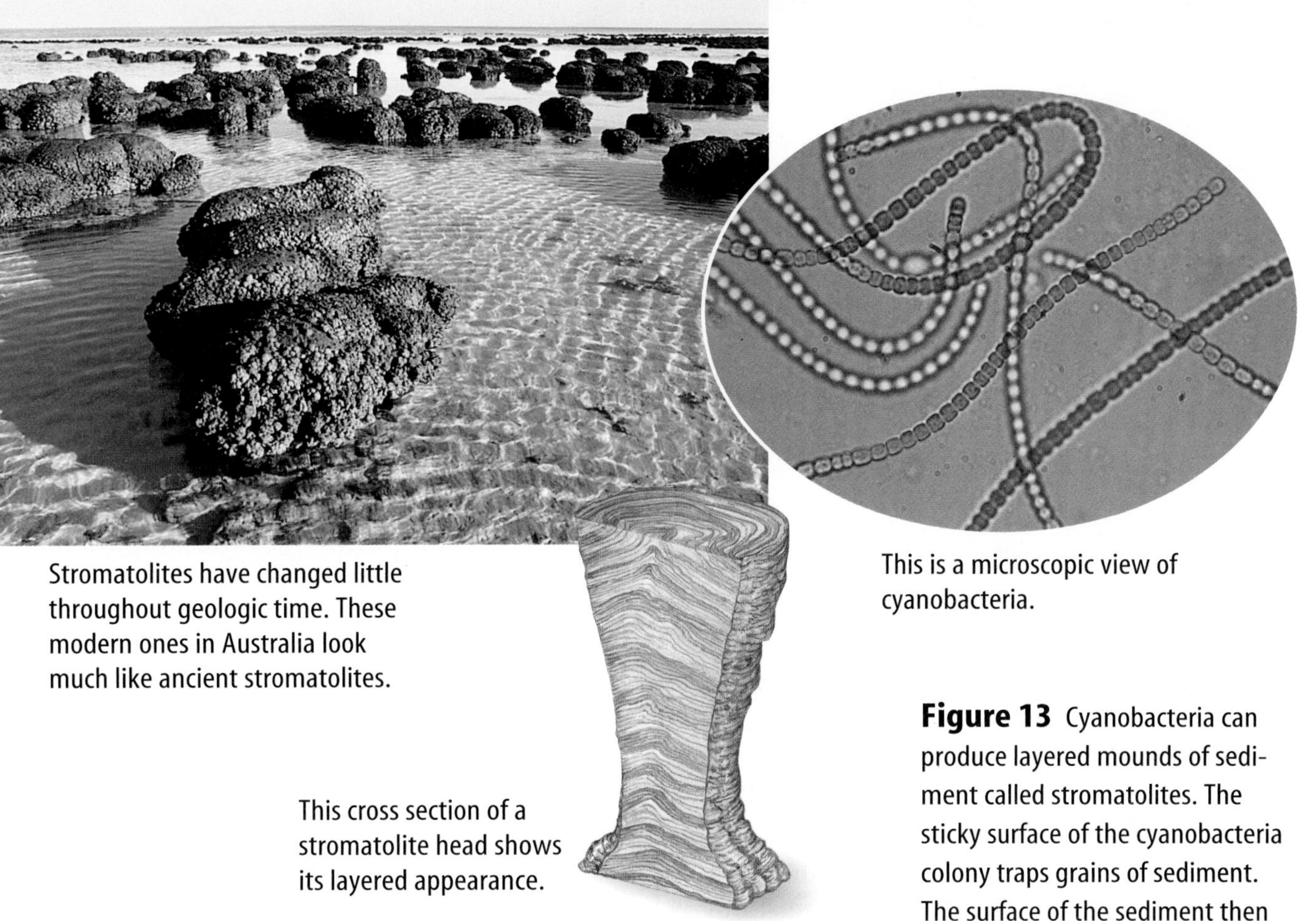

Stromatolites have changed little throughout geologic time. These modern ones in Australia look much like ancient stromatolites.

This cross section of a stromatolite head shows its layered appearance.

This is a microscopic view of cyanobacteria.

Figure 13 Cyanobacteria can produce layered mounds of sediment called stromatolites. The sticky surface of the cyanobacteria colony traps grains of sediment. The surface of the sediment then becomes colonized with cyanobacteria again, and the cycle repeats, producing the layers inside the stromatolite.

Early Life Many studies of the early history of life involve ancient stromatolites (stroh MA tuh lites). **Figure 13** shows stromatolites, which are layered mats formed by cyanobacteria colonies. **Cyanobacteria** are blue-green algae thought to be one of the earliest forms of life on Earth. Cyanobacteria first appeared about 3.5 billion years ago. They contained chlorophyll and used photosynthesis. This is important because during photosynthesis, they produced oxygen, which helped change Earth's atmosphere. Following the appearance of cyanobacteria, oxygen became a major atmospheric gas. Also of importance was that the ozone layer in the atmosphere began to develop, shielding Earth from ultraviolet rays. It is hypothesized that these changes allowed species of single-celled organisms to evolve into more complex organisms.

Reading Check *What atmospheric gas is produced by photosynthesis?*

Animals without backbones, called invertebrates (ihn VUR tuh brayts), appeared toward the end of Precambrian time. Imprints of invertebrates have been found in late Precambrian rocks, but because these early invertebrates were soft bodied, they weren't often preserved as fossils. Because of this, many Precambrian fossils are trace fossils.

Earth's First Air
Cyanobacteria are thought to have been one of the mechanisms by which Earth's early atmosphere became richer in oxygen. Research the composition of Earth's early atmosphere and where these gases probably came from. Record your findings in your Science Journal.

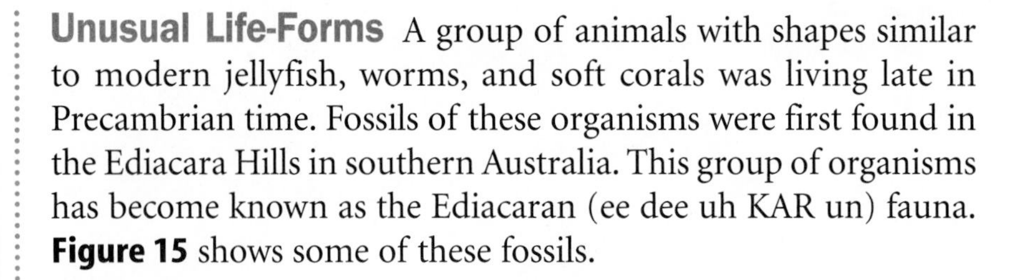

Unusual Life-Forms A group of animals with shapes similar to modern jellyfish, worms, and soft corals was living late in Precambrian time. Fossils of these organisms were first found in the Ediacara Hills in southern Australia. This group of organisms has become known as the Ediacaran (ee dee uh KAR un) fauna. **Figure 15** shows some of these fossils.

What modern organisms do some Ediacaran organisms resemble?

Ediacaran animals were bottom dwellers and might have had tough outer coverings like air mattresses. Trilobites and other invertebrates might have outcompeted the Ediacarans and caused their extinction, but nobody knows for sure why these creatures disappeared.

The Paleozoic Era

As you have learned, fossils are unlikely to form if organisms have only soft parts. An abundance of organisms with hard parts, such as shells, marks the beginning of the Paleozoic (pay lee uh ZOH ihk) Era. The **Paleozoic Era,** or era of ancient life, began about 544 million years ago and ended about 248 million years ago. Traces of life are much easier to find in Paleozoic rocks than in Precambrian rocks.

Paleozoic Life Because warm, shallow seas covered large parts of the continents during much of the Paleozoic Era, many of the life-forms scientists know about were marine, meaning they lived in the ocean. Trilobites were common, especially early in the Paleozoic. Other organisms developed shells that were easily preserved as fossils. Therefore, the fossil record of this era contains abundant shells. However, invertebrates were not the only animals to live in the shallow, Paleozoic seas.

Vertebrates, or animals with backbones, also evolved during this era. The first vertebrates were fishlike creatures without jaws. Armoured fish with jaws such as the one shown in **Figure 14** lived during the Devonian Period. Some of these fish were so huge that they could eat large sharks with their powerful jaws. By the Devonian Period, forests had appeared and vertebrates began to adapt to land environments, as well.

Dating Rock Layers with Fossils

Procedure

1. Draw three rock layers.
2. Number the layers 1 to 3, bottom to top.
3. Layer 1 contains fossil A. Layer 2 contains fossils A and B. Layer 3 contains fossil C.
4. Fossil A lived from the Cambrian through the Ordovician. Fossil B lived from the Ordovician through the Silurian. Fossil C lived in the Silurian and Devonian.

Analysis

1. Which layers were you able to date to a specific period?
2. Why isn't it possible to determine during which specific period the other layers formed?

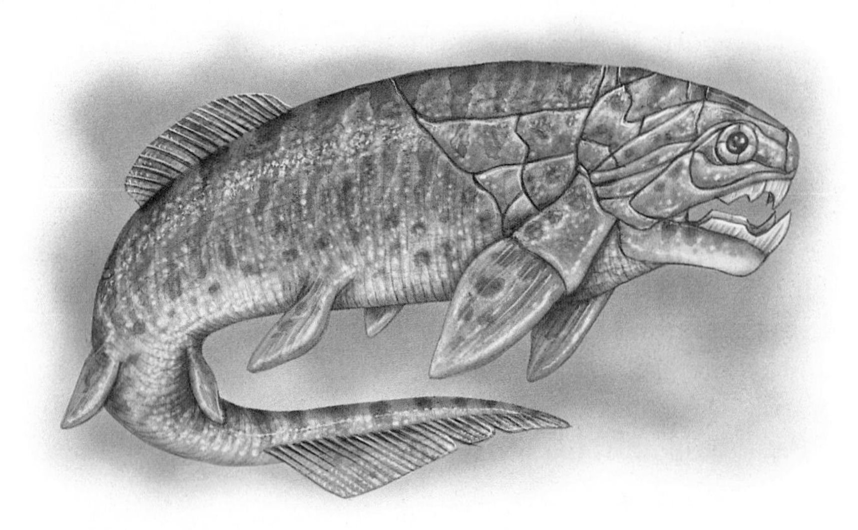

Figure 14 This giant predatory fish lived in seas that were present in North America during the Devonian Period. It grew to about 6 m in length.

NATIONAL GEOGRAPHIC **VISUALIZING UNUSUAL LIFE-FORMS**

Figure 15

A variety of 600-million-year-old fossils—known as Ediacaran (eed ee uh KAR un) fauna—have been found on every continent except Antarctica. These unusual organisms were originally thought to be descendants of early animals such as jellyfish, worms, and coral. Today, paleontologists debate whether these organisms were part of the animal kingdom or belonged to an entirely new kingdom whose members became extinct about 545 million years ago.

DICKENSONIA (dihk un suh NEE uh) Impressions of *Dickensonia,* a bottom-dwelling wormlike creature, have been discovered. Some are nearly one meter long.

RANGEA (rayn JEE uh) As it lay rooted in sea-bottom sediments, *Rangea* may have snagged tiny bits of food by filtering water through its body.

SPRIGGINA (sprih GIHN uh) Some scientists hypothesize that the four-centimeter-long *Spriggina* was a type of crawling, segmented organism. Others suggest that it sat upright while attached to the seafloor.

CYCLOMEDUSA (si kloh muh DEW suh) Although it looks a lot like a jellyfish, *Cyclomedusa* may have had more in common with modern sea anemones. Some paleontologists, however, hypothesize that it is unrelated to any living organism.

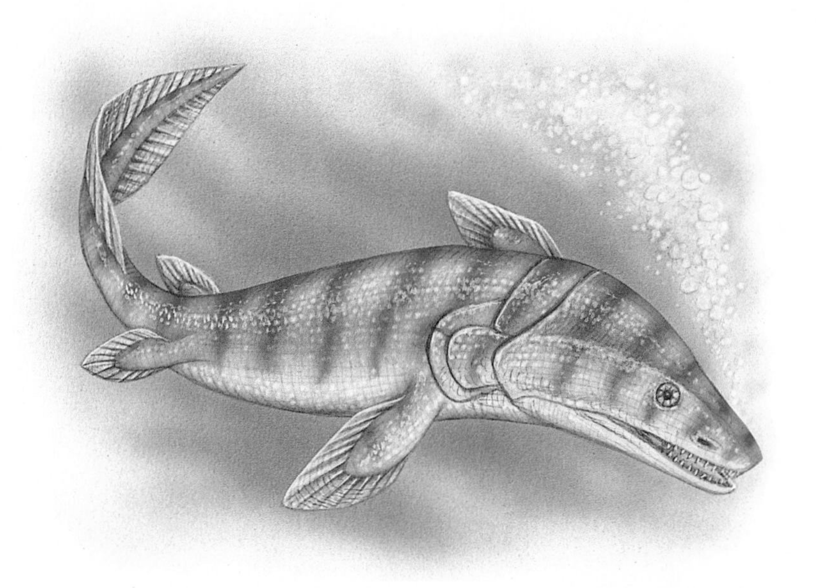

Figure 16 Amphibians probably evolved from fish like *Panderichthys* (pan dur IHK theez), which had leglike fins and lungs.

Life on Land Based on their structure, paleontologists know that many ancient fish had lungs as well as gills. Lungs enabled these fish to live in water with low oxygen levels—when needed they could swim to the surface and breathe air. Today's lungfish also can get oxygen from the water through gills and from the air through lungs.

One kind of ancient fish had lungs and leglike fins, which were used to swim and crawl around on the ocean bottom. Paleontologists hypothesize that amphibians might have evolved from this kind of fish, shown in **Figure 16.** The characteristics that helped animals survive in oxygen-poor waters also made living on land possible. Today, amphibians live in a variety of habitats in water and on land. They all have at least one thing in common, though. They must lay their eggs in water or moist places.

Reading Check ***What are some characteristics of the fish from which amphibians might have evolved?***

By the Pennsylvanian Period, some amphibians evolved an egg with a membrane that protected it from drying out. Because of this, these animals, called reptiles, no longer needed to lay eggs in water. Reptiles also have skin with hard scales that prevent loss of body fluids. This adaptation enables them to survive farther from water and in relatively dry climates, as shown in **Figure 17,** where many amphibians cannot live.

Topic: Paleozoic Life
Visit earth.msscience.com for Web links to information about Paleozoic life.

Activity Prepare a presentation on the organisms of one period of the Paleozoic Era. Describe a few animals from different groups, including how and where they lived. Are any of these creatures alive today, and if not, when did they become extinct?

Figure 17 Reptiles have scaly skins that allow them to live in dry places.

More than 375 million years ago, volcanic island chains formed in the ocean and were pushed against the coast as Africa moved toward North America.

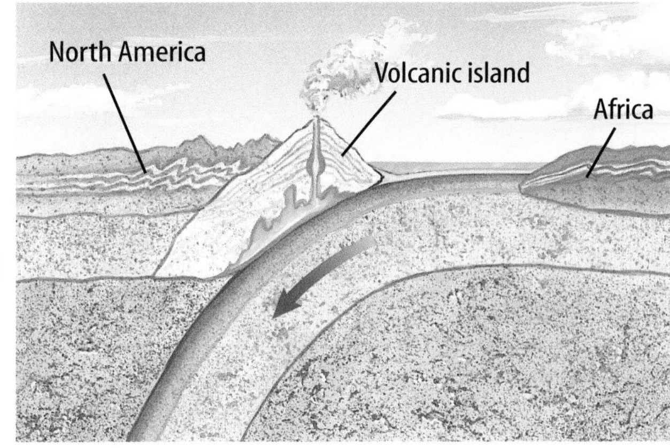

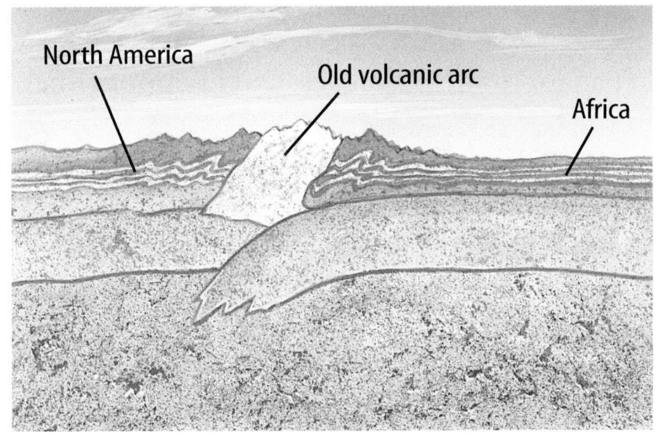

About 375 million years ago, the African plate collided with the North American plate, forming mountains on both continents.

About 200 million years ago, the Atlantic Ocean opened up, separating the two continents.

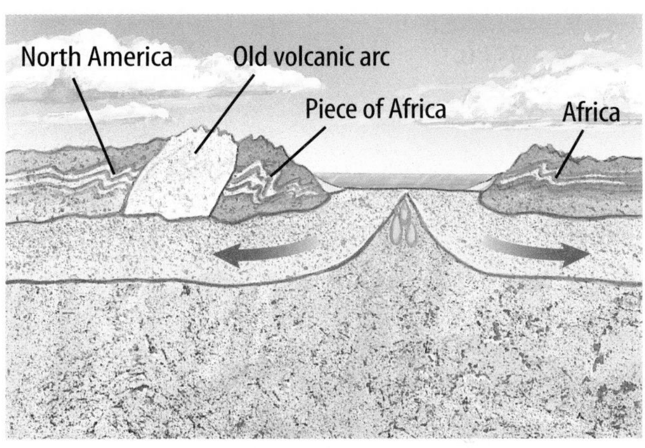

Figure 18 The Appalachian Mountains formed in several stages.
Infer *how these movements affected species in the Appalachians.*

Mountain Building Several mountain-building episodes occurred during the Paleozoic Era. The Appalachian Mountains, for example, formed during this time. This happened in several stages, as shown in **Figure 18.** The first mountain-building episode occurred as the ocean separating North America from Europe and Africa closed. Several volcanic island chains that had formed in the ocean collided with the North American Plate, as shown in the top picture of **Figure 18.** The collision of the island chains generated high mountains.

The next mountain-building episode was a result of the African Plate colliding with the North American Plate, as shown in the left picture of **Figure 18.** When Africa and North America collided, rock layers were folded and faulted. Some rocks originally deposited near the eastern coast of the North American Plate were pushed along faults as much as 65 km westward by the collision. Sediments were uplifted to form an immense mountain belt, part of which still remains today.

Figure 19 Hyoliths were organisms that became extinct at the end of the Paleozoic Era.

End of an Era At the end of the Paleozoic Era, more than 90 percent of all marine species and 70 percent of all land species died off. **Figure 19** shows one such animal. The cause of these extinctions might have been changes in climate and a lowering of sea level.

Near the end of the Permian Period, the continental plates came together and formed the supercontinent Pangaea. Glaciers formed over most of its southern part. The slow, gradual collision of continental plates caused mountain building. Mountain-building processes caused seas to close and deserts to spread over North America and Europe. Many species, especially marine organisms, couldn't adapt to these changes, and became extinct.

Other Hypotheses Other explanations also have been proposed for this mass extinction. During the late Paleozoic Era, volcanoes were extremely active. If the volcanic activity was great enough, it could have affected the entire globe. Another recent theory is similar to the one proposed to explain the extinction of dinosaurs. Perhaps a large asteroid or comet collided with Earth some 248 million years ago. This event could have caused widespread extinctions just as many paleontologists suggest happened at the end of the Mesozoic Era, 65 million years ago. Perhaps the extinction at the end of the Paleozoic Era was caused by several or all of these events happening at about the same time.

section 2 review

Summary

Precambrian Time

- Precambrian time covers almost 4 billion years of Earth history, but little is known about the organisms of this time.
- Cyanobacteria were among the earliest life-forms.

The Paleozoic Era

- Invertebrates developed shells and other hard parts, leaving a rich fossil record.
- Vertebrates—animals with backbones—appeared during this era.
- Plants and amphibians first moved to land during the Paleozoic Era.
- Adaptations in reptiles allow them to move away from water for reproduction.
- Geologic events at the end of the Paleozoic Era led to a mass extinction.

Self Check

1. **List** the geologic events that ended the Paleozoic Era.
2. **Infer** how geologic events at the end of the Paleozoic Era might have caused extinctions.
3. **Discuss** the advance that allowed reptiles to reproduce away from water. Why was this an advantage?
4. **Identify** the major change in life-forms that occurred at the end of Precambrian time.
5. **Think Critically** How did cyanobacteria aid the evolution of complex life on land? Do you think cyanobacteria are as significant to this process today as they were during Precambrian time?

Applying Skills

6. **Use a Database** Research trilobites and describe these organisms and their habitats in your Science Journal. Include hand-drawn illustrations and compare them with the illustrations in your references.

Science Online earth.msscience.com/self_check_quiz

Changing Species

In this lab, you will observe how adaptation within a species might cause the evolution of a particular trait, leading to the development of a new species.

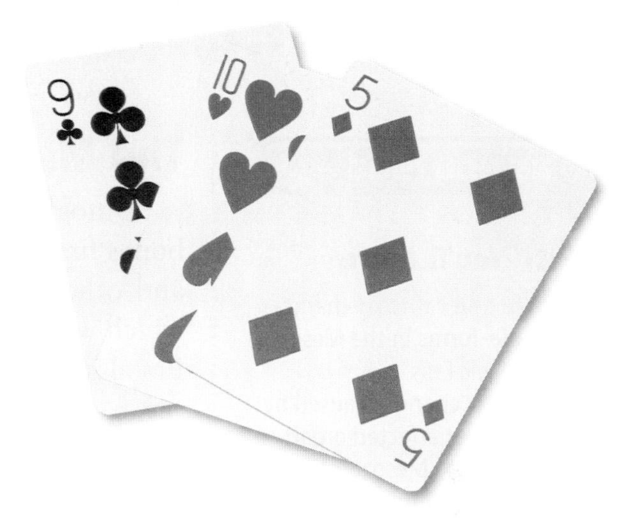

Real-World Question

How might adaptation within a species cause the evolution of a particular trait?

Goals

- **Model** adaptation within a species.

Materials

deck of playing cards

Procedure

1. **Remove** all of the kings, queens, jacks, and aces from a deck of playing cards.
2. Each remaining card represents an individual in a population of animals called "varimals." The number on each card represents the height of the individual.
3. **Calculate** the average height of the population of varimals represented by your cards.
4. Suppose varimals eat grass, shrubs, and leaves from trees. A drought causes many of these plants to die. All that's left are a few tall trees. Only varimals at least 6 units tall can reach the leaves on these trees.
5. All the varimals under 6 units leave the area or die from starvation. Discard all of the cards with a number less than 6. Calculate the new average height of the varimals.
6. Shuffle the deck of remaining cards.
7. **Draw** two cards at a time. Each pair represents a pair of varimals that will mate.
8. The offspring of each pair reaches the average height of its parents. Calculate and record the height of each offspring.
9. Discard all parents and offspring under 8 units tall and repeat steps 6–8. Now calculate the new average height of varimals. Include both the parents and offspring in your calculation.

Conclude and Apply

1. **Describe** how the height of the population changed.
2. **Explain** If you hadn't discarded the shortest varimals, would the average height of the population have changed as much?
3. Suppose the offspring grew to the height of one of its parents. How would the results change in each of the following scenarios?
   a. The height value for the offspring is chosen by coin toss.
   b. The height value for the offspring is whichever parent is tallest.
4. **Explain** If there had been no variation in height before the droughts occurred, would the species have been able to evolve?

section 3

Middle and Recent Earth History

as you read

What **You'll Learn**

- **Compare and contrast** characteristic life-forms in the Mesozoic and Cenozoic Eras.
- **Explain** how changes caused by plate tectonics affected organisms during the Mesozoic Era.
- **Identify** when humans first appeared on Earth.

Why **It's Important**

Many important groups of animals, like birds and mammals, appeared during the Mesozoic Era.

Review Vocabulary
dinosaur: a reptile from one of two orders that dominated the Mesozoic Era

New Vocabulary
- Mesozoic Era
- Cenozoic Era

The Mesozoic Era

Dinosaurs have captured people's imaginations since their bones first were unearthed more than 150 years ago. Dinosaurs and other interesting animals lived during the Mesozoic Era, which was between 248 and 65 million years ago. The Mesozoic Era also was marked by rapid movement of Earth's plates.

The Breakup of Pangaea The **Mesozoic** (meh zuh ZOH ihk) **Era,** or era of middle life, was a time of many changes on Earth. At the beginning of the Mesozoic Era, all continents were joined as a single landmass called Pangaea, as shown in **Figure 11.**

Pangaea separated into two large landmasses during the Triassic Period, as shown in **Figure 20.** The northern mass was Laurasia (law RAY zhuh), and Gondwanaland (gahn DWAH nuh land) was the southern landmass. As the Mesozoic Era continued, Laurasia and Gondwanaland broke apart and eventually formed the present-day continents.

Species that had adapted to the new environments survived the mass extinction at the end of the Paleozoic Era. Recall that a reptile's skin helps it retain bodily fluids. This characteristic, along with their shelled eggs, enabled reptiles to adapt readily to the drier climate of the Mesozoic Era. Reptiles became the most conspicuous animals on land by the Triassic Period.

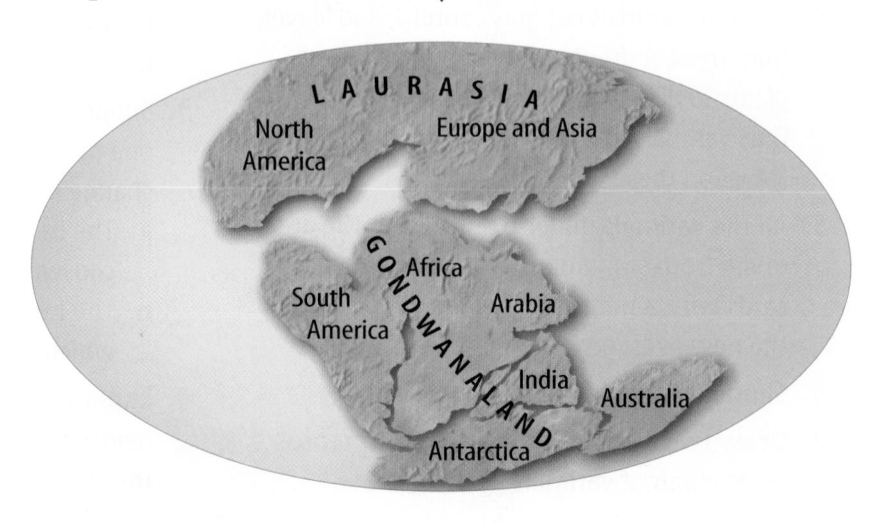

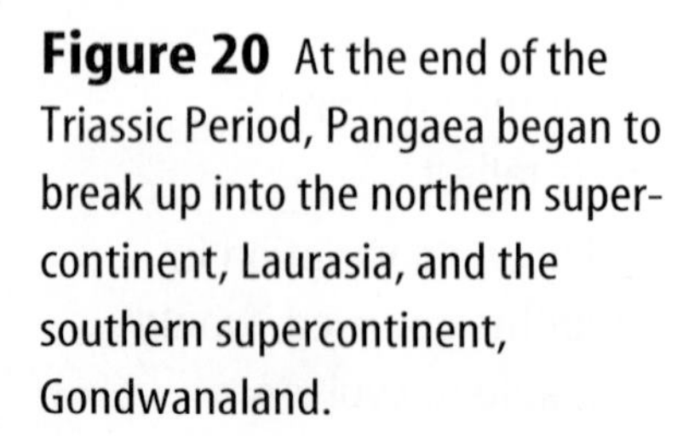

Figure 20 At the end of the Triassic Period, Pangaea began to break up into the northern supercontinent, Laurasia, and the southern supercontinent, Gondwanaland.

Dinosaurs What were the dinosaurs like? Dinosaurs ranged in height from less than 1 m to enormous creatures like *Apatosaurus* and *Tyrannosaurus.* The first small dinosaurs appeared during the Triassic Period. Larger species appeared during the Jurassic and Cretaceous Periods. Throughout the Mesozoic Era, new species of dinosaurs evolved and other species became extinct.

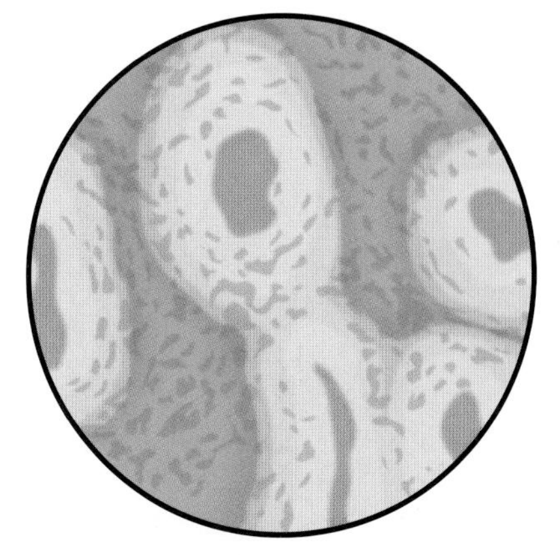

Dinosaur bone

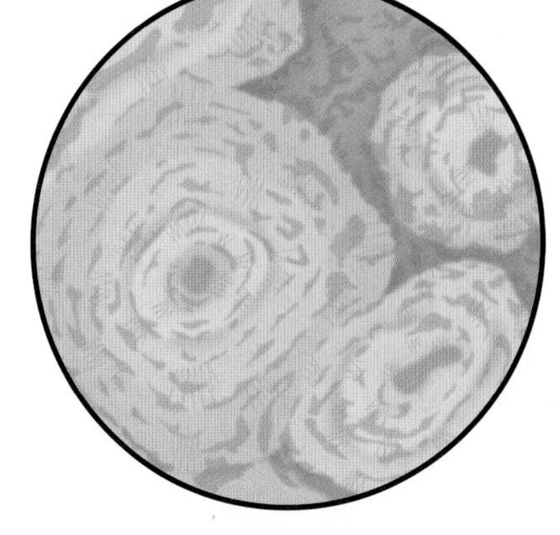

Mammal bone

Figure 21 Some dinosaur bones show structural features that are like mammal bones, leading some paleontologists to think that dinosaurs were warm blooded like mammals.

Dinosaurs Were Active Studying fossil footprints sometimes allows paleontologists to calculate how fast animals walked or ran. Some dinosaur tracks indicate that these animals were much faster runners than you might think. *Gallimimus* was 4 m long and could reach speeds of 65 km/h—as fast as a modern racehorse.

Some studies also indicate that dinosaurs might have been warm blooded, not cold blooded like present-day reptiles. The evidence that leads to this conclusion has to do with their bone structure. Slices through some cold-blooded animal bones show rings similar to growth rings in trees. The bones of some dinosaurs don't show this ring structure. Instead, they are similar to bones found in modern mammals, as you can see in **Figure 21.**

Reading Check ***Why do some paleontologists think that dinosaurs were warm blooded?***

These observations indicate that some dinosaurs might have been warm-blooded, fast-moving animals somewhat like present-day mammals and birds. They might have been quite different from present-day reptiles.

Good Mother Dinosaurs The fossil record also indicates that some dinosaurs nurtured their young and traveled in herds in which the adults surrounded their young.

One such dinosaur is *Maiasaura.* This dinosaur built nests in which it laid its eggs and raised its offspring. Nests have been found in relatively close clusters, indicating that more than one family of dinosaurs built in the same area. Some fossils of hatchlings have been found near adult animals, leading paleontologists to think that some dinosaurs nurtured their young. In fact, *Maiasaura* hatchlings might have stayed in the nest while they grew in length from about 35 cm to more than 1 m.

Topic: Warm Versus Cold
Visit earth.msscience.com for Web links to information about dinosaurs.

Activity Work with a partner to research the debate on warm-blooded versus cold-blooded dinosaurs. Present your finding to the class in the form of a debate. Be sure to cover the main points of disagreement between the two sides.

Figure 22 Birds might have evolved from dinosaurs.

A *Bambiraptor feinberger*, above, is a 75-million-year-old member of a family of meat-eating dinosaurs thought by some paleontologists to be closely related to birds.

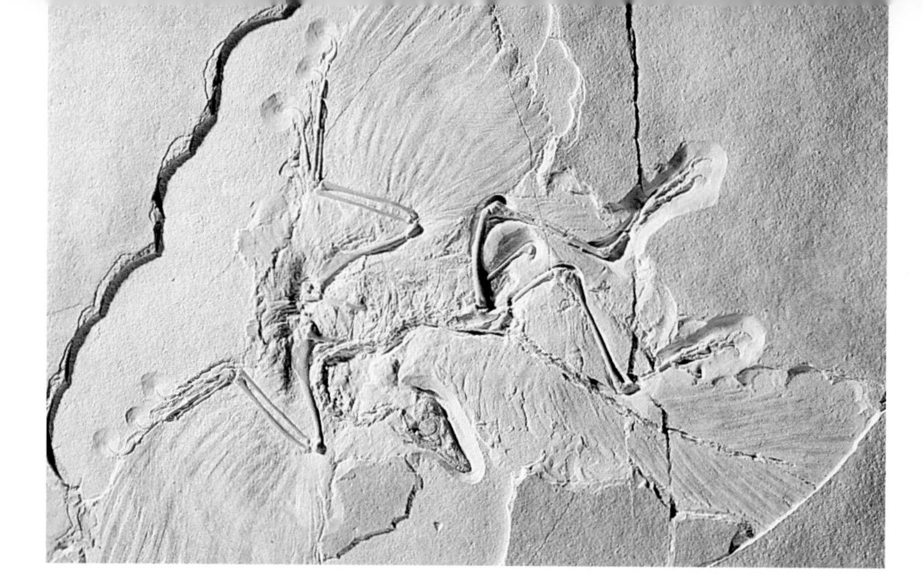

B Considered one of the world's most priceless fossils, *Archaeopteryx*, above, was first found in a limestone quarry in Germany in 1861.

Birds Birds appeared during the Jurassic Period. Some paleontologists think that birds evolved from small, meat-eating dinosaurs much like *Bambiraptor feinberger* in **Figure 22A.** The earliest bird, *Archaeopteryx*, shown in **Figure 22B,** had wings and feathers. However, because *Archaeopteryx* had features not shared with modern birds, scientists know it was not a direct ancestor of today's birds.

Mammals Mammals first appeared in the Triassic Period. The earliest mammals were small, mouselike creatures, as shown in **Figure 23.** Mammals are warm-blooded vertebrates that have hair covering their bodies. The females produce milk to feed their young. These two characteristics have enabled mammals to survive in many changing environments.

Gymnosperms During most of the Mesozoic Era, gymnosperms (JIHM nuh spurmz), which first appeared in the Paleozoic Era, dominated the land. Gymnosperms are plants that produce seeds but not flowers. Many gymnosperms are still around today. These include pines and ginkgo trees.

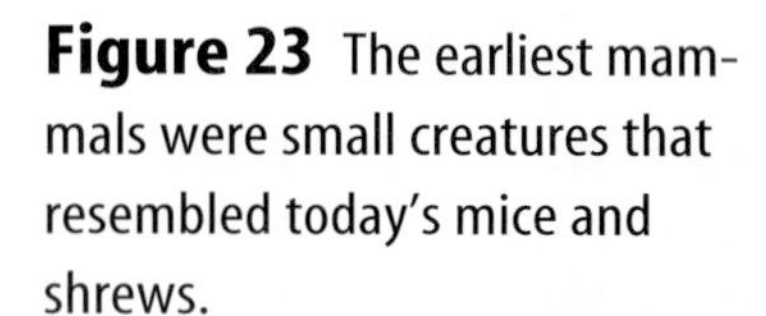

Figure 23 The earliest mammals were small creatures that resembled today's mice and shrews.

Angiosperms Angiosperms (AN jee uh spurmz), or flowering plants, first evolved during the Cretaceous Period. Angiosperms produce seeds with hard outer coverings.

Because their seeds are enclosed and protected, angiosperms can live in many environments. Angiosperms are the most diverse and abundant land plants today. Present-day angiosperms that evolved during the Mesozoic Era include magnolia and oak trees.

End of an Era The Mesozoic Era ended about 65 million years ago with a major extinction of land and marine species. Many groups of animals, including the dinosaurs, disappeared suddenly at this time. Many paleontologists hypothesize that a comet or asteroid collided with Earth, causing a huge cloud of dust and smoke to rise into the atmosphere, blocking out the Sun. Without sunlight the plants died, and all the animals that depended on these plants also died. Not everything died, however. All the organisms that you see around you today are descendants of the survivors of the great extinction at the end of the Mesozoic Era.

Applying Math — Calculate Percentages

CALCULATING EXTINCTION BY USING PERCENTAGES At the end of the Cretaceous Period, large numbers of plants and animals became extinct. Scientists still are trying to understand why some types of plants and animals survived while others died out. Looking at data about amphibians, reptiles, and mammals that lived during the Cretaceous Period, can you determine what percentage of amphibians survived this mass extinction?

Solution

1 *This is what you know:*

Animal Extinctions

Animal Type	Groups Living Before Extinction Event (n)	Groups Left After Extinction Event (t)
Amphibians	12	4
Reptiles	63	30
Mammals	24	8

2 *This is what you need to find out:* p = the percentage of amphibian groups that survived the Cretaceous extinction

3 *This is the equation you need to use:*
- $p = t / n \times 100$
- Both t and n are shown on the above chart.

4 *Substitute the known values:* $p = 4 / 12 \times 100 = 33.3\%$

Practice Problems

1. Using the same equation as demonstrated above, calculate the percentage of reptiles and then the percentage of mammals that survived. Which type of animal was least affected by the extinction?
2. What percentage of all groups survived?

Science Online — For more practice, visit earth.msscience.com/math_practice

Calculating the Age of the Atlantic Ocean

Procedure

1. On a **world map** or **globe,** measure the distance in kilometers between a point on the east coast of South America and a point on the west coast of Africa.
2. Measure in SI several times and take the average of your results.
3. Assuming that Africa has been moving away from South America at a rate of 3.5 cm per year, calculate how many years it took to create the Atlantic Ocean.

Analysis

1. Did the values used to obtain your average value vary much?
2. How close did your age come to the accepted estimate for the beginning of the breakup of Pangaea in the Triassic Period?

Try at Home

The Cenozoic Era

The **Cenozoic** (se nuh ZOH ihk) **Era,** or era of recent life, began about 65 million years ago and continues today. Many mountain ranges in North and South America and Europe began to form in the Cenozoic Era. In the late Cenozoic, the climate became much cooler and ice ages occurred. The Cenozoic Era is subdivided into two periods. The first of these is the Tertiary Period. The present-day period is the Quaternary Period. It began about 1.8 million years ago.

Reading Check *What happened to the climate during the late Cenozoic Era?*

Times of Mountain Building Many mountain ranges formed during the Cenozoic Era. These include the Alps in Europe and the Andes in South America. The Himalaya, shown in **Figure 24,** formed as India moved northward and collided with Asia. The collision crumpled and thickened Earth's crust, raising the highest mountains presently on Earth. Many people think the growth of these mountains has helped create cooler climates worldwide.

Figure 24 The Himalaya extend along the India-Tibet border and contain some of the world's tallest mountains. India drifted north and finally collided with Asia, forming the Himalaya.

Further Evolution of Mammals

Throughout much of the Cenozoic Era, expanding grasslands favored grazing plant eaters like horses, camels, deer, and some elephants. Many kinds of mammals became larger. Horses evolved from small, multi-toed animals into the large, hoofed animals of today. However, not all mammals remained on land. Ancestors of the present-day whales and dolphins evolved to live in the sea.

Figure 25 The wombat is one of many Australian marsupials. As a result of human activities, the number and range of wombats have diminished.

As Australia and South America separated from Antarctica during the continuing breakup of the continents, many species became isolated. They evolved separately from life-forms in other parts of the world. Evidence of this can be seen today in Australia's marsupials. Marsupials are mammals such as kangaroos, koalas, and wombats (shown in **Figure 25**) that carry their young in a pouch.

Your species, *Homo sapiens,* probably appeared about 140,000 years ago. Some people suggest that the appearance of humans could have led to the extinction of many other mammals. As their numbers grew, humans competed for food that other animals relied upon. Also, fossil bones and other evidence indicate that early humans were hunters.

section 3 review

Summary

The Mesozoic Era

- During the Triassic Period, Pangaea split into two continents.
- Dinosaurs were the dominant land animals of the Mesozoic Era.
- Birds, mammals, and flowering plants all appeared during this era.
- The Mesozoic Era ended 65 million years ago with a mass extinction.

The Cenozoic Era

- The Cenozoic Era has been a mountain-building period with cooler climates.
- Mammals became dominant with many new life-forms appearing after the dinosaurs disappeared.
- Humans also appeared in the Cenozoic Era, probably about 140,000 years ago.

Self Check

1. **List** the era, period, and epoch in which *Homo sapiens* first appeared.
2. **Discuss** whether mammals became more or less abundant after the extinction of the dinosaurs, and explain why.
3. **Infer** how seeds with a hard outer covering enabled angiosperms to survive in a wide variety of climates.
4. **Explain** why some paleontologists hypothesize that dinosaurs were warm-blooded animals.
5. **Think Critically** How could two species that evolved on separate continents have many similarities?

Applying Math

6. **Convert Units** A fossil mosasaur, a giant marine reptile, measured 9 m in length and had a skull that measured 45 cm in length. What fraction of the mosasaur's total length did the skull account for? Compare your length with the mosasaur's length.

Use the Internet

Discovering the Past

Goals

- **Gather** information about fossils found in your area.
- **Communicate** details about fossils found in your area.
- **Synthesize** information from sources about the fossil record and the changes in your area over time.

Data Source

Science Online

Visit earth.msscience.com/internet_lab for more information about fossils and changes over geologic time and for data collected by other students.

Real-World Question

Imagine what your state was like millions of years ago. What animals might have been roaming around the spot where you now sit? Can you picture a *Tyrannosaurus rex* roaming the area that is now your school? The animals and plants that once inhabited your region might have left some clues to their identity—fossils. Scientists use fossils to piece together what Earth looked like in the geologic past. Fossils can help determine whether an area used to be dry land or underwater. Fossils can help uncover clues about how plants and animals have evolved over the course of time. Using the resources of the Internet and by sharing data with your peers, you can start to discover how North America has changed through time. How has your area changed over geologic time? How might the area where you are now living have looked thousands or millions of years ago? Do you think that the types of animals and plants have changed much over time? Form a hypothesis concerning the change in organisms and geography from long ago to the present day in your area.

Fossils in Your Area

Fossil Name	Plant or Animal Fossil	Age of Fossils	Details About Plant or Animal Fossil	Location of Fossil	Additional Information
		Do not write in this book.			

Make a Plan

1. **Determine** the age of the rocks that make up your area. Were they formed during Precambrian time, the Paleozoic Era, the Mesozoic Era, or the Cenozoic Era?
2. Gather information about the plants and animals found in your area during one of the above geologic time intervals. Find specific information on when, where, and how the fossil organisms lived. If no fossils are known from your area, find out information about the fossils found nearest your area.

Follow Your Plan

1. Make sure your teacher approves your plan before you start.
2. Go to earth.msscience.com/internet_lab to post your data in the table. Add any additional information you think is important to understanding the fossils found in your area.

Analyze Your Data

1. What present-day relatives of prehistoric animals or plants exist in your area?
2. How have the organisms in your area changed over time? Is your hypothesis supported? Why or why not?
3. What other information did you discover about your area's climate or environment from the geologic time period you investigated?

Conclude and Apply

1. **Describe** the plant and animal fossils that have been discovered in your area. What clues did you discover about the environment in which these organisms lived? How do these compare to the environment of your area today?
2. **Infer** from the fossil organisms found in your area what the geography and climate were like during the geologic time period you chose.

Find this lab using the link below.

Science Online

earth.msscience.com/internet_lab

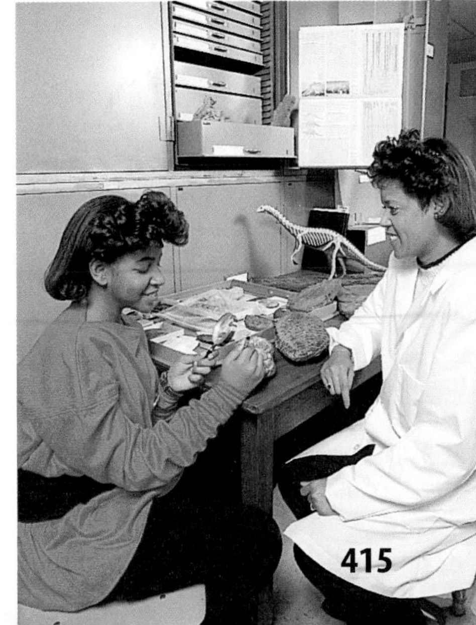

Science Stats

Extinct!

Did you know...

Saber-toothed cat

...The saber-toothed cat lived in the Americas from about 1.6 million to 8,000 years ago. *Smilodon,* the best-known saber-toothed cat, was among the most ferocious carnivores. It had large canine teeth, about 15 cm long, which it used to pierce the flesh of its prey.

Applying Math How many years did *Smilodon* live in the Americas before it became extinct?

Woolly mammoth

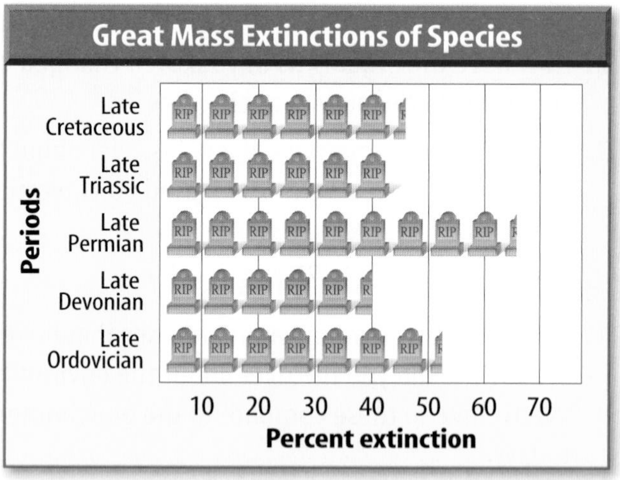
Great Mass Extinctions of Species
Late Cretaceous
Late Triassic
Late Permian
Late Devonian
Late Ordovician
Periods
RIP
10
20
30
40
50
60
70
Percent extinction

...The woolly mammoth lived in the cold tundra regions during the Ice Age. It looked rather like an elephant with long hair, had a mass between 5,300 kg and 7,300 kg, and was between 3 m and 4 m tall.

Write About It

Visit earth.msscience.com/science_stats to research extinct animals. Trace the origins of each of the species and learn how long its kind existed on Earth.

Reviewing Main Ideas

Section 1 Life and Geologic Time

1. Geologic time is divided into eons, eras, periods, and epochs.
2. Divisions within the geologic time scale are based largely on major evolutionary changes in organisms.
3. Plate movements affect organic evolution.

Section 2 Early Earth History

1. Cyanobacteria evolved during Precambrian time. Trilobites, fish, and corals were abundant during the Paleozoic Era.
2. Plants and animals began to move onto land during the middle of the Paleozoic Era.
3. The Paleozoic Era was a time of mountain building. The Appalachian Mountains formed when several islands and finally Africa collided with North America.
4. At the end of the Paleozoic Era, many marine invertebrates became extinct.

Section 3 Middle and Recent Earth History

1. Reptiles and gymnosperms were dominant land life-forms in the Mesozoic Era. Mammals and angiosperms began to dominate the land in the Cenozoic Era.
2. Pangaea broke apart during the Mesozoic Era. Many mountain ranges formed during the Cenozoic Era.

Visualizing Main Ideas

Copy and complete the concept map on geologic time using the following choices: Cenozoic, Trilobites in oceans, Mammals common, Paleozoic, Dinosaurs roam Earth, *and* Abundant gymnosperms.

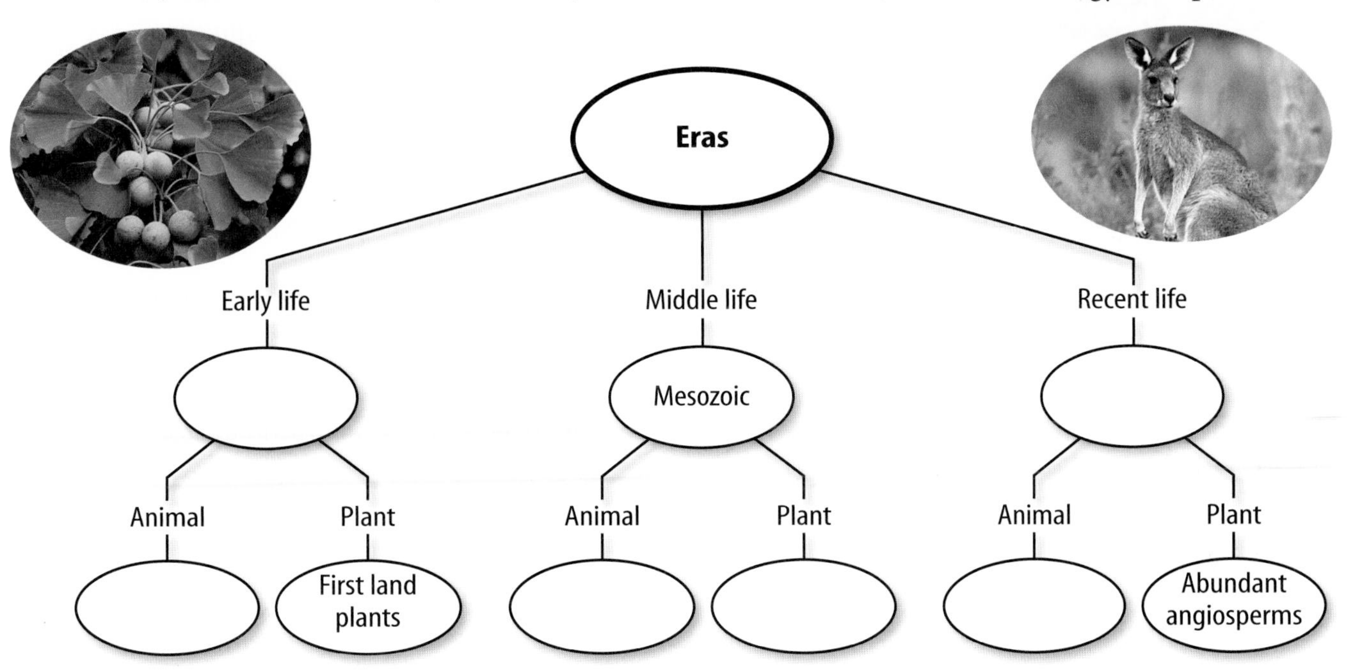

chapter 14 Review

Using Vocabulary

Cenozoic Era p. 412
cyanobacteria p. 401
eon p. 393
epoch p. 393
era p. 393
geologic time scale p. 392
Mesozoic Era p. 408
natural selection p. 395
organic evolution p. 394
Paleozoic Era p. 402
Pangaea p. 399
period p. 393
Precambrian time p. 400
species p. 394
trilobite p. 397

Fill in the blank with the correct word or words.

1. A change in the hereditary features of a species over a long period is ________.
2. A record of events in Earth history is the ________.
3. The largest subdivision of geologic time is the ________.
4. The process by which the best-suited individuals survive in their environment is ________.
5. A group of individuals that normally breed only among themselves is a(n) ________.

Checking Concepts

Choose the word or phrase that best completes the sentence.

6. How many millions of years ago did the era in which you live begin?
   A) 650 **C)** 1.6
   B) 245 **D)** 65
7. What is the process by which better-suited organisms survive and reproduce?
   A) endangerment **C)** gymnosperm
   B) extinction **D)** natural selection
8. During what period did the most recent ice age occur?
   A) Pennsylvanian **C)** Tertiary
   B) Triassic **D)** Quaternary
9. What is the next smaller division of geologic time after the era?
   A) period **C)** epoch
   B) stage **D)** eon
10. What was one of the earliest forms of life?
    A) gymnosperm **C)** angiosperm
    B) cyanobacterium **D)** dinosaur

Use the illustration below to answer question 11.

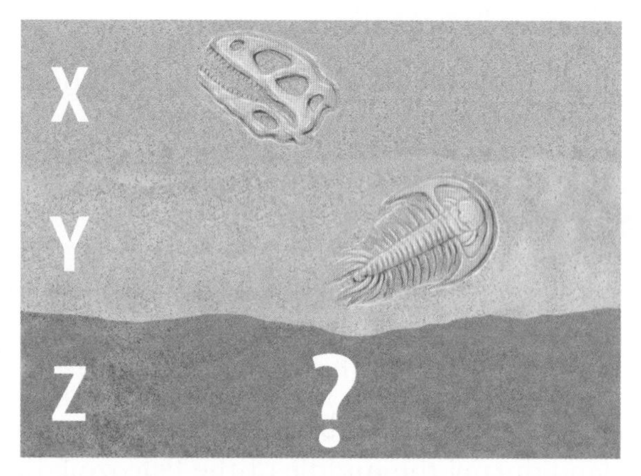

11. Consider the undisturbed rock layers in the figure above. If fossil X were a *Tyrannosaurus rex* bone, and fossil Y were a trilobite; then fossil Z could be which of the following?
    A) stromatolite **C)** angiosperm
    B) sabre-tooth cat **D)** *Homo sapiens*
12. During which era did the dinosaurs live?
    A) Mesozoic **C)** Miocene
    B) Paleozoic **D)** Cenozoic
13. Which type of plant has seeds without protective coverings?
    A) angiosperms **C)** gymnosperms
    B) apples **D)** magnolias
14. Which group of plants evolved during the Mesozoic Era and is dominant today?
    A) gymnosperms **C)** ginkgoes
    B) angiosperms **D)** algae
15. In which era did the Ediacaran fauna live?
    A) Precambrian **C)** Mesozoic
    B) Paleozoic **D)** Cenozoic

Science Online earth.msscience.com/vocabulary_puzzlemaker

Thinking Critically

16. **Infer** why plants couldn't move onto land until an ozone layer formed.
17. **Discuss** why trilobites are classified as index fossils.
18. **Compare and contrast** the most significant difference between Precambrian life-forms and Paleozoic life-forms.
19. **Describe** how natural selection is related to organic evolution.
20. **Explain** In the early 1800s, a naturalist proposed that the giraffe species has a long neck as a result of years of stretching their necks to reach leaves in tall trees. Why isn't this true?
21. **Infer** Use the outlines of the present-day continents to make a sketch of Pangaea.
22. **Form Hypotheses** Suggest some reasons why trilobites might have become extinct at the end of the Paleozoic Era.

23. **Interpret Data** A student found what she thought was a piece of dinosaur bone in Pleistocene sediment. How likely is it that she is right? Explain.
24. **Infer** why mammals didn't become dominant until after the dinosaurs disappeared.

Performance Activities

25. **Make a Model** In the Section 2 Lab, you learned how a particular characteristic might evolve within a species. Modify the experimental model by using color instead of height as a characteristic. Design your activity with the understanding that varimals live in a dark-colored forest environment.
26. **Make a Display** Certain groups of animals have dominated the land throughout geologic time. Use your textbook and other references to discover some of the dominant species of each era. Make a display that illustrates some animals from each era. Be sure to include appropriate habitats.

Applying Math

Use the graph below to answer questions 27 and 28.

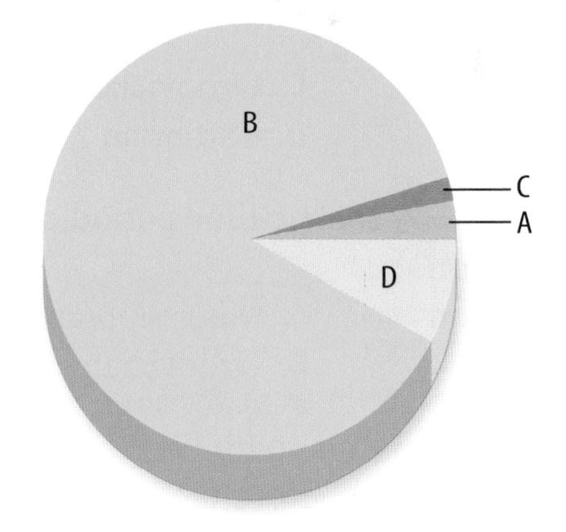

27. **Modeling Geologic Time** The circle graph above represents geologic time. Determine which interval of geologic time is represented by each portion of the graph. Which interval was longest? Which do we know the least about? Which of these intervals is getting larger?
28. **Interpret Data** The Cenozoic Era has lasted 65 million years. What percentage of Earth's 4.5-billion-year history is that?

Part 1 Multiple Choice

Record your answers on the answer sheet provided by your teacher or on a sheet of paper.

Examine the diagram below. Then answer questions 1–3.

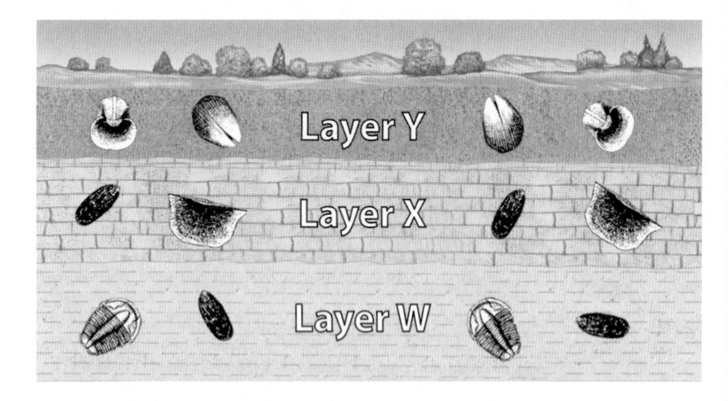

1. During which geologic time period did layer W form?
   A. Cambrian
   B. Ordovician
   C. Devonian
   D. Silurian

2. During which geologic time period did layer X form?
   A. Devonian
   B. Silurian
   C. Ordovician
   D. Cambrian

3. During which geologic time period did layer Y form?
   A. Cambrian
   B. Silurian
   C. Mississippian
   D. Ordovician

4. When did dinosaurs roam Earth?
   A. Precambrian time
   B. Paleozoic Era
   C. Mesozoic Era
   D. Cenozoic Era

5. What is the name of the supercontinent that formed at the end of the Paleozoic Era?
   A. Gondwanaland
   B. Eurasia
   C. Laurasia
   D. Pangaea

6. During which geologic period did modern humans evolve?
   A. Quaternary
   B. Triassic
   C. Ordovician
   D. Tertiary

7. How many body lobes did trilobites have?
   A. one
   B. two
   C. three
   D. four

8. Which mountain range formed because India collided with Asia?
   A. Alps
   B. Andes
   C. Ural
   D. Himalaya

Use the diagram below to answer questions 9–11.

Cenozoic Era	Quaternary Period	Holocene Epoch
		Pleistocene Epoch
	Tertiary Period	Pliocene Epoch
		Miocene Epoch
		Oligocene Epoch
		Eocene Epoch
		Paleocene Epoch

9. What is the oldest epoch in the Cenozoic Era?
   A. Pleistocene
   B. Paleocene
   C. Miocene
   D. Holocene

10. What is the youngest epoch in the Cenozoic Era?
    A. Miocene
    B. Holocene
    C. Paleocene
    D. Eocene

11. Which epoch is part of the Quaternary Period?
    A. Oligocene
    B. Eocene
    C. Pleistocene
    D. Pliocene

Part 2 Short Response/Grid In

Record your answers on the answer sheet provided by your teacher or on a sheet of paper.

12. Who was Charles Darwin? How did he contribute to science?
13. Explain one hypothesis about why dinosaurs might have become extinct.
14. Describe *Archaeopteryx*. Why is this an important fossil?
15. Why do many scientists think that dinosaurs were warm-blooded?
16. What are stromatolites? How do they form?
17. Define the term *species*.

Select one of the equations below to help you answer questions 18–20.

time = distance ÷ speed

or

speed = distance ÷ time

18. It recently was estimated that *T. rex* could run no faster than about 11 m/s. At this speed, how long would it take *T. rex* to run 200 m?
19. A typical ornithopod (plant-eating dinosaur that walked on two legs) probably moved at a speed of about 2 m/s. How long would it take this dinosaur to run 200 m?
20. In 1996, Michael Johnson ran 200 m in 19.32 s. What was his average speed? How does this compare with *T. rex?*

Test-Taking Tip

Show Your Work For constructed response questions, show all of your work and any calculations on your answer sheet.

Part 3 Open Ended

Record your answers on a sheet of paper.

Use the diagram below to answer questions 21 and 22. It shows the time ranges of various types of organisms on Earth. When a bar is wider, there were more species of that type of organism (higher diversity).

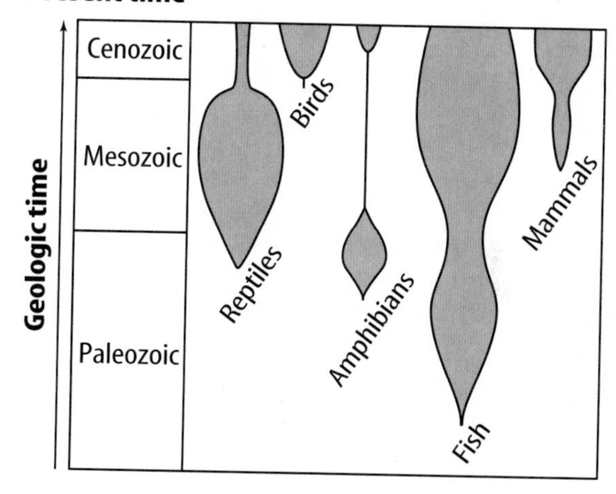

21. Describe how the diversity of reptiles changed through time.
22. How has the diversity of mammals changed through time? Do you see any relationship with how reptile diversity has changed?
23. Why might mammals in Australia be so much different than mammals on other continents?
24. Describe how natural selection might cause a species to change through time.
25. How did early photosynthetic organisms change the conditions on Earth to allow more advanced organisms to flourish?
26. What are mass extinctions? How have they affected life on Earth?
27. Write a description of what Earth was like during Precambrian time. Summarize how Earth was different than it is now.

How Are Bats & Tornadoes Connected?

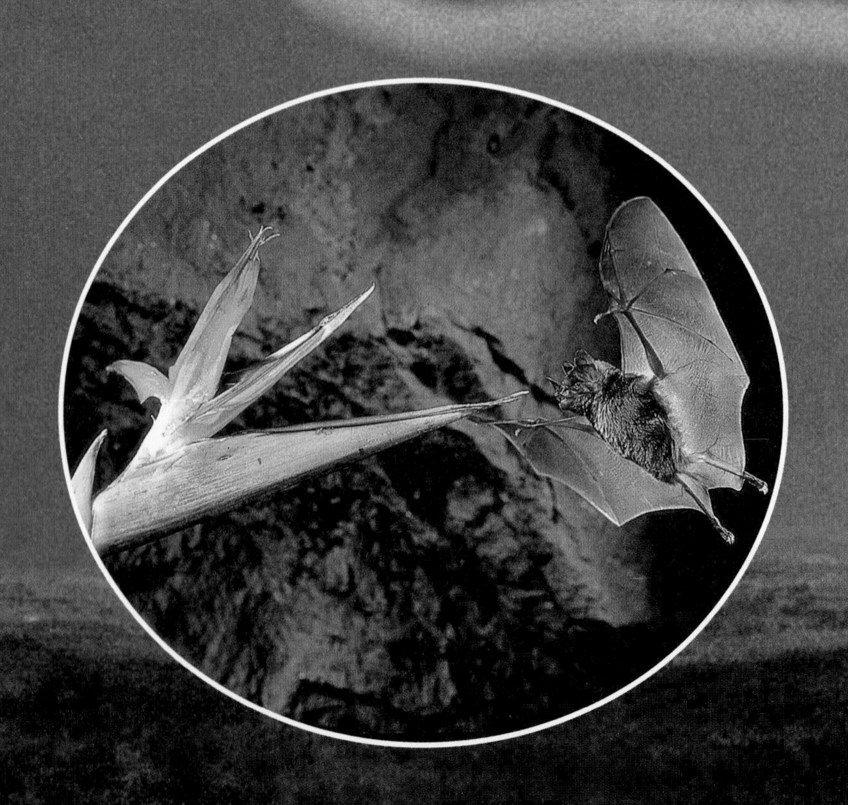

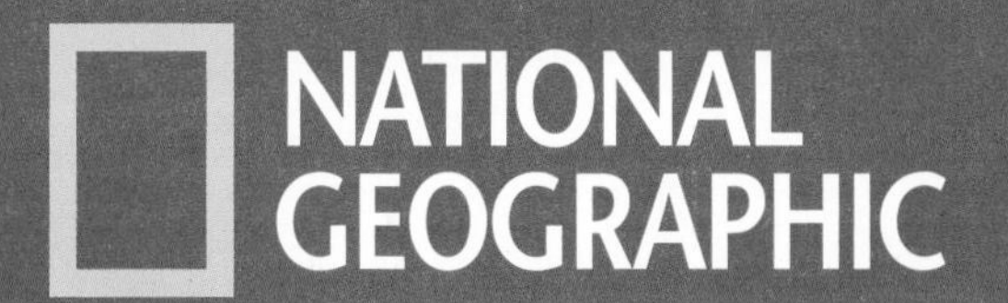

Bats are able to find food and avoid obstacles without using their vision. They do this by producing high-frequency sound waves which bounce off objects and return to the bat. From these echoes, the bat is able to locate obstacles and prey. This process is called echolocation. If the reflected waves have a higher frequency than the emitted waves, the bat senses the object is getting closer. If the reflected waves have a lower frequency, the object is moving away. This change in frequency is called the Doppler effect. Like echolocation, sonar technology uses sound waves and the Doppler effect to determine the position and motion of objects. Doppler radar also uses the Doppler effect, but with radar waves instead of sound waves. Higher frequency waves indicate if an object, such as a storm, is coming closer, while lower frequencies indicate if it is moving away. Meteorologists use frequency shifts indicated by Doppler radar to detect the formation of tornadoes and to predict where they will strike.

unit projects

Visit earth.msscience.com/unit_project to find project ideas and resources.
Projects include:

- **Technology** Predict and track the weather of a city in a different part of the world, and compare it to your local weather pattern.
- **Career** Explore weather-related careers while investigating different types of storms. Compare and contrast career characteristics and history.
- **Model** Research animal behavior to discover if animals are able to predict the weather. Present your samples of weather-predicting proverbs as a collection, or use them in a folklore tale.

WebQuest *Hurricanes!* investigates a variety of tropical storms, their source of energy, classifications, and destructive forces.

chapter 15

Atmosphere

The BIG Idea

Earth's atmosphere helps regulate the absorption and distribution of energy received from the Sun.

SECTION 1
Earth's Atmosphere
Main Idea Earth's atmosphere is a thin layer of air that forms a protective covering around the planet.

SECTION 2
Energy Transfer in the Atmosphere
Main Idea Earth's atmosphere helps control how much of the Sun's radiation is absorbed or lost to space.

SECTION 3
Air Movement
Main Idea Uneven heating of Earth's surface leads to a change in pressure that causes air to move.

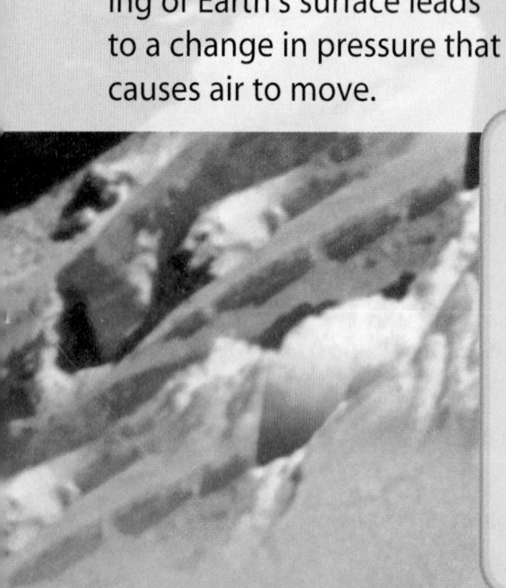

Fresh mountain air?

On top of Mt. Everest the air is a bit thin. Without breathing equipment, an average person quickly would become dizzy, then unconscious, and eventually would die. In this chapter you'll learn what makes the atmosphere at high altitudes different from the atmosphere we are used to.

Science Journal Write a short article describing how you might prepare to climb Mt. Everest.

Start-Up Activities

Observe Air Pressure

The air around you is made of billions of molecules. These molecules are constantly moving in all directions and bouncing into every object in the room, including you. Air pressure is the result of the billions of collisions of molecules into these objects. Because you usually do not feel molecules in air hitting you, do the lab below to see the effect of air pressure.

1. Cut out a square of cardboard about 10 cm from the side of a cereal box.
2. Fill a glass to the brim with water.
3. Hold the cardboard firmly over the top of the glass, covering the water, and invert the glass.
4. Slowly remove your hand holding the cardboard in place and observe.
5. **Think Critically** Write a paragraph in your Science Journal describing what happened to the cardboard when you inverted the glass and removed your hand. How does air pressure explain what happened?

Earth's Atmospheric Layers Make the following Foldable to help you visualize the five layers of Earth's atmosphere.

STEP 1 **Collect** 3 sheets of paper and layer them about 1.25 cm apart vertically. Keep the edges level.

STEP 2 **Fold** up the bottom edges of the paper to form 6 equal tabs.

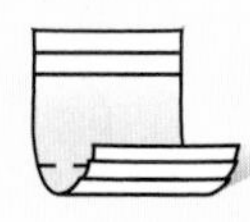

STEP 3 **Fold** the paper and crease well to hold the tabs in place. Staple along the fold. **Label** each tab.

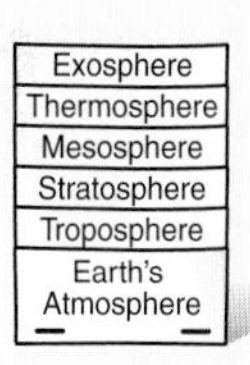

Find Main Ideas Label the tabs *Earth's Atmosphere, Troposphere, Stratosphere, Mesosphere, Thermosphere,* and *Exosphere* from bottom to top as shown. As you read the chapter, write information about each layer of Earth's atmosphere under the appropriate tab.

Science Online | **Preview this chapter's content and activities at** earth.msscience.com

Get Ready to Read

Identify the Main Idea

1 Learn It! Main ideas are the most important ideas in a paragraph, section, or chapter. Supporting details are facts or examples that explain the main idea. Understanding the main idea allows you to grasp the whole picture.

2 Practice It! Read the following paragraph. Draw a graphic organizer like the one below to show the main idea and supporting details.

> In addition to gases, Earth's atmosphere contains small, solid particles such as dust, salt, and pollen. Dust particles get into the atmosphere when wind picks them up off the ground and carries them along. Salt is picked up from ocean spray. Plants give off pollen that becomes mixed throughout part of the atmosphere.
>
> *—from page 427*

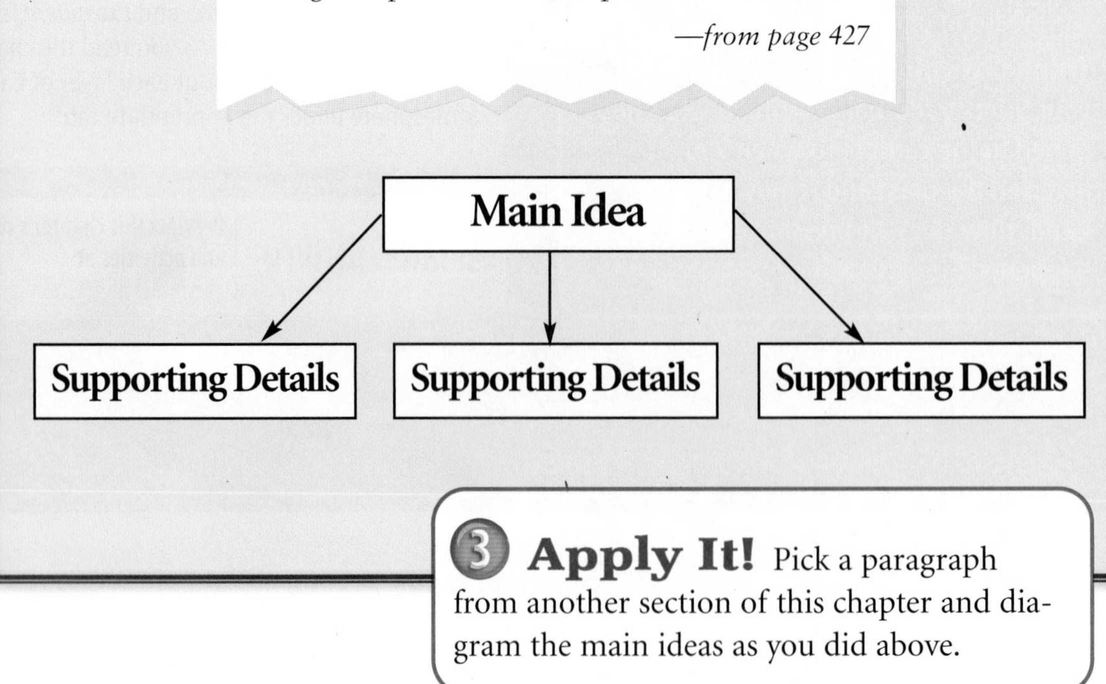

3 Apply It! Pick a paragraph from another section of this chapter and diagram the main ideas as you did above.

Reading Tip

The main idea is often the first sentence in a paragraph but not always.

Target Your Reading

Use this to focus on the main ideas as you read the chapter.

1. **Before you read** the chapter, respond to the statements below on your worksheet or on a numbered sheet of paper.
   - Write an **A** if you **agree** with the statement.
   - Write a **D** if you **disagree** with the statement.

2. **After you read** the chapter, look back to this page to see if you've changed your mind about any of the statements.
   - If any of your answers changed, explain why.
   - Change any false statements into true statements.
   - Use your revised statements as a study guide.

Before You Read A or D	Statement		After You Read A or D
	1	Earth's atmosphere is mostly oxygen.	
	2	Air pressure is greater near Earth's surface and decreases higher in the atmosphere.	
	3	The ozone layer absorbs most of the harmful infrared radiation that enters the atmosphere.	
	4	Conduction is the transfer of heat by the flow of material.	
	5	In the atmosphere, cold, dense air sinks, causing hot, less dense air to rise.	
	6	Wind is the movement of air from an area of lower pressure to an area of higher pressure.	
	7	Earth's surface is heated evenly by the Sun.	
	8	Earth's rotation affects the direction in which air and water move.	
	9	Jet streams are legally defined zones in the atmosphere where only jets are allowed to travel.	

Science Online

Print out a worksheet of this page at earth.msscience.com

section 1

Earth's Atmosphere

as you read

***What* You'll Learn**

- **Identify** the gases in Earth's atmosphere.
- **Describe** the structure of Earth's atmosphere.
- **Explain** what causes air pressure.

***Why* It's Important**

The atmosphere makes life on Earth possible.

Review Vocabulary
pressure: force exerted on an area

New Vocabulary
- atmosphere
- troposphere
- ionosphere
- ozone layer
- ultraviolet radiation
- chlorofluorocarbon

Importance of the Atmosphere

Earth's **atmosphere,** shown in **Figure 1,** is a thin layer of air that forms a protective covering around the planet. If Earth had no atmosphere, days would be extremely hot and nights would be extremely cold. Earth's atmosphere maintains a balance between the amount of heat absorbed from the Sun and the amount of heat that escapes back into space. It also protects lifeforms from some of the Sun's harmful rays.

Makeup of the Atmosphere

Earth's atmosphere is a mixture of gases, solids, and liquids that surrounds the planet. It extends from Earth's surface to outer space. The atmosphere is much different today from what it was when Earth was young.

Earth's early atmosphere, produced by erupting volcanoes, contained nitrogen and carbon dioxide, but little oxygen. Then, more than 2 billon years ago, Earth's early organisms released oxygen into the atmosphere as they made food with the aid of sunlight. These early organisms, however, were limited to layers of ocean water deep enough to be shielded from the Sun's harmful rays, yet close enough to the surface to receive sunlight. Eventually, a layer rich in ozone (O_3) that protects Earth from the Sun's harmful rays formed in the upper atmosphere. This protective layer eventually allowed green plants to flourish all over Earth, releasing even more oxygen. Today, a variety of life forms, including you, depends on a certain amount of oxygen in Earth's atmosphere.

Figure 1 Earth's atmosphere, as viewed from space, is a thin layer of gases. The atmosphere keeps Earth's temperature in a range that can support life.

Gases in the Atmosphere Today's atmosphere is a mixture of the gases shown in **Figure 2.** Nitrogen is the most abundant gas, making up 78 percent of the atmosphere. Oxygen actually makes up only 21 percent of Earth's atmosphere. As much as four percent of the atmosphere is water vapor. Other gases that make up Earth's atmosphere include argon and carbon dioxide.

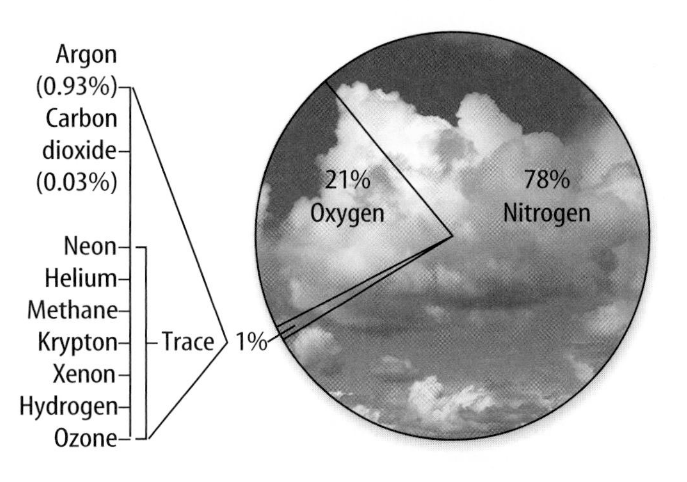

Figure 2 This circle graph shows the percentages of the gases, excluding water vapor, that make up Earth's atmosphere.
Determine *Approximately what fraction of Earth's atmosphere is oxygen?*

The composition of the atmosphere is changing in small but important ways. For example, car exhaust emits gases into the air. These pollutants mix with oxygen and other chemicals in the presence of sunlight and form a brown haze called smog. Humans burn fuel for energy. As fuel is burned, carbon dioxide is released as a by-product into Earth's atmosphere. Increasing energy use may increase the amount of carbon dioxide in the atmosphere.

Solids and Liquids in Earth's Atmosphere In addition to gases, Earth's atmosphere contains small, solid particles such as dust, salt, and pollen. Dust particles get into the atmosphere when wind picks them up off the ground and carries them along. Salt is picked up from ocean spray. Plants give off pollen that becomes mixed throughout part of the atmosphere.

The atmosphere also contains small liquid droplets other than water droplets in clouds. The atmosphere constantly moves these liquid droplets and solids from one region to another. For example, the atmosphere above you may contain liquid droplets and solids from an erupting volcano thousands of kilometers from your home, as illustrated in **Figure 3.**

Figure 3 Solids and liquids can travel large distances in Earth's atmosphere, affecting regions far from their source.

On June 12, 1991, Mount Pinatubo in the Philippines erupted, causing liquid droplets to form in Earth's atmosphere.

Droplets of sulfuric acid from volcanoes can produce spectacular sunrises.

Layers of the Atmosphere

What would happen if you left a glass of chocolate milk on the kitchen counter for a while? Eventually, you would see a lower layer with more chocolate separating from upper layers with less chocolate. Like a glass of chocolate milk, Earth's atmosphere has layers. There are five layers in Earth's atmosphere, each with its own properties, as shown in **Figure 4.** The lower layers include the troposphere and stratosphere. The upper atmospheric layers are the mesosphere, thermosphere, and exosphere. The troposphere and stratosphere contain most of the air.

Topic: Earth's Atmospheric Layers

Visit earth.msscience.com for Web links to information about layers of Earth's atmosphere.

Activity Locate data on recent ozone layer depletion. Graph your data.

Lower Layers of the Atmosphere You study, eat, sleep, and play in the **troposphere** which is the lowest of Earth's atmospheric layers. It contains 99 percent of the water vapor and 75 percent of the atmospheric gases. Rain, snow, and clouds occur in the troposphere, which extends up to about 10 km.

The stratosphere, the layer directly above the troposphere, extends from 10 km above Earth's surface to about 50 km. As **Figure 4** shows, a portion of the stratosphere contains higher levels of a gas called ozone. Each molecule of ozone is made up of three oxygen atoms bonded together. Later in this section you will learn how ozone protects Earth from the Sun's harmful rays.

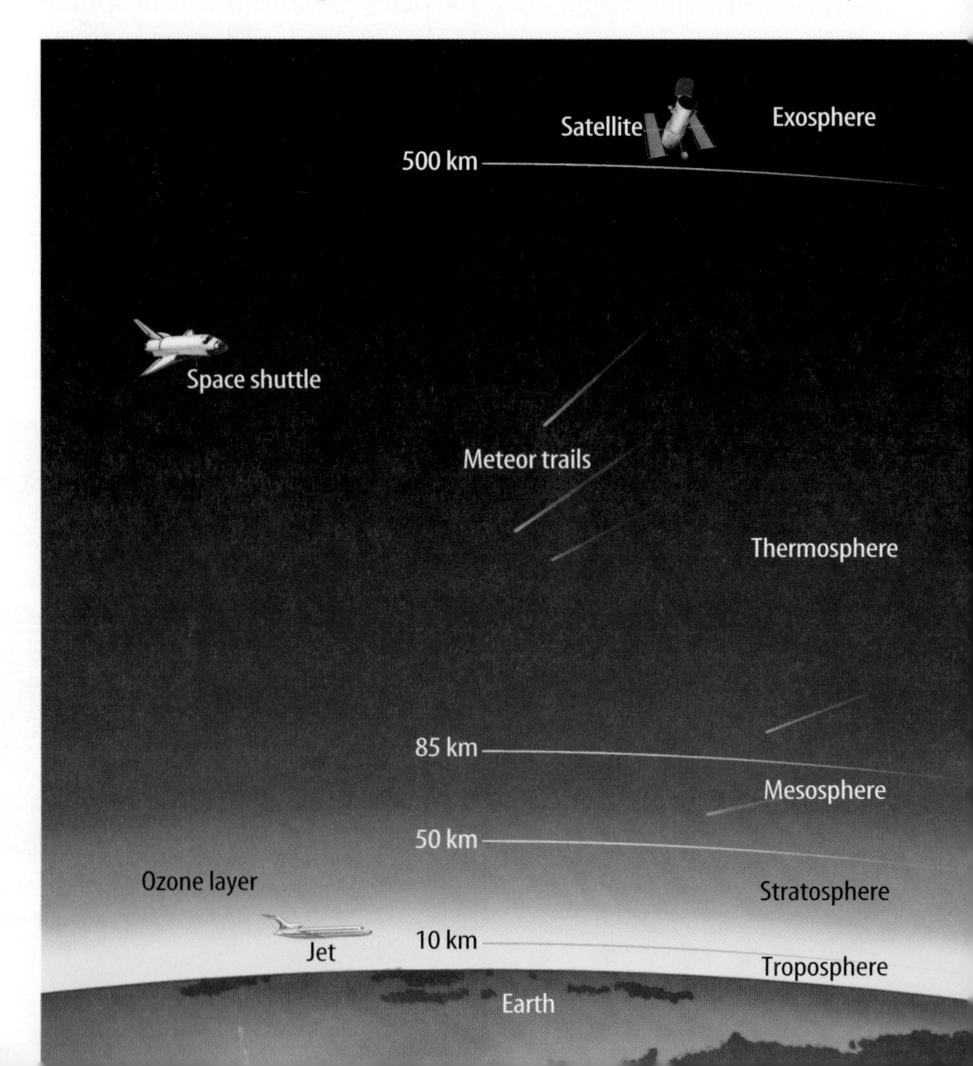

Figure 4 Earth's atmosphere is divided into five layers.
Describe *the layer of the atmosphere in which you live.*

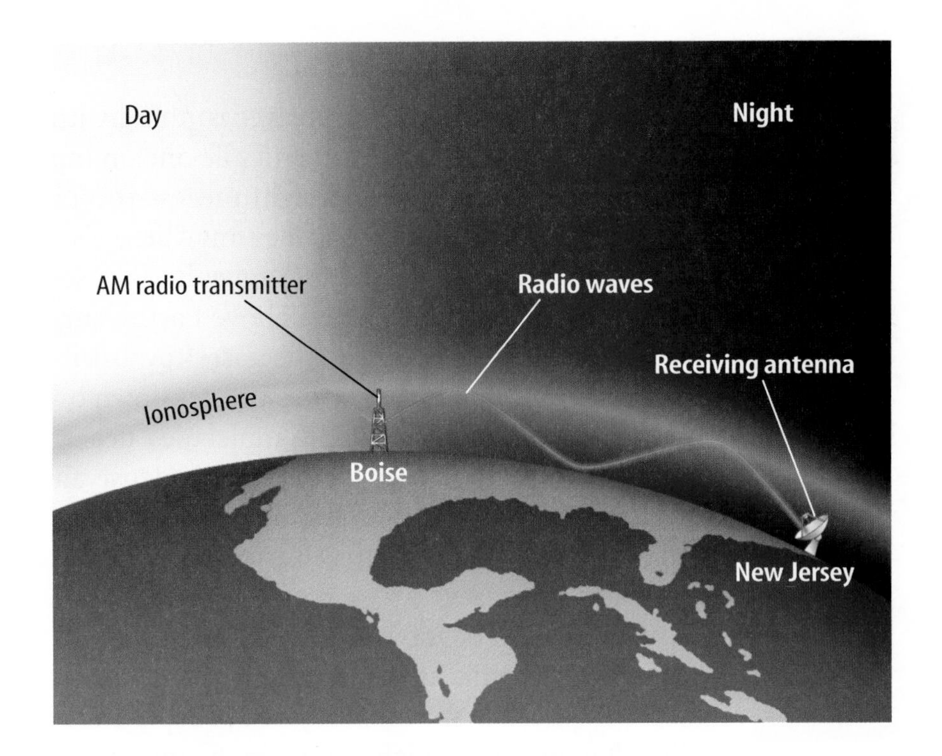

Figure 5 During the day, the ionosphere absorbs radio transmissions. This prevents you from hearing distant radio stations. At night, the ionosphere reflects radio waves. The reflected waves can travel to distant cities.
Describe *what causes the ionosphere to change between day and night.*

Upper Layers of the Atmosphere Beyond the stratosphere are the mesosphere, thermosphere, and exosphere. The mesosphere extends from the top of the stratosphere to about 85 km above Earth. If you've ever seen a shooting star, you might have witnessed a meteor in the mesosphere.

The thermosphere is named for its high temperatures. This is the thickest atmospheric layer and is found between 85 km and 500 km above Earth's surface.

Within the mesosphere and thermosphere is a layer of electrically charged particles called the **ionosphere** (i AH nuh sfihr). If you live in New Jersey and listen to the radio at night, you might pick up a station from Boise, Idaho. The ionosphere allows radio waves to travel across the country to another city, as shown in **Figure 5.** During the day, energy from the Sun interacts with the particles in the ionosphere, causing them to absorb AM radio frequencies. At night, without solar energy, AM radio transmissions reflect off the ionosphere, allowing radio transmissions to be received at greater distances.

The space shuttle in **Figure 6** orbits Earth in the exosphere. In contrast to the troposphere, the layer you live in, the exosphere has so few molecules that the wings of the shuttle are useless. In the exosphere, the spacecraft relies on bursts from small rocket thrusters to move around. Beyond the exosphere is outer space.

Reading Check *How does the space shuttle maneuver in the exosphere?*

Figure 6 Wings help move aircraft in lower layers of the atmosphere. The space shuttle can't use its wings to maneuver in the exosphere because so few molecules are present.

Figure 7 Air pressure decreases as you go higher in Earth's atmosphere.

Atmospheric Pressure

Imagine you're a football player running with the ball. Six players tackle you and pile one on top of the other. Who feels the weight more—you or the player on top? Like molecules anywhere else, atmospheric gases have mass. Atmospheric gases extend hundreds of kilometers above Earth's surface. As Earth's gravity pulls the gases toward its surface, the weight of these gases presses down on the air below. As a result, the molecules nearer Earth's surface are closer together. This dense air exerts more force than the less dense air near the top of the atmosphere. Force exerted on an area is known as pressure.

Like the pile of football players, air pressure is greater near Earth's surface and decreases higher in the atmosphere, as shown in **Figure 7.** People find it difficult to breathe in high mountains because fewer molecules of air exist there. Jets that fly in the stratosphere must maintain pressurized cabins so that people can breathe.

Where is air pressure greater—in the exosphere or in the troposphere?

Applying Science

How does altitude affect air pressure?

Atmospheric gases extend hundreds of kilometers above Earth's surface, but the molecules that make up these gases are fewer and fewer in number as you go higher. This means that air pressure decreases with altitude.

Identifying the Problem

The graph on the right shows these changes in air pressure. Note that altitude on the graph goes up only to 50 km. The troposphere and the stratosphere are represented on the graph, but other layers of the atmosphere are not. By examining the graph, can you understand the relationship between altitude and pressure?

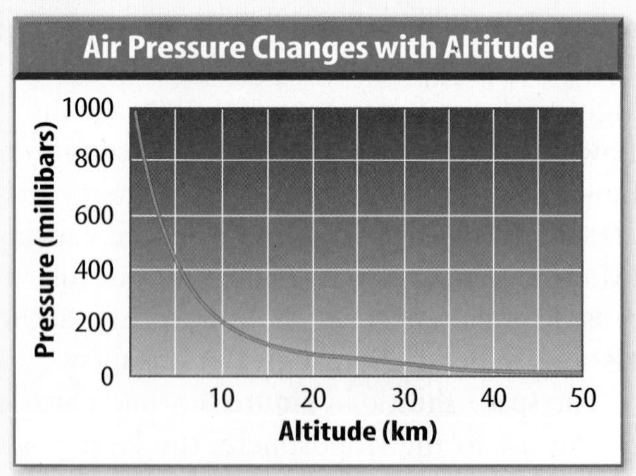

Solving the Problem

1. Estimate the air pressure at an altitude of 5 km.
2. Does air pressure change more quickly at higher altitudes or at lower altitudes?

Temperature in Atmospheric Layers

The Sun is the source of most of the energy on Earth. Before it reaches Earth's surface, energy from the Sun must pass through the atmosphere. Because some layers contain gases that easily absorb the Sun's energy while other layers do not, the various layers have different temperatures, illustrated by the red line in **Figure 8.**

Molecules that make up air in the troposphere are warmed mostly by heat from Earth's surface. The Sun warms Earth's surface, which then warms the air above it. When you climb a mountain, the air at the top is usually cooler than the air at the bottom. Every kilometer you climb, the air temperature decreases about 6.5°C.

Molecules of ozone in the stratosphere absorb some of the Sun's energy. Energy absorbed by ozone molecules raises the temperature. Because more ozone molecules are in the upper portion of the stratosphere, the temperature in this layer rises with increasing altitude.

Like the troposphere, the temperature in the mesosphere decreases with altitude. The thermosphere and exosphere are the first layers to receive the Sun's rays. Few molecules are in these layers, but each molecule has a great deal of energy. Temperatures here are high.

Determining if Air Has Mass

Procedure

1. On a **pan balance,** find the mass of an **inflatable ball** that is completely deflated.
2. Hypothesize about the change in the mass of the ball when it is inflated.
3. Inflate the ball to its maximum recommended inflation pressure.
4. Determine the mass of the fully inflated ball.

Analysis

1. What change occurs in the mass of the ball when it is inflated?
2. Infer from your data whether air has mass.

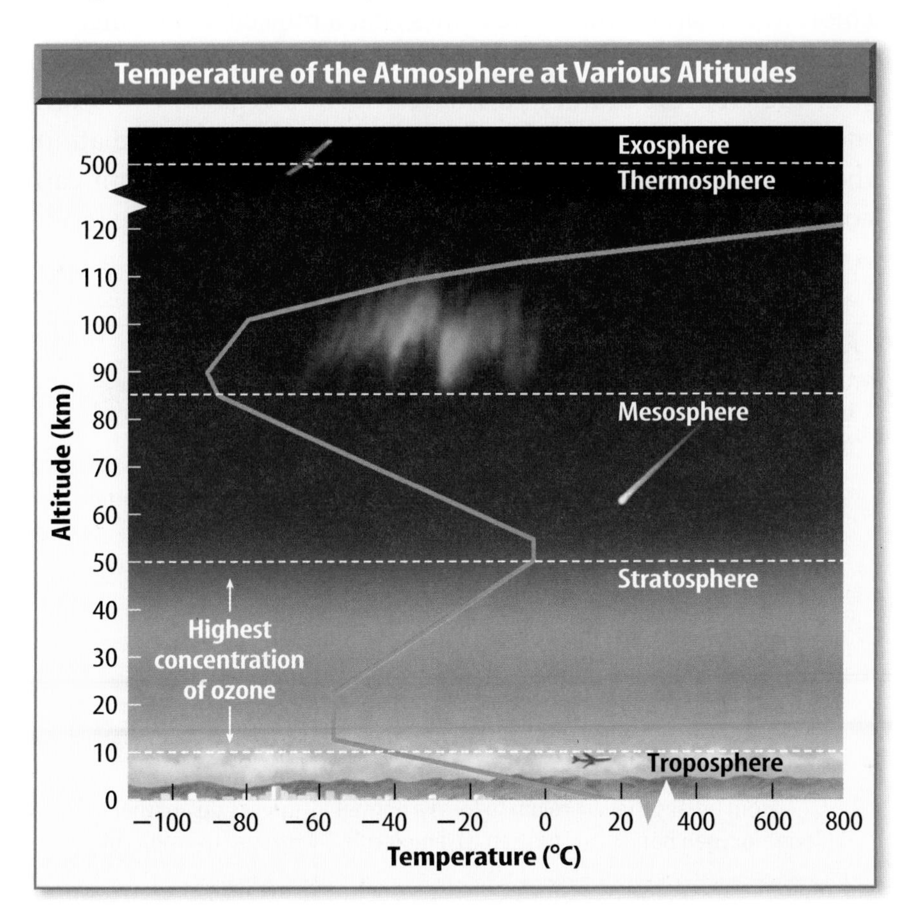

Figure 8 The division of the atmosphere into layers is based mainly on differences in temperature.
Determine *Does the temperature increase or decrease with altitude in the mesosphere?*

INTEGRATE
Life Science

Effects of UV Light on Algae Algae are organisms that use sunlight to make their own food. This process releases oxygen to Earth's atmosphere. Some scientists suggest that growth is reduced when algae are exposed to ultraviolet radiation. Infer what might happen to the oxygen level of the atmosphere if increased ultraviolet radiation damages some algae.

The Ozone Layer

Within the stratosphere, about 19 km to 48 km above your head, lies an atmospheric layer called the **ozone layer.** Ozone is made of oxygen. Although you cannot see the ozone layer, your life depends on it.

The oxygen you breathe has two atoms per molecule, but an ozone molecule is made up of three oxygen atoms bound together. The ozone layer contains a high concentration of ozone and shields you from the Sun's harmful energy. Ozone absorbs most of the ultraviolet radiation that enters the atmosphere. **Ultraviolet radiation** is one of the many types of energy that come to Earth from the Sun. Too much exposure to ultraviolet radiation can damage your skin and cause cancer.

CFCs Evidence exists that some air pollutants are destroying the ozone layer. Blame has fallen on **chlorofluorocarbons** (CFCs), chemical compounds used in some refrigerators, air conditioners, and aerosol sprays, and in the production of some foam packaging. CFCs can enter the atmosphere if these appliances leak or if they and other products containing CFCs are improperly discarded.

Recall that an ozone molecule is made of three oxygen atoms bonded together. Chlorofluorocarbon molecules, shown in **Figure 9,** destroy ozone. When a chlorine atom from a chlorofluorocarbon molecule comes near a molecule of ozone, the ozone molecule breaks apart. One of the oxygen atoms combines with the chlorine atom, and the rest form a regular, two-atom molecule. These compounds don't absorb ultraviolet radiation the way ozone can. In addition, the original chlorine atom can continue to break apart thousands of ozone molecules. The result is that more ultraviolet radiation reaches Earth's surface.

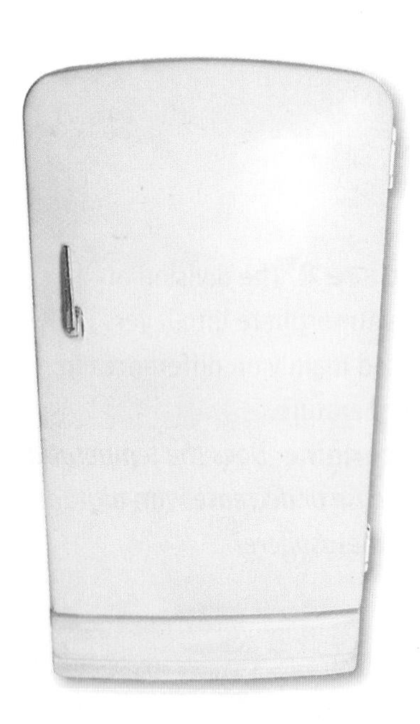

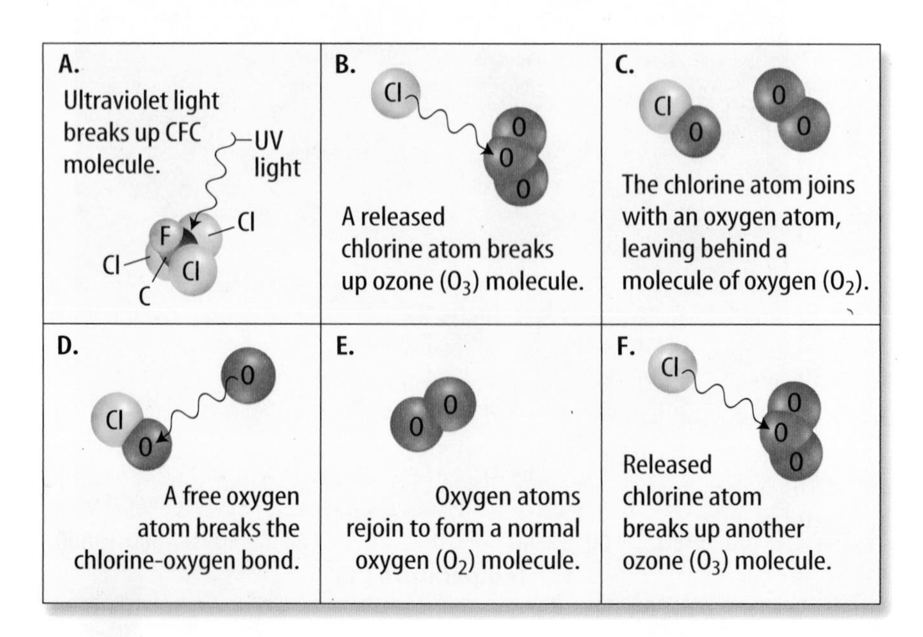
A.
Ultraviolet light breaks up CFC molecule.
UV light
F
Cl
Cl
Cl
C
B.
Cl
O
O
O
A released chlorine atom breaks up ozone (O_3) molecule.
C.
Cl
O
O
O
The chlorine atom joins with an oxygen atom, leaving behind a molecule of oxygen (O_2).
D.
Cl
O
O
A free oxygen atom breaks the chlorine-oxygen bond.
E.
O
O
Oxygen atoms rejoin to form a normal oxygen (O_2) molecule.
F.
Cl
O
O
O
Released chlorine atom breaks up another ozone (O_3) molecule.

Figure 9 Chlorofluorocarbon (CFC) molecules once were used in refrigerators and air conditioners. Each CFC molecule has three chlorine atoms. One atom of chlorine can destroy approximately 100,000 ozone molecules.

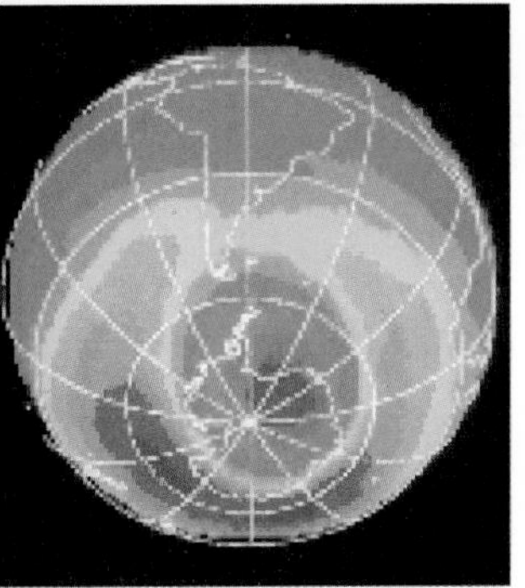

October 1980

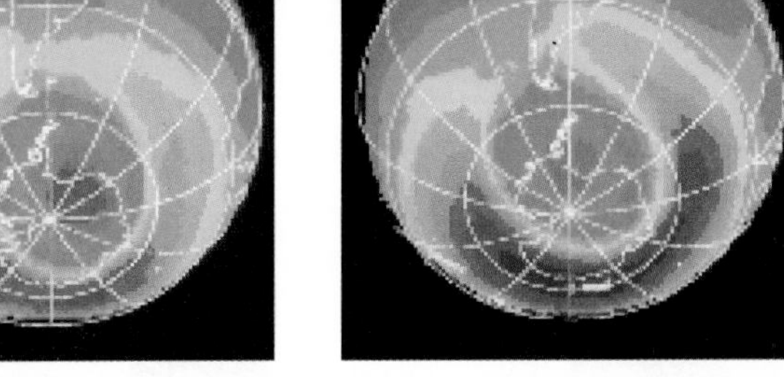

October 1988

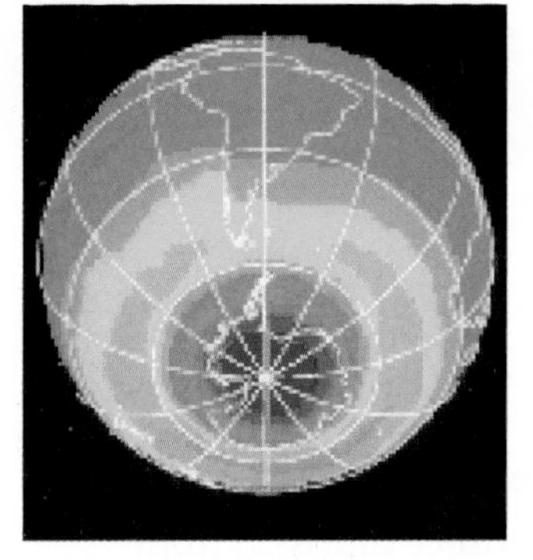

October 1990

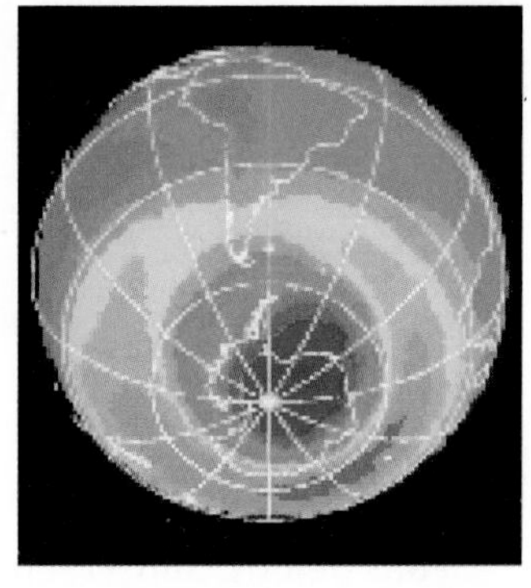

September 1999

INTEGRATE Health

The Ozone Hole The destruction of ozone molecules by CFCs seems to cause a seasonal reduction in ozone over Antarctica called the ozone hole. Every year beginning in late August or early September the amount of ozone in the atmosphere over Antarctica begins to decrease. By October, the ozone concentration reaches its lowest values and then begins to increase again. By December, the ozone hole disappears. **Figure 10** shows how the ozone hole over Antarctica has changed. In the mid-1990s, many governments banned the production and use of CFCs. Since then, the concentration of CFCs in the atmosphere has started to decrease.

Figure 10 These images of Antarctica were produced using data from a NASA satellite. The lowest values of ozone concentration are shown in dark blue and purple. These data show that the size of the seasonal ozone hole over Antarctica has grown larger over time.

section 1 review

Summary

Layers of the Atmosphere

- The atmosphere is a mixture of gases, solids, and liquids.
- The atmosphere has five layers—troposphere, stratosphere, mesosphere, thermosphere, and exosphere.
- The ionosphere is made up of electrically charged particles.

Atmospheric Pressure and Temperature

- Atmospheric pressure decreases with distance from Earth.
- Because some layers absorb the Sun's energy more easily than others, the various layers have different temperatures.

Ozone Layer

- The ozone layer absorbs most UV light.
- Chlorofluorocarbons (CFCs) break down the ozone layer.

Self Check

1. **Describe** How did oxygen come to make up 21 percent of Earth's present atmosphere?
2. **Infer** While hiking in the mountains, you notice that it is harder to breathe as you climb higher. Explain.
3. **State** some effects of a thinning ozone layer.
4. **Think Critically** Explain why, during the day, the radio only receives AM stations from a nearby city, while at night, you're able to hear a distant city's stations.

Applying Skills

5. **Interpret Scientific Illustrations** Using **Figure 2,** determine the total percentage of nitrogen and oxygen in the atmosphere. What is the total percentage of argon and carbon dioxide?
6. **Communicate** The names of the atmospheric layers end with the suffix *-sphere,* a word that means "ball." Find out what *tropo-, meso-, thermo-,* and *exo-* mean. Write their meanings in your Science Journal and explain if the layers are appropriately named.

Evaluating Sunscreens

Without protection, sun exposure can damage your health. Sunscreens protect your skin from UV radiation. In this lab, you will draw inferences using different sunscreen labels.

Real-World Question

How effective are various brands of sunscreens?

Goals

- **Draw inferences** based on labels on sunscreen brands.
- **Compare** the effectiveness of different sunscreen brands for protection against the Sun.
- **Compare** the cost of several sunscreen brands.

Materials

variety of sunscreens of different brand names

Safety Precautions

Procedure

1. Make a data table in your Science Journal using the following headings: *Brand Name, SPF, Cost per Milliliter,* and *Misleading Terms.*
2. The Sun Protection Factor (SPF) tells you how long the sunscreen will protect you. For example, an SPF of 4 allows you to stay in the Sun four times longer than if you did not use sunscreen. Record the SPF of each sunscreen on your data table.
3. **Calculate** the cost per milliliter of each sunscreen brand.
4. Government guidelines say that terms like *sunblock* and *waterproof* are misleading because sunscreens can't block the Sun's rays, and they do wash off in water. List misleading terms in your data table for each brand.

Sunscreen Assessment

Brand Name			
SPF			
Cost per Milliliter	Do not write in this book.		
Misleading Terms			

Conclude and Apply

1. **Explain** why you need to use sunscreen.
2. **Evaluate** A minimum of SPF 15 is considered adequate protection for a sunscreen. An SPF greater than 30 is considered by government guidelines to be misleading because sunscreens wash or wear off. Evaluate the SPF of each sunscreen brand.
3. **Discuss** Considering the cost and effectiveness of all the sunscreen brands, discuss which you consider to be the best buy.

Communicating Your Data

Create a poster on the proper use of sunscreens, and provide guidelines for selecting the safest product.

section 2

Energy Transfer in the Atmosphere

Energy from the Sun

The Sun provides most of Earth's energy. This energy drives winds and ocean currents and allows plants to grow and produce food, providing nutrition for many animals. When Earth receives energy from the Sun, three different things can happen to that energy, as shown in **Figure 11.** Some energy is reflected back into space by clouds, particles, and Earth's surface. Some is absorbed by the atmosphere or by land and water on Earth's surface.

Heat

Heat is energy that flows from an object with a higher temperature to an object with a lower temperature. Energy from the Sun reaches Earth's surface and heats it. Heat then is transferred through the atmosphere in three ways—radiation, conduction, and convection, as shown in **Figure 12.**

as you read

What **You'll Learn**

- **Describe** what happens to the energy Earth receives from the Sun.
- **Compare and contrast** radiation, conduction, and convection.
- **Explain** the water cycle and its effect on weather patterns and climate.

Why **It's Important**

The Sun provides energy to Earth's atmosphere, allowing life to exist.

Review Vocabulary

evaporation: when a liquid changes to a gas at a temperature below the liquid's boiling point

New Vocabulary

- radiation
- conduction
- convection
- hydrosphere
- condensation

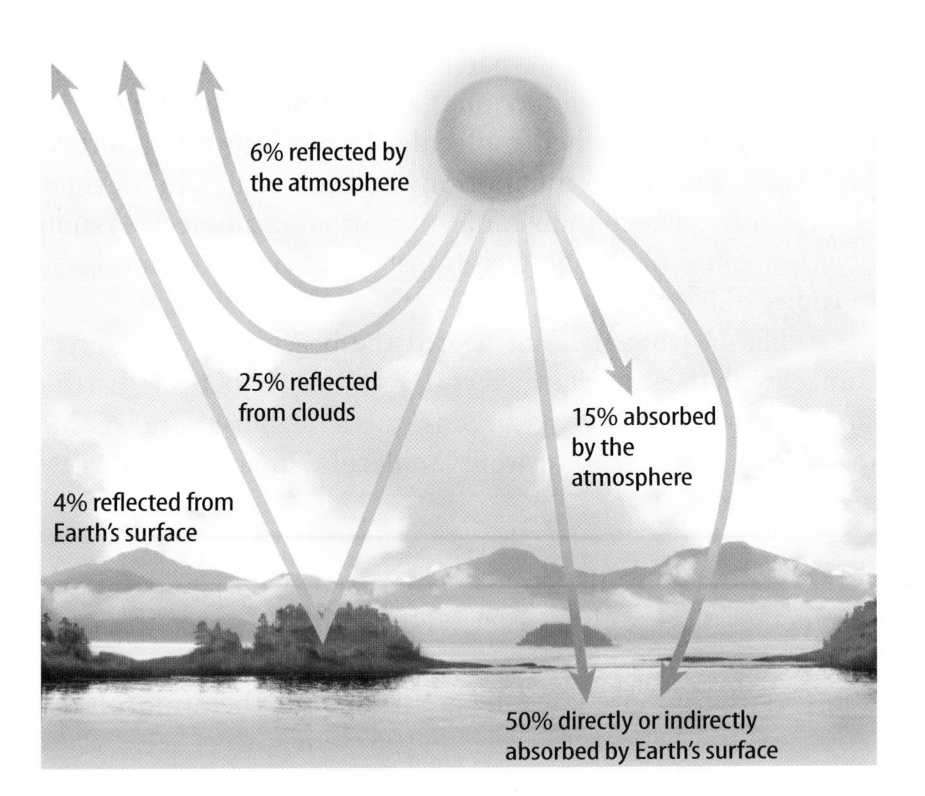

Figure 11 The Sun is the source of energy for Earth's atmosphere. Thirty-five percent of incoming solar radiation is reflected back into space.
Infer *how much is absorbed by Earth's surface and atmosphere.*

Radiation warms the surface.
The air near Earth's surface is heated by conduction.
Cooler air pushes warm air upward, creating a convection current.

Figure 12 Heat is transferred within Earth's atmosphere by radiation, conduction, and convection.

Radiation Sitting on the beach, you feel the Sun's warmth on your face. How can you feel the Sun's heat even though you aren't in direct contact with it? Energy from the Sun reaches Earth in the form of radiant energy, or radiation. **Radiation** is energy that is transferred in the form of rays or waves. Earth radiates some of the energy it absorbs from the Sun back toward space. Radiant energy from the Sun warms your face.

Reading Check *How does the Sun warm your skin?*

INTEGRATE
Physics

Specific Heat Specific heat is the amount of heat required to raise the temperature of one kilogram of a substance one degree Celsius. Substances with high specific heat absorb a lot of heat for a small increase in temperature. Land warms faster than water does. Infer whether soil or water has a higher specific heat value.

Conduction If you walk barefoot on a hot beach, your feet heat up because of conduction. **Conduction** is the transfer of energy that occurs when molecules bump into one another. Molecules are always in motion, but molecules in warmer objects move faster than molecules in cooler objects. When objects are in contact, energy is transferred from warmer objects to cooler objects.

Radiation from the Sun heated the beach sand, but direct contact with the sand warmed your feet. In a similar way, Earth's surface conducts energy directly to the atmosphere. As air moves over warm land or water, molecules in air are heated by direct contact.

Convection After the atmosphere is warmed by radiation or conduction, the heat is transferred by a third process called convection. **Convection** is the transfer of heat by the flow of material. Convection circulates heat throughout the atmosphere. How does this happen?

When air is warmed, the molecules in it move apart and the air becomes less dense. Air pressure decreases because fewer molecules are in the same space. In cold air, molecules move closer together. The air becomes more dense and air pressure increases. Cooler, denser air sinks while warmer, less dense air rises, forming a convection current. As **Figure 12** shows, radiation, conduction, and convection together distribute the Sun's heat throughout Earth's atmosphere.

The Water Cycle

Hydrosphere is a term that describes all the waters of Earth. The constant cycling of water within the atmosphere and the hydrosphere, as shown in **Figure 13,** plays an important role in determining weather patterns and climate types.

Energy from the Sun causes water to change from a liquid to a gas by a process called evaporation. Water that evaporates from lakes, streams, and oceans enters Earth's atmosphere. If water vapor in the atmosphere cools enough, it changes back into a liquid. This process of water vapor changing to a liquid is called **condensation.**

Clouds form when condensation occurs high in the atmosphere. Clouds are made up of tiny water droplets that can collide to form larger drops. As the drops grow, they fall to Earth as precipitation. This completes the water cycle within the hydrosphere. Classification of world climates is commonly based on annual and monthly averages of temperature and precipitation that are strongly affected by the water cycle.

Modeling Heat Transfer

Procedure

1. Cover the outside of an empty **soup can,** with **black construction paper.**
2. Fill the can with **cold water** and feel it with your fingers.
3. Place the can in sunlight for 1 h, then pour the water over your fingers.

Analysis

1. Does the water in the can feel warmer or cooler after placing the can in sunlight?
2. What types of heat transfer did you model?

Try at Home

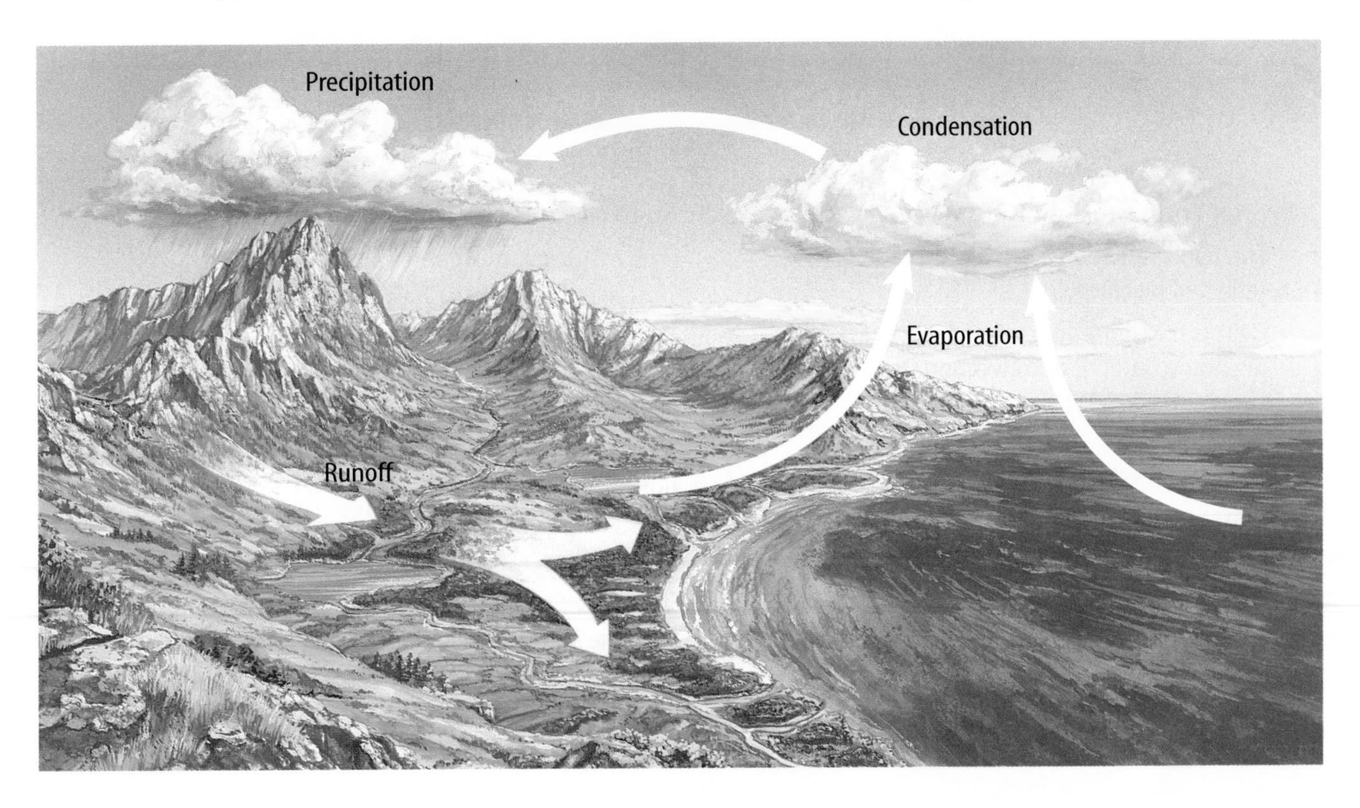

Figure 13 In the water cycle, water moves from Earth to the atmosphere and back to Earth again.

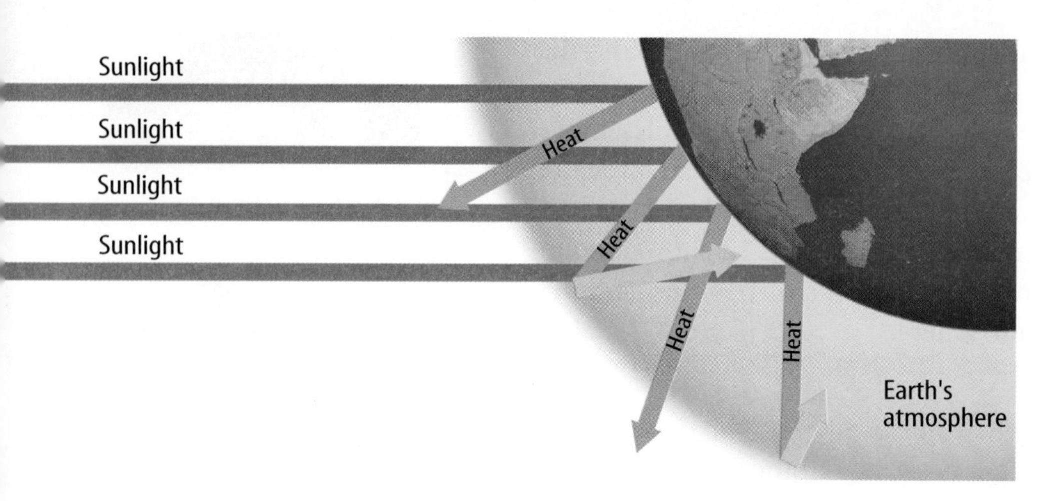

Figure 14 Earth's atmosphere creates a delicate balance between energy received and energy lost. **Infer** *What could happen if the balance is tipped toward receiving more energy than it does now?*

Earth's Atmosphere is Unique

On Earth, radiation from the Sun can be reflected into space, absorbed by the atmosphere, or absorbed by land and water. Once it is absorbed, heat can be transferred by radiation, conduction, or convection. Earth's atmosphere, shown in **Figure 14,** helps control how much of the Sun's radiation is absorbed or lost.

Reading Check ***What helps control how much of the Sun's radiation is absorbed on Earth?***

Why doesn't life exist on Mars or Venus? Mars is a cold, lifeless world because its atmosphere is too thin to support life or to hold much of the Sun's heat. Temperatures on the surface of Mars range from 35°C to −170°C. On the other hand, gases in Venus's dense atmosphere trap heat coming from the Sun. The temperature on the surface of Venus is 470°C. Living things would burn instantly if they were placed on Venus's surface. Life on Earth exists because the atmosphere holds just the right amount of the Sun's energy.

section 2 review

Summary

Energy From the Sun

- The Sun's radiation is either absorbed or reflected by Earth.
- Heat is transferred by radiation (waves), conduction (contact), or convection (flow).

The Water Cycle

- The water cycle affects climate.
- Water moves between the hydrosphere and the atmosphere through a continual process of evaporation and condensation.

Earth's Atmosphere is Unique

- Earth's atmosphere controls the amount of solar radiation that reaches Earth's surface.

Self Check

1. **State** how the Sun transfers energy to Earth.
2. **Contrast** the atmospheres of Earth and Mars.
3. **Describe** briefly the steps included in the water cycle.
4. **Explain** how the water cycle is related to weather patterns and climate.
5. **Think Critically** What would happen to temperatures on Earth if the Sun's heat were not distributed throughout the atmosphere?

Applying Math

6. **Solve One-Step Equations** Earth is about 150 million km from the Sun. The radiation coming from the Sun travels at 300,000 km/s. How long does it take for radiation from the Sun to reach Earth?

section 3

Air Movement

Forming Wind

Earth is mostly rock or land, with three-fourths of its surface covered by a relatively thin layer of water, the oceans. These two areas strongly influence global wind systems. Uneven heating of Earth's surface by the Sun causes some areas to be warmer than others. Recall that warmer air expands, becoming lower in density than the colder air. This causes air pressure to be generally lower where air is heated. Wind is the movement of air from an area of higher pressure to an area of lower pressure.

Heated Air Areas of Earth receive different amounts of radiation from the Sun because Earth is curved. **Figure 15** illustrates why the equator receives more radiation than areas to the north or south. The heated air at the equator is less dense, so it is displaced by denser, colder air, creating convection currents.

This cold, denser air comes from the poles, which receive less radiation from the Sun, making air at the poles much cooler. The resulting dense, high-pressure air sinks and moves along Earth's surface. However, dense air sinking as less-dense air rises does not explain everything about wind.

as you read

What You'll Learn

- **Explain** why different latitudes on Earth receive different amounts of solar energy.
- **Describe** the Coriolis effect.
- **Explain** how land and water surfaces affect the overlying air.

Why It's Important

Wind systems determine major weather patterns on Earth.

Review Vocabulary
density: mass per unit volume

New Vocabulary
- Coriolis effect
- jet stream
- sea breeze
- land breeze

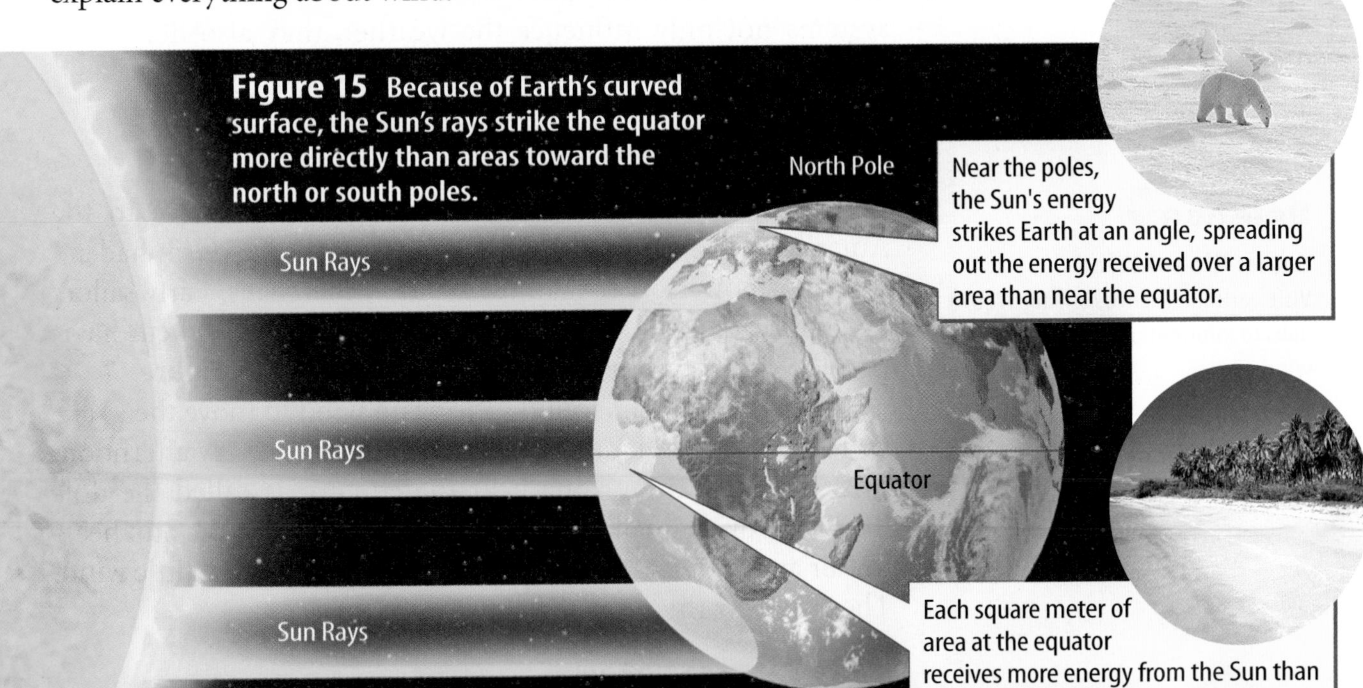
North Pole
Sun Rays
Sun Rays
Sun Rays
Equator
South Pole
Near the poles, the Sun's energy strikes Earth at an angle, spreading out the energy received over a larger area than near the equator.
Each square meter of area at the equator receives more energy from the Sun than each square meter at the poles does.

Figure 15 Because of Earth's curved surface, the Sun's rays strike the equator more directly than areas toward the north or south poles.

Figure 16 The Coriolis effect causes moving air to turn to the right in the northern hemisphere and to the left in the southern hemisphere.
Explain *What causes this to happen?*

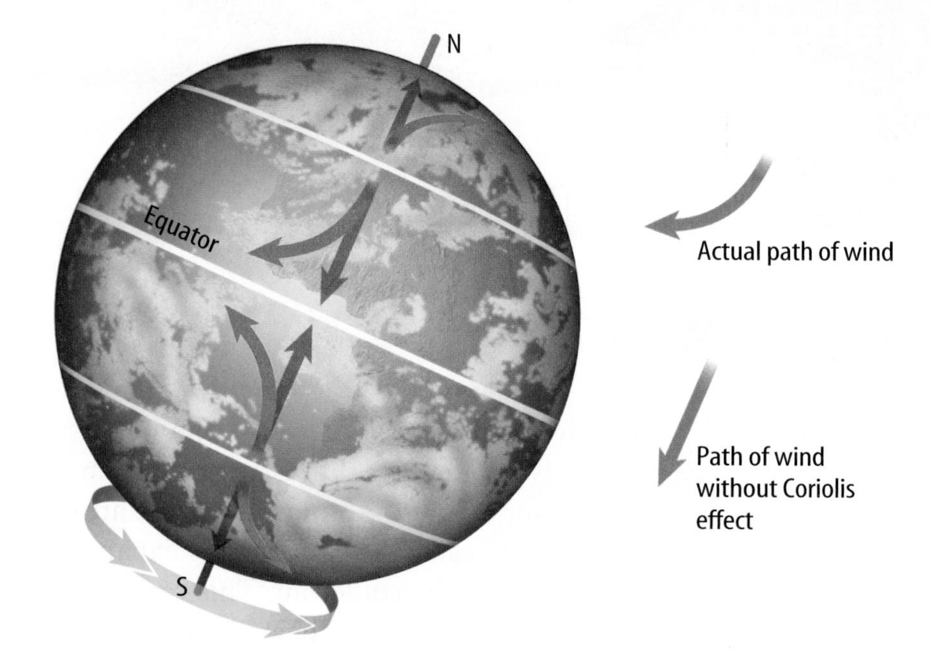

The Coriolis Effect What would happen if you threw a ball to someone sitting directly across from you on a moving merry-go-round? Would the ball go to your friend? By the time the ball got to the opposite side, your friend would have moved and the ball would appear to have curved.

Like the merry-go-round, the rotation of Earth causes moving air and water to appear to turn to the right north of the equator and to the left south of the equator. This is called the **Coriolis** (kohr ee OH lus) **effect.** It is illustrated in **Figure 16.** The flow of air caused by differences in the amount of solar radiation received on Earth's surface and by the Coriolis effect creates distinct wind patterns on Earth's surface. These wind systems not only influence the weather, they also determine when and where ships and planes travel most efficiently.

Topic: Global Winds
Visit earth.msscience.com for Web links to information about global winds.

Activity Make a model of Earth showing the locations of global wind patterns.

Global Winds

How did Christopher Columbus get from Spain to the Americas? The *Nina,* the *Pinta,* and the *Santa Maria* had no source of power other than the wind in their sails. Early sailors discovered that the wind patterns on Earth helped them navigate the oceans. These wind systems are shown in **Figure 17.**

Sometimes sailors found little or no wind to move their sailing ships near the equator. It also rained nearly every afternoon. This windless, rainy zone near the equator is called the doldrums. Look again at **Figure 17.** Near the equator, the Sun heats the air and causes it to rise, creating low pressure and little wind. The rising air then cools, causing rain.

Reading Check *What are the doldrums?*

NATIONAL GEOGRAPHIC VISUALIZING GLOBAL WINDS

Figure 17

The Sun's uneven heating of Earth's surface forms giant loops, or cells, of moving air. The Coriolis effect deflects the surface winds to the west or east, setting up belts of prevailing winds that distribute heat and moisture around the globe.

A WESTERLIES Near 30° north and south latitude, Earth's rotation deflects air from west to east as air moves toward the polar regions. In the United States, the westerlies move weather systems, such as this one along the Oklahoma-Texas border, from west to east.

B DOLDRUMS Along the equator, heating causes air to expand, creating a zone of low pressure. Cloudy, rainy weather, as shown here, develops almost every afternoon.

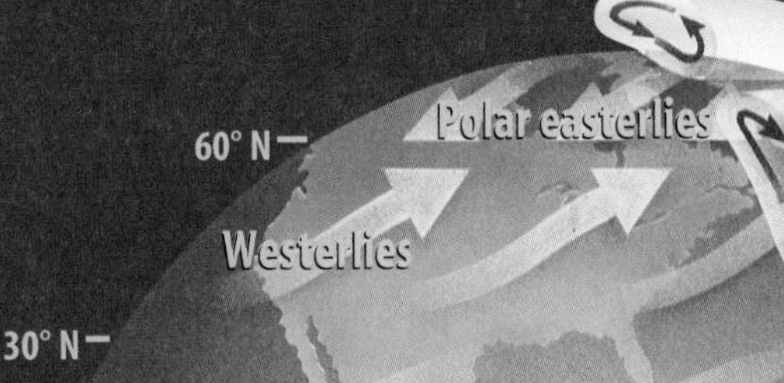

C TRADE WINDS Air warmed near the equator travels toward the poles but gradually cools and sinks. As the air flows back toward the low pressure of the doldrums, the Coriolis effect deflects the surface wind to the west. Early sailors, in ships like the one above, relied on these winds to navigate global trade routes.

D POLAR EASTERLIES In the polar regions, cold, dense air sinks and moves away from the poles. Earth's rotation deflects this wind from east to west.

Surface Winds Air descending to Earth's surface near 30° north and south latitude creates steady winds that blow in tropical regions. These are called trade winds because early sailors used their dependability to establish trade routes.

Between 30° and 60° latitude, winds called the prevailing westerlies blow in the opposite direction from the trade winds. Prevailing westerlies are responsible for much of the movement of weather across North America.

Polar easterlies are found near the poles. Near the north pole, easterlies blow from northeast to southwest. Near the south pole, polar easterlies blow from the southeast to the northwest.

Winds in the Upper Troposphere Narrow belts of strong winds, called **jet streams,** blow near the top of the troposphere. The polar jet stream forms at the boundary of cold, dry polar air to the north and warmer, more moist air to the south, as shown in **Figure 18.** The jet stream moves faster in the winter because the difference between cold air and warm air is greater. The jet stream helps move storms across the country.

Jet pilots take advantage of the jet streams. When flying eastward, planes save time and fuel. Going west, planes fly at different altitudes to avoid the jet streams.

Local Wind Systems

Global wind systems determine the major weather patterns for the entire planet. Smaller wind systems affect local weather. If you live near a large body of water, you're familiar with two such wind systems—sea breezes and land breezes.

Figure 18 The polar jet stream affecting North America forms along a boundary where colder air lies to the north and warmer air lies to the south. It is a swiftly flowing current of air that moves in a wavy west-to-east direction and is usually found between 10 km and 15 km above Earth's surface.

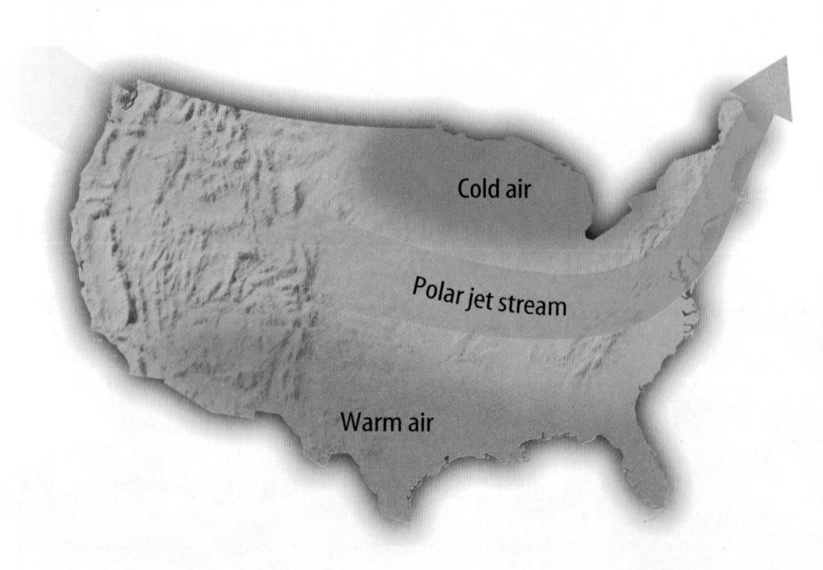

Flying from Boston to Seattle may take 30 min longer than flying from Seattle to Boston.

Think Critically *Why would it take longer to fly from east to west than it would from west to east?*

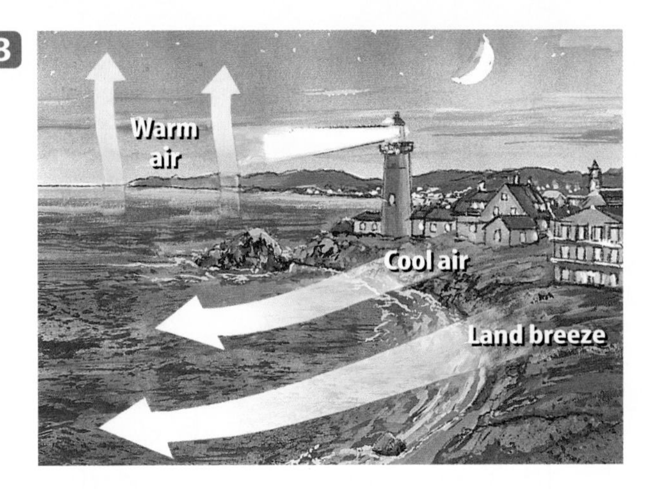

Sea and Land Breezes Convection currents over areas where the land meets the sea can cause wind. A **sea breeze,** shown in **Figure 19,** is created during the day because solar radiation warms the land more than the water. Air over the land is heated by conduction. This heated air is less dense and has lower pressure. Cooler, denser air over the water has higher pressure and flows toward the warmer, less dense air. A convection current results, and wind blows from the sea toward the land. The reverse occurs at night, when land cools much more rapidly than ocean water. Air over the land becomes cooler than air over the ocean. Cooler, denser air above the land moves over the water, as the warm air over the water rises. Movement of air toward the water from the land is called a **land breeze.**

Reading Check *How does a sea breeze form?*

Figure 19 These daily winds occur because land heats up and cools off faster than water does. **A** During the day, cool air from the water moves over the land, creating a sea breeze. **B** At night, cool air over the land moves toward the warmer air over the water, creating a land breeze.

section 3 review

Summary

Forming Wind

- Warm air is less dense than cool air.
- Differences in density and pressure cause air movement and wind.
- The Coriolis effect causes moving air to appear to turn right north of the equator and left south of the equator.

Wind Systems

- Wind patterns are affected by latitude.
- High-altitude belts of wind, called jet streams, can be found near the top of the troposphere.
- Sea breezes blow from large bodies of water toward land, while land breezes blow from land toward water.

Self Check

1. **Conclude** why some parts of Earth's surface, such as the equator, receive more of the Sun's heat than other regions.
2. **Explain** how the Coriolis effect influences winds.
3. **Analyze** why little wind and much afternoon rain occur in the doldrums.
4. **Infer** which wind system helped early sailors navigate Earth's oceans.
5. **Think Critically** How does the jet stream help move storms across North America?

Applying Skills

6. **Compare and contrast** sea breezes and land breezes.

Design Your Own

The Heat Is On

Goals

- **Design** an experiment to compare heat absorption and release for soil and water.
- **Observe** how heat release affects the air above soil and above water.

Possible Materials

ring stand
soil
metric ruler
water
masking tape
clear-plastic boxes (2)
overhead light
with reflector
thermometers (4)
colored pencils (4)

Safety Precautions

WARNING: *Be careful when handling the hot overhead light. Do not let the light or its cord make contact with water.*

Real-World Question

Sometimes, a plunge in a pool or lake on a hot summer day feels cool and refreshing. Why does the beach sand get so hot when the water remains cool? A few hours later, the water feels warmer than the land does. How do soil and water compare in their abilities to absorb and emit heat?

Form a Hypothesis

Form a hypothesis about how soil and water compare in their abilities to absorb and release heat. Write another hypothesis about how air temperatures above soil and above water differ during the day and night.

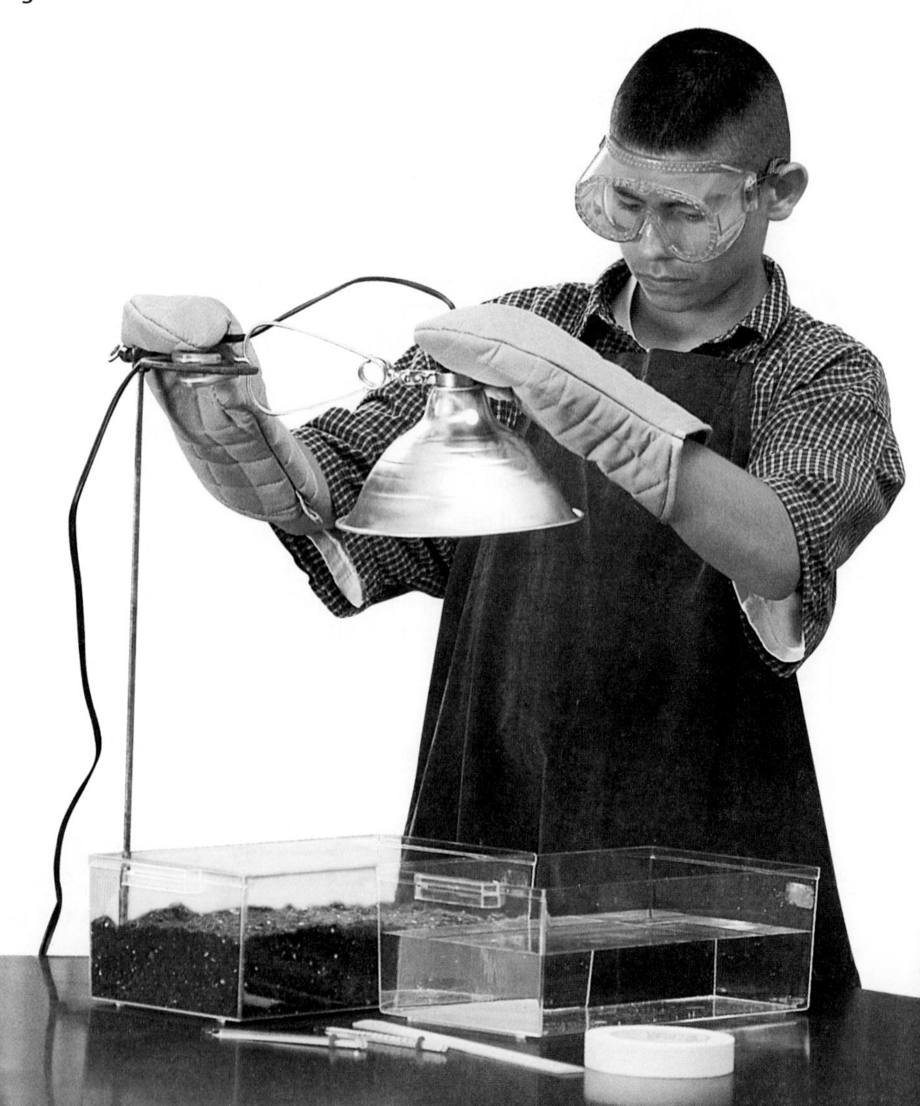

Test Your Hypothesis

Make a Plan

1. As a group, agree upon and write your hypothesis.
2. **List** the steps that you need to take to test your hypothesis. Include in your plan a description of how you will use your equipment to compare heat absorption and release for water and soil.
3. **Design** a data table in your Science Journal for both parts of your experiment—when the light is on and energy can be absorbed and when the light is off and energy is released to the environment.

Follow Your Plan

1. Make sure your teacher approves your plan and your data table before you start.
2. Carry out the experiment as planned.
3. During the experiment, record your observations and complete the data table in your Science Journal.
4. Include the temperatures of the soil and the water in your measurements. Also compare heat release for water and soil. Include the temperatures of the air immediately above both of the substances. Allow 15 min for each test.

Analyze Your Data

1. Use your colored pencils and the information in your data tables to make line graphs. Show the rate of temperature increase for soil and water. Graph the rate of temperature decrease for soil and water after you turn the light off.
2. **Analyze** your graphs. When the light was on, which heated up faster—the soil or the water?
3. **Compare** how fast the air temperature over the water changed with how fast the temperature over the land changed after the light was turned off.

Conclude and Apply

1. Were your hypotheses supported or not? Explain.
2. **Infer** from your graphs which cooled faster—the water or the soil.
3. **Compare** the temperatures of the air above the water and above the soil 15 minutes after the light was turned off. How do water and soil compare in their abilities to absorb and release heat?

Make a poster showing the steps you followed for your experiment. Include graphs of your data. Display your poster in the classroom.

Science and Language Arts

Song of the Sky Loom[1]

Brian Swann, ed.

This Native American prayer probably comes from the Tewa-speaking Pueblo village of San Juan, New Mexico. The poem is actually a chanted prayer used in ceremonial rituals.

Mother Earth Father Sky

we are your children
With tired backs we bring you gifts you love
Then weave for us a garment of brightness
its warp[2] the white light of morning,
weft[3] the red light of evening,
fringes the falling rain,
its border the standing rainbow.
Thus weave for us a garment of brightness
So we may walk fittingly where birds sing,
So we may walk fittingly where grass is green.

Mother Earth Father Sky

1 a machine or device from which cloth is produced

2 threads that run lengthwise in a piece of cloth

3 horizontal threads interlaced through the warp in a piece of cloth

Understanding Literature

Metaphor A metaphor is a figure of speech that compares seemingly unlike things. Unlike a simile, a metaphor does not use the connecting words *like* or *as*. Why does the song use the image of a garment to describe Earth's atmosphere?

Respond to Reading

1. What metaphor does the song use to describe Earth's atmosphere?
2. Why do the words *Mother Earth* and *Father Sky* appear on either side and above and below the rest of the words?
3. **Linking Science and Writing** Write a four-line poem that uses a metaphor to describe rain.

In this chapter, you learned about the composition of Earth's atmosphere. The atmosphere maintains the proper balance between the amount of heat absorbed from the Sun and the amount of heat that escapes back into space. The water cycle explains how water evaporates from Earth's surface back into the atmosphere. Using metaphor instead of scientific facts, the Tewa song conveys to the reader how the relationship between Earth and its atmosphere is important to all living things.

Reviewing Main Ideas

Section 1 Earth's Atmosphere

1. Earth's atmosphere is made up mostly of gases, with some suspended solids and liquids. The unique atmosphere allows life on Earth to exist.
2. The atmosphere is divided into five layers with different characteristics.
3. The ozone layer protects Earth from too much ultraviolet radiation, which can be harmful.
4. Unlike the atmosphere on Mars or Venus, Earth's unique atmosphere maintains a balance between energy received and energy lost that keeps temperatures mild. This delicate balance allows life on Earth to exist.

Section 2 Energy Transfer in the Atmosphere

1. Earth receives its energy from the Sun. Some of this energy is reflected back into space, and some is absorbed.
2. Heat is distributed in Earth's atmosphere by radiation, conduction, and convection.
3. Energy from the Sun powers the water cycle between the atmosphere and Earth's surface.

Section 3 Air Movement

1. Because Earth's surface is curved, not all areas receive the same amount of solar radiation. This uneven heating causes temperature differences at Earth's surface.
2. Convection currents modified by the Coriolis effect produce Earth's global winds.
3. The polar jet stream is a strong current of wind found in the upper troposphere. It forms at the boundary between cold, polar air and warm, tropical air.
4. Land breezes and sea breezes occur near the ocean.

Visualizing Main Ideas

Copy and complete the following cycle map on the water cycle.

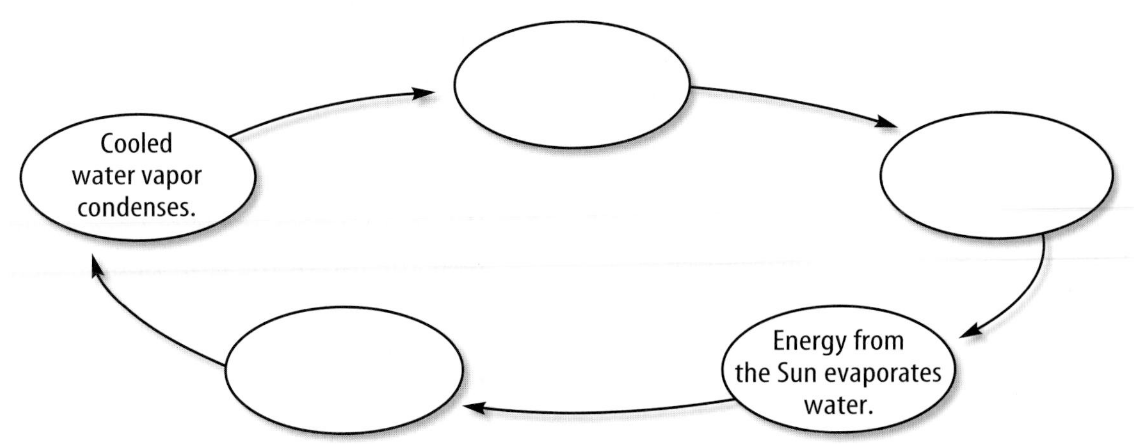

chapter 15 Review

Using Vocabulary

atmosphere p. 426	jet stream p. 442
chlorofluorocarbon p. 432	land breeze p. 443
condensation p. 437	ozone layer p. 432
conduction p. 436	radiation p. 436
convection p. 436	sea breeze p. 443
Coriolis effect p. 440	troposphere p. 428
hydrosphere p. 437	ultraviolet radiation p. 432
ionosphere p. 429	

Fill in the blanks below with the correct vocabulary word or words.

1. Chlorofluorocarbons are dangerous because they destroy the ________.

2. Narrow belts of strong winds called ________ blow near the top of the troposphere.

3. The thin layer of air that surrounds Earth is called the ________.

4. Heat energy transferred in the form of waves is called ________.

5. The ozone layer helps protect us from ________.

Checking Concepts

Choose the word or phrase that best answers the question.

6. Nitrogen makes up what percentage of the atmosphere?
A) 21% **C)** 78%
B) 1% **D)** 90%

7. What causes a brown haze near cities?
A) conduction
B) mud
C) car exhaust
D) wind

8. Which is the uppermost layer of the atmosphere?
A) troposphere **C)** exosphere
B) stratosphere **D)** thermosphere

9. What layer of the atmosphere has the most water?
A) troposphere **C)** mesosphere
B) stratosphere **D)** exosphere

10. What protects living things from too much ultraviolet radiation?
A) the ozone layer **C)** nitrogen
B) oxygen **D)** argon

11. Where is air pressure least?
A) troposphere **C)** exosphere
B) stratosphere **D)** thermosphere

12. How is energy transferred when objects are in contact?
A) trade winds **C)** radiation
B) convection **D)** conduction

13. Which surface winds are responsible for most of the weather movement across the United States?
A) polar easterlies
B) sea breeze
C) prevailing westerlies
D) trade winds

14. What type of wind is a movement of air toward water?
A) sea breeze
B) polar easterlies
C) land breeze
D) trade winds

15. What are narrow belts of strong winds near the top of the troposphere called?
A) doldrums
B) jet streams
C) polar easterlies
D) trade winds

Science Online earth.msscience.com/vocabulary_puzzlemaker

Thinking Critically

16. **Explain** why there are few or no clouds in the stratosphere.

17. **Describe** It is thought that life could not have existed on land until the ozone layer formed about 2 billion years ago. Why does life on land require an ozone layer?

18. **Diagram** Why do sea breezes occur during the day but not at night?

19. **Describe** what happens when water vapor rises and cools.

20. **Explain** why air pressure decreases with an increase in altitude.

21. **Concept Map** Copy and complete the cycle concept map below using the following phrases to explain how air moves to form a convection current: *Cool air moves toward warm air, warm air is lifted and cools, and cool air sinks.*

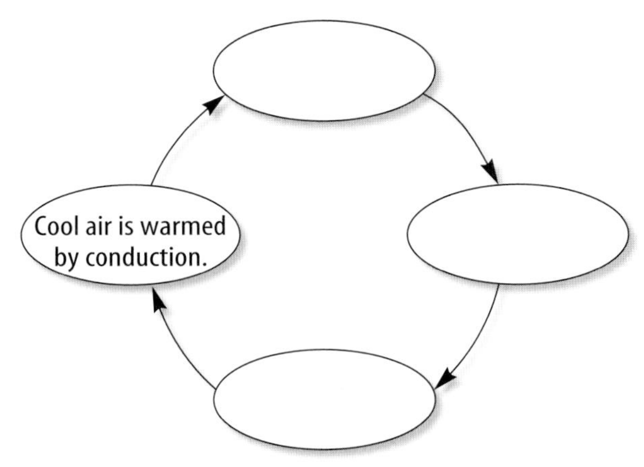

22. **Form Hypotheses** Carbon dioxide in the atmosphere prevents some radiation from Earth's surface from escaping to space. Hypothesize how the temperature on Earth might change if more carbon dioxide were released from burning fossil fuels.

23. **Identify and Manipulate Variables and Controls** Design an experiment to find out how plants are affected by differing amounts of ultraviolet radiation. In the design, use filtering film made for car windows. What is the variable you are testing? What are your constants? Your controls?

24. **Recognize Cause and Effect** Why is the inside of a car hotter than the outdoor temperature on a sunny summer day?

Performance Activities

25. **Make a Poster** Find newspaper and magazine photos that illustrate how the water cycle affects weather patterns and climate around the world.

26. **Experiment** Design and conduct an experiment to find out how different surfaces such as asphalt, soil, sand, and grass absorb and reflect solar energy. Share the results with your class.

Applying Math

Use the graph below to answer questions 27–28.

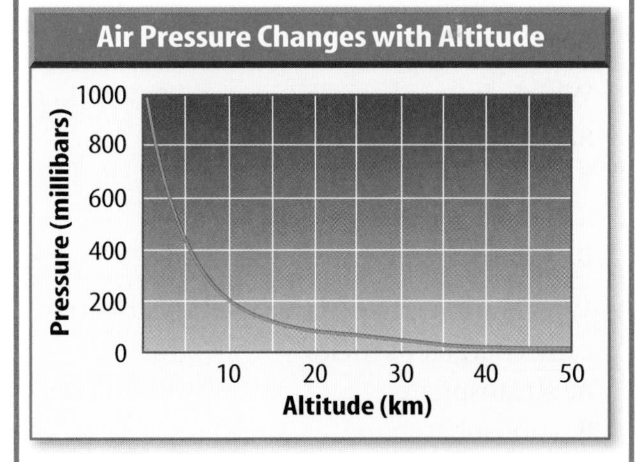

27. **Altitude and Air Pressure** What is the altitude at which air pressure is about 1,000 millibars? What is it at 200 millibars?

28. **Mt. Everest** Assume the altitude on Mt. Everest is about 10 km high. How many times greater is air pressure at sea level than on top of Mt. Everest?

Standardized Test Practice

Part 1 Multiple Choice

Record your answers on the answer sheet provided by your teacher or on a sheet of paper.

Use the illustration below to answer questions 1–3.

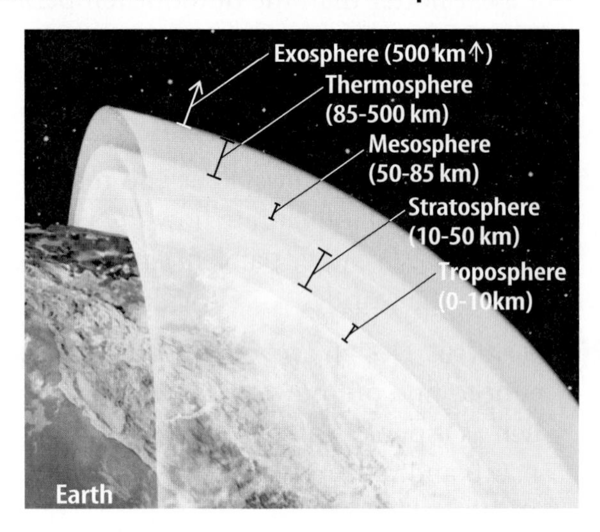

1. Which layer of the atmosphere contains the ozone layer?
A. exosphere
B. mesosphere
C. stratosphere
D. troposphere

2. Which atmospheric layer contains weather?
A. mesosphere
B. stratosphere
C. thermosphere
D. troposphere

3. Which atmospheric layer contains electrically charged particles?
A. stratosphere
B. ionosphere
C. exosphere
D. troposphere

4. What process changes water vapor to a liquid?
A. condensation
B. evaporation
C. infiltration
D. precipitation

5. Which process transfers heat by contact?
A. conduction
B. convection
C. evaporation
D. radiation

6. Which global wind affects weather in the U.S.?
A. doldrums
B. easterlies
C. trade winds
D. westerlies

Use the illustration below to answer question 7.

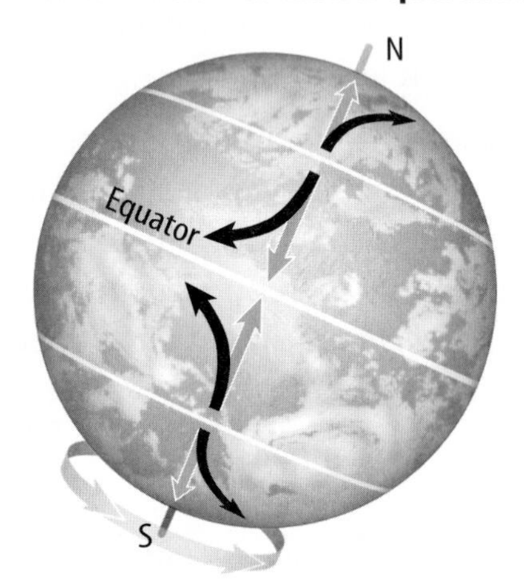

7. Which deflects winds to the west or east?
A. convection
B. Coriolis effect
C. jet stream
D. radiation

8. Which forms during the day because water heats slower than land?
A. easterlies
B. westerlies
C. land breeze
D. sea breeze

9. Which is the most abundant gas in Earth's atmosphere?
A. carbon dioxide
B. nitrogen
C. oxygen
D. water vapor

Part 2 Short Response/Grid In

Record your answers on the answer sheet provided by your teacher or on a sheet of paper.

10. Why does pressure drop as you travel upward from Earth's surface?

11. Why does the equator receive more radiation than areas to the north or south?

12. Why does a land breeze form at night?

13. Why does the jet stream move faster in the winter?

14. Why is one global wind pattern known as the trade winds?

Use the illustration below to answer questions 15–17.

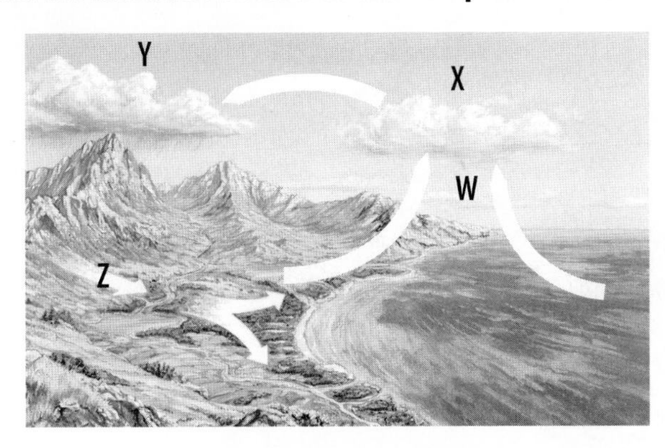

15. What process is illustrated?

16. Explain how this cycle affects weather patterns and climate.

17. What happens to water that falls as precipitation and does not runoff and flow into streams?

18. How do solid particles become part of Earth's atmosphere?

19. Why can flying from Seattle to Boston take less time than flying from Boston to Seattle in the same aircraft?

Part 3 Open Ended

Record your answers on a sheet of paper.

20. Explain how ozone is destroyed by chlorofluorocarbons.

21. Explain how Earth can heat the air by conduction.

22. Explain how humans influence the composition of Earth's atmosphere.

23. Draw three diagrams to demonstrate radiation, convection, and conduction.

24. Explain why the doldrums form over the equator.

Use the graph below to answer question 25.

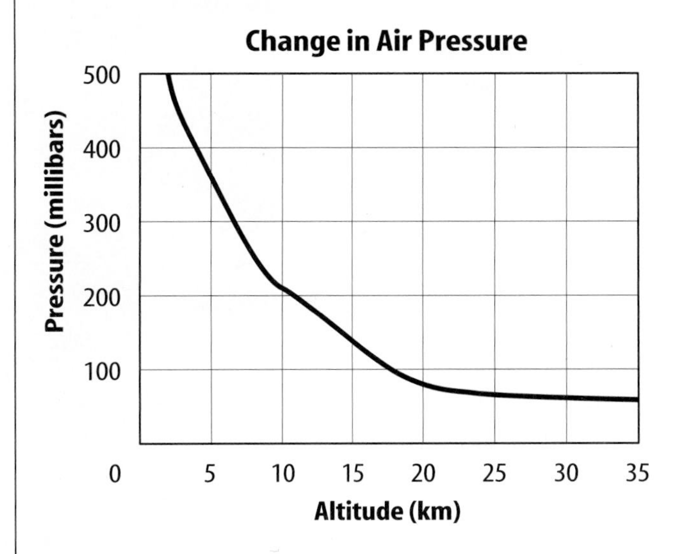

25. As you increase in altitude what happens to the air pressure? How might this affect people who move to the mountains?

Test-Taking Tip

Trends in Graphs When analyzing data in a table or graph, look for a trend. Questions about the pattern may use words like *increase, decrease, hypothesis,* or *summary.*

chapter 16

Weather

The BIG Idea

Weather refers to the state of the atmosphere at a specific time and place.

SECTION 1
What is weather?
Main Idea Weather describes conditions such as air pressure, wind, temperature, and the amount of moisture in the air.

SECTION 2
Weather Patterns
Main Idea Because of the movement of air and moisture in the atmosphere, weather constantly changes.

SECTION 3
Weather Forecasts
Main Idea Meteorologists gather information about current weather to make predictions about future weather patterns.

To play or not to play?

Will this approaching storm be over before the game begins? New weather technology can provide information that allows us to make plans based on predicted weather conditions, such as whether or not to delay the start of a baseball game.

Science Journal Write three questions you would ask a meteorologist about weather.

Start-Up Activities

What causes rain?

How can it rain one day and be sunny the next? Powered by heat from the Sun, the air that surrounds you stirs and swirls. This constant mixing produces storms, calm weather, and everything in between. What causes rain and where does the water come from? Do the lab below to find out. **WARNING:** *Boiling water and steam can cause burns.*

1. Bring a pan of water to a boil on a hot plate.
2. Carefully hold another pan containing ice cubes about 20 cm above the boiling water. Be sure to keep your hands and face away from the steam.
3. Keep the pan with the ice cubes in place until you see drops of water dripping from the bottom.
4. **Think Critically** In your Science Journal, describe how the droplets formed. Infer where the water on the bottom of the pan came from.

Weather When information is grouped into clear categories, it is easier to make sense of what you are learning. Make the following Foldable to help you organize your thoughts about weather.

STEP 1 **Collect** 2 sheets of paper and layer them about 1.25 cm apart vertically. Keep the edges level.

STEP 2 **Fold** up the bottom edges of the paper to form 4 equal tabs.

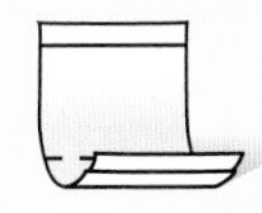

STEP 3 **Fold** the papers and crease well to hold the tabs in place. Staple along the fold.

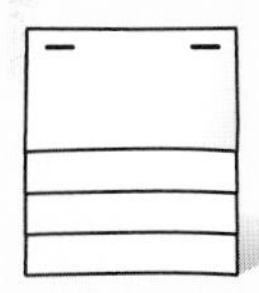

STEP 4 Label the tabs *Weather, What is weather?, Weather Patterns,* and *Forecasting Weather* as shown.

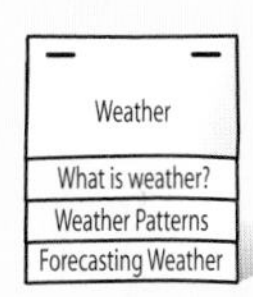

Summarize As you read the chapter, summarize what you learn under the appropriate tabs.

Preview this chapter's content and activities at earth.msscience.com

Get Ready to Read

New Vocabulary

1 Learn It! What should you do if you find a word you don't know or understand? Here are some suggested strategies:

1. Use context clues (from the sentence or the paragraph) to help you define it.
2. Look for prefixes, suffixes, or root words that you already know.
3. Write it down and ask for help with the meaning.
4. Guess at its meaning.
5. Look it up in the glossary or a dictionary.

2 Practice It! Look at the phrase *air mass* in the following passage. See how context clues can help you understand its meaning.

. . . an air mass that develops over land is dry compared with one that develops over water. An air mass that develops in the tropics is warmer than one that develops over northern regions. An air mass can cover thousands of square kilometers. When you observe a change in the weather from one day to the next, it is due to the movement of air masses.

—*from page 462*

Context Clue
Air masses can vary in the amount of moisture content and can vary in temperature.

Context Clue
Air masses can be very large.

Context Clue
Air masses can move.

3 Apply It! Make a vocabulary bookmark with a strip of paper. As you read, keep track of words you do not know or want to learn more about.

Target Your Reading

Use this to focus on the main ideas as you read the chapter.

1. **Before you read** the chapter, respond to the statements below on your worksheet or on a numbered sheet of paper.
   - Write an **A** if you **agree** with the statement.
   - Write a **D** if you **disagree** with the statement.

2. **After you read** the chapter, look back to this page to see if you've changed your mind about any of the statements.
   - If any of your answers changed, explain why.
   - Change any false statements into true statements.
   - Use your revised statements as a study guide.

Reading Tip

Read a paragraph containing a vocabulary word from beginning to end. Then, go back to determine the meaning of the word.

Before You Read A or D		Statement	After You Read A or D
	1	Heat from the Sun is absorbed by Earth's surface, which then heats the air above it.	
	2	Clouds are made of frozen air.	
	3	Dew point is a specific temperature and is not related to the amount of moisture in the air.	
	4	Clouds are classified by shape and height.	
	5	Cold fronts occur only in northern climates, and warm fronts occur only in southern climates.	
	6	Lightning can occur within a cloud, between clouds, or between a cloud and the ground.	
	7	Thunder is caused by clouds crashing into each other.	
	8	Hurricanes gain strength from heat and moisture of warm ocean water.	
	9	A meteorologist studies weather.	
	10	An isobar is a cold, icy, iron bar.	

Science Online
Print out a worksheet of this page at earth.msscience.com

section 1

What is weather?

as you read

What You'll Learn

- **Explain** how solar heating and water vapor in the atmosphere affect weather.
- **Discuss** how clouds form and how they are classified.
- **Describe** how rain, hail, sleet, and snow develop.

Why It's Important

Weather changes affect your daily activities.

Review Vocabulary
factor: something that influences a result

New Vocabulary
- weather
- humidity
- relative humidity
- dew point
- fog
- precipitation

Weather Factors

It might seem like small talk to you, but for farmers, truck drivers, pilots, and construction workers, the weather can have a huge impact on their livelihoods. Even professional athletes, especially golfers, follow weather patterns closely. You can describe what happens in different kinds of weather, but can you explain how it happens?

Weather refers to the state of the atmosphere at a specific time and place. Weather describes conditions such as air pressure, wind, temperature, and the amount of moisture in the air.

The Sun provides almost all of Earth's energy. Energy from the Sun evaporates water into the atmosphere where it forms clouds. Eventually, the water falls back to Earth as rain or snow. However, the Sun does more than evaporate water. It is also a source of heat energy. Heat from the Sun is absorbed by Earth's surface, which then heats the air above it. Differences in Earth's surface lead to uneven heating of Earth's atmosphere. Heat is eventually redistributed by air and water currents. Weather, as shown in **Figure 1,** is the result of heat and Earth's air and water.

Figure 1 The Sun provides the energy that drives Earth's weather.
Identify *storms in this image.*

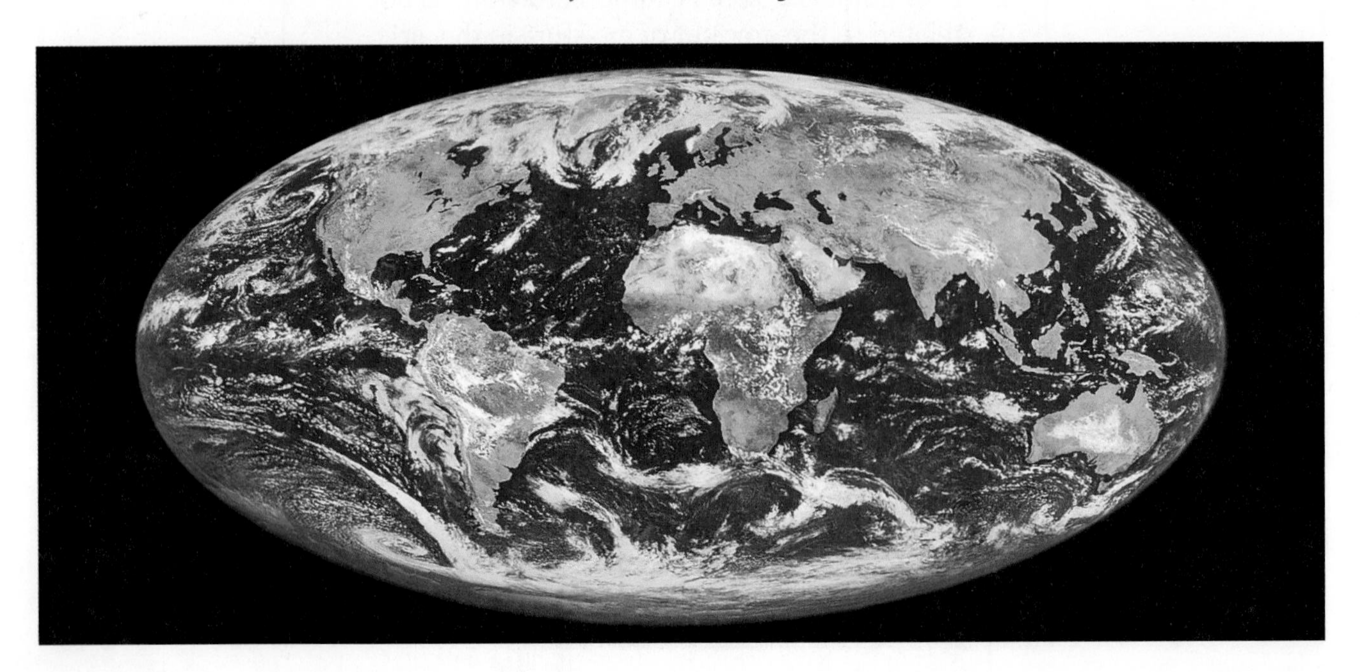

When air is heated, it expands and becomes less dense. This creates lower pressure.

Molecules making up air are closer together in cooler temperatures, creating high pressure. Wind blows from higher pressure toward lower pressure.

Figure 2 The temperature of air can affect air pressure. Wind is air moving from high pressure to low pressure.

Infer *In the above picture, which way would the wind move at night if the land cooled?*

Air Temperature During the summer when the Sun is hot and the air is still, a swim can be refreshing. But would a swim seem refreshing on a cold, winter day? The temperature of air influences your daily activities.

Air is made up of molecules that are always moving randomly, even when there's no wind. Temperature is a measure of the average amount of motion of molecules. When the temperature is high, molecules in air move rapidly and it feels warm. When the temperature is low, molecules in air move less rapidly, and it feels cold.

Wind Why can you fly a kite on some days but not others? Kites fly because air is moving. Air moving in a specific direction is called wind. As the Sun warms Earth's surface, air near the surface is heated by conduction. The air expands, becomes less dense, and rises. Warm, rising air has low atmospheric pressure. Cool, dense air tends to sink, bringing about high atmospheric pressure. Wind results because air moves from areas of high pressure to areas of low pressure. You may have experienced this if you've ever spent time along a beach, as in **Figure 2.**

Many instruments are used to measure wind direction and speed. Wind direction can be measured using a wind vane. A wind vane has an arrow that points in the direction from which the wind is blowing. A wind sock has one open end that catches the wind, causing the sock to point in the direction toward which the wind is blowing. Wind speed can be measured using an anemometer (a nuh MAH muh tur). Anemometers have rotating cups that spin faster when the wind is strong.

Body Temperature Birds and mammals maintain a fairly constant internal temperature, even when the temperature outside their bodies changes. On the other hand, the internal temperature of fish and reptiles changes when the temperature around them changes. Infer from this which group is more likely to survive a quick change in the weather.

Figure 3 Warmer air can have more water vapor than cooler air can because water vapor doesn't easily condense in warm air.

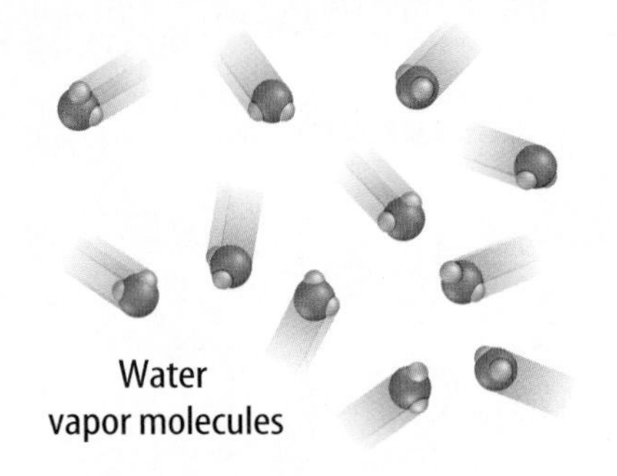

Water vapor molecules in warm air move rapidly. The molecules can't easily come together and condense.

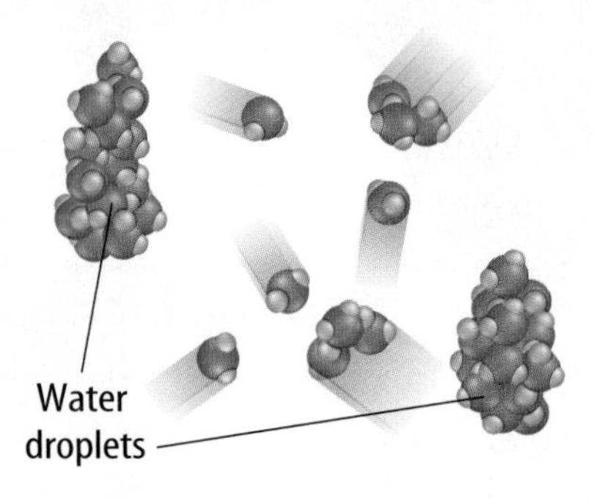

As air cools, water molecules in air move closer together. Some of them collide, allowing condensation to take place.

Determining Dew Point

Procedure

1. Partially fill a **metal can** with **room-temperature water**. Dry the outer surface of the can.
2. Place a **stirring rod** in the water.
3. Slowly stir the water and add small amounts of **ice.**
4. Make a data table in your **Science Journal.** With a **thermometer,** note the exact water temperature at which a thin film of moisture first begins to form on the outside of the metal can.
5. Repeat steps 1 through 4 two more times.
6. The average of the three temperatures at which the moisture begins to appear is the dew point temperature of the air surrounding the metal container.

Analysis

1. What determines the dew point temperature?
2. Will the dew point change with increasing temperature if the amount of moisture in the air doesn't change? Explain.

Humidity Heat evaporates water into the atmosphere. Where does the water go? Water vapor molecules fit into spaces among the molecules that make up air. The amount of water vapor present in the air is called **humidity.**

Air doesn't always contain the same amount of water vapor. As you can see in **Figure 3,** more water vapor can be present when the air is warm than when it is cool. At warmer temperatures, the molecules of water vapor in air move quickly and don't easily come together. At cooler temperatures, molecules in air move more slowly. The slower movement allows water vapor molecules to stick together and form droplets of liquid water. The formation of liquid water from water vapor is called condensation. When enough water vapor is present in air for condensation to take place, the air is saturated.

Why can more water vapor be present in warm air than in cold air?

Relative Humidity On a hot, sticky afternoon, the weather forecaster reports that the humidity is 50 percent. How can the humidity be low when it feels so humid? Weather forecasters report the amount of moisture in the air as relative humidity. **Relative humidity** is a measure of the amount of water vapor present in the air compared to the amount needed for saturation at a specific temperature.

If you hear a weather forecaster say that the relative humidity is 50 percent, it means that the air contains 50 percent of the water needed for the air to be saturated.

As shown in **Figure 4,** air at 25°C is saturated when it contains 22 g of water vapor per cubic meter of air. The relative humidity is 100 percent. If air at 25°C contains 11 g of water vapor per cubic meter, the relative humidity is 50 percent.

Dew Point

When the temperature drops, less water vapor can be present in air. The water vapor in air will condense to a liquid or form ice crystals. The temperature at which air is saturated and condensation forms is the dew point. The **dew point** changes with the amount of water vapor in the air.

You've probably seen water droplets form on the outside of a glass of cold milk. The cold glass cooled the air next to it to its dew point. The water vapor in the surrounding air condensed and formed water droplets on the glass. In a similar way, when air near the ground cools to its dew point, water vapor condenses and forms dew. Frost may form when temperatures are near 0°C.

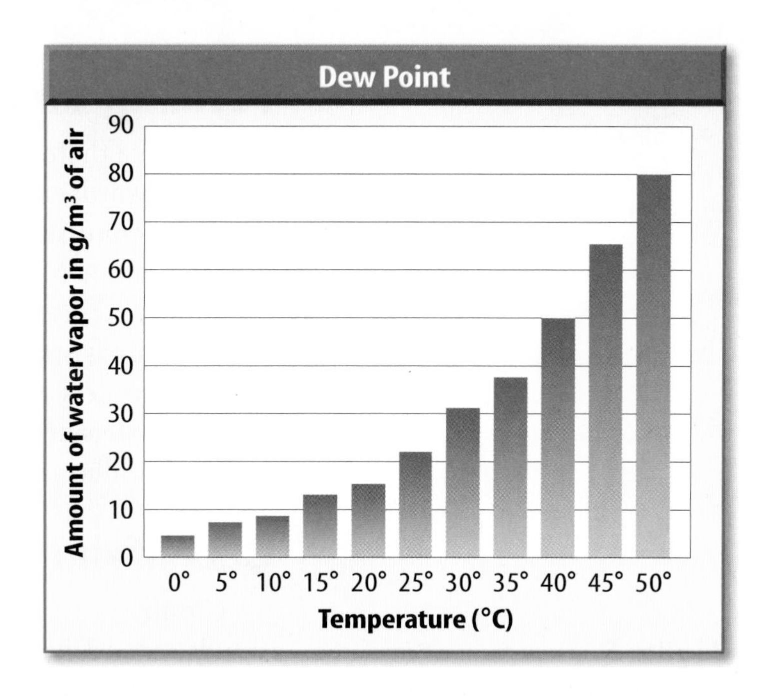

Figure 4 This graph shows that as the temperature of air increases, more water vapor can be present in the air.

Applying Math — Calculate Percent

DEW POINT One summer day, the relative humidity is 80 percent and the temperature is 35°C. Use **Figure 4** to find the dew point reached if the temperature falls to 25°C?

Solution

1 *This is what you know:*

Air Temperature (°C)	Amount of Water Vapor Needed for Saturation (g/m^3)
35	37
25	22
15	14

2 *This is what you need to find out:*

x = amount of water vapor in 35°C air at 80 percent relative humidity. Is $x > 22$ g/m^3 or is $x < 22$ g/m^3?

3 *This is how you solve the problem:*

$x = .80\ (37\ \text{g/m}^3)$

$x = 29.6\ \text{g/m}^3$ of water vapor

$29.6\ \text{g/m}^3 > 22\ \text{g/m}^3$, so the dew point is reached and dew will form.

Practice Problems

1. If the relative humidity is 50 percent and the air temperature is 35°C, will the dew point be reached if the temperature falls to 20°C?
2. If the air temperature is 25°C and the relative humidity is 30 percent, will the dew point be reached if the temperature drops to 15°C?

Science Online — For more practice, visit earth.msscience.com/math_practice

Forming Clouds

Why are there clouds in the sky? Clouds form as warm air is forced upward, expands, and cools. **Figure 5** shows several ways that warm, moist air forms clouds. As the air cools, the amount of water vapor needed for saturation decreases and the relative humidity increases. When the relative humidity reaches 100 percent, the air is saturated. Water vapor soon begins to condense in tiny droplets around small particles such as dust and salt. These droplets of water are so small that they remain suspended in the air. Billions of these droplets form a cloud.

Classifying Clouds

Clouds are classified mainly by shape and height. Some clouds extend high into the sky, and others are low and flat. Some dense clouds bring rain or snow, while thin, wispy clouds appear on mostly sunny days. The shape and height of clouds vary with temperature, pressure, and the amount of water vapor in the atmosphere.

Figure 5 Clouds form when moist air is lifted and cools. This occurs where air is heated, at mountain ranges, and where cold air meets warm air.

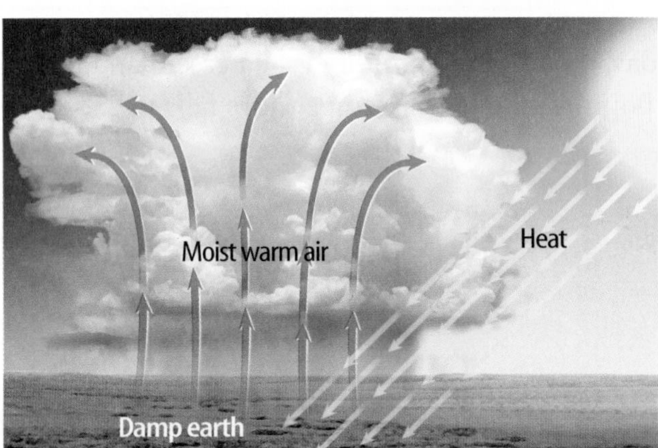

Rays from the Sun heat the ground and the air next to it. The warm air rises and cools. If the air is moist, some water vapor condenses and forms clouds.

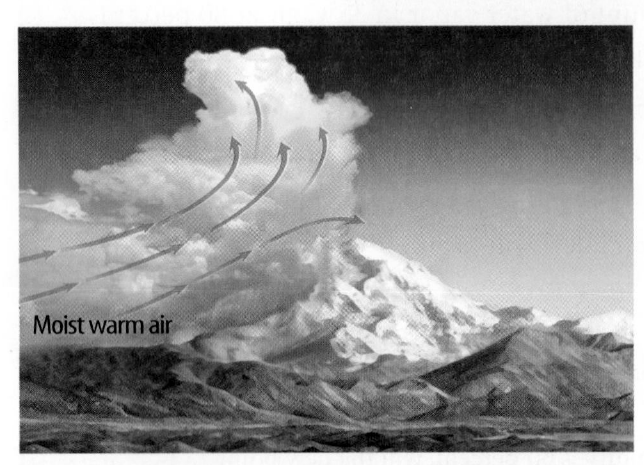

As moist air moves over mountains, it is lifted and cools. Clouds formed in this way can cover mountains for long periods of time.

When cool air meets warm, moist air, the warm air is lifted and cools.
Explain *what happens to the water vapor when the dew point is reached.*

Shape The three main cloud types are stratus, cumulus, and cirrus. Stratus clouds form layers, or smooth, even sheets in the sky. Stratus clouds usually form at low altitudes and may be associated with fair weather or rain or snow. When air is cooled to its dew point near the ground, it forms a stratus cloud called **fog,** as shown in **Figure 6.**

Figure 6 Fog surrounds the Golden Gate Bridge, San Francisco. Fog is a stratus cloud near the ground.
Think Critically *Why do you think fog is found in San Francisco Bay?*

Cumulus (KYEW myuh lus) clouds are masses of puffy, white clouds, often with flat bases. They sometimes tower to great heights and can be associated with fair weather or thunderstorms.

Cirrus (SIHR us) clouds appear fibrous or curly. They are high, thin, white, feathery clouds made of ice crystals. Cirrus clouds are associated with fair weather, but they can indicate approaching storms.

Height Some prefixes of cloud names describe the height of the cloud base. The prefix *cirro-* describes high clouds, *alto-* describes middle-elevation clouds, and *strato-* refers to clouds at low elevations. Some clouds' names combine the altitude prefix with the term *stratus* or *cumulus.*

Cirrostratus clouds are high clouds, like those in **Figure 7.** Usually, cirrostratus clouds indicate fair weather, but they also can signal an approaching storm. Altostratus clouds form at middle levels. If the clouds are not too thick, sunlight can filter through them.

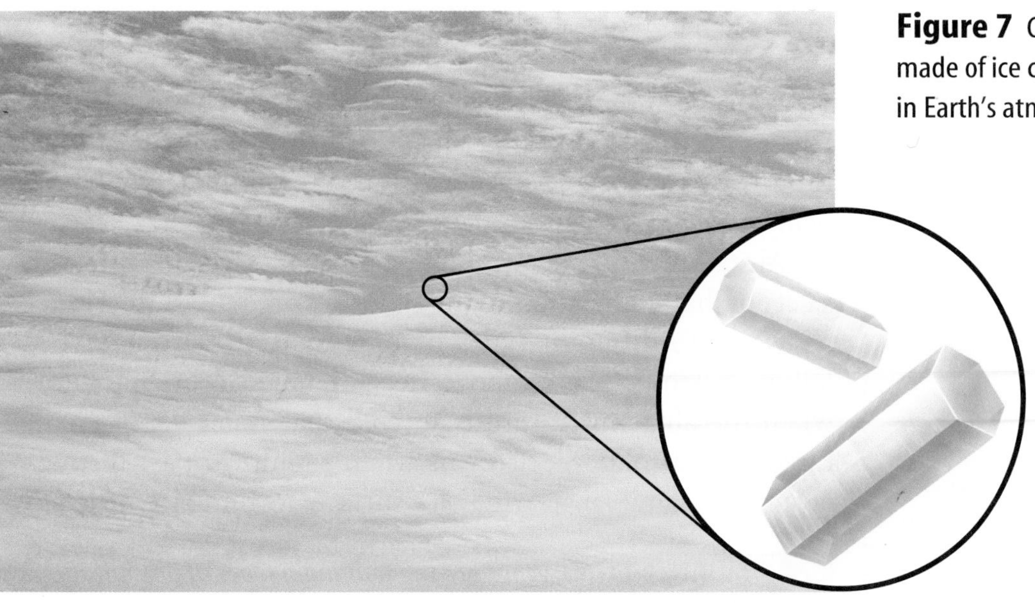

Figure 7 Cirrostratus clouds are made of ice crystals and form high in Earth's atmosphere.

Figure 8 Water vapor in air collects on particles to form water droplets or ice crystals. The type of precipitation that is received on the ground depends on the temperature of the air.

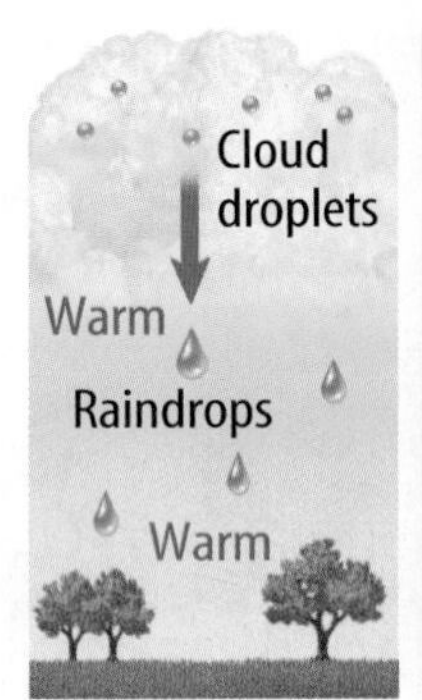

When the air is warm, water vapor forms raindrops that fall as rain.

When the air is cold, water vapor forms snowflakes.

Rain- or Snow-Producing Clouds Clouds associated with rain or snow often have the word nimbus attached to them. The term *nimbus* is Latin for "dark rain cloud" and this is a good description, because the water content of these clouds is so high that little sunlight can pass through them. When a cumulus cloud grows into a thunderstorm, it is called a cumulonimbus (kyew myuh loh NIHM bus) cloud. These clouds can tower to nearly 18 km. Nimbostratus clouds are layered clouds that can bring long, steady rain or snowfall.

Precipitation

Water falling from clouds is called **precipitation.** Precipitation occurs when cloud droplets combine and grow large enough to fall to Earth. The cloud droplets form around small particles, such as salt and dust. These particles are so small that a puff of smoke can contain millions of them.

You might have noticed that raindrops are not all the same size. The size of raindrops depends on several factors. One factor is the strength of updrafts in a cloud. Strong updrafts can keep drops suspended in the air where they can combine with other drops and grow larger. The rate of evaporation as a drop falls to Earth also can affect its size. If the air is dry, the size of raindrops can be reduced or they can completely evaporate before reaching the ground.

Air temperature determines whether water forms rain, snow, sleet, or hail—the four main types of precipitation. **Figure 8** shows these different types of precipitation. Drops of water falling in temperatures above freezing fall as rain. Snow forms when the air temperature is so cold that water vapor changes directly to a solid. Sleet forms when raindrops pass through a layer of freezing air near Earth's surface, forming ice pellets.

Reading Check *What are the four main types of precipitation?*

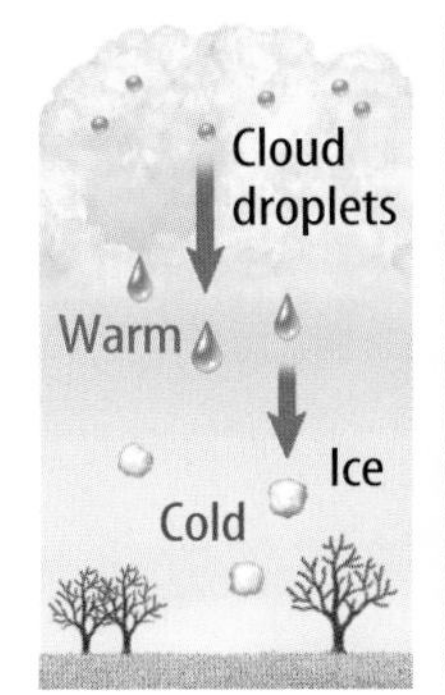

When the air near the ground is cold, sleet, which is made up of many small ice pellets, falls.

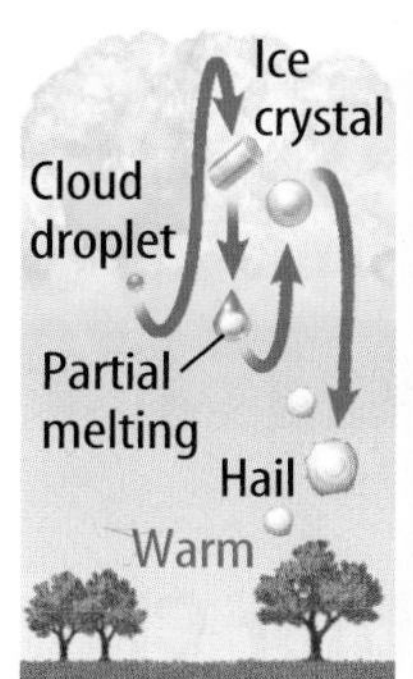

Hailstones are pellets of ice that form inside a cloud.

Hail Hail is precipitation in the form of lumps of ice. Hail forms in cumulonimbus clouds of a thunderstorm when water freezes in layers around a small nucleus of ice. Hailstones grow larger as they're tossed up and down by rising and falling air. Most hailstones are smaller than 2.5 cm but can grow larger than a softball. Of all forms of precipitation, hail produces the most damage immediately, especially if winds blow during a hailstorm. Falling hailstones can break windows and destroy crops.

If you understand the role of water vapor in the atmosphere, you can begin to understand weather. The relative humidity of the air helps determine whether a location will have a dry day or experience some form of precipitation. The temperature of the atmosphere determines the form of precipitation. Studying clouds can add to your ability to forecast weather.

section 1 review

Summary

Weather Factors

- Weather is the state of the atmosphere at a specific time and place.
- Temperature, wind, air pressure, dew point, and humidity describe weather.

Clouds

- Warm, moist air rises, forming clouds.
- The main types of clouds are stratus, cumulus, and cirrus.

Precipitation

- Water falling from clouds is called precipitation.
- Air temperature determines whether water forms rain, snow, sleet, or hail.

Self Check

1. **Explain** When does water vapor in air condense?
2. **Compare and contrast** humidity and relative humidity.
3. **Summarize** how clouds form.
4. **Describe** How does precipitation occur and what determines the type of precipitation that falls to Earth?
5. **Think Critically** Cumulonimbus clouds form when warm, moist air is suddenly lifted. How can the same cumulonimbus cloud produce rain and hail?

Applying Math

6. **Use Graphs** If the air temperature is 30°C and the relative humidity is 60 percent, will the dew point be reached if the temperature drops to 25°C? Use the graph in **Figure 4** to explain your answer.

section 2

Weather Patterns

as you read

What You'll Learn

- **Describe** how weather is associated with fronts and high- and low-pressure areas.
- **Explain** how tornadoes develop from thunderstorms.
- **Discuss** the dangers of severe weather.

Why It's Important

Air masses, pressure systems, and fronts cause weather to change.

Review Vocabulary

barometer: instrument used to measure atmospheric pressure

New Vocabulary

- air mass
- front
- tornado
- hurricane
- blizzard

Weather Changes

When you leave for school in the morning, the weather might be different from what it is when you head home in the afternoon. Because of the movement of air and moisture in the atmosphere, weather constantly changes.

Air Masses An **air mass** is a large body of air that has properties similar to the part of Earth's surface over which it develops. For example, an air mass that develops over land is dry compared with one that develops over water. An air mass that develops in the tropics is warmer than one that develops over northern regions. An air mass can cover thousands of square kilometers. When you observe a change in the weather from one day to the next, it is due to the movement of air masses. **Figure 9** shows air masses that affect the United States.

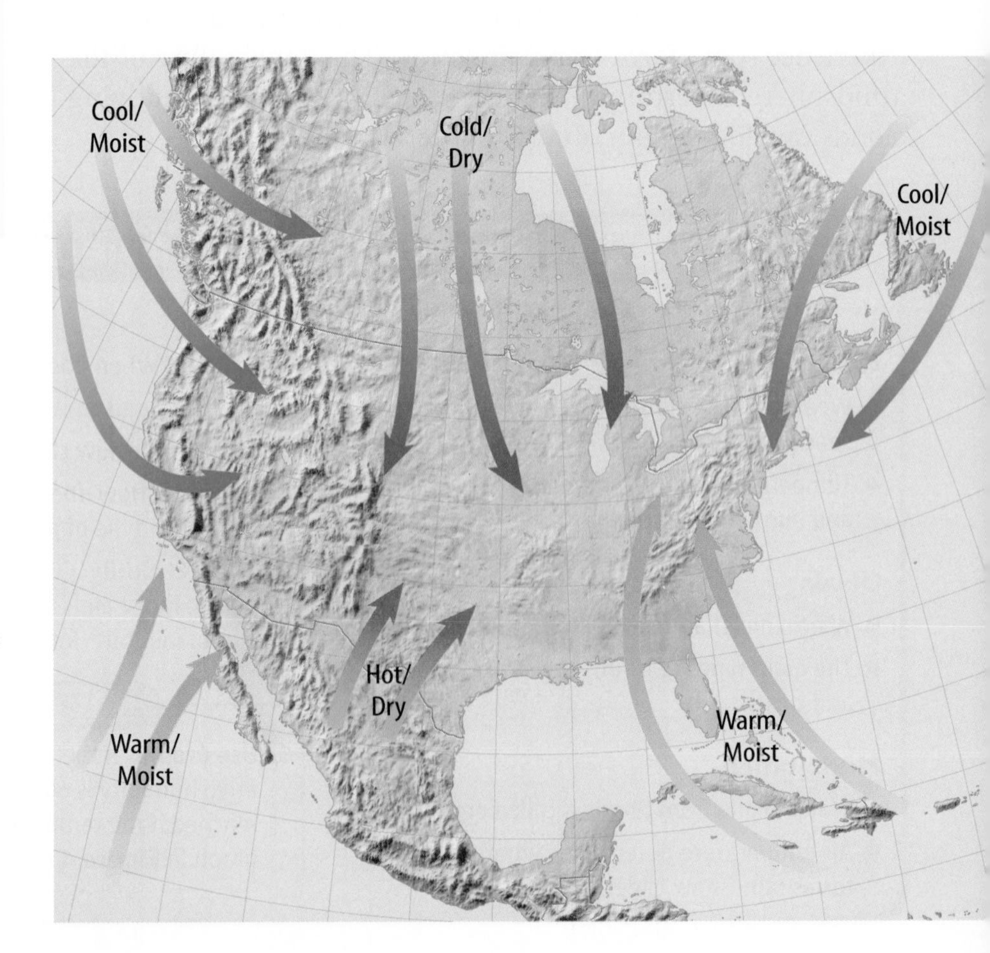

Figure 9 Six major air masses affect weather in the United States. Each air mass has the same characteristics of temperature and moisture content as the area over which it formed.

INTEGRATE Physics **Highs and Lows** Atmospheric pressure varies over Earth's surface. Anyone who has watched a weather report on television has heard about high- and low-pressure systems. Recall that winds blow from areas of high pressure to areas of low pressure. As winds blow into a low-pressure area in the northern hemisphere, Earth's rotation causes these winds to swirl in a counterclockwise direction. Large, swirling areas of low pressure are called cyclones and are associated with stormy weather.

Figure 10 A barometer measures atmospheric pressure. The red pointer points to the current pressure. Watch how atmospheric pressure changes over time when you line up the white pointer to the one indicating the current pressure each day.

Reading Check *How do winds move in a cyclone?*

Winds blow away from a center of high pressure. Earth's rotation causes these winds to spiral clockwise in the northern hemisphere. High-pressure areas are associated with fair weather and are called anticyclones. Air pressure is measured using a barometer, like the one shown in **Figure 10.**

Variation in atmospheric pressure affects the weather. Low pressure systems at Earth's surface are regions of rising air. Clouds form when air is lifted and cools. Areas of low pressure usually have cloudy weather. Sinking motion in high-pressure air masses makes it difficult for air to rise and clouds to form. That's why high pressure usually means good weather.

Fronts

A boundary between two air masses of different density, moisture, or temperature is called a **front.** If you've seen a weather map in the newspaper or on the evening news, you've seen fronts represented by various types of curving lines.

Cloudiness, precipitation, and storms sometimes occur at frontal boundaries. Four types of fronts include cold, warm, occluded, and stationary.

Cold and Warm Fronts A cold front, shown on a map as a blue line with triangles ▲▲▲, occurs when colder air advances toward warm air. The cold air wedges under the warm air like a plow. As the warm air is lifted, it cools and water vapor condenses, forming clouds. When the temperature difference between the cold and warm air is large, thunderstorms and even tornadoes may form.

Warm fronts form when lighter, warmer air advances over heavier, colder air. A warm front is drawn on weather maps as a red line with red semicircles ◢◣◢◣.

Topic: Atmospheric Pressure
Visit earth.msscience.com for Web links to information about the current atmospheric pressure of your town or nearest city.

Activity Look up the pressure of a city west of your town and the pressure of a city to the east. Compare the pressures to local weather conditions. Share your information with the class.

Occluded and Stationary Fronts An occluded front involves three air masses of different temperatures—colder air, cool air, and warm air. An occluded front may form when a cold air mass moves toward cool air with warm air between the two. The colder air forces the warm air upward, closing off the warm air from the surface. Occluded fronts are shown on maps as purple lines with triangles and semicircles ▲●▲.

A stationary front occurs when a boundary between air masses stops advancing. Stationary fronts may remain in the same place for several days, producing light wind and precipitation. A stationary front is drawn on a weather map as an alternating red and blue line. Red semicircles point toward the cold air and blue triangles point toward the warm air ●▼●. **Figure 11** summarizes the four types of fronts.

Figure 11 Cold, warm, occluded, and stationary fronts occur at the boundaries of air masses.
Describe *what type of weather occurs at front boundaries.*

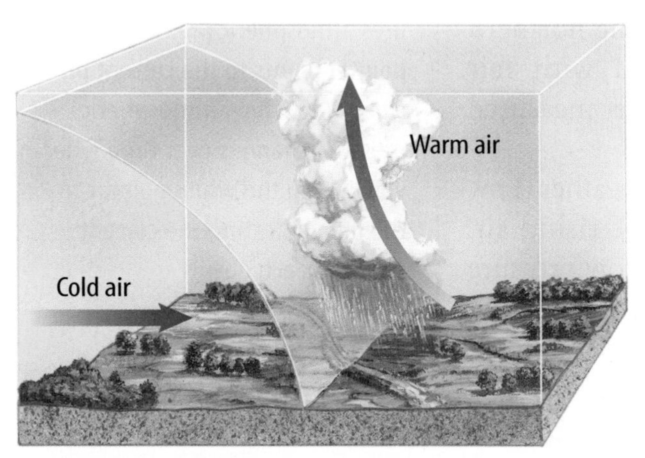

A cold front can advance rapidly. Thunderstorms often form as warm air is suddenly lifted up over the cold air.

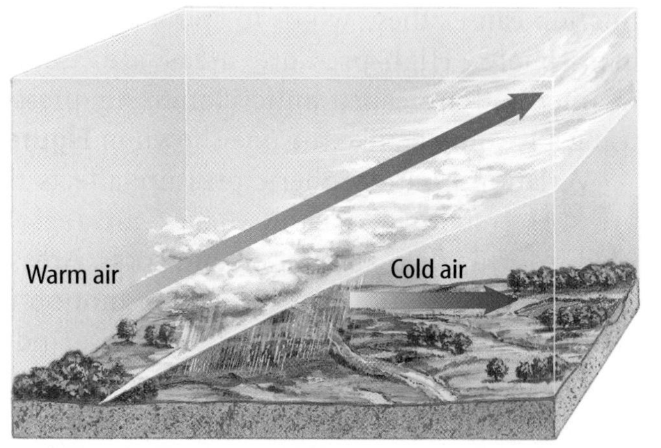

Warm air slides over colder air along a warm front, forming a boundary with a gentle slope. This can lead to hours, if not days, of wet weather.

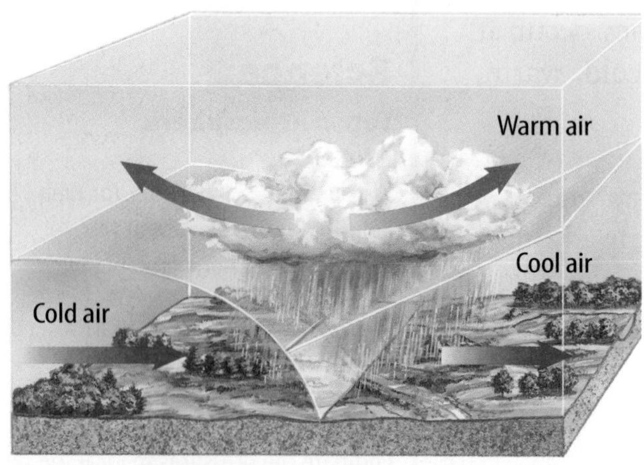

The term *occlusion* means "closure." Colder air forces warm air upward, forming an occluded front that closes off the warm air from the surface.

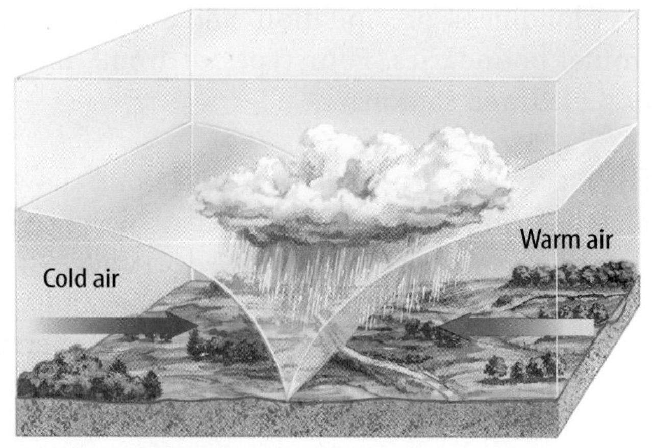

A stationary front results when neither cold air nor warm air advances.

Severe Weather

Despite the weather, you usually can do your daily activities. If it's raining, you still go to school. You can still get there even if it snows a little. However, some weather conditions, such as those caused by thunderstorms, tornadoes, and blizzards, prevent you from going about your normal routine. Severe weather poses danger to people, structures, and animals.

Thunderstorms In a thunderstorm, heavy rain falls, lightning flashes, thunder roars, and hail might fall. What forces cause such extreme weather conditions?

Thunderstorms occur in warm, moist air masses and along fronts. Warm, moist air can be forced upward where it cools and condensation occurs, forming cumulonimbus clouds that can reach heights of 18 km, like the one in **Figure 12.** When rising air cools, water vapor condenses into water droplets or ice crystals. Smaller droplets collide to form larger ones, and the droplets fall through the cloud toward Earth's surface. The falling droplets collide with still more droplets and grow larger. Raindrops cool the air around them. This cool, dense air then sinks and spreads over Earth's surface. Sinking, rain-cooled air and strong updrafts of warmer air cause the strong winds associated with thunderstorms. Hail also may form as ice crystals alternately fall to warmer layers and are lifted into colder layers by the strong updrafts inside cumulonimbus clouds.

Thunderstorm Damage Sometimes thunderstorms can stall over a region, causing rain to fall heavily for a period of time. When streams cannot contain all the water running into them, flash flooding can occur. Flash floods can be dangerous because they occur with little warning.

Strong winds generated by thunderstorms also can cause damage. If a thunderstorm is accompanied by winds traveling faster than 89 km/h, it is classified as a severe thunderstorm. Hail from a thunderstorm can dent cars and the aluminum siding on houses. Although rain from thunderstorms helps crops grow, hail has been known to flatten and destroy entire crops in a matter of minutes.

Figure 12 Tall cumulonimbus clouds may form quickly as warm, moist air rapidly rises.
Identify *some things these clouds are known to produce.*

Figure 13 This time-elapsed photo shows a thunderstorm over Arizona.

Lightning and Thunder What are lightning and thunder? Inside a storm cloud, warm air is lifted rapidly as cooler air sinks. This movement of air can cause different parts of a cloud to become oppositely charged. When current flows between regions of opposite electrical charge, lightning flashes. Lightning, as shown in **Figure 13,** can occur within a cloud, between clouds, or between a cloud and the ground.

Thunder results from the rapid heating of air around a bolt of lightning. Lightning can reach temperatures of about 30,000°C, which is more than five times the temperature of the surface of the Sun. This extreme heat causes air around the lightning to expand rapidly. Then it cools quickly and contracts. The rapid movement of the molecules forms sound waves heard as thunder.

Tornadoes Some of the most severe thunderstorms produce tornadoes. A **tornado** is a violently rotating column of air in contact with the ground. In severe thunderstorms, wind at different heights blows in different directions and at different speeds. This difference in wind speed and direction, called wind shear, creates a rotating column parallel to the ground. A thunderstorm's updraft can tilt the rotating column upward into the thunderstorm creating a funnel cloud. If the funnel comes into contact with Earth's surface, it is called a tornado.

Reading Check *What causes a tornado to form?*

A tornado's destructive winds can rip apart buildings and uproot trees. High winds can blow through broken windows. When winds blow inside a house, they can lift off the roof and blow out the walls, making it look as though the building exploded. The updraft in the center of a powerful tornado can lift animals, cars, and even houses into the air. Although tornadoes rarely exceed 200 m in diameter and usually last only a few minutes, they often are extremely destructive. In May 1999, multiple thunderstorms produced more than 70 tornadoes in Kansas, Oklahoma, and Texas. This severe tornado outbreak caused 40 deaths, 100 injuries, and more than $1.2 billion in property damage.

Topic: Lightning
Visit earth.msscience.com for Web links to research the number of lightning strikes in your state during the last year.

Activity Compare your findings with data from previous years. Communicate to your class what you learn.

NATIONAL GEOGRAPHIC VISUALIZING TORNADOES

Figure 14

Tornadoes are extremely rapid, rotating winds that form at the base of cumulonimbus clouds. Smaller tornadoes may even form inside larger ones. Luckily, most tornadoes remain on the ground for just a few minutes. During that time, however, they can cause considerable—and sometimes strange—damage, such as driving a fork into a tree.

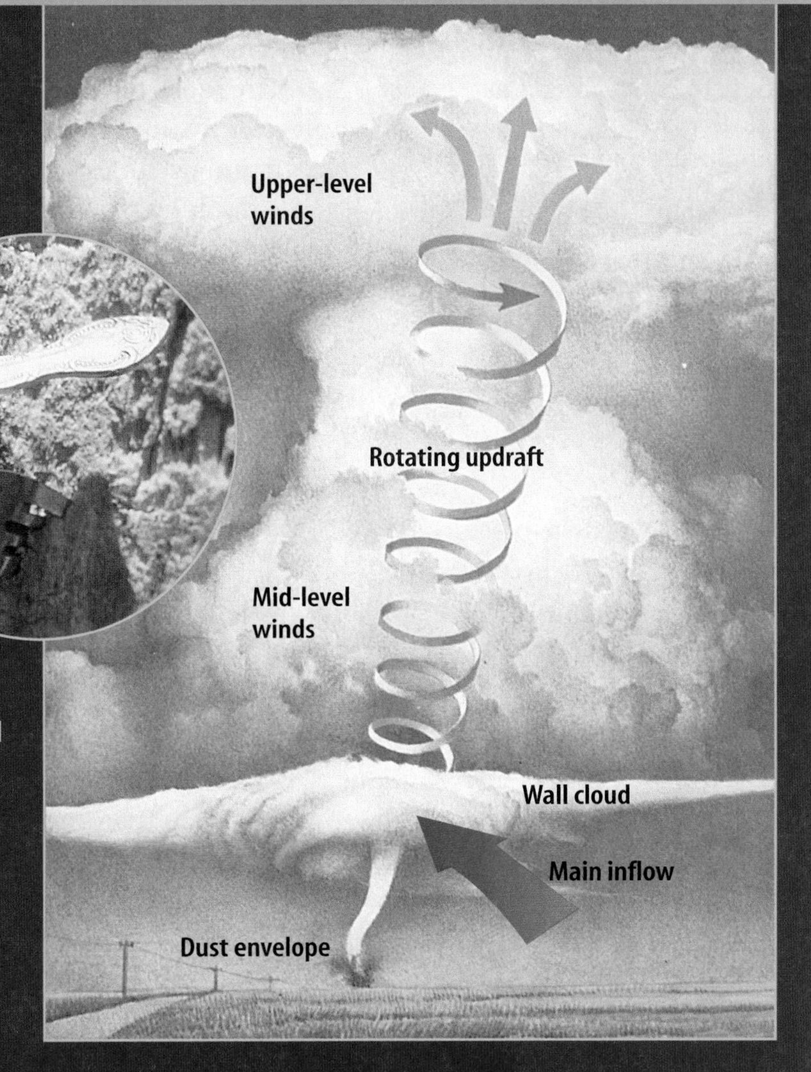

Tornadoes often form from a type of cumulonimbus cloud called a wall cloud. Strong, spiraling updrafts of warm, moist air may form in these clouds. As air spins upward, a low-pressure area forms, and the cloud descends to the ground in a funnel. The tornado sucks up debris as it moves along the ground, forming a dust envelope.

The Fujita Scale

	Wind speed (km/h)	Damage
F0	<116	Light: broken branches and chimneys
F1	116–180	Moderate: roofs damaged, mobile homes upturned
F2	181–253	Considerable: roofs torn off homes, large trees uprooted
F3	254–332	Severe: trains overturned, roofs and walls torn off
F4	333–419	Devastating: houses completely destroyed, cars picked up and carried elsewhere
F5	420–512	Incredible: total demolition

The Fujita scale, named after tornado expert Theodore Fujita, ranks tornadoes according to how much damage they cause. Fortunately, only one percent of tornadoes are classified as violent (F4 and F5).

Global Warming Some scientists hypothesize that Earth's ocean temperatures are increasing due to global warming. In your Science Journal, predict what might happen to the strength of hurricanes if Earth's oceans become warmer.

Hurricanes The most powerful storm is the hurricane. A **hurricane,** illustrated in **Figure 15,** is a large, swirling, low-pressure system that forms over the warm Atlantic Ocean. It is like a machine that turns heat energy from the ocean into wind. A storm must have winds of at least 119 km/h to be called a hurricane. Similar storms are called typhoons in the Pacific Ocean and cyclones in the Indian Ocean.

Hurricanes are similar to low-pressure systems on land, but they are much stronger. In the Atlantic and Pacific Oceans, low pressure sometimes develops near the equator. In the northern hemisphere, winds around this low pressure begin rotating counterclockwise. The strongest hurricanes affecting North America usually begin as a low-pressure system west of Africa. Steered by surface winds, these storms can travel west, gaining strength from the heat and moisture of warm ocean water.

When a hurricane strikes land, high winds, tornadoes, heavy rains, and high waves can cause a lot of damage. Floods from the heavy rains can cause additional damage. Hurricane weather can destroy crops, demolish buildings, and kill people and other animals. As long as a hurricane is over water, the warm, moist air rises and provides energy for the storm. When a hurricane reaches land, however, its supply of energy disappears and the storm loses power.

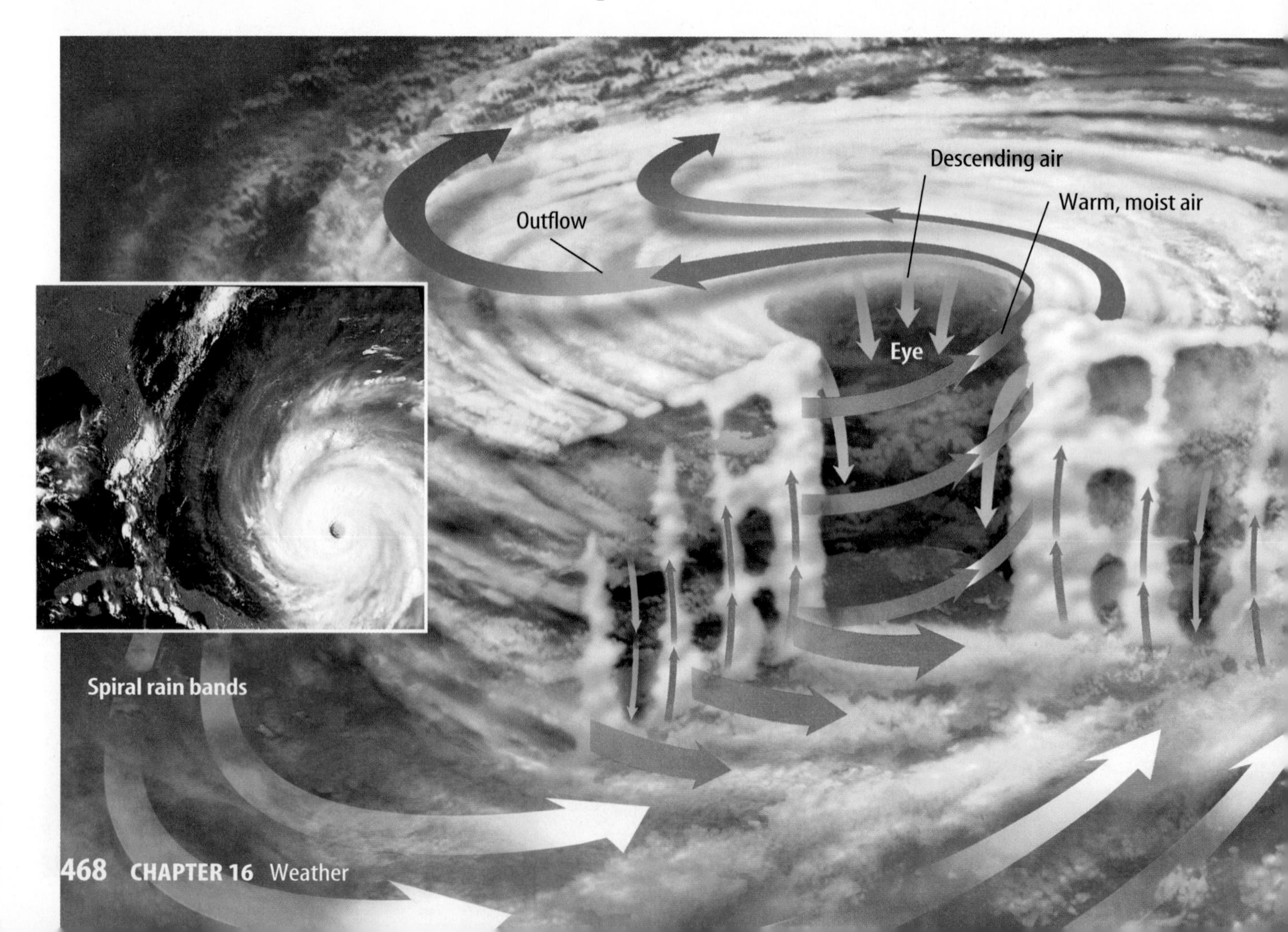

Figure 15 In this hurricane cross section, the small, red arrows indicate rising, warm, moist air. This air forms cumulus and cumulonimbus clouds in bands around the eye. The green arrows indicate cool, dry air sinking in the eye and between the cloud bands.

Blizzards Severe storms also can occur in winter. If you live in the northern United States, you may have awakened from a winter night's sleep to a cold, howling wind and blowing snow, like the storm in **Figure 16.** The National Weather Service classifies a winter storm as a **blizzard** if the winds are 56 km/h, the temperature is low, the visibility is less than 400 m in falling or blowing snow, and if these conditions persist for three hours or more.

Figure 16 Blizzards can be extremely dangerous because of their high winds, low temperatures, and poor visibility.

Severe Weather Safety When severe weather threatens, the National Weather Service issues a watch or warning. Watches are issued when conditions are favorable for severe thunderstorms, tornadoes, floods, blizzards, and hurricanes. During a watch, stay tuned to a radio or television station reporting the weather. When a warning is issued, severe weather conditions already exist. You should take immediate action. During a severe thunderstorm or tornado warning, take shelter in the basement or a room in the middle of the house away from windows. When a hurricane or flood watch is issued, be prepared to leave your home and move farther inland.

Blizzards can be blinding and have dangerously low temperatures with high winds. During a blizzard, stay indoors. Spending too much time outside can result in severe frostbite.

section 2 review

Summary

Weather Changes

- Air masses tend to have temperature and moisture properties similar to Earth's surface.
- Winds blow from areas of high pressure to areas of lower pressure.

Fronts

- A boundary between different air masses is called a front.

Severe Weather

- The National Weather Service issues watches or warnings, depending on the severity of the storm, for people's safety.

Self Check

1. **Draw Conclusions** Why is fair weather common during periods of high pressure?
2. **Describe** how a cold front affects weather.
3. **Explain** what causes lightning and thunder.
4. **Compare and contrast** a watch and a warning. How can you keep safe during a tornado warning?
5. **Think Critically** Explain why some fronts produce stronger storms than others.

Applying Skills

6. **Recognize Cause and Effect** Describe how an occluded front may form over your city and what effects it can have on the weather.

section 3

Weather Forecasts

as you read

What You'll Learn

- **Explain** how data are collected for weather maps and forecasts.
- **Identify** the symbols used in a weather station model.

Why It's Important

Weather observations help you predict future weather events.

Review Vocabulary

forecast: to predict a condition or event on the basis of observations

New Vocabulary

- meteorologist
- station model
- isotherm
- isobar

Weather Observations

You can determine current weather conditions by checking the thermometer and looking to see whether clouds are in the sky. You know when it's raining. You have a general idea of the weather because you are familiar with the typical weather where you live. If you live in Florida, you don't expect snow in the forecast. If you live in Maine, you assume it will snow every winter. What weather concerns do you have in your region?

A **meteorologist** (mee tee uh RAH luh jist) is a person who studies the weather. Meteorologists take measurements of temperature, air pressure, winds, humidity, and precipitation. Computers, weather satellites, Doppler radar shown in **Figure 17,** and instruments attached to balloons are used to gather data. Such instruments improve meteorologists' ability to predict the weather. Meteorologists use the information provided by weather instruments to make weather maps. These maps are used to make weather forecasts.

Forecasting Weather

Meteorologists gather information about current weather and use computers to make predictions about future weather patterns. Because storms can be dangerous, you do not want to be unprepared for threatening weather. However, meteorologists cannot always predict the weather exactly because conditions can change rapidly.

The National Weather Service depends on two sources for its information—data collected from the upper atmosphere and data collected on Earth's surface. Meteorologists of the National Weather Service collect information recorded by satellites, instruments attached to weather balloons, and from radar. This information is used to describe weather conditions in the atmosphere above Earth's surface.

Figure 17 A meteorologist uses Doppler radar to track a tornado. Since the nineteenth century, technology has greatly improved weather forecasting.

Station Models When meteorologists gather data from Earth's surface, it is recorded on a map using a combination of symbols, forming a **station model.** A station model, like the one in **Figure 18,** shows the weather conditions at a specific location on Earth's surface. Information provided by station models and instruments in the upper atmosphere is entered into computers and used to forecast weather.

Temperature and Pressure In addition to station models, weather maps have lines that connect locations of equal temperature or pressure. A line that connects points of equal temperature is called an **isotherm** (I suh thurm). *Iso* means "same" and *therm* means "temperature." You probably have seen isotherms on weather maps on TV or in the newspaper.

An **isobar** is a line drawn to connect points of equal atmospheric pressure. You can tell how fast wind is blowing in an area by noting how closely isobars are spaced. Isobars that are close together indicate a large pressure difference over a small area. A large pressure difference causes strong winds. Isobars that are spread apart indicate a smaller difference in pressure. Winds in this area are gentler. Isobars also indicate the locations of high- and low-pressure areas.

Reading Check *How do isobars indicate wind speed?*

Measuring Rain

Procedure

1. You will need a **straight-sided container,** such as a soup or coffee can, **duct tape,** and a **ruler.**
2. Tape the ruler to the inner wall of your container.
3. Place the container on a level surface outdoors, away from buildings or plants.
4. Measure the amount of water in your container after it rains. Continue to take measurements for a week.

Analysis

1. What was the average daily rainfall?
2. Why is it necessary to use containers with straight sides?

Try at Home

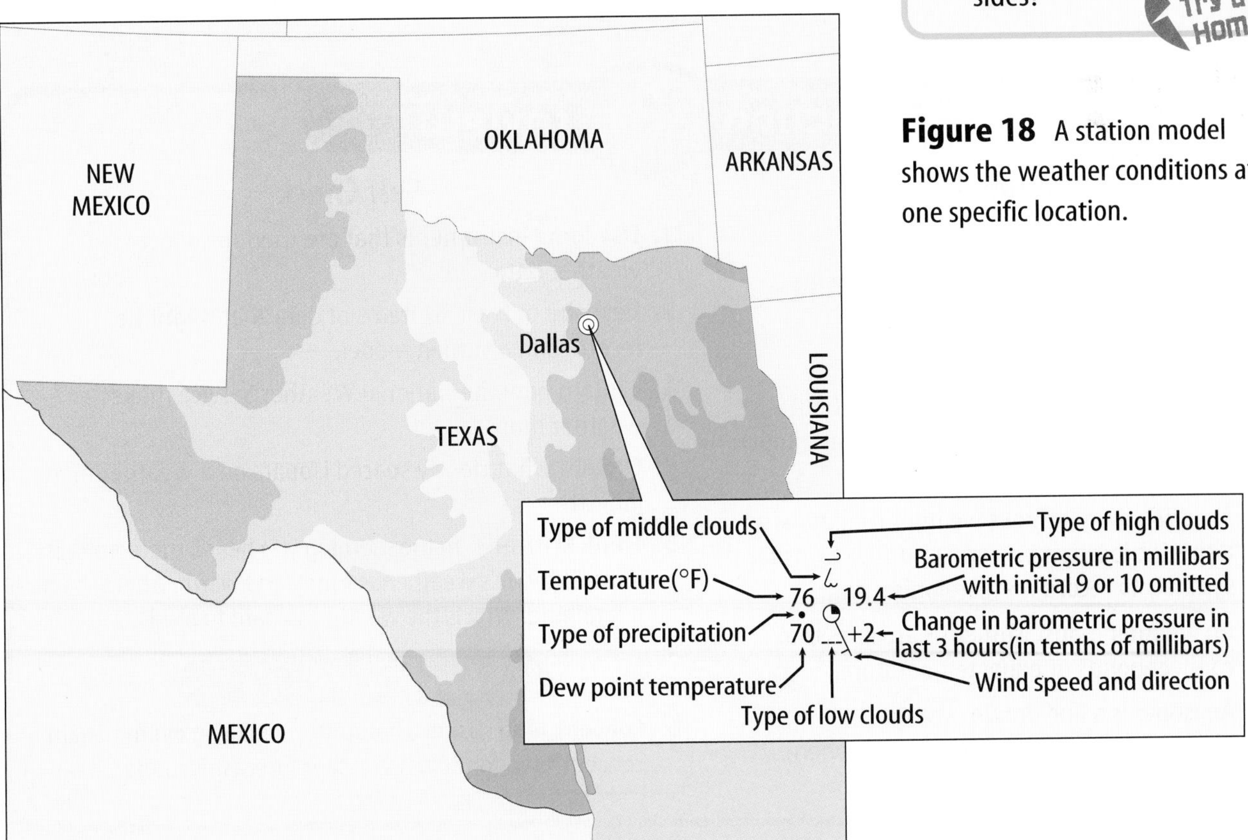

Figure 18 A station model shows the weather conditions at one specific location.

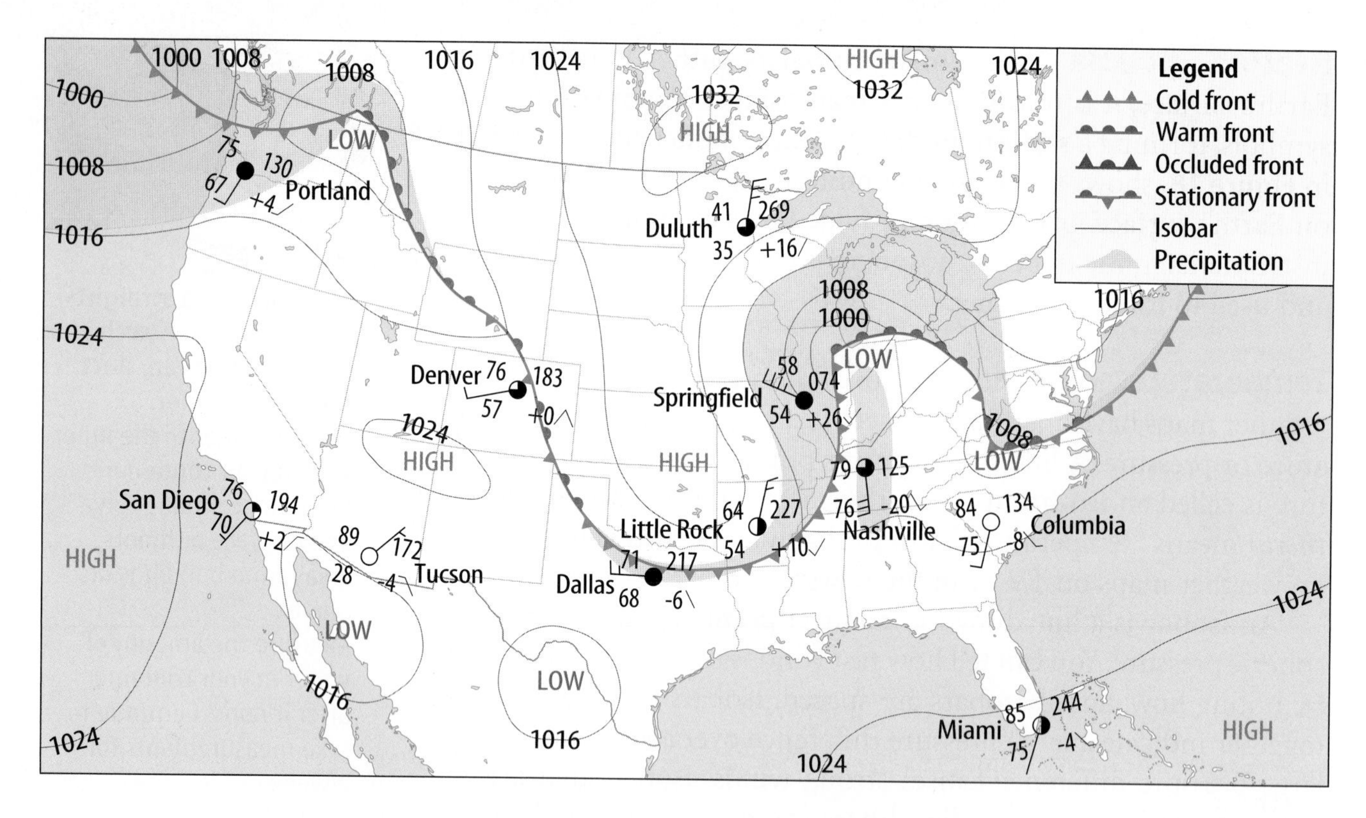

Figure 19 Highs, lows, isobars, and fronts on this weather map help meteorologists forecast the weather.

Weather Maps On a weather map like the one in **Figure 19,** pressure areas are drawn as circles with the word High or Low in the middle of the circle. Fronts are drawn as lines and symbols. When you watch weather forecasts on television, notice how weather fronts move from west to east. This is a pattern that meteorologists depend on to forecast weather.

section 3 review

Summary

Weather Observations

- Meteorologists are people who study the weather and make weather maps.

Forecasting Weather

- Meteorologists gather information about current weather and make predictions about future weather patterns.
- A station model shows weather conditions at a specific location on Earth's surface by using symbols to record meteorological data.
- On weather maps, isotherms are lines that connect points of equal temperature.
- An isobar is a line drawn on a weather map that connects points of equal atmospheric pressure.

Self Check

1. **List** some instruments that are used to collect weather data.
2. **Describe** at least six items of data that might be recorded in a station model.
3. **Explain** how the National Weather Service makes weather maps.
4. **Explain** what closely spaced isobars on a weather map indicate.
5. **Think Critically** In the morning you hear a meteorologist forecast today's weather as sunny and warm. After school, it is raining. Why is the weather so hard to predict?

Applying Skills

6. **Concept Map** Using a computer, make an events-chain concept map for how a weather forecast is made.

Science Online earth.msscience.com/self_check_quiz

Reading a Weather Map

Meteorologists use a series of symbols to provide a picture of local and national weather conditions. With what you know, can you interpret weather information from weather map symbols?

Real-World Question

How do you read a weather map?

Materials

magnifying lens
Weather Map Symbols Appendix
Figure 19 (Weather Map)

Goals

- **Learn** how to read a weather map.
- **Use** information from a station model and a weather map to forecast weather.

Procedure

Use the information provided in the questions below and the Weather Map Symbols Appendix to learn how to read a weather map.

1. Find the station models on the map for Portland, Oregon, and Miami, Florida. Find the dew point, wind direction, barometric pressure, and temperature at each location.
2. Looking at the placement of the isobars, determine whether the wind would be stronger at Springfield, Illinois, or at San Diego, California. Record your answer. What is another way to determine the wind speed at these locations?
3. **Determine** the type of front near Dallas, Texas. Record your answer.
4. The triangles or half-circles are on the side of the line toward the direction the front is moving. In which direction is the cold front located over Washington state moving?

Conclude and Apply

1. Locate the pressure system over southeast Kansas. Predict what will happen to the weather of Nashville, Tennessee, if this pressure system moves there.
2. Prevailing westerlies are winds responsible for the movement of much of the weather across the United States. Based on this, would you expect Columbia, South Carolina, to continue to have clear skies? Explain.
3. The direction line on the station model indicates the direction from which the wind blows. The wind is named for that direction. Infer from this the name of the wind blowing at Little Rock, Arkansas.

Communicating Your Data

Pretend you are a meteorologist for a local TV news station. Make a poster of your weather data and present a weather forecast to your class.

Model and Invent

Measuring Wind Speed

Goals

- **Invent** an instrument or devise a system for measuring wind speeds using common materials.
- **Devise** a method for using your invention or system to compare different wind speeds.

Possible Materials

paper
scissors
confetti
grass clippings
meterstick
**measuring tape*
**Alternate materials*

Safety Precautions

Data Source

Refer to Section 1 for more information about anemometers and other wind speed instruments. Consult the data table for information about Beaufort's wind speed scale.

Real-World Question

When you watch a gust of wind blow leaves down the street, do you wonder how fast the wind is moving? For centuries, people could only guess at wind speeds, but in 1805, Admiral Beaufort of the British navy invented a method for estimating wind speeds based on their effect on sails. Later, Beaufort's system was modified for use on land. Meteorologists use a simple instrument called an anemometer to measure wind speeds, and they still use Beaufort's system to estimate the speed of the wind. What type of instrument or system can you invent to measure wind speed? How could you use simple materials to invent an instrument or system for measuring wind speeds? What observations do you use to estimate the speed of the wind?

Make a Model

1. Scan the list of possible materials and choose the materials you will need to devise your system.
2. **Devise** a system to measure different wind speeds. Be certain the materials you use are light enough to be moved by slight breezes.

Check the Model Plans

1. **Describe** your plan to your teacher. Provide a sketch of your instrument or system and ask your teacher how you might improve its design.
2. Present your idea for measuring wind speed to the class in the form of a diagram or poster. Ask your classmates to suggest improvements in your design that will make your system more accurate or easier to use.

Test Your Model

1. Confetti or grass clippings that are all the same size can be used to measure wind speed by dropping them from a specific height. Measuring the distances they travel in different strength winds will provide data for devising a wind speed scale.
2. Different sizes and shapes of paper also could be dropped into the wind, and the strength of the wind would be determined by measuring the distances traveled by these different types of paper.

Analyze Your Data

1. **Develop** a scale for your method.
2. **Compare** your results with Beaufort's wind speed scale.
3. **Analyze** what problems may exist in the design of your system and suggest steps you could take to improve your design.

Conclude and Apply

1. **Explain** why it is important for meteorologists to measure wind speeds.
2. **Evaluate** how well your system worked in gentle breezes and strong winds.

Beaufort's Wind Speed Scale

Description	Wind Speed (km/h)
calm—smoke drifts up	less than 1
light air—smoke drifts with wind	1–5
light breeze—leaves rustle	6–11
gentle breeze—leaves move constantly	12–19
moderate breeze—branches move	20–29
fresh breeze—small trees sway	30–39
strong breeze—large branches move	40–50
moderate gale—whole trees move	51–61
fresh gale—twigs break	62–74
strong gale—slight damage to houses	75–87
whole gale—much damage to houses	88–101
storm—extensive damage	102–120
hurricane—extreme damage	more than 120

Communicating Your Data

Demonstrate your system for the class. Compare your results and measurements with the results of other classmates.

TIME SCIENCE AND Society

SCIENCE ISSUES THAT AFFECT YOU!

Rainmakers

Cloud seeding is an inexact science

You listen to a meteorologist give the long-term weather forecast. Another week with no rain in sight. As a farmer, you are concerned that your crops are withering in the fields. Home owners' lawns are turning brown. Wildfires are possible. Cattle are starving. And, if farmers' crops die, there could be a shortage of food and prices will go up for consumers.

Meanwhile, several states away, another farmer is listening to the weather report calling for another week of rain. Her crops are getting so water soaked that they are beginning to rot.

Weather. Can't scientists find a way to better control it? The answer is...not exactly. Scientists have been experimenting with methods to control our weather since the 1940s. And nothing really works.

Cloud seeding is one such attempt. It uses technology to enhance the natural rainfall process. The idea has been used to create rain where it is needed or to reduce hail damage. Government officials also use cloud seeding or weather modification to try to reduce the force of a severe storm.

Flares are lodged under a plane. The pilot will drop them into potential rain clouds.

Some people seed a cloud by flying a plane above it and releasing highway-type flares with chemicals, such as silver iodide. Another method is to fly beneath the cloud and spray a chemical that can be carried into the cloud by air currents.

Cloud seeding doesn't work with clouds that have little water vapor or are not near the dew point. Seeding chemicals must be released into potential rain clouds. The chemicals provide nuclei for water molecules to cluster around. Water then falls to Earth as precipitation.

Cloud seeding does have its critics. If you seed clouds and cause rain for your area, aren't you preventing rain from falling in another area? Would that be considered "rain theft" by people who live in places where the cloudburst would naturally occur? What about those cloud-seeding agents? Could the cloud-seeding chemicals, such as silver iodide and acetone, affect the environment in a harmful way? Are humans meddling with nature and creating problems in ways that haven't been determined?

Debate Learn more about cloud seeding and other methods of changing weather. Then debate whether or not cloud seeding can be considered "rain theft."

For more information, visit earth.msscience.com/time

Reviewing Main Ideas

Section 1 What is weather?

1. Factors that determine weather include air pressure, wind, temperature, and the amount of moisture in the air.
2. More water vapor can be present in warm air than in cold air. Water vapor condenses when the dew point is reached. Clouds are formed when warm, moist air rises and cools to its dew point.
3. Rain, hail, sleet, and snow are types of precipitation.

Section 2 Weather Patterns

1. Fronts form when air masses with different characteristics meet. Types of fronts include cold, warm, occluded, and stationary fronts.
2. High atmospheric pressure at Earth's surface usually means good weather. Cloudy and stormy weather occurs under low pressure.
3. Tornadoes, thunderstorms, hurricanes, and blizzards are examples of severe weather.

Section 3 Weather Forecasts

1. Meteorologists use information from radar, satellites, computers, and other weather instruments to forecast the weather.
2. Weather maps include information about temperature and air pressure. Station models indicate weather at a particular location.

Visualizing Main Ideas

Copy and complete the following concept map about air temperature, water vapor, and pressure.

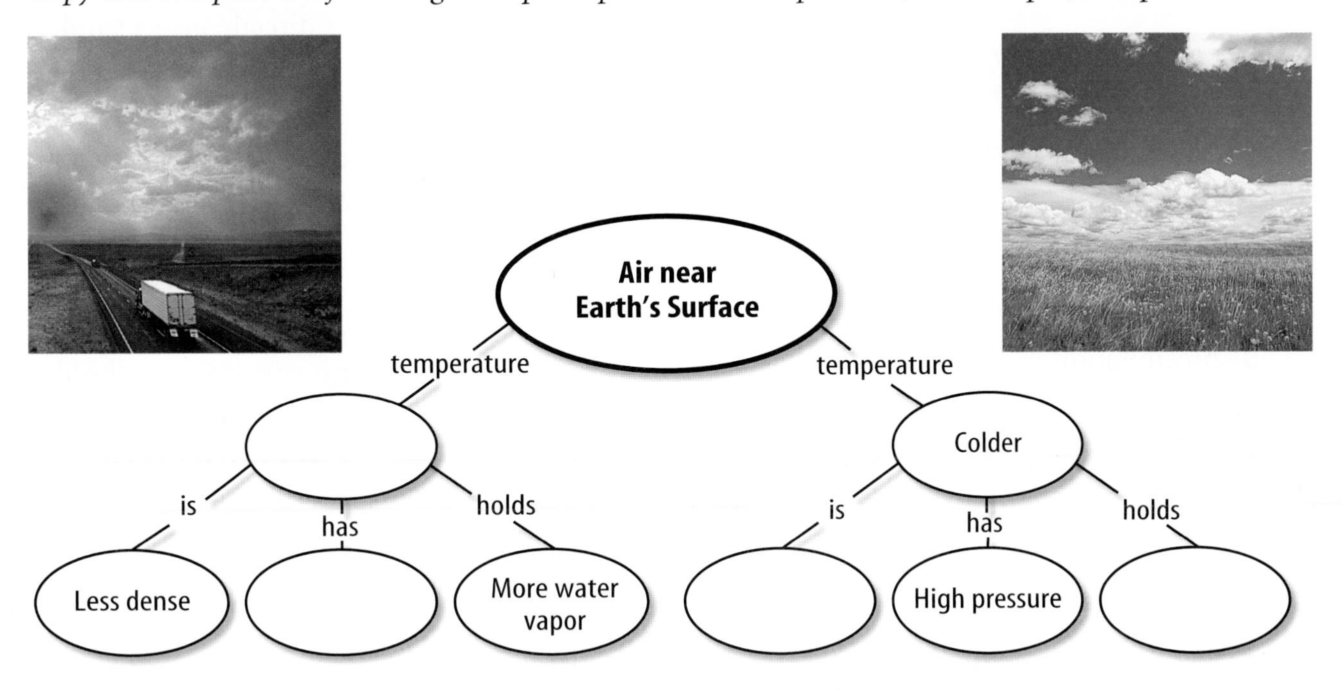

chapter 16 Review

Using Vocabulary

air mass p. 462	isotherm p. 471
blizzard p. 469	meteorologist p. 470
dew point p. 457	precipitation p. 460
fog p. 459	relative humidity p. 456
front p. 463	station model p. 471
humidity p. 456	tornado p. 466
hurricane p. 468	weather p. 454
isobar p. 471	

Explain the differences between the vocabulary words in each of the following sets.

1. air mass—front
2. humidity—relative humidity
3. relative humidity—dew point
4. dew point—precipitation
5. hurricane—tornado
6. blizzard—fog
7. meteorologist—station model
8. precipitation—fog
9. isobar—isotherm
10. isobar—front

Checking Concepts

Choose the word or phrase that best answers the question.

11. Which term refers to the amount of water vapor in the air?
   A) dew point **C)** humidity
   B) precipitation **D)** relative humidity

12. What does an anemometer measure?
   A) wind speed **C)** air pressure
   B) precipitation **D)** relative humidity

13. Which type of air has a relative humidity of 100 percent?
   A) humid **C)** dry
   B) temperate **D)** saturated

Use the photo below to answer question 14.

14. Which type of the following clouds are high feathery clouds made of ice crystals?
   A) cirrus **C)** cumulus
   B) nimbus **D)** stratus

15. What is a large body of air that has the same properties as the area over which it formed called?
   A) air mass **C)** front
   B) station model **D)** isotherm

16. At what temperature does water vapor in air condense?
   A) dew point **C)** front
   B) station model **D)** isobar

17. Which type of precipitation forms when water vapor changes directly into a solid?
   A) rain **C)** sleet
   B) hail **D)** snow

18. Which type of front may form when cool air, cold air, and warm air meet?
   A) warm **C)** stationary
   B) cold **D)** occluded

19. Which is issued when severe weather conditions exist and immediate action should be taken?
   A) front **C)** station model
   B) watch **D)** warning

20. What is a large, swirling storm that forms over warm, tropical water called?
   A) hurricane **C)** blizzard
   B) tornado **D)** hailstorm

Science Online earth.msscience.com/vocabulary_puzzlemaker

Thinking Critically

21. **Explain** the relationship between temperature and relative humidity.
22. **Describe** how air, water, and the Sun interact to cause weather.
23. **Explain** why northwest Washington often has rainy weather and southwest Texas is dry.
24. **Determine** What does it mean if the relative humidity is 79 percent?
25. **Infer** Why don't hurricanes form in Earth's polar regions?
26. **Compare and contrast** the weather at a cold front and the weather at a warm front.
27. **Interpret Scientific Illustrations** Use the cloud descriptions in this chapter to describe the weather at your location today. Then try to predict tomorrow's weather.
28. **Compare and contrast** tornadoes and thunderstorms. Include information about wind location and direction.
29. **Concept Map** Copy and complete the sequence map below showing how precipitation forms.

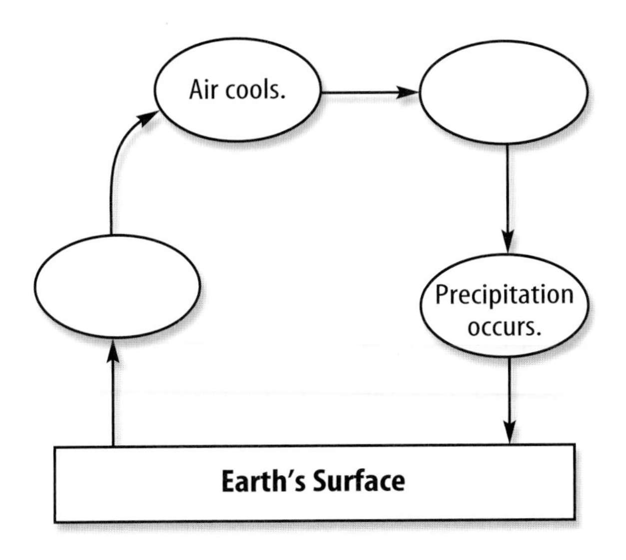

30. **Observe and Infer** You take a hot shower. The mirror in the bathroom fogs up. Infer from this information what has happened.

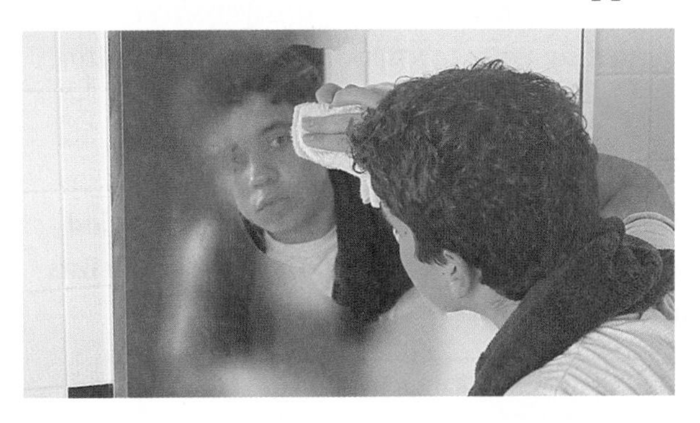

Performance Activities

31. **Board Game** Make a board game using weather terms. You could make cards to advance or retreat a token.
32. **Design** your own weather station. Record temperature, precipitation, and wind speed for one week.

Applying Math

Use the table below to answer question 33.

Air Temperature (°C)	Amount of Water Vapor Needed for Saturation (g/m^3)
25	22
20	15

33. **Dew Point** If the air temperature is 25°C and the relative humidity is 55 percent, will the dew point be reached if the temperature drops to 20°C?
34. **Rising Temperature** If the air temperature is 30°C and the relative humidity is 60 percent, will the dew point be reached if the temperature rises to 35°C? Use the graph in **Figure 4** to explain your answer.

Standardized Test Practice

Part 1 Multiple Choice

Record your answers on the answer sheet provided by your teacher or on a sheet of paper.

Use the table and paragraph below to answer questions 1 and 2.

Hurricanes are rated on a scale based on their wind speed and barometric pressure. The table below lists hurricane categories.

Hurricane Rating Scale		
Category	Wind Speed (km/h)	Barometric Pressure (millibars)
1	119–154	>980
2	155–178	965–980
3	179–210	945–964
4	211–250	920–944
5	>250	<920

1. Hurricane Mitch, with winds of 313 km/h and a pressure of 907 mb, struck the east coast of Central America in 1998. What category was Hurricane Mitch?
   A. 2
   B. 3
   C. 4
   D. 5

2. Which of the following is true when categorizing a hurricane?
   A. Storm category increases as wind increases and pressure decreases.
   B. Storm category increases as wind decreases and pressure increases.
   C. Storm category increases as wind and pressure increase.
   D. Storm category decreases as wind and pressure decrease.

Test-Taking Tip

Fill In All Blanks Never leave any answer blank.

3. Which of the following instruments is used to measure air pressure?
   A. anemometer
   B. thermometer
   C. barometer
   D. rain gauge

4. Which of the following is a description of a tornado?
   A. a large, swirling, low-pressure system that forms over the warm Atlantic Ocean
   B. a winter storm with winds at least 56 km/h and low visibility
   C. a violently rotating column of air in contact with the ground
   D. a boundary between two air masses of different density, moisture, or temperature

Use the figure below to answer questions 5 and 6.

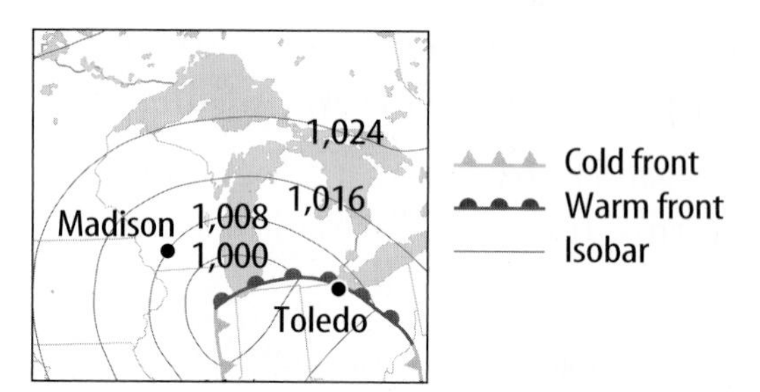

5. What is the atmospheric pressure in the city of Madison, Wisconsin?
   A. 1024 mb
   B. 1016 mb
   C. 1008 mb
   D. 1000 mb

6. What type of front is near Toledo, Ohio?
   A. cold front
   B. warm front
   C. stationary front
   D. occluded front

7. Which of the following terms is used to describe a person studies the weather?
   A. meteorologist
   B. geologist
   C. biologist
   D. paleontologist

Part 2 Short Response/Grid In

Record your answer on the answer sheet provided by your teacher or on a sheet of paper.

8. Compare and contrast the formation of stratus clouds, cumulus clouds and cirrus clouds.

Use the graph below to answer questions 9 and 10.

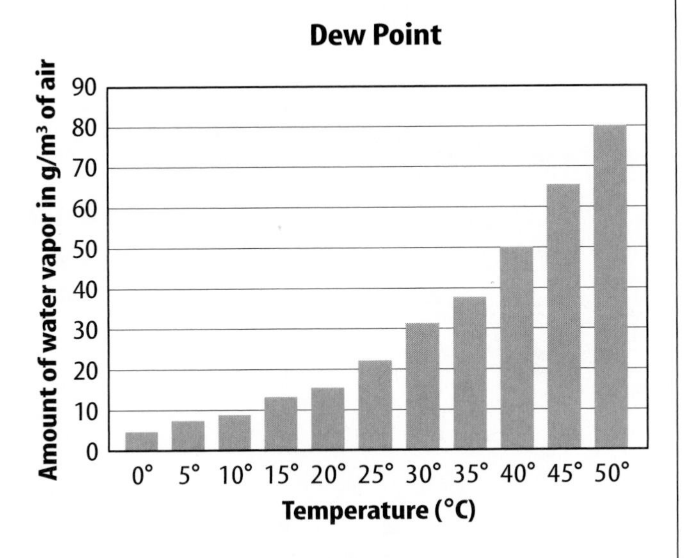

9. What amount of water vapor is in air at a temperature of 45°C?
10. On a fall day, the relative humidity is 72 percent and the temperature is 30°C. Will the dew point be reached if the temperature drops to 20°C? Why or why not?
11. Describe weather conditions during which hailstones form and the process by which they form.
12. What effects do high-pressure systems have on air circulation and weather? What effects do low-pressure systems have on weather?
13. Explain the relationship between differences in atmospheric pressure and wind speed.

Part 3 Open Ended

Record your answers on a sheet of paper.

14. Explain the relationship between lightning and thunder.
15. Describe how a hurricane in the Northern Hemisphere forms.
16. Explain why hurricanes lose power once they reach land.

Use the figure below to answer question 17.

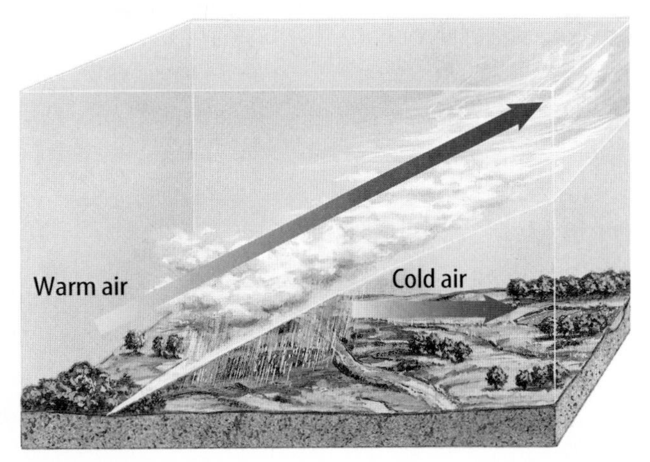

17. What type of front is shown? How does this type of front form?
18. Explain what type of weather occurs at front boundaries.
19. List the safety precautions you should take during a severe weather alert, including tornado warnings, flood watches, and blizzards, respectively.
20. Explain how the Sun's heat energy creates Earth's weather.
21. What are the four main types of precipitation? Describe the differences between each type.

chapter 17

Climate

The BIG Idea

Climate is the pattern of weather that occurs in an area over many years.

SECTION 1
What is climate?

Main Idea Some factors that affect the climate of a region include latitude, landforms, location of lakes and oceans, and ocean currents.

SECTION 2
Climate Types

Main Idea World climates can be classified by using averages of temperature and precipitation and the vegetation that is adapted to an area.

SECTION 3
Climate Changes

Main Idea The causes of climatic change can operate over short periods of time or very long periods of time.

Why do seasons change?

Why do some places have four distinct seasons, while others have only a wet and dry season? In this chapter, you will learn what climate is and how climates are classified. You will also learn what causes climate changes and how humans and animals adapt to different climates.

Science Journal Write a paragraph explaining what you already know about the causes of seasons.

Start-Up Activities

Tracking World Climates

You wouldn't go to Alaska to swim or to Jamaica to snow ski. You know the climates in these places aren't suited for these sports. In this lab, you'll explore the climates in different parts of the world.

1. Obtain a world atlas, globe, or large classroom map. Select several cities from different parts of the world.
2. Record the longitude and latitude of your cities. Note if they are near mountains or an ocean.
3. Research the average temperature of your cities. In what months are they hottest? Coldest? What is the average yearly rainfall? What kinds of plants and animals live in the region? Record your findings.
4. Compare your findings with those of the rest of your class. Can you see any relationship between latitude and climate? Do cities near an ocean or a mountain range have different climatic characteristics?
5. **Think Critically** Keep track of the daily weather conditions in your cities. Are these representative of the kind of climates your cities are supposed to have? Suggest reasons why day-to-day weather conditions may vary.

Classifying Climates Make the following Foldable to help you compare climatic types.

STEP 1 **Fold** two pieces of paper lengthwise into thirds.

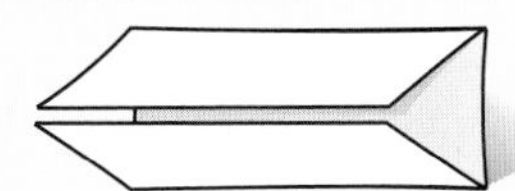

STEP 2 **Fold** the papers widthwise into fourths.

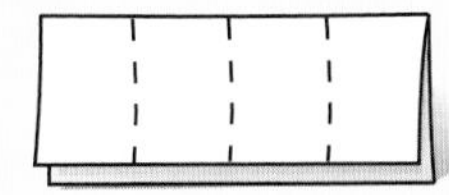

STEP 3 **Unfold,** lay the papers lengthwise, and draw lines along the folds as shown.

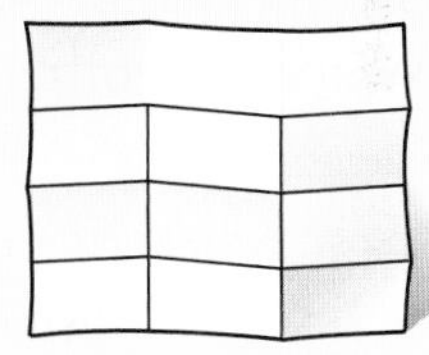

STEP 4 **Label** your tables as shown.

Climate Classification		
Tropical		
Mild		
Dry		

Climate Classification		
Continental		
Polar		
High elevation		

Make a Table As you read the chapter, define each type of climate and write notes on its weather characteristics.

Preview this chapter's content and activities at earth.msscience.com

Get Ready to Read

Monitor

1 Learn It! An important strategy to help you improve your reading is monitoring, or finding your reading strengths and weaknesses. As you read, monitor yourself to make sure the text makes sense. Discover different monitoring techniques you can use at different times, depending on the type of test and situation.

2 Practice It! The paragraph below appears in Section 2. Read the passage and answer the questions that follow. Discuss your answers with other students to see how they monitor their reading.

> Climatologists—people who study climates—usually use a system developed in 1918 by Wladimir Köppen to classify climates. Köppen observed that the types of plants found in a region depended on the climate of the area.... He classified world climates by using the annual and monthly averages of temperature and precipitation of different regions. He then related the types and distribution of native vegetation to the various climates.
>
> *—from page 488*

- What questions do you still have after reading?
- Do you understand all of the words in the passage?
- Did you have to stop reading often? Is the reading level appropriate for you?

3 Apply It! Identify one paragraph that is difficult to understand. Discuss it with a partner to improve your understanding.

Target Your Reading

Use this to focus on the main ideas as you read the chapter.

1. **Before you read** the chapter, respond to the statements below on your worksheet or on a numbered sheet of paper.
   - Write an **A** if you **agree** with the statement.
   - Write a **D** if you **disagree** with the statement.

2. **After you read** the chapter, look back to this page to see if you've changed your mind about any of the statements.
   - If any of your answers changed, explain why.
   - Change any false statements into true statements.
   - Use your revised statements as a study guide.

Reading Tip

Monitor your reading by slowing down or speeding up depending on your understanding of the text.

Before You Read A or D	Statement	After You Read A or D
	1 Climate is determined by averaging the weather of a region over a long period of time.	
	2 Latitude affects the climate of an area.	
	3 Ocean currents do not affect weather and climate.	
	4 The area on the leeward (downwind) side of a mountain experiences high rainfall.	
	5 There is only one type of climate in North America.	
	6 Over the past 100 years, Earth's average global surface temperature has increased.	
	7 El Niño can affect weather patterns, leading to droughts in some areas and flooding in others.	
	8 During the past century, atmospheric carbon dioxide has decreased.	
	9 Deforestation affects the amount of carbon dioxide in the atmosphere.	

Science Online
Print out a worksheet of this page at earth.msscience.com

section 1

What is climate?

as you read

What You'll Learn

- **Describe** what determines climate.
- **Explain** how latitude, oceans, and other factors affect the climate of a region.

Why It's Important

Climate affects the way you live.

Review Vocabulary
latitudes: distance in degrees north or south of the equator

New Vocabulary
- climate
- tropics
- polar zone
- temperate zone

Climate

If you wandered through a tropical rain forest, you would see beautiful plants flowering in shades of pink and purple beneath a canopy of towering trees. A variety of exotic birds and other animals would dart among the tree branches and across the forest floor. The sounds of singing birds and croaking frogs would surround you. All of these organisms thrive in hot temperatures and abundant rainfall. Rain forests have a hot, wet climate. **Climate** is the pattern of weather that occurs in an area over many years. It determines the types of plants or animals that can survive, and it influences how people live.

Climate is determined by averaging the weather of a region over a long period of time, such as 30 years. Scientists average temperature, precipitation, air pressure, humidity, and number of days of sunshine to determine an area's climate. Some factors that affect the climate of a region include latitude, landforms, location of lakes and oceans, and ocean currents.

Latitude and Climate

As you can see in **Figure 1,** regions close to the equator receive the most solar radiation. Latitude, a measure of distance north or south of the equator, affects climate. **Figure 2** compares cities at different latitudes. The **tropics**—the region between latitudes 23.5°N and 23.5°S—receive the most solar radiation because the Sun shines almost directly over these areas. The tropics have temperatures that are always hot, except at high elevations. The **polar zones** extend from 66.5°N and 66.5°S latitude to the poles. Solar radiation hits these zones at a low angle, spreading energy over a large area. During winter, polar regions receive little or no solar radiation. Polar regions are never warm.

Figure 1 The tropics are warmer than the temperate zones and the polar zones because the tropics receive the most direct solar energy.

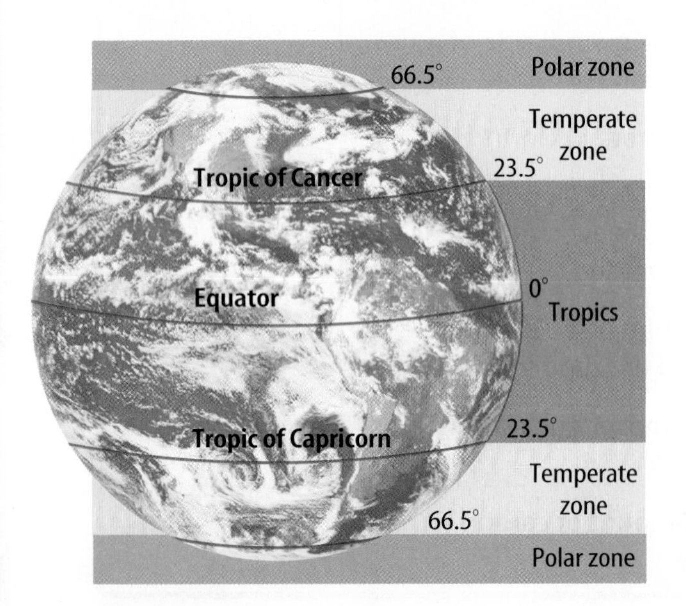

Reading Check *How does latitude affect climate?*

Between the tropics and the polar zones are the **temperate zones.** Temperatures here are moderate. Most of the United States is in a temperate zone.

Other Factors

In addition to the general climate divisions of polar, temperate, and tropical, natural features such as large bodies of water, ocean currents, and mountains affect climate within each zone. Large cities also change weather patterns and influence the local climate.

Large Bodies of Water If you live or have vacationed near an ocean, you may have noticed that water heats up and cools down more slowly than land does. This is because it takes a lot more heat to increase the temperature of water than it takes to increase the temperature of land. In addition, water must give up more heat than land does for it to cool. Large bodies of water can affect the climate of coastal areas by absorbing or giving off heat. This causes many coastal regions to be warmer in the winter and cooler in the summer than inland areas at similar latitude. Look at **Figure 2** again. You can see the effect of an ocean on climate by comparing the average temperatures in a coastal city and an inland city, both located at 37°N latitude.

Mini
LAB

Observing Solar Radiation

Procedure

1. Darken the room.
2. Hold a **flashlight** about 30 cm from a **globe.** Shine the light directly on the equator. With your finger, trace around the light.
3. Now, tilt the flashlight to shine on 30°N latitude. The size of the lighted area should increase. Repeat at 60°N latitude.

Analysis

1. How did the size and shape of the light beam change as you directed the light toward higher latitudes?
2. How does Earth's tilt affect the solar radiation received by different latitudes?

Figure 2 This map shows average daily low temperatures in four cities during January and July. It also shows average yearly precipitation.

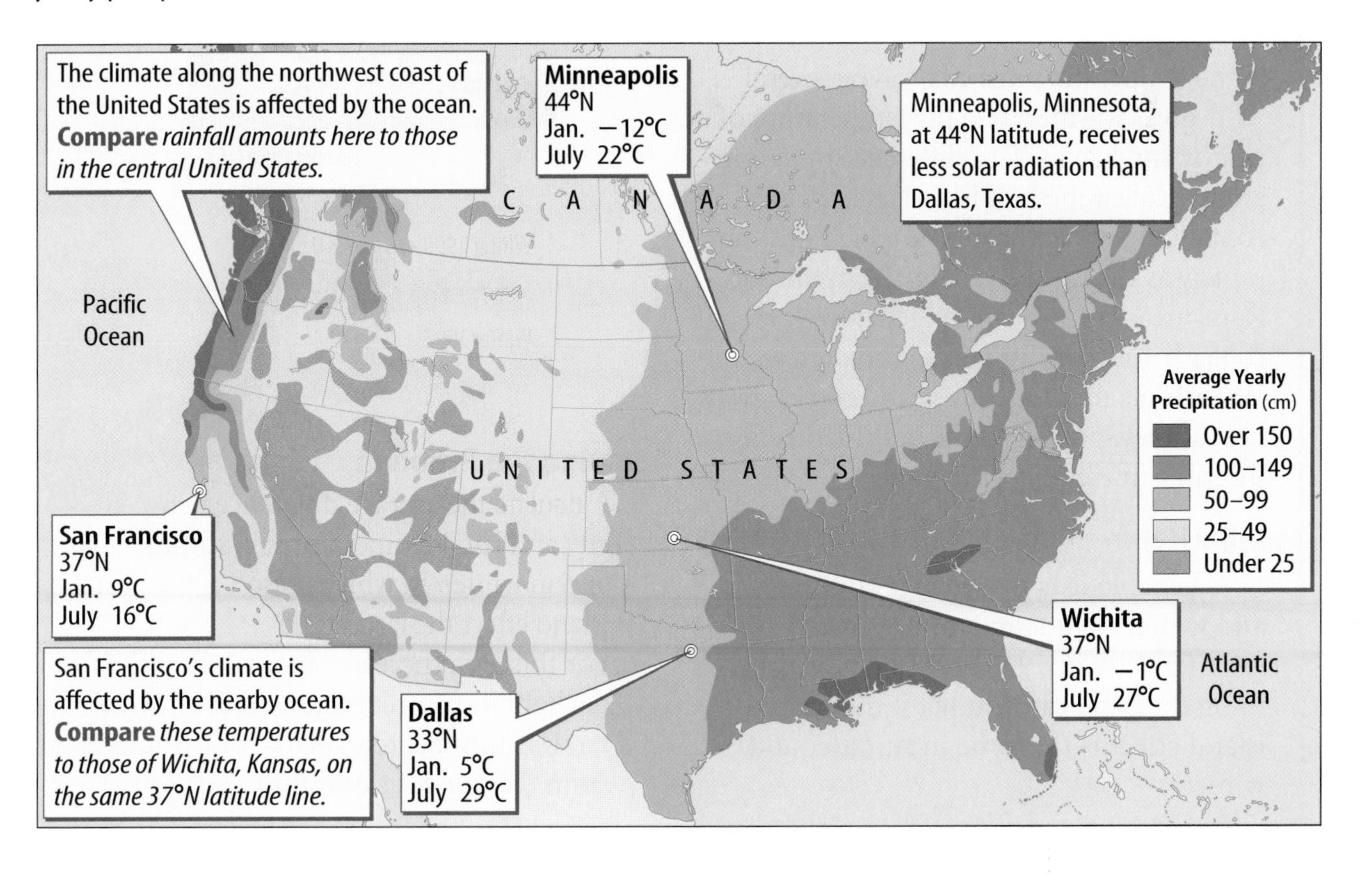
The climate along the northwest coast of the United States is affected by the ocean. **Compare** *rainfall amounts here to those in the central United States.*
Minneapolis
44°N
Jan. −12°C
July 22°C
Minneapolis, Minnesota, at 44°N latitude, receives less solar radiation than Dallas, Texas.
C A N A D A
Pacific Ocean
U N I T E D S T A T E S
Average Yearly Precipitation (cm)
Over 150
100–149
50–99
25–49
Under 25
San Francisco
37°N
Jan. 9°C
July 16°C
San Francisco's climate is affected by the nearby ocean. **Compare** *these temperatures to those of Wichita, Kansas, on the same 37°N latitude line.*
Dallas
33°N
Jan. 5°C
July 29°C
Wichita
37°N
Jan. −1°C
July 27°C
Atlantic Ocean

Mountain Air When air rises over a mountain, the air expands and its temperature decreases, causing water vapor to condense and form rain. Temperature changes caused by air expanding or contracting also occur in some machines. Why does the air coming out of a bicycle pump feel cold?

Ocean Currents Ocean currents affect coastal climates. Warm currents begin near the equator and flow toward higher latitudes, warming the land regions they pass. When the currents cool off and flow back toward the equator, they cool the air and climates of nearby land.

Reading Check *How do ocean currents affect climate?*

Winds blowing from the sea are often moister than those blowing from land. Therefore, some coastal areas have wetter climates than places farther inland. Look at the northwest coast of the United States shown in **Figure 2.** The large amounts of precipitation in Washington, Oregon, and northern California can be explained by this moist ocean air.

Mountains At the same latitude, the climate is colder in the mountains than at sea level. When radiation from the Sun is absorbed by Earth's surface, it heats the land. Heat from Earth then warms the atmosphere. Because Earth's atmosphere gets thinner at higher altitudes, the air in the mountains has fewer molecules to absorb heat.

Applying Science

How do cities influence temperature?

The temperature in a city can be several degrees warmer than the temperature of nearby rural areas. This difference in temperature is called the heat island effect. Cities contain asphalt and concrete which heat up rapidly as they absorb energy from the Sun. Rural areas covered with vegetation stay cooler because plants and soil contain water. Water heats up more slowly and carries away heat as it evaporates. Is the heat island effect the same in summer and winter?

Average Seasonal Temperatures

Season	Temperature (°C)	
	City	Rural
Winter 1996	−3.0	−4.4
Summer 1996	23.5	20.9
Winter 1997	−0.1	−1.8
Summer 1997	23.6	21.2

Identifying the Problem

The table lists the average summer and winter high temperatures in and around a city in 1996 and 1997. By examining the data, can you tell if the heat island effect is the same in summer and winter?

Solving the Problem

1. Calculate the average difference between city and rural temperatures in summer and in winter. In which season is the heat island effect the largest?
2. For this area there are about 15 hours of daylight in summer and 9 hours in winter. Use this fact to explain your results from the previous question.

Rain Shadows Mountains also affect regional climates, as shown in **Figure 3.** On the windward side of a mountain range, air rises, cools, and drops its moisture. On the leeward side of a mountain range air descends, heats up, and dries the land. Deserts are common on the leeward sides of mountains.

Figure 3 Large mountain ranges can affect climate by forcing air to rise over the windward side, cooling and bringing precipitation. The air descends with little or no moisture, creating desertlike conditions on the leeward side.

Cities Large cities affect local climates. Streets, parking lots, and buildings heat up, in turn heating the air. Air pollution traps this heat, creating what is known as the heat-island effect. Temperatures in a city can be 5°C higher than in surrounding rural areas.

section 1 review

Summary

Latitude and Climate

- Climate is the pattern of weather that occurs in an area over many years.
- The tropics receive the most solar radiation because the Sun shines most directly there.
- The polar zones receive the least solar energy due to the low-angled rays.
- Temperate zones, located between the tropics and the polar zones, have moderate temperatures.

Other Factors

- Natural features such as large bodies of water, ocean currents, and mountains can affect local and regional climates.
- Large cities can change weather patterns and influence local climates.

Self Check

1. **Explain** how two cities located at the same latitude can have different climates.
2. **Describe** how mountains affect climate.
3. **Define** the heat island effect.
4. **Compare and contrast** tropical and polar climates.
5. **Think Critically** Explain why plants found at different elevations on a mountain might differ. How can latitude affect the elevation at which some plants are found?

Applying Math

6. **Solve One-Step Equations** The coolest average summer temperature in the United States is 2°C at Barrow, Alaska, and the warmest is 37°C at Death Valley, California. Calculate the range of average summer temperatures in the United States.

section 2

Climate Types

as you read

What **You'll Learn**

- **Describe** a climate classification system.
- **Explain** how organisms adapt to particular climates.

Why **It's Important**

Many organisms can survive only in climates to which they are adapted.

Review Vocabulary
regions: places united by specific characteristics

New Vocabulary
- adaptation
- hibernation

Classifying Climates

What is the climate like where you live? Would you call it generally warm? Usually wet and cold? Or different depending on the time of year? How would you classify the climate in your region? Life is full of familiar classification systems—from musical categories to food groups. Classifications help to organize your thoughts and to make your life more efficient. That's why Earth's climates also are classified and are organized into the various types that exist. Climatologists—people who study climates—usually use a system developed in 1918 by Wladimir Köppen to classify climates. Köppen observed that the types of plants found in a region depended on the climate of the area. **Figure 4** shows one type of region Köppen might have observed. He classified world climates by using the annual and monthly averages of temperature and precipitation of different regions. He then related the types and distribution of native vegetation to the various climates.

The climate classification system shown in **Figure 5** separates climates into six groups—tropical, mild, dry, continental, polar, and high elevation. These groups are further separated into types. For example, the dry climate classification is separated into semiarid and arid.

Figure 4 The type of vegetation in a region depends on the climate. **Describe** *what these plants tell you about the climate shown here.*

Adaptations

Climates vary around the world, and as Köppen observed, the type of climate that exists in an area determines the vegetation found there. Fir trees aren't found in deserts, nor are cacti found in rain forests. In fact, all organisms are best suited for certain climates. Organisms are adapted to their environment. An **adaptation** is any structure or behavior that helps an organism survive in its environment. Structural adaptations are inherited. They develop in a population over a long period of time. Once adapted to a particular climate, organisms may not be able to survive in other climates.

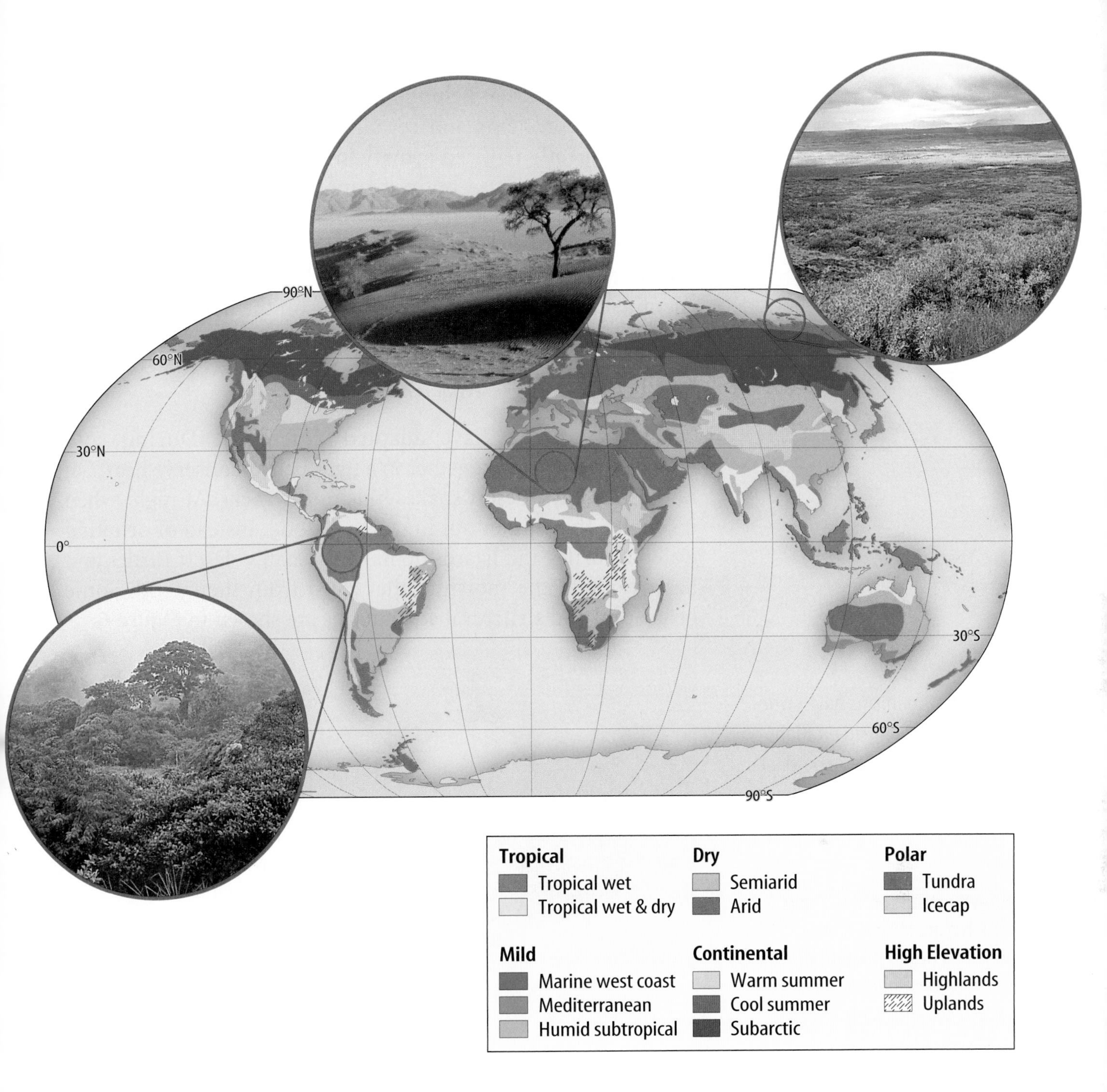

Figure 5 This map shows a climate classification system similar to the one developed by Köppen. **Describe** *the patterns you can see in the locations of certain climate types.*

INTEGRATE Life Science

Structural Adaptations Some organisms have body structures that help them survive in certain climates. The fur of mammals is really hair that insulates them from cold temperatures. A cactus has a thick, fleshy stem. This structural adaptation helps a cactus hold water. The waxy stem covering prevents water inside the cactus from evaporating. Instead of broad leaves, these plants have spiny leaves, called needles, that further reduce water loss.

Reading Check *How do cacti conserve water?*

Behavioral Adaptations Some organisms display behavioral adaptations that help them survive in a particular climate. For example, rodents and certain other mammals undergo a period of greatly reduced activity in winter called **hibernation.** During hibernation, body temperature drops and body processes are reduced to a minimum. Some of the factors thought to trigger hibernation include cooler temperatures, shorter days, and lack of adequate food. The length of time that an animal hibernates varies depending on the particular species of animal and the environmental conditions.

Reading Check *What is hibernation?*

Other animals have adapted differently. During cold weather, bees cluster together in a tight ball to conserve heat. On hot, sunny days, desert snakes hide under rocks. At night when it's cooler, they slither out in search of food. Instead of drinking water as turtles and lizards do in wet climates, desert turtles and lizards obtain the moisture they need from their food. Some behavioral and structural adaptations are shown in **Figure 6.**

Figure 6 Organisms have structural and behavioral adaptations that help them survive in particular climates.

These hibernating bats have adapted their behavior to survive winter.

The needles and the waxy skin of a cactus are structural adaptations to a desert climate. **Infer** *how these adaptations help cacti conserve water.*

Polar bears have structural adaptations to keep them warm. The hairs of their fur trap air and heat.

Figure 7 Lungfish survive periods of intense heat and drought by going into an inactive state called estivation. During the dry season when water evaporates, lungfish dig into the mud and curl up in a small chamber they make at the lake's bottom. During the wet season, lungfish reemerge to live in small lakes and pools.

Estivation Lungfish, shown in **Figure 7,** survive periods of intense heat by entering an inactive state called estivation (es tuh VAY shun). As the weather gets hot and water evaporates, the fish burrows into mud and covers itself in a leathery mixture of mud and mucus. It lives this way until the warm, dry months pass.

Like other organisms, you have adaptations that help you adjust to climate. In hot weather, your sweat glands release water onto your skin. The water evaporates, taking some heat with it. As a result, you become cooler. In cold weather, you may shiver to help your body stay warm. When you shiver, the rapid muscle movements produce some heat. What other adaptations to climate do people have?

section 2 review

Summary

Classifying Climates

- Climatologists classify climates into six main groups: tropical, mild, dry, continental, polar, and high elevation.

Adaptations

- Adaptations are any structures or behaviors that help an organism to survive.
- Structural adaptations such as fur, hair, and spiny needles help an organism to survive in certain climates.
- Behavioral adaptations include hibernation, a period of greatly reduced activity in winter; estivation, an inactive state during intense heat; clustering together in the cold; and obtaining water from food when water is not found elsewhere.

Self Check

1. **List** Use **Figure 5** and a world map to identify the climate type for each of the following locations: Cuba, North Korea, Egypt, and Uruguay.
2. **Compare and contrast** hibernation and estivation.
3. **Think Critically** What adaptations help dogs keep cool during hot weather?

Applying Skills

4. **Form Hypotheses** Some scientists have suggested that Earth's climate is getting warmer. What effects might this have on vegetation and animal life in various parts of the United States?
5. **Communicate** Research the types of vegetation found in the six climate regions shown in **Figure 5.** Write a paragraph in your Science Journal describing why vegetation can be used to help define climate boundaries.

section 3

Climatic Changes

as you read

What You'll Learn

- **Explain** what causes seasons.
- **Describe** how El Niño affects climate.
- **Explore** possible causes of climatic change.

Why It's Important

Changing climates could affect sea level and life on Earth.

Review Vocabulary

solar radiation: energy from the Sun transferred by waves or rays

New Vocabulary

- season
- El Niño
- greenhouse effect
- global warming
- deforestation

Earth's Seasons

In temperate zones, you can play softball under the summer Sun and in the winter go sledding with friends. Weather changes with the season. **Seasons** are short periods of climatic change caused by changes in the amount of solar radiation an area receives. **Figure 8** shows Earth revolving around the Sun. Because Earth is tilted, different areas of Earth receive changing amounts of solar radiation throughout the year.

Seasonal Changes Because of fairly constant solar radiation near the equator, the tropics do not have much seasonal temperature change. However, they do experience dry and rainy seasons. The middle latitudes, or temperate zones, have warm summers and cool winters. Spring and fall are usually mild.

Reading Check *What are seasons like in the tropics?*

Figure 8 As Earth revolves around the Sun, different areas of Earth are tilted toward the Sun, which causes different seasons.
Identify *During which northern hemisphere season is Earth closer to the Sun?*

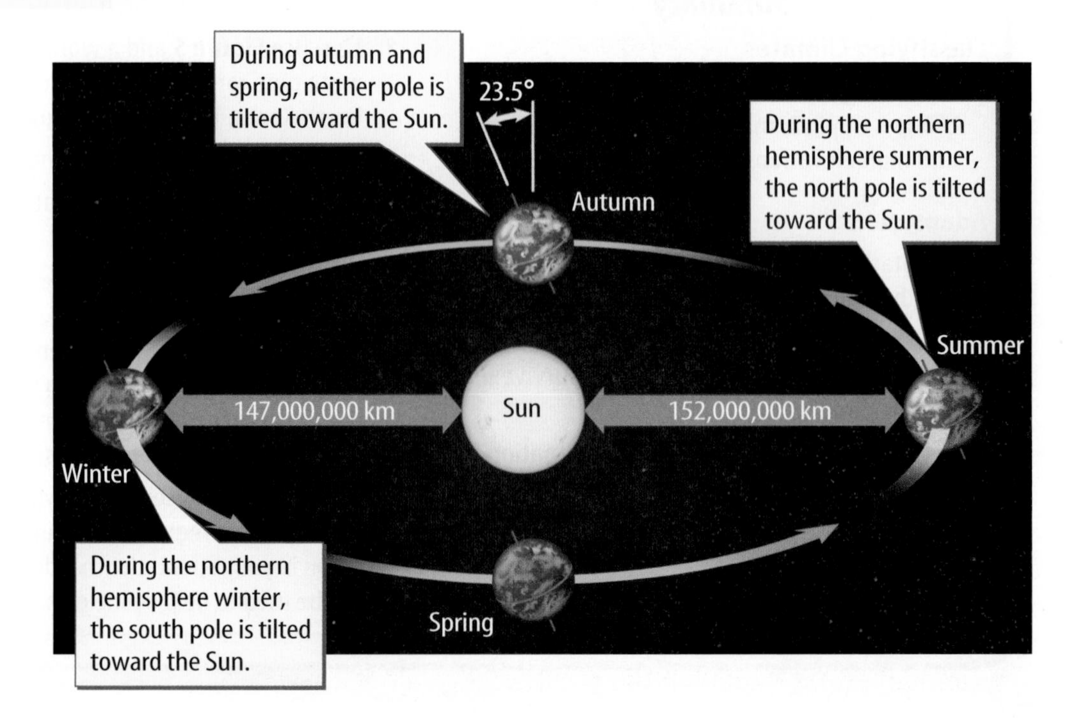
During autumn and spring, neither pole is tilted toward the Sun.
23.5°
Autumn
During the northern hemisphere summer, the north pole is tilted toward the Sun.
Summer
147,000,000 km
Sun
152,000,000 km
Winter
During the northern hemisphere winter, the south pole is tilted toward the Sun.
Spring

Figure 9 A strong El Niño, like the one that occurred in 1998, can affect weather patterns around the world.

A severe drought struck Indonesia, contributing to forest fires.

California was plagued by large storms that produced pounding surf and shoreline erosion.

High Latitudes During the year, the high latitudes near the poles have great differences in temperature and number of daylight hours. As shown in **Figure 8,** during summer in the northern hemisphere, the north pole is tilted toward the Sun. During summer at the north pole, the Sun doesn't set for nearly six months. During that same time, the Sun never rises at the south pole. At the equator days are about the same length all year long.

El Niño and La Niña

El Niño (el NEEN yoh) is a climatic event that involves the tropical Pacific Ocean and the atmosphere. During normal years, strong trade winds that blow east to west along the equator push warm surface water toward the western Pacific Ocean. Cold, deep water then is forced up from below along the coast of South America. During El Niño years, these winds weaken and sometimes reverse. The change in the winds allows warm, tropical water in the upper layers of the Pacific to flow back eastward to South America. Cold, deep water is no longer forced up from below. Ocean temperatures increase by 1°C to 7°C off the coast of Peru.

El Niño can affect weather patterns. It can alter the position and strength of one of the jet streams. This changes the atmospheric pressure off California and wind and precipitation patterns around the world. This can cause drought in Australia and Africa. This also affects monsoon rains in Indonesia and causes storms in California, as shown in **Figure 9.**

The opposite of El Niño is La Niña, shown in **Figure 10.** During La Niña, the winds blowing across the Pacific are stronger than normal, causing warm water to accumulate in the western Pacific. The water in the eastern Pacific near Peru is cooler than normal. La Niña may cause droughts in the southern United States and excess rainfall in the northwestern United States.

Mini LAB

Modeling El Niño

Procedure

1. During El Niño, trade winds blowing across the Pacific Ocean from east to west slacken or even reverse. Surface waters move back toward the coast of Peru.
2. Add **warm water** to a **9-in × 13-in baking pan** until it is two-thirds full. Place the pan on a smooth countertop.
3. Blow as hard as you can across the surface of the water along the length of the pan. Next, blow with less force. Then, blow in the opposite direction.

Analysis

1. What happened to the water as you blew across its surface? What was different when you blew with less force and when you blew from the opposite direction?
2. Explain how this is similar to what happens during an El Niño event.

Try at Home

VISUALIZING EL NIÑO AND LA NIÑA

NATIONAL GEOGRAPHIC

Figure 10

Weather in the United States can be affected by changes that occur thousands of kilometers away. Out in the middle of the Pacific Ocean, periodic warming and cooling of a huge mass of seawater—phenomena known as El Niño and La Niña, respectively—can impact weather across North America. During normal years (right), when neither El Niño nor La Niña is in effect, strong winds tend to keep warm surface waters contained in the western Pacific while cooler water wells up to the surface in the eastern Pacific.

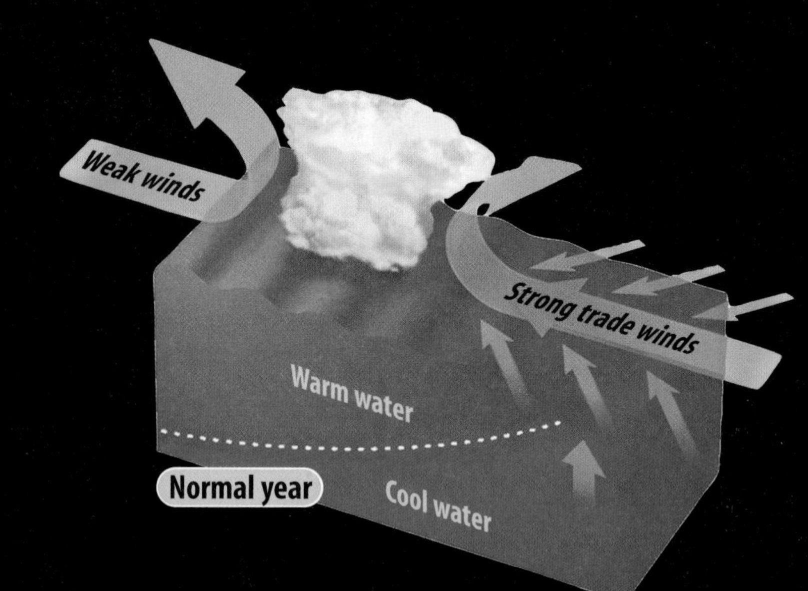

EL NIÑO During El Niño years, winds blowing west weaken and may even reverse. When this happens, warm waters in the western Pacific move eastward, preventing cold water from upwelling. These changes can alter global weather patterns and trigger heavier-than-normal precipitation across much of the United States.

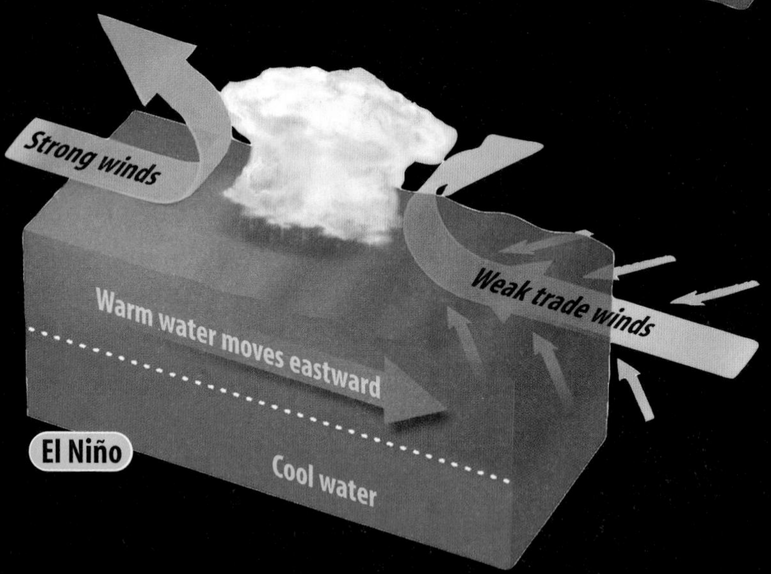

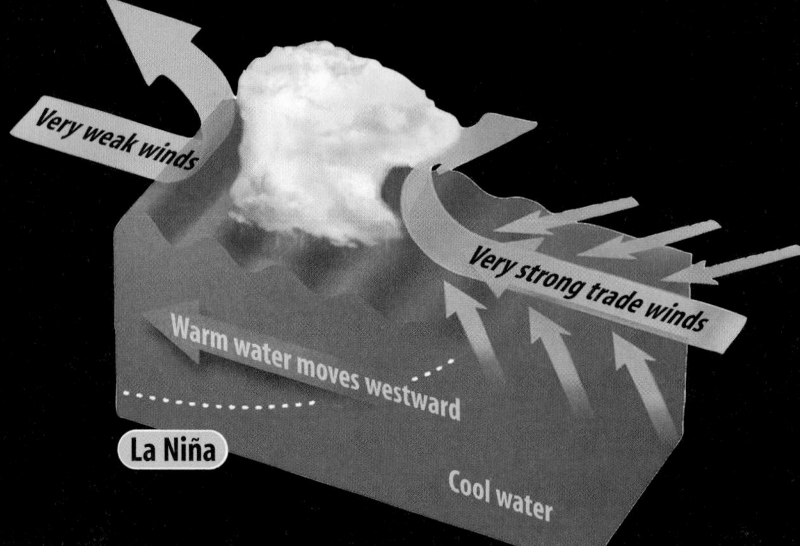

LA NIÑA During La Niña years, stronger-than-normal winds push warm Pacific waters farther west, toward Asia. Cold, deep-sea waters then well up strongly in the eastern Pacific, bringing cooler and often drier weather to many parts of the United States.

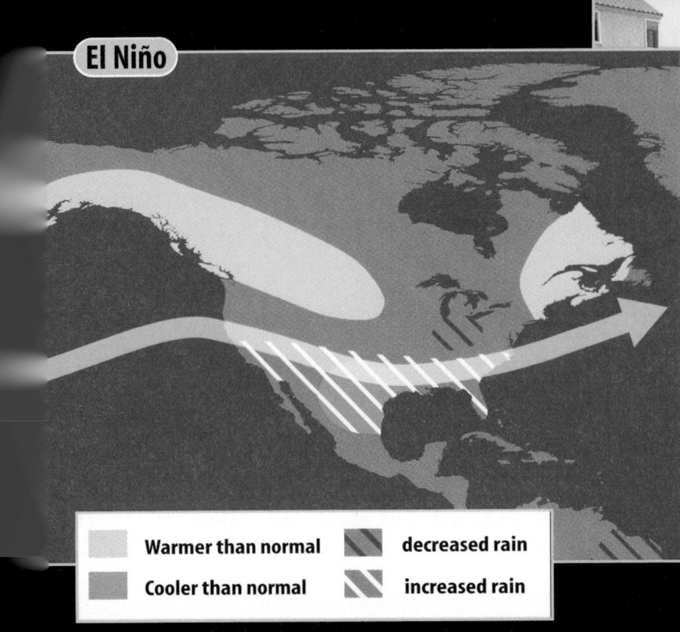

Sun-warmed surface water spans the Pacific Ocean during El Niño years. Clouds form above the warm ocean, carrying moisture aloft. The jet stream, shown by the white arrow above, helps bring some of this warm, moist air to the United States.

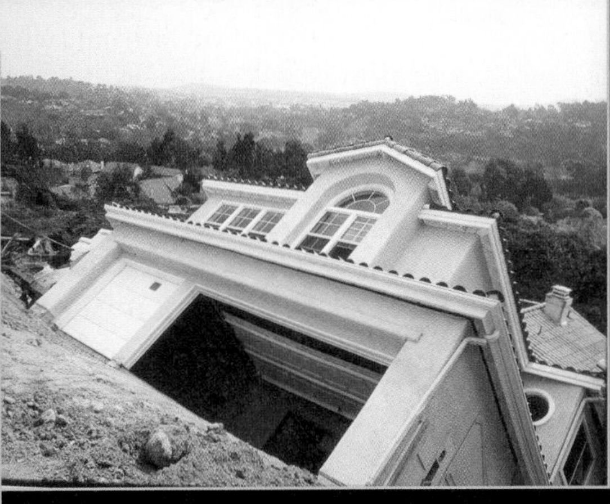

▲ LANDSLIDE Heavy rains in California resulting from El Niño can lead to landslides. This upended house in Laguna Niguel, California, took a ride downhill during the El Niño storms of 1998.

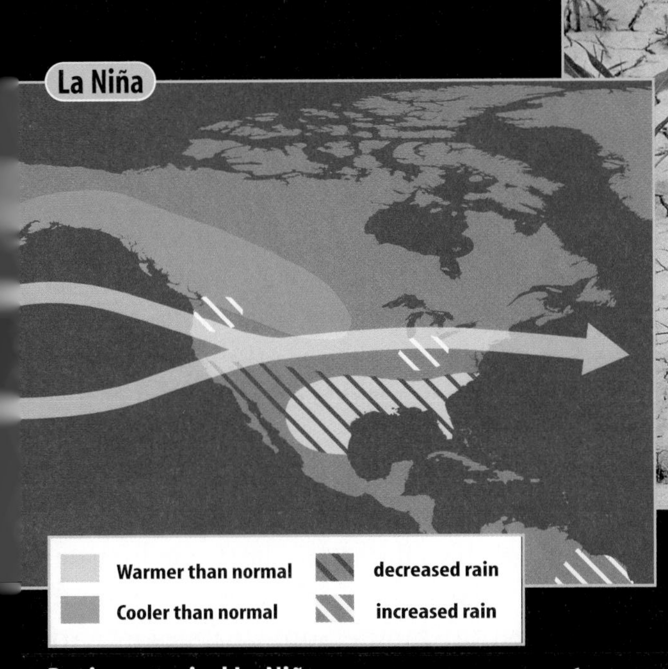

During a typical La Niña year, warm ocean waters, clouds, and moisture are pushed away from North America. A weaker jet stream often brings cooler weather to the northern parts of the continent and hot, dry weather to southern areas.

▲ PARCHED LAND The Southeast may experience drought conditions, like those that struck the cornfields of Montgomery County, Maryland, during the La Niña summer of 1988.

Climatic Change

If you were exploring in Antarctica near Earth's south pole and found a 3-million-year-old fossil of a warm-weather plant or animal, what would it tell you? You might conclude that the climate of that region changed because Antarctica is much too cold for similar plants and animals to survive today. Some warm-weather fossils found in polar regions indicate that at times in Earth's past, worldwide climate was much warmer than at present. At other times Earth's climate has been much colder than it is today.

Sediments in many parts of the world show that at several different times in the past 2 million years, glaciers covered large parts of Earth's surface. These times are called ice ages. During the past 2 million years, ice ages have alternated with warm periods called interglacial intervals. Ice ages seem to last 60,000 to 100,000 years. Most interglacial periods are shorter, lasting 10,000 to 15,000 years. We are now in an interglacial interval that began about 11,500 years ago. Additional evidence suggests that climate can change even more quickly. Ice cores record climate in a way similar to tree rings. Cores drilled in Greenland show that during the last ice age, colder times lasting 1,000 to 2,000 years changed quickly to warmer spells that lasted about as long. **Figure 11** shows a scientist working with ice cores.

Figure 11 Some ice cores consist of layers of ice that record detailed climate information for individual years. These ice cores can cover more than 300,000 years. **Describe** *how this is helpful.*

What causes climatic change?

Climatic change has many varied causes. These causes of climatic change can operate over short periods of time or very long periods of time. Catastrophic events, including meteorite collisions and large volcanic eruptions, can affect climate over short periods of time, such as a year or several years. These events add solid particles and liquid droplets to the upper atmosphere, which can change climate. Another factor that can alter Earth's climate is short- or long-term changes in solar output, which is the amount of energy given off by the Sun. Changes in Earth's movements in space affect climate over many thousands of years, and movement of Earth's crustal plates can change climate over millions of years. All of these things can work separately or together to alter Earth's climate.

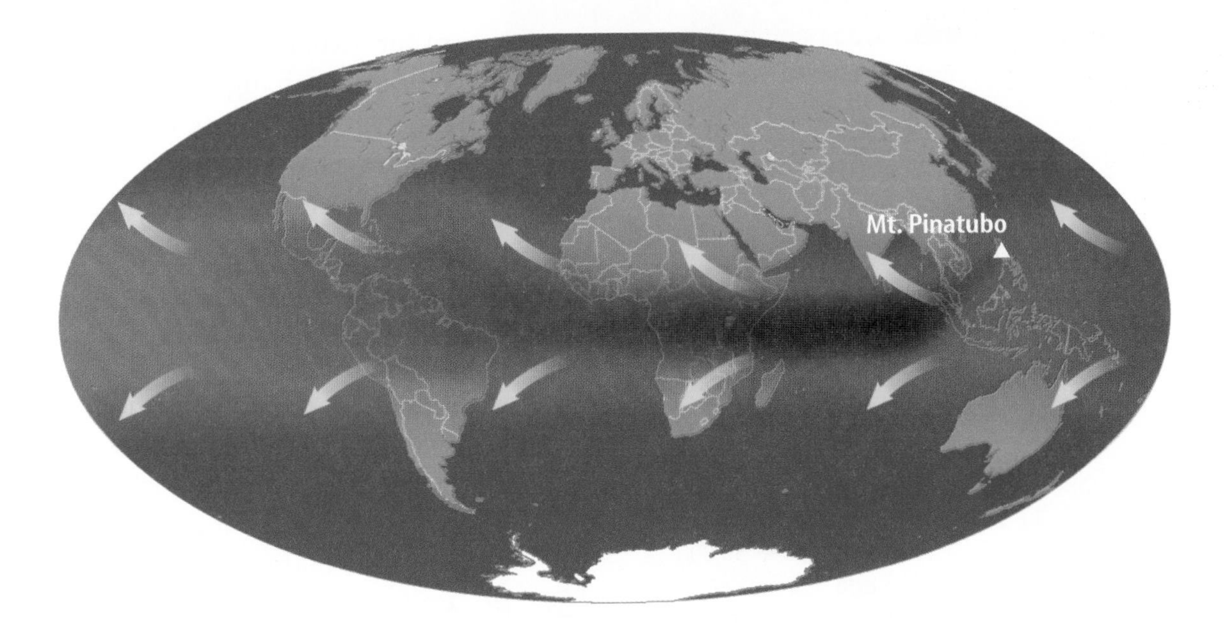

Atmospheric Solids and Liquids Small solid and liquid particles always are present in Earth's atmosphere. These particles can enter the atmosphere naturally or be added to the atmosphere by humans as pollution. Some ways that particles enter the atmosphere naturally include volcanic eruptions, soot from fires, and wind erosion of soil particles. Humans add particles to the atmosphere through automobile exhaust and smokestack emissions. These small particles can affect climate.

Catastrophic events such as meteorite collisions and volcanic eruptions put enormous volumes of dust, ash, and other particles into the atmosphere. These particles block so much solar radiation that they can cool the planet. **Figure 12** shows how a major volcanic eruption affected Earth's atmosphere.

In cities, particles put into the atmosphere as pollution can change the local climate. These particles can increase the amount of cloud cover downwind from the city. Some studies have even suggested that rainfall amounts can be reduced in these areas. This may happen because many small cloud droplets form rather than larger droplets that could produce rain.

Energy from the Sun Solar radiation provides Earth's energy. If the output of radiation from the Sun varies, Earth's climate could change. Some changes in the amount of energy given off by the Sun seem to be related to the presence of sunspots. Sunspots are dark spots on the surface of the Sun. **WARNING:** *Never look directly at the Sun.* Evidence supporting the link between sunspots and climate includes an extremely cold period in Europe between 1645 and 1715. During this time, very few sunspots appeared on the Sun.

Figure 12 Mount Pinatubo in the Philippines erupted in 1991. During the eruption, particles were spread high into the atmosphere and circled the globe. Over time, particles spread around the world, blocking some of the Sun's energy from reaching Earth. The gray areas show how particles from the eruption moved around the world.

Air Quality Control/Monitor
Atmospheric particles from pollution can affect human health as well as climate. These small particles, often called particulates, can enter the lungs and cause tissue damage. The Department of Environmental Protection employs people to monitor air pollution and its causes. Research what types of laws air quality control monitors must enforce.

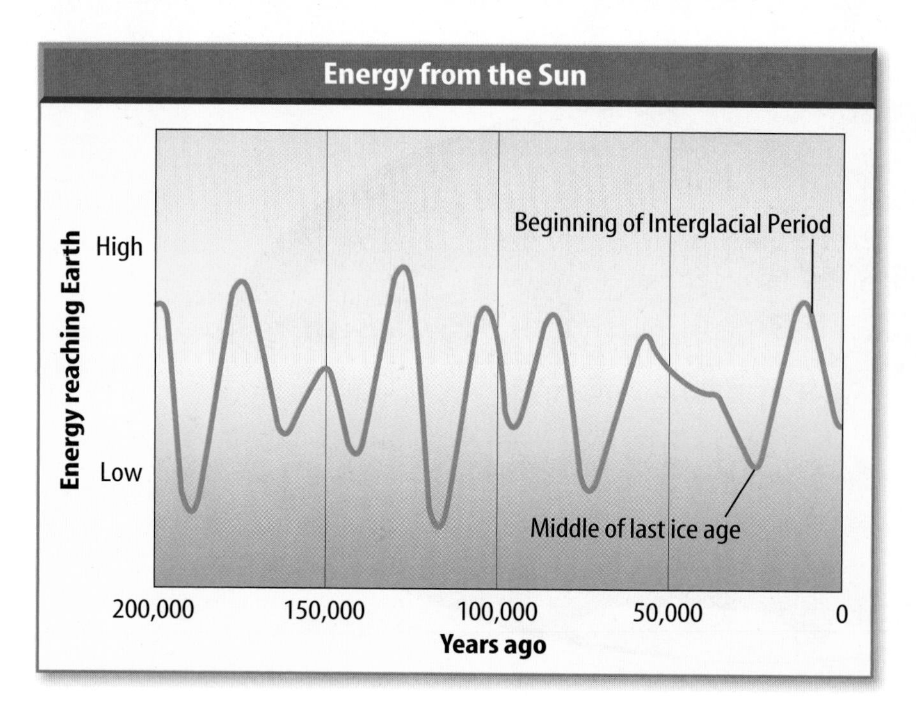

Figure 13 The curving line shows how the amount of the Sun's energy that strikes the northern hemisphere changed over the last 200,000 years.
Describe *the amount of energy that reached the northern hemisphere during the last ice age.*

Earth Movements Another explanation for some climatic changes involves Earth's movements in space. Earth's axis currently is tilted 23.5° from perpendicular to the plane of its orbit around the Sun. In the past, this tilt has increased to 24.5° and has decreased to 21.5°. When this tilt is at its maximum, the change between summer and winter is probably greater. Earth's tilt changes about every 41,000 years. Some scientists hypothesize that the change in tilt affects climate.

Two additional Earth movements also cause climatic change. Earth's axis wobbles in space just like the axis of a top wobbles when it begins to spin more slowly. This can affect the amount of solar energy received by different parts of Earth. Also, the shape of Earth's orbit changes. Sometimes it is more circular than at present and sometimes it is more flattened. The shape of Earth's orbit changes over a 100,000-year cycle.

Amount of Solar Energy These movements of Earth cause the amount of solar energy reaching different parts of Earth to vary over time, as shown in **Figure 13.** These changes might have caused glaciers to grow and shrink over the last few million years. However, they do not explain why glaciers have occurred so rarely over long spans of geologic time.

Crustal Plate Movement Another explanation for major climatic change over tens or hundreds of millions of years concerns the movement of Earth's crustal plates. The movement of continents and oceans affects the transfer of heat on Earth, which in turn affects wind and precipitation patterns. Through time, these altered patterns can change climate. One example of this is when movement of Earth's plates created the Himalaya about 40 million years ago. The growth of these mountains changed climate over much of Earth.

As you've learned, many theories attempt to answer questions about why Earth's climate has changed through the ages. Probably all of these things play some role in changing climates. More study needs to be done before all the factors that affect climate will be understood.

Climatic Changes Today

Beginning in 1992, representatives from many countries have met to discuss the greenhouse effect and global climate change. These subjects also have appeared frequently in the headlines of newspapers and magazines. Some people are concerned that the greenhouse effect could be responsible for some present-day warming of Earth's atmosphere and oceans.

The **greenhouse effect** is a natural heating process that occurs when certain gases in Earth's atmosphere trap heat. Radiation from the Sun strikes Earth's surface and causes it to warm. Some of this heat then is radiated back toward space. Some gases in the atmosphere, known as greenhouse gases, absorb a portion of this heat and then radiate heat back toward Earth, as shown in **Figure 14.** This keeps Earth warmer than it would be otherwise.

There are many natural greenhouse gases in Earth's atmosphere. Water vapor, carbon dioxide, and methane are some of the most important ones. Without these greenhouse gases, life would not be possible on Earth. Like Mars, Earth would be too cold. However, if the greenhouse effect is too strong, Earth could get too warm. High levels of carbon dioxide in its atmosphere indicate that this has happened on the planet Venus.

Topic: Greenhouse Effect
Visit earth.msscience.com for Web links to information about the greenhouse effect.

Activity Research changes in the greenhouse effect over the last 200 years. Infer what might be causing the changes.

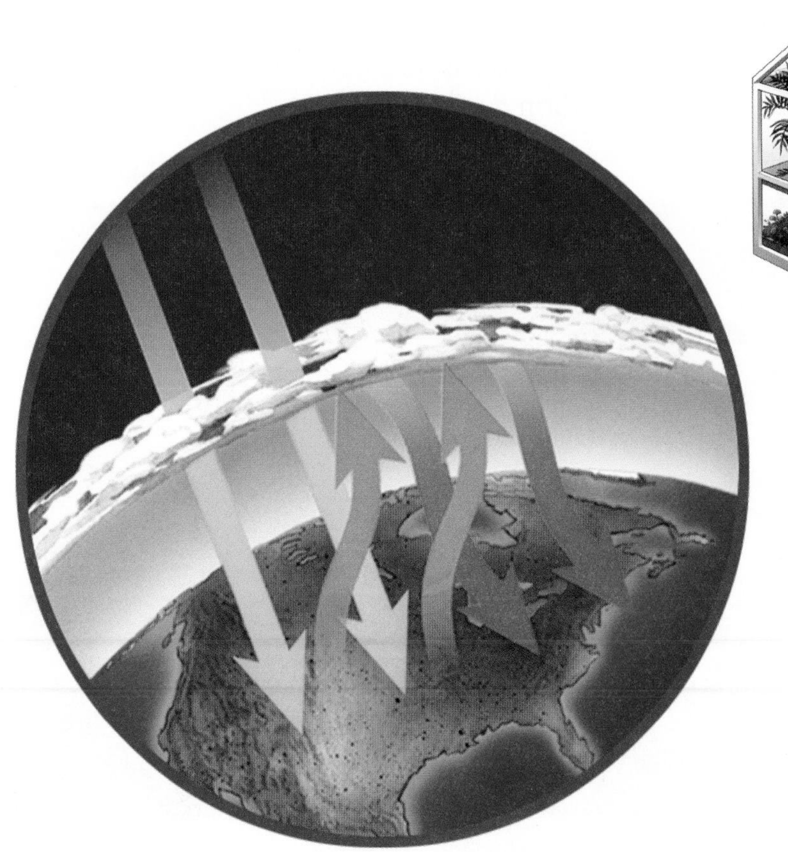

Figure 14 The Sun's radiation travels through Earth's atmosphere and heats the surface. Gases in our atmosphere trap the heat. **Compare and contrast** *this to the way a greenhouse works.*

Global Warming

Over the past 100 years, the average global surface temperature on Earth has increased by about 0.6°C. This increase in temperature is known as **global warming.** Over the same time period, atmospheric carbon dioxide has increased by about 20 percent. As a result, researchers hypothesize that the increase in global temperatures may be related to the increase in atmospheric carbon dioxide. Other hypotheses include the possibility that global warming might be caused by changes in the energy emitted by the Sun.

If Earth's average temperature continues to rise, many glaciers could melt. When glaciers melt, the extra water causes sea levels to rise. Low-lying coastal areas could experience increased flooding. Already some ice caps and small glaciers are beginning to melt and recede, as shown in **Figure 15.** Sea level is rising in some places. Some scientific studies show that these events are related to Earth's increased temperature.

You learned in the previous section that organisms are adapted to their environments. When environments change, can organisms cope? In some tropical waters around the world, corals are dying. Many people think these deaths are caused by warmer water to which the corals are not adapted.

Some climate models show that in the future, Earth's temperatures will increase faster than they have in the last 100 years. However, these predictions might change because of uncertainties in the climate models and in estimating future increases in atmospheric carbon dioxide.

Figure 15 This glacier in Greenland might have receded from its previous position because of global warming. The pile of sediment in front shows how far the glacier once reached.

Figure 16 When forests are cleared or burned, carbon dioxide levels increase in the atmosphere.

Human Activities

Human activities affect the air in Earth's atmosphere. Burning fossil fuels and removing vegetation increase the amount of carbon dioxide in the atmosphere. Because carbon dioxide is a greenhouse gas, it might contribute to global warming. Each year, the amount of carbon dioxide in the atmosphere continues to increase.

Burning Fossil Fuels When natural gas, oil, and coal are burned for energy, the carbon in these fossil fuels combines with atmospheric oxygen to form carbon dioxide. This increases the amount of carbon dioxide in Earth's atmosphere. Studies indicate that humans have increased carbon dioxide levels in the atmosphere by about 25 percent over the last 150 years.

Deforestation Destroying and cutting down trees, called **deforestation**, also affects the amount of carbon dioxide in the atmosphere. Forests, such as the one shown in **Figure 16,** are cleared for mining, roads, buildings, and grazing cattle. Large tracts of forest have been cleared in every country on Earth. Tropical forests have been decreasing at a rate of about one percent each year for the past two decades.

As trees grow, they take in carbon dioxide from the atmosphere. Trees use this carbon dioxide to produce wood and leaves. When trees are cut down, the carbon dioxide they could have removed from the atmosphere remains in the atmosphere. Cut-down trees often are burned for fuel or to clear the land. Burning trees produces even more carbon dioxide.

Reading Check *What can humans do to slow carbon dioxide increases in the atmosphere?*

Topic: Deforestation
Visit earth.msscience.com for Web links to information about deforestation.

Activity Collect data on the world's decline in forests. Infer what the world's forests will be like in 100 years.

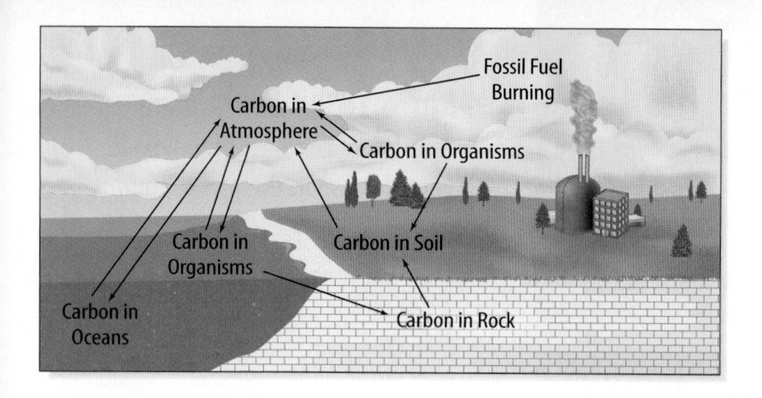

Figure 17 Carbon constantly is cycled among the atmosphere, oceans, solid earth, and biosphere.

The Carbon Cycle

Carbon, primarily as carbon dioxide, is constantly recycled in nature among the atmosphere, Earth's oceans, and organisms that inhabit the land. Organisms that undergo photosynthesis on land and in the water take in carbon dioxide and produce and store carbon-based food. This food is consumed by non-photosynthetic organisms. Carbon dioxide is released as food is broken down to release energy. When organisms die and decay, some carbon is stored as humus in soil and some carbon is released as carbon dioxide. This carbon cycle is illustrated in **Figure 17.**

Some carbon dioxide in the atmosphere dissolves in the oceans, and is used by algae and other photosynthetic, aquatic organisms. Just as on land, aquatic organisms give off carbon dioxide. However, Earth's oceans currently absorb more carbon dioxide from the atmosphere than they give off.

When Earth's climate changes, the amount of carbon dioxide that cycles among atmosphere, ocean, and land also can change. Some people hypothesize that if Earth's climate continues to warm, more carbon dioxide may be absorbed by oceans and land. Scientists continue to collect data to study any changes in the global carbon cycle.

section 3 review

Summary

Earth's Seasons

- Seasons are short periods of climatic changes due to Earth's tilt on its axis while revolving around the Sun, causing differing amounts of solar energy to reach areas of Earth.

El Niño and La Niña

- El Niño begins in the tropical Pacific Ocean when trade winds weaken or reverse directions, disrupting the normal temperature and precipitation patterns around the globe.

Climatic Changes Today

- The greenhouse effect is a natural heating process that occurs when certain gases in Earth's atmosphere trap heat.
- Burning fossil fuels increases the amount of carbon dioxide in the air.
- Deforestation increases the amount of carbon dioxide in the atmosphere.

Self Check

1. **Explain** how Earth's tilted axis is responsible for seasons.
2. **Compare and contrast** El Niño and La Niña. What climate changes do they demonstrate?
3. **List** factors that can cause Earth's climate to change.
4. **Explain** how people are adding carbon dioxide to the atmosphere.
5. **Think Critically** If Earth's climate continues to warm, how might your community be affected?

Applying Skills

6. **Use Models** Using a globe, model the three movements of Earth in space that can cause climatic change.
7. **Use a word processor** to make a table that lists the different processes that might cause Earth's climate to change. Include in your table a description of the process and how it causes climate to change.

earth.msscience.com/self_check_quiz

The Greenhouse Effect

Do you remember climbing into the car on a warm, sunny day? Why was it so hot inside the car when it wasn't that hot outside? It was hotter in the car because the car functioned like a greenhouse. You experienced the greenhouse effect.

Real-World Question

How can you demonstrate the greenhouse effect?

Goals

- **Model** the greenhouse effect.
- **Measure and graph** temperature changes.

Materials

identical large, empty glass jars (2)
lid for one jar
nonmercury thermometers (3)

Safety Precautions

WARNING: *Be careful when you handle glass thermometers. If a thermometer breaks, do not touch it. Have your teacher dispose of the glass safely.*

Procedure

1. Lay a thermometer inside each jar.
2. Place the jars next to each other by a sunny window. Lay the third thermometer between the jars.
3. **Record** the temperatures of the three thermometers. They should be the same.
4. Place the lid on one jar.
5. **Record** the temperatures of all three thermometers at the end of 5, 10, and 15 min.
6. Make a line graph that shows the temperatures of the three thermometers for the 15 min of the experiment.

Conclude and Apply

1. **Explain** why you placed a thermometer between the two jars.
2. **List** the constants in this experiment. What was the variable?
3. **Identify** which thermometer showed the greatest temperature change during your experiment.
4. **Analyze** what occurred in this experiment. How was the lid in this experiment like the greenhouse gases in the atmosphere?
5. **Infer** from this experiment why you should never leave a pet inside a closed car in warm weather.

Communicating Your Data

Give a brief speech describing your conclusions to your class.

MICROCLIMATES

Goals

- **Observe** temperature, wind speed, relative humidity, and precipitation in areas outside your school.
- **Identify** local microclimates.

Materials

thermometers
psychrometer
paper strip or wind sock
large cans (4 or 5)
** beakers or rain gauges (4 or 5)*
unlined paper
**Alternate materials*

Safety Precautions

WARNING: *If a thermometer breaks, do not touch it. Have your teacher dispose of the glass safely.*

Real-World Question

A microclimate is a localized climate that differs from the main climate of a region. Buildings in a city, for instance, can affect the climate of the surrounding area. Large buildings, such as the Bank of America Plaza in Dallas, Texas, can create microclimates by blocking the Sun or changing wind patterns. Does your school create microclimates?

Procedure

1. Select four or five sites around your school building. Also, select a control site well away from the school.
2. Attach a thermometer to an object near each of the locations you selected. Set up a rain gauge, beaker, or can to collect precipitation.
3. Visit each site at two predetermined times, one in the morning and one in the afternoon, each day for a week. Record the temperature and measure any precipitation that might have fallen. Use a wind sock or paper strip to determine wind direction.

Relative Humidity

Dry Bulb Temperature (°C)	Dry Bulb Temperature Minus Wet Bulb Temperature (°C)									
	1	2	3	4	5	6	7	8	9	10
14	90	79	70	60	51	42	34	26	18	10
15	90	80	71	61	53	44	36	27	20	13
16	90	81	71	63	54	46	38	30	23	15
17	90	81	72	64	55	47	40	32	25	18
18	91	82	73	65	57	49	41	34	27	20
19	91	82	74	65	58	50	43	36	29	22
20	91	83	74	66	59	51	44	37	31	24
21	91	83	75	67	60	53	46	39	32	26
22	92	83	76	68	61	54	47	40	34	28
23	92	84	76	69	62	55	48	42	36	30
24	92	84	77	69	62	56	49	43	37	31
25	92	84	77	70	63	57	50	44	39	33

4. To find relative humidity, you'll need to use a psychrometer. A psychrometer is an instrument with two thermometers—one wet and one dry. As moisture from the wet thermometer evaporates, it takes heat energy from its environment, and the environment immediately around the wet thermometer cools. The thermometer records a lower temperature. Relative humidity can be found by finding the difference between the wet thermometer and the dry thermometer and by using the chart on the previous page. Record all of your weather data.

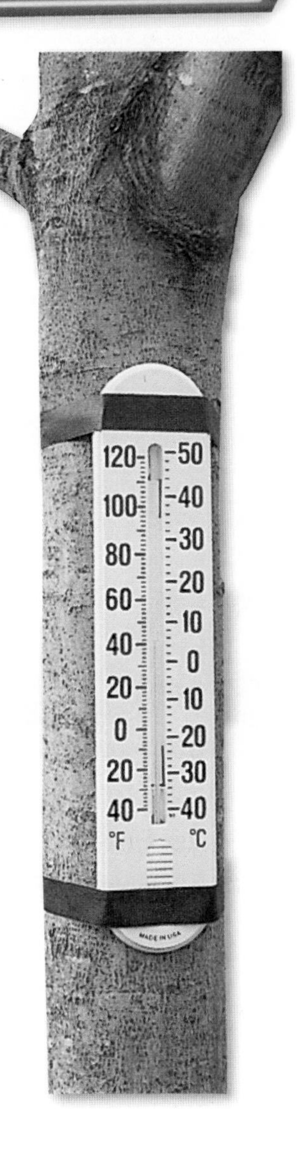

Analyze Your Data

1. Make separate line graphs for temperature, relative humidity, and precipitation for your morning and afternoon data. Make a table showing wind direction data.
2. **Compare and contrast** weather data for each of your sites. What microclimates did you identify around your school building? How did these climates differ from the control site? How did they differ from each other?

Conclude and Apply

1. **Explain** Why did you take weather data at a control site away from the school building? How did the control help you analyze and interpret your data?
2. **Infer** what conditions could have caused the microclimates that you identified. Are your microclimates similar to those that might exist in a large city? Explain.

Communicating Your Data

Use your graphs to make a large poster explaining your conclusions. Display your poster in the school building. **For more help, refer to the Science Skill Handbook.**

TIME SCIENCE AND HISTORY

SCIENCE CAN CHANGE THE COURSE OF HISTORY!

The Year there was No Summer

You've seen pictures of erupting volcanoes. One kind of volcano sends smoke, rock, and ash high into the air above the crater. Another kind of volcano erupts with fiery, red-hot rivers of lava snaking down its sides. Erupting volcanoes are nature's forces at their mightiest, causing destruction and death. But not everyone realizes how far-reaching the destruction can be. Large volcanic eruptions can affect people thousands of kilometers away. In fact, major volcanic eruptions can have effects that reach around the globe.

An erupting volcano can temporarily change Earth's climate. The ash a volcano ejects into the atmosphere can create day after day without sunshine. Other particles move high into the atmosphere and are carried all the way around Earth, sometimes causing global temperatures to drop for several months.

The Summer That Never Came

An example of a volcanic eruption with wide-ranging effects occurred in 1783 in Iceland, an island nation in the North Atlantic Ocean. Winds carried a black cloud of ash from an erupting volcano in Iceland westward across northern Canada, Alaska, and across the Pacific Ocean to Japan. The summer turned bitterly cold in these places. Water froze, and heavy snowstorms pelted the land. Sulfurous gases from the erupting volcano combined with water to form particles of acid that reflected solar energy back into space. This "blanket" in the atmosphere kept the Sun's rays from heating up part of Earth.

The most tragic result of this eruption was the death of many Kauwerak people, who lived in western Alaska. Only a handful of Kauwerak survived the summer that never came. They had no opportunity to catch needed foods to keep them alive through the following winter.

Locate Using an atlas, locate Indonesia and Iceland. Using reference materials, find five facts about each place. Make a map of each nation and illustrate the map with your five facts.

For more information, visit earth.msscience.com/time

Reviewing Main Ideas

Section 1 What is climate?

1. An area's climate is the average weather over a long period of time, such as 30 years.
2. The three main climate zones are tropical, polar, and temperate.
3. Features such as oceans, mountains, and even large cities affect climate.

Section 2 Climate Types

1. Climates can be classified by various characteristics, such as temperature, precipitation, and vegetation. World climates commonly are separated into six major groups.
2. Organisms have structural and behavioral adaptations that help them survive in particular climates. Many organisms can survive only in the climate they are adapted to.
3. Adaptations develop in a population over a long period of time.

Section 3 Climatic Changes

1. Seasons are caused by the tilt of Earth's axis as Earth revolves around the Sun.
2. El Niño disrupts normal temperature and precipitation patterns around the world.
3. Geological records show that over the past few million years, Earth's climate has alternated between ice ages and warmer periods.
4. The greenhouse effect occurs when certain gases trap heat in Earth's atmosphere.
5. Carbon dioxide enters the atmosphere when fossil fuels such as oil and coal are burned.

Visualizing Main Ideas

Copy and complete the following concept map on climate.

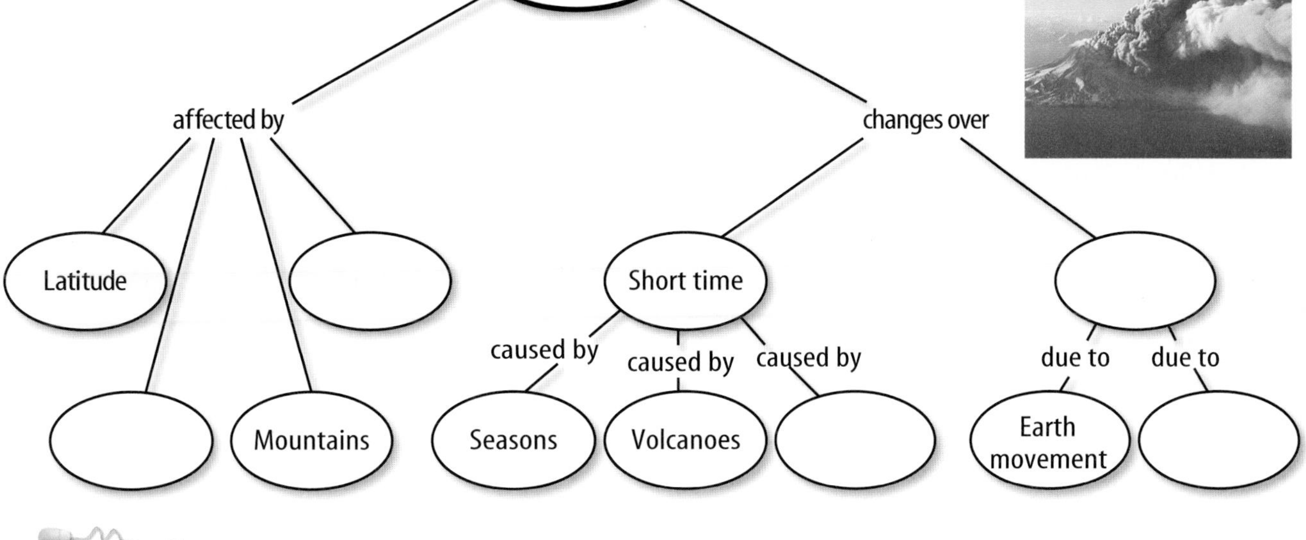

Climate
affected by
changes over
Latitude
Short time
caused by
caused by
caused by
due to
due to
Mountains
Seasons
Volcanoes
Earth movement

chapter 17 Review

Using Vocabulary

adaptation p. 488	hibernation p. 490
climate p. 484	polar zone p. 484
deforestation p. 501	season p. 492
El Niño p. 493	temperate zone p. 484
global warming p. 500	tropics p. 484
greenhouse effect p. 499	

Fill in the blanks with the correct vocabulary word or words.

1. Earth's north pole is in the ____________.
2. ____________ causes the Pacific Ocean to become warmer off the coast of Peru.
3. During ____________, an animal's body temperature drops.
4. ____________ is the pattern of weather that occurs over many years.
5. ____________ means global temperatures are rising.

Checking Concepts

Choose the word or phrase that best answers the question.

6. Which of the following is a greenhouse gas in Earth's atmosphere?
   A) helium **C)** hydrogen
   B) carbon dioxide **D)** oxygen

7. During which of the following is the eastern Pacific warmer than normal?
   A) El Niño **C)** summer
   B) La Niña **D)** spring

8. Which latitude receives the most direct rays of the Sun year-round?
   A) 60°N **C)** 30°S
   B) 90°N **D)** 0°

9. What happens as you climb a mountain?
   A) temperature decreases
   B) temperature increases
   C) air pressure increases
   D) air pressure remains constant

10. Which of the following is true of El Niño?
    A) It cools the Pacific Ocean near Peru.
    B) It causes flooding in Australia.
    C) It cools the waters off Alaska.
    D) It may occur when the trade winds slacken or reverse.

11. What do changes in Earth's orbit affect?
    A) Earth's shape **C)** Earth's rotation
    B) Earth's climate **D)** Earth's tilt

12. The Köppen climate classification system includes categories based on precipitation and what other factor?
    A) temperature **C)** winds
    B) air pressure **D)** latitude

13. Which of the following is an example of structural adaptation?
    A) hibernation **C)** fur
    B) migration **D)** estivation

14. Which of these can people do in order to help reduce global warming?
    A) burn coal **C)** conserve energy
    B) remove trees **D)** produce methane

Use the illustration below to answer question 15.

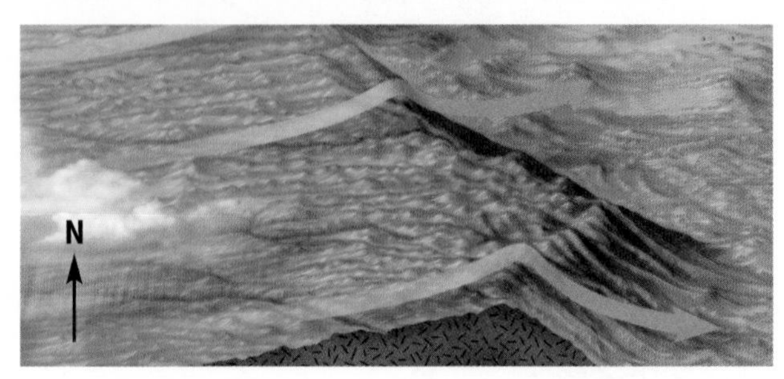

15. What would you most likely find on the leeward side of this mountain range?
    A) lakes **C)** deserts
    B) rain forests **D)** glaciers

Science Online earth.msscience.com/vocabulary_puzzlemaker

Thinking Critically

16. **Draw a Conclusion** How could climate change cause the types of organisms in an area to change?
17. **Infer** What might you infer if you find fossils of tropical plants in a desert?
18. **Describe** On a summer day, why would a Florida beach be cooler than an orange grove that is 2 km inland?
19. **Infer** what would happen to global climates if the Sun emitted more energy.
20. **Explain** why it will be cooler if you climb to a higher elevation in a desert.
21. **Communicate** Explain how atmospheric pressure over the Pacific Ocean might affect how the trade winds blow.
22. **Predict** Make a chain-of-events chart to explain the effect of a major volcanic eruption on climate.
23. **Form Hypotheses** A mountain glacier in South America has been getting smaller over several decades. What hypotheses should a scientist consider to explain why this is occurring?
24. **Concept Map** Copy and complete the concept map using the following: *tropics, 0°–23.5° latitude, polar, temperate, 23.5°–66.5° latitude,* and *66.5° latitude to poles.*

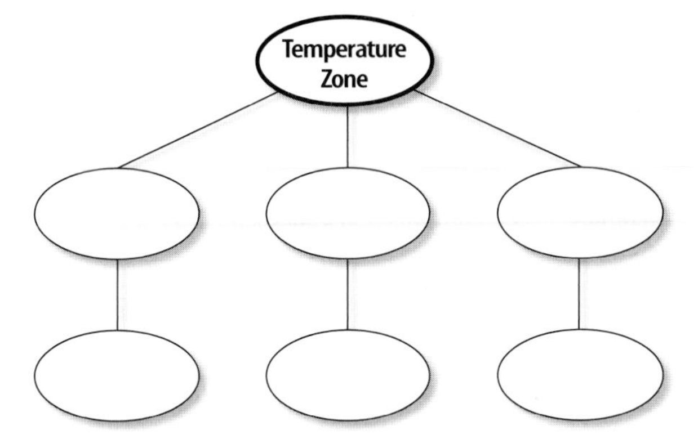

25. **Explain** how global warming might lead to the extinction of some organisms.
26. **Describe** how dust and ash from large volcanoes can change the atmosphere.
27. **Explain** how heat energy carried by ocean currents influences climate.
28. **Describe** how sediments, fossils, and ice cores record Earth's geologic history.
29. **Describe** how volcanic eruptions or meteorite collisions have changed past climates.

Performance Activities

30. **Science Display** Make a display illustrating different factors that can affect climate. Be sure to include detailed diagrams and descriptions for each factor in your display. Present your display to the class.

Applying Math

Use the table below to answer questions 31 and 32.

Precipitation in Phoenix, Arizona

Season	Precipitation (cm)
Winter	5.7
Spring	1.2
Summer	6.7
Autumn	5.9
Total	19.5

31. **Precipitation Amounts** The following table gives average precipitation amounts for Phoenix, Arizona. Make a bar graph of these data. Which climate type do you think Phoenix represents?
32. **Local Precipitation** Use the table above to help estimate seasonal precipitation for your city or one that you choose. Create a bar graph for that data.

Part 1 Multiple Choice

Record your answers on the answer sheet provided by your teacher or on a sheet of paper.

Use the graph below to answer questions 1 and 2.

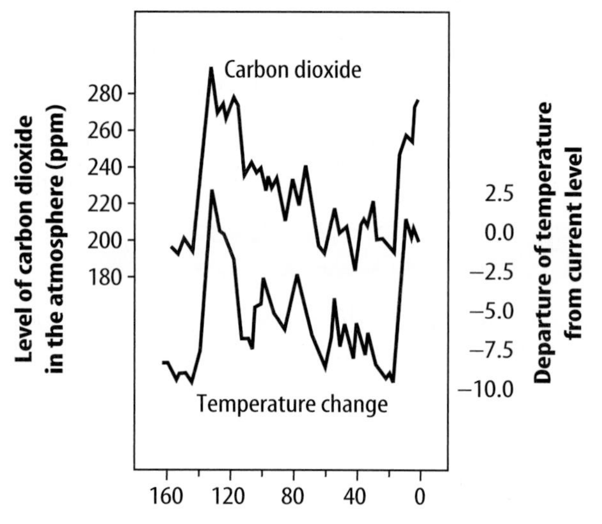

1. Which of these statements is true according to the graph?

A. Earth's mean temperature has never been hotter than it is today.
B. The level of CO_2 has never been higher than today.
C. The mean global temperature 60,000 years ago was less than today.
D. The level of CO_2 in the atmosphere 80,000 years ago was 280 parts per million.

2. Which of the following statements best describes this graph?

A. As CO_2 levels have increased, so has global temperature.
B. As CO_2 levels have increased, global temperature has decreased.
C. As global temperature has increased, CO_2 levels have decreased.
D. No relationship exists between CO_2 and global temperatures.

Use the table below to answer question 3.

Apparent Temperature Index					
	Relative Humidity (%)				
Air Temperature (°F)		80	85	90	95
	85	97	99	102	105
	80	86	87	88	89
	75	78	78	79	79
	70	71	71	71	71

3. The National Weather Service created the Apparent Temperature Index to show the temperature the human body feels when heat and humidity are combined. If the relative humidity is 85% and the temperature is 75°F, what is the apparent temperature?

A. 78°F
B. 79°F
C. 88°F
D. 89°F

4. What is the most likely reason that the air temperature is warmest at the tropical latitudes?

A. These latitudes receive the most solar radiation because there are no clouds.
B. These latitudes receive the most solar radiation because the sun's angle is high.
C. These latitudes receive the least solar radiation because the sun's angle is low.
D. These latitudes receive the least solar radiation because of heavy cloud cover.

Test-Taking Tip

Qualifiers Look for quantifiers in a question. Such questions are not looking for absolute answers. Qualifiers would be words such as *most likely, most common,* or *least common.*

Question 2 Look for the most likely scientific explanation.

Part 2 Short Response/Grid In

Record your answers on the answer sheet provided by your teacher or on a sheet of paper.

5. Explain how a large body of water can affect the climate of a nearby area.
6. Describe the relationship between ocean currents and precipitation in a coastal region.
7. The city of Redmond, Oregon is near the Cascade Mountain Range. The average annual rainfall for the Redmond, OR area is about 8 inches. Infer whether Redmond, OR is located on the windward side or leeward side of the mountain range. Explain your answer.

Use the figure below to answer question 8.

8. What is the greenhouse effect?
9. List three greenhouse gases.
10. How does the greenhouse effect positively affect life on Earth? How could it negatively affect life on Earth?
11. Explain why the temperature of a city can be up to 5°C warmer than the surrounding rural areas.
12. What are the different ways in which solid and liquid particles enter the Earth's atmosphere?

Part 3 Open Ended

Record your answers on a sheet of paper.

Use the figure below to answer questions 13–15.

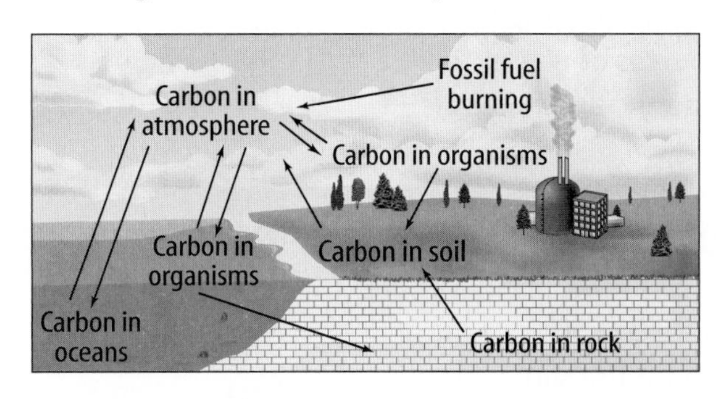

13. Describe the carbon cycle. Explain how carbon is transferred from organisms to soil.
14. How does the burning of fossil fuels affect the amount of carbon dioxide entering the carbon cycle?
15. How does deforestation affect the amount of carbon dioxide entering the carbon cycle?
16. In 1991 Mt. Pinatubo erupted, releasing volcanic particulates into the atmosphere. Temperatures around the world fell by as much as 0.7°C below average during 1992. How was this global temperature change related to the volcanic eruption?
17. What is global warming? What hypotheses help explain global warming? Explain the relationship between global warming and the level of seawater.
18. A scientist analyzes the pollen of ancient plants found preserved in lake sediments. The pollen is determined to be from a plant that needs moisture and year-round warm temperatures to grow. Make an inference about the type of climate that area experienced during the time the plant lived.

chapter 18

Ocean Motion

The BIG Idea

Oceans are important sources of water, food, energy, and minerals.

SECTION 1
Ocean Water
Main Idea Elements are added to the oceans at about the same rate that they are removed.

SECTION 2
Ocean Currents
Main Idea An ocean current is a mass movement, or flow, of ocean water.

SECTION 3
Ocean Waves and Tides
Main Idea In the ocean, waves and tides move through seawater.

The Power of Waves

Surfers from around the world experience firsthand the enormous power of moving water. Wind blowing across the ocean surface can create small ripples, the wave shown above in Fiji, and even the giant waves of hurricanes.

Science Journal Record in your Science Journal some facts you know about ocean currents, waves, or tides. Include some pictures to show your ideas.

Start-Up Activities

Explore How Currents Work

Surface currents are caused by wind. Deep-water currents are created by differences in the density of ocean water. Several factors affect water density. One is temperature. Do the lab below to see how temperature differences create deep-water currents.

1. In a bowl, mix ice and cold water to make ice water.
2. Fill a beaker with warm tap water.
3. Add a few drops of food coloring to the ice water and stir the mixture.
4. Use a dropper to place some of this ice water on top of the warm water.
5. **Think Critically** In your Science Journal, describe what happened. Did adding cold water on top produce a current? Look up the word *convection* in a dictionary. Infer why the current you created is called a convection current.

Preview this chapter's content and activities at earth.msscience.com

Ocean Motion Make the following Foldable to help you understand the cause-and-effect relationship of ocean motion.

STEP 1 **Fold** a vertical sheet of paper in half from top to bottom.

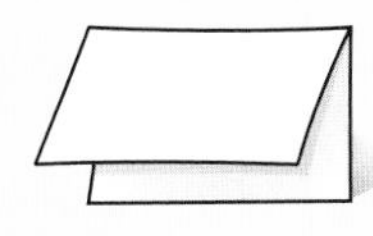

STEP 2 **Fold** in half from side to side with the previous fold at the top.

STEP 3 **Unfold** the paper once. **Cut** only the fold of the top flap to make two tabs.

STEP 4 **Turn** the paper vertically and **label** the front tabs as shown.

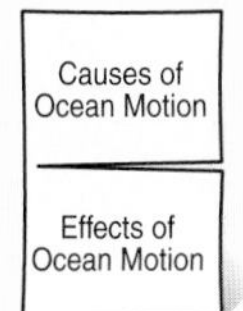

Read and Write As you read the chapter, write what you learn about why the ocean moves and the effects of ocean motion under the appropriate tabs.

Get Ready to Read

Visualize

1 Learn It! Visualize by forming mental images of the text as you read. Imagine how the text descriptions look, sound, feel, smell, or taste. Look for any pictures or diagrams on the page that may help you add to your understanding.

2 Practice It! Read the following paragraph. As you read, use the underlined details to form a picture in your mind.

> Notice that waves look like hills and valleys. The crest is the highest point of the wave. The trough (TRAWF) is the lowest point of a wave. Wavelength is the horizontal distance between the crests or between the troughs of two adjacent waves. Wave height is the vertical distance between crest and trough.
>
> *—from page 524*

Based on the description above, try to visualize the parts of a wave. Now look at the illustration on page 524.

- How closely does it match your mental picture?
- Reread the passage and look at the picture again. Did your ideas change?
- Compare your image with what others in your class visualized.

3 Apply It! Read the chapter and list three subjects you were able to visualize. Make a rough sketch showing what you visualized.

Forming your own mental images will help you remember what you read.

Target Your Reading

Use this to focus on the main ideas as you read the chapter.

1. **Before you read** the chapter, respond to the statements below on your worksheet or on a numbered sheet of paper.
   - Write an **A** if you **agree** with the statement.
   - Write a **D** if you **disagree** with the statement.

2. **After you read** the chapter, look back to this page to see if you've changed your mind about any of the statements.
   - If any of your answers changed, explain why.
   - Change any false statements into true statements.
   - Use your revised statements as a study guide.

Before You Read A or D	Statement		After You Read A or D
	1	Approximately 50 percent of Earth's surface is covered by ocean water.	
	2	Ocean water contains dissolved gases and salts.	
	3	Salinity is a measure of the amount of gases dissolved in seawater.	
	4	Surface currents are powered by gravity.	
	5	Upwelling is a vertical circulation in the ocean that brings deep, cold water to the ocean surface.	
	6	Deep in the ocean, waters circulate because of density differences.	
	7	Density currents circulate water rapidly.	
	8	Wavelength is the vertical distance between the crest and the trough of a wave.	
	9	A tide is caused by a giant wave produced by the gravitational pull of the Sun and the Moon.	
	10	As Earth rotates, different locations on its surface pass through high and low tide.	

Science Online
Print out a worksheet of this page at earth.msscience.com

section 1

Ocean Water

as you read

What You'll Learn

- **Identify** the origin of the water in Earth's oceans.
- **Explain** how dissolved salts and other substances get into seawater.
- **Describe** the composition of seawater.

Why It's Important

Oceans are a reservoir of valuable food, energy, and mineral resources.

Review Vocabulary
resource: a reserve source of supply, such as a material or mineral

New Vocabulary
- basin
- salinity

Importance of Oceans

Imagine yourself lying on a beach and listening to the waves gently roll onto shore. A warm breeze blows off the water, making it seem as if you're in a tropical paradise. It's easy to appreciate the oceans under these circumstances, but the oceans affect your life in other ways, too.

Varied Resources Oceans are important sources of food, energy, and minerals. **Figure 1** shows two examples of food resources collected from oceans. Energy sources such as oil and natural gas are found beneath the ocean floor. Oil wells often are drilled in shallow water. Mineral resources including copper and gold are mined in shallow waters as well. Approximately one-third of the world's table salt is extracted from seawater through the process of evaporation. Oceans also allow for the efficient transportation of goods. For example, millions of tons of oil, coal, and grains are shipped over the oceans each year.

Reading Check *What resources come from oceans?*

Figure 1 People depend on the oceans for many resources.

Krill are tiny, shrimplike animals that live in the Antarctic Ocean. Some cultures use krill in noodles and rice cakes.

Kelp is a fast-growing seaweed that is a source of algin, used in making ice cream, salad dressing, medicines, and cosmetics.

Origin of Oceans

During Earth's first billion years, its surface, shown in the top portion of **Figure 2,** was much more volcanically active than it is today. When volcanoes erupt, they spew lava and ash, and they give off water vapor, carbon dioxide, and other gases. Scientists hypothesize that about 4 billion years ago, this water vapor began to be stored in Earth's early atmosphere. Over millions of years, it cooled enough to condense into storm clouds. Torrential rains began to fall. Shown in the bottom portion of **Figure 2,** oceans were formed as this water filled low areas on Earth called **basins.** Today, approximately 70 percent of Earth's surface is covered by ocean water.

Figure 2 Earth's oceans formed from water vapor.

Water vapor was released into the atmosphere by volcanoes that also gave off other gases, such as carbon dioxide and nitrogen.

Condensed water vapor formed storm clouds. Oceans formed when basins filled with water from torrential rains.

Composition of Oceans

Ocean water contains dissolved gases such as oxygen, carbon dioxide, and nitrogen. Oxygen is the gas that almost all organisms need for respiration. It enters the oceans in two ways—directly from the atmosphere and from organisms that photosynthesize. Carbon dioxide enters the ocean from the atmosphere and from organisms when they respire. The atmosphere is the only important source of nitrogen gas. Bacteria combine nitrogen and oxygen to create nitrates, which are important nutrients for plants.

If you've ever tasted ocean water, you know that it is salty. Ocean water contains many dissolved salts. Chloride, sodium, sulfate, magnesium, calcium, and potassium are some of the ions in seawater. An ion is a charged atom or group of atoms. Some of these ions come from rocks that are dissolved slowly by rivers and groundwater. These include calcium, magnesium, and sodium. Rivers carry these chemicals to the oceans. Erupting volcanoes add other ions, such as bromide and chloride.

How do sodium and chloride ions get into seawater?

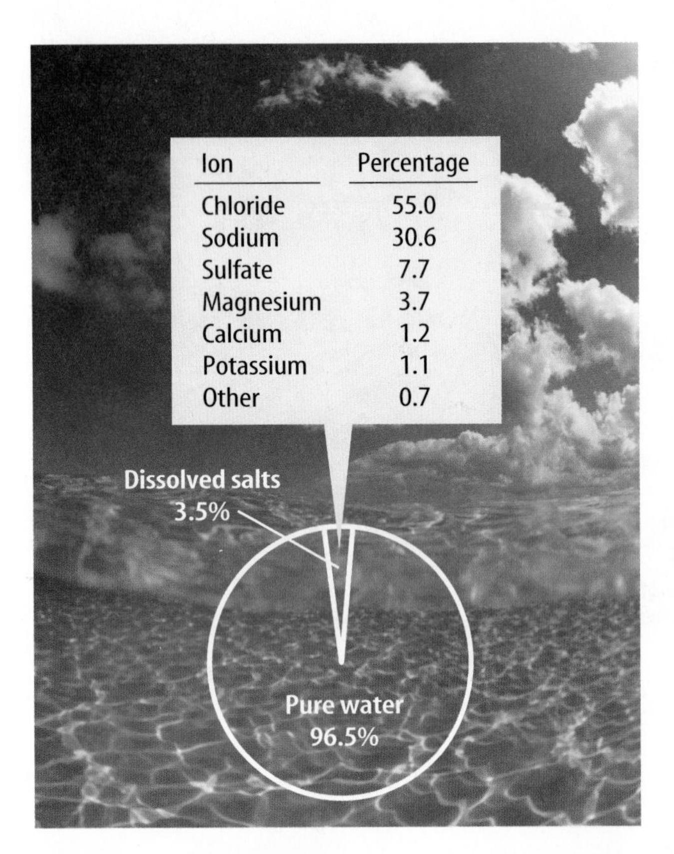

Ion
Percentage
Chloride 55.0
Sodium 30.6
Sulfate 7.7
Magnesium 3.7
Calcium 1.2
Potassium 1.1
Other 0.7
Dissolved salts
3.5%
Pure water
96.5%

Figure 3 Ocean water contains about 3.5 percent dissolved salts.
Calculate *If you evaporated 1,000 g of seawater, how many grams of salt would be left?*

Salts The most abundant elements in sea water are the hydrogen and oxygen that make up water. Many other ions are found dissolved in seawater. When seawater is evaporated, these ions combine to form materials called salts. Sodium and chloride make up most of the ions in seawater. If seawater evaporates, the sodium and chloride ions combine to form a salt called halite. Halite is the common table salt you use to season food. It is this dissolved salt and similar ones that give ocean water its salty taste.

Salinity (say LIH nuh tee) is a measure of the amount of salts dissolved in seawater. It usually is measured in grams of dissolved salt per kilogram of seawater. One kilogram of ocean water contains about 35 g of dissolved salts, or 3.5 percent. The chart in **Figure 3** shows the most abundant ions in ocean water. The proportion and amount of dissolved salts in seawater remain nearly constant and have stayed about the same for hundreds of millions of years. This tells you that the composition of the oceans is in balance. Evidence that scientists have gathered indicates that Earth's oceans are not growing saltier.

INTEGRATE Life Science

Removal of Elements Although rivers, volcanoes, and the atmosphere constantly add material to the oceans, the oceans are considered to be in a steady state. This means that elements are added to the oceans at about the same rate that they are removed. Dissolved salts are removed when they precipitate out of ocean water and become part of the sediment. Some marine organisms use dissolved salts to make body parts. Some remove calcium ions from the water to form bones. Other animals, such as oysters, use the dissolved calcium to form shells. Some algae, called diatoms, have silica shells. Because many organisms use calcium and silicon, these elements are removed more quickly from seawater than elements such as chlorine or sodium.

Desalination Salt can be removed from ocean water by a process called desalination (dee sa luh NAY shun). If you have ever swum in the ocean, you know what happens when your skin dries. The white, flaky substance on your skin is salt. As seawater evaporates, salt is left behind. As demand for freshwater increases throughout the world, scientists are working on technology to remove salt to make seawater drinkable.

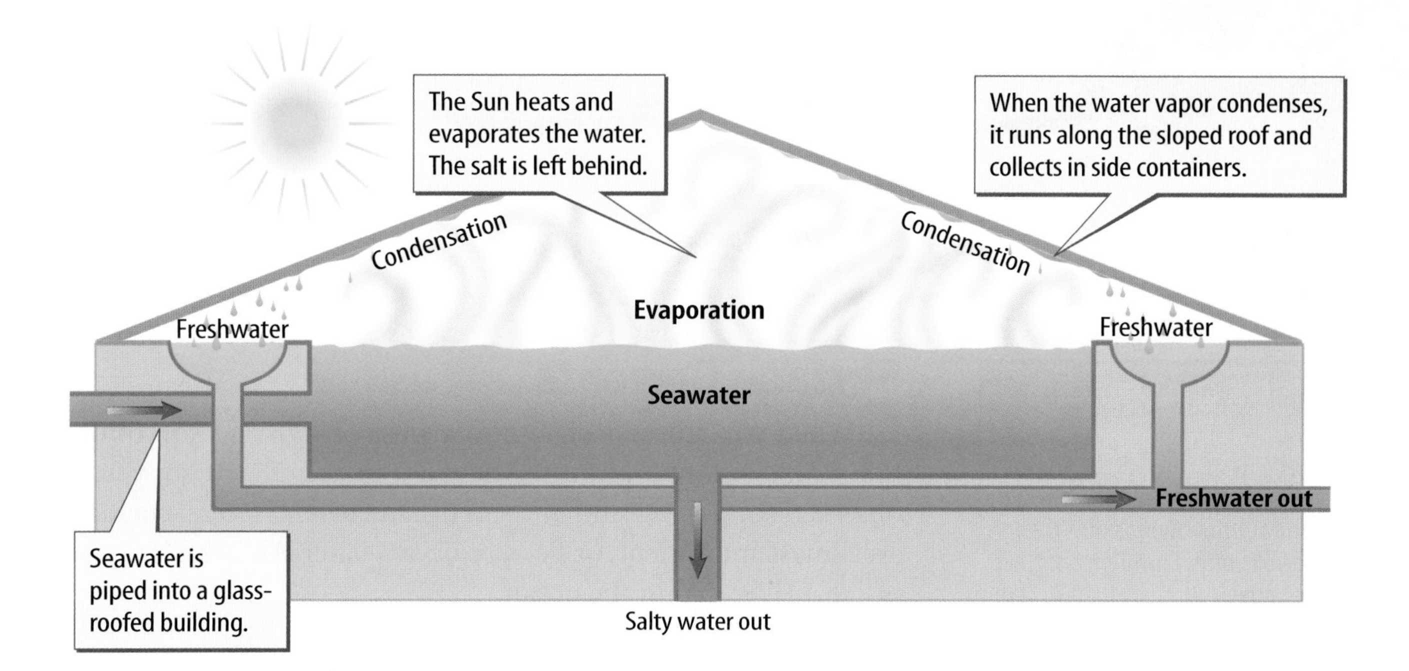
The Sun heats and evaporates the water. The salt is left behind.
When the water vapor condenses, it runs along the sloped roof and collects in side containers.
Condensation
Condensation
Evaporation
Freshwater
Freshwater
Seawater
Freshwater out
Seawater is piped into a glass-roofed building.
Salty water out

Desalination Plants Some methods of desalination include evaporating seawater and collecting the freshwater as it condenses on a glass roof. **Figure 4** shows how a desalination plant that uses solar energy works. Other plants desalinate water by passing it through a membrane that removes the dissolved salts. Freshwater also can be obtained by melting frozen seawater. As seawater freezes, the ice crystals that form contain much less salt than the remaining water. The salty, unfrozen water then can be separated from the ice. The ice can be washed and melted to produce freshwater.

Figure 4 This desalination plant uses solar energy to produce freshwater.

section 1 review

Summary

Importance of Oceans

- Oceans are a source of food, energy, and minerals.
- Oceans allow for the efficient transportation of goods such as oil, coal, and grains.

Origin of Oceans

- Scientists hypothesize that about 4 billion years ago, water vapor from volcanic eruptions cooled and condensed into storm clouds. Oceans formed as water from torrential rains filled Earth's basins.

Composition of Oceans

- Ocean water contains dissolved gases and salts.
- Oceans are considered to be in a steady state.

Self Check

1. **Describe** five ways Earth's oceans affect your life.
2. **Explain** the relationship between volcanic activity and the origin of Earth's oceans.
3. **Identify** the components of seawater. How do dissolved salts enter oceans? How does oxygen enter oceans?
4. **Think Critically** Organisms in the oceans are important sources of food and medicine. What steps can humans take to ensure that these resources are available for future generations?

Applying Math

5. **Use Proportions** If the average salinity of seawater is 35 parts per thousand, how many grams of dissolved salts will 500 g of seawater contain?

section 2

Ocean Currents

as you read

What You'll Learn

- **Explain** how winds and the Coriolis effect influence surface currents.
- **Discuss** the temperatures of coastal waters.
- **Describe** density currents.

Why It's Important

Ocean currents and the atmosphere transfer heat that affects the climate you live in.

Review Vocabulary
circulation: a water current flow occurring in an area in a closed, circular pattern

New Vocabulary
- surface current
- Coriolis effect
- upwelling
- density current

Surface Currents

When you stir chocolate into a glass of milk, do you notice the milk swirling around in the glass in a circle? If so, you've observed something similar to an ocean current. Ocean currents are a mass movement, or flow, of ocean water. An ocean current is like a river within the ocean.

Surface currents move water horizontally—parallel to Earth's surface. These currents are powered by wind. The wind forces the ocean to move in huge, circular patterns. **Figure 5** shows these major surface currents. Notice that some currents are shown with red arrows and some are shown with blue arrows. Red arrows indicate warm currents. Blue arrows indicate cold currents. The currents on the ocean's surface are related to the general circulation of winds on Earth.

Surface currents move only the upper few hundred meters of seawater. Some seeds and plants are carried between continents by surface currents. Sailors take advantage of these currents along with winds to sail more efficiently from place to place.

Figure 5 These are the major surface currents of Earth's oceans.

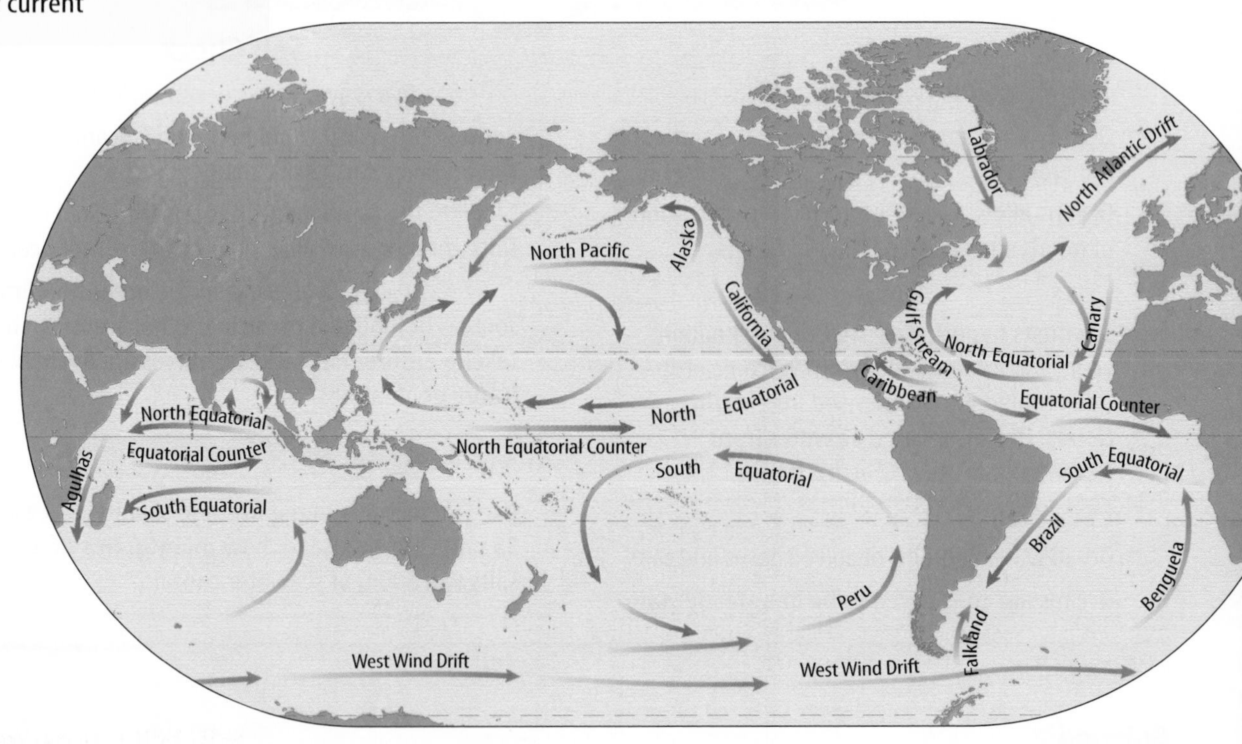

How Surface Currents Form Surface ocean currents and surface winds are affected by the Coriolis (kor ee OH lus) effect. The **Coriolis effect** is the shifting of winds and surface currents from their expected paths that is caused by Earth's rotation. Imagine that you try to draw a line straight out from the center of a disk to the edge of the disk. You probably could do that with no problem. But what would happen if the disk were slowly spinning like the one in **Figure 6?** As the student tried to draw a straight line, the disk rotated and, as shown in **Figure 6,** the line curved.

Figure 6 The student draws a line straight out from the center while spinning the disk. Because the disk was spinning, the line is curved.
Describe *how the Coriolis effect influences surface currents in the southern hemisphere.*

A similar thing happens to wind and surface currents. Because Earth rotates toward the east, winds appear to curve to the right in the northern hemisphere and to the left in the southern hemisphere. These surface winds can cause water to pile up in certain parts of the ocean. When gravity pulls water off the pile, the Coriolis effect turns the water. This causes surface water in the oceans to spiral around the piles of water. The Coriolis effect causes currents north of the equator to turn to the right. Currents south of the equator are turned to the left. Look again at the map of surface currents in **Figure 5** to see the results of the Coriolis effect.

The Gulf Stream Although satellites provide new information about ocean movements, much of what is known about surface currents comes from records that were kept by sailors of the nineteenth century. Sailors always have used surface currents to help them travel quickly. Sailing ships depend on some surface currents to carry them to the west and others to carry them east. During the American colonial era, ships floated on the 100-km-wide Gulf Stream current to go quickly from North America to England. Find the Gulf Stream current in the Atlantic Ocean on the map in **Figure 5.**

In the late 1700s, Deputy Postmaster General Benjamin Franklin received complaints about why it took longer to receive a letter from England than it did to send one there. Upon investigation, Franklin found that a Nantucket whaling captain's map furnished the answer. Going against the Gulf Stream delayed ships sailing west from England by up to 110 km per day.

Topic: Ocean Currents
Visit earth.msscience.com for Web links to information about ocean currents.

Activity Choose an area of the world's oceans and make a map of the currents in that area. In which direction do they travel? What are the currents named?

Figure 7 Bottles and other floating objects that enter the ocean are used to gain information about surface currents.

Tracking Surface Currents Items that wash up on beaches, such as the bottle shown in **Figure 7,** provide information about ocean currents. Drift bottles containing messages and numbered cards are released from a variety of coastal locations. The bottles are carried by surface currents and might end up on a beach. The person who finds a bottle writes down the date and the location where the bottle was found. Then the card is sent back to the institution that launched the bottle. By doing this, valuable information is provided about the current that carried the bottle.

Warm and Cold Surface Currents Notice in **Figure 5** that currents on the west coasts of continents begin near the poles where the water is colder. The California Current that flows along the west coast of the United States is a cold surface current. East-coast currents originate near the equator where the water is warmer. Warm surface currents, such as the Gulf Stream, distribute heat from equatorial regions to other areas of Earth. **Figure 8** shows the warm water of the Gulf Stream in red and orange. Cooler water appears in blue and green.

As warm water flows away from the equator, heat is released to the atmosphere. The atmosphere is warmed. This transfer of heat influences climate.

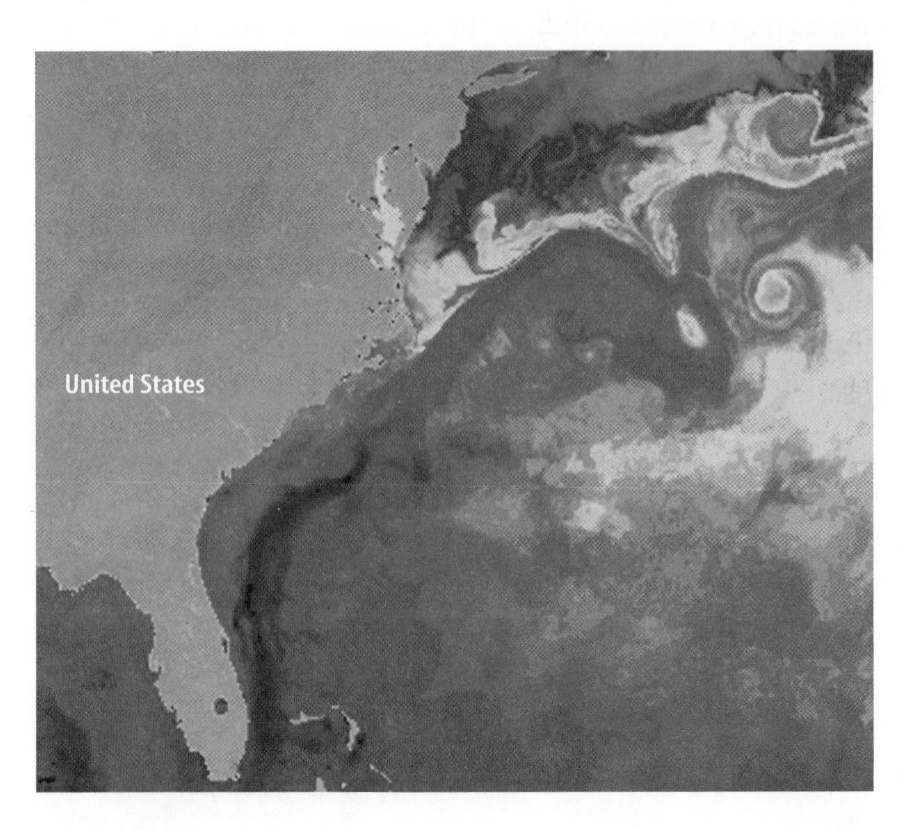

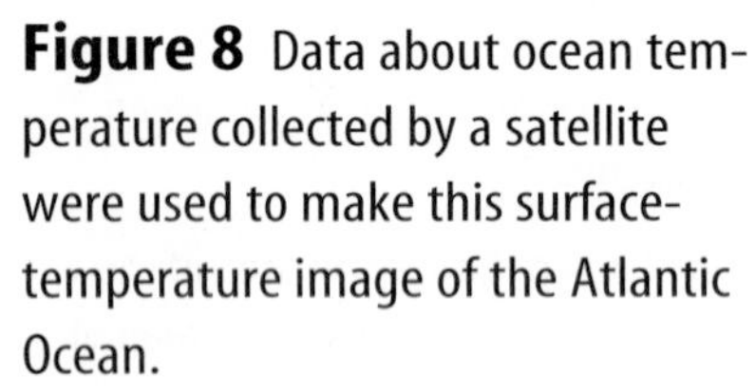

Figure 8 Data about ocean temperature collected by a satellite were used to make this surface-temperature image of the Atlantic Ocean.
Infer *Where does the Gulf Stream originate?*

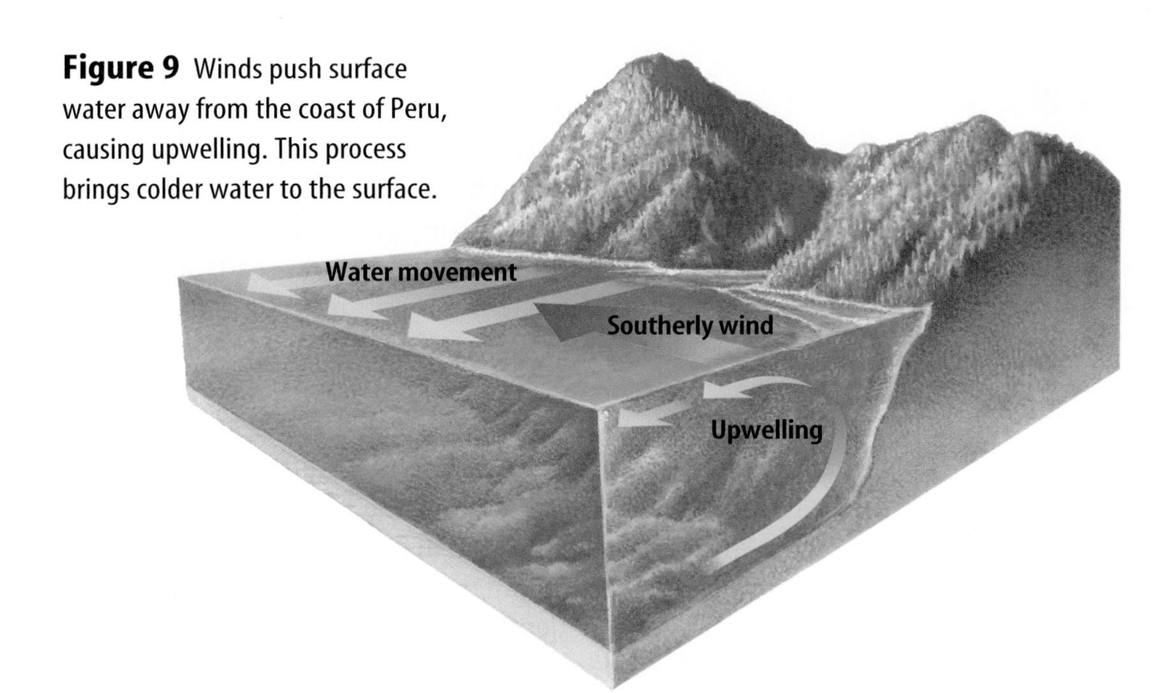

Figure 9 Winds push surface water away from the coast of Peru, causing upwelling. This process brings colder water to the surface.

Upwelling

Upwelling is a vertical circulation in the ocean that brings deep, cold water to the ocean surface. Along some coasts of continents, wind blowing parallel to the coast carries water away from the land because of the Coriolis effect, as shown in **Figure 9.** Cold, deep ocean water rises to the surface and replaces water that has moved away from shore. This water contains high concentrations of nutrients from organisms that died, sank to the bottom, and decayed. Nutrients promote primary production and plankton growth, which attracts fish. Areas of upwelling occur along the coasts of Oregon, Washington, and Peru and create important fishing grounds.

Density Currents

Deep in the ocean, waters circulate not because of wind but because of density differences. A **density current** forms when a mass of seawater becomes more dense than the surrounding water. Gravity causes more dense seawater to sink beneath less dense seawater. This deep, dense water then slowly spreads to the rest of the ocean.

The density of seawater increases if salinity increases, as you can see if you perform the MiniLAB on this page. It also increases when temperature decreases. In the Launch Lab at the beginning of the chapter, the cold water was more dense than the warm water in the beaker. The cold water sank to the bottom. This created a density current that moved the food coloring.

Changes in temperature and salinity work together to create density currents. Density currents circulate ocean water slowly.

Mini LAB

Modeling a Density Current

Procedure

1. Fill a **clear-plastic storage box** (shoe-box size) with room-temperature **water.**
2. Mix several spoonfuls of **table salt** into a **glass of water** at room temperature.
3. Add a few drops of **food coloring** to the saltwater solution. Pour the solution slowly into the freshwater in the large container.

Analysis

1. Describe what happened when you added salt water to freshwater.
2. How does this lab model density currents?

Try at Home

Oceanographer An oceanographer studies the composition of ocean water; ocean currents, waves, and tides; and marine life. Dr. Robert D. Ballard is an oceanographer who revolutionized deep-sea archaeology. He developed several high-tech vessels that can explore ocean bottoms previously out of reach.

Deep Waters An important density current begins in Antarctica where the most dense ocean water forms during the winter. As ice forms, seawater freezes, but the salt is left behind in the unfrozen water. This extra salt increases the salinity and, therefore, the density of the ocean water until it is very dense. This dense water sinks and slowly spreads along the ocean bottom toward the equator, forming a density current. In the Pacific Ocean, this water could take 1,000 years to reach the equator.

In the North Atlantic Ocean, cold, dense water forms around Norway, Greenland, and Labrador. These waters sink, forming North Atlantic Deep Water. In about the northern one-third to one-half of the Atlantic Ocean, North Atlantic Deep Water forms the bottom layer of ocean water. In the southern part of the Atlantic Ocean, it flows at depths of about 3,000 m, just above the denser water formed near Antarctica. The dense waters circulate more quickly in the Atlantic Ocean than in the Pacific Ocean. In the Atlantic, a density current could circulate in 275 years.

Applying Math — Calculate Density

DENSITY OF SALT WATER You have an aquarium full of freshwater in which you have dissolved salt. If the mass of the salt water is 123,000 g and its volume is 120,000 cm^3, what is the density of the salt water?

Solution

1 *This is what you know:*
- volume: $v = 120{,}000 \text{ cm}^3$
- mass of salt water: $m = 123{,}000 \text{ g}$

2 *This is what you need to find:* density of water: d

3 *This is the equation you need to use:* $d = m/v$

4 *Substitute the known values:* $d = 123{,}000\text{g} / 120{,}000\text{cm}^3 = 1.025 \text{ g/cm}^3$

5 *Check your answer:* Multiply your answer by the volume. Do you calculate the same mass of salt water that was given?

Practice Problems

1. Calculate the density of 78,000 cm^3 of salt water with a mass of 79,000 g.
2. If a sample of ocean water has a density of 1.03 g/cm^3 and a mass of 50,000 g, what is the volume of the water?

For more practice, visit earth.msscience.com/math_practice

Intermediate Waters A density current also occurs in the Mediterranean Sea, a nearly enclosed body of water. The warm temperatures and dry air in the region cause large amounts of water to evaporate from the surface of the sea. This evaporation increases the salinity and density of the water. This dense water from the Mediterranean flows through the narrow Straits of Gibraltar into the Atlantic Ocean at a depth of about 320 m. When it reaches the Atlantic, it flows to depths of 1,000 m to 2,000 m because it is more dense than the water in the upper parts of the North Atlantic Ocean. However, the water from the Mediterranean is less dense than the very cold, salty water flowing from the North Atlantic Ocean around Greenland, Norway, and Labrador. Therefore, as shown in **Figure 10,** the Mediterranean water forms a middle layer of water—the Mediterranean Intermediate Water.

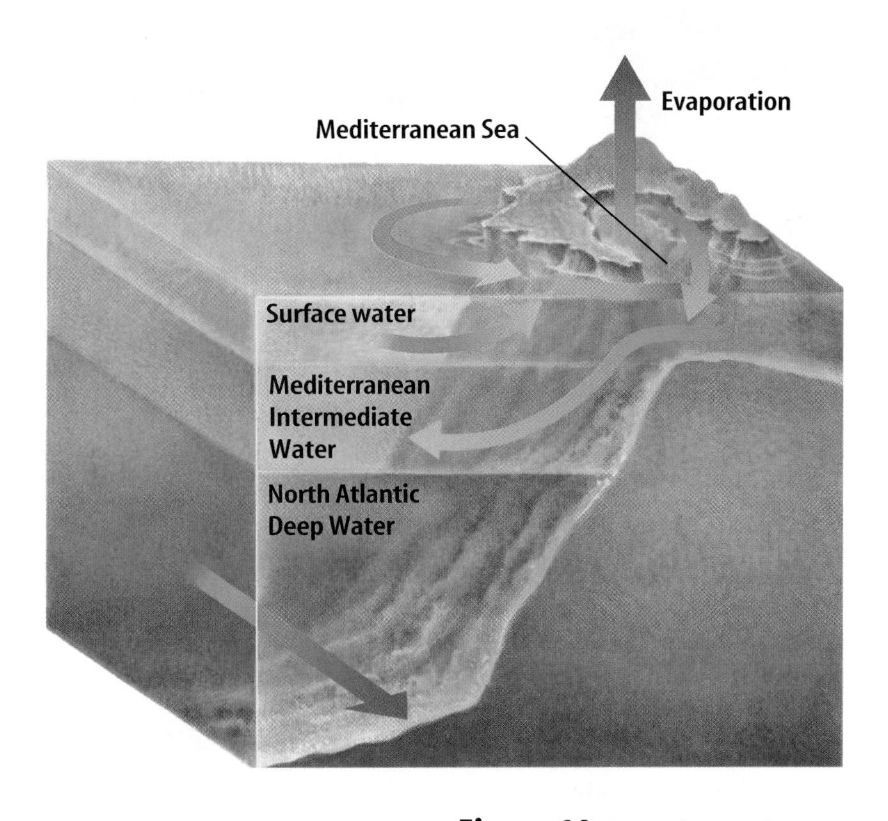

Figure 10 Dense layers of North Atlantic Deep Water form in the Greenland, Labrador, and Norwegian Seas. This water flows southward along the North Atlantic seafloor. Less dense water from the Mediterranean Sea forms Mediterranean Intermediate Water.

Reading Check *What causes the Mediterranean Intermediate Water to form?*

section 2 review

Summary

Surface Currents

- Surface currents are wind-powered ocean currents that move water horizontally.

Upwelling

- Upwelling is vertical circulation that brings deep, cold water to the ocean surface.

Density Currents

- Gravity acts on masses of seawater that are denser than surrounding water, causing the denser water to sink.
- Density currents slowly circulate deep ocean water.

Self Check

1. **Explain** how winds and the Coriolis effect influence surface currents.
2. **Summarize** why upwelling is important.
3. **Describe** how density currents circulate water.
4. **Think Critically** The latitudes of San Diego, California, and Charleston, South Carolina, are exactly the same. However, the average yearly water temperature in the ocean off Charleston is much higher than the water temperature off San Diego. Explain why.

Applying Skills

5. **Predict** what will happen to a layer of freshwater as it flows into the ocean. Explain your prediction.

section 3

Ocean Waves and Tides

as you read

What You'll Learn

- **Describe** wave formation.
- **Distinguish** between the movement of water particles in a wave and the movement of the wave.
- **Explain** how ocean tides form.

Why It's Important

Waves and tides affect life and property in coastal areas.

Review Vocabulary
energy: the ability to cause change

New Vocabulary
- wave
- crest
- trough
- breaker
- tide
- tidal range

Waves

If you've been to the seashore or seen a beach on TV, you've watched waves roll in. There is something hypnotic about ocean waves. They keep coming and coming, one after another. But what is an ocean wave? A **wave** is a rhythmic movement that carries energy through matter or space. In the ocean, waves like those in **Figure 11** move through seawater.

Describing Waves Several terms are used to describe waves, as shown in **Figure 11.** Notice that waves look like hills and valleys. The **crest** is the highest point of the wave. The **trough** (TRAWF) is the lowest point of the wave. Wavelength is the horizontal distance between the crests or between the troughs of two adjacent waves. Wave height is the vertical distance between crest and trough.

Half the distance of the wave height is called the amplitude (AM pluh tewd) of the wave. The amplitude squared is proportional to the amount of energy the wave carries. For example, a wave with twice the amplitude of the wave in **Figure 11** carries four times ($2 \times 2 = 4$) the energy. On a calm day, the amplitude of ocean waves is small. But during a storm, wave amplitude increases and the waves carry a lot more energy. Large waves can damage ships and coastal property.

Figure 11 Ocean waves carry energy through seawater.

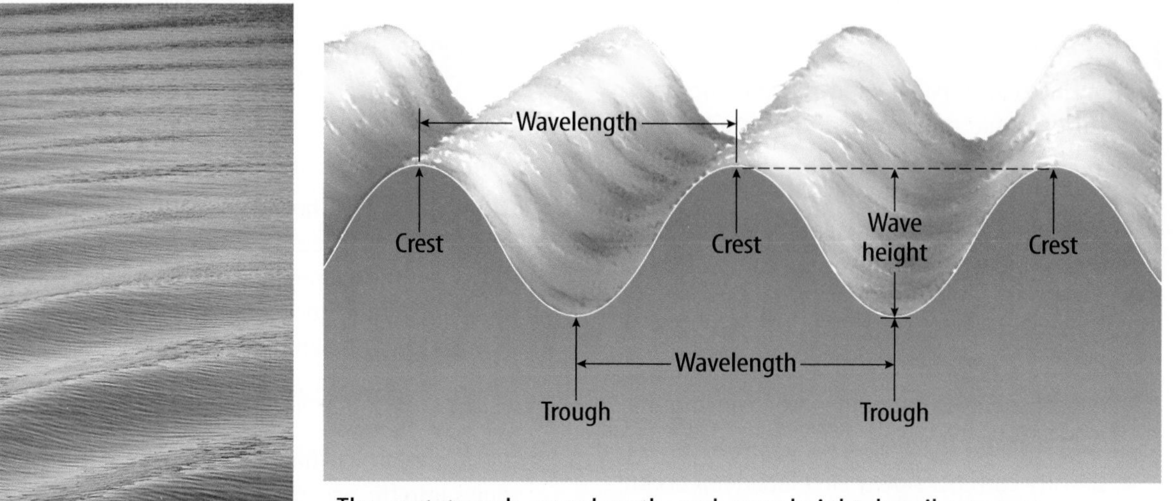

The crest, trough, wavelength, and wave height describe a wave.
Identify *the crests and troughs in the photo on the left.*

Figure 12 As a wave passes, only energy moves forward. The water particles and the bobber nearly return to their original positions after the wave has passed. **Describe** *what happens to water movement below a depth equal to about half the wavelength.*

Wave Movement You might have noticed that if you throw a pebble into a pond, a circular wave moves outward from where the pebble entered the water, as shown in **Figure 12.** A bobber on a fishing line floating in the water will bob up and down as the wave passes, but it will not move outward with the wave. Notice that the bobber returns to near its original position.

When you watch an ocean wave, it looks as though the water is moving forward. But unless the wave is breaking onto shore, the water does not move forward. Each molecule of water returns to near its original position after the wave passes. **Figure 13** shows this. Water molecules in a wave move around in circles. Only the energy moves forward while the water molecules remain in about the same place. Below a depth equal to about half the wavelength, water movement stops. Below that depth, water is not affected by waves. Submarines that travel below this level usually are not affected by surface storms.

Breakers A wave changes shape in the shallow area near shore. Near the shoreline, friction with the ocean bottom slows water at the bottom of the wave. As the wave slows, its crest and trough come closer together. The wave height increases. The top of a wave, not slowed by friction, moves faster than the bottom. Eventually, the top of the wave outruns the bottom and it collapses. The wave crest falls as water tumbles over on itself. The wave breaks onto the shore. **Figure 13** also shows this process. This collapsing wave is a **breaker.** It is the collapse of this wave that propels a surfer and surfboard onto shore. After a wave breaks onto shore, gravity pulls the water back into the sea.

What causes an ocean wave to slow down?

Modeling Water Particle Movement

Procedure

1. Put a piece of **tape** on the outside bottom of a **rectangular, clear-plastic storage box.** Fill the box with **water.**
2. Float a **cork** in the container above the piece of tape.
3. Use a **spoon** to make gentle waves in the container.
4. Observe the movement of the waves and the cork.

Analysis

1. Describe the movement of the waves and the motion of the cork.
2. Compare the movement of the cork in the water with the movement of water molecules in a wave.

NATIONAL GEOGRAPHIC

VISUALIZING WAVE MOVEMENT

Figure 13

As ocean waves move toward the shore, they seem to be traveling in from a great distance, hurrying toward land. Actually, the water in waves moves relatively little, as shown here. It's the energy in the waves that moves across the ocean surface. Eventually that energy is transferred—in a crash of foam and spray—to the land.

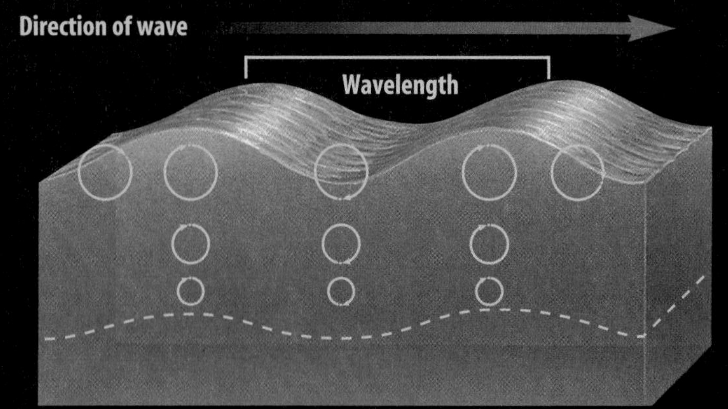

A Particles of water move around in circles rather than forward. Near the water's surface, the circles are relatively large. Below the surface, the circles become progressively smaller. Little water movement occurs below a depth equal to about one-half of a wave's length.

B The energy in waves, however, does move forward. One way to visualize this energy movement is to imagine a line of dominoes. Knock over the first domino, and the others fall in sequence. As they fall, individual dominoes—like water particles in waves—remain close to where they started. But each transfers its energy to the next one down the line.

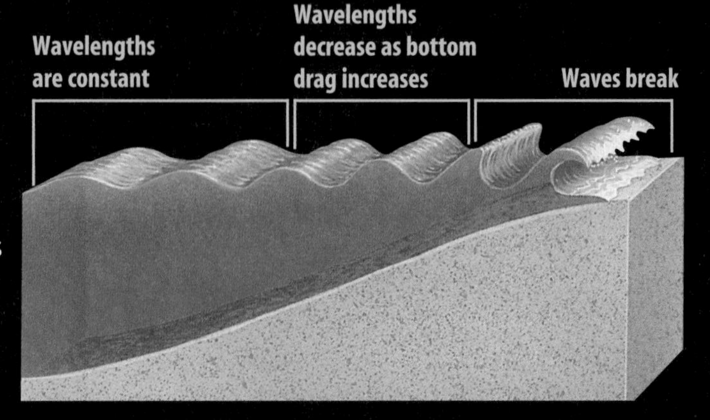

C As waves approach shore, wavelength decreases and wave height increases. This causes breakers to form. Where ocean floor rises steeply to beach, incoming waves break quickly at a great height, forming huge arching waves.

Figure 14 Waves formed by storm winds can reach heights of 20 m to 30 m.
List *the three factors that affect wave height.*

How Water Waves Form On a windy day, waves form on a lake or ocean. When wind blows across a body of water, wind energy is transferred to the water. If the wind speed is great enough, the water begins to pile up, forming a wave. As the wind continues to blow, the wave increases in height. Some waves reach tremendous heights, as shown in **Figure 14.** Storm winds have been known to produce waves more than 30 m high—taller than a six-story building.

The height of waves depends on the speed of the wind, the distance over which the wind blows, and the length of time the wind blows. When the wind stops blowing, waves stop forming. But once set in motion, waves continue moving for long distances, even if the wind stops. The waves you see lapping at a beach could have formed halfway around the world.

Reading Check *What factors affect the height of waves?*

Topic: Tides
Visit earth.msscience.com for Web links to information about tides.

Activity There are three types of tides: diurnal, semidiurnal, and mixed. Define each type and give an example of where each type occurs.

Tides

When you go to a beach, you probably notice the level of the sea rise and fall during the day. This rise and fall in sea level is called a **tide.** A tide is caused by a giant wave produced by the gravitational pull of the Sun and the Moon. This wave has a wave height of only 1 m or 2 m, but it has a wavelength that is thousands of kilometers long. As the crest of this wave approaches the shore, sea level appears to rise. This rise in sea level is called high tide. Later, as the trough of the wave approaches, sea level appears to drop. This drop in sea level is referred to as low tide.

Figure 15 A large difference between high tide and low tide can be seen at Mont-Saint-Michel off the northwestern coast of France.

Mont-Saint-Michel lies about 1.6 km offshore and is connected to the mainland at low tide.

Incoming tides move very quickly, making Mont-Saint-Michel an island at high tide.

Tidal Range As Earth rotates, different locations on Earth's surface pass through the high and low positions. Many coastal locations, such as the Atlantic and Pacific coasts of the United States, experience two high tides and two low tides each day. One low-tide/high-tide cycle takes 12 h, 25 min. A daily cycle of two high tides and two low tides takes 24 h, 50 min—slightly more than a day. But because ocean basins vary in size and shape, some coastal locations, such as many along the Gulf of Mexico, have only one high and one low tide each day. The **tidal range** is the difference between the level of the ocean at high tide and low tide. Notice the tidal range in the photos in **Figure 15.**

Figure 16 The Bay of Fundy has the greatest tidal range in the world. **Infer** *Was this picture taken at high tide or low tide?*

Extreme Tidal Ranges The shape of the seacoast and the shape of the ocean floor affect the ranges of tides. Along a smooth, wide beach, the incoming water can spread over a large area. There the water level might rise only a few centimeters at high tide. In a narrow gulf or bay, however, the water might rise many meters at high tide.

Most shorelines have tidal ranges between 1 m and 2 m. Some places, such as those on the Mediterranean Sea, have tidal ranges of only about 30 cm. Other places have large tidal ranges. Mont-Saint-Michel, shown in **Figure 15,** lies in the Gulf of Saint-Malo off the northwestern coast of France. There the tidal range reaches about 13.5 m.

The dock shown in **Figure 16** is in Digby, Nova Scotia in the Bay of Fundy. This bay is extremely narrow, which contributes to large tidal ranges. The difference between water levels at high tide and low tide can be as much as 15 m.

Tidal Bores In some areas when a rising tide enters a shallow, narrow river from a wide area of the sea, a wave called a tidal bore forms. A tidal bore can have a breaking crest or it can be a smooth wave. Tidal bores tend to be found in places with large tidal ranges. The Amazon River in Brazil, the Tsientang River in China, and rivers that empty into the Bay of Fundy in Nova Scotia have tidal bores.

When a tidal bore enters a river, it causes water to reverse its flow. In the Amazon River, the tidal bore rushes 650 km upstream at speeds of 65 km/h, causing a wave more than 5 m in height. Four rivers that empty into the Bay of Fundy have tidal bores. In those rivers, bore rafting is a popular sport.

The Gravitational Effect of the Moon For the most part, tides are caused by the interaction of gravity in the Earth-Moon system. The Moon's gravity exerts a strong pull on Earth. Earth and the water in Earth's oceans respond to this pull. The water bulges outward as Earth and the Moon revolve around a common center of mass. These events are explained in **Figure 17.**

Two bulges of water form, one on the side of Earth closest to the Moon and one on the opposite side of Earth. The bulge on the side of Earth closest to the Moon is caused by the gravitational attraction of the Moon on Earth. The force of gravity here is greater than another, opposing force generated by the motion of Earth and the Moon. As a result, surface water is pulled in the direction of the Moon. The bulge on the opposite side of Earth is caused by the same opposing force that, here, is greater than the force of gravity. The imbalance in forces results in surface water being pulled away from the Moon. The ocean bulges are the high tides, and the areas of Earth's oceans that are not toward or away from the Moon are the low tides. As Earth rotates, different locations on its surface pass through high and low tide.

Life in the Tidal Zone
Limpets are sea snails that live on rocky shores. When the tide comes in, they glide over the rocks to graze on seaweed. When the tide goes out, they use strong muscles to pull their shells tight against the rocks. Find out how other organisms survive in the zone between high and low tides.

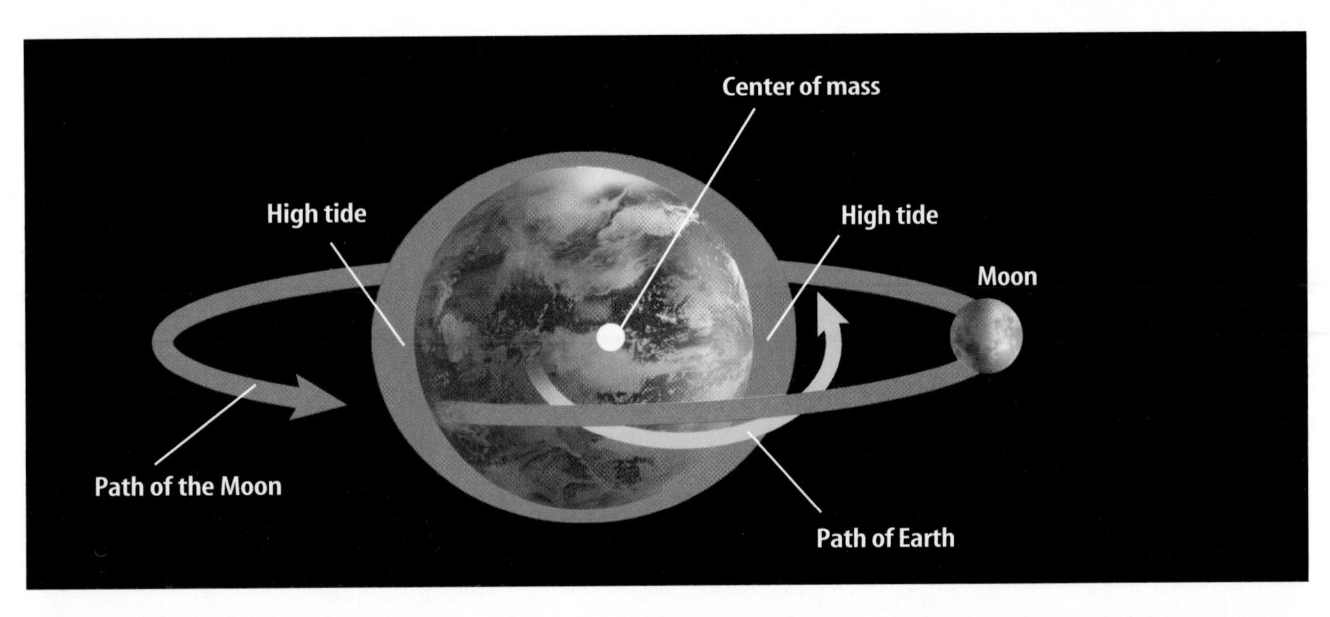

Figure 17 The Moon and Earth revolve around a common center of mass. Because the Moon's gravity pulls harder on parts of Earth closer to the Moon, a bulge of water forms on the side of Earth facing the Moon and the side of Earth opposite the Moon.

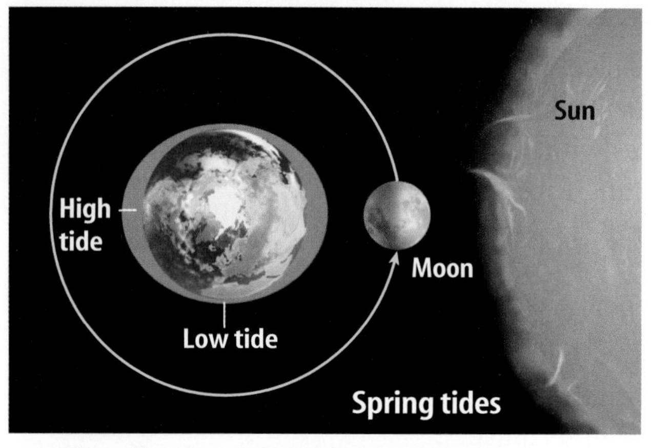

When the Sun, the Moon, and Earth are aligned, spring tides occur.

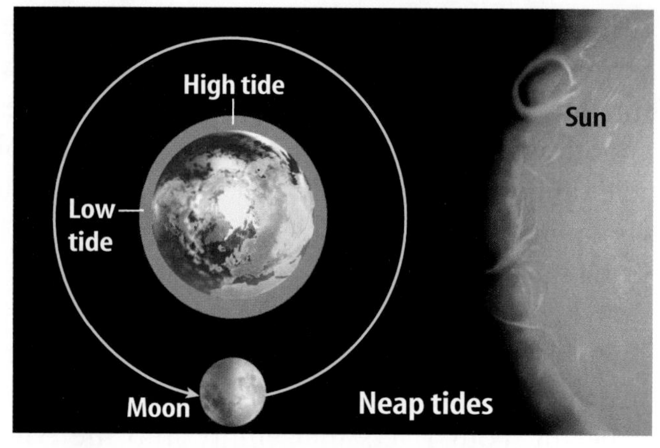

When the Sun, Earth, and the Moon form a right angle, neap tides occur.

Figure 18 The gravitational attraction of the Sun causes spring tides and neap tides.

The Gravitational Effect of the Sun The Sun also affects tides. The Sun can strengthen or weaken the Moon's effects. When the Moon, Earth, and the Sun are lined up together, the combined pull of the Sun and the Moon causes spring tides, shown in **Figure 18.** During spring tides, high tides are higher and low tides are lower than normal. The name *spring tide* has nothing to do with the season of spring. It comes from the German word *springen*, which means "to jump." When the Sun, Earth, and the Moon form a right angle, also shown in **Figure 18,** high tides are lower and low tides are higher than normal. These are called neap tides.

section 3 review

Summary

Waves

- A wave is a rhythmic movement that carries energy through matter or space.
- Water waves form as wind blows across a body of water.
- The height of a wave depends on the speed of the wind, the distance over which the wind blows, and the length of time the wind blows.

Tides

- Tides result from the gravitational pull of the Moon and the Sun on Earth.
- The shape of the seafloor and coast affect the range of tides in an area.
- Depending on the alignment of the Sun, the Moon, and Earth, spring tides or neap tides occur.

Self Check

1. **Identify** the parts of an ocean wave.
2. **Explain** how wind creates water waves.
3. **Describe** what causes high tides. Describe what causes spring tides.
4. **Summarize** the movement of water molecules in a wave and wave movement.
5. **Think Critically** At the ocean, you spot a wave about 200 m from shore. A few seconds later, the wave breaks on the beach. Explain why the water in the breaker is not the same water that was in the wave 200 m away.

Applying Skills

6. **Compare and contrast** the effects of the Sun and the Moon on Earth's tides.

earth.msscience.com/self_check_quiz

Wave Properties

Ocean wave energy impacts coastlines around the world. Understanding wave properties helps scientists predict the movement and effects of waves.

Real-World Question

How are wave characteristics related to each other and to the energy source that causes waves?

Goals

- **Test** statements about wave properties.
- **Summarize** the relationship between wave properties and the energy source which causes waves.

Materials

rectangular, clear-plastic box
water
straw
metric ruler
3-cm chalk piece
3-cm ball aluminum foil

Safety Precautions 

Procedure

1. Copy the data table above.
2. Fill the clear, plastic box with water to a depth of about 5.5 cm.
3. Test statement 1. Hold the straw just above the water. Blow through the straw. Record your observations in your data table.
4. Test statement 2. Hold the straw just above the water at one end of the box. Blow gently and continuously. Use the metric ruler to compare the wavelengths close to the straw and on the other end of the box. Record your observations.

Statement	Observations
1. Wind causes waves.	
2. Wavelength increases as the distance from the energy sources increases.	
3. The effects of wave motion are felt relatively close to the surface only.	Do not write in this book.
4. Wave energy is transferred through the water: the water itself does not move forward with the wave.	

5. Test statement 3. Sink the chalk piece in the middle of the box. Hold the straw just above the water at one end of the box. Blow gently and continuously. Observe any movement of the chalk and record your observations.
6. Test statement 4. Float the aluminum foil ball in the middle of the box. Hold the straw just above the water at one end of the box. Blow gently and continuously. Observe and record any movement of the aluminum foil ball.

Conclude and Apply

1. **Explain** how wind causes waves to form.
2. **Describe** Did your results support the statement *Wavelength increases as the distance from the energy source increases?* Why or why not?
3. **Explain** how you know the effects of wave motion are not felt below a certain depth.
4. **Infer** How did you prove that the water in a wave does not move forward with the wave? Did your observations surprise you? Why or why not?

Design Your Own

Sink or Float?

Goals

- **Design** an experiment to identify how increasing salinity affects the ability of a potato to float in water.

Possible Materials

small, uncooked potato
teaspoon
salt
large glass bowl
water
balance
large graduated cylinder
metric ruler

Safety Precautions

Real-World Question

As you know, ocean water contains many dissolved salts. How does this affect objects within the oceans? Why do certain objects float on top of the ocean's waves, while others sink directly to the bottom? Density is a measurement of mass per volume. You can use density to determine whether an object will float within a certain volume of water of a specific salinity. Based on what you know so far about salinity, why things float or sink, and the density of a potato, plus what it looks and feels like, formulate a hypothesis. Do you think the salinity of water has any effect on objects that are floating in water? What kind of effect? Will they float or sink? How would a dense object like a potato be different from a less dense object like a cork?

Test Your Hypothesis

Make a Plan

1. As a group, agree upon and write your hypothesis statement.
2. Devise a method to test how salinity affects whether a potato floats in water.
3. List the steps you need to take to test your hypothesis. Be specific, describing exactly what you will do at each step.
4. Read over your plan for testing your hypothesis.
5. How will you determine the densities of the potato and the different water samples? How will you measure the salinity of the water? How will you change the salinity of the water? Will you add teaspoons of salt one at a time?

6. How will you measure the ability of an object to float? Could you somehow measure the displacement of the water? Perhaps you could draw a line somewhere on your bowl and see how the position of the potato changes.
7. **Design** a data table where you can record your results. Include columns/rows for the salinity and float/sink measurements. What else should you include?

Follow Your Plan

1. Make sure your teacher approves your plan before you start.
2. Carry out the experiment.
3. While conducting the experiment, record your data and any observations that you or other group members make in your Science Journal.

Analyze Your Data

1. **Compare** how the potato floated in water with different salinities.
2. How does the ability of an object to float change with changing salinity?

Conclude and Apply

1. Did your experiment support the hypothesis you made?
2. A heavily loaded ship barely floats in the Gulf of Mexico. Based on what you learned, infer what might happen to the ship if it travels into the freshwater of the Mississippi River.

Communicating Your Data

Prepare a large copy of your data table and share the results of your experiment with members of your class. For more help, refer to the **Science Skill Handbook.**

Science and Language Arts

"The Jungle of Ceylon"

from Passions and Impressions

by Pablo Neruda

The following passage is part of a travel chronicle describing the Chilean poet Pablo Neruda's visit to the island of Ceylon, now called Sri Lanka, which is located southeast of India. The author considered himself so connected to Earth that he wrote in green ink.

Felicitous[1] shore! A coral reef stretches parallel to the beach; there the ocean interposes in its blues the perpetual white of a rippling ruff[2] of feathers and foam; the triangular red sails of sampans[3]; the unmarred line of the coast on which the straight trunks of the coconut palms rise like explosions, their brilliant green Spanish combs nearly touching the sky.

... In the deep jungle, there is a silence like that of libraries: abstract and humid.

1 Happy
2 round collar made of layers of lace
3 East Asian boats

Respond to the Reading

1. What were his impressions of the island on arrival?
2. What words does the author choose to describe waves?
3. **Linking Science and Writing** Write a weather report for fishers and others who work at sea.

Linking Science and Writing

Imagery Imagery is a series of words that evoke pictures to the reader. Poets use imagery to connect images to abstract concepts. The poet, here, wants to capture a particular feature of the reef and does so by describing it as a "ruff of feathers and foam," invoking the image of a gentle place, without the author saying so.
Where else in the poem does the poet use imagery to convey a mood or feeling?

INTEGRATE Earth Science Sri Lanka often is plagued by monsoons, which affect ocean conditions and local climate. Monsoons are seasonal reversals of the regional winds. During the wet season, moist winds blow in from the sea, causing storms and producing waves. During the dry season, winds blow from the land and sunny days are common.

Reviewing Main Ideas

Section 1 Ocean Water

1. Earth's ocean water might have originated from water vapor released from volcanoes. Over millions of years, the water condensed and rain fell, filling basins.
2. The oceans are a mixture of water, dissolved salts, and dissolved gases.
3. Ions are added to ocean water by rivers, volcanic eruptions, and the atmosphere. When seawater is evaporated, these ions combine to form salts.

Section 2 Ocean Currents

1. Wind causes surface currents. Surface currents are affected by the Coriolis effect.
2. Cool currents off western coasts originate far from the equator. Warmer currents along eastern coasts begin near the equator.
3. Differences in temperature and salinity between water masses in the oceans set up circulation patterns called density currents.

Section 3 Ocean Waves and Tides

1. A wave is a rhythmic movement that carries energy.
2. In a wave, energy moves forward while water molecules move around in small circles.
3. Wind causes water to pile up and form waves. Tides are caused by gravitational forces.

Visualizing Main Ideas

Copy and complete the following concept map on ocean motions.

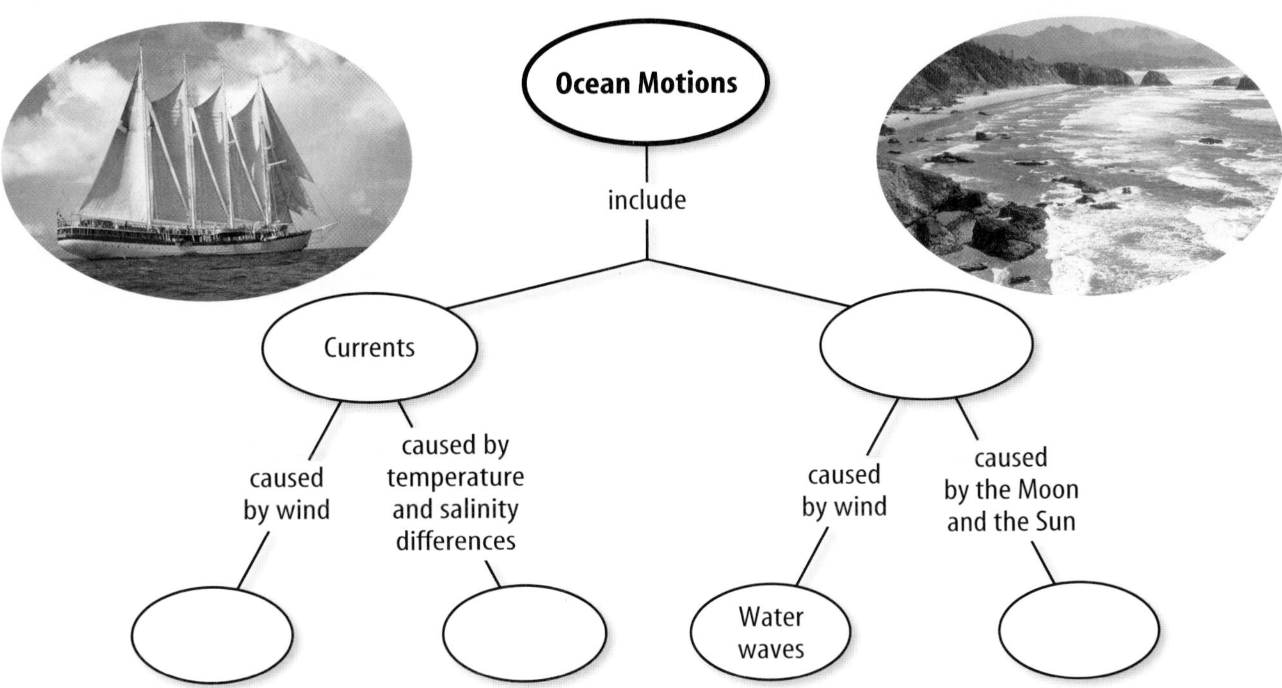

chapter 18 Review

Using Vocabulary

basin p. 515
breaker p. 525
Coriolis effect p. 519
crest p. 524
density current p. 521
salinity p. 516
surface current p. 518
tidal range p. 528
tide p. 527
trough p. 524
upwelling p. 521
wave p. 524

Fill in the blanks with the correct vocabulary word or words.

1. The ________ of seawater has stayed about the same for hundreds of millions of years.
2. An area of ________ is a good place to catch fish.
3. A(n) ________ is created when a mass of more dense water sinks beneath less dense water.
4. Along most ocean beaches, a rise and fall of the ocean related to gravitational pull, or a(n) ________, is easy to see.
5. The difference between the level of the ocean at high tide and low tide is ________.

Checking Concepts

Choose the word or phrase that best answers the question.

6. Where might ocean water have originated?
   A) salt marshes **C)** basins
   B) volcanoes **D)** surface currents
7. How does chloride enter the oceans?
   A) volcanoes **C)** density currents
   B) rivers **D)** groundwater
8. What is the most common ion found in ocean water?
   A) chloride **C)** boron
   B) calcium **D)** sulfate
9. What causes most surface currents?
   A) density differences
   B) the Gulf Stream
   C) salinity
   D) wind

Use the illustration below to answer question 10.

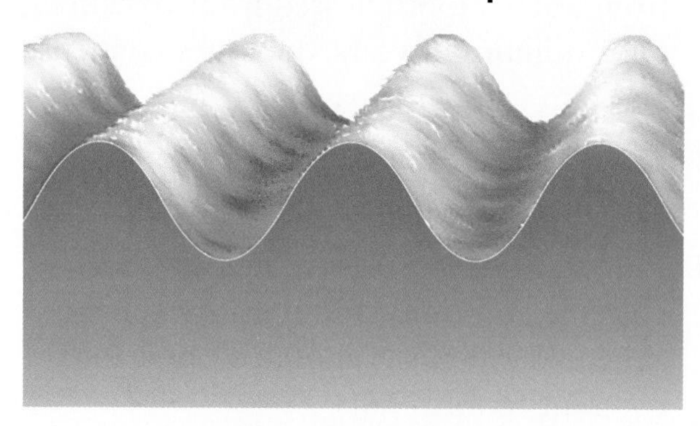

10. What is the highest point on a wave called?
    A) wave height **C)** crest
    B) trough **D)** wavelength
11. In the ocean, what is the rhythmic movement that carries energy through seawater?
    A) current **C)** crest
    B) wave **D)** upwelling
12. Which of the following causes the density of seawater to increase?
    A) a decrease in temperature
    B) a decrease in salinity
    C) an increase in temperature
    D) a decrease in pressure
13. In which direction does the Coriolis effect cause currents in the northern hemisphere to turn?
    A) east **C)** counterclockwise
    B) south **D)** clockwise
14. Tides are affected by the positions of which celestial bodies?
    A) Earth and the Moon
    B) Earth, the Moon, and the Sun
    C) Venus, Earth, and Mars
    D) the Sun, Earth, and Mars

Science Online earth.msscience.com/vocabulary_puzzlemaker

Thinking Critically

15. **Infer** If a sealed bottle is dropped into the ocean off the coast of California, where do you think it might wash up?

16. **Compare and contrast** the density of seawater at the mouth of the Mississippi River and in the Mediterranean Sea.

17. **Recognize Cause and Effect** What causes upwelling? What effect does it have? What can happen when upwelling stops?

18. **Compare and contrast** ocean waves and ocean currents.

Use the figure below to answer question 19.

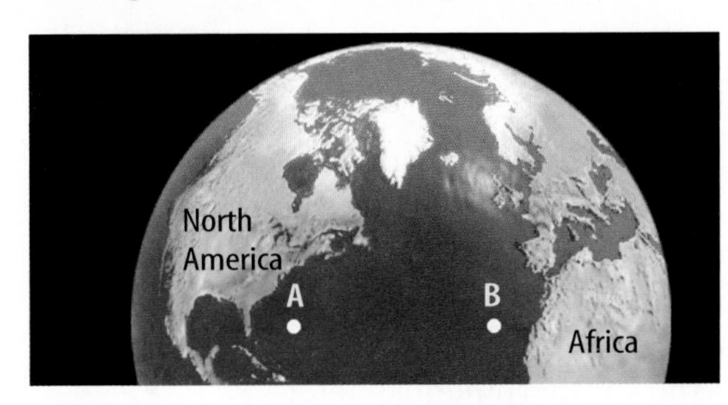

19. **Predict** how drift bottles dropped into the ocean at points A and B will move. Explain.

20. **Recognize Cause and Effect** In the Mediterranean Sea, a density current forms because of the high rate of evaporation of water from the surface. How can evaporation cause a density current?

21. **Evaluate** One water mass has a temperature of 5°C and a salinity of 37 parts per thousand. Another water mass has a temperature of 10°C and a salinity of 35 parts per thousand. Which water mass will sit on top of the other? Why?

22. **Infer** In some areas tidal energy is used as an alternative energy source. What are some advantages and disadvantages of doing this?

Performance Activities

23. **Invention** Design a method for desalinating water that does not use solar energy. Draw it, and display it for your class.

24. **Design and Perform an Experiment** Create an experiment to test the density of water at different temperatures.

Applying Math

25. **Wave Speed** Wave speed of deep water waves is calculated using the formula $S = L/T$, where S represents wave speed, L represents the wavelength, and T represents the period of the wave. What is the speed of a wave if $L = 100$ m and $T = 11$ s?

26. **Wave Steepness** The steepness of a wave is represented by the formula Steepness $= H/L$, where $H =$ wave height and $L =$ wavelength. When the steepness of a wave reaches 1/7, the wave becomes unstable and breaks. If $L = 50$ m, at what height will the wave break?

Use the graph below to answer question 27.

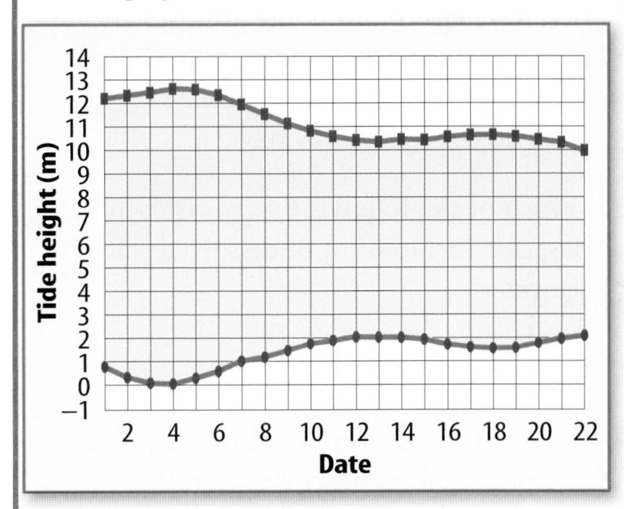

27. **Tides** The graph above shows the tidal ranges for each day. The maximum tidal range is called a spring tide. The minimum tidal range is referred to as a neap tide. Calculate the tidal range for the spring tide and the neap tide. On which date does each occur?

Standardized Test Practice

Part 1 Multiple Choice

Record your answers on the answer sheet provided by your teacher or on a sheet of paper.

Use the table below to answer question 1.

Ions in Seawater	
Ion	**Percentage**
Chloride	55.0
Sodium	30.6
Sulfate	7.7
Magnesium	3.7
Calcium	1.2
Potassium	1.1
Other	0.7

1. Which ion makes up 7.7 percent of the ions in seawater?
A. calcium
B. sulfate
C. chloride
D. sodium

2. Which of the following dissolved gases enters ocean water both from the atmosphere and from organisms that photosynthesize?
A. carbon
B. nitrogen
C. hydrogen
D. oxygen

3. Which of the following terms is used to describe the amount of dissolved salts in seawater?
A. density
B. temperature
C. salinity
D. buoyancy

4. Which of the following describes upwelling?
A. horizontal ocean circulation that brings deep, cold water to the surface
B. vertical ocean circulation that brings deep, warm water to the surface
C. horizontal ocean circulation that brings deep, warm water to the surface
D. vertical ocean circulation that brings deep, cold water to the surface

5. What is the lowest point on a wave called?
A. trough
B. wavelength
C. crest
D. wave height

Use the illustration below to answer questions 6 and 7.

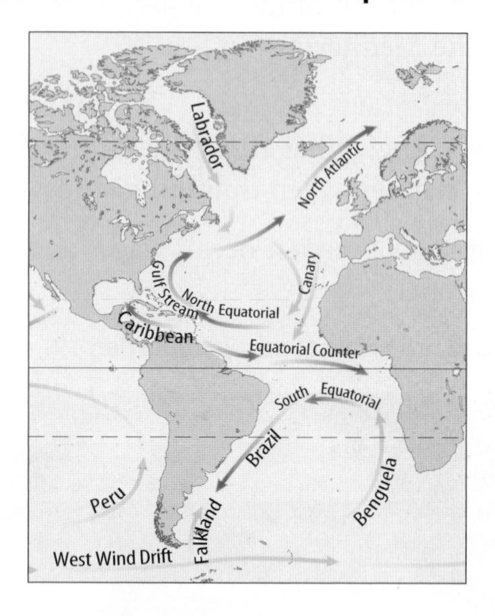

6. What is the direction of ocean currents in the southern hemisphere?
A. counterclockwise
B. north to south only
C. clockwise
D. east to west only

7. Which of the following is a reasonable conclusion based on the information in the figure?
A. The oceans' currents only flow in one direction.
B. The oceans' waters are constantly in motion.
C. The Gulf Stream flows east to west.
D. The Atlantic Ocean is deep.

8. What affects surface currents?
A. crests
B. upwellings
C. the Coriolis effect
D. tides

Part 2 | Short Response/Grid In

Record your answers on the answer sheet provided by your teacher or on a sheet of paper.

9. Explain the difference between surface currents and density currents in the ocean.

Use the illustration below to answer questions 10 and 11.

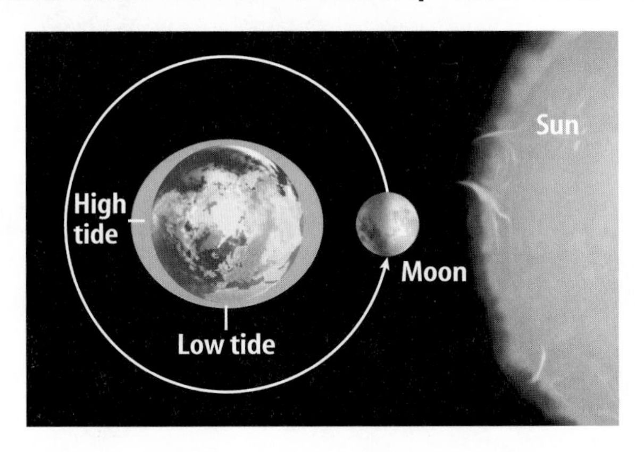

10. Which type of tide occurs when the Sun, the Moon, and Earth are aligned?

11. Describe how the Sun, the Moon, and Earth are positioned relative to each other during a neap tide.

12. What is tidal range?

13. On June 17th 2003, in Santa Barbara, California, the morning low tide was measured at −0.365 m. High tide was measured at 1.12 m. Calculate the tidal range between these tides.

14. Explain what the term *steady state* means in relation to ocean salinity. What processes keep ocean salinity in a steady state?

15. Explain how the ocean can influence the climate of an area.

Part 3 | Open Ended

Record your answers on a sheet of paper.

16. Draw a diagram that explains the process of upwelling. An area of upwelling exists off of the western coast of South America. During El Niño events, upwelling does not occur and surface water is warm and nutrient-poor. What effect could this change have on the marine organisms in this area?

Use the illustration below to answer question 17.

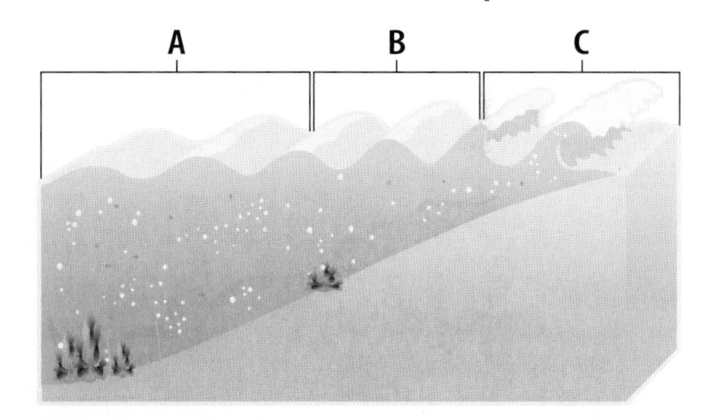

17. Describe the changes that occur as a wave approaches shore. Explain how wavelength is affected at each stage—A, B, and C—on the diagram.

18. What is the Coriolis effect? Explain how it affects ocean surface currents.

19. Compare and contrast the formation of North Atlantic Deep Water and Mediterranean Intermediate Water.

Test-Taking Tip

Answer All Parts Make sure each part of the question is answered when listing discussion points. For example, if the question asks you to compare and contrast, make sure you list both similarities and differences.

Question 19 Be sure to list the similarities and differences between the formation of the two masses of water.

chapter 19

Oceanography

The BIG Idea

Many organisms, including humans, depend on the ocean for their survival.

SECTION 1
The Seafloor

Main Idea Ocean basins, which are low areas of Earth that are filled with water, have many different features.

SECTION 2
Life in the Ocean

Main Idea Nearly all of the energy used by organisms in the ocean ultimately comes from the Sun.

SECTION 3
Ocean Pollution

Main Idea Pollutants from land eventually will reach the ocean.

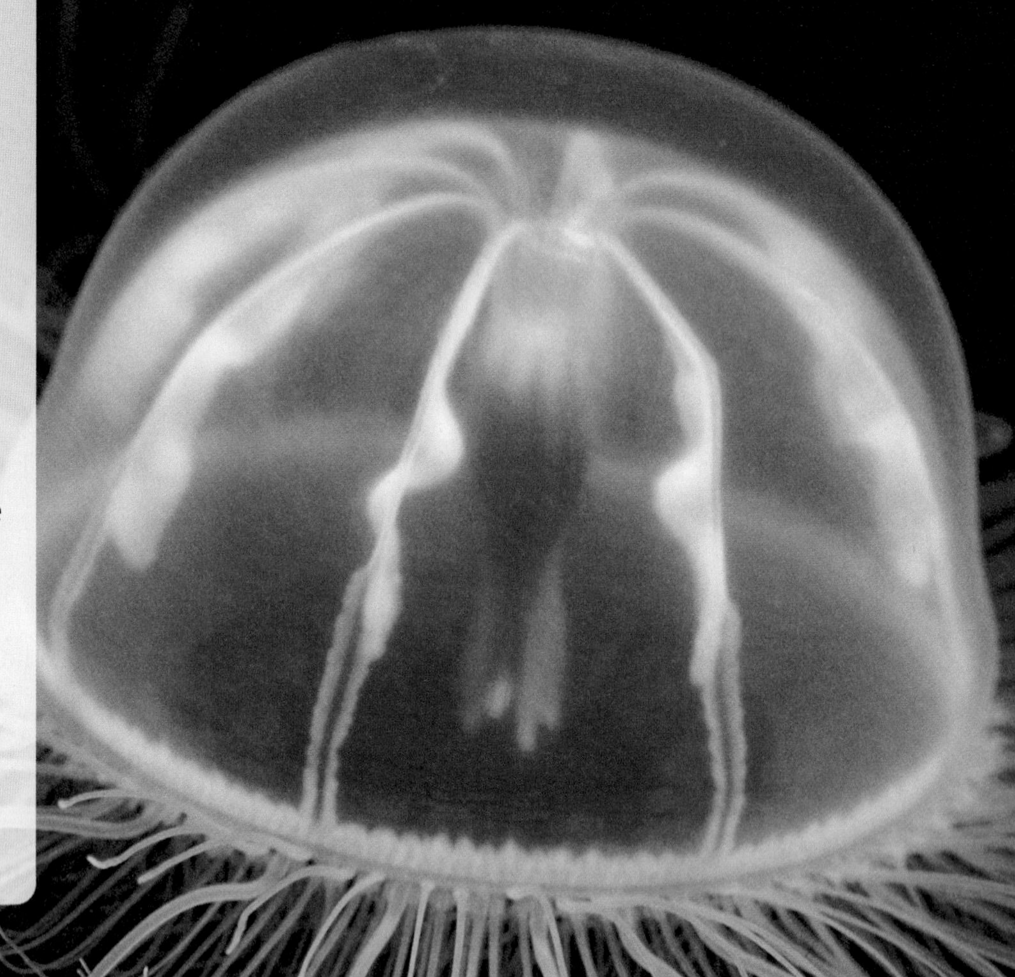

Marine Life

This inch-long jelly drifts with ocean currents and uses stinging tentacles to capture small prey. Through a chemical process called bioluminescence it is able to produce light and glow in the dark.

Science Journal Describe characteristics of three marine organisms you are familiar with.

Start-Up Activities

How deep is the ocean?

Sonar is used to measure ocean depth. You will model sonar in this lab.

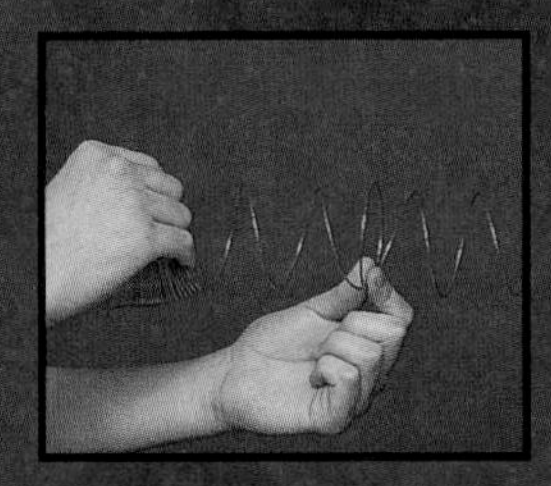

1. With one person holding each end, stretch a spring until it is taut. Measure the distance between the ends.
2. Pinch two coils together. When the spring is steady, release the coils to create a wave.
3. Record the time it takes the wave to travel back and forth five times. Divide this number by five to calculate the time of one round trip.
4. Calculate the speed of the wave by multiplying the distance by two and dividing this number by the time.
5. Move closer to your partner. Take in coils to keep the spring at the same tension. Repeat steps 2 and 3.
6. Calculate the new distance by multiplying the new time by the speed from step 4, and then dividing this number by two.
7. **Think Critically** Write a paragraph in your Science Journal that describes how this lab models sonar.

The Seafloor Make the following Foldable to help you identify the features of the seafloor.

STEP 1 **Fold** a sheet of paper in half lengthwise. Make the back edge about 1.25 cm longer than the front edge.

STEP 2 **Fold** in half, then fold in half again to make three folds.

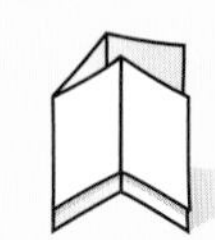

STEP 3 **Unfold and cut** only the top layer along the three folds to make four tabs.

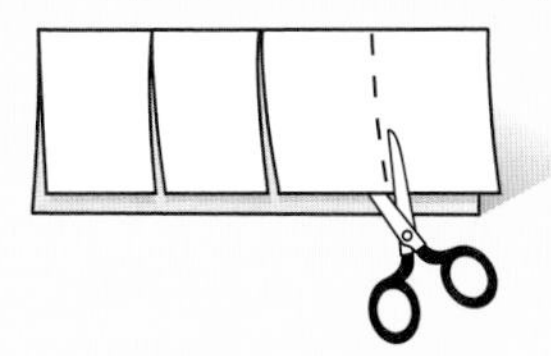

STEP 4 **Label** the tabs as shown.

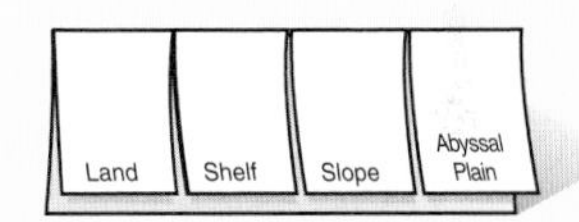

Find Main Ideas As you read the chapter, draw seafloor features on the front of the tabs and write information about them under the tabs.

Preview this chapter's content and activities at earth.msscience.com

Get Ready to Read

Questioning

1 Learn It! Asking questions helps you to understand what you read. As you read, think about the questions you'd like answered. Often you can find the answer in the next paragraph or section. Learn to ask good questions by asking *who, what, when, where, why,* and *how.*

2 Practice It! Read the following passage from Section 2.

> An estuary is an area where the mouth of a river opens into an ocean. ... Rivers bring nutrients to estuaries. Areas with many nutrients usually have many phytoplankton, which form the base of the food chain. ... Estuaries are an important habitat to many marine organisms. Newly hatched fish, shrimps, crabs, and other animals enter estuaries as microscopic organisms and remain there until adulthood. For these vulnerable animals, fewer predators and more food are found in estuaries.
>
> *—from page 556*

Here are some questions you might ask about this paragraph:

- What is an estuary?
- What are phytoplankton?
- Why are estuaries important?

3 Apply It! As you read the chapter, look for answers to section headings that are in the form of questions.

Test yourself. Create questions and then read to find answers to your own questions.

Target Your Reading

Use this to focus on the main ideas as you read the chapter.

1. **Before you read** the chapter, respond to the statements below on your worksheet or on a numbered sheet of paper.
   - Write an **A** if you **agree** with the statement.
   - Write a **D** if you **disagree** with the statement.

2. **After you read** the chapter, look back to this page to see if you've changed your mind about any of the statements.
   - If any of your answers changed, explain why.
   - Change any false statements into true statements.
   - Use your revised statements as a study guide.

Before You Read A or D		Statement	After You Read A or D
	1	The continental shelf is always wider than 100 km.	
	2	A mid-ocean ridge is the area in an ocean basin where ocean floor is formed.	
	3	Mid-ocean ridges can be found at the bottom of all ocean basins.	
	4	Most ocean trenches are found in the Pacific Basin.	
	5	Chemosynthesis requires sunlight to change carbon dioxide and water into sugar and oxygen.	
	6	Most marine organisms live in the waters of the deep ocean or on the floor of the abyssal plains.	
	7	An estuary is an area where the mouth of a river opens into an ocean.	
	8	Air pollution does not affect ocean pollution.	
	9	Less than 5 percent of oil pollution that reaches the ocean comes from land.	
	10	Today, there is not a single area of the ocean that is not polluted in some way.	

Science Online
Print out a worksheet of this page at earth.msscience.com

section 1

The Seafloor

as you read

What You'll Learn

- **Differentiate** between a continental shelf and a continental slope.
- **Describe** a mid-ocean ridge, an abyssal plain, and an ocean trench.
- **Identify** the mineral resources found on the continental shelf and in the deep ocean.

Why It's Important

Oceans cover nearly three fourths of Earth's surface.

Review Vocabulary
magma: hot, melted rock beneath Earth's surface

New Vocabulary
- continental shelf
- continental slope
- abyssal plain
- mid-ocean ridge
- trench

The Ocean Basins

Imagine yourself driving a deep-sea submersible along the ocean floor. Surrounded by cold, black water, the lights of your vessel reflect off of what looks like a mountain range ahead. As you continue, you find a huge opening in the seafloor—so deep you can't even see the bottom. What other ocean floor features can you find in **Figure 1?**

Ocean basins, which are low areas of Earth that are filled with water, have many different features. Beginning at the ocean shoreline is the continental shelf. The **continental shelf** is the gradually sloping end of a continent that extends under the ocean. On some coasts, the continental shelf extends a long distance. For instance, on North America's Atlantic and Gulf coasts, it extends 100 km to 350 km into the sea. On the Pacific Coast, where the coastal range mountains are close to the shore, the shelf is only 10 km to 30 km wide. The ocean covering the continental shelf can be as deep as 350 m.

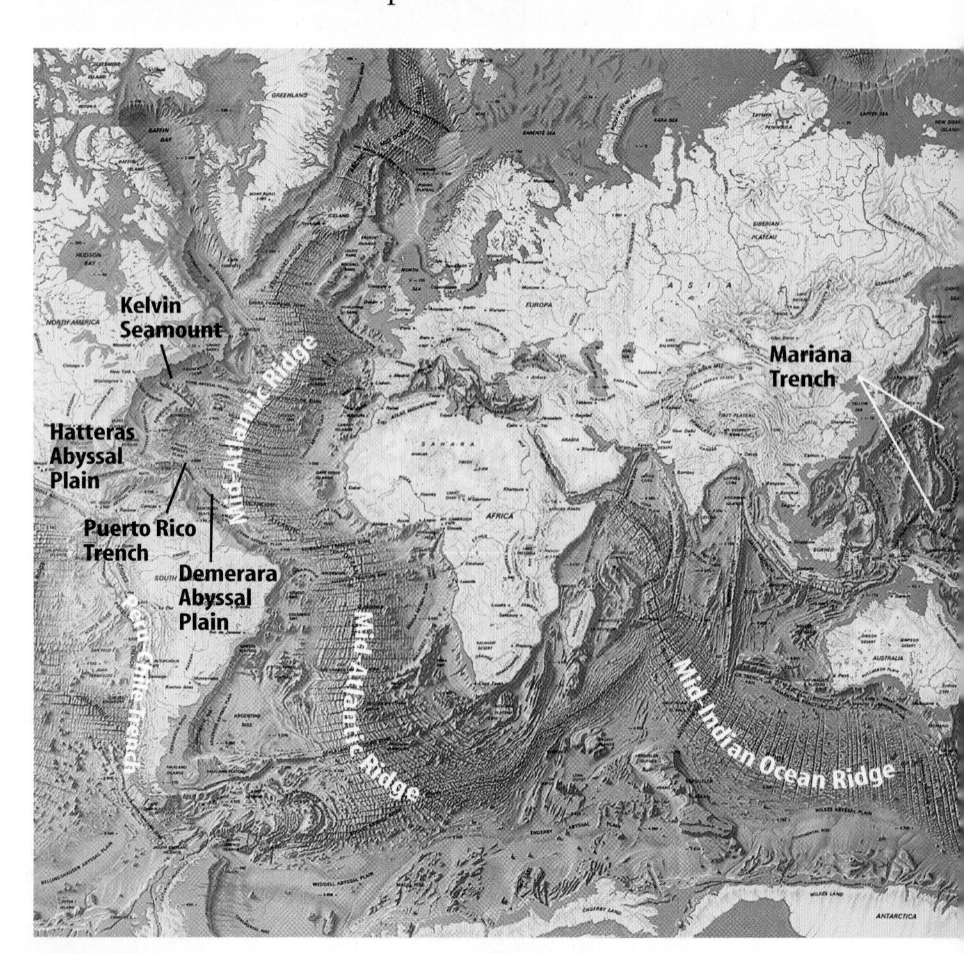

Figure 1 This map shows features of the ocean basins. Locate a trench and a mid-ocean ridge.

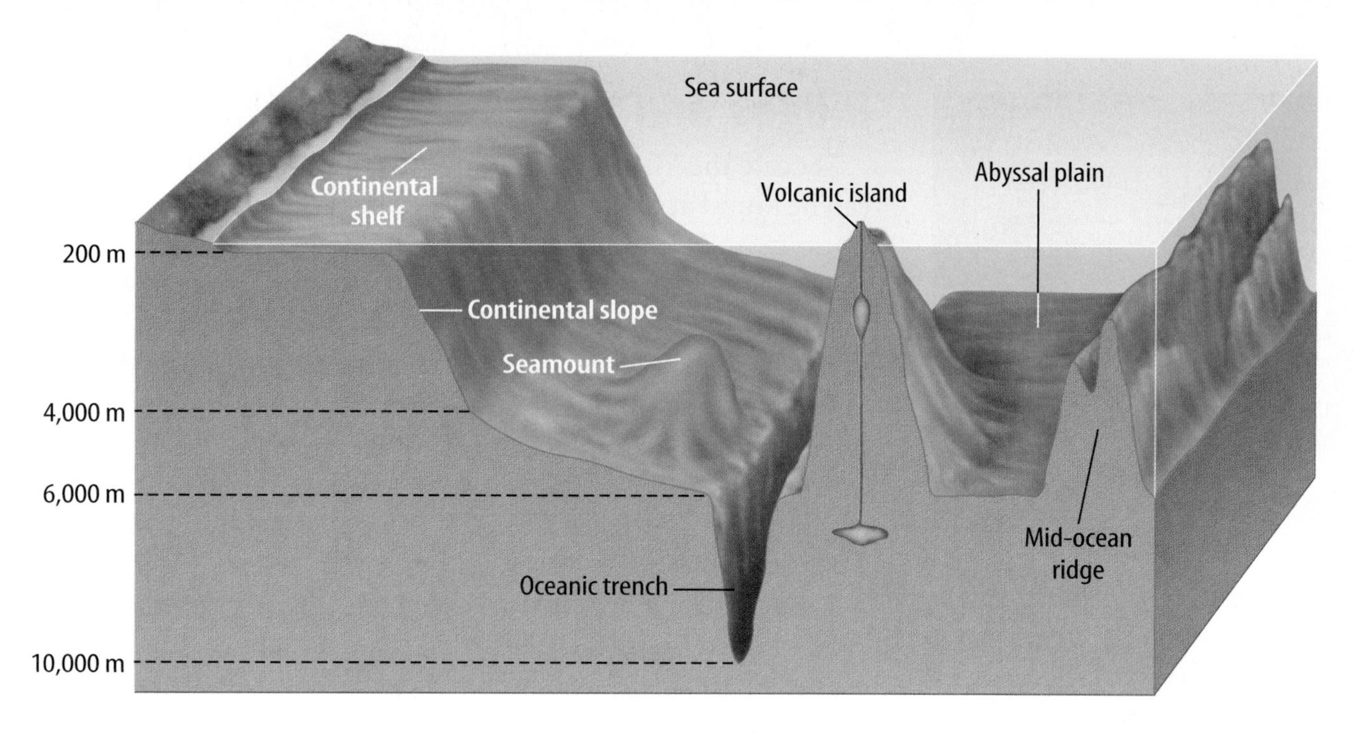

Figure 2 shows that beyond the shelf, the ocean floor drops more steeply, forming the continental slope. The **continental slope** extends from the outer edge of the continental shelf down to the ocean floor. Beyond the continental slope lie the trenches, valleys, plains, mountains, and ridges of the ocean basin.

In the deep ocean, sediment, derived mostly from land, settles constantly on the ocean floor. These deposits fill in valleys and create flat seafloor areas called **abyssal** (uh BIH sul) **plains.** Abyssal plains are from 4,000 m to 6,000 m below the ocean surface. Can you locate the abyssal plain shown in **Figure 2?**

In the Atlantic Ocean, areas of extremely flat abyssal plains can be large. One example is the Canary Abyssal Plain, which has an area of approximately 900,000 km^2. Other abyssal plains found in the Atlantic Ocean include the Hatteras and Demerara Abyssal Plains, both shown in **Figure 1.** Some areas of abyssal plains have small hills and seamounts. Seamounts are under-water, inactive volcanic peaks. They most commonly are found in the Pacific Ocean. Can you locate a seamount in **Figure 1?**

What are seamounts?

Figure 2 Ocean basin features are continuous from shore to shore. (Features in this diagram are not to scale.)
Describe *where the continental shelf ends and the continental slope begins.*

Topic: Ocean Basins
Visit earth.msscience.com for Web links to information about ocean basins.

Activity Find five fun facts about ocean basins, such as the location of the deepest known point in the ocean. Present your facts to the class.

Figure 3 New seafloor forms at mid-ocean ridges. A type of lava called pillow lava lies newly formed at this ridge on the ocean floor.

Ridges and Trenches

Locate the Mid-Atlantic Ridge in **Figure 1.** Mid-ocean ridges can be found at the bottom of all ocean basins. They form a continuous underwater ridge approximately 70,000 km long. A **mid-ocean ridge** is the area in an ocean basin where new ocean floor is formed. Crustal plates, which are large sections of Earth's uppermost mantle and crust, are moving constantly. As they move, the ocean floor changes. When ocean plates separate, hot magma from Earth's interior forms new ocean crust. This is the process of seafloor spreading. New ocean floor is being formed at a rate of approximately 2.5 cm per year along the Mid-Atlantic Ridge.

New ocean floor forms along mid-ocean ridges as lava erupts through cracks in Earth's crust. **Figure 3** shows newly erupted lava on the seafloor. When the lava hits the water, it cools quickly into solid rock, forming new seafloor. While seafloor is being formed in some parts of the oceans, it is being destroyed in others. Areas where old ocean floor slides beneath another plate and descends into Earth's mantle are called subduction zones.

Reading Check *How does new ocean floor form?*

Applying Math — Find the Slope

CALCULATING A FEATURE'S SLOPE If the width of a continental shelf is 320 km and it increases in depth a total of 300 m in that distance, what is its slope?

Solution

1. *This is what you know:*
   - width = 320 km
   - increase in depth = 300 m
2. *This is what you need to find:* slope: s
3. *This is the equation you need to use:* s = increase in depth ÷ width
4. *Solve the equation by substituting in known values:* $s = 300 \text{ m} \div 320 \text{ km} = 0.94 \text{ m/km}$

Practice Problems

1. The width of a continental slope is 40 km. It increases in depth by 2,000 m. What is the slope of the continental slope?
2. If the depth of a continental slope increases by 3,700 m and the slope is 74 m/km, what is the width of the slope?

Science Online For more practice, visit earth.msscience.com/math_practice

Figure 4 Located at subduction zones, trenches are important ocean basin features.

In 1960, the world's deepest dive was made in the Mariana Trench. The *Trieste* carried Jacque Piccard and Donald Walsh to a depth of almost 11 km.

Height of Mt. Everest

11,000 m

Depth of trench

If Earth's tallest mountain, Mount Everest, were set in the bottom of the Mariana Trench of the Pacific Basin, it would be covered with more than 2,000 m of water.

Subduction Zones On the ocean floor, subduction zones are marked by deep ocean trenches. A **trench** is a long, narrow, steep-sided depression where one crustal plate sinks beneath another. Most trenches are found in the Pacific Basin. Ocean trenches are usually longer and deeper than any valley on any continent. One trench, famous for its depth, is the Mariana Trench. It is located to the south and east of Japan in the Pacific Basin. This trench reaches 11 km below the surface of the water, and it is the deepest place in the Pacific. The photo in **Figure 4** shows the deep-sea vessel, the *Trieste,* that descended into the trench in 1960. **Figure 4** also illustrates that the Mariana Trench is so deep that Mount Everest could easily fit into it.

Mineral Resources from the Seafloor

Resources can be found in many places in the ocean. Some deposits on the continental shelf are relatively easy to extract. Others can be found only in the deep abyssal regions on the ocean floor. People still are trying to figure out how to get these valuable resources to the surface. As you read, suggest some methods that could be used to retrieve hard-to-reach resources.

Modeling the Mid-Atlantic Ridge

Procedure

1. Set two **tray tables** 2 cm apart.
2. Gather ten **paper towels** that are still connected. Lay one end of the paper towels on each table so the towels hang down into the space between the tables.
3. Slowly pull each end of the paper towels away from each other.

Analysis

1. Explain how this models the Mid-Atlantic Ridge.
2. How long does it take for 2.5 cm of new ocean crust to form at the Mid-Atlantic Ridge? How long does it take 25 cm to form?

Try at Home

Figure 5 The ocean is rich with mineral resources.
Determine *In which areas of the world can phosphorite be found? Where can diamonds be found?*

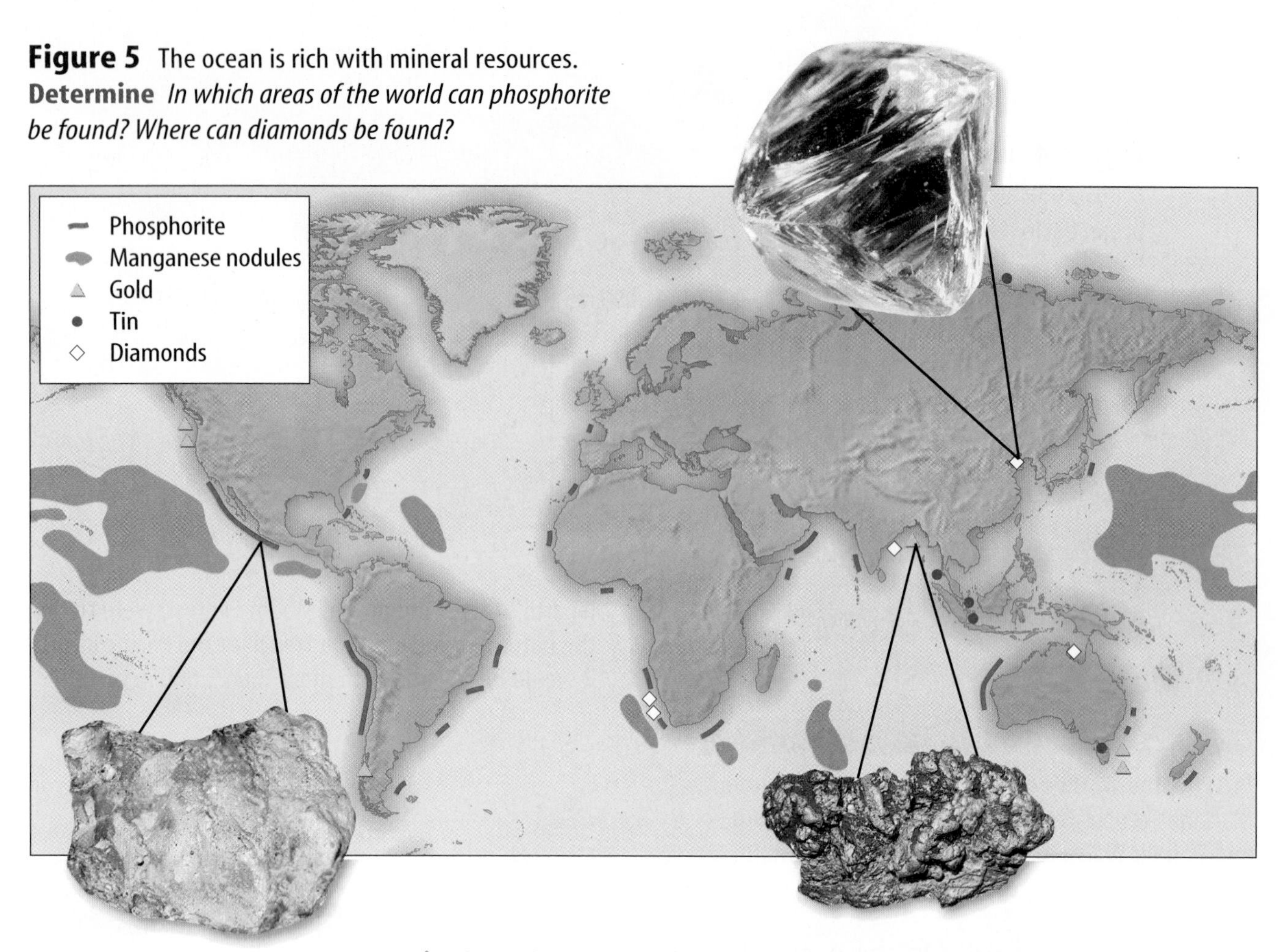

Continental Shelf Deposits A high amount of organic activity occurs in the waters above the continental shelf, and sediment accumulates to great thickness on the ocean floor. This is why many different kinds of resources can be found there, such as petroleum and natural gas deposits. Approximately 20 percent of the world's oil comes from under the seabed. To extract these substances, wells are drilled into the seafloor from floating vessels and fixed platforms.

Other deposits on the continental shelf include phosphorite, which is used to make fertilizer, and limestone, which is used to make cement. Sand and gravel, both economically important, also can be dredged from the continental shelf.

Rivers that flow into oceans transport important minerals to the continental shelf from land. Sometimes the energy of ocean waves and currents can cause denser mineral grains that have been brought in by rivers to concentrate in one place. These deposits, called placer (PLAHS ur) deposits, can occur in coastal regions where rivers entering the ocean suddenly lose energy, slow down, and drop their sediment. Metals such as gold and titanium and gemstones such as diamonds are mined from placer deposits in some coastal regions. **Figure 5** shows where some resources in the ocean can be found.

Deep-Water Deposits Through the holes and cracks along mid-ocean ridges, plumes of hot water billow out into surrounding seawater. As the superheated water cools, mineral deposits sometimes form. As a result, elements such as sulfur and metals like iron, copper, zinc, and silver can be concentrated in these areas. Today, no one is mining these valuable materials from the depths because it would be too expensive to recover them. However, in the future, these deposits could become important.

Other mineral deposits can precipitate from seawater. In this process, minerals that are dissolved in ocean water come out of solution and form solids on the ocean floor. Manganese nodules are small, darkly colored lumps strewn across large areas of the ocean basins. **Figure 6** shows these nodules. Manganese nodules form by a chemical process that is not fully understood. They form around nuclei such as discarded sharks' teeth, growing slowly, perhaps as little as 1 mm to 10 mm per million years. These nodules are rich in manganese, copper, iron, nickel, and cobalt, which are used in the manufacture of steel, paint, and batteries. Most of the nodules lie thousands of meters deep in the ocean and are not currently being mined, although suction devices similar to huge vacuum cleaners have been tested to collect them.

Figure 6 These manganese nodules were found on the floor of the Pacific Ocean.
Think Critically *Can you think of an efficient way to gather the nodules from a depth of 4 km?*

section 1 review

Summary

The Ocean Basins

- Ocean basins have many different features, including the continental shelf, continental slope, and abyssal plains.

Ridges and Trenches

- New ocean floor forms along mid-ocean ridges.
- Trenches mark areas of ocean floor where one crustal plate is sinking beneath another.

Mineral Resources from the Seafloor

- Many mineral deposits, such as petroleum and natural gas, can be found on the continental shelf.
- Other mineral deposits, such as manganese nodules, can be found in deep water.

Self Check

1. **Compare and contrast** continental shelves and continental slopes.
2. **Contrast** mid-ocean ridges and trenches.
3. **Describe** how an abyssal plain looks and how it forms.
4. **Think Critically** Why is the formation of continental shelf deposits different from that of deep-water deposits? Name two examples of each type of deposit.

Applying Skills

5. **Infer** Depth soundings, taken as a ship moves across an ocean, are consistently between 4,000 m and 4,500 m. Over which area of seafloor is the ship passing?

Science Online earth.msscience.com/self_check_quiz

Mapping the Ocean Floor

In this lab you will use sonar data from the Atlantic Ocean to make a profile of the ocean bottom.

Real-World Question

What does the ocean floor look like?

Goals

- **Make** a profile of the ocean floor.
- **Identify** seafloor structures.

Materials

graph paper

Procedure

1. Copy and complete a graph like the one shown.
2. **Plot** each data point and connect the points with a smooth line.

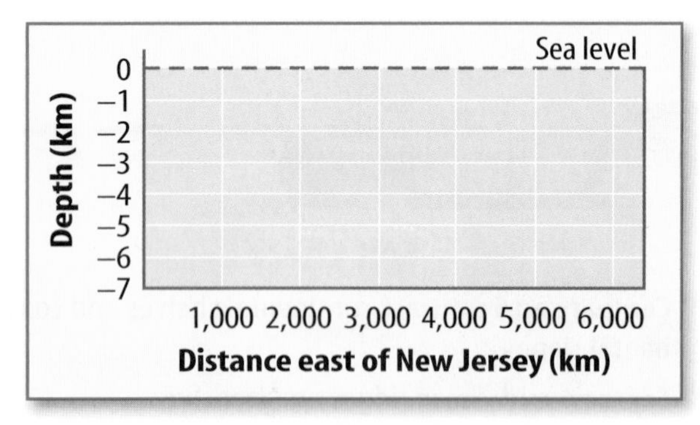

3. Color water blue and the seafloor brown.

Ocean Floor Data

Station Number	Distance from New Jersey (km)	Depth to Ocean Floor (m)
1	0	0
2	160	165
3	200	1,800
4	500	3,500
5	1,050	5,450
6	1,450	5,100
7	1,800	5,300
8	2,000	5,600
9	2,300	4,750
10	2,400	3,500
11	2,600	3,100
12	3,000	4,300
13	3,200	3,900
14	3,450	3,400
15	3,550	2,100
16	3,700	1,275
17	3,950	1,000
18	4,000	0
19	4,100	1,800
20	4,350	3,650
21	4,500	5,100
22	5,000	5,000
23	5,300	4,200
24	5,450	1,800
25	5,500	920
26	5,650	0

Conclude and Apply

1. What ocean floor structures occur between 160 km and 1,050 km east of New Jersey? Between 2,000 km and 4,500 km? Between 5,300 km and 5,500 km?
2. When a profile of a feature is drawn to scale, the horizontal and vertical scales must be the same. Does your profile give an accurate picture of the ocean floor? Explain.

section 2

Life in the Ocean

Life Processes

Life processes such as breathing oxygen, digesting food, making new cells, and growing take place in your body every day. It takes energy to do this, plus walk between classrooms or play soccer. Organisms that live in the ocean also carry out life processes every day. The octopus shown in **Figure 7** will get the oxygen it needs from the water. It will have to eat, and it will use energy to capture prey and to escape predators. It will make new cells and eventually reproduce. Like other marine organisms, it is adapted to accomplish these processes in the salty water of the ocean.

One of the most important processes in the ocean, as it is on land, is that organisms obtain food to use for energy. Obtaining the food necessary to survive can be done in several ways.

INTEGRATE Life Science **Photosynthesis** Nearly all of the energy used by organisms in the ocean ultimately comes from the Sun. Radiant energy from the Sun penetrates seawater to an average depth of 100 m. Marine organisms such as plants and algae use energy from the Sun to build their tissues and produce their own food. This process of making food is called **photosynthesis.** During photosynthesis, carbon dioxide and water are changed to sugar and oxygen in the presence of sunlight. Organisms that undergo photosynthesis are called producers. Producers also need nutrients, such as nitrogen and phosphorus, in order to produce organic matter. These and other nutrients are obtained from the surrounding water. Marine producers include sea grasses, seaweeds, and microscopic algae. Although they might seem unimportant because they are small, microscopic algae are responsible for approximately 90 percent of all marine production. Organisms that feed on producers are called consumers. Consumers in the marine environment include shrimp, fish, dolphins, whales, and sharks.

Figure 7 Hunting at night, this Pacific octopus feeds on snails and crabs. It uses camouflage, ink, and speed to avoid predators.

as you read

What You'll Learn

- **Describe** photosynthesis and chemosynthesis in the oceans.
- **List** the key characteristics of plankton, nekton, and benthos.
- **Compare and contrast** ocean margin habitats.

Why It's Important

The ocean environment is fragile, and many organisms, including humans, depend on it for their survival.

Review Vocabulary

nutrient: a substance needed for the production of organic matter

New Vocabulary

- photosynthesis
- chemosynthesis
- plankton
- nekton
- benthos
- estuary
- reef

Ecologist An ecologist is a scientist who studies the interactions between organisms and their environment. Ecologists may specialize in areas such as marine ecosystems or tropical rain forests. They may also study how energy is transferred from one organism to another, such as through a food web.

Energy Relationships Energy from the Sun is transferred through food chains. Although the organisms of the ocean capture only a small part of the Sun's energy, this energy is passed from producer to consumer, then to other consumers. In **Figure 8,** notice that in one food chain, a large whale shark consumes small, shrimplike organisms as its basic food. In the other chain, microscopic algae found in water are eaten by microscopic animals called copepods (KOH puh pahdz). The copepods are, in turn, eaten by herring. Cod eat the herring, seals eat the cod, and eventually great white sharks eat the seals. At each stage in the food chain, energy obtained by one organism is used by other organisms to move, grow, repair cells, reproduce, and eliminate waste.

Reading Check *What is passed on at each stage in a food chain?*

In an ecosystem—a community of organisms and their environment—many complex feeding relationships exist. Most organisms depend on more than one species for food. For example, herring eat more than copepods, cod eat more than herring, seals eat more than cod, and white sharks eat more than seals. In an ecosystem, food chains overlap and are connected much like the threads of a spider's web. These highly complex systems are called food webs.

Figure 8 Numerous food chains exist in the ocean. Some food chains are simple and some are complex.

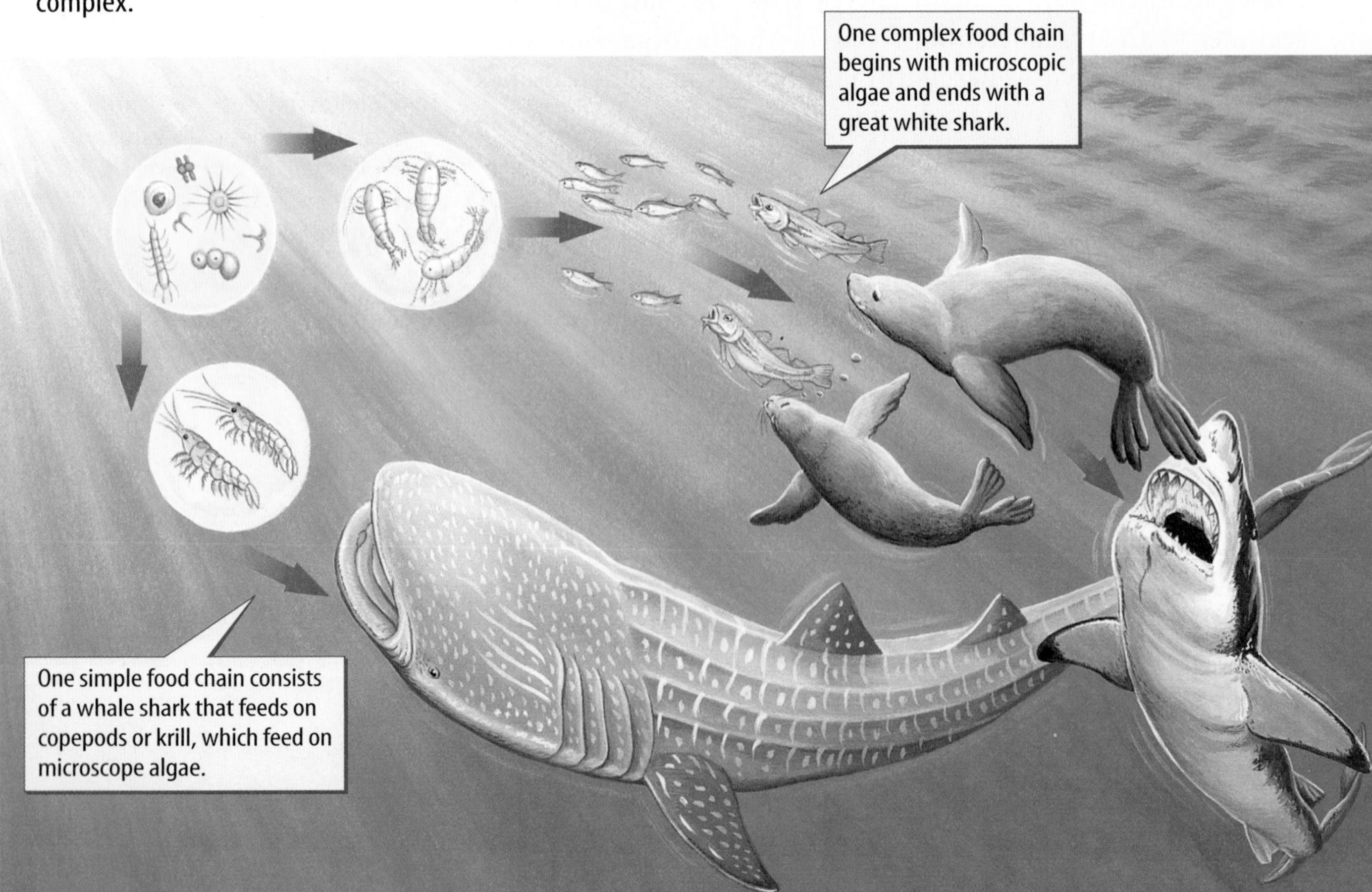

One complex food chain begins with microscopic algae and ends with a great white shark.
One simple food chain consists of a whale shark that feeds on copepods or krill, which feed on microscope algae.

Chemosynthesis Other types of food webs do not depend on the Sun and photosynthesis. These food webs depend on bacteria that perform chemosynthesis. **Chemosynthesis** (kee moh SIHN thuh sus) involves using sulfur or nitrogen compounds as an energy source, instead of light from the Sun, to produce food. Bacteria that perform chemosynthesis using sulfur compounds live along mid-ocean ridges near hydrothermal vents where no light is available. Recall that superheated water from the crust contains high amounts of sulfur. The bacteria found here form the base of a food chain and support a host of highly specialized organisms such as giant tube worms, clams, crabs, and shrimp.

Figure 9 Because sponges live attached to the ocean bottom, they depend on currents to carry their reproductive cells to nearby sponges.
Infer *what would happen if a sponge settled in an area without strong currents.*

Other Life Processes Reproduction also is a vital life process. Some organisms, such as corals and sponges, depend on ocean currents for successful reproduction. Shown in **Figure 9,** these organisms release reproductive cells into the water where they unite to form more organisms of the same type. Other organisms, such as salmon and the Atlantic eel, travel long distances across the ocean in order to reproduce in a specific location. One important aspect of successful reproduction is finding a safe place for eggs and newly hatched larvae to develop. You will learn later in this section that some places in the ocean are used by marine organisms for this purpose.

Ocean Life

Many varieties of plants and animals live in the ocean. Although some organisms live in the open ocean or on the deep ocean floor, most marine organisms live in the waters above or on the floor of the continental shelf. In this relatively shallow water, the Sun penetrates to the bottom, allowing for photosynthesis. Because light is available for photosynthesis, large numbers of producers live in the waters above the continental shelf. These waters also contain many nutrients that producers use to carry out life processes. As a result, the greatest source of food is located in the waters of the continental shelf.

Diatoms are phytoplankton that live in freshwater and ocean water.

The zooplankton shown here is a copepod. Although it has reached its adult size, it is still microscopic.

Figure 10 Some plankton are producers, others are consumers.

Plankton Marine organisms that drift with the currents are called **plankton.** Plankton range from microscopic algae and animals to organisms as large as jellyfish. Most phytoplankton—plankton that are producers—are one-celled organisms that float in the upper layers of the ocean where light needed for photosynthesis is available. One abundant form of phytoplankton is a one-celled organism called a diatom. Diatoms are shown in **Figure 10.** Diatoms and other phytoplankton are the source of food for zooplankton, animals that drift with ocean currents.

Zooplankton includes newly hatched fish and crabs, jellies, and tiny adults of some organisms like the one shown in **Figure 10,** which feed on phytoplankton and are usually the second step in ocean food chains. Most animal plankton depend on surface currents to move them, but some can swim short distances.

Nekton Animals that actively swim, rather than drift with the currents in the ocean, are called **nekton.** Nekton include all swimming forms of fish and other animals, from tiny herring to huge whales. Nekton can be found from polar regions to the tropics and from shallow water to the deepest parts of the ocean. In **Figure 11,** the Greenland shark, the manatee, and the deep-ocean fish are all nekton. As nekton move throughout the oceans, it is important that they are able to control their buoyancy, or how easily they float or sink. What happens when you hold your breath underwater, then let all of the air out of your lungs at once? The air held in your lungs provides buoyancy and helps you float. As the air is released, you sink. Many fish have a special organ filled with gas that helps them control their buoyancy. By changing their buoyancy, organisms can change their depth in the ocean. The ability to move between different depths allows animals to search more areas for food.

Reading Check *What are nekton?*

Some deep-dwelling nekton are adapted with special light-generating organs. The light has several uses for these organisms. The deep-sea fish, shown in **Figure 11,** dangles a luminous lure from beneath its jaw. When prey attracted by the lure are close enough, they are swallowed quickly. Some deep-sea organisms use this light to momentarily blind predators so they can escape. Others use it to attract mates.

INTEGRATE Chemistry

Bioluminescence Some marine organisms, including types of bacteria, one-celled algae, and fish, can make their own light through a process called bioluminescence. The main molecule involved in producing light is luciferin. In the process of a chemical reaction involving luciferin, a burst of light is produced.

Figure 11 Nekton are found living in all areas of the ocean, warm or cold, shallow or deep.

This Greenland shark lives in the cold waters of the North Atlantic.

Manatees are found in tropical regions around the world.

This deep-sea fish is adapted to living under pressure at a depth of 4 km.

Bottom Dwellers The plants and animals living on or in the seafloor are the **benthos** (BEN thahs). Benthic animals include crabs, snails, sea urchins, and bottom-dwelling fish such as flounder. These organisms move or swim across the bottom to find food. Other benthic animals that live permanently attached to the bottom, such as sea anemones and sponges, filter out food particles from seawater. Certain types of worms live burrowed in the sediment of the ocean floor. Bottom-dwelling animals can be found living from the shallow water of the continental shelf to the deepest areas of the ocean. Benthic plants and algae, however, are limited to the shallow areas of the ocean where enough sunlight penetrates the water to allow for photosynthesis. One example of a benthic algae is kelp, which is anchored to the bottom and grows toward the surface from depths of up to 30 m.

Ocean Margin Habitats

The area of the environment where a plant or animal normally lives is called a habitat. Along the near-shore areas of the continental shelf, called ocean margins, a variety of habitats exist. Beaches, rocky shores, estuaries, and coral reefs are some examples of the different habitats found along ocean margins.

Observing Plankton

Procedure

1. Place one or two drops of **pond, lake, or ocean water** onto a **microscope slide.**
2. Use a **microscope** to observe your sample. Look for microscopic life.
3. Find at least three different types of plankton.

Analysis

1. Draw detailed pictures of three types of plankton.
2. Classify the plankton as phytoplankton or zooplankton.

Beaches At the edge of a sandy beach where the waves splash, you can find some microscopic organisms and worms that spend their entire lives between moist grains of sand. Burrowing animals such as small clams and mole crabs make holes in the sand. When water covers the holes, these animals rise to the surface to filter food from the water. Where sand is covered constantly by water, larger animals like horseshoe crabs, snails, fish, turtles, and sand dollars reside. **Figure 12** shows some of the organisms that are found living on sandy beaches.

Although the beach is great fun for people, it is a very stressful environment for the plants and animals that live there. They constantly deal with waves, changing tides, and storms, all of which redistribute large amounts of sand. Large waves produced by storms, such as hurricanes, can cause damage to beaches as they crash onto shore. These organisms must adapt to natural changes as well as changes created by humans. Damming rivers, building harbors, and constructing homes and hotels near the shoreline disrupts natural processes on the beach.

Topic: Beach Erosion
Visit earth.msscience.com for Web links to information about beach erosion.

Activity Find two examples of beaches that are having problems with erosion. What is being done to help solve the problem?

Rocky Shore Areas In some regions the shoreline is rocky, as shown in **Figure 13.** Algae, sea anemones, mussels, and barnacles encrust submerged rocks. Sea stars, sea urchins, octopuses, and hermit crabs crawl along the rock surfaces, looking for food.

Tide pools are formed when water remains onshore, trapped by the rocks during low tide. Tide pools are an important habitat for many marine organisms. They serve as protected areas where many animals, such as octopuses and fish, can develop safely from juveniles to adults. Tide pools contain an abundance of food and offer protection from larger predators.

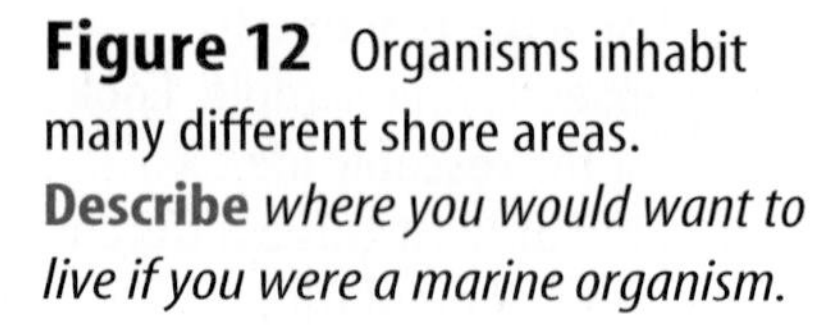

Figure 12 Organisms inhabit many different shore areas.
Describe *where you would want to live if you were a marine organism.*

NATIONAL GEOGRAPHIC VISUALIZING THE ROCKY SHORE HABITAT

Figure 13

Life is tough in the intertidal zone—the coastal area between the highest high tide and the lowest low tide. Organisms here are pounded by waves and alternately covered and uncovered by water as tides rise and fall. These organisms tend to cluster into three general zones along the shore. Where they live depends on how well they tolerate being washed by waves, submerged at high tide, or exposed to air and sunlight when the tide is low.

Wavy turban snail

Stone crab

Periwinkle

Algae

UPPER INTERTIDAL ZONE This part of the intertidal zone is splashed by high waves and is usually covered by water only during the highest tides each month. It is home to crabs that scuttle among periwinkle snails, limpets, and a few kinds of algae that can withstand long periods of dryness.

Gooseneck barnacles

Lined chiton

Blue mussels

MID-INTERTIDAL ZONE Submerged at most high tides and exposed at most low tides, this zone is populated by brown algae, sponges, barnacles, mussels, chitons, snails, and sea stars. These creatures are resistant to drying out and good at clinging to slippery surfaces.

LOWER INTERTIDAL ZONE This section of the intertidal zone is exposed only during the lowest tides each month. It contains the most diverse collection of living things. Here you find sea urchins, large sea stars, brittle stars, nudibranchs, sea cucumbers, anemones, and many kinds of fish.

Sea lemon nudibranch

African sea star

Figure 14 Estuaries are called the nurseries of the oceans because many creatures spend their early lives there.

Estuaries An **estuary** is an area where the mouth of a river opens into an ocean. Because estuaries receive freshwater from rivers, they are not as salty as the ocean. Rivers also bring nutrients to estuaries. Areas with many nutrients usually have many phytoplankton, which form the base of the food chain. Shown in **Figure 14,** estuaries are full of life from salt-tolerant grasses to oysters, clams, shrimps, fish, and even manatees.

Estuaries are an important habitat to many marine organisms. Newly hatched fish, shrimps, crabs, and other animals enter estuaries as microscopic organisms and remain there until adulthood. For these vulnerable animals, fewer predators and more food are found in estuaries.

Coral Reefs Corals thrive in clear, warm water that receives a lot of sunlight. This means that they generally live in warm latitudes, between 30°N and 30°S, and in water that is no deeper than 40 m. Each coral animal builds a hard capsule around its body from the calcium it removes from seawater. Each capsule is cemented to others to form a large colony called a reef. A **reef** is a rigid, wave-resistant structure built by corals from skeletal material. As a coral reef forms, other benthos such as sea stars and sponges and nekton such as fish and turtles begin living on it.

In all ocean margin habitats, nutrients, food, and energy are cycled among organisms in complex food webs. Plankton, nekton, and benthos depend on each other for survival.

section 2 review

Summary

Life Processes

- Organisms in the ocean obtain food to use for energy in several ways.
- Photosynthesis and chemosynthesis are processes used by producers to make food. Other organisms are consumers.
- Reproduction is also a vital life process.

Ocean Life

- Organisms in the ocean can be classified as plankton, nekton, or benthos depending on where they live.

Ocean Margin Habitats

- Ocean margin habitats such as beaches, rocky shore areas, estuaries, and coral reefs exist along the near-shore areas of the continental shelf.

Self Check

1. **Describe** the processes of photosynthesis and chemosynthesis.
2. **Identify** the key characteristics of plankton, nekton, and benthos.
3. **Compare and contrast** the characteristics of coral reef and estuary habitats.
4. **Think Critically** The amount of nutrients in the water decreases as the distance from the continental shelf increases. What effect does this have on open-ocean food chains?

Applying Skills

5. **Use Graphics Software** Design a creative poster that shows energy relationships in a food chain. Begin with photosynthesis. Use clip art, scanned photographs, or computer graphics.

Science Online earth.msscience.com/self_check_quiz

section 3

Ocean Pollution

Sources of Pollution

How would you feel if someone came into your bedroom; spilled oil on your carpet; littered your room with plastic bags, cans, bottles, and newspapers; then sprayed insect killer and scattered sand all over? Organisms in the ocean experience these things when people pollute seawater.

Pollution is the introduction of harmful waste products, chemicals, and other substances not native to an environment. A pollutant is a substance that causes damage to organisms by interfering with life processes.

Pollutants from land eventually will reach the ocean in one of four main ways. They can be dumped deliberately and directly into the ocean. Material can be lost overboard accidentally during storms or shipwrecks. Some pollutants begin in the air and enter the ocean through rain. Other pollutants will reach the ocean by being carried in rivers that empty into the ocean. **Figure 15** illustrates how pollutants from land enter the oceans.

as you read

What You'll Learn

- **List** five types of ocean pollution.
- **Explain** how ocean pollution affects the entire world.
- **Describe** how ocean pollution can be controlled.

Why It's Important

Earth's health depends on the oceans being unpolluted.

Review Vocabulary
runoff: water that does not soak into the ground or evaporate but instead flows over Earth's surface, eventually entering streams, lakes, or oceans

New Vocabulary
- pollution

Figure 15 Ocean pollution comes from many sources.

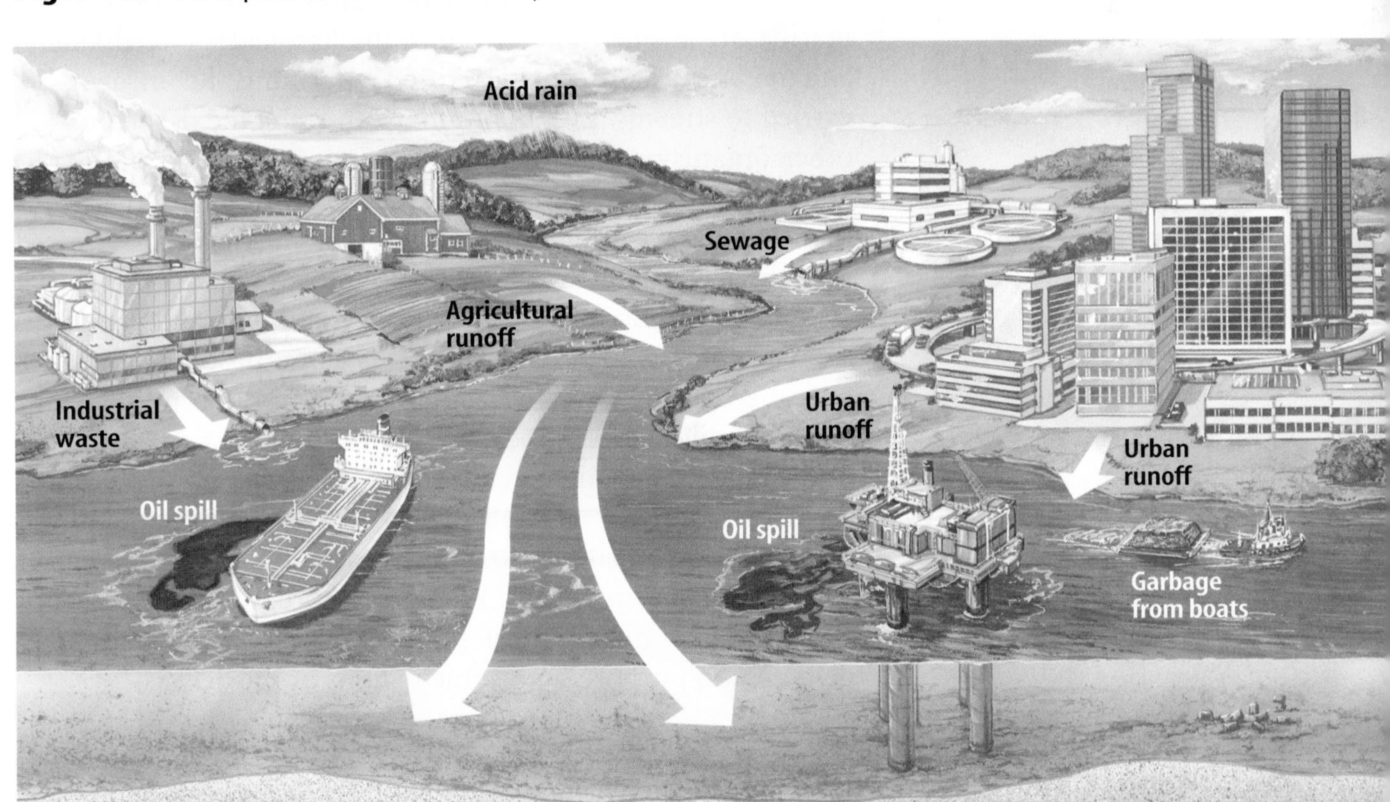

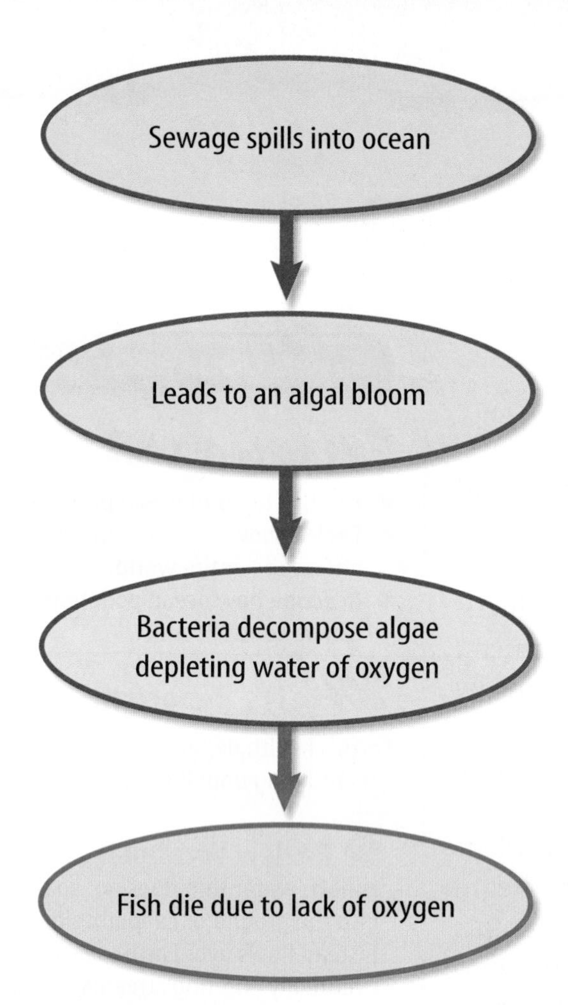

Figure 16 Fish kills occur when the oxygen supply is low.
Infer *how fish kills affect the food web.*

Sewage In some regions, human sewage leaks from septic tanks or is pumped directly into oceans or into rivers leading to an ocean. The introduction of sewage to an area of the ocean can cause immediate changes in the ecosystem, as shown by the following example. Sewage is a pollutant that acts like fertilizer. It is rich in nutrients that cause some types of algae to reproduce rapidly, creating what is called a bloom. The problem occurs when the algae die. As huge numbers of bacteria reproduce and decompose the algae, much of the oxygen in the water is used up. Other organisms, such as fish, cannot get enough oxygen. As a result, fish die in a phenomenon called a fish kill, as illustrated in **Figure 16.**

Reading Check *What is an algal bloom?*

When sewage is dumped routinely into the same area year after year, changes take place. Entire ecosystems have been altered drastically as a result of long-term, repeated exposure to sewage and fertilizer runoff. In some areas of the world, sewage is dumped directly onto coral reefs. When this happens algae can outgrow the coral because the sewage acts like a fertilizer. Eventually, the coral organisms die. If this occurs, other organisms that depend on the reef for food and shelter also can be affected.

Chemical Pollution Industrial wastes from land can harm marine organisms. When it rains, the herbicides (weed killers) and insecticides (insect killers) used in farming and on lawns are carried to streams. Eventually, they can reach the ocean and kill other organisms far from where they were applied originally. Sometimes industrial wastes are released directly into streams that eventually empty into oceans. Other chemicals are released into the air, where they later settle into the ocean. Industrial chemicals include metals like mercury and lead and chemicals like polychlorinated biphenyls (PCBs). In a process called biological amplification (am plah fah KAY shun), harmful chemicals can build up in the tissues of organisms that are at the top of the food chain. Higher consumers like dolphins and seabirds accumulate greater amounts of a toxin as they continue to feed on smaller organisms. At high concentrations, some chemicals can damage an organism's immune and reproductive systems. Explosives and nuclear wastes also have been dumped, by accident and on purpose, into some regions of the oceans.

Oil Pollution Although oil spills from tankers that have collided or are leaking are usually highly publicized, they are not the biggest source of oil pollution in the ocean. As much as 44 percent of oil that reaches the ocean comes from land. Oil that washes from cars and streets, or that is poured down drains or into soil, flows into streams. Eventually, this oil reaches the ocean. Other sources of oil pollution are leaks at offshore oil wells and oil mixed with wastewater that is pumped out of ships. **Figure 17** shows the percentage of different sources of oil entering the oceans each year.

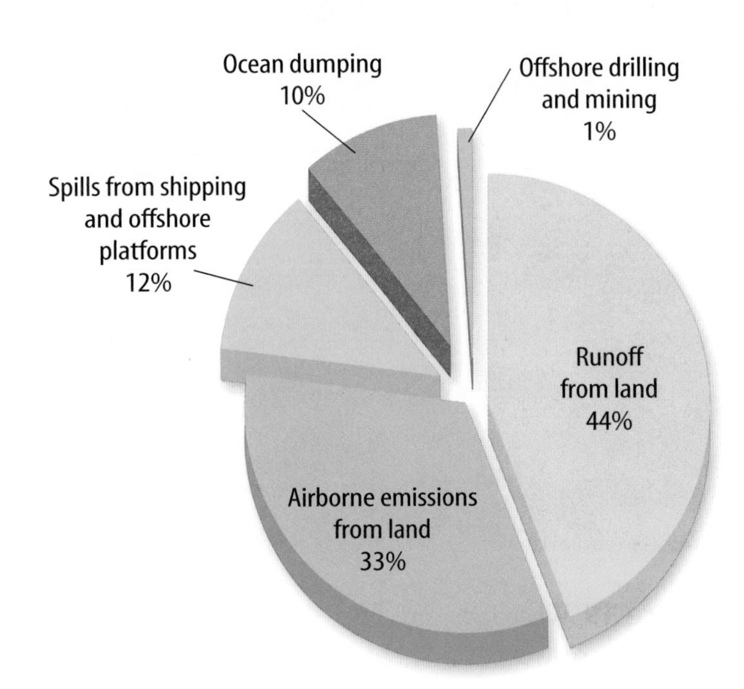

Figure 17 Although oil spills are highly publicized and tragic, the same harmful oil enters the ocean every day from many other sources.
Think Critically *What can be done to reduce the amount of oil entering the oceans?*

Solid-Waste Pollution Even in the most remote areas of the world, such as uninhabited islands that are thousands of miles from any major city, large amounts of trash wash up on the beach. **Figure 18** shows the amount of debris collected by a scientist on an island in the Pacific Ocean, 8,000 km east of Australia in just one day. The presence of trash ruins a beautiful beach, and solid wastes, such as plastic bags and fishing line, can entangle animals. Animals such as sea turtles mistakenly eat plastic bags, because they look so much like their normal prey, floating jellyfish. Illegally dumped medical waste such as needles, plastic tubing, and bags also are a threat to humans and other animals.

Figure 18 These items are like the ones found washed ashore on one of the Pitcairn Islands in the South Pacific. The number of each item found is shown below the figure. Also among the rubble were broken toys, two pairs of gloves, and an asthma inhaler.

Sediment Silt also can pollute the ocean. Human activities such as agriculture, deforestation, and construction tear up the soil. Rain washes soil into streams and eventually into an ocean nearby, as shown in **Figure 19.** This causes huge amounts of silt to accumulate in many coastal areas. Coral reefs and saltwater marshes are safe, protected places where young marine organisms grow to adults. When large amounts of silt cover coral reefs and fill marshes, these habitats are destroyed. Without a safe place to grow larger, many organisms will not survive.

Figure 19 When large amounts of silt enter seawater, the filter-feeding systems of animals such as oysters and clams can be clogged.

Effects of Pollution

You already have learned some examples of how pollution affects the ocean and the organisms that live there. Today, there is not a single area of the ocean that is not polluted in some way. As pollution from land continues to reach the ocean, scientists are recording dramatic changes in this environment.

Estuaries and the rivers that feed into them from Delaware to North Carolina have suffered from toxic blooms of *Pfiesteria* since the late 1980s. These blooms have killed billions of fish. *Pfiesteria,* a type of plankton, also has caused rashes, nausea, memory loss, and the weakening of the immune system in humans. The cause of these blooms is thought to be runoff contaminated by fertilizers and other waste materials. In Florida, toxic red tides kill fish and manatees. Some people also blame these red tides on sewage releases and fertilizer runoff. **Figure 20** shows an increase in the number of harmful algal blooms since the early 1970s.

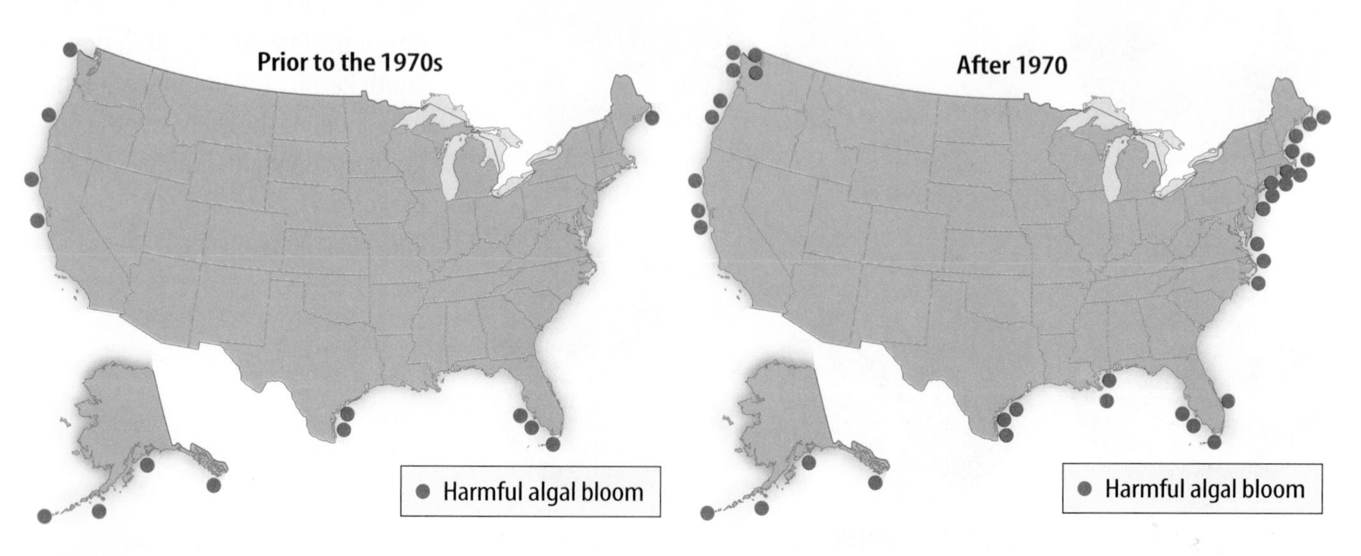

Figure 20 Some scientists hypothesize that a relationship exists between increased pollution in the ocean and the number of harmful algal blooms in the last 30 years.

Controlling Pollution

Some people believe that oceans take care of themselves because they are large. However, other people view ocean pollution as a serious problem. Many international organizations have met to develop ways of reducing ocean pollution. Treaties prohibit the dumping of some kinds of hazardous wastes from vessels, aircraft, and platforms. One treaty requires that some ships and operators of offshore platforms have oil pollution emergency plans. This includes having the proper equipment to combat oil spills and practicing what to do if a spill takes place. Recall that a large amount of pollution enters the ocean from land. Although the idea of reducing land pollution to better protect the ocean has been discussed, no international agreement exists to prevent and control land-based activities that affect the oceans.

What has been done to help control ocean pollution?

What You Can Do Current international and U.S. laws aren't effective enough. Further cooperation is needed to reduce ocean pollution. You can help by disposing of wastes properly and volunteering for beach or community cleanups, like the one shown in **Figure 21.** You can recycle materials such as newspapers, glass, and plastics and never dump chemicals like oil or paint onto soil or into water. One of the best things you can do is continue to learn about marine pollution and how people affect the oceans. What other things will help reduce ocean pollution?

Figure 21 Under careful supervision, picking up trash is an easy way to help reduce ocean pollution.

section 3 review

Summary

Sources of Pollution

- Sewage, industrial wastes, oil, solid waste, and sediment are the main types of pollution entering ocean water.

Effects of Pollution

- As pollution from land continues to reach the ocean, ecosystems and organisms are negatively affected.

Controlling Pollution

- International treaties and U.S. laws have been made to help reduce ocean pollution.
- Everyone can help reduce ocean pollution.

Self Check

1. **Identify** five human activities that pollute the oceans. Suggest a solution to each.
2. **Explain** how pollution of the oceans affects the world.
3. **Describe** the ways international treaties have helped reduce pollution.
4. **Think Critically** To widen beaches, some cities pump offshore sediment onto them. How might this affect organisms that live in coastal waters?

Applying Skills

5. **Concept Map** Make an events-chain concept map that describes how runoff can reach the ocean. Include examples of pollution that could be in the runoff.

Use the Internet

Resources from the Oceans

Goals

- **Research and identify** organisms that are used to make products.
- **Explain** why it is important to keep oceans clean.

Data Source

Science Online

Visit earth.msscience.com/internet_lab for Web links to more information about resources from the oceans, hints on which products come from the oceans, and data from other students.

Real-World Question

Oceans cover most of Earth's surface. Humans get many things from oceans, such as seafood, medicines, oil, and diamonds. Humans also use oceans for recreation and to transport materials from place to place. What else comes from oceans? Scientists continue to discover and research new uses for ocean resources. You might not realize that you probably use many products every day that are made from organisms that live in oceans. Think about the plants and animals that live in the oceans. How could these organisms be used to make everyday products? Form a hypothesis about the types of products that could be manufactured from these organisms.

Make a Plan

1. **Identify** Web links shown in the Data Source section above and identify other resources that will help you complete the data table shown on the right.
2. **Observe** that to complete the table you must identify products made from marine organisms, where the organisms are collected or harvested, and alternative products.
3. **Plan** how and when you will locate the information.

Ocean Conservation Organizations Data

Organization	Location Where Conservation Takes Place	Main Goal	Affiliation
	Do not write in this book.		

Follow Your Plan

1. Make sure your teacher approves your plan and your resource list before you begin.
2. **Describe** at least three ocean organisms that are used to make products you use every day.
3. **Identify** the name and any uses of the product.
4. **Research** where each organism lives and the method by which it is collected or harvested.
5. **Identify** alternative products.

Analyze Your Data

1. **Describe** the different ways in which ocean organisms are useful to humans.
2. **Explain** Are there any substitutes or alternatives available for the ocean organisms in the products?

Conclude and Apply

1. **Infer** How might the activities of humans affect any of the ocean organisms you researched?
2. **Determine** Are the substitute or alternative products more or less expensive?
3. **Describe** Can you tell whether the ocean-made product is better than the substitute product?
4. **Explain** why it is important to conserve ocean resources and keep oceans clean.

Communicating Your Data

Find this lab using the link below. Post your data in the table provided. Compare your data to those of other students.

Science Online

earth.msscience.com/internet_lab

Strange Creatures from the Ocean Floor

In 1977, the *Alvin,* a small submersible craft specially designed to explore the ocean depths, took three geologists down about 2,200 m below the sea surface. They wanted to be the first to observe and study the formations of the Galápagos Rift deep in the Pacific Ocean. What they saw was totally unexpected. Instead of barren rock, the geologists found life—a lot of life. And they had never even considered having a life scientist as part of the research team!

The crew of the *Alvin* discovered hydrothermal vents—underwater openings where hot water (400°C) spurts from cracks in the rocks on the ocean floor. Some organisms thrive there because of the hydrogen sulfide that exists at the vents. Many of these organisms are like nothing humans had ever seen before. They are organisms that live in extremely hot temperatures and use hydrogen sulfide as their food supply.

The discovery and study of hydrothermal vents almost has been overshadowed by the amazing variety of life that was found there. But scientists think these openings on the ocean floor (many located along the Mid-Atlantic Ridge) control the temperature and movement of nearby ocean waters, as well as have a significant effect on the ocean's chemical content. These vents also act as outlets for Earth's inner heat.

Scientists also have discovered that the vent communities are temporary. Each vent eventually shuts down and the organisms somehow disperse to other vents. Exactly how this happens is an area of ongoing research.

Blood-red tube worms live deep beneath the sea.

Creative Writing **Imagine you were a passenger on the *Alvin.* Write about your adventure as you came upon the hydrothermal communities. Describe and draw in detail some of the unique creatures you saw.**

Science Online
For more information, visit earth.msscience.com/oops

Reviewing Main Ideas

Section 1 The Seafloor

1. The continental shelf is a gently sloping part of the continent that extends into the oceans. The continental slope extends from the continental shelf to the ocean floor. The abyssal plain is a flat area of the ocean floor.
2. Along mid-ocean ridges, new seafloor forms. Seafloor slips beneath another crustal plate at a trench.
3. Petroleum, natural gas, and placer deposits are mined from continental shelves. Manganese nodules and other mineral deposits can be found in deep water.

Section 2 Life in the Ocean

1. Marine organisms are specially adapted to live in salt water. They produce or consume food, and reproduce in the oceans.
2. Photosynthesis is the basis of most of the food chains in the ocean. Chemosynthesis is a process of making food using chemical energy. Energy is transferred through food webs.
3. Organisms that drift in ocean currents are called plankton. Nekton are marine organisms that swim. Benthos are plants and animals that live on or near the ocean floor.
4. Ocean margin habitats, found along the continental shelf, include sandy beaches, rocky shores, estuaries, and coral reefs.

Section 3 Ocean Pollution

1. Sources of pollution include sewage, chemical pollution, oil spills, solid waste pollution, and sediment.
2. Ocean pollution can disrupt food webs and threaten marine organisms.
3. International treaties and U.S. laws have been made to help reduce ocean pollution.

Visualizing Main Ideas

Copy and complete the following chart on ocean-margin organisms.

Ocean-Margin Organisms

Organism	Ocean-Margin Environments			
	Sandy Beach	Rocky Shore	Estuary	Coral Reef
Plankton	phytoplankton zooplankton		phytoplankton zooplankton	
Nekton		octopuses, fish		fish, turtles
Benthos			grasses, snails, clams	

chapter 19 Review

Using Vocabulary

abyssal plain p. 543	nekton p. 552
benthos p. 553	photosynthesis p. 549
chemosynthesis p. 551	plankton p. 552
continental shelf p. 542	pollution p. 557
continental slope p. 543	reef p. 556
estuary p. 556	trench p. 545
mid-ocean ridge p. 544	

Fill in the blanks with the correct words.

1. Animals such as whales, sea turtles, and fish are examples of ________.
2. ________ occurs in areas where organisms use sulfur as energy to produce their own food.
3. The ________ drops from the edge of a continent out to the deep abyssal plains of the ocean floor.
4. New ocean floor is formed at ________.
5. An area where the mouth of a river opens into an ocean is a(n) ________.

Checking Concepts

Choose the word or phrase that best answers the question.

6. What are formed along subduction zones?
   A) mid-ocean ridges
   B) continental shelves
   C) trenches
   D) density currents
7. What are ocean organisms that drift in ocean currents called?
   A) nekton **C)** benthos
   B) pollutants **D)** plankton
8. How do some deep-water bacteria in the ocean make food?
   A) photosynthesis **C)** respiration
   B) chemosynthesis **D)** rifting

Use the illustration below to answer question 9.

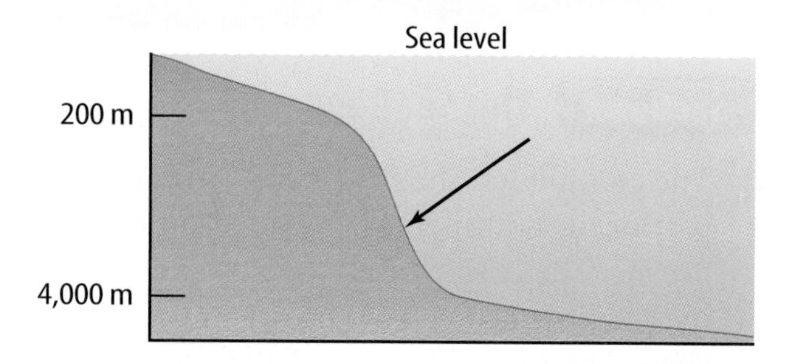

9. Which feature of the ocean floor is the arrow pointing to?
   A) rift valley **C)** abyssal plain
   B) seamount **D)** continental slope
10. What might be found in areas where rivers enter oceans?
    A) rift valleys
    B) manganese nodules
    C) abyssal plains
    D) placer deposits
11. Which organisms reproduce rapidly, resulting eventually in a lack of oxygen?
    A) fish **C)** algae
    B) corals **D)** animal plankton
12. In which area of the ocean is the greatest source of food found?
    A) on abyssal plains
    B) in trenches
    C) along continental shelves
    D) along the mid-ocean ridge
13. Where does most oil pollution originate?
    A) tanker collisions
    B) runoff from land
    C) leaks at offshore wells
    D) in wastewater pumped from ships
14. Where does new seafloor form?
    A) trenches
    B) mid-ocean ridges
    C) abyssal plains
    D) continental shelves

Science Online earth.msscience.com/vocabulary_puzzlemaker

Thinking Critically

15. **Infer** why some industries might be interested in mining manganese nodules.
16. **Explain** why ocean pollution is considered to be a serious international problem.
17. **Summarize** How can agricultural chemicals kill marine organisms?
18. **Draw Conclusions** Would you expect coral reefs to grow around the bases of underwater volcanoes off the coast of Alaska?
19. **Think Critically** Scientists currently are researching the use of chemicals produced by marine organisms to help fight diseases including certain types of cancer. How would ocean pollution affect the ability to discover and research new drugs?
20. **Use Scientific Illustrations** Use **Figure 8** to determine which organisms would starve if phytoplankton became extinct.
21. **Classify** each of these sea creatures as plankton, nekton, or benthos: shrimps, dolphins, sea stars, krill, coral, manatees, and algae.

Use the illustration below to answer question 22.

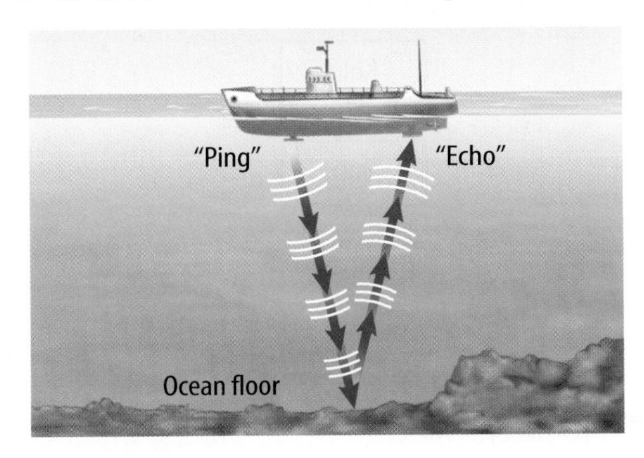

22. **Draw Conclusions** At point A an echo, a sound wave bounced off the ocean floor, took 2 s to reach the ship. It took 2.4 s at point B. Which point is deeper?

Performance Activities

23. **Graph a Profile** Previously, you made a profile of the ocean bottom along the 38° N latitude line, but it was not drawn to scale. Make a scale profile of the area between 3,600 km and 4,100 km from New Jersey. Use the scale 1 mm = 1,000 m.
24. **Poster** Choose a sea animal and research its life processes. Classify it as plankton, nekton, or benthos. Design a poster that includes all of this information.

Applying Math

Use the graph below to answer questions 25 and 26.

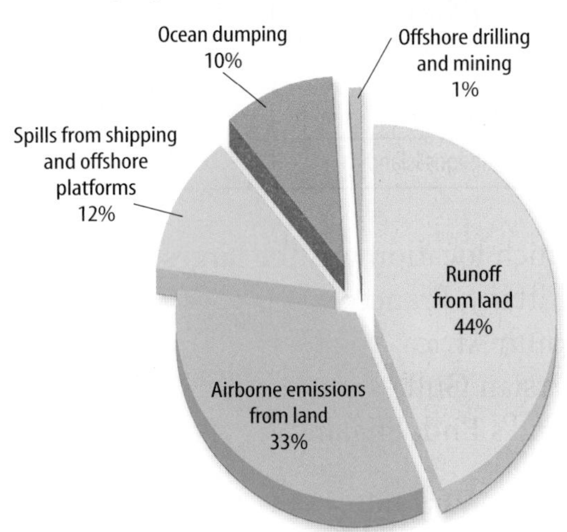

25. **Oil Production** If 6,600,000 tons of oil enter the world's oceans in one year, approximately what amount (in tons) is from runoff from land?
26. **Sources of Pollution** How many more times are airborne emissions from land a source of oil pollution than ocean dumping?
27. **Kelp Growth** If kelp grows at a steady rate of 30 cm per day, how long would it take to reach a length of 25 m?
28. **Seafloor Spreading** The distance between two locations across an ocean basin increases by 1.8 cm, 4.1 cm, and 3.2 cm, each year respectively. What is the average rate of separation of these locations during this time?

Part 1 Multiple Choice

Record your answers on the answer sheet provided by your teacher or on a sheet of paper.

Use the table below to answer questions 1 and 2.

Oil Spills Around the World		
Year	Location	Spill Size (millions of liters)
1967	Land's End, England	144.7
1972	Gulf of Oman, Oman	143.5
1978	Brittany, France	260.2
1979	Bay of Campeche, Mexico	530.3
1983	South Africa	297.3
1988	Newfoundland, Canada	163.2
1991	Persian Gulf	909.0
2001	Galápagos Islands	0.6

1. At which location was the largest spill?
- **A.** Brittany, France
- **B.** South Africa
- **C.** Persian Gulf
- **D.** Land's End, England

2. Approximately how many more liters were spilled in the Persian Gulf than in the Bay of Campeche, Mexico?
- **A.** 530 million liters
- **B.** 470 million liters
- **C.** 379 million liters
- **D.** 279 million liters

3. Why is oil entering the ocean a concern?
- **A.** The presence of oil can reduce water quality.
- **B.** The presence of oil in the water is not harmful to marine life.
- **C.** Large spills can be easy to clean up.
- **D.** The presence of oil can improve water quality.

4. A rigid, wave-resistant structure built by corals from skeletal materials is
- **A.** an estuary.
- **B.** a reef.
- **C.** a beach.
- **D.** a rocky shore.

5. Some bacteria undergo chemosynthesis. Chemosynthesis is
- **A.** a process that involves using sulfur or nitrogen compounds as an energy source to produce food.
- **B.** a process by which other organisms are consumed as a source of energy.
- **C.** a process by which reproductive cells are released into the water.
- **D.** a process that involves using light from the Sun as an energy source to produce food.

Use the illustration of a simple food chain below to answer questions 6 and 7.

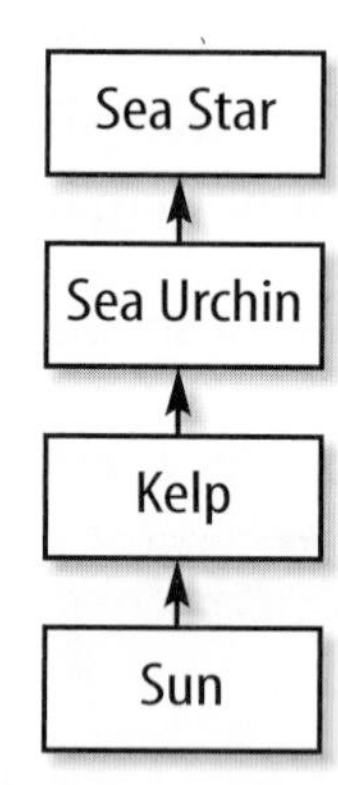

6. Which is a producer?
- **A.** the Sun
- **B.** kelp
- **C.** sea urchin
- **D.** sea star

7. Producers undergo which process in order to make food?
- **A.** bioluminescence
- **B.** respiration
- **C.** reproduction
- **D.** photosynthesis

Part 2 | Short Response/Grid In

Record your answers on the answer sheet provided by your teacher or on a sheet of paper.

8. Define the words *producer* and *consumer*. Give two examples of each that can be found in the ocean.

9. Compare and contrast rocky shore and beach habitats.

Use the illustration below to answer question 10.

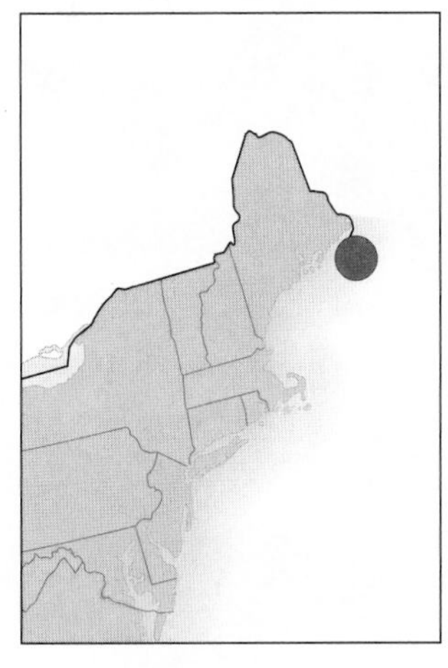

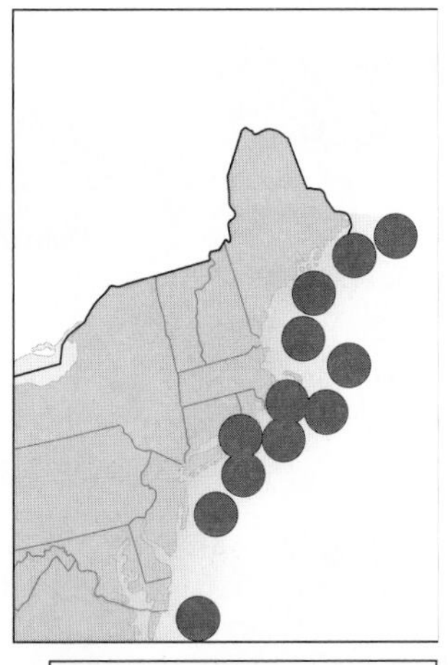

10. What changes have occurred in regard to harmful algal blooms since 1970? What do some scientists hypothesize is the cause of these changes?

11. Explain the process of seafloor spreading.

12. Ten percent of the total available energy is stored by a consumer at each level of the food chain. If 2,533 energy units are passed on to a salmon feeding on zooplankton, how many energy units will the salmon store?

13. Why are estuaries referred to as nurseries? What other marine habitat could also be referred to this way? Why?

Part 3 | Open Ended

Record your answers on a sheet of paper.

14. Compare and contrast the Atlantic Ocean Basin with the Pacific Ocean Basin. Which basin contains many deep-sea trenches? Which basin is getting larger with time?

15. Describe how you could set up an experiment to test the effects of different amounts of light on marine producers.

16. Write a paragraph that explains why ocean pollution is a problem that people can help prevent. List examples of things people can do to help.

Use the illustration below to answer question 17.

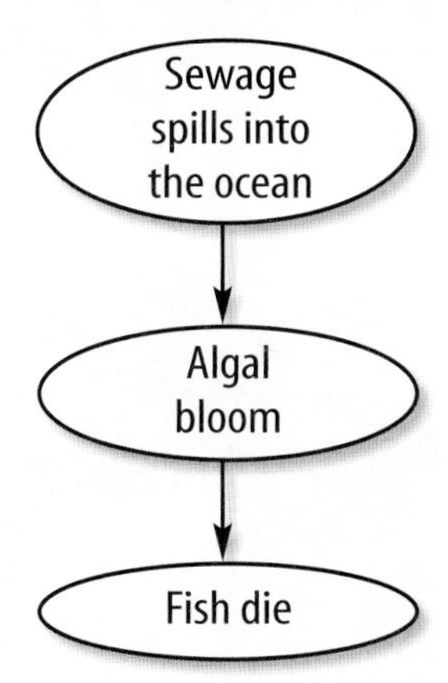

17. Explain in detail how the events are related.

18. Explain the relationship between hydrothermal vents and minerals on the ocean floor. Why are these minerals not being mined?

Test-Taking Tip

Organize Main Ideas For essay questions, spend a few minutes listing and organizing the main ideas on your scratch paper.

Question 14 Make two columns, titled *Atlantic Ocean* and *Pacific Ocean*, on your scratch paper. Fill in information about each topic in the columns.

How Are Cotton & Cookies Connected?

In the 1800s, the economy of the South depended heavily on cotton and tobacco—two crops that rob the soil of nutrients, especially nitrogen. By the late 1800s, the soil was in poor shape. A scientist named George Washington Carver set out to change that. He promoted the technique of crop rotation—alternating soil-depleting crops such as cotton with soil-enriching crops such as peanuts. Many farmers listened to Carver and began planting peanuts. However, there was little market for the crop. So Carver poured his energy into developing uses for peanuts. Ultimately, he came up with more than 300 products made from peanuts—everything from soap to axle grease. He also created the first recipe for peanut butter cookies, which have become an American favorite.

unit projects

Visit earth.msscience.com/unit_project to find project ideas and resources. Projects include:

- **History** Investigate land use in the U.S. during the past 200 years. Design flip charts to compare, and then predict, the land use of the future.
- **Technology** Research how legumes "fix" nitrogen and why crop rotation keeps farm land more productive.
- **Model** Explore peanut inventions. Design your own creative use for peanuts, write directions, and build a prototype for a class peanut invention fair.

WebQuest *Recycling Plastics* explores the seven classes of plastics and their uses, as well as the chemistry of plastic. Discover what it takes to recycle plastic, glass, paper, and aluminum.

chapter 20

Our Impact on Land

The BIG Idea

As the world population continues to rise, the strain on the environment may worsen.

SECTION 1
Population Impact on the Environment

Main Idea As human population continues to grow, more resources are used and more waste is created.

SECTION 2
Using Land

Main Idea Agriculture, logging, garbage disposal, and urban development often impact Earth's land resources.

SECTION 3
Conserving Resources

Main Idea Ways to conserve resources include reducing the use of materials and reusing and recycling materials.

How many people live on your street?

There are a lot of people on Earth and more are added every second. Each person uses land for food, shelter, transportation, and waste disposal. Fortunately, scientists and community leaders are discovering many ways to protect the land we live on.

Science Journal Write three ways you can reduce the amount of trash you throw in the garbage.

Start-Up Activities

What happens as the human population grows?

You're the first one on the school bus in the morning. After a few more stops, you notice that the bus is rather noisy. By the time you get to school, every seat is taken. Like the school bus, space on Earth is limited.

1. On a piece of paper, draw a square that is 10 cm on each side. This square represents 1 km^2 of land.
2. In 1965, an average of 22 people lived on 1 km^2 of land. Draw 22 small circles inside your square to represent this.
3. In 1990, the average was 35. Add 13 circles to illustrate this increase.
4. In 2025, the estimated number of people will be 52. Add enough circles to represent this increase.
5. Prepare a bar graph that shows population density for these years.
6. **Think Critically** Use your bar graph to explain how Earth's human population has changed over time.

Preview this chapter's content and activities at earth.msscience.com

How Human Acivities Impact Land Make the following Foldable to help identify what you know, what you want to know, and what you learned about how humans impact land.

STEP 1 **Fold** a vertical sheet of paper from side to side. Make the front edge about 1.25 cm shorter than the back edge.

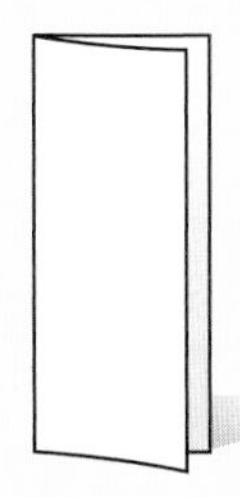

STEP 2 **Turn** lengthwise and **fold** into thirds.

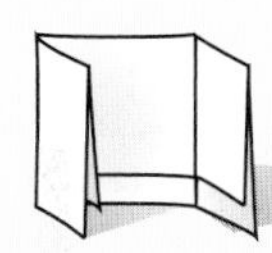

STEP 3 **Unfold** and cut only the top layer along both folds to make three tabs. **Label** each tab as shown.

Identify Questions Before you read the chapter, write what you already know about the impact of human activities on land under the left tab of your Foldable, and write questions about what you'd like to know under the center tab. As you read the chapter, list what you learned under the right tab.

Get Ready to Read

Make Predictions

1 Learn It! A prediction is an educated guess based on what you already know. One way to predict while reading is to guess what you believe the author will tell you next. As you are reading, each new topic should make sense because it is related to the previous paragraph or passage.

2 Practice It! Read the excerpt below from Section 1. Based on what you have read, make predictions about what you will read in the rest of the lesson. After you read Section 1, go back to your predictions to see if they were correct.

Think about how people compete for resources today.

Predict whether Earth will have the same, more, or fewer natural resources in the future than there are today.

Predict what living conditions might be like in the future.

By 2050, the human population is predicted to be about 9 billion—one and a half times what it is now. **Imagine the effect** such a large human population will have on the environment. Will the things that have helped the population grow, such as improved health care, agriculture, and clean water, be maintainable? Will Earth have **enough natural resources** to **support such a large population**?

—from page 575

3 Apply It! Before you read, skim the questions in the Chapter Review. Choose three questions and predict the answers.

As you read, check the predictions you made to see if they were correct.

Target Your Reading

Use this to focus on the main ideas as you read the chapter.

1. **Before you read** the chapter, respond to the statements below on your worksheet or on a numbered sheet of paper.
   - Write an **A** if you **agree** with the statement.
   - Write a **D** if you **disagree** with the statement.

2. **After you read** the chapter, look back to this page to see if you've changed your mind about any of the statements.
   - If any of your answers changed, explain why.
   - Change any false statements into true statements.
   - Use your revised statements as a study guide.

Before You Read A or D		Statement	After You Read A or D
	1	Each day, the number of humans on Earth increases by approximately 10,000.	
	2	Earth is now experiencing a population explosion.	
	3	Earth's resources are unlimited.	
	4	As human population continues to grow, more resources are used and more waste is created.	
	5	Deforestation is the clearing of forested land for agriculture, grazing, development, or logging.	
	6	Deforestation does not affect climate.	
	7	Land is a natural resource.	
	8	Paving land has no effect on underground water supplies.	
	9	Developed countries such as the United States use fewer natural resources than other regions.	
	10	Recycling is a method of conserving resources.	

Science Online
Print out a worksheet of this page at earth.msscience.com

section 1

Population Impact on the Environment

as you read

What You'll Learn

- **Describe** how fast the human population is increasing.
- **Identify** reasons for Earth's rapid increase in human population.
- **List** several ways each person can affect the environment.

Why It's Important

As the human population grows, resources are depleted and more waste is produced.

Review Vocabulary
natural resource: materials supplied by nature that are necessary or useful for life

New Vocabulary
- population
- carrying capacity
- pollutant

Population and Carrying Capacity

Look around and identify the kinds of living things you see. You might see students, fish in an aquarium, or squirrels in the trees. Perhaps plants are on the windowsill. These are examples of populations. A **population** is all of the individuals of one species occupying a particular area. As you can see in **Figure 1,** the area can be small or large. For example, a human population can be of one community, such as Los Angeles, or the entire planet.

Earth's Increasing Population Do you ever wonder how many people live on Earth? The global population in 2000 was 6.1 billion. Each day, the number of humans increases by approximately 200,000. Earth is now experiencing a population explosion. The word explosion is used because the rate at which the population is growing has increased rapidly in recent history.

Reading Check *Why is the increasing number of humans on Earth called a population explosion?*

Figure 1 A population is the number of individuals occupying an area. The population of Cranford includes the population of the classroom.

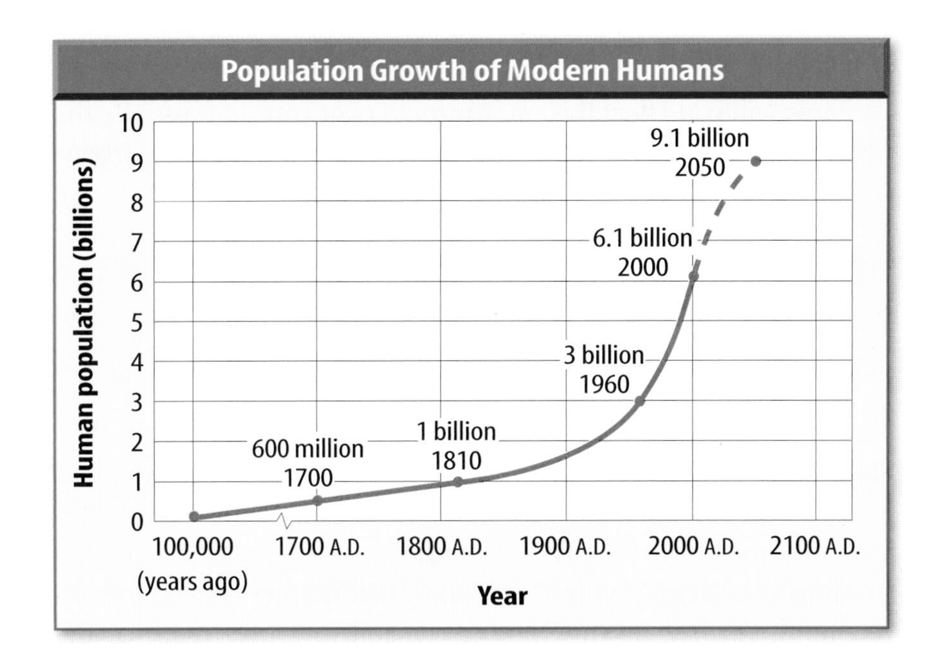

Figure 2 Human population growth remained relatively steady until the beginning of the nineteenth century. The growth rate then began to increase rapidly. **Infer** *why the human population has experienced a sharp increase in growth rate since 1800.*

Population Growth Many years ago, few people lived on Earth. You can see in **Figure 2** that it took thousands of years for the population to reach one billion people. After the mid 1800's, the population increased much faster. The human population has increased because modern medicine, clean water, and better nutrition have decreased the death rate. This means that more people are living longer. In addition, the number of births has increased because more people survive to the age at which they can have children.

By 2050, the population is predicted to be about 9 billion—one and a half times what it is now. Imagine the effect such a large human population will have on the environment. Will the things that have helped the population grow, such as improved health care, agriculture, and clean water, be maintainable? Will Earth have enough natural resources to support such a large population?

Population Limits Each person uses space and resources. Population size depends on the amount of available resources and how members of the population use them. If resources become scarce or if the environment is damaged, members of the population can suffer and population size could decrease.

People once thought that Earth had an endless supply of resources such as fossil fuels, metals, rich soils, and clean water. It's now known that this isn't true. Earth's resources are limited. The planet has a carrying capacity. **Carrying capacity** is the largest number of individuals of a particular species that the environment can support. Unless Earth's resources are treated with care, they could disappear and the human population might reach its carrying capacity.

Topic: Human Population
Visit earth.msscience.com for Web links to information about population updates.

Activity Compare the rate of people added every 30 seconds in the U.S. to the global rate.

Pesticides Bald eagles are fish-eating birds whose population in the United States declined rapidly during the 1950s and early 1960s. One of the reasons this occurred was the use of pesticides, which affected reproduction. Researchers who study how pesticides applied to land can end up in an eagle's body include biologists, chemists, mathematicians, and toxicologists.

People and the Environment

How will you affect the environment over your lifetime? By the time you're 75 years old, you will have produced enough garbage to equal the mass of eleven African elephants (53,000 kg). You will have consumed enough water to fill 68,000 bathtubs (18 million L). If you live in the United States, you will have used several times as much energy as an average person living elsewhere in the world.

Daily Activities Every day you affect the environment. The electricity you use might be generated by burning fossil fuels. The environment changes when fuels are mined, and again later when they are burned. The water that you use must be treated to make it as clean as possible before being returned to the environment. You eat food, which needs soil to grow. Much of the food you eat is grown using chemical substances, such as pesticides and herbicides, to kill insects and weeds. These chemicals can get into water supplies and threaten the health of living things if the chemicals become too concentrated. How else do you and other people affect the environment?

As you can see in **Figure 3,** many of the products you use are made of plastic and paper. Plastic begins as oil. The process of refining oil can produce **pollutants**—substances that contaminate the environment. In the process of changing trees to paper, several things happen that impact the environment. Trees are cut down. Oil is used to transport the trees to the paper mill, and water and air pollutants are given off in the papermaking process.

Reading Check *How do the products you use affect the environment?*

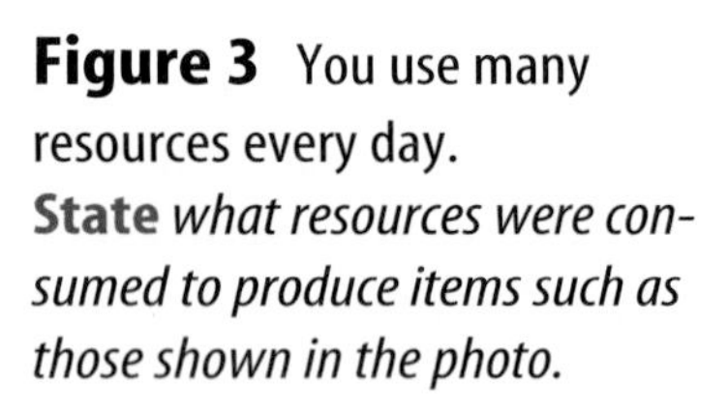

Figure 3 You use many resources every day.
State *what resources were consumed to produce items such as those shown in the photo.*

Packaging Produces Waste

The land is changed when resources are removed from it. The environment is further impacted when those resources are shaped into usable products. After the products are produced and consumed, they must be discarded. Look at **Figure 4.** Unnecessary packaging is only one of the problems associated with waste disposal.

Figure 4 Packaging foods for single servings uses more paper and plastic than buying food in bulk does.

The Future As the population continues to grow, more resources are used and more waste is created. If these resources are not used wisely and if waste is not managed properly, environmental problems are possible. What can be done to prevent these problems? As you learn more about how you affect the environment, you'll discover what you can do to help make the future world one that everyone can live in and enjoy. An important step that you can take is to think carefully about your use of natural resources. If you conserve resources, you can lessen the impact on the environment.

section 1 review

Summary

Population and Carrying Capacity

- Population growth rate has rapidly increased since 1800.
- The human population might reach Earth's carrying capacity if resources are not used wisely.

People and the Environment

- A person's daily activities use resources and produce waste.
- Less packaging produces less waste.
- Conserving resources can lessen our impact on land.

Self Check

1. **Use Graphs** Using **Figure 2,** estimate the human population increase from 1800 to 1960.
2. **State** three reasons why the human population is increasing rapidly.
3. **Infer** what might happen if the human population reaches its carrying capacity.
4. **Think Critically** How do your daily activities affect Earth's available resources?

Applying Skills

5. **Research Information** Some areas of the world are experiencing a decrease in population. Find out where and some reasons for the decrease.

section 2

Using Land

as you read

What You'll Learn

- **Identify** ways that land is used.
- **Explain** how land use creates environmental problems.
- **Identify** things you can do to help protect the environment.

Why It's Important

Using land responsibly will help conserve this natural resource.

Review Vocabulary
erosion: a process that wears away surface materials and moves them from one place to another

New Vocabulary
- stream discharge
- sanitary landfill
- hazardous waste
- enzyme

Land Usage

You may not think of land as a natural resource. Yet it is as important to people as oil, clean air, and clean water. We use land for agriculture, logging, garbage disposal, and urban development. These activities often impact Earth's land resources.

Agriculture About 16 million km^2 of Earth's total land surface is used as farmland. To feed the growing world population, some farmers use higher-yielding seeds and chemical fertilizers. These methods help increase the amount of food grown on each km^2 of land. Herbicides and pesticides also are used to reduce weeds, insects, and other pests that can damage crops.

Organic farming techniques, as shown in **Figure 5,** use natural fertilizers, crop rotation, and biological pest controls. These methods help crops grow without using chemicals. However, organic farming cannot currently produce enough food to feed all of Earth's people.

Whenever vegetation is removed from an area, such as a construction site or tilled farmland, soil is exposed. Without plant roots to hold soil in place, nothing prevents the soil from being carried away by running water and wind. Several centimeters of topsoil may be lost in one year. In some places, it can take more than 1,000 years for new topsoil to develop.

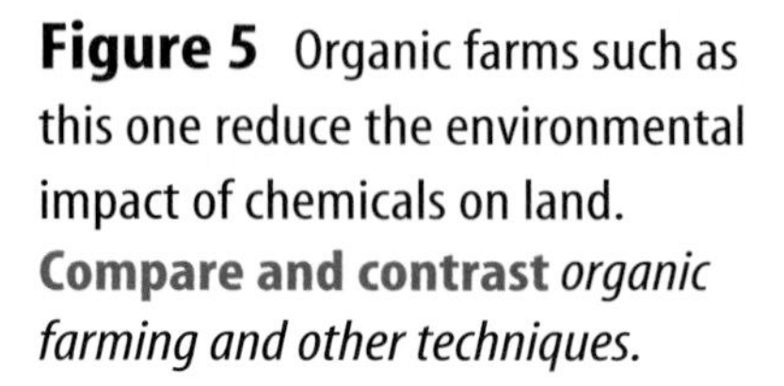

Figure 5 Organic farms such as this one reduce the environmental impact of chemicals on land.
Compare and contrast *organic farming and other techniques.*

Reducing Erosion Some farmers practice no-till farming, as shown in **Figure 6.** They don't plow the soil from harvest until planting. Instead, farmers plant seed between the stubble left from the previous year.

Other methods also are used to reduce soil loss. One method is contour plowing. The rows are tilled across hills and valleys. When it rains, water and soil are captured by the plowed rows, reducing erosion. Other techniques include planting trees in rows along fields. The trees slow the wind, which reduces the amount of soil blown from the land. Cover crops, crops that are not harvested, also can be planted to reduce erosion.

Feeding Livestock Land also is used for feeding livestock. Animals such as cattle eat vegetation and then are used as food for humans. About sixty-five percent of the farmland in Texas is used for grazing cattle. Other regions of the United States such as the west and midwest also set aside land as pasture. Other land is used to grow crops to be fed to cattle. Many farmers raise corn and hay for this purpose. These crops provide cattle with a variety of nutrients and can improve the quality of the meat.

Modeling Earth's Farmland

Procedure

1. Cut an **apple** into four quarters and set aside three. One quarter of Earth's surface is land. The remaining 3/4 is covered with water.
2. Slice the remaining quarter into thirds.
3. Set aside two of the three pieces, because 2/3 of Earth's land is too hot, too cold, or too mountainous to farm or live on.
4. Carefully peel the remaining piece. This represents the usable land surface that must support the entire human population.

Analysis

What may happen if available farmland is converted to other uses?

Figure 6 No-till farming can reduce erosion of topsoil.
Describe *other techniques that can be used to reduce soil erosion.*

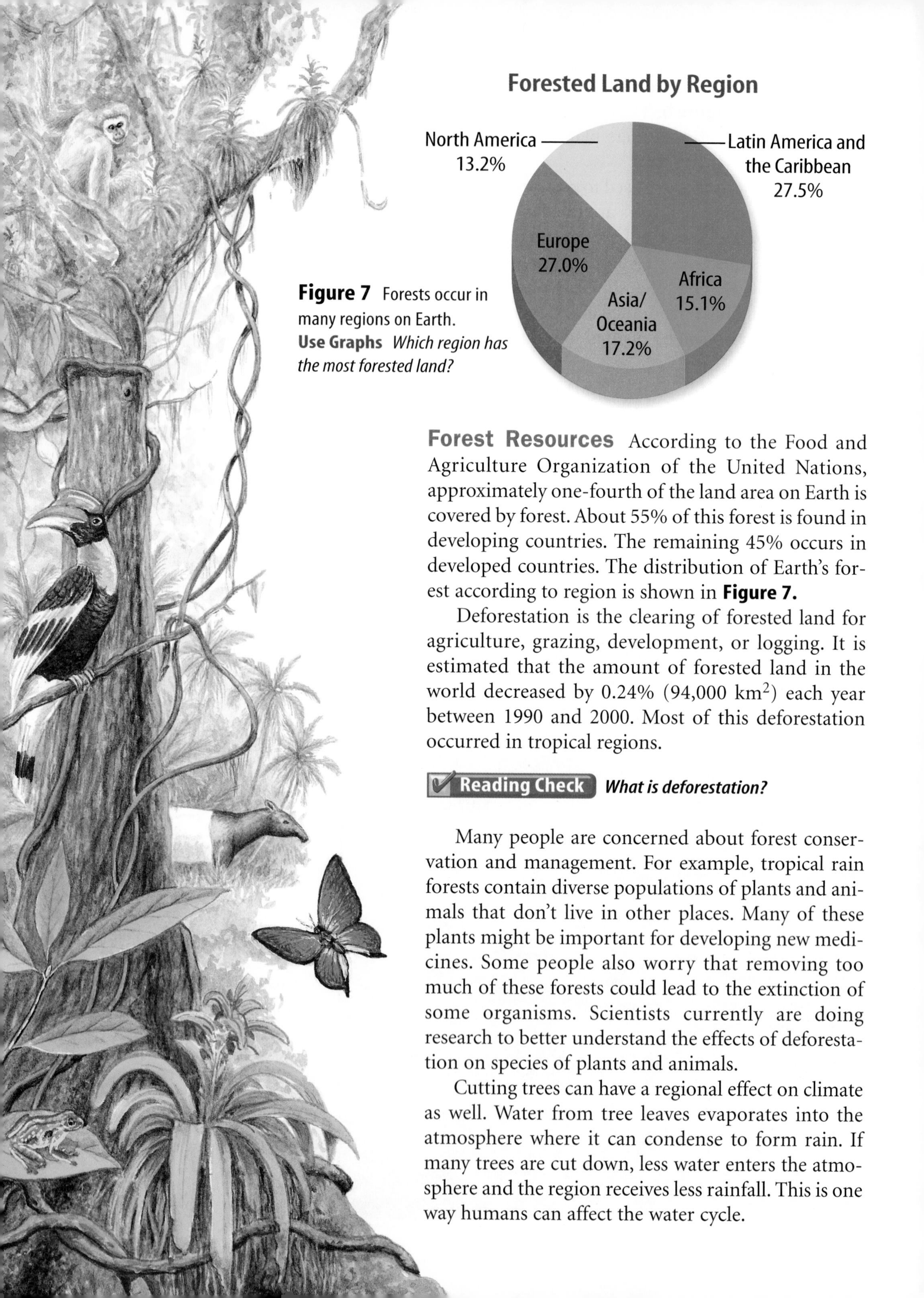

Figure 7 Forests occur in many regions on Earth.
Use Graphs *Which region has the most forested land?*

Forest Resources According to the Food and Agriculture Organization of the United Nations, approximately one-fourth of the land area on Earth is covered by forest. About 55% of this forest is found in developing countries. The remaining 45% occurs in developed countries. The distribution of Earth's forest according to region is shown in **Figure 7.**

Deforestation is the clearing of forested land for agriculture, grazing, development, or logging. It is estimated that the amount of forested land in the world decreased by 0.24% (94,000 km^2) each year between 1990 and 2000. Most of this deforestation occurred in tropical regions.

Reading Check *What is deforestation?*

Many people are concerned about forest conservation and management. For example, tropical rain forests contain diverse populations of plants and animals that don't live in other places. Many of these plants might be important for developing new medicines. Some people also worry that removing too much of these forests could lead to the extinction of some organisms. Scientists currently are doing research to better understand the effects of deforestation on species of plants and animals.

Cutting trees can have a regional effect on climate as well. Water from tree leaves evaporates into the atmosphere where it can condense to form rain. If many trees are cut down, less water enters the atmosphere and the region receives less rainfall. This is one way humans can affect the water cycle.

Development From 1990 to 2000, the number of kilometers of urban roadways in the United States increased by more than 13 percent. Highway building often leads to more paving as office buildings, stores, and parking lots are constructed.

Paving land prevents water from soaking into the soil. Instead, it runs off into sewers or streams. A stream's discharge increases when more water enters its channel. **Stream discharge** is the volume of water flowing past a point per unit of time. During heavy rainstorms in paved areas, rainwater flows directly into streams, increasing stream discharge and the risk of flooding.

Many communities use underground water supplies for drinking. Covering land with roads, sidewalks, and parking lots reduces the amount of rainwater that soaks into the ground to refill underground water supplies.

Some communities, businesses, and private groups preserve areas rather than pave them. Land is set aside for environmental protection, as shown in **Figure 8.** Preserving space beautifies the environment, increases the area into which water can soak, and provides space for recreation and other outdoor activities.

Figure 8 Some land in urban areas is preserved, such as this area near Portland, Oregon.
Discuss *how preserving green space near cities helps protect the environment.*

Applying Science

How does land use affect stream discharge?

It's not unusual for streams and rivers to flood after heavy rain. The amount of water flowing quickly into waterways may be more than streams and rivers can carry. Land use can affect how much runoff enters a waterway. Would changing the landscape increase flooding? Use your ability to interpret a data table to find out.

Rainfall Runoff Percentages

Land Use	Runoff to Streams (%)
Commercial (offices and stores)	75
Residential (houses)	40
Natural areas (forest and grassland)	29

Identifying the Problem

The table above lists the percentage of rainfall that runs off land. Compare the amount of runoff for each of the land uses listed. Assume that all of the regions are the same size and have the same slope. Looking at the table, do you see a relationship between what is on the land and how much water runs off of it?

Solving the Problem

1. Two years after construction of a commercial development near a stream, houses downstream flooded after a heavy rain. What contributed to the flooding?
2. What are some ways that developers can help reduce the risk of flooding?

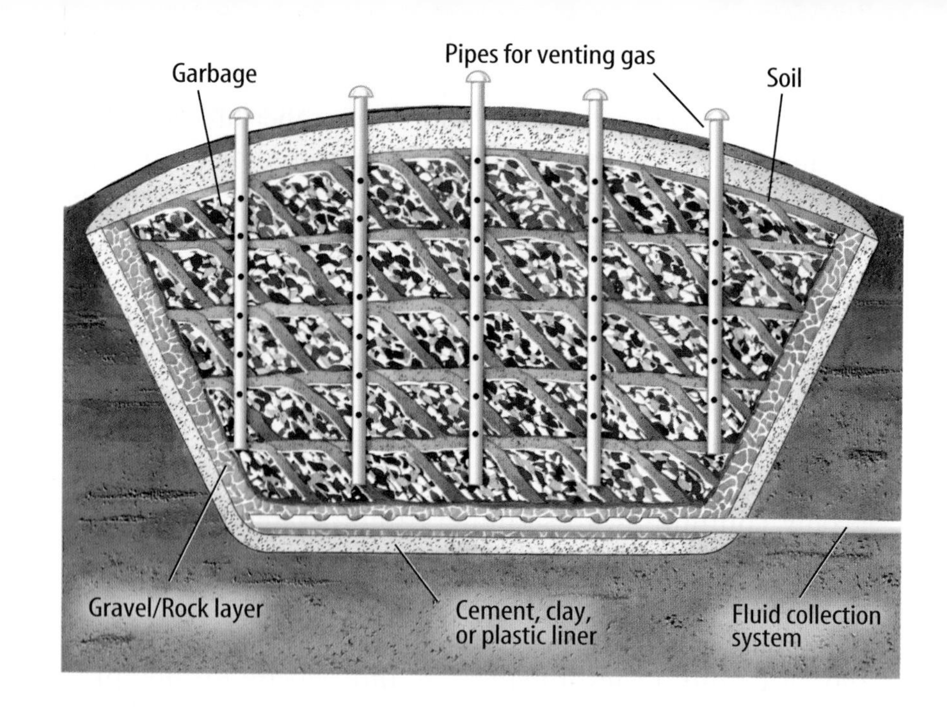

Figure 9 The majority of garbage is deposited in sanitary landfills designed to contain garbage and prevent contamination of the surrounding land and water.
Describe *the problems associated with landfill disposal.*

Sanitary Landfills Land also is used when consumed products are thrown away. About 60 percent of our garbage goes into sanitary landfills. A **sanitary landfill,** like the one illustrated in **Figure 9,** is an area where each day's garbage is deposited and covered with soil. The soil prevents the deposit from blowing away, helps decompose some materials, and reduces the odor produced by the decaying waste.

Sanitary landfills also are designed to prevent liquid wastes from draining into the soil and groundwater below. New sanitary landfills are lined with plastic, concrete, or clay-rich soils that trap the liquid waste. Because of these linings, sanitary landfills greatly reduce the chance that pollutants will leak into the surrounding soil and groundwater.

Since many materials do not decompose in landfills, or they decompose slowly, landfills fill with garbage, and new ones must be built. Locating an acceptable area to build a landfill can be difficult. Type of soil, the depth to groundwater, and neighborhood concerns must be considered.

Nuclear Waste Wastes from nuclear power plants must be stored safely because radioactivity is dangerous. The U.S. government is currently studying a site in Nevada for nuclear waste disposal because the area is remote, little rain falls, and the underground water supply is far below the proposed storage facility. What is radioactivity and how can it harm the environment?

Hazardous Wastes

Some of the wastes that are thrown away are dangerous to organisms. Wastes that are poisonous, that cause cancer, or that can catch fire are called **hazardous wastes.** Previously, everyone—industries and individuals alike—put hazardous wastes into landfills, along with household garbage. In the 1980s, many states passed environmental laws that prohibit industries from disposing of hazardous wastes in sanitary landfills. New technologies which help recycle hazardous wastes have decreased the need to dispose of them.

Household Hazardous Waste Unlike most industries, individuals discard hazardous wastes such as insect sprays, batteries, drain cleaners, bleaches, medicines, and paints in the trash. It may seem that when you throw something in the garbage, it's gone and you don't need to be concerned with it anymore. Unfortunately, some garbage can remain unchanged in a landfill for hundreds of years. You can help by disposing of hazardous wastes at special hazardous waste-collection sites. Contact your local government to find out about collections in your area.

Phytoremediation Hazardous substances can contaminate soil. These contaminants may come from nearby industries or leaking landfills. Water contaminated from such a source can filter into the ground and leave the toxic substances in the soil. Some plants can help fix this problem in a method called phytoremediation (FI toh ruh mee dee AY shun). *Phyto* means "plant" and *remediation* means "to fix or remedy a problem."

During phytoremediation, roots of certain plants such as alfalfa, grasses, and pine trees can absorb metals, including copper, lead, and zinc from contaminated soil just as they absorb other nutrients. **Figure 10** shows how metals are absorbed from the soil and taken into plant tissue.

What happens to these plants after they absorb metals? If livestock were to eat contaminated alfalfa, the harmful metals could end up in your milk or meat. Plants that become concentrated with metals from soil eventually must be harvested and either composted to recycle the metals or burned. If these plants are destroyed by burning, the ash residue contains the hazardous waste that was in the plant tissue and must be disposed of at a hazardous waste site.

INTEGRATE Chemistry

Breaking Down Organic Pollutants

Living things also can clean up pollutants other than metals. Substances that contain carbon and other elements like hydrogen, oxygen, and nitrogen, are called organic compounds. Examples of organic pollutants are gasoline, oil, and solvents.

Organic pollutants can be broken down into simpler, harmless substances, some of which plants use for growth. Some plant roots release enzymes (EN zimez) into the soil. **Enzymes** are substances that make chemical reactions go faster. Enzymes from plant roots increase the rate at which organic pollutants are broken down into simpler substances. Plants use these substances for growth.

Reading Check *How do enzymes affect organic pollutants in soil?*

Figure 10 Metals such as copper can be removed from soil and be absorbed by plant tissues.
State *why this vegetation can't be fed to livestock.*

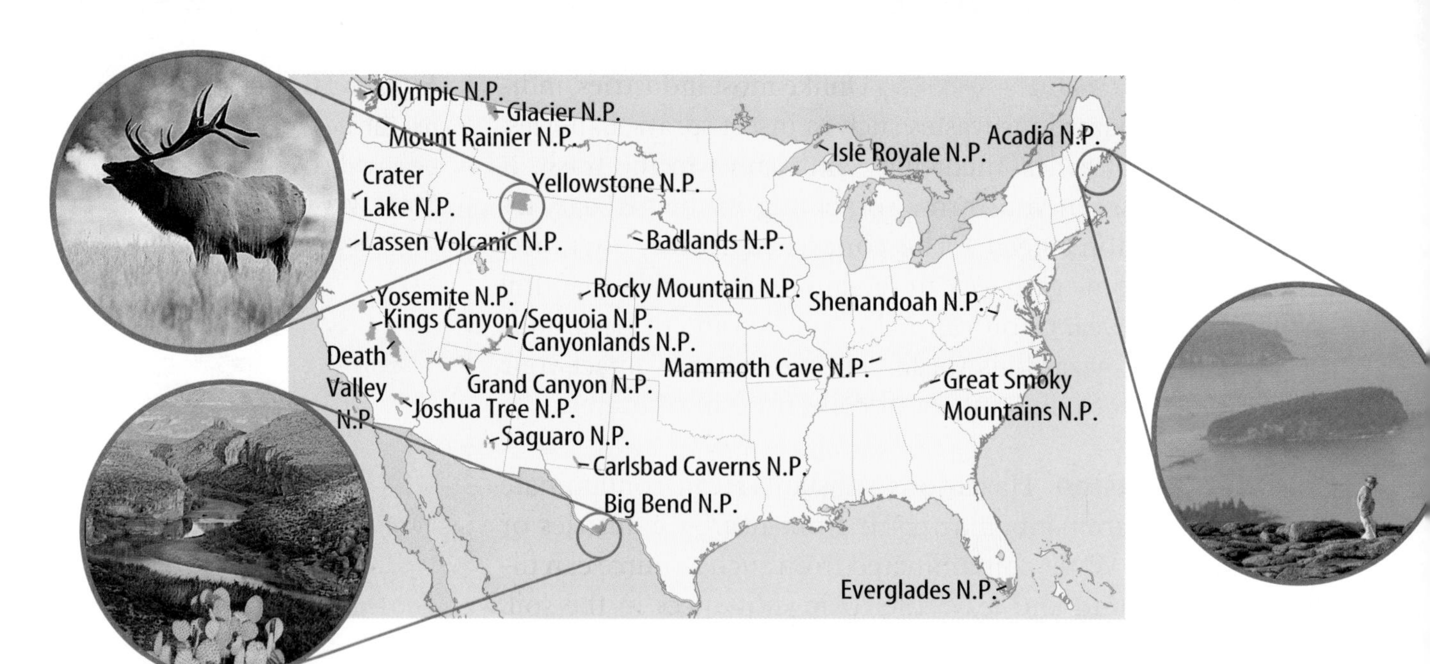

Figure 11 Many countries set aside land in the form of national parks as natural preserves. **Discuss** *how natural preserves might benefit humans and other living things.*

Natural Preserves

Not all land on Earth is being utilized to produce usable materials or for storing waste. As shown in **Figure 11,** some land remains mostly uninhabited by people. National forestlands, grasslands, and national parks in the United States are protected from many problems that you've read about in this section. In many other countries throughout the world, land also is set aside for natural preserves. As the world population continues to rise, the strain on the environment may worsen. Preserving some land in its natural state will benefit future generations.

section 2 review

Summary

Land Usage

- Pesticides and herbicides may be used on farms to grow more food per km^2.
- No-till farming can reduce erosion.
- Development can increase runoff.
- A sanitary landfill is designed to protect soil and groundwater.

Hazardous Wastes

- New technologies help recycle hazardous wastes.
- Hazardous wastes can be broken down by enzymes or phytoremediation.

Natural Preserves

- Many countries set aside land for protection.

Self Check

1. **List** six ways that people use land.
2. **Discuss** environmental problems that can be created by agriculture and trash disposal.
3. **Infer** what you can do that would benefit the environment.
4. **Describe** how development can increase flooding.
5. **Think Critically** Preserving land beautifies the environment, provides recreational space, and benefits future generations. Are there any disadvantages to setting aside large areas of land as natural preserves?

Applying Skills

6. **Form a Hypothesis** Develop a hypothesis about how migrating birds might be affected by cutting down forests.

Science Online earth.msscience.com/self_check_quiz

What to Wear?

What items in your house will end up in a landfill? You might think about milk jugs or food scraps. What about old clothing? In this lab, you'll observe what happens to different types of clothes that are buried in a landfill.

Real-World Question

Do materials decompose at the same rate?

Goals

- **Compare** the decomposing rates of natural and artificial clothing materials.
- **Infer** the effect of these materials on landfills.

Materials

identical baking trays (2)
garden soil
clothing made of natural fibers (linen, cotton, wool, silk)
clothing made of artificial materials (fleece, polyester, acrylic, rayon)
toothpicks
transparent tape
scissors
spray bottle filled with water

Safety Precautions

Procedure

1. Collect several articles of clothing and separate those made with natural fibers from those made from artificial materials.
2. Cut 3-cm squares of each type of clothing.
3. Cut 1-cm × 3-cm labels from a sheet of notebook paper, and write one label for each of your clothing squares. Tape each label to the tip of a toothpick.
4. Fill each tray halfway with garden soil. Lay your artificial cloth squares in one tray and your natural cloth squares in the other tray. Be certain the squares don't overlap. Thoroughly moisten all squares using the spray bottle.
5. Identify each clothing square by attaching a toothpick label.
6. Cover your squares with soil. Moisten the soil and place the trays in a dark place. Keep the soil moist for three weeks.
7. After three weeks, dig up your samples and observe each square. Record your observations in your Science Journal.

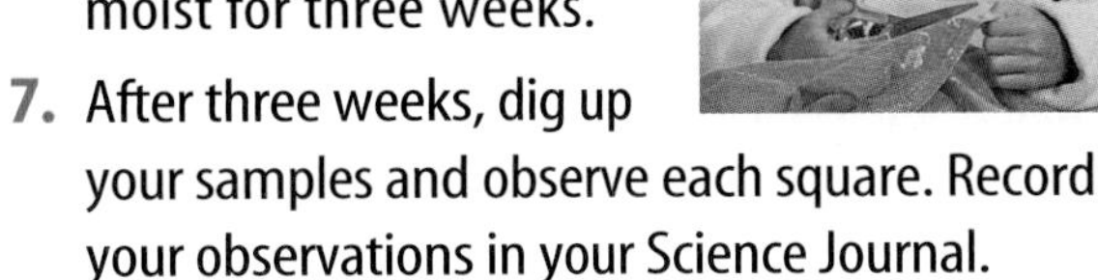

Conclude and Apply

1. **Compare** the amount of decomposition of the two types of materials.
2. **Infer** the effects of clothing made with natural materials on landfills.
3. **Infer** the effects of clothing made with artificial materials on landfills.
4. **Research** materials used to manufacture clothing. Determine if the material is made from recycled products such as plastic bottles.

Communicating Your Data

Compare the types of clothing worn by your classmates with the types you used in your experiment. Contrast the results of their experiments with your observations. **For more help, refer to the** Science Skill Handbook.

section 3

Conserving Resources

as you read

What You'll Learn

- **Identify** three ways to conserve resources.
- **Explain** the advantages of recycling.

Why It's Important

Conserving resources helps reduce solid waste.

Review Vocabulary
consumption: using up materials

New Vocabulary
- conservation
- composting
- recycling

Resource Use

Resources such as petroleum and metals are important for making the products you use every day at home and in school. For example, petroleum is used to produce plastics and fuel. Minerals are used to make automobiles and bicycles. However, if these resources are not used carefully, the environment can be damaged. **Conservation** is the careful use of earth materials to reduce damage to the environment. Conservation can prevent future shortages of some materials, such as certain metals.

Reduce, Reuse, Recycle

Developed countries such as the United States use more natural resources than other regions, as shown in **Figure 12.** Ways to conserve resources include reducing the use of materials, and reusing and recycling materials. You can reduce the consumption of materials in simple ways, such as using both sides of notebook paper or carrying lunch to school in a nondisposable container. Reusing an item means finding another use for it instead of throwing it away. You can reuse old clothes by giving them to someone else or by cutting them into rags. The rags can be used in place of paper towels for cleaning jobs around your home.

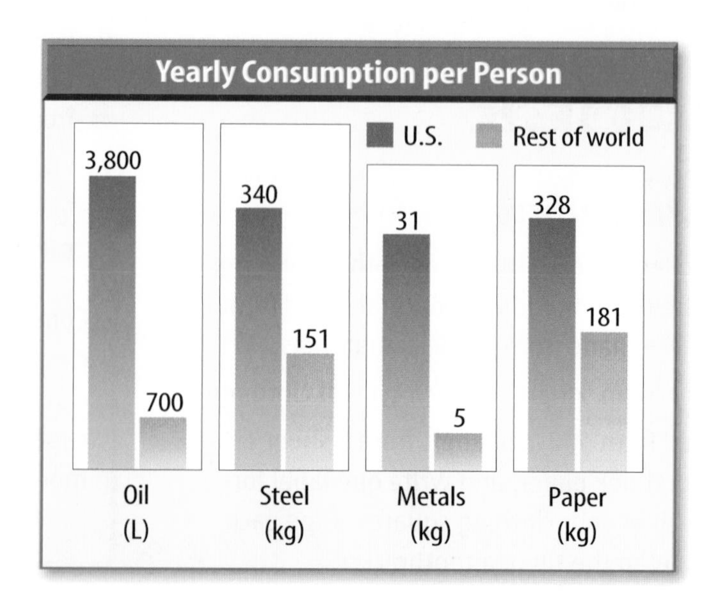

Figure 12 A person in the United States uses more resources than the average person elsewhere.

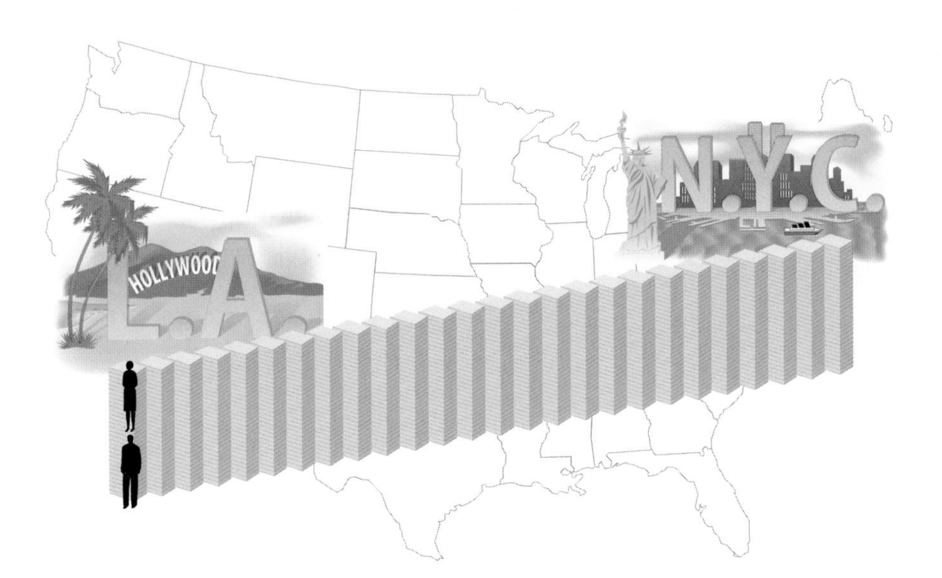

Figure 13 People in the United States throw away enough office and writing paper each year to build a wall 3.6 m high stretching from New York City to Los Angeles.

Reusing Yard Waste Outdoors, you can do helpful things, too. If you cut grass or rake leaves, you can compost these items instead of putting them into the trash. **Composting** means piling yard wastes where they can decompose gradually. Decomposed material provides needed nutrients for your garden or flower bed. Some cities no longer pick up yard waste to take to landfills. In these places, composting is common. If everyone in the United States composted, it would reduce the trash put into landfills by 20 percent.

Recycling Materials Using materials again is called **recycling.** When you recycle wastes such as glass, paper, plastic, steel, or tires, you help conserve Earth's resources, energy, and landfill space.

Paper makes up about 40 percent of the mass of trash. As shown in **Figure 13,** Americans throw away a large amount of paper each year. Recycling this paper would use 58 percent less water and generate 74 percent less air pollution than producing new paper from trees. The paper shown in the figure doesn't even include newspapers. More than 500,000 trees are cut every week just to print newspapers.

Companies have found that recycling can be good for business. For example, companies can recover part of the cost of many materials by recycling the waste. Some businesses use scrap materials such as steel to make new products. These practices save money, benefit the environment, and reduce the amount of waste sent to landfills.

Figure 14 shows that the amount of material deposited in landfills has decreased since 1980. In addition to saving landfill space, reducing, reusing and recycling can reduce energy use and minimize the need to extract raw materials from Earth.

Classifying Your Trash for One Day

Procedure

1. Label a table with the following columns: *Paper, Plastic, Glass, Metal,* and *Food Waste.*
2. Record items you throw out in one day. At the end of the day, count the number of trash items in each column.
3. Rank each column by number from the fewest trash items to the most trash items.

Analysis

1. Compare your rankings with those of others in your household.
2. What activities can you change to decrease the amount of trash you produce?

Try at Home

NATIONAL GEOGRAPHIC **VISUALIZING TRASH DISPOSAL**

Figure 14

Although trash production in the United States is increasing, the amount of trash deposited in landfills is decreasing. In 1980, 82 percent of discarded trash ended up in a landfill. Today, only 55 percent is taken to the dump—thanks to the use of waste-reducing methods such as those shown below.

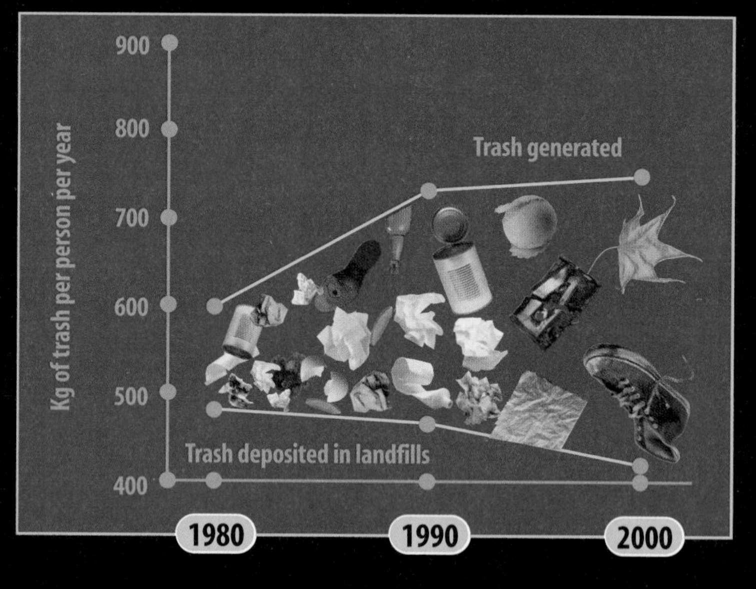

▲ RECYCLING In 1980, about nine percent of trash was recycled. Now nearly 30 percent of America's trash is reused.

◀ COMPOSTING Yard trimmings placed in a pile will decompose and form a substance called compost. Compost then can be used on flowers and vegetables to help them grow.

▶ WASTE TO ENERGY Some waste material can be burned to produce electricity. This plant in Rochester, Massachusetts, burns trash to generate electricity for a local paper company.

Recycling Methods What types of recycling programs does your state have? Many states or cities have some form of recycling laws. For example, in some places people who recycle pay lower trash-collection fees. In other places a refundable deposit is made on all beverage containers. This means paying extra money at the store for a drink, but you get your money back if you return the container to the store for recycling.

How have states and cities encouraged people to recycle?

There are several disadvantages to recycling. More people and trucks are needed to haul materials separately from your trash. The materials then must be separated at special facilities like the one shown in **Figure 15.** In addition, demand for things made from recycled materials must exist, and items made from recycled materials often cost more.

Figure 15 In recycling facilities like this one, materials must be separated before they can be reused.

The Population Outlook The human population explosion already has had an effect on the environment and the organisms that inhabit Earth. It's unlikely that the population will begin to decline in the near future. To make up for this, resources must be used wisely. Conserving resources by reducing, reusing, and recycling is an important way that you can make a difference.

section 3 review

Summary

Resource Use

- Earth's resources are used to make products.
- Conservation of resources can help prevent future shortages.

Reduce, Reuse, Recycle

- There are many simple ways to reduce the amount of materials you use.
- Composting yard waste reduces trash in landfills and provides nutrients for plants.
- Recycling materials can save money, benefit the environment, and save landfill space.

Self Check

1. **List** four advantages and two disadvantages of recycling.
2. **Compare and contrast** reducing and reusing materials.
3. **List** two simple ways that you and your classmates can reduce your consumption of Earth materials.
4. **Think Critically** Why is it more important to conserve resources as the human population increases?

Applying Skills

5. **Research Information** Contact a sanitary landfill near you. Find out how long it will take for your community's landfill to be full. How will waste be disposed of after the landfill is full?

A World Full of People

Goals

- **Demonstrate** the world's human population increase in the next decade.
- **Predict** the world's population in 50 years.
- **Record, graph,** and **interpret** population data.

Materials

small objects such as popcorn kernels or dried beans (1,000)
large map of the world (the map must show the countries of the world)
clock or watch
calculator

Safety Precautions

Never eat or taste anything in the lab, even if you are confident that you know what it is.

Real-World Question

Every second, five people are born on Earth and two or three people die. As a result, there is a net increase of two or three people in the world every second of every day. That amounts to about 81 million new people every year. This is nearly equal to the population of Central Africa. What effects will this rapid increase in human population have on Earth? How crowded will different regions of Earth become in the next ten years?

Procedure

1. Copy the data table below in your Science Journal.
2. Lay the map out on a table. The map represents Earth and the people already living here.
3. Each minute of time will represent one year. During your first minute, place 78 popcorn kernels on the continents of your map. Each kernel represents 1 million new people.
4. Place one kernel inside the borders of developed countries such as the United States, Canada, Japan, Australia, and countries in Europe. Place 77 kernels inside the borders of developing nations located in South America, Africa, and Asia.
5. Continue adding 78 kernels to your map in the same fashion each minute for 10 min. Record the total population increase for each year (each minute of the lab) in your data table.

Population Data

Time (in years)	Total Population Increase
1	78 million
2	
3	
4	
5	Do not write in this book.
6	
7	
8	
9	
10	

Analyze Your Data

1. **Draw and label** a graph of your data showing the time in years on the horizontal axis and the world population on the vertical axis.
2. **Calculate** the world's population in 50 years by using an average rate of 71 million people per year.
3. **Determine** world population in ten years if only 4.5 million people are added each year.

Conclude and Apply

1. **Infer** how many new people will be added to Earth in the next 10 years. Determine the world's population in 10 years.
2. **Compare** the population growth in developed countries to the growth of developing countries.
3. **Discuss** ways the increase in the human population will affect Earth's resources in the future.

Communicating Your Data

Draw your graph on a computer and present your findings to the class. **For more help, refer to the** Science Skill Handbook.

TIME SCIENCE AND Society

SCIENCE ISSUES THAT AFFECT YOU!

It causes health risks, but how do we safely get rid of it?

A danger sign in a garbage dump alerts visitors to the presence of hazardous waste.

Hazardous Waste

Danger: Hazardous Waste Area. Unauthorized Persons Keep Out.

During much of the 1980s, this sign greeted visitors to Love Canal, a housing project in Niagara Falls, New York. The housing project was closed because it had been built on a hazardous waste dump and people were getting sick. Exposure to hazardous waste can cause nerve damage, birth defects, and lowered resistance to disease.

The Environmental Protection Agency (EPA) estimates that U.S. industries produce about 265 million metric tons of hazardous wastes each year. Much of this waste is recycled or converted to harmless substances. About 60 million tons of hazardous waste, however, must be disposed of in a safe manner. Incineration, or burning, is one way to dispose of hazardous wastes. However, the safety of this method is hotly debated.

For Incineration

People in favor of incineration note that, if done correctly, it destroys 99.99 percent of toxic materials. Although the remaining ash must still be disposed of, it is often less hazardous than the original waste material. Supporters also note that incineration is safer than storing the hazardous wastes or dumping them in landfills.

Against Incineration

Other people say that incinerators fail to destroy all hazardous wastes and that some toxins are released in the process. They also note that new substances are generated during incineration, and that scientists don't yet know how these new substances will impact the environment or human health. Lastly, they say that incineration may reduce efforts to reuse or recycle hazardous wastes.

While the debate goes on, scientists continue to develop better methods for dealing with hazardous wastes. As Roberta Crowell Barbalace, an environmental scientist, wrote in an article, "In an ideal environment there would be no hazardous waste facilities. The problem is that we don't live in an ideal environment ... Until some new technology is found for dealing with or eliminating hazardous waste, disposal facilities will be necessary to protect both humans and the environment."

Research Find out more about incineration. Then use this feature and your research to conduct a class debate about the advantages and disadvantages of incineration.

For more information, visit earth.msscience.com/time

Reviewing Main Ideas

Section 1 Population Impact on the Environment

1. Modern medicine, clean water, and better nutrition have contributed to the human population explosion on Earth.
2. Earth's resources are limited.
3. Our daily activities use resources and produce waste.

Section 2 Using Land

1. Land is used for farming, grazing livestock, lumber, development, and disposal.
2. Farming and development are ways that using land can impact the environment.
3. Using forest resources can impact organisms and Earth's climate.
4. Plants are sometimes used to break down and absorb pollutants from contaminated land.
5. New technologies have reduced greatly the need for hazardous waste disposal.
6. One way to preserve our land is to set aside natural areas.

Section 3 Conserving Resources

1. Recycling, reducing, and reusing materials are important ways to conserve natural resources.
2. Reducing, reusing, and recycling has decreased the amount of trash deposited in landfills since 1980.
3. Different methods can be used to encourage recycling.

Visualizing Main Ideas

Copy and complete the concept map about using land.

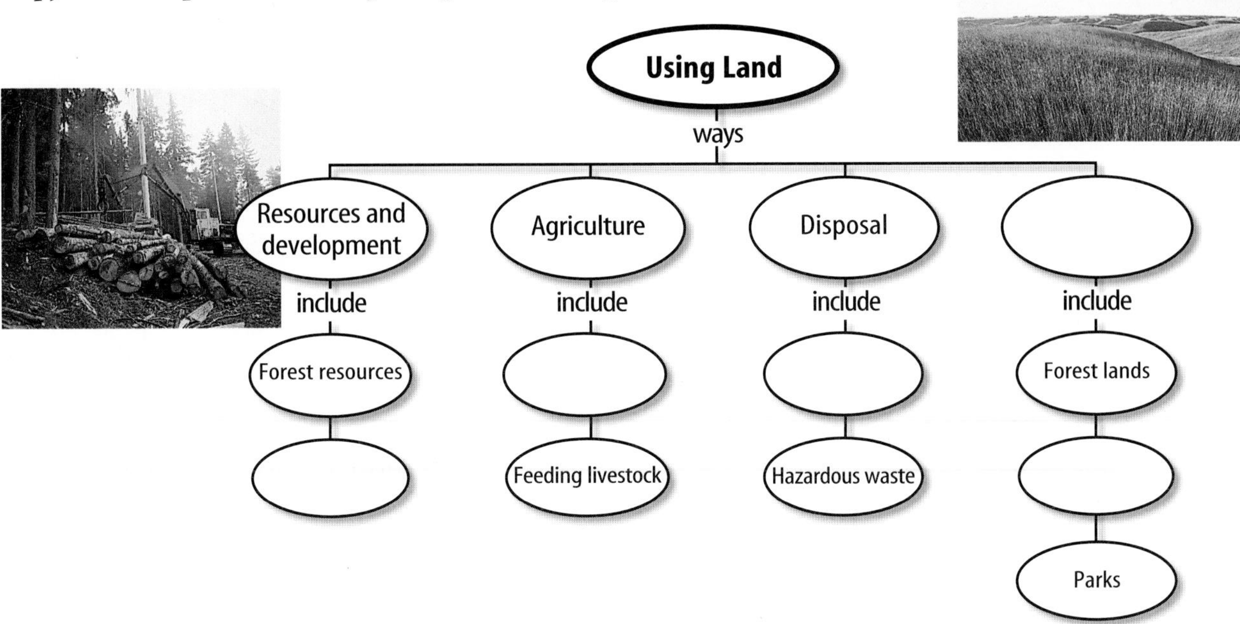

chapter 20 Review

Using Vocabulary

carrying capacity p. 575	pollutant p. 576
composting p. 587	population p. 574
conservation p. 586	recycling p. 587
enzyme p. 583	sanitary landfill p. 582
hazardous waste p. 582	stream discharge p. 581

Fill in the blanks with the correct words.

1. The total number of individuals of a particular species in an area is called ________.
2. ________ means using resources carefully to reduce damage to the environment.
3. ________ means using materials again.
4. The maximum number of individuals of a particular species that the environment can support is called ________.
5. The volume of river water flowing past a point per unit of time is called ________.

Checking Concepts

Choose the word or phrase that best answers the question.

6. Where is most of the trash in the United States disposed of?
   A) recycling centers
   B) landfills
   C) hazardous waste sites
   D) compost piles

7. Between 1960 and 2000, world population increased by how many billions of people?
   A) 5.9 **C)** 1.0
   B) 4.2 **D)** 3.1

8. Which of the following is a substance that contaminates the environment?
   A) compost **C)** pollutant
   B) development **D)** groundwater

9. Which of the following might be poisonous, cause cancer, or catch fire?
   A) enzyme
   B) compost
   C) metals
   D) hazardous waste

10. What do we call an object that can be processed so that it can be used again?
    A) trash **C)** disposable
    B) recyclable **D)** hazardous

11. What is about 40 percent of the mass of our trash made up of?
    A) glass **C)** yard waste
    B) aluminum **D)** paper

12. What term is used to describe using plants to clean up contaminated soil?
    A) recycling
    B) composting
    C) phytoremediation
    D) sanitary landfill

Use the illustration below to answer question 13.

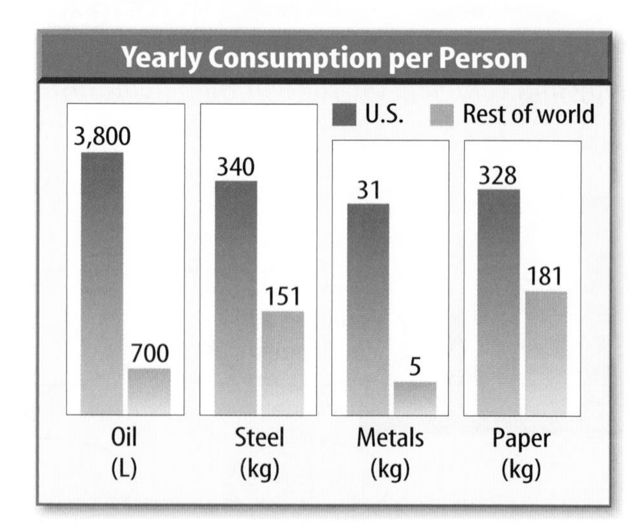

13. How much more oil does each person use in the U.S. compared to the average use per person elsewhere?
    A) 310 L **C)** 189 L
    B) 3,100 L **D)** 27 L

Thinking Critically

14. **Explain** how reducing materials used for packaging products would affect our disposal of solid wastes.

15. **Discuss** ways that land can be developed without changing stream discharge.

16. **Infer** Although land is farmable in several developing countries, hunger is a major problem in many of these places. Give some reasons why this might be so.

17. **Form a Hypothesis** Forests in Germany are dying due to acid rain. What effects might this loss of trees have on the environment?

18. **Describe** how you could encourage your neighbors to recycle their aluminum cans.

19. **Classify** Group the following materials as hazardous or nonhazardous: gasoline, newspaper, leaves, lead, can of paint, glass.

20. **Concept Map** Copy and complete this concept map about phytoremediation.

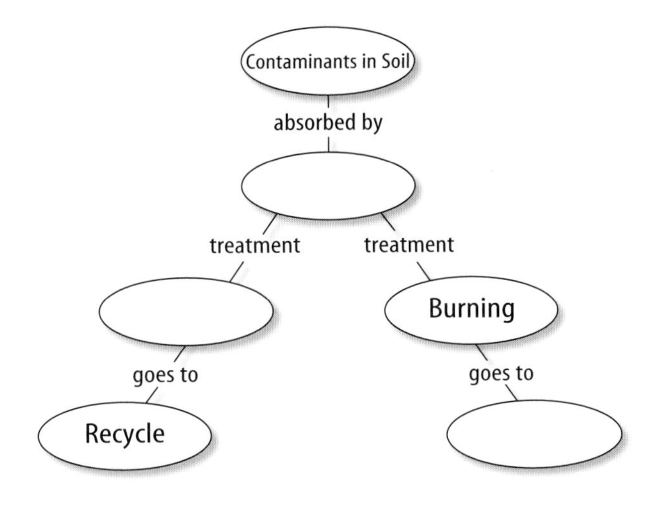

21. **Compare and contrast** farming and developing land. How do these activities affect stream discharge?

22. **Research** Find out whether your community excludes yard waste from landfills.

Performance Activities

23. **Evaluate a Hypothesis** Design an experiment to determine factors that decrease the time it takes for newspapers or yard wastes to decompose.

24. **Display** Make a display showing how paving land can increase stream discharge.

Applying Math

25. **Junk Mail** Collect your family's junk mail for one week and weigh it. Divide this weight by the number of people in your home. Multiply this number by 300 million (the U.S. population). If 17 trees are cut to make each metric ton of paper, calculate how many trees are cut each year to make junk mail for the entire U.S. population.

26. **Interpret Scientific Illustrations** One hectare, shown here, is a square of land measuring 100 meters by 100 meters. How many hectares are in 50,000 m^2 of land?

100 m

100 m

1 hectare or 10,000 m^2

Use the table below to answer question 27.

World Population

Year	Population (Billions)
1960	3.0
1980	4.4
2000	6.1

27. **Growth Rate** Calculate the percent increase in world population from 1960 to 1980 and from 1980 to 2000. Infer what is happening to the rate of population growth on Earth.

Part 1 Multiple Choice

Record your answers on the answer sheet provided by your teacher or on a sheet of paper.

Use the photo below to answer question 1.

1. Why do some farmers use no-till farming?
A. to improve crop yields
B. to reduce soil erosion
C. to save fuel
D. to reduce the use of fertilizer

2. Which term means decreasing the amount of material used?
A. recycle **C.** reduce
B. compost **D.** reuse

3. Which term describes using land to grow crops and raise farm animals?
A. agriculture **C.** development
B. landfill **D.** remediation

4. Which of the following is a substance that makes chemical reactions go faster?
A. enzyme
B. compost
C. pollutant
D. hazardous waste

Test-Taking Tip

Keep Moving If you don't know the answer to a multiple choice question, mark your best guess and move on.

5. Which term describes all the individuals of one species that occupy an area?
A. population explosion
B. carrying capacity
C. population
D. population limit

6. Which of the following describes how the human population on Earth changed during the last 50 years?
A. It decreased.
B. It increased.
C. It didn't change.
D. It first increased, then decreased.

Use the circle graph below to answer questions 7 and 8.

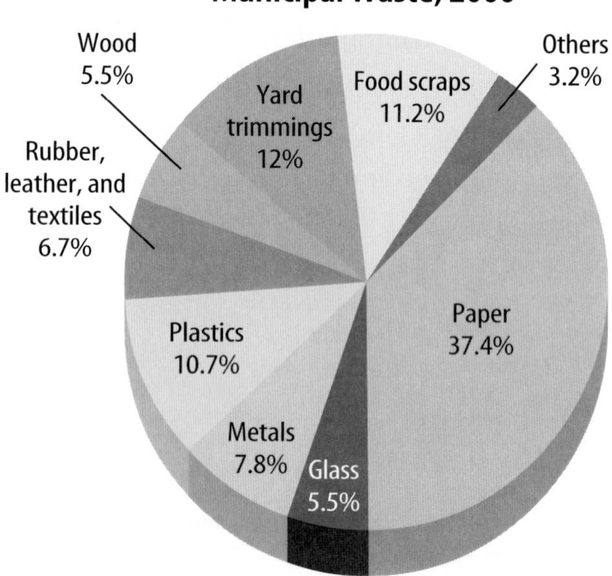

7. Which material made up the largest portion of the municipal solid waste produced in the year 2000?
A. plastics **C.** metals
B. paper **D.** wood

8. What percentage of waste was food scraps?
A. 11.2 **C.** 5.5
B. 6.7 **D.** 37.4

Part 2 Short Response/Grid In

Record your answers on the answer sheet provided by your teacher or on a sheet of paper.

9. List three ways that people affect the environment.
10. How does unnecessary packaging increase the amount of solid waste? How can this waste be reduced?
11. Why are many people concerned about the loss of tropical forest?
12. What are hazardous wastes? How are they different from other types of waste?
13. Should household hazardous waste be included with the normal trash? Explain.
14. What is phytoremediation? Why can't plants used in phytoremediation be used to feed livestock?

Use the graph below to answer questions 15–17.

Municipal Solid Waste Recycled

Percent of municipal solid waste recycled

50.0%
40.0%
30.0%
20.0%
10.0%
0.0%

5.6
8.0
9.6
16.2
30.1

1960 1970 1980 1990 2000

Year

Source: U.S. Environmental Protection Agency

15. What percentage of municipal wastes were recycled in 2000?
16. How much higher is the percentage of waste recycled in 2000 than in 1960?
17. Describe the trend shown on the graph.

Part 3 Open Ended

Record yours answers on a sheet of paper.

Use the diagram below to answer questions 18–21.

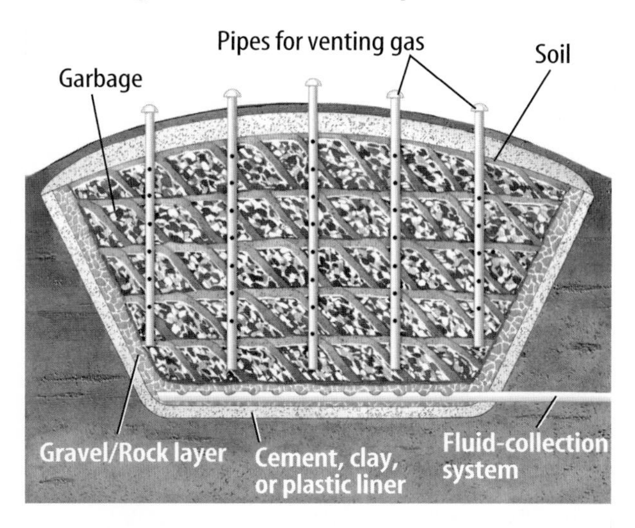

18. Why are the gas venting pipes shown on the diagram necessary?
19. Why is a liner placed at the bottom of the landfill?
20. What is a fluid collection system?
21. Explain why it is important to design landfills this way.
22. Why have natural preserves been established in the United States and in other countries?
23. What is organic farming? How is it different from conventional farming?
24. Why have some communities preserved unpaved areas where rainwater can soak into the ground?
25. Are there any disadvantages to recycling? Explain.
26. Why is human population growth currently called a population explosion?
27. Why has human population increased so dramatically in the past century?

chapter 21

Our Impact on Water and Air

The BIG Idea

When states and nations cooperate, pollution problems can be reduced.

SECTION 1
Water Pollution

Main Idea Polluted water contains chemicals and organisms that can cause disease.

SECTION 2
Air Pollution

Main Idea Millions of people in the United States still breathe unhealthy air.

Do you enjoy the outdoors?

At one time, all the water on Earth was like this. Clean water and air help create a pleasant outdoor experience. Too many substances released into air and water from human activity may damage these resources.

Science Journal Hypothesize what happens to the water in your home after the water goes down the drain.

Start-Up Activities

Is pollution always obvious?

Some water pollution is easy to see. The water can be discolored, have an odor, or contain dead fish. Suppose the water appears to be clean. Does that mean it's free of pollution? You'll find out during this lab.

1. Pour 125 mL of water into a large jar.
2. Add one drop (0.05 mL) of food coloring to the water and stir.
3. Add an additional 125 mL of water to the jar and stir.
4. Repeat step 3 until you cannot see the food coloring.
5. **Think Critically** Calculate the concentration of food coloring in your jar with each 125-mL addition of water. Will the concentration of food coloring ever become zero by diluting the solution?

Preview this chapter's content and activities at earth.msscience.com

Pollution Make the following Foldable to compare and contrast the characteristics of water pollution and air pollution.

STEP 1 **Fold** one sheet of paper lengthwise.

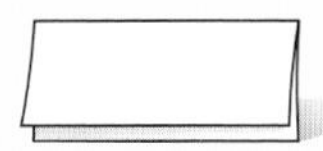

STEP 2 **Fold** into thirds.

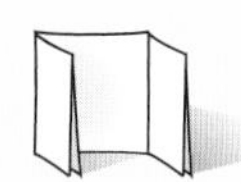

STEP 3 **Unfold** and draw overlapping ovals. **Cut** the top sheet along the folds.

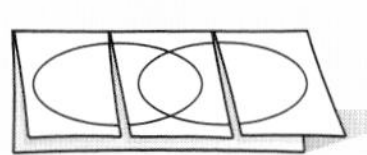

STEP 4 **Label** the ovals as shown.

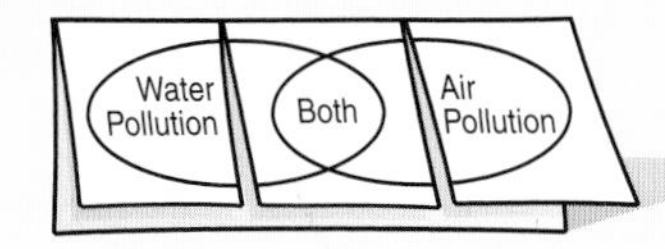

Construct a Venn Diagram As you read this chapter, list the characteristics unique to water pollution under the left tab, those unique to air pollution under the right tab, and those characteristics common to both under the middle tab.

Get Ready to Read

Identify Cause and Effect

1 Learn It! A cause is the reason something happens. The result of what happens is called an effect. Learning to identify causes and effects helps you understand why things happen. By using graphic organizers, you can sort and analyze causes and effects as you read.

2 Practice It! Read the following paragraph. Then use the graphic organizer below to show what happened when fertilizers wash into streams.

> Fertilizers are chemicals that help plants grow. However, rain washes away as much as 25 percent of the fertilizers applied to farms and yards into ponds, streams, and rivers. Fertilizers contain nitrogen and phosphorus that algae, living in water, use to grow and multiply. Lakes and ponds with high nitrogen and phosphorus levels, such as the one shown in Figure 3, can be choked with algae. When algae die and decompose, oxygen in the lake is used up more rapidly. This can cause fish and other organisms to die.
>
> —*from page 601*

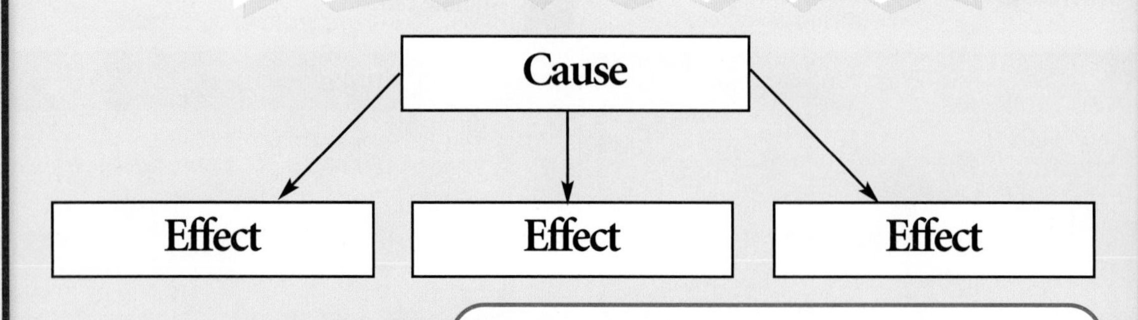

3 Apply It! As you read the chapter, be aware of causes and effects of air and water pollution. Find five causes and their effects.

Reading Tip

Graphic organizers such as the Cause-Effect organizer help you organize what you are reading so you can remember it later.

Target Your Reading

Use this to focus on the main ideas as you read the chapter.

1. **Before you read** the chapter, respond to the statements below on your worksheet or on a numbered sheet of paper.
   - Write an **A** if you **agree** with the statement.
   - Write a **D** if you **disagree** with the statement.

2. **After you read** the chapter, look back to this page to see if you've changed your mind about any of the statements.
   - If any of your answers changed, explain why.
   - Change any false statements into true statements.
   - Use your revised statements as a study guide.

Before You Read A or D		Statement	After You Read A or D
	1	All organisms need water.	
	2	Fertilizers, which are chemicals that help plants grow, can pollute water.	
	3	Metals such as mercury, lead, nickel, and cadmium are poisonous only in large amounts.	
	4	Only a few streams in the United States are polluted.	
	5	Clear, sparkling streams are safe for drinking.	
	6	*Smog* is a term that was originally used to describe the combination of smoke and fog.	
	7	Major sources of smog include cars, factories, and power plants.	
	8	The smaller particles in air pollution are less dangerous than the larger particles.	
	9	Electric power plants that burn fossil fuels emit particulates into the atmosphere.	
	10	Pollutants moving through the atmosphere stop when they reach state borders.	

Science Online

Print out a worksheet of this page at earth.msscience.com

section 1

Water Pollution

as you read

What You'll Learn
- **Identify** types of water pollutants and their effects.
- **Discuss** ways to reduce water pollution.
- **List** ways that you can help reduce water pollution.

Why It's Important
All organisms on Earth depend on water for life.

Review Vocabulary
pollution: the introduction of harmful substances to the environment

New Vocabulary
- point source pollution
- nonpoint source pollution
- pesticide
- fertilizer
- sewage

Importance of Clean Water

All organisms need water. Plants need water to make food from sunlight. Some animals such as fish, frogs, and whales live in water. What about you? You cannot live without drinking water. What happens if water isn't clean? Polluted water contains chemicals and organisms that can cause disease or bring death to many living things. Water also can be polluted with sediments, such as silt and clay.

Sources of Water Pollution

If you were hiking along a stream or lake and became thirsty, would it be safe to drink the water? Many streams and lakes in the United States are polluted in some way. Even streams that look clear and sparkling might not be safe for drinking.

Point source pollution is pollution that enters water from a specific location, such as drainpipes or ditches, as shown in **Figure 1.** Pollution from point sources can be controlled or treated before the water is released to a body of water.

However, many times bodies of water become polluted and no one knows exactly where the pollution comes from. Pollution that enters a body of water from a large area, such as lawns, construction sites, and roads, is called **nonpoint source pollution.** Nonpoint sources also include pollutants in rain or snow. Nonpoint source pollution is the largest source of water quality problems in the United States.

Figure 1 Water can be polluted in two ways.

Nonpoint sources cannot be traced to a single location.

Point sources include industrial wastes from outfalls.

Sediment The largest source of water pollution in the United States is sediment. Sediment is loose material, such as rock fragments and mineral grains, that is moved by erosion. Rivers always have carried sediment to oceans, but human activities can increase the amount of sediment in rivers, lakes, and oceans. Each year, about 25 billion metric tons of sediment are carried from farm fields to bodies of water on Earth. At least 50 billion additional tons run off of construction sites, cleared forests, and land used to graze livestock. Sediment makes water cloudy and blocks sunlight that underwater plants need to make food. Sediment also covers the eggs of organisms that live in water, preventing organisms from receiving the oxygen they need to develop.

Figure 2 Research suggests that some pesticides in the environment could lead to deformities in frogs, such as missing legs.

Agriculture and Lawn Care Farmers and home owners apply **pesticides,** which are substances that destroy pests, to keep insects and weeds from destroying their crops and lawns. When farmers and home owners apply pesticides to their crops and lawns, some of the chemicals run off into water. These chemicals might be harmful to people and other organisms, such as the frog in **Figure 2.**

Fertilizers are chemicals that help plants grow. However, rain washes away as much as 25 percent of the fertilizers applied to farms and yards into ponds, streams, and rivers. Fertilizers contain nitrogen and phosphorus that algae, living in water, use to grow and multiply. Lakes or ponds with high nitrogen and phosphorous levels, such as the one shown in **Figure 3,** can be choked with algae. When algae die and decompose, oxygen in the lake is used up more rapidly. This can cause fish and other organisms to die. Earth's nitrogen cycle is modified when fertilizers enter the water system.

Reading Check *How do fertilizers cause water pollution?*

Without enough oxygen, fish may die.
Algae grow and multiply.
Algae die and decay, using up oxygen.
Fertilizer applied to lawns or farms runs off.

Figure 3 Nitrogen and phosphorus in fertilizer cause algae to grow and multiply. Fish can die when algae decompose, using up oxygen.

Human Waste When you flush a toilet or take a shower, the water that goes into drains, called **sewage,** contains human waste, household detergents, and soaps. Human waste contains harmful organisms that can make people sick.

In most cities and towns in the United States, underground pipes take the water you use from your home to a sewage treatment plant. Sewage treatment plants, such as the one in **Figure 4,** remove pollution using several steps. These steps purify the water by removing solid materials from the sewage, killing harmful bacteria, and reducing the amount of nitrogen and phosphorus.

Applying Math — Calculate Percentages

SURFACE WATER POLLUTION This table shows the number of sampling stations that have an increased or a decreased level of pollution in a 10-year period. What percent of stations has shown an increase in nitrogen over a 10-year period?

Trends in River and Stream Water Quality

Measured Pollutant	Total No. of Stations Examined	No. of Stations With Decrease in Pollutant Level	No. of Stations With Increase in Pollutant Level
Sediments	324	36	6
Bacteria from sewage	313	41	9
Total phosphorus	410	90	21
Nitrogen	344	27	21

Solution

1. *This is what you know:* Nitrogen: number of stations with an increase = 21; total number of stations examined = 344
2. *This is what you need to find:* percentage: ______%
3. *This is the equation you need to use:*
   - % = (stations with increase)/(total stations) × 100
   - (21)/(344) × 100 = 6.1%
4. *Check your answer:* Multiply the total stations by the percent in decimal form to obtain the number of stations with an increase.

Practice Problems

1. What percent of stations has shown a decrease in bacteria?
2. What percent of stations has shown an increase in sediment?

For more practice, visit earth.msscience.com/math_practice

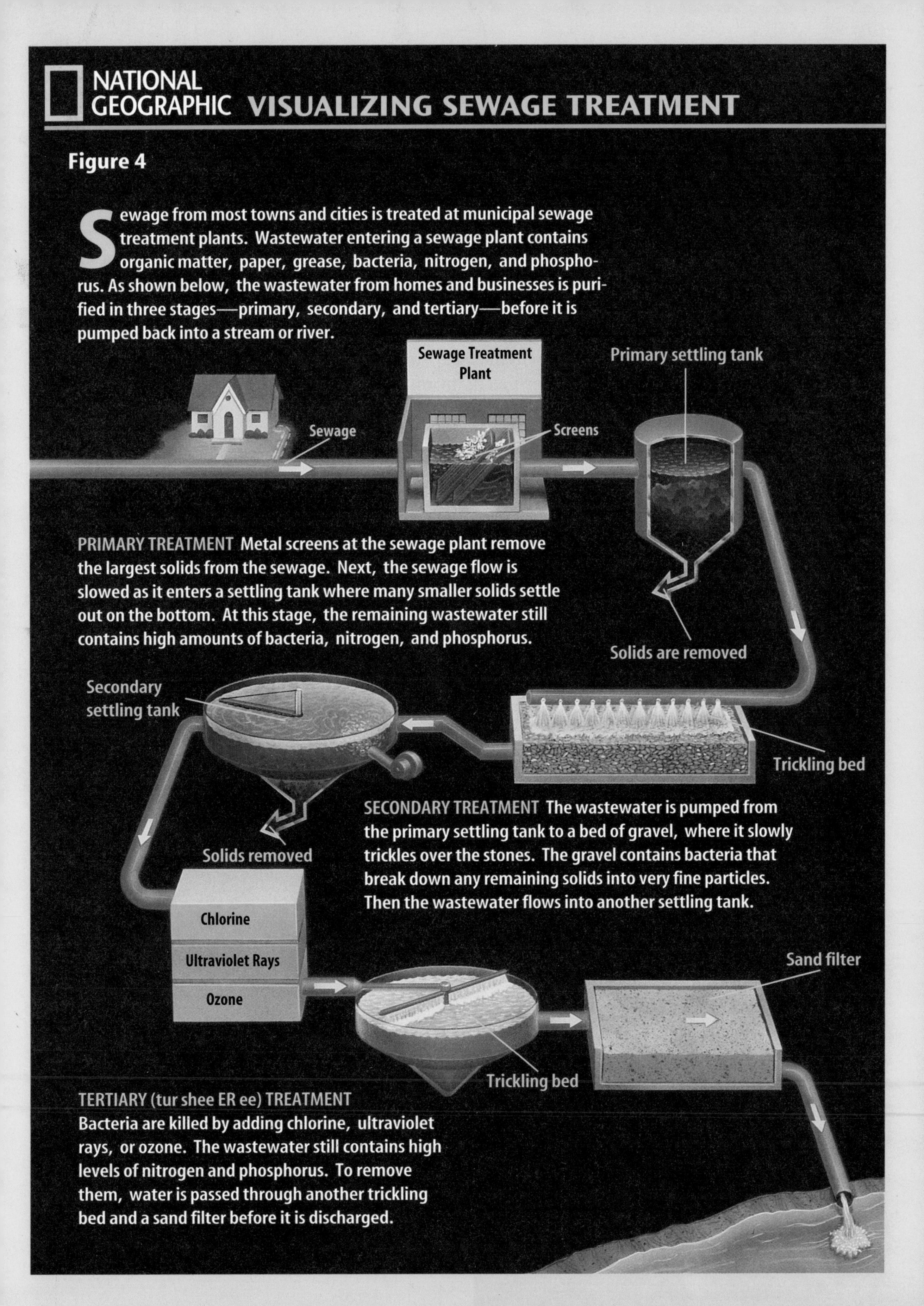
NATIONAL GEOGRAPHIC VISUALIZING SEWAGE TREATMENT
Figure 4
Sewage from most towns and cities is treated at municipal sewage treatment plants. Wastewater entering a sewage plant contains organic matter, paper, grease, bacteria, nitrogen, and phosphorus. As shown below, the wastewater from homes and businesses is purified in three stages—primary, secondary, and tertiary—before it is pumped back into a stream or river.
Sewage Treatment Plant
Primary settling tank
Sewage
Screens
PRIMARY TREATMENT Metal screens at the sewage plant remove the largest solids from the sewage. Next, the sewage flow is slowed as it enters a settling tank where many smaller solids settle out on the bottom. At this stage, the remaining wastewater still contains high amounts of bacteria, nitrogen, and phosphorus.
Solids are removed
Secondary settling tank
Trickling bed
SECONDARY TREATMENT The wastewater is pumped from the primary settling tank to a bed of gravel, where it slowly trickles over the stones. The gravel contains bacteria that break down any remaining solids into very fine particles. Then the wastewater flows into another settling tank.
Solids removed
Chlorine
Ultraviolet Rays
Ozone
Sand filter
Trickling bed
TERTIARY (tur shee ER ee) TREATMENT
Bacteria are killed by adding chlorine, ultraviolet rays, or ozone. The wastewater still contains high levels of nitrogen and phosphorus. To remove them, water is passed through another trickling bed and a sand filter before it is discharged.

Environmental Engineering Earth's atmosphere and oceans have a limited capacity to absorb wastes and recycle materials naturally. However, overabundance of pollutants has negative effects. Environmental engineering offers opportunities to work in environmental protection. Major areas include air pollution control, water supply, wastewater management, and storm water management.

Metals Many metals such as mercury, lead, nickel, and cadmium can be poisonous, even in small amounts. For example, lead and mercury in drinking water can damage the nervous system. However, metals such as these are valuable in making items you use such as paints and stereos. Before environmental laws were written, a large amount of metals was released with wastewater from factories. Today, laws control how much metal can be released. Because metals remain in the environment for a long time, metals released many years ago still are polluting bodies of water today.

Mining also releases metals into water. For example, in the state of Tennessee, more than 43 percent of all streams and lakes contain metals from mining activities. In the mid 1980s, gold was found near the Amazon River in South America. Miners use mercury to trap the gold and separate it from sediments. Each year, more than 130 tons of mercury end up in the Amazon River.

Oil and Gasoline Oil and gasoline run off roads and parking lots into streams and rivers when it rains. These compounds contain pollutants that might cause cancer. Gasoline is stored at gas stations in tanks below the ground. In the past, the tanks were made of steel. Some of these tanks rusted and leaked gasoline into the surrounding soil and groundwater. As little as one gallon of gasoline can make an entire city's water supply unsafe for drinking.

Federal laws passed in 1988 require all new gasoline tanks to have a double layer of steel or fiberglass. In addition, by 1998, all new and old underground tanks must have had equipment installed that detects spills and must be made of materials that will not develop holes. These laws help protect soil and groundwater from gasoline and oil stored in underground tanks.

Heat When a factory makes a product, heat often is released. Sometimes, cool water from a nearby ocean, river, lake, or underground supply is used to cool factory machines. The heated water then is released. This water can pollute because it contains less oxygen than cool water does. In addition, organisms that live in water are sensitive to changes in temperature. A sudden release of heated water can kill a large number of fish in a short time. Water can be cooled before it is released into a river by using a cooling tower or pond, as shown in **Figure 5.**

Figure 5 During the manufacture of many products, such as electricity from this power plant, water is needed for cooling the machinery. Heated water remains in large towers and ponds until it has cooled to a temperature that is safe for fish and other organisms.

Reducing Water Pollution

One way to reduce water pollution is by treating water before it enters a stream, lake, or river. In 1972, the United States Congress amended the Water Pollution Control Act. This law provided funds to build sewage-treatment facilities. It required industries to remove or treat pollution in water discharged to a lake or stream. The Clean Water Act of 1987 made additional money available for sewage treatment and set goals for reducing point source and nonpoint source pollution.

Another law, the Safe Drinking Water Act of 1996, strengthens health standards for drinking water. This legislation also protects rivers, lakes, and streams that are sources of drinking water.

Figure 6 Because of laws passed since 1972, Lake Erie's water has improved.

A fire on the Cuyahoga River in Cleveland, Ohio, alerted people in the United States to water pollution problems.

Today, millions of people enjoy this natural resource.

International Cooperation Several countries have worked together to reduce water pollution. Lake Erie is on the border between the United States and Canada. Prior to the 1970s, phosphorus and nitrogen from sewage, soaps, and fertilizers entered Lake Erie from homes, yards, and farms, causing algae to grow and reproduce. The lake became a green, soupy mess. In the summer, the algae died and sank to the lake bottom. As the dead algae decayed, large areas of the lake bottom no longer had oxygen and, therefore, no life.

Pollutants also were discharged from many steel, automobile, and other factories along Lake Erie. **Figure 6** shows how on June 22, 1969, greasy debris on a large river flowing through Cleveland, Ohio, caught fire. This event was a wake-up call for everyone concerned about the quality of water in the United States and around the world.

In the 1970s, the United States and Canada made two water-quality agreements. These agreements set goals for reducing pollution in the Great Lakes. As a result of these agreements, limits were placed on the amount of phosphorus and other pollutants allowed into Lake Erie.

Today, the green slime is gone and the fish are back. However, more than 300 human-made chemicals still can be found in Lake Erie, and some of them are hazardous. The United States and Canada are studying ways to remove them from the lake.

Reading Check ***Which countries worked together to control water pollution in Lake Erie?***

Hazardous Wastes
Some wastes are called hazardous because they are carcinogenic (kar sih nuh JEH nik). What does carcinogenic mean? What may happen if hazardous wastes seeped into a drinking water supply?

How can you help?

Through laws and regulations, the quality of many streams, rivers, and lakes in the United States has improved. However, as **Figure 7** shows, much remains to be done. Individuals and industries alike need to continue to work to reduce water pollution. You easily can help by keeping contaminants out of Earth's water supply and by conserving water.

Dispose of Wastes Safely When you dispose of household chemicals such as paint and motor oil, don't pour them onto the ground or down the drain. Hazardous wastes that are poured directly onto the ground move through the soil and eventually might reach the groundwater below. Pouring them down the drain is no better because they flow through the sewer, through the wastewater-treatment plant, and into a stream or river where they can harm the organisms living there.

What should you do with these wastes? First, read the label on the container for instructions on disposal. Don't throw the container into the trash if the label tells you not to. Store chemical wastes so that they can't leak. Call your local government officials and ask how to dispose of these wastes in your area safely. Many communities have specific times each year when they collect hazardous wastes. These wastes then are disposed of at special hazardous waste sites.

Topic: Water Conservation
Visit earth.msscience.com for Web links to information about water conservation.

Activity Turn on a faucet until it drips. Collect the water for 10 min. Calculate how much water goes down the drain each day.

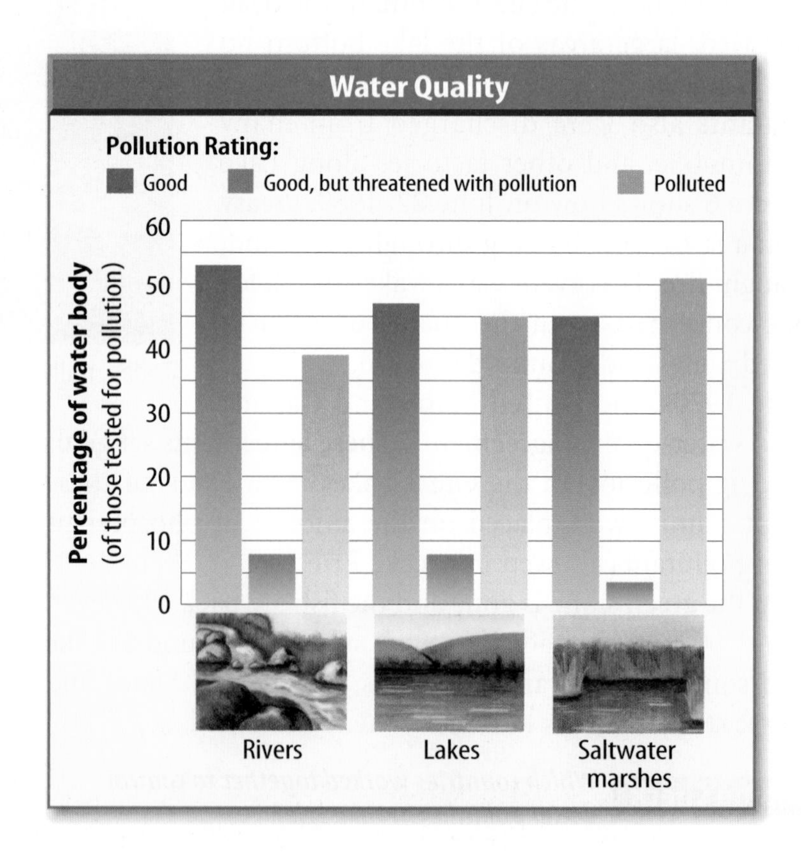

Figure 7 This graph shows that water pollution is still a problem in the United States.
Determine *the percentage of rivers that are listed as polluted.*

Figure 8 Water pollution can be reduced if less water is used.

One drip every 5 s from a leaky faucet will waste nearly 7,500 L of water per year.

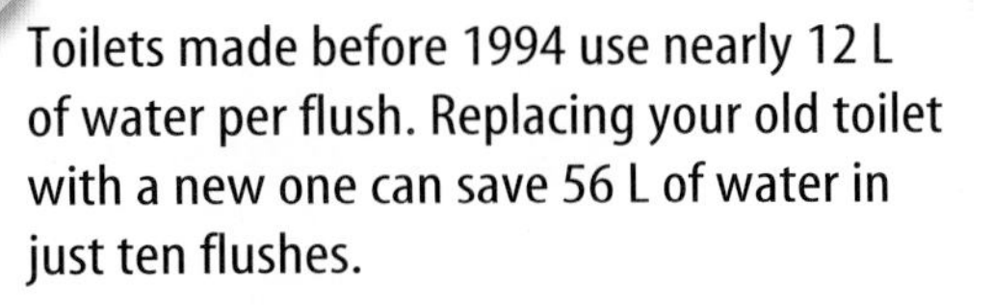

Toilets made before 1994 use nearly 12 L of water per flush. Replacing your old toilet with a new one can save 56 L of water in just ten flushes.

Turning off the water while brushing your teeth will save more than 19 L per day.

Conserve Water How much water do you use every day? You use water every time you flush a toilet, take a bath, clean your clothes, wash dishes, wash a car, or use a hose or lawn sprinkler. A typical U.S. citizen uses an average of 375 L of water per day. Unless it comes from a home well, this water must be purified before it reaches your home. After you use it, it must be treated again. **Figure 8** shows how using simple conservation methods can save water. Conserving water reduces the need for water treatment and reduces water pollution.

section 1 review

Summary

Importance of Clean Water

- All life on Earth needs water.
- Water pollution can harm living things.

Sources of Water Pollution

- Pollution enters water from point and non-point sources.
- Lawn and farm chemicals, sewage, metals, oil and heat all contribute to water pollution.

Reducing Water Pollution

- Federal laws and international agreements have helped reduce water pollution.

How can you help?

- Conserving water helps reduce pollution.

Self Check

1. **Compare and contrast** point source and nonpoint source pollution.
2. **Infer** how U.S. laws have helped reduce water pollution.
3. **Describe** ways you can conserve water.
4. **Think Critically** Southern Florida has many dairy farms and sugarcane fields. It also contains Everglades National Park—a shallow river system. What kinds of pollutants might be in the Everglades? How did they get there?

Applying Skills

5. **Use graphics software** to design a pamphlet that informs people how to reduce the amount of water they use.

Elements in Water

When you look at water, it is often clear and looks as if there is not much in it. However, there are many compounds, microscopic organisms, and other substances that can be in the water, but aren't easily visible. How can you find out what else might be in the water? Can you also find out how much of it is in the water?

Real-World Question

What is the nitrate and phosphate content of water?

Goals

- **Determine** the nitrate and phosphate content of two samples of water.
- **Compare** the levels and explain any differences you find.

Materials

beakers (2)
tap water
plant fertilizer
teaspoon
nitrate test kit
phosphate test kit
stirrer

Safety Precautions

Never eat or drink anything in the lab. Use gloves and goggles when handling fertilizer.

Procedure

1. Half-fill two large beakers with tap water.
2. Add a teaspoon of plant fertilizer to one of the beakers and stir well.
3. **Predict** which beaker might have a greater level of nitrate.
4. Using an appropriate kit, measure the nitrate content of each beaker of water.

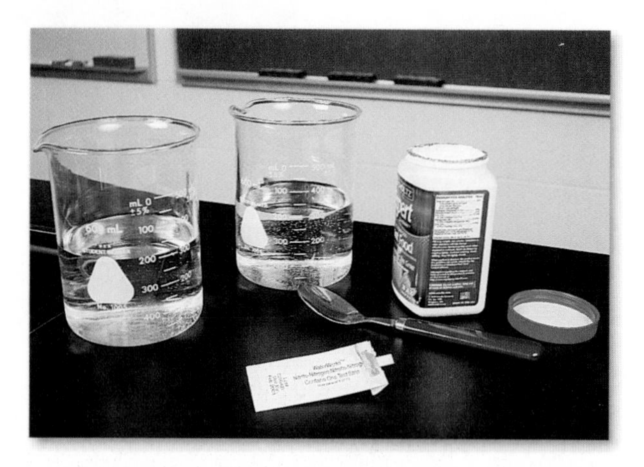

5. Clean the test kit between measurements. Record your measurements.
6. **Predict** which beaker might have a greater level of phosphates.
7. Using an appropriate kit, measure the phosphate content of each beaker of water. Be sure to clean the kit between measurements. Record your measurements.

Conclude and Apply

1. **Describe** your results. Were the levels of each compound you measured the same in both samples?
2. **Infer** if your predictions were correct.
3. **Explain** any differences that you found.
4. **Explain** how the use of fertilizers can cause problems in lakes and streams.

Communicating Your Data

Compare your results with those of others in your class. **Discuss** any differences found in your measurements.

section 2

Air Pollution

Causes of Air Pollution

Cities can be exciting because they are centers of business, culture, and entertainment. Unfortunately, cities also have many cars, buses, and trucks that burn fuel for energy. The brown haze you sometimes see forms from the exhaust of these vehicles. Air pollution also comes from burning fuels in factories, generating electricity, and burning trash. Dust from plowed fields, construction sites, and mines also contributes to air pollution.

Natural sources add pollutants to the air, too. For example, radon is a naturally occurring gas given off by certain kinds of rock. This gas can seep into basements of homes built on these rocks. Exposure to radon can increase the risk of lung cancer. Natural sources of pollution also include particles and gases emitted into air from erupting volcanoes and fires.

What is smog?

One type of air pollution found in urban areas is called smog, a term originally used to describe the combination of smoke and fog. Major sources of smog, shown in **Figure 9,** include cars, factories, and power plants.

as you read

What You'll Learn

- **List** the different sources of air pollutants.
- **Describe** how air pollution affects people and the environment.
- **Discuss** how air pollution can be reduced.

Why It's Important

Air pollution can adversely affect your health and the health of others.

Review Vocabulary

ozone layer: a layer of the stratosphere that absorbs most of the Sun's ultraviolet radiation

New Vocabulary

- photochemical smog
- acid rain
- pH scale
- acid
- base
- carbon monoxide
- particulate matter
- scrubber

Figure 9 Cars are one of the main sources of air pollution in the United States.
Calculate *the percentage of smog that comes from power plants and industry combined.*

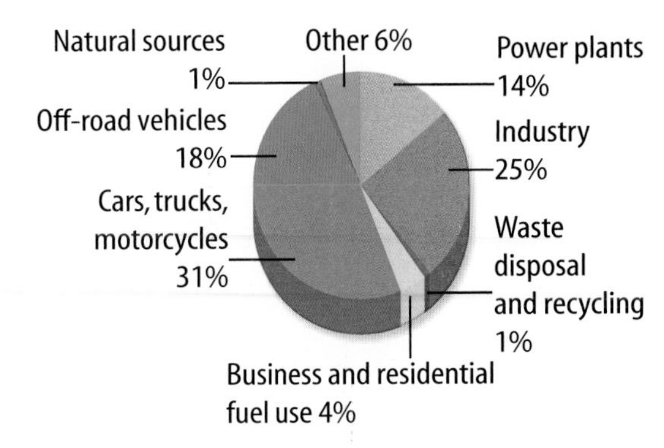

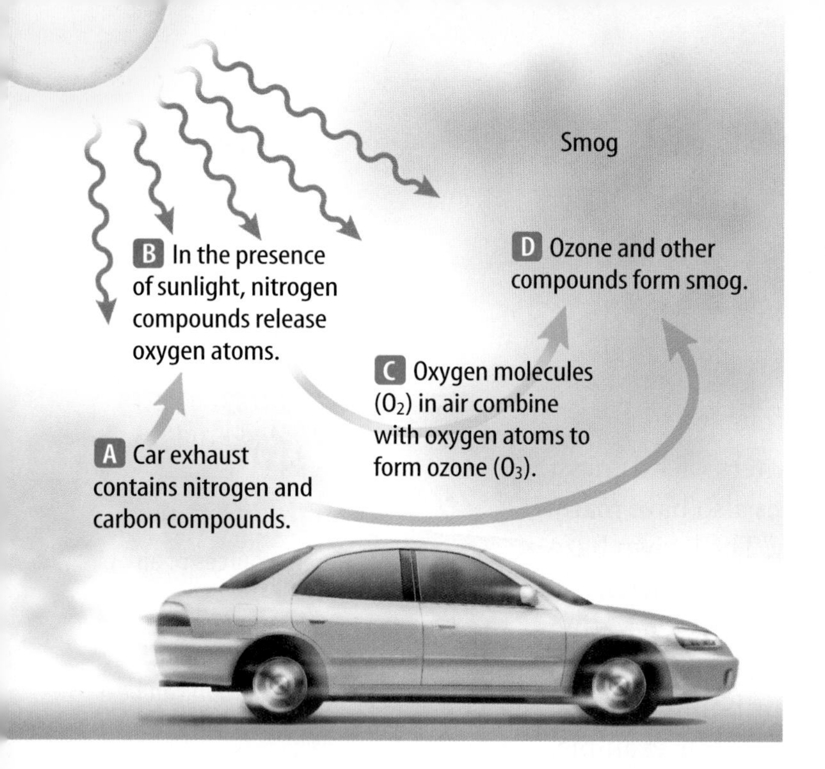
Smog
B In the presence of sunlight, nitrogen compounds release oxygen atoms.
D Ozone and other compounds form smog.
C Oxygen molecules (O_2) in air combine with oxygen atoms to form ozone (O_3).
A Car exhaust contains nitrogen and carbon compounds.

How Smog Forms The hazy, yellowish brown blanket of smog that is sometimes found over cities is called **photochemical smog** because it forms with the help of sunlight. Pollutants get into the air when gasoline is burned, releasing nitrogen and carbon compounds. These compounds, as shown in **Figure 10,** react in the presence of sunlight to produce other substances. One of the substances produced is ozone. Ozone high in the atmosphere protects you from the Sun's ultraviolet radiation. However, ozone near Earth's surface is a major component of smog. Smog can damage sensitive tissues, like plants or your lungs.

Figure 10 Exhaust from cars can form smog in the presence of sunlight.

Nature and Smog Certain natural conditions contribute to smoggy air. For example, some cities do not have serious smog problems because their pollutants often are dispersed by winds. In other areas, landforms add to smog development. The mountains surrounding Los Angeles, for example, can prevent smog from being carried away by winds.

Figure 11 shows how the atmosphere also can influence the formation of smog. Normally, warmer air is found near Earth's surface. However, sometimes warm air traps cool air near the ground. This is called a temperature inversion, and it reduces the capacity of the atmosphere to mix materials, causing pollutants to accumulate near Earth's surface.

Figure 11 Conditions in the atmosphere can worsen air pollution.

Normal Conditions

Cooler air
Cool air
Warm air

Usually, air temperature decreases with distance above Earth's surface. Air pollutants can be carried far away from their source.

Temperature Inversion

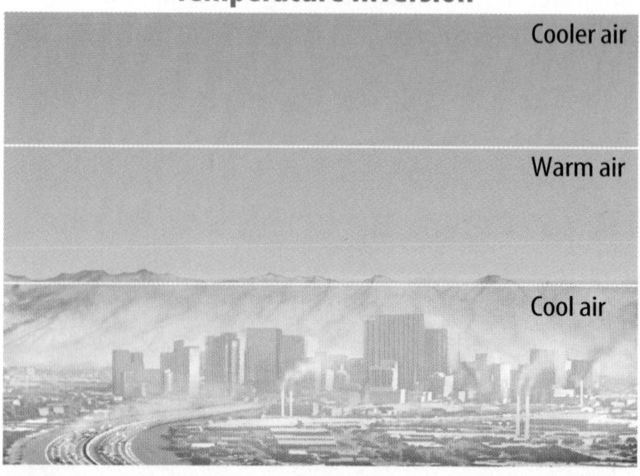
Cooler air
Warm air
Cool air

During a temperature inversion, warm air overlies cool air. Air pollutants can't be dispersed and can accumulate to unhealthy levels.

Acid Rain

INTEGRATE Chemistry

When sulfur oxides from coal-burning power plants and nitrogen oxides from cars combine with moisture in the air, they form acids. When acidic moisture falls to Earth as rain or snow, it is called **acid rain.** Acid rain can corrode structures, damage forests, and harm organisms. The amount of acid is measured using the **pH scale.** A lower number means greater acidity. Substances with a pH lower than 7 are **acids.** Substances with a pH above 7 are **bases.**

Natural lakes and streams have a pH between 6 and 8. Acid rain is precipitation with a pH below 5.6. When rain is acidic, the pH of streams and lakes may decrease. As **Figure 12** shows, certain organisms, like snails, can't live in acidic water.

CFCs

About 20 km above Earth's surface is a layer of atmosphere that contains a higher concentration of ozone, called the ozone layer. Recall that ozone is a molecule made of three oxygen atoms and is found in smog. However, unlike smog, the ozone that exists at high altitudes helps Earth's organisms by absorbing some of the Sun's harmful ultraviolet (UV) rays. Chlorofluorocarbons (CFCs) from air conditioners and refrigerators might be destroying this ozone layer. Each CFC molecule can destroy thousands of ozone molecules. Even though the use of CFCs has been declining worldwide, these compounds can remain in the upper atmosphere for many decades.

Identifying Acid Rain

Procedure

1. Use a clean **glass or plastic container** to collect a **sample of precipitation.**
2. Use **pH paper** or a **pH computer probe** to determine the acidity level of your sample. If you have collected snow, allow it to melt before measuring its pH.
3. Record the indicated pH of your sample and compare it with the results of other classmates who have followed the same procedure.

Analysis

1. What is the average pH of the samples obtained from this precipitation?
2. Compare and contrast the pH of your samples with those of the substances shown on a pH scale.

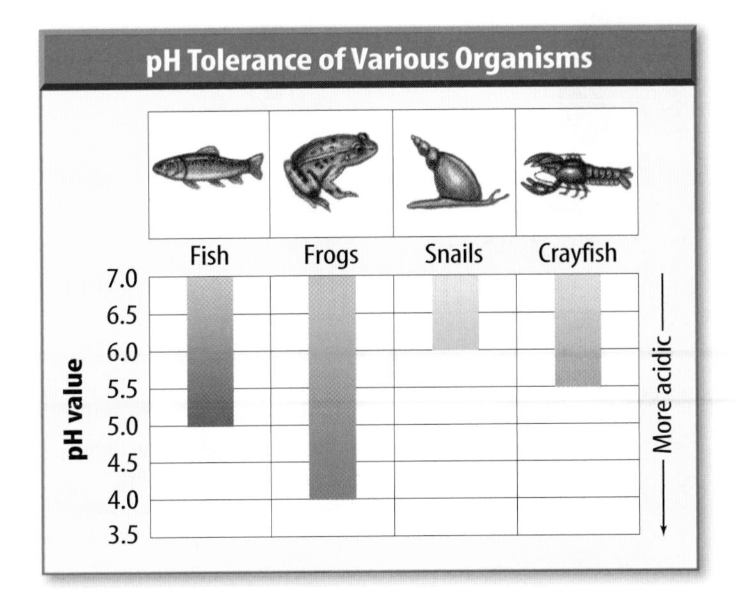

Figure 12 This chart shows water acidity levels where different organisms can live. **Infer** *which types of organisms you might find in a pond with a pH of 5.5 or greater. Which organisms would be in the pond if the pH dropped to 4.5?*

Topic: Air Quality
Visit earth.msscience.com for Web links to information about air quality indices.

Activity Look up the air quality index of a city near you each day for 10 days. Graph the results.

Air Pollution and Your Health

Suppose you're an athlete in a large city training for a competition. You might have to get up at 4:30 A.M. to exercise. Later in the day, the smog levels might be so high that it wouldn't be safe for you to exercise outdoors. In some large cities, athletes adjust their training schedules to avoid exposure to ozone and other pollutants. Schools schedule football games for Saturday afternoons when smog levels are lower. Parents are warned to keep their children indoors when smog exceeds certain levels.

Health Disorders How hazardous is dirty air? Approximately 250,000 people in the United States suffer from pollution-related breathing disorders. About 70,000 deaths each year in the United States are blamed on air pollution. **Figure 13** illustrates some of the health problems caused by air pollution. Ozone damages lung tissue, making people more susceptible to diseases such as pneumonia and asthma. Less severe symptoms of ozone include burning eyes, dry throat, and headache.

How do you know if ozone levels in your community are safe? You may have seen the Air Quality Index reported in your newspaper. **Table 1** shows the index along with ways to protect your health when ozone levels are high.

Carbon monoxide, a colorless, odorless gas found in car exhaust, also contributes to air pollution. This gas can make people ill, even in small concentrations because it replaces oxygen in your blood.

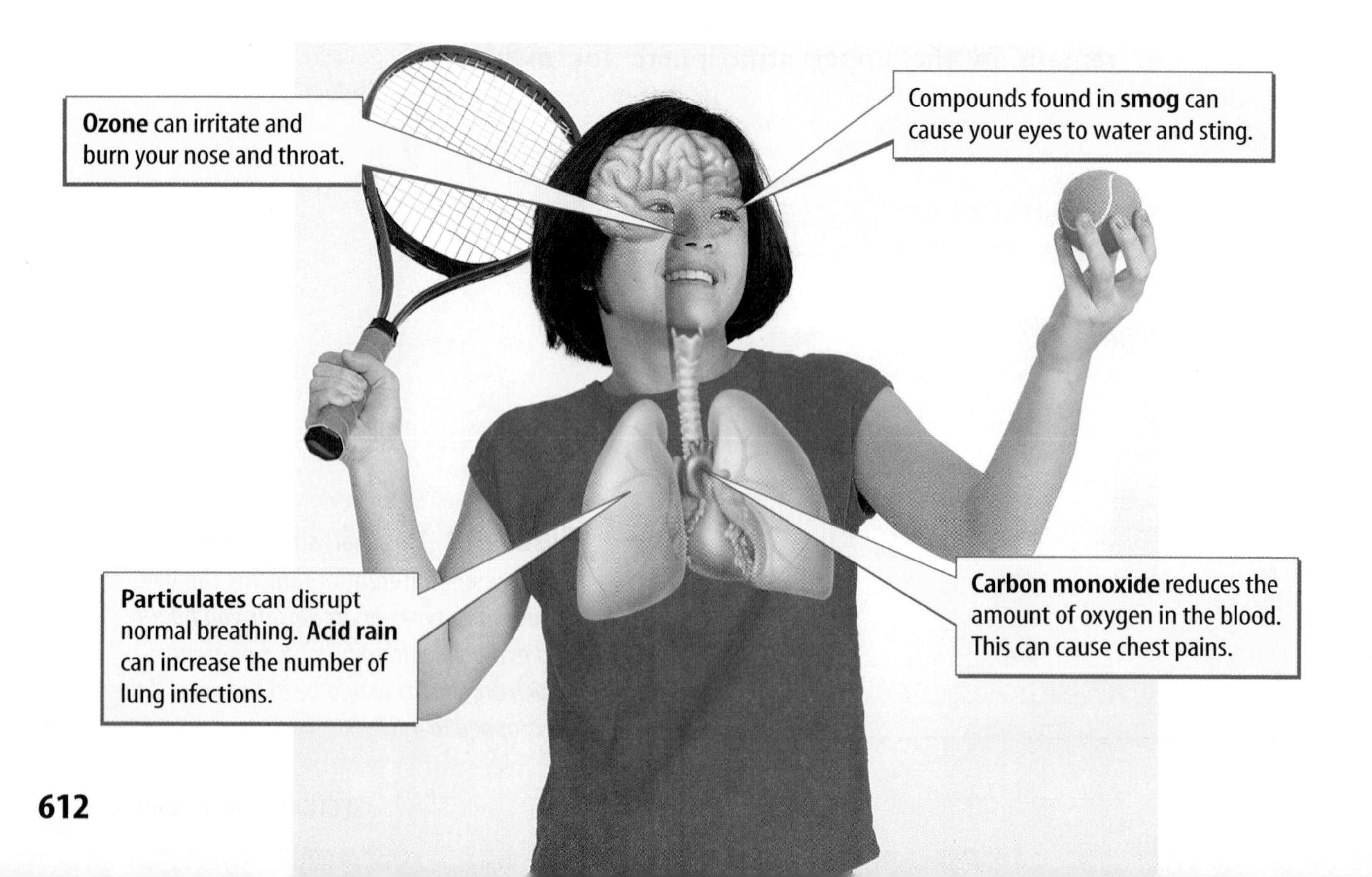

Ozone can irritate and burn your nose and throat.
Compounds found in **smog** can cause your eyes to water and sting.
Particulates can disrupt normal breathing. **Acid rain** can increase the number of lung infections.
Carbon monoxide reduces the amount of oxygen in the blood. This can cause chest pains.

Figure 13 Air pollution can be a health hazard. Compounds in the air can affect your body.

Acid Rain What do you suppose happens when you inhale the humid air from acid rain? Acid is breathed deep inside your lungs. This may cause irritation and reduce your ability to fight respiratory infections. When you breathe, oxygen travels from the air to your lungs. Lungs damaged by acid rain cannot move oxygen to the blood easily. This puts stress on your heart.

Table 1 Air Quality Index

Air Quality	Air Quality Index	Protect Your Health
Good	0–50	No health impacts occur.
Moderate	51–100	People with breathing problems should limit outdoor exercise.
Unhealthy for certain people	101–150	Everyone, especially children and elderly, should not exercise outside for long periods of time.
Unhealthy	151–200	People with breathing problems should avoid outdoor activities.

Particulates Thick, black smoke from a forest fire, exhaust from school buses and large trucks, smoke billowing from a factory, and dust picked up by the wind all contain particulate (par TIH kyuh luht) matter. **Particulate matter** consists of fine particles such as dust, pollen, mold, ash, and soot that are in the air.

Particulate matter ranges in size from large, visible solids like dust and soil particles to microscopic particles that form when substances are burned. Smaller particles are more dangerous, because they can travel deeper into the lungs. When particulate matter is breathed in, it can irritate and damage the lungs, causing breathing problems.

Reading Check *Why are small particles dangerous to your health?*

Reducing Air Pollution

Pollutants moving through the atmosphere don't stop when they reach the borders between states and countries. They go wherever the wind carries them. This makes them difficult to control. Even if one state or country reduces its air pollution levels, pollutants from another state or country can blow across the border. For example, burning coal in midwestern states might cause acid rain in the northeast and Canada.

When states and nations cooperate, pollution problems can be reduced. People from around the world have met on several occasions to try to eliminate some kinds of air pollution. At one meeting in Montreal, Canada, an agreement called the Montreal Protocol was written to phase out the manufacture and use of CFCs by 2000. In 1989, 29 countries that consumed 82 percent of CFCs signed the agreement. By 1999, 184 countries signed it.

Examining the Content of Air

WARNING: *Use caution when reaching high places. Students with dust allergies should not perform this lab.*

Procedure

1. Find a **high shelf or the top of a tall cabinet** in your home—someplace that hasn't been cleaned for a while.
2. Using a **white cloth,** thoroughly dust the surface.
3. Observe the cloth under a **magnifying lens.**

Analysis

1. What did you see on your cloth? Where did these particles come from?
2. Explain what you think happens when you breathe in these particles.

Try at Home

Table 2 Clean Air Regulations

Urban air pollution	All cars manufactured since 1996 must reduce nitrogen oxide emissions by 60 percent and hydrocarbons by 35 percent from their 1990 levels.
Acid rain	Sulfur dioxide emissions had to be reduced by 14 million tons from 1990 levels by the year 2000.
Airborne toxins	Industries must limit the emission of 200 compounds that cause cancer and birth defects.
Ozone-depleting chemicals	Industries were required to immediately cease production of many ozone-depleting substances in 1996.

Air Pollution in the United States The United States Congress passed several laws to protect the air. The Clean Air Act of 1990, summarized in **Table 2,** addressed some air pollution problems by regulating emissions from cars, energy production, and other industries. In 1997, new levels for ozone and particulate matter were proposed.

Since the passage of the Clean Air Act, the amount of some pollutants released into the air has decreased, as shown in **Figure 14.** However, millions of people in the United States still breathe unhealthy air.

Reducing Emissions More than 80 percent of sulfur dioxide emissions comes from coal-burning power plants. Coal from some parts of the United States contains a lot of sulfur. When this coal is burned, sulfur dioxides combine with moisture in the air to form sulfuric acid, causing acid rain. Sulfur dioxide can be removed by passing the smoke through a scrubber. A **scrubber** lets the gases react with a limestone and water mixture. Another way to decrease the amount of sulfur dioxide is by burning low-sulfur coal.

Electric power plants that burn fossil fuels emit particulates into the atmosphere. Particulate matter in smoke from power plants can be removed with an electrostatic separator, shown in **Figure 15.** Plates in the separator give the smoke particles a positive charge. As the smoke particles move past negatively-charged plates, the positively-charged particles adhere to the negatively-charged plates.

Figure 14 This graph shows that some air pollutants have decreased since the passage of the Clean Air Act.
Determine *how many tons of particulates were released to the air in 1994.*

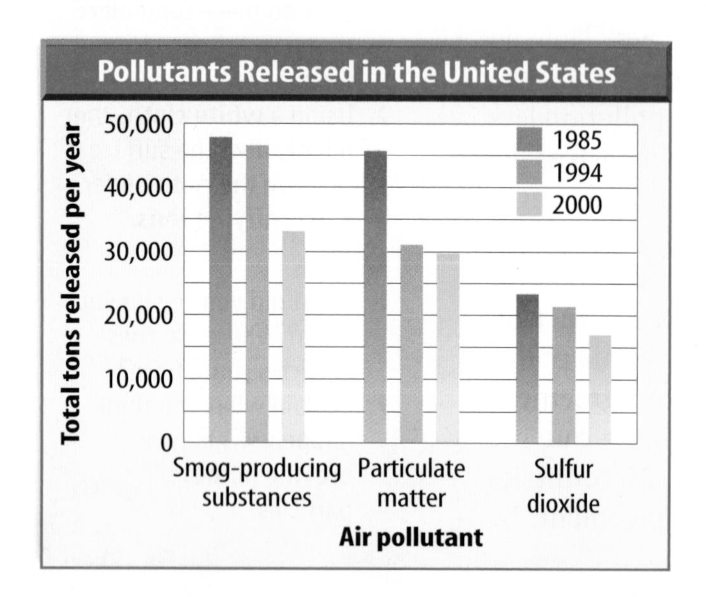

Getting Around Recent improvements in vehicle design and in how gasoline is made, as well as the use of emissions-control devices such as catalytic converters, have reduced automobile emissions significantly. Future advances in technology might reduce emissions further. Why is this important? Americans are driving more today than they did in the past. More time spent driving leads to more traffic congestion. Cars and trucks produce more pollution when they are stopped in traffic.

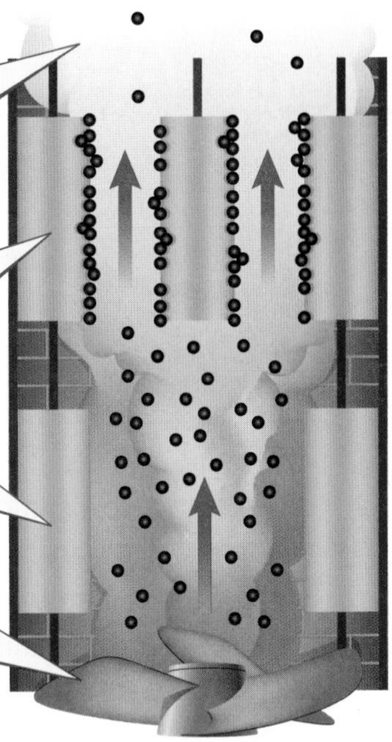

The Clean Air Act can work only if we all cooperate. Cleaning the air takes money, time, and effort. How might you take part in this cleanup? You can change your lifestyle. For example, you can walk, ride a bike, or use public transportation to get to a friend's house instead of asking for a ride. You also can set the thermostat in your house lower in the winter and higher in the summer.

The smoke, with up to 99 percent of the particulates removed, is released through the smokestack.

The positively charged particulates move past negatively charged plates. The particulates are attracted to and held by the plates.

The plates give the particles a positive charge.

A fan blows the smoke with particulates past electronically charged plates.

Figure 15 Electrostatic separators can remove almost all of the particulates in industrial smoke.

Reading Check *What can you do to prevent air pollution?*

section 2 review

Summary

Causes of Air Pollution

- Vehicles, electric generation, and dust from human activity contribute to air pollution.

What is smog?

- Smog forms when compounds react in the presence of sunlight to form ozone.
- Temperature inversions can worsen smog.

Acid Rain

- Exhaust from burning coal and gasoline can form acid rain.

CFCs

- Chlorofluorcarbons can damage Earth's ozone layer.

Air Pollution and Your Health

- Air pollution can cause breathing problems.

Self Check

1. **List** three pollutants released into the air when fuels are burned.
2. **Explain** how smog forms.
3. **Infer** how people can reduce air pollution.
4. **Think Critically** Laws were passed in 1970 requiring coal-burning power plants to use tall smokestacks to disperse pollutants. Power plants in the midwestern states complied with that law, and people in eastern Canada began complaining about acid rain. Explain the connection.

Applying Skills

5. **Classify** Use the information in **Table 1** to classify the following air quality indices: 43, 152, 7, 52, 147, and 98. Explain why it is important to have limits on pollutants from cars and factories as the U.S. population grows and people drive more.

Design Your Own

WHAT'S IN THE AIR?

Goals

- **Design** an experiment to collect and analyze particulate matter in the air in your community.
- **Observe and describe** the particulate matter you collect.

Possible Materials

small box of plain gelatin
hot plate
pan or pot
water
marker
refrigerator
plastic lids (4)
microscope
**magnifying lens*
**Alternate materials*

Safety Precautions

Wear a thermal mitt, safety goggles, and an apron while working with a hot plate and while pouring the gelatin from the pan or pot into the lids. Never eat anything in the lab.

Real-World Question

When you dust items in your household, you are cleaning up particles that settled out of the air. How often do you have to dust to keep your furniture clean? Just imagine how many pieces of particulate matter the air must hold. Do some areas of your environment have more particulates than other areas?

Form a Hypothesis

Based on your knowledge of your neighborhood, form a hypothesis to explain whether all areas in your community contain the same types and amounts of particulate matter.

Test Your Hypothesis

Make a Plan

1. As a group, agree upon your hypothesis and decide how you will test it.
2. **List** the steps you need to take to test your hypothesis. Describe exactly what you will do at each step. List your materials.
3. Prepare a data table in your Science Journal to record your observations.
4. Label your lids with the location where you decide to place them.
5. Mix the gelatin according to the directions on the box. Carefully pour a thin layer of gelatin into each lid. Use this to collect air particulate matter.
6. Read over your entire experiment to make sure that all steps are in a logical order.
7. Identify any constants, variables, and controls of the experiment.

Follow Your Plan

1. Make sure your teacher approves your plan.
2. Carry out the experiment as planned.
3. **Record** any observations that you make and complete the data table in your Science Journal.

Analyze Your Data

1. **Describe** the types of materials you collected in each lid.
2. **Calculate** the number of particles on each lid.
3. **Compare and contrast** your controls and your variables in this experiment.
4. **Graph** your results using a bar graph. Place the number of particulates on the *y*-axis and the test site location on the *x*-axis.

Conclude and Apply

1. **Determine** if the results support your hypothesis.
2. **Explain** why different sizes of particulate matter may be found at different locations.
3. **Infer** why some test-site locations showed more particulates than other sites did.

Communicating Your Data

Develop Multimedia Presentations
Give an oral presentation of your experiment on air pollution in your community to another class. **For more help, refer to the Science Skill Handbook.**

TIME SCIENCE AND HISTORY

SCIENCE CAN CHANGE THE COURSE OF HISTORY!

A biologist and writer who made people aware of the fragility of nature

In 1958, retired biologist Rachel Carson (1907–1964) received a letter from a worried friend. Several songbirds had died immediately after the pesticide DDT was sprayed over an area of woods. In the 1940s and 1950s, DDT was sprayed over large areas of land to kill insects that caused crop damage and to eliminate diseases such as malaria. DDT was considered to be a scientific miracle. The letter Carson received, however, indicated something she had long suspected—there was a downside to the miracle.

After four years of research, interviews, and analysis, Carson wrote her famous book, *Silent Spring.* In it, she stated her findings that pesticides were killing birds and fish, and poisoning human food supplies. She wrote that unless action was taken, an eerie stillness would settle over the world, a world without songbirds—a silent spring.

The publication of *Silent Spring* led to a heated debate in the United States over the use of pesticides. But it also led to a change in how people thought about the natural world. Before Carson's book, few people thought about nature and how human activities might affect Earth's organisms. Thanks to *Silent Spring,* many people began to realize that Earth and the organisms living on Earth are closely connected.

Carson's findings were verified and DDT was banned. Many species of birds owe their continuing existence to her efforts. The most famous example is the national symbol of the United States—the bald eagle. DDT caused the bald eagles' eggs to weaken and break, bringing the species close to extinction. Since the ban on DDT, bald eagles have been making a comeback.

Make Posters Research an environmental issue you are concerned about. Make posters to educate others in your school or community about the issue. Look for quotes you'd like to use to help illustrate your poster. You might find some in Carson's book.

For more information, visit earth.msscience.com/time

Reviewing Main Ideas

Section 1 Water Pollution

1. Water pollution comes from industrial discharge; runoff of pesticides and fertilizers from lawns and farms; and sewage.
2. Sewage from homes and businesses is purified before it is released back into a stream or river.
3. National and international cooperation is necessary to reduce water pollution. In the United States, the 1990 Clean Water Act set up standards for sewage and wastewater-treatment facilities and for nonpoint sources.
4. Conserving water in your daily activities can help reduce water pollution.

Section 2 Air Pollution

1. Exhaust from vehicles pollutes the air. Other sources of air pollution include power plants, fires and volcanoes.
2. Natural conditions, such as landforms and temperature inversions, can affect the ability of the atmosphere to disperse air pollutants.
3. Polluted air can affect human health. Breathing particles, ozone, and acid rain can damage your lungs. Carbon monoxide can replace oxygen in your blood.
4. Air pollutants don't have boundaries. They drift between states and countries. National and international cooperation is necessary to reduce the problem.

Visualizing Main Ideas

Copy and complete the following concept map on types of pollution.

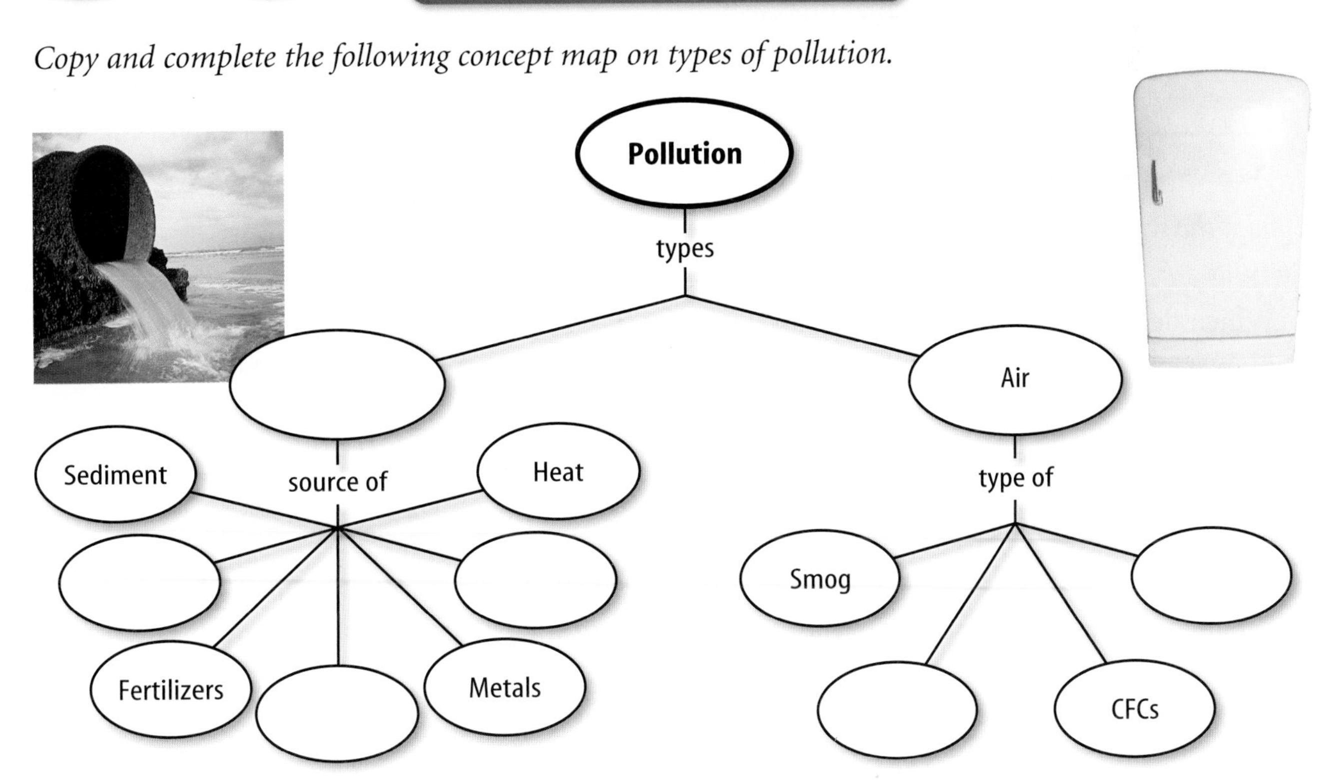

chapter 21 Review

Using Vocabulary

acid p. 611	pesticide p. 601
acid rain p. 611	pH scale p. 611
base p. 611	photochemical smog p. 610
carbon monoxide p. 612	point source pollution p. 600
fertilizer p. 601	scrubber p. 614
nonpoint source pollution p. 600	sewage p. 602
particulate matter p. 613	

Fill in the blanks with the correct word or words.

1. A type of pollution that forms when nitrogen and carbon compounds are exposed to sunlight is called ________.
2. ________ can be controlled or treated because it enters water from a specific location.
3. The water that goes into drains, called ________ contains wastes and detergent.
4. Substance with a pH higher than 7 is called a(n)________.
5. ________ are fine solids such as dust, ash and soot.

Checking Concepts

Choose the word or phrase that best answers the question.

6. Which describes warm air over cool air?
   A) inversion
   B) CFCs
   C) pollution
   D) scrubber
7. Which of the following describes substances with a low pH?
   A) neutral
   B) acidic
   C) dense
   D) basic
8. What combines with moisture in the air to form acid rain?
   A) ozone
   B) sulfur oxides
   C) lead
   D) oxygen
9. Which of the following is a nonpoint source?
   A) runoff from a golf course
   B) discharge from a sewage treatment plant
   C) wastewater from industry
   D) discharge from a ditch into a river
10. What is the largest source of water pollution in the United States?
    A) sediment
    B) metals
    C) heat
    D) gasoline
11. What is the pH of acid rain?
    A) less than 5.6
    B) between 5.6 and 7.0
    C) greater than 7.0
    D) greater than 9.5
12. What kind of pollution are airborne solids that range in size from large grains to microscopic?
    A) pH
    B) ozone
    C) particulate matter
    D) acid rain
13. Which of the following causes algae to grow?
    A) pesticides
    B) sediment
    C) metals
    D) fertilizers

Use the table below to answer question 14.

Phosphorus Entering Lake Erie

Year	Metric Tons
1976	15,000
1982	12,000
1988	8,000
1995	7,000

14. Which of the following is the best estimate of the decrease in phosphorous entering Lake Erie from 1976 to 1995?
    A) 10%
    B) 20%
    C) 75%
    D) 50%

Science Online earth.msscience.com/vocabulary_puzzlemaker

Thinking Critically

15. **Describe** how cities with smog problems might lessen the dangers to people who live and work in the cities.
16. **List** some ways to control nonpoint pollution sources.
17. **Recognize Cause and Effect** Why is it important to place stricter limits on pollutants from cars and factories as the U.S. population grows and people drive more?
18. **Draw Conlusions** Pollution occurs when heated water is released into a nearby body of water. What effects does this type of pollution have on organisms living in the water?
19. **Infer** how a community in a desert might cope with water-supply problems.
20. **Concept Map** Copy and complete this concept map about sewage treatment.

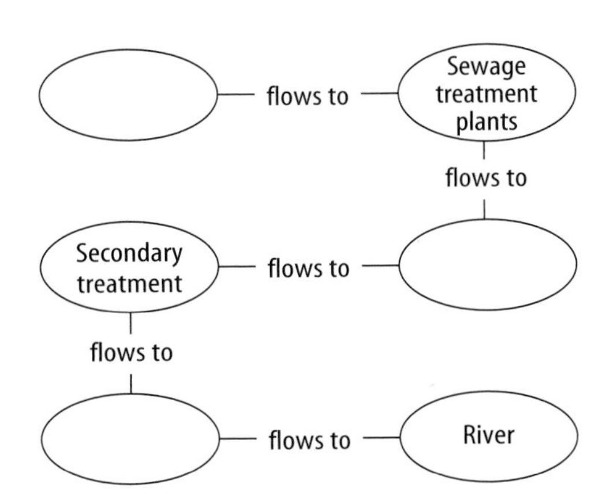

21. **Communicate** Explain what you personally can do to reduce air pollution.
22. **List** three ways air pollution can affect your health.
23. **Recognize Cause and Effect** Your community is downstream from a large metropolitan area. Explain why it might cost more money to produce clean drinking water for your community than a similar community upstream.

Performance Activities

24. **Design and Perform an Experiment** to test the effects of acid rain on vegetation. You might choose to use different types of vegetation as your variable and use acidity level as your constant or you might want to use the pH of the solution as your variable and use the type of vegetation as your constant. Remember to test one variable at a time.
25. **Letter to the Editor** Survey your town for evidence of air and water pollution. Write a letter to the editor of your local newspaper communicating what you have observed. Include suggestions for reducing pollution.

Applying Math

Use the figure below to answer questions 26 and 27.

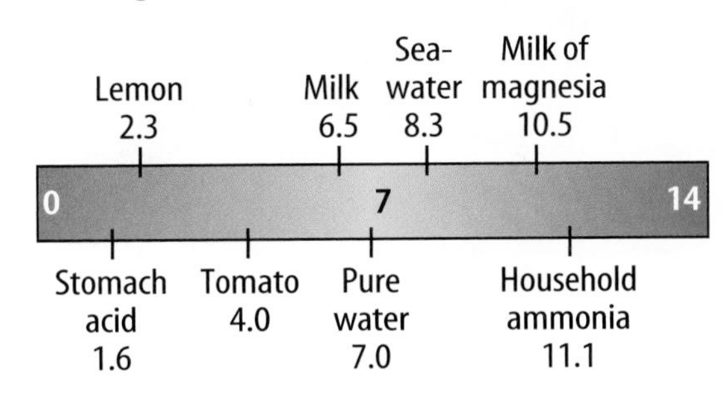

26. **pH Scale** A decrease of one pH unit on the pH scale means that the solution is ten times more acidic. A decrease of two means the solution is 100 times more acidic. How much more acidic is tomato juice than pure water?
27. **Estimate** How much more acidic is milk than milk of magnesia?
28. **Travel** Your family car travels 20 miles on one gallon of gas. You visit your friend who lives two miles away three times a week. Calculate how much gas you would save in five weeks if you walked or rode your bike to your friend's house. Estimate how much gas you would save in one year.

Part 1 Multiple Choice

Record your answers on the answer sheet provided by your teacher or on a sheet of paper.

1. This term describes pollution from industrial outfalls or ditches.
   - **A.** point source pollution
   - **B.** nonpoint source pollution
   - **C.** runoff pollution
   - **D.** chemical pollution

Use the photo below to answer question 2.

2. Which type of pollution might come from this site?
   - **A.** point source pollution
   - **B.** nonpoint source pollution
   - **C.** acid rain
   - **D.** smog

3. Which is the largest source of water pollution in the United States?
   - **A.** sewage
   - **B.** metals
   - **C.** oil
   - **D.** sediment

4. Which substance in fertilizers can cause an overgrowth of algae?
   - **A.** carbon
   - **B.** nitrogen
   - **C.** hydrogen
   - **D.** oil

5. Which areas have the most smog?
   - **A.** urban areas
   - **B.** mountains
   - **C.** deserts
   - **D.** saltwater marshes

Use the graph below to answer questions 6 and 7.

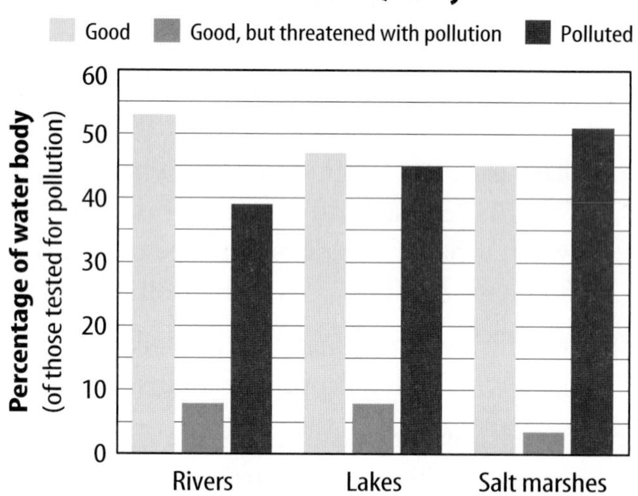

6. What is the estimated percentage of rivers with a pollution rating of good?
   - **A.** 17%
   - **B.** 38%
   - **C.** 45%
   - **D.** 53%

7. What percentage of lakes are polluted?
   - **A.** 8%
   - **B.** 45%
   - **C.** 53%
   - **D.** 50%

8. What might happen if heated water is released to a body of water?
   - **A.** oxygen levels decrease
   - **B.** oxygen levels increase
   - **C.** pH decreases
   - **D.** pH increases

9. Acid rain has what affect on the pH of natural lakes and streams?
   - **A.** it increases the pH
   - **B.** it decreases the pH
   - **C.** it stabilizes the pH
   - **D.** it doesn't affect the pH

Part 2 Short Response/Grid In

Record your answers on the answer sheet provided by your teacher or on a sheet of paper.

10. List three pollutants that can be removed by a sewage treatment plant.

Use the illustration below to answer questions 11 and 12.

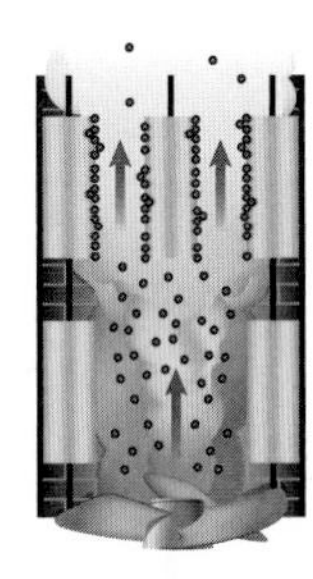

11. What type of pollution is removed by this device?

12. Explain how this device removes pollutants from smoke.

13. Why is it better for the environment for countries to work together to reduce pollution?

14. What type of damage does acid rain cause?

15. What is the ozone layer and how is it being damaged?

16. How do particulates get into the air and what type of damage do they cause to your lungs?

17. What pollution sources contribute to smog?

18. What pollutant was addressed by the Montreal Protocol?

19. What device can lower sulfur dioxide emissions from coal-burning power plants?

Test-Taking Tip

Diagrams Study a diagram carefully, being sure to read all labels and captions.

Part 3 Open Ended

Record your answers on a sheet of paper.

Use the table below to answer question 20.

Source of Water Pollution	Example of Water Pollution from Source
Agriculture and lawn care	pesticide
Sewage	human waste
Highway runoff	oil and gasoline
Factories using water to cool equipment	heat

20. List one method that might be used to reduce each example of water pollution.

21. In a year, a family can save 2,400 L of water by fixing a leaky faucet, 82,000 L by replacing an old toilet, and 7,000 L by turning the water off while brushing their teeth. How much water would be saved if 50 families did these water conservation efforts for 5 years? Show your results in a bar graph showing the total amount saved for each conservation measure.

22. The United States Department of Commerce Census Bureau reported that construction rates raised 1.6 percent during the first five months of 2003. How might this rate effect water pollution? What type of damages might be caused?

23. How did the old gasoline storage tanks at gas stations contribute to water pollution? How is this problem being solved?

24. How did Lake Erie become polluted? What measures have been taken to reduce the amount of pollution?

25. Explain how sunny days and other conditions in the atmosphere can worsen air pollution and smog.

unit 7 Astronomy

How Are Thunderstorms & Neutron Stars Connected?

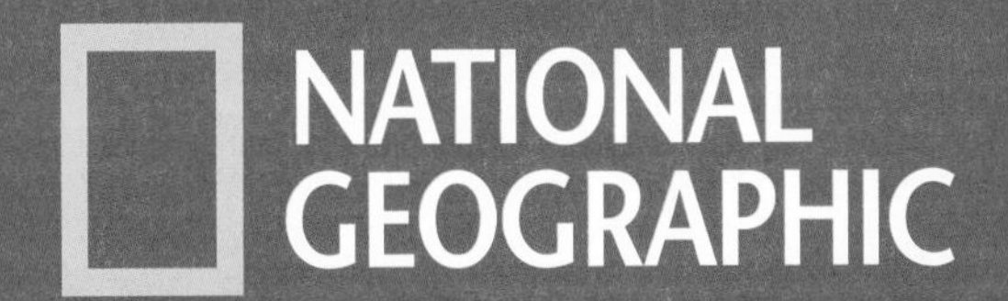

In 1931, an engineer built an antenna to study thunderstorm static that was interfering with radio communication. The antenna did detect static from storms, but it also picked up something else: radio signals coming from beyond our solar system. That discovery marked the birth of radio astronomy. By the 1960s, radio astronomy was thriving. In 1967, astronomer Jocelyn Bell Burnell detected a peculiar series of radio pulses coming from far out in space. At first, she and her colleagues theorized that the signals might be a message from a distant civilization. Soon, however, scientists determined that the signals must be coming from something called a neutron star (below)—a rapidly spinning star that gives off a radio beam from its magnetic pole, similar to the rotating beam of a lighthouse.

Axis

Magnetic pole

Direction of spin

Radio beam

unit projects

Visit earth.msscience.com/unit_project to find project ideas and resources. Projects include:

- **Career** Discover what is required to have a career in space exploration or space technology. Design an interview and present your findings to the class.
- **Technology** Research interesting information describing the planet of your choice in a newspaper article.
- **Model** After a thorough investigation of an existing neutron star, design a creative mobile to display your research information.

WebQuest *Sun and Energy* uses online resources to explore the composition of the Sun and the possibilities of harnessing its energy for everyday use.

chapter 22

Exploring Space

The BIG Idea

Even though scientists have learned a great deal about the Moon and planets from telescopes, they want to learn more by sending spacecraft.

SECTION 1
Radiation from Space

Main Idea The light and other energy leaving a star are forms of radiation.

SECTION 2
Early Space Missions

Main Idea The space age began in 1957 when the former Soviet Union used a rocket to send *Sputnik I* into space.

SECTION 3
Current and Future Space Missions

Main Idea Many people have benefited from research done for space programs.

Fiery end or new beginning?

These colorful streamers are the remains of a star that exploded in a nearby galaxy thousands of years ago. Eventually, new stars and planets may form from this material, just as our Sun and planets formed from similar debris billions of years ago.

Science Journal Do you think space exploration is worth the risk and expense? Explain why.

Start-Up Activities

An Astronomer's View

You might think exploring space with a telescope is easy because the stars seem so bright and space is dark. But starlight passing through Earth's atmosphere, and differences in temperature and density of the atmosphere can distort images.

1. Cut off a piece of clear plastic wrap about 15 cm long.
2. Place an opened book in front of you and observe the clarity of the text.
3. Hold the piece of plastic wrap close to your eyes, keeping it taut using both hands.
4. Look at the same text through the plastic wrap.
5. Fold the plastic wrap in half and look at the text again through both layers.
6. **Think Critically** Write a paragraph in your Science Journal comparing reading text through plastic wrap to an astronomer viewing stars through Earth's atmosphere. Predict what might occur if you increased the number of layers.

Preview this chapter's content and activities at earth.msscience.com

FOLDABLES™ Study Organizer

Exploring Space Make the following Foldable to help identify what you already know, what you want to know, and what you learned about exploring space.

STEP 1 **Fold** a vertical sheet of paper from side to side with the front edge about 1.25 cm shorter than the back.

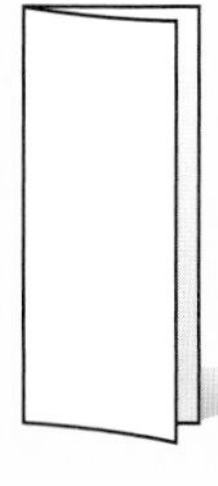

STEP 2 **Turn** lengthwise and **fold** into thirds.

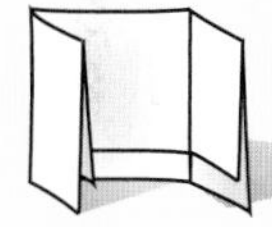

STEP 3 **Unfold and cut** only the top layer along both folds to make three tabs. **Label** each tab.

Identify Questions Before you read the chapter, write what you already know about exploring space under the left tab of your Foldable, and write questions about what you'd like to know under the center tab. After you read the chapter, list what you learned under the right tab.

Get Ready to Read

Make Connections

1 Learn It! Make connections between what you read and what you already know. Connections can be based on personal experiences (text-to-self), what you have read before (text-to-text), or events in other places (text-to-world).

As you read, ask connecting questions. Are you reminded of a personal experience? Have you read about the topic before? Did you think of a person, a place, or an event in another part of the world?

2 Practice It! Read the excerpt below and make connections to your own knowledge and experience.

Text-to-text:
What is electromagnetic radiation? Where have you read about electromagnetic radiation before?

Text-to-self:
What type of radiation are you experiencing when you peer down a microscope?

Text-to-world:
How many ways can you recall in which people use electromagnetic radiation?

Electromagnetic radiation is everywhere around you. When you turn on the radio, peer down a microscope, or have an X ray taken—you're using various forms of electromagnetic radiation.

—from page 628

3 Apply It! As you read this chapter, choose five words or phrases that make a connection to something you already know.

Reading Tip

Make connections with memorable events, places, or people in your life. The better the connection, the more likely you will be to remember it.

Target Your Reading

Use this to focus on the main ideas as you read the chapter.

1. **Before you read** the chapter, respond to the statements below on your worksheet or on a numbered sheet of paper.
   - Write an **A** if you **agree** with the statement.
   - Write a **D** if you **disagree** with the statement.

2. **After you read** the chapter, look back to this page to see if you've changed your mind about any of the statements.
   - If any of your answers changed, explain why.
   - Change any false statements into true statements.
   - Use your revised statements as a study guide.

Before You Read A or D		Statement	After You Read A or D
	1	Light is a form of radiation.	
	2	Sound waves can travel through empty space.	
	3	All electromagnetic waves travel at the same speed in a vacuum.	
	4	It sometimes takes millions of years for starlight to reach Earth.	
	5	A satellite is any object that revolves around another object.	
	6	In 1961, John Glenn became the first U.S. citizen in space.	
	7	In 1969, *Apollo 13* landed on the Moon's surface.	
	8	Edwin Aldrin was the first human to set foot on the Moon.	
	9	The *International Space Station* transports astronauts and materials to and from space.	
	10	After a mission, the space shuttle glides back to Earth and lands like an airplane.	

Science Online
Print out a worksheet of this page at earth.msscience.com

section 1

Radiation from Space

as you read

What You'll Learn

- **Explain** the electromagnetic spectrum.
- **Identify** the differences between refracting and reflecting telescopes.
- **Recognize** the differences between optical and radio telescopes.

Why It's Important

Learning about space can help us better understand our own world.

Review Vocabulary

telescope: an instrument that can magnify the size of distant objects

New Vocabulary

- electromagnetic spectrum
- refracting telescope
- reflecting telescope
- observatory
- radio telescope

Electromagnetic Waves

As you have read, we have begun to explore our solar system and beyond. With the help of telescopes like the *Hubble,* we can see far into space, but if you've ever thought of racing toward distant parts of the universe, think again. Even at the speed of light it would take many years to reach even the nearest stars.

Light from the Past When you look at a star, the light that you see left the star many years ago. Although light travels fast, distances between objects in space are so great that it sometimes takes millions of years for the light to reach Earth.

The light and other energy leaving a star are forms of radiation. Radiation is energy that is transmitted from one place to another by electromagnetic waves. Because of the electric and magnetic properties of this radiation, it's called electromagnetic radiation. Electromagnetic waves carry energy through empty space and through matter.

Electromagnetic radiation is everywhere around you. When you turn on the radio, peer down a microscope, or have an X ray taken—you're using various forms of electromagnetic radiation.

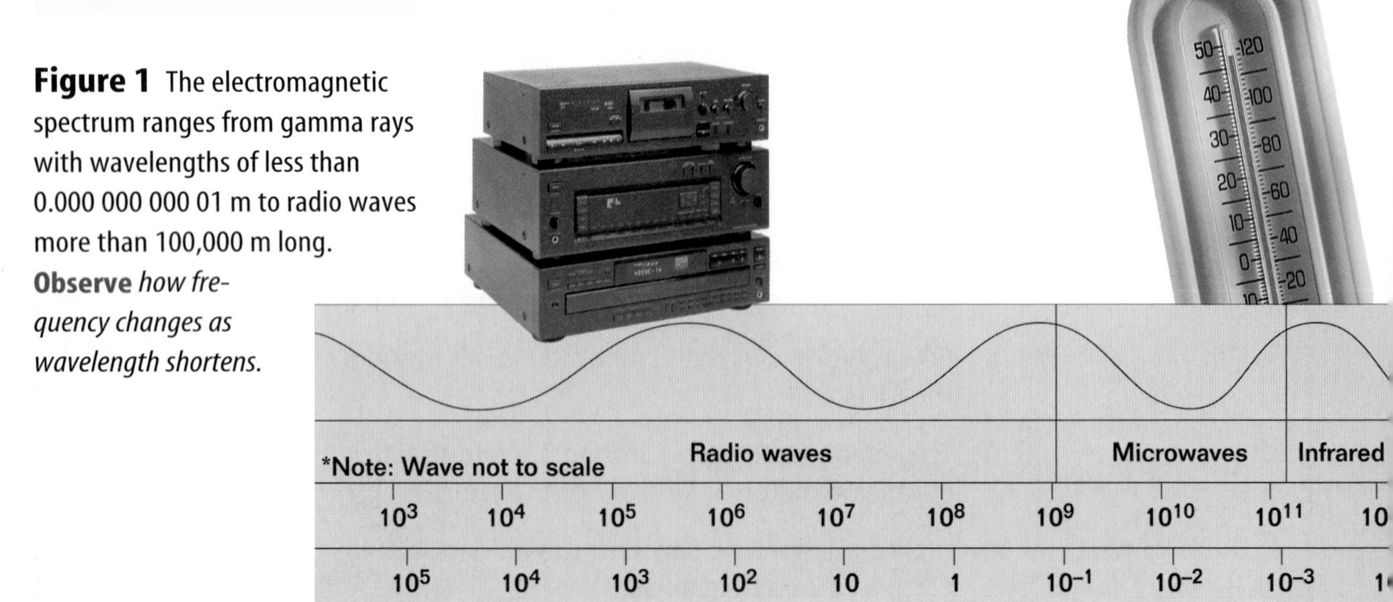

Figure 1 The electromagnetic spectrum ranges from gamma rays with wavelengths of less than 0.000 000 000 01 m to radio waves more than 100,000 m long. **Observe** *how frequency changes as wavelength shortens.*

Electromagnetic Radiation Sound waves, which are a type of mechanical wave, can't travel through empty space. How, then, do we hear the voices of the astronauts while they're in space? When astronauts speak into a microphone, the sound waves are converted into electromagnetic waves called radio waves. The radio waves travel through space and through Earth's atmosphere. They're then converted back into sound waves by electronic equipment and audio speakers.

Radio waves and visible light from the Sun are just two types of electromagnetic radiation. Other types include gamma rays, X rays, ultraviolet waves, infrared waves, and microwaves. **Figure 1** shows these forms of electromagnetic radiation arranged according to their wavelengths. This arrangement of electromagnetic radiation is called the **electromagnetic spectrum.** Forms of electromagnetic radiation also differ in their frequencies. Frequency is the number of wave crests that pass a given point per unit of time. The shorter the wavelength is, the higher the frequency, as shown in **Figure 1.**

Ultraviolet Light Many newspapers include an ultraviolet (UV) index to urge people to minimize their exposure to the Sun. Compare the wavelengths and frequencies of red and violet light, shown below in **Figure 1.** Infer what properties of UV light cause damage to tissues of organisms.

Speed of Light Although the various electromagnetic waves differ in their wavelengths, they all travel at 300,000 km/s in a vacuum. This is called the speed of light. Visible light and other forms of electromagnetic radiation travel at this incredible speed, but the universe is so large that it takes millions of years for the light from some stars to reach Earth.

When electromagnetic radiation from stars and other objects reaches Earth, scientists use it to learn about its source. One tool for studying such electromagnetic radiation is a telescope.

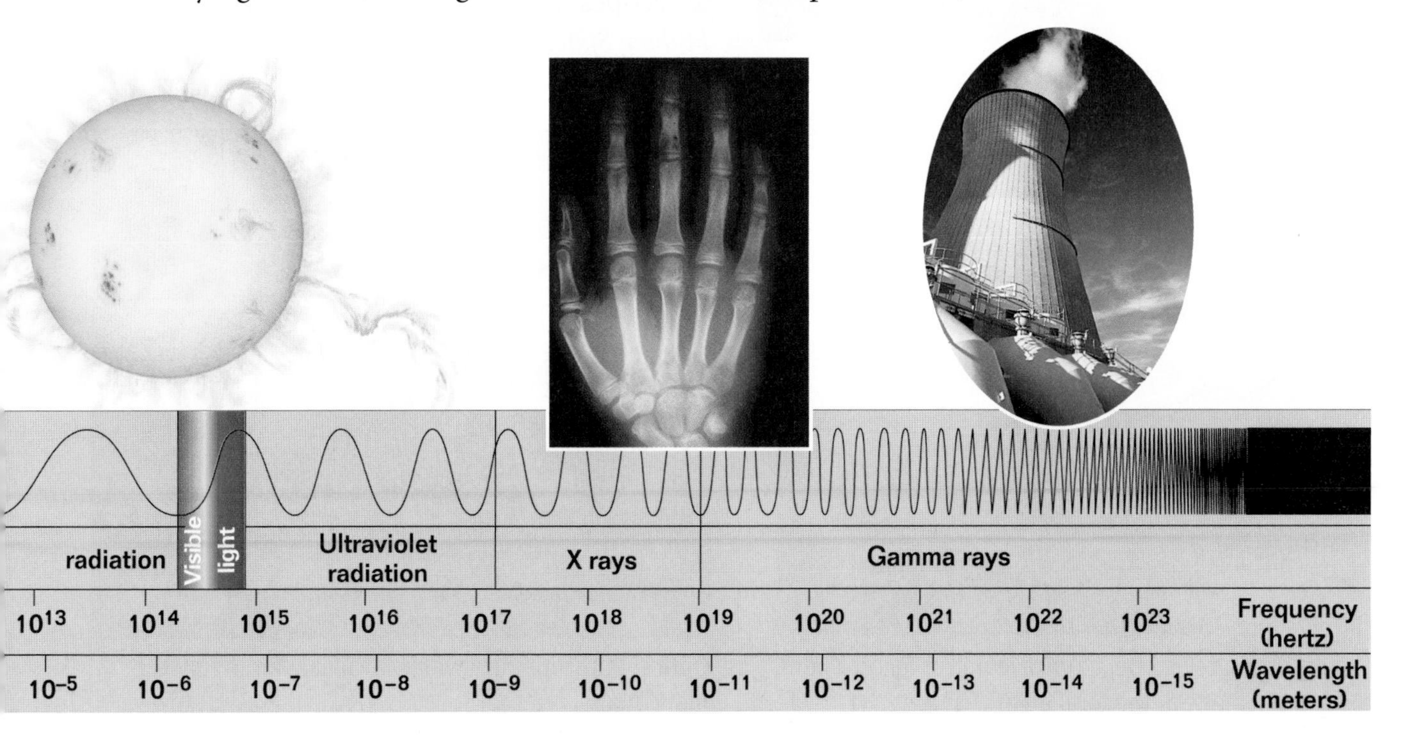

Optical Telescopes

Optical telescopes use light, which is a form of electromagnetic radiation, to produce magnified images of objects. Light is collected by an objective lens or mirror, which then forms an image at the focal point of the telescope. The focal point is where light that is bent by the lens or reflected by the mirror comes together to form an image. The eyepiece lens then magnifies the image. The two types of optical telescopes are shown in **Figure 2.**

A **refracting telescope** uses convex lenses, which are curved outward like the surface of a ball. Light from an object passes through a convex objective lens and is bent to form an image at the focal point. The eyepiece magnifies the image.

A **reflecting telescope** uses a curved mirror to direct light. Light from the object being viewed passes through the open end of a reflecting telescope. This light strikes a concave mirror, which is curved inward like a bowl and located at the base of the telescope. The light is reflected off the interior surface of the bowl to the focal point where it forms an image. Sometimes, a smaller mirror is used to reflect light into the eyepiece lens, where it is magnified for viewing.

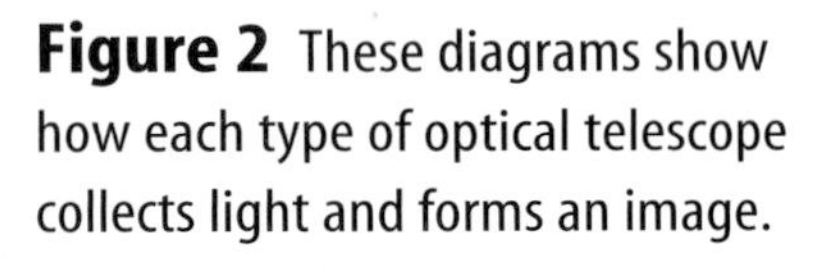

Figure 2 These diagrams show how each type of optical telescope collects light and forms an image.

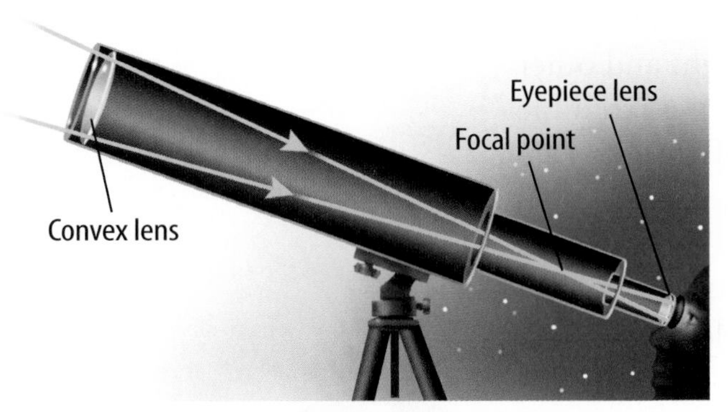

In a refracting telescope, a convex lens focuses light to form an image at the focal point.

Using Optical Telescopes

Most optical telescopes used by professional astronomers are housed in buildings called **observatories.** Observatories often have dome-shaped roofs that can be opened up for viewing. However, not all telescopes are located in observatories. The *Hubble Space Telescope* is an example.

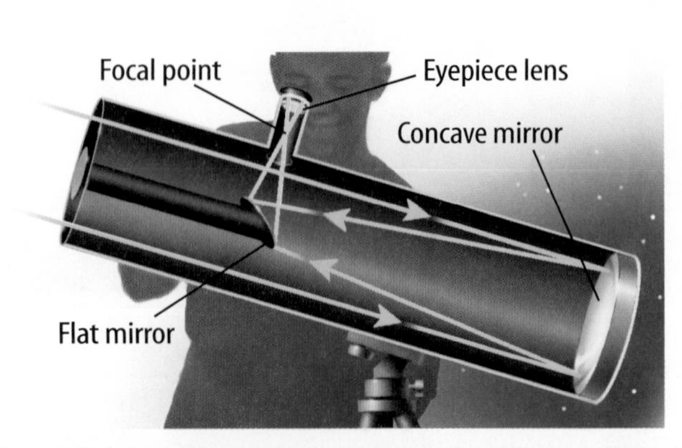

In a reflecting telescope, a concave mirror focuses light to form an image at the focal point.

Optical telescopes are widely available for use by individuals.

Hubble Space Telescope The *Hubble Space Telescope* was launched in 1990 by the space shuttle *Discovery.* Because *Hubble* is located outside Earth's atmosphere, which absorbs and distorts some of the energy received from space, it should have produced clear images. However, when the largest mirror of this reflecting telescope was shaped, a mistake was made. As a result, images obtained by the telescope were not as clear as expected. In December 1993, a team of astronauts repaired the *Hubble Space Telescope* by installing a set of small mirrors designed to correct images obtained by the faulty mirror. Two more missions to service *Hubble* were carried out in 1997 and 1999, shown in **Figure 3.** Among the objects viewed by *Hubble* after it was repaired in 1999 was a large cluster of galaxies known as Abell 2218.

Reading Check *Why is* **Hubble** *located outside Earth's atmosphere?*

Figure 3 The *Hubble Space Telescope* was serviced at the end of 1999. Astronauts replaced devices on *Hubble* that are used to stabilize the telescope.

Observing Effects of Light Pollution

Procedure

1. Obtain a **cardboard tube** from an empty roll of paper towels.
2. Go outside on a clear night about two hours after sunset. Look through the cardboard tube at a specific constellation decided upon ahead of time.
3. Count the number of stars you can see without moving the observing tube. Repeat this three times.
4. Calculate the average number of observable stars at your location.

Analysis

1. Compare and contrast the number of stars visible from other students' homes.
2. Explain the causes and effects of your observations.

Large Reflecting Telescopes Since the early 1600s, when the Italian scientist Galileo Galilei first turned a telescope toward the stars, people have been searching for better ways to study what lies beyond Earth's atmosphere. For example, the twin Keck reflecting telescopes, shown in **Figure 4,** have segmented mirrors 10 m wide. Until 2000, these mirrors were the largest reflectors ever used. To cope with the difficulty of building such huge mirrors, the Keck telescope mirrors are built out of many small mirrors that are pieced together. In 2000, the European Southern Observatory's telescope, in Chile, consisted of four 8.2-m reflectors, making it the largest optical telescope in use.

Reading Check *About how long have people been using telescopes?*

Active and Adaptive Optics The most recent innovations in optical telescopes involve active and adaptive optics. With active optics, a computer corrects for changes in temperature, mirror distortions, and bad viewing conditions. Adaptive optics is even more ambitious. Adaptive optics uses a laser to probe the atmosphere and relay information to a computer about air turbulence. The computer then adjusts the telescope's mirror thousands of times per second, which lessens the effects of atmospheric turbulence. Telescope images are clearer when corrections for air turbulence, temperature changes, and mirror-shape changes are made.

Figure 4 The twin Keck telescopes on Mauna Kea in Hawaii can be used together, more than doubling their ability to distinguish objects. A Keck reflector is shown in the inset photo. Currently, plans include using these telescopes, along with four others to obtain images that will help answer questions about the origin of planetary systems.

Radio Telescopes

As shown in the spectrum illustrated in **Figure 1,** stars and other objects radiate electromagnetic energy of various types. Radio waves are an example of long-wavelength energy in the electromagnetic spectrum. A **radio telescope,** such as the one shown in **Figure 5,** is used to study radio waves traveling through space. Unlike visible light, radio waves pass freely through Earth's atmosphere. Because of this, radio telescopes are useful 24 hours per day under most weather conditions.

Radio waves reaching Earth's surface strike the large, concave dish of a radio telescope. This dish reflects the waves to a focal point where a receiver is located. The information allows scientists to detect objects in space, to map the universe, and to search for signs of intelligent life on other planets.

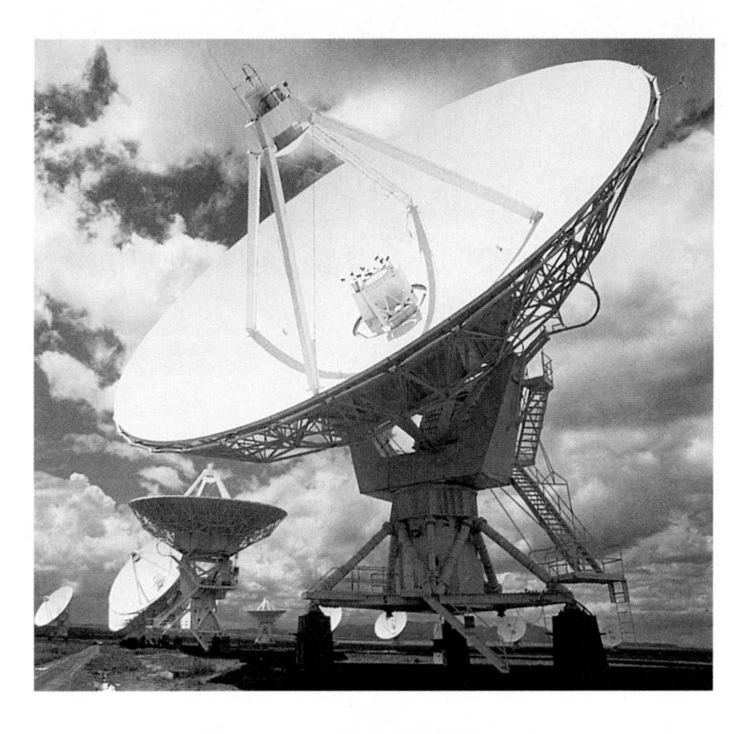

Figure 5 This radio telescope is used to study radio waves traveling through space.

section 1 review

Summary

Electromagnetic Waves

- Light is a form of electromagnetic radiation.
- Electromagnetic radiation includes radio waves, microwaves, X rays, gamma rays, and infrared and ultraviolet radiation.
- Light travels at 300,000 km/s in a vacuum.

Optical Telescopes

- A refracting telescope uses lenses to collect, focus, and view light.
- A reflecting telescope uses a mirror to collect and focus light and a lens to view the image.
- Computers and lasers are used to reduce problems caused by looking through Earth's atmosphere.
- These telescopes are housed in domed buildings called observatories.
- Placing a telescope in space avoids problems caused by Earth's atmosphere.

Radio Telescopes

- Radio telescopes collect and measure radio waves coming from stars and other objects.

Self Check

1. **Identify** one advantage of radio telescopes over optical telescopes.
2. **Infer** If red light has a longer wavelength than blue light, which has a greater frequency?
3. **Explain** the difference between sound waves and radio waves.
4. **Describe** how adaptive optics in a telescope help solve problems caused by atmospheric turbulence.
5. **Think Critically** It takes light from the closest star to Earth (other than the Sun) about four years to reach Earth. If intelligent life were on a planet circling that star, how long would it take for scientists on Earth to send them a radio transmission and for the scientists to receive their reply?

Applying Math

6. **Calculate** how long it takes for a radio signal to reach the Moon, which is about 380,000 km away.
7. **Use Numbers** If an X ray has a frequency of 10^{18} hertz and a gamma ray has a frequency of 10^{21} hertz, how many times greater is the frequency of the gamma ray?

Building a Reflecting Telescope

Nearly four hundred years ago, Galileo Galilei saw what no human had ever seen. Using the telescope he built, he saw moons around Jupiter, details of lunar craters, and sunspots. What was it like to make these discoveries? Find out as you make your own reflecting telescope.

Real-World Question

How do you construct a reflecting telescope?

Goals

- **Construct** a reflecting telescope.
- **Observe** magnified images using the telescope and different magnifying lenses.

Materials

flat mirror
shaving or cosmetic mirror (a curved, concave mirror)
magnifying lenses of different magnifications (3–4)

Safety Precautions

WARNING: *Never observe the Sun directly or with mirrors.*

Procedure

1. Position the cosmetic mirror so that you can see the reflection of the object you want to look at. Choose an object such as the Moon, a planet, or an artificial light source.
2. Place the flat mirror so that it is facing the cosmetic mirror.
3. Adjust the position of the flat mirror until you can see the reflection of the object in it.
4. View the image of the object in the flat mirror with one of your magnifying lenses. Observe how the lens magnifies the image.
5. Use your other magnifying lenses to view the image of the object in the flat mirror. Observe how the different lenses change the image of the object.

Analyze Your Data

1. **Describe** how the image changed when you used different magnifying lenses.
2. **Identify** the part or parts of your telescope that reflected the light of the image.
3. **Identify** the parts of your telescope that magnified the image.

Conclude and Apply

1. **Explain** how the three parts of your telescope worked to reflect and magnify the light of the object.
2. **Infer** how the materials you used would have differed if you had constructed a refracting instead of a reflecting telescope.

Communicating Your Data

Write an instructional pamphlet for amateur astronomers about how to construct a reflecting telescope.

section 2

Early Space Missions

The First Missions into Space

You're offered a choice—front-row-center seats for this weekend's rock concert, or a copy of the video when it's released. Wouldn't you rather be right next to the action? Astronomers feel the same way about space. Even though telescopes have taught them a great deal about the Moon and planets, they want to learn more by going to those places or by sending spacecraft where humans can't go.

Rockets The space program would not have gotten far off the ground using ordinary airplane engines. To break free of gravity and enter Earth's orbit, spacecraft must travel at speeds greater than 11 km/s. The space shuttle and several other spacecrafts are equipped with special engines that carry their own fuel. **Rockets,** like the one in **Figure 6,** are engines that have everything they need for the burning of fuel. They don't even require air to carry out the process. Therefore, they can work in space, which has no air. The simplest rocket engine is made of a burning chamber and a nozzle. More complex rockets have more than one burning chamber.

Rocket Types Two types of rockets are distinguished by the type of fuel they use. One type is the liquid-propellant rocket and the other is the solid-propellant rocket. Solid-propellant rockets are generally simpler but they can't be shut down after they are ignited. Liquid-propellant rockets can be shut down after they are ignited and can be restarted. For this reason, liquid-propellant rockets are preferred for use in long-term space missions. Scientists on Earth can send signals that start and stop the spacecraft's engines whenever they want to modify its course or adjust its orbit. Liquid propellants successfully powered many space probes, including the two *Voyagers* and *Galileo.*

as you read

What You'll Learn

- **Compare and contrast** natural and artificial satellites.
- **Identify** the differences between artificial satellites and space probes.
- **Explain** the history of the race to the Moon.

Why It's Important

Early missions that sent objects and people into space began a new era of human exploration.

Review Vocabulary
thrust: the force that propels an aircraft or missile

New Vocabulary
- rocket
- satellite
- orbit
- space probe
- Project Mercury
- Project Gemini
- Project Apollo

Figure 6 Rockets differ according to the types of fuel used to launch them. Liquid oxygen is used often to support combustion.

Figure 7 The space shuttle uses both liquid and solid fuels. Here the red liquid fuel tank is visible behind a white, solid rocket booster.

Rocket Launching Solid-propellant rockets use a rubberlike fuel that contains its own oxidizer. The burning chamber of a rocket is a tube that has a nozzle at one end. As the solid propellant burns, hot gases exert pressure on all inner surfaces of the tube. The tube pushes back on the gas except at the nozzle where hot gases escape. Thrust builds up and pushes the rocket forward.

Liquid-propellant rockets use a liquid fuel and an oxidizer, such as liquid oxygen, stored in separate tanks. To ignite the rocket, the oxidizer is mixed with the liquid fuel in the burning chamber. As the mixture burns, forces are exerted and the rocket is propelled forward. **Figure 7** shows the space shuttle, with both types of rockets, being launched.

Applying Math — Make and Use Graphs

DRAWING BY NUMBERS Points are defined by two coordinates, called an ordered pair. To plot an ordered pair, find the first number on the horizontal x-axis and the second on the vertical y-axis. The point is placed where these two coordinates intersect. Line segments are drawn to connect points. Draw a symmetrical house by using an x-y grid and these coordinates: (1,1), (5,1), (5,4), (3,6), (1,4)

Solution

1 *On a piece of graph paper, label and number the x-axis 0 to 6 and the y-axis 0 to 6, as shown here.*

2 *Plot the above points and connect them with straight line segments, as shown here.*

Section	Points
1	(1, −8) (3,−13) (6, −21) (9,−21) (9,−17) (8,−15) (8,−12) (6,−8) (5,−4) (4,−3) (4,−1) (5,1) (6,3) (8,3) (9,4) (9,7) (7,11) (4,14) (4,22) (−9,22) (−9,10) (−10,5) (−11,−1) (−11,−7) (−9,−8) (−8,−7) (−8,−1) (−6,3) (−6,−3) (−6,−9) (−7,−20) (−8,−21) (−4,−21) (−4,−18) (−3,−14) (−1,−8)
2	(0,11) (2,13) (2,17) (0,19) (−4,19) (−6,17) (−6,13) (−4,11)
3	(−4,9) (1,9) (1,5) (−1,5) (−2,6) (−4,6)

Practice Problems

1. Label and number the x-axis –12 to 10 and the y-axis –22 to 23. Draw an astronaut by plotting and connecting the points in each section. Do not draw segments to connect points in different sections.
2. Make your own drawing on graph paper and write its coordinates as ordered pairs. Then give it to a classmate to solve.

Science Online — For more practice, visit earth.msscience.com/math_practice

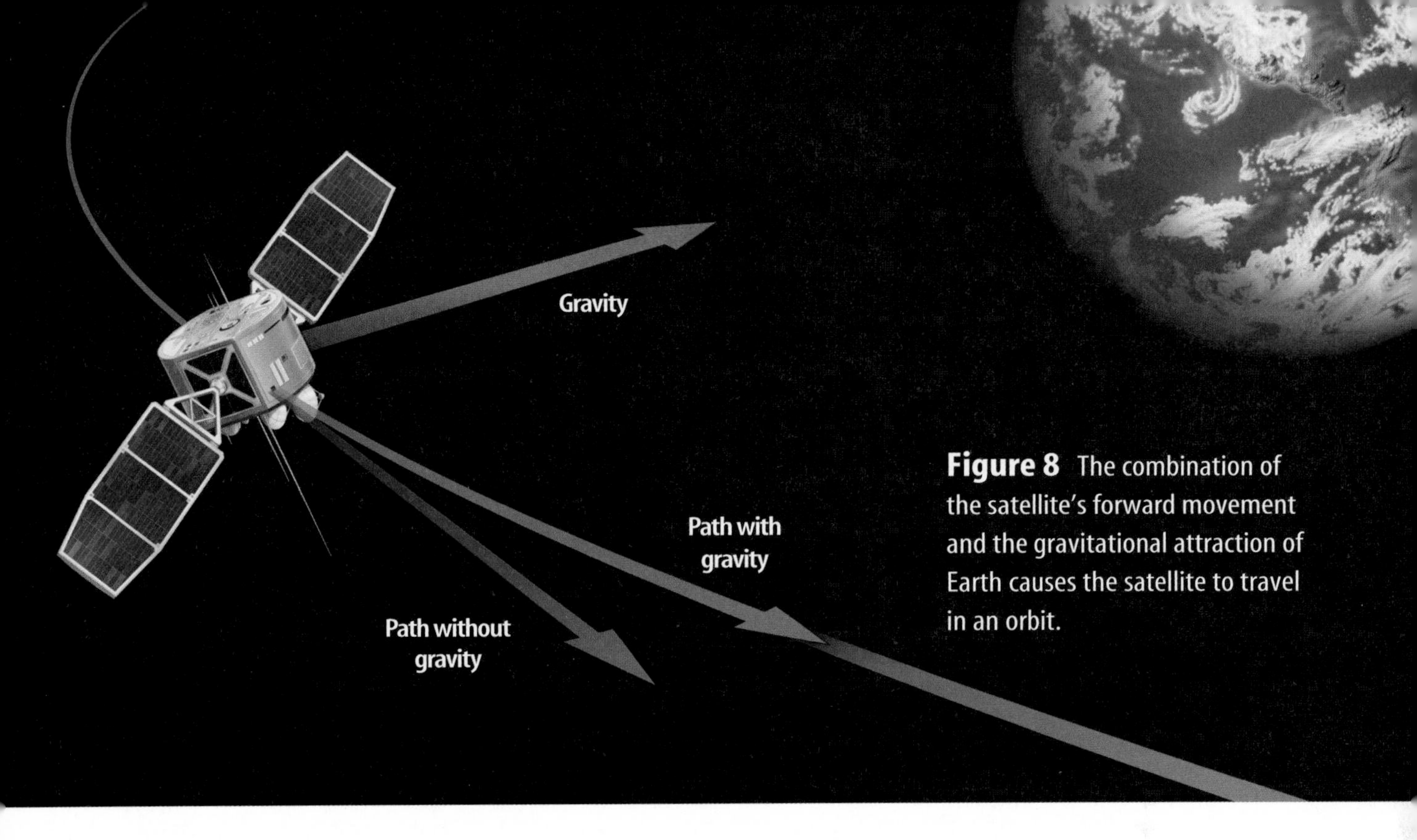

Figure 8 The combination of the satellite's forward movement and the gravitational attraction of Earth causes the satellite to travel in an orbit.

Satellites The space age began in 1957 when the former Soviet Union used a rocket to send *Sputnik I* into space. *Sputnik I* was the first artificial satellite. A **satellite** is any object that revolves around another object. When an object enters space, it travels in a straight line unless a force, such as gravity, makes it turn. Earth's gravity pulls a satellite toward Earth. The result of the satellite traveling forward while at the same time being pulled toward Earth is a curved path, called an **orbit,** around Earth. This is shown in **Figure 8.** *Sputnik I* orbited Earth for 57 days before gravity pulled it back into the atmosphere, where it burned up.

Satellite Uses *Sputnik I* was an experiment to show that artificial satellites could be made and placed into orbit around Earth.

Today, thousands of artificial satellites orbit Earth. Communication satellites transmit radio and television programs to locations around the world. Other satellites gather scientific data, like those shown in **Figure 9,** which can't be obtained from Earth, and weather satellites constantly monitor Earth's global weather patterns.

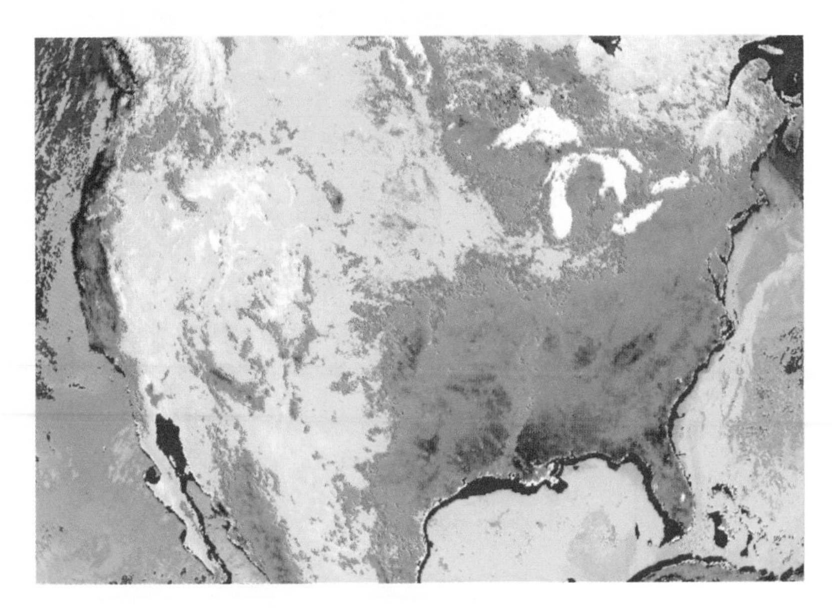

Figure 9 Data obtained from the satellite *Terra,* launched in 1999, illustrates the use of space technology to study Earth. This false-color image includes data on spring growth, sea-surface temperature, carbon monoxide concentrations, and reflected sunlight, among others.

Astronomy Astronomers today have more choices than ever before. Some still use optical telescopes to study stars and galaxies. Others explore the universe using the radio, X-ray, infrared, or even gamma-ray regions of the electromagnetic spectrum. Still others deal with theory and work with physicists to understand the big bang and the nature of matter in the universe. Government, universities, and private industry offer jobs for astronomers.

Space Probes

Not all objects carried into space by rockets become satellites. Rockets also can be used to send instruments into space to collect data. A **space probe** is an instrument that gathers information and sends it back to Earth. Unlike satellites that orbit Earth, space probes travel into the solar system as illustrated in **Figure 10.** Some even have traveled to the edge of the solar system. Among these is *Pioneer 10,* launched in 1972. Although its transmitter failed in 2003, it continues on through space. Also, both *Voyager* spacecrafts should continue to return data on the outer reaches of the solar system until about 2020.

Space probes, like many satellites, carry cameras and other data-gathering equipment, as well as radio transmitters and receivers that allow them to communicate with scientists on Earth. **Table 1** shows some of the early space probes launched by the National Aeronautics and Space Administration (NASA).

Table 1 Some Early Space Missions

Mission Name	Date Launched	Destination	Data Obtained
Mariner 2	August 1962	Venus	verified high temperatures in Venus's atmosphere
Pioneer 10	March 1972	Jupiter	sent back photos of Jupiter—first probe to encounter an outer planet
Viking 1	August 1975	Mars	orbiter mapped the surface of Mars; lander searched for life on Mars
Magellan	May 1989	Venus	mapped Venus's surface and returned data on the composition of Venus's atmosphere

NATIONAL GEOGRAPHIC VISUALIZING SPACE PROBES

Figure 10

Probes have taught us much about the solar system. As they travel through space, these car-size craft gather data with their onboard instruments and send results back to Earth via radio waves. Some data collected during these missions are made into pictures, a selection of which is shown here.

A In 1974, *Mariner 10* obtained the first good images of the surface of Mercury.

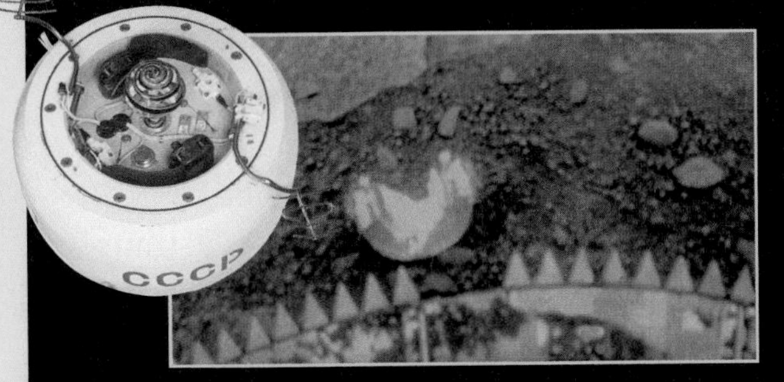

B A Soviet *Venera* probe took this picture of the surface of Venus on March 1, 1982. Parts of the spacecraft's landing gear are visible at the bottom of the photograph.

D In 1990, *Magellan* imaged craters, lava domes, and great rifts, or cracks, on the surface of Venus.

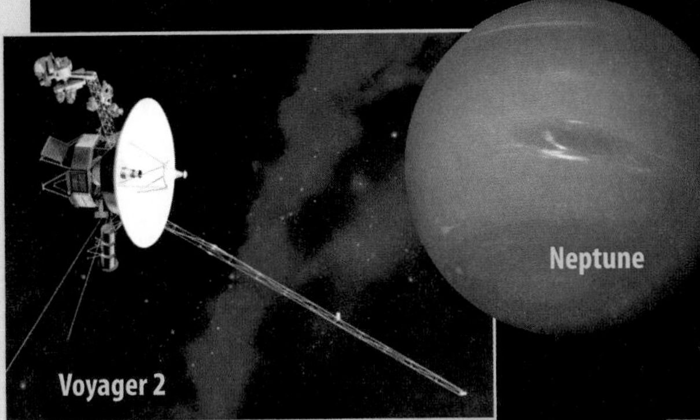

C The *Voyager 2* mission included flybys of the outer planets Jupiter, Saturn, Uranus, and Neptune. *Voyager* took this photograph of Neptune in 1989 as the craft sped toward the edge of the solar system.

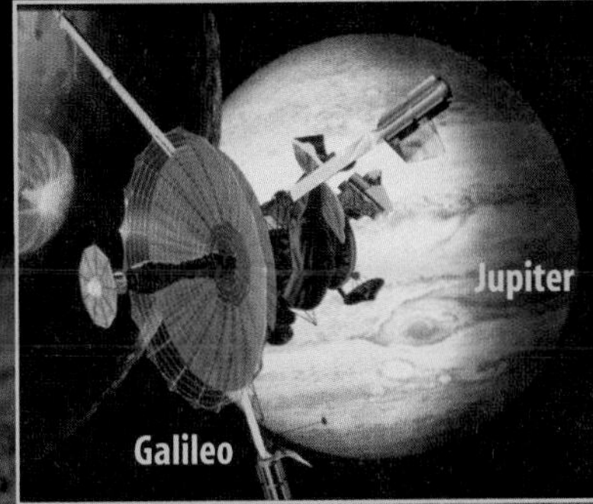

E NASA's veteran space traveler *Galileo* nears Jupiter in this artist's drawing. The craft arrived at Jupiter in 1995 and sent back data, including images of Europa, one of Jupiter's 61 moons, seen below in a color-enhanced view.

Europa

Topic: Galileo Mission
Go to earth.msscience.com for Web links to informaiton about the Galileo mission.

Activity Make a list of the places the *Galileo* spacecraft visited on its mission.

***Voyager* and *Pioneer* Probes** Space probes *Voyager 1* and *Voyager 2* were launched in 1977 and now are heading toward deep space. *Voyager 1* flew past Jupiter and Saturn. *Voyager 2* flew past Jupiter, Saturn, Uranus, and Neptune. These probes will explore beyond the solar system as part of the Voyager Interstellar Mission. Scientists expect these probes to continue to transmit data to Earth for at least 20 more years.

Pioneer 10, launched in 1972, was the first probe to survive a trip through the asteroid belt and encounter an outer planet, Jupiter. As of 2003, *Pioneer 10* was more than 12 billion km from Earth, and will continue beyond the solar system. The probe carries a gold medallion with an engraving of a man, a woman, and Earth's position in the galaxy.

Galileo Launched in 1989, *Galileo* reached Jupiter in 1995. In July 1995, *Galileo* released a smaller probe that began a five-month approach to Jupiter. The small probe took a parachute ride through Jupiter's violent atmosphere in December 1995.

Before being crushed by the atmospheric pressure, it transmitted information about Jupiter's composition, temperature, and pressure to the satellite orbiting above. *Galileo* studied Jupiter's moons, rings, and magnetic fields and then relayed this information to scientists who were waiting eagerly for it on Earth.

Studies of Jupiter's moon Europa by *Galileo* indicate that an ocean of water may exist under the surface of Europa. A cracked outer layer of ice makes up Europa's surface, shown in **Figure 11.** The cracks in the surface may be caused by geologic activity that heats the ocean underneath the surface. Sunlight penetrates these cracks, further heating the ocean and setting the stage for the possible existence of life on Europa. *Galileo* ended its study of Europa in 2000. More advanced probes will be needed to determine whether life exists on this icy moon.

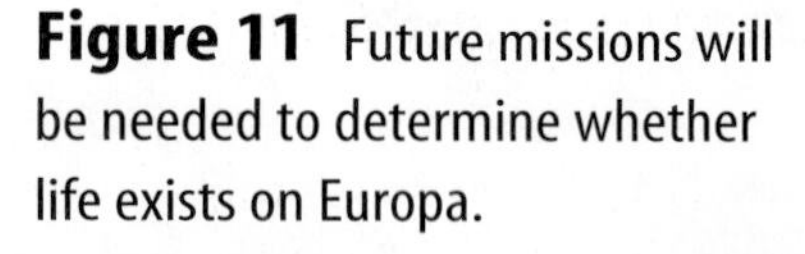

Figure 11 Future missions will be needed to determine whether life exists on Europa.

Reading Check ***What features on Europa suggest the possibility of life existing on this moon?***

In October and November of 1999, *Galileo* approached Io, another one of Jupiter's moons. It came within 300 km and took photographs of a volcanic vent named Loki, which emits more energy than all of Earth's volcanoes combined. *Galileo* also discovered eruption plumes that shoot gas made of sulfur and oxygen.

Moon Quest

Throughout the world, people were shocked when they turned on their radios and television sets in 1957 and heard the radio transmissions from *Sputnik I* as it orbited Earth. All that *Sputnik I* transmitted was a sort of beeping sound, but people quickly realized that launching a human into space wasn't far off.

In 1961, Soviet cosmonaut Yuri A. Gagarin became the first human in space. He orbited Earth and returned safely. Soon, President John F. Kennedy called for the United States to send humans to the Moon and return them safely to Earth. His goal was to achieve this by the end of the 1960s. The race for space was underway.

Figure 12 An important step in the attempt to reach the Moon was John Glenn's first orbit around Earth.

The U.S. program to reach the Moon began with **Project Mercury.** The goals of Project Mercury were to orbit a piloted spacecraft around Earth and to bring it back safely. The program provided data and experience in the basics of space flight. On May 5, 1961, Alan B. Shepard became the first U.S. citizen in space. In 1962, *Mercury* astronaut John Glenn became the first U.S. citizen to orbit Earth. **Figure 12** shows Glenn preparing for liftoff.

Reading Check *What were the goals of Project Mercury?*

Project Gemini The next step in reaching the Moon was called **Project Gemini.** Teams of two astronauts in the same *Gemini* spacecraft orbited Earth. One *Gemini* team met and connected with another spacecraft in orbit—a skill that would be needed on a voyage to the Moon.

The *Gemini* spacecraft was much like the *Mercury* spacecraft, except it was larger and easier for the astronauts to maintain. It was launched by a rocket known as a *Titan II*, which was a liquid fuel rocket.

In addition to connecting spacecraft in orbit, another goal of Project Gemini was to investigate the effects of space travel on the human body.

Along with the *Mercury* and *Gemini* programs, a series of robotic probes was sent to the Moon. *Ranger* proved that a spacecraft could be sent to the Moon. In 1966, *Surveyor* landed gently on the Moon's surface, indicating that the Moon's surface could support spacecraft and humans. The mission of *Lunar Orbiter* was to take pictures of the Moon's surface that would help determine the best future lunar landing sites.

Modeling a Satellite

WARNING: *Stand a safe distance away from classmates.*

Procedure

1. Tie one end of a strong, 50-cm-long **string** to a small **cork.**
2. Hold the other end of the string tightly with your arm fully extended.
3. Move your hand back and forth so that the cork swings in a circular motion.
4. Gradually decrease the speed of the cork.

Analysis

1. What happened as the cork's motion slowed?
2. How does the motion of the cork resemble that of a satellite in orbit?

Figure 13 The Lunar Rover vehicle was first used during the *Apollo 15* mission. Riding in the moon buggy, *Apollo 15, 16,* and *17* astronauts explored the lunar surface.

Project Apollo The final stage of the U.S. program to reach the Moon was **Project Apollo.** On July 20, 1969, *Apollo 11* landed on the Moon's surface. Neil Armstrong was the first human to set foot on the Moon. His first words as he stepped onto its surface were, "That's one small step for man, one giant leap for mankind." Edwin Aldrin, the second of the three *Apollo 11* astronauts, joined Armstrong on the Moon, and they explored its surface for two hours. While they were exploring, Michael Collins remained in the Command Module; Armstrong and Aldrin then returned to the Command Module before beginning the journey home. A total of six lunar landings brought back more than 2,000 samples of moon rock and soil for study before the program ended in 1972. **Figure 13** shows an astronaut exploring the Moon's surface from the Lunar Rover vehicle.

section 2 review

Summary

First Missions into Space

- Rockets are engines that have everything they need to burn fuel.
- Rockets may be fueled with liquid or solid propellants.
- A satellite is any object that revolves around another object.

Space Probes

- A space probe is an instrument that gathers information and sends it back to Earth.
- *Voyager* and *Pioneer* are probes designed to explore the solar system and beyond.
- *Galileo* is a space probe that explored Jupiter and its moons.

Moon Quest

- Project Mercury sent the first piloted spacecraft around Earth.
- *Ranger* and *Surveyor* probes explored the Moon's surface.
- *Gemini* orbited teams of two astronauts.
- Project Apollo completed six lunar landings.

Self Check

1. **Explain** why Neptune has eleven satellites even though it is not orbited by human-made objects.
2. **Explain** why *Galileo* was considered a space probe as it traveled to Jupiter. However, once there, it became an artificial satellite.
3. **List** several discoveries made by the *Voyager 1* and *Voyager 2* space probes.
4. **Sequence** Draw a time line beginning with *Sputnik* and ending with Project Apollo. Include descriptions of important missions.
5. **Think Critically** Is Earth a satellite of any other body in space? Explain.

Applying Math

6. **Solve Simple Equations** A standard unit of measurement in astronomy is the astronomical unit, or AU. It equals is about 150,000,000,000 (1.5×10^{11}) m. In 2000, *Pioneer 10* was more than 11 million km from Earth. How many AUs is this?
7. **Convert Units** A spacecraft is launched at a velocity of 40,200 km/h. Express this speed in kilometers per second. Show your work.

section 3

Current and Future Space Missions

as you read

What You'll Learn

- **Explain** the benefits of the space shuttle.
- **Identify** the usefulness of orbital space stations.
- **Explore** future space missions.
- **Identify** the applications of space technology to everyday life.

Why It's Important

Experiments performed on future space missions may benefit you.

Review Vocabulary
cosmonaut: astronaut of the former Soviet Union or present-day Russian space program

New Vocabulary
- space shuttle
- space station

The Space Shuttle

Imagine spending millions of dollars to build a machine, sending it off into space, and watching its 3,000 metric tons of metal and other materials burn up after only a few minutes of work. That's exactly what NASA did with the rocket portions of spacecraft for many years. The early rockets were used only to launch a small capsule holding astronauts into orbit. Then sections of the rocket separated from the rest and burned when reentering the atmosphere.

A Reusable Spacecraft NASA administrators, like many others, realized that it would be less expensive and less wasteful to reuse resources. The reusable spacecraft that transports astronauts, satellites, and other materials to and from space is called the **space shuttle**, shown in **Figure 14,** as it is landing.

At launch, the space shuttle stands on end and is connected to an external liquid-fuel tank and two solid-fuel booster rockets. When the shuttle reaches an altitude of about 45 km, the emptied, solid-fuel booster rockets drop off and parachute back to Earth. These are recovered and used again. The external liquid-fuel tank separates and falls back to Earth, but it isn't recovered.

Work on the Shuttle After the space shuttle reaches space, it begins to orbit Earth. There, astronauts perform many different tasks. In the cargo bay, astronauts can conduct scientific experiments and determine the effects of spaceflight on the human body. When the cargo bay isn't used as a laboratory, the shuttle can launch, repair, and retrieve satellites. Then the satellites can be returned to Earth or repaired onboard and returned to space. After a mission, the shuttle glides back to Earth and lands like an airplane. A large landing field is needed as the gliding speed of the shuttle is 335 km/h.

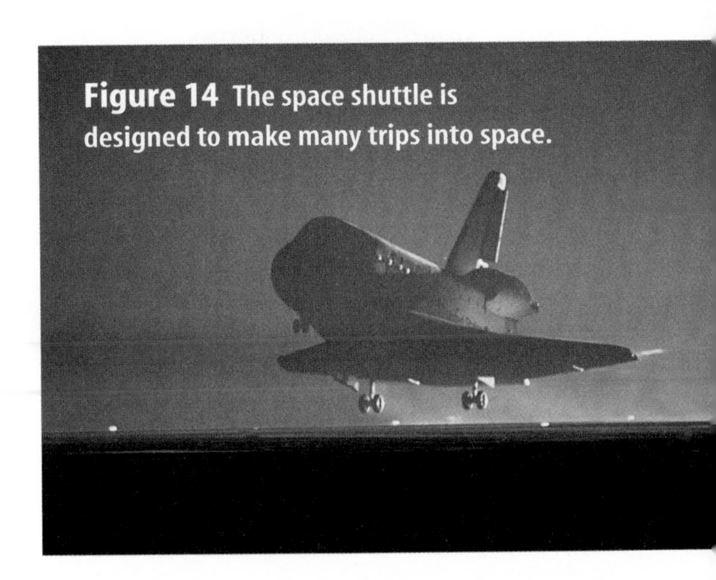

Figure 14 The space shuttle is designed to make many trips into space.

Space Stations

Astronauts can spend only a short time living in the space shuttle. Its living area is small, and the crew needs more room to live, exercise, and work. A **space station** has living quarters, work and exercise areas, and all the equipment and support systems needed for humans to live and work in space.

In 1973, the United States launched the space station *Skylab,* shown in **Figure 15.** Crews of astronauts spent up to 84 days there, performing experiments and collecting data on the effects on humans of living in space. In 1979, the abandoned *Skylab* fell out of orbit and burned up as it entered Earth's atmosphere.

Figure 15 Astronauts performed a variety of tasks while living and working in space onboard *Skylab.*

Crews from the former Soviet Union have spent more time in space, onboard the space station *Mir,* than crews from any other country. Cosmonaut Dr. Valery Polyakov returned to Earth after 438 days in space studying the long-term effects of weightlessness.

Cooperation in Space

In 1995, the United States and Russia began an era of cooperation and trust in exploring space. Early in the year, American Dr. Norman Thagard was launched into orbit aboard the Russian *Soyuz* spacecraft, along with two Russian cosmonaut crewmates. Dr. Thagard was the first U.S. astronaut launched into space by a Russian booster and the first American resident of the Russian space station *Mir.*

Figure 16 Russian and American scientists have worked together to further space exploration. **Explain** *why the docking of the space shuttle with* Mir *was so important.*

In June 1995, Russian cosmonauts rode into orbit onboard the space shuttle *Atlantis,* America's 100th crewed launch. The mission of *Atlantis* involved, among other studies, a rendezvous and docking with the space station *Mir.* The cooperation that existed on this mission, as shown in **Figure 16,** continued through eight more space shuttle-*Mir* docking missions. Each of the eight missions was an important step toward building and operating the *International Space Station.* In 2001, the abandoned *Mir* space station fell out of orbit and burned up upon reentering the atmosphere. Cooperation continued as the *International Space Station* began to take form.

The International Space Station The *International Space Station (ISS)* will be a laboratory designed for long-term research projects. Diverse topics will be studied, including research on the growth of protein crystals. This particular project will help scientists determine protein structure and function, which is expected to enhance work on drug design and the treatment of many diseases.

The *ISS* will draw on the resources of more than 15 nations. These nations will build units for the space station, which then will be transported into space onboard the space shuttle and Russian launch rockets. The station will be constructed in space. **Figure 17** shows what the completed station will look like.

Figure 17 This is a picture of what the proposed *International Space Station* will look like when it is completed.

***What is the purpose of the* International Space Station?**

Phases of *ISS* NASA is planning the *ISS* program in phases. Phase One, now concluded, involved the space shuttle-*Mir* docking missions. Phase Two began in 1998 with the launch of the Russian-built *Zarya Module,* also known as the Functional Cargo Block. In December 1998, the first assembly of *ISS* occurred when a space shuttle mission attached the Unity module to *Zarya.* During this phase, crews of three people were delivered to the space station. Phase Two ended in 2001 with the addition of a U.S. laboratory.

Living in Space The project will continue with Phase Three when the Japanese Experiment Module, the European Columbus Orbiting Facility, and another Russian lab will be delivered.

It is hoped that the *International Space Station* will be completed in 2010. Eventually, a seven-person crew should be able to work comfortably onboard the station. A total of 80 U.S. and Russian flights will be required to take all the components of the *ISS* into space and prepare it for continuous habitation. NASA plans for crews of astronauts to stay onboard the station for several months at a time. NASA already has conducted numerous tests to prepare crews of astronauts for extended space missions. One day, the station could be a construction site for ships that will travel to the Moon and Mars.

Topic: *International Space Station*

Visit earth.msscience.com for Web links to information about the *International Space Station.*

Activity You can see the station travel across the sky with an unaided eye. Find out the schedule and try to observe it.

Figure 18 Gulleys, channels, and aprons of sediment imaged by the *Mars Global Surveyor* are similar to features on Earth known to be caused by flowing water. This water is thought to seep out from beneath the surface of Mars.

Exploring Mars

Two of the most successful missions in recent years were the 1996 launchings of the *Mars Global Surveyor* and the *Mars Pathfinder. Surveyor* orbited Mars, taking high-quality photos of the planet's surface as shown in **Figure 18.** *Pathfinder* descended to the Martian surface, using rockets and a parachute system to slow its descent. Large balloons absorbed the shock of landing. *Pathfinder* carried technology to study the surface of the planet, including a remote-controlled robot rover called Sojourner. Using information gathered by studying photographs taken by *Surveyor,* scientists determined that water recently had seeped to the surface of Mars in some areas.

Reading Check ***What type of data were obtained by the* Mars Global Surveyor?**

Another orbiting spacecraft, the *Mars Odyssey* began mapping the surface of Mars in 2002. Soon after, its data confirmed the findings of *Surveyor*—that Martian soil contains frozen water in the southern polar area. The next step was to send robots to explore the surface of Mars. Twin rovers named *Spirit* and *Opportunity* were launched in 2003 and arrived at their separate destinations on Mars in January 2004. They analyzed Martian rocks and soils, which told scientists more about Martian geology and provided clues about the role of water on Mars. Future plans include *Phoenix* in 2008, a robot lander capable of digging over a meter into the surface.

New Millennium Program

To continue space missions into the future, NASA has created the New Millennium Program (NMP). The goal of the NMP is to develop advanced technology that will let NASA send smart spacecraft into the solar system. This will reduce the amount of ground control needed. They also hope to reduce the size of future spacecraft to keep the cost of launching them under control. NASA's challenge is to prove that certain cutting-edge technologies, as well as mission concepts, work in space.

Exploring the Moon

Does water exist in the craters of the Moon's poles? This is one question NASA intends to explore with data gathered from the *Lunar Prospector* spacecraft shown in **Figure 19.** Launched in 1998, the *Lunar Prospector's* one-year mission was to orbit the Moon, mapping its structure and composition. Data obtained from the spacecraft indicate that water ice might be present in the craters at the Moon's poles. Scientists first estimated up to 300 million metric tons of water may be trapped as ice, and later estimates are much higher. In the permanently shadowed areas of some craters, the temperature never exceeds −230°C. Therefore water delivered to the Moon by comets or meteorites early in its history could remain frozen indefinitely.

At the end of its mission, *Lunar Prospector* was deliberately crashed into a lunar crater. Using special telescopes, scientists hoped to see evidence of water vapor thrown up by the collision. None was seen, however scientists still believe that much water ice is there. If so, this water would be useful if a colony is ever built on the Moon.

Topic: New Millennium Program

Visit earth.msscience.com for Web links to information about NASA's New Millennium Program.

Activity Prepare a table listing proposed missions, projected launch dates, and what they will study.

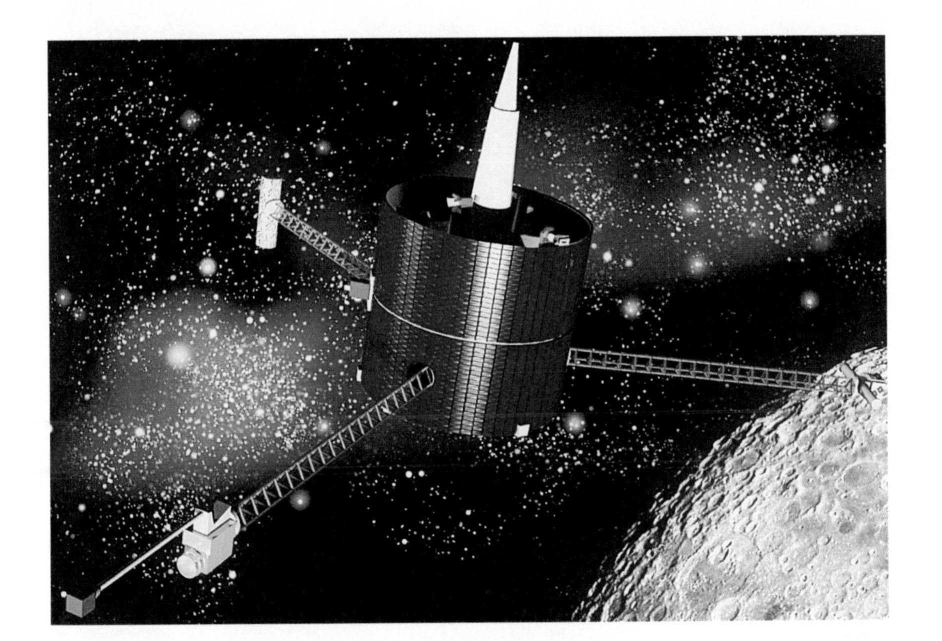

Figure 19 The *Lunar Prospector* analyzed the Moon's composition during its one-year mission. **Explain** *why* Lunar Prospector *was deliberately crashed on the Moon.*

Figure 20 *Cassini* is currently on its way to Saturn. After it arrives, it will spend four years studying Saturn and its surrounding area.

Cassini

In October 1997, NASA launched the space probe *Cassini.* This probe's destination was Saturn. *Cassini,* shown in **Figure 20,** reached its goal in 2004. The space probe will explore Saturn and surrounding areas for four years. As part of its mission, it delivered the European Space Agency's *Huygens* probe to Saturn's largest moon, Titan. Some scientists theorize that Titan's atmosphere may be similar to the atmosphere of early Earth.

The Next Generation Space Telescope Not all space missions involve sending astronauts or probes into space. Plans are being made to launch a new space telescope that is capable of observing the first stars and galaxies in the universe. The *James Webb Space Telescope,* shown in **Figure 21,** will be the successor to the *Hubble Space Telescope.* As part of the Origins project, it will provide scientists with the opportunity to study the evolution of galaxies, the production of elements by stars, and the process of star and planet formation. To accomplish these tasks, the telescope will have to be able to see objects 400 times fainter than those currently studied with ground-based telescopes such as the twin Keck telescopes. NASA intends to launch the *James Webb Space Telescope* no earlier than 2013.

Figure 21 The *James Webb Space Telescope* honors the NASA administrator who contributed greatly to the Apollo Program. It will help scientists learn more about how galaxies form.

Everyday Space Technology Many people have benefited from research done for space programs. Medicine especially has gained much from space technology. Space medicine led to better ways to diagnose and treat heart disease here on Earth and to better heart pacemakers. A screening system that works on infants is helping eye doctors spot vision problems early. Cochlear implants that help thousands of deaf people hear were developed using knowledge gained during the space shuttle program.

Space technology can even help catch criminals and prevent accidents. For example, a method to sharpen images that was devised for space studies is being used by police to read numbers on blurry photos of license plates. Equipment using space technology can be placed on emergency vehicles. This equipment automatically changes traffic signals as an emergency vehicle approaches intersections, so that crossing vehicles have time to stop safely. A hand-held device used for travel directions is shown in **Figure 22.**

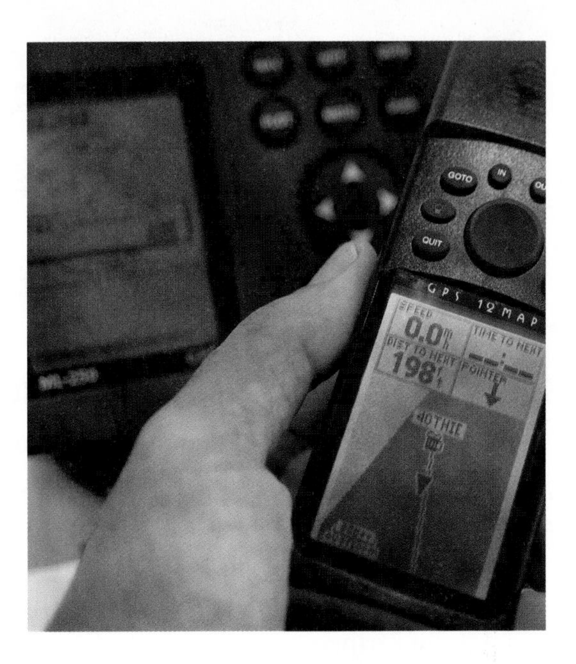

Figure 22 Global Positioning System (GPS) technology uses satellites to determine location on Earth's surface.

How have research and technology developed for space benefited people here on Earth?

section 3 review

Summary

The Space Station

- A space station is an orbiting laboratory.
- The new *International Space Station (ISS)* is being built with the aid of 16 nations.
- The space shuttle transports astronauts, satellites, and other materials to and from the *ISS*.

Exploring Mars and the Moon

- The *Mars Global Surveyor* orbited Mars and the *Mars Pathfinder* studied its surface.
- *Lunar Prospector* orbited the Moon, mapping its structure and composition.
- Recent data indicate that water ice crystals may exist in shadows of lunar craters.

Future Missions

- The *Cassini* probe is scheduled to explore Saturn and its moons.
- The successor to the *Hubble* will be the *James Webb Space Telescope*.

Self Check

1. **Identify** the main advantage of the space shuttle.
2. **Describe** the importance of space shuttle-*Mir* docking missions.
3. **Explain** how *International Space Station* is used.
4. **Identify** three ways that space technology is a benefit to everyday life.
5. **Think Critically** What makes the space shuttle more versatile than earlier spacecraft?

Applying Math

6. **Solve One-Step Equations** *Voyager 1* had about 30 kg of hydrazine fuel left in 2003. If it uses about 500 g per year, how long will this fuel last?
7. **Use Percentages** Suppose you're in charge of assembling a crew of 50 people. Decide how many to assign each task, such as farming, maintenance, scientific experimentation, and so on. Calculate the percent of the crew assigned to each task. Justify your decisions.

Use the Internet

Star Sightings

Goals

- **Record** your sightings of Polaris.
- **Share** the data with other students to calculate the circumference of Earth.

Data Source

Science Online

Go to earth.msscience.com/internet_lab to obtain instructions on how to make an astrolabe. Also visit the Web site for more information about the location of Polaris, and for data from other students.

Safety Precautions

WARNING: *Do not use the astrolabe during the daytime to observe the Sun.*

Real-World Question

For thousands of years, people have measured their position on Earth using the position of Polaris, the North Star. At any given observation point, it always appears at the same angle above the horizon. For example, at the north pole, Polaris appears directly overhead, and at the equator it is just above the northern horizon. Other locations can be determined by measuring the height of Polaris above the horizon using an instrument called an astrolabe. Could you use Polaris to determine the size of Earth? You know that Earth is round. Knowing this, do you think you can estimate the circumference of Earth based on star sightings?

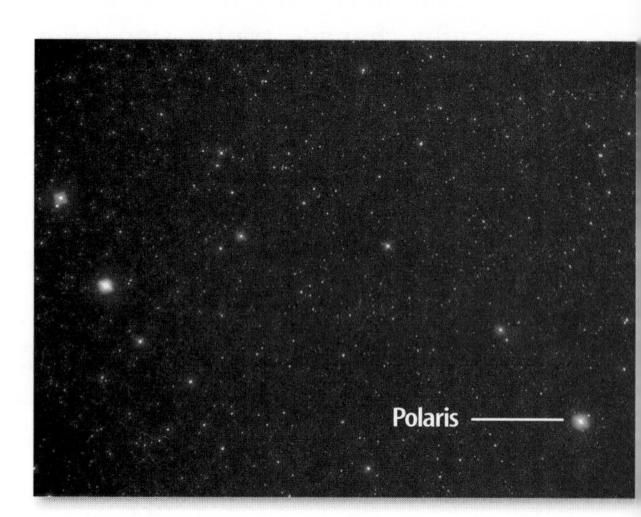

Make a Plan

1. Obtain an astrolabe or construct one using the instructions posted by visiting the link above.
2. **Design** a data table in your Science Journal similar to the one below.

Polaris Observations

Your Location:

Date	Time	Astrolabe Reading
Do not write in this book.		

3. Decide as a group how you will make your observations. Does it take more than one person to make each observation? When will it be easiest to see Polaris?

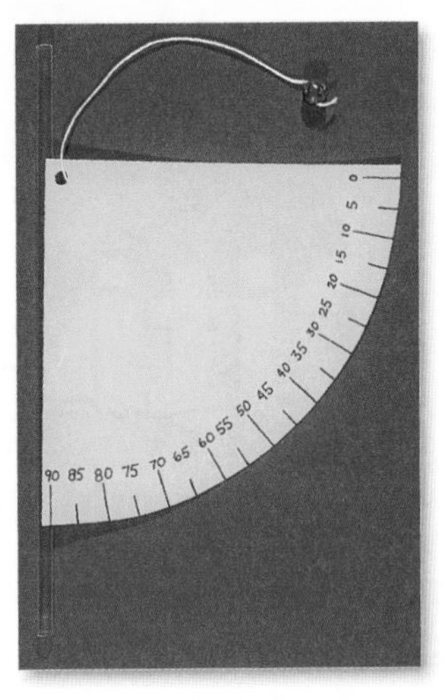

Follow Your Plan

1. Make sure your teacher approves your plan before you start.
2. Carry out your observations.
3. **Record** your observations in your data table.
4. Average your readings and post them in the table provided at the link shown below.

Analyze Your Data

1. **Research** the names of cities that are at approximately the same longitude as your hometown. Gather astrolabe readings from students in one of those cities at the link shown below.
2. **Compare** your astrolabe readings. Subtract the smaller reading from the larger one.
3. **Determine** the distance between your star sighting location and the other city.
4. **Calculate** the circumference of Earth using the following relationship.

   Circumference = (360°) × (distance between locations)/difference between readings

Conclude and Apply

1. **Analyze** how the circumference of Earth that you calculated compares with the accepted value of 40,079 km.
2. **Determine** some possible sources of error in this method of establishing the size of Earth. What improvements would you suggest?

Communicating Your Data

Find this lab using the link below. **Create** a poster that includes a table of your data and data from students in other cities. **Perform** a sample circumference calculation for your class.

Science Online

earth.msscience.com/internet_lab

TIME SCIENCE AND Society

SCIENCE ISSUES THAT AFFECT YOU!

Should the U.S. spend money to colonize space?

Humans have landed on the Moon, and spacecrafts have landed on Mars. But these space missions are just small steps that may lead to a giant new space program. As technology improves, humans may be able to visit and even live on other planets. But is it worth the time and money involved?

Those in favor of living in space point to the International Space Station that already is orbiting Earth. It's an early step toward establishing floating cities where astronauts can live and work. As Earth's population continues to increase and there is less room on this planet, why not expand to other planets or build a floating city in space? Also, the fact that there is little pollution in space makes the idea appealing to many.

Critics of colonizing space think we should spend the hundreds of billions of dollars that it would cost to colonize space on projects to help improve people's lives here on Earth. Building better housing, developing ways to feed the hungry, finding cures for diseases, and increasing funds for education should come first, these people say. And, critics continue, if people want to explore, why not explore right here on Earth, for example, the ocean floor.

Moon or Mars? If humans were to move permanently to space, the two most likely destinations would be Mars and the Moon, both bleak places. But those in favor of moving to these places say humans could find a way to make them livable as they have made homes in harsh climates and in many rugged areas here on Earth.

Water may be locked in lunar craters, and photos suggest that Mars once had liquid water on its surface. If that water is frozen underground, humans may be able to access it. NASA is studying whether it makes sense to send astronauts and scientists to explore Mars.

Transforming Mars into an Earthlike place with breathable air and usable water will take much longer, but some small steps are being taken. Experimental plants are being developed that could absorb Mars's excess carbon dioxide and release oxygen. Solar mirrors that could warm Mars's surface are available.

Those for and against colonizing space agree on one thing—it will take large amounts of money, research, and planning. It also will take the same spirit of adventure that has led history's pioneers into so many bold frontiers—deserts, the poles, and the sky.

Debate with your class the pros and cons of colonizing space. Do you think the United States should spend money to create space cities or use the money now to improve lives of people on Earth?

For more information, visit earth.msscience.com/time

Reviewing Main Ideas

Section 1 Radiation from Space

1. The arrangement of electromagnetic waves according to their wavelengths is the electromagnetic spectrum.
2. Optical telescopes produce magnified images of objects.
3. Radio telescopes collect and record radio waves given off by some space objects.

Section 2 Early Space Missions

1. A satellite is an object that revolves around another object. The moons of planets are natural satellites. Artificial satellites are those made by people.
2. A space probe travels into the solar system, gathers data, and sends them back to Earth.
3. American piloted space programs included the Gemini, Mercury, and Apollo Projects.

Section 3 Current and Future Space Missions

1. Space stations provide the opportunity to conduct research not possible on Earth. The *International Space Station* is being constructed in space with the cooperation of more than a dozen nations.
2. The space shuttle is a reusable spacecraft that carries astronauts, satellites, and other cargo to and from space.
3. Space technology is used to solve problems on Earth, too. Advances in engineering related to space travel have aided medicine, environmental sciences, and other fields.

Visualizing Main Ideas

Copy and complete the following concept map about the race to the Moon. Use the phrases: first satellite, Project Gemini, Project Mercury, team of two astronauts orbits Earth, Project Apollo.

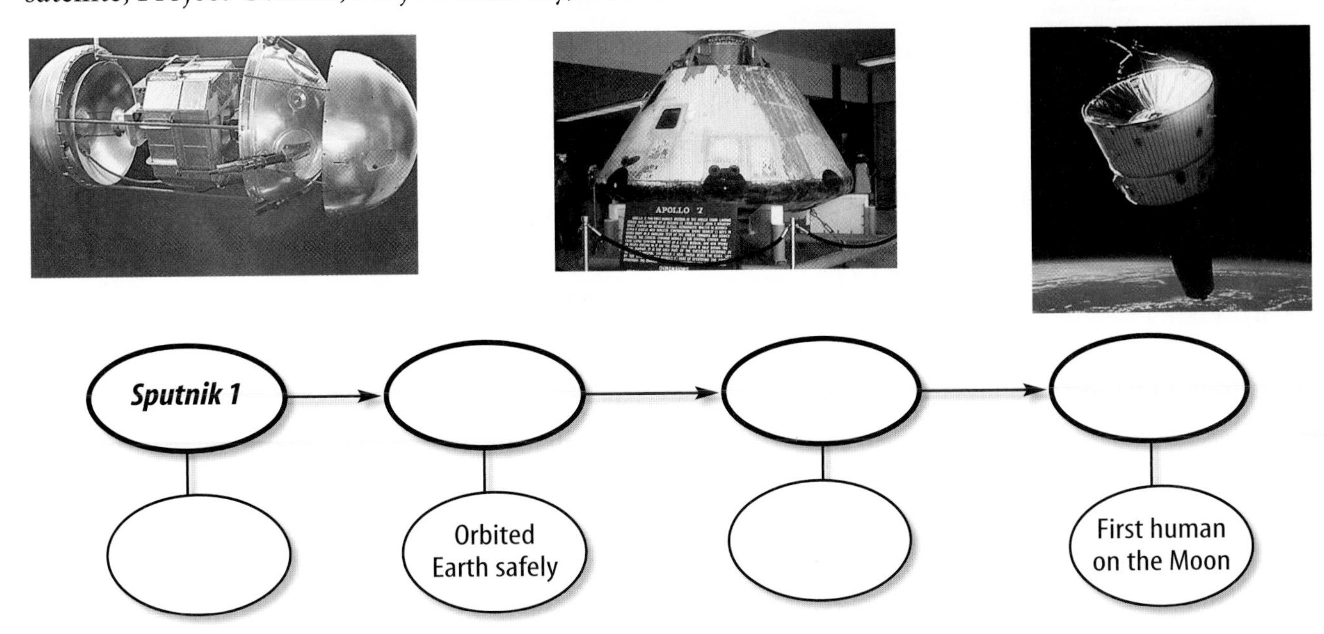

chapter 22 Review

Using Vocabulary

electromagnetic spectrum p. 629
observatory p. 630
orbit p. 637
Project Apollo p. 642
Project Gemini p. 641
Project Mercury p. 641
radio telescope p. 633
reflecting telescope p. 630
refracting telescope p. 630
rocket p. 635
satellite p. 637
space probe p. 638
space shuttle p. 643
space station p. 644

Fill in the blanks with the correct vocabulary word(s).

1. A(n) ________ telescope uses lenses to bend light.

2. A(n) ________ is an object that revolves around another object in space.

3. ________ was the first piloted U.S. space program.

4. A(n) ________ carries people and tools to and from space.

5. In the ________, electromagnetic waves are arranged, in order, according to their wavelengths.

Checking Concepts

Choose the word or phrase that best answers the question.

6. Which spacecraft has sent images of Venus to scientists on Earth?
A) *Voyager*
B) *Viking*
C) *Apollo 11*
D) *Magellan*

7. Which kind of telescope uses mirrors to collect light?
A) radio
B) electromagnetic
C) refracting
D) reflecting

8. What was *Sputnik I?*
A) the first telescope
B) the first artificial satellite
C) the first observatory
D) the first U.S. space probe

9. Which kind of telescope can be used during the day or night and during bad weather?
A) radio
B) electromagnetic
C) refracting
D) reflecting

10. When fully operational, what is the maximum number of people who will crew the *International Space Station?*
A) 3
B) 7
C) 15
D) 50

11. Which space mission's goal was to put a spacecraft into orbit and bring it back safely?
A) Project Mercury
B) Project Apollo
C) Project Gemini
D) *Viking I*

12. Which of the following is a natural satellite of Earth?
A) *Skylab*
B) the space shuttle
C) the Sun
D) the Moon

13. What does the space shuttle use to place a satellite into space?
A) liquid-fuel tank
B) booster rocket
C) mechanical arm
D) cargo bay

14. What part of the space shuttle is reused?
A) liquid-fuel tanks
B) *Gemini* rockets
C) booster engines
D) Saturn rockets

Science Online earth.msscience.com/vocabulary_puzzlemaker

Thinking Critically

15. **Compare and contrast** the advantages of a moon-based telescope with an Earth-based telescope.

16. **Infer** how sensors used to detect toxic chemicals in the space shuttle could be beneficial to a factory worker.

17. **Drawing Conclusions** Which do you think is a wiser method of exploration—space missions with people onboard or robotic space probes? Why?

18. **Explain** Suppose two astronauts are outside the space shuttle orbiting Earth. The audio speaker in the helmet of one astronaut quits working. The other astronaut is 1 m away and shouts a message. Can the first astronaut hear the message? Support your reasoning.

19. **Make and Use Tables** Copy and complete the table below. Use information from several resources.

United States Space Probes

Probe	Launch Date(s)	Planets or Objects Visited
Vikings 1 and *2*		
Galileo	Do not write in this book.	
Lunar Prospector		
Pathfinder		

20. **Classify** the following as a satellite or a space probe: *Cassini, Sputnik I, Hubble Space Telescope,* space shuttle, and *Voyager 2.*

21. **Compare and contrast** space probes and artificial satellites.

Performance Activities

22. **Display** Make a display showing some of the images obtained from the *Hubble Space Telescope.* Include samples of three types of galaxies, nebulae, and star clusters.

Applying Math

23. **Space Communications** In May 2003 *Voyager 1* was 13 billion km from the Sun. Calculate how long it takes for a signal to travel this far assuming it travels at 3×10^8 m/s.

Use the graph below to answer question 24.

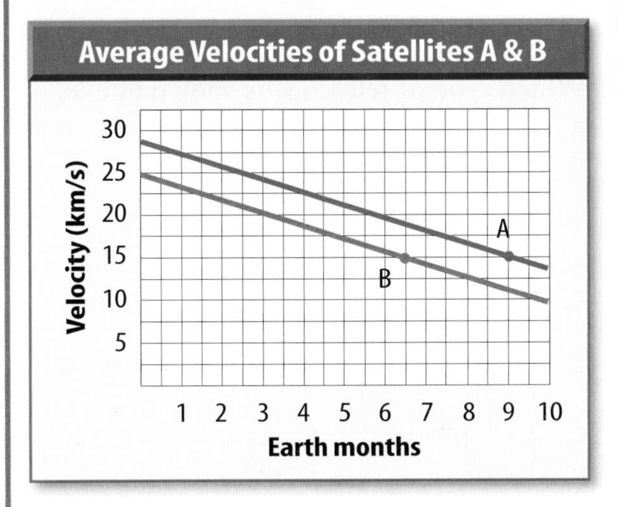

24. **Satellite Orbits** The graph above predicts the average velocities of satellites A and B in orbit around a hypothetical planet. Because of contact with the planet's atmosphere, their velocities are decreasing. At a velocity of 15 km/s their orbits will decay and they will spiral downwards to the surface. Using the graph, determine how long will it take for each satellite to reach this point?

25. **Calculate Fuel** A spacecraft carries 30 kg of hydrazine fuel and uses and average of 500 g/y. How many years could this fuel last?

26. **Space Distances** Find the distance in AUs to a star 68 light-years (LY) distant. (1 LY $= 6.3 \times 10^4$ AUs)

Part 1 Multiple Choice

Record your answers on the answer sheet provided by your teacher or on a sheet of paper.

Use the figure below to answer question 1.

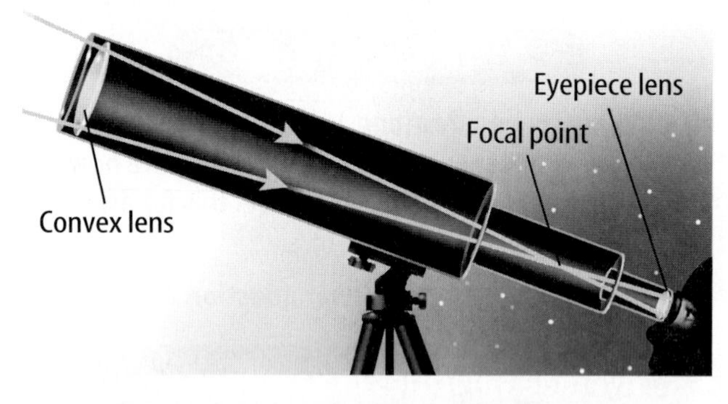

1. Which type of telescope is shown above?
   A. refracting **C.** reflecting
   B. radio **D.** space
2. Who was the first human in space?
   A. Edwin Aldrin **C.** Neil Armstrong
   B. John Glenn **D.** Yuri Gagarin
3. Which is an engine that can launch an object into space?
   A. space probe **C.** capsule
   B. shuttle **D.** rocket
4. Which is the speed of light in a vacuum?
   A. 300 km/s **C.** 3,000 km/s
   B. 300,000 km/s **D.** 30,00 km/s
5. Which of the following is an advantage of space telescopes?
   A. They are cheaper to build.
   B. They have fewer technical problems.
   C. They obtain higher quality images.
   D. They can be repaired easily.

Test-Taking Tip

Making Answers Do not mark the test booklet when taking the test. Be sure to mark ALL answers on your answer sheet and leave no blanks.

6. Which type of radiation has a shorter wavelength than visible light does?
   A. ultraviolet **C.** infrared
   B. microwaves **D.** radio waves
7. Which space probe visited Mars?
   A. *Viking 1* **C.** *Magellan*
   B. *Mariner 2* **D.** *Pioneer 10*
8. Which United States space program included several lunar landings?
   A. Gemini **C.** Apollo
   B. Mercury **D.** space shuttle

Examine the diagram below. Then answer questions 9–11.

9. What is the name of the curved path that the satellite follows?
   A. an orbit **C.** a revolution
   B. a rotation **D.** a track
10. Which force pulls the satellite toward Earth?
    A. the Moon's gravity
    B. Earth's gravity
    C. the Sun's gravity
    D. Earth's magnetic field
11. Imagine that the satellite in the diagram above started to orbit at a slower speed. Which of the following probably would happen to the satellite?
    A. It would fly off into space.
    B. It would crash into the Moon.
    C. It would crash into the Sun.
    D. It would crash into Earth.

Part 2 Short Response/Grid In

Record your answers on the answer sheet provided by your teacher or on a sheet of paper.

12. Explain the difference between a space probe and a satellite that is orbiting Earth.

13. Why was the flight of *Sputnik 1* important?

14. List four ways that satellites are useful.

15. How are radio telescopes different from optical telescopes?

Use the table below to answer questions 16–19. The table includes data collected by *Mars Pathfinder* on the third Sol, or Martian day, of operation.

Sol 3 Temperature Data from Mars Pathfinder			
Proportion of Sol	Temperature (°C)		
	1.0 m above surface	0.5 m above surface	0.25 m above surface
3.07	−70.4	−70.7	−73.4
3.23	−74.4	−74.9	−75.9
3.33	−53.0	−51.9	−46.7
3.51	−22.3	−19.2	−15.7
3.60	−15.1	−12.5	−8.9
3.70	−26.1	−25.7	−24.0
3.92	−63.9	−64.5	−65.8

16. Which proportion of sol value corresponds to the warmest temperatures at all three heights?

17. Which proportion of sol value corresponds to the coldest temperatures at all three heights?

18. What is the range of the listed temperature values for each distance above the surface?

19. Explain the data in the table. Why do the temperatures vary in this way?

Part 3 Open Ended

Record your answers on a sheet of paper.

Use the diagram below to answer question 20.

20. Explain the purpose of each of the labeled objects.

21. List four advancements in technology directly attributable to space exploration and how they have impacted everyday life on Earth.

22. What are the advantages of having reusable spacecraft? Are there any disadvantages? Explain.

23. What is the *International Space Station?* How is it used?

24. What are the advantages of international cooperation during space exploration? Are there disadvantages?

25. Explain how the voices of astronauts onboard the space shuttle can be heard on Earth.

26. List several benefits and costs of space exploration. Do you think that the benefits of space exploration outweigh the costs? Explain why you do or do not.

chapter 23

The Sun-Earth-Moon System

The BIG Idea

Many common observations, such as seasons, eclipses, and lunar phases, are caused by interactions between the Sun, Earth, and the Moon.

SECTION 1
Earth
Main Idea Earth is a sphere that rotates on a tilted axis and revolves around the Sun.

SECTION 2
The Moon—Earth's Satellite
Main Idea Eclipses and phases of the Moon occur as the Moon moves in relation to the Sun and Earth.

SECTION 3
Exploring Earth's Moon
Main Idea Knowledge of the Moon's structure and composition has been increased by many spacecraft missions to the Moon.

Full Moon Rising—The Real Story

Why does the Moon's appearance change throughout the month? Do the Sun and Moon really rise? You will find the answers to these questions and also learn why we have summer and winter.

Science Journal Rotation or revolution—which motion of Earth brings morning and which brings summer?

Start-Up Activities

Model Rotation and Revolution

The Sun rises in the morning; at least, it seems to. Instead, it is Earth that moves. The movements of Earth cause day and night, as well as the seasons. In this lab, you will explore Earth's movements.

1. Hold a basketball with one finger at the top and one at the bottom. Have a classmate gently spin the ball.
2. Explain how this models Earth's rotation.
3. Continue to hold the basketball and walk one complete circle around another student in your class.
4. How does this model Earth's revolution?
5. **Think Critically** Write a paragraph in your Science Journal describing how these movements of the basketball model Earth's rotation and revolution.

Preview this chapter's content and activities at earth.msscience.com

Earth and the Moon All on Earth can see and feel the movements of Earth and the Moon as they circle the Sun. Make the following Foldable to organize what you learn about these movements and their effects.

STEP 1 **Fold** a sheet of paper in half lengthwise.

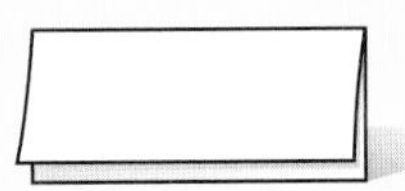

STEP 2 **Fold** paper down 2.5 cm from the top. (Hint: From the tip of your index finger to your middle knuckle is about 2.5 cm.)

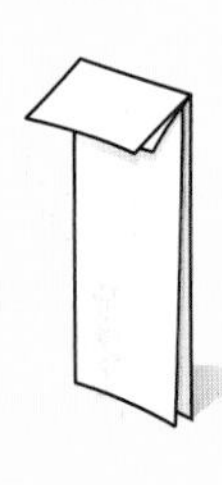

STEP 3 **Open and draw** lines along the 2.5-cm fold. **Label** as shown.

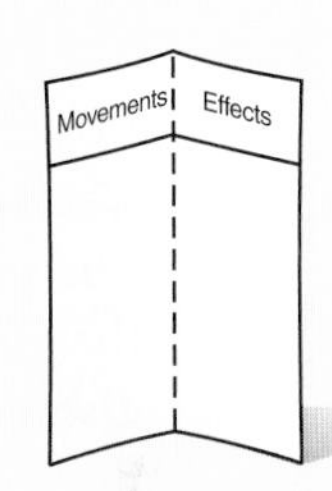

Summarize in a Table As you read the chapter, summarize the movements of Earth and the Moon in the left column and the effects of these movements in the right column.

Get Ready to Read

Summarize

1 Learn It!

Summarizing helps you organize information, focus on main ideas, and reduce the amount of information to remember. To summarize, restate the important facts in a short sentence or paragraph. Be brief and do not include too many details.

2 Practice It!

Read the text on page 664 labeled *Solstices*. Then read the summary below and look at the important facts from that passage.

Summary

The solstice is the day when the Sun reaches its greatest distance north or south of the equator.

Important Facts

- In the northern hemisphere, the summer solstice occurs in June, and the winter solstice occurs in December.
- In the southern hemisphere, the winter solstice occurs in June, and the summer solstice occurs in December.
- Summer solstice is about the longest period of daylight of the year.
- Winter solstice is about the shortest period of daylight of the year.

3 Apply It!

Practice summarizing as you read this chapter. Stop after each section and write a brief summary.

Reading Tip

Reread your summary to make sure you didn't change the author's original meaning or ideas.

Target Your Reading

Use this to focus on the main ideas as you read the chapter.

1. **Before you read** the chapter, respond to the statements below on your worksheet or on a numbered sheet of paper.
   - Write an **A** if you **agree** with the statement.
   - Write a **D** if you **disagree** with the statement.

2. **After you read** the chapter, look back to this page to see if you've changed your mind about any of the statements.
   - If any of your answers changed, explain why.
   - Change any false statements into true statements.
   - Use your revised statements as a study guide.

Before You Read A or D		Statement	After You Read A or D
	1	Earth's revolution around the Sun causes day and night to occur.	
	2	Earth's magnetic poles are aligned on Earth's rotational axis.	
	3	Summer occurs in the northern hemisphere when Earth is closest to the Sun.	
	4	During an equinox, the number of daylight hours is nearly equal with the number of night-time hours all over the world.	
	5	When observing the phases of the Moon, the Moon's lighted surface area is daylight on the Moon and the dark portion is nighttime on the Moon.	
	6	The length of one Moon day is about the same amount of time as the length of one Earth day.	
	7	A lunar eclipse occurs when the Moon comes between Earth and the Sun.	
	8	Humans first walked on the Moon during the *Apollo* spacecraft missions.	

Science Online

Print out a worksheet of this page at earth.msscience.com

section 1

Earth

as you read

What You'll Learn

- **Examine** Earth's physical characteristics.
- **Differentiate** between rotation and revolution.
- **Discuss** what causes seasons to change.

Why It's Important

Your life follows the rhythm of Earth's movements.

Review Vocabulary
orbit: the path taken by an object revolving around another

New Vocabulary
- sphere
- axis
- rotation
- revolution
- ellipse
- solstice
- equinox

Figure 1 For many years, sailors have observed that the tops of ships coming across the horizon appear first. This suggests that Earth is spherical, not flat, as was once widely believed.

Properties of Earth

You awaken at daybreak to catch the Sun "rising" from the dark horizon. Then it begins its daily "journey" from east to west across the sky. Finally the Sun "sinks" out of view as night falls. Is the Sun moving—or are you?

It wasn't long ago that people thought Earth was the center of the universe. It was widely believed that the Sun revolved around Earth, which stood still. It is now common knowledge that the Sun only appears to be moving around Earth. Because Earth spins as it revolves around the Sun, it creates the illusion that the Sun is moving across the sky.

Another mistaken idea about Earth concerned its shape. Even as recently as the days of Christopher Columbus, many people believed Earth to be flat. Because of this, they were afraid that if they sailed far enough out to sea, they would fall off the edge of the world. How do you know this isn't true? How have scientists determined the true shape of Earth?

Spherical Shape A round, three-dimensional object is called a **sphere** (SFIHR). Its surface is the same distance from its center at all points. Some common examples of spheres are basketballs and tennis balls.

In the late twentieth century, artificial satellites and space probes sent back pictures showing that Earth is spherical. Much earlier, Aristotle, a Greek astronomer and philosopher who lived around 350 B.C., suspected that Earth was spherical. He observed that Earth cast a curved shadow on the Moon during an eclipse.

In addition to Aristotle, other individuals made observations that indicated Earth's spherical shape. Early sailors, for example, noticed that the tops of approaching ships appeared first on the horizon and the rest appeared gradually, as if they were coming over the crest of a hill, as shown in **Figure 1.**

Additional Evidence Sailors also noticed changes in how the night sky looked. As they sailed north or south, the North Star moved higher or lower in the sky. The best explanation was a spherical Earth.

Today, most people know that Earth is spherical. They also know all objects are attracted by gravity to the center of a spherical Earth. Astronauts have clearly seen the spherical shape of Earth. However, it bulges slightly at the equator and is somewhat flattened at the poles, so it is not a perfect sphere.

Rotation Earth's **axis** is the imaginary vertical line around which Earth spins. This line cuts directly through the center of Earth, as shown in the illustration accompanying **Table 1.** The poles are located at the north and south ends of Earth's axis. The spinning of Earth on its axis, called **rotation,** causes day and night to occur. Here is how it works. As Earth rotates, you can see the Sun come into view at daybreak. Earth continues to spin, making it seem as if the Sun moves across the sky until it sets at night. During night, your area of Earth has rotated so that it is facing away from the Sun. Because of this, the Sun is no longer visible to you. Earth continues to rotate steadily, and eventually the Sun comes into view again the next morning. One complete rotation takes about 24 h, or one day. How many rotations does Earth complete during one year? As you can infer from **Table 1,** it completes about 365 rotations during its one-year journey around the Sun.

Reading Check *Why does the Sun seem to rise and set?*

Earth's Rotation
Suppose that Earth's rotation took twice as long as it does now. In your Science Journal, predict how conditions such as global temperatures, work schedules, plant growth, and other factors might change under these circumstances.

Table 1 Physical Properties of Earth

Diameter (pole to pole)	12,714 km
Diameter (equator)	12,756 km
Circumference (poles)	40,008 km
Circumference (equator)	40,075 km
Mass	5.98×10^{24} kg
Average density	5.52 g/cm^3
Average distance to the Sun	149,600,000 km
Period of rotation (1 day)	23 h, 56 min
Period of revolution (1 year)	365 days, 6 h, 9 min

Axis

Rotation

Figure 2 Earth's magnetic field is similar to that of a bar magnet, almost as if Earth contained a giant magnet. Earth's magnetic axis is angled 11.5 degrees from its rotational axis.

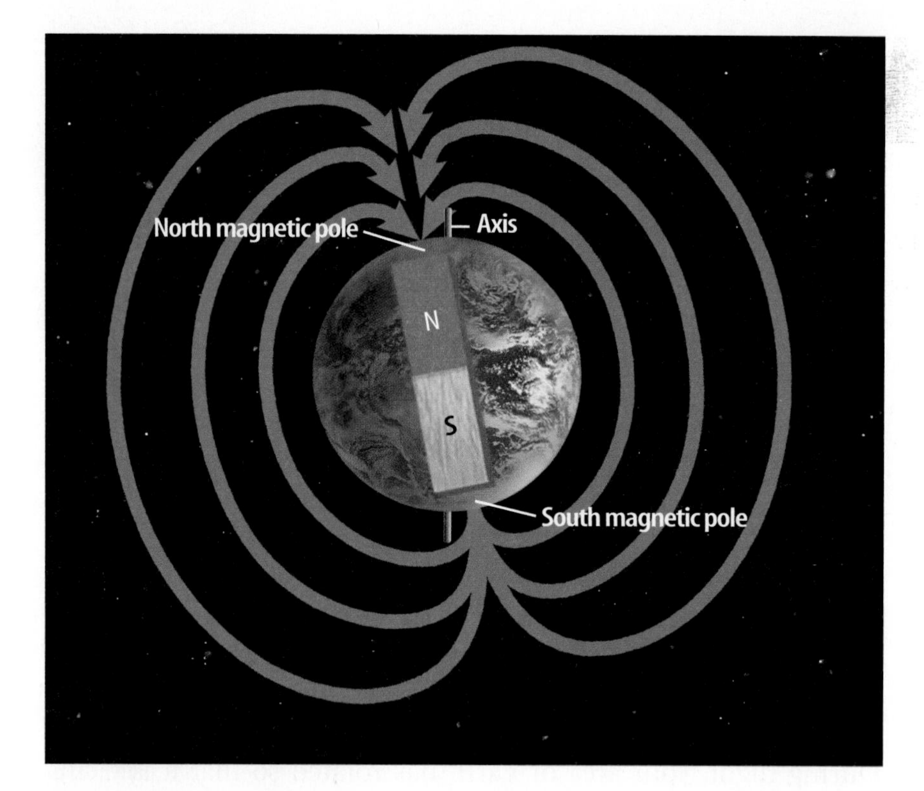

Mini LAB

Making Your Own Compass

Procedure

WARNING: *Use care when handling sharp objects.*

1. Cut off the bottom of a **plastic foam cup** to make a polystyrene disk.
2. Magnetize a **sewing needle** by continuously stroking the needle in the same direction with a **magnet** for 1 min.
3. **Tape** the needle to the center of the foam disk.
4. Fill a **plate** with **water** and float the disk, needle side up, in the water.

Analysis

1. What happened to the needle and disk when you placed them in the water? Why did this happen?
2. Infer how ancient sailors might have used magnets to help them navigate on the open seas.

Magnetic Field

INTEGRATE Physics

Scientists hypothesize that the movement of material inside Earth's core, along with Earth's rotation, generates a magnetic field. This magnetic field is much like that of a bar magnet. Earth has a north and a south magnetic pole, just as a bar magnet has opposite magnetic poles at each of its ends. When you sprinkle iron shavings over a bar magnet, the shavings align with the magnetic field of the magnet. As you can see in **Figure 2,** Earth's magnetic field is similar—almost as if Earth contained a giant bar magnet. Earth's magnetic field protects you from harmful solar radiation by trapping many charged particles from the Sun.

Magnetic Axis When you observe a compass needle pointing north, you are seeing evidence of Earth's magnetic field. Earth's magnetic axis, the line joining its north and south magnetic poles, does not align with its rotational axis. The magnetic axis is inclined at an angle of 11.5° to the rotational axis. If you followed a compass needle, you would end up at the magnetic north pole rather than the rotational north pole.

The location of the magnetic poles has been shown to change slowly over time. The magnetic poles move around the rotational (geographic) poles in an irregular way. This movement can be significant over decades. Many maps include information about the position of the magnetic north pole at the time the map was made. Why would this information be important?

What causes changing seasons?

Flowers bloom as the days get warmer. The Sun appears higher in the sky, and daylight lasts longer. Spring seems like a fresh, new beginning. What causes these wonderful changes?

Orbiting the Sun You learned earlier that Earth's rotation causes day and night. Another important motion is **revolution,** which is Earth's yearly orbit around the Sun. Just as the Moon is Earth's satellite, Earth is a satellite of the Sun. If Earth's orbit were a circle with the Sun at the center, Earth would maintain a constant distance from the Sun. However, this is not the case. Earth's orbit is an **ellipse** (ee LIHPS)—an elongated, closed curve. The Sun is not at the center of the ellipse but is a little toward one end. Because of this, the distance between Earth and the Sun changes during Earth's yearlong orbit. Earth gets closest to the Sun—about 147 million km away—around January 3. The farthest Earth gets from the Sun is about 152 million km away. This happens around July 4 each year.

Reading Check *What is an ellipse?*

Does this elliptical orbit cause seasonal temperatures on Earth? If it did, you would expect the warmest days to be in January. You know this isn't the case in the northern hemisphere, something else must cause the change.

Even though Earth is closest to the Sun in January, the change in distance is small. Earth is exposed to almost the same amount of Sun all year. But the amount of solar energy any one place on Earth receives varies greatly during the year. Next, you will learn why.

Topic: Ellipses
Visit earth.msscience.com for Web links to information about orbits and ellipses.

Activity Scientists compare orbits by how close they come to being circular. To do this, they use a measurement called eccentricity. A circle has an eccentricity of zero. Ellipses have eccentricities that are greater than zero, but less than one. The closer the eccentricity is to one, the more elliptical the orbit. Compare the orbits of the four inner planets. List them in order of increasing eccentricity.

A Tilted Axis Earth's axis is tilted 23.5° from a line drawn perpendicular to the plane of its orbit. It is this tilt that causes seasons. The number of daylight hours is greater for the hemisphere, or half of Earth, that is tilted toward the Sun. Think of how early it gets dark in the winter compared to the summer. As shown in **Figure 3,** the hemisphere that is tilted toward the Sun receives more hours of sunlight each day than the hemisphere that is tilted away from the Sun. The longer period of sunlight is one reason summer is warmer than winter, but it is not the only reason.

Figure 3 In summer, the northern hemisphere is tilted toward the Sun. Notice that the north pole is always lit during the summer.
Observe *Why is there a greater number of daylight hours in the summer than in the winter?*

Radiation from the Sun Earth's tilt also causes the Sun's radiation to strike the hemispheres at different angles. Sunlight strikes the hemisphere tilted towards the Sun at a higher angle, that is, closer to 90 degrees, than the hemisphere tilted away. Thus it receives more total solar radiation than the hemisphere tilted away from the Sun, where sunlight strikes at a lower angle.

Summer occurs in the hemisphere tilted toward the Sun, when its radiation strikes Earth at a higher angle and for longer periods of time. The hemisphere receiving less radiation experiences winter.

Solstices

The **solstice** is the day when the Sun reaches its greatest distance north or south of the equator. In the northern hemisphere, the summer solstice occurs on June 21 or 22, and the winter solstice occurs on December 21 or 22. Both solstices are illustrated in **Figure 4.** In the southern hemisphere, the winter solstice is in June and the summer solstice is in December. Summer solstice is about the longest period of daylight of the year. After this, the number of daylight hours become less and less, until the winter solstice, about the shortest period of daylight of the year. Then the hours of daylight start to increase again.

Figure 4 During the summer solstice in the northern hemisphere, the Sun is directly over the tropic of Cancer, the latitude line at 23.5° N latitude. During the winter solstice, the Sun is directly over the tropic of Capricorn, the latitude line at 23.5° S latitude. At fall and spring equinoxes, the Sun is directly over the equator.

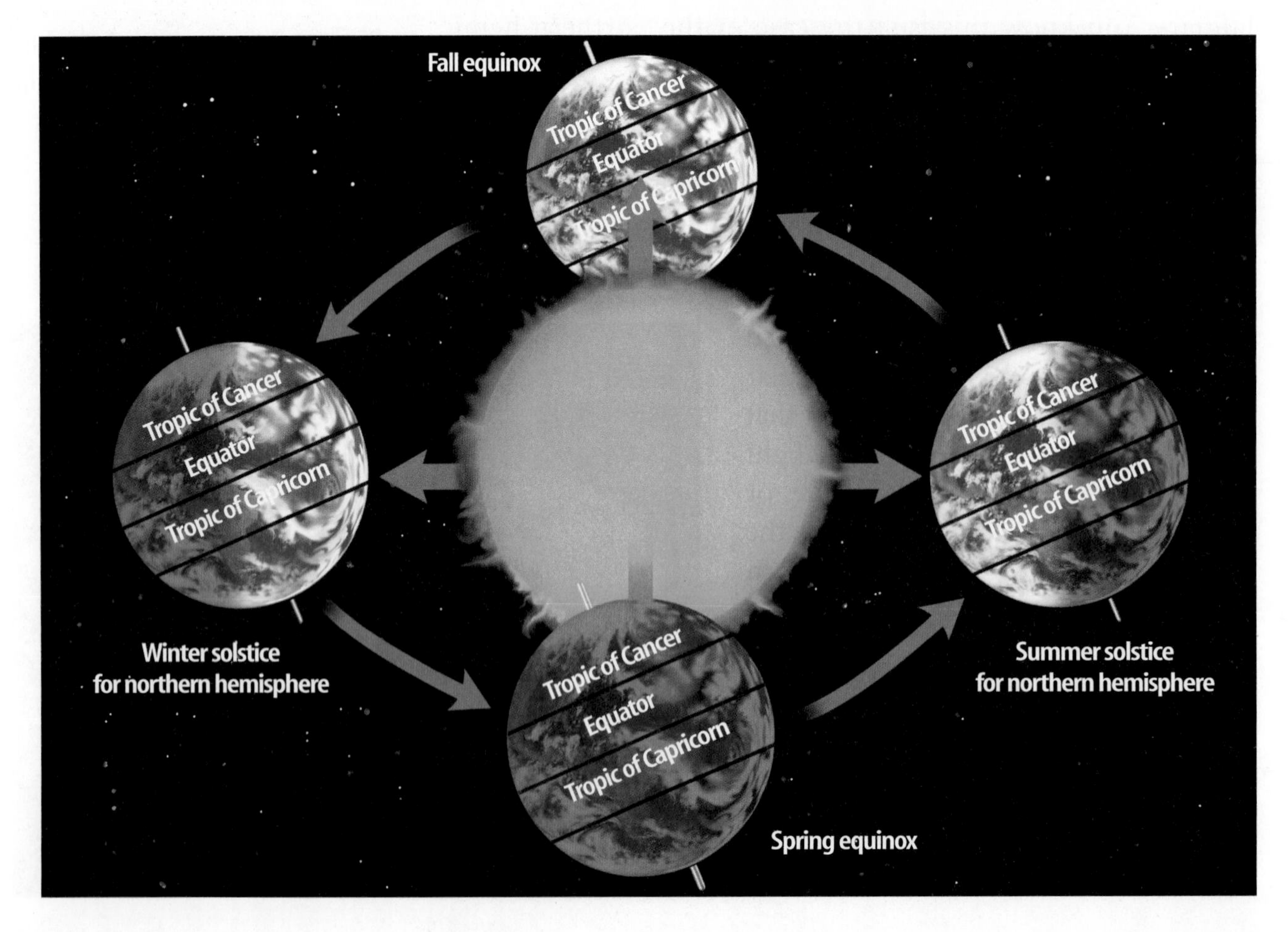

Equinoxes

An **equinox** (EE kwuh nahks) occurs when the Sun is directly above Earth's equator. Because of the tilt of Earth's axis, the Sun's position relative to the equator changes constantly. Most of the time, the Sun is either north or south of the equator, but two times each year it is directly over it, resulting in the spring and fall equinoxes. As you can see in **Figure 4,** at an equinox the Sun strikes the equator at the highest possible angle, 90°.

During an equinox, the number of daylight hours and nighttime hours is nearly equal all over the world. Also at this time, neither the northern hemisphere nor the southern hemisphere is tilted toward the Sun.

In the northern hemisphere, the Sun reaches the spring equinox on March 20 or 21, and the fall equinox occurs on September 22 or 23. In the southern hemisphere, the equinoxes are reversed. Spring occurs in September and fall occurs in March.

Earth Data Review As you have learned, Earth is a sphere that rotates on a tilted axis. This rotation causes day and night. Earth's tilted axis and its revolution around the Sun cause the seasons. One Earth revolution takes one year. In the next section, you will read how the Moon rotates on its axis and revolves around Earth.

Topic: Seasons
Visit earth.msscience.com for Web links to information about the seasons.

Activity Make a poster describing how the seasons differ in other parts of the world. Show how holidays might be celebrated differently and how farming might vary between hemispheres.

section 1 review

Summary

Properties of Earth

- Earth is a slightly flattened sphere that rotates around an imaginary line called an axis.
- Earth has a magnetic field, much like a bar magnet.
- The magnetic axis of Earth differs from its rotational axis.

Seasons

- Earth revolves around the Sun in an elliptical orbit.
- The tilt of Earth's axis and its revolution cause the seasons.
- Solstices are days when the Sun reaches its farthest points north or south of the equator.
- Equinoxes are the points when the Sun is directly over the equator.

Self Check

1. **Explain** why Aristotle thought Earth was spherical.
2. **Compare and contrast** rotation and revolution.
3. **Describe** how Earth's distance from the Sun changes throughout the year. When is Earth closest to the Sun?
4. **Explain** why it is summer in Earth's northern hemisphere at the same time it is winter in the southern hemisphere.
5. **Think Critically** **Table 1** lists Earth's distance from the Sun as an average. Why isn't an exact measurement available for this distance?

Applying Skills

6. **Classify** The terms *clockwise* and *counterclockwise* are used to indicate the direction of circular motion. How would you classify the motion of the Moon around Earth as you view it from above Earth's north pole? Now try to classify Earth's movement around the Sun.

Science Online earth.msscience.com/self_check_quiz

section 2

The Moon—Earth's Satellite

as you read

What **You'll Learn**

- **Identify** phases of the Moon and their cause.
- **Explain** why solar and lunar eclipses occur.
- **Infer** what the Moon's surface features may reveal about its history.

Why **It's Important**

Learning about the Moon can teach you about Earth.

Review Vocabulary
mantle: portion of the interior of a planet or moon lying between the crust and core

New Vocabulary

- moon phase
- new moon
- waxing
- full moon
- waning
- solar eclipse
- lunar eclipse
- maria

Motions of the Moon

Just as Earth rotates on its axis and revolves around the Sun, the Moon rotates on its axis and revolves around Earth. The Moon's revolution around Earth is responsible for the changes in its appearance. If the Moon rotates on its axis, why can't you see it spin around in space? The reason is that the Moon's rotation takes 27.3 days—the same amount of time it takes to revolve once around Earth. Because these two motions take the same amount of time, the same side of the Moon always faces Earth, as shown in **Figure 5.**

You can demonstrate this by having a friend hold a ball in front of you. Direct your friend to move the ball in a circle around you while keeping the same side of it facing you. Everyone else in the room will see all sides of the ball. You will see only one side. If the moon didn't rotate, we would see all of its surface during one month.

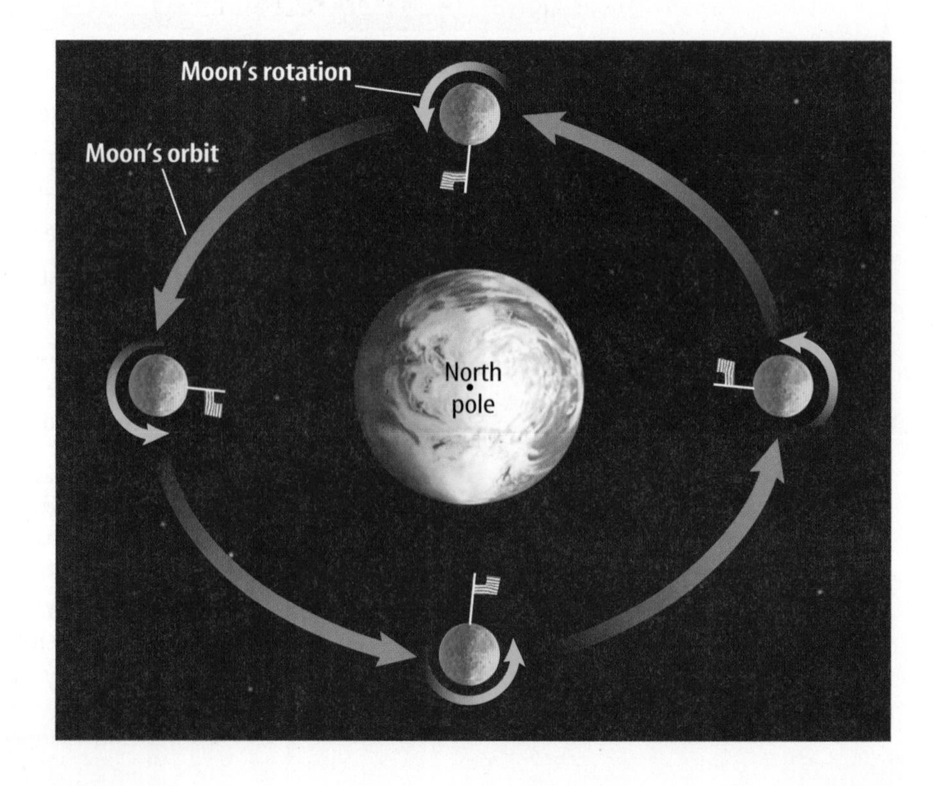

Figure 5 In about 27.3 days, the Moon orbits Earth. It also completes one rotation on its axis during the same period. **Think Critically** *Explain how this affects which side of the Moon faces Earth.*

Reflection of the Sun The Moon seems to shine because its surface reflects sunlight. Just as half of Earth experiences day as the other half experiences night, half of the Moon is lighted while the other half is dark. As the Moon revolves around Earth, you see different portions of its lighted side, causing the Moon's appearance to change.

Phases of the Moon

Moon phases are the different forms that the Moon takes in its appearance from Earth. The phase depends on the relative positions of the Moon, Earth, and the Sun, as seen in **Figure 6** on the next page. A **new moon** occurs when the Moon is between Earth and the Sun. During a new moon, the lighted half of the Moon is facing the Sun and the dark side faces Earth. The Moon is in the sky, but it cannot be seen. The new moon rises and sets with the Sun.

Reading Check *Why can't you see a new moon?*

Waxing Phases After a new moon, the phases begin waxing. **Waxing** means that more of the illuminated half of the Moon can be seen each night. About 24 h after a new moon, you can see a thin slice of the Moon. This phase is called the waxing crescent. About a week after a new moon, you can see half of the lighted side of the Moon, or one quarter of the Moon's surface. This is the first quarter phase.

The phases continue to wax. When more than one quarter is visible, it is called waxing *gibbous* after the Latin word for "humpbacked." A **full moon** occurs when all of the Moon's surface facing Earth reflects light.

Waning Phases After a full moon, the phases are said to be waning. When the Moon's phases are **waning,** you see less of its illuminated half each night. Waning gibbous begins just after a full moon. When you can see only half of the lighted side, it is the third-quarter phase. The Moon continues to appear to shrink. Waning crescent occurs just before another new moon. Once again, you can see only a small slice of the Moon.

It takes about 29.5 days for the Moon to complete its cycle of phases. Recall that it takes about 27.3 days for the Moon to revolve around Earth. The discrepancy between these two numbers is due to Earth's revolution. The roughly two extra days are what it takes for the Sun, Earth, and Moon to return to their same relative positions.

Comparing the Sun and the Moon

Procedure

1. Find an area where you can make a chalk mark on **pavement or similar surface.**
2. Tie a piece of **chalk** to one end of a 200-cm-long **string.**
3. Hold the other end of the string to the pavement.
4. Have a friend pull the string tight and walk around you, drawing a circle (the Sun) on the pavement.
5. Draw a 1-cm-diameter circle in the middle of the larger circle (the Moon).

Analysis

1. How big is the Sun compared to the Moon?
2. The diameter of the Sun is 1.39 million km. The diameter of Earth is 12,756 km. Draw two new circles modeling the sizes of the Sun and Earth. What scale did you use?

Try at Home

Figure 6 The phases of the Moon change during a cycle that lasts about 29.5 days.

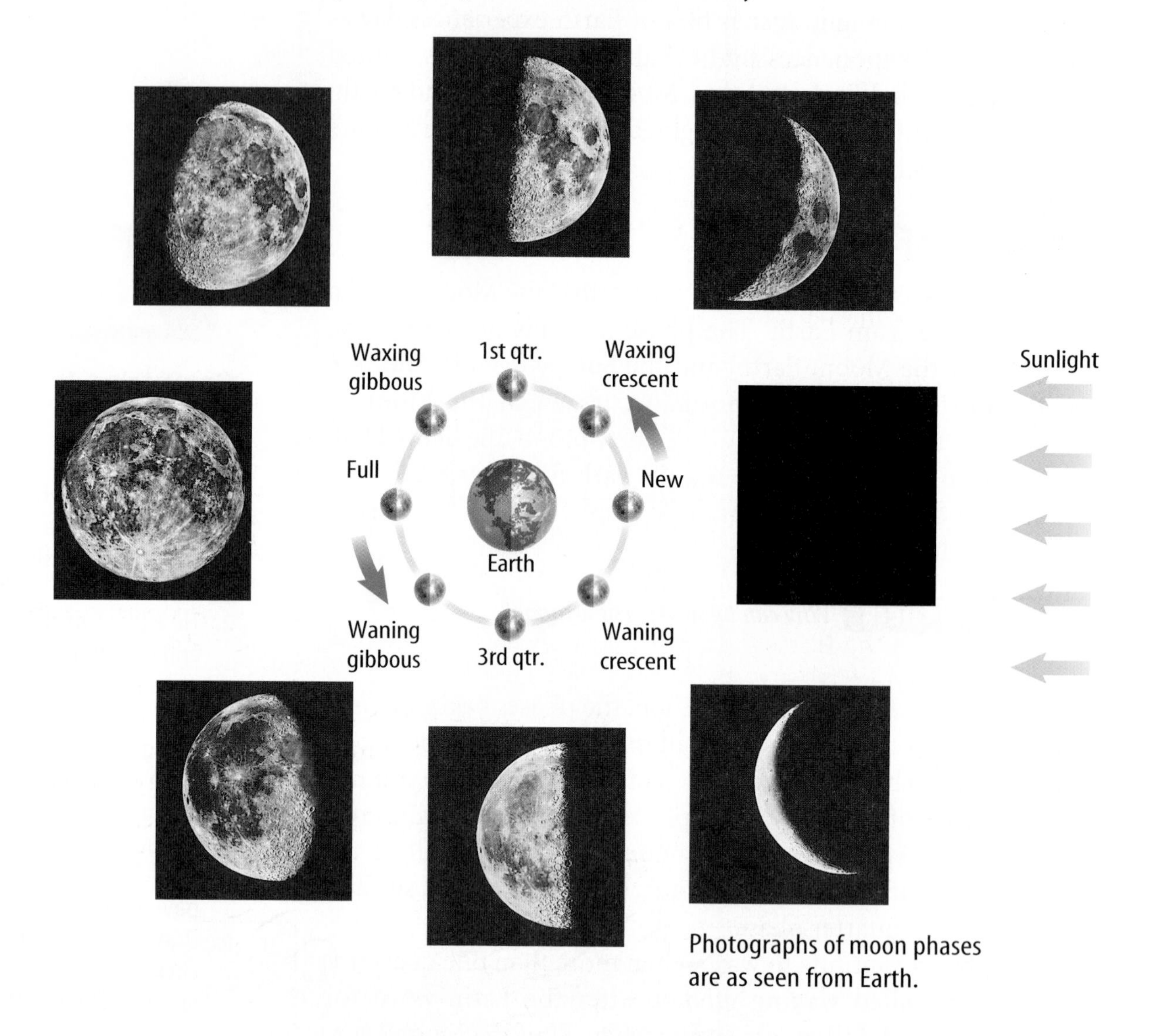

Figure 7 The outer portion of the Sun's atmosphere is visible during a total solar eclipse. It looks like a halo around the Moon.

Eclipses

Imagine living 10,000 years ago. You are foraging for nuts and fruit when unexpectedly the Sun disappears from the sky. The darkness lasts only a short time, and the Sun soon returns to full brightness. You know something strange has happened, but you don't know why. It will be almost 8,000 years before anyone can explain what you just experienced.

The event just described was a total solar eclipse (ih KLIPS), shown in **Figure 7.** Today, most people know what causes such eclipses, but without this knowledge, they would have been terrifying events. During a solar eclipse, many animals act as if it is nighttime. Cows return to their barns and chickens go to sleep. What causes the day to become night and then change back into day?

Reading Check *What happens during a total solar eclipse?*

What causes an eclipse? The revolution of the Moon causes eclipses. Eclipses occur when Earth or the Moon temporarily blocks the sunlight from reaching the other. Sometimes, during a new moon, the Moon's shadow falls on Earth and causes a solar eclipse. During a full moon, Earth's shadow can be cast on the Moon, resulting in a lunar eclipse.

An eclipse can occur only when the Sun, the Moon, and Earth are lined up perfectly. Because the Moon's orbit is not in the same plane as Earth's orbit around the Sun, lunar eclipses occur only a few times each year.

Topic: Eclipses
Visit earth.msscience.com for Web links to information about solar and lunar eclipses.

Activity Make a chart showing the dates when lunar and solar eclipses will be visible in your area. Include whether the eclipses will be total or partial.

Eclipses of the Sun A **solar eclipse** occurs when the Moon moves directly between the Sun and Earth and casts its shadow over part of Earth, as seen in **Figure 8.** Depending on where you are on Earth, you may experience a total eclipse or a partial eclipse. The darkest portion of the Moon's shadow is called the umbra (UM bruh). A person standing within the umbra experiences a total solar eclipse. During a total solar eclipse, the only visible portion of the Sun is a pearly white glow around the edge of the eclipsing Moon.

Surrounding the umbra is a lighter shadow on Earth's surface called the penumbra (puh NUM bruh). Persons standing in the penumbra experience a partial solar eclipse. **WARNING:** *Regardless of which eclipse you view, never look directly at the Sun. The light can permanently damage your eyes.*

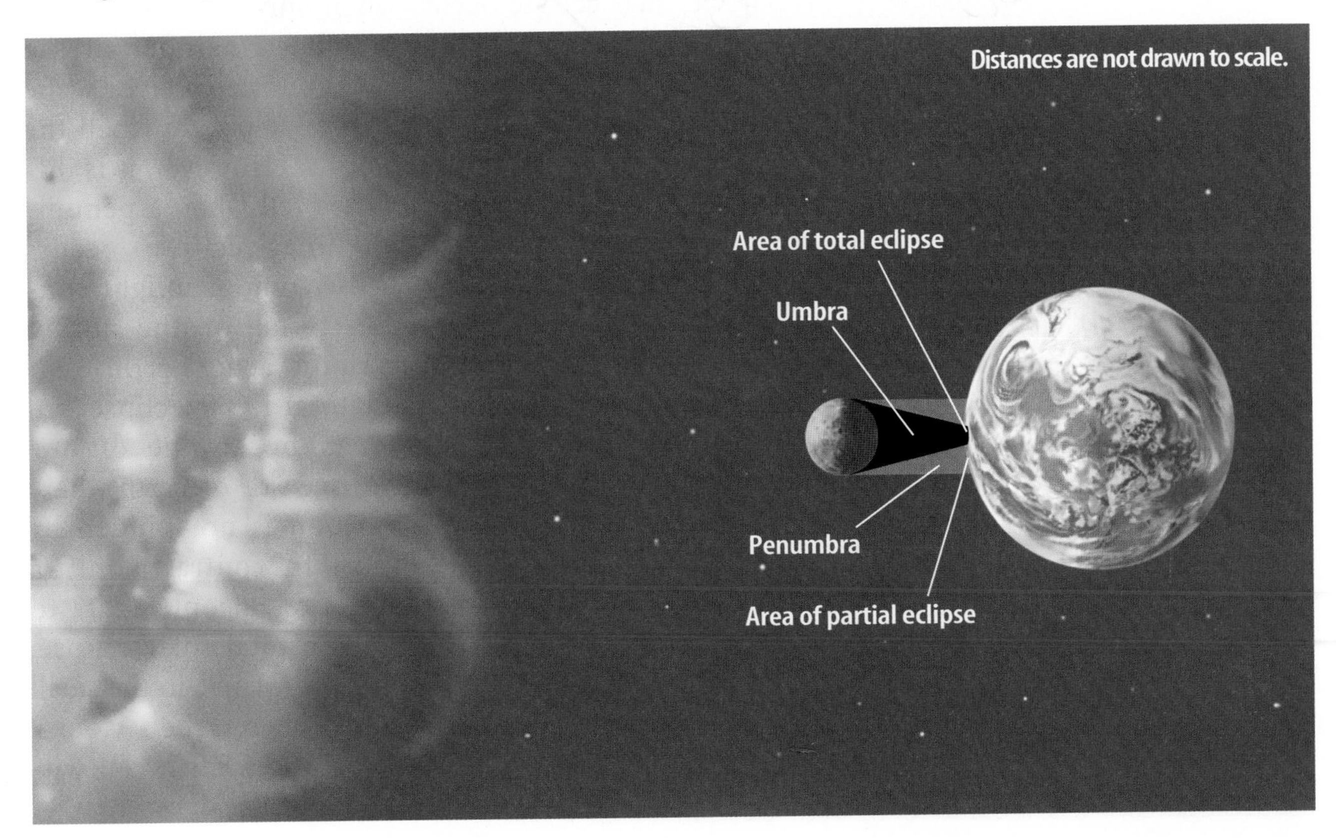

Figure 8 Only a small area of Earth experiences a total solar eclipse during the eclipse event.

Figure 9 These photographs show the Moon moving from right to left into Earth's umbra, then out again.

Eclipses of the Moon When Earth's shadow falls on the Moon, a **lunar eclipse** occurs. A lunar eclipse begins when the Moon moves into Earth's penumbra. As the Moon continues to move, it enters Earth's umbra and you see a curved shadow on the Moon's surface, as in **Figure 9.** Upon moving completely into Earth's umbra, as shown in **Figure 10,** the Moon goes dark, signaling that a total lunar eclipse has occurred. Sometimes sunlight bent through Earth's atmosphere causes the eclipsed Moon to appear red.

A partial lunar eclipse occurs when only a portion of the Moon moves into Earth's umbra. The remainder of the Moon is in Earth's penumbra and, therefore, receives some direct sunlight. A penumbral lunar eclipse occurs when the Moon is totally within Earth's penumbra. However, it is difficult to tell when a penumbral lunar eclipse occurs because some sunlight continues to fall on the side of the Moon facing Earth.

A total lunar eclipse can be seen by anyone on the nighttime side of Earth where the Moon is not hidden by clouds. In contrast, only a lucky few people get to witness a total solar eclipse. Only those people in the small region where the Moon's umbra strikes Earth can witness one.

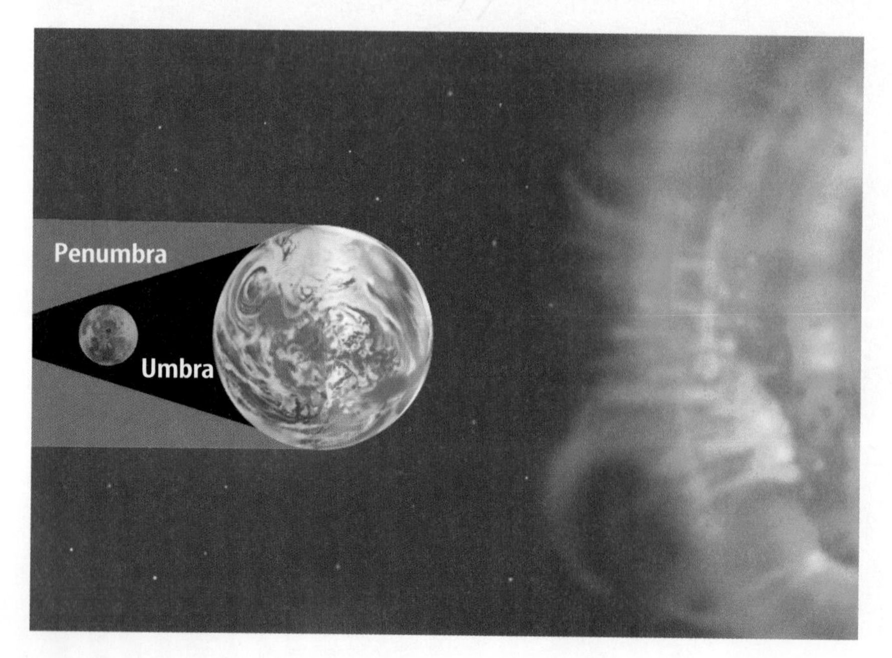

Figure 10 During a total lunar eclipse, Earth's shadow blocks light coming from the Sun.

The Moon's Surface

When you look at the Moon, as shown in **Figure 12** on the next page, you can see many depressions called craters. Meteorites, asteroids, and comets striking the Moon's surface created most of these craters, which formed early in the Moon's history. Upon impact, cracks may have formed in the Moon's crust, allowing lava to reach the surface and fill up the large craters. The resulting dark, flat regions are called **maria** (MAHR ee uh). The igneous rocks of the maria are 3 billion to 4 billion years old. So far, they are the youngest rocks to be found on the Moon. This indicates that craters formed after the Moon's surface originally cooled. The maria formed early enough in the Moon's history that molten material still remained in the Moon's interior. The Moon once must have been as geologically active as Earth is today. Before the Moon cooled to the current condition, the interior separated into distinct layers.

Seismology A seismologist is an Earth scientist who studies the propagation of seismic waves in geological materials. Usually this means studying earthquakes, but some seismologists apply their knowledge to studies of the Moon and planets. Seismologists usually study geology, physics, and applied mathematics in college and later specialize in seismology for an advanced degree.

Inside the Moon

Earthquakes allow scientists to learn about Earth's interior. In a similar way, scientists use instruments such as the one in **Figure 11** to study moonquakes. The data they have received have led to the construction of several models of the Moon's interior. One such model, shown in **Figure 11,** suggests that the Moon's crust is about 60 km thick on the side facing Earth. On the far side, it is thought to be about 150 km thick. Under the crust, a solid mantle may extend to a depth of 1,000 km. A partly molten zone of the mantle may extend even farther down. Below this mantle may lie a solid, iron-rich core.

Figure 11 Equipment, such as the seismograph left on the Moon by the *Apollo 12* mission, helps scientists study moonquakes.

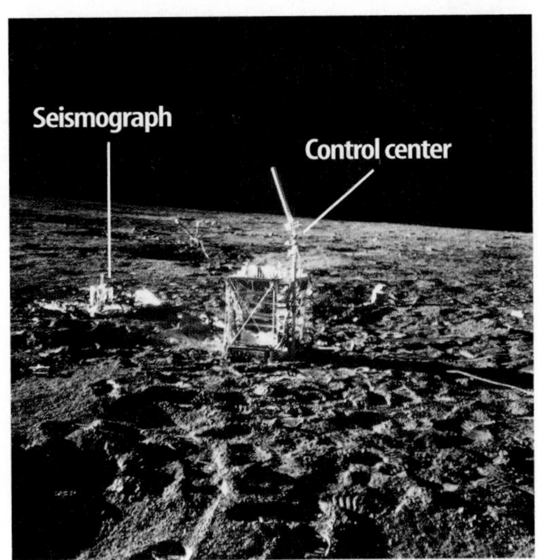

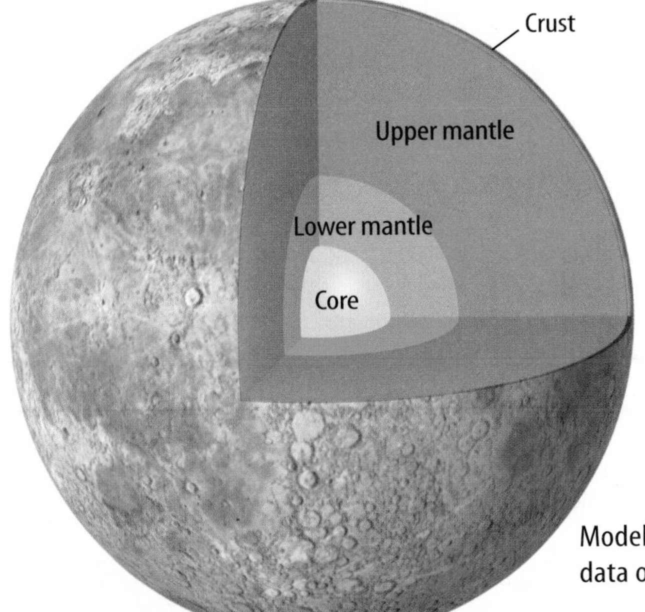

Models of the Moon's interior were created from data obtained by scientists studying moonquakes.

NATIONAL GEOGRAPHIC **VISUALIZING THE MOON'S SURFACE**

Figure 12

By looking through binoculars, you can see many of the features on the surface of the Moon. These include craters that are hundreds of kilometers wide, light-colored mountains, and darker patches that early astronomers called maria (Latin for "seas"). However, as the NASA Apollo missions discovered, these so-called seas do not contain water. In fact, maria (singular, mare) are flat, dry areas formed by ancient lava flows. Some of the Moon's geographic features are shown below, along with the landing sites of Apollo missions sent to investigate Earth's closest neighbor in space.

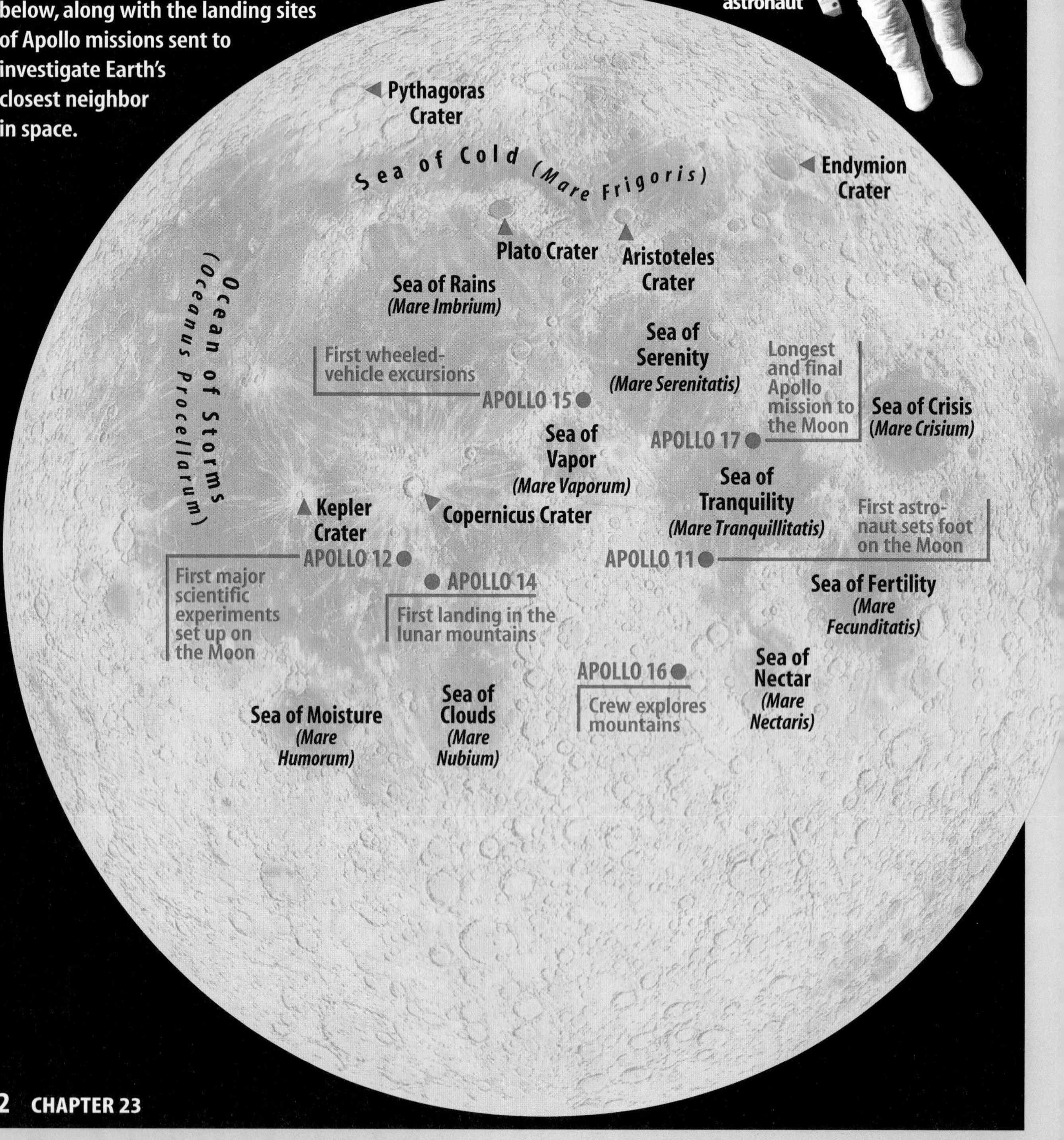

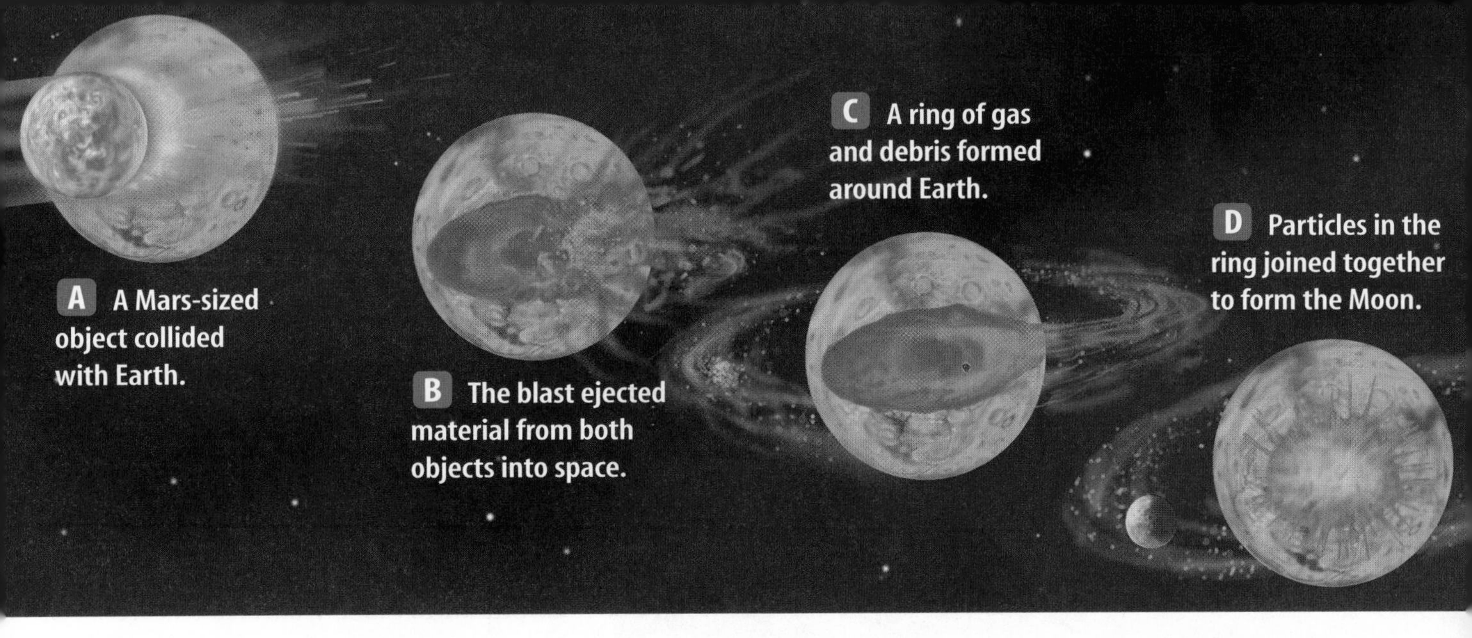

A A Mars-sized object collided with Earth.
B The blast ejected material from both objects into space.
C A ring of gas and debris formed around Earth.
D Particles in the ring joined together to form the Moon.

Figure 13 According to the impact theory, a Mars-sized object collided with Earth around 4.6 billion years ago. Vaporized materials ejected by the collision began orbiting Earth and quickly consolidated into the Moon.

The Moon's Origin

Before the *Apollo* space missions in the 1960s and 1970s, there were three leading theories about the Moon's origin. According to one theory, the Moon was captured by Earth's gravity. Another held that the Moon and Earth condensed from the same cloud of dust and gas. An alternative theory proposed that Earth ejected molten material that became the Moon.

The Impact Theory The data gathered by the *Apollo* missions have led many scientists to support a new theory, known as the impact theory. It states that the Moon formed billions of years ago from condensing gas and debris thrown off when Earth collided with a Mars-sized object as shown in **Figure 13.**

Applying Science

What will you use to survive on the Moon?

You have crash-landed on the Moon. It will take one day to reach a moon colony on foot. The side of the Moon that you are on will be facing away from the Sun during your entire trip. You manage to salvage the following items from your wrecked ship: food, rope, solar-powered heating unit, battery-operated heating unit, oxygen tanks, map of the constellations, compass, matches, water, solar-powered radio transmitter, three flashlights, signal mirror, and binoculars.

Identifying the Problem

The Moon lacks a magnetic field and has no atmosphere. How do the Moon's physical properties and the lack of sunlight affect your decisions?

Solving the Problem

1. Which items will be of no use to you? Which items will you take with you?
2. Describe why each of the salvaged items is useful or not useful.

Figure 14 Moon rocks collected by astronauts provide scientists with information about the Moon and Earth.

The Moon in History Studying the Moon's phases and eclipses led to the conclusion that both Earth and the Moon were in motion around the Sun. The curved shadow Earth casts on the Moon indicated to early scientists that Earth was spherical. When Galileo first turned his telescope toward the Moon, he found a surface scarred by craters and maria. Before that time, many people believed that all planetary bodies were perfectly smooth and lacking surface features. Now, actual moon rocks are available for scientists to study, as seen in **Figure 14.** By doing so, they hope to learn more about Earth.

Reading Check *How has observing the Moon been important to science?*

section 2 review

Summary

Motions of the Moon

- The Moon rotates on its axis about once each month.
- The Moon also revolves around Earth about once every 27.3 days.
- The Moon shines because it reflects sunlight.

Phases of the Moon

- During the waxing phases, the illuminated portion of the Moon grows larger.
- During waning phases, the illuminated portion of the Moon grows smaller.
- Earth passing directly between the Sun and the Moon causes a lunar eclipse.
- The Moon passing between Earth and the Sun causes a solar eclipse.

Structure and Origin of the Moon

- The Moon's surface is covered with depressions called impact craters.
- Flat, dark regions within craters are called maria.
- The Moon may have formed as the result of a collision between Earth and a Mars-sized object.

Self Check

1. **Explain** how the Sun, Moon, and Earth are positioned relative to each other during a new moon and how this alignment changes to produce a full moon.
2. **Describe** what phase the Moon must be in to have a lunar eclipse. A solar eclipse?
3. **Define** the terms *umbra* and *penumbra* and explain how they relate to eclipses.
4. **Explain** why lunar eclipses are more common than solar eclipses and why so few people ever have a chance to view a total solar eclipse.
5. **Think Critically** What do the surface features and their distribution on the Moon's surface tell you about its history?

Applying Math

6. **Solve Simple Equations** The Moon travels in its orbit at about 3,400 km/h. Therefore, during a solar eclipse, its shadow sweeps at this speed from west to east. However, Earth rotates from west to east at about 1,670 km/h near the equator. At what speed does the shadow really move across this part of Earth's surface?

Moon Phases and Eclipses

In this lab, you will demonstrate the positions of the Sun, the Moon, and Earth during certain phases and eclipses. You also will see why only a small portion of the people on Earth witness a total solar eclipse during a particular eclipse event.

Real-World Question

Can a model be devised to show the positions of the Sun, the Moon, and Earth during various phases and eclipses?

Goals

- **Model** moon phases.
- **Model** solar and lunar eclipses.

Materials

light source (unshaded)
globe
polystyrene ball
pencil

Safety Precautions

Procedure

1. Review the illustrations of moon phases and eclipses shown in Section 2.
2. Use the light source as a Sun model and a polystyrene ball on a pencil as a Moon model. Move the Moon around the globe to duplicate the exact position that would have to occur for a lunar eclipse to take place.
3. Move the Moon to the position that would cause a solar eclipse.
4. Place the Moon at each of the following phases: first quarter, full moon, third quarter, and new moon. Identify which, if any, type of eclipse could occur during each phase. Record your data.

Moon Phase Observations

Moon Phase	Observations
First quarter	
Full moon	
Third quarter	Do not write in this book.
New moon	

5. Place the Moon at the location where a lunar eclipse could occur. Move it slightly toward Earth, then away from Earth. Note the amount of change in the size of the shadow.
6. Repeat step 5 with the Moon in a position where a solar eclipse could occur.

Conclude and Apply

1. **Identify** which phase(s) of the Moon make(s) it possible for an eclipse to occur.
2. **Describe** the effect of a small change in distance between Earth and the Moon on the size of the umbra and penumbra.
3. **Infer** why a lunar and a solar eclipse do not occur every month.
4. **Explain** why only a few people have experienced a total solar eclipse.
5. **Diagram** the positions of the Sun, Earth, and the Moon during a first-quarter moon.
6. **Infer** why it might be better to call a full moon a half moon.

Communicate your answers to other students.

section 3

Exploring Earth's Moon

as you read

What You'll Learn

- **Describe** recent discoveries about the Moon.
- **Examine** facts about the Moon that might influence future space travel.

Why It's Important

Continuing moon missions may result in discoveries about Earth's origin.

Review Vocabulary
comet: space object orbiting the Sun formed from dust and rock particles mixed with frozen water, methane, and ammonia

New Vocabulary
- impact basin

Missions to the Moon

The Moon has always fascinated humanity. People have made up stories about how it formed. Children's stories even suggested it was made of cheese. Of course, for centuries astronomers also have studied the Moon for clues to its makeup and origin. In 1959, the former Soviet Union launched the first *Luna* spacecraft, enabling up-close study of the Moon. Two years later, the United States began a similar program with the first *Ranger* spacecraft and a series of *Lunar Orbiters.* The spacecraft in these early missions took detailed photographs of the Moon.

The next step was the *Surveyor* spacecraft designed to take more detailed photographs and actually land on the Moon. Five of these spacecraft successfully touched down on the lunar surface and performed the first analysis of lunar soil. The goal of the *Surveyor* program was to prepare for landing astronauts on the Moon. This goal was achieved in 1969 by the astronauts of *Apollo 11.* By 1972, when the *Apollo* missions ended, 12 U.S. astronauts had walked on the Moon. A time line of these important moon missions can be seen in **Figure 15.**

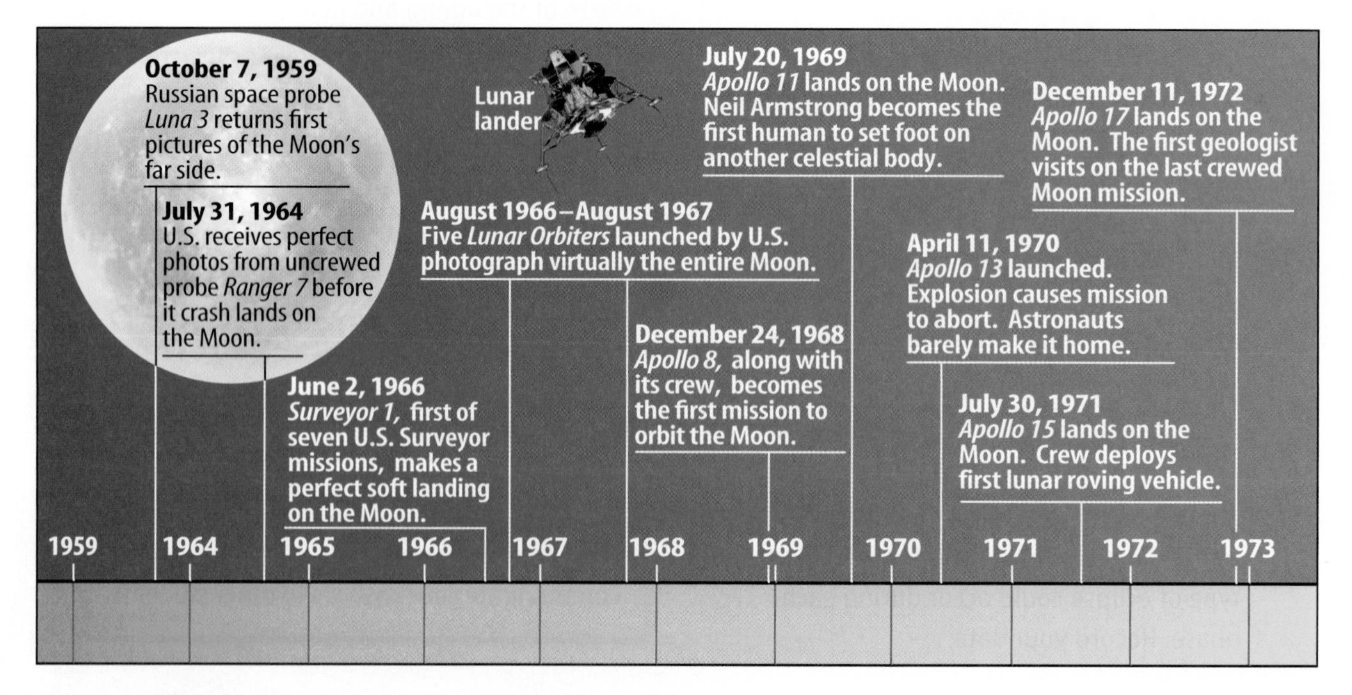

October 7, 1959
Russian space probe Luna 3 returns first pictures of the Moon's far side.
July 31, 1964
U.S. receives perfect photos from uncrewed probe Ranger 7 before it crash lands on the Moon.
June 2, 1966
Surveyor 1, first of seven U.S. Surveyor missions, makes a perfect soft landing on the Moon.
Lunar lander
August 1966–August 1967
Five Lunar Orbiters launched by U.S. photograph virtually the entire Moon.
December 24, 1968
Apollo 8, along with its crew, becomes the first mission to orbit the Moon.
July 20, 1969
Apollo 11 lands on the Moon. Neil Armstrong becomes the first human to set foot on another celestial body.
April 11, 1970
Apollo 13 launched. Explosion causes mission to abort. Astronauts barely make it home.
July 30, 1971
Apollo 15 lands on the Moon. Crew deploys first lunar roving vehicle.
December 11, 1972
Apollo 17 lands on the Moon. The first geologist visits on the last crewed Moon mission.
1959
1964
1965
1966
1967
1968
1969
1970
1971
1972
1973

Figure 15 This time line illustrates some of the most important events in the history of moon exploration.

Surveying the Moon There is still much to learn about the Moon and, for this reason, the United States resumed its studies. In 1994, the *Clementine* was placed into lunar orbit. Its goal was to conduct a two-month survey of the Moon's surface. An important aspect of this study was collecting data on the mineral content of Moon rocks. In fact, this part of its mission was instrumental in naming the spacecraft. Clementine was the daughter of a miner in the ballad *My Darlin' Clementine.* While in orbit, *Clementine* also mapped features on the Moon's surface, including huge impact basins.

Reading Check *Why was* **Clementine** *placed in lunar orbit?*

Impact Basins When meteorites and other objects strike the Moon, they leave behind depressions in the Moon's surface. The depression left behind by an object striking the Moon is known as an **impact basin,** or impact crater. The South Pole-Aitken Basin is the oldest identifiable impact feature on the Moon's surface. At 12 km in depth and 2,500 km in diameter, it is also the largest and deepest impact basin in the solar system.

Impact basins at the poles were of special interest to scientists. Because the Sun's rays never strike directly, the crater bottoms remain always in shadow. Temperatures in shadowed areas, as shown in **Figure 16,** would be extremely low, probably never more than −173°C. Scientists hypothesize that any ice deposited by comets impacting the Moon throughout its history would remain in these shadowed areas. Indeed, early signals from *Clementine* indicated the presence of water. This was intriguing, because it could be a source of water for future moon colonies.

Science Online

Topic: The Far Side

Visit earth.msscience.com for Web links to information about the far side of the Moon.

Activity Compare the image of the far side of the Moon with that of the near side shown in **Figure 12.** Make a list of all the differences you note and then compare them with lists made by other students.

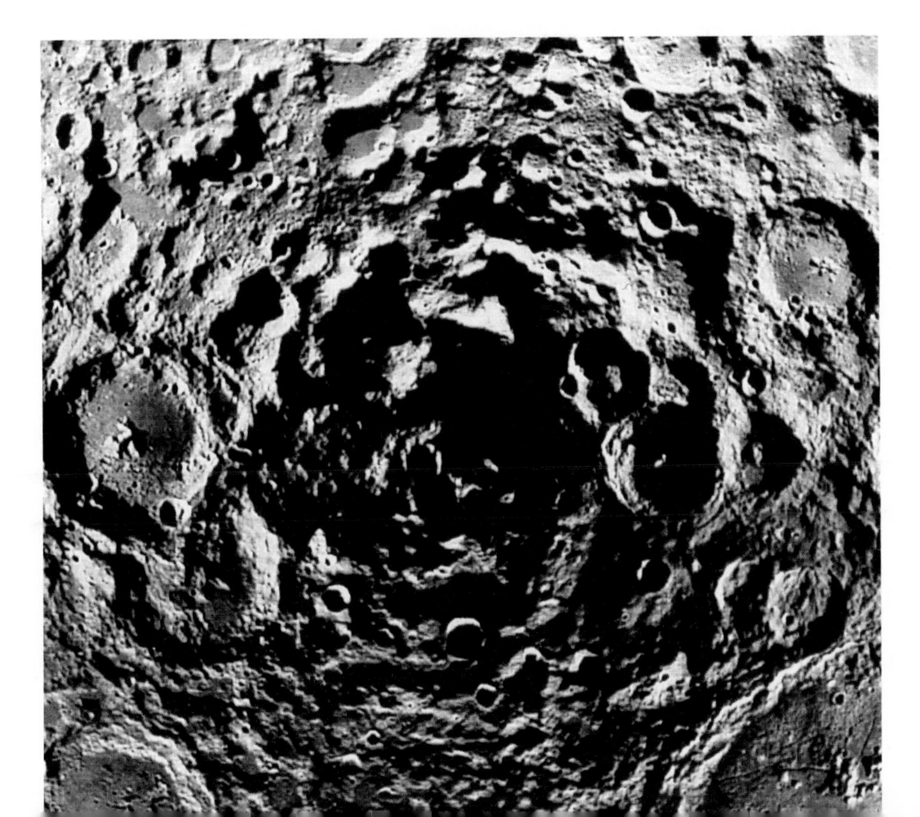

Figure 16 The South Pole-Aitken Basin is the largest of its kind found anywhere in the solar system. The deepest craters in the basin stay in shadow throughout the Moon's rotation. Ice deposits from impacting comets are thought to have collected at the bottom of these craters.

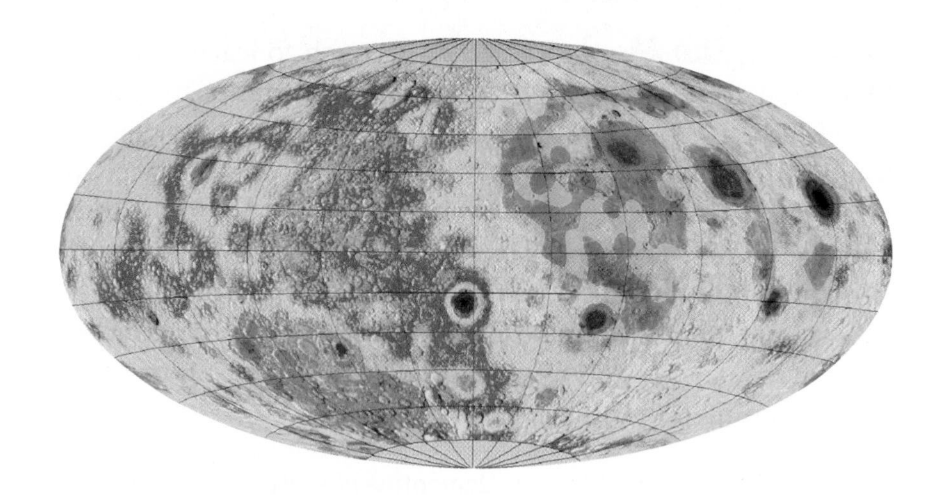

Figure 17 This computer-enhanced map based on *Clementine* data indicates the thickness of the Moon's crust. The crust of the side of the Moon facing Earth, shown mostly in red, is thinner than the crust on the far side of the Moon.

Mapping the Moon

A large part of *Clementine's* mission included taking high-resolution photographs so a detailed map of the Moon's surface could be compiled. *Clementine* carried cameras and other instruments to collect data at wavelengths ranging from infrared to ultraviolet. One camera could resolve features as small as 20 m across. One image resulting from *Clementine* data is shown in **Figure 17.** It shows that the crust on the side of the Moon that faces Earth is much thinner than the crust on the far side. Additional information shows that the Moon's crust is thinnest under impact basins. Based on analysis of the light data received from *Clementine,* a global map of the Moon also was created that shows its composition, as seen in **Figure 18.**

Reading Check ***What information about the Moon did scientists learn from*** **Clementine?**

The Lunar Prospector The success of Clementine opened the door for further moon missions. In 1998, NASA launched the desk-sized *Lunar Prospector,* shown in **Figure 18,** into lunar orbit. The spacecraft spent a year orbiting the Moon from pole to pole, once every two hours. The resulting maps confirmed the *Clementine* data. Also, data from *Lunar Prospector* confirmed that the Moon has a small, iron-rich core about 600 km in diameter. A small core supports the impact theory of how the Moon formed—only a small amount of iron could be blasted away from Earth.

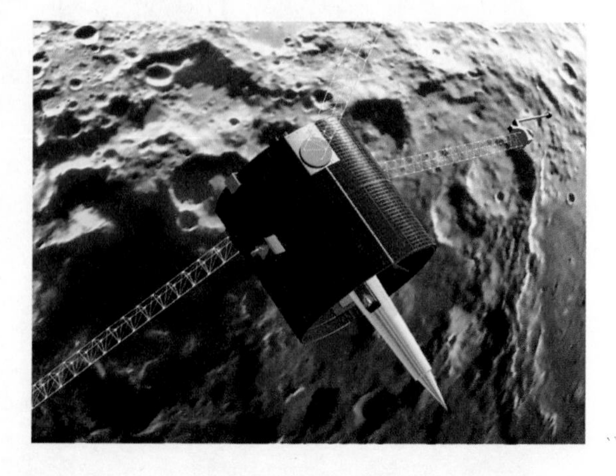

Figure 18 *Lunar Prospector* performed high-resolution mapping of the lunar surface and had instruments that detected water ice at the lunar poles.

Icy Poles In addition to photographing the surface, *Lunar Prospector* carried instruments designed to map the Moon's gravity, magnetic field, and the abundances of 11 elements in the lunar crust. This provided scientists with data from the entire lunar surface rather than just the areas around the Moon's equator, which had been gathered earlier. Also, *Lunar Prospector* confirmed the findings of *Clementine* that water ice was present in deep craters at both lunar poles.

Later estimates concluded that as much as 3 billion metric tons of water ice was present at the poles, with a bit more at the north pole. Using data from *Lunar Prospector,* scientists prepared maps showing the location of water ice at each pole. **Figure 19** shows how water may be distributed at the Moon's north pole. At first it was thought that ice crystals were mixed with lunar soil, but most recent results suggest that the ice may be in the form of more compact deposits.

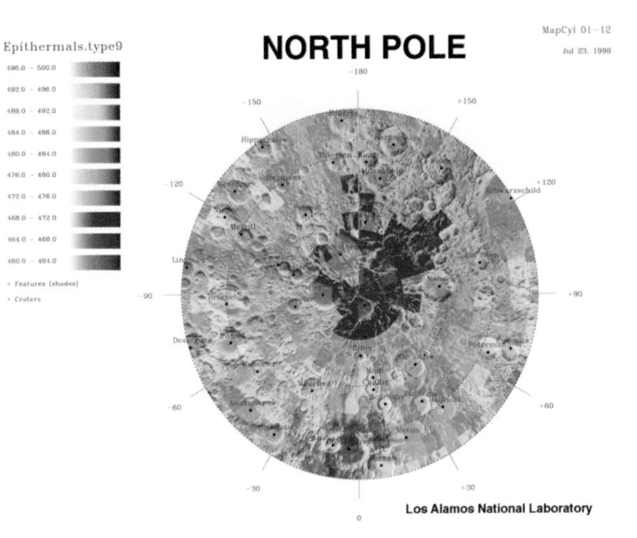

Figure 19 The *Lunar Prospector* data indicates that ice exists in crater shadows at the Moon's poles.

section 3 review

Summary

Missions to the Moon

- The first lunar surveys were done by *Luna,* launched by the former Soviet Union, and U.S.-launched *Ranger* and *Lunar Orbiters.*
- Five *Surveyor* probes landed on the Moon.
- U.S. Astronauts landed on and explored the Moon in the *Apollo* program.
- *Clementine,* a lunar orbiter, mapped the lunar surface and collected data on rocks.
- *Clementine* found that the lunar crust is thinner on the side facing Earth.
- Data from *Clementine* indicated that water ice could exist in shaded areas of impact basins.

Mapping the Moon

- *Lunar Prospector* orbited the Moon from pole to pole, collecting data that confirm *Clementine* results and that the Moon has a small iron-rich core.
- Data from *Lunar Prospector* indicate the presence of large quantities of water ice in craters at the lunar poles.

Self Check

1. **Name** the first U.S. spacecraft to successfully land on the Moon. What was the major purpose of this program?
2. **Explain** why scientists continue to study the Moon long after the *Apollo* program ended and list some of the types of data that have been collected.
3. **Explain** how water ice might be preserved in portions of deep impact craters.
4. **Describe** how the detection of a small iron-rich core supports the theory that the Moon was formed from a collision between Earth and a Mars-sized object.
5. **Think Critically** Why might the discovery of ice in impact basins at the Moon's poles be important to future space flights?

Applying Skills

6. **Infer** why it might be better to build a future moon base on a brightly lit plateau near a lunar pole in the vicinity of a deep crater. Why not build a base in the crater itself?

TILT AND TEMPERATURE

Goals

- **Measure** the temperature change in a surface after light strikes it at different angles.
- **Describe** how the angle of light relates to seasons on Earth.

Materials

tape
black construction paper (one sheet)
gooseneck lamp with 75-watt bulb
Celsius thermometer
watch
protractor

Safety Precautions

WARNING: *Do not touch the lamp without safety gloves. The lightbulb and shade can be hot even when the lamp has been turned off. Handle the thermometer carefully. If it breaks, do not touch anything. Inform your teacher immediately.*

If you walk on blacktop pavement at noon, you can feel the effect of solar energy. The Sun's rays hit at the highest angle at midday. Now consider the fact that Earth is tilted on its axis. How does this tilt affect the angle at which light rays strike an area on Earth? How is the angle of the light rays related to the amount of heat energy and the changing seasons?

Real-World Question

How does the angle at which light strikes Earth affect the amount of heat energy received by any area on Earth?

Procedure

1. Choose three angles that you will use to aim the light at the paper.
2. **Determine** how long you will shine the light at each angle before you measure the temperature. You will measure the temperature at two times for each angle. Use the same time periods for each angle.
3. Copy the following data table into your Science Journal to record the temperature the paper reaches at each angle and time.

Temperature Data

Angle of Lamp	Initial Temperature	Temperature at ___ Minutes/Seconds	Temperature at ___ Minutes/Seconds
First angle			
Second angle	Do not write in this book.		
Third angle			

4. Form a pocket out of a sheet of black construction paper and tape it to a desk or the floor.
5. Using the protractor, set the gooseneck lamp so that it will shine on the paper at one of the angles you chose.

6. Place the thermometer in the paper pocket. Turn on the lamp. Use the thermometer to measure the temperature of the paper at the end of the first time period. Continue shining the lamp on the paper until the second time period has passed. Measure the temperature again. Record your data in your data table.
7. Turn off the lamp until the paper cools to room temperature. Repeat steps 5 and 6 using your other two angles.

Conclude and Apply

1. **Describe** your experiment. Identify the variables in your experiment. Which were your independent and dependent variables?
2. **Graph** your data using a line graph. Describe what your graph tells you about the data.
3. **Describe** what happened to the temperature of the paper as you changed the angle of light.
4. **Predict** how your results might have been different if you used white paper. Explain why.
5. **Describe** how the results of this experiment apply to seasons on Earth.

Communicating Your Data

Compare your results with those of other students in your class. **Discuss** how the different angles and time periods affected the temperatures.

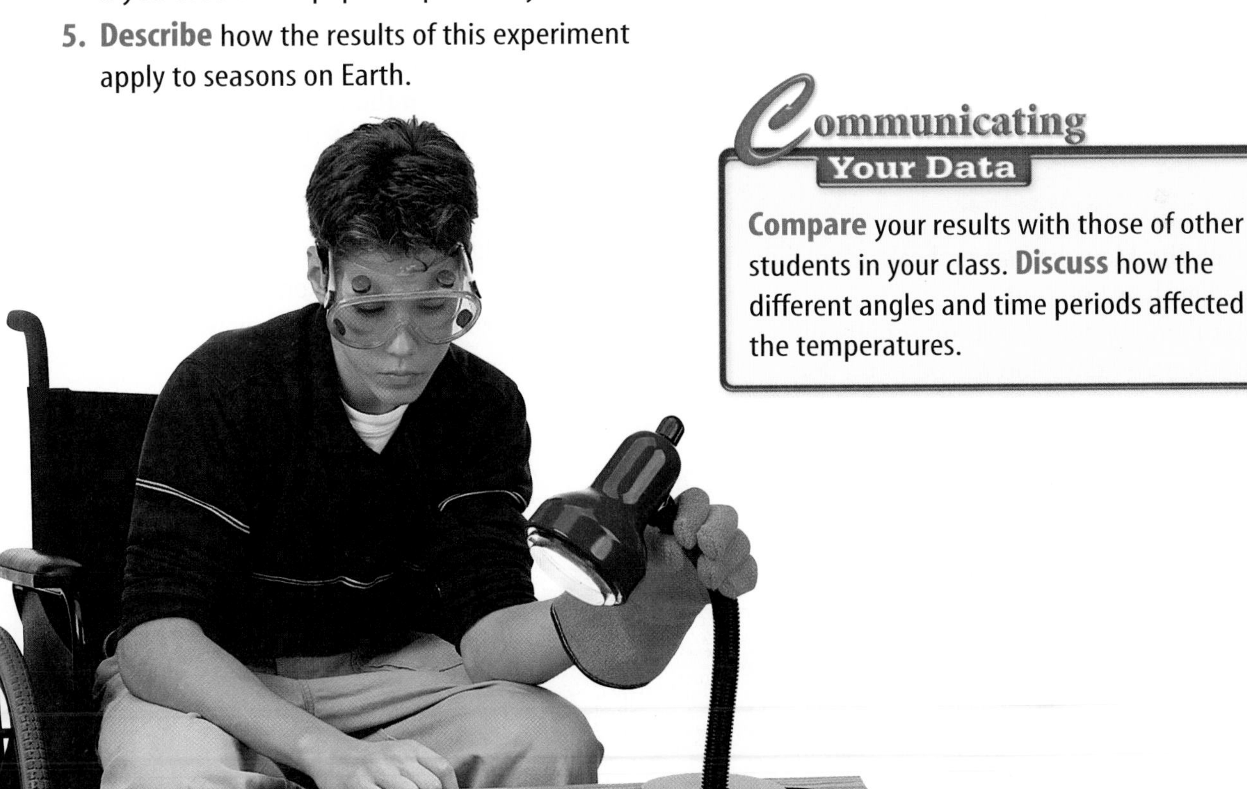

THE Mayan Calendar

Most people take for granted that a week is seven days, and that a year is 12 months. However there are other ways to divide time into useful units. Roughly 1,750 years ago, in what is now south Mexico and Central America, the Mayan people invented a calendar system based on careful observations of sun and moon cycles.

These glyphs represent four different days of the Tzolkin calendar.

In fact, the Maya had several calendars that they used at the same time.

Two calendars were most important—one was based on 260 days and the other on 365 days. The calendars were so accurate and useful that later civilizations, including the Aztecs, adopted them.

The 260-Day Calendar

This Mayan calendar, called the *Tzolkin* (tz uhl KIN), was used primarily to time planting, harvesting, drying, and storing of corn—their main crop. Each day of the *Tzolkin* had one of 20 names, as well as a number from 1 to 13 and a Mayan god associated with it.

The 365-Day Calendar

Another Mayan calendar, called the *Haab* (HAHB), was based on the orbit of Earth around the Sun. It was divided into 18 months with 20 days each, plus five extra days at the end of each year.

These calendars were used together making the Maya the most accurate reckoners of time before the modern period. In fact, they were only one day off every 6,000 years.

The Kukulkan, built around the year 1050 A.D., in what is now Chichén Itzá, Mexico, was used by the Maya as a calendar. It had four stairways, each with 91 steps, a total of 365 including the platform on top.

Drawing Symbols The Maya created picture symbols for each day of their week. Historians call these symbols glyphs. Collaborate with another student to invent seven glyphs—one for each weekday. Compare them with other glyphs at msscience.com/time.

For more information, visit earth.msscience.com/time

Reviewing Main Ideas

Section 1 Earth

1. Earth is spherical and bulges slightly at its equator.
2. Earth rotates once per day and orbits the Sun in a little more than 365 days.
3. Earth has a magnetic field.
4. Seasons on Earth are caused by the tilt of Earth's axis as it orbits the Sun.

Section 2 The Moon—Earth's Satellite

1. Earth's Moon goes through phases that depend on the relative positions of the Sun, the Moon, and Earth.
2. Eclipses occur when Earth or the Moon temporarily blocks sunlight from reaching the other.
3. The Moon's maria are the result of ancient lava flows. Craters on the Moon's surface formed from impacts with meteorites, asteroids, and comets.

Section 3 Exploring Earth's Moon

1. The *Clementine* spacecraft took detailed photographs of the Moon's surface and collected data indicating the presence of water in deep craters.
2. NASA's *Lunar Prospector* spacecraft found additional evidence of ice.

Visualizing Main Ideas

Copy and complete the following concept map on the impact theory of the Moon's formation.

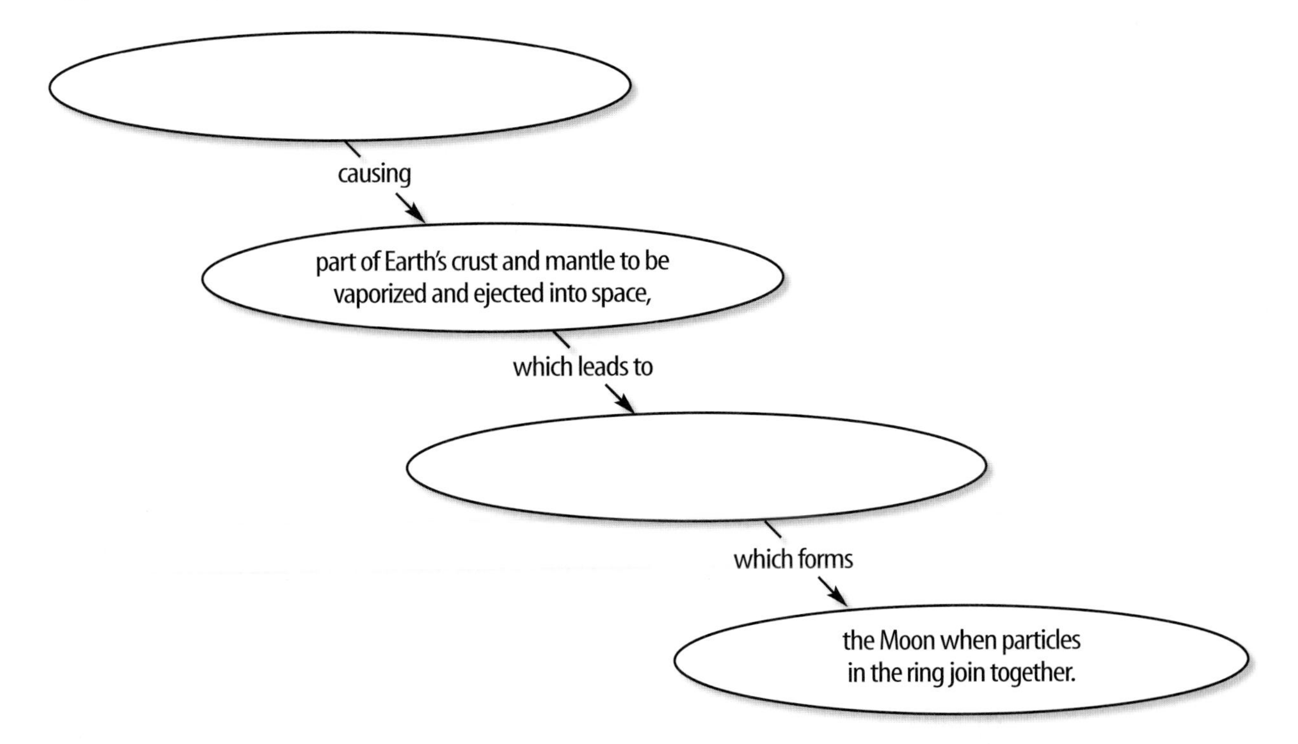
causing
part of Earth's crust and mantle to be vaporized and ejected into space,
which leads to
which forms
the Moon when particles in the ring join together.

Using Vocabulary

axis p. 661	new moon p. 667
ellipse p. 663	revolution p. 663
equinox p. 665	rotation p. 661
full moon p. 667	solar eclipse p. 669
impact basin p. 677	solstice p. 664
lunar eclipse p. 670	sphere p. 660
maria p. 671	waning p. 667
moon phase p. 667	waxing p. 667

Fill in the blanks with the correct vocabulary word or words.

1. The spinning of Earth around its axis is called ________.
2. The ________ is the point at which the Sun reaches its greatest distance north or south of the equator.
3. The Moon is said to be ________ when less and less of the side facing Earth is lighted.
4. The depression left behind by an object striking the Moon is called a(n) ________.
5. Earth's orbit is a(n) ________.

Checking Concepts

Choose the word or phrase that best answers the question.

6. How long does it take for the Moon to rotate once?
   A) 24 hours **C)** 27.3 hours
   B) 365 days **D)** 27.3 days
7. Where is Earth's circumference greatest?
   A) equator **C)** poles
   B) mantle **D)** axis
8. Earth is closest to the Sun during which season in the northern hemisphere?
   A) spring **C)** winter
   B) summer **D)** fall
9. What causes the Sun to appear to rise and set?
   A) Earth's revolution
   B) the Sun's revolution
   C) Earth's rotation
   D) the Sun's rotation

Use the photo below to answer question 10.

10. What phase of the Moon is shown in the photo above?
    A) waning crescent **C)** third quarter
    B) waxing gibbous **D)** waning gibbous
11. How long does it take for the Moon to revolve once around Earth?
    A) 24 hours **C)** 27.3 hours
    B) 365 days **D)** 27.3 days
12. What is it called when the phases of the Moon appear to get larger?
    A) waning **C)** rotating
    B) waxing **D)** revolving
13. What kind of eclipse occurs when the Moon blocks sunlight from reaching Earth?
    A) solar **C)** full
    B) new **D)** lunar
14. What is the darkest part of the shadow during an eclipse?
    A) waxing gibbous **C)** waning gibbous
    B) umbra **D)** penumbra
15. What is the name for a depression on the Moon caused by an object striking its surface?
    A) eclipse **C)** phase
    B) moonquake **D)** impact basin

Science Online earth.msscience.com/vocabulary_puzzlemaker

Thinking Critically

16. **Predict** how the Moon would appear to an observer in space during its revolution. Would phases be observable? Explain.

17. **Predict** what the effect would be on Earth's seasons if the axis were tilted at 28.5° instead of 23.5°.

18. **Infer** Seasons in the two hemispheres are opposite. Explain how this supports the statement that seasons are NOT caused by Earth's changing distance from the Sun.

19. **Draw Conclusions** How would solar eclipses be different if the Moon were twice as far from Earth? Explain.

20. **Predict** how the information gathered by moon missions could be helpful in the future for people wanting to establish a colony on the Moon.

21. **Use Variables, Constants, and Controls** Describe a simple activity to show how the Moon's rotation and revolution work to keep the same side facing Earth at all times.

22. **Draw Conclusions** Gravity is weaker on the Moon than it is on Earth. Why might more craters be present on the far side of the Moon than on the side of the Moon facing Earth?

23. **Recognize Cause and Effect** During a new phase of the Moon, we cannot see it because no sunlight reaches the side facing Earth. Yet sometimes when there is a thin crescent visible, we do see a faint image of the rest of the Moon. Explain what might cause this to happen.

24. **Describe** Earth's magnetic field. Include an explanation of how scientists believe it is generated and two ways in which it helps people on Earth.

Performance Activities

25. **Display** Draw a cross section of the Moon. Include the crust, outer and inner mantles, and possible core based on the information in this chapter. Indicate the presence of impact craters and show how the thickness of the crust varies from one side of the Moon to the other.

26. **Poem** Write a poem in which you describe the various surface features of the Moon. Be sure to include information on how these features formed.

Applying Math

27. **Orbital Tilt** The Moon's orbit is tilted at an angle of 5° to Earth's orbit around the sun. Using a protractor, draw the Moon's orbit around Earth. What fraction of a full circle (360°) is 5°?

Use the illustration below to answer question 28.

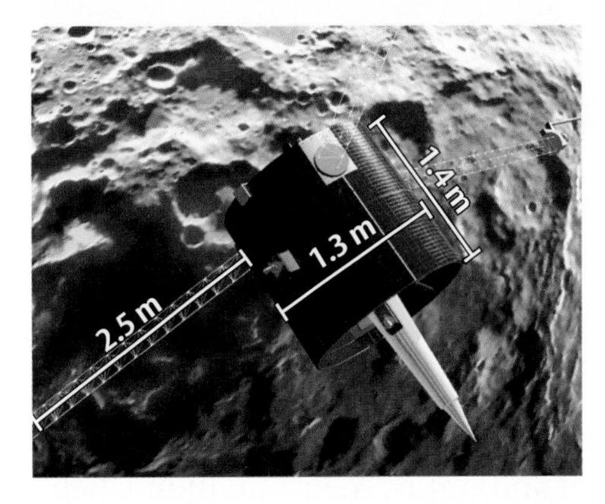

28. **Model to Scale** You are planning to make a scale model of the *Lunar Prospector* spacecraft, shown above. Assuming that the three instrument masts are of equal length, draw a labeled diagram of your model using a scale of 1 cm equals 30 cm.

29. **Spacecraft Velocity** The *Lunar Prospector* spacecraft shown above took 105 hours to reach the Moon. Assuming that the average distance from Earth to Moon is 384,000 km, calculate its average velocity on the trip.

Part 1 Multiple Choice

Record your answers on the answer sheet provided by your teacher or on a sheet of paper.

1. Which of the following terms would you use to describe the spinning of Earth on its axis?

A. revolution
B. ellipse
C. rotation
D. solstice

Use the illustration below to answer questions 2 and 3.

2. Which season is beginning for the southern hemisphere when Earth is in this position?

A. spring
B. summer
C. fall
D. winter

3. Which part of Earth receives the greatest total amount of solar radiation when Earth is in this position?

A. northern hemisphere
B. South Pole
C. southern hemisphere
D. equator

4. Which term describes the dark, flat areas on the Moon's surface which are made of cooled, hardened lava?

A. spheres
B. moonquakes
C. highlands
D. maria

Use the illustration below to answer questions 5 and 6.

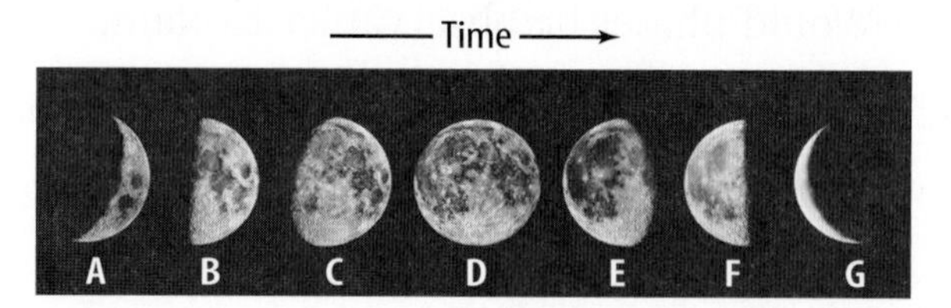

5. Which letter corresponds to the moon phase waning gibbous?

A. G
B. C
C. E
D. A

6. The Moon phase cycle lasts about 29.5 days. Given this information, about how long does it take the Moon to wax from new moon to full moon?

A. about 3 days
B. about 1 week
C. about 2 weeks
D. about 4 weeks

7. Where have large amounts of water been detected on the Moon?

A. highlands
B. lunar equator
C. maria
D. lunar poles

8. In what month is Earth closest to the Sun?

A. March
B. September
C. July
D. January

9. So far, where on the Moon have the youngest rocks been found?

A. lunar highlands
B. maria
C. lunar poles
D. lunar equator

Test-Taking Tip

Eliminate Choices If you don't know the answer to a multiple-choice question, eliminate as many incorrect choices as possible. Mark your best guess from the remaining answers before moving to the next question.

Question 5 Eliminate those phases that you know are not gibbous.

Part 2 Short Response/Grid In

Record your answers on the answer sheet provided by your teacher or on a sheet of paper.

10. Explain why the North Pole is always in sunlight during summer in the northern hemisphere.

11. Describe one positive effect of Earth's magnetic field.

12. Explain the difference between a solstice and an equinox. Give the dates of these events on Earth.

13. Explain how scientists know that the Moon was once geologically active.

Use the illustration below to answer questions 14 and 15.

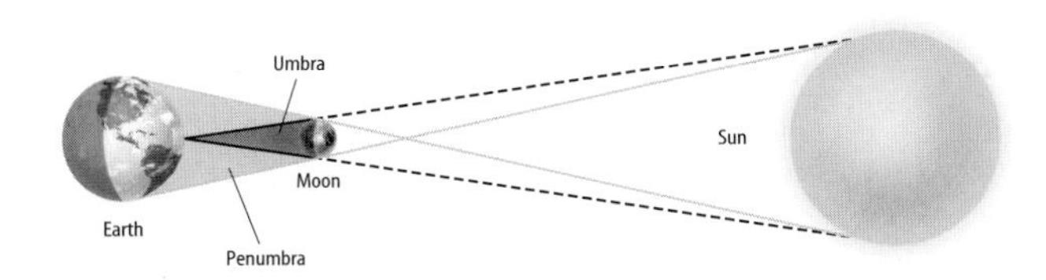

14. What type of eclipse is shown above?

15. Describe what a person standing in the Moon's umbra would see if he or she looked at the sky wearing protective eyewear.

16. The tilt of Earth on its axis causes seasons. Give two reasons why this tilt causes summer to be warmer than winter.

17. When the Apollo missions ended in 1972, 12 astronauts had visited the Moon and brought back samples of moon soil and rock. Explain why we continue to send orbiting spacecraft to study the Moon.

18. Define the term *impact basin*, and name the largest one known in the solar system.

Part 3 Open Ended

Record your answers on a sheet of paper.

Use the illustrations below to answer questions 19 and 20.

19. As a ship comes into view over the horizon, the top appears before the rest of the ship. How does this demonstrate that Earth is spherical?

20. If Earth were flat, how would an approaching ship appear differently?

21. Explain why eclipses of the Sun occur only occasionally despite the fact that the Moon's rotation causes it to pass between Earth and the Sun every month.

22. Recent data from the spacecraft *Lunar Prospector* indicate the presence of large quantities of water in shadowed areas of lunar impact basins. Describe the hypothesis that scientists have developed to explain how this water reached the moon and how it might be preserved.

23. Compare the impact theory of lunar formation with one of the older theories proposed before the *Apollo* mission.

24. Describe how scientists study the interior of the Moon and what they have learned so far.

25. Explain why Earth's magnetic north poles must be mapped and why these maps must be kept up-to-date.

chapter 24

The Solar System

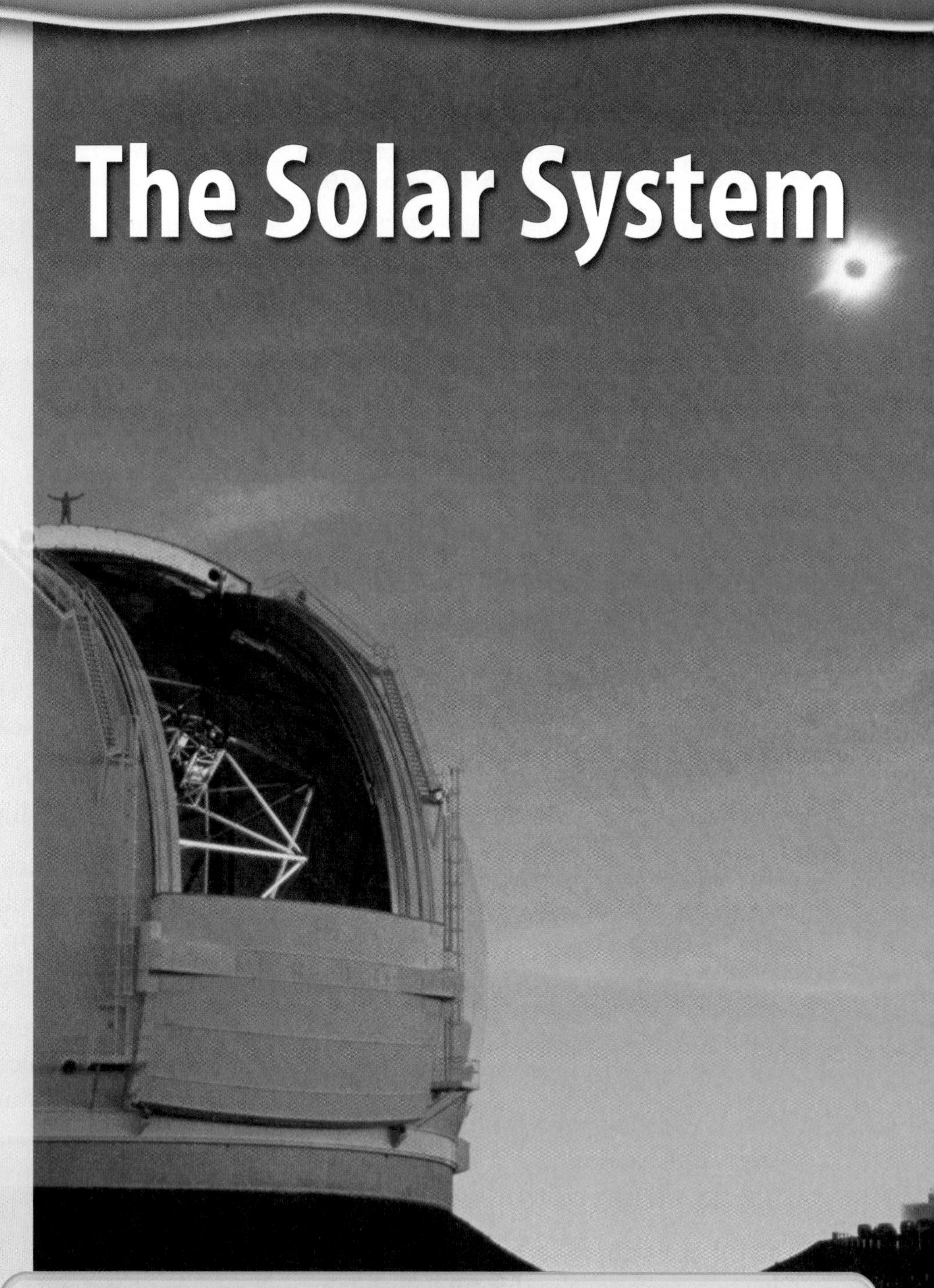

The BIG Idea

The solar system consists of planets and their moons, comets, meteoroids, and asteroids that all orbit the Sun.

SECTION 1
The Solar System

Main Idea Gravity assisted in the formation of the solar system and continues to hold the planets in their places as they orbit the Sun.

SECTION 2
The Inner Planets

Main Idea The inner planets—Mercury, Venus, Earth, and Mars—are the closest planets to the Sun.

SECTION 3
The Outer Planets

Main Idea The outer planets—Jupiter, Saturn, Uranus, and Neptune—are the farthest planets from the Sun.

SECTION 4
Other Objects in the Solar System

Main Idea Comets, meteoroids, and asteroids are smaller than planets but also orbit the Sun and are part of the solar system.

How is space explored?

You've seen the Sun and the Moon. You also might have observed some of the planets. But to get a really good look at the solar system from Earth, telescopes are needed. The optical telescope at the Keck Observatory in Hawaii allows scientists a close-up view.

Science Journal If you could command the Keck telescope, what would you view? Describe what you would see.

Start-Up Activities

Model Crater Formation

Some objects in the solar system have many craters. The Moon is covered with them. The planet Mercury also has a cratered landscape. Even Earth has some craters. All of these craters formed when rocks from space hit the surface of the planet or moon. In this lab, you'll explore crater formation.

1. Place white flour into a metal cake pan to a depth of 3 cm.
2. Cover the flour with 1 cm of colored powdered drink mix or different colors of gelatin powder.
3. From different heights, ranging from 10 cm to 25 cm, drop various-sized marbles into the pan.
4. **Think Critically** Make drawings in your Science Journal that show what happened to the surface of the powder when marbles were dropped from different heights.

Preview this chapter's content and activities at earth.msscience.com

The Solar System Make the following Foldable to help you identify what you already know, what you want to know, and what you learned about the solar system.

STEP 1 **Fold** a vertical sheet of paper from side to side. Make the front edge about 1.25 cm shorter than the back edge.

STEP 2 **Turn** lengthwise and **fold** into thirds.

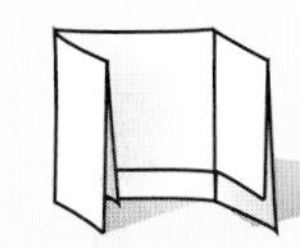

STEP 3 **Unfold and cut** only the top layer along both folds to make three tabs.

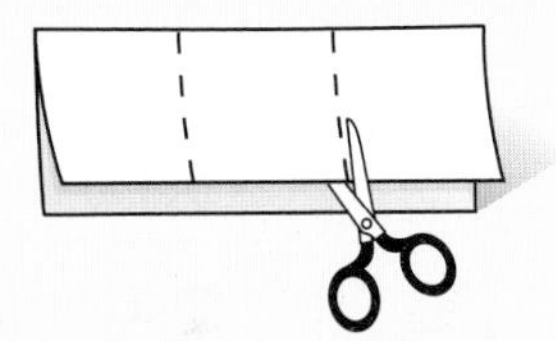

STEP 4 **Label** each tab.

Identify Questions Before you read the chapter, write what you already know about the solar system under the left tab of your Foldable. Write questions about what you'd like to know under the center tab. After you read the chapter, list what you learned under the right tab.

Get Ready to Read

Compare and Contrast

1 Learn It! Good readers compare and contrast information as they read. This means they look for similarities and differences to help them to remember important ideas. Look for signal words in the text to let you know when the author is comparing or contrasting.

Compare and Contrast Signal Words	
Compare	**Contrast**
as	but
like	or
likewise	unlike
similarly	however
at the same time	although
in a similar way	on the other hand

2 Practice It! Read the excerpt below and notice how the author uses contrast signal words to describe the differences between types of planets.

> You can see that the planets **closer** to the Sun travel **faster** than planets **farther** away from the Sun. Because of their **slower** speeds and the **longer** distances they must travel, the outer planets take much **longer** to orbit the Sun than the inner planets do.
>
> —*from page 694*

3 Apply It! Compare and contrast Earth's characteristics on page 698 to the other planets.

Reading Tip

As you read, use other skills, such as summarizing and connecting, to help you understand comparisons and contrasts.

Target Your Reading

Use this to focus on the main ideas as you read the chapter.

1. **Before you read** the chapter, respond to the statements below on your worksheet or on a numbered sheet of paper.
   - Write an **A** if you **agree** with the statement.
   - Write a **D** if you **disagree** with the statement.

2. **After you read** the chapter, look back to this page to see if you've changed your mind about any of the statements.
   - If any of your answers changed, explain why.
   - Change any false statements into true statements.
   - Use your revised statements as a study guide.

Before You Read A or D	Statement	After You Read A or D
	1 The Sun contains more than 99 percent of the mass of the entire solar system.	
	2 Venus is sometimes called Earth's twin because its size and mass are similar to Earth's.	
	3 Mars is the third planet from the Sun and is nicknamed the blue planet.	
	4 Earth is the most volcanically active object in the solar system.	
	5 Jupiter, Saturn, Uranus, and Neptune all have rings that orbit these planets.	
	6 Most asteroids are located in an area called the asteroid belt, which is located between the orbits of Jupiter and Saturn.	
	7 Asteroids, meteoroids, and comets do not contain water.	

Science Online

Print out a worksheet of this page at earth.msscience.com

section 1

The Solar System

as you read

What You'll Learn

- **Compare** models of the solar system.
- **Explain** that gravity holds planets in orbits around the Sun.

Why It's Important

New technology has come from exploring the solar system.

Review Vocabulary
system: a portion of the universe and all of its components, processes, and interactions

New Vocabulary
- solar system

Ideas About the Solar System

People have been looking at the night sky for thousands of years. Early observers noted the changing positions of the planets and developed ideas about the solar system based on their observations and beliefs. Today, people know that objects in the solar system orbit the Sun. People also know that the Sun's gravity holds the solar system together, just as Earth's gravity holds the Moon in its orbit around Earth. However, our understanding of the solar system changes as scientists make new observations.

Earth-Centered Model Many early Greek scientists thought the planets, the Sun, and the Moon were fixed in separate spheres that rotated around Earth. The stars were thought to be in another sphere that also rotated around Earth. This is called the Earth-centered model of the solar system. It included Earth, the Moon, the Sun, five planets—Mercury, Venus, Mars, Jupiter, and Saturn—and the sphere of stars.

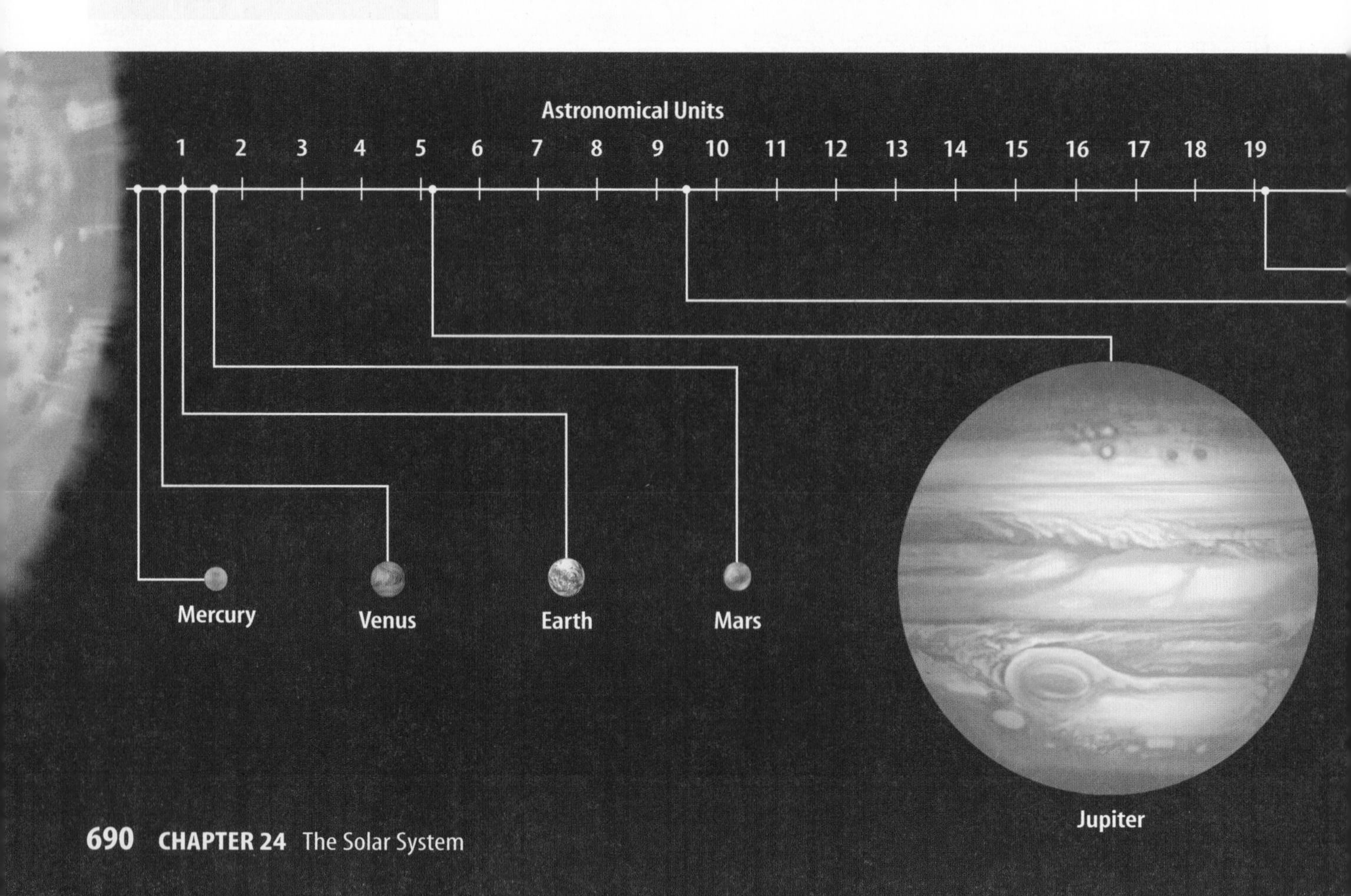

Topic: Solar System
Visit earth.msscience.com for Web links to information about planets.

Sun-Centered Model People believed the idea of an Earth-centered solar system for centuries. Then in 1543, Nicholas Copernicus published a different view. Copernicus stated that the Moon revolved around Earth and that Earth and the other planets revolved around the Sun. He also stated that the daily movement of the planets and the stars was caused by Earth's rotation. This is the Sun-centered model of the solar system.

Using his telescope, Galileo Galilei observed that Venus went through a full cycle of phases like the Moon's. He also observed that the apparent diameter of Venus was smallest when the phase was near full. This only could be explained if Venus were orbiting the Sun. Galileo concluded that the Sun is the center of the solar system.

Modern View of the Solar System As of 2006, the **solar system** is made up of eight planets, including Earth, and many smaller objects that orbit the Sun. The eight planets and the Sun are shown in **Figure 1.** Notice how small Earth is compared with some of the other planets and the Sun.

The solar system includes a huge volume of space that stretches in all directions from the Sun. Because the Sun contains 99.86 percent of the mass of the solar system, its gravity is immense. The Sun's gravity holds the planets and other objects in the solar system in their orbits.

Figure 1 Each of the eight planets in the solar system is unique. The distances between the planets and the Sun are shown on the scale. One astronomical unit (AU) is the average distance between Earth and the Sun.

Rotational Motion You might have noticed that when a twirling ice skater pulls in her arms, she spins faster. The same thing occurs when a cloud of gas, ice, and dust in a nebula contracts. As mass moves toward the center of the cloud, the cloud rotates faster.

How the Solar System Formed

Scientists hypothesize that the solar system formed from part of a nebula of gas, ice, and dust, like the one shown in **Figure 2.** Follow the steps shown in **Figures 3A** through **3D** to learn how this might have happened. A nearby star might have exploded and the shock waves produced by these events could have caused the cloud to start contracting. As it contracted, the nebula likely fragmented into smaller and smaller pieces. The density in the cloud fragments became greater, and the attraction of gravity pulled more gas and dust toward several centers of contraction. This in turn caused them to flatten into disks with dense centers. As the cloud fragments continued to contract, they began to rotate faster and faster.

As each cloud fragment contracted, its temperature increased. Eventually, the temperature in the core of one of these cloud fragments reached about 10 million degrees Celsius. Nuclear fusion began when hydrogen atoms started to fuse and release energy. A star was born—the beginning of the Sun.

It is unlikely that the Sun formed alone. A cluster of stars like the Sun likely formed from parts of the original cloud. The Sun, which is one of many stars in our galaxy, probably escaped from this cluster and has since revolved around the galaxy many times.

Reading Check *What is nuclear fusion?*

Planet Formation Not all of the nearby gas, ice, and dust was drawn into the core of the cloud fragment. The matter that did not get pulled into the center collided and stuck together to form the planets and asteroids. Close to the Sun, the temperature was hot, and the easily vaporized elements could not condense into solids. This is why lighter elements are scarcer in the planets near the Sun than in planets farther out in the solar system.

The inner planets of the solar system—Mercury, Venus, Earth, and Mars—are small, rocky planets with iron cores. The outer planets are Jupiter, Saturn, Uranus, and Neptune. The outer planets are much larger and are made mostly of lighter substances such as hydrogen, helium, methane, and ammonia.

Figure 2 Systems of planets, such as the solar system, form in areas of space like this, called a nebula.

NATIONAL GEOGRAPHIC

VISUALIZING THE SOLAR SYSTEM'S FORMATION

Figure 3

Through careful observations, astronomers have found clues that help explain how the solar system may have formed. A More than 4.6 billion years ago, the solar system was a cloud fragment of gas, ice, and dust. B Gradually, this cloud fragment contracted into a large, tightly packed, spinning disk. The disk's center was so hot and dense that nuclear fusion reactions began to occur, and the Sun was born. C Eventually, the rest of the material in the disk cooled enough to clump into scattered solids. D Finally, these clumps collided and combined to become the eight planets that make up the solar system today.

Planet	Average Orbital Speed (km/s)
Mercury	48
Venus	35
Earth	30
Mars	24
Jupiter	13
Saturn	9.7
Uranus	6.8
Neptune	5.4

Table 1 Average Orbital Speed

Johannes Kepler

Motions of the Planets

When Nicholas Copernicus developed his Sun-centered model of the solar system, he thought that the planets orbited the Sun in circles. In the early 1600s, German mathematician Johannes Kepler began studying the orbits of the planets. He discovered that the shapes of the orbits are not circular. They are oval shaped, or elliptical. His calculations further showed that the Sun is not at the center of the orbits but is slightly offset.

Kepler also discovered that the planets travel at different speeds in their orbits around the Sun, as shown in **Table 1.** You can see that the planets closer to the Sun travel faster than planets farther away from the Sun. Because of their slower speeds and the longer distances they must travel, the outer planets take much longer to orbit the Sun than the inner planets do.

Copernicus's ideas, considered radical at the time, led to the birth of modern astronomy. Early scientists didn't have technology such as space probes to learn about the planets. Nevertheless, they developed theories about the solar system that still are used today.

section 1 review

Summary

Ideas About the Solar System

- The planets in the solar system revolve around the Sun.
- The Sun's immense gravity holds the planets in their orbits.

How the Solar System Formed

- The solar system formed from a piece of a nebula of gas, ice, and dust.
- As the piece of nebula contracted, nuclear fusion began at its center and the Sun was born.

Motion of the Planets

- The planets' orbits are elliptical.
- Planets that are closer to the Sun revolve faster than those that are farther away from the Sun.

Self Check

1. **Describe** the Sun-centered model of the solar system. What holds the solar system together?
2. **Explain** how the planets in the solar system formed.
3. **Infer** why life is unlikely on the outer planets in spite of the presence of water, methane, and ammonia—materials needed for life to develop.
4. **List** two reasons why the outer planets take longer to orbit the Sun than the inner planets do.
5. **Think Critically** Would a year on the planet Neptune be longer or shorter than an earth year? Explain.

Applying Skills

6. **Concept Map** Make a concept map that compares and contrasts the Earth-centered model with the Sun-centered model of the solar system.

earth.msscience.com/self_check_quiz

Planetary Orbits

Planets travel around the Sun along paths called orbits. As you construct a model of a planetary orbit, you will observe that the shape of planetary orbits is an ellipse.

Real-World Question

How can you model planetary orbits?

Goals

- **Model** planetary orbits.
- **Calculate** the eccentricity of ellipses.

Materials

thumbtacks or pins (2)
cardboard (23 cm × 30 cm)
paper (21.5 cm × 28 cm)
metric ruler
string (25 cm)
pencil

Safety Precautions 

Procedure

1. Place a blank sheet of paper on top of the cardboard and insert two thumbtacks or pins about 3 cm apart.
2. Tie the string into a circle with a circumference of 15 cm to 20 cm. Loop the string around the thumbtacks. With someone holding the tacks or pins, place your pencil inside the loop and pull it tight.
3. Moving the pencil around the tacks and keeping the string tight, mark a line until you have completed a smooth, closed curve.
4. Repeat steps 1 through 3 several times. First, vary the distance between the tacks, then vary the length of the string. However, change only one of these each time. Make a data table to record the changes in the sizes and shapes of the ellipses.
5. Eccentricity is a measure of how an orbit varies from a perfect circle. Eccentricity, *e*, is determined by dividing the distance, *d*, between the foci (fixed points—here, the tacks) by the length, *l*, of the major axis.

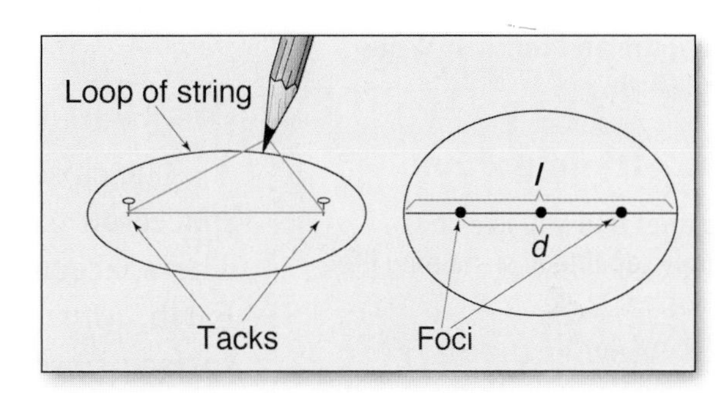

6. **Calculate** and record the eccentricity of the ellipses that you constructed.
7. **Research** the eccentricities of planetary orbits. Construct an ellipse with the same eccentricity as Earth's orbit.

Conclude and Apply

1. **Analyze** the effect that a change in the length of the string or the distance between the tacks has on the shape of the ellipse.
2. **Hypothesize** what must be done to the string or placement of tacks to decrease the eccentricity of a constructed ellipse.
3. **Describe** the shape of Earth's orbit. Where is the Sun located within the orbit?

Compare your results with those of other students. **For more help, refer to the** Science Skill Handbook.

section 2

The Inner Planets

as you read

What You'll Learn

- **List** the inner planets in order from the Sun.
- **Describe** each inner planet.
- **Compare and contrast** Venus and Earth.

Why It's Important

The planet that you live on is uniquely capable of sustaining life.

Review Vocabulary
space probe: an instrument that is sent to space to gather information and send it back to Earth

New Vocabulary
- Mercury
- Venus
- Earth
- Mars

Inner Planets

Today, people know more about the solar system than ever before. Better telescopes allow astronomers to observe the planets from Earth and space. In addition, space probes have explored much of the solar system. Prepare to take a tour of the solar system through the eyes of some space probes.

Mercury

The closest planet to the Sun is **Mercury.** The first American spacecraft mission to Mercury was in 1974–1975 by *Mariner 10.* The spacecraft flew by the planet and sent pictures back to Earth. *Mariner 10* photographed only 45 percent of Mercury's surface, so scientists don't know what the other 55 percent looks like. What they do know is that the surface of Mercury has many craters and looks much like Earth's Moon. It also has cliffs as high as 3 km on its surface. These cliffs might have formed at a time when Mercury shrank in diameter, as seen in **Figure 4.**

Why would Mercury have shrunk? *Mariner 10* detected a weak magnetic field around Mercury. This indicates that the planet has an iron core. Some scientists hypothesize that Mercury's crust solidified while the iron core was still hot and molten. As the core started to solidify, it contracted. The cliffs resulted from breaks in the crust caused by this contraction.

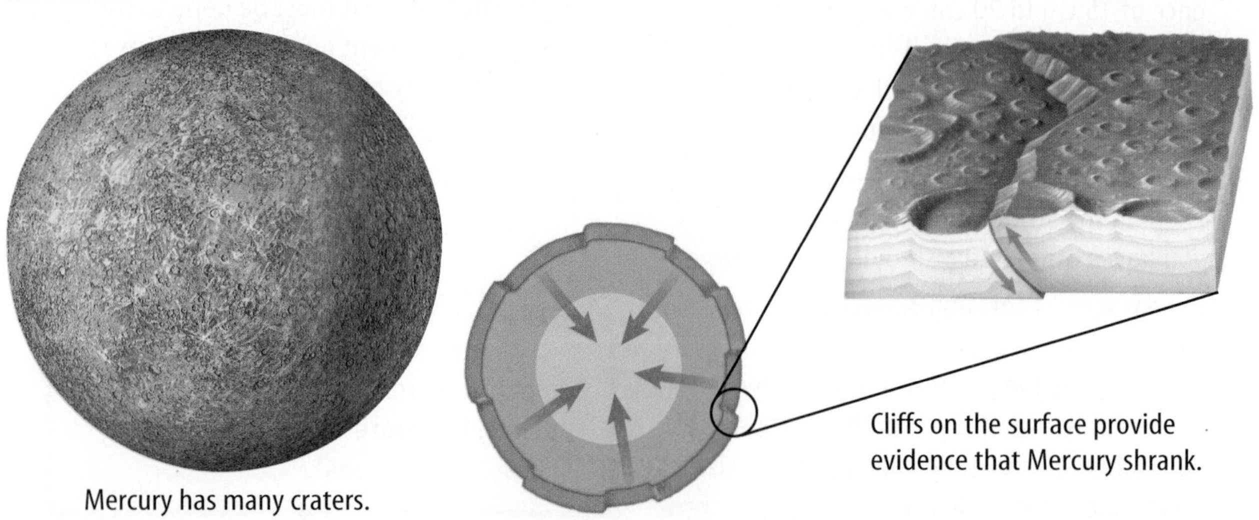

Mercury has many craters.
Cliffs on the surface provide evidence that Mercury shrank.

Figure 4 Large cliffs on Mercury might have formed when the crust of the planet broke as the planet contracted.

Does Mercury have an atmosphere? Because of Mercury's low gravitational pull and high daytime temperatures, most gases that could form an atmosphere escape into space. *Mariner 10* found traces of hydrogen and helium gas that were first thought to be an atmosphere. However, these gases are now known to be temporarily taken from the solar wind.

The lack of atmosphere and its nearness to the Sun cause Mercury to have great extremes in temperature. Mercury's temperature can reach 425°C during the day, and it can drop to −170°C at night.

Future Mission Launched in 2004, *Messenger* is the next mission to Mercury. This space probe will fly by the planet in 2008 and orbit it in 2011. The probe will photograph and map the entire surface.

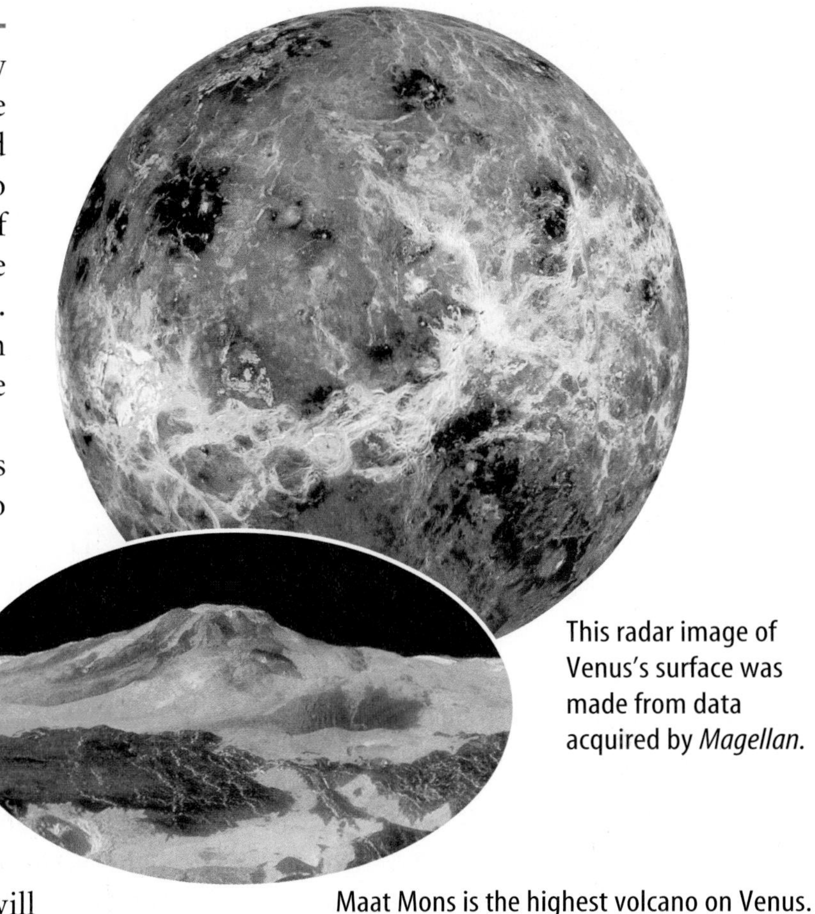

This radar image of Venus's surface was made from data acquired by *Magellan.*

Maat Mons is the highest volcano on Venus. Lava flows extend for hundreds of kilometers across the plains.

Figure 5 Venus is the second planet from the Sun.

Venus

The second planet from the Sun is **Venus,** shown in **Figure 5.** Venus is sometimes called Earth's twin because its size and mass are similar to Earth's. In 1962, *Mariner 2* flew past Venus and sent back information about Venus's atmosphere and rotation. The former Soviet Union landed the first probe on the surface of Venus in 1970. *Venera 7,* however, stopped working in less than an hour because of the high temperature and pressure. Additional *Venera* probes photographed and mapped the surface of Venus. Between 1990 and 1994, the U.S. *Magellan* probe used its radar to make the most detailed maps yet of Venus's surface. It collected radar images of 98 percent of Venus's surface. Notice the huge volcano in **Figure 5.**

Clouds on Venus are so dense that only a small percentage of the sunlight that strikes the top of the clouds reaches the planet's surface. The sunlight that does get through warms Venus's surface, which then gives off heat to the atmosphere. Much of this heat is absorbed by carbon dioxide gas in Venus's atmosphere. This causes a greenhouse effect similar to, but more intense than, Earth's greenhouse effect. Due to this intense greenhouse effect, the temperature on the surface of Venus is between 450°C and 475°C.

Earth

Figure 6 shows **Earth,** the third planet from the Sun. The average distance from Earth to the Sun is 150 million km, or one astronomical unit (AU). Unlike other planets, Earth has abundant liquid water and supports life. Earth's atmosphere causes most meteors to burn up before they reach the surface, and its ozone layer protects life from the effects of the Sun's intense radiation.

Mars

Look at **Figure 7.** Can you guess why **Mars,** the fourth planet from the Sun, is called the red planet? Iron oxide in soil on its surface gives it a reddish color. Other features visible from Earth are Mars's polar ice caps and changes in the coloring of the planet's surface. The ice caps are made of frozen water covered by a layer of frozen carbon dioxide.

Figure 6 More than 70 percent of Earth's surface is covered by liquid water.
Explain *how Earth is unique.*

Most of the information scientists have about Mars came from *Mariner 9,* the *Viking* probes, *Mars Pathfinder, Mars Global Surveyor, Mars Odyssey,* and the Mars Exploration Rovers. *Mariner 9* orbited Mars in 1971 and 1972. It revealed long channels on the planet that might have been carved by flowing water. *Mariner 9* also discovered the largest volcano in the solar system, Olympus Mons, shown in **Figure 7.** Olympus Mons is probably extinct. Large rift valleys in the Martian crust also were discovered. One such valley, Valles Marineris, is shown in **Figure 7.**

Figure 7 Many features on Mars are similar to those on Earth.

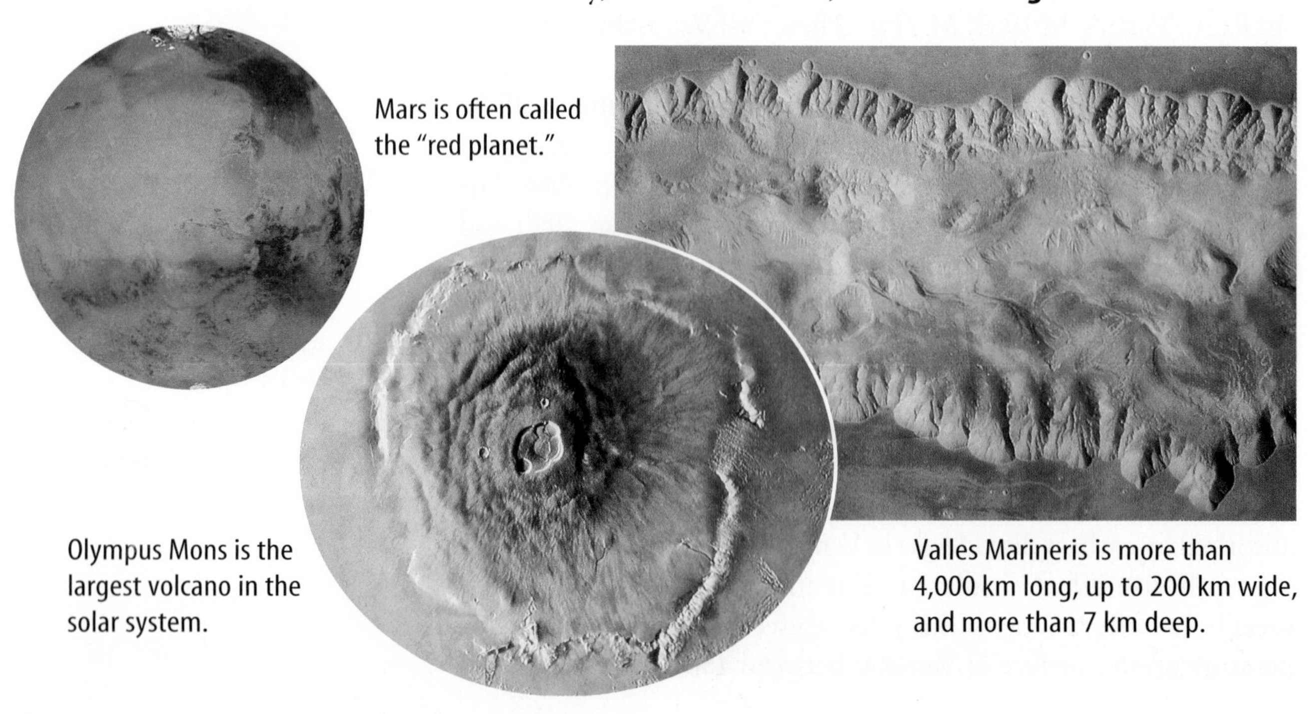

Mars is often called the "red planet."

Olympus Mons is the largest volcano in the solar system.

Valles Marineris is more than 4,000 km long, up to 200 km wide, and more than 7 km deep.

The *Viking* Probes The *Viking 1* and *2* probes arrived at Mars in 1976. Each probe consisted of an orbiter and a lander. The orbiters photographed the entire planet from their orbits, while the landers touched down on the surface. Instruments on the landers attempted to detect possible life by analyzing gases in the Martian soil. The tests found no conclusive evidence of life.

Pathfinder* and *Global Surveyor The *Mars Pathfinder* carried a robot rover named *Sojourner* to test samples of Martian rocks and soil. The data showed that iron in the crust might have been leached out by groundwater. Cameras onboard *Global Surveyor* showed features that looked like sediment gullies and deposits formed by flowing water. These features, shown in **Figure 8,** seem to indicate that groundwater might exist on Mars and that it reached the surface. The features are similar to those formed by flash floods on Earth, such as on Mount St. Helens.

***Odyssey* and Mars Exploration Rovers** In 2002, *Mars Odyssey* began orbiting Mars. It measured elements in Mars's crust and searched for signs of water. Instruments on *Odyssey* detected high levels of hematite, a mineral that forms in water, and subsurface ice near the poles.

Odyssey also relayed data to Earth from the Mars Exploration Rovers *Spirit* and *Opportunity* in 2004. These robot rovers analyzed Martian geology. Data from *Opportunity* confirmed that there were once bodies of water on Mars's surface.

Reading Check *What evidence indicates that Mars has water?*

Inferring Effects of Gravity

Procedure

1. Suppose you are a crane operator who is sent to Mars to help build a colony.
2. Your crane can lift 4,500 kg on Earth, but the force due to gravity on Mars is only 40 percent as large as that on Earth.
3. Determine how much mass your crane could lift on Earth and Mars.

Analysis

1. How can what you have discovered be an advantage over construction on Earth?
2. How might construction advantages change the overall design of the Mars colony?

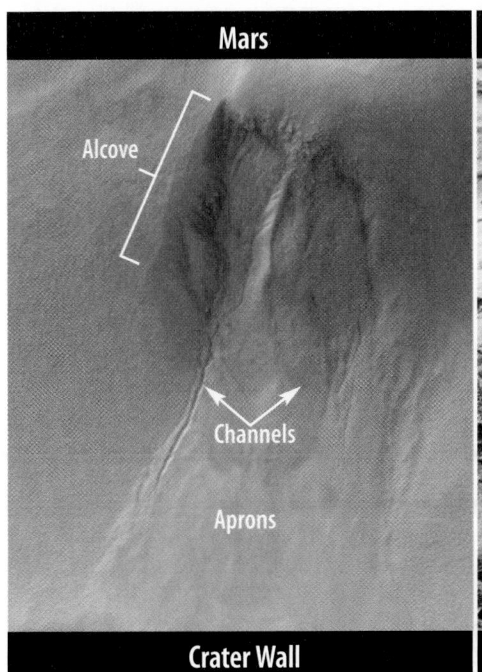

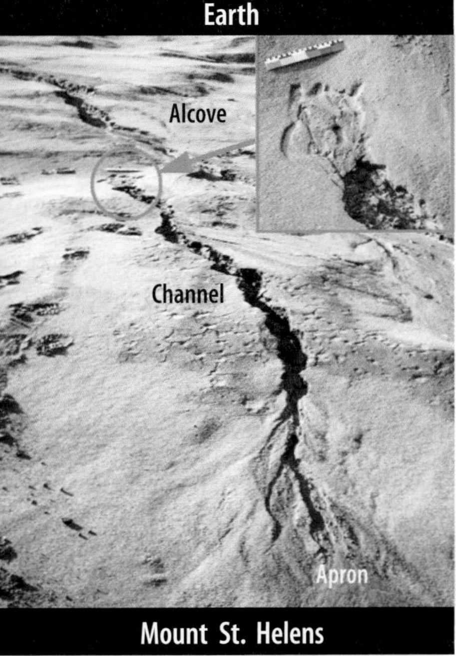

Figure 8 Compare the features found on Mars with those found on an area of Mount St. Helens in Washington state that experienced a flash flood.

Topic: Mars Exploration

Visit earth.msscience.com for Web links to information about future missions to Mars.

Activity Make a timeline that shows when each probe is scheduled to reach Mars. Include the mission objectives for each probe on your timeline.

Mars's Atmosphere The *Viking* and *Global Surveyor* probes analyzed gases in the Martian atmosphere and determined atmospheric pressure and temperature. They found that Mars's atmosphere is much thinner than Earth's. It is composed mostly of carbon dioxide, with some nitrogen and argon. Surface temperatures range from −125°C to 35°C. The temperature difference between day and night results in strong winds on the planet, which can cause global dust storms during certain seasons. This information will help in planning possible human exploration of Mars in the future.

Martian Seasons Mars's axis of rotation is tilted 25°, which is close to Earth's tilt of 23.5°. Because of this, Mars goes through seasons as it orbits the Sun, just like Earth does. The polar ice caps on Mars change with the season. During winter, carbon dioxide ice accumulates and makes the ice cap larger. During summer, carbon dioxide ice changes to carbon dioxide gas and the ice cap shrinks. As one ice cap gets larger, the other ice cap gets smaller. The color of the ice caps and other areas on Mars also changes with the season. The movement of dust and sand during dust storms causes the changing colors.

Applying Math — Use Percentages

DIAMETER OF MARS The diameter of Earth is 12,756 km. The diameter of Mars is 53.3 percent of the diameter of Earth. Calculate the diameter of Mars.

Solution

1 *This is what you know:*
- diameter of Earth: 12,756 km
- percent of Earth's diameter: 53.3%
- decimal equivalent: 0.533 (53.3% ÷ 100)

2 *This is what you need to find:* diameter of Mars

3 *This is the procedure you need to use:* Multiply the diameter of Earth by the decimal equivalent.
$(12{,}756 \text{ km}) \times (0.533) = 6{,}799 \text{ km}$

Practice Problems

1. Use the same procedure to calculate the diameter of Venus. Its diameter is 94.9 percent of the diameter of Earth.
2. Calculate the diameter of Mercury. Its diameter is 38.2 percent of the diameter of Earth.

For more practice, visit earth.msscience.com/math_practice

Martian Moons Mars has two small, irregularly shaped moons that are heavily cratered. Phobos, shown in **Figure 9,** is about 25 km in length, and Deimos is about 13 km in length. Deimos orbits Mars once every 31 h, while Phobos speeds around Mars once every 7 h.

Phobos has many interesting surface features. Grooves and chains of smaller craters seem to radiate out from the large Stickney Crater. Some of the grooves are 700 m across and 90 m deep. These features probably are the result of the large impact that formed the Stickney Crater.

Deimos is the outer of Mars's two moons. It is among the smallest known moons in the solar system. Its surface is smoother in appearance than that of Phobos because some of its craters have partially filled with soil and rock.

As you toured the inner planets through the eyes of the space probes, you saw how each planet is unique. Refer to **Table 3** following Section 3 for a summary of the planets. Mercury, Venus, Earth, and Mars are different from the outer planets, which you'll explore in the next section.

Figure 9 Phobos orbits Mars once every 7 h.
Infer *why Phobos has so many craters.*

section 2 review

Summary

Mercury
- Mercury is extremely hot during the day and extremely cold at night.
- Its surface has many craters.

Venus
- Venus's size and mass are similar to Earth's.
- Temperatures on Venus are between 450°C and 475°C.

Earth
- Earth is the only planet known to support life.

Mars
- Mars has polar ice caps, channels that might have been carved by water, and the largest volcano in the solar system, Olympus Mons.

Self Check

1. **Explain** why Mercury's surface temperature varies so much from day to night.
2. **List** important characteristics for each inner planet.
3. **Infer** why life is unlikely on Venus.
4. **Identify** the inner planet that is farthest from the Sun. Identify the one that is closest to the Sun.
5. **Think Critically** Aside from Earth, which inner planet could humans visit most easily? Explain.

Applying Math

6. **Use Statistics** The inner planets have the following average densities: Mercury, 5.43 g/cm^3; Venus, 5.24 g/cm^3; Earth, 5.52 g/cm^3; and Mars, 3.94 g/cm^3. Which planet has the highest density? Which has the lowest? Calculate the range of these data.

section 3

The Outer Planets

as you read

What You'll Learn

- **Describe** the characteristics of Jupiter, Saturn, Uranus, and Neptune.
- **Describe** the largest moons of each of the outer planets.

Why It's Important

Studying the outer planets will help scientists understand Earth.

Review Vocabulary

moon: a natural satellite of a planet that is held in its orbit around the planet by the planet's gravitational pull

New Vocabulary

- Jupiter
- Great Red Spot
- Saturn
- Uranus
- Neptune
- Pluto

Outer Planets

You might have heard about *Voyager, Galileo,* and *Cassini.* They were not the first probes to the outer planets, but they gathered a lot of new information about them. Follow the spacecrafts as you read about their journeys to the outer planets.

Jupiter

In 1979, *Voyager 1* and *Voyager 2* flew past **Jupiter,** the fifth planet from the Sun. *Galileo* reached Jupiter in 1995, and *Cassini* flew past Jupiter on its way to Saturn in 2000. The spacecrafts gathered new information about Jupiter. The *Voyager* probes revealed that Jupiter has faint dust rings around it and that one of its moons has active volcanoes on it.

Jupiter's Atmosphere Jupiter is composed mostly of hydrogen and helium, with some ammonia, methane, and water vapor. Scientists hypothesize that the atmosphere of hydrogen and helium changes to an ocean of liquid hydrogen and helium toward the middle of the planet. Below this liquid layer might be a rocky core. The extreme pressure and temperature, however, would make the core different from any rock on Earth.

You've probably seen pictures from the probes of Jupiter's colorful clouds. In **Figure 10,** you can see bands of white, red, tan, and brown clouds in its atmosphere. Continuous storms of swirling, high-pressure gas have been observed on Jupiter. The **Great Red Spot** is the most spectacular of these storms.

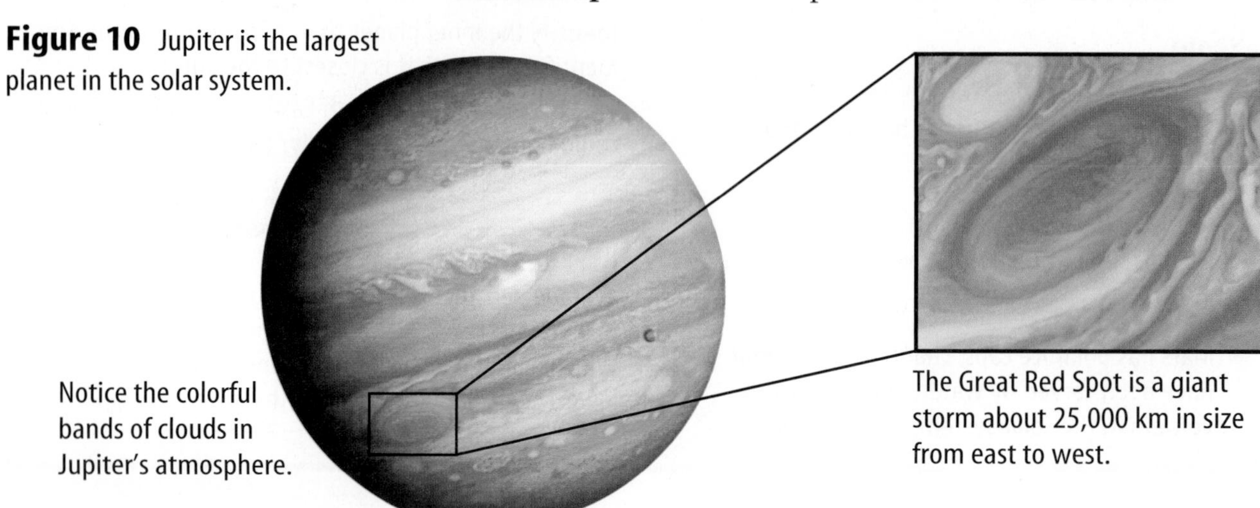
Figure 10 Jupiter is the largest planet in the solar system.

Notice the colorful bands of clouds in Jupiter's atmosphere.

The Great Red Spot is a giant storm about 25,000 km in size from east to west.

Table 2 Large Moons of Jupiter	
Io The most volcanically active object in the solar system; sulfurous compounds give it its distinctive reddish and orange colors; has a thin atmosphere of sulfur dioxide .	
Europa Rocky interior is covered by a smooth 5-km-thick crust of ice, which has a network of cracks; a 50-km-deep ocean might exist under the ice crust; has a thin oxygen atmosphere.	
Ganymede Has a heavily cratered crust of ice covered with grooves; has a rocky interior surrounding a molten iron core and a thin oxygen atmosphere.	
Callisto Has a heavily cratered crust with a mixture of ice and rock throughout the interior; has a rock core and a thin atmosphere of carbon dioxide.	

Moons of Jupiter At least 63 moons orbit Jupiter. In 1610, the astronomer Galileo Galilei was the first person to see Jupiter's four largest moons, shown in **Table 2.** Io (I oh) is the closest large moon to Jupiter. Jupiter's tremendous gravitational force and the gravity of Europa, Jupiter's next large moon, pull on Io. This force heats up Io, causing it to be the most volcanically active object in the solar system. You can see a volcano erupting on Io in **Figure 11.** Europa is composed mostly of rock with a thick, smooth crust of ice. Under the ice might be an ocean as deep as 50 km. If this ocean of water exists, it will be the only place in the solar system, other than Earth, where liquid water exists in large quantities. Next is Ganymede, the largest moon in the solar system—larger even than the planet Mercury. Callisto, the last of Jupiter's large moons, is composed mostly of ice and rock. Studying these moons adds to knowledge about the origin of Earth and the rest of the solar system.

Figure 11 *Voyager 2* photographed the eruption of this volcano on Io in July 1979.

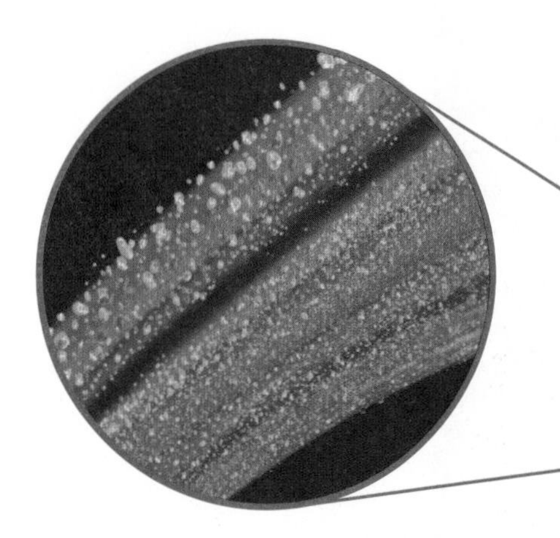

Figure 12 Saturn's rings are composed of pieces of rock and ice.

Modeling Planets

Procedure

1. Research the planets to determine how the sizes of the planets in the solar system compare with Earth's size.
2. Select a scale for the diameter of Earth.
3. Make a model by drawing a circle with this diameter on **paper.**
4. Using Earth's diameter as 1.0 unit, draw each of the other planets to scale.

Analysis

1. Which planet is largest? Which is smallest?
2. Which scale diameter did you select for Earth? Was this a good choice? Why or why not?

Try at Home

Saturn

The *Voyager* probes next surveyed Saturn in 1980 and 1981. *Cassini* reached Saturn on July 1, 2004. **Saturn** is the sixth planet from the Sun. It is the second-largest planet in the solar system, but it has the lowest density.

Saturn's Atmosphere Similar to Jupiter, Saturn is a large, gaseous planet. It has a thick outer atmosphere composed mostly of hydrogen and helium. Saturn's atmosphere also contains ammonia, methane, and water vapor. As you go deeper into Saturn's atmosphere, the gases gradually change to liquid hydrogen and helium. Below its atmosphere and liquid layer, Saturn might have a small, rocky core.

Rings and Moons The *Voyager* and *Cassini* probes gathered information about Saturn's ring system. The probes showed that there are several broad rings. Each large ring is composed of thousands of thin ringlets. **Figure 12** shows that Saturn's rings are composed of countless ice and rock particles. These particles range in size from a speck of dust to tens of meters across. Saturn's ring system is the most complex one in the solar system.

At least 47 moons orbit Saturn. Saturn's gravity holds these moons in their orbits around Saturn, just like the Sun's gravity holds the planets in their orbits around the Sun. The largest moon, Titan, is larger than the planet Mercury. It has a thick atmosphere of nitrogen, argon, and methane. *Cassini* delivered the *Huygens* probe to analyze Titan's atmosphere in 2005.

Uranus

Beyond Saturn, *Voyager 2* flew by Uranus in 1986. **Uranus** (YOOR uh nus) is the seventh planet from the Sun and was discovered in 1781. It is a large, gaseous planet with at least 27 moons and a system of thin, dark rings. Uranus's largest moon, Titania, has many craters and deep valleys. The valleys on this moon indicate that some process reshaped its surface after it formed. Uranus's 11 rings surround the planet's equator.

Figure 13 The atmosphere of Uranus gives the planet its distinct bluish-green color.

Uranus's Characteristics The atmosphere of Uranus is composed of hydrogen, helium, and some methane. Methane gives the planet the bluish-green color that you see in **Figure 13.** Methane absorbs the red and yellow light, and the clouds reflect the green and blue. Few cloud bands and storm systems can be seen on Uranus. Evidence suggests that under its atmosphere, Uranus is composed primarily of rock and various ices. There is no separate core.

Figure 14 shows one of the most unusual features of Uranus. Its axis of rotation is tilted on its side compared with the other planets. The axes of rotation of the other planets are nearly perpendicular to the planes of their orbits. However, Uranus's axis of rotation is nearly parallel to the plane of its orbit. Some scientists believe a collision with another object tipped Uranus on its side.

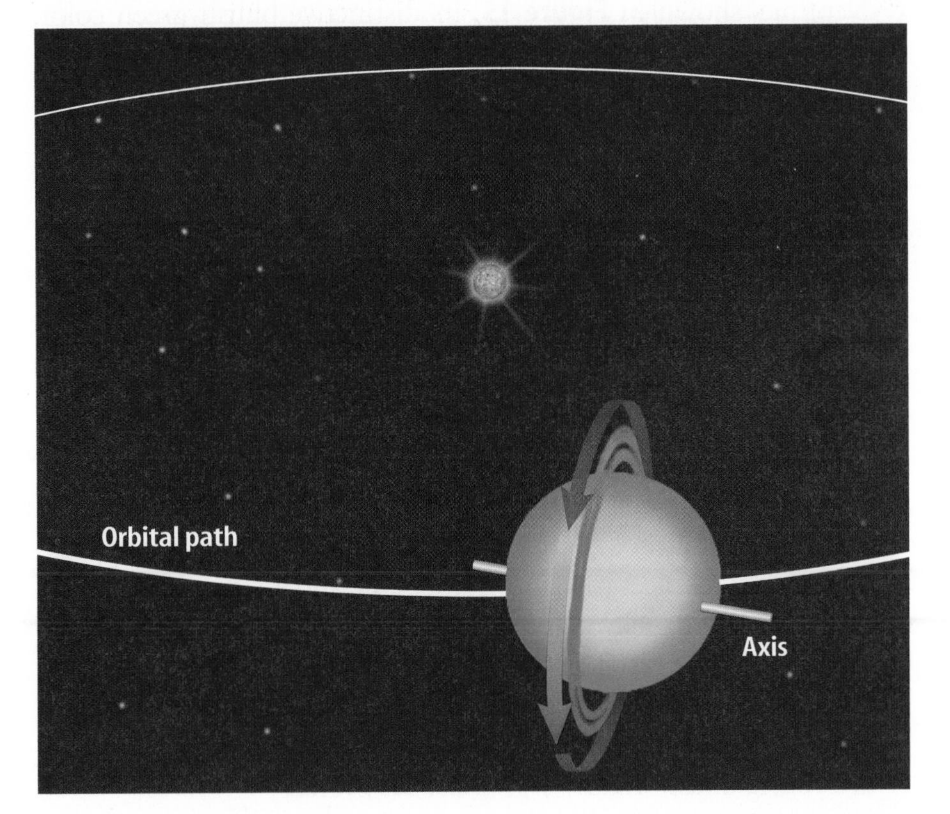

Figure 14 Uranus's axis of rotation is nearly parallel to the plane of its orbit. During its revolution around the Sun, each pole, at different times, points almost directly at the Sun.

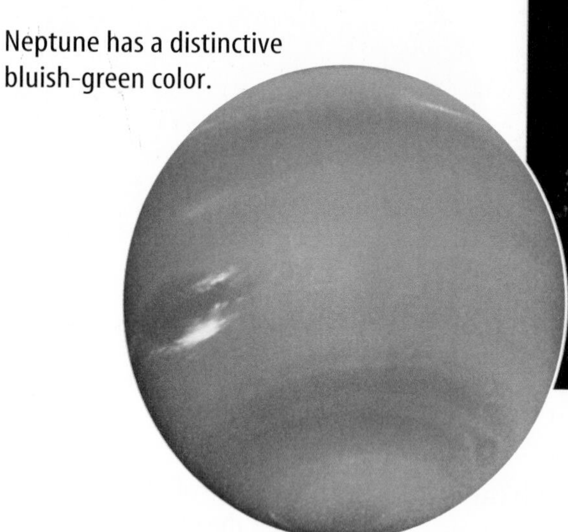
Neptune has a distinctive bluish-green color.

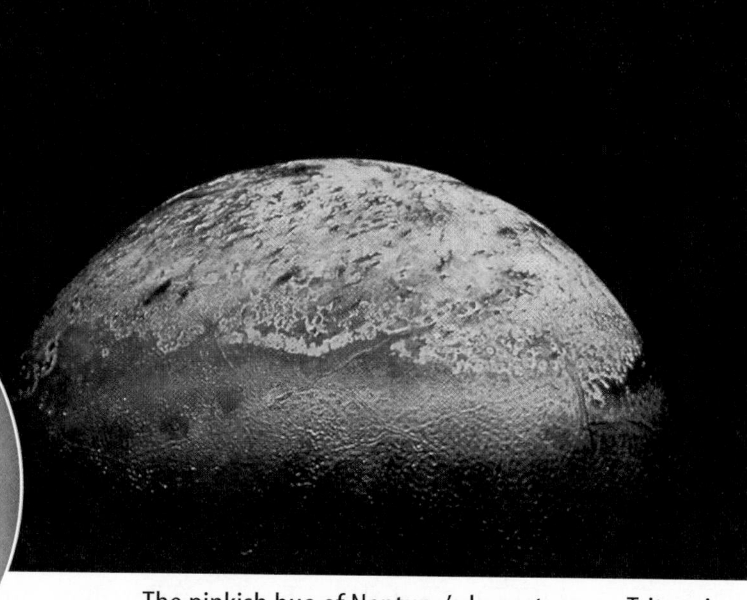
The pinkish hue of Neptune's largest moon, Triton, is thought to come from an evaporating layer of nitrogen and methane ice.

Figure 15 Neptune is the eighth planet from the Sun.

Neptune

Passing Uranus, *Voyager 2* traveled to Neptune, another large, gaseous planet. Discovered in 1846, **Neptune** is the eighth planet from the Sun.

Neptune's Characteristics Like Uranus's atmosphere, Neptune's atmosphere is made up of hydrogen and helium, with smaller amounts of methane. The methane content gives Neptune, shown in **Figure 15,** its distinctive bluish-green color, just as it does for Uranus.

Reading Check *What gives Neptune its bluish-green color?*

Neptune has dark-colored storms in its atmosphere that are similar to the Great Red Spot on Jupiter. One discovered by *Voyager 2* in 1989 was called the Great Dark Spot. It was about the size of Earth with windspeeds higher than any other planet. Observations by the *Hubble Space Telescope* in 1994 showed that the Great Dark Spot disappeared and then reappeared. Bright clouds also form and then disappear. Scientists don't know what causes these changes, but they show that Neptune's atmosphere is active and changes rapidly.

Under its atmosphere, Neptune has a mixture of rock and various types of ices made from methane and ammonia. Neptune probably has a rocky core.

Neptune has at least 13 moons and several rings. Triton, shown in **Figure 15,** is Neptune's largest moon. It has a thin atmosphere composed mostly of nitrogen. Neptune's dark rings are young and probably won't last very long.

Names of Planets The names of most of the planets in the solar system come from Roman or Greek mythology. For example, Neptune was the Roman god of the sea. Research to learn about the names of the other planets. Write a paragraph in your Science Journal that summarizes what you learn.

Dwarf Planets

From the time of its discovery in 1930 until 2006 Pluto was considered the ninth planet in the solar system. But with the discovery of Eris (EE rihs), which is larger than Pluto, the International Astronomical Union decided to define the term *planet.* Now, scientists call Pluto a dwarf planet.

Ceres Ceres was discovered in 1801. It has an average diameter of about 940 km and is located within the asteroid belt at an average distance of about 2.7 AU from the Sun. Ceres orbits the Sun about once every 4.6 years.

Pluto 1930 Pluto has a diameter of 2,300 km. It is an average distance of 39.2 AU from the Sun and takes 248 years to complete one orbit. It is surrounded by only a thin atmosphere and it has a solid, icy-rock surface. Pluto has three moons: Charon, Hydra, and Nix. The largest moon, Charon, has a diameter of about 1,200 km and orbits Pluto at a distance of about 19,500 km.

Eris Astronomers discovered Eris in 2005, originally calling it UB313. With a diameter of about 2,400 km, Eris is slightly larger than Pluto. Eris has an elliptical orbit that varies from between about 38 AU to 98 AU from the Sun. Eris orbits the Sun once every 557 years and has one moon, named Dysnomia (dihs NOH mee uh).

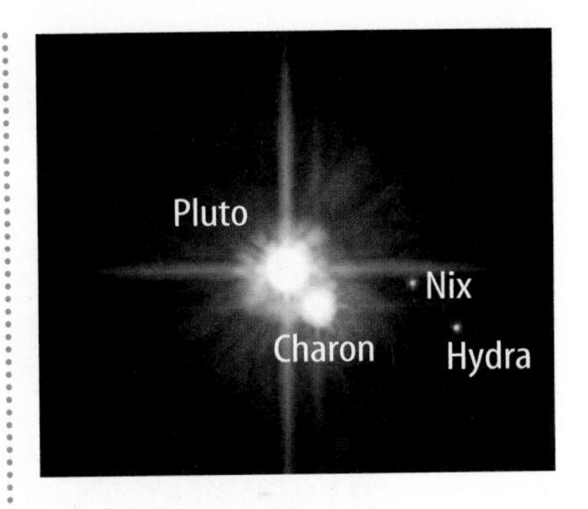

Figure 16 Hydra and Nix are about three times farther from Pluto than Charon is.

section 3 review

Summary

Jupiter
- Jupiter is the largest planet in the solar system.
- The Great Red Spot is a huge storm on Jupiter.

Saturn
- Saturn has a complex system of rings.

Uranus
- Uranus has a bluish-green color caused by methane in its atmosphere.

Neptune
- Like Uranus, Neptune has a bluish-green color.
- Neptune's atmosphere can change rapidly.

Dwarf Planets
- Pluto is made of ice and rock.
- Ceres is a dwarf planet within the asteroid belt.

Self Check

1. **Describe** the differences between the outer planets and the inner planets.
2. **Describe** what Saturn's rings are made from.
3. **Compare** Pluto to the eight planets.
4. **Explain** how Uranus's axis of rotation differs from those of most other planets.
5. **Think Critically** What would seasons be like on Uranus? Explain.

Applying Skills

6. **Identify a Question** When a probe lands on Pluto, so many questions will be answered. Think of a question about Pluto that you'd like to have answered. Then, explain why the answer is important to you.

Table 3 Planets

Mercury		■ closest to the Sun ■ second-smallest planet ■ surface has many craters and high cliffs ■ no atmosphere ■ temperatures range from 425°C during the day to −170°C at night ■ has no moons
Venus		■ similar to Earth in size and mass ■ thick atmosphere made mostly of carbon dioxide ■ droplets of sulfuric acid in atmosphere give clouds a yellowish color ■ surface has craters, faultlike cracks, and volcanoes ■ greenhouse effect causes surface temperatures of 450°C to 475°C ■ has no moons
Earth		■ atmosphere protects life ■ surface temperatures allow water to exist as solid, liquid, and gas ■ only planet where life is known to exist ■ has one large moon
Mars		■ surface appears reddish-yellow because of iron oxide in soil ■ ice caps are made of frozen carbon dioxide and water ■ channels indicate that water had flowed on the surface; has large volcanoes and valleys ■ has a thin atmosphere composed mostly of carbon dioxide ■ surface temperatures range from −125°C to 35°C ■ huge dust storms often blanket the planet ■ has two small moons

Table 3 Planets

Jupiter		■ largest planet ■ has faint rings ■ atmosphere is mostly hydrogen and helium; continuous storms swirl on the planet—the largest is the Great Red Spot ■ has four large moons and at least 59 smaller moons; one of its moons, Io, has active volcanoes
Saturn		■ second-largest planet ■ thick atmosphere is mostly hydrogen and helium ■ has a complex ring system ■ has at least 47 moons—the largest, Titan, is larger than Mercury
Uranus		■ large, gaseous planet with thin, dark rings ■ atmosphere is hydrogen, helium, and methane ■ axis of rotation is nearly parallel to plane of orbit ■ has at least 27 moons
Neptune	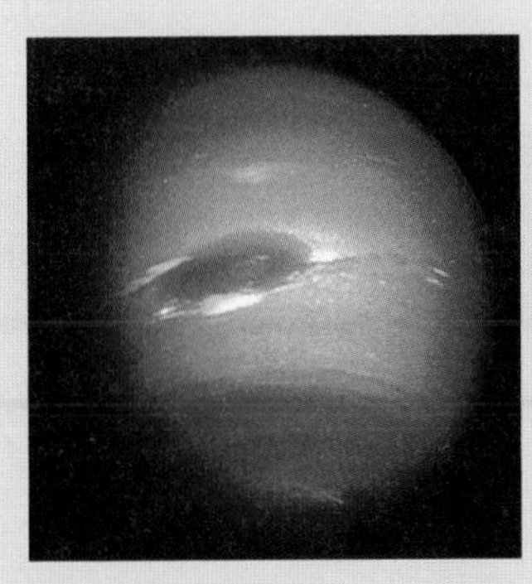	■ large, gaseous planet with rings that vary in thickness ■ is sometimes farther from the Sun than Pluto is ■ methane atmosphere causes its bluish-green color ■ has dark-colored storms in atmosphere ■ has at least 13 moons

section 4

Other Objects in the Solar System

as you read

What You'll Learn

- **Describe** how comets change when they approach the Sun.
- **Distinguish** among comets, meteoroids, and asteroids.
- **Explain** that objects from space sometimes impact Earth.

Why It's Important

Comets, asteroids, and most meteorites are very old. Scientists can learn about the early solar system by studying them.

Review Vocabulary
crater: a nearly circular depression in a planet, moon, or asteroid that formed when an object from space hit its surface

New Vocabulary
- comet
- meteor
- meteorite
- asteroid

Comets

The planets and their moons are the most noticeable members of the Sun's family, but many other objects also orbit the Sun. Comets, meteoroids, and asteroids are other important objects in the solar system.

You might have heard of Halley's comet. A **comet** is composed of dust and rock particles mixed with frozen water, methane, and ammonia. Halley's comet was last seen from Earth in 1986. English astronomer Edmund Halley realized that comet sightings that had taken place about every 76 years were really sightings of the same comet. This comet, which takes about 76 years to orbit the Sun, was named after him.

Oort Cloud Astronomer Jan Oort proposed the idea that billions of comets surround the solar system. This cloud of comets, called the Oort Cloud, is located beyond the orbit of Pluto. Oort suggested that the gravities of the Sun and nearby stars interact with comets in the Oort Cloud. Comets either escape from the solar system or get captured into smaller orbits.

Comet Hale-Bopp On July 23, 1995, two amateur astronomers made an exciting discovery. A new comet, Comet Hale-Bopp, was headed toward the Sun. Larger than most that approach the Sun, it was the brightest comet visible from Earth in 20 years. Shown in **Figure 17,** Comet Hale-Bopp was at its brightest in March and April 1997.

Figure 17 Comet Hale-Bopp was most visible in March and April 1997.

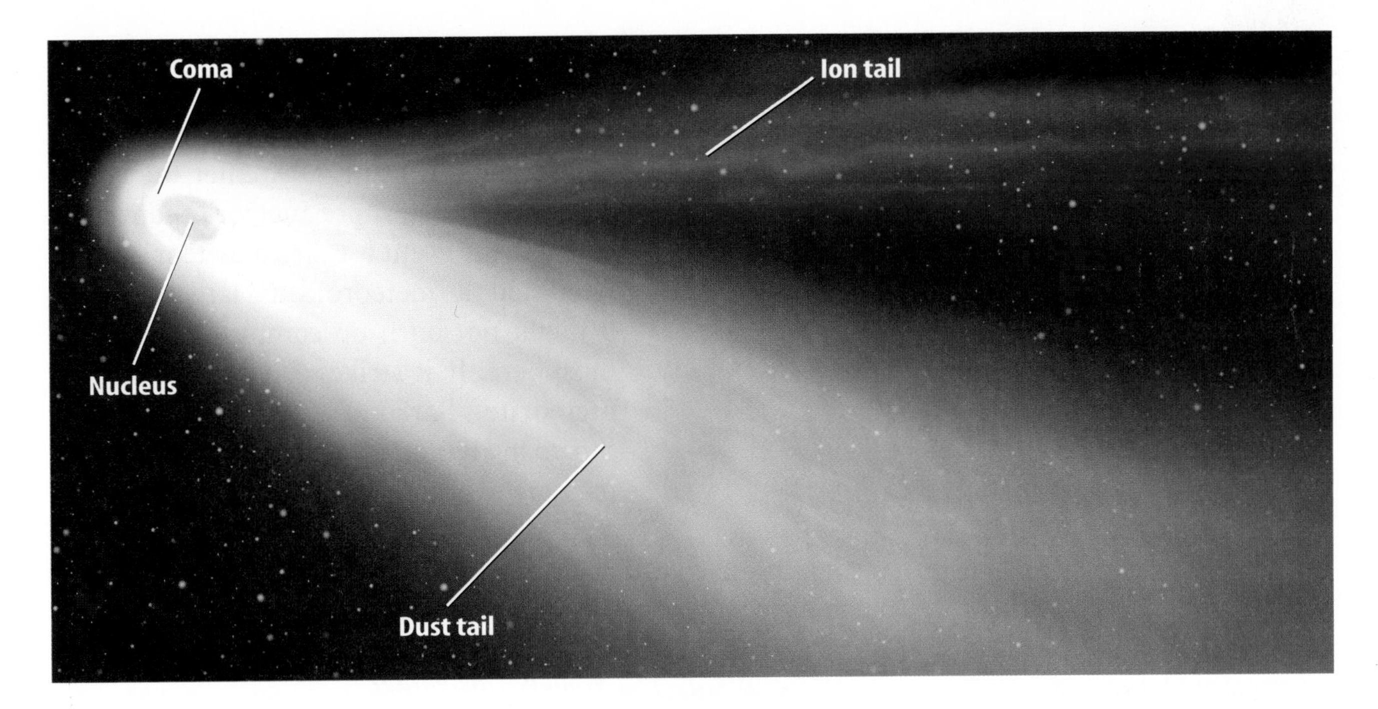

Structure of Comets The *Hubble Space Telescope* and spacecrafts such as the *International Cometary Explorer* have gathered information about comets. In 2006, a spacecraft called *Stardust* will return a capsule to Earth containing samples of dust from a comet's tail. Notice the structure of a comet shown in **Figure 18.** It is a mass of frozen ice and rock.

As a comet approaches the Sun, it changes. Ices of water, methane, and ammonia vaporize because of the heat from the Sun. This releases dust and jets of gas. The gases and released dust form a bright cloud called a coma around the nucleus, or solid part, of the comet. The solar wind pushes on the gases and dust in the coma, causing the particles to form separate tails that point away from the Sun.

After many trips around the Sun, most of the ice in a comet's nucleus has vaporized. All that's left are dust and rock, which are spread throughout the orbit of the original comet.

Figure 18 A comet consists of a nucleus, a coma, a dust tail, and an ion tail. Pictures of the comet Wild 2 from *Stardust* show that the comet has a rocky, cratered surface.

Meteoroids, Meteors, and Meteorites

You learned that comets vaporize and break up after they have passed close to the Sun many times. The small pieces from the comet's nucleus spread out into a loose group within the original orbit of the comet. These pieces of dust and rock, along with those derived from other sources, are called meteoroids.

Sometimes the path of a meteoroid crosses the position of Earth, and it enters Earth's atmosphere at speeds of 15 km/s to 70 km/s. Most meteoroids are so small that they completely burn up in Earth's atmosphere. A meteoroid that burns up in Earth's atmosphere is called a **meteor,** shown in **Figure 19.**

Figure 19 A meteoroid that burns up in Earth's atmosphere is called a meteor.

Figure 20 Meteorites occasionally strike Earth's surface. A large meteorite struck Arizona, forming a crater about 1.2 km in diameter and about 200 m deep.

Meteor Showers Each time Earth passes through the loose group of particles within the old orbit of a comet, many small particles of rock and dust enter the atmosphere. Because more meteors than usual are seen, the event is called a meteor shower.

When a meteoroid is large enough, it might not burn up completely in the atmosphere. If it strikes Earth, it is called a **meteorite.** Barringer Crater in Arizona, shown in **Figure 20,** was formed when a large meteorite struck Earth about 50,000 years ago. Most meteorites are probably debris from asteroid collisions or broken-up comets, but some originate from the Moon and Mars.

Reading Check *What is a meteorite?*

Asteroids

An **asteroid** is a piece of rock similar to the material that formed into the planets. Most asteroids are located in an area between the orbits of Mars and Jupiter called the asteroid belt. Find the asteroid belt in **Figure 21.** Why are they located there? The gravity of Jupiter might have kept a planet from forming in the area where the asteroid belt is located now.

Other asteroids are scattered throughout the solar system. They might have been thrown out of the belt by Jupiter's gravity. Some of these asteroids have orbits that cross Earth's orbit. Scientists monitor the positions of these asteroids. However, it is unlikely that an asteroid will hit Earth in the near future.

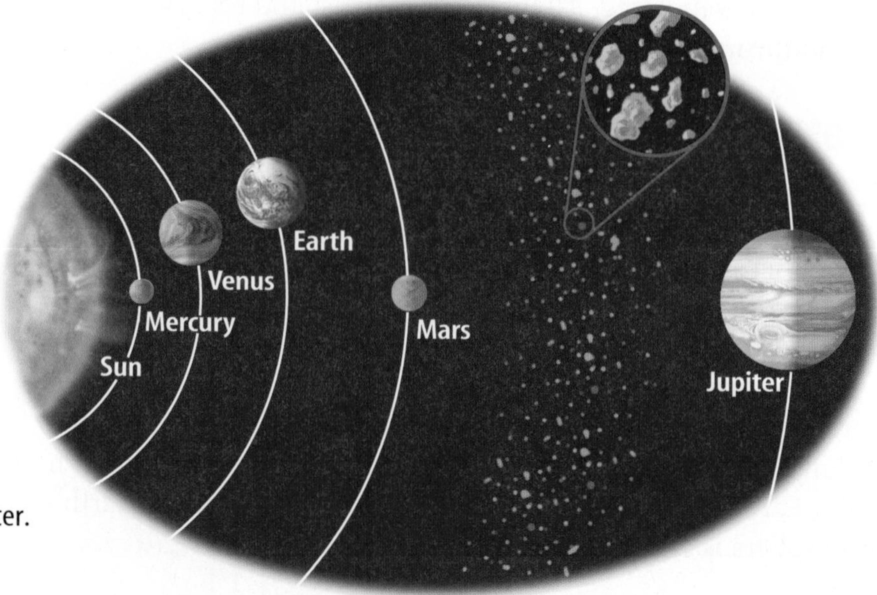

Figure 21 The asteroid belt lies between the orbits of Mars and Jupiter.

Asteroid Sizes The sizes of the asteroids in the asteroid belt range from tiny particles to objects 940 km in diameter. Ceres is the largest and the first one discovered. The next three in order of size are Vesta (530 km), Pallas (522 km), and 10 Hygiea (430 km). The asteroid Gaspra, shown in **Figure 22,** was photographed by *Galileo* on its way to Jupiter.

Figure 22 The asteroid Gaspra is about 20 km long.

Exploring Asteroids On February 14, 2000, the *Near Earth Asteroid Rendezvous (NEAR)* spacecraft went into orbit around the asteroid 433 Eros and later completed its one-year mission of gathering data. Data from the probe show that Eros has many craters and is similar to meteorites on Earth. The Japanese space probe *Hayabusa* arrived at the asteroid Itokawa in November 2005. Its mission is to collect samples and return them to Earth in a capsule in June 2010.

Comets, asteroids, and most meteorites formed early in the history of the solar system. Scientists study these space objects to learn what the solar system might have been like long ago. Understanding this could help scientists better understand how Earth formed.

section 4 review

Summary

Comets

- Comets consist of dust, rock, and different types of ice.
- The Oort Cloud was proposed as a source of comets in the solar system.

Meteoroids, Meteors, Meteorites

- When meteoroids burn up in the atmosphere, they are called meteors.
- Meteor showers occur when Earth crosses the orbital path of a comet.

Asteroids

- Many asteroids occur between the orbits of Mars and Jupiter. This region is called the asteroid belt.

Self Check

1. **Describe** how a comet changes when it comes close to the Sun.
2. **Explain** how craters form.
3. **Summarize** the differences between comets and asteroids.
4. **Describe** the mission of the *NEAR* space probe.
5. **Think Critically** A meteorite found in Antarctica is thought to have come from Mars. How could a rock from Mars get to Earth?

Applying Math

6. **Use Proportions** During the 2001 Leonid Meteor Shower, some people saw 20 meteors each minute. Assuming a constant rate, how many meteors did these people see in one hour?

Model and Invent

Solar System Distance Model

Goals

- **Design** a table of scale distances and model the distances between and among the Sun and the planets.

Possible Materials

meterstick
scissors
pencil
colored markers
string (several meters)
notebook paper (several sheets)

Safety Precautions

Use care when handling scissors.

Real-World Question

Distances between the Sun and the planets of the solar system are large. These large distances can be difficult to visualize. Can you design and create a model that will demonstrate the distances in the solar system?

Make a Model

1. **List** the steps that you need to take to make your model. Describe exactly what you will do at each step.
2. **List** the materials that you will need to complete your model.
3. **Describe** the calculations that you will use to get scale distances from the Sun for all nine planets.
4. **Make** a table of scale distances that you will use in your model. Show your calculations in your table.
5. **Write** a description of how you will build your model. Explain how it will demonstrate relative distances between and among the Sun and planets of the solar system.

Planetary Distances

Planet	Distance to Sun (km)	Distance to Sun (AU)	Scale Distance (1 AU = 10 cm)	Scale Distance (1 AU = 2 cm)
Mercury	5.97×10^7	0.39		
Venus	1.08×10^8	0.72		
Earth	1.50×10^8	1.00		
Mars	2.28×10^8	1.52		Do not write in this book.
Jupiter	7.78×10^8	5.20		
Saturn	1.43×10^9	9.54		
Uranus	2.87×10^9	19.19		
Neptune	4.50×10^9	30.07		

Test Your Model

1. **Compare** your scale distances with those of other students. Discuss why each of you chose the scale that you did.
2. Make sure your teacher approves your plan before you start.
3. **Construct** the model using your scale distances.
4. While constructing the model, write any observations that you or other members of your group make, and complete the data table in your Science Journal. Calculate the scale distances that would be used in your model if 1 AU = 2 m.

Analyze Your Data

1. **Explain** how a scale distance is determined.
2. Was it possible to work with your scale? Explain why or why not.
3. How much string would be required to construct a model with a scale distance of 1 AU = 2 m?
4. Proxima Centauri, the closest star to the Sun, is about 270,000 AU from the Sun. Based on your scale, how much string would you need to place this star on your model?

Conclude and Apply

1. **Summarize** your observations about distances in the solar system. How are distances between the inner planets different from distances between the outer planets?
2. Using your scale distances, determine which planet orbits closest to Earth. Which planet's orbit is second closest?

ASTRONOMY
Plate LXXXVII
Fig. 207 The GRAND ORRERY by Rowley
Fig. 208

Communicating Your Data

Compare your scale model with those of other students. Discuss any differences. **For more help, refer to the Science Skill Handbook.**

Oops! Accidents in SCIENCE

SOMETIMES GREAT DISCOVERIES HAPPEN BY ACCIDENT!

IT CAME FROM OUTER SPACE!

On September 4, 1990, Frances Pegg was unloading bags of groceries in her kitchen in Burnwell, Kentucky. Suddenly, she heard a loud crashing sound. Her husband Arthur heard the same sound. The sound frightened the couple's goat and horse. The noise had come from an object that had crashed through the Pegg's roof, their ceiling, and the floor of their porch. They couldn't see what the object was, but the noise sounded like a gunshot, and pieces of wood from their home flew everywhere. The next day the couple looked under their front porch and found the culprit—a chunk of rock from outer space. It was a meteorite.

For seven years, the Peggs kept their "space rock" at home, making them local celebrities. The rock appeared on TV, and the couple was interviewed by newspaper reporters. In 1997, the Peggs sold the meteorite to the National Museum of Natural History in Washington, D.C., which has a collection of more than 9,000 meteorites. Scientists there study meteorites to learn more about the solar system. One astronomer explained, "Meteorites formed at about the same time as the solar system, about 4.6 billion years ago, though some are younger."

Scientists especially are interested in the Burnwell meteorite because its chemical make up is different from other meteorites previously studied. The Burnwell meteorite is richer in metallic iron and nickel than other known meteorites and is less rich in some metals such as cobalt. Scientists are comparing the rare Burnwell rock with other data to find out if there are more meteorites like the one that fell on the Peggs' roof. But so far, it seems the Peggs' visitor from outer space is one-of-a-kind.

The Burnwell meteorite crashed into the Peggs' home and landed in their basement on the right.

Research Do research to learn more about meteorites. How do they give clues to how our solar system formed? Report to the class.

For more information, visit earth.msscience.com/oops

Reviewing Main Ideas

Section 1 The Solar System

1. Early Greek scientists thought that Earth was at the center of the solar system. They thought that the planets and stars circled Earth.
2. Today, people know that objects in the solar system revolve around the Sun.

Section 2 The Inner Planets

1. The inner planets are Mercury, Venus, Earth, and Mars.
2. The inner planets are small, rocky planets.

Section 3 The Outer Planets

1. The outer planets are Jupiter, Saturn, Uranus, and Neptune.
2. Pluto is a small icy dwarf planet. Other dwarf planets include Ceres and Eris.

Section 4 Other Objects in the Solar System

1. Comets are masses of ice and rock. When a comet approaches the Sun, some ice turns to gas and the comet glows brightly.
2. Meteors occur when small pieces of rock enter Earth's atmosphere and burn up.

Visualizing Main Ideas

Copy and complete the following concept map about the solar system.

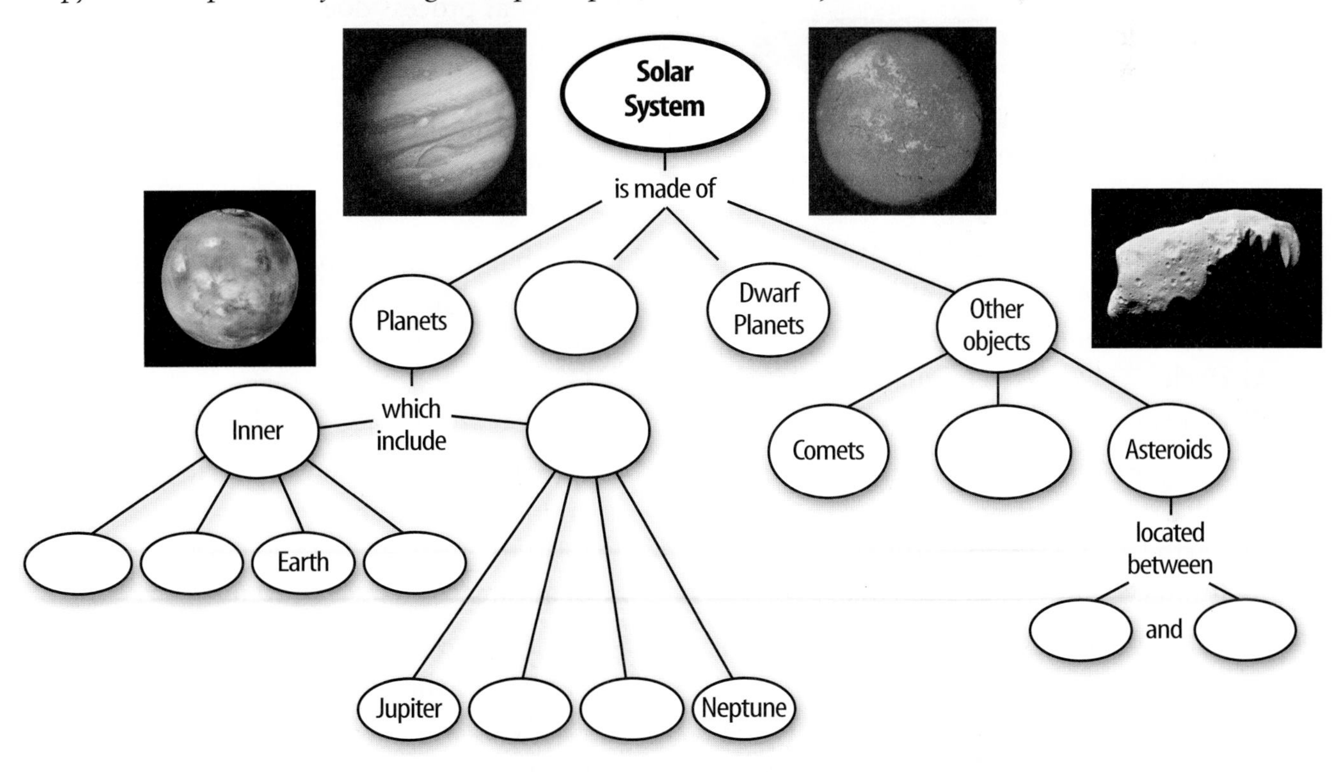

Using Vocabulary

asteroid p. 712	meteorite p. 712
comet p. 710	Neptune p. 706
Earth p. 698	Pluto p. 707
Great Red Spot p. 702	Saturn p. 704
Jupiter p. 702	solar system p. 691
Mars p. 698	Uranus p. 705
Mercury p. 696	Venus p. 697
meteor p. 711	

Fill in the blanks with the correct words.

1. A meteoroid that burns up in Earth's atmosphere is called a(n) ________.
2. The Great Red Spot is a giant storm on ________.
3. ________ is the second largest planet.
4. The *Viking* landers tested for life on ________.
5. The ________ includes the Sun, planets, moons, and other objects.

Checking Concepts

Choose the word or phrase that best answers the question.

6. Who proposed a Sun-centered solar system?
   A) Ptolemy **C)** Galileo
   B) Copernicus **D)** Oort
7. What is the shape of planetary orbits?
   A) circles **C)** squares
   B) ellipses **D)** rectangles
8. Which planet has extreme temperatures because it has no atmosphere?
   A) Earth **C)** Saturn
   B) Jupiter **D)** Mercury
9. Where is the largest volcano in the solar system?
   A) Earth **C)** Mars
   B) Jupiter **D)** Uranus

Use the photo below to answer question 10.

10. Which planet has a complex ring system consisting of thousands of ringlets?
    A) Jupiter **C)** Uranus
    B) Saturn **D)** Mars
11. What is a rock from space that strikes Earth's surface?
    A) asteroid **C)** meteorite
    B) meteoroid **D)** meteor
12. By what process does the Sun produce energy?
    A) magnetism
    B) nuclear fission
    C) nuclear fusion
    D) gravity
13. In what direction do comet tails point?
    A) toward the Sun
    B) away from the Sun
    C) toward Earth
    D) away from the Oort Cloud
14. Which planet has abundant surface water and is known to have life?
    A) Mars **C)** Earth
    B) Jupiter **D)** Venus
15. Which planet has the highest temperatures because of the greenhouse effect?
    A) Mercury **C)** Saturn
    B) Venus **D)** Earth

Science Online earth.msscience.com/vocabulary_puzzlemaker

Thinking Critically

16. **Infer** Why are probe landings on Jupiter not possible?

17. **Concept Map** Copy and complete the concept map on this page to show how a comet changes as it travels through space.

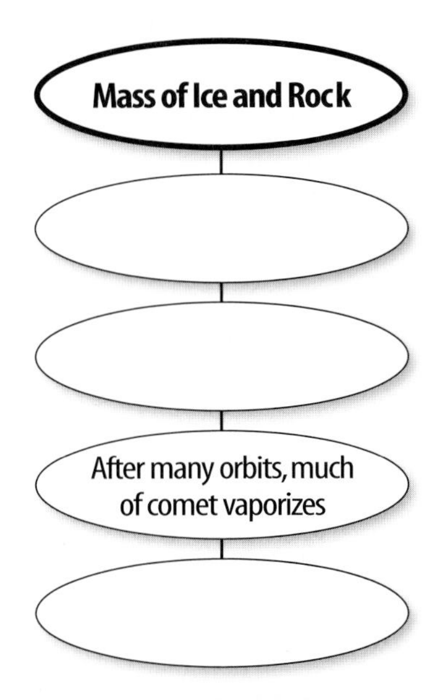

18. **Recognize Cause and Effect** What evidence suggests that liquid water is or once was present on Mars?

19. **Venn Diagram** Create a Venn diagram for Earth and Venus. Create a second Venn diagram for Uranus and Neptune. Which two planets do you think are more similar?

20. **Recognize Cause and Effect** Mercury is closer to the Sun than Venus, yet Venus has higher temperatures. Explain.

21. **Make Models** Make a model that includes the Sun, Earth, and the Moon. Use your model to demonstrate how the Moon revolves around Earth and how Earth and the Moon revolve around the Sun.

22. **Form Hypotheses** Why do Mars's two moons look like asteroids?

Performance Activities

23. **Display** Mercury, Venus, Mars, Jupiter, and Saturn can be observed with the unaided eye. Research when and where in the sky these planets can be observed during the next year. Make a display illustrating your findings. Take some time to observe some of these planets.

24. **Short Story** Select one of the planets or a moon in the solar system. Write a short story from the planet's or moon's perspective. Include scientifically correct facts and concepts in your story.

Applying Math

25. **Saturn's Atmosphere** Saturn's atmosphere consists of 96.3% hydrogen and 3.25% helium. What percentage of Saturn's atmosphere is made up of other gases?

26. **Length of Day on Pluto** A day on Pluto lasts 6.39 times longer than a day on Earth. If an Earth day lasts 24 h, how many hours is a day on Pluto?

Use the graph below to answer question 27.

Weight on Several Planets

Planet	Proportion of Earth's Gravity	Melissa's Weight (kg)
Mercury	0.378	
Venus	0.903	
Earth	1.000	31.8
Mars	0.379	
Jupiter	2.54	

27. **Gravity and Weight** Melissa weighs 31.8 kg on Earth. Multiply Melissa's weight by the proportion of Earth's gravity for each planet to find out how much Melissa would weigh on each.

Part 1 Multiple Choice

Record your answers on the answer sheet provided by your teacher or on a sheet of paper.

Use the photo below to answer question 1.

1. What is shown in the photo above?
   A. asteroids **C.** meteors
   B. comets **D.** meteorites

2. Which is the eighth planet from the Sun?
   A. Earth **C.** Jupiter
   B. Mars **D.** Neptune

3. Which is Pluto's largest moon?
   A. Hydra **C.** Charon
   B. Nix **D.** Triton

4. Which object's gravity holds the planets in their orbits?
   A. Gaspra **C.** Mercury
   B. Earth **D.** the Sun

5. Which of the following occurs in a cycle?
   A. appearance of Halley's comet
   B. condensation of a nebula
   C. formation of a crater
   D. formation of a planet

Test-Taking Tip

No Peeking During the test, keep your eyes on your own paper. If you need to rest them, close them or look up at the ceiling.

6. Which planet likely will be visited by humans in the future?
   A. Jupiter **C.** Mars
   B. Venus **D.** Neptune

7. Between which two planets' orbits does the asteroid belt occur?
   A. Mercury and Venus
   B. Earth and Mars
   C. Uranus and Neptune
   D. Mars and Jupiter

8. Who discovered that planets have elliptical orbits?
   A. Galileo Galilei
   B. Johannes Kepler
   C. Albert Einstein
   D. Nicholas Copernicus

Use the illustration below to answer question 9.

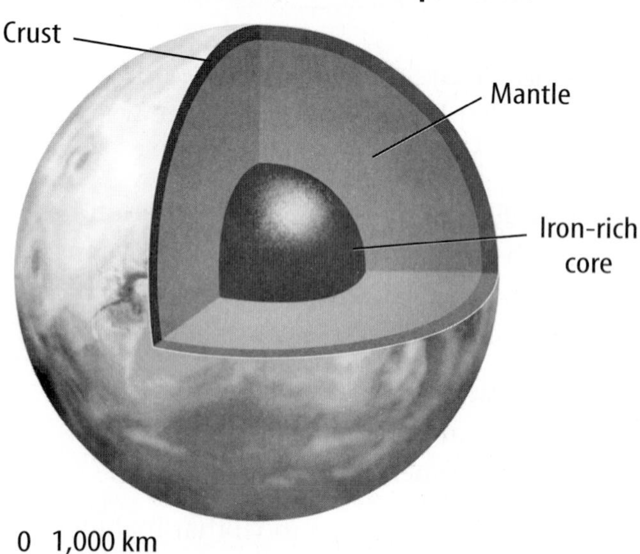

9. Which of the following answers is a good estimate for the diameter of Mars?
   A. 23,122 km **C.** 1,348 km
   B. 6,794 km **D.** 12,583 km

Part 2 Short Response/Grid In

Record your answers on the answer sheet provided by your teacher or on a sheet of paper.

10. Why does a moon remain in orbit around a planet?

11. Compare and contrast the inner planets and the outer planets.

12. Describe the difference between Pluto and the eight planets.

13. Describe Saturn's rings. What are they made of?

14. What is the Great Red Spot?

15. How is Earth different from the other planets in the solar system?

Use the graph below to answfer questions 16–19.

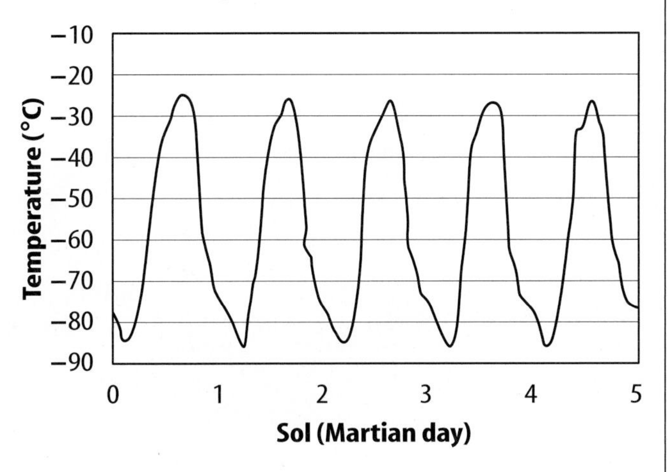

16. Why do the temperatures in the graph vary in a pattern?

17. Approximate the typical high temperature value measured by *Viking I.*

18. Approximate the typical low temperature value measured by *Viking I.*

19. What is the range of these temperature values?

Part 3 Open Ended

Record your answers on a sheet of paper.

20. How might near-Earth-asteroids affect life on Earth? Why do astronomers search for them and monitor their positions?

Use the illustration below to answer question 21.

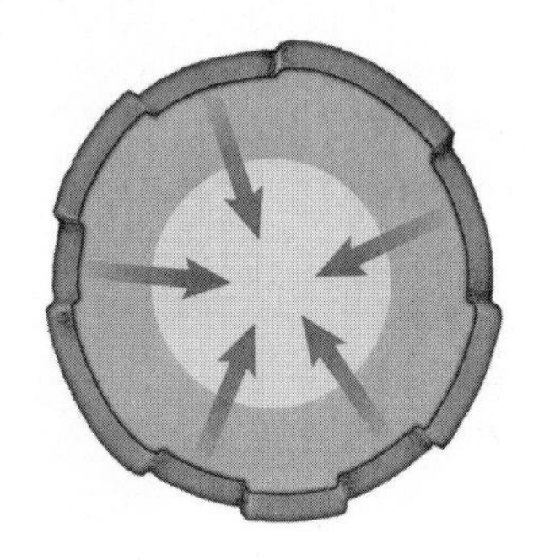

21. Explain how scientists hypothesize that the large cliffs on Mercury formed.

22. Describe the Sun-centered model of the solar system. How is it different from the Earth-centered model?

23. What is an astronomical unit? Why is it useful?

24. Compare and contrast the distances between the planets in the solar system. Which planets are relatively close together? Which planets are relatively far apart?

25. Summarize the current hypothesis about how the solar system formed.

26. Explain how Earth's gravity affects objects that are on or near Earth.

27. Describe the shape of planets' orbits. What is the name of this shape? Where is the Sun located?

28. Describe Jupiter's atmosphere. What characteristics can be observed in images acquired by space probes?

chapter 25

Stars and Galaxies

The BIG Idea

The universe is made up of stars and galaxies.

SECTION 1
Stars

Main Idea For many years, people have been learning about stars by observing their locations in the sky and by studying their light.

SECTION 2
The Sun

Main Idea The Sun is an enormous ball of gas which produces energy by fusing hydrogen into helium.

SECTION 3
Evolution of Stars

Main Idea Stars pass through several stages as they evolve.

SECTION 4
Galaxies and the Universe

Main Idea By studying galaxies, scientists have observed that the universe is expanding.

What's your address?

You know your address at home. You also know your address at school. But do you know your address in space? You live on a planet called Earth that revolves around a star called the Sun. Earth and the Sun are part of a galaxy called the Milky Way. It looks similar to galaxy M83, shown in the photo.

Science Journal Write a description in your Science Journal of the galaxy shown on this page.

Start-Up Activities

Why do clusters of galaxies move apart?

Astronomers know that most galaxies occur in groups of galaxies called clusters. These clusters are moving away from each other in space. The fabric of space is stretching like an inflating balloon.

1. Partially inflate a balloon. Use a piece of string to seal the neck.
2. Draw six evenly spaced dots on the balloon with a felt-tipped marker. Label the dots A through F.
3. Use string and a ruler to measure the distance, in millimeters, from dot A to each of the other dots.
4. Inflate the balloon more.
5. Measure the distances from dot A again.
6. Inflate the balloon again and make new measurements.
7. **Think Critically** Imagine that each dot represents a cluster of galaxies and that the balloon represents the universe. Describe the motion of the clusters in your Science Journal.

Preview this chapter's content and activities at earth.msscience.com

FOLDABLES™ Study Organizer

Stars, Galaxies, and the Universe Make the following Foldable to show what you know about stars, galaxies, and the universe.

STEP 1 **Fold** a sheet of paper from side to side. Make the front edge about 1.25 cm shorter than the back edge.

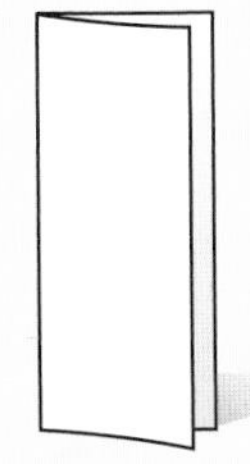

STEP 2 **Turn** lengthwise and **fold** into thirds.

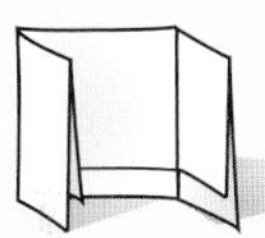

STEP 3 **Unfold and cut** only the top layer along both folds to make three tabs.

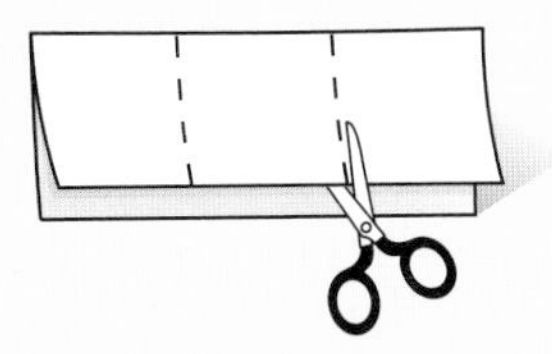

STEP 4 **Label** the tabs *Stars, Galaxies,* and *Universe.*

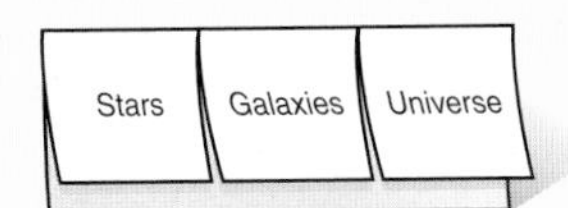

Read and Write Before you read the chapter, write what you already know about stars, galaxies, and the universe. As you read the chapter, add to or correct what you have written under the tabs.

Get Ready to Read

Make Inferences

1 Learn It! When you make inferences, you draw conclusions that are not directly stated in the text. This means you "read between the lines." You interpret clues and draw upon prior knowledge. Authors rely on a reader's ability to infer because all the details are not always given.

2 Practice It! Read the excerpt below and pay attention to highlighted words as you make inferences. Use this Think-Through chart to help you make inferences.

> Notice how the front two stars of the Big Dipper point almost directly at Polaris, which often is called the North Star. Polaris is located at the end of the Little Dipper in the constellation Ursa Minor.
>
> —*from page 725*

Text	Question	Inferences
the front two stars of the Big Dipper	Which are the "front" two stars?	The two "bowl" stars which are farthest from the "handle?"
point almost directly at Polaris	How do the two stars "point?"	Visualize a straight line through the two stars toward the Little Dipper?
located at the end of the Little Dipper	Which is the "end" of the Little Dipper?	The last star in the handle away from the bowl?

3 Apply It! As you read this chapter, practice your skill at making inferences by making connections and asking questions.

Reading Tip

Sometimes you make inferences by using other reading skills, such as questioning and predicting.

Target Your Reading

Use this to focus on the main ideas as you read the chapter.

① **Before you read** the chapter, respond to the statements below on your worksheet or on a numbered sheet of paper.

- Write an **A** if you **agree** with the statement.
- Write a **D** if you **disagree** with the statement.

② **After you read** the chapter, look back to this page to see if you've changed your mind about any of the statements.

- If any of your answers changed, explain why.
- Change any false statements into true statements.
- Use your revised statements as a study guide.

Before You Read A or D		Statement	After You Read A or D
	1	A constellation is a group of stars which are close together in space.	
	2	A light-year is a measurement of time.	
	3	The color of a star indicates its temperature.	
	4	The Sun is the closest star to Earth.	
	5	Light from the Sun reaches Earth in about eight minutes.	
	6	Most of the heat energy produced by the Sun is caused by the fission, or radioactive decay, of helium into hydrogen in the Sun's core.	
	7	A black hole is a location in space where there is no mass, gravity, or light.	
	8	The Milky Way Galaxy is located within the solar system.	
	9	A red shift in the light spectrum coming from a star means that the star is becoming hotter.	
	10	Most scientific evidence currently suggests that the universe is expanding.	

Science Online
Print out a worksheet of this page at earth.msscience.com

section 1

Stars

as you read

What You'll Learn

- **Explain** why some constellations are visible only during certain seasons.
- **Distinguish** between absolute magnitude and apparent magnitude.

Why It's Important

The Sun is a typical star.

Review Vocabulary

star: a large, spherical mass of gas that gives off light and other types of radiation

New Vocabulary

- constellation
- absolute magnitude
- apparent magnitude
- light-year

Constellations

It's fun to look at clouds and find ones that remind you of animals, people, or objects that you recognize. It takes more imagination to play this game with stars. Ancient Greeks, Romans, and other early cultures observed patterns of stars in the night sky called **constellations.** They imagined that the constellations represented mythological characters, animals, or familiar objects.

From Earth, a constellation looks like spots of light arranged in a particular shape against the dark night sky. **Figure 1** shows how the constellation of the mythological Greek hunter Orion appears from Earth. It also shows that the stars in a constellation often have no relationship to each other in space.

Stars in the sky can be found at specific locations within a constellation. For example, you can find the star Betelgeuse (BEE tul jooz) in the shoulder of the mighty hunter Orion. Orion's faithful companion is his dog, Canis Major. Sirius, the brightest star that is visible from the northern hemisphere, is in Canis Major.

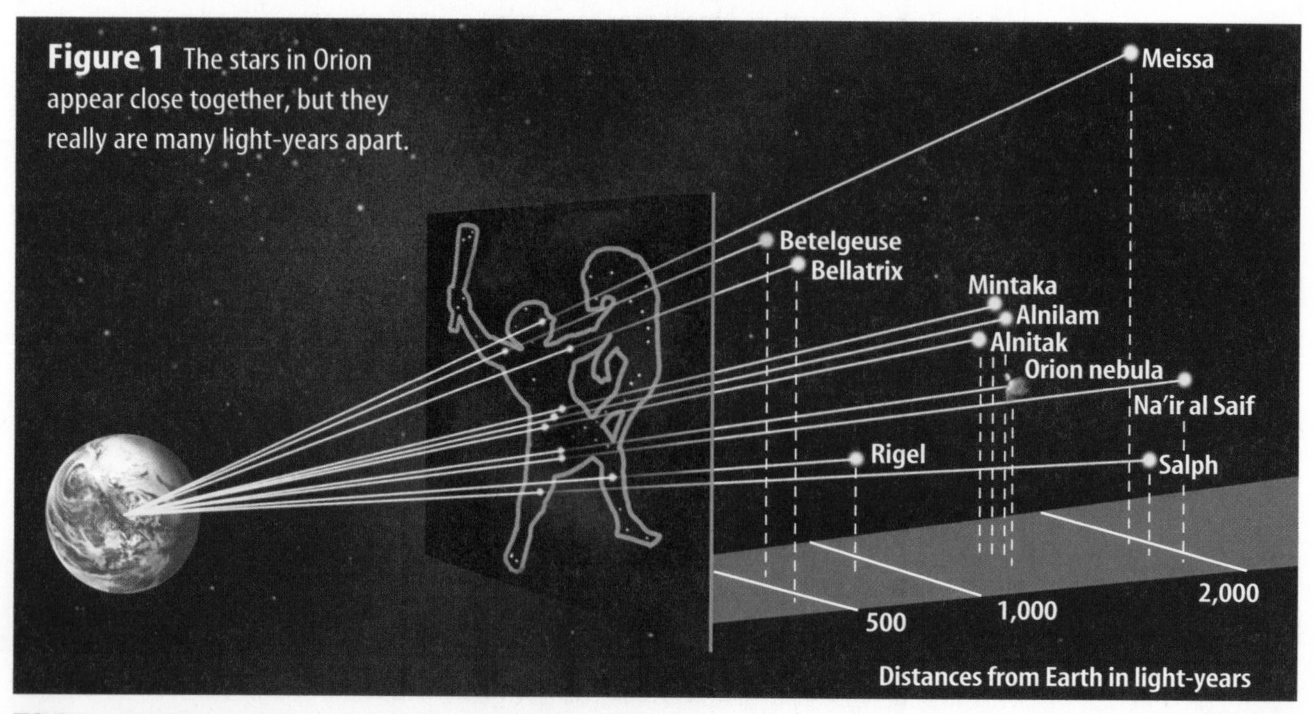

Figure 1 The stars in Orion appear close together, but they really are many light-years apart.

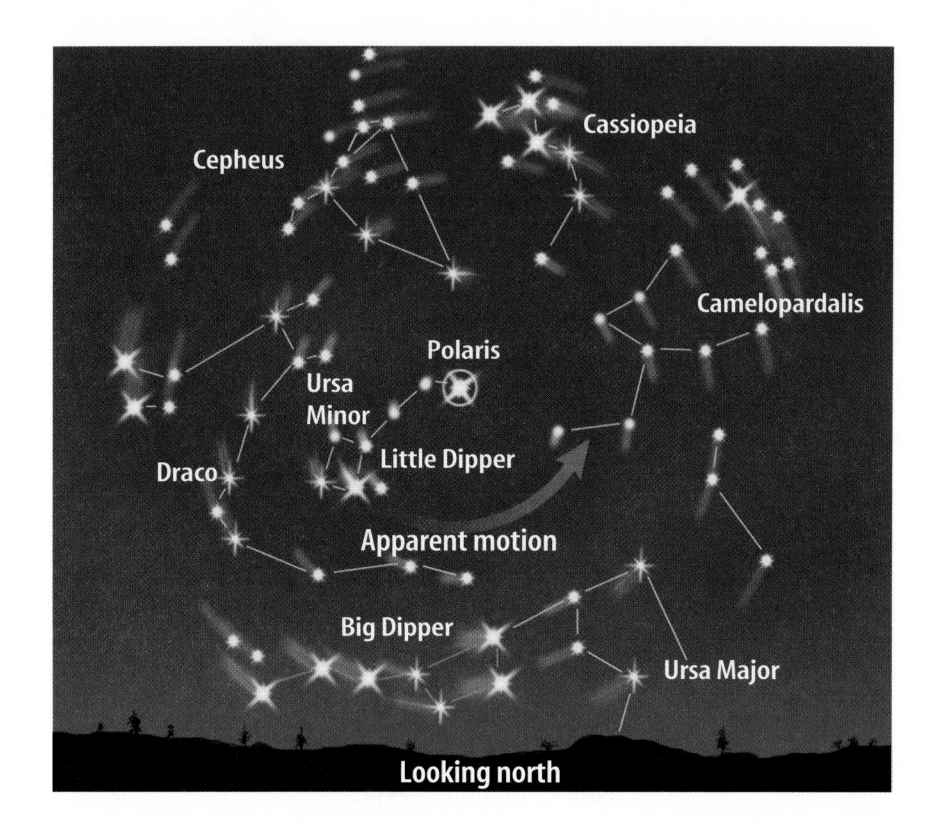

Figure 2 The Big Dipper, in red, is part of the constellation Ursa Major. It is visible year-round in the northern hemisphere. Constellations close to Polaris rotate around Polaris, which is almost directly over the north pole.

Modern Constellations Modern astronomy divides the sky into 88 constellations, many of which were named by early astronomers. You probably know some of them. Can you recognize the Big Dipper? It's part of the constellation Ursa Major, shown in **Figure 2.** Notice how the front two stars of the Big Dipper point almost directly at Polaris, which often is called the North Star. Polaris is located at the end of the Little Dipper in the constellation Ursa Minor. It is positioned almost directly over Earth's north pole.

Circumpolar Constellations As Earth rotates, Ursa Major, Ursa Minor, and other constellations in the northern sky circle around Polaris. Because of this, they are called circumpolar constellations. The constellations appear to move, as shown in **Figure 2,** because Earth is in motion. The stars appear to complete one full circle in the sky in about 24 h as Earth rotates on its axis. One circumpolar constellation that's easy to find is Cassiopeia (ka see uh PEE uh). You can look for five bright stars that form a big W or a big M in the northern sky, depending on the season.

As Earth orbits the Sun, different constellations come into view while others disappear. Because of their unique position, circumpolar constellations are visible all year long. Other constellations are not. Orion, which is visible in the winter in the northern hemisphere, can't be seen there in the summer because the daytime side of Earth is facing it.

Observing Star Patterns

Procedure

1. On a clear night, go outside after dark and study the stars. Take an adult with you.
2. Let your imagination flow to find patterns of stars that look like something familiar.
3. Draw the stars you see, note their positions, and include a drawing of what you think each star pattern resembles.

Analysis

1. Which of your constellations match those observed by your classmates?
2. How can recognizing star patterns be useful?

Try at Home

Absolute and Apparent Magnitudes

When you look at constellations, you'll notice that some stars are brighter than others. For example, Sirius looks much brighter than Rigel. Is Sirius a brighter star, or is it just closer to Earth, making it appear to be brighter? As it turns out, Sirius is 100 times closer to Earth than Rigel is. If Sirius and Rigel were the same distance from Earth, Rigel would appear much brighter in the night sky than Sirius would.

When you refer to the brightness of a star, you can refer to its absolute magnitude or its apparent magnitude. The **absolute magnitude** of a star is a measure of the amount of light it gives off. A measure of the amount of light received on Earth is the **apparent magnitude.** A star that's dim can appear bright in the sky if it's close to Earth, and a star that's bright can appear dim if it's far away. If two stars are the same distance away, what might cause one of them to be brighter than the other?

What is the difference between absolute and apparent magnitude?

Applying Science

Are distance and brightness related?

The apparent magnitude of a star is affected by its distance from Earth. This activity will help you determine the relationship between distance and brightness.

Identifying the Problem

Luisa conducted an experiment to determine the relationship between distance and the brightness of stars. She used a meterstick, a light meter, and a lightbulb. She placed the bulb at the zero end of the meterstick, then placed the light meter at the 20-cm mark and recorded the distance and the light-meter reading in her data table. Readings are in luxes, which are units for measuring light intensity. Luisa then increased the distance from the bulb to the light meter and took more readings. By examining the data in the table, can you see a relationship between the two variables?

Effect of Distance on Light

Distance (cm)	Meter Reading (luxes)
20	4150.0
40	1037.5
60	461.1
80	259.4

Solving the Problem

1. What happened to the amount of light recorded when the distance was increased from 20 cm to 40 cm? When the distance was increased from 20 cm to 60 cm?
2. What does this indicate about the relationship between light intensity and distance? What would the light intensity be at 100 cm? Would making a graph help you visualize the relationship?

Measurement in Space

How do scientists determine the distance from Earth to nearby stars? One way is to measure parallax—the apparent shift in the position of an object when viewed from two different positions. Extend your arm and look at your thumb first with your left eye closed and then with your right eye closed, as the girl in **Figure 3A** is doing. Your thumb appears to change position with respect to the background. Now do the same experiment with your thumb closer to your face, as shown in **Figure 3B.** What do you observe? The nearer an object is to the observer, the greater its parallax is.

Astronomers can measure the parallax of relatively close stars to determine their distances from Earth. **Figure 4** shows how a close star's position appears to change. Knowing the angle that the star's position changes and the size of Earth's orbit, astronomers can calculate the distance of the star from Earth.

Because space is so vast, a special unit of measure is needed to record distances. Distances between stars and galaxies are measured in light-years. A **light-year** is the distance that light travels in one year. Light travels at 300,000 km/s, or about 9.5 trillion km in one year. The nearest star to Earth, other than the Sun, is Proxima Centauri. Proxima Centauri is a mere 4.3 light-years away, or about 40 trillion km.

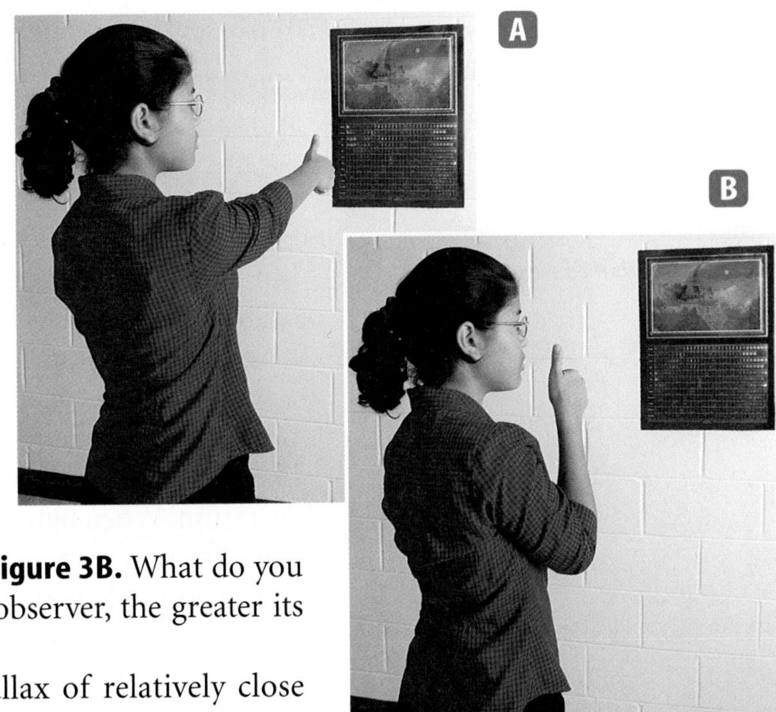

Figure 3 A Your thumb appears to move less against the background when it is farther away from your eyes. B It appears to move more when it is closer to your eyes.

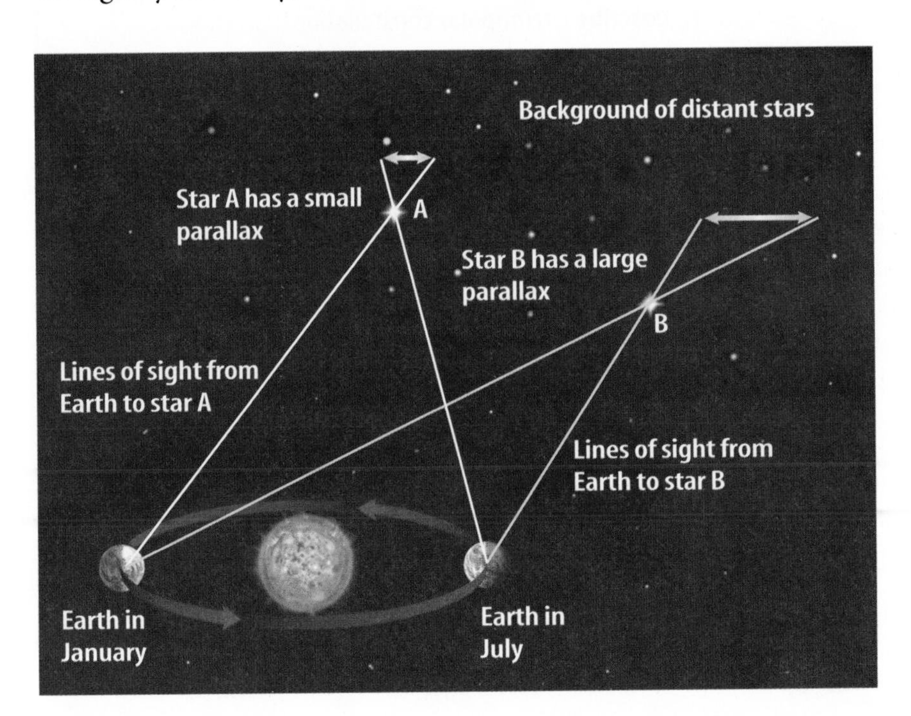

Figure 4 Parallax is determined by observing the same star when Earth is at two different points in its orbit around the Sun. The star's position relative to more distant background stars will appear to change.
Infer *whether star* **A** *or* **B** *is farther from Earth.*

Figure 5 This star spectrum was made by placing a diffraction grating over a telescope's objective lens. A diffraction grating produces a spectrum by causing interference of light waves. **Explain** *what causes the lines in spectra.*

Properties of Stars

The color of a star indicates its temperature. For example, hot stars are a blue-white color. A relatively cool star looks orange or red. Stars that have the same temperature as the Sun have a yellow color.

Astronomers study the composition of stars by observing their spectra. When fitted into a telescope, a spectroscope acts like a prism. It spreads light out in the rainbow band called a spectrum. When light from a star passes through a spectroscope, it breaks into its component colors. Look at the spectrum of a star in **Figure 5.** Notice the dark lines caused by elements in the star's atmosphere. Light radiated from a star passes through the star's atmosphere. As it does, elements in the atmosphere absorb some of this light.The wavelengths of visible light that are absorbed appear as dark lines in the spectrum. Each element absorbs certain wavelengths, producing a unique pattern of dark lines. Like a fingerprint, the patterns of lines can be used to identify the elements in a star's atmosphere.

section 1 review

Summary

Constellations

- Constellations are patterns of stars in the night sky.
- The stars in a constellation often have no relationship to each other in space.

Absolute and Apparent Magnitudes

- Absolute magnitude is a measure of how much light is given off by a star.
- Apparent magnitude is a measure of how much light from a star is received on Earth.

Measurement in Space

- Distances between stars are measured in light-years.

Properties of Stars

- Astronomers study the composition of stars by observing their spectra.

Self Check

1. **Describe** circumpolar constellations.
2. **Explain** why some constellations are visible only during certain seasons.
3. **Infer** how two stars could have the same apparent magnitude but different absolute magnitudes.
4. **Explain** how a star is similar to the Sun if it has the same absorption lines in its spectrum that occur in the Sun's spectrum.
5. **Think Critically** If a star's parallax angle is too small to measure, what can you conclude about the star's distance from Earth?

Applying Skills

6. **Recognize Cause and Effect** Suppose you viewed Proxima Centauri, which is 4.3 light-years from Earth, through a telescope. How old were you when the light that you see left this star?

earth.msscience.com/self_check_quiz

section 2

The Sun

The Sun's Layers

The Sun is an ordinary star, but it's important to you. The Sun is the center of the solar system, and the closest star to Earth. Almost all of the life on Earth depends on energy from the Sun.

Notice the different layers of the Sun, shown in **Figure 6,** as you read about them. Like other stars, the Sun is an enormous ball of gas that produces energy by fusing hydrogen into helium in its core. This energy travels outward through the radiation zone and the convection zone. In the convection zone, gases circulate in giant swirls. Finally, energy passes into the Sun's atmosphere.

The Sun's Atmosphere

The lowest layer of the Sun's atmosphere and the layer from which light is given off is the **photosphere.** The photosphere often is called the surface of the Sun, although the surface is not a smooth feature. Temperatures there are about 6,000 K. Above the photosphere is the **chromosphere.** This layer extends upward about 2,000 km above the photosphere. A transition zone occurs between 2,000 km and 10,000 km above the photosphere. Above the transition zone is the **corona.** This is the largest layer of the Sun's atmosphere and extends millions of kilometers into space. Temperatures in the corona are as high as 2 million K. Charged particles continually escape from the corona and move through space as solar wind.

as you read

What You'll Learn

- **Explain** that the Sun is the closest star to Earth.
- **Describe** the structure of the Sun.
- **Describe** sunspots, prominences, and solar flares.

Why It's Important

The Sun is the source of most energy on Earth.

Review Vocabulary

cycle: a repeating sequence of events, such as the sunspot cycle

New Vocabulary

- photosphere
- chromosphere
- corona
- sunspot

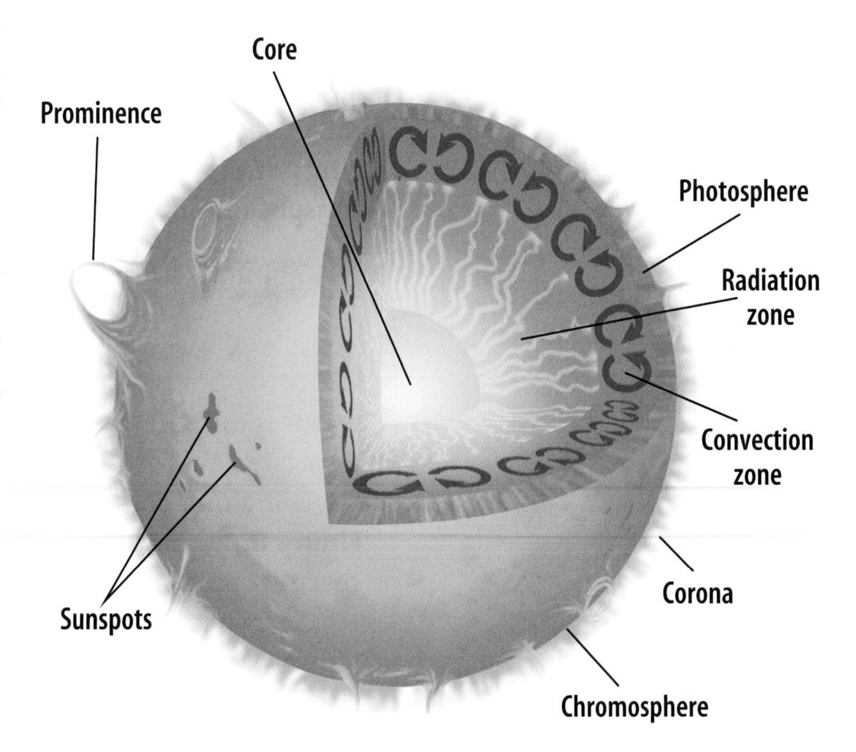

Figure 6 Energy produced in the Sun's core by fusion travels outward by radiation and convection. The Sun's atmosphere shines by the energy produced in the core.

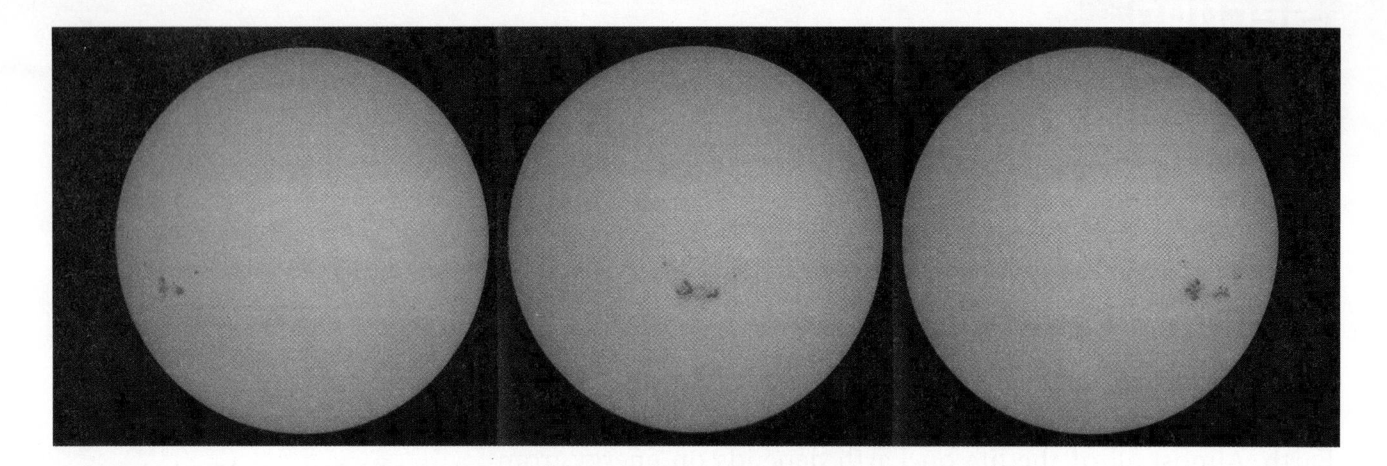

Figure 7 Sunspots are bright, but when viewed against the rest of the photosphere, they appear dark. Notice how these sunspots move as the Sun rotates. **Describe** *the Sun's direction of rotation.*

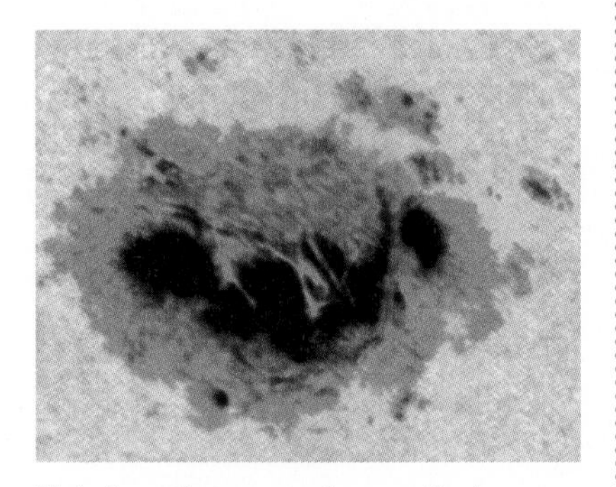

This is a close-up photo of a large sunspot.

Surface Features

From the viewpoint that you observe the Sun, its surface appears to be a smooth layer. But the Sun's surface has many features, including sunspots, prominences, flares, and CMEs.

Sunspots Areas of the Sun's surface that appear dark because they are cooler than surrounding areas are called **sunspots.** Ever since Galileo Galilei made drawings of sunspots, scientists have been studying them. Because scientists could observe the movement of individual sunspots, shown in **Figure 7,** they concluded that the Sun rotates. However, the Sun doesn't rotate as a solid body, as Earth does. It rotates faster at its equator than at its poles. Sunspots at the equator take about 25 days to complete one rotation. Near the poles, they take about 35 days.

Sunspots aren't permanent features on the Sun. They appear and disappear over a period of several days, weeks, or months. The number of sunspots increases and decreases in a fairly regular pattern called the sunspot, or solar activity, cycle. Times when many large sunspots occur are called sunspot maximums. Sunspot maximums occur about every 10 to 11 years. Periods of sunspot minimum occur in between.

Reading Check *What is a sunspot cycle?*

Prominences and Flares Sunspots are related to several features on the Sun's surface. The intense magnetic fields associated with sunspots might cause prominences, which are huge, arching columns of gas. Notice the huge prominence in **Figure 8.** Some prominences blast material from the Sun into space at speeds ranging from 600 km/s to more than 1,000 km/s.

Gases near a sunspot sometimes brighten suddenly, shooting outward at high speed. These violent eruptions are called solar flares. You can see a solar flare in **Figure 8.**

CMEs Coronal mass ejections (CMEs) occur when large amounts of electrically-charged gas are ejected suddenly from the Sun's corona. CMEs can occur as often as two or three times each day during a sunspot maximum.

CMEs present little danger to life on Earth, but they do have some effects. CMEs can damage satellites in orbit around Earth. They also can interfere with radio and power distribution equipment. CMEs often cause auroras. High-energy particles contained in CMEs and the solar wind are carried past Earth's magnetic field. This generates electric currents that flow toward Earth's poles. These electric currents ionize gases in Earth's atmosphere. When these ions recombine with electrons, they produce the light of an aurora, shown in **Figure 8.**

Topic: Space Weather
Visit earth.msscience.com for Web links to information about space weather and its effects.

Activity Record space weather conditions for several weeks. How does space weather affect Earth?

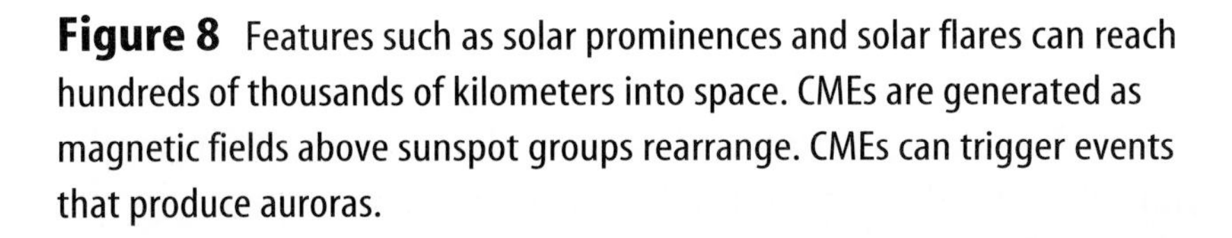

Figure 8 Features such as solar prominences and solar flares can reach hundreds of thousands of kilometers into space. CMEs are generated as magnetic fields above sunspot groups rearrange. CMEs can trigger events that produce auroras.

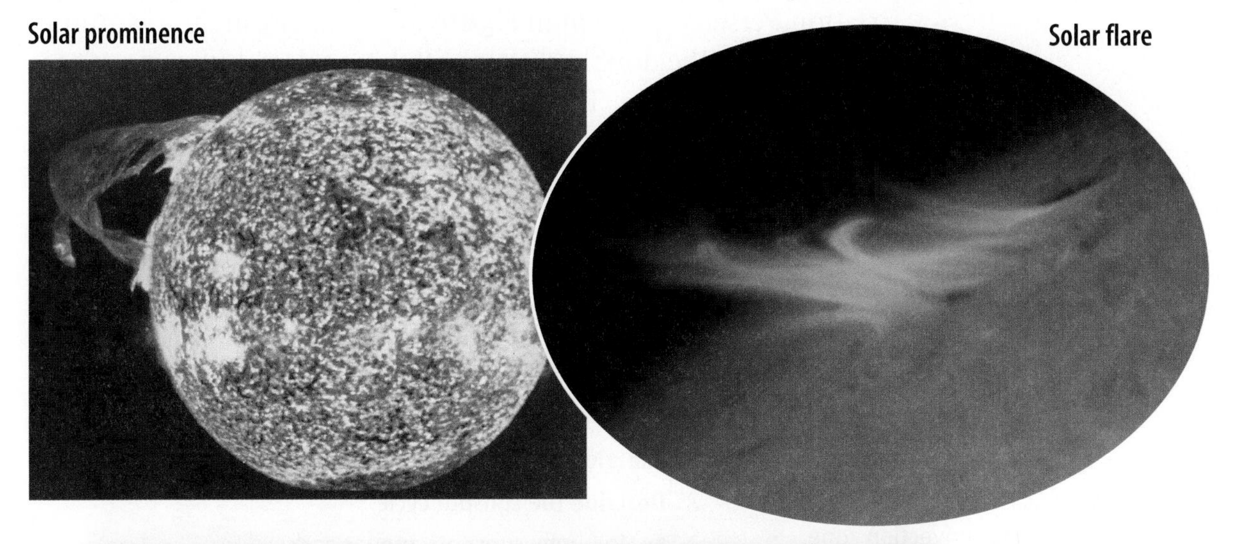

Aurora borealis, or northern lights

Figure 9 Most stars originally formed in large clusters containing hundreds, or even thousands, of stars.
Draw and label *a sketch of the double cluster.*

The Sun—An Average Star

The Sun is an average star. It is middle-aged, and its absolute magnitude is about average. It shines with a yellow light. Although the Sun is an average star, it is much closer to Earth than other stars. Light from the Sun reaches Earth in about eight minutes. Light from other stars takes from many years to many millions of years to reach Earth.

The Sun is unusual in one way. It is not close to any other stars. Most stars are part of a system in which two or more stars orbit each other. When two stars orbit each other, it is called a binary system. When three stars orbit each other, it is called a triple star system. The closest star system to the Sun—the Alpha Centauri system, including Proxima Centauri—is a triple star.

Stars also can move through space as a cluster. In a star cluster, many stars are relatively close, so the gravitational attraction among the stars is strong. Most star clusters are far from the solar system. They sometimes appear as a fuzzy patch in the night sky. The double cluster in the northern part of the constellation Perseus is shown in **Figure 9.** On a dark night in autumn, you can see the double cluster with binoculars, but you can't see its individual stars. The Pleiades star cluster can be seen in the constellation of Taurus in the winter sky. On a clear, dark night, you might be able to see seven of the stars in this cluster.

section 2 review

Summary

The Sun's Layers

- The Sun's interior has layers that include the core, radiation zone, and convection zone.

The Sun's Atmosphere

- The Sun's atmosphere includes the photosphere, chromosphere, and corona.

Surface Features

- The number of sunspots on the Sun varies in a 10- to 11-year cycle.
- Auroras occur when charged particles from the Sun interact with Earth's magnetic field.

The Sun—An Average Star

- The Sun is an average star, but it is much closer to Earth than any other star.

Self Check

1. **Explain** why the Sun is important for life on Earth.
2. **Describe** the sunspot cycle.
3. **Explain** why sunspots appear dark.
4. **Explain** why the Sun, which is an average star, appears so much brighter from Earth than other stars do.
5. **Think Critically** When a CME occurs on the Sun, it takes a couple of days for effects to be noticed on Earth. Explain.

Applying Skills

6. **Communicate** Make a sketch that shows the Sun's layers in your Science Journal. Write a short description of each layer.

Science Online earth.msscience.com/self_check_quiz

Sunspots

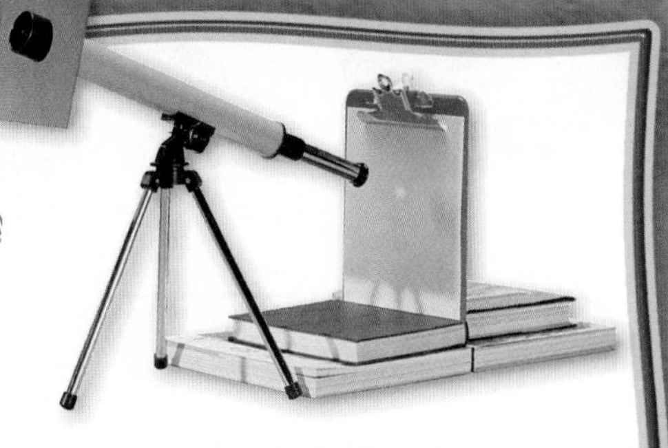

Sunspots can be observed moving across the face of the Sun as it rotates. Measure the movement of sunspots, and use your data to determine the Sun's period of rotation.

Real-World Question

Can sunspot motion be used to determine the Sun's period of rotation?

Goals

- **Observe** sunspots and estimate their size.
- **Estimate** the rate at which sunspots move across the face of the Sun.

Materials

several books
piece of cardboard
drawing paper
refracting telescope
clipboard
small tripod
scissors

Safety Precautions

WARNING: *Handle scissors with care.*

Procedure

1. Find a location where the Sun can be viewed at the same time of day for a minimum of five days. **WARNING:** *Do not look directly at the Sun. Do not look through the telescope at the Sun. You could damage your eyes.*
2. If the telescope has a small finder scope attached, remove it or keep it covered.
3. Set up the telescope with the eyepiece facing away from the Sun, as shown. Align the telescope so that the shadow it casts on the ground is the smallest size possible. Cut and attach the cardboard as shown in the photo.
4. Use books to prop the clipboard upright. Point the eyepiece at the drawing paper.
5. Move the clipboard back and forth until you have the largest image of the Sun on the paper. Adjust the telescope to form a clear image. Trace the outline of the Sun on the paper.
6. Trace any sunspots that appear as dark areas on the Sun's image. Repeat this step at the same time each day for a week.
7. Using the Sun's diameter (approximately 1,390,000 km), estimate the size of the largest sunspots that you observed.
8. **Calculate** how many kilometers the sunspots move each day.
9. **Predict** how many days it will take for the same group of sunspots to return to the same position in which they appeared on day 1.

Conclude and Apply

1. What was the estimated size and rate of motion of the largest sunspots?
2. **Infer** how sunspots can be used to determine that the Sun's surface is not solid like Earth's surface.

Communicating Your Data

Compare your conclusions with those of other students in your class. **For more help, refer to the** Science Skill Handbook.

section 3

Evolution of Stars

as you read

What You'll Learn

- **Describe** how stars are classified.
- **Compare** the Sun to other types of stars on the H-R diagram.
- **Describe** how stars evolve.

Why It's Important

Earth and your body contain elements that were made in stars.

Review Vocabulary

gravity: an attractive force between objects that have mass

New Vocabulary

- nebula
- giant
- white dwarf
- supergiant
- neutron star
- black hole

Classifying Stars

When you look at the night sky, all stars might appear to be similar, but they are quite different. Like people, they vary in age and size, but stars also vary in temperature and brightness.

In the early 1900s, Ejnar Hertzsprung and Henry Russell made some important observations. They noticed that, in general, stars with higher temperatures also have brighter absolute magnitudes.

Hertzsprung and Russell developed a graph, shown in **Figure 10,** to show this relationship. They placed temperatures across the bottom and absolute magnitudes up one side. A graph that shows the relationship of a star's temperature to its absolute magnitude is called a Hertzsprung-Russell (H-R) diagram.

The Main Sequence As you can see, stars seem to fit into specific areas of the graph. Most stars fit into a diagonal band that runs from the upper left to the lower right of the graph. This band, called the main sequence, contains hot, blue, bright stars in the upper left and cool, red, dim stars in the lower right. Yellow main sequence stars, like the Sun, fall in between.

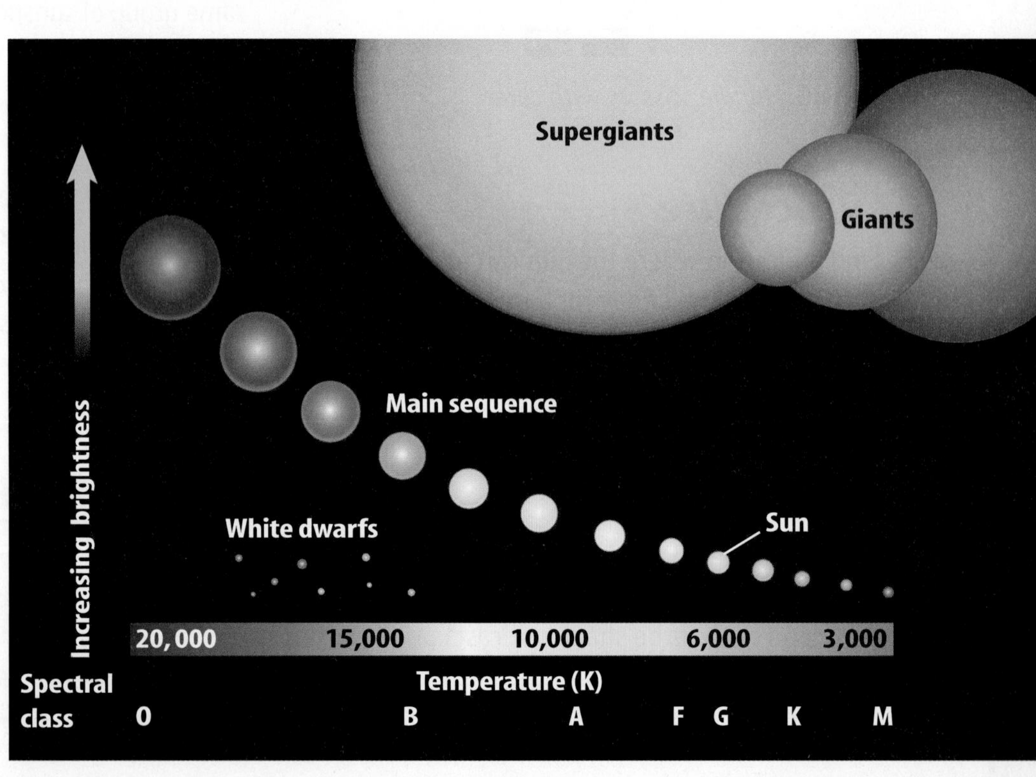

Figure 10 The relationships among a star's color, temperature, and brightness are shown in this H-R diagram. Stars in the upper left are hot, bright stars, and stars in the lower right are cool, dim stars.
Classify *Which type of star shown in the diagram is the hottest, dimmest star?*

Dwarfs and Giants About 90 percent of all stars are main sequence stars. Most of these are small, red stars found in the lower right of the H-R diagram. Among main sequence stars, the hottest stars generate the most light and the coolest ones generate the least. What about the ten percent of stars that are not part of the main sequence? Some of these stars are hot but not bright. These small stars are located on the lower left of the H-R diagram and are called white dwarfs. Other stars are extremely bright but not hot. These large stars on the upper right of the H-R diagram are called giants, or red giants, because they are usually red in color. The largest giants are called supergiants. **Figure 11** shows the supergiant, Antares—a star 300 times the Sun's diameter—in the constellation Scorpius. It is more than 11,000 times as bright as the Sun.

Figure 11 Antares is a bright supergiant located 400 light-years from Earth. Although its temperature is only about 3,500 K, it is the 16th brightest star in the sky.

Reading Check *What kinds of stars are on the main sequence?*

How do stars shine?

For centuries, people were puzzled by the questions of what stars were made of and how they produced light. Many people had estimated that Earth was only a few thousand years old. The Sun could have been made of coal and shined for that long. However, when people realized that Earth was much older, they wondered what material possibly could burn for so many years. Early in the twentieth century, scientists began to understand the process that keeps stars shining for billions of years.

Generating Energy In the 1930s, scientists discovered reactions between the nuclei of atoms. They hypothesized that temperatures in the center of the Sun must be high enough to cause hydrogen to fuse to make helium. This reaction releases tremendous amounts of energy. Much of this energy is emitted as different wavelengths of light, including visible, infrared, and ultraviolet light. Only a tiny fraction of this light comes to Earth. During the fusion reaction, four hydrogen nuclei combine to create one helium nucleus. The mass of one helium nucleus is less than the mass of four hydrogen nuclei, so some mass is lost in the reaction.

Years earlier, in 1905, Albert Einstein had proposed a theory stating that mass can be converted into energy. This was stated as the famous equation $E = mc^2$. In this equation, E is the energy produced, m is the mass, and c is the speed of light. The small amount of mass "lost" when hydrogen atoms fuse to form a helium atom is converted to a large amount of energy.

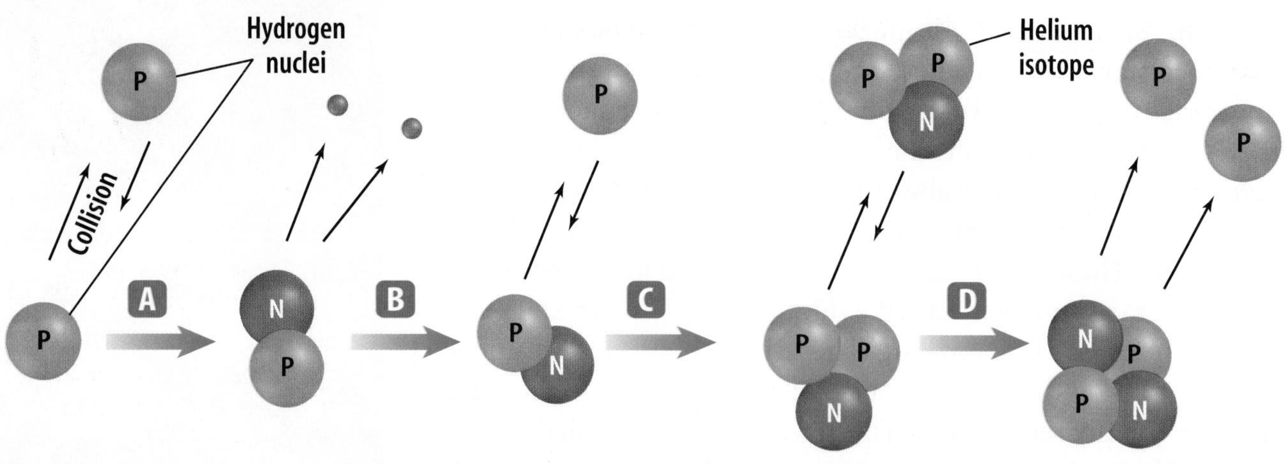

A Two protons (hydrogen nuclei) collide. One proton decays to a neutron, releasing subatomic particles and some energy.

B Another proton fuses with a proton and neutron to form an isotope of helium. Energy is given off again.

C Two helium isotopes collide with enough energy to fuse.

D A helium nucleus (two protons and two neutrons) forms as two protons break away. During the process, still more energy is released.

Figure 12 Fusion of hydrogen into helium occurs in a star's core. **Infer** *what happens to the "lost" mass during this process.*

Fusion Shown in **Figure 12,** fusion occurs in the cores of stars. Only in the core are temperatures high enough to cause atoms to fuse. Normally, they would repel each other, but in the core of a star where temperatures can exceed 15,000,000 K, atoms can move so fast that some of them fuse upon colliding.

Evolution of Stars

INTEGRATE Physics The H-R diagram explained a lot about stars. However, it also led to more questions. Many wondered why some stars didn't fit in the main sequence group and what happened when a star depleted its supply of hydrogen fuel. Today, scientists have theories about how stars evolve, what makes them different from one another, and what happens when they die. **Figure 13** illustrates the lives of different types of stars.

When hydrogen fuel is depleted, a star loses its main sequence status. This can take less than 1 million years for the brightest stars to many billions of years for the dimmest stars. The Sun has a main sequence life span of about 10 billion years. Half of its life span is still in the future.

Topic: Evolution of Stars
Visit earth.msscience.com for Web links to information about the evolution of stars.

Activity Make a three-circle Venn diagram to compare and contrast white dwarfs, neutron stars, and black holes.

Nebula Stars begin as a large cloud of gas and dust called a **nebula.** As the particles of gas and dust exert a gravitational force on each other, the nebula begins to contract. Gravitational forces cause instability within the nebula. The nebula can break apart into smaller and smaller pieces. Each piece eventually might collapse to form a star.

A Star Is Born As the particles in the smaller pieces of nebula move closer together, the temperatures in each nebula piece increase. When the temperature inside the core of a nebula piece reaches 10 million K, fusion begins. The energy released radiates outward through the condensing ball of gas. As the energy radiates into space, stars are born.

Reading Check *How are stars born?*

Main Sequence to Giant Stars In the newly formed star, the heat from fusion causes pressure to increase. This pressure balances the attraction due to gravity. The star becomes a main sequence star. It continues to use its hydrogen fuel.

When hydrogen in the core of the star is depleted, a balance no longer exists between pressure and gravity. The core contracts, and temperatures inside the star increase. This causes the outer layers of the star to expand and cool. In this late stage of its life cycle, a star is called a **giant.**

After the core temperature reaches 100 million K, helium nuclei fuse to form carbon in the giant's core. By this time, the star has expanded to an enormous size, and its outer layers are much cooler than they were when it was a main sequence star. In about 5 billion years, the Sun will become a giant.

White Dwarfs After the star's core uses much of its helium, it contracts even more and its outer layers escape into space. This leaves behind the hot, dense core. At this stage in a star's evolution, it becomes a **white dwarf.** A white dwarf is about the size of Earth. Eventually, the white dwarf will cool and stop giving off light.

White Dwarf Matter The matter in white dwarf stars is more than 500,000 times as dense as the matter in Earth. In white dwarf matter, there are free electrons and atomic nuclei. The resistance of the electrons to pack together more provides pressure that keeps the star from collapsing. This state of matter is called electron degeneracy.

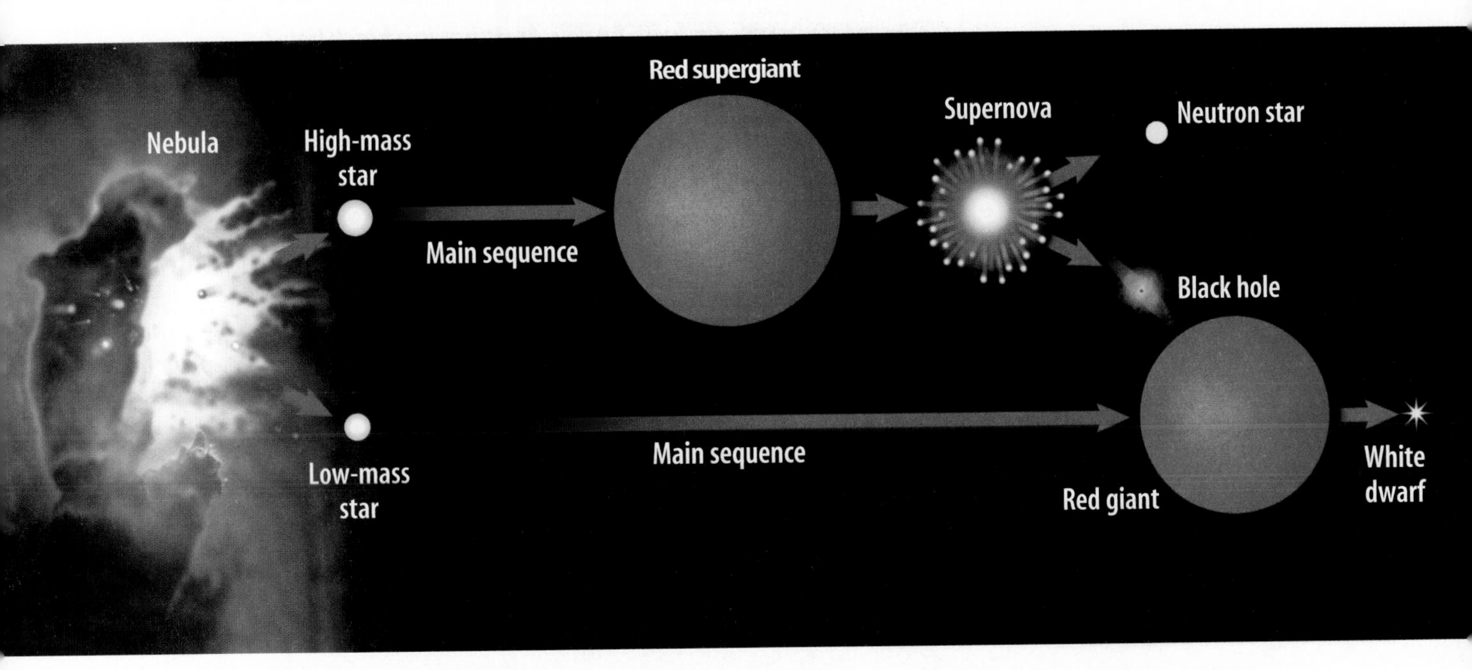

Figure 13 The life of a star depends on its mass. Massive stars eventually become neutron stars or black holes.
Explain *what happens to stars that are the size of the Sun.*

386 Supernova In 386 A.D., Chinese observers described a new star—a supernova—in the night sky. More recently, astronomers using the *Chandra X-ray Observatory* found evidence of a spinning neutron star, called a pulsar, in exactly the same location. Because of the Chinese account, astronomers better understand how neutron stars form and evolve.

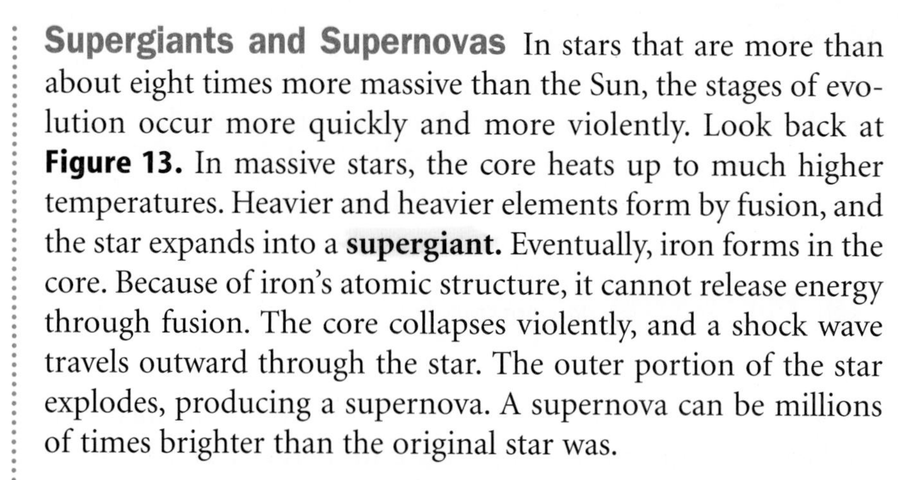

Supergiants and Supernovas In stars that are more than about eight times more massive than the Sun, the stages of evolution occur more quickly and more violently. Look back at **Figure 13.** In massive stars, the core heats up to much higher temperatures. Heavier and heavier elements form by fusion, and the star expands into a **supergiant.** Eventually, iron forms in the core. Because of iron's atomic structure, it cannot release energy through fusion. The core collapses violently, and a shock wave travels outward through the star. The outer portion of the star explodes, producing a supernova. A supernova can be millions of times brighter than the original star was.

Neutron Stars If the collapsed core of a supernova is between about 1.4 and 3 times as massive as the Sun, it will shrink to approximately 20 km in diameter. Only neutrons can exist in the dense core, and it becomes a **neutron star.** Neutron stars are so dense that a teaspoonful would weigh more than 600 million metric tons in Earth's gravity. As dense as neutron stars are, they can contract only so far because the neutrons resist the inward pull of gravity.

Black Holes If the remaining dense core from a supernova is more than about three times more massive than the Sun, probably nothing can stop the core's collapse. Under these conditions, all of the core's mass collapses to a point. The gravity near this mass is so strong that nothing can escape from it, not even light. Because light cannot escape, the region is called a **black hole.** If you could shine a flashlight on a black hole, the light simply would disappear into it.

Figure 14 The black hole at the center of galaxy M87 pulls matter into it at extremely high velocities. Some matter is ejected to produce a jet of gas that streams away from the center of the galaxy at nearly light speed.

What is a black hole?

Black holes, however, are not like giant vacuum cleaners, sucking in distant objects. A black hole has an event horizon, which is a region inside of which nothing can escape. If something—including light—crosses the event horizon, it will be pulled into the black hole. Beyond the event horizon, the black hole's gravity pulls on objects just as it would if the mass had not collapsed. Stars and planets can orbit around a black hole.

The photograph in **Figure 14** was taken by the *Hubble Space Telescope.* It shows a jet of gas streaming out of the center of galaxy M87. This jet of gas formed as matter flowed toward a black hole, and some of the gas was ejected along the polar axis.

Recycling Matter A star begins its life as a nebula, such as the one shown in **Figure 15.** Where does the matter in a nebula come from? Nebulas form partly from the matter that was once in other stars. A star ejects enormous amounts of matter during its lifetime. Some of this matter is incorporated into nebulas, which can evolve to form new stars. The matter in stars is recycled many times.

What about the matter created in the cores of stars and during supernova explosions? Are elements such as carbon and iron also recycled? These elements can become parts of new stars. In fact, spectrographs have shown that the Sun contains some carbon, iron, and other heavier elements. Because the Sun is an average, main sequence star, it is too young and its mass is too small to have formed these elements itself. The Sun condensed from material that was created in stars that died many billions of years ago.

Some elements condense to form planets and other bodies rather than stars. In fact, your body contains many atoms that were fused in the cores of ancient stars. Evidence suggests that the first stars formed from hydrogen and helium and that all the other elements have formed in the cores of stars or as stars explode.

Figure 15 Stars are forming in the Orion Nebula and other similar nebulae.
Describe *a star-forming nebula.*

section 3 review

Summary

Classifying Stars

- Most stars plot on the main sequence of an H-R diagram.
- As stars near the end of their lives, they move off of the main sequence.

How do stars shine?

- Stars shine because of a process called fusion.
- During fusion, nuclei of a lighter element merge to form a heavier element.

Evolution of Stars

- Stars form in regions of gas and dust called nebulae.
- Stars evolve differently depending on how massive they are.

Self Check

1. **Explain** how the Sun is different from other stars on the main sequence. How is it different from a giant star? How is it different from a white dwarf?
2. **Describe** how stars release energy.
3. **Outline** the past and probable future of the Sun.
4. **Define** a black hole.
5. **Think Critically** How can white dwarf stars be both hot and dim?

Applying Math

6. **Convert Units** A neutron star has a diameter of 20 km. One kilometer equals 0.62 miles. What is the neutron star's diameter in miles?

section 4

Galaxies and the Universe

as you read

What You'll Learn

- **Describe** the Sun's position in the Milky Way Galaxy.
- **Explain** that the same natural laws that apply to our solar system also apply in other galaxies.

Why It's Important

Studying the universe could determine whether life exists elsewhere.

Review Vocabulary
universe: the space that contains all known matter and energy

New Vocabulary
- galaxy
- big bang theory

Galaxies

If you enjoy science fiction, you might have read about explorers traveling through the galaxy. On their way, they visit planets around other stars and encounter strange alien beings. Although this type of space exploration is futuristic, it is possible to explore galaxies today. Using a variety of telescopes, much is being learned about the Milky Way and other galaxies.

A **galaxy** is a large group of stars, gas, and dust held together by gravity. Earth and the solar system are in a galaxy called the Milky Way. It might contain as many as one trillion stars. Countless other galaxies also exist. Each of these galaxies contains the same elements, forces, and types of energy that occur in Earth's solar system. Galaxies are separated by huge distances—often millions of light-years.

In the same way that stars are grouped together within galaxies, galaxies are grouped into clusters. The cluster that the Milky Way belongs to is called the Local Group. It contains about 45 galaxies of various sizes and types. The three major types of galaxies are spiral, elliptical, and irregular.

Spiral Galaxies Spiral galaxies are galaxies that have spiral arms that wind outward from the center. The arms consist of bright stars, dust, and gas. The Milky Way Galaxy, shown in **Figure 16,** is a spiral galaxy. The Sun and the rest of the solar system are located near the outer edge of the Milky Way Galaxy.

Spiral galaxies can be normal or barred. Arms in a normal spiral start close to the center of the galaxy. Barred spirals have spiral arms extending from a large bar of stars and gas that passes through the center of the galaxy.

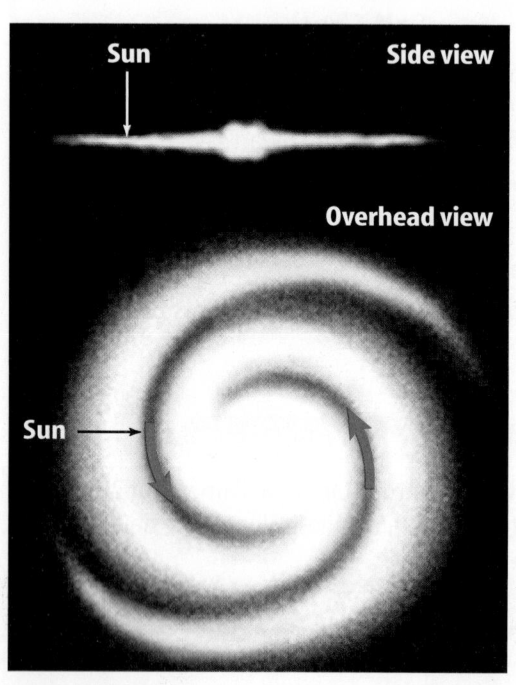

Figure 16 This illustration shows a side view and an overhead view of the Milky Way.
Describe *where the Sun is in the Milky Way.*

Elliptical Galaxies A common type of galaxy is the elliptical galaxy. **Figure 17** shows an elliptical galaxy in the constellation Andromeda. These galaxies are shaped like large, three-dimensional ellipses. Many are football shaped, but others are round. Some elliptical galaxies are small, while others are so large that several galaxies the size of the Milky Way would fit inside one of them.

Irregular Galaxies The third type—an irregular galaxy—includes most of those galaxies that don't fit into the other categories. Irregular galaxies have many different shapes. They are smaller than the other types of galaxies. Two irregular galaxies called the Clouds of Magellan orbit the Milky Way. The Large Magellanic Cloud is shown in **Figure 18.**

How do the three different types of galaxies differ?

Figure 17 This photo shows an example of an elliptical galaxy. **Identify** *the two other types of galaxies.*

The Milky Way Galaxy

The Milky Way might contain one trillion stars. The visible disk of stars shown in **Figure 16** is about 100,000 light-years across. Find the location of the Sun. Notice that it is located about 26,000 light-years from the galaxy's center in one of the spiral arms. In the galaxy, all stars orbit around a central region, or core. It takes about 225 million years for the Sun to orbit the center of the Milky Way.

The Milky Way often is classified as a normal spiral galaxy. However, recent evidence suggests that it might be a barred spiral. It is difficult to know for sure because astronomers have limited data about how the galaxy looks from the outside.

You can't see the shape of the Milky Way because you are located within one of its spiral arms. You can, however, see the Milky Way stretching across the sky as a misty band of faint light. You can see the brightest part of the Milky Way if you look low in the southern sky on a moonless summer night. All the stars you can see in the night sky belong to the Milky Way.

Like many other galaxies, the Milky Way has a supermassive black hole at its center. This black hole might be more than 2.5 million times as massive as the Sun. Evidence for the existence of the black hole comes from observing the orbit of a star near the galaxy's center. Additional evidence includes X-ray emissions detected by the *Chandra X-ray Observatory.* X rays are produced when matter spirals into a black hole.

Figure 18 The Large Magellanic Cloud is an irregular galaxy. It's a member of the Local Group, and it orbits the Milky Way.

Measuring Distance in Space

Procedure

1. On a large sheet of **paper,** draw an overhead view of the Milky Way. If necessary, refer to **Figure 16.** Choose a scale to show distance in light-years.
2. Mark the approximate location of the solar system, which is about two-thirds of the way out on one of the spiral arms.
3. Now, draw a side view of the Milky Way Galaxy. Mark the position of the solar system.

Analysis

1. What scale did you use to represent distance on your model of the Milky Way?
2. The Andromeda Galaxy is about 2.9 million light-years from Earth. What scale distance would this represent?

Origin of the Universe

People long have wondered how the universe formed. Several models of its origin have been proposed. One model is the steady state theory. It suggests that the universe always has been the same as it is now. The universe always existed and always will. As the universe expands, new matter is created to keep the overall density of the universe the same or in a steady state. However, evidence indicates that the universe was much different in the past.

A second idea is called the oscillating model. In this model, the universe began with expansion. Over time, the expansion slowed and the universe contracted. Then the process began again, oscillating back and forth. Some scientists still hypothesize that the universe expands and contracts in a cycle.

A third model of how the universe formed is called the big bang theory. The universe started with a big bang and has been expanding ever since. This theory will be described later.

Expansion of the Universe

What does it sound like when a train is blowing its whistle while it travels past you? The whistle has a higher pitch as the train approaches you. Then the whistle seems to drop in pitch as the train moves away. This effect is called the Doppler shift. The Doppler shift occurs with light as well as with sound. **Figure 19** shows how the Doppler shift causes changes in the light coming from distant stars and galaxies. If a star is moving toward Earth, its wavelengths of light are compressed. If a star is moving away from Earth, its wavelengths of light are stretched.

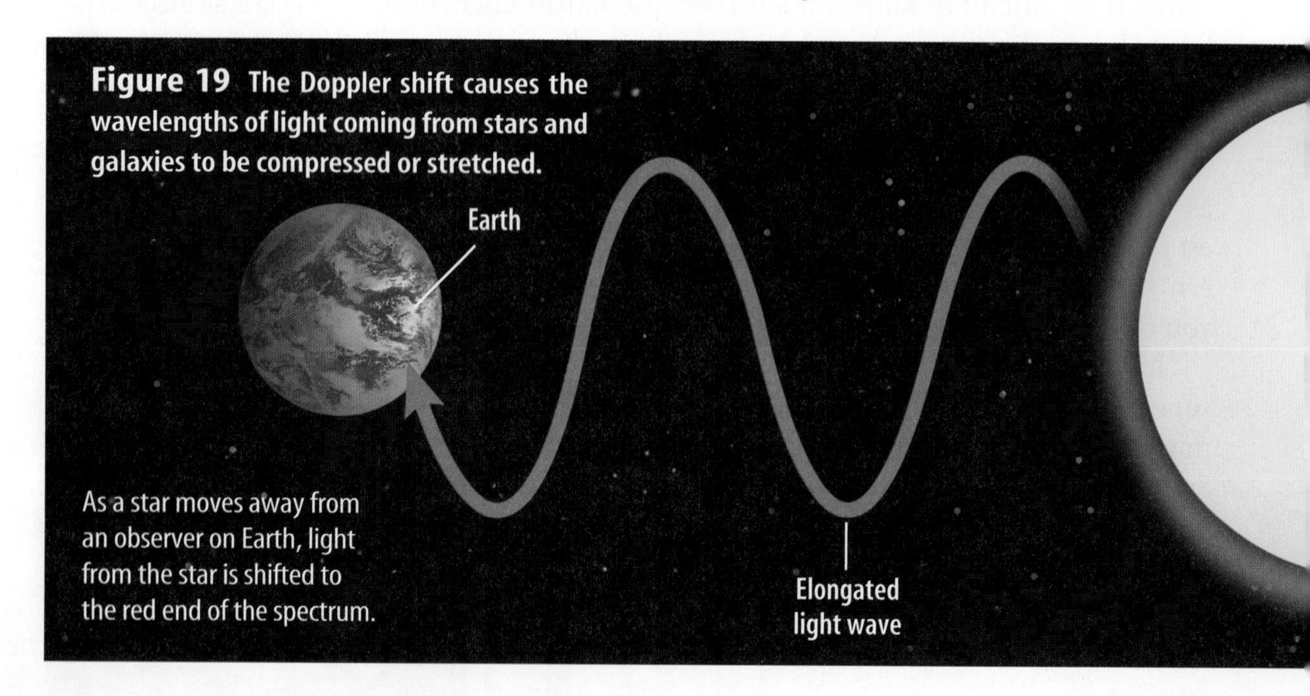

Earth
As a star moves away from an observer on Earth, light from the star is shifted to the red end of the spectrum.
Elongated light wave

Figure 19 The Doppler shift causes the wavelengths of light coming from stars and galaxies to be compressed or stretched.

The Doppler Shift Look at the spectrum of a star in **Figure 20A.** Note the position of the dark lines. How do they compare with the lines in **Figures 20B** and **20C?** They have shifted in position. What caused this shift? As you just read, when a star is moving toward Earth, its wavelengths of light are compressed, just as the sound waves from the train's whistle are. This causes the dark lines in the spectrum to shift toward the blue-violet end of the spectrum. A red shift in the spectrum occurs when a star is moving away from Earth. In a red shift, the dark lines shift toward the red end of the spectrum.

A
B
C

Figure 20 **A** This spectrum shows dark absorption lines. **B** The dark lines shift toward the blue-violet end for a star moving toward Earth. **C** The lines shift toward the red end for a star moving away from Earth.

Red Shift In 1929, Edwin Hubble published an interesting fact about the light coming from most galaxies. When a spectrograph is used to study light from galaxies beyond the Local Group, a red shift occurs in the light. What does this red shift tell you about the universe?

Because all galaxies beyond the Local Group show a red shift in their spectra, they must be moving away from Earth. If all galaxies outside the Local Group are moving away from Earth, then the entire universe must be expanding. Remember the Launch Lab at the beginning of the chapter? The dots on the balloon moved apart as the model universe expanded. Regardless of which dot you picked, all the other dots moved away from it. In a similar way, galaxies beyond the Local Group are moving away from Earth.

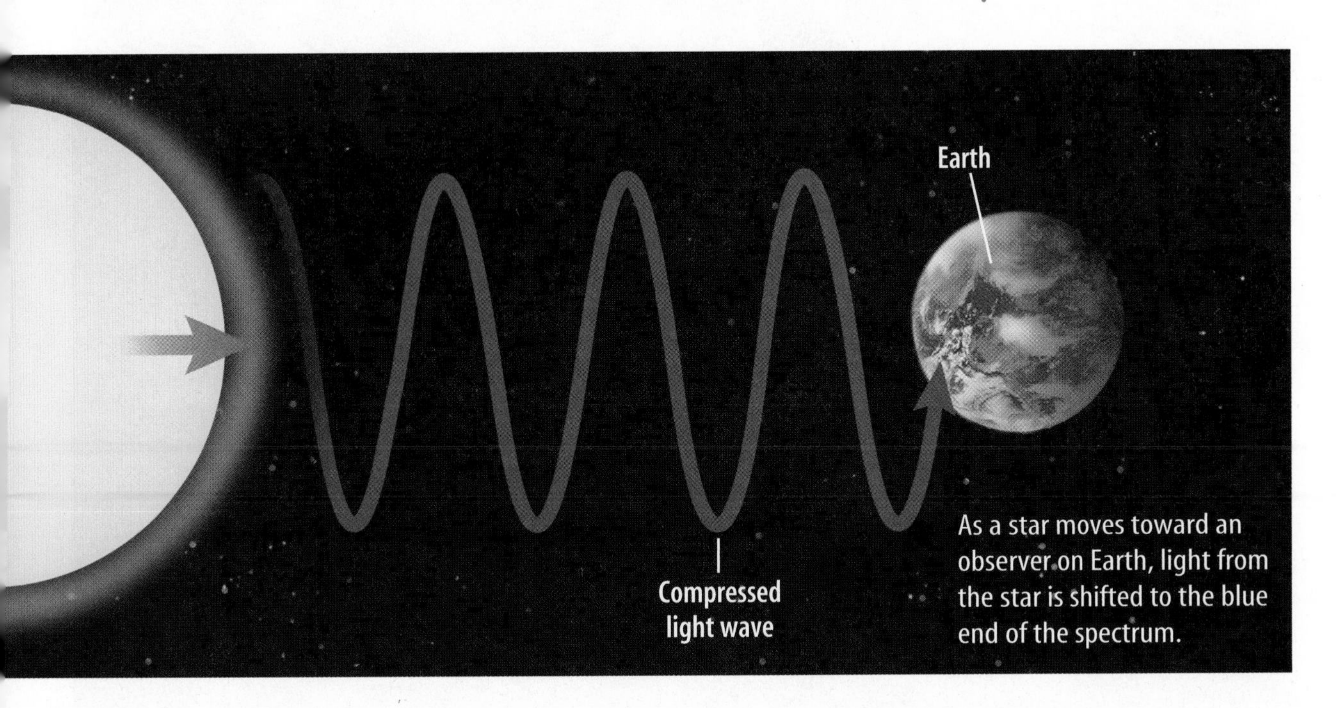

Earth
Compressed light wave
As a star moves toward an observer on Earth, light from the star is shifted to the blue end of the spectrum.

NATIONAL GEOGRAPHIC

VISUALIZING THE BIG BANG THEORY

Figure 21

The big bang theory states that the universe probably began about 13.7 billion years ago with an enormous explosion. Even today, galaxies are rushing apart from this explosion.

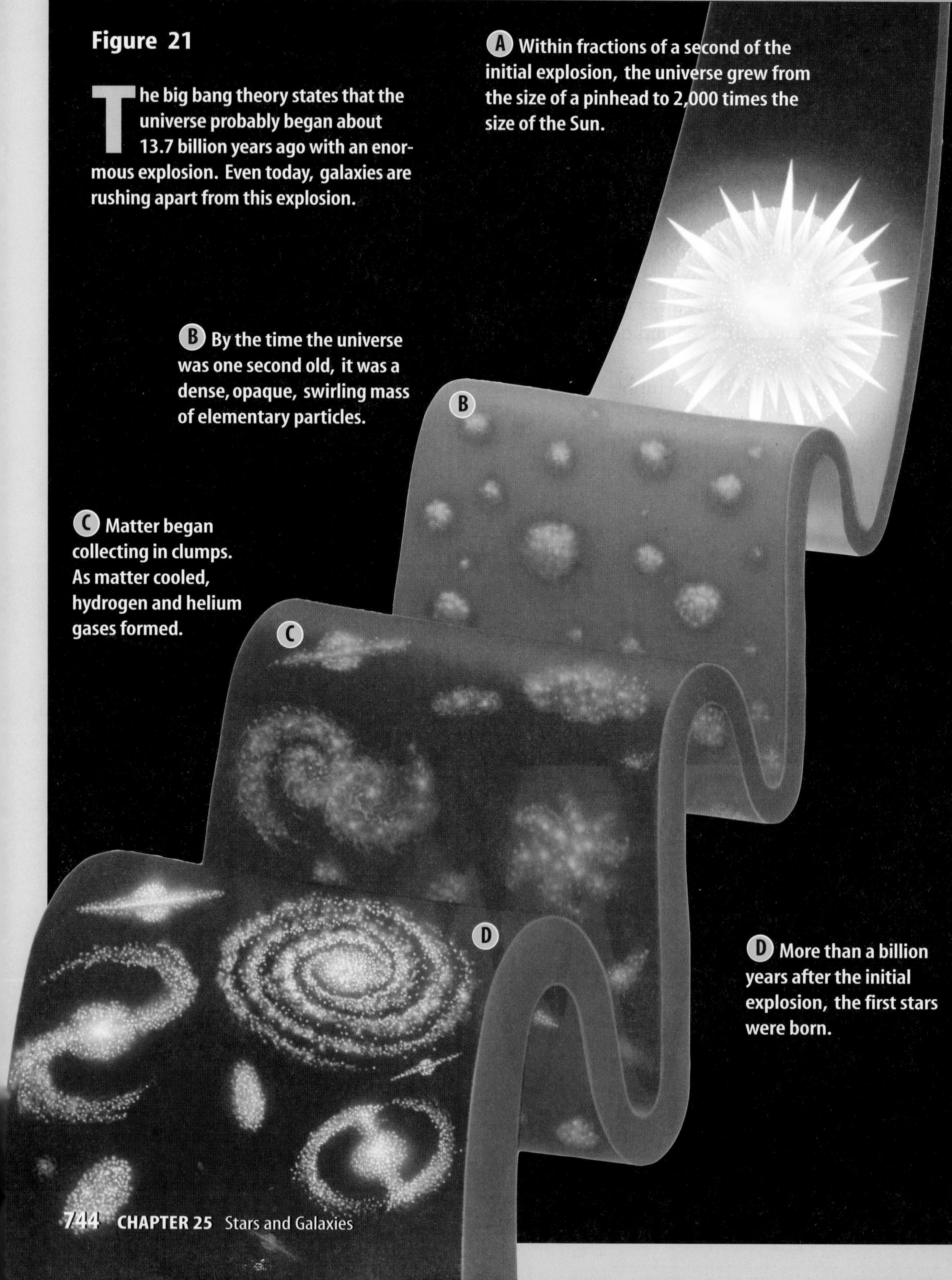

A Within fractions of a second of the initial explosion, the universe grew from the size of a pinhead to 2,000 times the size of the Sun.

B By the time the universe was one second old, it was a dense, opaque, swirling mass of elementary particles.

C Matter began collecting in clumps. As matter cooled, hydrogen and helium gases formed.

D More than a billion years after the initial explosion, the first stars were born.

The Big Bang Theory

When scientists determined that the universe was expanding, they developed a theory to explain their observations. The leading theory about the formation of the universe is called the **big bang theory.** **Figure 21** illustrates the big bang theory. According to this theory, approximately 13.7 billion years ago, the universe began with an enormous explosion. The entire universe began to expand everywhere at the same time.

Figure 22 The light from the galaxies in this photo mosaic took billions of years to reach Earth.

Looking Back in Time The time-exposure photograph shown in **Figure 22** was taken by the *Hubble Space Telescope.* It shows more than 1,500 galaxies at distances of more than 10 billion light-years. These galaxies could date back to when the universe was no more than 1 billion years old. The galaxies are in various stages of development. One astronomer says that humans might be looking back to a time when the Milky Way was forming.

Whether the universe will expand forever or stop expanding is still unknown. If enough matter exists, gravity might halt the expansion, and the universe will contract until everything comes to a single point. However, studies of distant supernovae indicate that an energy, called dark energy, is causing the universe to expand faster. Scientists are trying to understand how dark energy might affect the fate of the universe.

section 4 review

Summary

Galaxies

- The three main types of galaxies are spiral, elliptical, and irregular.

The Milky Way Galaxy

- The Milky Way is a spiral galaxy and the Sun is about 26,000 light-years from its center.

Origin of the Universe

- Theories about how the universe formed include the steady state theory, the oscillating universe theory, and the big bang theory.

The Big Bang Theory

- This theory states that the universe began with an explosion about 13.7 billion years ago.

Self Check

1. **Describe** elliptical galaxies. How are they different from spiral galaxies?
2. **Identify** the galaxy that you live in.
3. **Explain** the Doppler shift.
4. **Explain** how all galaxies are similar.
5. **Think Critically** All galaxies outside the Local Group show a red shift. Within the Local Group, some show a red shift and some show a blue shift. What does this tell you about the galaxies in the Local Group?

Applying Skills

6. **Compare and contrast** the theories about the origin of the universe.

Design Your Own

Measuring Parallax

Real-World Question

Parallax is the apparent shift in the position of an object when viewed from two locations. How can you build a model to show the relationship between distance and parallax?

Goals

- **Design** a model to show how the distance from an observer to an object affects the object's parallax shift.
- **Describe** how parallax can be used to determine the distance to a star.

Possible Materials

meterstick
masking tape
metric ruler
pencil

Safety Precautions

WARNING: *Be sure to wear goggles to protect your eyes.*

Form a Hypothesis

State a hypothesis about how parallax varies with distance.

Test Your Hypothesis

Make a Plan

1. As a group, agree upon and write your hypothesis statement.
2. **List** the steps you need to take to build your model. Be specific, describing exactly what you will do at each step.
3. **Devise** a method to test how distance from an observer to an object, such as a pencil, affects the parallax of the object.
4. **List** the steps you will take to test your hypothesis. Be specific, describing exactly what you will do at each step.
5. Read over your plan for the model to be used in this experiment.

6. How will you determine changes in observed parallax? Remember, these changes should occur when the distance from the observer to the object is changed.
7. You should measure shifts in parallax from several different positions. How will these positions differ?
8. How will you measure distances accurately and compare relative position shift?

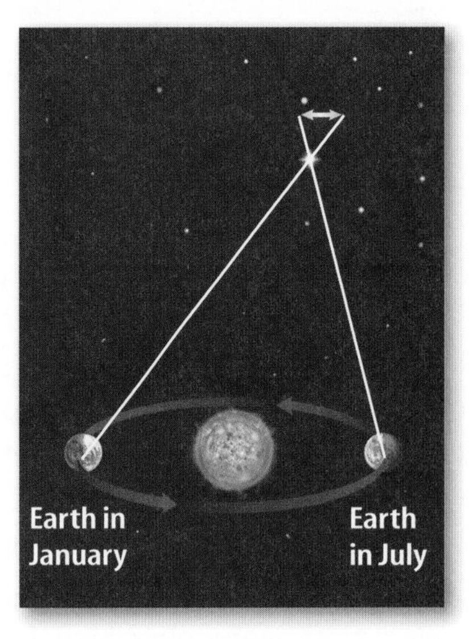

Follow Your Plan

1. Make sure your teacher approves your plan before you start.
2. **Construct** the model your team has planned.
3. Carry out the experiment as planned.
4. While conducting the experiment, record any observations that you or other members of your group make in your Science Journal.

Analyze Your Data

1. **Compare** what happened to the object when it was viewed with one eye closed, then the other.
2. At what distance from the observer did the object appear to shift the most?
3. At what distance did it appear to shift the least?

Conclude and Apply

1. **Infer** what happened to the apparent shift of the object's location as the distance from the observer was increased or decreased.
2. **Describe** how astronomers might use parallax to study stars.

Prepare a chart showing the results of your experiment. Share the chart with members of your class. **For more help, refer to the** Science Skill Handbook.

SCIENCE Stats

Stars and Galaxies

Did you know...

. . . A star in Earth's galaxy explodes as a supernova about once a century. The most famous supernova of this galaxy occurred in 1054 and was recorded by the ancient Chinese and Koreans. The explosion was so powerful that it could be seen during the day, and its brightness lasted for weeks. Other major supernovas in the Milky Way that were observed from Earth occurred in 185, 386, 1006, 1181, 1572, and 1604.

. . . The large loops of material called solar prominences can extend more than 320,000 km above the Sun's surface. This is so high that two Jupiters and three Earths could fit under the arch.

. . . The red giant star Betelgeuse has a diameter larger than that of Earth's Sun. This gigantic star measures 450,520,000 km in diameter, while the Sun's diameter is a mere 1,390,176 km.

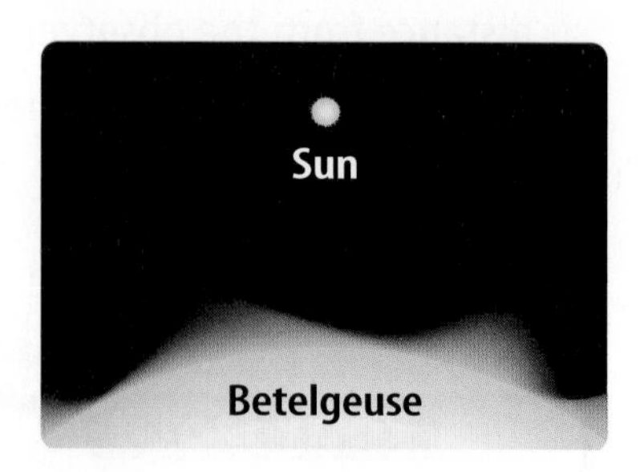

Applying Math Use words to express the number 450,520,000.

Write About It

Visit earth.msscience.com/science_stats to learn whether it might be possible for Earth astronauts to travel to the nearest stars. How long would such a trip take? What problems would have to be overcome? Write a brief report about what you find.

Reviewing Main Ideas

Section 1 Stars

1. Constellations are patterns of stars in the night sky. Some constellations can be seen all year. Other constellations are visible only during certain seasons.
2. Parallax is the apparent shift in the position of an object when viewed from two different positions. Parallax is used to find the distance to nearby stars.

Section 2 The Sun

1. The Sun is the closest star to Earth.
2. Sunspots are areas on the Sun's surface that are cooler and less bright than surrounding areas.

Section 3 Evolution of Stars

1. Stars are classified according to their position on the H-R diagram.
2. Low-mass stars end their lives as white dwarfs. High-mass stars become neutron stars or black holes.

Section 4 Galaxies and the Universe

1. A galaxy consists of stars, gas, and dust held together by gravity.
2. Earth's solar system is in the Milky Way, a spiral galaxy.
3. The universe is expanding. Scientists don't know whether the universe will expand forever or contract to a single point.

Visualizing Main Ideas

Copy and complete the following concept map that shows the evolution of a main sequence star with a mass similar to that of the Sun.

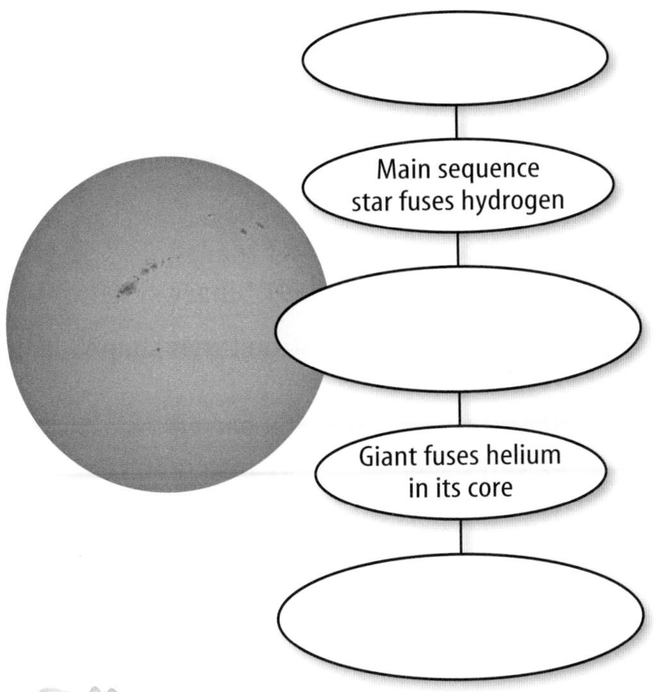

Using Vocabulary

absolute magnitude p. 726	giant p. 737
apparent magnitude p. 726	light-year p. 727
big bang theory p. 745	nebula p. 736
black hole p. 738	neutron star p. 738
chromosphere p. 729	photosphere p. 729
constellation p. 724	sunspot p. 730
corona p. 729	supergiant p. 738
galaxy p. 740	white dwarf p. 737

Explain the difference between the terms in each of the following sets.

1. absolute magnitude—apparent magnitude
2. galaxy—constellation
3. giant—supergiant
4. chromosphere—photosphere
5. black hole—neutron star

Checking Concepts

Choose the word or phrase that best answers the question.

6. What is a measure of the amount of a star's light that is received on Earth?
   A) absolute magnitude
   B) apparent magnitude
   C) fusion
   D) parallax
7. What is higher for closer stars?
   A) absolute magnitude
   B) red shift
   C) parallax
   D) blue shift
8. What happens after a nebula contracts and its temperature increases to 10 million K?
   A) a black hole forms
   B) a supernova occurs
   C) fusion begins
   D) a white dwarf forms
9. Which of these has an event horizon?
   A) giant **C)** black hole
   B) white dwarf **D)** neutron star
10. What forms when the Sun fuses hydrogen?
    A) carbon **C)** iron
    B) oxygen **D)** helium

Use the illustration below to answer question 11.

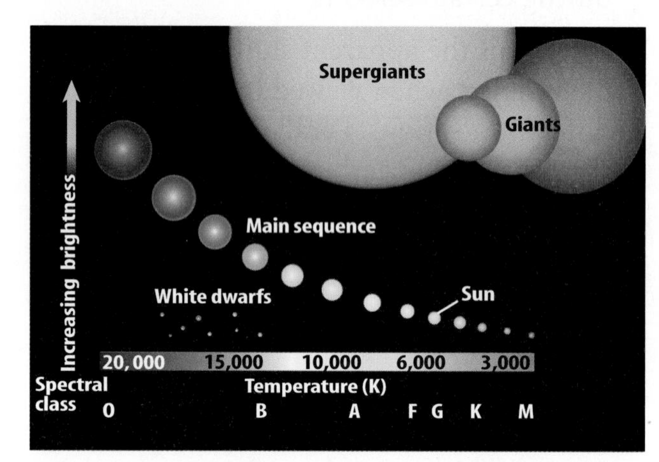

11. Which of the following best describes giant stars?
    A) hot, dim stars
    B) cool, dim stars
    C) hot, bright stars
    D) cool, bright stars
12. Which of the following are loops of matter flowing from the Sun?
    A) sunspots **C)** coronas
    B) auroras **D)** prominences
13. What are groups of galaxies called?
    A) clusters **C)** giants
    B) supergiants **D)** binary systems
14. Which galaxies are sometimes shaped like footballs?
    A) spiral **C)** barred
    B) elliptical **D)** irregular
15. What do scientists study to determine shifts in wavelengths of light?
    A) spectrum **C)** corona
    B) parallax **D)** nebula

 Science Online earth.msscience.com/vocabulary_puzzlemaker

Thinking Critically

Use the table below to answer question 16.

Magnitude and Distance of Stars

Star	Apparent Magnitude	Absolute Magnitude	Distance in Light-Years
A	−26	4.8	0.00002
B	−1.5	1.4	8.7
C	0.1	4.4	4.3
D	0.1	−7.0	815
E	0.4	−5.9	520
F	1.0	−0.6	45

16. **Interpret Data** Use the table above to answer the following questions. *Hint: lower magnitude values are brighter than higher magnitude values.*
    a. Which star appears brightest from Earth?
    b. Which star would appear brightest from a distance of 10 light-years?
    c. Infer which star in the table above is the Sun.

17. **Infer** How do scientists know that black holes exist if these objects don't emit visible light?

18. **Recognize Cause and Effect** Why can parallax only be used to measure distances to stars that are relatively close to Earth?

19. **Compare and contrast** the Sun with other stars on the H-R diagram.

20. **Concept Map** Make a concept map showing the life history of a very large star.

21. **Make Models** Make a model of the Sun. Include all of the Sun's layers in your model.

Performance Activities

22. **Story** Write a short science-fiction story about an astronaut traveling through the universe. In your story, describe what the astronaut observes. Use as many vocabulary words as you can.

23. **Photomontage** Gather photographs of the aurora borealis from magazines and other sources. Use the photographs to create a photomontage. Write a caption for each photo.

Applying Math

24. **Travel to Vega** Vega is a star that is 26 light-years away. If a spaceship could travel at one-tenth the speed of light, how long would it take to reach this star?

Use the illustration below to answer question 25.

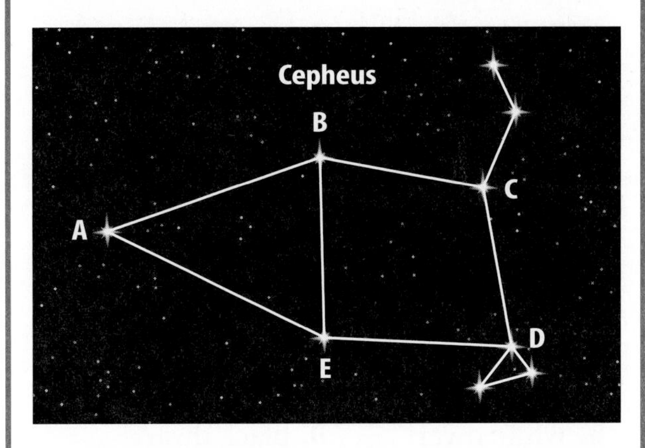

25. **Constellation Cepheus** The illustration above shows the constellation Cepheus. Answer the following questions about this contellation.
    a. Which of the line segments are nearly parallel?
    b. Which line segments are nearly perpendicular?
    c. Which angles are oblique?
    d. What geometric shape do the three stars at the left side of the drawing form?

Part 1 Multiple Choice

Record your answers on the answer sheet provided by your teacher or on a sheet of paper.

Use the illustration below to answer question 1.

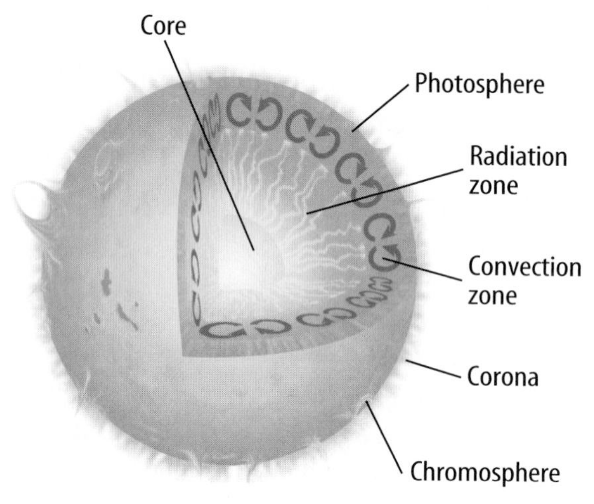

1. The illustration above shows the interior of which object?
   A. Earth
   B. Saturn
   C. the Sun
   D. the Moon

2. Which is a group of stars, gas, and dust held together by gravity?
   A. constellation
   B. supergiant
   C. black hole
   D. galaxy

3. The most massive stars end their lives as which type of object?
   A. black hole
   B. white dwarf
   C. neutron star
   D. black dwarf

4. In which galaxy does the Sun exist?
   A. Arp's galaxy
   B. Barnard's galaxy
   C. Milky Way galaxy
   D. Andromeda galaxy

Test-Taking Tip

Process of Elimination If you don't know the answer to a multiple-choice question, eliminate as many incorrect choices as possible. Mark your best guess from the remaining answers before moving on to the next question.

5. Which is the closest star to Earth?
   A. Sirius
   B. the Sun
   C. Betelgeuse
   D. the Moon

6. In which of the following choices are the objects ordered from smallest to largest?
   A. stars, galaxies, galaxy clusters, universe
   B. galaxy clusters, galaxies, stars, universe
   C. universe, galaxy clusters, galaxies, stars
   D. universe, stars, galaxies, galaxy clusters

Use the graph below to answer questions 7 and 8.

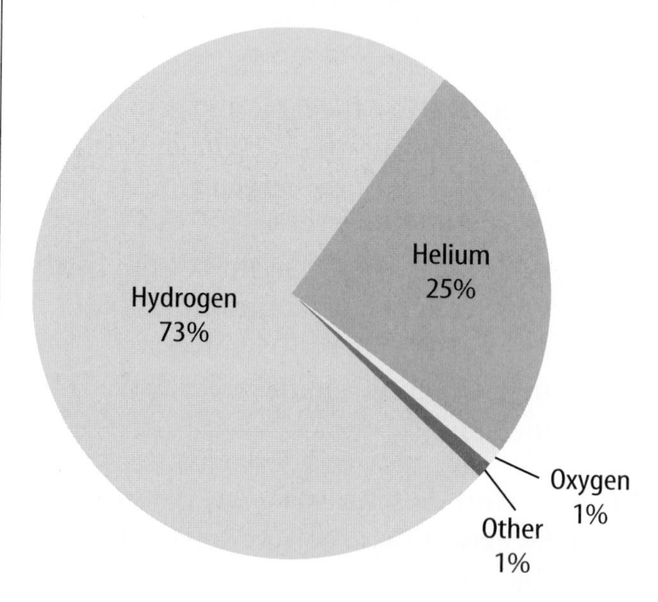

7. Which is the most abundant element in the Sun?
   A. helium
   B. hydrogen
   C. oxygen
   D. carbon

8. How will this circle graph change as the Sun ages?
   A. The hydrogen slice will get smaller.
   B. The hydrogen slice will get larger.
   C. The helium slice will get smaller.
   D. The circle graph will not change.

Part 2 | Short Response/Grid In

Record your answers on the answer sheet provided by your teacher or on a sheet of paper.

9. How can events on the Sun affect Earth? Give one example.
10. How does a red shift differ from a blue shift?
11. How do astronomers know that the universe is expanding?
12. What is the main sequence?
13. What is a constellation?

Use the illustration below to answer questions 14–16.

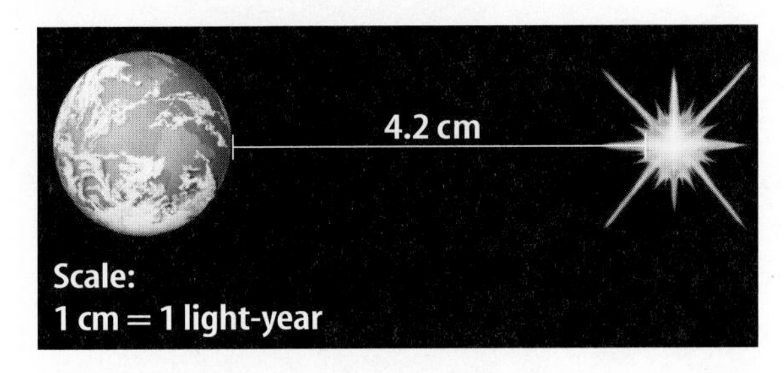

14. According to the illustration, how many light-years from Earth is Proxima Centauri?
15. How many years would it take for light from Proxima Centauri to get to Earth?
16. At this scale, how many centimeters would represent the distance to a star that is 100 light-years from Earth?
17. How can a star's color provide information about its temperature?
18. Approximately how long does it take light from the Sun to reach Earth? In general, how does this compare to the amount of time it takes light from all other stars to reach Earth?
19. How does the size, temperature, age, and brightness of the Sun compare to other stars in the Milky Way Galaxy?

Part 3 | Open Ended

Record your answers on a sheet of paper.

Use the graph below to answer question 20.

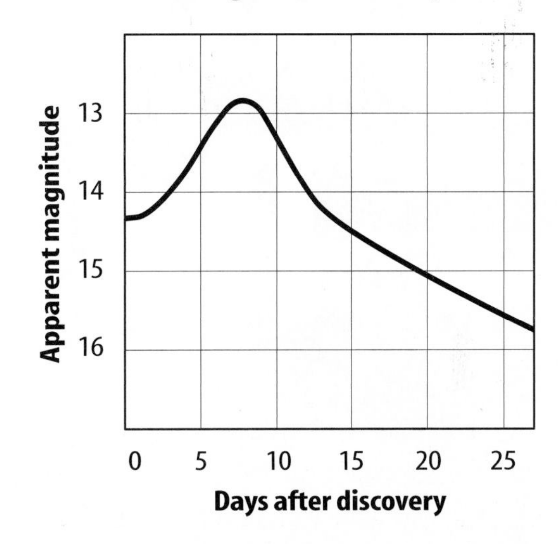

20. The graph above shows the brightness of a supernova that was observed from Earth in 1987. Describe how the brightness of this supernova changed through time. When was it brightest? What happened before May 20? What happened after May 20? How much did the brightness change?
21. Compare and contrast the different types of galaxies.
22. Write a detailed description of the Sun. What is it? What is it like?
23. Explain how parallax is used to measure the distance to nearby stars.
24. Why are some constellations visible all year? Why are other constellations only visible during certain seasons?
25. What are black holes? How do they form?
26. Explain the big bang theory.
27. What can be learned by studying the dark lines in a star's spectrum?

Student Resources

CONTENTS

Scientific Methods

Scientists use an orderly approach called the scientific method to solve problems. This includes organizing and recording data so others can understand them. Scientists use many variations in this method when they solve problems.

Identify a Question

The first step in a scientific investigation or experiment is to identify a question to be answered or a problem to be solved. For example, you might ask which gasoline is the most efficient.

Figure 1 The Internet can be a valuable research tool.

Gather and Organize Information

After you have identified your question, begin gathering and organizing information. There are many ways to gather information, such as researching in a library, interviewing those knowledgeable about the subject, testing and working in the laboratory and field. Fieldwork is investigations and observations done outside of a laboratory.

Researching Information Before moving in a new direction, it is important to gather the information that already is known about the subject. Start by asking yourself questions to determine exactly what you need to know. Then you will look for the information in various reference sources, like the student is doing in **Figure 1.** Some sources may include textbooks, encyclopedias, government documents, professional journals, science magazines, and the Internet. Always list the sources of your information.

Evaluate Sources of Information Not all sources of information are reliable. You should evaluate all of your sources of information, and use only those you know to be dependable. For example, if you are researching ways to make homes more energy efficient, a site written by the U.S. Department of Energy would be more reliable than a site written by a company that is trying to sell a new type of weatherproofing material. Also, remember that research always is changing. Consult the most current resources available to you. For example, a 1985 resource about saving energy would not reflect the most recent findings.

Sometimes scientists use data that they did not collect themselves, or conclusions drawn by other researchers. This data must be evaluated carefully. Ask questions about how the data were obtained, if the investigation was carried out properly, and if it has been duplicated exactly with the same results. Would you reach the same conclusion from the data? Only when you have confidence in the data can you believe it is true and feel comfortable using it.

Interpret Scientific Illustrations As you research a topic in science, you will see drawings, diagrams, and photographs to help you understand what you read. Some illustrations are included to help you understand an idea that you can't see easily by yourself, like the tiny particles in an atom in **Figure 2.** A drawing helps many people to remember details more easily and provides examples that clarify difficult concepts or give additional information about the topic you are studying. Most illustrations have labels or a caption to identify or to provide more information.

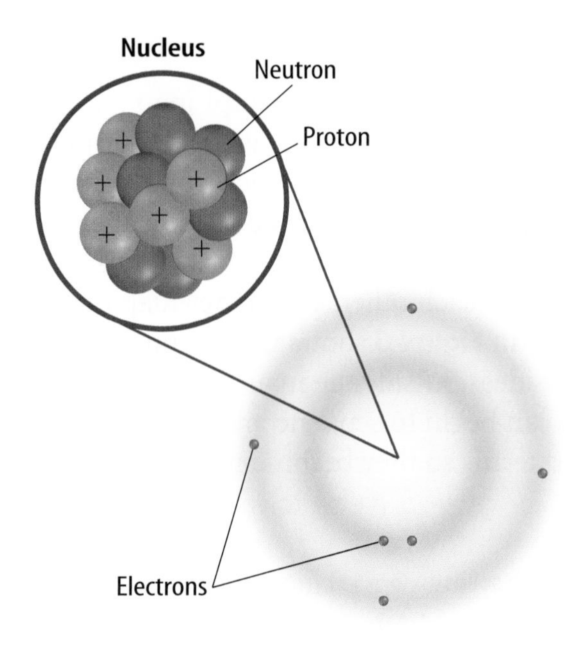

Figure 2 This drawing shows an atom of carbon with its six protons, six neutrons, and six electrons.

Concept Maps One way to organize data is to draw a diagram that shows relationships among ideas (or concepts). A concept map can help make the meanings of ideas and terms more clear, and help you understand and remember what you are studying. Concept maps are useful for breaking large concepts down into smaller parts, making learning easier.

Network Tree A type of concept map that not only shows a relationship, but how the concepts are related is a network tree, shown in **Figure 3.** In a network tree, the words are written in the ovals, while the description of the type of relationship is written across the connecting lines.

When constructing a network tree, write down the topic and all major topics on separate pieces of paper or notecards. Then arrange them in order from general to specific. Branch the related concepts from the major concept and describe the relationship on the connecting line. Continue to more specific concepts until finished.

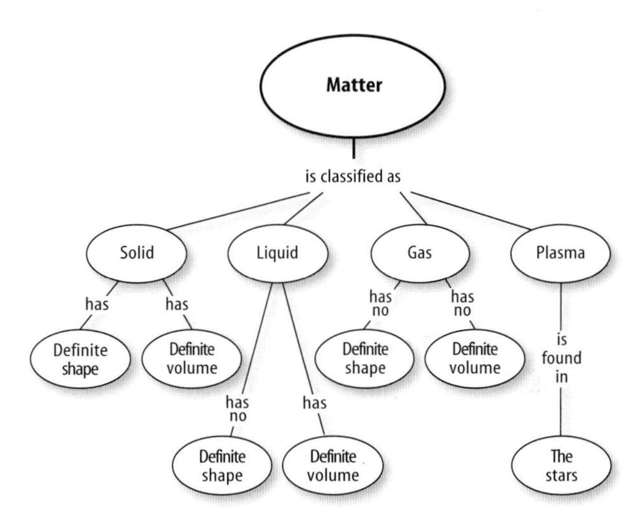

Figure 3 A network tree shows how concepts or objects are related.

Events Chain Another type of concept map is an events chain. Sometimes called a flow chart, it models the order or sequence of items. An events chain can be used to describe a sequence of events, the steps in a procedure, or the stages of a process.

When making an events chain, first find the one event that starts the chain. This event is called the initiating event. Then, find the next event and continue until the outcome is reached, as shown in **Figure 4.**

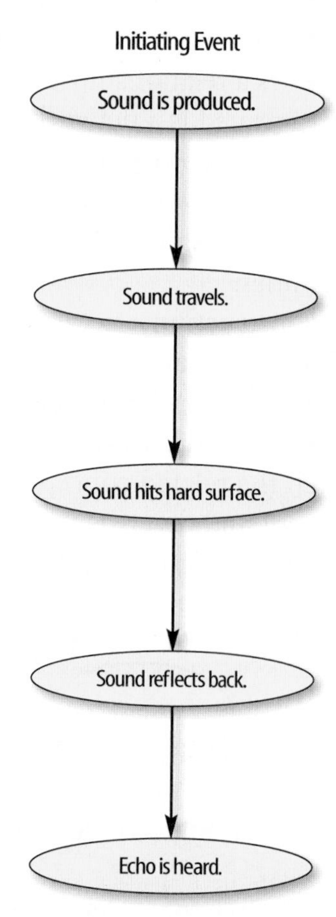

Figure 4 Events-chain concept maps show the order of steps in a process or event. This concept map shows how a sound makes an echo.

Cycle Map A specific type of events chain is a cycle map. It is used when the series of events do not produce a final outcome, but instead relate back to the beginning event, such as in **Figure 5.** Therefore, the cycle repeats itself.

To make a cycle map, first decide what event is the beginning event. This is also called the initiating event. Then list the next events in the order that they occur, with the last event relating back to the initiating event. Words can be written between the events that describe what happens from one event to the next. The number of events in a cycle map can vary, but usually contain three or more events.

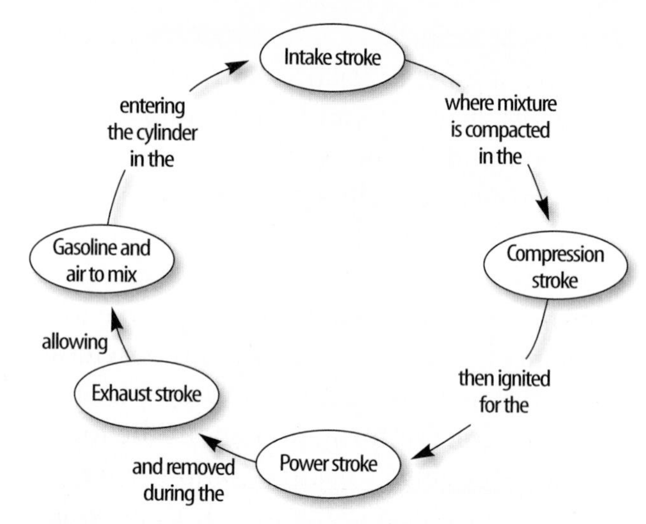

Figure 5 A cycle map shows events that occur in a cycle.

Spider Map A type of concept map that you can use for brainstorming is the spider map. When you have a central idea, you might find that you have a jumble of ideas that relate to it but are not necessarily clearly related to each other. The spider map on sound in **Figure 6** shows that if you write these ideas outside the main concept, then you can begin to separate and group unrelated terms so they become more useful.

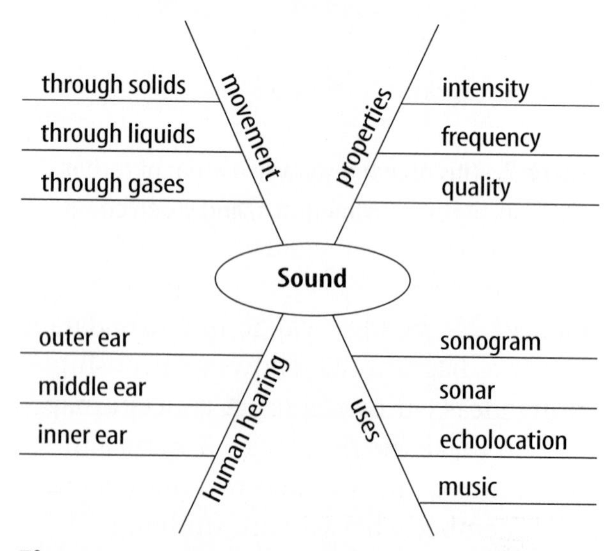

Figure 6 A spider map allows you to list ideas that relate to a central topic but not necessarily to one another.

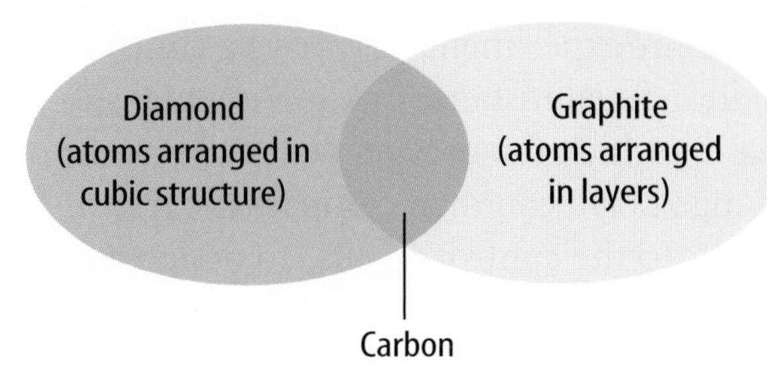

Figure 7 This Venn diagram compares and contrasts two substances made from carbon.

Venn Diagram To illustrate how two subjects compare and contrast you can use a Venn diagram. You can see the characteristics that the subjects have in common and those that they do not, shown in **Figure 7.**

To create a Venn diagram, draw two overlapping ovals that that are big enough to write in. List the characteristics unique to one subject in one oval, and the characteristics of the other subject in the other oval. The characteristics in common are listed in the overlapping section.

Make and Use Tables One way to organize information so it is easier to understand is to use a table. Tables can contain numbers, words, or both.

To make a table, list the items to be compared in the first column and the characteristics to be compared in the first row. The title should clearly indicate the content of the table, and the column or row heads should be clear. Notice that in **Table 1** the units are included.

Table 1 Recyclables Collected During Week

Day of Week	Paper (kg)	Aluminum (kg)	Glass (kg)
Monday	5.0	4.0	12.0
Wednesday	4.0	1.0	10.0
Friday	2.5	2.0	10.0

Make a Model One way to help you better understand the parts of a structure, the way a process works, or to show things too large or small for viewing is to make a model. For example, an atomic model made of a plastic-ball nucleus and pipe-cleaner electron shells can help you visualize how the parts of an atom relate to each other. Other types of models can by devised on a computer or represented by equations.

Form a Hypothesis

A possible explanation based on previous knowledge and observations is called a hypothesis. After researching gasoline types and recalling previous experiences in your family's car you form a hypothesis—our car runs more efficiently because we use premium gasoline. To be valid, a hypothesis has to be something you can test by using an investigation.

Predict When you apply a hypothesis to a specific situation, you predict something about that situation. A prediction makes a statement in advance, based on prior observation, experience, or scientific reasoning. People use predictions to make everyday decisions. Scientists test predictions by performing investigations. Based on previous observations and experiences, you might form a prediction that cars are more efficient with premium gasoline. The prediction can be tested in an investigation.

Design an Experiment A scientist needs to make many decisions before beginning an investigation. Some of these include: how to carry out the investigation, what steps to follow, how to record the data, and how the investigation will answer the question. It also is important to address any safety concerns.

Test the Hypothesis

Now that you have formed your hypothesis, you need to test it. Using an investigation, you will make observations and collect data, or information. This data might either support or not support your hypothesis. Scientists collect and organize data as numbers and descriptions.

Follow a Procedure In order to know what materials to use, as well as how and in what order to use them, you must follow a procedure. **Figure 8** shows a procedure you might follow to test your hypothesis.

Procedure

1. Use regular gasoline for two weeks.
2. Record the number of kilometers between fill-ups and the amount of gasoline used.
3. Switch to premium gasoline for two weeks.
4. Record the number of kilometers between fill-ups and the amount of gasoline used.

Figure 8 A procedure tells you what to do step by step.

Identify and Manipulate Variables and Controls In any experiment, it is important to keep everything the same except for the item you are testing. The one factor you change is called the independent variable. The change that results is the dependent variable. Make sure you have only one independent variable, to assure yourself of the cause of the changes you observe in the dependent variable. For example, in your gasoline experiment the type of fuel is the independent variable. The dependent variable is the efficiency.

Many experiments also have a control—an individual instance or experimental subject for which the independent variable is not changed. You can then compare the test results to the control results. To design a control you can have two cars of the same type. The control car uses regular gasoline for four weeks. After you are done with the test, you can compare the experimental results to the control results.

Collect Data

Whether you are carrying out an investigation or a short observational experiment, you will collect data, as shown in **Figure 9.** Scientists collect data as numbers and descriptions and organize it in specific ways.

Observe Scientists observe items and events, then record what they see. When they use only words to describe an observation, it is called qualitative data. Scientists' observations also can describe how much there is of something. These observations use numbers, as well as words, in the description and are called quantitative data. For example, if a sample of the element gold is described as being "shiny and very dense" the data are qualitative. Quantitative data on this sample of gold might include "a mass of 30 g and a density of 19.3 g/cm^3."

Figure 9 Collecting data is one way to gather information directly.

Figure 10 Record data neatly and clearly so it is easy to understand.

When you make observations you should examine the entire object or situation first, and then look carefully for details. It is important to record observations accurately and completely. Always record your notes immediately as you make them, so you do not miss details or make a mistake when recording results from memory. Never put unidentified observations on scraps of paper. Instead they should be recorded in a notebook, like the one in **Figure 10.** Write your data neatly so you can easily read it later. At each point in the experiment, record your observations and label them. That way, you will not have to determine what the figures mean when you look at your notes later. Set up any tables that you will need to use ahead of time, so you can record any observations right away. Remember to avoid bias when collecting data by not including personal thoughts when you record observations. Record only what you observe.

Estimate Scientific work also involves estimating. To estimate is to make a judgment about the size or the number of something without measuring or counting. This is important when the number or size of an object or population is too large or too difficult to accurately count or measure.

Sample Scientists may use a sample or a portion of the total number as a type of estimation. To sample is to take a small, representative portion of the objects or organisms of a population for research. By making careful observations or manipulating variables within that portion of the group, information is discovered and conclusions are drawn that might apply to the whole population. A poorly chosen sample can be unrepresentative of the whole. If you were trying to determine the rainfall in an area, it would not be best to take a rainfall sample from under a tree.

Measure You use measurements everyday. Scientists also take measurements when collecting data. When taking measurements, it is important to know how to use measuring tools properly. Accuracy also is important.

Length To measure length, the distance between two points, scientists use meters. Smaller measurements might be measured in centimeters or millimeters.

Length is measured using a metric ruler or meter stick. When using a metric ruler, line up the 0-cm mark with the end of the object being measured and read the number of the unit where the object ends. Look at the metric ruler shown in **Figure 11.** The centimeter lines are the long, numbered lines, and the shorter lines are millimeter lines. In this instance, the length would be 4.50 cm.

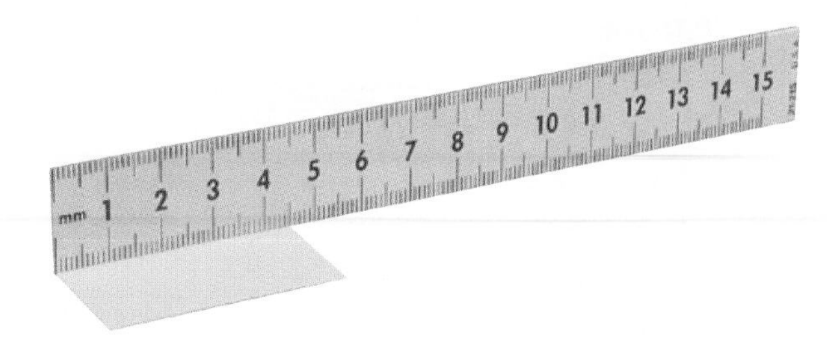

Figure 11 This metric ruler has centimeter and millimeter divisions.

Mass The SI unit for mass is the kilogram (kg). Scientists can measure mass using units formed by adding metric prefixes to the unit gram (g), such as milligram (mg). To measure mass, you might use a triple-beam balance similar to the one shown in **Figure 12.** The balance has a pan on one side and a set of beams on the other side. Each beam has a rider that slides on the beam.

When using a triple-beam balance, place an object on the pan. Slide the largest rider along its beam until the pointer drops below zero. Then move it back one notch. Repeat the process for each rider proceeding from the larger to smaller until the pointer swings an equal distance above and below the zero point. Sum the masses on each beam to find the mass of the object. Move all riders back to zero when finished.

Instead of putting materials directly on the balance, scientists often take a tare of a container. A tare is the mass of a container into which objects or substances are placed for measuring their masses. To mass objects or substances, find the mass of a clean container. Remove the container from the pan, and place the object or substances in the container. Find the mass of the container with the materials in it. Subtract the mass of the empty container from the mass of the filled container to find the mass of the materials you are using.

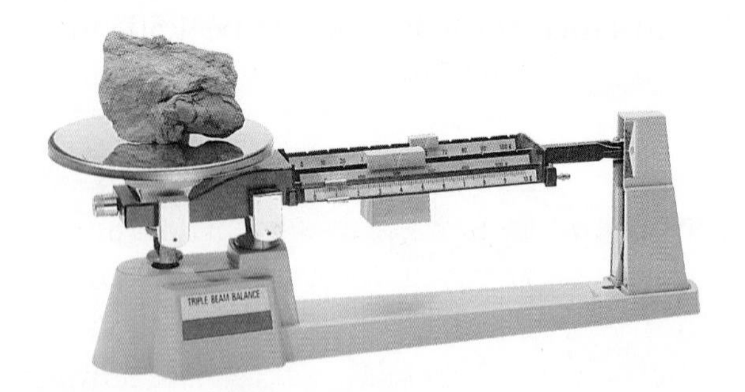

Figure 12 A triple-beam balance is used to determine the mass of an object.

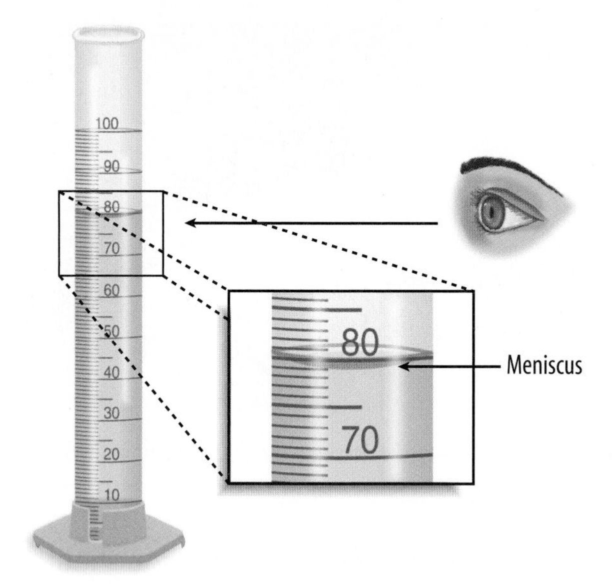

Figure 13 Graduated cylinders measure liquid volume.

Liquid Volume To measure liquids, the unit used is the liter. When a smaller unit is needed, scientists might use a milliliter. Because a milliliter takes up the volume of a cube measuring 1 cm on each side it also can be called a cubic centimeter ($cm^3 = cm \times cm \times cm$).

You can use beakers and graduated cylinders to measure liquid volume. A graduated cylinder, shown in **Figure 13,** is marked from bottom to top in milliliters. In lab, you might use a 10-mL graduated cylinder or a 100-mL graduated cylinder. When measuring liquids, notice that the liquid has a curved surface. Look at the surface at eye level, and measure the bottom of the curve. This is called the meniscus. The graduated cylinder in **Figure 13** contains 79.0 mL, or 79.0 cm^3, of a liquid.

Temperature Scientists often measure temperature using the Celsius scale. Pure water has a freezing point of 0°C and boiling point of 100°C. The unit of measurement is degrees Celsius. Two other scales often used are the Fahrenheit and Kelvin scales.

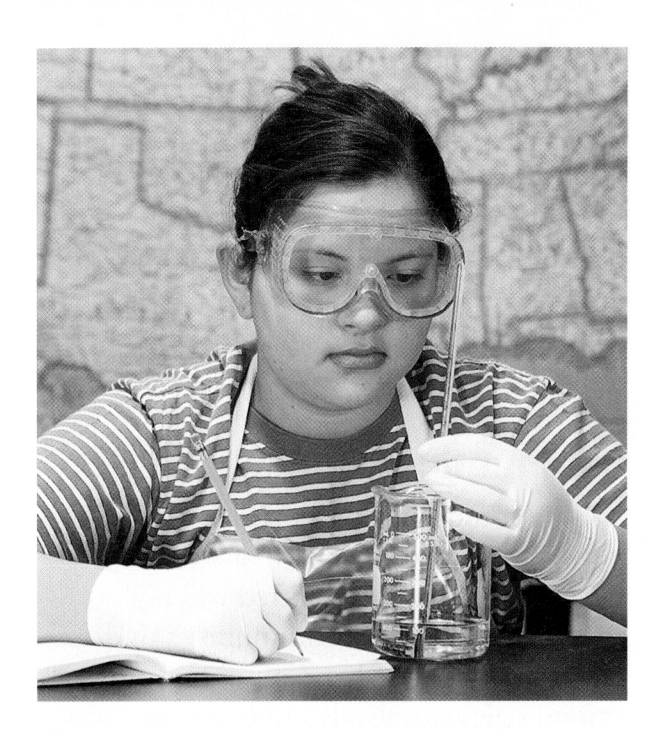

Figure 14 A thermometer measures the temperature of an object.

Scientists use a thermometer to measure temperature. Most thermometers in a laboratory are glass tubes with a bulb at the bottom end containing a liquid such as colored alcohol. The liquid rises or falls with a change in temperature. To read a glass thermometer like the thermometer in **Figure 14,** rotate it slowly until a red line appears. Read the temperature where the red line ends.

Form Operational Definitions An operational definition defines an object by how it functions, works, or behaves. For example, when you are playing hide and seek and a tree is home base, you have created an operational definition for a tree.

Objects can have more than one operational definition. For example, a ruler can be defined as a tool that measures the length of an object (how it is used). It can also be a tool with a series of marks used as a standard when measuring (how it works).

Analyze the Data

To determine the meaning of your observations and investigation results, you will need to look for patterns in the data. Then you must think critically to determine what the data mean. Scientists use several approaches when they analyze the data they have collected and recorded. Each approach is useful for identifying specific patterns.

Interpret Data The word *interpret* means "to explain the meaning of something." When analyzing data from an experiment, try to find out what the data show. Identify the control group and the test group to see whether or not changes in the independent variable have had an effect. Look for differences in the dependent variable between the control and test groups.

Classify Sorting objects or events into groups based on common features is called classifying. When classifying, first observe the objects or events to be classified. Then select one feature that is shared by some members in the group, but not by all. Place those members that share that feature in a subgroup. You can classify members into smaller and smaller subgroups based on characteristics. Remember that when you classify, you are grouping objects or events for a purpose. Keep your purpose in mind as you select the features to form groups and subgroups.

Compare and Contrast Observations can be analyzed by noting the similarities and differences between two more objects or events that you observe. When you look at objects or events to see how they are similar, you are comparing them. Contrasting is looking for differences in objects or events.

Recognize Cause and Effect A cause is a reason for an action or condition. The effect is that action or condition. When two events happen together, it is not necessarily true that one event caused the other. Scientists must design a controlled investigation to recognize the exact cause and effect.

Draw Conclusions

When scientists have analyzed the data they collected, they proceed to draw conclusions about the data. These conclusions are sometimes stated in words similar to the hypothesis that you formed earlier. They may confirm a hypothesis, or lead you to a new hypothesis.

Infer Scientists often make inferences based on their observations. An inference is an attempt to explain observations or to indicate a cause. An inference is not a fact, but a logical conclusion that needs further investigation. For example, you may infer that a fire has caused smoke. Until you investigate, however, you do not know for sure.

Apply When you draw a conclusion, you must apply those conclusions to determine whether the data supports the hypothesis. If your data do not support your hypothesis, it does not mean that the hypothesis is wrong. It means only that the result of the investigation did not support the hypothesis. Maybe the experiment needs to be redesigned, or some of the initial observations on which the hypothesis was based were incomplete or biased. Perhaps more observation or research is needed to refine your hypothesis. A successful investigation does not always come out the way you originally predicted.

Avoid Bias Sometimes a scientific investigation involves making judgments. When you make a judgment, you form an opinion. It is important to be honest and not to allow any expectations of results to bias your judgments. This is important throughout the entire investigation, from researching to collecting data to drawing conclusions.

Communicate

The communication of ideas is an important part of the work of scientists. A discovery that is not reported will not advance the scientific community's understanding or knowledge. Communication among scientists also is important as a way of improving their investigations.

Scientists communicate in many ways, from writing articles in journals and magazines that explain their investigations and experiments, to announcing important discoveries on television and radio. Scientists also share ideas with colleagues on the Internet or present them as lectures, like the student is doing in **Figure 15.**

Figure 15 A student communicates to his peers about his investigation.

SAFETY SYMBOLS	HAZARD	EXAMPLES	PRECAUTION	REMEDY
DISPOSAL	Special disposal procedures need to be followed.	certain chemicals, living organisms	Do not dispose of these materials in the sink or trash can.	Dispose of wastes as directed by your teacher.
BIOLOGICAL	Organisms or other biological materials that might be harmful to humans	bacteria, fungi, blood, unpreserved tissues, plant materials	Avoid skin contact with these materials. Wear mask or gloves.	Notify your teacher if you suspect contact with material. Wash hands thoroughly.
EXTREME TEMPERATURE	Objects that can burn skin by being too cold or too hot	boiling liquids, hot plates, dry ice, liquid nitrogen	Use proper protection when handling.	Go to your teacher for first aid.
SHARP OBJECT	Use of tools or glassware that can easily puncture or slice skin	razor blades, pins, scalpels, pointed tools, dissecting probes, broken glass	Practice common-sense behavior and follow guidelines for use of the tool.	Go to your teacher for first aid.
FUME	Possible danger to respiratory tract from fumes	ammonia, acetone, nail polish remover, heated sulfur, moth balls	Make sure there is good ventilation. Never smell fumes directly. Wear a mask.	Leave foul area and notify your teacher immediately.
ELECTRICAL	Possible danger from electrical shock or burn	improper grounding, liquid spills, short circuits, exposed wires	Double-check setup with teacher. Check condition of wires and apparatus.	Do not attempt to fix electrical problems. Notify your teacher immediately.
IRRITANT	Substances that can irritate the skin or mucous membranes of the respiratory tract	pollen, moth balls, steel wool, fiberglass, potassium permanganate	Wear dust mask and gloves. Practice extra care when handling these materials.	Go to your teacher for first aid.
CHEMICAL	Chemicals can react with and destroy tissue and other materials	bleaches such as hydrogen peroxide; acids such as sulfuric acid, hydrochloric acid; bases such as ammonia, sodium hydroxide	Wear goggles, gloves, and an apron.	Immediately flush the affected area with water and notify your teacher.
TOXIC	Substance may be poisonous if touched, inhaled, or swallowed.	mercury, many metal compounds, iodine, poinsettia plant parts	Follow your teacher's instructions.	Always wash hands thoroughly after use. Go to your teacher for first aid.
FLAMMABLE	Flammable chemicals may be ignited by open flame, spark, or exposed heat.	alcohol, kerosene, potassium permanganate	Avoid open flames and heat when using flammable chemicals.	Notify your teacher immediately. Use fire safety equipment if applicable.
OPEN FLAME	Open flame in use, may cause fire.	hair, clothing, paper, synthetic materials	Tie back hair and loose clothing. Follow teacher's instruction on lighting and extinguishing flames.	Notify your teacher immediately. Use fire safety equipment if applicable.

Eye Safety
Proper eye protection should be worn at all times by anyone performing or observing science activities.

Clothing Protection
This symbol appears when substances could stain or burn clothing.

Animal Safety
This symbol appears when safety of animals and students must be ensured.

Handwashing
After the lab, wash hands with soap and water before removing goggles.

Safety in the Science Laboratory

The science laboratory is a safe place to work if you follow standard safety procedures. Being responsible for your own safety helps to make the entire laboratory a safer place for everyone. When performing any lab, read and apply the caution statements and safety symbol listed at the beginning of the lab.

General Safety Rules

1. Obtain your teacher's permission to begin all investigations and use laboratory equipment.
2. Study the procedure. Ask your teacher any questions. Be sure you understand safety symbols shown on the page.
3. Notify your teacher about allergies or other health conditions which can affect your participation in a lab.
4. Learn and follow use and safety procedures for your equipment. If unsure, ask your teacher.

5. Never eat, drink, chew gum, apply cosmetics, or do any personal grooming in the lab. Never use lab glassware as food or drink containers. Keep your hands away from your face and mouth.
6. Know the location and proper use of the safety shower, eye wash, fire blanket, and fire alarm.

Prevent Accidents

1. Use the safety equipment provided to you. Goggles and a safety apron should be worn during investigations.
2. Do NOT use hair spray, mousse, or other flammable hair products. Tie back long hair and tie down loose clothing.
3. Do NOT wear sandals or other open-toed shoes in the lab.
4. Remove jewelry on hands and wrists. Loose jewelry, such as chains and long necklaces, should be removed to prevent them from getting caught in equipment.
5. Do not taste any substances or draw any material into a tube with your mouth.
6. Proper behavior is expected in the lab. Practical jokes and fooling around can lead to accidents and injury.
7. Keep your work area uncluttered.

Laboratory Work

1. Collect and carry all equipment and materials to your work area before beginning a lab.
2. Remain in your own work area unless given permission by your teacher to leave it.

3. Always slant test tubes away from yourself and others when heating them, adding substances to them, or rinsing them.
4. If instructed to smell a substance in a container, hold the container a short distance away and fan vapors towards your nose.
5. Do NOT substitute other chemicals/substances for those in the materials list unless instructed to do so by your teacher.
6. Do NOT take any materials or chemicals outside of the laboratory.
7. Stay out of storage areas unless instructed to be there and supervised by your teacher.

Laboratory Cleanup

1. Turn off all burners, water, and gas, and disconnect all electrical devices.
2. Clean all pieces of equipment and return all materials to their proper places.
3. Dispose of chemicals and other materials as directed by your teacher. Place broken glass and solid substances in the proper containers. Never discard materials in the sink.
4. Clean your work area.
5. Wash your hands with soap and water thoroughly BEFORE removing your goggles.

Emergencies

1. Report any fire, electrical shock, glassware breakage, spill, or injury, no matter how small, to your teacher immediately. Follow his or her instructions.
2. If your clothing should catch fire, STOP, DROP, and ROLL. If possible, smother it with the fire blanket or get under a safety shower. NEVER RUN.
3. If a fire should occur, turn off all gas and leave the room according to established procedures.
4. In most instances, your teacher will clean up spills. Do NOT attempt to clean up spills unless you are given permission and instructions to do so.
5. If chemicals come into contact with your eyes or skin, notify your teacher immediately. Use the eyewash or flush your skin or eyes with large quantities of water.
6. The fire extinguisher and first-aid kit should only be used by your teacher unless it is an extreme emergency and you have been given permission.
7. If someone is injured or becomes ill, only a professional medical provider or someone certified in first aid should perform first-aid procedures.

From Your Kitchen, Junk Drawer, or Yard

1 The Strongest Bag

Real-World Question

How can you test the strength of self-sealing bags?

Possible Materials

- several self-sealing bags of different brand names
- several generic-brand self-sealing bags
- marbles or round stones
- plastic basin

Procedure

1. Place a handful of marbles or stones in one of the self-sealing bags and seal the bag tight.
2. Invert the bag and hold it upside down over a plastic basin for 5 s.
3. If the seal does not break, open the bag, and add more stones or marbles. Reseal the bag, and invert it for 5 s over the plastic basin.
4. Continue adding weights to the bag until the seal can no longer hold them, and they spill into the basin. Record the maximum number of marbles or stones the bag held in your science journal.
5. Repeat steps 1–4 for the other self-sealing bags to test their strengths.

Conclude and Apply

1. Identify the strongest brand of self-sealing bag. Identify the weakest bag.
2. Infer whether or not brand-name bags are worth the extra cost.

2 Kitchen Compounds

Real-World Question

How can we make new compounds?

Possible Materials

- baking soda
- white vinegar
- bowl
- glass
- water
- iron nail
- jar
- dry yeast
- hydrogen peroxide
- jar with lid
- measuring cup
- safety goggles

Procedure

1. Pour 25 mL of baking soda into a bowl until it covers the bottom.
2. Pour 25 mL of white vinegar into the bowl with the baking soda and observe the reaction.
3. Pour a package of dry yeast into a jar. Pour 25 mL of hydrogen peroxide into the jar, seal the jar with a lid, and observe the reaction.
4. Fill a jar with water and drop an iron nail into the water. Observe the nail over the next two weeks.

Conclude and Apply

1. Describe the three reactions you caused.
2. Infer what compounds were made from these reactions.

Adult supervision required for all labs.

3 Panning Minerals

Real-World Question

How can minerals be separated from sand?

Possible Materials

- large, aluminum pie pan
- gallon jug filled with water
- empty gallon jug
- clean sand
- funnel
- coffee filter
- squirt bottle of water
- magnifying lens
- white paper
- hand magnet

Procedure

1. Conduct this lab outdoors.
2. Line the funnel with a coffee filter. Insert the funnel stem into an empty gallon jug.
3. Add a small amount of sand to the pie pan. Add some water and swirl the pan.
4. Continue to shake and swirl the pan until only black sand is left in the pan.
5. Use the squirt bottle to wash the black sand into the coffee filter. Repeat steps 3–5 until you have a good sample of black sand.
6. Let the black sand dry. Then observe it with a magnifying lens. Test the sand with a magnet.

Conclude and Apply

1. Why was black sand left in the pan after swirling it?
2. Describe how the sand looked under the lens. Did you see any well-shaped crystals?
3. What happened when you tested the sand with a magnet? Explain.

4 Changing Rocks

Real-World Question

How can the change of metamorphic rock be modeled?

Possible Materials

- soil
- water
- measuring cup
- bowl
- spoon
- clothes apron
- shale sample
- slate sample
- schist sample
- gneiss sample
- hornfel sample

Procedure

1. Mix equal parts of soil and water in a measuring cup or bowl. Stir the mixture until you make mud.
2. Place the bowl of mud on the table near the top edge.
3. Lay a sample of shale below the mud, a sample of slate below the shale, a sample of schist below the slate, a sample of gneiss below the schist, and a sample of hornfel below the gneiss.
4. Observe the different stages of sedimentary and metamorphic rocks that are formed by heat and pressure over long periods of time.

Conclude and Apply

1. Identify which rock sample(s) are sedimentary rock and which sample(s) are metamorphic rock.
2. Infer which type of rock is found at the greatest depth beneath the surface of Earth.

Adult supervision required for all labs.

5 Why Recycle?

Real-World Question

What are the effects of throwing out aluminum cans instead of recycling them?

Possible Materials

- calculator
- aluminum can

Procedure

1. An aluminum can has a mass of about 13 g.
2. Convert the can's mass from grams to kilograms by dividing the mass by 1,000.
3. Find the volume of the can (in mL) on the label.
4. Convert the volume from milliliters to liters by dividing it by 1,000.

Conclude and Apply

1. Calculate the mass of aluminum cans thrown out by Americans each year by multiplying the mass of the can in kilograms times 50,000,000,000.
2. Calculate the amount of fuel needed to remake the cans thrown out by Americans each year by multiplying the volume of the can in liters times 50,000,000,000 and dividing your total by 2.
3. Infer the environmental effects of throwing out aluminum cans instead of recycling them.

6 3-D Maps

Real-World Question

How can you make a topographical map of your room?

Possible Materials

- meterstick
- metric ruler
- metric tape measure
- poster board
- black marker
- construction paper
- transparent tape

Procedure

1. Measure the length and width of your room in meters. Include the measurements of any odd shapes or angles in the room.
2. Decide upon a scale for your map.
3. Using your scale, draw the outline of your room on the poster board.
4. Measure the length, width, and height of a piece of furniture.
5. Using your scale, measure and cut out the sides for a model of the furniture piece from construction paper. Tape the pieces of the model together.
6. Place your furniture model on your map to match the actual piece's location in your room.
7. Construct two or three other models of furniture for your map.

Conclude and Apply

1. What scale did you use for your map?
2. Infer how a biologist might use a topographical map.

Adult supervision required for all labs.

7 Rock and Roll

Real-World Question
How can we model the weathering of rock?

Possible Materials
- white glue
- sand
- plastic bowl
- plastic spoon
- cookie tray
- barbecue brush
- aluminum foil
- cooking oil
- empty coffee can with lid
- water
- measuring cup
- transparent packing tape
- safety goggles

Procedure
1. Make your own sedimentary rocks by adding equal amounts of white glue and sand to a bowl. Stir the sand and glue together until you make several small lumps.
2. Lay aluminum foil on the bottom of a cookie tray and coat the foil with cooking oil.
3. Lay your rocks on the tray in direct sunlight for 3 days until they dry.
4. Place your rocks in a coffee can and pour 50 mL of water into the can.
5. Place the lid on the can and secure the lid with thick transparent tape.
6. Shake the contents of the can for 4 minutes, open the lid, and observe your rocks.

Conclude and Apply
1. Describe what happened to your sedimentary rocks.
2. Infer how this lab modeled the weathering of rocks.

8 Modeling Mudslides

Real-World Question
How can mudslides be prevented?

Possible Materials
- deep, rectangular basin
- bowl
- measuring cup
- hose
- water
- soil (potting or garden)
- blocks of wood or bricks
- protractor
- sod
- several houseplants
- clothing aprons

Procedure
1. Pour soil into one-half of a deep, rectangular basin. Lightly pack the soil down.
2. Prop the container up on wooden blocks until the end with the soil in it is raised to a 60° angle.
3. Fill a bowl with water and slowly pour the water over the soil near the edge of the basin.
4. Continue pouring water onto the soil until a mudslide is created.
5. Clean the basin and repeat the lab, but plant sod or houseplants in the soil before adding water.

Conclude and Apply
1. Describe what happened to the soil without plants and with plants.
2. Explain how this lab modeled a mudslide.
3. Describe the relationship between vegetation and mudslides.

Adult supervision required for all labs.

9 In Deep Water

Real-World Question

How can the water table and a well be modeled?

Possible Materials

- clear-plastic drink bottle (500 mL)
- aquarium gravel
- water
- blue food dye
- measuring cup
- long dropper

Procedure

1. Fill a clear-plastic bottle with aquarium gravel.
2. Pour 450 mL of water into the the measuring cup and add several drops of blue food dye.
3. Pour 300 mL of the blue water into the bottle with the gravel. Insert the dropper down into the gravel and try to suck out some of the water.
4. Pour another 150 mL of water into the bottle and try to suck out some of the water with the dropper.

Conclude and Apply

1. Describe how this lab models the water table and a well.
2. Infer how deep a well must be dug for it to yield water.
3. Infer why some wells only yield water at certain times of the year.

10 Measuring Movement

Real-World Question

How can we model continental drift?

Possible Materials

- flashlight, nail, rubber band or tape, thick circle of paper
- protractor
- mirror
- stick-on notepad paper
- marker
- metric ruler
- calculator

Procedure

1. Cut a circle of paper to fit around the lens of the flashlight. Use a nail to make a hole in the paper. Fasten the paper with the rubber band or tape. You should now have a flashlight that shines a focused beam of light.
2. Direct the light beam of the flashlight on a protractor held horizontally so that the beam lines up to the 90° mark.
3. Darken a room and aim the light beam at a mirror from an angle. Measure the angle. Observe where the reflected beam hits the wall.
4. Have a partner place a stick on the wall and mark the location of the beam with a marker.
5. Move the flashlight to a 100° angle and mark the beam's location on the wall with a second note.
6. Measure the distance between the two points on the wall and divide by 10 to determine the distance per degree.

Conclude and Apply

1. What was the distance per degree of your measurements?
2. Calculate what the distance would be between the first spot and a third spot marking the location of the flashlight at a 40° angle.
3. Explain how this lab models measuring continental drift.

Adult supervision required for all labs.

11 Making Waves

Real-World Question
What do earthquake waves look like?

Possible Materials

- rope (3-m length)
- garden hose
- coiled spring
- safety goggles

Procedure
1. With a partner, stretch a coiled spring out on the floor. Firmly push your side of the spring in and out and observe the waves you created.
2. With a partner, stretch the rope out on the floor. Quickly wave your end back and forth and observe the waves you created.
3. Stand with a partner, stretch the rope out, and hold it waist high. Quickly move your hand up and down and observe the waves you created.

Conclude and Apply
1. Infer the type of seismic wave you modeled with the coiled spring.
2. Infer the type of seismic waves you modeled with the rope on the floor.
3. Infer the type of seismic waves you modeled with the rope in the air.

12 Mini Eruptions

Real-World Question
How can we model the eruptions of shield and cinder cone volcanoes?

Possible Materials

- tube of toothpaste
- metal straight pin
- unopened bottle of carbonated soda
- newspaper
- paper towels
- safety goggles
- clothes apron

Procedure
1. Lay down newspaper or paper towels.
2. Press down on the back end of a full tube of toothpaste to move all the paste to the front of the tube.
3. Have a partner press a long pin into the center of the tube. Observe what happens to the toothpaste.
4. Go outside and vigorously shake a bottle of carbonated soda for 1 min.
5. Point the bottle away from other people and quickly remove the cap. Observe what happens to the soda.

Conclude and Apply
1. Describe what happened to the toothpaste and soda.
2. Infer how you modeled a shield volcano eruption.
3. Infer how you modeled a cinder cone volcano eruption.

13 Making Burrows

Real-World Question
How does burrowing affect sediment layers?

Possible Materials
- clear-glass bowl
- white flour
- colored gelatin powder (3 packages)
- paintbrush
- pencil

Procedure
1. Add 3 cm of white flour to the bowl. Flatten the top of the flour layer.
2. Carefully sprinkle gelatin powder over the flour to form a colored layer about 1/4 cm thick.
3. The two layers represent two different layers of sediment.
4. Use a paintbrush or pencil to make "burrows" in the "sediment."
5. Make sure to make some of the burrows at the edge of the bowl so that you can see how it affects the sediment.
6. Continue to make more burrows and observe the effect on the two layers.

Conclude and Apply
1. How did the two layers of powder change as you continued to make burrows?
2. Were the "trace fossils" easy to recognize at first? How about after a lot of burrowing?
3. How do you think burrowing animals affect layers of sediment on the ocean floor? How could this burrowing be recognized in rock?

14 History in a Bottle

Real-World Question
What does the geologic column look like?

Possible Materials
- clear-plastic 2-L bottle
- scissors
- 3-in × 5-in index cards
- colored markers
- permanent marker
- transparent tape
- metric ruler
- sand (3 different colors)
- aquarium gravel (3 different colors)

Procedure
1. Cut the top 5 cm off a clear-plastic 2-liter soda bottle. Remove the label.
2. Cut 12 square cards measuring 2 cm × 2 cm.
3. Draw a picture of a trilobite, coral, fish, amphibian, insect, reptile, mouse, conifer tree, dinosaur, bird, flower, large mammal, and human on the 12 cards.
4. Starting at the bottom, inside of the bottle and working up, tape the trilobite, coral, fish, amphibian, insect, and reptile pictures face out in that order. The reptile picture should be about halfway up the bottle.
5. Pour red sand into the bottle until it covers your reptile picture.
6. Tape the mouse, conifer tree, dinosaur, bird, and flower pictures above the red sand in that order. Pour in blue sand until it covers the flower picture.
7. Tape the large mammal and human pictures above the blue sand. Pour green sand into the bottle until it covers the person.

Conclude and Apply
1. Research what era each color of sand represents.
2. Infer why few fossils of organisms living before the Paleozoic Era are found.

Adult supervision required for all labs.

15 The Pressure's On

Real-World Question

How can atmospheric air pressure changes be modeled?

Possible Materials

- large pot
- stove or hot plate
- tongs
- oven mitt
- empty aluminum soda can
- water
- cold water
- small aquarium or large bowl
- large jar
- measuring cup
- safety goggles

Procedure

1. Fill a small aquarium or large bowl with cold water.
2. Pour water into a large pot and boil it.
3. Pour 25 mL of water into an empty aluminum can.
4. Using an oven mitt and tongs, hold the bottom of the can in the boiling water for 1 minute.
5. Remove the can from the pot and immediately submerge it upside-down in the cold water in the aquarium or large bowl.

Conclude and Apply

1. Describe what happened to the can in the cold water.
2. Infer why the can changed in the cold water.

16 Bottling a Tornado

Real-World Question

How can you model a tornado?

Possible Materials

- two 2-L soda bottles
- dish soap
- masking tape
- duct tape
- metric measuring cup
- towel
- clothing apron
- safety goggles

Procedure

1. Remove the labels from two 2-L soda bottles.
2. Fill one bottle with 1.5 L of water.
3. Add two drops of dish soap to the bottle with the water.
4. Invert the second bottle and connect the openings of the bottles.
5. Attach the two bottles together with duct tape.
6. Flip the bottles upside-down and quickly swirl the top bottle with a smooth motion. Observe the tornado pattern made in the water.

Conclude and Apply

1. Describe how you modeled a tornado.
2. Research how a real tornado forms.

Adult supervision required for all labs.

Extra Try at Home Labs

17 Getting Warmer

Real-World Question

How do artificial surfaces affect temperature?

Possible Materials

- thermometer
- moist leaf litter
- large self-sealing bag

Procedure

1. Collect moist leaf litter in a large self-sealing bag.
2. Pile the leaf litter on a patch of grass that is exposed to direct sunlight.
3. Set the thermometer in the center of the leaf litter, wait 3 min, and measure the temperature.
4. Place the thermometer on the grass in direct sunlight, wait 3 min, and measure the temperature.
5. Place the thermometer on a cement surface in direct sunlight, wait 3 min, and measure the temperature.
6. Place the thermometer on an asphalt surface in direct sunlight, wait 3 min, and measure the temperature.

Conclude and Apply

1. Compare the temperatures of the different surfaces.
2. Explain how you measured the heat-island effect.

18 That's Cold!

Real-World Question

How does salt affect the freezing temperature of water?

Possible Materials

- drinking glasses (2)
- ice cubes
- salt
- clock

Procedure

1. Place the same number of ice cubes into two different glasses. Make sure that the ice cubes have equal volume.
2. Sprinkle about 10 g of salt into one of the glasses.
3. Observe both glasses every 10 min until the ice is melted. Record the results.

Conclude and Apply

1. In which glass did the ice melt more quickly? Describe what you observed.
2. In which glass did the water have the lower freezing point? How do you know?
3. Why does ocean water get colder than freshwater before freezing?

Adult supervision required for all labs.

19 Water Pressure

Real-World Question
How does water pressure change with depth?

Possible Materials
- plastic gallon milk jug
- two 40-cm lengths of 1/4-in. plastic tubing
- nail
- duct tape
- metric ruler
- water

Procedure
1. Use a nail to make two holes in one side of a plastic jug. One hole should be near the bottom of the jug, and one should be closer to the top.
2. Insert a piece of plastic tubing into each hole and seal with duct tape.
3. While holding the ends of the tubing in the air, fill the jug with water.
4. Measure how high the water rises above the hole in each piece of tubing. Also measure the depth of water in the jug at each hole.

Conclude and Apply
1. How high was the water column in each piece of tubing? Was the water column higher for the top hole or the bottom hole? Explain.
2. How do you think pressure changes with depth in the ocean? How might this affect the way submersible vehicles are designed?

20 Pack It Up

Real-World Question
Which sizes and shapes of packaging contain the least waste?

Possible Materials
- several types of packaging

Procedure
1. Separate each type of packaging from its contents. Set all of the packages in a row on a table so you can compare them.
2. Make a data table with the following headings: *Name of Product, Mass of Product in the Package,* and *Amount of Packaging.*
3. In the last column, describe each type of packaging. Try to guess what percent of the mass of product the packaging would be.
4. Express the amount of packaging a second way by describing how much packaging there is per usage of the product (e.g. one load of laundry or one serving of food).
5. Compare the amount of packaging in each case. Make a list in order from most packaging to least.

Conclude and Apply
1. Which sizes and shapes of packaging do you think contained the least waste per ounce of product?
2. How could consumers and companies use this knowledge to reduce the amount of household waste?
3. Do you think that some packaging materials are better for the environment than others? Explain your answer.

Adult supervision required for all labs.

21 Conserving Water

Real-World Question

How can you save water when using the bathroom faucet?

Possible Materials

- measuring cup
- watch with second hand or stopwatch
- calculator

Procedure

1. Turn on the faucet on your bathroom sink as you normally would.
2. Place the measuring cup in the stream of water for 5 s. Record how much water is in the measuring cup.
3. Now, slow the water. Place the empty measuring cup in the stream of water for 5 s. Record how much water is in the cup.
4. Divide the amount of water measured in Step 2 by the amount of water measured in Step 3. Record this value.
5. Subtract the value you recorded from 1 and multiply by 100. This is the percent of water that you could save by opening the faucet less.

Conclude and Apply

1. What percent of water could you save?
2. A typical person in the United States uses about 4,055 L of water from indoor faucets. How many liters per year could you save?

22 Space Probe Flights

Real-World Question

How can we compare the distances traveled by space probes to their destinations?

Possible Materials

- polystyrene balls (5)
- toothpicks (5)
- small stick-on labels (5)
- tennis ball
- meterstick

Procedure

1. Write the names *Mariner 2, Pioneer 10, Mariner 10, Viking 1,* and *Voyager 2* on the five labels and stick each label on a toothpick. Stick a labeled toothpick into each of the polystyrene balls to represent these five United States space probes.
2. Place the tennis ball in an open space such as a basketball court or field.
3. Measure a distance of 0.42 m from the tennis ball and place the *Mariner 2* probe in that spot. Place the *Pioneer 10* probe 6.28 m away, the *Mariner 10* probe 0.92 m from the ball, the *Viking 1* probe 0.78 m away, and the *Voyager 2* probe 43.47 m from the tennis ball.

Conclude and Apply

1. Create a time line showing the year each probe was launched and its destination, and relate this information to the distance traveled.
2. Mercury is 58 million km from the Sun and Earth is 150 million km. Use this information to calculate the scale used for this lab.

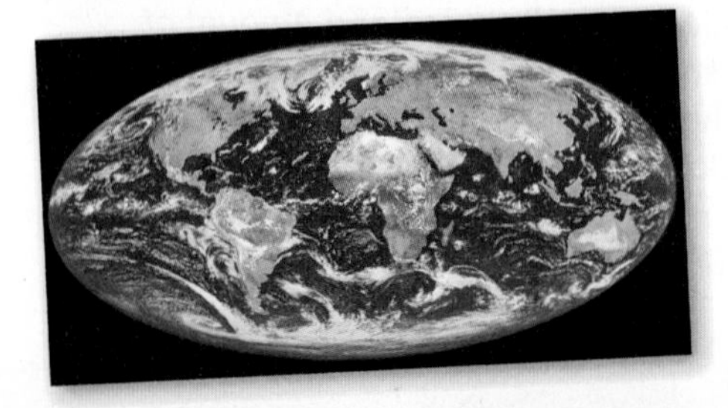

Adult supervision required for all labs.

23 Creating Craters

Real-World Question

Why does the Moon have craters?

Possible Materials

- drink mix or powdered baby formula
- black pepper
- paprika
- large, deep cooking tray
- large bowl
- marbles
- small, round candies
- aquarium gravel
- tweezers
- bag of cotton balls

Procedure

1. Pour a 3-cm layer of powder over the bottom of a large, deep cooking tray.
2. Sprinkle a fine layer of black pepper over the powder.
3. Lay a 2–3 cm layer of cotton over half of the powder.
4. Drop marbles and other small objects into the powder not covered by the cotton. Carefully remove the objects with tweezers and observe the craters and impact patterns they make.
5. Drop objects on to the half of the tray covered by cotton.
6. Remove the objects and cotton and observe the marks made by objects in the powder.

Conclude and Apply

1. Compare the impacts made by the objects in the powder not covered by cotton with the impacts in the powder covered by cotton.
2. Infer why the Moon has many craters on its surface but Earth does not.

24 Many Moons

Real-World Question

How do the number of moons of the nine planets compare?

Possible Materials

- golf balls (5)
- softballs (4)
- colored construction paper
- hole puncher
- pennies (10)
- quarters (8)
- meterstick

Procedure

1. Lay the golf balls and softballs on the floor in a row about a meter apart to represent the nine planets. The golf balls should represent the terrestrial planets and the softballs the gas planets.
2. Next to the golf ball representing Earth, place one quarter. The quarter represents a moon with a diameter greater than 1,000 km. Research which planets have moons this size and place quarters next to them.
3. Use pennies to represent moons with a diameter between 200–1,000 km. Place pennies next to the planets with moons this size.
4. Use a hole punch to punch out holes from colored construction paper. These holes represent moons smaller than 200 km in diameter. Research which planets have moons this size and place the holes next to them.

Conclude and Apply

1. Infer why terrestrial planets have fewer moons than gas planets.
2. Infer why astronomers do not believe all the moons in the solar system have been discovered.

Adult supervision required for all labs.

25 Big Stars

Real-World Question

How does the size of Earth compare to the size of stars?

Possible Materials

- metric ruler
- meterstick
- tape measure
- masking tape
- white paper
- black marker

Procedure

1. Tape a sheet of white paper to the floor.
2. Draw a dot in the center to the paper. Measure a 1-mm distance from the dot and draw a second dot. This distance represents the diameter of Earth.
3. Measure a distance of 10.9 cm from the first dot and draw a third dot. This distance represents the diameter of the Sun.
4. Measure a distance of 5 m from the first dot and mark the location on the floor with a piece of masking tape. This distance represents the average diameter of a red giant star.
5. Measure a distance of 30 m from the first dot and mark the location on the floor with a piece of masking tape. This distance represents the diameter of the supergiant star Antares.

Conclude and Apply

1. The diameter of Earth is 12,756 km. What is the diameter of the Sun?
2. What is the diameter of an average red giant?

Computer Skills

People who study science rely on computers, like the one in **Figure 16,** to record and store data and to analyze results from investigations. Whether you work in a laboratory or just need to write a lab report with tables, good computer skills are a necessity.

Using the computer comes with responsibility. Issues of ownership, security, and privacy can arise. Remember, if you did not author the information you are using, you must provide a source for your information. Also, anything on a computer can be accessed by others. Do not put anything on the computer that you would not want everyone to know. To add more security to your work, use a password.

Use a Word Processing Program

A computer program that allows you to type your information, change it as many times as you need to, and then print it out is called a word processing program. Word processing programs also can be used to make tables.

Figure 16 A computer will make reports neater and more professional looking.

Learn the Skill To start your word processing program, a blank document, sometimes called "Document 1," appears on the screen. To begin, start typing. To create a new document, click the *New* button on the standard tool bar. These tips will help you format the document.

- The program will automatically move to the next line; press *Enter* if you wish to start a new paragraph.
- Symbols, called non-printing characters, can be hidden by clicking the *Show/Hide* button on your toolbar.
- To insert text, move the cursor to the point where you want the insertion to go, click on the mouse once, and type the text.
- To move several lines of text, select the text and click the *Cut* button on your toolbar. Then position your cursor in the location that you want to move the cut text and click *Paste.* If you move to the wrong place, click *Undo.*
- The spell check feature does not catch words that are misspelled to look like other words, like "cold" instead of "gold." Always reread your document to catch all spelling mistakes.
- To learn about other word processing methods, read the user's manual or click on the *Help* button.
- You can integrate databases, graphics, and spreadsheets into documents by copying from another program and pasting it into your document, or by using desktop publishing (DTP). DTP software allows you to put text and graphics together to finish your document with a professional look. This software varies in how it is used and its capabilities.

Use a Database

A collection of facts stored in a computer and sorted into different fields is called a database. A database can be reorganized in any way that suits your needs.

Learn the Skill A computer program that allows you to create your own database is a database management system (DBMS). It allows you to add, delete, or change information. Take time to get to know the features of your database software.

- Determine what facts you would like to include and research to collect your information.
- Determine how you want to organize the information.
- Follow the instructions for your particular DBMS to set up fields. Then enter each item of data in the appropriate field.
- Follow the instructions to sort the information in order of importance.
- Evaluate the information in your database, and add, delete, or change as necessary.

Use the Internet

The Internet is a global network of computers where information is stored and shared. To use the Internet, like the students in **Figure 17,** you need a modem to connect your computer to a phone line and an Internet Service Provider account.

Learn the Skill To access internet sites and information, use a "Web browser," which lets you view and explore pages on the World Wide Web. Each page is its own site, and each site has its own address, called a URL. Once you have found a Web browser, follow these steps for a search (this also is how you search a database).

Figure 17 The Internet allows you to search a global network for a variety of information.

- Be as specific as possible. If you know you want to research "gold," don't type in "elements." Keep narrowing your search until you find what you want.
- Web sites that end in *.com* are commercial Web sites; *.org, .edu,* and *.gov* are non-profit, educational, or government Web sites.
- Electronic encyclopedias, almanacs, indexes, and catalogs will help locate and select relevant information.
- Develop a "home page" with relative ease. When developing a Web site, NEVER post pictures or disclose personal information such as location, names, or phone numbers. Your school or community usually can host your Web site. A basic understanding of HTML (hypertext mark-up language), the language of Web sites, is necessary. Software that creates HTML code is called authoring software, and can be downloaded free from many Web sites. This software allows text and pictures to be arranged as the software is writing the HTML code.

Use a Spreadsheet

A spreadsheet, shown in **Figure 18,** can perform mathematical functions with any data arranged in columns and rows. By entering a simple equation into a cell, the program can perform operations in specific cells, rows, or columns.

Learn the Skill Each column (vertical) is assigned a letter, and each row (horizontal) is assigned a number. Each point where a row and column intersect is called a cell, and is labeled according to where it is located—Column A, Row 1 (A1).

- Decide how to organize the data, and enter it in the correct row or column.
- Spreadsheets can use standard formulas or formulas can be customized to calculate cells.
- To make a change, click on a cell to make it activate, and enter the edited data or formula.
- Spreadsheets also can display your results in graphs. Choose the style of graph that best represents the data.

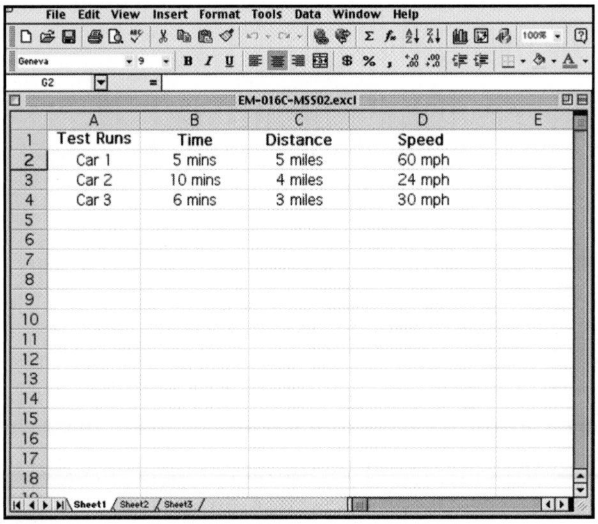

	A	B	C	D	E
1	Test Runs	Time	Distance	Speed	
2	Car 1	5 mins	5 miles	60 mph	
3	Car 2	10 mins	4 miles	24 mph	
4	Car 3	6 mins	3 miles	30 mph	

Figure 18 A spreadsheet allows you to perform mathematical operations on your data.

Use Graphics Software

Adding pictures, called graphics, to your documents is one way to make your documents more meaningful and exciting. This software adds, edits, and even constructs graphics. There is a variety of graphics software programs. The tools used for drawing can be a mouse, keyboard, or other specialized devices. Some graphics programs are simple. Others are complicated, called computer-aided design (CAD) software.

Learn the Skill It is important to have an understanding of the graphics software being used before starting. The better the software is understood, the better the results. The graphics can be placed in a word-processing document.

- Clip art can be found on a variety of internet sites, and on CDs. These images can be copied and pasted into your document.
- When beginning, try editing existing drawings, then work up to creating drawings.
- The images are made of tiny rectangles of color called pixels. Each pixel can be altered.
- Digital photography is another way to add images. The photographs in the memory of a digital camera can be downloaded into a computer, then edited and added to the document.
- Graphics software also can allow animation. The software allows drawings to have the appearance of movement by connecting basic drawings automatically. This is called in-betweening, or tweening.
- Remember to save often.

Presentation Skills

Develop Multimedia Presentations

Most presentations are more dynamic if they include diagrams, photographs, videos, or sound recordings, like the one shown in **Figure 19.** A multimedia presentation involves using stereos, overhead projectors, televisions, computers, and more.

Learn the Skill Decide the main points of your presentation, and what types of media would best illustrate those points.

- Make sure you know how to use the equipment you are working with.
- Practice the presentation using the equipment several times.
- Enlist the help of a classmate to push play or turn lights out for you. Be sure to practice your presentation with him or her.
- If possible, set up all of the equipment ahead of time, and make sure everything is working properly.

Figure 19 These students are engaging the audience using a variety of tools.

Computer Presentations

There are many different interactive computer programs that you can use to enhance your presentation. Most computers have a compact disc (CD) drive that can play both CDs and digital video discs (DVDs). Also, there is hardware to connect a regular CD, DVD, or VCR. These tools will enhance your presentation.

Another method of using the computer to aid in your presentation is to develop a slide show using a computer program. This can allow movement of visuals at the presenter's pace, and can allow for visuals to build on one another.

Learn the Skill In order to create multimedia presentations on a computer, you need to have certain tools. These may include traditional graphic tools and drawing programs, animation programs, and authoring systems that tie everything together. Your computer will tell you which tools it supports. The most important step is to learn about the tools that you will be using.

- Often, color and strong images will convey a point better than words alone. Use the best methods available to convey your point.
- As with other presentations, practice many times.
- Practice your presentation with the tools you and any assistants will be using.
- Maintain eye contact with the audience. The purpose of using the computer is not to prompt the presenter, but to help the audience understand the points of the presentation.

Math Review

Use Fractions

A fraction compares a part to a whole. In the fraction $\frac{2}{3}$, the 2 represents the part and is the numerator. The 3 represents the whole and is the denominator.

Reduce Fractions To reduce a fraction, you must find the largest factor that is common to both the numerator and the denominator, the greatest common factor (GCF). Divide both numbers by the GCF. The fraction has then been reduced, or it is in its simplest form.

Example Twelve of the 20 chemicals in the science lab are in powder form. What fraction of the chemicals used in the lab are in powder form?

Step 1 Write the fraction.

$\frac{\text{part}}{\text{whole}} = \frac{12}{20}$

Step 2 To find the GCF of the numerator and denominator, list all of the factors of each number.

Factors of 12: 1, 2, 3, 4, 6, 12 (the numbers that divide evenly into 12)

Factors of 20: 1, 2, 4, 5, 10, 20 (the numbers that divide evenly into 20)

Step 3 List the common factors.

1, 2, 4.

Step 4 Choose the greatest factor in the list.

The GCF of 12 and 20 is 4.

Step 5 Divide the numerator and denominator by the GCF.

$\frac{12 \div 4}{20 \div 4} = \frac{3}{5}$

In the lab, $\frac{3}{5}$ of the chemicals are in powder form.

Practice Problem At an amusement park, 66 of 90 rides have a height restriction. What fraction of the rides, in its simplest form, has a height restriction?

Add and Subtract Fractions To add or subtract fractions with the same denominator, add or subtract the numerators and write the sum or difference over the denominator. After finding the sum or difference, find the simplest form for your fraction.

Example 1 In the forest outside your house, $\frac{1}{8}$ of the animals are rabbits, $\frac{3}{8}$ are squirrels, and the remainder are birds and insects. How many are mammals?

Step 1 Add the numerators.

$\frac{1}{8} + \frac{3}{8} = \frac{(1 + 3)}{8} = \frac{4}{8}$

Step 2 Find the GCF.

$\frac{4}{8}$ (GCF, 4)

Step 3 Divide the numerator and denominator by the GCF.

$\frac{4}{4} = 1,\ \frac{8}{4} = 2$

$\frac{1}{2}$ of the animals are mammals.

Example 2 If $\frac{7}{16}$ of the Earth is covered by freshwater, and $\frac{1}{16}$ of that is in glaciers, how much freshwater is not frozen?

Step 1 Subtract the numerators.

$\frac{7}{16} - \frac{1}{16} = \frac{(7 - 1)}{16} = \frac{6}{16}$

Step 2 Find the GCF.

$\frac{6}{16}$ (GCF, 2)

Step 3 Divide the numerator and denominator by the GCF.

$\frac{6}{2} = 3,\ \frac{16}{2} = 8$

$\frac{3}{8}$ of the freshwater is not frozen.

Practice Problem A bicycle rider is going 15 km/h for $\frac{4}{9}$ of his ride, 10 km/h for $\frac{2}{9}$ of his ride, and 8 km/h for the remainder of the ride. How much of his ride is he going over 8 km/h?

Unlike Denominators To add or subtract fractions with unlike denominators, first find the least common denominator (LCD). This is the smallest number that is a common multiple of both denominators. Rename each fraction with the LCD, and then add or subtract. Find the simplest form if necessary.

Example 1 A chemist makes a paste that is $\frac{1}{2}$ table salt (NaCl), $\frac{1}{3}$ sugar ($C_6H_{12}O_6$), and the rest water (H_2O). How much of the paste is a solid?

Step 1 Find the LCD of the fractions.

$\frac{1}{2} + \frac{1}{3}$ (LCD, 6)

Step 2 Rename each numerator and each denominator with the LCD.

$1 \times 3 = 3,\ 2 \times 3 = 6$

$1 \times 2 = 2,\ 3 \times 2 = 6$

Step 3 Add the numerators.

$\frac{3}{6} + \frac{2}{6} = \frac{(3 + 2)}{6} = \frac{5}{6}$

$\frac{5}{6}$ of the paste is a solid.

Example 2 The average precipitation in Grand Junction, CO, is $\frac{7}{10}$ inch in November, and $\frac{3}{5}$ inch in December. What is the total average precipitation?

Step 1 Find the LCD of the fractions.

$\frac{7}{10} + \frac{3}{5}$ (LCD, 10)

Step 2 Rename each numerator and each denominator with the LCD.

$7 \times 1 = 7,\ 10 \times 1 = 10$

$3 \times 2 = 6,\ 5 \times 2 = 10$

Step 3 Add the numerators.

$\frac{7}{10} + \frac{6}{10} = \frac{(7 + 6)}{10} = \frac{13}{10}$

$\frac{13}{10}$ inches total precipitation, or $1\frac{3}{10}$ inches.

Practice Problem On an electric bill, about $\frac{1}{8}$ of the energy is from solar energy and about $\frac{1}{10}$ is from wind power. How much of the total bill is from solar energy and wind power combined?

Example 3 In your body, $\frac{7}{10}$ of your muscle contractions are involuntary (cardiac and smooth muscle tissue). Smooth muscle makes up $\frac{3}{15}$ of your muscle contractions. How many of your muscle contractions are made by cardiac muscle?

Step 1 Find the LCD of the fractions.

$\frac{7}{10} - \frac{3}{15}$ (LCD, 30)

Step 2 Rename each numerator and each denominator with the LCD.

$7 \times 3 = 21,\ 10 \times 3 = 30$

$3 \times 2 = 6,\ 15 \times 2 = 30$

Step 3 Add the numerators.

$\frac{21}{30} - \frac{6}{30} = \frac{(21 - 6)}{30} = \frac{15}{30}$

Step 4 Find the GCF.

$\frac{15}{30}$ (GCF, 15)

$\frac{1}{2}$

$\frac{1}{2}$ of all muscle contractions are cardiac muscle.

Example 4 Tony wants to make cookies that call for $\frac{3}{4}$ of a cup of flour, but he only has $\frac{1}{3}$ of a cup. How much more flour does he need?

Step 1 Find the LCD of the fractions.

$\frac{3}{4} - \frac{1}{3}$ (LCD, 12)

Step 2 Rename each numerator and each denominator with the LCD.

$3 \times 3 = 9,\ 4 \times 3 = 12$

$1 \times 4 = 4,\ 3 \times 4 = 12$

Step 3 Add the numerators.

$\frac{9}{12} - \frac{4}{12} = \frac{(9 - 4)}{12} = \frac{5}{12}$

$\frac{5}{12}$ of a cup of flour.

Practice Problem Using the information provided to you in Example 3 above, determine how many muscle contractions are voluntary (skeletal muscle).

Multiply Fractions To multiply with fractions, multiply the numerators and multiply the denominators. Find the simplest form if necessary.

Example Multiply $\frac{3}{5}$ by $\frac{1}{3}$.

Step 1 Multiply the numerators and denominators.

$$\frac{3}{5} \times \frac{1}{3} = \frac{(3 \times 1)}{(5 \times 3)} = \frac{3}{15}$$

Step 2 Find the GCF.

$\frac{3}{15}$ (GCF, 3)

Step 3 Divide the numerator and denominator by the GCF.

$\frac{3}{3} = 1,\ \frac{15}{3} = 5$

$\frac{1}{5}$

$\frac{3}{5}$ multiplied by $\frac{1}{3}$ is $\frac{1}{5}$.

Practice Problem Multiply $\frac{3}{14}$ by $\frac{5}{16}$.

Find a Reciprocal Two numbers whose product is 1 are called multiplicative inverses, or reciprocals.

Example Find the reciprocal of $\frac{3}{8}$.

Step 1 Inverse the fraction by putting the denominator on top and the numerator on the bottom.

$\frac{8}{3}$

The reciprocal of $\frac{3}{8}$ is $\frac{8}{3}$.

Practice Problem Find the reciprocal of $\frac{4}{9}$.

Divide Fractions To divide one fraction by another fraction, multiply the dividend by the reciprocal of the divisor. Find the simplest form if necessary.

Example 1 Divide $\frac{1}{9}$ by $\frac{1}{3}$.

Step 1 Find the reciprocal of the divisor.

The reciprocal of $\frac{1}{3}$ is $\frac{3}{1}$.

Step 2 Multiply the dividend by the reciprocal of the divisor.

$$\frac{\frac{1}{9}}{\frac{1}{3}} = \frac{1}{9} \times \frac{3}{1} = \frac{(1 \times 3)}{(9 \times 1)} = \frac{3}{9}$$

Step 3 Find the GCF.

$\frac{3}{9}$ (GCF, 3)

Step 4 Divide the numerator and denominator by the GCF.

$\frac{3}{3} = 1,\ \frac{9}{3} = 3$

$\frac{1}{3}$

$\frac{1}{9}$ divided by $\frac{1}{3}$ is $\frac{1}{3}$.

Example 2 Divide $\frac{3}{5}$ by $\frac{1}{4}$.

Step 1 Find the reciprocal of the divisor.

The reciprocal of $\frac{1}{4}$ is $\frac{4}{1}$.

Step 2 Multiply the dividend by the reciprocal of the divisor.

$$\frac{\frac{3}{5}}{\frac{1}{4}} = \frac{3}{5} \times \frac{4}{1} = \frac{(3 \times 4)}{(5 \times 1)} = \frac{12}{5}$$

$\frac{3}{5}$ divided by $\frac{1}{4}$ is $\frac{12}{5}$ or $2\frac{2}{5}$.

Practice Problem Divide $\frac{3}{11}$ by $\frac{7}{10}$.

Use Ratios

When you compare two numbers by division, you are using a ratio. Ratios can be written 3 to 5, 3:5, or $\frac{3}{5}$. Ratios, like fractions, also can be written in simplest form.

Ratios can represent probabilities, also called odds. This is a ratio that compares the number of ways a certain outcome occurs to the number of outcomes. For example, if you flip a coin 100 times, what are the odds that it will come up heads? There are two possible outcomes, heads or tails, so the odds of coming up heads are 50:100. Another way to say this is that 50 out of 100 times the coin will come up heads. In its simplest form, the ratio is 1:2.

Example 1 A chemical solution contains 40 g of salt and 64 g of baking soda. What is the ratio of salt to baking soda as a fraction in simplest form?

Step 1 Write the ratio as a fraction.

$$\frac{\text{salt}}{\text{baking soda}} = \frac{40}{64}$$

Step 2 Express the fraction in simplest form.

The GCF of 40 and 64 is 8.

$$\frac{40}{64} = \frac{40 \div 8}{64 \div 8} = \frac{5}{8}$$

The ratio of salt to baking soda in the sample is 5:8.

Example 2 Sean rolls a 6-sided die 6 times. What are the odds that the side with a 3 will show?

Step 1 Write the ratio as a fraction.

$$\frac{\text{number of sides with a 3}}{\text{number of sides}} = \frac{1}{6}$$

Step 2 Multiply by the number of attempts.

$$\frac{1}{6} \times 6 \text{ attempts} = \frac{6}{6} \text{ attempts} = 1 \text{ attempt}$$

1 attempt out of 6 will show a 3.

Practice Problem Two metal rods measure 100 cm and 144 cm in length. What is the ratio of their lengths in simplest form?

Use Decimals

A fraction with a denominator that is a power of ten can be written as a decimal. For example, 0.27 means $\frac{27}{100}$. The decimal point separates the ones place from the tenths place.

Any fraction can be written as a decimal using division. For example, the fraction $\frac{5}{8}$ can be written as a decimal by dividing 5 by 8. Written as a decimal, it is 0.625.

Add or Subtract Decimals When adding and subtracting decimals, line up the decimal points before carrying out the operation.

Example 1 Find the sum of 47.68 and 7.80.

Step 1 Line up the decimal places when you write the numbers.

$$\begin{array}{r} 47.68 \\ +\ 7.80 \\ \hline \end{array}$$

Step 2 Add the decimals.

$$\begin{array}{r} 47.68 \\ +\ 7.80 \\ \hline 55.48 \end{array}$$

The sum of 47.68 and 7.80 is 55.48.

Example 2 Find the difference of 42.17 and 15.85.

Step 1 Line up the decimal places when you write the number.

$$\begin{array}{r} 42.17 \\ -15.85 \\ \hline \end{array}$$

Step 2 Subtract the decimals.

$$\begin{array}{r} 42.17 \\ -15.85 \\ \hline 26.32 \end{array}$$

The difference of 42.17 and 15.85 is 26.32.

Practice Problem Find the sum of 1.245 and 3.842.

Multiply Decimals To multiply decimals, multiply the numbers like any other number, ignoring the decimal point. Count the decimal places in each factor. The product will have the same number of decimal places as the sum of the decimal places in the factors.

Example Multiply 2.4 by 5.9.

Step 1 Multiply the factors like two whole numbers.
$24 \times 59 = 1416$

Step 2 Find the sum of the number of decimal places in the factors. Each factor has one decimal place, for a sum of two decimal places.

Step 3 The product will have two decimal places.
14.16

The product of 2.4 and 5.9 is 14.16.

Practice Problem Multiply 4.6 by 2.2.

Divide Decimals When dividing decimals, change the divisor to a whole number. To do this, multiply both the divisor and the dividend by the same power of ten. Then place the decimal point in the quotient directly above the decimal point in the dividend. Then divide as you do with whole numbers.

Example Divide 8.84 by 3.4.

Step 1 Multiply both factors by 10.
$3.4 \times 10 = 34$, $8.84 \times 10 = 88.4$

Step 2 Divide 88.4 by 34.

$$\begin{array}{r} 2.6 \\ 34\overline{)88.4} \\ -68 \\ \hline 20\,4 \\ -20\,4 \\ \hline 0 \end{array}$$

8.84 divided by 3.4 is 2.6.

Practice Problem Divide 75.6 by 3.6.

Use Proportions

An equation that shows that two ratios are equivalent is a proportion. The ratios $\frac{2}{4}$ and $\frac{5}{10}$ are equivalent, so they can be written as $\frac{2}{4} = \frac{5}{10}$. This equation is a proportion.

When two ratios form a proportion, the cross products are equal. To find the cross products in the proportion $\frac{2}{4} = \frac{5}{10}$, multiply the 2 and the 10, and the 4 and the 5. Therefore $2 \times 10 = 4 \times 5$, or $20 = 20$.

Because you know that both proportions are equal, you can use cross products to find a missing term in a proportion. This is known as solving the proportion.

Example The heights of a tree and a pole are proportional to the lengths of their shadows. The tree casts a shadow of 24 m when a 6-m pole casts a shadow of 4 m. What is the height of the tree?

Step 1 Write a proportion.
$$\frac{\text{height of tree}}{\text{height of pole}} = \frac{\text{length of tree's shadow}}{\text{length of pole's shadow}}$$

Step 2 Substitute the known values into the proportion. Let h represent the unknown value, the height of the tree.
$$\frac{h}{6} = \frac{24}{4}$$

Step 3 Find the cross products.
$h \times 4 = 6 \times 24$

Step 4 Simplify the equation.
$4h = 144$

Step 5 Divide each side by 4.
$$\frac{4h}{4} = \frac{144}{4}$$
$h = 36$

The height of the tree is 36 m.

Practice Problem The ratios of the weights of two objects on the Moon and on Earth are in proportion. A rock weighing 3 N on the Moon weighs 18 N on Earth. How much would a rock that weighs 5 N on the Moon weigh on Earth?

Use Percentages

The word *percent* means "out of one hundred." It is a ratio that compares a number to 100. Suppose you read that 77 percent of the Earth's surface is covered by water. That is the same as reading that the fraction of the Earth's surface covered by water is $\frac{77}{100}$. To express a fraction as a percent, first find the equivalent decimal for the fraction. Then, multiply the decimal by 100 and add the percent symbol.

Example Express $\frac{13}{20}$ as a percent.

Step 1 Find the equivalent decimal for the fraction.

$$\begin{array}{r} 0.65 \\ 20\overline{)13.00} \\ \underline{12\,0} \\ 1\,00 \\ \underline{1\,00} \\ 0 \end{array}$$

Step 2 Rewrite the fraction $\frac{13}{20}$ as 0.65.

Step 3 Multiply 0.65 by 100 and add the % sign.

$0.65 \times 100 = 65 = 65\%$

So, $\frac{13}{20} = 65\%$.

This also can be solved as a proportion.

Example Express $\frac{13}{20}$ as a percent.

Step 1 Write a proportion.

$\frac{13}{20} = \frac{x}{100}$

Step 2 Find the cross products.

$1300 = 20x$

Step 3 Divide each side by 20.

$\frac{1300}{20} = \frac{20x}{20}$

$65\% = x$

Practice Problem In one year, 73 of 365 days were rainy in one city. What percent of the days in that city were rainy?

Solve One-Step Equations

A statement that two things are equal is an equation. For example, $A = B$ is an equation that states that A is equal to B.

An equation is solved when a variable is replaced with a value that makes both sides of the equation equal. To make both sides equal the inverse operation is used. Addition and subtraction are inverses, and multiplication and division are inverses.

Example 1 Solve the equation $x - 10 = 35$.

Step 1 Find the solution by adding 10 to each side of the equation.

$x - 10 = 35$

$x - 10 + 10 = 35 + 10$

$x = 45$

Step 2 Check the solution.

$x - 10 = 35$

$45 - 10 = 35$

$35 = 35$

Both sides of the equation are equal, so $x = 45$.

Example 2 In the formula $a = bc$, find the value of c if $a = 20$ and $b = 2$.

Step 1 Rearrange the formula so the unknown value is by itself on one side of the equation by dividing both sides by b.

$a = bc$

$\frac{a}{b} = \frac{bc}{b}$

$\frac{a}{b} = c$

Step 2 Replace the variables a and b with the values that are given.

$\frac{a}{b} = c$

$\frac{20}{2} = c$

$10 = c$

Step 3 Check the solution.

$a = bc$

$20 = 2 \times 10$

$20 = 20$

Both sides of the equation are equal, so $c = 10$ is the solution when $a = 20$ and $b = 2$.

Practice Problem In the formula $h = gd$, find the value of d if $g = 12.3$ and $h = 17.4$.

Use Statistics

The branch of mathematics that deals with collecting, analyzing, and presenting data is statistics. In statistics, there are three common ways to summarize data with a single number—the mean, the median, and the mode.

The **mean** of a set of data is the arithmetic average. It is found by adding the numbers in the data set and dividing by the number of items in the set.

The **median** is the middle number in a set of data when the data are arranged in numerical order. If there were an even number of data points, the median would be the mean of the two middle numbers.

The **mode** of a set of data is the number or item that appears most often.

Another number that often is used to describe a set of data is the range. The **range** is the difference between the largest number and the smallest number in a set of data.

A **frequency table** shows how many times each piece of data occurs, usually in a survey. **Table 2** below shows the results of a student survey on favorite color.

Table 2 Student Color Choice

Color	Tally	Frequency
red	||||	4
blue	~~||||~~	5
black	||	2
green	|||	3
purple	~~||||~~ ||	7
yellow	~~||||~~ |	6

Based on the frequency table data, which color is the favorite?

Example The speeds (in m/s) for a race car during five different time trials are 39, 37, 44, 36, and 44.

To find the mean:

Step 1 Find the sum of the numbers.
$39 + 37 + 44 + 36 + 44 = 200$

Step 2 Divide the sum by the number of items, which is 5.
$200 \div 5 = 40$

The mean is 40 m/s.

To find the median:

Step 1 Arrange the measures from least to greatest.
36, 37, 39, 44, 44

Step 2 Determine the middle measure.
36, 37, <u>39</u>, 44, 44

The median is 39 m/s.

To find the mode:

Step 1 Group the numbers that are the same together.
44, 44, 36, 37, 39

Step 2 Determine the number that occurs most in the set.
<u>44, 44</u>, 36, 37, 39

The mode is 44 m/s.

To find the range:

Step 1 Arrange the measures from largest to smallest.
44, 44, 39, 37, 36

Step 2 Determine the largest and smallest measures in the set.
<u>44</u>, 44, 39, 37, <u>36</u>

Step 3 Find the difference between the largest and smallest measures.
$44 - 36 = 8$

The range is 8 m/s.

Practice Problem Find the mean, median, mode, and range for the data set 8, 4, 12, 8, 11, 14, 16.

Use Geometry

The branch of mathematics that deals with the measurement, properties, and relationships of points, lines, angles, surfaces, and solids is called geometry.

Perimeter The **perimeter** (P) is the distance around a geometric figure. To find the perimeter of a rectangle, add the length and width and multiply that sum by two, or $2(l + w)$. Find perimeters of irregular figures by adding the length of the sides.

Example 1 Find the perimeter of a rectangle that is 3 m long and 5 m wide.

Step 1 You know that the perimeter is 2 times the sum of the width and length.
$P = 2(3\text{ m} + 5\text{ m})$

Step 2 Find the sum of the width and length.
$P = 2(8\text{ m})$

Step 3 Multiply by 2.
$P = 16\text{ m}$

The perimeter is 16 m.

Example 2 Find the perimeter of a shape with sides measuring 2 cm, 5 cm, 6 cm, 3 cm.

Step 1 You know that the perimeter is the sum of all the sides.
$P = 2 + 5 + 6 + 3$

Step 2 Find the sum of the sides.
$P = 2 + 5 + 6 + 3$
$P = 16$

The perimeter is 16 cm.

Practice Problem Find the perimeter of a rectangle with a length of 18 m and a width of 7 m.

Practice Problem Find the perimeter of a triangle measuring 1.6 cm by 2.4 cm by 2.4 cm.

Area of a Rectangle The **area** (A) is the number of square units needed to cover a surface. To find the area of a rectangle, multiply the length times the width, or $l \times w$. When finding area, the units also are multiplied. Area is given in square units.

Example Find the area of a rectangle with a length of 1 cm and a width of 10 cm.

Step 1 You know that the area is the length multiplied by the width.
$A = (1\text{ cm} \times 10\text{ cm})$

Step 2 Multiply the length by the width. Also multiply the units.
$A = 10\text{ cm}^2$

The area is 10 cm^2.

Practice Problem Find the area of a square whose sides measure 4 m.

Area of a Triangle To find the area of a triangle, use the formula:

$$A = \frac{1}{2}(\text{base} \times \text{height})$$

The base of a triangle can be any of its sides. The height is the perpendicular distance from a base to the opposite endpoint, or vertex.

Example Find the area of a triangle with a base of 18 m and a height of 7 m.

Step 1 You know that the area is $\frac{1}{2}$ the base times the height.
$A = \frac{1}{2}(18\text{ m} \times 7\text{ m})$

Step 2 Multiply $\frac{1}{2}$ by the product of 18×7. Multiply the units.
$A = \frac{1}{2}(126\text{ m}^2)$
$A = 63\text{ m}^2$

The area is 63 m^2.

Practice Problem Find the area of a triangle with a base of 27 cm and a height of 17 cm.

Circumference of a Circle The **diameter** (d) of a circle is the distance across the circle through its center, and the **radius** (r) is the distance from the center to any point on the circle. The radius is half of the diameter. The distance around the circle is called the **circumference** (C). The formula for finding the circumference is:

$C = 2\pi r$ *or* $C = \pi d$

The circumference divided by the diameter is always equal to 3.1415926... This nonterminating and nonrepeating number is represented by the Greek letter π (pi). An approximation often used for π is 3.14.

Example 1 Find the circumference of a circle with a radius of 3 m.

Step 1 You know the formula for the circumference is 2 times the radius times π.
$C = 2\pi(3)$

Step 2 Multiply 2 times the radius.
$C = 6\pi$

Step 3 Multiply by π.
$C = 19$ m

The circumference is 19 m.

Example 2 Find the circumference of a circle with a diameter of 24.0 cm.

Step 1 You know the formula for the circumference is the diameter times π.
$C = \pi(24.0)$

Step 2 Multiply the diameter by π.
$C = 75.4$ cm

The circumference is 75.4 cm.

Practice Problem Find the circumference of a circle with a radius of 19 cm.

Area of a Circle The formula for the area of a circle is:
$A = \pi r^2$

Example 1 Find the area of a circle with a radius of 4.0 cm.

Step 1 $A = \pi(4.0)^2$

Step 2 Find the square of the radius.
$A = 16\pi$

Step 3 Multiply the square of the radius by π.
$A = 50\ cm^2$

The area of the circle is 50 cm^2.

Example 2 Find the area of a circle with a radius of 225 m.

Step 1 $A = \pi(225)^2$

Step 2 Find the square of the radius.
$A = 50625\pi$

Step 3 Multiply the square of the radius by π.
$A = 158962.5$

The area of the circle is 158,962 m^2.

Example 3 Find the area of a circle whose diameter is 20.0 mm.

Step 1 You know the formula for the area of a circle is the square of the radius times π, and that the radius is half of the diameter.
$A = \pi\left(\frac{20.0}{2}\right)^2$

Step 2 Find the radius.
$A = \pi(10.0)^2$

Step 3 Find the square of the radius.
$A = 100\pi$

Step 4 Multiply the square of the radius by π.
$A = 314\ mm^2$

The area is 314 mm^2.

Practice Problem Find the area of a circle with a radius of 16 m.

Volume The measure of space occupied by a solid is the **volume** (V). To find the volume of a rectangular solid multiply the length times width times height, or $V = l \times w \times h$. It is measured in cubic units, such as cubic centimeters (cm^3).

Example Find the volume of a rectangular solid with a length of 2.0 m, a width of 4.0 m, and a height of 3.0 m.

Step 1 You know the formula for volume is the length times the width times the height.

$V = 2.0\ m \times 4.0\ m \times 3.0\ m$

Step 2 Multiply the length times the width times the height.

$V = 24\ m^3$

The volume is 24 m^3.

Practice Problem Find the volume of a rectangular solid that is 8 m long, 4 m wide, and 4 m high.

To find the volume of other solids, multiply the area of the base times the height.

Example 1 Find the volume of a solid that has a triangular base with a length of 8.0 m and a height of 7.0 m. The height of the entire solid is 15.0 m.

Step 1 You know that the base is a triangle, and the area of a triangle is $\frac{1}{2}$ the base times the height, and the volume is the area of the base times the height.

$V = \left[\frac{1}{2}(b \times h)\right] \times 15$

Step 2 Find the area of the base.

$V = \left[\frac{1}{2}(8 \times 7)\right] \times 15$

$V = \left(\frac{1}{2} \times 56\right) \times 15$

Step 3 Multiply the area of the base by the height of the solid.

$V = 28 \times 15$

$V = 420\ m^3$

The volume is 420 m^3.

Example 2 Find the volume of a cylinder that has a base with a radius of 12.0 cm, and a height of 21.0 cm.

Step 1 You know that the base is a circle, and the area of a circle is the square of the radius times π, and the volume is the area of the base times the height.

$V = (\pi r^2) \times 21$

$V = (\pi 12^2) \times 21$

Step 2 Find the area of the base.

$V = 144\pi \times 21$

$V = 452 \times 21$

Step 3 Multiply the area of the base by the height of the solid.

$V = 9490\ cm^3$

The volume is 9490 cm^3.

Example 3 Find the volume of a cylinder that has a diameter of 15 mm and a height of 4.8 mm.

Step 1 You know that the base is a circle with an area equal to the square of the radius times π. The radius is one-half the diameter. The volume is the area of the base times the height.

$V = (\pi r^2) \times 4.8$

$V = \left[\pi\left(\frac{1}{2} \times 15\right)^2\right] \times 4.8$

$V = (\pi 7.5^2) \times 4.8$

Step 2 Find the area of the base.

$V = 56.25\pi \times 4.8$

$V = 176.63 \times 4.8$

Step 3 Multiply the area of the base by the height of the solid.

$V = 847.8$

The volume is 847.8 mm^3.

Practice Problem Find the volume of a cylinder with a diameter of 7 cm in the base and a height of 16 cm.

Science Applications

Measure in SI

The metric system of measurement was developed in 1795. A modern form of the metric system, called the International System (SI), was adopted in 1960 and provides the standard measurements that all scientists around the world can understand.

The SI system is convenient because unit sizes vary by powers of 10. Prefixes are used to name units. Look at **Table 3** for some common metric prefixes and their meanings.

Table 3 Common SI Prefixes

Prefix	Symbol	Meaning	
kilo-	k	1,000	thousand
hecto-	h	100	hundred
deka-	da	10	ten
deci-	d	0.1	tenth
centi-	c	0.01	hundredth
milli-	m	0.001	thousandth

Example How many grams equal one kilogram?

Step 1 Find the prefix *kilo* in **Table 3.**

Step 2 Using **Table 3,** determine the meaning of *kilo.* According to the table, it means 1,000. When the prefix *kilo* is added to a unit, it means that there are 1,000 of the units in a "*kilo*unit."

Step 3 Apply the prefix to the units in the question. The units in the question are grams. There are 1,000 grams in a kilogram.

Practice Problem Is a milligram larger or smaller than a gram? How many of the smaller units equal one larger unit? What fraction of the larger unit does one smaller unit represent?

Dimensional Analysis

Convert SI Units In science, quantities such as length, mass, and time sometimes are measured using different units. A process called dimensional analysis can be used to change one unit of measure to another. This process involves multiplying your starting quantity and units by one or more conversion factors. A conversion factor is a ratio equal to one and can be made from any two equal quantities with different units. If 1,000 mL equal 1 L then two ratios can be made.

$$\frac{1{,}000 \text{ mL}}{1 \text{ L}} = \frac{1 \text{ L}}{1{,}000 \text{ mL}} = 1$$

One can covert between units in the SI system by using the equivalents in **Table 3** to make conversion factors.

Example 1 How many cm are in 4 m?

Step 1 Write conversion factors for the units given. From **Table 3,** you know that 100 cm = 1 m. The conversion factors are

$$\frac{100 \text{ cm}}{1 \text{ m}} \text{ and } \frac{1 \text{ m}}{100 \text{ cm}}$$

Step 2 Decide which conversion factor to use. Select the factor that has the units you are converting from (m) in the denominator and the units you are converting to (cm) in the numerator.

$$\frac{100 \text{ cm}}{1 \text{ m}}$$

Step 3 Multiply the starting quantity and units by the conversion factor. Cancel the starting units with the units in the denominator. There are 400 cm in 4 m.

$$4 \cancel{\text{m}} \times \frac{100 \text{ cm}}{1 \cancel{\text{m}}} = 400 \text{ cm}$$

Practice Problem How many milligrams are in one kilogram? (Hint: You will need to use two conversion factors from **Table 3.**)

Table 4 Unit System Equivalents

Type of Measurement	Equivalent
Length	1 in = 2.54 cm 1 yd = 0.91 m 1 mi = 1.61 km
Mass and Weight*	1 oz = 28.35 g 1 lb = 0.45 kg 1 ton (short) = 0.91 tonnes (metric tons) 1 lb = 4.45 N
Volume	1 in^3 = 16.39 cm^3 1 qt = 0.95 L 1 gal = 3.78 L
Area	1 in^2 = 6.45 cm^2 1 yd^2 = 0.83 m^2 1 mi^2 = 2.59 km^2 1 acre = 0.40 hectares
Temperature	$°C = \frac{(°F - 32)}{1.8}$ K = °C + 273

*Weight is measured in standard Earth gravity.

Convert Between Unit Systems **Table 4** gives a list of equivalents that can be used to convert between English and SI units.

Example If a meterstick has a length of 100 cm, how long is the meterstick in inches?

Step 1 Write the conversion factors for the units given. From **Table 3,** 1 in = 2.54 cm.

$$\frac{1 \text{ in}}{2.54 \text{ cm}} \text{ and } \frac{2.54 \text{ cm}}{1 \text{ in}}$$

Step 2 Determine which conversion factor to use. You are converting from cm to in. Use the conversion factor with cm on the bottom.

$$\frac{1 \text{ in}}{2.54 \text{ cm}}$$

Step 3 Multiply the starting quantity and units by the conversion factor. Cancel the starting units with the units in the denominator. Round your answer based on the number of significant figures in the conversion factor.

$$100 \text{ cm} \times \frac{1 \text{ in}}{2.54 \text{ cm}} = 39.37 \text{ in}$$

The meterstick is 39.4 in long.

Practice Problem A book has a mass of 5 lbs. What is the mass of the book in kg?

Practice Problem Use the equivalent for in and cm (1 in = 2.54 cm) to show how 1 in^3 = 16.39 cm^3.

Precision and Significant Digits

When you make a measurement, the value you record depends on the precision of the measuring instrument. This precision is represented by the number of significant digits recorded in the measurement. When counting the number of significant digits, all digits are counted except zeros at the end of a number with no decimal point such as 2,050, and zeros at the beginning of a decimal such as 0.03020. When adding or subtracting numbers with different precision, round the answer to the smallest number of decimal places of any number in the sum or difference. When multiplying or dividing, the answer is rounded to the smallest number of significant digits of any number being multiplied or divided.

Example The lengths 5.28 and 5.2 are measured in meters. Find the sum of these lengths and record your answer using the correct number of significant digits.

Step 1 Find the sum.

5.28 m	2 digits after the decimal
+ 5.2 m	1 digit after the decimal
10.48 m	

Step 2 Round to one digit after the decimal because the least number of digits after the decimal of the numbers being added is 1.

The sum is 10.5 m.

Practice Problem How many significant digits are in the measurement 7,071,301 m? How many significant digits are in the measurement 0.003010 g?

Practice Problem Multiply 5.28 and 5.2 using the rule for multiplying and dividing. Record the answer using the correct number of significant digits.

Scientific Notation

Many times numbers used in science are very small or very large. Because these numbers are difficult to work with scientists use scientific notation. To write numbers in scientific notation, move the decimal point until only one non-zero digit remains on the left. Then count the number of places you moved the decimal point and use that number as a power of ten. For example, the average distance from the Sun to Mars is 227,800,000,000 m. In scientific notation, this distance is 2.278×10^{11} m. Because you moved the decimal point to the left, the number is a positive power of ten.

The mass of an electron is about 0.000 000 000 000 000 000 000 000 000 000 911 kg. Expressed in scientific notation, this mass is 9.11×10^{-31} kg. Because the decimal point was moved to the right, the number is a negative power of ten.

Example Earth is 149,600,000 km from the Sun. Express this in scientific notation.

Step 1 Move the decimal point until one non-zero digit remains on the left.
1.496 000 00

Step 2 Count the number of decimal places you have moved. In this case, eight.

Step 3 Show that number as a power of ten, 10^8.

The Earth is 1.496×10^8 km from the Sun.

Practice Problem How many significant digits are in 149,600,000 km? How many significant digits are in 1.496×10^8 km?

Practice Problem Parts used in a high performance car must be measured to 7×10^{-6} m. Express this number as a decimal.

Practice Problem A CD is spinning at 539 revolutions per minute. Express this number in scientific notation.

Make and Use Graphs

Data in tables can be displayed in a graph—a visual representation of data. Common graph types include line graphs, bar graphs, and circle graphs.

Line Graph A line graph shows a relationship between two variables that change continuously. The independent variable is changed and is plotted on the x-axis. The dependent variable is observed, and is plotted on the y-axis.

Example Draw a line graph of the data below from a cyclist in a long-distance race.

Table 5 Bicycle Race Data

Time (h)	Distance (km)
0	0
1	8
2	16
3	24
4	32
5	40

Step 1 Determine the x-axis and y-axis variables. Time varies independently of distance and is plotted on the x-axis. Distance is dependent on time and is plotted on the y-axis.

Step 2 Determine the scale of each axis. The x-axis data ranges from 0 to 5. The y-axis data ranges from 0 to 40.

Step 3 Using graph paper, draw and label the axes. Include units in the labels.

Step 4 Draw a point at the intersection of the time value on the x-axis and corresponding distance value on the y-axis. Connect the points and label the graph with a title, as shown in **Figure 20.**

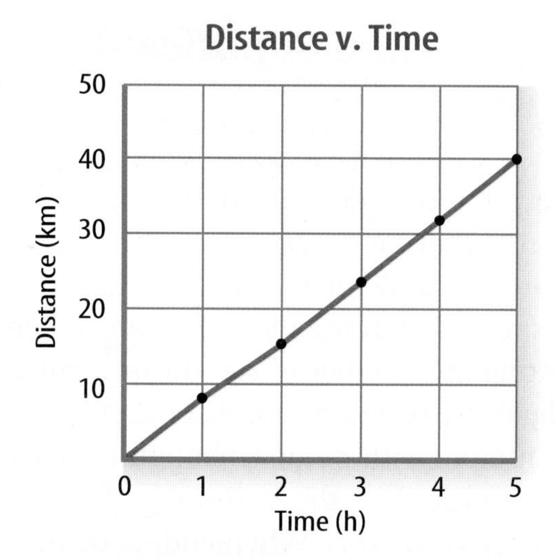

Figure 20 This line graph shows the relationship between distance and time during a bicycle ride.

Practice Problem A puppy's shoulder height is measured during the first year of her life. The following measurements were collected: (3 mo, 52 cm), (6 mo, 72 cm), (9 mo, 83 cm), (12 mo, 86 cm). Graph this data.

Find a Slope The slope of a straight line is the ratio of the vertical change, rise, to the horizontal change, run.

$$\text{Slope} = \frac{\text{vertical change (rise)}}{\text{horizontal change (run)}} = \frac{\text{change in } y}{\text{change in } x}$$

Example Find the slope of the graph in **Figure 20.**

Step 1 You know that the slope is the change in y divided by the change in x.

$$\text{Slope} = \frac{\text{change in } y}{\text{change in } x}$$

Step 2 Determine the data points you will be using. For a straight line, choose the two sets of points that are the farthest apart.

$$\text{Slope} = \frac{(40-0)\text{ km}}{(5-0)\text{ hr}}$$

Step 3 Find the change in y and x.

$$\text{Slope} = \frac{40\text{ km}}{5\text{h}}$$

Step 4 Divide the change in y by the change in x.

$$\text{Slope} = \frac{8\text{ km}}{\text{h}}$$

The slope of the graph is 8 km/h.

Bar Graph To compare data that does not change continuously you might choose a bar graph. A bar graph uses bars to show the relationships between variables. The x-axis variable is divided into parts. The parts can be numbers such as years, or a category such as a type of animal. The y-axis is a number and increases continuously along the axis.

Example A recycling center collects 4.0 kg of aluminum on Monday, 1.0 kg on Wednesday, and 2.0 kg on Friday. Create a bar graph of this data.

Step 1 Select the x-axis and y-axis variables. The measured numbers (the masses of aluminum) should be placed on the y-axis. The variable divided into parts (collection days) is placed on the x-axis.

Step 2 Create a graph grid like you would for a line graph. Include labels and units.

Step 3 For each measured number, draw a vertical bar above the x-axis value up to the y-axis value. For the first data point, draw a vertical bar above Monday up to 4.0 kg.

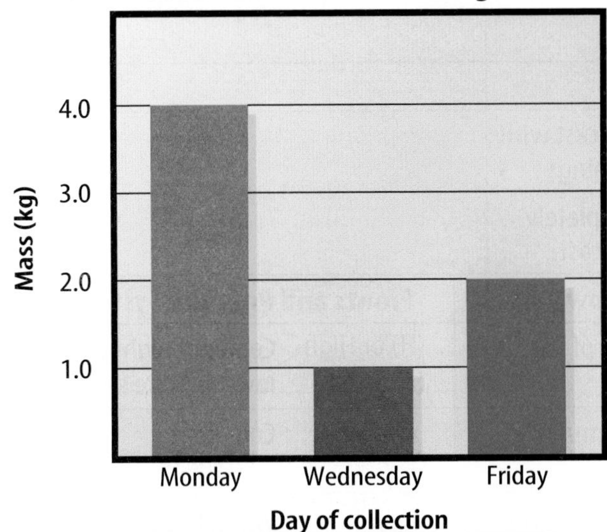

Practice Problem Draw a bar graph of the gases in air: 78% nitrogen, 21% oxygen, 1% other gases.

Circle Graph To display data as parts of a whole, you might use a circle graph. A circle graph is a circle divided into sections that represent the relative size of each piece of data. The entire circle represents 100%, half represents 50%, and so on.

Example Air is made up of 78% nitrogen, 21% oxygen, and 1% other gases. Display the composition of air in a circle graph.

Step 1 Multiply each percent by 360° and divide by 100 to find the angle of each section in the circle.

$$78\% \times \frac{360^\circ}{100} = 280.8^\circ$$

$$21\% \times \frac{360^\circ}{100} = 75.6^\circ$$

$$1\% \times \frac{360^\circ}{100} = 3.6^\circ$$

Step 2 Use a compass to draw a circle and to mark the center of the circle. Draw a straight line from the center to the edge of the circle.

Step 3 Use a protractor and the angles you calculated to divide the circle into parts. Place the center of the protractor over the center of the circle and line the base of the protractor over the straight line.

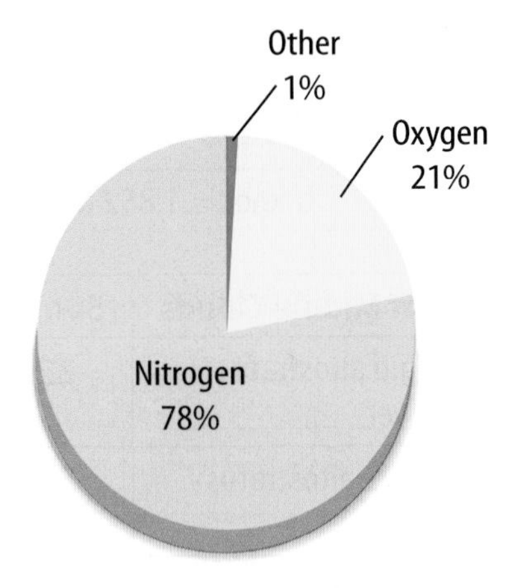

Practice Problem Draw a circle graph to represent the amount of aluminum collected during the week shown in the bar graph to the left.

Weather Map Symbols

Sample Station Model

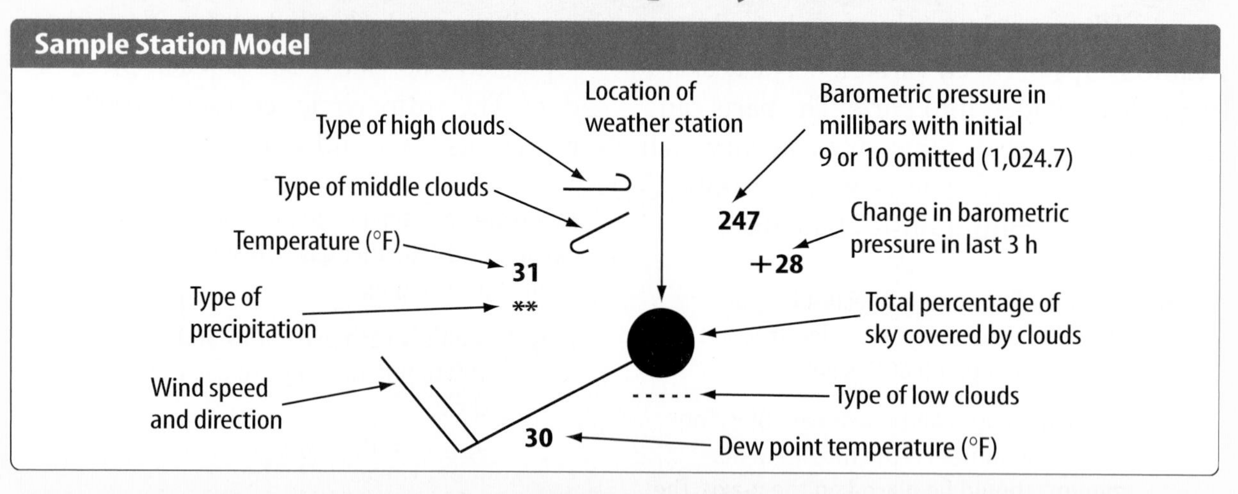

Sample Plotted Report at Each Station

Precipitation	Wind Speed and Direction	Sky Coverage	Some Types of High Clouds
≡ Fog	○ 0 calm	○ No cover	Scattered cirrus
★ Snow	1–2 knots	1/10 or less	Dense cirrus in patches
● Rain	3–7 knots	2/10 to 3/10	Veil of cirrus covering entire sky
Thunderstorm	8–12 knots	4/10	Cirrus not covering entire sky
, Drizzle	13–17 knots	–	
▽ Showers	18–22 knots	6/10	
	23–27 knots	7/10	
	48–52 knots	Overcast with openings	
	1 knot = 1.852 km/h	● Completely overcast	

Some Types of Middle Clouds	Some Types of Low Clouds	Fronts and Pressure Systems	
Thin altostratus layer	Cumulus of fair weather	(H) or High (L) or Low	Center of high- or low-pressure system
Thick altostratus layer	Stratocumulus		Cold front
Thin altostratus in patches	----- Fractocumulus of bad weather		Warm front
Thin altostratus in bands	— Stratus of fair weather		Occluded front
			Stationary front

Rocks

Rock Type	Rock Name	Characteristics
Igneous (intrusive)	Granite	Large mineral grains of quartz, feldspar, hornblende, and mica. Usually light in color.
	Diorite	Large mineral grains of feldspar, hornblende, and mica. Less quartz than granite. Intermediate in color.
	Gabbro	Large mineral grains of feldspar, augite, and olivine. No quartz. Dark in color.
Igneous (extrusive)	Rhyolite	Small mineral grains of quartz, feldspar, hornblende, and mica, or no visible grains. Light in color.
	Andesite	Small mineral grains of feldspar, hornblende, and mica or no visible grains. Intermediate in color.
	Basalt	Small mineral grains of feldspar, augite, and possibly olivine or no visible grains. No quartz. Dark in color.
	Obsidian	Glassy texture. No visible grains. Volcanic glass. Fracture looks like broken glass.
	Pumice	Frothy texture. Floats in water. Usually light in color.
Sedimentary (detrital)	Conglomerate	Coarse grained. Gravel or pebble-size grains.
	Sandstone	Sand-sized grains 1/16 to 2 mm.
	Siltstone	Grains are smaller than sand but larger than clay.
	Shale	Smallest grains. Often dark in color. Usually platy.
Sedimentary (chemical or organic)	Limestone	Major mineral is calcite. Usually forms in oceans and lakes. Often contains fossils.
	Coal	Forms in swampy areas. Compacted layers of organic material, mainly plant remains.
Sedimentary (chemical)	Rock Salt	Commonly forms by the evaporation of seawater.
Metamorphic (foliated)	Gneiss	Banding due to alternate layers of different minerals, of different colors. Parent rock often is granite.
	Schist	Parallel arrangement of sheetlike minerals, mainly micas. Forms from different parent rocks.
	Phyllite	Shiny or silky appearance. May look wrinkled. Common parent rocks are shale and slate.
	Slate	Harder, denser, and shinier than shale. Common parent rock is shale.
Metamorphic (nonfoliated)	Marble	Calcite or dolomite. Common parent rock is limestone.
	Soapstone	Mainly of talc. Soft with greasy feel.
	Quartzite	Hard with interlocking quartz crystals. Common parent rock is sandstone.

Minerals

Minerals

Mineral (formula)	Color	Streak	Hardness	Breakage Pattern	Uses and Other Properties
Graphite (C)	black to gray	black to gray	1–1.5	basal cleavage (scales)	pencil lead, lubricants for locks, rods to control some small nuclear reactions, battery poles
Galena (PbS)	gray	gray to black	2.5	cubic cleavage perfect	source of lead, used for pipes, shields for X rays, fishing equipment sinkers
Hematite (Fe_2O_3)	black or reddish-brown	reddish-brown	5.5–6.5	irregular fracture	source of iron; converted to pig iron, made into steel
Magnetite (Fe_3O_4)	black	black	6	conchoidal fracture	source of iron, attracts a magnet
Pyrite (FeS_2)	light, brassy, yellow	greenish-black	6–6.5	uneven fracture	fool's gold
Talc ($Mg_3Si_4O_{10}(OH)_2$)	white, greenish	white	1	cleavage in one direction	used for talcum powder, sculptures, paper, and tabletops
Gypsum ($CaSO_4 \cdot 2H_2O$)	colorless, gray, white, brown	white	2	basal cleavage	used in plaster of paris and dry wall for building construction
Sphalerite (ZnS)	brown, reddish-brown, greenish	light to dark brown	3.5–4	cleavage in six directions	main ore of zinc; used in paints, dyes, and medicine
Muscovite ($KAl_3Si_3O_{10}(OH)_2$)	white, light gray, yellow, rose, green	colorless	2–2.5	basal cleavage	occurs in large, flexible plates; used as an insulator in electrical equipment, lubricant
Biotite ($K(Mg,Fe)_3(AlSi_3O_{10})(OH)_2$)	black to dark brown	colorless	2.5–3	basal cleavage	occurs in large, flexible plates
Halite ($NaCl$)	colorless, red, white, blue	colorless	2.5	cubic cleavage	salt; soluble in water; a preservative

Minerals

Minerals					
Mineral (formula)	Color	Streak	Hardness	Breakage Pattern	Uses and Other Properties
Calcite ($CaCO_3$)	colorless, white, pale blue	colorless, white	3	cleavage in three directions	fizzes when HCl is added; used in cements and other building materials
Dolomite ($CaMg(CO_3)_2$)	colorless, white, pink, green, gray, black	white	3.5–4	cleavage in three directions	concrete and cement; used as an ornamental building stone
Fluorite (CaF_2)	colorless, white, blue, green, red, yellow, purple	colorless	4	cleavage in four directions	used in the manufacture of optical equipment; glows under ultraviolet light
Hornblende ($(CaNa)_{2-3}(Mg,Al,Fe)_5$-$(Al,Si)_2Si_6O_{22}(OH)_2$)	green to black	gray to white	5–6	cleavage in two directions	will transmit light on thin edges; 6-sided cross section
Feldspar ($(KAlSi_3O_8)$ $(NaAlSi_3O_8)$, $(CaAl_2Si_2O_8)$)	colorless, white to gray, green	colorless	6	two cleavage planes meet at 90° angle	used in the manufacture of ceramics
Augite ($((Ca,Na)(Mg,Fe,Al)(Al,Si)_2O_6)$)	black	colorless	6	cleavage in two directions	square or 8-sided cross section
Olivine ($((Mg,Fe)_2SiO_4)$)	olive, green	none	6.5–7	conchoidal fracture	gemstones, refractory sand
Quartz ((SiO_2))	colorless, various colors	none	7	conchoidal fracture	used in glass manufacture, electronic equipment, radios, computers, watches, gemstones

PERIODIC TABLE OF THE ELEMENTS

Columns of elements are called groups. Elements in the same group have similar chemical properties.

Element — Hydrogen
Atomic number — 1
Symbol — H
Atomic mass — 1.008
State of matter

Gas
Liquid
Solid
Synthetic

The first three symbols tell you the state of matter of the element at room temperature. The fourth symbol identifies elements that are not present in significant amounts on Earth. Useful amounts are made synthetically.

	1	2	3	4	5	6	7	8	9
1	Hydrogen 1 H 1.008								
2	Lithium 3 Li 6.941	Beryllium 4 Be 9.012							
3	Sodium 11 Na 22.990	Magnesium 12 Mg 24.305							
4	Potassium 19 K 39.098	Calcium 20 Ca 40.078	Scandium 21 Sc 44.956	Titanium 22 Ti 47.867	Vanadium 23 V 50.942	Chromium 24 Cr 51.996	Manganese 25 Mn 54.938	Iron 26 Fe 55.845	Cobalt 27 Co 58.933
5	Rubidium 37 Rb 85.468	Strontium 38 Sr 87.62	Yttrium 39 Y 88.906	Zirconium 40 Zr 91.224	Niobium 41 Nb 92.906	Molybdenum 42 Mo 95.94	Technetium 43 Tc (98)	Ruthenium 44 Ru 101.07	Rhodium 45 Rh 102.906
6	Cesium 55 Cs 132.905	Barium 56 Ba 137.327	Lanthanum 57 La 138.906	Hafnium 72 Hf 178.49	Tantalum 73 Ta 180.948	Tungsten 74 W 183.84	Rhenium 75 Re 186.207	Osmium 76 Os 190.23	Iridium 77 Ir 192.217
7	Francium 87 Fr (223)	Radium 88 Ra (226)	Actinium 89 Ac (227)	Rutherfordium 104 Rf (261)	Dubnium 105 Db (262)	Seaborgium 106 Sg (266)	Bohrium 107 Bh (264)	Hassium 108 Hs (277)	Meitnerium 109 Mt (268)

The number in parentheses is the mass number of the longest-lived isotope for that element.

Rows of elements are called periods. Atomic number increases across a period.

The arrow shows where these elements would fit into the periodic table. They are moved to the bottom of the table to save space.

Lanthanide series	Cerium 58 Ce 140.116	Praseodymium 59 Pr 140.908	Neodymium 60 Nd 144.24	Promethium 61 Pm (145)	Samarium 62 Sm 150.36
Actinide series	Thorium 90 Th 232.038	Protactinium 91 Pa 231.036	Uranium 92 U 238.029	Neptunium 93 Np (237)	Plutonium 94 Pu (244)

Metal
Metalloid
Nonmetal

The color of an element's block tells you if the element is a metal, nonmetal, or metalloid.

Science Online
Visit earth.msscience.com for updates to the periodic table.

10	11	12	13	14	15	16	17	18
								Helium 2 He 4.003
			Boron 5 B 10.811	Carbon 6 C 12.011	Nitrogen 7 N 14.007	Oxygen 8 O 15.999	Fluorine 9 F 18.998	Neon 10 Ne 20.180
			Aluminum 13 Al 26.982	Silicon 14 Si 28.086	Phosphorus 15 P 30.974	Sulfur 16 S 32.065	Chlorine 17 Cl 35.453	Argon 18 Ar 39.948
Nickel 28 Ni 58.693	Copper 29 Cu 63.546	Zinc 30 Zn 65.409	Gallium 31 Ga 69.723	Germanium 32 Ge 72.64	Arsenic 33 As 74.922	Selenium 34 Se 78.96	Bromine 35 Br 79.904	Krypton 36 Kr 83.798
Palladium 46 Pd 106.42	Silver 47 Ag 107.868	Cadmium 48 Cd 112.411	Indium 49 In 114.818	Tin 50 Sn 118.710	Antimony 51 Sb 121.760	Tellurium 52 Te 127.60	Iodine 53 I 126.904	Xenon 54 Xe 131.293
Platinum 78 Pt 195.078	Gold 79 Au 196.967	Mercury 80 Hg 200.59	Thallium 81 Tl 204.383	Lead 82 Pb 207.2	Bismuth 83 Bi 208.980	Polonium 84 Po (209)	Astatine 85 At (210)	Radon 86 Rn (222)
Darmstadtium 110 Ds (281)	Roentgenium 111 Rg (272)	Ununbium * 112 Uub (285)		Ununquadium * 114 Uuq (289)				

* The names and symbols for elements 112 and 114 are temporary. Final names will be selected when the elements' discoveries are verified.

Europium 63 Eu 151.964	Gadolinium 64 Gd 157.25	Terbium 65 Tb 158.925	Dysprosium 66 Dy 162.500	Holmium 67 Ho 164.930	Erbium 68 Er 167.259	Thulium 69 Tm 168.934	Ytterbium 70 Yb 173.04	Lutetium 71 Lu 174.967
Americium 95 Am (243)	Curium 96 Cm (247)	Berkelium 97 Bk (247)	Californium 98 Cf (251)	Einsteinium 99 Es (252)	Fermium 100 Fm (257)	Mendelevium 101 Md (258)	Nobelium 102 No (259)	Lawrencium 103 Lr (262)

Topographic Map Symbols

Topographic Map Symbols	
Primary highway, hard surface	Index contour
Secondary highway, hard surface	Supplementary contour
Light-duty road, hard or improved surface	Intermediate contour
Unimproved road	Depression contours
Railroad: single track	
Railroad: multiple track	Boundaries: national
Railroads in juxtaposition	State
	County, parish, municipal
Buildings	Civil township, precinct, town, barrio
cem Schools, church, and cemetery	Incorporated city, village, town, hamlet
Buildings (barn, warehouse, etc.)	Reservation, national or state
Wells other than water (labeled as to type)	Small park, cemetery, airport, etc.
Tanks: oil, water, etc. (labeled only if water)	Land grant
Located or landmark object; windmill	Township or range line, U.S. land survey
Open pit, mine, or quarry; prospect	
	Township or range line, approximate location
Marsh (swamp)	
Wooded marsh	Perennial streams
Woods or brushwood	Elevated aqueduct
Vineyard	Water well and spring
Land subject to controlled inundation	Small rapids
Submerged marsh	Large rapids
Mangrove	Intermittent lake
Orchard	Intermittent stream
Scrub	Aqueduct tunnel
Urban area	Glacier
	Small falls
x7369 Spot elevation	Large falls
670 Water elevation	Dry lake bed

Cómo usar el glosario en español:
1. Busca el término en inglés que desees encontrar.
2. El término en español, junto con la definición, se encuentran en la columna de la derecha.

Pronunciation Key

Use the following key to help you sound out words in the glossary.

a .... back (BAK)
ay .... day (DAY)
ah .... father (FAH thur)
ow .... flower (FLOW ur)
ar .... car (CAR)
e .... less (LES)
ee .... leaf (LEEF)
ih .... trip (TRIHP)
i (i + con + e) .. idea (i DEE uh)
oh .... go (GOH)
aw .... soft (SAWFT)
or .... orbit (OR buht)
oy .... coin (COYN)
oo .... foot (FOOT)
ew .... food (FEWD)
yoo .... pure (PYOOR)
yew .... few (FYEW)
uh .... comma (CAH muh)
u (+ con) .... rub (RUB)
sh .... shelf (SHELF)
ch .... nature (NAY chur)
g .... gift (GIHFT)
j .... gem (JEM)
ing .... sing (SING)
zh .... vision (VIH zhun)
k .... cake (KAYK)
s .... seed, cent (SEED, SENT)
z .... zone, raise (ZOHN, RAYZ)

English — A — Español

abrasion: a type of erosion that occurs when wind-blown sediments strike rocks and sediments, polishing and pitting their surface. (p. 222)

abrasión: tipo de erosión que ocurre cuando los sedimentos arrastrados por el viento golpean las rocas y los sedimentos, puliendo y llenando de hoyos su superficie. (p. 222)

absolute age: age, in years, of a rock or other object; can be determined by using properties of the atoms that make up materials. (p. 377)

edad absoluta: edad, en años, de una roca u otro objeto; puede determinarse utilizando las propiedades de los átomos de los materiales. (p. 377)

absolute magnitude: measure of the amount of light a star actually gives off. (p. 726)

magnitud absoluta: medida de la cantidad real de luz que genera una estrella. (p. 726)

abyssal (uh BIH sul) plain: flat seafloor area from 4,000 m to 6,000 m below the ocean surface, formed by the deposition of sediments. (p. 543)

planicie abisal: área plana del suelo marino entre 4000 y 6000 metros por debajo de la superficie del océano, formada por deposición de sedimentos. (p. 543)

acid: substance with a pH lower than 7. (p. 611)

ácido: sustancia con un pH menor de 7. (p. 611)

acid rain: acidic moisture, with a pH below 5.6, that falls to Earth as rain or snow and can damage forests, harm organisms, and corrode structures. (p. 611)

lluvia ácida: humedad ácida con un pH menor de 5.6 y que cae a la Tierra en forma de lluvia o nieve; puede dañar bosques y organismos o corroer estructuras. (p. 611)

adaptation: any structural or behavioral change that helps an organism survive in its particular environment. (p. 488)

adaptación: cualquier cambio de estructura o comportamiento que ayude a un organismo a sobrevivir en su medio ambiente particular. (p. 488)

air mass: large body of air that has the same characteristics of temperature and moisture content as the part of Earth's surface over which it formed. (p. 462)

masa de aire: gran cuerpo de aire que tiene las mismas características de temperatura y contenido de humedad que la parte de la superficie terrestre sobre la cual se formó. (p. 462)

apparent magnitude: measure of the amount of light from a star that is received on Earth. (p. 726)

magnitud aparente: medida de la cantidad de luz recibida en la Tierra desde una estrella. (p. 726)

aquifer (AK wuh fur): layer of permeable rock that allows water to flow through. (p. 250)

asteroid: a piece of rock or metal made up of material similar to that which formed the planets; mostly found in the asteroid belt between the orbits of Mars and Jupiter. (p. 712)

asthenosphere (as THE nuh sfihr): plasticlike layer of Earth on which the lithospheric plates float and move around. (p. 280)

atmosphere: Earth's air, which is made up of a thin layer of gases, solids, and liquids; forms a protective layer around the planet and is divided into five distinct layers. (p. 426)

atom: tiny building block of matter, made up of protons, neutrons, and electrons. (p. 34)

atomic number: the number of protons in an atom. (p. 37)

axis: imaginary vertical line that cuts through the center of Earth and around which Earth spins. (p. 661)

acuífero: capa de roca permeable que permite que el agua fluya a través de ella. (p. 250)

asteroide: pedazo de roca o metal formado de material similar al que forma los planetas; se encuentran principalmente en el cinturón de asteroides entre las órbitas de Marte y Júpiter. (p. 712)

astenosfera: capa flexible de la Tierra en la que las placas litosféricas flotan y se mueven de un lugar a otro. (p. 280)

atmósfera: el aire de la Tierra; está compuesta por una capa fina de gases, sólidos y líquidos, forma una capa protectora alrededor del planeta y está dividida en cinco capas distintas. (p. 426)

átomo: bloque diminuto de construcción de la materia, formado por protones, neutrones y electrones. (p. 34)

número atómico: el número de protones en un átomo. (p. 37)

eje: línea vertical imaginaria que atraviesa el centro de la Tierra y alrededor de la cual gira ésta. (p. 661)

B

basaltic: describes dense, dark-colored igneous rock formed from magma rich in magnesium and iron and poor in silica. (p. 97)

base: substance with a pH above 7. (p. 611)

basin: low area on Earth in which an ocean formed when the area filled with water from torrential rains. (p. 515)

batholith: largest intrusive igneous rock body that forms when magma being forced upward toward Earth's crust cools slowly and solidifies underground. (p. 346)

beach: deposit of sediment whose materials vary in size, color, and composition and is most commonly found on a smooth, gently sloped shoreline. (p. 257)

benthos: marine plants and animals that live on or in the ocean floor. (p. 553)

bias: personal opinion. (p. 21)

big bang theory: states that about 13.7 billion years ago, the universe began with a huge, fiery explosion. (p. 745)

biomass energy: renewable energy derived from burning organic materials such as wood and alcohol. (p. 133)

black hole: final stage in the evolution of a very massive star, where the core's mass collapses to a point that it's gravity is so strong that not even light can escape. (p. 738)

basáltica: describe roca ígnea densa de color oscuro que se forma a partir de magma rico en magnesio y hierro pero pobre en sílice. (p. 97)

base: sustancia con un pH mayor de 7. (p. 611)

depresión: área baja de la Tierra en la que se forma un océano cuando el área es llenada con agua proveniente de lluvias torrenciales. (p. 515)

batolito: gran cuerpo rocoso ígneo intrusivo que se forma cuando el magma es forzado a salir a la superficie de la corteza terrestre, se enfría lentamente y se solidifica en el subsuelo. (p. 346)

playa: depósito de sedimentos cuyos materiales varían en tamaño, color y composición y que comúnmente se encuentran en las líneas costeras planas y poco inclinadas. (p. 257)

bentos: plantas y animales marinos que subsisten o viven en el suelo del océano. (p. 553)

sesgo: opinión personal. (p. 21)

teoría de la Gran Explosión: establece que hace aproximadamente 13.7 billones de años el universo se originó con una enorme explosión. (p. 745)

energía de biomasa: energía renovable derivada de la combustión de materiales orgánicos tales como la madera y el alcohol. (p. 133)

agujero negro: etapa final en la evolución de una estrella masiva, en donde la masa del núcleo se colapsa hasta el punto de que su gravedad es tan fuerte que ni siquiera la luz puede escapar. (p. 738)

blizzard: winter storm that lasts at least three hours with temperatures of −12°C or below, poor visibility, and winds of at least 51 km/h. (p. 469)

breaker: collapsing ocean wave that forms in shallow water and breaks onto the shore. (p. 525)

nevasca: tormenta invernal que dura por lo menos tres horas con temperaturas de −12°C o menores, escasa visibilidad y vientos de por lo menos 51 km/h. (p. 469)

rompiente: ola oceánica colapsante que se forma en aguas poco profundas y rompe en la orilla. (p. 525)

C

caldera: large, circular-shaped opening formed when the top of a volcano collapses. (p. 348)

carbon film: thin film of carbon residue preserved as a fossil. (p. 364)

carbon monoxide: colorless, odorless gas that reduces the oxygen content in the blood, is found in car exhaust, and contributes to air pollution. (p. 612)

carrying capacity: maximum number of individuals of a given species that the environment will support. (p. 575)

cast: a type of body fossil that forms when crystals fill a mold or sediments wash into a mold and harden into rock. (p. 365)

cave: underground opening that can form when acidic groundwater dissolves limestone. (p. 253)

cementation: sedimentary rock-forming process in which sediment grains are held together by natural cements that are produced when water moves through rock and soil. (p. 105)

Cenozoic (sen uh ZOH ihk) Era: era of recent life that began about 66 million years ago and continues today; includes the first appearance of *Homo sapiens* about 400,000 years ago. (p. 412)

channel: groove created by water moving down the same path. (p. 240)

chemical weathering: occurs when chemical reactions dissolve the minerals in rocks or change them into different minerals. (p. 185)

chemosynthesis (kee moh SIHN thuh sihs): food-making process using sulfur or nitrogen compounds, rather than light energy from the Sun, that is used by bacteria living near hydrothermal vents. (p. 551)

chlorofluorocarbons (CFCs): group of chemical compounds used in refrigerators, air conditioners, foam packaging, and aerosol sprays that may enter the atmosphere and destroy ozone. (p. 432)

caldera: apertura grande circular que se crea cuando la cima de un volcán se colapsa. (p. 348)

película de carbono: capa delgada de residuos de carbono preservada como un fósil. (p. 364)

monóxido de carbono: gas inodoro e incoloro que reduce el contenido de oxígeno en la sangre; se emite a través del escape de los automóviles y contribuye a la contaminación del aire. (p. 612)

capacidad de carga: máximo número de individuos de una especie determinada que es capaz de albergar el medio ambiente. (p. 575)

vaciado: tipo de cuerpo fósil que se forma cuando los cristales llenan un molde o los sedimentos son lavados hacia un molde y se endurecen convirtiéndose en roca. (p. 365)

cueva: apertura subterránea que puede formarse cuando el agua subterránea acidificada disuelve la piedra caliza. (p. 253)

cementación: proceso de formación de la roca sedimentaria en el que las partículas de sedimento están unidas por cementos naturales producidos cuando el agua se mueve a través de la roca y el suelo. (p. 105)

Era Cenozoica: era de vida reciente que comenzó hace aproximadamente 66 millones de años y continúa hasta hoy; incluye la aparición del Homo sapiens cerca de 400,000 años atrás. (p. 412)

canal: surco creado por el agua cuando se mueve cuesta abajo por el mismo curso. (p. 240)

erosión química: ocurre cuando las reacciones químicas disuelven los minerales en las rocas o los convierten en diferentes minerales. (p. 185)

quimiosíntesis: proceso de producción de alimentos que utiliza compuestos de azufre o nitrógeno en lugar de energía solar; este proceso es utilizado por las bacterias que viven cerca de los conductos hidrotérmicos. (p. 551)

clorofluorocarbonos (CFCs): grupo de compuestos químicos usados en refrigeradores, acondicionadores de aire, espumas de empaque y aerosoles; pueden entrar en la atmósfera y destruir el ozono. (p. 432)

chromosphere: layer of the Sun's atmosphere above the photosphere. (p. 729)

cinder cone volcano: steep-sided, loosely packed volcano formed when tephra falls to the ground. (p. 340)

cleavage: physical property of some minerals that causes them to break along smooth, flat surfaces. (p. 71)

climate: average weather pattern in an area over a long period of time; can be classified by temperature, humidity, precipitation, and vegetation. (pp. 186, 484)

coal: sedimentary rock formed from decayed plant material; the world's most abundant fossil fuel. (p. 121)

comet: space object made of dust and rock particles mixed with frozen water, methane, and ammonia that forms a bright coma as it approaches the Sun. (p. 710)

compaction: process that forms sedimentary rocks when layers of sediments are compressed by the weight of the layers above them. (p. 104)

composite volcano: volcano built by alternating explosive and quiet eruptions that produce layers of tephra and lava; found mostly where Earth's plates come together and one plate sinks below the other. (p. 341)

composting: conservation method in which yard wastes such as cut grass, pulled weeds, and raked leaves are piled and left to decompose gradually. (p. 587)

compound: atoms of more than one type of element that are chemically bonded together. (p. 40)

condensation: process in which water vapor changes to a liquid. (p. 437)

conduction: transfer of energy that occurs when molecules bump into each other. (p. 436)

conic projection: map made by projecting points and lines from a globe onto a cone. (p. 165)

conservation: careful use of resources to reduce damage to the environment though such methods as composting and recycling materials. (p. 586)

constant: variable that does not change in an experiment. (p. 10)

constellation: group of stars that forms a pattern in the sky that looks like a familiar object (Libra), animal (Pegasus), or character (Orion). (p. 724)

cromosfera: capa de la atmósfera del sol que se encuentra sobre la fotosfera. (p. 729)

volcán de cono de ceniza: volcán de laderas inclinadas, poco compactado, que se forma cuando la tefra cae al suelo. (p. 340)

exfoliación: propiedad física de algunos minerales que causa que se rompan junto a superficies planas y lisas. (p. 71)

clima: modelo meteorológico en un área durante un periodo de tiempo largo; puede clasificarse por temperatura, humedad, precipitación y vegetación. (pp. 186, 484)

carbón mineral: roca sedimentaria formada a partir de material vegetal descompuesto; es el combustible fósil más abundante en el mundo. (p. 121)

cometa: objeto espacial formado por partículas de polvo y roca mezcladas con agua congelada, metano y amoníaco que forman una cola brillante cuando se aproxima al sol. (p. 710)

compactación: proceso que forma rocas sedimentarias cuando las capas de sedimento son comprimidas por el peso de las capas superiores. (p. 104)

volcán compuesto: volcán formado por explosiones alternantes y erupciones de baja intensidad que producen capas de tefra y lava; se encuentran principalmente donde se unen las placas continentales y una se sumerge bajo la otra. (p. 341)

compostaje: método de conservación en el que los desechos de jardinería tales como el césped cortado, las malezas arrancadas y las hojas barridas se apilan y dejan para que se descompongan gradualmente. (p. 587)

compuesto: átomos de más de un tipo de elemento que están químicamente unidos. (p. 40)

condensación: proceso mediante el cual el vapor de agua cambia a su forma líquida. (p. 437)

conducción: transferencia de energía que ocurre cuando las moléculas chocan unas con otras. (p. 436)

proyección cónica: mapa hecho por la proyección de puntos y líneas desde un globo a un cono. (p. 165)

conservación: uso consciente de los recursos naturales para reducir el daño ambiental a través de métodos como el compostaje y el reciclaje de materiales. (p. 586)

constante: variable que no cambia en un experimento. (p. 10)

constelación: grupo de estrellas que forma un patrón en el cielo y que semeja un objeto (Libra), un animal (Pegaso) o un personaje familiar (Orión). (p. 724)

continental drift: Wegener's hypothesis that all continents were once connected in a single large landmass that broke apart about 200 million years ago and drifted slowly to their current positions. (p. 272)

continental shelf: gradually sloping end of a continent that extends beneath the ocean. (p. 542)

continental slope: ocean basin feature that dips steeply down from the continental shelf. (p. 543)

contour farming: planting along the natural contours of the land to reduce soil erosion. (p. 199)

contour line: line on a map that connects points of equal elevation. (p. 166)

control: standard for comparison in an experiment. (p. 10)

convection: transfer of heat by the flow of material. (p. 436)

convection current: current in Earth's mantle that transfers heat in Earth's interior and is the driving force for plate tectonics. (p. 285)

Coriolis (kor ee OH lus) effect: causes moving air and water to turn left in the southern hemisphere and turn right in the northern hemisphere due to Earth's rotation. (pp. 440, 519)

corona: outermost, largest layer of the Sun's atmosphere; extends millions of kilometers into space and has temperatures up to 2 million K. (p. 729)

crater: steep-walled depression around a volcano's vent. (p. 332)

creep: a type of mass movement in which sediments move down-slope very slowly; is common in areas of freezing and thawing, and can cause walls, trees, and fences to lean downhill. (p. 212)

crest: highest point of a wave. (p. 524)

crystal: solid in which the atoms are arranged in an orderly, repeating pattern. (p. 63)

cyanobacteria: chlorophyll-containing, photosynthetic bacteria thought to be one of Earth's earliest life-forms. (p. 401)

deriva continental: hipótesis de Wegener respecto a que todos los continentes estuvieron alguna vez conectados en una gran masa terrestre única que se fraccionó cerca de 200 millones de años atrás y sus trozos se han movilizado lentamente a la deriva hasta sus posiciones actuales. (p. 272)

plataforma continental: extremo gradualmente inclinado de un continente que se extiende por debajo del océano. (p. 542)

talud continental: depresión oceánica característica que se sumerge abruptamente desde la plataforma continental. (p. 543)

cultivo de contorno: plantación a lo largo de los contornos naturales de la tierra para reducir la erosión de los suelos. (p. 199)

curva de nivel: línea en un mapa que conecta puntos de la misma elevación. (p. 166)

control: modelo de comparación en un experimento. (p. 10)

convección: transferencia de calor mediante flujo de material. (p. 436)

corriente de convección: corriente en el manto de la Tierra que transfiere calor en el interior de la Tierra y es la causa de la tectónica de placas. (p. 285)

efecto de Coriolis: causa el movimiento del aire y el agua hacia la izquierda en el hemisferio sur y hacia la derecha en el hemisferio norte, debido a la rotación de la Tierra. (pp. 440, 519)

corona: capa más externa y más grande de la atmósfera solar; se extiende millones de kilómetros dentro del espacio y tiene una temperatura hasta de 2 millones de grados Kelvin. (p. 729)

cráter: depresión con paredes pronunciadas alrededor de la apertura volcánica. (p. 332)

reptación: tipo de movimiento en masa en el que los sedimentos se mueven hacia abajo muy lentamente; es común en áreas sujetas a congelación y descongelación y puede causar que los muros, los árboles y los cercos se inclinen hacia abajo. (p. 212)

cresta: el punto más alto de una ola. (p. 524)

cristal: sólido en el que los átomos están alineados en forma ordenada y repetitiva. (p. 63)

cianobacteria: bacteria fotosintética que contiene clorofila; se cree que es una de las primeras formas de vida que surgió en la tierra. (p. 401)

D

deflation: a type of erosion that occurs when wind blows over loose sediments, removes small particles, and leaves coarser sediments behind. (p. 222)

deflación: tipo de erosión que ocurre cuando el viento sopla sobre los sedimentos sueltos, retira partículas pequeñas y deja los sedimentos grandes. (p. 222)

deforestation: destruction and cutting down of forests—often to clear land for mining, roads, and grazing of cattle—resulting in increased atmospheric CO_2 levels. (p. 501)

deforestación: destrucción y tala de los bosques—a menudo el despeje de la tierra para minería, carreteras y ganadería—resultando en el aumento de los niveles atmosféricos de dióxido de carbono. (p. 501)

density: measurement of the mass of an object divided by its volume. (p. 46)

densidad: medida de la masa de un objeto dividida entre su volumen. (p. 46)

density current: circulation pattern in the ocean that forms when a mass of more dense seawater sinks beneath less dense seawater. (p. 521)

corriente de densidad: patrón de circulación en el océano que se forma cuando una masa de agua marina más densa se hunde por debajo del agua marina menos densa. (p. 521)

dependent variable: factor being measured in an experiment. (p. 10)

variable dependiente: factor que se mide en un experimento. (p. 10)

deposition: dropping of sediments that occurs when an agent of erosion, such as gravity, a glacier, wind, or water, loses its energy and can no longer carry its load. (p. 211)

deposición: caída de sedimentos que ocurre cuando un agente erosivo como la gravedad, un glaciar, el viento o el agua, pierde su energía y ya no puede continuar con su carga. (p. 211)

dew point: temperature at which air is saturated and condensation forms. (p. 457)

punto de condensación: temperatura a la que el aire se satura y se genera la condensación. (p. 457)

dike: igneous rock feature formed when magma is squeezed into a vertical crack that cuts across rock layers and hardens underground. (p. 347)

dique: característica de la roca ígnea formada cuando el magma es comprimido en una grieta vertical que cruza capas rocosas y se endurece en el subsuelo. (p. 347)

drainage basin: land area from which a river or stream collects runoff. (p. 242)

cuenca de drenaje: terreno del que un río o arroyo recolecta sus aguas. (p. 242)

dune (DOON): mound formed when windblown sediments pile up behind an obstacle; common landform in desert areas. (p. 225)

duna: amontonamiento de tierra formado cuando los sedimentos arrastrados por el aire se apilan detrás de un obstáculo; forma de terreno común en las áreas desérticas. (p. 225)

E

Earth: third planet from the Sun; has an atmosphere that protects life and surface temperatures that allow water to exist as a solid, liquid, and gas. (p. 698)

Tierra: tercer planeta más cercano al sol; tiene una atmósfera que protege la vida y temperaturas en su superficie que permiten la presencia de agua en estado sólido, líquido y gaseoso. (p. 698)

earthquake: vibrations produced when rocks break along a fault. (p. 301)

terremoto: vibraciones producidas cuando las rocas se rompen a lo largo de una falla. (p. 301)

Earth science: study of Earth and space, including rocks, fossils, climate, volcanoes, land use, ocean water, earthquakes, and objects in space. (p. 9)

ciencia de la Tierra: estudio de la Tierra y el espacio, incluyendo rocas, fósiles, clima, volcanes, uso del suelo, aguas oceánicas, terremotos y objetos en el espacio. (p. 9)

electromagnetic spectrum: arrangement of electromagnetic waves according to their wavelengths. (p. 629)

electron: particle with a negative charge. (p. 36)

element: substance that is made of only one type of atom. (p. 35)

ellipse (ee LIHPS): elongated, closed curve that describes Earth's yearlong orbit around the Sun. (p. 663)

El Niño (el NEEN yoh): climatic event that begins in the tropical Pacific Ocean; may occur when trade winds weaken or reverse, and can disrupt normal temperature and precipitation patterns around the world. (p. 493)

enzyme: substance that causes chemical reactions to happen more quickly. (p. 583)

eon: longest subdivision in the geologic time scale that is based on the abundance of certain types of fossils and is subdivided into eras, periods, and epochs. (p. 393)

epicenter (EP ih sen tur): point on Earth's surface directly above an earthquake's focus. (p. 305)

epoch: next-smaller division of geologic time after the period; is characterized by differences in life-forms that may vary regionally. (p. 393)

equator: imaginary line that wraps around Earth at 0° latitude, halfway between the north and south poles. (p. 160)

equinox (EE kwuh nahks): twice-yearly time—each spring and fall—when the Sun is directly over the equator and the number of daylight and nighttime hours are equal worldwide. (p. 665)

era: second-longest division of geologic time; is subdivided into periods and is based on major worldwide changes in types of fossils. (p. 393)

erosion: process in which surface materials are worn away and transported from one place to another by agents such as gravity, water, wind, and glaciers. (p. 210)

estuary: area where a river meets the ocean that contains a mixture of freshwater and ocean water and provides an important habitat to many marine organisms. (p. 556)

ethics: study of moral values about what is good or bad. (p. 20)

extrusive: describes fine-grained igneous rock that forms when magma cools quickly at or near Earth's surface. (p. 95)

espectro electromagnético: ordenamiento de las ondas electromagnéticas de acuerdo con su longitud de onda. (p. 629)

electrón: partícula con carga negativa. (p. 36)

elemento: sustancia formada por un solo tipo de átomo. (p. 35)

elipse: curva cerrada y elongada que describe la órbita anual de la Tierra alrededor del sol. (p. 663)

El Niño: evento climático que comienza en el Océano Pacífico tropical; puede ocurrir cuando los vientos alisios se debilitan o se invierten; puede desestabilizar los patrones normales de precipitación y temperatura del mundo. (p. 493)

enzima: sustancia que acelera las reacciones químicas. (p. 583)

eón: la más grande subdivisión en la escala del tiempo geológico; se basa en la abundancia de cierto tipo de fósiles y está dividida en eras, periodos y épocas. (p. 393)

epicentro: punto de la superficie terrestre directamente encima del foco del terremoto. (p. 305)

época: la siguiente división más pequeña del tiempo geológico después del periodo; está caracterizada por diferencias en las formas de vida que pueden variar regionalmente. (p. 393)

ecuador: línea imaginaria que rodea a la Tierra en el punto de latitud 0°, a la mitad de la distancia entre el polo norte y el polo sur. (p. 160)

equinoccio: dos veces al año—en primavera y otoño—cuando el sol está posicionado directamente sobre el ecuador y el número de horas del día y de la noche son iguales en todo el mundo. (p. 665)

era: la segunda división más grande del tiempo geológico; está subdividida en periodos y se basa en cambios mayores en todo el mundo con respecto a los tipos de fósiles. (p. 393)

erosión: proceso mediante el cual los materiales de la superficie son desgastados y transportados de un lugar a otro por agentes como la gravedad, el agua, el viento o los glaciares. (p. 210)

estuario: área donde un río desemboca en el océano, contiene una mezcla de agua dulce y agua salada y proporciona un hábitat importante para muchos organismos marinos. (p. 556)

ética: estudio de los valores morales sobre lo que es bueno o malo. (p. 20)

extrusivo: describe rocas ígneas de grano fino que se forman cuando el magma se enfría rápidamente en o cerca de la superficie terrestre. (p. 95)

fault: surface along which rocks move when they pass their elastic limit and break. (p. 300)

fault-block mountains: mountains formed from huge, tilted blocks of rock that are separated from surrounding rocks by faults. (p. 158)

fertilizer: chemical that helps plants and other organisms grow. (p. 601)

focus: in an earthquake, the point below Earth's surface where energy is released in the form of seismic waves. (p. 304)

fog: a stratus cloud that forms when air is cooled to its dew point near the ground. (p. 459)

folded mountains: mountains formed when horizontal rock layers are squeezed from opposite sides, causing them to buckle and fold. (p. 157)

foliated: describes metamorphic rock, such as slate and gneiss, whose mineral grains line up in parallel layers. (p. 101)

fossil fuel: nonrenewable energy resource, such as oil and coal, formed over millions of years from the remains of dead plants and other organisms. (p. 120)

fossils: remains, imprints, or traces of prehistoric organisms that can tell when and where organisms once lived and how they lived. (p. 363)

fracture: physical property of some minerals that causes them to break with uneven, rough, or jagged surfaces. (p. 71)

front: boundary between two air masses with different temperatures, density, or moisture; can be cold, warm, occluded, and stationary. (p. 463)

full moon: phase that occurs when all of the Moon's surface facing Earth reflects light. (p. 667)

falla: área a lo largo de la cual las rocas se mueven cuando sobrepasan su límite elástico y se rompen. (p. 300)

montañas de fallas: montañas formadas por bloques rocosos grandes e inclinados separados de las rocas circundantes por fracturas. (p. 158)

fertilizante: agente químico que ayuda a crecer a las plantas y a otros organismos. (p. 601)

foco: en un terremoto, el punto bajo la superficie terrestre donde se libera la energía en forma de ondas sísmicas. (p. 304)

niebla: nube de estrato que se forma cuando el aire se enfría a su punto de condensación cerca del suelo. (p. 459)

montañas de plegamiento: montañas formadas cuando las capas rocosas horizontales son comprimidas desde lados opuestos, causando que se colapsen y plieguen. (p. 157)

foliado: describe rocas metamórficas, como pizarra y gneis, cuyas vetas minerales se alinean en capas paralelas. (p. 101)

combustible fósil: recurso energético no renovable, como el petróleo y el carbón mineral, formado durante millones de años a partir de restos de plantas y otros organismos muertos. (p. 120)

fósiles: restos, huellas o trazas de organismos prehistóricos que pueden informar cuándo, dónde y cómo vivieron tales organismos. (p. 363)

fractura: propiedad física de algunos minerales que causa que se rompan formando superficies irregulares, ásperas o dentadas. (p. 71)

frente: límite entre dos masas de aire con temperatura, densidad o humedad diferentes; puede ser frío, caliente, ocluido o estacionario. (p. 463)

luna llena: fase que ocurre cuando toda la superficie de la luna frente a la Tierra refleja la luz del sol. (p. 667)

G

galaxy: large group of stars, dust, and gas held together by gravity; can be elliptical, spiral, or irregular. (p. 740)

gem: beautiful, rare, highly prized mineral that can be worn in jewelry. (p. 73)

geologic time scale: division of Earth's history into time units based largely on the types of life-forms that lived only during certain periods. (p. 392)

galaxia: grupo grande de estrellas, polvo y gas en donde todo está unido por gravedad; puede ser elíptica, espiral o irregular. (p. 740)

gema: mineral hermoso, raro y altamente valorado que puede usarse como joya. (p. 73)

escala del tiempo geológico: división de la historia de la Tierra en unidades de tiempo; se basa en los tipos de formas de vida que vivieron sólo durante ciertos periodos. (p. 392)

geothermal energy: inexhaustible energy resource that uses hot magma or hot, dry rocks from below Earth's surface to generate electricity. (p. 132)

energía geotérmica: recurso energético inagotable que utiliza el magma caliente o las piedras secas calientes encontradas debajo de la superficie terrestre para producir electricidad. (p. 132)

geyser: hot spring that erupts periodically and shoots water and steam into the air—for example, Old Faithful in Yellowstone National Park. (p. 253)

géiser: aguas termales que erupcionan periódicamente arrojando agua y vapor al aire—por ejemplo, Old Faithful en el Parque Nacional Yellowstone. (p. 253)

giant: late stage in the life of comparatively low-mass main sequence star in which hydrogen in the core is deleted, the core contracts and temperatures inside the star increase, causing its outer layers to expand and cool. (p. 737)

gigante: etapa tardía en la vida de una estrella de secuencia principal, de relativamente poca masa, en la que el hidrógeno en el núcleo está agotado, el núcleo se contrae y la temperatura en el interior de la estrella aumenta, causando que las capas externas se expandan y enfríen. (p. 737)

glaciers: large, moving masses of ice and snow that change large areas of Earth's surface through erosion and deposition. (p. 215)

glaciares: grandes masas de hielo y nieve en movimiento que cambian extensas áreas de la superficie terrestre a través de la erosión y la deposición. (p. 215)

global warming: increase in the average global temperature of Earth. (p. 500)

calentamiento global: incremento del promedio de la temperatura global. (p. 500)

granitic: describes generally light-colored, silica-rich igneous rock that is less dense than basaltic rock. (p. 97)

granítica: describe roca ígnea rica en sílice, generalmente de color claro y menos densa que la rocas basáltica. (p. 97)

Great Red Spot: giant, high-pressure storm in Jupiter's atmosphere. (p. 702)

La Gran Mancha Roja: tormenta gigante de alta presión en la atmósfera de Júpiter. (p. 702)

greenhouse effect: natural heating that occurs when certain gases in Earth's atmosphere, such as methane, CO_2, and water vapor, trap heat. (p. 499)

efecto invernadero: calentamiento natural que ocurre cuando ciertos gases en la atmósfera terrestre, como el metano, el dióxido de carbono y el vapor de agua atrapan el calor. (p. 499)

groundwater: water that soaks into the ground and collects in pores and empty spaces and is an important source of drinking water. (p. 249)

agua subterránea: agua que se difunde en el suelo y se acumula en poros y espacios vacíos siendo una fuente importante de agua potable. (p. 249)

H

half-life: time it takes for half the atoms of an isotope to decay. (p. 378)

vida media: tiempo que le toma a la mitad de los átomos de un isótopo para desintegrarse. (p. 378)

hardness: measure of how easily a mineral can be scratched. (p. 69)

dureza: medida de la facilidad con que un mineral puede ser rayado. (p. 69)

hazardous waste: poisonous, ignitable, or cancer-causing waste. (p. 582)

desechos peligrosos: desechos venenosos, cancerígenos o inflamables. (p. 582)

heterogeneous mixture: a mixture which is not mixed evenly and each component retains its own properties. (p. 43)

mezcla heterogénea: mezcla no uniforme en la que cada componente conserva sus propiedades. (p. 43)

hibernation: behavioral adaptation for winter survival in which an animal's activity is greatly reduced, its body temperature drops, and body processes slow down. (p. 490)

hibernación: adaptación del comportamiento para sobrevivir durante el invierno en la cual la actividad del animal se ve fuertemente reducida, su temperatura corporal se reduce y los procesos corporales disminuyen su ritmo. (p. 490)

homogeneous mixture: a mixture which is evenly mixed throughout. (p. 43)

mezcla homogénea: mezcla uniforme. (p. 43)

horizon: each layer in a soil profile—horizon A (top layer of soil), horizon B (middle layer), and horizon C (bottom layer). (p. 190)

hot spot: the result of an unusually hot area at the boundary between Earth's mantle and core that forms volcanoes when melted rock is forced upward and breaks through the crust. (p. 334)

humidity: amount of water vapor held in the air. (p. 456)

humus (HYEW mus): dark-colored, decayed organic matter that supplies nutrients to plants and is found mainly in topsoil. (p. 190)

hurricane: large, severe storm that forms over tropical oceans, has winds of at least 120 km/h, and loses power when it reaches land. (p. 468)

hydroelectric energy: electricity produced by waterpower using large dams in a river. (p. 132)

hydrosphere: all the water on Earth's surface. (p. 437)

hypothesis: an educated guess. (p. 7)

horizonte: cada capa en un perfil de suelos: horizonte A (la capa superior del suelo), horizonte B (la capa media) y horizonte C (la capa inferior). (p. 190)

punto caliente: el resultado de un área extraordinariamente caliente en los límites entre el manto y el núcleo de la Tierra; forma volcanes cuando la roca fundida es empujada hacia arriba y se abre paso hacia la corteza. (p. 334)

humedad: cantidad de vapor de agua suspendido en el aire. (p. 456)

humus: materia orgánica en descomposición, de color oscuro, que suministra nutrientes a las plantas y se encuentra principalmente en la parte superior del suelo. (p. 190)

huracán: tormenta grande y severa que se forma sobre los océanos tropicales, tiene vientos de por lo menos 120 km/h y pierde su fuerza cuando alcanza la costa. (p. 468)

energía hidroeléctrica: electricidad producida por la energía hidráulica generada mediante represas grandes construidas en los ríos. (p. 132)

hidrosfera: toda el agua en la superficie terrestre. (p. 437)

hipótesis: una suposición fundamentada. (p. 7)

I

ice wedging: mechanical weathering process that occurs when water freezes in the cracks of rocks and expands, causing the rock to break apart. (p. 184)

igneous rock: rock formed when magma or lava cools and hardens. (p. 94)

impact basin: a hollow left on the surface of the Moon caused by an object striking its surface. (p. 677)

impermeable: describes materials that water cannot pass through. (p. 250)

independent variable: factor that changes in an experiment. (p. 10)

index fossils: remains of species that existed on Earth for a relatively short period of time, were abundant and widespread geographically, and can be used by geologists to assign the ages of rock layers. (p. 367)

intrusive: describes a type of igneous rock that generally contains large crystals and forms when magma cools slowly beneath Earth's surface. (p. 95)

ion: electrically-charged atom whose charge results from an atom losing or gaining electrons. (p. 41)

gelifracción: proceso de erosión mecánica que ocurre cuando el agua se congela en las grietas de las rocas y luego se expande, causando que la roca de fraccione. (p. 184)

roca ígnea: roca formada cuando se enfría y endurece el magma o la lava. (p. 94)

cráter de impacto: un hueco dejado en la superficie de la luna causada por un objeto que chocó contra su superficie. (p. 677)

impermeable: describe materiales que impiden el paso del agua a través de ellos. (p. 250)

variable independiente: factor que cambia en un experimento. (p. 10)

fósiles índice: restos de especies que existieron sobre la Tierra durante un periodo de tiempo relativamente corto y que fueron abundantes y ampliamente diseminadas geográficamente; los geólogos pueden usarlos para inferir las edades de las capas rocosas. (p. 367)

intrusivo: describe un tipo de roca ígnea que generalmente contiene cristales grandes y se forma cuando el magma se enfría lentamente por debajo de la superficie terrestre. (p. 95)

ión: átomo con carga eléctrica cuya carga es el resultado de la pérdida o ganancia de electrones por parte de un átomo. (p. 41)

ionosphere: layer of electrically charged particles in the thermosphere that absorbs AM radio waves during the day and reflects them back at night. (p. 429)

ionosfera: capa de partículas con carga eléctrica presentes en la termosfera, la cual absorbe las ondas de radio AM durante el día y las refleja durante la noche. (p. 429)

isobars: lines drawn on a weather map that connect points having equal atmospheric pressure; also indicate the location of high- and low-pressure areas and can show wind speed. (p. 471)

isobaras: líneas dibujadas en un mapa meteorológico que conectan los puntos que tienen una presión atmosférica similar; también indican la ubicación de las áreas de baja y alta presión y pueden mostrar la velocidad del viento. (p. 471)

isotherm (I suh thurm): line drawn on a weather map that connects points having equal temperature. (p. 471)

isoterma: línea dibujada en un mapa meteorológico que conecta los puntos que tienen la misma temperatura. (p. 471)

isotopes: atoms of the same element that have different numbers of neutrons. (p. 37)

isótopos: átomos del mismo elemento que tienen diferente número de neutrones. (p. 37)

J

jet stream: narrow belt of strong winds that blows near the top of the troposphere. (p. 442)

corriente de chorro: faja angosta de vientos fuertes que soplan cerca de la parte superior de la troposfera. (p. 442)

Jupiter: largest and fifth planet from the Sun; contains more mass than all the other planets combined, has continuous storms of high-pressure gas, and an atmosphere mostly of hydrogen and helium. (p. 702)

Júpiter: el quinto planeta más cercano al sol, y también el más grande; contiene más masa que todos los otros planetas en conjunto, tiene tormentas continuas de gas a alta presión y una atmósfera compuesta principalmente por hidrógeno y helio. (p. 702)

L

land breeze: movement of air from land to sea at night, created when cooler, denser air from the land forces up warmer air over the sea. (p. 443)

brisa terrestre: movimiento de aire nocturno de la tierra al mar, generado cuando el aire denso y frío proveniente de la tierra empuja hacia arriba al aire caliente que está sobre el mar. (p. 443)

latitude: distance in degrees north or south of the equator. (p. 160)

latitud: distancia en grados al norte o sur del ecuador. (p. 160)

lava: molten rock that flows from volcanoes onto Earth's surface. (p. 94)

lava: roca derretida que fluye de los volcanes hacia la superficie terrestre. (p. 94)

leaching: removal of minerals that have been dissolved in water. (p. 191)

lixiviación: remoción de minerales que han sido disueltos en el agua. (p. 191)

light-year: unit representing the distance light travels in one year—about 9.5 trillion km—used to record distances between stars and galaxies. (p. 727)

año luz: unidad que representa la distancia que la luz viaja en un año—cerca de 9.5 trillones de kilómetros—usada para registrar las distancias entre las estrellas y las galaxias. (p. 727)

liquefaction: occurs when wet soil acts more like a liquid during an earthquake. (p. 315)

licuefacción: ocurre cuando el suelo húmedo se comporta como un líquido durante un terremoto. (p. 315)

lithosphere (LIH thuh sfihr): rigid layer of Earth about 100 km thick, made of the crust and a part of the upper mantle. (p. 280)

litosfera: capa rígida de la Tierra de unos 100 kilómetros de profundidad, comprende la corteza y una parte del manto superior. (p. 280)

litter: twigs, leaves, and other organic matter that help prevent erosion and hold water and may eventually be changed into humus by decomposing organisms. (p. 191)

loess (LOOS): windblown deposit of tightly packed, fine-grained sediments. (p. 225)

longitude: distance in degrees east or west of the prime meridian. (p. 161)

longshore current: current that runs parallel to the shoreline, is caused by waves colliding with the shore at slight angles, and moves tons of loose sediment. (p. 256)

lunar eclipse: occurs when Earth's shadow falls on the Moon. (p. 670)

luster: describes the way a mineral reflects light from its surface; can be metallic or nonmetallic. (p. 70)

hojarasca: ramitas, hojas y demás material orgánico que ayuda a prevenir la erosión y a mantener el agua, y que eventualmente puede ser transformado en humus por los organismos descomponedores. (p. 191)

loes: depósito arrastrado por el viento que se compone de sedimentos de partículas finas y se encuentran muy compactados. (p. 225)

longitud: distancia en grados al este u oeste del meridiano inicial. (p. 161)

corriente costera: corriente que corre paralela a la línea costera, es causada por olas que chocan contra la orilla en ángulos tenues y mueve toneladas de sedimentos sueltos. (p. 256)

eclipse lunar: ocurre cuando la sombra de la Tierra cubre la luna. (p. 670)

brillo: describe la forma en que un mineral refleja la luz desde su superficie; puede ser metálicos o no metálicos. (p. 70)

M

magma: hot, melted rock material beneath Earth's surface. (p. 65)

magnitude: measure of the energy released during an earthquake. (p. 314)

map legend: explains the meaning of symbols used on a map. (p. 168)

map scale: relationship between distances on a map and distances on Earth's surface that can be represented as a ratio or as a small bar divided into sections. (p. 168)

maria (MAHR ee uh): dark-colored, relatively flat regions of the Moon formed when ancient lava reached the surface and filled craters on the Moon's surface. (p. 671)

Mars: fourth planet from the Sun; has polar ice caps, a thin atmosphere, and a reddish appearance caused by iron oxide in weathered rocks and soil. (p. 698)

mass movement: any type of erosion that occurs as gravity moves materials down-slope. (p. 211)

mass number: the number of protons plus the number of neutrons in an atom. (p. 37)

matter: anything that has mass and takes up space. (p. 34)

magma: material rocoso fundido y caliente que se encuentra por debajo de la superficie terrestre. (p. 65)

magnitud: medida de la energía liberada durante un terremoto. (p. 314)

leyenda del mapa: explica el significado de los símbolos utilizados en un mapa. (p. 168)

escala del mapa: relación entre las distancias en un mapa y las distancias sobre la superficie terrestre, que puede representarse como una relación o como una barra pequeña dividida en secciones. (p. 168)

mares: regiones relativamente planas, de color oscuro, que se encuentran en la luna y que fueron formadas cuando la lava antigua alcanzó la superficie y llenó los cráteres sobre la superficie lunar. (p. 671)

Marte: cuarto planeta más cercano al sol; tiene casquetes de hielo polar, una atmósfera delgada y una apariencia rojiza causada por el óxido de hierro presente en las rocas y suelo de su superficie. (p. 698)

movimiento en masa: cualquier tipo de erosión que ocurre cuando la gravedad mueve materiales cuesta abajo. (p. 211)

número de masa: el número de protones más el número de neutrones en un átomo. (p. 37)

materia: cualquier cosa que tenga masa y ocupe un lugar en el espacio. (p. 34)

meander (mee AN dur): broad, c-shaped curve in a river or stream, formed by erosion of its outer bank. (p. 243)

mechanical weathering: physical processes that break rock apart without changing its chemical makeup; can be caused by ice wedging, animals, and plant roots. (p. 183)

Mercury: smallest planet, closest to the Sun; does not have a true atmosphere; has a surface with many craters and high cliffs. (p. 696)

Mesozoic (mez uh ZOH ihk) Era: middle era of Earth's history, during which Pangaea broke apart, dinosaurs appeared, and reptiles and gymnosperms were the dominant land life-forms. (p. 408)

metamorphic rock: forms when heat, pressure, or fluids act on igneous, sedimentary, or other metamorphic rock to change its form or composition, or both. (p. 99)

meteor: a meteoroid that burns up in Earth's atmosphere. (p. 711)

meteorite: a meteoroid that strikes the surface of a moon or planet. (p. 712)

meteorologist (meet ee uh RAHL uh just): studies weather and uses information from Doppler radar, weather satellites, computers and other instruments to make weather maps and provide forecasts. (p. 470)

mid-ocean ridge: area where new ocean floor is formed when lava erupts through cracks in Earth's crust. (p. 544)

mineral: naturally occurring inorganic solid that has a definite chemical composition and an orderly internal atomic structure. (p. 62)

mineral resources: resources from which metals are obtained. (p. 137)

mixture: composed of two or more substances that are not chemically combined. (p. 43)

mold: a type of body fossil that forms in rock when an organism with hard parts is buried, decays or dissolves, and leaves a cavity in the rock. (p. 365)

molecule: a group of atoms connected by covalent bonds. (p. 41)

moon phase: change in appearance of the Moon as viewed from the Earth, due to the relative positions of the Moon, Earth, and Sun. (p. 667)

moraine: large ridge of rocks and soil deposited by a glacier when it stops moving forward. (p. 217)

meandro: curva amplia en forma de C en un río o arroyo, formada por la erosión de su rivera externa. (p. 243)

erosión mecánica: procesos físicos que fraccionan la roca sin cambiar su composición química; puede ser causada por gelifracción, animales y raíces de las plantas. (p. 183)

Mercurio: el planeta más pequeño y más cercano al sol; no tiene una atmósfera verdadera; tiene una superficie con muchos cráteres y grandes acantilados. (p. 696)

Era Mesozoica: era media de la historia de la Tierra durante la cual se escindió la Pangea y aparecieron los dinosaurios; los reptiles y gimnospermas fueron las formas de vida que dominaron la tierra. (p. 408)

roca metamórfica: se forma cuando el calor, la presión o los fluidos actúan sobre una roca ígnea, sedimentaria u otra roca metamórfica para cambiar su forma, composición o ambas. (p. 99)

meteoro: un meteoroide que se incinera en la atmósfera de la Tierra. (p. 711)

meteorito: un meteoroide que choca contra la superficie de la luna o de algún planeta. (p. 712)

meteorólogo: persona que estudia el clima y usa información del radar Doppler, satélites meteorológicos, computadoras y otros instrumentos para elaborar mapas del estado del tiempo y hacer pronósticos. (p. 470)

surco en mitad del océano: área donde se forma el nuevo suelo oceánico cuando la lava brota a través de grietas en la corteza terrestre. (p. 544)

mineral: sólido inorgánico que se encuentra en la naturaleza, tiene una composición química definida y una estructura atómica ordenada. (p. 62)

recursos minerales: recursos a partir de los cuales pueden obtenerse metales. (p. 137)

mezcla: compuesto de dos o más sustancias que no están combinadas químicamente. (p. 43)

moldura: tipo de cuerpo fósil que se formó en la roca cuando un organismo con partes duras fue enterrado, descompuesto o disuelto, dejando una cavidad en la roca. (p. 365)

molécula: grupo de átomos unidos por enlaces covalentes. (p. 41)

fase lunar: cambio en la apariencia de la luna según es vista desde la Tierra; se debe a las posiciones relativas de la luna, la Tierra y el sol. (p. 667)

morrena: grandes cúmulos de rocas y suelo depositados por un glaciar cuando deja de moverse hacia adelante. (p. 217)

N

natural gas: fossil fuel formed from marine organisms that is often found in tilted or folded rock layers and is used for heating and cooking. (p. 123)

gas natural: combustible fósil formado a partir de organismos marinos y que a menudo se encuentra en capas rocosas inclinadas o plegadas; se usa para calefacción y para cocinar. (p. 123)

natural selection: process by which organisms that are suited to a particular environment are better able to survive and reproduce than organisms that are not. (p. 395)

selección natural: proceso mediante el cual los organismos que están adaptados a un ambiente particular están mejor capacitados para sobrevivir y reproducirse que los organismos que no están adaptados. (p. 395)

nebula: large cloud of gas and dust that contracts under gravitational force and breaks apart into smaller pieces, each of which might collapse to form a star. (p. 736)

nebulosa: nube grande de polvo y gas que se contrae bajo la fuerza gravitacional y se descompone en pedazos más pequeños, cada uno de los cuales se puede colapsar para formar una estrella. (p. 736)

nekton: marine organisms that actively swim in the ocean. (p. 552)

necton: organismos marinos que nadan activamente en el océano. (p. 552)

Neptune: usually the eighth planet from the Sun; is large and gaseous, has rings that vary in thickness, and is bluish-green in color. (p. 706)

Neptuno: el octavo planeta desde el sol; es grande y gaseoso, tiene anillos que varían en espesor y tiene un color verde-azulado. (p. 706)

neutron: particle without an electrical charge. (p. 36)

neutrón: partícula sin carga eléctrica. (p. 36)

neutron star: collapsed core of a supernova that can shrink to about 20 km in diameter and contains only neutrons in the dense core. (p. 738)

estrella de neutrones: núcleo colapsado de una supernova que puede contraerse hasta tener un diámetro de 20 kilómetros y contiene sólo neutrones en su denso núcleo. (p. 738)

new moon: moon phase that occurs when the Moon is between Earth and the Sun, at which point the Moon cannot be seen because its lighted half is facing the Sun and its dark side faces Earth. (p. 667)

luna nueva: fase lunar que ocurre cuando la luna se encuentra entre la Tierra y el sol, punto en el cual la luna no puede verse porque su mitad iluminada está frente al sol y su lado oscuro frente a la Tierra. (p. 667)

nonfoliated: describes metamorphic rock, such as quartzite or marble, whose mineral grains grow and rearrange but generally do not form layers. (p. 102)

no foliado: describe rocas metamórficas, como la cuarcita o el mármol, cuyas vetas minerales se acumulan y reestructuran pero rara vez forman capas. (p. 102)

nonpoint source pollution: pollution that enters water from a large area and cannot be traced to a single location. (p. 600)

contaminación sin fuente establecida: contaminación que entra en el agua desde un área grande y no puede ser rastreada hasta una ubicación de origen. (p. 600)

normal fault: break in rock caused by tension forces, where rock above the fault surface moves down relative to the rock below the fault surface. (p. 302)

falla normal: ruptura en la roca causada por fuerzas de tensión, donde la roca sobre la superficie de la falla se mueve hacia abajo con respecto a la roca debajo de la superficie de la falla. (p. 302)

no-till farming: method for reducing soil erosion; plant stalks are left in the field after harvesting and the next year's crop is planted within the stalks without plowing. (p. 198)

cultivo sin labranza: método para reducir la erosión del suelo; los tallos de las plantas se dejan en el terreno después de la cosecha y el cultivo del siguiente año se siembra entre los tallos sin hacer labranza alguna. (p. 198)

nuclear energy: alternative energy source that is based on atomic fission. (p. 127)

energía nuclear: fuente de energía alternativa que se basa en la fisión atómica. (p. 127)

O

observatory: building that can house an optical telescope; often has a dome-shaped roof that can be opened for viewing. (p. 630)

observatorio: edificación que puede albergar un telescopio óptico; a menudo tiene un techo en forma de domo que puede abrirse para la observación. (p. 630)

oil: liquid fossil fuel formed from marine organisms that is burned to obtain energy and used in the manufacture of plastics. (p. 123)

petróleo: combustible fósil líquido formado a partir de organismos marinos; es quemado para obtener energía y se usa en la manufactura de plásticos. (p. 123)

orbit: curved path followed by a satellite as it revolves around an object. (p. 637)

órbita: trayectoria curva seguida por un satélite conforme gira alrededor de un objeto. (p. 637)

ore: deposit in which a mineral exists in large enough amounts to be mined at a profit. (pp. 77, 137)

mena: depósito en el que existe un mineral en cantidades suficientes para la explotación minera. (pp. 77, 137)

organic evolution: change of organisms over geologic time. (p. 394)

evolución orgánica: cambio de los organismos a través del tiempo geológico. (p. 394)

outwash: material deposited by meltwater from a glacier. (p. 217)

derrubio: material depositado por la corriente de agua del hielo derretido de un glaciar. (p. 217)

oxidation (ahk sih DAY shun): chemical weathering process that occurs when some minerals are exposed to oxygen and water over time. (p. 186)

oxidación: proceso de erosión química que ocurre cuando algunos minerales son expuestos al oxígeno y al agua. (p. 186)

ozone layer: layer of the stratosphere with a high concentration of ozone; absorbs most of the Sun's harmful ultraviolet radiation. (p. 432)

capa de ozono: capa de la estratosfera con una concentración alta de ozono y que absorbe la mayor parte de la radiación ultravioleta dañina del sol. (p. 432)

P

Paleozoic Era: era of ancient life, which began about 544 million years ago, when organisms developed hard parts, and ended with mass extinctions about 245 million years ago. (p. 402)

Era Paleozoica: era de la vida antigua que comenzó hace 544 millones de años, cuando los organismos desarrollaron partes duras; terminó con extinciones en masa hace unos 245 millones de años. (p. 402)

Pangaea (pan JEE uh): large ancient landmass that was composed of all the continents joined together. (pp. 272, 399)

Pangea: masa terrestre antigua que estaba compuesta por todos los continentes unidos. (pp. 272, 399)

particulate (par TIHK yuh layt) matter: fine solids such as pollen, dust, mold, ash, and soot as well as liquid droplets in the air that can irritate and damage lungs when breathed in. (p. 613)

material en polvo: sólidos finos como el polen, el polvo, el moho, las cenizas y el hollín, así como las gotas líquidas en el aire, que pueden irritar y dañar los pulmones cuando son inhaladas. (p. 613)

period: third-longest division of geologic time; is subdivided into epochs and is characterized by the types of life that existed worldwide. (p. 393)

periodo: la tercera división más grande del tiempo geológico; está subdividido en épocas y se caracteriza por los tipos de vida que existieron en todo el mundo. (p. 393)

permeable (PUR mee uh bul): describes soil and rock with connecting pores through which water can flow. (p. 250)

permeable: describe el suelo y la roca con poros conectados a través de los cuales el agua puede fluir. (p. 250)

permineralized remains: fossils in which the spaces inside are filled with minerals from groundwater. (p. 364)

restos permineralizados: fósiles en los que los espacios interiores son llenados con minerales de aguas subterráneas. (p. 364)

pesticide: substance used to keep insects and weeds from destroying crops and lawns. (p. 601)

photochemical smog: hazy, yellow-brown blanket of smog found over cities that is formed with the help of sunlight, contains ozone near Earth's surface, and can damage lungs and plants. (p. 610)

photosphere: lowest layer of the Sun's atmosphere; gives off light and has temperatures of about 6,000 K. (p. 729)

photosynthesis: food-making process using light energy from the Sun, carbon dioxide, and water. (p. 549)

pH scale: scale used to measure how acidic or basic something is. (p. 611)

plain: large, flat landform that often has thick, fertile soil and is usually found in the interior region of a continent. (p. 154)

plankton: marine organisms that drift in ocean currents. (p. 552)

plate: a large section of Earth's oceanic or continental crust and rigid upper mantle that moves around on the asthenosphere. (p. 280)

plateau (pla TOH): flat, raised landform made up of nearly horizontal rocks that have been uplifted. (p. 156)

plate tectonics: theory that Earth's crust and upper mantle are broken into plates that float and move around on a plasticlike layer of the mantle. (p. 280)

plucking: process that adds gravel, sand, and boulders to a glacier's bottom and sides as water freezes and thaws, breaking off pieces of surrounding rock. (p. 216)

point source pollution: pollution that enters water from a specific location and can be controlled or treated before it enters a body of water. (p. 600)

polar zones: climate zones that receive solar radiation at a low angle, extend from 66°N and S latitude to the poles, and are never warm. (p. 484)

pollutant: any substance that contaminates the environment. (p. 576)

pollution: introduction of wastes to an environment, such as sewage and chemicals, that can damage organisms. (p. 557)

population: total number of individuals of one species occupying the same area. (p. 574)

pesticida: sustancia utilizada para evitar que los insectos y malezas destruyan los cultivos y los prados. (p. 601)

smog fotoquímico: cubierta brumosa de color amarillo-marrón que se encuentra sobre las ciudades; se forma con ayuda de la luz solar, contiene ozono cerca de la superficie terrestre y puede dañar los pulmones y las plantas. (p. 610)

fotosfera: capa más interna de la atmósfera del sol; emite luz y tiene temperaturas de cerca de 6,000 grados Kelvin. (p. 729)

fotosíntesis: proceso de producción de alimentos usando la energía luminosa del sol, dióxido de carbono y agua. (p. 549)

escala de pH: escala usada para medir el grado de acidez o alcalinidad de una sustancia. (p. 611)

planicie: formación de terreno extenso y plano que a menudo tiene suelos gruesos y fértiles; generalmente se encuentra en la región interior de un continente. (p. 154)

plancton: organismos marinos que se desplazan a la deriva en las corrientes oceánicas. (p. 552)

placa: gran sección de la corteza terrestre u oceánica y del manto rígido superior que se mueve sobre la astenosfera. (p. 280)

meseta: formación de terreno plano y elevado constituida por rocas casi horizontales que han sido levantadas. (p. 156)

tectónica de placas: teoría respecto a que la corteza terrestre y el manto superior están fraccionados en placas que flotan y se mueven sobre una capa plástica del manto. (p. 280)

gelivación: proceso que agrega grava, arena y cantos a la parte inferior y lateral de un glaciar conforme el agua se congela y descongela, fraccionando las piezas de las rocas circundantes. (p. 216)

contaminación de fuente establecida: contaminación que entra en el agua desde una ubicación específica y puede controlarse o tratarse antes de que entre en la masa de agua. (p. 600)

zonas polares: zonas climáticas que reciben radiación solar a un ángulo reducido, se extienden desde los 66° de latitud norte y sur hasta los polos y nunca son cálidas. (p. 484)

contaminante: cualquier sustancia que contamine el medio ambiente. (p. 576)

contaminación: introducción de desechos al medio ambiente, como aguas residuales y químicos, que pueden causar daño a los organismos. (p. 557)

población: número total de individuos de una especie que ocupan la misma área. (p. 574)

Precambrian (pree KAM bree un) time: longest part of Earth's history, lasting from 4.0 billion to about 544 million years ago. (p. 400)

precipitation: water falling from clouds—including rain, snow, sleet, and hail—whose form is determined by air temperature. (p. 460)

primary wave: seismic wave that moves rock particles back-and-forth in the same direction that the wave travels. (p. 305)

prime meridian: imaginary line that represents 0° longitude and runs from the north pole through Greenwich, England, to the south pole. (p. 161)

principle of superposition: states that in undisturbed rock layers, the oldest rocks are on the bottom and the rocks become progressively younger toward the top. (p. 370)

Project Apollo: final stage in the U.S. program to reach the Moon in which Neil Armstrong was the first human to step onto the Moon's surface. (p. 642)

Project Gemini: second stage in the U.S. program to reach the Moon in which an astronaut team connected with another spacecraft in orbit. (p. 641)

Project Mercury: first step in the U.S. program to reach the Moon that orbited a piloted spacecraft around Earth and brought it back safely. (p. 641)

proton: particle that has a positive electric charge. (p. 36)

tiempo precámbrico: la parte más duradera de la historia de la Tierra; duró desde hace 4.0 billones de años hasta hace aproximadamente 544 millones de años. (p. 400)

precipitación: agua que cae de las nubes—incluyendo lluvia, nieve, aguanieve y granizo—cuya forma está determinada por la temperatura del aire. (p. 460)

onda primaria: onda sísmica que mueve partículas rocosas en la misma dirección en que viaja la onda. (p. 305)

meridiano inicial: línea imaginaria que representa los cero grados de longitud y va desde el polo norte pasando por Greenwich, Inglaterra, hasta el polo sur. (p. 161)

principio de superposición: establece que en las capas rocosas no perturbadas, las rocas más antiguas están en la parte inferior y las rocas son más jóvenes conforme están más cerca de la superficie. (p. 370)

Proyecto Apolo: etapa final en el proyecto norteamericano para llegar a la luna en el que Neil Armstrong fue el primer ser humano en caminar sobre la superficie lunar. (p. 642)

Proyecto Géminis: segunda etapa del proyecto norteamericano para llegar a la luna en el que un grupo de astronautas se conectó con otra nave espacial en órbita. (p. 641)

Proyecto Mercurio: primera etapa del proyecto norteamericano para llegar a la luna en el que una nave espacial tripulada recorrió la órbita de la Tierra y regresó de manera segura. (p. 641)

protón: partícula con carga positiva. (p. 36)

R

radiation: energy transferred by waves or rays. (p. 436)

radioactive decay: process in which some isotopes break down into other isotopes and particles. (p. 377)

radiometric dating: process used to calculate the absolute age of rock by measuring the ratio of parent isotope to daughter product in a mineral and knowing the half-life of the parent. (p. 379)

radio telescope: collects and records radio waves traveling through space; can be used day or night under most weather conditions. (p. 633)

recycling: conservation method in which old materials are processed to make new ones. (pp. 141, 587)

radiación: energía transmitida por ondas o rayos. (p. 436)

desintegración radiactiva: proceso en el que algunos isótopos se desintegran en otros isótopos y partículas. (p. 377)

fechado radiométrico: proceso utilizado para calcular la edad absoluta de las rocas midiendo la relación isótopo parental a producto derivado en un mineral y conociendo la vida media del parental. (p. 379)

radiotelescopio: recolecta y registra ondas de radio que viajan a través del espacio; puede usarse de día o de noche en la mayoría de condiciones climáticas. (p. 633)

reciclaje: método de conservación en el cual los materiales usados son procesados para fabricar otros nuevos. (pp. 141, 587)

reef: rigid, wave-resistant, ocean margin habitat built by corals from skeletal materials and calcium. (p. 556)

reflecting telescope: optical telescope that uses a concave mirror to focus light and form an image at the focal point. (p. 630)

refracting telescope: optical telescope that uses a double convex lens to bend light and form an image at the focal point. (p. 630)

relative age: the age of something compared with other things. (p. 371)

relative humidity: measure of the amount of moisture held in the air compared with the amount it can hold at a given temperature; can range from 0 percent to 100 percent. (p. 456)

reserve: amount of a fossil fuel that can be extracted from Earth at a profit using current technology. (p. 125)

reverse fault: break in rock caused by compressive forces, where rock above the fault surface moves upward relative to the rock below the fault surface. (p. 302)

revolution: Earth's yearlong elliptical orbit around the Sun. (p. 663)

rock: mixture of one or more minerals, rock fragments, volcanic glass, organic matter, or other natural materials; can be igneous, metamorphic, or sedimentary. (p. 90)

rock cycle: model that describes how rocks slowly change from one form to another through time. (p. 91)

rocket: special engine that can work in space and burns liquid or solid fuel. (p. 635)

rotation: spinning of Earth on its imaginary axis, which takes about 24 hours to complete and causes day and night to occur. (p. 661)

runoff: any rainwater that does not soak into the ground or evaporate but flows over Earth's surface; generally flows into streams and has the ability to erode and carry sediments. (p. 238)

arrecife: hábitat de los márgenes oceánicos, rígido y resistente a las olas; es generado por los corales a partir de materiales esqueléticos y calcio. (p. 556)

telescopio reflectante: telescopio óptico que utiliza un espejo cóncavo para enfocar la luz y formar una imagen en el punto focal. (p. 630)

telescopio de refracción: telescopio óptico que utiliza un lente doble convexo para formar una imagen en el punto focal. (p. 630)

edad relativa: la edad de algo comparado con otras cosas. (p. 371)

humedad relativa: medida de la cantidad de humedad suspendida en el aire en comparación con la cantidad que puede contener a una temperatura determinada; puede variar del cero al cien por ciento. (p. 456)

reserva: depósito de un combustible fósil que puede extraerse de la Tierra y del cual, utilizando la tecnología actual, se obtienen utilidades. (p. 125)

falla inversa: ruptura en la roca causada por fuerzas de compresión, donde la roca sobre la superficie de la falla se mueve hacia arriba con respecto a la roca debajo de la superficie de la falla. (p. 302)

revolución: órbita elíptica de un año de duración que la Tierra recorre alrededor del sol. (p. 663)

roca: mezcla de uno o más minerales, fragmentos de roca, obsidiana, materia orgánica u otros materiales naturales; puede ser ígnea, metamórfica o sedimentaria. (p. 90)

ciclo de la roca: modelo que describe cómo cambian lentamente las rocas de una forma a otra a través del tiempo. (p. 91)

cohete: máquina especial que puede funcionar en el espacio y quema combustible sólido o líquido. (p. 635)

rotación: rotación de la Tierra sobre su eje imaginario, lo cual toma cerca de 24 horas para completarse y causa la alternancia entre el día y la noche. (p. 661)

escorrentía: agua de lluvia que no se difunde en el suelo ni se evapora pero que fluye sobre la superficie terrestre; generalmente fluye hacia los arroyos y tiene la capacidad de causar erosión y transportar sedimentos. (p. 238)

S

salinity (suh LIH nuh tee): a measure of the amount of salts dissolved in seawater. (p. 516)

sanitary landfill: area where garbage is deposited and covered with soil and that is designed to prevent contamination of land and water. (p. 582)

salinidad: medida de la cantidad de sales disueltas en el agua marina. (p. 516)

relleno sanitario: área donde las basuras son depositadas y cubiertas con suelo y que están diseñadas para prevenir la contaminación del suelo y del agua. (p. 582)

satellite: any natural or artificial object that revolves around another object. (p. 637)

Saturn: second-largest and sixth planet from the Sun; has a complex ring system, at least 31 moons, and a thick atmosphere made mostly of hydrogen and helium. (p. 704)

science: process of looking at and studying things in the world in order to gain knowledge. (p. 8)

scientific law: rule that describes the behavior of something in nature; usually describes what will happen in a situation but not why it happens. (p. 19)

scientific methods: problem-solving procedures that can include identifying the problem or question, gathering information, developing a hypothesis, testing the hypothesis, analyzing the results, and drawing conclusions. (p. 8)

scientific theory: explanation that is supported by results from repeated experimentation or testing. (p. 18)

scrubber: device that lowers sulfur emissions from coal-burning power plants. (p. 614)

sea breeze: movement of air from sea to land during the day when cooler air from above the water moves over the land, forcing the heated, less dense air above the land to rise. (p. 443)

seafloor spreading: Hess's theory that new seafloor is formed when magma is forced upward toward the surface at a mid-ocean ridge. (p. 277)

season: short period of climate change in an area caused by the tilt of Earth's axis as Earth revolves around the Sun. (p. 492)

secondary wave: seismic wave that moves rock particles at right angles to the direction of the wave. (p. 305)

sedimentary rock: forms when sediments are compacted and cemented together or when minerals form from solutions. (p. 103)

sediments: loose materials, such as rock fragments, mineral grains, and the remains of once-living plants and animals, that have been moved by wind, water, ice, or gravity. (p. 103)

seismic (SIZE mihk) wave: wave generated by an earthquake. (p. 304)

seismograph: instrument used to register earthquake waves and record the time that each arrived. (p. 307)

satélite: cualquier objeto natural o artificial que gire alrededor de otro objeto. (p. 637)

Saturno: además de ser el sexto planeta más cercano al sol, también es el segundo en tamaño; tiene un sistema de anillos complejo, por lo menos 31 lunas y una atmósfera gruesa compuesta principalmente de hidrógeno y helio. (p. 704)

ciencia: proceso de observación y estudio de las cosas en el mundo con el propósito de adquirir conocimientos. (p. 8)

ley científica: regla que describe el comportamiento de algo en la naturaleza; usualmente describe qué sucederá en una situación pero no el porqué sucedería. (p. 19)

métodos científicos: procedimientos para solucionar problemas que pueden incluir la identificación del problema o pregunta, la recopilación de información, el desarrollo de una hipótesis, la prueba de la hipótesis, el análisis de los resultados y la extracción de conclusiones. (p. 8)

teoría científica: explicación apoyada por los resultados de la experimentación o de pruebas repetidas. (p. 18)

filtro de fricción: dispositivo que disminuye las emisiones de sulfuro provenientes de plantas eléctricas que funcionan con carbón. (p. 614)

brisa marina: movimiento de aire del mar a la tierra durante el día, cuando el aire frío que está sobre el mar empuja al aire caliente y menos denso que está sobre la tierra. (p. 443)

expansión del suelo oceánico: teoría de Hess respecto a que se forma un nuevo suelo oceánico cuando el magma es empujado hacia la superficie a través de un surco en la mitad del océano. (p. 277)

estación: periodo corto de cambio climático en un área, causado por la inclinación del eje de la Tierra conforme gira alrededor del sol. (p. 492)

onda secundaria: onda sísmica que mueve partículas rocosas en ángulos rectos respecto a la dirección de la onda. (p. 305)

roca sedimentaria: se forma cuando los sedimentos son compactados y cementados o cuando se forman minerales a partir de soluciones. (p. 103)

sedimentos: materiales sueltos, como fragmentos de roca, granos minerales y restos de animales y plantas, que han sido arrastrados por el viento, el agua, el hielo o la gravedad. (p. 103)

onda sísmica: onda generada por un terremoto. (p. 304)

sismógrafo: instrumento utilizado para registrar las ondas sísmicas y la hora a la que llega cada una. (p. 307)

sewage: water that goes into drains and contains human waste, household detergents, and soaps. (p. 602)

aguas residuales: agua que entra a los desagües y contiene desechos humanos, detergentes de uso doméstico y jabones. (p. 602)

sheet erosion: a type of surface water erosion caused by runoff that occurs when water flowing as sheets picks up sediments and carries them away. (p. 241)

erosión laminar: tipo de erosión causada por las corrientes de agua de lluvia; ocurre cuando el agua fluye laminarmente recogiendo sedimentos y llevándolos a otro lugar. (p. 241)

shield volcano: broad, gently sloping volcano formed by quiet eruptions of basaltic lava. (p. 340)

volcán de escudo: volcán levemente inclinado y de gran extensión, formado por erupciones de baja intensidad de lava basáltica. (p. 340)

silicate: describes a mineral that contains silicon and oxygen and usually one or more other elements. (p. 66)

silicato: describe mineral que contiene sílice y oxígeno y generalmente uno o varios elementos distintos. (p. 66)

sill: igneous rock feature formed when magma is squeezed into a horizontal crack between layers of rock and hardens underground. (p. 347)

alféizar: roca ígnea característica formada cuando el magma es comprimido en una grieta horizontal entre capas de roca y se endurece en el subsuelo. (p. 347)

slump: a type of mass movement that occurs when a mass of material moves down a curved slope. (p. 211)

desprendimiento: tipo de movimiento en masa que ocurre cuando un volumen de material se mueve hacia abajo de una cuesta curvada. (p. 211)

soil: mixture of weathered rock and mineral fragments, decayed organic matter, mineral fragments, water, and air that can take thousands of years to develop. (p. 188)

suelo: mezcla de roca erosionada y fragmentos minerales, materia orgánica en descomposición, fragmentos minerales, agua y aire, y que puede tardar miles de años para formarse. (p. 188)

soil profile: vertical section of soil layers, each of which is a horizon. (p. 190)

perfil de suelos: sección vertical de las capas del suelo, cada una de las cuales es un horizonte. (p. 190)

solar eclipse: occurs when the Moon passes directly between the Sun and Earth and casts a shadow over part of Earth. (p. 669)

eclipse solar: ocurre cuando la luna pasa directamente entre el sol y la Tierra y se genera una sombra sobre una parte de la Tierra. (p. 669)

solar energy: energy from the Sun that is clean, inexhaustible, and can be transformed into electricity by solar cells. (p. 130)

energía solar: energía del sol, la cual es limpia e inagotable y puede transformarse en electricidad a través de celdas solares. (p. 130)

solar system: system of eight planets, including Earth, and other objects that revolve around the Sun. (p. 691)

sistema solar: sistema de ocho planetas, incluyendo a la Tierra y otros objetos que giran alrededor del sol. (p. 691)

solstice: twice-yearly point at which the Sun reaches its greatest distance north or south of the equator. (p. 664)

solsticio: punto en el cual dos veces al año el sol alcanza su mayor distancia al norte o al sur del ecuador. (p. 664)

solution: a mixture which is evenly mixed throughout; also known as a homogeneous mixture. (p. 43)

solución: una mezcla combinada uniformemente; también conocida como mezcla homogénea. (p. 43)

space probe: instrument that travels far into the solar system and gathers data that it sends them back to Earth. (p. 638)

sonda espacial: instrumento que viaja grandes distancias en el sistema solar, recopila datos y los envía a la Tierra. (p. 638)

space shuttle: reusable spacecraft that can carry cargo, astronauts, and satellites to and from space. (p. 643)

trasbordador espacial: nave espacial reutilizable que puede llevar carga, astronautas y satélites hacia y desde el espacio. (p. 643)

space station: large facility with living quarters, work and exercise areas, and equipment and support systems for humans to live and work in space and conduct research. (p. 644)

estación espacial: instalación grande con áreas para hospedarse, trabajar y hacer ejercicio; tiene equipos y sistemas de apoyo para que los seres humanos vivan, trabajen y lleven a cabo investigaciones en el espacio. (p. 644)

species: group of organisms that reproduces only with other members of their own group. (p. 394)

specific gravity: ratio of a mineral's weight compared with the weight of an equal volume of water. (p. 70)

sphere (SFIHR): a round, three-dimensional object whose surface is the same distance from its center at all points; Earth is a sphere that bulges somewhat at the equator and is slightly flattened at the poles. (p. 660)

spring: forms when the water table meets Earth's surface; often found on hillsides and used as a freshwater source. (p. 253)

station model: indicates weather conditions at a specific location, using a combination of symbols on a map. (p. 471)

streak: color of a mineral when it is in powdered form. (p. 71)

stream discharge: volume of water that flows past a specific point per unit of time. (p. 581)

strike-slip fault: break in rock caused by shear forces, where rocks move past each other without much vertical movement. (p. 303)

sunspots: areas on the Sun's surface that are cooler and less bright than surrounding areas, are caused by the Sun's magnetic field, and occur in cycles. (p. 730)

supergiant: late stage in the life cycle of a massive star in which the core heats up, heavy elements form by fusion, and the star expands; can eventually explode to form a supernova. (p. 739)

surface current: wind-powered ocean current that moves water horizontally, parallel to Earth's surface, and moves only the upper few hundred meters of seawater. (p. 518)

surface wave: seismic wave that moves rock particles up-and-down in a backward rolling motion and side-to-side in a swaying motion. (p. 305)

especie: grupo de organismos que se reproduce sólo entre los miembros de su mismo grupo. (p. 394)

gravedad específica: cociente del peso de un mineral comparado con el peso de un volumen igual de agua. (p. 70)

esfera: un objeto tridimensional y redondo donde cualquier punto de su superficie está a la misma distancia del centro; la Tierra es una esfera algo abultada en el ecuador y ligeramente achatada en los polos. (p. 660)

manantial: se forma cuando el nivel freático alcanza la superficie terrestre; a menudo se encuentran en las laderas y se usan como fuente de agua potable. (p. 253)

modelo estacional: indica las condiciones del estado del tiempo en una ubicación específica, utilizando una combinación de símbolos en un mapa. (p. 471)

veta: color de un mineral en forma de polvo. (p. 71)

descarga de corriente: volumen de agua que fluye a través de un punto específico por unidad de tiempo. (p. 581)

falla deslizante: ruptura en la roca causada por fuerzas opuestas, donde las rocas se mueven una tras otra sin mucho movimiento vertical. (p. 303)

manchas solares: áreas en la superficie solar que son más frías y menos brillantes que las áreas circundantes, son causadas por el campo magnético solar y ocurren en ciclos. (p. 730)

supergigante: etapa tardía en el ciclo de vida de una estrella masiva en la que el núcleo se calienta, se forman elementos pesados por fusión y la estrella se expande; eventualmente puede explotar para formar una supernova. (p. 739)

corriente de superficie: corriente oceánica empujada por el viento que mueve el agua horizontalmente, paralela a la superficie de la Tierra, y mueve sólo unos cientos de metros de la parte superior del agua marina. (p. 518)

onda de superficie: onda sísmica que mueve partículas rocosas en forma ascendente y descendente en un movimiento circular en retroceso y de un lado a otro en un movimiento oscilante. (p. 305)

technology: use of scientific discoveries for practical purposes, making people's lives easier and better. (p. 12)

tecnología: uso de descubrimientos científicos para propósitos prácticos, haciendo que la vida de las personas sea mejor y más fácil. (p. 12)

temperate zones: climate zones with moderate temperatures that are located between the tropics and the polar zones. (p. 484)

tephra (TEFF ruh): bits of rock or solidified lava dropped from the air during an explosive volcanic eruption; ranges in size from volcanic ash to volcanic bombs and blocks. (p. 340)

terracing: farming method used to reduce erosion on steep slopes. (p. 199)

tidal range: the difference between the level of the ocean at high tide and the level at low tide. (p. 528)

tide: daily rise and fall in sea level caused, for the most part, by the interaction of gravity in the Earth-Moon system. (p. 527)

till: mixture of different-sized sediments that is dropped from the base of a retreating glacier and can cover huge areas of land. (p. 216)

topographic map: map that shows the changes in elevation of Earth's surface and indicates such features as roads and cities. (p. 166)

tornado: violent, whirling windstorm that crosses land in a narrow path and can result from wind shears inside a thunderhead. (p. 466)

trench: long, narrow, steep-sided depression in the seafloor formed where one crustal plate sinks beneath another. (p. 545)

trilobite (TRI luh bite): organism with a three-lobed exoskeleton that was abundant in Paleozoic oceans and is considered to be an index fossil. (p. 393)

tropics: climate zone that receives the most solar radiation, is located between latitudes 23°N and 23°S, and is always hot, except at high elevations. (p. 484)

troposphere: layer of Earth's atmosphere that is closest to the ground, contains 99 percent of the water vapor and 75 percent of the atmospheric gases, and is where clouds and weather occur. (p. 428)

trough (TRAWF): lowest point of a wave. (p. 524)

tsunami (soo NAH mee): seismic sea wave that begins over an earthquake focus and can be highly destructive when it crashes on shore. (p. 316)

zonas templadas: zonas climáticas con temperaturas moderadas que están localizadas entre los trópicos y las zonas polares. (p. 484)

tefra: trozos de roca o lava solidificada que caen del aire durante una erupción volcánica explosiva; su tamaño oscila desde la ceniza volcánica hasta las bombas o bloques volcánicos. (p. 340)

terraceo: método de siembra usado para reducir la erosión en cuestas inclinadas. (p. 199)

rango de la marea: la diferencia entre el nivel del océano en marea alta y marea baja. (p. 528)

marea: elevación y disminución diaria del nivel del mar causada, en su mayor parte, por la interacción de la gravedad en el sistema Tierra-Luna. (p. 527)

tillita: mezcla de sedimentos de diferentes tamaños que ha caído de la base de un glaciar en retroceso y puede cubrir grandes extensiones de terreno. (p. 216)

mapa topográfico: mapa que muestra los cambios en la elevación de la superficie terrestre que puede ser representado como una relación e indica características como carreteras y ciudades. (p. 166)

tornado: tormenta de viento en forma de remolino que cruza la tierra en un curso estrecho y puede resultar de vientos que se entrecruzan en direcciones opuestas dentro del frente de una tormenta. (p. 466)

fosa: depresión estrecha, alargada y de bordes pronunciados en el suelo marino; se forma cuando una placa de la corteza se hunde por debajo de otra. (p. 545)

trilobite: organismo con un exoesqueleto trilobulado que fue abundante en los océanos del Paleozoico y es considerado como un fósil índice. (p. 393)

trópicos: zonas climáticas que reciben la mayor parte de la radiación solar, están localizadas entre los 23° de latitud norte y 23° de latitud sur y siempre son cálidas excepto a grandes alturas. (p. 484)

troposfera: capa de la atmósfera terrestre que se encuentra cerca del suelo, contiene el 99 por ciento del vapor de agua y el 75 por ciento de los gases atmosféricos; es donde se forman las nubes y las condiciones meteorológicas. (p. 428)

seno: el punto más bajo de una ola. (p. 524)

maremoto: onda sísmica marina que comienza sobre el foco del terremoto y que puede ser altamente destructiva cuando se estrella en la costa. (p. 316)

ultraviolet radiation: a type of energy that comes to Earth from the Sun, can damage skin and cause cancer, and is mostly absorbed by the ozone layer. (p. 432)

unconformity (un kun FOR mih tee): gap in the rock layer that is due to erosion or periods without any deposition. (p. 372)

uniformitarianism: principle stating that Earth processes occurring today are similar to those that occurred in the past. (p. 381)

upwarped mountains: mountains formed when blocks of Earth's crust are pushed up by forces inside Earth. (p. 158)

upwelling: vertical circulation in the ocean that brings deep, cold water to the ocean surface. (p. 521)

Uranus (YOOR uh nus): seventh planet from the Sun; is large and gaseous, has a distinct bluish-green color, and rotates on an axis nearly parallel to the plane of its orbit. (p. 705)

radiación ultravioleta: tipo de energía que llega a la Tierra desde el sol y que puede dañar la piel y causar cáncer; la mayor parte de esta radiación es absorbida por la capa de ozono. (p. 432)

discordancia: brecha en la capa rocosa que es debida a la erosión o a periodos sin deposición. (p. 372)

uniformitarianismo: principio que establece que los procesos de la Tierra que ocurren actualmente son similares a los que ocurrieron en el pasado. (p. 381)

montañas de levantamiento: montañas que se forman cuando los bloques de la corteza terrestre son empujados hacia arriba por fuerzas del interior de la Tierra. (p. 158)

solevantamiento: circulación vertical en el océano que trae el agua fría de las profundidades a la superficie del océano. (p. 521)

Urano: séptimo planeta desde el sol; es grande y gaseoso, tiene un color verde-azulado distintivo y gira sobre un eje casi paralelo al plano de su órbita. (p. 705)

variables: different factors that can be changed in an experiment. (p. 10)

vent: opening where magma is forced up and flows out onto Earth's surface as lava, forming a volcano. (p. 332)

Venus: second planet from the Sun; similar to Earth in mass and size; has a thick atmosphere and a surface with craters, faultlike cracks, and volcanoes. (p. 697)

volcanic mountains: mountains formed when molten material reaches Earth's surface through a weak crustal area and piles up into a cone-shaped structure. (p. 159)

volcanic neck: solid igneous core of a volcano left behind after the softer cone has been eroded. (p. 347)

volcano: opening in Earth's surface that erupts sulfurous gases, ash, and lava; can form at Earth's plate boundaries, where plates move apart or together, and at hot spots. (p. 330)

variables: diferentes factores que pueden cambiarse en un experimento. (p. 10)

chimenea: apertura donde el magma es empujado hacia arriba y fluye sobre la superficie terrestre como lava, formando un volcán. (p. 332)

Venus: segundo planeta más cercano al sol; similar a la Tierra en masa y tamaño; tiene una atmósfera gruesa y una superficie con cráteres, grietas similares a fallas y volcanes. (p. 697)

montañas volcánicas: montañas formadas cuando material derretido alcanza la superficie a través de un área débil de la corteza terrestre y se acumula formando una estructura en forma de cono. (p. 159)

cuello volcánico: núcleo ígneo sólido de un volcán que queda después de que el cono más blando ha sido erosionado. (p. 347)

volcán: apertura en la superficie terrestre que arroja gases sulfurosos, ceniza y lava; puede formarse en los límites de las placas continentales, donde las placas se separan o encuentran y en los puntos calientes. (p. 330)

waning: describes phases that occur after a full moon, as the visible lighted side of the Moon grows smaller. (p. 10)

water table: upper surface of the zone of saturation; drops during a drought. (p. 250)

wave: rhythmic movement that carries energy through matter or space; can be described by its crest, trough, wavelength, and wave height. (p. 524)

waxing: describes phases following a new moon, as more of the Moon's lighted side becomes visible. (p. 10)

weather: state of the atmosphere at a specific time and place, determined by factors including air pressure, amount of moisture in the air, temperature, wind, and precipitation. (p. 454)

weathering: mechanical or chemical surface processes that break rock into smaller and smaller pieces. (p. 182)

white dwarf: late stage in the life cycle of a comparatively low-mass main sequence star; formed when its core depletes its helium and its outer layers escape into space, leaving behind a hot, dense core. (p. 738)

wind farm: area where many windmills use wind to generate electricity. (p. 131)

menguante: describe las fases posteriores a la luna llena, de manera que el lado iluminado de la luna es cada vez menos visible. (p. 10)

nivel freático: parte superior de la zona de saturación; desciende durante las sequías. (p. 250)

ola: movimiento rítmico que lleva energía a través de la materia o del espacio; puede describirse por su cresta, valle, longitud de la ola y altura de la ola. (p. 524)

creciente: describe las fases posteriores a la luna nueva, de manera que el lado iluminado de la luna es cada vez más visible. (p. 10)

estado del tiempo: estado de la atmósfera en un momento y lugar específicos, determinado por factores que incluyen la presión del aire, cantidad de humedad en el aire, temperatura, viento y precipitación. (p. 454)

erosión: proceso superficial químico o mecánico que fracciona la roca en trozos cada vez más pequeños. (p. 182)

enana blanca: etapa tardía en el ciclo de vida de una estrella de secuencia principal, de relativamente poca masa, formada cuando el núcleo agota su helio y sus capas externas escapan al espacio, dejando atrás un núcleo denso y caliente. (p. 738)

granja de energía eólica: área en donde muchos molinos usan el viento para generar electricidad. (p. 131)

Index

Italic numbers = illustration/photo **Bold numbers = vocabulary term**
lab *= indicates a page on which the entry is used in a lab*
act *= indicates a page on which the entry is used in an activity*

Index

B

Index

K

L

M

N

O

Q

R

S

Magnification Key: Magnifications listed are the magnifications at which images were originally photographed.
LM–Light Microscope
SEM–Scanning Electron Microscope
TEM–Transmission Electron Microscope

Acknowledgments: Glencoe would like to acknowledge the artists and agencies who participated in illustrating this program: Absolute Science Illustration; Andrew Evansen; Argosy; Articulate Graphics; Craig Attebery represented by Frank & Jeff Lavaty; CHK America; John Edwards and Associates; Gagliano Graphics; Pedro Julio Gonzalez represented by Melissa Turk & The Artist Network; Robert Hynes represented by Mendola Ltd.; Morgan Cain & Associates; JTH Illustration; Laurie O'Keefe; Matthew Pippin represented by Beranbaum Artist's Representative; Precision Graphics; Publisher's Art; Rolin Graphics, Inc.; Wendy Smith represented by Melissa Turk & The Artist Network; Kevin Torline represented by Berendsen and Associates, Inc.; WILDlife ART; Phil Wilson represented by Cliff Knecht Artist Representative; Zoo Botanica.

Photo Credits

Cover George Steinmetz/CORBIS; **i ii** PhotoDisc; **vii** Aaron Haupt; **viii** John Evans; **ix** (t)PhotoDisc, (b)John Evans; **x** (l)John Evans, (r)Geoff Butler; **xi** (l)John Evans, (r)PhotoDisc; **xii** PhotoDisc; **xiii** Doug Martin; **xiv** David Muench/CORBIS; **xvii** Larry Lee/CORBIS; **xviii** (t)Darryl Torckler/Stone/Getty Images, (b)Herb Segars/Animals Animals; **xix** Chuck Place/Stock Boston; **xx** (t)USGS/TSADO/Tom Stack & Assoc., (c)NASA/Photo Researchers, (b)USGS/Science Photo Library/Photo Researchers; **xxi** Science Photo Library/Photo Researchers; **xxii** CORBIS/PictureQuest; **xxv** David J. Cross/Peter Arnold, Inc.; **xxvi** Matt Meadows; **xxvii** Fred Bavendam/Minden Pictures; **xxviii** Louis Psihoyos/Matrix; **1** (t)NASA/Science Source/Photo Researchers, (bl)Arthur R. Hill/Visuals Unlimited, (br)Mark Schneider/Visuals Unlimited; **2–3** Mike Zens/CORBIS; **3** (l r)Mark A. Schneider/Visuals Unlimited; **4–5** Dutheil Didier/CORBIS Sygma; **7** (t)Michael Habicht/Earth Scenes, (b)Michael Wilhelm/ENP Images; **8** Richard Cummins/CORBIS; **9** John Heseltine/Science Photo Library/Photo Researchers; **10 11** Aaron Haupt; **12** (t)Michael Dwyer/Stock Boston, (c)Tim Courlas, (b)Mark Segal/Stock Boston; **13** (t)Science Museum/Science & Society Picture Library, (cl)reprinted by permission of Parks Canada and Newfoundland Museum, (c cr)Dorling Kindersley, (b)NASA; **14** Russ Underwood/Lockheed Martin Space Systems; **15** Todd Gustafson/Danita Delimont, Agent; **16** Smithsonian Institution; **18** NASA/MSFC; **19** European Space Agency/Science Photo Library/Photo Researchers; **20** (l)Frans Lanting/Minden Pictures, (c)Ted Levin/Animals Animals, (r)Al & Linda Bristor/Visuals Unlimited; **21** Matt Meadows; **22** Bates Littlehales/National Geographic Image Collection; **23** Aaron Haupt; **24** Mark Burnett; **25** Timothy Fuller; **26** AP/Wide World Photos; **27** (l)CABISCO/Visuals Unlimited, (r)Jan Hinsch/Science Photo Libray/Photo Researchers; **29** Paul Silverman/Fundamental Photographs; **32–33** John Coletti/Index Stock; **35** (tl)Stephan Frisch/Stock Boston, (tcl)Dane S. Johnson/Visuals Unlimited, (tcr)Ken Lucas/Visuals Unlimited, (tr)Mark A. Schneider/Photo Researchers, (bl br)Aaron Haupt, (bcl)Amanita Pictures, (bcr)Charles D. Winters/Photo Researchers; **36** John Evans; **39** (l)Herbert Kehrer/OKAPIA/Photo Researchers, (c)Doug Martin, (r)Bruce Hands/Getty Images; **40** Kenji Kerins; **42** Ken Whitmore/Getty Images; **43** Mark Steinmetz; **44** Stuart Westmorland/Danita Delimont, Agent; **46** John S. Lough/Visuals Unlimited; **48** CORBIS; **49** (tl)Breck P. Kent/Earth Scenes, (tr)Storm Pirate Productions/Artville/PictureQuest, (bl)CORBIS/PictureQuest, (br)E.R. Degginger/Earth Scenes; **50** (t)Paul Chesley/Getty Images, (b)David Muench/CORBIS; **51** NASA/JPL/Malin Space Science Systems; **52** Matt Meadows; **53** (t)Tim Courlas, (b)Royalty-Free/CORBIS; **54** (t)KS Studios, (b)StudiOhio; **55** (l)Doug Martin, (r)Matt Meadows; **58** Ed Young/CORBIS; **60–61** SuperStock; **62** Matt Meadows; **63** (inset)John R. Foster/Photo Researchers, (l)Mark A. Schneider/Visuals Unlimited; **64** (tr br)Mark A. Schneider/Visuals Unlimited, (cl)A.J. Copley/Visuals Unlimited, (cr bl)Harry Taylor/DK Images, (bc)Mark A. Schneider/Photo Researchers; **65** (inset)Patricia K. Armstrong/Visuals Unlimited, (r)Dennis Flaherty Photography/Photo Researchers; **67** KS Studios; **68** (l)Mark Burnett/Photo Researchers, (c)Dan Suzio/Photo Researchers, (r)Breck P. Kent/Earth Scenes; **69** (t)Bud Roberts/Visuals Unlimited, (inset)Icon Images, (b)Charles D. Winters/Photo Researchers; **70** (l)Andrew McClenaghan/Science Photo Library/Photo Researchers, (r)Charles D. Winters/Photo Researchers; **71** (t)Goeff Butler, (bl)Doug Martin, (br)Photo Researchers; **72** Matt Meadows; **73** Reuters NewMedia, Inc./CORBIS; **74** (Beryl, Spinel)Biophoto Associates/Photo Researchers, (Emerald, Topaz)H. Stern/Photo Researchers, (Ruby Spinel, Tanzanite)A.J. Copley/Visuals Unlimited, (Zoisite)Visuals Unlimited, (Uncut Topaz) Mark A. Schneider/Visuals Unlimited; **75** (Olivine) University of Houston, (Peridot)Charles D. Winters/Photo Researchers, (Garnet)Arthur R. Hill/Visuals Unlimited, (Almandine)David Lees/CORBIS, (Quartz, Corundum)Doug Martin, (Amethyst)A.J. Copley/Visuals Unlimited, (Blue Sapphire) Vaughan Fleming/Science Photo Library/Photo Researchers; **76** (l)Francis G. Mayer/CORBIS, (r)Smithsonian Institution; **77** (l)Fred Whitehead/Earth Scenes, (inset)Doug Martin; **78** (t)Matt Meadows, (bl)Paul Silverman/Fundamental Photographs, (br)Biophoto Associates/Photo Researchers; **79** Jim Cummins/Getty Images; **80** Matt Meadows; **81** (t)Doug Martin, (inset)José Manuel Sanchis Calvete/CORBIS, (bl)Andrew J. Martinez/Photo Researchers, (br)Charles D. Winter/Photo Researchers; **82** (bkgd)Science Photo Library/Custom Medical Stock Photo, (bl)Bettmann/CORBIS; **83** José Manuel Sanchis Calvete/CORBIS; **84** R. Weller/Cochise College; **86** José Manuel Sanchis Calvete/CORBIS; **87** Breck P. Kent/Earth Scenes; **88–89** Michael T. Sedam/CORBIS; **90** (l)CORBIS, (inset) Doug Martin; **91** (tl)Steve Hoffman, (tr)Breck P. Kent/Earth Scenes, (cl)Brent Turner/BLT Productions, (cr)Breck P. Kent/Earth Scenes; **92** (bkgd)CORBIS/ PictureQuest, (t)CORBIS, (bl)Martin Miller, (bc)Jeff Gnass, (br)Doug Sokell/Tom Stack & Assoc.; **93** Russ Clark; **94** USGS/HVO; **95** (t)Breck P. Kent/Earth Scenes, (b)Doug Martin; **96** (Basalt)Mark Steinmetz, (Andesite, Granite), Doug Martin, (Pumice)Tim Courlas, (others)Breck P. Kent/Earth Scenes; **98** (l)Breck P. Kent/Earth Scenes, (r)Doug Martin/Photo Researchers; **99** (l r) Breck P. Kent/Earth Scenes, (c)Courtesy Kent Ratajeski & Dr. Allen Glazner, University of NC, (inset)Alfred Pasieka/Photo Researchers; **101** (inset)Aaron Haupt, (r)Robert Estall/CORBIS; **102** Paul Rocheleau/Index Stock; **103** (l)Timothy Fuller, (r)Steve McCutcheon/Visuals Unlimited; **105** (l)Icon Images, (cl)Doug Martin, (cr)Andrew Martinez/Photo

Researchers, (r)John R. Foster/Photo Researchers; **106** (l)Breck P. Kent/Earth Scenes, (r)Aaron Haupt; **107** Georg Gerster/Photo Researchers, (inset)Icon Images; **109** Beth Davidow/Visuals Unlimited; **110** (l)con Images, (r)Breck P. Kent/Earth Scenes; **111** (l)Jack Sekowski, (r)Tim Courlas; **112** (bkgd)Y. Kawasaki/Photonica, (inset)Matt Turner/Liaison Agency; **114** Breck P. Kent/Earth Scenes; **115** Jeremy Woodhouse/DRK Photo; **118–119** Bill Ross/CORBIS; **121** Visuals Unlimited, (l)George Lepp/CORBIS; **124** (r)Carson Baldwin Jr./Earth Scenes; **125** Paul A. Souders/CORBIS; **126** (l)Emory Kristof, (r)National Energy Technology Laboratory, (bkgd)Ian R. MacDonald/Texas A&M University; **127** Hal Beral/Visuals Unlimited; **129** Roger Ressmeyer/CORBIS; **130** Spencer Grant/PhotoEdit, Inc.; **131** Inga Spence/Visuals Unlimited; **132** Robert Cameron/Stone/Getty Images; **133** Vince Streano/CORBIS; **134** (t)David Young-Wolff/PhotoEdit, Inc., (b)Earl Young/Archive Photos; **135** Peter Holden/Visuals Unlimited; **137** Aaron Haupt; **138** Joseph Nettis/Photo Researchers; **139** (t)Mark Joseph/Stone/Getty Images, (bl)Aaron Haupt, (br)Wyoming Mining Association; **142** (t)Aaron Haupt, (b)Joel W. Rogers/CORBIS; **143** Aaron Haupt; **144** (t)Ed Clark, (bl)Brown Brothers, (br)Shell Oil Co.; **145** (l)Andrew J. Martinez/Photo Researchers, (r)Coco McCoy/Rainbow; **149** Mark Joseph/Stone/Getty Images; **150–151** Thierry Borredon/Stone/Getty Images; **151** (inset)Robert Caputo/Aurora/PictureQuest; **152–153** GSFC/NASA; **155** (tl)Alan Maichrowicz/Peter Arnold, Inc., (tr)Carr Clifton/Minden Pictures, (b)Stephen G. Maka/DRK Photo; **156** Ron Mellot; **157** John Lemker/Earth Scenes; **158** (t)John Kieffer/Peter Arnold, Inc., (b)Carr Clifton/Minden Pictures; **159** David Muench/CORBIS; **162** Dominic Oldershaw; **167** (t)Rob & Ann Simpson, (b)courtesy Maps a la Carte, Inc. and TopoZone.com; **170** CORBIS; **172** (t)Layne Kennedy/CORBIS, (b)John Evans; **173** John Evans; **174** (tl)Culver Pictures, (tcl b)PhotoDisc, (c)Pictor, (tcr)William Manning/The Stock Market/CORBIS, (tr)Kunio Owaki/The Stock Market/CORBIS; **175** (l)Tom Bean/DRK Photo, (r)Marc Muench; **176** William Weber; **178** Aaron Haupt; **180–181** Andrew Brown, Ecoscene/CORBIS; **183** (l)StudiOhio, (r)Tom Bean/DRK Photo; **184** W. Perry Conway/CORBIS; **185** Hans Strand/Stone; **186** (tl)Craig Kramer, (tr)A.J. Copley/Visuals Unlimited, (bl br)John Evans; **187** (l)William Johnson/Stock Boston, (r)Runk/Schoenberger from Grant Heilman; **189** (bkgd) Stephen R. Wagner, (t)James D. Balog, (c)Martin Miller, (b)Steven C. Wilson/Entheos; **190** (l)Bonnie Heidel/Visuals Unlimited, (r)John Bova/Photo Researchers; **196** (l)Gary Braasch/CORBIS, (r)Donna Ikenberry/Earth Scenes; **197** Chip & Jill Isenhart/Tom Stack & Associates; **198** (t)Dr. Russ Utgard, (b)Denny Eilers from Grant Heilman; **199** Georg Custer/ Photo Researchers; **200** (t)George H. Harrison from Grant Heilman, (b)Bob Daemmrich; **201** KS Studios; **202** Larry Hamill; **203** (l)Tom Bean/DRK Photo, (r)David M. Dennis/ Earth Scenes; **205** Matt Meadows; **206** Georg Custer/Photo Researchers; **208–209** Paul A. Souders/CORBIS; **210** Robert L. Schuster/USGS; **211** Martin G. Miller/Visuals Unlimited; **212** (t)John D. Cunningham/Visuals Unlimited, (bl)Sylvester Allred/Visuals Unlimited, (br)Tom Uhlman/Visuals Unlimited; **213** AP/Wide World Photos; **214** Martin G. Miller/Visuals Unlimited; **216** James N. Westwater; **217** (t)Tom Bean/Stone/Getty Images, (b)Tom Bean/CORBIS; **218** John Gerlach /Visuals Unlimited; **219** Gregory G. Dimijian/Photo Researchers; **220** Mark E. Gibson/Visuals Unlimited; **221** Timothy Fuller; **222** Galen Rowell/CORBIS; **224** Fletcher & Baylis/Photo Researchers; **225** (t)John D. Cunningham/Visuals Unlimited, (b)file photo; **226** (tl)Stephen J. Krasemann/Photo Researchers, (tr)Steve McCurry, (b)Wyman P. Meinzer, (bkgd)Breck P. Kent/Earth Scenes; **227** John Giustina/FPG/Getty Images; **228** (t)Greg Vaughn/ Tom Stack & Assoc., (b)Matt Meadows; **230** (t)World Class Images, (c)Yann Arthus-Bertrand/CORBIS, (b)AP/Wide World Photos; **232** John D. Cunningham/Visuals Unlimited; **236–237** William Manning/The Stock Market/CORBIS; **237** Aaron Haupt; **238** (l)Michael Busselle/Stone/Getty Images, (r)David Woodfall/DRK Photo; **239** Tim Davis/Stone/Getty Images; **240** (t)Grant Heilman Photography, (b)KS Studios; **242** Mel Allen/ICL/Panoramic Images; **244** CORBIS/PictureQuest; **245** (l)Harald Sund/The Image Bank/Getty Images, (r)Loren McIntyre; **246** C. Davidson/Comstock; **247** James L. Amos/CORBIS; **248** (l)Wolfgang Kaehler, (r)Nigel Press/Stone/Getty Images; **249** First Image; **251** CORBIS; **252** file photo; **253** Barbara Filet; **254** Chad Ehlers/Stone/Getty Images; **256** Macduff Everton/The Image Bank/Getty Images; **257** (tl)Steve Bentsen, (tr)SuperStock, (bl)Runk/Schoenberger from Grant Heilman, (br)Breck P. Kent/Earth Scenes; **258** Bruce Roberts/Photo Researchers; **260 261** KS Studios; **262** Gary Bogdon/CORBIS Sygma; **263** (l)Todd Powell/Index Stock, (r)J. Wengle/DRK Photo; **266** Barbara Filet; **267** Grant Heilman Photography; **268–269** James Watt/Earth Scenes; **268** (l r)Ken Lucas/TCL/Masterfile; **269** (l)Patrice Ceisel/Stock Boston/PictureQuest, (r)Hal Beral/Photo Network/PictureQuest; **270–271** Bourseiller/Durieux/Photo Researchers; **274** Martin Land/Science Source/Photo Researchers; **277** Ralph White/CORBIS; **283** Davis Meltzer; **284** Craig Aurness/CORBIS; **286** Craig Brown/Index Stock; **287** Ric Ergenbright/CORBIS; **288** Roger Ressmeyer/CORBIS; **290** AP/Wide World Photos; **292** L. Lauber/Earth Scenes; **298–299** Chuck Nacke/TimeLife Pictures/Getty Images; **300** Tom & Therisa Stack; **302** (t)Tom Bean/DRK Photo, (b)Lysbeth Corsi/Visuals Unlimited; **303** David Parker/Photo Researchers; **304** Tom & Therisa Stack; **306** Robert W. Tope/Natural Science Illustrations; **313** (l)Steven D. Starr/Stock Boston, (r)Berkeley Seismological Laboratory; **314** AP/Wide World Photos; **315** David J. Cross/Peter Arnold, Inc.; **318** James L. Stanfield/National Geographic Image Collection; **319** David Young-Wolff/PhotoEdit, Inc.; **321** Reuters NewMedia Inc./CORBIS; **322** (tr)Richard Cummins/CORBIS, (l)Bettmann/CORBIS, (br)RO-MA Stock/Index Stock; **323** (l)Science VU/Visuals Unlimited, (r)Peter Menzel/Stock Boston; **326** Vince Streano/CORBIS; **328–329** Art Wolfe/Getty Images; **329** KS Studios; **330** Sigurjon Sindrason; **331** (t)John Cancalosi/DRK Photo, (b)Deborah Brosnan, Sustainable Ecosystems Institute; **335** Image courtesy NASA/GSFC/JPL, MISR Team; **336** Gary Rosenquist; **337** Gary Rosenquist; **338** (bkgd)API/Explorer/Photo Researchers, (t)Krafft/HOA-QUI/Photo Researchers, (bl)Robert Hessler/Planet Earth Pictures, (br)Paul Chesley; **339** (l)Steve Kaufman/DRK Photo, (r)Dee Breger/Photo Researchers; **341** (t)Krafft/Explorer/Science Source/Photo Researchers, (b)Darrell Gulin/DRK Photo; **346** (tl)Joyce Photo/Photo Researchers, (tr)Doug Martin, (b)Brent Turner; **347** (t)Dick Canby/DRK Photo, (b)Tom Bean/DRK Photo; **349** Larry Ulrich/DRK Photo; **350** Amanita Pictures; **351** (t)Spencer Grant/PhotoEdit, Inc., (b)Darrell Gulin/DRK Photo; **352** Mimmo Jodice/CORBIS; **353** (l)Soames Summerhays/Photo Researchers, (r)Photri/The Stock Market/CORBIS; **354** Krafft/Explorer/Science Source/Photo Researchers; **356** Kerrick James/Getty Images;

358–359 Coco McCoy from Rainbow/PictureQuest; **359** (inset)Mary Evans Picture Library; **360–361** Hugh Sitton/Getty Images; **362** (t)Mark E. Gibson/Visuals Unlimited, (b)D.E. Hurlbert & James DiLoreto/Smithsonian Institution; **363** Jeffrey Rotman/CORBIS; **364** (t)Dr. John A. Long, (b)A.J. Copley/Visuals Unlimited; **366** (t)PhotoTake, NYC/PictureQuest, (b)Louis Psihoyos/Matrix; **368** David M. Dennis; **369** (l)Gary Retherford/Photo Researchers, (r)Lawson Wood/CORBIS; **370** Aaron Haupt; **373** (l)IPR/12-18 T. Bain, British Geological Survey/NERC. All rights reserved, (r)Tom Bean/CORBIS, (bkgd)Lyle Rosbotham; **374** Jim Hughes/PhotoVenture/Visuals Unlimited; **375** (l)Michael T. Sedam/CORBIS, (r)Pat O'Hara/CORBIS; **377** Aaron Haupt; **379** James King-Holmes/Science Photo Library/Photo Researchers; **380** Kenneth Garrett; **381** WildCountry/CORBIS; **382** (t)A.J. Copley/Visuals Unlimited, (b)Lawson Wood/CORBIS; **383** Matt Meadows; **384** Jacques Bredy; **385** (tl)François Gohier/Photo Researchers, (tr)Sinclair Stammers/Photo Researchers, (b)Mark E. Gibson/DRK Photo; **388** Tom Bean/CORBIS; **390–391** Roger Garwood & Trish Ainslie/CORBIS; **391** KS Studios; **392** Tom & Therisa Stack/Tom Stack & Assoc.; **394** (l)Gerald & Buff Corsi/Visuals Unlimited, (r)John Gerlach/Animals Animals; **396** (tl)Mark Boulron/Photo Researchers, (others)Walter Chandoha; **397** Jeff Lepore/Photo Researchers; **401** (l)Mitsuaki Iwago/Minden Pictures, (r)R. Calentine/Visuals Unlimited; **403** J.G. Gehling/Discover Magazine; **404** Gerry Ellis/ENP Images; **407** Matt Meadows; **410** (l)David Burnham/Fossilworks, Inc., (r)François Gohier/Photo Researchers; **412** Michael Andrews/Earth Scenes; **413** Tom J. Ulrich/Visuals Unlimited; **414** David M. Dennis; **415** Mark Burnett; **417** (l)E. Webber/Visuals Unlimited, (r)Len Rue, Jr./Animals Animals; **419** John Cancalosi/Stock Boston; **422–423** S.P. Gillette/CORBIS; **424** NASA; **425** (l)Frank Rossotto/The Stock Market/CORBIS, (r)Larry Lee/CORBIS; **428** Laurence Fordyce/CORBIS; **430** Doug Martin; **431** NASA/GSFC; **432** Michael Newman/PhotoEdit, Inc.; **437** (t)Dan Guravich/Photo Researchers, (b)Bill Brooks/Masterfile; **439** (cw from top)Gene Moore/PhotoTake NYC/PictureQuest, Phil Schermeister/CORBIS, Joel W. Rogers, Kevin Schafer/CORBIS, Stephen R. Wagner; **440** Bill Brooks/Masterfile; **442 443** David Young-Wolff/PhotoEdit, Inc.; **444** Bob Rowan/CORBIS; **452–453** Reuters NewMedia Inc./CORBIS; **451** KS Studios; **454** Kevin Horgan/Stone/Getty Images; **455** Fabio Colombini/Earth Scenes; **459** (t)Charles O'Rear/CORBIS, (b)Joyce Photographics/Photo Researchers; **460** (l)Roy Morsch/The Stock Market/CORBIS, (r)Mark McDermott; **461** (l)Mark E. Gibson/Visuals Unlimited, (r)EPI Nancy Adams/Tom Stack & Assoc.; **463** Van Bucher/Science Source/Photo Researchers; **465** Jeffrey Howe/Visuals Unlimited; **466** Roy Johnson/Tom Stack & Assoc.; **467** (l)Warren Faidley/Weatherstock, (r)Robert Hynes; **468** NASA/Science Photo Library/Photo Researchers; **469** Fritz Pölking/Peter Arnold, Inc; **470** Howard Bluestein/Science Source/Photo Researchers; **473** Mark Burnett; **474** (t)Marc Epstein/DRK Photo, (b)Timothy Fuller, (bkgd)Erik Rank/Photonica; **476** courtesy Weather Modification Inc.; **477** (l)George D. Lepp/Photo Researchers, (r)Janet Foster/Masterfile; **478** Ruth Dixon; **479** Bob Daemmrich; **482–483** Andrew Wenzel/Masterfile; **487** (l)William Leonard/DRK Photo, (r)Bob Rowan, Progressive Image/CORBIS; **488** John Shaw/Tom Stack & Assoc.; **489** (tl)David Hosking/CORBIS, (tr)Yva Momatiuk & John Eastcott/Photo Researchers, (b)Michael Melford/The Image Bank/Getty Images; **490** (t)S.R. Maglione/Photo Researchers, (c)Jack Grove/Tom Stack & Assoc., (b)Fritz Pölking/Visuals Unlimited; **491** Zig Leszczynski/Animals Animals; **493** (l)Jonathan Head/AP/Wide World Photos, (r)Jim Corwin/Index Stock; **495** (t)A. Ramey/PhotoEdit, Inc., (b)Peter Beck/Pictor; **496** Galen Rowell/Mountain Light; **500** John Bolzan; **501** Chip & Jill Isenhart/Tom Stack & Assoc.; **503** Matt Meadows; **505** Doug Martin; **506** Alberto Garcia/Saba; **507** Steve Kaufman/DRK Photo; **510** (inset)Stephen Dalton/Animals Animals; **510–511** A.T. Willett/Image Bank/Getty Images; **512–513** Warren Bolster/Getty Images; **514** (l)Norbert Wu/Peter Arnold, Inc., (r)Darryl Torckler/Stone/Getty Images; **516** Cathy Church/Picturesque/PictureQuest; **519** Bob Daemmrich; **520** (t)Darryl Torckler/Stone/Getty Images, (b)Raven/Explorer/Photo Researchers; **524** Jack Fields/Photo Researchers; **525** Tom & Therisa Stack; **526** (t b)Stephen R. Wagner, (cl)Spike Mafford/PhotoDisc, (cr)Douglas Peebles/CORBIS; **527** Arnulf Husmo/Stone/Getty Images; **528** (l)Groenendyk/Photo Researchers, (r)Patrick Ingrand/Stone/Getty Images, (b)Kent Knudson/Stock Boston; **532** (t)Mark E. Gibson/Visuals Unlimited, (b)Timothy Fuller; **533** Timothy Fuller; **534** (bkgd)Chris Lisle/CORBIS, Sovfoto/Eastfoto/PictureQuest; **535** (l)Carl R. Sams II/Peter Arnold, Inc., (r)Edna Douthat; **540–541** Bill Curtsinger/Getty Images; **542–543** The Floor of the Oceans by Bruce C. Heezen and Marie Tharp, ©1977 by Marie Tharp. Reproduced by permission of Marie Tharp; **541** Mark Burnett; **544** Woods Hole Oceanographic Institution; **545** Thomas J. Abercrombie/National Geographic Society; **546** (t)J. & L. Weber/Peter Arnold, Inc., (bl)Arthur Hill/Visuals Unlimited, (br)John Cancalosi/Peter Arnold, Inc.; **547** (t)Instutute of Oceanographic Sciences/NERC/Science Photo Library/Photo Researchers, (b)Biophoto Associates/Photo Researchers; **549** Fred Bavendam/Minden Pictures; **551** Nanct Sefton/Photo Researchers; **552** (t)Manfred Kage/Peter Arnold, Inc., (b)M.I. Walker/Science Source/Photo Researchers; **553** (l)Nick Caloyianis/National Geographic Society, (c)Herb Segars/Animals Animals, (r)Norbert Wu; **554** Fred Bavendam/Peter Arnold, Inc.; **555** (1)Lloyd K. Townsend, (2)Fred Whitehead/Animals Animals, (3)Andrew J. Martinez/Photo Researchers, (4)Andrew J. Martinez/Photo Researchers, (5)Hal Beral/Visuals Unlimited, (6)Michael Abbey/Photo Researchers, (7)Gerald & Buff Corsi/Visuals Unlimited, (8)Anne W. Rosenfeld/Animals Animals, (9)Andrew J. Martinez/Photo Researchers, (10)Zig Leszczynski/Animals Animals, (11)Gregory Ochocki/Photo Researchers, (12)Peter Skinner/Photo Researchers; **556** James H. Robinson/Photo Researchers; **559** (tl)C.C. Lockwood/Earth Scenes, (bl)C.C. Lockwood/DRK Photo, (others)David Young-Wolff/PhotoEdit, Inc.; **560** NASA; **561** David Young-Wolff/PhotoEdit, Inc.; **562** (t)Jim Nilsen/Stone/Getty Images, (b)Fred Bavendam/Minden Pictures; **563** (t)Fred Bavendam/Minden Pictures, (b)Jeff Rotman/Peter Arnold, Inc.; **564** (t)Rick Price/CORBIS, (b)Emory Kristof/National Geographic; **565** Fred Bavendam/Minden Pictures; **570–571** Joseph Sohm/ChromoSohm, Inc./CORBIS; **571** (l r)Andrew A. Wagner; **572–573** George D. Lepp/Photo Researchers; **574** (inset)Joseph Sohm/ChromoSohm Inc./CORBIS, (r)Bob Daemmrich/Stock Boston; **576** Timothy Fuller; **577** Aaron Haupt; **578** Paul Bousquet; **579** Gene Alexander/Soil Conservation Service;

Credits

581 Rich Iwasaki; **582** Simon Fraser/Northumbrian Environmental Management Ltd./Science Photo Library/Photo Researchers; **584** (tl)Gloria H. Chomica/Masterfile, (r)Raymond Gehman/CORBIS, (bl)David Muench/Getty Images;
585 John Evans; **588** (tl)Philip James Corwin/CORBIS, (cr)Skiold/PhotoEdit/PictureQuest, (bl)Bill Gallery/Stock Boston/PictureQuest, (br)Aerials Only, (included in graph)Digital Stock, Image Ideas, PhotoDisc; **589** David Young-Wolff/PhotoEdit, Inc.; **591** (t)SuperStock, (b)Matt Meadows; **592** VCG/FPG; **593** (l)Stacy Pick/Stock Boston, (r)Tom Bean/Stone/Getty Images; **596** Gene Alexander/Soil Conservation Service; **598–599** Russell Borden/Index Stock Imagery; **600** (l)Michael J. Pettypool/Pictor, (r)Visuals Unlimited; **601** (t)Bob Child/AP/Wide World Photos, (b)David Hoffman; **604** Colin Raw/Stone/Getty Images; **605** (t)Cleveland Public Library Photograph Collection, (b)Jim Baron/The Image Finders; **607** (l)C. Squared Studios/PhotoDisc, (c)Larry Lefever from Grant Heilman, (r)Dominic Oldershaw; **608** Mark Burnett; **609** Alan Pitcairn from Grant Heilman; **612** John Evans; **617** (t)file photo, (b)Dominic Oldershaw; **618** Eric Hartmann/Magnum Photos; **619** (l)David Woodfall/Stone/Getty Images, (r)Doug Martin; **622** Scott T. Smith/CORBIS; **624–625** Steve Murray/PictureQuest; **625** (inset)Davis Meltzer; **626–627** TSADO/NASA/Tom Stack & Assoc.; **628** (l)Weinberg-Clark/The Image Bank/Getty Images, (r)Stephen Marks/The Image Bank/Getty Images; **629** (l)PhotoEdit, Inc., (r)Wernher Krutein/Liaison Agency/Getty Images; **630** Chuck Place/Stock Boston; **631** NASA; **632** (t)Roger Ressmeyer/CORBIS, (b)Simon Fraser/Science Photo Library/Photo Researchers; **633** Raphael Gaillarde/Liaison Agency/Getty Images; **634** (t)Icon Images, (b)Diane Graham-Henry & Kathleen Culbert-Aguilar; **635** NASA; **636** NASA/Science Photo Library/Photo Researchers; **637** NASA; **638** (t tc)NASA/Science Source/Photo Researchers, (bc)M. Salaber/Liaison Agency/Getty Images, (b)Julian Baum/Science Photo Library/Photo Researchers; **639** (Venera 8)Dorling Kindersley Images, (Surface of Venus)TASS from Sovfoto, (Mercury, Venus)NASA/JPL, (Voyager 2, Neptune)NASA/JPL/Caltech, (others)NASA; **640** AFP/CORBIS; **641** NASA; **642** NASA/Science Source/Photo Researchers; **643** NASA/Liaison Agency/Getty Images; **644** (t)NASA, (b)NASA/Liaison Agency/Getty Images; **645** NASA/Science Source/Photo Researchers; **646** NASA/JPL/Malin Space Science Systems; **647** NASA/JPL/ Liaison Agency/Getty Images; **648** (t)David Ducros/Science Photo Library/Photo Researchers, (b)NASA; **649** The Cover Story/CORBIS; **650** Roger Ressmeyer/CORBIS; **651** Doug Martin; **652** Robert McCall; **653** (l)Novosti/Science Photo Library/Photo Researchers, (c)Roger K. Burnard, (r)NASA; **656** Tom Steyer/Getty Images; **657** NASA/Science Photo Library/Photo Researchers; **658–659** Chad Ehlers/Stone/Getty Images.; **668** (bl)Richard J. Wainscoat/Peter Arnold, Inc., (others)Lick Observatory; **670** Dr. Fred Espenak/Science Photo Library/Photo Researchers; **671** Bettmann/CORBIS; **672** NASA; **674** Roger Ressmeyer/CORBIS;
677 BMDO/NRL/LLNL/Science Photo Library/Photo Researchers; **678** (t)Zuber et al/Johns Hopkins University/NASA/Photo Researchers, (b)NASA; **679** NASA; **681** Matt Meadows; **682** Cosmo Condina/Stone; **684** Lick Observatory; **685** NASA; **688–689** Roger Ressmeyer/CORBIS; **689** Matt Meadows; **692** European Southern Observatory/Photo Researchers; **694** Bettmann/CORBIS; **696** USGS/Science Photo Library/Photo Researchers; **697** (t)NASA/Photo Researchers, (inset)JPL/TSADO/Tom Stack & Assoc.; **698** (t)Science Photo Library/Photo Researchers, (bl)USGS/TSADO/Tom Stack & Assoc., (inset)USGS/Tom Stack & Assoc., (br)USGS/Tom Stack & Assoc.; **699** NASA/JPL/Malin Space Science Systems; **701** Science Photo Library/Photo Researchers; **702** (l)NASA/ Science Photo Library/Photo Researchers, (r)CORBIS; **703** (t to b)USGS/TSADO/Tom Stack & Assoc., NASA /JPL/Photo Researchers, NASA/TSADO/Tom Stack & Assoc., JPL, NASA; **704** JPL; **705** Heidi Hammel/NASA; **706** (inset)NASA/Science Source/Photo Researchers, (r)NASA/JPL/TSADO/Tom Stack & Assoc.; **707** NASA, ESA, H.Weaver (JHU/APL), A.Stern (SWRI), and the HST Pluto Companion Search Team; **708** (t to b)NASA/JPL/TSADO/Tom Stack & Assoc., NASA/Science Source/Photo Researchers, CORBIS, NASA/USGS/TSADO/Tom Stack & Assoc.; **709** CORBIS, (t to b)NASA/Science Photo Library/Photo Researchers, NASA/Science Source/Photo Researchers, ASP/Science Source/Photo Researchers, W. Kaufmann/JPL/Science Source/Photo Researchers; **710** Pekka Parviainen/Science Photo Library/Photo Researchers; **711** Pekka Parviainen/ Science Photo Library/Photo Researchers; **712** Georg Gerster/Photo Researchers; **713** JPL/TSADO/Tom Stack & Assoc.; **715** Bettmann/CORBIS; **716** (t)Museum of Natural History/Smithsonian Institution, (b)Museum of Natural History/Smithsonian Institution; **717** (l r)NASA, (cl)JPL/NASA, (cr)file photo; **718** NASA/Science Source/Photo Researchers; **720** John R. Foster/Photo Researchers; **722–723** TSADO/ESO/Tom Stack & Assoc.; **727** Bob Daemmrich; **730** (t)Carnegie Institution of Washington, (b)NSO/SEL/Roger Ressmeyer/CORBIS; **731** (l)NASA, (r)Picture Press/CORBIS, (b)Bryan & Cherry Alexander/Photo Researchers; **732** Celestial Image Co./Science Photo Library/Photo Researchers; **733** Tim Courlas; **735** Luke Dodd/Science Photo Library/Photo Researchers; **738** AFP/CORBIS; **739** NASA; **741** (t)Kitt Peak National Observatory, (b)CORBIS; **745** R. Williams (ST Scl)/NASA; **746** Matt Meadows; **748** Dennis Di Cicco/Peter Arnold, Inc.; **749** (l)file photo, (r)AFP/CORBIS; **754** PhotoDisc; **756** Tom Pantages; **760** Michell D. Bridwell/PhotoEdit, Inc.; **761** (t)Mark Burnett, (b)Dominic Oldershaw; **762** StudiOhio; **763** Timothy Fuller; **764** Aaron Haupt; **766** KS Studios; **767** Matt Meadows; **769** (t)Matt Meadows, (b)Doug Martin; **770** (t)Doug Martin, (b)PhotoDisc; **771** Giboux/Gamma-Liaison/Getty Images; **775** (t)Matt Meadows, (b)Mary Lou Uttermohlen; **776** Doug Martin; **777** Jeff Smith/Fotosmith; **778** (t)Mark Burnett, (b)Kevin Horgan/Stone/Getty Images; **781** Amanita Pictures; **782** Bob Daemmrich; **784** Davis Barber/PhotoEdit, Inc.

PERIODIC TABLE OF THE ELEMENTS

Columns of elements are called groups. Elements in the same group have similar chemical properties.

Gas
Liquid
Solid
Synthetic

Element — Hydrogen
Atomic number — 1
Symbol — H
Atomic mass — 1.008
State of matter

The first three symbols tell you the state of matter of the element at room temperature. The fourth symbol identifies elements that are not present in significant amounts on Earth. Useful amounts are made synthetically.

Period	1	2	3	4	5	6	7	8	9
1	Hydrogen 1 H 1.008								
2	Lithium 3 Li 6.941	Beryllium 4 Be 9.012							
3	Sodium 11 Na 22.990	Magnesium 12 Mg 24.305							
4	Potassium 19 K 39.098	Calcium 20 Ca 40.078	Scandium 21 Sc 44.956	Titanium 22 Ti 47.867	Vanadium 23 V 50.942	Chromium 24 Cr 51.996	Manganese 25 Mn 54.938	Iron 26 Fe 55.845	Cobalt 27 Co 58.933
5	Rubidium 37 Rb 85.468	Strontium 38 Sr 87.62	Yttrium 39 Y 88.906	Zirconium 40 Zr 91.224	Niobium 41 Nb 92.906	Molybdenum 42 Mo 95.94	Technetium 43 Tc (98)	Ruthenium 44 Ru 101.07	Rhodium 45 Rh 102.906
6	Cesium 55 Cs 132.905	Barium 56 Ba 137.327	Lanthanum 57 La 138.906	Hafnium 72 Hf 178.49	Tantalum 73 Ta 180.948	Tungsten 74 W 183.84	Rhenium 75 Re 186.207	Osmium 76 Os 190.23	Iridium 77 Ir 192.217
7	Francium 87 Fr (223)	Radium 88 Ra (226)	Actinium 89 Ac (227)	Rutherfordium 104 Rf (261)	Dubnium 105 Db (262)	Seaborgium 106 Sg (266)	Bohrium 107 Bh (264)	Hassium 108 Hs (277)	Meitnerium 109 Mt (268)

The number in parentheses is the mass number of the longest-lived isotope for that element.

Rows of elements are called periods. Atomic number increases across a period.

The arrow shows where these elements would fit into the periodic table. They are moved to the bottom of the table to save space.

Lanthanide series	Cerium 58 Ce 140.116	Praseodymium 59 Pr 140.908	Neodymium 60 Nd 144.24	Promethium 61 Pm (145)	Samarium 62 Sm 150.36
Actinide series	Thorium 90 Th 232.038	Protactinium 91 Pa 231.036	Uranium 92 U 238.029	Neptunium 93 Np (237)	Plutonium 94 Pu (244)